Tribology and dynamics of engine and powertrain

Related titles:

Vehicle noise and vibration refinement
(ISBN 978-1-84569-497-5)
High standards of noise, vibration and harshness performance are expected in vehicle design. Refinement is therefore one of the main engineering/design attributes to be addressed when developing new vehicle models and components. This book provides a review of noise and vibration refinement principles and methods, advanced experimental and modelling techniques, and palliative treatments necessary in the process of vehicle design, development and integration in order to meet noise and vibration standards. Case studies from the collective experience of specialists working for major automotive companies are included to form an important reference for engineers in the motor industry who seek to overcome the technological challenges faced in developing quieter, more comfortable cars.

Materials, design and manufacturing for lightweight vehicles
(ISBN 978-1-84569-463-0)
Research into the manufacture of lightweight automobiles has led to the consideration of a variety of materials, such as high-strength steels, aluminium alloys, magnesium alloys, plastics and composites. This research is driven by a need to reduce fuel consumption to preserve dwindling hydrocarbon resources without compromising other attributes such as safety, performance, recyclability and cost. This important book will make it easier for engineers not only to learn about the materials being considered for lightweight automobiles, but also to compare their characteristics and properties. It also covers issues such as crashworthiness and recycling.

Diesel engine system design
(ISBN 978-1-84569-715-0)
Diesel engine design is highly complex, involving many individuals and companies from original equipment manufacturers to suppliers. A system design approach for setting up the right engine performance specifications is essential to streamline the processes. This important book links everything a diesel engineer needs to know about engine performance and system design in order to master all the essential topics quickly; the focus is on how to use advanced analysis methods to solve practical design problems. Numerous case studies and examples illustrate advanced design approaches using engine cycle simulation tools. The central theme is how to design a good engine system performance specification at an early stage of the product development cycle.

Details of these and other Woodhead Publishing books can be obtained by:

- visiting our web site at www.woodheadpublishing.com
- contacting Customer Services (e-mail: sales@woodheadpublishing.com; fax: +44 (0) 1223 893694; tel.: +44 (0) 1223 891358 ext. 130; address: Woodhead Publishing Limited, Abington Hall, Granta Park, Great Abington, Cambridge CB21 6AH, UK)

If you would like to receive information on forthcoming titles, please send your address details to: Francis Dodds (address, tel. and fax as above; e-mail: francis.dodds@woodheadpublishing.com). Please confirm which subject areas you are interested in.

Tribology and dynamics of engine and powertrain

Fundamentals, applications and future trends

Edited by

Homer Rahnejat

WOODHEAD
PUBLISHING

Oxford   Cambridge   Philadelphia   New Delhi

Published by Woodhead Publishing Limited, Abington Hall, Granta Park, Great Abington, Cambridge CB21 6AH, UK
www.woodheadpublishing.com

Woodhead Publishing, 525 South 4th Street #241, Philadelphia, PA 19147, USA

Woodhead Publishing India Private Limited, G-2, Vardaan House, 7/28 Ansari Road, Daryaganj, New Delhi – 110002, India
www.woodheadpublishingindia.com

First published 2010, Woodhead Publishing Limited
© Woodhead Publishing Limited, 2010
The authors have asserted their moral rights.

British Library Cataloguing in Publication Data
A catalogue record for this book is available from the British Library.

ISBN 978-0-08-101435-6 (print)
ISBN 978-1-84569-993-2 (online)

The publisher's policy is to use permanent paper from mills that operate a sustainable forestry policy, and which has been manufactured from pulp which is processed using acid-free and elemental chlorine-free practices. Furthermore, the publisher ensures that the text paper and cover board used have met acceptable environmental accreditation standards.

Typeset by Replika Press Pvt Ltd, India
Printed by TJI Digital, Padstow, Cornwall, UK

Contents

xiv   Contents

Part III Micro-systems and nano-conjunctions

Contributor contact details

(* = main contact)

Chapter 1

H. Rahnejat
Wolfson School of Mechanical and
    Manufacturing Engineering
Loughborough University
Loughborough
UK

E-mail: H.Rahnejat@Lboro.ac.uk

Chapter 2

D. Arnell
Jost Institute for Tribotechnology
University of Central Lancashire
Preston PR1 2HE
UK

E-mail: rdarnell@uclan.ac.uk

Chapter 3

P. Prokopovich
Wolfson School of Mechanical and
    Manufacturing Engineering
Loughborough University
Loughborough
UK

E-mail: P.P.Prokopovich@lboro.ac.uk

M. Teodorescu
Department of Automotive
    Engineering
School of Engineering
Cranfield University
Cranfield MK43 0AL
UK

E-mail: m.s.teodorescu@cranfield.ac.uk

H. Rahnejat*
Wolfson School of Mechanical and
    Manufacturing Engineering
Loughborough University
Loughborough
UK

E-mail: H.Rahnejat@Lboro.ac.uk

Chapter 4

M. Teodorescu
Department of Automotive
    Engineering
School of Engineering
Cranfield University
Cranfield MK43 0AL
UK

E-mail: m.s.teodorescu@cranfield.ac.uk

V. Votsios
Atos
Spain

E-mail: vasilis.votsios@atosorigin.com

P. M. Johns-Rahnejat
Formerly Imperial College London
London
UK

H. Rahnejat*
Wolfson School of Mechanical and
    Manufacturing Engineering
Loughborough University
Loughborough
UK

E-mail: H.Rahnejat@Lboro.ac.uk

Chapters 5 and 7

R. Gohar
c/o H. Rahnejat
Wolfson School of Mechanical and
    Manufacturing Engineering
Loughborough University
Loughborough
UK

E-mail: c.pereagohar@imperial.ac.uk

M. M. A. Safa
22 Alexandra Drive
Surbiton KT59AB
UK

E-mail: m.safa@kingston.ac.uk

Chapter 6

F. Sadeghi
School of Mechanical Engineering
Purdue University
West Lafayette, IN 47907
USA

E-mail: sadeghi@ecn.purdue.edu

Chapter 8

Dr D. R. Adams
Engine Engineering
Dagenham Development Centre
Dagenham RM9 6SA
UK

E-mail: dadams69@ford.com

Chapter 9

Dr I. McLuckie
AIES Ltd
PO Box 7784
Market Harborough LE16 7YH
UK

E-mail: ian.mcluckie@aiesl.co.uk

Chapter 10

Dr A. Senatore* and Professor V.
    D'Agostino
Department of Mechanical
    Engineering, University of
    Salerno
Via Ponte don Melillo, 1
I 84084 Fisciano (SA)
Italy

E-mail: a.senatore@unisa.it
E-mail: dagostino@unisa.it

Chapter 11

I. Sherrington
Jost Institute for Tribotechnology
University of Central Lancashire
Preston PR1 2HE
UK

E-mail: isherrington@uclan.ac.uk

Chapter 12

R. S. Dwyer-Joyce
Department of Mechanical
    Engineering
University of Sheffield
Mappin Street
Sheffield S1 3JD
UK

E-Mail: r.dwyerjoyce@sheffield.ac.uk

Chapter 13

I. Etsion
Department of Mechanical
    Engineering
Technion
Haifa 32000
Israel

E-mail: etsion@technion.ac.il

Chapter 14

R. Rahmani
Dynamics Research Group
Wolfson School of Mechanical and
    Manufacturing Engineering
Loughborough University
Loughborough
Leicestershire LE11 3TU
UK

E-mail: r.rahmani@lboro.ac.uk

A. Shirvani and H. Shirvani
Faculty of Science and Technology
Anglia Ruskin University
Bishop Hall Lane
Chelmsford CM1 1SQ
UK

Email: Ayoub.Shirvani@anglia.ac.uk
        Hassan.Shirvani@anglia.ac.uk

Chapter 15

P. C. Mishra, H. Rahnejat* and P.
    King
Wolfson School of Mechanical and
    Manufacturing Engineering
Loughborough University
Loughborough
UK

E-mail: H.Rahnejat@Lboro.ac.uk

Chapter 16

M. Kushwaha
Litens
Canada

E-mail: manu.kushwaha@litens.com

Chapter 17

M. Teodorescu
Department of Automotive
    Engineering
School of Engineering
Cranfield University
Cranfield MK43 0AL
UK

E-mail: m.s.teodorescu@cranfield.ac.uk

Chapter 18

S. Balakrishnan
Mercedes Benz High Performance
    Engines
Brixworth
UK

E-mail: Sashi.Balakrishnan@Mercedes-
    benz-hpe.com

C. McMinn
Ford Motor Company
Dagenham
UK

E-mail:cmcminn@ford.com

C. E. Baker and H. Rahnejat*
Wolfson School of Mechanical and
    Manufacturing Engineering
Loughborough University
Loughborough
UK

E-mail: c.e.baker@lboro.ac.uk
E-mail: H.Rahnejat@Lboro.ac.uk

Chapter 19

S. Boedo
Department of Mechanical
    Engineering
Rochester Institute of Technology
76 Lomb Memorial Drive
Rochester, NY 14623-5604
USA

E-mail: sxbeme@rit.edu

Chapter 20

P. C. Mishra and H. Rahnejat*
Wolfson School of Mechanical and
    Manufacturing Engineering
Loughborough University
Loughborough
UK

E-mail: H.Rahnejat@Lboro.ac.uk

Chapter 21

M. Menday
Loughborough University
UK

E-mail: michael_menday@yahoo.co.uk

Chapter 22

M. Menday and H. Rahnejat*
Loughborough University
UK

E-mail: michael_menday@yahoo.co.uk

Chapter 23

G. Mavros
Aeronautical and Automotive
    Engineering
Loughborough University
Epinal Way
Loughborough
Leicestershire LE11 3TU
UK

E-Mail: G.Mavros@lboro.ac.uk

Chapter 24

S. K. Mohan
Magna Powertrain
6600 New Venture Gear Drive
East Syracuse, NY 1307
USA

E-mail: sankar.mohan@gmail.com

Chapter 25

S. Natsiavas* and D. Giagopoulos
Department of Mechanical
    Engineering
Aristotle University
54 124 Thessoloniki
Greece

E-mail: natsiava@auth.gr

Chapter 26

S. N. Doğan
Daimler AG
Mercedesstr.122
70372 Stuttgart
Germany

E-mail: suereyya_nejat.dogan@daimler.
    com

Chapter 27

P. Kelly*
Ford Werke GmbH
Cologne
Germany

E-mail: pkelly7@ford.com

M. Menday
Loughborough University
UK

E-mail: michael_menday@yahoo.co.uk

Chapter 28

P. Kelly*
Ford Werke GmbH
Cologne
Germany

E-mail: pkelly7@ford.com

B. Pennec, R. Seebacher, B. Tlatlik
    and M. Mueller
LuK GmbH & Co. oHG
Industriestrasse 3
77815
Germany

E-mail: Roland.Seebacher@schaeffler.
    com

Chapter 29

S. Theodossiades, O. Tangasawi
    and H. Rahnejat*
Wolfson School of Mechanical and
    Manufacturing Engineering
Loughborough University
Loughborough
UK

E-mail: s.theodossiades@lboro.ac.uk
        H.Rahnejat@Lboro.ac.uk

Chapter 30

M. Gnanakumarr
Wolfson School of Mechanical and
   Manufacturing Engineering
Loughborough University
Loughborough
UK

E-mail: M.M.Gnanakumarr@lboro.ac.uk

Chapter 31

M. S. M. Perera, S. Theodossiades
   and H. Rahnejat*
Wolfson School of Mechanical and
   Manufacturing Engineering
Loughborough University
Loughborough
UK

E-mail: M.S.M.Perera@lboro.ac.uk
       s.theodossiades@lboro.ac.uk
       H.Rahnejat@Lboro.ac.uk

Chapter 32

M. Teodorescu
School of Engineerng
Cranfield University
Cranfield
UK

H. Rahnejat* and S. Theodossiades
Wolfson School of Mechanical and
   Manufacturing Engineering
Loughborough University
Loughborough
UK

E-mail: H.Rahnejat@Lboro.ac.uk

Chapter 33

Frank W. DelRio
Ceramics Division, Materials
   Science and Engineering
   Laboratory
National Institute of Standards and
   Technology
100 Bureau Drive, Stop 8520
Gaithersburg, MD 20899
USA

Email: frank.delrio@nist.gov

Carlo Carraro and Professor Roya
   Maboudian
Department of Chemical
   Engineering
University of California Berkeley
B78 Tan Hall
Berkeley, CA 94720
USA

E-mail: carraro@yahoo.com
       maboudia@berkeley.edu

Preface

Future developments in engines and powertrain systems will necessarily be sympathetic to nature. There is a growing need to preserve the ever-diminishing natural sources of energy; mainly fossil fuels. The rate by which these resources are used far exceeds their natural recurrence. For the foreseeable future the transport sector in all its variety will use these diminishing resources to some extent, even with the emergence of alternative sources of energy.

The efficient use of these precious natural assets has never been as important as it is today. The quest to reduce the rate of depletion of these fuels would directly affect every facet of science and engineering. One key discipline is tribology; the science of friction, wear and lubrication. Its efficient and often empirical application to means of transport has been a subject interwoven with human civilisation from its very outset. As our knowledge has grown, the seemingly empirical nature of rules governing the science of interacting surfaces has advanced towards a more fundamental understanding. Tribology is now one of the cornerstones in the quest for efficiency, conservation of resources and the protection of ecosystems. These noble causes have driven us from relatively large-scale observed phenomena such as friction down to the scale of minutiae to seek answers to many questions.

Tribology cannot be viewed as an isolated discipline impervious to a host of other phenomena that interact with it and determine its outcome. It may, therefore, be regarded as a facet of physics of motion within a limited scale. At its upper boundary we encounter the behaviour of contiguous surfaces and beyond that the structure of solids themselves, and yet further on, the conglomerate of bodies that are subject to some form of constrained motions. At the lower bound of the tribological world we strive to understand the interaction of small matter; molecules of lubricant with each other and with the rough terrain on the surfaces themselves. These interactions, minute or relatively large, are parts of the physics of motion at all physical scales and are currently described by a plethora of rules and laws. Yet in the absence of a unique law, these various facets act as our microscope pointing to one end of the scale to search for underlying causes and conversely as a telescope to the other end to observe the effects. In engineering this amounts to down-cascading to the root causes and up-cascading to a remedial solution. This approach is best seen in problems associated with noise, vibration and harshness (NVH) refinement, a key concern in industry.

Vibration and noise are parts of nature itself, as static equilibrium is only a mathematical convenience. Vibration and noise appear to follow the principle of least action in all facets of dynamics. Like water taking the path of least resistance to conserve energy, all other matter from atoms to stars move just sufficiently to maintain the status quo. We should learn from nature in our inventions, including engines and powertrains, conserve energy and strive to maintain peace and tranquillity. The affinity to nature translates to adherence to the principle of parsimony and a quest for the principle of least action (optimal conservation of energy).

In the absence of an adequate understanding of the laws and rules of nature, we should adopt our evolving knowledge to multi-physics, multi-scale problem-solving. This is the approach used throughout this volume. I would like to take this opportunity of thanking my contributing colleagues, who have enthusiastically strived to share their experiences with the readers of this work.

Homer Rahnejat

Foreword

Since applications of several remarkable developments in science and engineering are eye-catching and attract much publicity, it is easy to overlook other equally impressive developments of mechanical systems because they are so familiar to us in everyday life. The automobile is one such system. Just a few decades ago, the reliability of automobiles was far from satisfactory, fuel consumption was unacceptably high and frequent servicing was essential. Engines now run smoothly and much more efficiently with pleasing ease and negligible wear. Noise levels have been greatly reduced and the demands of frequent servicing much diminished.

Advances in tribology have played a major role in this progress. Modelling, analysis, experimentation, design, manufacture, materials and lubricants have all contributed to tremendous improvements in the performance, reliability and efficiency of modern engines. How often do motorists experience problems with cylinder, piston ring, or valve train wear, or bearing failure?

Analysis and design of linked engine components and a complete powertrain present fascinating challenges to the engineer, particularly in relation to tribology and multi-body dynamics. The editor of this extensive volume is well placed to bring the reader up to date on these matters, having intimate experience of both fields. The fundamentals of each topic are presented and followed by contributions by experienced engineers on analysis, experimental procedures and design.

The reader is introduced to significant progress in both tribology and dynamics. Effective performance and reliability of major engine components rely upon extensive analysis embracing current concepts of transient thermo-elastohydrodynamic lubrication, terminology that was unfamiliar a few decades ago. A fascinating feature of the text is the mixture and interchange of such engineering science vocabulary with practical terminology such as clonk, clatter, rattle, judder, shuffle and whoop!

The 33 chapters in this major text (each with many valuable references) are arranged in three parts, with the bulk of the work being presented in Part II:

I Introduction to *dynamics* and tribology within the *multi-physics* environment.
II Engine and powertrain technologies and applications.
III Micro-systems and nano-conjunctions.

Basic aspects of dynamics, friction, wear, and lubrication, surface interactions and impact dynamics in Part I, contributed by authors from five UK universities, Spain and the United States, provide the foundations for subsequent topics. Conventional approaches to bearing design, based upon the governing, classical Reynolds equation of fluid film lubrication, are presented and followed by detailed accounts of elastohydrodynamic lubrication.

Intriguing discussion of the approach to engine and powertrain analysis, experimentation and design are presented in Part II. Piston systems, valve trains and engine bearings receive special attention in Sections II.II, II.III and II.IV, while Section II.V is devoted to drivetrain systems. The latter topic, which reflects long-standing research interests at Loughborough, the editor's home university, provides a fascinating demonstration of the combined power of multi-body dynamics and tribology. Problems with airborne and structural noise transmission, clutch take-up judder, tyre–road interactions, differential gears, traction control devices and flywheels are all addressed. The authors of chapters in this balanced central core of the book are based in industry in Germany, the UK and the USA and in universities in Greece, Israel, Italy and the UK.

Special problems associated with the miniaturisation of mechanical drives have attracted much attention in recent years. The two chapters in Part III introduce readers to developing surface interaction concepts in micro-systems and small-scale surface engineering.

This fascinating and comprehensive book on current developments in the fields of tribology and multi-body dynamics related to vehicle problems will be of interest to both the industrial and academic communities.

Duncan Dowson
School of Mechanical Engineering
The University of Leeds
Leeds LS2 9JT
and
Visiting Professor
Loughborough University

Introduction

Fuel efficiency, environmental sustainability, reliability, quality and noise, vibration and harshness (NVH) refinement are key targets for the automotive industry into the future. All these attributes require design and development, driven by fundamental and applied research in a multi-disciplinary environment. Fuel efficiency is an important economic objective which also impinges on the longer-term environmental sustainability of the vehicle industry in terms of conservation of energy resources. Other aspects of the environmental impact of transport systems include emissions, which is a major growing concern. This means that alternative methods of propulsion should be considered as well as better combustion strategies in internal combustion (IC) engines. Whichever method is finally employed, friction, as the main source of parasitic losses, will always be a major problem and methods to mitigate its effects are of paramount importance.

It is not surprising that with such a broad range of aforementioned ideals, downsizing of propulsion units has been favoured for some time now. In fact, the future points to ever more compact high output power – to lightweight power train units. However, one repercussion has been a plethora of NVH issues, which should be regarded as error states; or errant system dynamics. The palliation methods are often quite costly and sometimes unnecessarily complex.

The combination of parasitic losses and errant dynamic states are of concern in terms of fuel efficiency and represent undesired environmental impacts (emissions and noise pollution). Furthermore, these can be regarded as being of poor quality and as potential sources of unreliability.

The foregoing shows the importance of tribology and dynamics as key disciplines in design and analysis of future power trains. These are the two aspects covered in this volume. They interact with each other and other phenomena, also noted in this book, at all physical scales from the minutiae of contacting surfaces to relatively large displacement of pistons and the drive train.

The book covers many important practical engineering and technological issues that the industry faces in design and development today and into the future. Many of the authors are experienced technologists, specialists and component engineers from industry with many years of experience in the fields of their expertise. They are joined by established academics and promising researchers to make this a rather unique and comprehensive volume.

Professor Richard Parry-Jones
Former Vice President of Ford Global Development and
Chairman of Premier Automotive Group

Part I

Introduction to *dynamics* and tribology within the *multi-physics* environment

1

An introduction to multi-physics multi-scale approach

H. RAHNEJAT, Loughborough University, UK

Abstract: All natural phenomena or those which are instigated by people through devised mechanisms, machines or devices are subject to dynamics. Dynamics, as physics of motion, is the broadest of sciences. It happens that its many facets are explained by various laws of physics and theories, and methods of analysis. Therefore, one can view dynamics as a multi-physics science. The underlying principles and methods describing the multi-physics nature of dynamics have their roots in laws and rules which belie the development of the various branches of physics. These are often based on kinetic laws that seemingly govern interactions at various physical scales. Thus, dynamics is a multi-physics multi-scale science of interactions of all matter. Engines and powertrains are no exception, and perhaps very good examples of the interactive nature of multi-faceted dynamics. This chapter introduces the fundamentals of dynamics, the science that has engaged human inquisition and has arguably contributed to the advancement of civilisation for the longest.

Key words: Newtonian axioms, constraints, Lagrangian dynamics, multi-body dynamics, elasticity.

1.1   Introduction

Dynamics and tribology, described in this book, may be regarded as subsets of physics of motion (in a multi-physics perspective). Dynamics is the study of motion of entities caused by the underlying forces. Historically, in the discipline of dynamics and within engineering these entities have been considered to be an assembly of parts (a system), solid inertial elements (a component) and rigid particles. When the study of motion of a *material point* (a generic term used to describe these entities; a particle, a body: a conglomerate of such particles or a system: an assembly or cluster of bodies) is observation-based only (without regard to the underlying cause: force), then the field of investigation is referred to as *kinematics*. In the case of a multi-body or a many-body system, kinematics refer to studies with no degrees of freedom; relative motions between their constituent material points (their motion is prespecified).

When a system undergoes no displacements with respect to a specified frame of reference (a co-ordinate set; t, x, y, z), then it is regarded to be *static*.

3

The forces applied to such a system are said to be in a state of equilibrium (no net force). If a multitude of such equilibria can be assumed, then these various states of the system may be termed as *quasi-static*.

Real systems are not rigid. An example is the powertrain system, subject of this book, where hollow driveshaft tubes undergo small amplitude elastic deformation under load, while they undertake much larger inertial motions (see Chapter 30, explaining the clonk phenomenon). The same is true of all material points, although in many cases, such as molecules, the deformation would be infinitesimal and thus almost insensible. Therefore, flexible systems are subject to *elastodynamics*. Since the deformation amplitudes are different in scale of measurement to the overall inertial displacements, the problem is *multi-scale*. If one disregards the larger scale and the study is confined to small amplitude oscillations, then the problem at hand is regarded as one of *vibration*. In general all real material points, being compliant, can assume many forms when vibrating. These forms are known as modes. For example, Chapter 30 shows various modes of flexible driveshaft tubes. The same is true of very small material points such as electrons with their wavy motions with many spins. At the other end of the scale, it is surmised that even heavenly bodies pulsate or quiver, spreading waves on the fabric of space, rather similar to the wave propagation on the surface of driveshaft tubes, explained in Chapter 30.

When undertaking study of a problem in dynamics, the boundary of the system must be defined, because there is no generic system. The interactions between the defined system and those material points extraneous to it are then ignored. This is a fundamental rule of experimentation. Thus, for example, in vehicle engineering, problems are defined as those of the powertrain system or vehicle–road interactions, and not a vehicle within the universe! With the system boundaries defined, interaction of key material points are considered. These interactions are simply forces acting between them, causing motions in a multitude of physical scales. Therefore, the interaction scale(s) of interest should also be determined. For example, powertrain dynamics problems may be in the scale of large displacements (inertial dynamics: shuffle, see Chapters 21, 23 and 30) or structural response (modal behaviour of driveshaft tubes or the transmission case, see Chapter 30) or noise propagation (acoustic response of thin-walled structures). These may be regarded as wave motions from scales of metres to sub-millimetre and on to nanometres respectively, but they are all part of dynamics of a defined system (as are the usual micro-scale deflections of load-bearing conjunctions in tribology; see Chapters 4, 5 and 6). The environment outside the system boundary is considered to be rigid, to which a global frame of reference for measurement of multi-scale physics of motion is firmly attached (Rahnejat, 1998). In reality the extraneous environment is not rigid, nor is any place within the known or surmised universe. The experiment carried out in a laboratory within a defined system is positioned on Earth which moves around the Sun at 66730 miles/h (average), while the Solar system is dragged by the Sagitarius A* at the centre of the Galaxy at 45000 miles/h towards the constellation of Hercules. However, one can consider the dominant forces in the experiment to be because of material points of the defined system and in some cases (bodies of significant size) due to the Earth's gravitational pull only. Thus, in dynamics the motion of a material point is governed by all those within the same system. This is the essence of **Mach's principle** and is fundamental to the subject of dynamics. Now with this philosophical basis and within any system of any conglomerate of material points i, any one such point has an acceleration due to its interactions with others as:

$$a = \frac{\sum\limits_i F_i}{m} \qquad (1.1)$$

This is **Newton's second law** of motion, where m is the mass of any material point. Newton called this an axiom, because in his perspective it was a natural observation for which no proof was required at the time of his enunciation, same as Euclid's geometrical axioms. The second law is the foundation upon which all the field of dynamics resides. Later a fundamental proof for this axiom is provided through energy consideration: Lagrange's equation.

The notion of material points is not confined to those of a solid nature, but all matter in any physical state including fluids. Thus, a system may be defined as a volume of fluid bounded by solid surfaces such as a river and its impervious banks. The volume of fluid may be considered as a series of elemental volumes progressing through the system in the same manner (but not exactly) as the deformation wavefronts progress in the hollow driveshaft tubes in Chapter 30. The study is, therefore, one of continuous fluid flow due to a pressure gradient and velocity profile (both as a result of forces) through the assumed system. The subject is called **hydrodynamics** (see Chapter 5). This is also a multi-scale problem, like other forms of dynamics. For a large expanse of fluid, the elemental volumes may be considered large, but finite within which a state of equilibrium may be assumed relative to the interactions between any pair of such elements themselves. Large elemental volumes mean significant body forces (weight) and inertial forces. On the other hand, in tribological conjunctions, the narrowness of the gap between the boundary solids means small elemental volumes and their assumed uniform motion through the system (the conjunction). Thus, the problem simplifies to that of flow induced by changes in pressure gradient and any relative motion of bounding surfaces, forming a **wedge effect** (Gohar and Rahnejat, 2008; also see Chapter 5). Therefore, the complex flow dynamics is reduced to a manageable problem in hydrodynamics, with a fundamental equation:

$$\frac{\partial}{\partial x}\left(\frac{\rho h^3}{12\eta}\frac{\partial p}{\partial x}\right) + \frac{\partial}{\partial y}\left(\frac{\rho h^3}{12\eta}\frac{\partial p}{\partial y}\right) = u\,\frac{\partial \rho h}{\partial x} + v\,\frac{\partial \rho h}{\partial y} + \frac{\partial \rho h}{\partial t} \qquad (1.2)$$

Reynolds obtained this equation by ignoring the effect of surface, body and inertial forces for small incremental elemental volumes of fluid, confined by a pair of close solid boundaries (Chapter 5 describes the equation in detail). Therefore, as in the case of solids, a question of scale exists in the case of fluids as well. If the size of the system is reduced, the incremental *computational* volume must also decrease accordingly, where the conditions within such a volume may be considered to be in equilibrium. It is, therefore, clear that in the extreme cases (ultra-thin film tribology), with molecular interactions and surface energy effects, no bulk properties such as a computational elemental volume may be assumed (see Chapter 3).

Therefore, one may surmise that physical interactions regardless of the state of matter are functions of size of the assumed system and that of a material point considered, or the ratio ε/ℓ (Rahnejat, 2008). It turns out that the nature of physical interactions (force) changes according to scale (the same ratio). However, the size of the material point ε is explained by a host of physical attributes such as mass or charge and that of the system ℓ by density, viscosity, permittivity, elasticity, coefficient of friction, etc. This means that forces other than gravity are related to kinematic quantities (displacement, velocity and acceleration) by physical properties of material points and the environment of the system. The introduction to Chapter 3 describes the philosophical concern about the multiplicity of forces of nature. This means the current knowledge is based on acceptance of a multi-physics character for interactions of material points at multi-scale within defined systems, thus the increasingly used phrase: ***multi-scale multi-physics analysis***. Finally, this brief introduction has shown that both dynamics and tribology are subsets of physics of motion.

1.2   Newtonian mechanics

Kinematics, being the study of motion without regard to the underlying cause (force), is one of the oldest sciences. Its roots can be traced back to the ancient studies of heavenly bodies, such as Homer's Earth-centred universe in the *Iliad*. As an observation-based science, *kinematics* is concerned with measurement of the state of motion of a material point (displacement, velocity and acceleration) with respect to a frame of reference. As already discussed, it is particularly convenient to firmly attach this frame of reference to a fixed (static) object. Therefore, it was particularly convenient to assume Earth to be the fixed central entity about which all the heavens would revolve (presumably in the adoration of humanity!). The heavenly bodies would then describe curvilinear paths whose slope at a given position would yield their relative velocity with respect to the frame of measurement. Much later, Galileo understood that deviation from a straight-line motion corresponded to non-uniform velocity and the curvature was due to accelerated motion. Therefore, kinematics is the study of curves; their local slope and curvature. By the late sixteenth and early seventeenth centuries kinematics had finally attained the status only hitherto afforded to geometry as a fundamental science, because of the historical prominence of Ancient and Middle Age geometers such as Homer, Pythagoras, Archimedes, Ptolemy, Khayam, Tusi and Copernicus, among others. Using astronomical observations and a cursory understanding of non-uniform motion, Galileo and Kepler put an end to the concept of the Earth-centred universe and obviously put a sizeable dent in the human vanity! Kinematics had its greatest moment in history with the acceptance of the *heliocentric system*.

Using Kepler's observations and his laws of motion, Newton explained the elliptical orbit of planets around the Sun by a central force due to gravitational attraction. The cause belying kinematics was found: force. In the case of planetary motions, the force of gravity caused the non-uniform accelerated motion; curvature of the path. The law of universal gravitation states:

$$F = \frac{GMm}{r^2} = mg \tag{1.3}$$

where G is the *universal gravitational constant*, M *the* mass of a source (such as the Sun) and m that of a target body (such as Earth). Thus, the radius of curvature r at any position along the path is $r = \sqrt{(GM/g)}$ for a two-body system (Sun and Earth). Since the Earth's path is elliptical (only slightly) on the Ecliptic plane, then r is not a constant, which means that g varies accordingly.

The simple calculations here assume a two-body system, but the path of a body within a system (Earth in the solar system) is subject to all material points within it (other planets); remember Mach's principle. More comprehensive treatment of this problem is given by Chandrasekhar (1995).

Newton then stated that in general equation (1.3) can be extended to his second axiom; equation (1.1). If there is no net force; $F = 0$, a body at rest remains stationary, while one in motion pursues a straight-line path; *Newton's first axiom*. A straight-line path is an extremal path (shortest path) due to uniform motion of a material point relative to an observer. One can surmise from equation (1.3) that attraction between two bodies necessitates equal and opposite forces. This is *Newton's third axiom*; for every action there is an equal and opposite reaction.

The assertion of these axioms by Newton, in addition to the law of universal gravitation, resulted in scientific disputes, some of which persisted beyond his lifetime. One concerned a fundamental proof for the second axiom, to render the same as a law of physics. An *axiom* is an assertion which appeals to all observers who would all agree on the cause of a phenomenon. This definition does not put the onus of acceptance on a mathematical proof; such as the existence of the Sun. Some have proposed intangible proofs for certain axiomatic concepts such as Descartes' for 'life': *I think, therefore, I am.* Mathematical discourse has increasingly been viewed as a requirement for proof since the seventeenth century. In this respect, **Lagrange's equation** is the proof of Newton's second axiom as:

$$\frac{d}{dt}\left(\frac{\partial K}{\partial \dot{q}^j}\right) + \frac{\partial U}{\partial q^j} = F_{a^j} \tag{1.4}$$

where K is the kinetic energy (considered to be independent of displacement q with the right choice of co-ordinate system), U is the potential energy and F_{a^j} the component of net applied force in the co-ordinate direction q^j (a generalised proof is given in Section 1.3).

In general a completely unconstrained material point in space has six degrees of freedom, therefore, the generalised co-ordinate set: $q^j \in x, y, z,$ ψ, θ, φ. The **kinetic energy** has, therefore, components: $K_j = \frac{1}{2}m\dot{q}^{j^2}$ for the translational degrees of freedom and $K_j = \frac{1}{2}I_{q^j}\dot{q}^{j^2}$ for rotational ones. Therefore, it can be seen that the first term on the left-hand side of Lagrange's equation is the inertial force, for example for $q^j = x$:

$$\frac{d}{dt}\left(\frac{\partial K}{\partial \dot{x}}\right) = \frac{d}{dt}\left[\frac{\partial\left(\frac{1}{2}m\dot{x}^2\right)}{\partial \dot{x}}\right] = \frac{d(m\dot{x})}{dt} = m\ddot{x} \tag{1.5}$$

Now, the second term on the left-hand side of Lagrange's equation is **Euler's equation**, simply stating that the rate of change of **potential energy** with respect to displacement is the body or restoring force:

$$F_{q^j} = -\frac{\partial U}{\partial q^j} \tag{1.6}$$

Thus, a material point falling freely under the influence of gravity towards the centre of Earth from any height x, with the frame of reference q aligned with the direction of motion has a body force:

$$F_x = -\frac{\partial(mgx)}{\partial x} = -mg \tag{1.7}$$

Using Lagrange's equation and noting that there is no applied force (free fall); $F_{ax} = 0$, then: $F_{ax} = m\ddot{x} - mg$ $(\ddot{x} = g)$, which is Newton's second axiom.

Therefore, Lagrange's equation is essentially the determination of net force, causing an acceleration (same as equation (1.1)):

$$a_q = -\frac{\partial \phi}{\partial q} = \frac{1}{\Omega}\left(F_{aq} - \frac{\partial U}{\partial q}\right) \tag{1.8}$$

where $\Omega \in m, I$ according to the degree of freedom (translational or rotational). Thus, if a potential ϕ can be specified, then acceleration of all material points within such a field can be determined. One can now revisit the same example of the falling matter above, this time attaching the frame of reference to the material point itself, falling within a field, where $\phi = -(GM/x)$. In this case, $q = x$, $\Omega = m$, $F_{ax} = 0$ as before and $a_q = \ddot{x}$, then:

$$\ddot{x} = \frac{GM}{x^2} = g \tag{1.9}$$

which yields the same results as previously. Two important observations should be made. Firstly, the potential used is due to gravitation, thus equation (1.3) is proven from first principles. If the field is due to Earth's gravity then M represents its mass. $x = r + H$ is the distance to the centre of Earth, r its radius and h the height of the falling matter above the Earth's surface. Since usually $H << r$, g hardly changes near the surface of Earth. This justifies the use of a constant value for g in engineering. Secondly, the above alternative analyses yield the same result, indicating the equivalence of the two systems; one in a gravitational field and the other falling uniformly with an equivalent inertial acceleration. This was noted by Einstein as the ***equivalence principle***, the implication being that inertial acceleration produces gravitational action. There are many examples, such as a material point in curvilinear motion or planetary motion or a vehicle cornering. This means that motion on curves induces gravitational action. This became clear with Einstein's general relativity; after all physics of motion in all its forms could be reduced to study of curves and motion of material points upon them. This appears to be true apart from various electromagnetic phenomena in the scale of minutiae as described in Chapter 3. The problem is that general relativity is based on a theory for gravity (macroscopic material points), Thus, seemingly prevalent potentials at very small scale deviate from it. The next section discusses Lagrange's equation. Readers should note that inertial and body forces which are dominant in the equation play an insignificant role in the scale of minutiae (see also Chapter 3).

1.3 Lagrange's equation and reduced configuration space

Lagrange's equation (1.4) is for unconstrained systems, where any material point within the defined system enjoys six degrees of freedom as already indicated above. However, recall that acceleration of a material point within a defined system is due to all other such points present (Mach's principle). Therefore, the defined system may be regarded as a **reduced configuration space**, where motion of material points are restricted by their interactions (**constrained system dynamics**).

Newton's second axiom in any co-ordinate direction q may be presented in the form: $-F_{q^i} + m_{q^i}\ddot{q}^i = 0$. This form of equation simply states that any applied external force on a material point m_{q^i} must be balanced by its inertial response. This form of second axiom is known as the **D'Alembert's principle** (Rahnejat, 2008). Johannes Bernoulli extended D'Alembert's principle to the net **virtual work done** for an n cluster of unconstrained material points, having $3n$ co-ordinates as: $\sum_{i=1}^{3n}(-F_{q^i} + m_{q^i}\ddot{q}^i)\delta q_i = 0$. If now there exists l constrained co-ordinates, then the reduced configuration space is: $r = 3n - l$. A new set of co-ordinates ζ is chosen, last l of which are constrained. Then, the **holonomic constraints** are:

$$q^i\big|_{i=1,3n} = f(\xi^j\big|_{j=1,r},\ t) \quad \text{and} \quad \xi^j\big|_{j=r+1,3n} = f(q^i\big|_{i=1,3n},\ t) = 0$$

$$(1.10)$$

Then, the infinitesimal change in a co-ordinate is: $\delta q_i = \sum_j \dfrac{\partial q^i}{\partial \xi^j}\,\delta \xi_j$. Now replacing in the above expression for the virtual work done:

$$\sum_i \sum_j (-F_{q^i} + m_{q^i}\ddot{q}^i)\frac{\partial q^i}{\partial \xi^j}\,\delta \xi_j = 0 \qquad (1.11)$$

This can be written in the form:

$$\sum_i \sum_j \left(m_{q^i}\frac{\partial q^i}{\partial \xi^j}\ddot{q}^i - F_{q^i}\frac{\partial q^i}{\partial \xi^j}\right)\delta \xi_j = 0 \qquad (1.12)$$

Since the co-ordinates ξ^j, velocities $\dot{\xi}^j$ and time are considered as independent variables, the first term can be written as:

$$\sum_j m_{q^i}\frac{\partial q^i}{\partial \xi^j}\ddot{q}^i = \frac{d}{dt}\left(\sum_j m_{q^i}\frac{\partial q^i}{\partial \xi^j}\frac{dq^i}{dt}\right) - \sum_j m_{q^i}\frac{dq^i}{dt} - \frac{d}{dt}\left(\frac{\partial q^i}{\partial \xi^j}\right) \qquad (1.13)$$

and the second term as:

$$\sum_i F_{q^i}\frac{\partial q^i}{\partial \xi^j} = -\sum_i \frac{\partial U}{\partial q^i}\frac{\partial q^i}{\partial \xi^j} = -\frac{\partial U}{\partial \xi^j} \qquad (1.14)$$

To simplify the above terms one needs the derivatives: dq^i/dt and $\partial\dot{q}^i/\partial\dot{\xi}^j$.

The latter is the same as $\partial q^i/\partial \xi^j$ and is substituted by it in the first term of (1.13). The former can be obtained from the constraints in (1.10). If the constraints are considered as time-dependent for an evolving system as would be the general case, then the absolute derivative (**covariant vector for space-time**) is:

$$\frac{dq^i}{dt} = \frac{\partial q^i}{dt} + \sum_i \frac{\partial q^i}{\partial \xi^j} \frac{d\xi^j}{dt} \tag{1.15}$$

These can all be substituted back into (1.13) and (1.14), completing equation (1.12). One additional clarification is for the second term in (1.13), where:

$$\frac{d}{dt}\left(\frac{\partial q^i}{\partial \xi^j}\right) = \frac{\partial^2 q^i}{\partial \xi^j \partial t} + \sum_k \frac{\partial^2 q^k}{\partial \xi^j \partial \xi^k} \frac{d\xi^k}{dt} = \frac{\partial}{\partial \xi^j}\left(\frac{dq^i}{dt}\right)$$

After all the indicated substitutions and some manipulation, equation (1.12) becomes:

$$\sum_i \left[\frac{d}{dt} \sum_j m_{qi} \frac{\partial q^i}{\partial \xi^j} \underbrace{\left(\frac{\partial q^i}{\partial t} + \sum_i \frac{\partial q^i}{\partial \xi^j} \frac{d\xi^j}{dt}\right)}_{\frac{dq^i}{dt}} \right.$$

$$\left. - \sum_j m_{qi} \underbrace{\left(\frac{\partial q^i}{\partial t} + \sum_i \frac{\partial q^i}{\partial \xi^j} \frac{d\xi^j}{dt}\right)}_{\frac{dq^i}{dt}}\left(\frac{\partial \dot{q}^i}{\partial \xi^j}\right) + \frac{\partial U}{\partial \xi^j} \right] \delta q_i = 0 \tag{1.16}$$

Since $\delta q_i \neq 0$, then the term in the curly bracket must vanish. As kinetic energy $K = \frac{1}{2}\sum_i = m_{q^i}(d\dot{q}^i/dt)$, then it is clear that the first term in the brackets is $\partial K/\partial \dot{\xi}^j$ and the second term is $\partial K/\partial \xi^j$. Thus:

$$\frac{d}{dt}\left(\frac{\partial K}{\partial \dot{\xi}^j}\right) - \frac{\partial K}{\partial \xi^j} + \frac{\partial U}{\partial \xi^j} = 0 \tag{1.17}$$

This is a more general form of the **Lagrange's equation** than (1.4). The time dependent holonomic constraints are taken into account, which means that for such a system the set of equations are for $j = 1 \rightarrow r$. A solution for the defined system of n material points is thus obtained for this set of differential equations with the holonomic constraints $\xi^j|_{j=r+1,3n} = f(q^i|_{i=1,3n},t) = 0$ (as described above). As can be seen, the holonomic constraints are relationships that forbid displacements along or about certain defined co-ordinates. The derivation here assumes these constraints to be time dependent. In most engineering applications holonomic constraints are time independent. This means that the first term on the left-hand side of the velocity transform in (1.15) does not exist. This leads to a ***contravariant velocity vector***.

Constraints may be ***non-holonomic***, such as velocity dependency of pairs of co-ordinate systems. Readers should refer to specialist texts on multi-body dynamics (e.g. Rahnejat, 1998). If a co-ordinate system is fixed onto the ground in a gravitational field and another falls with respect to it, then the relationship between them can be described by non-holonomic constraints as co-ordinate functions of the gravitational field.

The simple derivation of Lagrange's equation here suffices, both for mathematical proof of Newton's second axiom and for use in mechanical problems. Many other formulations of Lagrange's equation can be arrived at, which suit particular systems or aid certain solution methods. Readers should refer to Orlandea (1999).

Multi-body mechanical systems as an assembly of parts are viewed as constrained system dynamics problems. Such systems comprise a number of components that, in general, are referred to as *parts*. Parts are joined together by constraint functions which provide relationships between co-ordinates attached to certain points on these neighbouring parts (similar relationships to equation (1.10)). These points are known as *markers*. In mechanical systems there are an assortment of joints, such as hinge/revolute, ball-in-socket/spherical, hook/universal, cylindrical, translational, and more complex joints such as various constant velocity joints, clevis joints, to name but a few. An introduction to mechanical joints may be found in Rahnejat (1998), Hunt (1973) and Gilmartin (1978).

In general, each mechanical joint introduces a number of constraints (i.e. a series of relationships between co-ordinates attached to the aforementioned *markers*). These provide a number of algebraic equations which must be satisfied simultaneously with differential equations of motion of *parts* (application of Lagrange's equation for various degrees of freedom of each part) within a multi-body system. Various solution methods for such sets of differential-algebraic equations in space–time exist. Again readers should refer to Rahnejat (1998) or Orlandea (2008).

1.4   Multi-body mechanical systems

1.4.1   Equations of motion

Mechanical systems, including engine and powertrain systems are constrained multi-body systems. At first sight many practical mechanical multi-body systems appear to be quite complex when a dynamic model, comprising a differential-algebraic set of equations is to be made. However, this task is made quite simple by commercial software, many of which generate the set of equations for the user in an automatic manner. In order to achieve this, Lagrange's equation for constrained systems may be stated in the form:

$$\frac{d}{dt}\left(\frac{\partial K}{\partial \dot{q}^i}\right) + \frac{\partial U}{\partial q^i} + \sum_m \lambda_m \frac{\partial C_{kl}}{\partial q^i} = F_{aq^i} \tag{1.18}$$

where the constraint functions C_{kl}, $k = 1, n$ and $l = 1, n$ with $k \neq l$ are functions of co-ordinates of *markers* on *parts* $(1, n)$ in the multi-body system and λ_m are **Lagrange multipliers** for constraints applied to a *part*, which are unknowns to be determined.

To obtain the equation set, it is usual to define a fixed co-ordinate system with respect to which all translational and rotational motions of *parts* within a multi-body system are measured. This fixed frame of reference is usually referred to as the **global frame of reference**. It is global in the sense that all the motions of other frames of reference attached to individual components of a multi-body system undergo transformations with respect to it. Recall from previous discussion that a generic global frame of reference cannot be assumed anywhere, unless within an isolated suitably defined system of investigation. The co-ordinates fixed to each *part* of the system are referred to as a **local part frame of reference**. If the *parts* are considered to be rigid and such frames of reference are attached suitably to their centres of mass/inertia, then the kinetic energy remains a function of $\dot{q}^i$ only and the form of Lagrange's equation (1.18) holds true for all parts in the multi-body system.

Transformation between a triad of axes in a local part frame of reference (LPRF) and the global frame of reference (GRF) is required in order to determine the kinematic attributes of the part in an instantaneous manner (in other words the GRF is the frame of observation). The kinematic observation model is, therefore, q^i, $\dot{q}^i$ for all parts i, while the dynamics model includes the underlying causes (forces); inertial dP_i/dt (P_i being the momentum conjugate to the co-ordinate q^i), the generalised body/restoring/resistive force F_{q^i} the constraint reactions $\sum_m \lambda_m \frac{\partial C_{kl}}{\partial q^i}$ and any applied forces. For known inertial properties and constraint functions (type of mechanical joints), the vector of unknowns in a multi-body mechanical dynamic system becomes: $\{q^i, \dot{q}^i, \lambda_j\}^T_{\in i,j}$. A noteworthy point is that while $F_{q^i} = -\partial U/\partial q^i$ may be regarded as a restraint (a resistance), the term $\sum \lambda_i \frac{\partial C_{kl}}{\partial q^i}$ is a constraint (a rigid restraint) determining the reduced configuration space. Thus, in mechanical systems stored energy in a linear spring $\frac{1}{2}k\delta^2$ by virtue of its deflection δ or that in a solid elastic sphere $\frac{2}{5}k\delta^{5/2}$ (see Chapter 4) account for resistance in certain co-ordinates, while constraints C_{kl} define the limit of a dynamic system in relative motion of parts k and l. Constraints, therefore, remove the working space of mechanisms (a reduced configuration space).

Although constraints are often used in multi-body formulation, they represent rigidity that is idealistic. Recall that there is no rigid location in space–time, to which one can attach a frame of reference. Constraints, therefore, have the dual purpose of problem simplification (when warranted) or to achieve kinematic conditions, where observation of articulation of a mechanism is deemed as a prelude to a later, more detailed, dynamic analysis (see Rahnejat, 1998).

To capture the position and orientation of an LPRF with respect to GRF it is usual to use roll–pitch–yaw transformations (common in aircraft and ship dynamics) or Euler's body 3–1–3 frame of reference (successive rotations of the embedded LPRF, a generalised form extensively used since Euler).

Now in a multi-body system the GRF is fixed on the ground, LPRFs are attached to the centres of mass/inertia of the moving bodies (the material points). The position vector of the origin of LPRFs (q^i, $i = 1,6)_{k=1,n}$ with respect to the GRF (ξ^j, $j = 1,6$) are given as:

$$\{R_k\}_{k=1,n} = \{||\,\xi^j\,||_k, j = 1,3\}^{\mathrm{T}} \{\xi^1\,\xi^2\,\xi^3\}$$

LPRFs assume varying orientations with respect to the GRF as system dynamics evolves in time. When the equations of motion are written for each part with respect to the respective embedded LPRF, it is necessary to have the Euler transformations from GRF to these:

$$\{q^i{}_k, i = 1,3\}^{\mathrm{T}} = [T]_k \{\xi^j, j = 1,3\}^{\mathrm{T}} \tag{1.19}$$

where the **Euler transformation** matrix is:

$$[T] = \begin{bmatrix} C\psi C\varphi - S\psi C\theta S\varphi & -C\psi S\varphi - S\psi C\theta C\varphi & S\psi S\theta \\ S\psi C\psi + C\psi C\theta S\varphi & -S\psi S\varphi + C\psi C\theta C\varphi & -C\psi S\theta \\ S\theta S\varphi & S\theta C\varphi & C\theta \end{bmatrix} \tag{1.20}$$

where $C \equiv \cos$, $S \equiv \sin$. Also note that $\xi^4 = \psi$, $\xi^5 = \theta$, $\xi^6 = \varphi$ are the **Euler angles** in the Euler body 3–1–3 frame of reference.

The translational components of velocity of any *part* are given as:

$$\{v_k\}_{k=1,n} = \left\{\frac{\partial R_k}{\partial t}\right\}_{k=1,n} = \left\{\left\|\frac{\partial \xi^j}{\partial t}\right\|_k, j = 1,3\right\}^{\mathrm{T}} \{\xi^1\xi^2\xi^3\}$$

In the Euler's frame of reference rotations of an LPRF relative to the GRF are given as: $\{\dot{\xi}^j, j = 4, 6\}^{\mathrm{T}} = \{\dot{\psi}, \dot{\theta}, \dot{\varphi}\}^{\mathrm{T}}$. Rotational kinetic energy is obtained in terms of derivatives of LPRF co-ordinates: $\dot{q}^i$, $i = 4, 6$. These are transformed to the global frame of reference as: $\{\dot{q}^i, i = 4, 6\}^{\mathrm{T}} = [T^*] \{\dot{\psi}, \dot{\theta}, \dot{\varphi}\}^{\mathrm{T}}$, where (Rahnejat, 1998):

$$[T^*] = \begin{bmatrix} S\theta S\phi & 0 & C\varphi \\ S\theta & 0 & -S\varphi \\ C\theta & 1 & 0 \end{bmatrix}$$

with co-directed ξ^3 and q^3 axes. Now simply replace q^i by ξ^j in equation (1.18) to find six equations of motion for each *part* (in Fig. 1.1) in the reduced configuration space defined by any algebraic constraint functions that join the *parts* of the multi-body system in terms of the global co-ordinates; GRF.

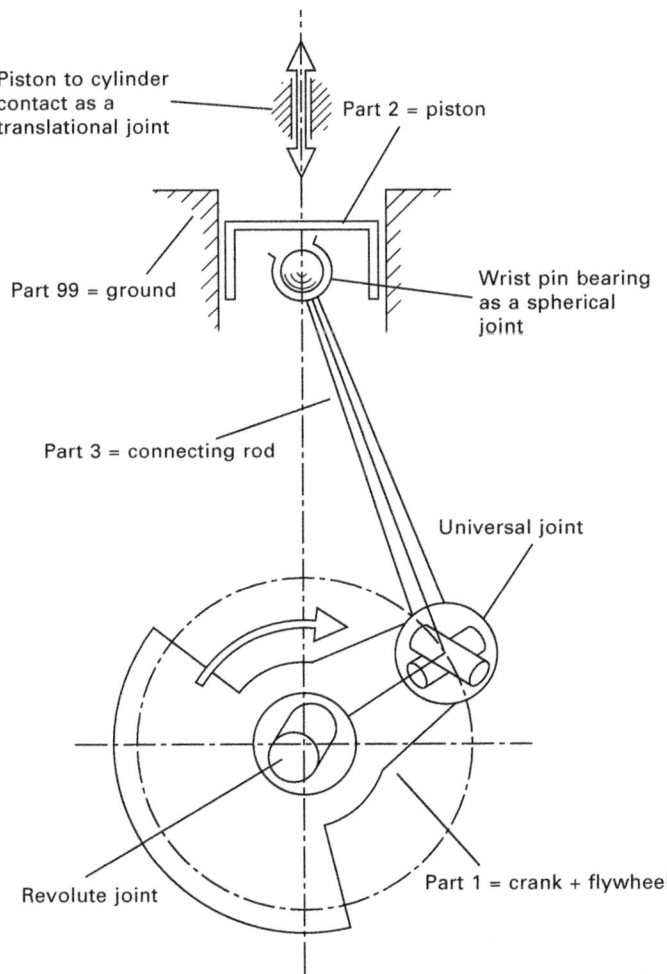

Piston to cylinder contact as a translational joint

Part 2 = piston

Part 99 = ground

Wrist pin bearing as a spherical joint

Part 3 = connecting rod

Universal joint

Revolute joint

Part 1 = crank + flywheel

1.1 A kinematic model for a single cylinder engine.

1.4.2   Constraint functions

The equations of motion for any constrained mechanical multi-body system require definition of the **constraint functions** C_{kl} in terms of the co-ordinates ξ^j. Mechanical joints, such as those mentioned above, consist of a number of basic algebraic functions, usually referred to as **primitive constraints**. Such constraints are usually imposed between two *parts* in a system at a specific geometric location, which is defined by points (*markers*) on a pair of assembled *parts*. Rahnejat (1998) provides detailed treatment of mechanical joint constraint formulation. Here a brief introduction to the subject is made.

The most common primitive constraint function is the **at-point** or **point coincident constraint**. If two *markers* on two *parts* k and l are compelled to remain coincident at all times irrespective of the orientation of their attached triad of axes, then the constraint function can be stated as:

$$C_{kl} = (R_k + r_k) - (R_l + r_l) = 0 \qquad (1.21)$$

where R is the global position vector of the centre of mass of the part upon which a marker is defined and r is the local position vector of the marker from its centre of mass. Clearly, the global position vectors are a function of the co-ordinates ξ^j, whereas the local position vectors are functions of q_k^i and need to be transformed to the former, using equation (1.19). Thus:

$$C_{kl} = \{\xi_k^j - \xi_l^j\}^T + \{[T]_k \xi_k^j - [T]_l \xi_l^j\}^T = 0 \quad \text{for} \quad j = 1,3 \qquad (1.22)$$

This yields three constraint functions as indicated. Therefore, three degrees of freedom are removed. A **spherical joint** or a **ball-in-socket joint** is simply described by an at-point constraint.

Some constraint functions are related to orientation of *parts* with respect to each other, such as co-directing the axes of two parts to form a hinge. If the nominated co-directed axes for a **hinge primitive constraint** on markers k and l of two *parts* are q_k^3 and q_l^3, then:

$$q_k^3 \bullet q_l^1 = 0 \quad \text{and} \quad q_k^3 \bullet q_l^2 = 0 \qquad (1.23)$$

where the dot product renders scalar constraint functions. Clearly, the co-ordinates q_k^i and q_l^i should be transformed to the global co-ordinates as before, thus an expression of the form $C_{kl} = [T]_k [T]_l \xi_k^j \xi_l^j = 0$ is obtained for both cases in (1.23), removing two degrees of freedom. The addition of the at-point and hinge primitive constraints at a location for markers k and l results in a **revolute joint**. Therefore, a revolute joint introduces five constraints.

In a mechanical multi-body system *parts* are joined together by various joints or couplers (these relate motion of two *parts* such as rack and pinion).

Once the appropriate constraint functions are determined, the relevant reactions $\sum\limits_{m} \lambda_m \dfrac{\partial C_{kl}}{\partial \xi^j}$ are obtained. Now a set of equations of motion is found. These equations need to be solved simultaneously with the constraint functions themselves as previously described in Section 1.2. Thus, the differential-algebraic equation set is:

$$\frac{\mathrm{d}}{\mathrm{d}t}\left(\frac{\partial K}{\partial \dot{\xi}^j}\right) + \frac{\partial U}{\partial \xi^j} + \sum_m \lambda_m \frac{\partial C_{kl}}{\partial \xi^j} = F_{a\xi^j} \quad \text{and} \quad C_{kl} = f(\xi^j) = 0 \qquad (1.24)$$

Rahnejat (1998) and Orlandea (1999, 2008) describe in detail the various methods of solution. However, a number of important points should be made since multi-body dynamic analysis codes are now readily available and increasingly used in industry and academia.

Firstly, in a dynamic analysis only suitably small changes in co-ordinates are permitted within a time marching integration method with small time steps δt. The solution is usually made by making substitutions: $\vartheta_k^j = \dot{\xi}_k^j$. Hence, the vector of unknowns to be determined is $\{\delta\vartheta_k^j, \delta\dot{\vartheta}_k^j, \delta\lambda_m\}^{\mathrm{T}}$. Often, users of multi-body codes choose an inappropriate time step size for their model's simulation. As a rough guideline this is related to the rate of change of system variables (the required solution vector). Small time steps of a few tenths of a second suffice for large displacement rigid body dynamics, but not when impulsive loading or impact forces are involved. Impact durations are usually of the order of a few tenths to several milliseconds, requiring very small time steps. All contact dynamic problems usually require time steps of the order of microseconds. Secondly, time marching simulation studies are usually iterative, requiring convergence criteria set to be satisfied. Various chapters in this book describe choice of criteria and size of error tolerated for a sensible solution vector to be found. For detailed discussion of this point readers are referred to Rahnejat (1998). Finally, one should note that in practice *rigid* constraints do not exist and all real joints are subject to deformation under load. Only if the applied loads are insufficient, a rigid constraint may be assumed. Joints such as ball-in-socket are also subject to friction, which is ignored when they are considered as rigid idealised constraints. The same is true of some couplers, such as assumed gearing constraint, where a common velocity marker on a meshing pair is assumed. Meshing pairs are subject to contact loads as a function of their lubricated separation and friction due to viscous shear of a lubricant film and potential asperity interactions (see, for example, the appendix to this chapter and Chapters 15, 20 and 29). Furthermore, a common velocity *marker* may only be assumed at the pitch point of a teeth pair contact, whilst elsewhere during meshing slide-roll motion occurs.

Nevertheless, for particular types of analysis one may choose a combination of joints to either eliminate certain degrees of freedom or, for example, render a kinematic analysis (no degrees of freedom; mechanism follows a prescribed motion; note that a motion function introduces a constraint). Two other problems can occur with use of constraints in order to create a model, representative of a multi-body system. One is the use of what physically represents a practical joint in a closest manner. This can render an over-constrained multi-body model. Note that rigid constraints do not exist in practical mechanisms, but are mathematically convenient. The other problem is to introduce in a model constraint functions which replicate one another (***repeat constraints***). If such constraint functions are parts of a defined joint (as in commercial software) repeat constraints lead to redundant equations which are then automatically removed. The ensuing analysis would now be different from that intended or expected.

The method for representation of multi-body dynamics presented here is based on ***constrained Lagrangian dynamics***, where a system is defined in a reduced configuration space with respect to a global frame of reference. An alternative and older method was developed by Euler, now referred to as the ***Newton–Euler method***. This is based on direct application of Newton's second axiom, as in equation (1.1). It is, therefore, clear that all the forces acting on an element (*part*) of a system in $\sum_i F_i$ must be known *a priori*. The ingenuity of Euler was to propose the use of free-body diagrams for constituent parts of a system, where reaction forces as in $F_{q^i} = -\partial U / \partial q^i$ can be specified to ensure the continuity of the system as a whole (Fig. 1.2). A series of equations (as in (1.1)) results for all parts of the system

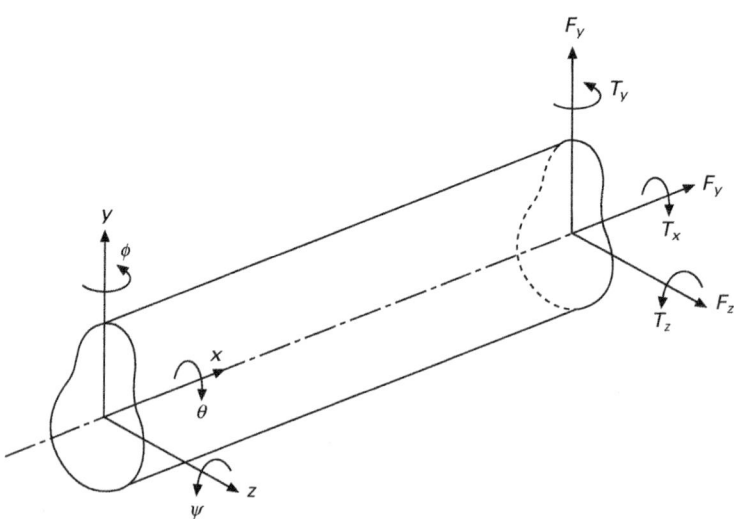

1.2 An Eulerian beam.

in the Euler frame of reference (previously described). It is clear that only the unconstrained degrees of freedom need to be modelled in this case, thus a smaller set of equations than those in constrained Lagrangian dynamics would result. This is the advantage of the Newton–Euler method. Its main disadvantage is that the omission of any reaction/restoring force can lead to an incorrect outcome. Furthermore, it is often difficult to visualise the degrees of freedom of a complex mechanism, prior knowledge of which is not required in the Lagrangian approach. Therefore, the Newton–Euler method demands from users a greater understanding of dynamics. All other methods developed since Lagrange and Euler are variations of the same with different interpretations or for manipulation of larger degrees of freedom systems, as a sparse matrix tableaux (Orlandea, 1999, 2008).

Although the use of constrained Lagrangian dynamics in multi-body software packages has made dynamics more accessible to the novice and uninitiated, certain considerations are essential and care is required in the development of models. Firstly, the approach becomes more useful with larger systems, and particularly inefficient with simpler systems (where Newton–Euler approach should be preferred). Secondly, it is necessary to determine the degrees of freedom of the eventually constructed model to ensure: (i) the model is not over-constrained, (ii) no repeat/redundant constraints exist and (iii) the remaining degrees of freedom can be identified and verified. The *Gruebler–Kutzbach expression* can be used to ascertain the degrees of freedom of a system model:

$$n\text{DOF} = 6(n-1) - \sum_{i}^{m} C \qquad (1.25)$$

where n is the number of *parts* in a multi-body system including the ground (the assumed extraneous rigid environment), m is the number of constraint functions and C is the number of constraints introduced by each constraining function. For some standard mechanical joints, the number of constraints introduced are: spherical, 3; universal/hook, 4; revolute, 5; translational, 5; and cylindrical, 4. Each specified motion introduces a constraint, so do the couplers. Thus, systems which yield zero degrees of freedom with one or a number of specified motions are *kinematic*. In such cases no solution to equations of motion is required (the underlying cause; forces, are of no concern). Thus, the solution is obtained by simultaneous solution of constraint functions, $C_{kl} = f(\xi^j) = 0$, which includes the specified motion(s).

For *static analysis*, let $\dot{\vartheta}_k^j = 0$ in equations (1.24). The solution vector $\{\vartheta_k^j, 0, \lambda_j\}^\text{T}$ corresponds to the equilibrium position, where $\sum_i F_j = 0$. If the solution vector has a unique outcome, then static condition is obtained. A multitude of such solutions corresponds to a *quasi-static analysis*.

1.5   Engine as a multi-body system

Perhaps the simplest multi-body model of an engine would be that of a single cylinder kinematic model. Such a model would be useful as a visualisation tool or simply to identify constraint functions required for a subsequent dynamics analysis. Figure 1.1 shows such a model. The *parts* in the engine are considered to be: flywheel, crankshaft, piston and connecting rod. The ground (the boundary of the system) is considered to be the cylinder block and beyond. The parts are numbered as $k = 1, 2, 3$ and 99 respectively as shown in the figure. Therefore, in the Gruebler–Kutzbach expression $n = 5$. The mechanical joints are so chosen in order to avoid repeat or redundant constraint functions. For example, the cylindrical joint, representing the big end bearing between the connecting rod and the crank-pin allows translation of crankshaft relative to the connecting rod. However, this is constrained by the revolute joint between the crankshaft and the ground. The universal joint at the position of the small end bearing allows the articulation of the connecting rod relative to the piston. However, it also enables the piston to tilt back and forth (in and out of the plane of paper), which is constrained by the cylindrical joint between the piston and the ground. Thus, there are no repeat/redundant constraints. In practice, the motion of the crankshaft is specified by the combustion gas force as $C_{kl} = C_{199} = 2 \pi Nt$, N being the rotational speed in rev/sec. Using the Gruebler–Kutzbach expression:

$$nDOF = 6(5-1) - \Sigma \ (1X \underset{\text{Fxd}}{6} + 1X \underset{\text{Rev}}{5} + 2X \underset{\text{Cyl}}{4} + 1X \underset{\text{Uni}}{4} + 1X \underset{\text{Mot}}{1} \ )$$

$$= 24 - 24 = 0$$

thus a kinematic model results, which follows the specified motion.

Kinematic models, using other combination of constraints may also be found. For example, an alternative choice is a translational joint between the piston and the ground, a revolute joint between the piston and the connecting rod, representing the wrist-pin bearing, an in-line joint between the connecting rod and crankshaft. The flywheel is considered fixed to the crankshaft, and the crankshaft has a revolute joint to the ground with the same specified motion as before. Thus:

$$nDOF = 6(4-1) - \Sigma \ (2X \underset{\text{Rev}}{5} + 1X \underset{\text{Tra}}{5} + 1X \underset{\text{Inl}}{2} + 1X \underset{\text{Mot}}{1} \ ) = 18 - 18 = 0$$

Note that the *in-line primitive constraint* introduces two constraint functions as:

$$C_{kl} = [(R_k + r_k) - (R_l + r_l)] \bullet q^l_{l|l=1,2} = \{(\xi^j_k - \xi^j_l)^T$$

$$+ \{[T]_k \xi^j_k - [T]_l \xi^j_l\}^T\} \bullet \{[T]_l \xi^j_{l|l=1,2}\}^T = 0$$

If k represents a marker on the crank/flywheel assembly and l the coincident marker on the connecting rod, then k can translate with respect to l with all its rotational freedoms intact. However, the translational motion is constrained by the revolute joint to the ground.

The motion constraint is in fact governed by the combustion process, with an initial condition, usually determined by the starter motor characteristics. Therefore, one can achieve a very basic dynamics model simply by removing the specified motion constraint and apply the gas force instead to the piston. A one degree of freedom system results, which couples the translational motion of the piston to rotation of the flywheel–crank assembly. Mass and inertial properties of the parts in the system should be specified. The choice of constraint functions in the assembly of parts can now be quite important, depending on the intended analysis.

For basic tribological studies, suitable constraint functions must be chosen to allow motions that are constrained in the previous examples. For instance, a piston has secondary motions as described in Chapters 8 and 10–15, involving piston lateral motion within the confine of its clearance with the cylinder liner or bore, as well as tilting motion about the axis of the wrist-pin bearing. Therefore, it is clear that choice of a translation joint between the piston and the ground (the engine block) prohibits these motions. The same is true of an in-line primitive constraint at the position of the big end bearing, constraining the lateral motions of crank-pin journal centre with respect to the bushing/sleeve fitted onto the connecting rod. The in-line primitive constraint can be replaced by a ***planar constraint***, which can best be described as a hockey puck sliding on ice. If the puck is regarded as *part k* and the ice surface as *part l*, then this joint primitive allows rotation with respect to their common orthogonal axis with no translation, while rotation about other axes are also constrained, so that contact is maintained at all times. These lateral axes of part k can translate with those of part l, allowing sliding motion of the puck. Thus, three constraint function are introduced. If k represents the crankshaft and l the connecting rod, then: $C_{kl} = [(R_k + r_k) - (R_l + r_l)] \bullet q_l^3 = 0$ and $q_k^3 \bullet q_l^i|_{i \in 1,2} = 0$, which introduces three constraints, which can be transformed in terms of the co-ordinates ξ_l^j, ξ_k^j. To avoid repeat constraints, the revolute joint between crank/flywheel assembly is replaced by a cylindrical joint as the planar joint already inhibits motion along the crank axis. The revolute joint between the connecting rod and the piston is also replaced by a spherical joint as the rotations of the connecting rod other than that about the wrist-pin axis are already constrained by the planar constraint functions. The rotation of the crankshaft is determined by the combustion curve (gas force). Hence, the motion constraint remains, this time with an applied gas force on the piston. A two degrees of freedom dynamic model results which correspond to the lateral motions of the crank/flywheel assembly relative to the connecting rod (the motion of the crank-pin journal relative to the bushing/sleeve). To restrict these (restraints), journal bearing forces should be used, similar to those described in Chapters 18–20 as functions of the eccentricity ratio. Hence:

$$n\text{DOF} = 6(4-1) - (1X \underset{\text{Cyl}}{4} + 1X \underset{\text{Tra}}{5} + 1X \underset{\text{Sph}}{3} + 1X \underset{\text{Pla}}{3} + 1X \underset{\text{Mot}}{1})$$

$$= 18 - 16 = 2$$

Therefore, multi-body models with suitable constraints can be developed to enable description of load-bearing conjunctions, where important tribological contributions to system dynamics can be included in the analysis.

Most multi-body dynamic analyses, however, were initially performed for low frequency phenomena such as suspension analysis, ride comfort and vehicle handling responses. Rahnejat (1998) provides some basic examples. Larger detailed vehicle models are commonly used in industry as a part of vehicle development programmes. Representative literature include a series of papers by Blundell (1999), Hegazy *et al.* (2000) and Hussain *et al.* (2007), which include tyre forces, aerodynamic forces and complex manoeuvres (see also Chapter 23, using the Newton–Euler approach).

Inclusion of component flexibility became possible later with integration of finite element techniques and multi-body methods through mode reduction and selection techniques such as component mode synthesis. This enabled representative analysis of systems as they are subject to deformation loads. A good example is the inclusion of anti-roll bars in vehicle models, where its structural compliance resists vehicle roll during cornering manoeuvres. Another example is the inclusion of structural resistive suspension elements such as leading or trailing arms, which restrict vehicle dive in braking or squat under sudden acceleration (see Azman *et al.*, 2007). These phenomena are still low to medium frequency events, dominated by large displacement dynamics (that of the sprung or unsprung masses).

In recent years multi-body dynamics approach has been used in conjunction with tribological studies (for example see Boysal and Rahnejat, 1997). Some detailed models, including component flexibility are presented by Kushwaha *et al.* (2002) and Perera *et al.* (2007) for engine and drivetrain dynamics with experimental validation. It is important to briefly describe the increasing need for inclusion of component flexibility in multi-body dynamics models, as well as different methods which can be employed to achieve this.

1.6   Elasto-multi-body dynamics analysis

Component flexibility plays an important role in dynamics of real systems. Its role has become more pronounced in recent years with increasing use of lighter components in an effort to reduce mechanical losses (mainly due to out-of-balances) and thus enhance fuel efficiency. However, lighter components, often made of materials of lower elastic modulus or with increasing use of hollow components are subject to elastic deformations and vibration. In the case of engines and powertrains a plethora of noise and vibration concerns have emerged in the past two decades due to component flexibility as well as increased output power (for example higher torques in diesel engines). Some of these noise and vibration problems are discussed in this book, such as transmission rattle (see Chapters 21, 26–29 and also the appendix in this chapter) and driveline clonk (see Chapters 21 and 30). There are many other such noise, vibration and harshness (NVH) concerns, such as vehicle body boom (see Chapter 21), engine roughness (Rahnejat, 1998) and clutch in-cycle vibration or whoop (Kushwaha *et al.*, 2002).

Rahnejat (1998) describes some of the basic modelling approaches to include component flexibility. These include the transfer matrix method (TMM) and dynamic stiffness matrix method. Other advanced methods are described by Shabana (2005) and Nikravesh (2008). Here a number of basic approaches are highlighted.

A simple approach to represent flexibility is by **Eulerian beams**. Compliance functions relate six restrained (elastic) degrees of freedom motion of one end of the beam (e.g. $q^i \in x, y, z, \theta, \phi, \psi$) with respect to the forces applied at the other (e.g. $F_{aq^i} \in F_x, F_y, F_z, T_x, T_y, T_z$) (see Fig. 1.2). The mass of the flexible part is discretised into two masses concentrated at the either ends of the beam. Therefore, these masses act as *parts* and would require inertial properties. In multi-body terminology they are referred to as dummy parts, because often they are a specific component of a mechanical system. For example, crankshaft flexibility may be represented by crank-pins as Eulerian beams and crank-webs as the point masses, generally referred to as concentrated inertial elements. The beams act as restraining elements which introduce forces between the dummy parts or in the case of the crankshaft between the successive crank-webs as:

$$\{F_{aq^i}\}^{\mathrm{T}} = \begin{bmatrix} \dfrac{EA}{L} & 0 & 0 & 0 & 0 & 0 \\[2mm] 0 & \dfrac{12EI_{q^3}}{L^3} & 0 & 0 & 0 & \dfrac{-6EI_{q^3}}{L^2} \\[2mm] 0 & 0 & \dfrac{12EI_{q^2}}{L^3} & 0 & \dfrac{6EI_{q^2}}{L^2} & 0 \\[2mm] 0 & 0 & 0 & \dfrac{GI_{q^1}}{L} & 0 & 0 \\[2mm] 0 & 0 & \dfrac{6EI_{q^2}}{L^2} & 0 & \dfrac{4EI_{q^2}}{L} & 0 \\[2mm] 0 & \dfrac{-6EI_{q^3}}{L^2} & 0 & 0 & 0 & \dfrac{4EI_{q^3}}{L} \end{bmatrix} \{\delta q^i\}^{\mathrm{T}} - [D]\{\delta \dot{q}^i\}^{\mathrm{T}}$$

$$(1.26)$$

where, these restraining/resistive forces are treated as applied forces. Note the longitudinal axis of the beam is designated as q^i in the stiffness matrix above. $[D]$ is the damping matrix, elements of which are usually quite difficult to specify. A simple approach is to specify its elements as a percentage of the stiffness matrix (usually around 1% for lightly damped powertrain components). The restraining forces are then transformed to the co-ordinates ξ^j, using equation (1.19) for $\{\xi q^i\}$ for $i \in 1,2,3$ and for $i \in 4,5,6$ the Euler transformation matrix $[T^*]$ is used, as described in Section 1.4.1. This is the basis of **transfer matrix method** (TMM) (see Rahnejat, 1998).

It is clear that if a crank-pin is considered as a single beam element, then many of its torsional-bending modes would be ignored. Thus, flexible parts may have to be discretised into more than a single beam interspersed with dummy parts between successive beams. The more flexible the structure or larger the deformation energy, a greater number of discretisations would be required (more mode shapes). This approach leads to ever smaller beams or in fact any regions, where certain stress–strain (load–deflection) relationships may be assumed. These regions are usually referred to as finite elements. Therefore, finite element analysis provides the right approach in determining the modal behaviour of flexible structures.

Unlike the bodies which are considered to be rigid, the elastic bodies undergo deformation while in motion. Therefore, a flexible body may itself be regarded as a constrained configuration space described by a set of co-ordinates, referred to as **elastic co-ordinates**. Its inertial behaviour remains the same as before; described by the global position vector of a reference point on it (e.g. centre of mass) with respect to the global frame of reference as already described. Therefore, the elastic deformation of the body is described by the elastic co-ordinates of its many points with respect to the local part frame of reference at the nominated reference point and through transformation to the global frame of reference.

$$R_k^p = R_k^0 + u_k^p = R_k^0 + f(r_k^B + \delta_k^p) \tag{1.27}$$

where the point B is an *initial* position of a point P prior to its deformation δ_k^p in a small time step δt. Clearly, the function $f(r_k^B + \delta_k^p)$ is a shape function, referred to as a mode shape or eigen-vector which should be determined. One general way of including this function in (1.27) is through finite element analysis.

As already described in finite element analysis (FEA) a continuous flexible body can be represented by a collection of interconnected elemental regions with specified constitutive (stress–strain) relations. The stiffness and mass matrices for the these elements can be obtained using assumed shape functions. This involves evaluation of mass and stiffness matrices to represent the elemental behaviour for given assumed shape functions as:

$$[m_\xi] = [T]^T [m_q][T] \quad \text{and} \quad [k_\xi] = [T]^T [k_q][T] \tag{1.28}$$

In the same manner the corresponding resistive forces associated with points P (nodes in the finite elements), F_{aq^i} can be transformed to the global co-ordinates: $F_{\xi j} = [T]^T F_{aq^i}$.

For Eulerian beams the localised stiffness matrix is given in (1.26) and the mass matrix in (1.28), $[m_\xi]$ is termed the *consistent mass matrix* as it is obtained for the same shape functions assumed for the calculation of the stiffness matrix. Now equations of motion for a flexible part can be written, using the same approach as before. Furthermore, there would be no need to concentrate the mass of a flexible member in certain locations.

However, with many finite elements a very large degree of freedom can result, corresponding to many mode shapes ϕ_i, all of which would contribute to δ_k^p as a linear combination in the local frame of reference q^k:

$$\delta_k^p = \sum_{i=1}^n \phi_i q^i = \Phi_k q^k \tag{1.29}$$

where n is the number of mode shapes and Φ_k is the modal matrix for the body k. Equation (1.29) effectively transforms a larger set of physical coordinates δ_k^p to a smaller set of modal co-ordinates q^k. Depending on component geometry and physical state and nature of loading, the modal matrix can be very large indeed. Hollow thin-walled tubes, for example (e.g. driveshaft tubes), have many coupled torsional-bending modes (see Chapter 30). These are usually represented by combination of circumferential and axial waves on the tubes, such as those shown in Chapter 30. Such structures are said to have a high modal density. On the other hand, short and stubby solid structures made of materials of high elastic modulus can potentially only undergo rigid body motions, even when subjected to moderately high loads, such as machine tool spindles. Depending on loading and purpose of investigation, a **reduced order model** may suffice, by considering certain modal responses to be constrained. A reduced order model can be achieved by solving only for $m < n$ mode shapes, because of the *modal superposition* approach in (1.29). Therefore, mode shapes should be selected in a manner that results in a good approximation for a predetermined frequency band of interest.

The basic technique used is **component mode synthesis** (Craig, 1995), where the physical co-ordinates δ_k^p for a component k is divided into boundary components which are regarded as not being subject to modal superposition (*elastically* constrained, dependent DOFs), δ_{dk}^p and those which are subject to deformation (independent DOFs), δ_{ik}^p. Now if $\delta_{dk}^p = 0$ because of the imposed constraints, then:

$$[m_{ik}]\{\ddot{\delta}_{ik}^{P}\} + [k_{ik}]\{\dot{\delta}_{ik}^{P}\} = \{f_{ik}\} \tag{1.30}$$

Combining this equation with (1.29) enables substitution for physical co-ordinates δ_{ik}^{p} with the reduced number of modal co-ordinates ϕ, using FEA. The detailed procedure is described in Gnanakumarr et al. (2005). With component flexibility included in a multi-body model other important features to be incorporated are usually contact/impact and frictional forces. This means that realistic models for all forms of machines and mechanisms can be constructed. Such models incorporate various physical characteristics; inertial, elasticity, tribological, etc.; thus may be regarded as multi-physics. They also comprise interactions at the various scales; large displacements, vibration and micro-scale tribology; thus multi-scale. The most appropriate way to demonstrate such an approach is through some examples. In the appendix to this chapter, one of Loughborough's researchers describes analysis of some of the most pertinent current concerns in the automobile industry.

1.7   References and further reading

Azman, M., King, P.D. and Rahnejat, H., 'Combined bounce, pitch, and roll dynamics of vehicles negotiating single speed bump events', *Proc. Instn. Mech. Engrs., Part K: J. Multi-body Dyn.*, **221(1)**, 2007, pp 33–40.

Blundell, M.V., 'The modelling and simulation of vehicle handling, Part 1: analysis methods', *Proc. Instn., Mech. Engrs., Part K: J. Multi-body Dyn.*, **213(2)**, 1999, pp. 103–118.

Boysal, A. and Rahnejat, H., 'Torsional vibration analysis of a multi-body single cylinder internal combustion engine model', *App. Math. Modelling*, **21**, 1997, pp. 481–493.

Chandrasekhar, S., *Newton's Principia*, Oxford University Press, Oxford, New York, 1995.

Craig, R.R. Jr, *Structural Dynamics: An Introduction to Computer Methods*, Soc. Exp. Mechs., Bethel, CT, 1995.

De la Cruz, M., Theodossiades, S., Rahnejat, H. and Kelly, P., 'Impact dynamic behaviour of meshing loaded teeth in transmission drive rattle', *6th Euromech, Non-linear dynamics Conference*, S Petersburg, Russia, 2008.

De la Cruz, M., Theodossiades, S., Rahnejat, H. and Kelly, P., 'The effect of thermohydrodynamics on manual automotive transmissions gear rattle', *Proc. ASME, IDETC/CIE 2009*, Pap. No. DETC2009-87226, San Diego, CA, 2009a.

De la Cruz, M., Theodossiades, S., Rahnejat, H. and Kelly, P., 'Numerical and experimental analysis of manual automotive transmissions – gear rattle', *SAE Int.*, Pap. No. 2009-01-0328, 2009b.

Gilmartin, G.M., 'Constant velocity joints', *Engng.*, Feb. 1978, I-VIII.

Gnanakumarr, G., Theodossiades, S., Rahnejat, H. and Menday, M., 'Impact-induced vibration in vehicular driveline systems: theoretical and experimental investigations', *Proc. Instn. Mech. Engrs., Part K: J. Multi-body Dyn.*, 2005, **219**, pp. 1–12.

Gohar, R., *Elastohydrodynamics*, Imperial College Press, London, 2001.

Gohar, R. and Rahnejat, H., *Fundamentals of Tribology*, Imperial College Press, London, 2008.

Greenwood, T.A. and Tripp, J.H., 'The elastic contact of rough surfaces', *Trans. ASME, J. Lubn. Tech.*, 1967, p. 417.

Grubin, A.N., 'Contact stresses in toothed gears and worm gears', *Book 30* CSRI for Tech. & Mech. Engng., Moscow, 1949, DSRI Trans., **37**.

Hegazy, S., Rahnejat, H. and Hussain, K., 'Multi-body dynamics in full-vehicle handling analysis under transient manoeuvre', *Vehicle System Dynamics*, **34**, 2000, pp 1–24.

Houpert, L., 'New results of traction force calculation in elastohydrodynamic contacts', *Trans. ASME, J. Trib.*, **107**, 1985, pp. 241–248.

Hunt, K.H., 'Constant velocity shaft couplings: A general theory', *Trans. ASME, J. Engng. Indust.*, 1973, pp. 455–464.

Hussain, K., Rahnejat, H. and Hegazy, S., 'Transient vehicle handling analysis with aerodynamic interactions', *Proc. Instn. Mech. Engrs., Part K: J. Multi-body Dyn.*, **221(1)**, 2007, pp. 21–32.

Kushwaha, M., Gupta, S., Kelly, P. and Rahnejat, H., 'Elasto-multi-body dynamics of a multi-cylinder internal combustion engine', *Proc. Instn. Mech. Engrs., Part K: Multi-body Dyn.*, **216**, 2002, pp 281–293.

Nikravesh, P.E., 'Newtonian-based methodologies in multi-body dynamics', *Proc. Instn., Mech. Engrs., Part K: J. Multi-body Dyn.*, **222(4)**, 2008, pp. 277–288.

Orlandea, N.V., 'A study of the effect of lower index methods on ADAMS sparse tableau formulation for computational dynamics of multi-body mechanical systems', *Proc. Instn., Mech. Engrs., Part K: J. Multi-body Dyn.*, **213(1)**, 1999, pp. 1–9.

Orlandea, N.V., 'From Newtonian dynamics to sparse tableaux formulation and multi-body dynamics', *Proc. Instn., Mech. Engrs., Part K: J. Multi-body Dyn.*, **222(4)**, 2008, pp. 301–314.

Perera, M.S.M., Theodossiades, S. and Rahnejat, H., 'A multi-physics multi-scale approach in engine design analysis', *Proc. Instn. Mech. Engrs., Part K: J. Multi-body Dyn.*, **221(3)**, 2007, pp. 335–348.

Rahnejat, H., *Multi-body Dynamics: Vehicles, Machines and Mechanisms*, Professional Engng. Publ. & SAE (Joint Publ.), Bury St Edmunds (UK) and Warrendale, PA, 1998.

Rahnejat, H., 'Physics of causality and continuum: questioning Nature', *Proc. Instn., Mech. Engrs., Part K: J. Multi-body Dyn.*, **222(4)**, 2008, pp. 255–264.

Shabana, A.A., *Dynamics of Multi-body Systems*, Cambridge University Press, Cambridge, 2005.

Tangasawi, O.A.M., Theodossiades, S. and Rahnejat, H., 'Gear teeth impacts in hydrodynamic conjunctions promoting idle gear rattle', *J. Sound and Vibration*, **303(3–5)**, 2007, pp 632–658.

1.8   Nomenclature

a	Acceleration
C_{ij}	Constraint function
E	Young's modulus of elasticity
F	Force
F_{a^j}	Applied forces
F_{q^j}	Generalised body forces
g	Gravitational acceleration

G	Universal gravitational constant
h	Conjunctional film thickness
I_{q^j}	Mass moments of inertia
ℓ	Characteristic 'size' of a system
L	Length of Eulerian beam
k	Stiffness
K	Kinetic energy
m	Mass of a material point
M	Mass of a source
p	Pressure
P	Momentum
q^j	Generalised co-ordinates (usually Eulerian 3-1-3 body-centred rotations)
r	Distance from a source or local position vector
R	Global position vector
u, v	Conjunctional velocities of flow
U	Potential energy
t	Time
$[T], [T^*]$	Transformation matrices between sets of co-ordinates
x, y, z	Cartesian co-ordinates
δ	Deflection/deformation
δq	Small change in co-ordinate q
ε	'Size' of a generic material point
ϕ	Potential
η	Dynamic viscosity
λ	Lagrange multiplier
ρ	Density
ξ^j	Global co-ordinates in reduced configuration space
ψ, θ, φ	Euler angles

Subscripts:

a^i	Applied in the co-ordinate i
q^j	Refers to a generalised co-ordinate

Superscripts:

i, j, k, l	Refer to sets of co-ordinates
T	Transposed
.	First time derivative
..	Second time derivative

Other symbols:

Σ	Summation
•	Vector dot product
$\in$	Inclusive of

1.9   Appendix: multi-physics analysis for investigation of manual transmission gear rattle – drive/creep rattle*

M. DE LA CRUZ, Loughborough University, Loughborough, UK

Nomenclature

b	Minor semi-half-width of contact
C	Backlash
C_p	Specific heat capacity
c	Clearance between loose wheel and retaining shaft
E^*	Effective elastic modulus (see Chapter 4)
F_b	Boundary friction
F_v	Viscous friction
$F_{fwi,j}$	Flank friction per teeth pair
F_{pi}	Petrof friction of loose wheels
h	Film thickness
I_2	Inertia of second gear
I_i	Inertia of loose wheels
I_{os}	Inertia of output shaft
I_7	Inertia of reverse pinion
k_t	Thermal conductivity of transmission fluid
l	Contact length
I_{brg}	Gear blank width
m	Equivalent mass
N	Relative rotation of loose wheel and retaining shaft in rev/s
p	Pressure
Q_s	Side leakage flow
R_{brgi}	Radius of supporting shaft of loose wheel
R_{bpi}	Base radius of pinion
R_{bwi}	Base radius of wheel
$r_{xi,j}$	Equivalent radius of a teeth pair in x–z plane (see Chapter 4)
T_{D2}	Resistive torque of differential referred to second gear
u	Speed of entraining motion (average speed of a meshing teeth pair)
Δu	Sliding velocity (relative surface speed of a meshing teeth pair)
u_{brg}	Speed of entraining motion in the loose wheel-to-shaft conjunction
v_e	Coefficient of thermal expansion
v_p	Velocity of approach of contacting/impacting gear teeth pair
v_x	Surface velocity in the x-direction
W_{brg}	Hydrodynamic load in the loose wheel-to-shaft conjunction
$W_{i,k}$	Contact load per teeth pair
α	Pressure–viscosity coefficient of the transmission fluid

α^* Temperature dependent pressure–viscosity coefficient
δ Contact elastic deflection
ε Eccentricity ratio
θ Temperature
φ_i Rotational displacement of wheel
φ_{os} Angular displacement of output shaft
φ_p Rotational displacement of pinion
η Effective dynamic viscosity at contact temperature
η_o Dynamic viscosity at bulk oil temperature and atmospheric pressure
ρ Density
ψ Attitude angle in the wheel-to-shaft conjunction

Mathematical formulation

Gear teeth impacts (**rattle**) are induced owing to oscillations of loose (unselected) gears within the confine of their clearances. These oscillations are caused by torsional fluctuations of the transmission input shaft caused by combination of engine inertial imbalance and combustion loading (referred to as **engine order vibrations**, see Rahnejat, 1998). These vibrations are particularly present in diesel engines output torque. Owing to various driving conditions, different types of rattle have been defined: idle, drive, creep and over-run rattles (see Chapters 21, 23–30). It is noteworthy, however, that rattle only originates within the unselected gear pairs. The distinction described above only changes the gearbox's input shaft excitations and the contribution of a given engaged gear to its counterparts. Figure A.1 shows a diagrammatic representation of a six speed forward plus reverse manual gearbox used in this analysis (also see Chapter 26).

In the multi-physics approach described in this chapter, the model comprises inertial dynamics and impact/contact tribological characteristics. There are a number of lubricated conjunctions, shown in Figure A.2.

For the loose rattling gear wheels, the regime of lubrication is assumed to be hydrodynamic (see Chapter 5). This assumption is supported by the low loads, W and a relatively large film thickness, h. In the case of engaged gear pair, this assumption is no longer true as moderate to high loads lead to elastohydrodynamic regime of lubrication (**EHL**) (see Chapter 6). These assumptions have been verified through use of a lubrication chart, such as that presented in Gohar and Rahnejat (2008). Typically, EHL conjunctions have film thicknesses in the sub-micrometre region.

Grubin (1949) provided the original analytical solution to the EHL problem. Equation A.1 presents Grubin's expression for film thickness:

$$\frac{h}{r_x} = 2.076 \left(\frac{\alpha\eta u}{r_x}\right)^{8/11} \left(\frac{E^* l r_x}{W}\right)^{1/11} \tag{A.1}$$

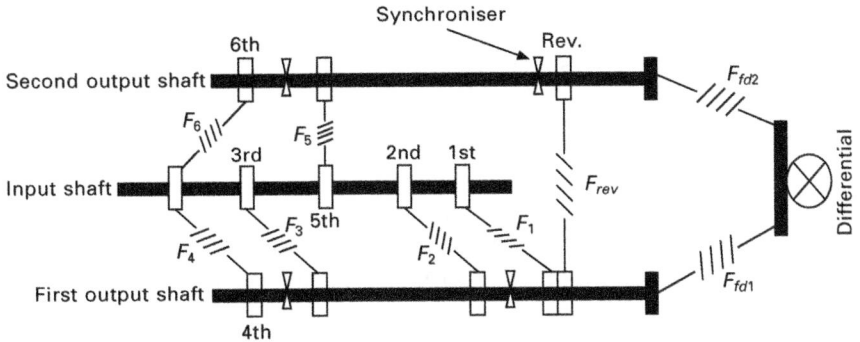

A.1 Front wheel drive, six speed + reverse gearbox under investigation.

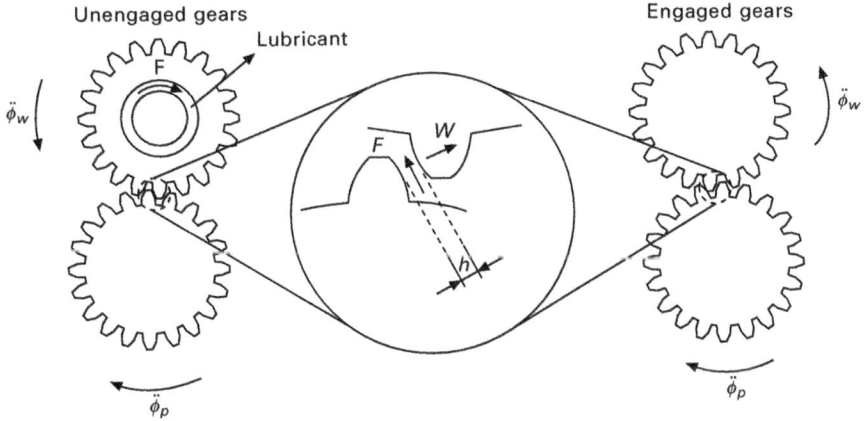

A.2 Forcing elements in conjunctions of engaged and unengaged gear pairs.

The contact load can be obtained through Hertzian analysis for an elastic line contact as:

$$W \approx \left\{ \frac{\pi l E^*}{2 \left[\ln \left(\frac{2l}{b} + \frac{1}{2} \right) \right]} \right\} \delta \tag{A.2}$$

where, under a totally elastic impact, the deflection is found as (Gohar and Rahnejat, 2008):

$$\delta = \left\{ \frac{\left[2\left(\ln\left(\frac{2l}{b} + \frac{1}{2}\right) mv_p^2 \right) \right]}{\pi l E^*} \right\}^{1/2}$$

(A.3)

With the aid of kinematics and Hertzian contact mechanics, all the parameters needed in equations (A.3) are found (De la Cruz *et al.*, 2008).

Friction for thin EHL films in gear pair contacts is due to a combination of viscous shear and boundary interactions, thus:

$$F = F_b + F_v$$

(A.4)

This is rather involved and readers should refer to Greenwood and Tripp (1967) model (see also Chapter 3).

Hydrodynamic reactions in loose gear teeth pairs are represented by the analytical relationship (Gohar and Rahnejat, 2008):

$$W = \frac{2bu\eta r_x}{h} - \frac{3\pi b\eta}{\sqrt{2}} \left(\frac{r_x}{h}\right)^{3/2} \frac{\partial h}{\partial t}$$

(A.5)

The expressions are also based on contact geometry and operating conditions. A very important parameter is the squeeze film velocity in approach and separation of gear teeth pairs (see Chapter 5); $\partial h/\partial t$. A negative value corresponds to the approach of the contacting solids (increasing the contact load in (A.5)). When a positive value is encountered, this is set to zero, i.e. entraining motion only, as the lubricant cannot sustain a tensile force.

The film thickness value for use in equation (A.5) is obtained as a function of gear pair dynamics (Tangasawi *et al.*, 2007):

$$h = C - |R_{bwi}\varphi_i - R_{bpi}\varphi_{pi}|$$

(A.6)

As can be seen in Fig. A.2, there are two distinct conjunctions in loose gear pairs. One conjunction corresponds to the impact/contact zones of gear teeth pairs, in which flank friction is generated due to viscous shear under hydrodynamic condition. A number of teeth pairs are usually in simultaneous action. The flank friction in each pair is a function of the contact sliding velocity, thus (Gohar, 2001):

$$F_{fw} = \begin{cases} \dfrac{l\pi\eta\Delta u}{\sqrt{2h}}\sqrt{r_x}, & \Delta u \geq 0 \\[3mm] \dfrac{-l\pi\eta\Delta u}{\sqrt{2h}}\sqrt{r_x}, & \Delta u < 0 \end{cases}$$

(A.7)

The other conjunction is the conforming contact between the output shaft supporting the loose wheel and the wheels inner (bore) surface. This contact may be approximated by a journal bearing with no eccentricity. The viscous friction in this conjunction is given as (Gohar and Rahnejat, 2008):

$$F_P = \frac{c\varepsilon W_{brg}}{2 R_{brg}} \sin \psi + \frac{2\pi \eta\, u_{brg}\, R_{brg}\, l_{brg}}{c(1 - \varepsilon^2)^{1/2}} \tag{A.8}$$

where $\varepsilon \approx 0$, thus:

$$F_P = \frac{2\pi \eta\, u_{brg} R_{brg} l_{brg}}{c}$$

Having stated all the required forcing functions, based on the tribological contacts, the dynamics of the problem can now be introduced. This multi-body system is solved using equations of motion following the Newton–Euler approach. Hence, for the seven degrees of freedom model presented in figure A.1, the equations of motion can be divided into three groups:

1   For the engaged pinion-wheel pair (in this case the second gear):

$$(I_2 + I_{os})\ddot{\varphi}_2 = \sum_{k=1,n'} W_{2,k}\, R_{bw2} - \sum_{i=1,3,4,7} \sum_{j=1,n} F_{i,j} r_{xi,j}$$

$$- \sum_{k=1,n'} F_{2,k} r_{x2,k} - T_{D2} \tag{A.9}$$

2   For the loose unselected pinion-wheel pairs $i \in 3 - 6$:

$$I_i \ddot{\varphi}_i = \sum_{j=1,n'} W_{i,j} R_{bwi} - \sum_{j=1,n'} F_{i,j} r_{xi,j} - F_{pi} R_{brgi} \tag{A.10}$$

3   Finally, for the reverse pinion-wheel pair (loose wheel on the second transmission output shaft, meshing with the reverse pinion mounted on the first output shaft):

$$I_7 \ddot{\varphi}_7 = \sum_{k=1,m'} W_{7,k} R_{bw7} - \sum_{k=1,m'} F_{7,k} r_{x7,k} - F_{p7} R_{brg7} \tag{A.11}$$

Under the EHL regime of lubrication, lubricant viscosity is dependent upon both the generated conjunctional pressures and temperature. Pressure causes a rise in viscosity and responsible lubricants almost incompressible behaviour (Chapters 5 and 6). Also, due to the viscous shear in the contact temperature rises, which in turn reduces the lubricant viscosity (see Chapter 5) In moderate to highly loaded concentrated contacts the conditions are referred to as *thermo-elastohydrodynamics* (see Chapter 6). Thus, together with the dynamics, the model is multi-physics, multi-level approach, dealing with variables from sub-micrometres (film thickness) to those of large displacements. For very small micro-electromechanical gear pairs (MEMS), the physics of conjunctions are in the nano-scale (see Gohar and Rahnejat, 2008, and Chapters 3 and 32).

Thermal effects must be accommodated in two major conjunctions, flank interactions and those of loose wheels-to-the retaining shaft. When dealing with heat generation in contacting flank pairs, the energy equation can be solved analytically (see Chapter 5). The energy equation is:

$$\underbrace{v_e v_x \theta \left(\frac{\partial p}{\partial x} \right)}_{\text{compressive heating}} + \underbrace{\eta \left(\frac{\partial v_x}{\partial z} \right)^2}_{\text{viscous heating}} = \underbrace{\rho v_x C_p \left(\frac{\partial \theta}{\partial x} \right)}_{\text{convection cooling}} - \underbrace{k_c \left(\frac{\partial^2 \theta}{\partial z^2} \right)}_{\text{conduction cooling}} \qquad (A.12)$$

For thin elastohydrodynamic films heat generation within the contact region is carried away by conduction through the contiguous surfaces. Thus, the convection cooling term may be ignored. Also, the compressive heating term is small compared with shear heating, particularly within the flat region of the EHL film. Thus an analytic solution is possible, where the temperature rise in the flat oil film is found as (Gohar and Rahnejat, 2008):

$$\Delta \theta = \frac{2 \eta \Delta u^2}{k_t} \qquad (A.13)$$

Under hydrodynamic conditions, relatively thicker films in the teeth flank conjunctions promote convection cooling and compressive heating can be ignored due to low pressures, thus:

$$\Delta \theta = \frac{8 b \eta \Delta u}{h^2 \rho C_p} \qquad (A.14)$$

For the conforming contact between the loose wheels and their retaining shafts, an analytic solution similar to that for a journal bearing may be used (see Chapter 8 and Gohar and Rahnejat, 2008). Thus:

$$\Delta \theta = \left(\frac{2 K_1 W_{brg}}{\rho C_p R_{brg} l_{brg}} \right) \left(\frac{\mu^*}{Q_s^*} \right) \qquad (A.15)$$

where $Q_s^* = \frac{Q_s}{\pi N l R c}$, $\mu^* = \frac{\mu R}{c}$ and $K_1 = \frac{k_1}{\rho c_p}$, where $k_1 < 1$ indicates that not all the heat is carried away by convection.

With $\Delta \theta$ obtained for each conjunction and the inlet bulk oil temperature known (that of engine operating condition), the effective viscosity of the lubricant in each conjunction can be evaluated using Houperts (1985) equation. This is given in Chapter 5. Thus, an iterative procedure is used which includes tribology, thermal effects and system dynamics (De la Cruz *et al.*, 2009a).

Results and discussion

Figure A.3 shows the predictions for partial loading creep rattle condition at the engine idling speed of 830 engine RPM with second gear engaged. Two different cases are presented; one for low and the other for high bulk oil temperature. Previous studies have indicated that temperature plays a key role in transmission rattle (see Tangasawi *et al.*, 2007, and Chapter 26). It is suggested that at higher temperatures loose gear pairs are more prone to rattle due to reduction in resistive (drag) torque. It has also been shown that at some *ideal temperature values* gear rattle is attenuated. It may be surmised that lubricant viscosity variation in different conjunctions, described above, may be the cause of this. However, this variation is different for each single gear pair and hence a unique solution may be difficult to find.

The result in figure A.3 relate to one meshing cycle, comprising simultaneous interactions of two to three gear teeth pairs of the selected second gear pair. The variations are for one set of teeth in order to demonstrate the effect of temperature, often ignored in all such analyses. As expected, because of EHL conditions the film thickness in general is rather insensitive to load, but profoundly sensitive to contact temperature and geometry. This affects the flank friction and thus the torque transmitted to the loose gear pairs, which themselves are affected by temperature as well.

The changes as the result of meshing conditions in the engaged gear pair, together with engine order vibration transmitted to the loose unselected gear pairs induce transmission creep rattle in the case studied here. Further analysis of loose gear results, typically by fast Fourier analysis of their dynamic behaviour sheds light on their rattle behaviour. De la Cruz *et al.* (2009b) used an ***impulsion ratio*** I_m, to define rattle conditions:

$$I_m = \frac{T_{\text{drive}}}{T_{\text{drag}}} \propto C \frac{C_{pet} \eta_f}{h \eta_{pet}} \tag{A.16}$$

This attempts to predict conditions that induce propensity to rattle. The ratio takes into account the inertial torque, inducing motion (due to impact forces of simultaneous meshing pairs) and the resistive or drag torque (because of frictional losses). A ratio of unity leads to uniform motion (no acceleration), while values exceeding unity indicate impulsive action and those below unity correspond to decaying oscillations (Fig. A.4).

It is noted that at a given temperature, high inertia gears show a lower energy content in their frequency spectrum, suggesting that rattle could be more noticeable in the low inertia gears. Even though the values of the impulsion ratio are higher for the first gear, it is the frequency at which the limiting value of unity is exceeded which fundamentally affects the rattle behaviour.

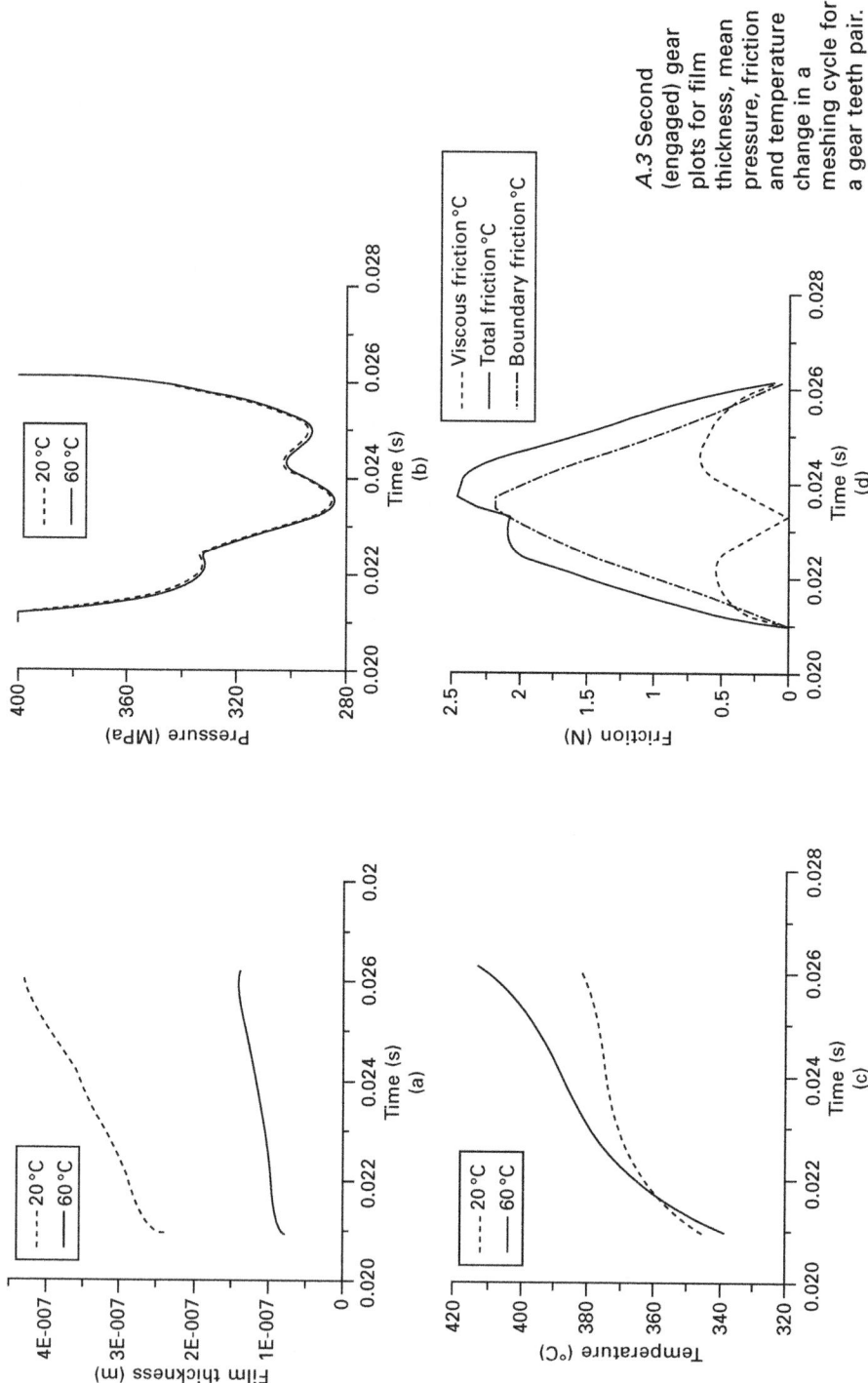

A.3 Second (engaged) gear plots for film thickness, mean pressure, friction and temperature change in a meshing cycle for a gear teeth pair.

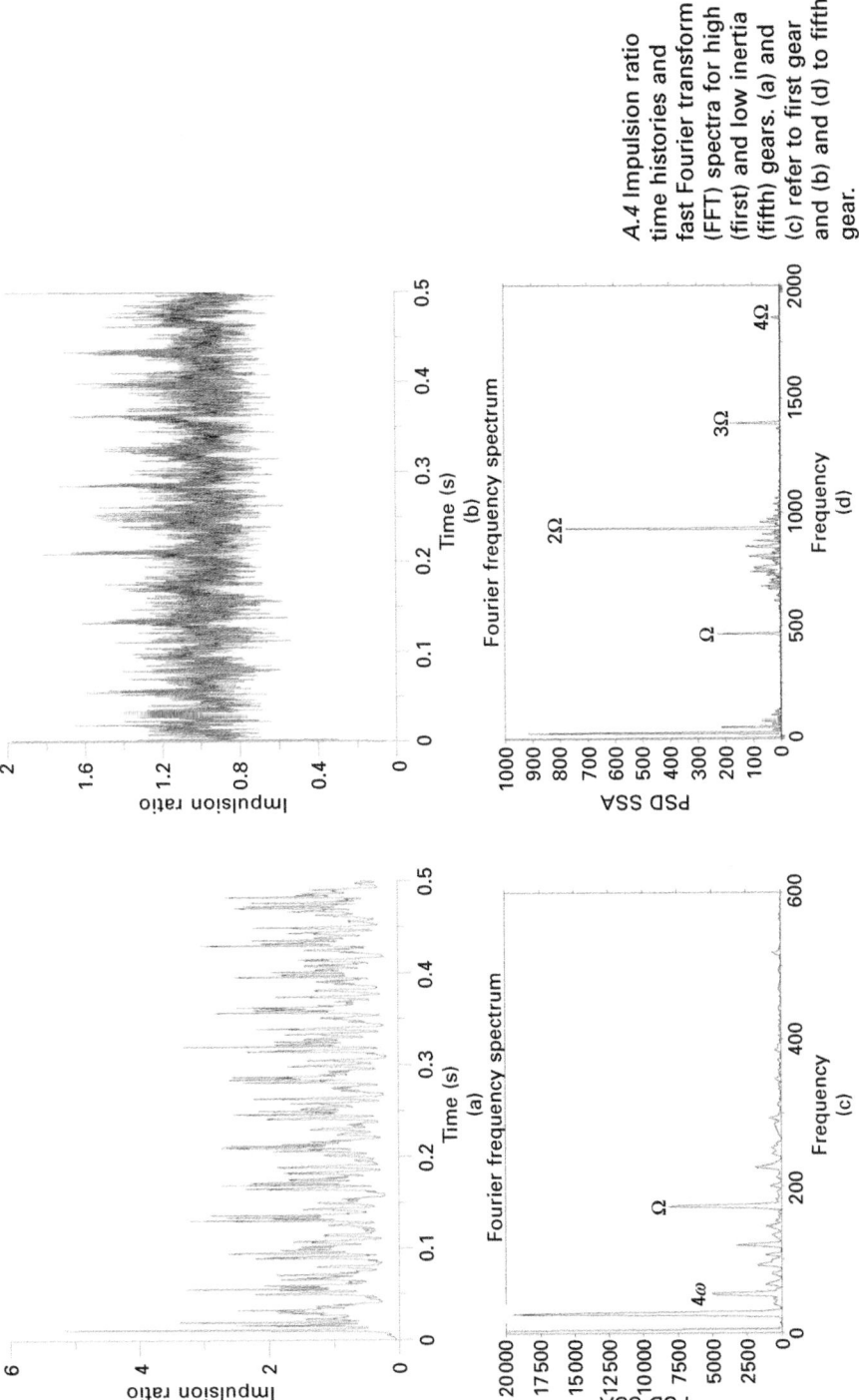

A.4 Impulsion ratio time histories and fast Fourier transform (FFT) spectra for high (first) and low inertia (fifth) gears. (a) and (c) refer to first gear and (b) and (d) to fifth gear.

References

De la Cruz, M., Theodossiades, S., Rahnejat, H. and Kelly, P., 'Impact dynamic behaviour of meshing loaded teeth in transmission drive rattle', *6th Euromech, Non-linear dynamics Conference*, S Petersburg, Russia, 2008.

De la Cruz, M., Theodossiades, S., Rahnejat, H. and Kelly, P., 'The effect of thermohydrodynamics on manual automotive transmissions gear rattle', *Proc. ASME, IDETC/CIE 2009*, Pap. No. DETC2009-87226, San Diego, CA, 2009a.

De la Cruz, M., Theodossiades, S., Rahnejat, H. and Kelly, P., 'Numerical and experimental analysis of manual automotive transmissions – gear rattle', *SAE Int.*, Pap. No. 2009-01-0328, 2009b.

Gohar, R., *Elastohydrodynamics*, Imperial College Press, London, 2001.

Gohar, R. and Rahnejat, H., *Fundamentals of Tribology*, Imperial College Press, London, 2008.

Greenwood, T.A. and Tripp, J.H., 'The elastic contact of rough surfaces', *Trans. ASME, J. Lubn. Tech.*, 1967, p. 417.

Grubin, A.N., 'Contact stresses in toothed gears and worm gears', *Book 30* CSRI for Tech. & Mech. Engng., Moscow, 1949, DSRI Trans., **37**.

Houpert, L., 'New results of traction force calculation in elastohydrodynamic contacts', *Trans. ASME, J. Trib.*, **107**, 1985, pp. 241–248.

Rahnejat, H., *Multi-body Dynamics: Vehicles, Machines and Mechanisms*, Professional Engng. Publ. & SAE (Joint Publ.), Bury St Edmunds (UK) and Warrendale, PA, 1998.

Tangasawi, O.A.M., Theodossiades, S. and Rahnejat, H., 'Gear teeth impacts in hydrodynamic conjunctions promoting idle gear rattle', *J. Sound and Vibration*, **303(3–5)**, 2007, pp 632–658.

Section I.I

Fundamentals of tribology and dynamics

2

Mechanisms and laws of friction and wear

D. ARNELL, University of Central Lancashire, UK

Abstract: This chapter initially introduces the concept of friction and wear arising from stresses at localised contact spots between opposing surfaces. It then discusses the nature and the description of the topography of engineering surfaces and the stresses at asperity contacts under normal and frictional forces. The development of theoretical descriptions of the tribological behaviour of various types of material are then critically reviewed and examples of wear in practice are described.

Key words: surface topography, friction, wear, contact of surfaces.

2.1   Introduction

It is the purpose of this chapter to describe and explain the phenomena of friction and wear that occur during the interaction of two surfaces in relative motion.

It is obvious that the nature of the surface will have a great influence on the observed behaviour and therefore Section 2.2 gives a brief description of what is understood, in engineering terms, by a solid surface. All engineering surfaces are found to be rough at some scale and, as a result, when two surfaces are placed in contact, they touch at only a few of their higher points, which must carry the total load, so that the local contact stresses would be very high even at modest loads. Therefore, Section 2.3 explains the nature and quantitative description of surface topography and the stresses that arise when two surfaces are made to contact. This prepares the ground for descriptions of the developments leading to the current understanding of friction (Section 2.5) and wear (Section 2.6). Section 2.7 outlines the current trends in the amelioration of friction and wear, which becomes ever more demanding as modern technology requires operation in increasingly hostile environments.

2.2   The nature of engineering surfaces

A conventional description of a surface would simply be the outer boundary of a body, and it would be envisaged as having length and breadth but no thickness. However, in describing *engineering surfaces* one must regard them as having significant depth, through which the physical, chemical

41

and functional properties can be substantially different from those of the underlying bulk material. Depending on the method of preparation and the operating environment, surfaces defined in this way can have depths varying from a few nanometres to many micrometres.

Above the bulk material, there is usually a ***work-hardened layer*** formed by the machining process. The outer part of this layer is often contaminated by materials from cutting or forming tools and their associated lubricants, and the superficial layer normally consists of reaction products formed by chemical reaction with the environment, usually oxides, together with contaminants such as dust and grease from the atmosphere.

In addition, the surface is characterised by irregularities having different amplitudes and frequencies. This property, known as the ***surface texture***, is of fundamental importance in the study of tribology and it is described in more detail in the following section.

2.3   Surface topography and contact

2.3.1   Measurement and description of surface topography

In manufacturing, any machining process used to finish a component leaves characteristic marks, collectively referred to as the '***surface topography***' or 'surface texture', which can influence the appearance and performance of the component in many ways. No engineering surface is ideally smooth. Even when very highly polished, such surfaces appear rough when viewed at a sufficiently high magnification, with height variations up to ~ 50 nm above and below the mean.

Surface topography is usually examined using an instrument known as a ***profilometer***. There are now many different types of profilometer, some using stylus measurements, and others using, for example, contactless, laser reflection methods. The different techniques can show the surface in two or three dimensions, but the principles of profilometry are best understood by describing the two-dimensional stylus technique, which is still the most commonly used.

In the stylus instrument a very finely pointed diamond stylus traverses the surface to be measured under a very light load. Its vertical movements are measured with a precision of nanometres and recorded as a function of distance along the surface. Figure 2.1 shows a true profile and the corresponding recorded profile, which is distorted due to the difference in horizontal and vertical magnification.

A machined surface can have different profile components. Traditionally, these components are qualitatively divided into the following types. ***Roughness*** refers to irregularities inherent in the production process left by the machining

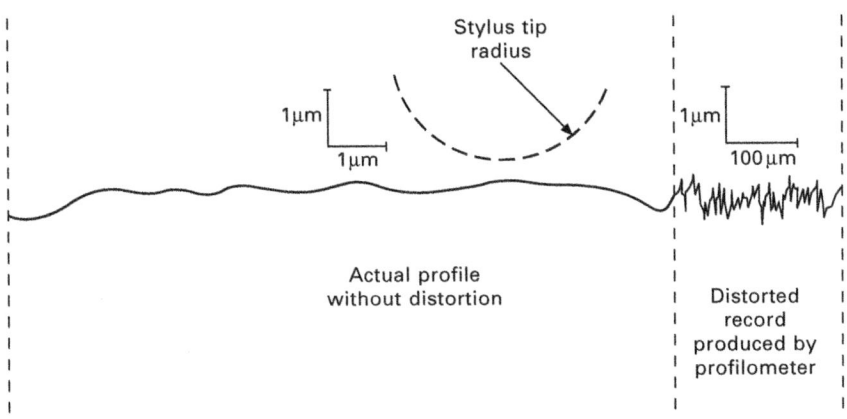

Stylus tip
radius

1μm

1μm

Actual profile
without distortion

1μm

100μm

Distorted
record
produced by
profilometer

2.1 Illustration of profile distortion caused by different vertical and horizontal magnifications of the profilometer trace. The true profile is shown on the left.

agent, e.g. tool tip, grit, spark; **waviness** refers to 'that component of the texture on which roughness is superimposed'; it may arise from machine/work deflection, tool vibration, etc.; form is 'the general shape of the surface neglecting variations due to roughness and waviness', with deviations from ideal being termed 'errors of form'; and lay refers to prominent directional machining marks.

Control of the surface topography of the product is essential for successful operation in many production processes and is also critical for the proper operation of all types of bearing components. As we shall see later, the stresses at the interactions between the individual peaks on surfaces need to be minimised to reduce friction and wear.

In many tribological situations, the roughness is the component of surface texture that is the most critical, so it is necessary to have a quantitative description of roughness. In modern instruments, such descriptions are based upon digitised surface profiles, as shown in Fig. 2.2. The surface height above an arbitrary datum is measured at equal intervals, the mean height of the readings is calculated, to define a centre line, and the height readings relative to the centre line are then calculated.

There are then several ways of describing the roughness in terms of deviations of the profile from the centre line, the two principal ones being the **roughness average**, Ra, and the **root mean square deviation** Rq:

$$Ra = \frac{1}{N} \sum_{n=1}^{N} |z_n|''$$ (2.1)

$$Rq = \sqrt{\frac{1}{N} \sum_{n=1}^{N} z_n^2}$$ (2.2)

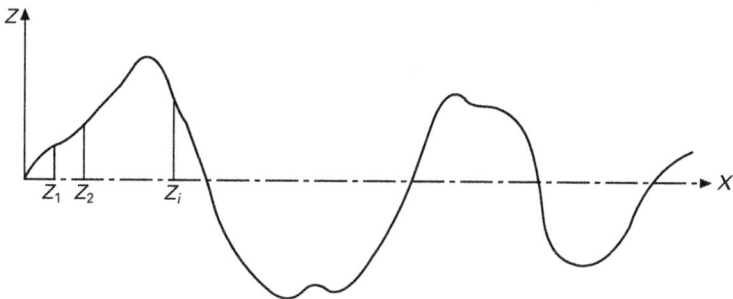

2.2 The digitisation of a surface profile.

2.3 The method for deriving the all-ordinate distribution.

where N is the total number of readings and z_n is the distance of the nth reading from the centre line.

Ra and *Rq* values have to be used with care, to compare surfaces prepared using the same process, e.g. grinding. It is possible for surfaces prepared by different processes to have very different shapes although they have similar roughness values.

The distribution of the surface height can also be described by an all-ordinate distribution, as shown in Fig. 2.3. This is obtained by taking measurements of the height at discrete intervals and summing the number of ordinates at any given height level. The distribution curve can then be drawn as the best smooth curve through the resulting histogram.

Skewed distributions can show up differences in profiles having the same *Ra* or *Rq* values, as illustrated in Fig. 2.4. Distribution curves can also be obtained for the peak and valley distributions. If these distributions can be approximated by mathematical functions, e.g. Gaussian distributions, they can then be used in the modelling of surface contacts.

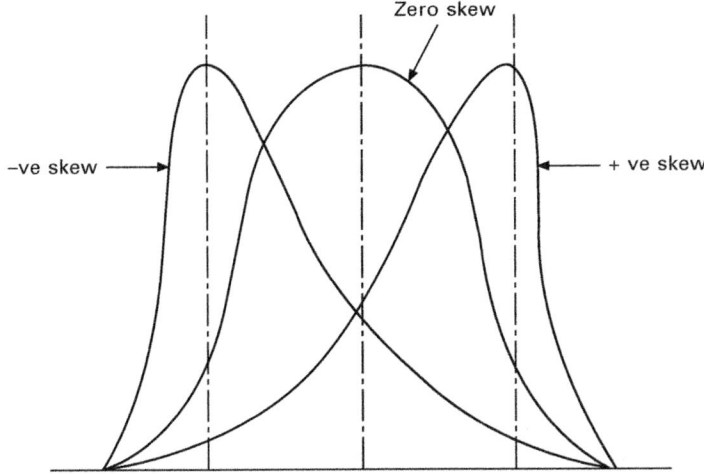

2.4 All-ordinate distributions for two surfaces having opposite skew.

2.3.2 Stresses at surface contacts

The peaks on a surface are known as asperities. When surfaces are placed in contact, they touch only at the tips of the higher asperities. The sum of the areas of these localised contacts is the real contact area of the surfaces and, in general, this is only a very small fraction of the apparent contact area. As an example, if two plates of low-carbon steel are placed together under a load of 100 N the total real contact area is only of the order of 0.1 mm^2, whatever the extent of the apparent area. Because the real area of contact is so small, the stresses at the localised contacts can be very high, even under modest loads. The stresses at these local contacts and the chemical nature of the contacting surfaces are largely responsible for observed friction and wear behaviour. Therefore, knowledge of the stresses and how they arise is important in understanding these phenomena.

The stresses at the contacting asperities depend on: the surface roughness; the asperity shapes; the elastic properties of the contacts; and the plastic properties of the contacts.

The *elastic properties* of the contacting materials are usually described by the their elastic modulus E and Poisson's ratio v, while their *plastic properties* are usually described by their hardness H. If the surfaces of two materials with elastic modulus and Poisson's ratio E_1, v_1 and E_2, v_2, respectively, are placed in contact under an applied load, both surfaces deform under the applied load and the effective elastic modulus E' at the contact is given by:

$$\frac{1}{E'} = \frac{1 - v_1^2}{E_1} + \frac{1 - v_2^2}{E_2}$$

There are several ways of describing hardness, but the most useful way to describe plastic behaviour in a tribological context is by the magnitude of the indentation hardness, *H*.

Indentation hardness is usually measured by pushing a square-based diamond pyramid (a Vickers diamond) into the material's surface under a known load and measuring the size of the indentation. The Vickers hardness number is given by the applied load divided by the area of the indentation. It has been traditionally measured in $kgf\,mm^{-2}$, but is now measured in $MN\,m^{-2}$ (or MPa).

Hardness, as measured by this method, is identical to the yield pressure of the material and this, in turn, is very closely related to the uniaxial yield stress σ_Y. This is because the indenter comes to equilibrium in the hardness test when the load is just supported elastically. For most metals:

$$H \approx 2.8\,\sigma_Y$$

Therefore, the **limiting elastic strain**, ε_p, i.e. the strain for the onset of plastic deformation, is given by

$$\varepsilon_p = \frac{H}{2.8E}$$

This is an important result for friction and wear.

When two extended engineering surfaces are placed in contact they initially touch only at the tips of the higher asperities. These asperities are unable to support the load and they plastically deform, so that the surfaces move closer together, bringing more asperities into contact. The contacting asperities are deformed by different amounts, with the higher ones being deformed to a greater extent, so that some are deformed elastically and others plastically. As the load is increased, more asperities come into contact and suffer different amounts of deformation, so that there are always populations of elastically and plastically deformed asperities. Plastically deformed asperities will make larger contributions to both friction and wear, so the relative proportions of elastically and plastically deforming asperities are important.

In modelling asperity contacts, it is usually assumed that they have spherical curvature at their tips, so that the contact is similar to that between two spheres of radii R_1 and R_2. However, the modelling is simplified if all the curvature is put on one surface, so that the model becomes a contact between a sphere of radius R and a plane surface.

These contacts are equivalent if:

$$1/R = 1/R_1 + 1/R_2$$

The problems of elastic contact between curved bodies were first solved in the nineteenth century by Hertz and are described in detail by Johnson (1985). Such contacts are known as **Hertzian contacts**.

Under a normal load P, the sphere and plane deform and approach each other by a distance δ to give a circular contact spot of radius a. The Hertz equations for this contact are:

$$a = (3PR/4E')^{1/3}$$

$$\delta = a^2/R = 1/2(9P^2/16RE'^2)^{1/3}$$

Maximum pressure, $p_0 = 3P/2\pi a^2 = (6PE'^2/\pi^3R^2)^{1/3}$

The distribution of the normal pressure over the contact surface is given by:

$$p = p_0 \left[1 - (r/a)^2\right]^{1/2}$$

as illustrated in Fig. 2.5. The distribution of shear stress in the contact zone is shown in Fig. 2.6. It has a maximum value of $\sim 0.31\, p_0$ at a depth of $0.48a$ (for $v = 0.3$).

As the load is increased, yielding first takes place below the surface when the maximum shear stress reaches a value of $\sim 0.5\,\sigma_Y$, where σ_Y is the uniaxial yield stress. At this point, the normal mean pressure on the surface is $\approx \sigma_Y$.

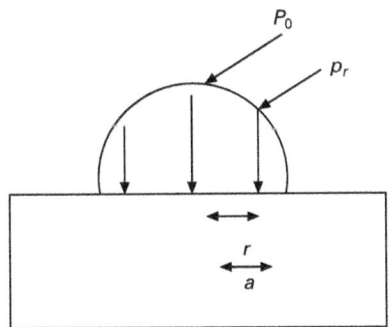

2.5 The normal pressure distribution for Hertzian contact of a ball and plane.

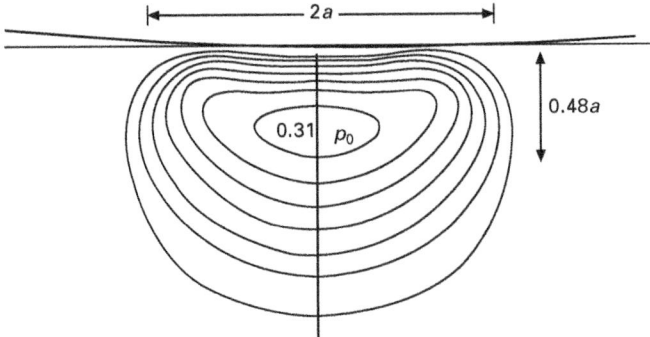

2.6 The distribution of shear stress beneath the Hertzian contact of Fig. 2.5.

However, the plastic zone is constrained by the surrounding elastic material and bulk yielding does not take place. As the load is increased further, the mean pressure increases and the plastic zone expands. When the mean pressure becomes $\sim 2.8\,\sigma_Y$, the plastic zone reaches the free surface, and at this point, the mean contact pressure is equal to the material's hardness.

2.4   The contact of rough surfaces

The contact between two surfaces is best visualised by showing the contact of a rough surface with an ideal plane surface, as shown in Fig. 2.7. It can be seen that the two central asperities in this figure have been heavily deformed while the outer ones are just in contact with the plane. In general, there will be asperities in each of the three states of deformation: purely elastic; contained elastic/plastic; and fully plastic.

The topography on most surfaces is such that, as the load increases, the number of contacting asperities increases, but the relative proportions of asperities in each deformation mode remain the same. The actual relationship between the load and the contact area depends on the height distribution of the deforming asperities which, in practice, means, at most, the outer 10%. If this distribution is exponential, then the area is proportional to the load, whatever the mode of deformation, and for distributions close to exponential this is almost true. In particular, for a Gaussian distribution of the heights of the outer asperities, which is very close to many actual distributions, the ratio of load to contact area only doubles while the load is increased by 10^5. It can, therefore, usually be taken that contact area is proportional to load, whatever the deformation mode of the asperities.

2.4.1   The plasticity index

The relative amounts of elastic and plastic deformation under normal load are related to the *plasticity index*, ψ (Greenwood and Williamson, 1966), where:

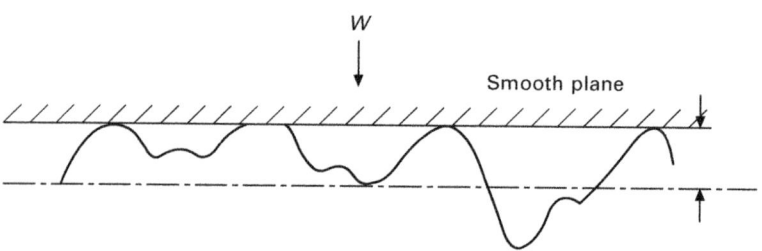

2.7 Asperity deformation during contact of a rough surface and a smooth plane.

$$\psi = (E'/H)(R/\beta)^{1/2}$$

and R is the standard deviation of the peak height distribution, β is the mean effective radius of the asperities and H is the hardness of the softer material.

If one examines the expression for the plasticity index, it can be seen that:

- E'/H is proportional to $1/\varepsilon_P$, so small E'/H corresponds to a high limit to elastic strain;
- small R correspond to low roughness;
- large β corresponds to a large asperity radius.

All three of these factors favour elastic rather than plastic behaviour. Thus, the plasticity index is a good indicator of the relative amounts of elastic and plastic deformation. For contact under normal load only, it has been shown by a recent finite element analysis (Kogut and Etsion, 2003) that:

- when ψ is less than 0.5, contact is entirely elastic;
- when $0.5 < \psi < 1.4$, contact is contained elastic/plastic;
- when $\psi \geq 8.0$, contact is entirely plastic, with the range between 1.4 and 8 covering the transition from contained elastic/plastic to fully plastic deformation.

Note that the expression for the plasticity index does not depend on load. As long as the asperities can deform independently and their deformation is not influenced by that of neighbouring asperities, a plasticity index of <0.5 will correspond to almost fully elastic behaviour, whatever the load. Similarly, a plasticity index of >8.0 will correspond to almost fully plastic behaviour, whatever the load. The effect of the load is simply to increase the number of contacting asperities in each state of deformation. The index can change during a wear process, e.g. during running in of an engine, R decreases and H increases as the surfaces become smoother and harder. As a result, after running in, contact can become almost entirely elastic, with very little wear.

During sliding, the asperities are subjected to both normal and tangential loads and the stress situation becomes more complicated. As the friction coefficient increases the initial plastic zone moves closer to the surface and unconstrained plastic flow occurs at a lower normal pressure. When the friction coefficient rises to ~0.3, the initial plastic zone is at the surface and local plastic deformation can occur at any significant normal pressure. However, knowledge of the plasticity index and the factors that underlie it are still important in explaining observed friction and wear behaviour.

2.5   Friction

2.5.1   Introduction

Friction and wear are easy to describe as phenomena but this is deceptive. Unlike many apparently more complex phenomena, they cannot be predicted from first principles and must be determined experimentally. They can also change with time due to changes in surface properties, composition and texture as sliding continues.

The most damaging tribological contacts, involving highest friction and wear, are those involving relative sliding, and this chapter will concentrate on this mode of contact. However, many contacts in engines and transmissions involve rolling or combinations of rolling and sliding, where both friction and wear are much lower than in sliding. For a full treatment or rolling, readers are referred to the excellent treatment by Johnson (1985). The descriptions given here for sliding are also relevant to the sliding contribution in rolling/ sliding contacts.

Friction coefficients in normal environments vary only in the range 0.1 to 1, and in the presence of a little lubricant the range is usually less than this. Also, friction only manifests itself as an energy cost, which in individual applications, is often very low. However, there are pressing economic needs to lower friction, even though possible savings in energy costs are usually only a few per cent. For example, such savings in fuel costs to individual motorists would be low but, summed over all operating engines, the saving would be of global importance.

2.5.2   The laws of friction

The normal load and friction force at a sliding contact are shown in Fig. 2.8. The *laws of friction* were first stated by Amontons and are known as Amontons' laws.

1.   Friction force is proportional to normal load, i.e.

$$F = \mu N \tag{2.3}$$

where the constant μ is known as the coefficient of friction.

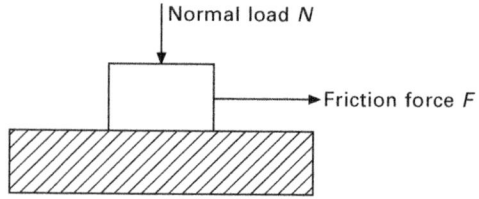

2.8 Normal and frictional forces in a sliding contact.

2. F is independent of the apparent area of contact.

These 'laws' are generally true, but there are exceptions. The force to initiate sliding is known as the **static friction** force and that to maintain sliding is known as the kinetic or dynamic friction force. The **kinetic friction** force is generally lower than the static one, so that one needs to define two friction coefficients, the static coefficient μ_s and the kinetic coefficient μ_k. Unless stated otherwise, μ is taken to mean the kinetic coefficient.

To explain Amontons' laws, one can make two assumptions.

1. During contact, the resistive force per unit area of contact is constant, i.e.

$$F = As \qquad (2.4)$$

where F is the friction force, A is the true contact area and s is the friction force per unit contact area. This assumption is easily justified as it simply implies that any part of the contact area is representative of the whole.

2. The **real area of contact** is proportional to the normal load, i.e.

$$A = qN \qquad (2.5)$$

where q is the constant of proportionality. This assumption could not always be justified but it is shown in Section 2.1 that it applies to most engineering surfaces as a result of the statistical distribution of asperity heights.

Eliminating A from these two equations gives:

$$F = qsN \qquad (2.6)$$

i.e. F is proportional to N. The purpose of any theory of friction is therefore to explain the actual magnitude of the friction coefficient.

There is, as yet, no theoretical method of predicting from first principles either the friction coefficient or the wear rate when any two materials are in sliding contact. However, conceptual understanding of the phenomena is advancing and what is given below is a brief overview of theories of friction.

It is known that friction arises from interactions between opposing asperities. Each asperity interaction will contribute and the total friction force at any time will be the sum of forces at the individual contacts. The observation that friction forces are often almost constant is due to the fact that the number of interactions taking place at any time is so large that the statistical distribution of the contact processes is almost constant.

It is conventional to describe two types of asperity interaction: adhesion and deformation.

2.5.3   Adhesion theories

Simple adhesion theory of friction

The first modern theory of friction, developed by Bowden and Tabor (1950) for ideal elastic/plastic metals is summarised as follows. When surfaces are loaded together they make contact only at asperity tips. The contact pressures are so high that the asperities of the softer metal deform plastically. This plastic deformation causes the contact area to increase, by both growth of individual contacts and initiation of new ones, until the real area of contact is just enough to support the load elastically.

Under these conditions, for an ideal elastic/plastic metal,

$$N = AH \tag{2.7}$$

where N is the normal load, A is the real area of contact and H is the hardness of the softer material.

As a result of the severe plastic deformation, the asperity junctions cold weld, so that strong adhesive bonds are formed. The specific friction force is then the force to cause shear failure of unit area of welded junction, so that

$$F = As \tag{2.8}$$

These two equations can be combined to give

$$F/N = \mu = s/H \tag{2.9}$$

This theory offered the first theoretical explanation of Amontons' laws.

For shear failure in the softer of the two materials, it is reasonable to take s as equal to k, the critical shear stress of the softer material, so that $\mu = k/H$. The ratio k/H is fairly constant for most metals, with a value of ~0.16, which gives friction coefficients of the right order for many material pairs in a normal environment. However, the theory is inadequate in several respects. First, the predicted friction coefficient depends only on the mechanical properties of the softer material, so one should expect any particular soft material to have similar friction coefficients against all harder materials, which is not observed. Second, actual values of μ in normal environments for many metal pairs are in the range 0.4 to 0.8, several times greater than the predicted value of 0.16. Third, many materials, particularly ductile metals, exhibit friction coefficients $>>1$ when their surfaces are perfectly clean, as in space or ultra-high vacuum. In some cases, when a shear force is applied to the contact between such surfaces, sliding does not occur at all and a weld is formed across much of the ***apparent contact area***. This led Bowden and Tabor to present a more realistic theory, as described in the following section.

Combined stress adhesion theory

The simple ***adhesion*** theory implicitly assumed that the normal and shear stresses on the asperities acted separately, with the normal stress forming the contact area and the shear stress shearing it. In fact both stresses act simultaneously and this must be taken into account.

Consider a simplified two-dimensional stress system with an axial pressure p. The simple Mohr's circle construction of Fig. 2.9 shows that, for this case, the maximum shear stress, $s_{max} = p/2$, and, when p is equal to the yield stress, p_Y, the ***maximum shear stress*** is equal to the shear stress for yield, k.

If now one considers the case where an external shear stress s is imposed, then the stress system changes, as shown in Fig. 2.10. The maximum shear stress in this system is the radius R, and, from the geometry of Mohr's circle, one can write

$$(p/2)^2 + s^2 = R^2 \qquad (2.10)$$

When R becomes equal to k the material yields and plastic deformation

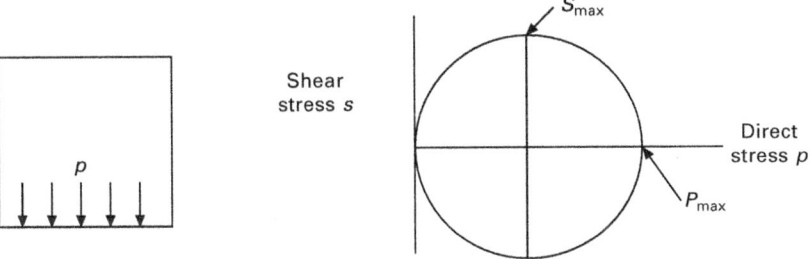

2.9 The maximum shear stress in a 2-dimensional junction.

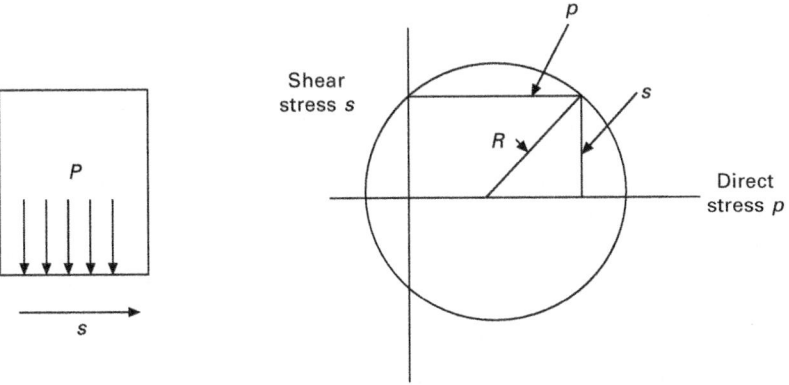

2.10 The maximum stress and maximum shear stress in a 2-dimensional junction subjected to pressure and shear.

takes place. Therefore, yielding is dependent on the action of the combined stresses and not on p alone. Even when p is less than the yield stress, the combined stresses can cause the material to yield.

One can now extend this to a 3-dimensional elastic/plastic asperity junction under a normal load N so that it has yielded and is at a pressure p_Y, with maximum shear stress k.

If an increment of tangential force F is now applied, its effect would be to increase the maximum shear stress to above the value of k. This cannot happen, so the material yields further until the contact area has grown to the point where the normal and shear stresses are again equal to p_Y and k, respectively.

The exact relationship between the stresses at this point is not known but, by analogy with the 2-dimensional case described above, it is assumed that it is of the form

$$P^2 + \alpha s^2 = K^2 \tag{2.11}$$

where α and K are constants yet to be determined. Therefore:

$$(N/A)^2 + \alpha(F/A)^2 = K^2 \tag{2.12}$$

where A is the area of the junction. To find the constants, boundary conditions are considered.

1. When s is zero, the pressure over the junction must be p_Y, so

    $$K = p_Y = H \tag{2.13}$$

2. When F becomes very large, N/A is very small compared with F/A, so that:

    $$\alpha s^2 \approx p_Y^2 \tag{2.14}$$

At this point $s = k$ and $p_Y = H$, so:

$$\alpha \approx (H/k)^2 \tag{2.15}$$

For most metals, H/k is ≈ 6, suggesting that $\alpha \approx 36$. However, experiment suggests that α should have a lower value, and Bowden and Tabor chose a value of 9. If further increments of tangential force are applied, the process described above is repeated so that the contact area continues to increase – a process known as **junction growth** – and no gross sliding takes place. This has been confirmed for clean surfaces of ductile metals in space and ultra-high vacuum, where no sliding occurs and the surfaces weld together over a large fraction of the apparent area.

Note that, although the above reasoning was applied to an asperity which was already plastically deformed, it applies equally well to asperities that have deformed elastically under the normal load. The addition of the high shear stress can again move such contacts into the plastic range.

The reason that such high friction is not observed in normal atmospheres is that metals are covered by oxides or other contaminant films, with the effects described below.

Adhesion theory of metals with contaminant films

One can assume that, at any junction, there is a film of contaminant of shear strength τ_f that is lower than k, the **shear strength** of the solid, so that we can write

$$\tau_f = ck \tag{2.16}$$

where $c < 1$.

While the frictional stress F/A is less than τ_f, junction growth will proceed as described above, but when $F/A = \tau_f$ the contaminating film will shear, junction growth will end and gross sliding will occur. Thus, the condition for sliding is:

$$P^2 + \alpha\tau_f^2 = H^2 \tag{2.17}$$

But it has already been shown that $H^2 = \alpha k^2$. Therefore:

$$P^2 + \alpha\tau_f^2 = \alpha k^2 \tag{2.18}$$

or

$$P^2 + \alpha\tau_f^2 = \frac{\alpha\tau_f^2}{c^2} \tag{2.19}$$

Hence,

$$\mu = \frac{\tau_f}{P} = \frac{c}{[\alpha(1-c^2)]^{1/2}} \tag{2.20}$$

Therefore, as c tends to 1, μ tends to infinity, as described earlier.

When c is small, i.e. when the surface film has low shear strength, $\mu \approx c/\alpha^{1/2}$, so one can obtain a low friction coefficient by having a film of low shear strength separating the surfaces. This is the principle underlying lubrication by soft metal films and boundary lubricants.

Figure 2.11 shows the variation of μ with c for various values of α. It can be seen that μ falls very rapidly as c reduces, so that a small amount of weakening of the interface causes a large reduction in friction. It can also be seen that the precise value of α is not important.

This modified theory also applies when the interface is inherently weaker than either of the asperity materials, even in the absence of contaminants. In this case, no junction growth is necessary and the interface can shear while both materials remain fully elastic.

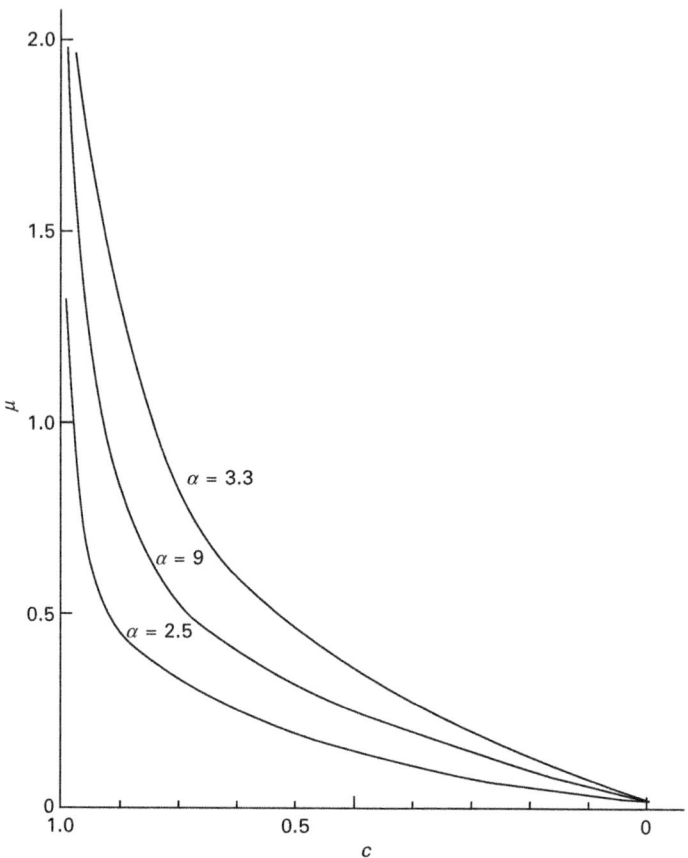

2.11 The variation of μ with c and α.

2.5.4   Deformation theories

Apart from adhesion, the only alternative resistance to motion would result from material being deformed or displaced during sliding. Two types of interaction are considered: ploughing and asperity deformation.

Ploughing

Ploughing occurs when the asperities on the harder surface plough grooves in the softer surface. It is illustrated in Fig. 2.12 for the example of a hard conical asperity of semi-angle θ.

During rubbing, only the front surface of the asperity is in contact with the softer material. The normal load N_i is supported by the horizontal projection of the contact, so that:

$$N_i = 0.5\,\pi r^2 H \tag{2.21}$$

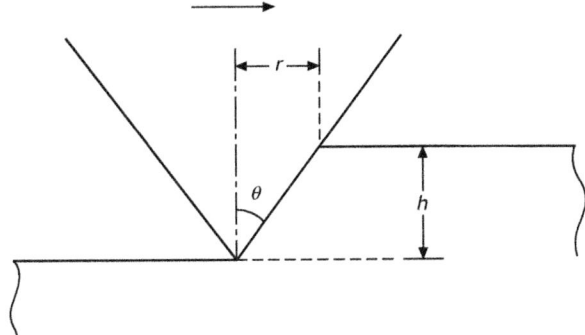

2.12 Ploughing of a softer surface by a hard conical asperity.

The friction force F_i is supported by the vertical projection, so that:

$$F_i = rhH \tag{2.22}$$

Therefore:

$$\mu = F_i/N_i = 2h/\pi r = 2 \cot \theta/\pi \tag{2.23}$$

Similar expressions can be calculated for other shapes. Ploughing contributions to friction are often negligible but abrasive material is usually very angular and can make a significant contribution to friction. According to simple Bowden and Tabor theory, the observed friction coefficient is simply the sum of the adhesive coefficient and the ploughing coefficient.

Asperity deformation theories of friction

In the adhesion theories, described above, the normal and shear stresses on a single asperity were taken to represent the stresses on all asperities. Deformation theories recognise that the normal and shear stresses on asperities will vary during the lifetime of a junction.

The basis of all deformation theories is that, in sliding of macroscopically flat surfaces, motion is parallel to the interface and the separation of the surfaces remains constant. This must be the case to maintain the real area of contact at the constant level that will support the constant load, and the consequence is that the contacting asperities must deform to allow movement to continue. There have been several models based on this principle. It is illustrated by the model due to Edwards and Halling (1968), who considered two *wedge-shaped asperities* of semi-angle θ moving as shown in Fig. 2.13.

By assuming that the material is ideal elastic/plastic and that the asperities deform plastically, they were able to calculate the instantaneous values of the normal and frictional forces throughout the life of the junction. They did

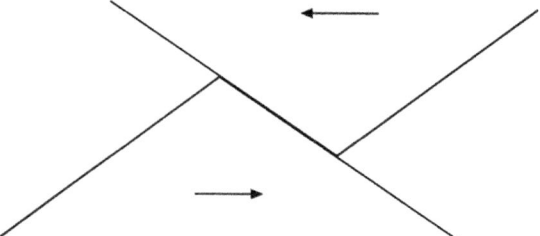

2.13 Idealised wedge-shaped asperities studied in the plastic interaction theory.

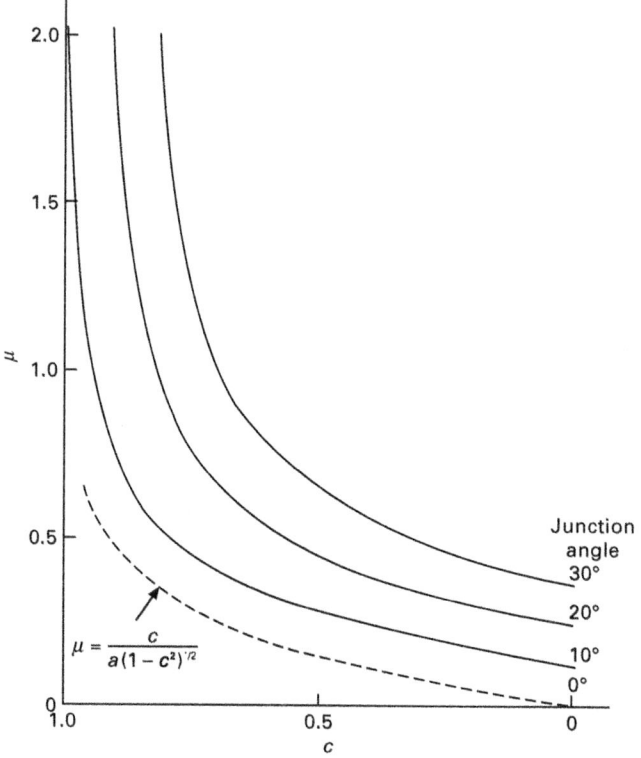

$$\mu = \frac{c}{a(1 - c^2)^{1/2}}$$

2.14 The variation of μ with c for various junction angles.

this for the case where the shear strength of the junction was taken to be k and for cases where the asperities were separated by a film of strength ck, where c and k have the same meanings as in the Bowden and Tabor theory. Calculations were made for various values of θ and c, and the results are shown in Fig. 2.14.

When the wedge angle is zero, the expression obtained for μ is identical to that from the Bowden and Tabor theory, but as wedge angle increases μ increases. Except when using abrasive media, asperity angles are normally <10°, so for most engineering surfaces, there is little difference in the predictions of the two theories.

2.5.5   Comparison of theories of sliding friction

At present, there is no generally accepted explanation of energy dissipation at asperity junctions. Many workers still maintain that the shearing of strong adhesion bonds between asperities is the main cause of energy dissipation. In what follows, it is shown that adhesion may be present, and, if it is, it will contribute to the observed friction force. However, it is also shown that adhesion is not necessarily present, and is likely to be a significant contribution only in high friction/high wear situations.

Energy dissipation at asperity junctions

There are only three possible mechanisms of energy dissipation at friction junctions: fracture, elastic deformation and plastic deformation. It will be shown later that many thousands of stress cycles are needed to create a wear particle, so the fraction of the total energy dissipation due to wear particle formation will be negligible. Also, as most metals are very close to being linear elastic, elastic hysteresis can also be ignored. Therefore, plastic deformation is the principal cause of metal friction.

This seems surprising as most asperities are flattened during the first few cycles, so that they then deform elastically. However, even when they are elastic, subsequent contacts will be close to the yield pressure, so that, on each contact, they can be approximated as highly loaded elastic Hertzian contacts. They will therefore be subjected to **sub-surface yield** on each subsequent contact, and the deformation involved will require energy dissipation. Furthermore, the resulting frictional stress will cause the sub-surface plastic zone to move closer to the surface, so that it will be less constrained, and a greater amount of plastic deformation could occur.

The interaction of deformation and adhesion

When two asperities make contact, force equilibrium demands that they must each be subjected to the same stress, and their behaviour during sliding is very dependent on their relative yield pressures. If the materials are identical, the opposing asperities will be strained in the same way, whether elastically or plastically.

If the yield pressures are significantly different, the softer material may deform elastically or plastically, but the harder one must remain nominally elastic. However, if the hardnesses do not differ by a factor of $> \sim2$, the harder material may undergo *sub-surface yield*.

If the hard material is very much harder than the softer one, stress similarity can only be obtained if the hard asperity penetrates the softer one, so that some ploughing occurs. In the discussion below, ploughing action is not considered.

Consider the *life of an asperity junction* in three stages: formation; shearing; and separation. This is an over-simplification, but it is helpful. The three stages can be characterised by the direction of the direct stress normal to the interface. During formation it is compressive; during shear it progressively falls to zero; and during separation, if adhesion occurs, it can become tensile.

During junction formation, the asperities must deform until they can pass at constant surface separation. Any plastic deformation can result in junction growth, disruption of protective films, and, possibly, adhesive bond formation.

If the interface is weak, sliding will take place at the interface and the only adhesive contribution to the friction will be the small force required to shear the interface. However, if the interface is strong, owing to the formation of strong adhesive bonds, plastic deformation can continue, initially under compression, then in pure shear and, finally, as the asperities begin to separate, in tension. Therefore, the adhesion and deformation theories are not competitive but complementary, with the energy being consumed in deformation being dependent on the strength of any adhesive bond.

2.6   Wear

2.6.1   Introduction to wear

Wear is defined as 'the progressive loss of substance from the operating surface of a body occurring as a result of relative motion at the surface', as shown by the Jost report (DES, 1966). This 'loss of substance' can be of enormous economic importance. In contrast to friction coefficients, wear rates vary by many orders of magnitude, and they can change catastrophically as the result of a small change in operating conditions, so the wrong choice of materials can have disastrous consequences.

Wear and friction arise from exactly the same asperity contacts, so, as with friction, there is no reliable way of predicting wear from first principles. The only safe way to select materials for wear resistance is on the basis of tests – realistic laboratory tests followed by field tests. However, there are now generally accepted qualitative explanations of wear behaviour, and they are described here. This chapter largely concentrates on wear of metals, with some mention later of other materials.

2.6.2   Mechanical wear processes

An excellent early survey of wear mechanisms (Burwell, 1957) listed four major mechanisms: adhesive wear, abrasive wear, surface fatigue and corrosion.

Adhesive wear

The theory of **adhesive wear** (again due to Bowden and Tabor) has the same basis as the adhesive theory of friction, i.e. strong cold welds are postulated to be formed between contacting asperities and these welds must be sheared for sliding to take place. The amount of wear then depends on where the junction is sheared. If it shears at the original interface, the contribution to wear is zero; if shear takes place away from the interface, a fragment of material is transferred from one surface to the other, usually from softer to harder but occasionally from harder to softer.

Note that the theory has only explained material transfer and not the formation of loose wear debris. In the original theory, it was simply stated that subsequent sliding causes some of the transferred particles to become detached.

Archard (1953) derived a theoretical expression for the rate of **adhesive wear**. It is assumed that the contact area consists of n circular spots of radius a. Each contact has an area of πa^2 and it supports a load of $\pi a^2 H$, where H is the hardness of the softer material. The opposing asperity passes over a softer asperity in a distance $2a$ and it is assumed that a hemispherical fragment is transferred of volume $2\pi a^3/3$. Then the **wear volume** δQ produced by one asperity contact in unit slid distance is given by

$$\delta Q = (2\pi a^3/3)/2a = \pi a^2/3 \tag{2.24}$$

and the total wear, Q, in unit slid distance is $n\pi a^2/3$.

Each contact, however, supports a load of $\pi a^2 H$, so the total load $N = n\pi a^2 H$, and:

$$Q = N/3H \tag{2.25}$$

This equation suggests three **laws of adhesive wear**:

1. The wear volume is proportional to the sliding distance.
2. The wear volume is proportional to the load.
3. The wear volume is inversely proportional to the hardness of the softer material.

The first of these 'laws' is found to be true over a wide range of conditions. The second law is generally true from low loads up to some load at which the wear increases catastrophically. The load where this occurs is at an apparent pressure of ~ $H/3$, i.e. at the point when the apparent pressure is equal to the **gross yield stress**. Therefore, at this point, the whole surface begins to deform plastically and the asperities are no longer independent. The third 'law' is often true, for similar materials, but there are other factors that can influence wear rates.

In one respect, Archard's equation is very misleading. It states that wear rate is equal to $N/3H$; whereas observed wear rates are much less than this, usually by several orders of magnitude.

The equation was derived on the assumption that every asperity contact produces a **wear particle**. Archard reconciled the theory with observed behaviour by suggesting that wear particles were only produced at a small fraction of contacts, so that:

$$Q = kN/3H \tag{2.26}$$

where $k \ll 1$, or:

$$Q = KN/H \tag{2.27}$$

where $K = k/3$. K is known as the **wear coefficient**, which can only be found by experiment. This concept is revisited later, when fatigue wear is discussed. Because wear rate is proportional to load, one can calculate a specific wear rate for any material under any chosen condition. This is also given the symbol k, and it should not be confused with the k used in the derivation of Archard equation.

The specific wear rate, k, is the volume worn per Newton load/per metre slid, so it has the units $m^3 N^{-1} m^{-1}$ or $m^2 N^{-1}$. The first set of units is preferable as it is a reminder of the meaning of the specific wear rate.

The **specific wear rate** of any material varies with factors such as the nature and roughness of the counterface, the temperature, the environment and, ultimately, the load, and it must be found by experiment. However, once it has been established for realistic conditions and over the allowable load range, it can then be used to predict wear within this range.

The predicted volume of wear for a load, N, and sliding distance L, is given by:

$$Q = kNL \ (m^3) \tag{2.28}$$

It is often more convenient to use $L = Vt$, where v is the sliding speed and t the time. So:

$$Q = kNVt \tag{2.29}$$

Again, it is often more useful to predict the **depth of wear** and this is simply Q/A, where A is the apparent area of the worn surface. So depth of wear, d, is given by:

$$d = kN/AVt = kPVt \qquad (2.30)$$

where P is the pressure over the apparent area. Therefore, depth wear rate d/t is proportional to pressure times sliding speed, the PV factor.

Abrasive wear

There are two types of **abrasive wear**, two-body abrasion and three-body abrasion, in each of which a soft material is ploughed by a relatively hard material. In **two-body abrasion** a hard surface rubs against a softer one; in **three-body abrasion**, hard particles, trapped between two surfaces, abrade one or both of them. One can obtain a semi-quantitative expression for the rate of abrasive wear by considering ploughing by a hard conical asperity of a softer surface, as shown earlier in Fig. 2.12. In travelling unit distance, the asperity displaces a volume $v = rd = r^2\tan\theta$. The normal load on the asperity, N_i, is equal to $0.5\,\pi r^2 H$, where H is the hardness of the soft material. So $v = 2N_i\tan\theta/\pi H$, for the single asperity. So, for all the asperities:

$$Q = 2N\tan\theta/\pi H \qquad (2.31)$$

where N is the total load.

As with adhesive wear, only a proportion k of contacts produce wear particles, thus:

$$Q = KN/H \qquad (2.32)$$

where $K = k\tan\theta/\pi$. This is similar to the equation for adhesive wear, with only the value of the constant being different, so the same wear 'laws' still apply. Not all the displaced material becomes loose wear debris; much of it simply piles up at the sides of the grooves, so typical values of k are ~0.1.

Two body abrasion is sufficiently well understood for it not to be a problem in engineering situations, although it is deliberately used in processes such as grinding and polishing. Three-body abrasion can still be an important cause of wear since many wear particles, such as oxides and work hardened metals are themselves abrasive. The solution to such problems is to flush away the abrasive particles and remove them by filtration.

Fatigue wear

Fatigue wear is important on two scales: macroscopic and microscopic. **Macroscopic fatigue wear** occurs at non-conforming loaded surfaces, such as those found in rolling contacts and sliding/rolling contacts such as gears; **microscopic fatigue wear** occurs at the contacts between sliding asperities.

Fatigue wear in rolling contacts

In well-lubricated rolling element bearings there is no progressive visible wear but the bearing life is limited by fatigue. Large wear particles are formed after a critical number of revolutions. Before this, there is no visible wear, but when the first fragments appear the bearing life is at its end. It is not useful to talk about the wear rate of a rolling bearing; the appropriate term is 'useful life'.

The *fatigue* failures that produce the wear fragments are due to repeated stress cycling with one cycle for each passage of a ball or roller. All sub-surface elements are subjected to shear stress cycles above the stress for sub-surface yield, and the accumulation of fatigue damage finally causes the propagation of fatigue crack and the formation of fragments.

The position of failure in a perfect material would be at the position of *maximum shear stress*. In practice, materials are not perfect, and the cracks are often nucleated at impurity particles. Bearing performance has been improved greatly in recent years by using increasingly high-quality materials with fewer and smaller defects.

The specified life of a rolling bearing is the minimum number of cycles that will be reached by 90% of similar bearings under identical conditions. It has been found that the life, L, is inversely proportional to the cube of the applied load W, i.e.

$$W^3L = \text{constant} \tag{2.33}$$

Fatigue wear in sliding contacts

Wear coefficients in engineering situations are usually in the range 10^{-3} to 10^{-7}. This raises two problems with the adhesion theory of wear:

1. Why are wear coefficients so low if all asperities are deformed plastically?
2. Why, when all gross plastic deformation of an asperity occurs in the first few cycles, do wear particles form after many thousands of further cycles of nominally elastic deformation?

Both these questions are answered if one assumes that, during the nominally elastic deformation, the asperities are subject to *sub-surface stresses* that give cyclic contained plastic deformation, and that the resulting damage accumulates to cause localised fatigue failures.

The wear coefficient can now be interpreted as the number of fatigue cycles needed, on average, to cause the detachment of a wear particle; if the wear coefficient is 10^{-6}, this means that, on average, it takes 10^6 fatigue cycles to form a wear particle. Also, although asperity deformation may be nominally elastic, cyclic sub-surface deformation can still lead to fatigue failures after many cycles. This explanation provides an explanation for the form of the wear laws, the formation of loose wear debris and the fact that hard materials can be worn by softer ones.

The form of the wear laws

It is known that the surface topography of most engineering surfaces is such that an increase in the load gives a proportionate increase in the true contact area, with the proportion of contacts in each stress state remaining constant. This means that one can write a general law of mechanical wear as:

$$Q = KA \qquad\qquad (2.34)$$

or:

$$Q = KN/\beta \qquad\qquad (2.35)$$

where β is a mechanical constant defining the **load-bearing capacity**. For plastic contacts, β is just the hardness H; for elastic contacts we know that many asperities are very close to yield, so β can be taken to be $\sim H$ for all practical situations. Therefore, as with Amonton's laws of friction, the wear laws are basically due to the surface topography of the wearing surface.

The formation of loose wear particles

Adhesion theory explains wear by stating that the adhesion between a harder material and a softer one is so strong that a welded junction breaks in the softer material, rather than at the interface. It then states that a loose particle is formed when the transferred material is detached during a subsequent contact. It is hard to see how this can be so. If the bond to the harder material is so strong that it is above the tensile strength of the soft material, how can the softer material be strong enough to break the bond between the transferred material and the hard asperity?

This problem does not arise with fatigue theory; the bond to the harder surface may be very weak, or even non-existent; the wear particle might have been finally detached just by collision with an opposing asperity.

Wear of dissimilar materials

If the hardnesses of two materials are such that, after running in, both can remain nominally elastic – a common situation in engineering – it is likely that each asperity of a contacting pair will be subjected to some sub-surface plasticity and, hence, fatigue. Both will be subjected to cycles of the same fatigue stress and repeated cycles will cause failure in both materials. However, at any particular stress, the harder material will require more stress cycles to form a wear particle, as illustrated in Fig. 2.15 and its wear rate will therefore be lower.

Therefore, fatigue wear explains the wear of a harder material by a softer one – which cannot be explained by adhesion theory.

Other wear processes

Corrosive wear

When rubbing takes place in a corrosive environment, reaction products can be formed on one or both surfaces. These products can then be removed by rubbing and the cycle of formation and removal can continue indefinitely. *Corrosive wear* can be much more severe than the effects of corrosion and wear acting separately.

The most common corrosive medium is oxygen in air. One may, therefore, describe *oxidative wear* but the principles would apply to any other corrosive environment. The first thing to note is that oxidation is usually beneficial as it prevents the metal to metal contact that causes severe adhesion.

As reactivity and film thickness are functions of both time and temperature, the effectiveness of an oxide film in protecting the metal can change dramatically with very small changes in rubbing conditions. The best known example of this is the mild-to-severe wear transition observed in the sliding of many unlubricated metal pairs. This is illustrated in Fig. 2.16, which shows a log–log plot of the variation of wear rate with load for carbon steel sliding on carbon steel. Below the first transition load, T_1, the wear rate is low, the surfaces remain smooth and the wear debris is finely divided oxide. At T_1, the wear rate increases by two orders of magnitude, the wear debris

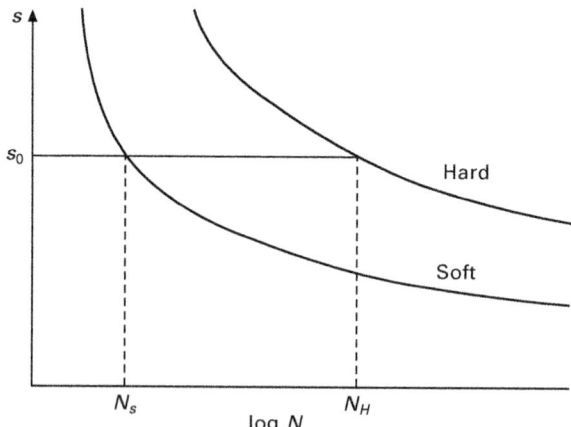

2.15 How dissimilar materials fatigued at the same stress will wear at different rates.

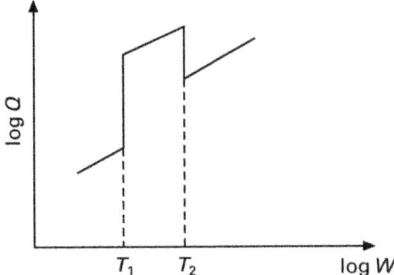

2.16 A mild-to-severe wear transition.

is relatively large metal particles and the surfaces are roughened. Over this range, there is insufficient time for the asperities to re-oxidise between contacts, so severe adhesion-enhanced wear takes place.

The rate of oxidation is very dependent on the local temperature, and this temperature depends on the input of frictional work, and therefore on the load. The rate of oxidation increases exponentially with temperature, and therefore approximately exponentially with the load, but the time between collisions decreases only linearly. Therefore, at T_2, the temperature effect predominates and the situation returns to one of mild oxidative wear. A similar effect to the one described above would also be seen as a result of increasing sliding speed.

Fretting

Fretting (Waterhouse, 1972) can take place whenever low-amplitude vibratory sliding takes place between two surfaces. It is a common occurrence because most machinery is subject to vibration, both in transit and in operation. Examples of vulnerable components are shrink fits, bolted parts and rolling bearings.

Fretting can combine many of the wear processes described earlier. The oscillatory motion causes fatigue wear, which can be enhanced by adhesion. The wear can also be combined with corrosion – principally oxidation – and the corrosion products can be abrasive. The fact that no macroscopic sliding takes place often means that wear debris cannot escape, but is trapped between the surfaces.

In fretting there may be no true macroscopic sliding. The surfaces may be in static contact in the central region of a contact, where the normal pressure is high, but subject to ***microslip*** at the periphery, where the pressure is low and the tangential traction is sufficiently high to overcome the static friction. Figure 2.17 shows ***stick and slip zones*** at such a fretting contact. In this situation, fretting takes place in the microslip region, fatigue cracks can be nucleated, and this can lead to gross component failure.

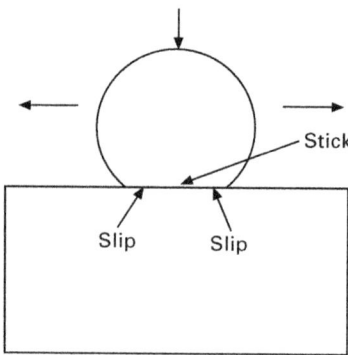

2.17 Stick and slip zones at a fretting contact.

Erosion

Erosion is damage experienced when liquid or solid particles impinge on a solid surface. There are two types of fluid erosion: liquid impact erosion and cavitation erosion

Liquid impact erosion: There are different damage mechanisms for brittle solids, ductile metals and polymers. On brittle materials, liquid impact generates momentary stresses that can be sufficiently high to cause cracking in initially unconnected ring cracks. With further impact, the cracks eventually join and material is removed in the form of chips. The erosion damage of ductile materials, such as most metals and thermoplastic polymers, initially takes the form of surface depressions with raised edges. With further impact the edges eventually fail, giving surface roughening. The erosion damage of brittle thermosetting polymers is as described for brittle materials. Fluid erosion damage can be very severe, for example on the leading edges of the wings of fast aircraft flying through rain.

Cavitation: Cavitation damage occurs when bubbles in a liquid implode against a solid surface. It is similar to liquid impact erosion but on a finer scale. It can cause severe damage to solids that are moving in a liquid, such as ships' propellers and pump impellers.

Solid erosion: When a stream of solid particles strikes a solid surface, the rate of erosive wear depends on the angle of incidence of the particles, and the wear rates for ductile and brittle materials follow different curves as shown in Fig. 2.18.

For ductile materials, it is thought that two damage processes are occurring: *cutting wear*, which predominates at low angles, and deformation wear, similar to that for liquid erosion, which occurs at high angles. The observed

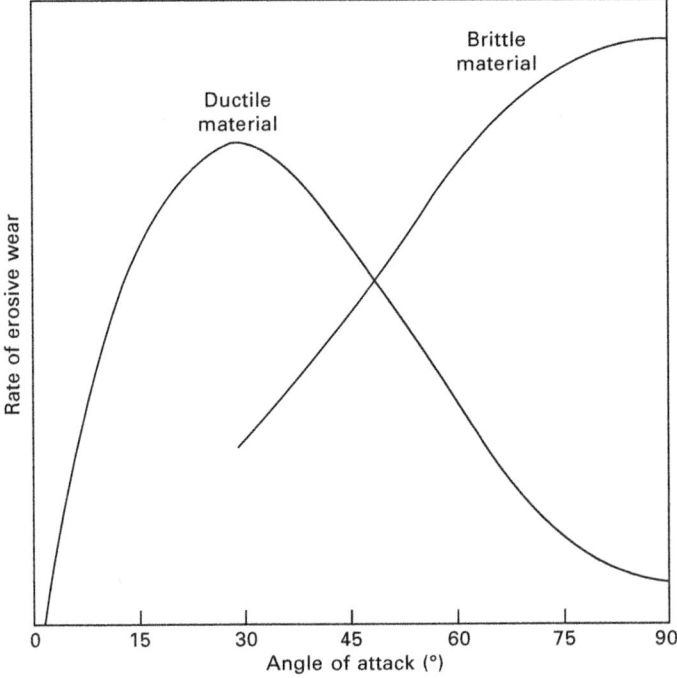

2.18 The dependence of the erosion rate on the impingement angle.

maximum wear rate, at ~20°, is where the combination of the two effects is at its maximum value. For brittle materials, erosion is similar to that with liquid impact, with wear particles being formed as a result of crack coalescence. There are no reliable methods of predicting rates of erosive wear.

Wear of lubricated surfaces

Fluid lubricants have three important effects in reducing both friction and wear at sliding surfaces.

1. They reduce, or even eliminate adhesion by providing a *low shear strength film* between the surfaces.
2. They cool the surfaces both by reducing frictional heating and by removing heat.
3. They carry wear debris away from the interface, thus preventing three-body abrasion by wear debris.

Wear coefficients are almost invariably reduced if surfaces are continuously lubricated with oil. However, the prediction of wear rates is even more difficult than in the case of dry rubbing. This is because, at any time, the lubricant is preventing solid contact over an unknown proportion of the surface, and there is one wear coefficient for the protected regions and a higher one for the regions that are unprotected.

Scuffing

Scuffing is a particularly damaging form of wear in highly loaded lubricated contacts such as piston rings and gear teeth. There is local penetration of the lubricant film, and breakdown of the boundary lubricants, leading to localised solid phase welding. When the weld is broken, the surfaces are roughened to such an extent that they continue to penetrate the oil film at each subsequent contact, with both the severity and the extent of the roughening increasing. This can lead to very rapid gross damage. It is a feature of scuffing failure that it occurs very rapidly. A mechanism that has been running smoothly can suddenly become noisy, and failure often ensues within minutes.

Wear of ceramics and cermets

Ceramics can exhibit very low wear rates, compared with metals, and they are being used increasingly in tribological situations. They also retain similar properties up to high temperatures and they are very unreactive chemically, so they will certainly find increasing use in hostile environments. Although they have similar elastic moduli to metals, they have very much higher hardness, so they are much more likely than metals to be subjected to fully elastic deformation. **Ceramics** also have low density and, as weight saving becomes increasingly important, the combination of high hardness, low wear and low density will favour increased use of ceramic tribological components.

Historically, bulk ceramics have contained many flaws and internal cracks, and this has made them prone to bulk fracture and lack of reliability. However, ceramic technology is improving rapidly, so that more homogeneous materials with fewer and smaller cracks are being produced. Also, problems of bulk fracture are being avoided by the use of cermets, i.e. materials that consist of mixtures of metal and ceramics.

Polymers and composites

Bearings made from polymers and composites have many advantages and are widely used. In automobiles, they find uses in such areas as steering joints, where their adoption has obviated the need for frequent re-greasing. However, they are not relevant to the operation of engines and powertrains and are not covered here. Readers requiring more information on polymer based bearings readers should consult the excellent ESDU Unit 96015: Design and material selection for dry rubbing bearings

2.7   Future trends

Many improvements in the tribological performance of engines and powertrains will no doubt be made by continuing development of advanced lubricants, and these topics will be covered in detail elsewhere in this book. Here, likely improvements to be made though advances in materials technology are discussed.

Advances will certainly be made due to the development of improved constructional materials, particularly through the increasing use of reliable, tough ceramics and cermets. However, a material that would be selected for its tribological properties might not be suitable for the bulk of the component. High hardness helps to resist wear, but hardness is often associated with brittleness, and this can often make a material that would be chosen for its wear resistance unsuitable for the bulk of a component. Also many materials with good tribological properties can be very expensive and difficult to fabricate, so the cost of using them could be very high, and prohibitive for mass market products such as automobiles. Therefore, many components to be used in tribological situations should ideally be made from two materials: a bulk material that provides the necessary strength and toughness, but is cheap and easy to fabricate, and a surface material that provides the necessary tribological properties. Technologies that allow this independent choice of bulk and surface materials are examples of surface engineering. Surface engineering techniques have been in use for many years, examples being *phosphating* of steel to improve running in, case hardening of gear teeth to give the required combination of a hard wear-resistant surface on a tough strong core, and diffusion processes such as *carburising* and *nitriding* to achieve similar effects. However, recent advances are likely to lead to rapidly increasing use of such techniques. These advances have largely been in the field of surface coating.

Highly adherent coatings of very high quality can now be deposited by a variety of techniques at thicknesses from < 1 to > 100 μm (Datta and Burnell-Gray, 1997). *Plasma spraying*, detonation gun plating and high-velocity-oxygen-fuel (HVOF) plating can be used to deposit metallic coatings and a wide range of ceramic and cermet coatings, typically of the order of 10 to 100 μm thick.

Finally, coatings only of the order of 1 μm thick can be deposited by physical vapour deposition (PVD) and chemical vapour deposition (CVD). These are rapidly developing technologies, which can produce spectacularly successful results in terms of wear resistance and low friction.

2.8   Sources of further information and advice

There are many textbooks available on the subject of friction and wear, with a variation in styles and approaches to the subject. A representative selection of established texts would include those by: Rabinowicz (ISBN 0-471-83084-4); Bowden and Tabor re-issue (ISBN 0-19-267011-5); Rigney (ISBN 0-871-70115-4); Hutchings (ISBN 0-849-37764-1); Halling (ISBN 0-333-24686-1); Williams (ISBN 0-198-56503-8); Arnell *et al.* (ISBN 0-387-91402-1); Bayer (ISBN 0-824-79027-8).

The principal journals that publish papers in the area of friction and wear are *Wear* (Elsevier); *Tribology International* (Elsevier); *Journal of Tribology* (ASME); *Tribology Transactions* (STLE); *Proc I Mech E, J Journal of Engineering Tribology* (Professional Engineering Publishing)

The worldwide web is also an invaluable data source.

2.9   References

Archard J F (1953), 'The contact and rubbing of flat surfaces', *J Appl Phys*, **24**, 981–988.

Bowden F P and Tabor D (1950), *The Friction and Lubrication of Solids,* Pt 2, Oxford University Press.

Burwell J T (1957), 'Survey of possible wear mechanisms', *Wear*, **1**, 119–141.

Datta P K and Burnell-Gray J S (1997), *Advances in Surface Engineering*, Royal Society of Chemistry.

DES (1966), *Lubrication (Tribology) Education and Research,* HMSO.

Edwards C M and Halling J (1968), 'An analysis of the plastic interaction of surface asperities and its relevance to the value of the coefficient of friction', *J Mech Eng Sci*, **10**, 101.

Greenwood J A and Williamson J B P (1966), 'Contact of nominally flat surfaces', *Proc R Soc*, **295A**, 300.

Johnson K L (1985), *Contact Mechanics*, Cambridge University Press.

Kogut L and Etsion I (2003), 'A finite element based elastic-plastic model for the contact of rough surfaces', *Trib Trans*, **46**, 3, 383–390.

Waterhouse R B (1972), *Fretting Corrosion*, Pergamon.

Surface phenomena in thin-film tribology

P. PROKOPOVICH and H. RAHNEJAT, Loughborough
University, UK and M. TEODORESCU,
Cranfield University, UK

Abstract: With miniaturisation of mechanisms and devices, load-bearing and transmitting conjunctions are progressively reducing in size. At the same time the gap size between contiguous solids is continually decreasing. These conjunctions are also subject to quite low loads. A whole new era in tribology has dawned which encompasses a wide area of interest from micro-electromechanical systems to data storage devices (ultra-smooth surfaces) to biological and bio-inspired systems, such as microfibres mimicking contacts of insects and geckos with very smooth surfaces. The field of study is broadly termed *nano-tribology*, but for biomimetics may be referred to as *nano-biotribology*. Relatively lightly loaded contacts also occur, for example in fairly rough elastomeric seals, prevalent in drug dispensing devices such as syringes and inhalers with low viscosity biocompatible compounds. Also, gels and cosmetic creams are applied with low load on fairly rough surfaces. Therefore, another new area of nano-tribology is emerging, which may be regarded as *pharmaceutical tribology*. In nearly all such conjunctions surface energy effects, adhesion, van der Waals interaction, electrostatic effect, meniscus action, hydration or solvation play more significant roles than the usual hydrodynamic viscous action as mechanisms of lubrication/tribology.

Key words: surface energy, meniscus action, van der Waals interaction, hydration, bio-mimetics, asperity adhesion.

3.1   Introduction

Dynamics of bodies and systems or tribology of contacts are ultimately described in terms of prevailing balance of forces. This has been the human understanding of nature and all the devised mechanisms and machines ever since proper description of what is understood to constitute a *force* in the seventeenth century by Newton (1687). The most important force of nature is force of gravity $F = (GMm')/r^2$. Although empirical in nature, it fitted the measured observations made previously by Kepler (1609). It is no wonder that, apart from some exceptions the successive pioneers tried to fit similar force laws to other interaction phenomena, such as Coulomb's electrostatic force between charged particles, which follows the same inverse distance squared law as that of universal gravitation. The elegance in the universality of gravitational constant, however, could not be retained for constants of

73

proportionality between the force and distance for these other force laws (i.e. K' in $\propto K'/r^2$). Nor have these empirical laws withstood the fundamental physical scrutiny in time, as noted by Teodorescu and Rahnejat (2008a).

Israelachvili (1992) shows that in the small scale the various force laws necessarily deviate from the Newtonian inverse distance squared force law in order to retain any physical sense vis-à-vis observations or intuition. For small particles and molecules, which one is concerned with in ultra-thin tribological conjunctions, the interactions follow inverse powers of distance exceeding 3. For example, **van der Waals** interactions follow $1/r^6$ and electrostatics: $1/r^{12}$ according to the Leonard–Jones potential (see Chapter 33; Atkins, 1986; Israelachvili, 1992; Gohar and Rahnejat, 2008). When a distribution of such particles or molecules is encountered within a conjunction (a system of finite size), then depending on geometry, forces are generated that deviate from the law of universal gravitation in their form. Some of these are described briefly later. The interactions described so far (particles and molecules) are at closer range than one is familiar with in traditional tribology (they are in nano-scale). In micro-scale (from tenths to tens of micrometres) the interaction forces also change in form. In fact, they bear no resemblance to those either in nanocosm or in the large scale. Take the case of Amontons' friction, described in Chapter 2. This is another empirical law, whose validity as shown in Chapter 2 can be reasoned in quite a simple manner.

Thus, one is left with a conundrum; either one should accept that force is when *size really matters*, thus nature has many complex facets or that Einstein's principle of parsimony holds, which means that the existing force laws are worryingly empirical and analyses may yet prove to be seriously misguided. A sobering thought, which one should always bear in mind when a plethora of empirical force laws are used to describe a conjunctional behaviour.

If one chooses, out of compulsion driven by incomplete knowledge, to accept the notion of a complex nature, then one would arrive at sciences used in this volume among others. Then, the theories at macrocosm differ from those at microcosm and are even further apart in nano-scale. This is the basis of this chapter too. If, on the other hand, a unified force law is seen as underlying, then a generic interaction potential must reside at the heart of nature, which only manifests itself according to *size* of *bodies* (or *feature on surfaces*) and that of an assumed *system* (see, for example, Rahnejat, 2008). It is almost certain that this *unified force law* would be neo-Newtonian.

Now, accepting the empirical nature of force laws, one can see that those applied in micro and nano-scales, and not necessarily in both, or at least in the same manner, form the basis of today's ultra-thin film tribology.

This chapter deals with surface interaction phenomena. In doing so, one should note that the problem of scale described thus far is also further extended to physical and chemical properties (i.e. medium of interactions).

Note that this dependence on the medium of interactions accounts for the deviation of force laws from universality (unlike the constant G in the law of universal gravitation). This is painfully manifested in the different regimes of lubrication between contiguous solids. One can see that conjunctional behaviour is different when wetted to varying degrees (see Teodorescu and Rahnejat, 2008b).

Thus, the established force laws are empirical, because they are discriminated by *size* and *wetness*, in other words by interaction scale and ***physical chemistry***. Therefore, the structure of this chapter follows this established classification of force laws.

3.2   A question of wetness

Chapter 2 describes dry contact of rough surfaces, which is generally referred to as ***boundary regime of lubrication***. This is shown in Fig. 3.1 (not to scale). This type of graph is referred to as the ***Stribeck curve*** (1907) (see also Chapter 6). When a film of fluid completely separates the surfaces, so that no direct surface-to-surface interactions can take place, then a ***viscous regime of lubrication*** exists in traditional micro-scale tribology (elastohydrodynamic, see Chapter 6, and hydrodynamics, see Chapter 5). The intervening regime is referred to as ***mixed lubrication***, where part of the surfaces is lubricated and part of them suffers direct contact. $\lambda_s = h/\sigma$, known as the ***oil film parameter***, on the horizontal axis is the ratio of the depth of the fluid film to that of the root mean square of surfaces roughness (described in Chapters 2 and 6).

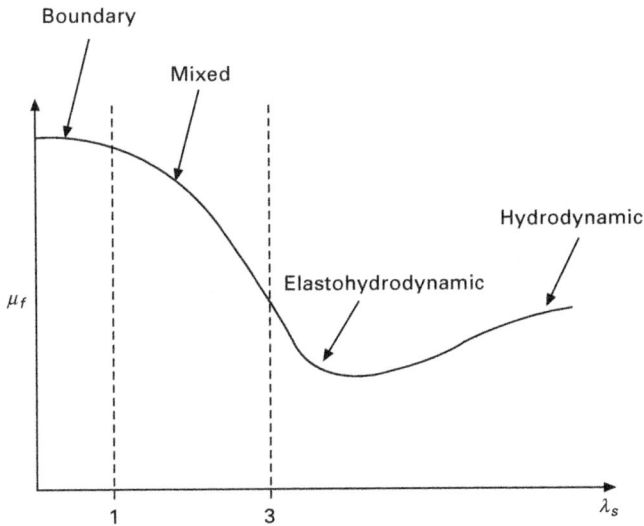

3.1 Stribeck curve.

The vertical axis is usually the *coefficient of friction*, μ_f, this representing the percentage of losses (i.e. ≈ 0.2–0.4 for direct interaction of dry metallic surfaces represents 20–40% wastage of input work, for example, in sliding motion). Finally, when such dry contacts occur, the regime of lubrication is considered to be boundary. Chapter 2 deals with boundary interactions, essentially for the micro-scale.

As the surface roughness is further reduced to nano-scale (almost smooth surfaces), *adhesion* is significantly enhanced. In fact, Gohar and Rahnejat (2008) observe that ideal smooth surfaces in vacuum would have very high adhesion. In practice, however, exposed surfaces have thin *low shear strength oxide films* formed upon them, which reduce the effect of adhesion and aid *boundary lubrication*. Solid lubricants/coatings are also used to reduce friction (see Chapter 4).

There is a tendency to reduce surface roughness in conjunctions where it is deemed that full fluid lubrication cannot be assured. This is increasingly achieved by hard wear-resistant coatings (refer to Chapter 2 and concept of plasticity index). However, this is not a generic solution and in fact in many instances rougher surfaces can entrap lubricant and reduce friction (see also Chapters 13, 14 and 20). In fact, a completely dry and idealised smooth surface is very unlikely, and usually some lubricant or moisture is adsorbed to the roughness features. Therefore, in practice additives are included in base lubricants to form low shear strength surface films to reduce friction in relative motion of contiguous surfaces, where no significant entraining motion of bulk lubricant is expected. These additives are referred to as *friction modifiers*. This has been the solution of choice in micro-scale tribology. In nano-scale the formation of surface films is due to quite complex mechanisms described later.

Thus, the Stribeck curve cannot be regarded as quantitative, but an indication of contact conditions, when a proper analysis is carried out (similar to those, for example, in Chapters 6 and 20). However, ideas behind the Stribeck curve are very useful for practitioners in industry, where a quick assessment of prevailing conditions is desired in the competitive environment which pervades the current global competition.

There are an assortment of mechanisms that play some role in wetting of surfaces and formation of desired surface films. These mechanisms depend on the physiochemical properties of the lubricant and its molecular composition, as well as surface topography and interaction energies between the fluid and adjacent solids, as well as between the solid surfaces themselves. These interactions are also dependent on the physics of scale, both in terms of gap and molecular size (gap; representing the size of the system, and molecule being the generic body: see Rahnejat, 2008). It turns out that the current knowledge for these various ultra-thin surface film-forming mechanisms is empirical and not one that either unifies them or distinguishes between them as a problem of scale only (i.e. still retains other physiochemical and thermodynamic properties). However, the prevailing knowledge is sufficient to note that the usual dominant viscous action in micro-scale (see Chapters 5 and 6) becomes insignificant in vanishingly small gaps with light load and also on free surfaces.

One mechanism wetting the surfaces, depending on its property and that of the fluid is known as *surface tension*. In short a volume of the fluid clings to the surface, and in the case of contacting surfaces in relative motion it forms a meniscus between them, drawing them together. Therefore, this action is sometimes referred to as *meniscus action*. Menisci can also form between opposing asperities of contiguous surfaces.

3.3   Meniscus action: surface tension

Molecules of a fluid in its bulk are *held* together by **cohesive forces** acting between them. This action ensures a uniform distribution of the fluid molecules in the stationary bulk. At the boundaries of the volume, the fluid molecules come into contact with that of another substance which constitutes the boundary of the volume. The molecules of the fluid now interact with those of the boundary with **adhesive forces**. This means that the distribution of the fluid molecules near the barriers differs from that in its bulk volume. Sometimes this is clearly visible, for example, the top layer of water in a glass, in touch with air. It looks as if there is a skin formed on the water surface. This is an interfacial layer formed by the water molecules. The force per unit length of the layer (N/m) required to form this interface is called **surface tension**, γ. This phenomenon is not confined to the water–air interface, but for any combination of liquid–vapour (γ_{lv}), solid–liquid (γ_{sl}) and solid–vapour (γ_{sv}). These surface tensions can be thought of as free energy exchanges that form interfaces between the different phases. Since contact conjunctions, being between macroscopic objects, such as a ball on a plane or a pair of opposing asperities on microscopic scale, comprise contiguous solids drawn together by a liquid in presence of some vapour (usually air), one needs to obtain the meniscus force. This is the force required to be overcome to separate the bodies. Section 3.5 describes methods of measurement of surface tension, as values are not always available.

When the contact of a hemispherical body of radius R against a flat plane is properly wetted, a meniscus of radii r_1 and r_2 (see Fig. 3.2) is formed between them. The same would be true of any pair of surfaces of ellipsoidal solids in contact, including, at small scale, a pair of assumed opposing asperities. The pressure inside the meniscus is less than that outside it, thus facilitating its formation. However, unlike Reynolds equation, there is no flow through the contact and a solid roof is not required (see Chapter 5) to form the conjunction. It turns out that another fundamental equation governs the

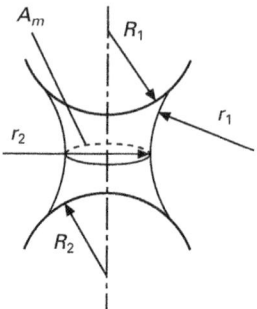

3.2 Geometry of a meniscus.

formation of interfaces in fluid mechanics (see Middleman, 1998; Gohar and Rahnejat, 2008). This is the **Young–Laplace equation** (not derived here):

$$\Delta p = -\gamma_{lv}\left(\frac{1}{r_1} + \frac{1}{r_2}\right) = -\frac{\gamma_{lv}}{r'} \tag{3.1}$$

In Fig. 3.2 the surface area of the sphere R immersed into the volume of meniscus is: $A_m \approx 2\pi Rd = 4\pi R r_1 \cos\theta$. Thus, the meniscus force is (Bowden and Tabor, 1950):

$$F_m = \frac{\Delta p}{A_m} = -4\pi R \gamma_{lv} \cos\theta \tag{3.2}$$

Note that Δp is negative indicating that the meniscus force draws the contacting surfaces together. θ is termed the **contact angle**, which is defined as: the angle formed by the solid–liquid or liquid–vapour interface, measured through the liquid (see Section 3.4). Equation (3.2) assumes counterface surfaces of same material. If dissimilar surfaces are considered the contact angle that a fluid makes with them may be quite different. Thus, the product $4\cos\theta$ is replaced by $2(\cos\theta_1 + \cos\theta_2)$. In fact, the contact angle is an indication of degree of success that a meniscus bridge would be formed between the contiguous surfaces (i.e. the effectiveness of the wetting of a surface by a given fluid). Note that when $\theta = 90°$, the meniscus force in equation (3.2) vanishes, and when $\theta > 90°$, $F_m > 0$ (i.e. a repulsive force). Physical interpretation of this is when a liquid fails to wet a surface (forms spherical droplets that easily fall from the surface). Good examples of this are the droplets formed on duck feathers that are easily dislodged or raindrops that fall readily from waxy leaves, such as that of lotus leaves. This means that the cohesive force between molecules of water within a raindrop represent stronger interactions than the adhesive forces which strive to adhere them to a waxy leaf surface. Nature is intrinsically intelligent, expending the minimum amount of energy required to form the spherical droplet shape (least surface area). In dynamics this concept is well established as Maupertis principle of least action (see for example Brillouin and Brennan, 1964). To describe the surfaces that water does not wet successfully the word ***hydrophobic*** was coined. This attribute is now readily used for all surfaces vis-à-vis all fluids. The converse of the foregoing is true of other surfaces, which are referred to as ***hydrophilic***. If the topography of such surfaces is investigated in some detail (see Chapter 2) it becomes clear that relative size of their surface features to that of water molecule (≈ 0.3 nm) determines their wettability. In the case of waxy leaves, their smallest roughness features are such that molecules of water cannot easily reside upon them. Thus, wettability is a function of fluid rheology and surface topography, a fact that is increasingly exploited in many applications, from inkjet printing, spray cooling of engines to crop spraying. However, there remains a long way to go in engineering of surfaces (see for example Chapters 13 and 14) and engineering the physical chemistry of coolants and lubricants. The discussions here should also have implied to the reader that formation of menisci is also ultimately a question of scale. Note that as $\theta \rightarrow 0$, F_m attains its maximum value. This is in fact the limit of adhesion due to meniscus action alone. However, as described later other forces would also contribute to the work of adhesion of rough surfaces in very close contiguity, while some others inhibit the same through repulsion.

3.4   Contact angle of liquids

Contact angle is an important thermodynamic parameter which is used as a measure of the hydrophobicity of a solid surface as described above (Fig. 3.3). The contact angle of a solid–liquid–vapour system is generally defined as the angle between the tangent to the liquid and the tangent to the solid surface, as shown in Fig. 3.4(a).

Depending on the nature of the solid and the liquid the three cases of *wetting* are distinguished: non-wetting (Fig. 3.4(a)), partial wetting (Fig. 3.4(b)) and complete wetting (Fig. 3.4(c)). A typical example of the non-wetting case is a drop of mercury deposited on a glass slide, while a drop of

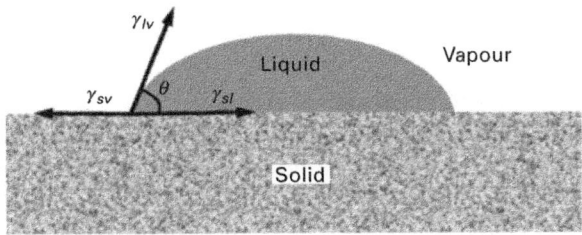

3.3 A sessile drop with contact angle θ.

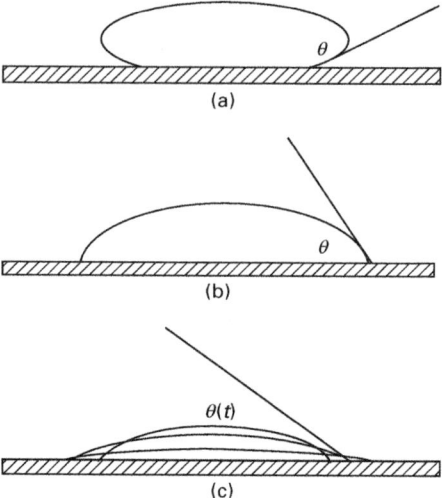

3.4 (a) Non-wetting case: $\theta > \pi/2$; (b) partial wetting case: contact angle $0 < \theta \leq 90$ (c) complete wetting case: the drop spreads out completely and only dynamic contact angle can be measured: $\theta(t) \rightarrow 0°$.

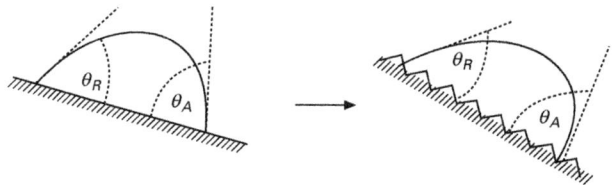

3.5 Receding and advancing contact angles.

water partially spreads on the same glass slide and finally, complete wetting is achieved when an oil drop is placed on a glass substrate.

For an ideal (smooth, homogeneous, rigid and insoluble) solid surface, the thermodynamic equilibrium condition, which is well known as the **Neumann–Young equation**, predicts a single contact angle:

$$\cos \theta_{eq} = \frac{(\gamma_{sv} - \gamma_{sl})}{\gamma_{lv}} \qquad (3.3)$$

where θ_{eq} is the equilibrium contact angle.

In practice, real surfaces are not smooth (see Chapter 2). Thus, the contact angle of a liquid drop advancing on a solid surface (θ_A) is usually different from that obtained when it is receding (θ_R) (see Fig. 3.5). The difference is known as **contact angle hysteresis** (θ_{hyst}). This phenomenon has been studied extensively both from theoretical and experimental viewpoints. Historically, contact angle hysteresis was attributed to surface roughness (see for example Johnson and Dettre, 1964; Lin *et al.*, 1993) and surface heterogeneity (Johnson and Dettre, 1965; Neumann and Good, 1972; Brandon and Marmur, 1996; Rangwalla *et al.*, 2004). The point to bear in mind is that the contact angle depends on surface texture and orientation, which itself depends on the manufacturing processes used to generate it. Thus, a unique value does not exist for a given fluid–surface material type combination.

Equation (3.3) does not take into account microscopic surface roughness or surface heterogeneity. If surfaces are not smooth, the microscopic surface roughness is known to have a significant influence on the contact angle, which is given by the ***Wenzel equation*** (1936):

$$\cos \theta = R_f \cos \theta_{eq} \tag{3.4}$$

where θ is the contact angle for a rough surface, θ_{eq} is the contact angle for a smooth surface and R_f is a roughness factor.

However, it is important to emphasise that hysteresis is not limited only to rough and heterogeneous surfaces. In recent years other causes for hysteresis have been identified. These include the following:

- Molecular-scale topography and rigidity of molecules (Fadeev and McCarthy 1999; McHale *et al.*, 2004). Note that the former brings again the question of physical scale, and the latter indicates that unlike the Newtonian slow viscous motion, where molecules are considered to be ideally rigid (see Chapter 5), molecules in real fluids can actually deform.
- Change in the organisation of molecules of the solid surface (Yasuda *et al.*, 1994): recall the equilibrium of cohesive and adhesive forces described above.
- Penetration of liquid and swelling of the solid (Sedev *et al.*, 1996; Hennig *et al.*, 2004).
- Liquid sorption and retention (Sedev *et al.*, 1996; Lam *et al.*, 2002).
- Size and shape of liquid molecules (Lam *et al.*, 2001).
- Strong interactions between molecules of solid and liquid (Extrand, 2004).
- Rate of motion of the three-phase line on the solid surface (Elliott and Riddiford, 1967; Johnson *et al.*, 1977; Blake, 1993).

3.5   Estimation of interfacial tension between a liquid and a solid

Direct measurement of ***surface tension*** of liquids is straightforward and several techniques can be used. These are (Wilhelmy, 1863; Rusanov *et al.*, 1996):

- Wilhelmy plate technique;
- drop weight method;
- oscillating jet method;
- capillary wave method;
- spinning drop method;
- static drop shape techniques.

In contrast, owing to the immobility of molecules in a solid phase, solid surface tensions cannot be measured directly. Hence, several indirect approaches, both experimental and theoretical have been used. They include contact angle measurements with different liquids (Zisman, 1964; van Oss *et al.*, 1988; Good and van Oss, 1992) and direct force measurements (Derjaguin *et al.*, 1975; Johnson *et al.*, 1971; Pashley *et al.*, 1985; Christenson, 1986).

3.5.1   Determination of free surface energy parameters

The total *free surface energy* γ^{TOT} consists of two components:

$$\gamma^{TOT} = \gamma^{LW} + \gamma^{AB} \tag{3.5}$$

γ^{LW} is the polar component of the free surface energy associated with Lifshitz's van der Waals interactions, γ^{AB}, which is the acid–base component of the free surface energy. γ^{AB} results from the electron-donor (γ^-) and electron-acceptor (γ^+) molecular interactions (i.e. *Lewis acid–base interactions*). The acid-base term is expressed as the product of the electron donor and electron acceptor parameters as:

$$\gamma^{AB} = 2\sqrt{\gamma^+ \gamma^-} \tag{3.6}$$

The interfacial energy, γ_{sl}, is defined as (van Oss *et al.*, 1988):

$$\gamma_{sl} = (\sqrt{\gamma_s^{LW}} - \sqrt{\gamma_l^{LW}})^2 + 2(\sqrt{\gamma_s^+ \gamma_s^-} + \sqrt{\gamma_l^+ \gamma_l^-} - \sqrt{\gamma_s^+ \gamma_l^-} - \sqrt{\gamma_s^- \gamma_l^+}) \tag{3.7}$$

where the subscripts s and l refer to the solid and liquid phases, respectively. Now the Young equation can be combined with that of Young–Dupree to yield (Middleman, 1998; Gohar and Rahnejat, 2008):

$$W_{svl} = \gamma_{lv}(1 + \cos\theta) = \gamma_{sv} + \gamma_{lv} - \gamma_{sl} = -\Delta\gamma \tag{3.8}$$

where W_{svl} is the *work of adhesion*.

Now substituting the appropriate expressions:

$$W_{svl} = \gamma_{lv}(1 + \cos\theta) = 2(\sqrt{\gamma_s^{LW} \gamma_l^{LW}} + \sqrt{\gamma_s^+ \gamma_l^-} + \sqrt{\gamma_s^- \gamma_l^+}) \tag{3.9}$$

It can be seen that a set of three simultaneous equations can be obtained for three liquids with known parameters.

An example: to determine surface energy of two biopolymer films: HPC (hydroxylpropylcellulose) and CMC (carboxylmethylcellulose), three liquids of different polarities (water, formamide and hexadecane) were used with their known surface energy parameters (see Table 3.1). Contact angles of these liquids on the HPC and CMC films were measured using a goniometer (Prokopovich and Perni, 2009) and the average contact angles recorded (Table 3.2). Solving equation (3.9) for the three liquids and for each biopolymer film gives: γ_s^{LW}, γ_s^+, γ_s^-, γ_s^{AB} and γ_s^{TOT} (Prokopovich and Perni, 2009) (Table 3.3).

3.5.2   Direct force measurements

A direct method is to obtain the ***work of adhesion***. This is the energy required to break through a junction formed between two surfaces, usually pairs of asperities. Since the tests are usually carried out in normal atmosphere, menisci may also be formed between a pair of rough contiguous surfaces. Such surfaces may be fully or partially wetted or indeed not wetted at all, although the last is unlikely unless perfectly dry surfaces are in contact in

Table 3.1 Surface free energy parameters of the liquids at 20 °C used for contact angle determination

Liquids	γ^{TOT} (mJ/m²)	γ^{LW} (mJ/m²)	γ^{AB} (mJ/m²)	γ^+ (mJ/m²)	γ^- (mJ/m²)
Water	72.8	21.8	51	25.5	25.5
Formamide	58	39	19	39.6	2.28
Hexadecane	27.2	27.2	0	0	0

Table 3.2 Contact angle in ° of liquids at 20 °C on synthetic polysaccharide-based films (Prokopovich and Perni, 2009)

Film material	Water	Formamide	Hexadecane
HPC	71	63	21.6
CMC	87	32	13

Table 3.3 Surface energy parameters for synthetic polysaccharide-based films (Prokopovich and Perni, 2009)

Film material	γ_S^{LW} (mJ/m²)	γ_S^+ (mJ/m²)	γ_S^- (mJ/m²)	γ_S^{AB} (mJ/m²)	γ_S^{TOT} (mJ/m²)
HPC	25.33	17.7	0.48	5.85	31.17
CMC	25.16	0.27	5.87	2.54	27.71

vacuum. In this method solid surface tension values are determined using an *atomic force microscope* (AFM). The method is based on measurement of the pull-off force of the AFM tip of radius R from a surface. For rough real surfaces, one would need an adhesion model. There are a number of such models, depending on average asperity attributes, such as compliance. The main models are referred to as the JKR model (Johnson *et al.*, 1971) and the DMT model (Derjaguin *et al.*, 1975). Both these models (described later) correlate the pull-off force to the work of adhesion as:

$$F_c = c \pi R W_{svl} \tag{3.10}$$

where F_c is the pull-off force and W_{svl} is the energy of adhesion (or solid surface tension). The constant c depends on the choice of model. For DMT model, $c = 2$, and for JKR model, $c = 1.5$. In general, the DMT model is more appropriate for systems with hard materials having low surface energy and a small radius of curvature for the AFM tip. The JKR model is more suited to softer asperities with higher surface energy and with use of larger AFM tips. However, interpretation of the results is not always straightforward. A detailed characterisation of a substrate material with an AFM probe is crucial in identifying asperity geometry; shape and size. The adhesion models are based on ideal contacts such as hemispherical asperities and perfectly hemispherical tips. In practice real systems deviate from these idealised geometries, causing problems in interpretation of measured AFM tip pull-off forces. As a result, calculated values of work of adhesion from equation (3.10) are prone to errors. Also note that the AFM tips are usually very delicate and subject to wear and damage.

3.6   Adhesion of rough surfaces

The mechanism of *adhesion* is described in Chapter 2. Briefly, any pair of bodies in loaded contact can *stick* by forming *adhesive junctions* through interaction of their rough topography. It turns out that adhesion is also a problem of scale. Smoother surfaces with smaller asperities form stronger *adhesive junctions* with many asperity pairs involved. The strength of these junctions depend on the compliance of asperities, thus various models have been developed, among them the JKR and DMT models mentioned above, all of which consider direct asperity interactions (i.e. dry contacts). The work of adhesion is the effort required to break off these asperity junctions and initiate motion. For fairly smooth surfaces the work of adhesion, therefore, can be quite significant relative to the applied load. Thus, ideal smooth surfaces would have very high coefficients of friction, which seem to suggest a contrary argument to those of Amontons (1699) and de Coulomb (1784). However, such ideal smooth surfaces do not exist in practice due to manufacturing processes. Also, apart from in vacuum, the exposed top layer of surfaces is normally covered with a thin film of oxides or by moisture. Therefore, an important point to note is that measured surface energy or work of adhesion is usually due to combination of surface films/oxides and solid surface.

One discerns the **work of adhesion** when a force is required to initiate relative motion of contacting surfaces. A tangible example is the effort required to pick an inverted glass from a draining board, some time after it was washed and left there to dry. Both the glass lip and the draining board surfaces are rough and moisture has also formed tiny menisci between some of the opposing asperities. This pull-off force is a measure of work of adhesion as described in the previous section. Nature has provided a range of insects and **geckos** with the mechanism of attachment to the surfaces by adherence of their rough (in nano-scale) *feet*, which are then *peeled* (**pull-off action**) rather than the mammalian *heel strike* and *toe-off* motion. Large numbers of **spatulae** at the ends of hair-like **setae**, many of which are in turn attached to lamellae structures provide adequate forces to more than balance their weight (see Fig. 3.6). Roughness features on a spatula interact with those of an assortment of surfaces as a problem of scale, but are generally in the nanometre range. Note that in humans, roughness of skin is in the micro-scale, with $6\,\mu m$ for the index finger, being the finest and around 20–$30\,\mu m$ on the heel. These are much rougher than most surfaces (within area of contact) that one comes into contact with or walks upon bare foot. Furthermore, the continual presence of a thin film of moisture or perspiration on human skin hydrates it and guards against adhesion. Therefore, it is clear that adhesion is a function of the roughness scale. To ascertain this, an appropriate model is required to include asperity deformation and *stretching* when a **pull-off force** is applied. It is then interesting to note that motion, based on a peel-off process, is governed by adhesion. Although this is an ingenious method for insects and geckos, it would be a poor solution for engineering surfaces, indicating that the common notion of achieving smooth surfaces in tribological conjunctions may be seriously misguided. However, very rough surfaces also result in boundary interactions. Thus, *optimum* topography is required for a given circumstance, a realisation that has led to the field of surface engineering (see Chapters 13 and 14, and Gohar and Rahnejat, 2008).

Returning to the issue of a model for asperity interaction, here the JKR model is chosen as an example. Later, the difference between this and other similar models is described.

One can assume idealised rough surfaces, meaning that roughness features can be assumed to be hemispherical such that the contact of a pair of these on opposing surfaces can be represented by an average equivalent radius R (see Chapters 4 and 6). Then, their deformation under normally loaded contact can be described by the Hertzian theory for a circular point contact (see Chapter 4) as:

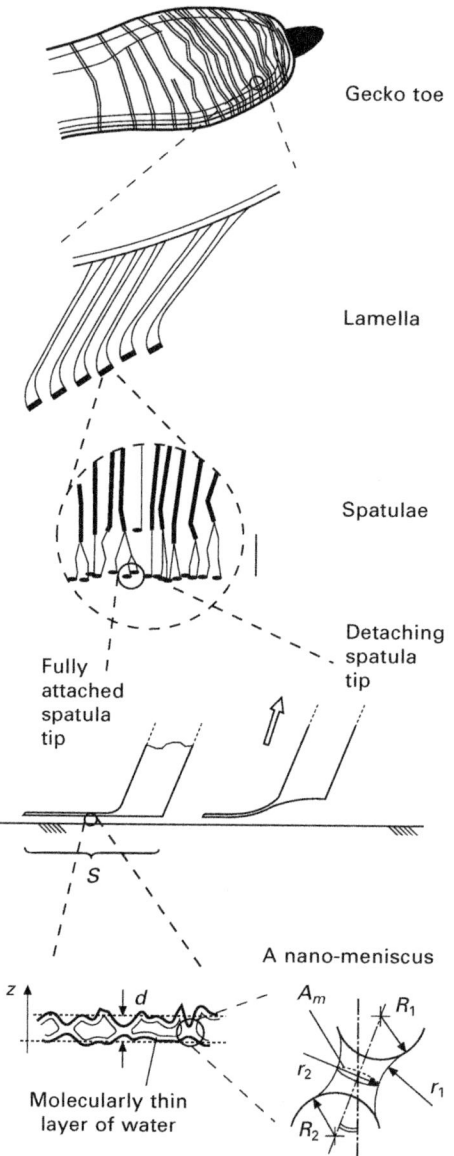

Gecko toe

Lamella

Spatulae

Detaching
spatula
tip

Fully
attached
spatula
tip

A nano-meniscus

Molecularly thin
layer of water

3.6 Hierarchical structure of gecko feet in wet adhesive contact.

$$F_H = \frac{4E^*a^3}{3R} \tag{3.11}$$

Equation (3.11) can be used for n asperities and an appropriate relationship between the total contact load W_e and the real contact area A_e (as opposed to the apparent contact area found through Hertzian theory, see Chapter 4, and Gohar and Rahnejat, 2008) is found as: $A_e \propto F_e^{2/3}$ (the suffix e indicates that the asperities are assumed to remain within the elastic limit). In reality the asperities may weld together and form junctions. When an effort is put to separate the surfaces, these junctions resist by stretching. This is due to adhesion. Thus, for a pair of asperities (Johnson *et al.*, 1971):

$$F = F_H + F_a = \frac{4E^* a^3}{3R} - \sqrt{8\pi a^3 \Delta \gamma E^*} \qquad (3.12)$$

Figure 3.7 is, therefore, a representation of rough surface contact, in which some asperities show penetration, some unaffected due to their smaller heights, and others stretched from their undeformed initial height, when surfaces commence to separate. This stretching of asperities before pull-off is due to adhesion. δ_c is the maximum stretching of the asperity tips prior to pull-off (contact separation). Note that the real contact area comprises such asperity tip contact diameters $2a$.

The second term on the right-hand side of (3.12) is the resistance of a pair of asperities to separate (thus, the negative sign). Equation (3.12) is the basis of the JKR model. When the contact load F is removed, a ***pull-off force*** is still needed to separate the asperities because owing to their compliance their welded junctions stretch before they break off. This is given as (Fuller and Tabor, 1975):

$$F_c = 4\sqrt{\frac{\beta}{3}} E^* \delta_c^{3/2} = 1.5\pi \beta \Delta \gamma \qquad (3.13)$$

where β is the representative average asperity tip radius for a real surface (not idealised) and can be measured using an AFM. Equation (3.13) is in fact (3.10), where as noted before $c = 1.5$ for the *JKR model*. Observe from Section 3.5.2b that for the *DMT model* $c = 2$, meaning that asperities are less compliant (more stiff, thus less adhesion) and a lower pull-off force

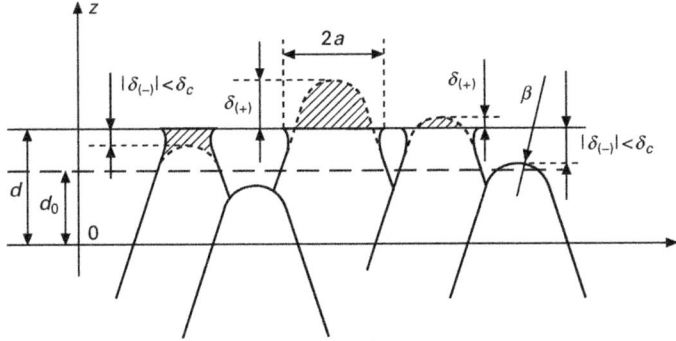

3.7 Representation of rough adhesive contacts.

is required. Therefore, each model suits a particular type of material and asperity geometry. Many real surfaces fall in between these models, where intermediate models have been proposed, as discussed later.

The adhesion phenomenon applies to all surfaces, but is clearly a function of β for surfaces of same material, but with different topography, where $\Delta \gamma = \gamma_1 + \gamma_2 + \gamma_{12}$, and $\gamma_1 = \gamma_2$ one may assume $\gamma_{12} = 0$. It is clear that with increasing β, representative of smoother surfaces, a greater pull-off force would be required. Therefore, adhesion plays a more prominent role in the sub-micrometre and nano-scales.

Equation (3.13) gives the maximum *asperity stretching* of:

$$\delta_c = \frac{1}{3} \left(\frac{3.375 \pi \Delta \gamma}{E^*} \right)^{2/3} \beta^{1/3} \qquad (3.14)$$

It is customary for the above equations to be put in non-dimensional forms, also using equations (3.13) and (3.14) as (see Fig. 3.8):

$$\begin{cases} \bar{F} = \bar{a}^3 - \sqrt{2\bar{a}^3} \\ \bar{\delta} = \bar{\delta}_0 - \bar{\delta}_a = 3^{\frac{1}{3}} \bar{a}^2 \left(1 - \frac{2\sqrt{2}}{3} \bar{a}^{-\frac{3}{2}} \right) \end{cases} \qquad (3.15)$$

where:

$$\bar{a} = a \left(\frac{4E^*}{9\pi \Delta \gamma \beta^2} \right)^{\frac{1}{3}}; \bar{\delta} = \delta \left(\frac{16E^{*2}}{9\pi^2 \Delta \gamma^2 \beta} \right)^{\frac{1}{3}} = \left(\frac{3}{4} \right)^{\frac{1}{3}} \frac{\delta}{\delta_c}; \bar{F} = \frac{F}{3\pi \Delta \gamma R} = \frac{F}{2F_c}$$

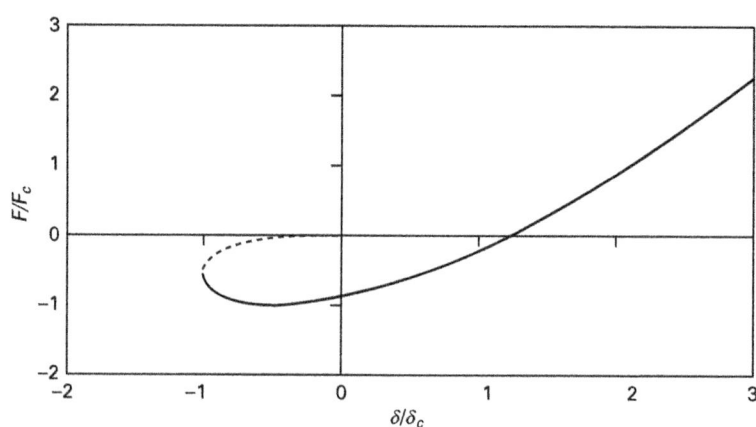

3.8 Adhesive contact of an asperity pair (Teodorescu and Rahnejat, 2008b).

From equation (3.15) it is clear that a rather complex relationship between load and deflection can be found: $\bar{F} = f(\bar{\delta})$, which is a replacement for the Hertzian relationship: $W = K\delta^{3/2}$ given in Chapters 2 and 4. In fact, Fig. 3.8 shows the variation.

The Hertzian load–deflection relationship follows a 3/2 law. When adhesion is included a pair of asperities are in compressive contact, the second term in (3.12) accounts for the deviation from Hertz. This is observed in Fig. 3.8 when $\bar{\delta} > 0$ and $\bar{F} < 0$, indicating a reduction in contact force due to adhesion. With $\bar{\delta} = 0$, there is a residual force, which must be overcome before the asperity-pair junction breaks off. The effort follows the locus $\bar{\delta} < 0$ as shown in the figure.

The analysis thus far deals with a pair of asperities. However, rough surfaces have many asperities. It is assumed that asperity height and tip radius vary according to Gaussian distributions, then the probability that an asperity has a height between z and $z + dz$ above the plane, defined by the **mean asperity height** is:

$$\phi(z) = \frac{1}{\sigma\sqrt{2\pi}} e^{-z^2/2\sigma^2} \tag{3.16}$$

Similarly for probability of an asperity having a radius between ρ and $\rho + d\rho$ can be given as:

$$\psi(\rho) = \frac{2}{\beta\sqrt{2\pi}} e^{-z^2/2\beta^2} \tag{3.17}$$

Combining these for a number of asperities with the height between any arbitrary limits z_1 and z_2 **asperity tip radius** between ρ_1 and ρ_2, then:

$$n'^{\rho_1 \div \rho_2}_{z_1 \div z_2} = N \int_{z_1}^{z_2} \phi(z)\,dz \int_{\rho_1}^{\rho_2} \psi(\rho)\,d\rho$$

which yields the total adhesion force as:

$$W_a = N\frac{F_c^B}{\sqrt{2\pi}} \int_{-L}^{\infty} f\left(\frac{\Delta}{\Delta_c}\right) e^{\left\{\left[-\frac{1}{2}(h+\Delta)^2\right] - \frac{\rho^2}{2\beta^2}\right\}} d\rho\,d\Delta \tag{3.18}$$

where

$$F_c^B = 1.5\pi B\Delta\gamma;\ B = \sqrt{\frac{2}{\pi}} \int_0^{\infty} \exp\left(-\frac{\rho^2}{2\beta^2}\right) d\rho;\ \Delta = \frac{\delta}{\sigma};\ \Delta_c = \frac{\delta_c^\beta}{\sigma};\ h = \frac{l}{\sigma}$$

see Fig. 3.7, where the gap h and the localised deformations are normalised with respect to the surface roughness Ra (σ).

It is clear that in normal atmosphere, moisture forms on exposed surfaces. In lubricated contacts a very thin low shear strength film forms near surfaces.

This is encouraged by addition of friction modifiers to the base lubricants. In all these cases the fluids wet the real rough surfaces as described in Section 3.3. There, it was shown that menisci bridges formed between the asperities of contiguous surfaces also contribute to the work of adhesion. Equation (3.2) gives the **meniscus force** between a pair of asperities, and to take into account the different contact angle that a fluid makes with the contiguous surfaces: $F_m = -2\pi R \gamma_{lv} (\cos\theta_1 + \cos\theta_2)$. Assuming the same Gaussian distributions as before, then (see Fig. 3.9):

$$W_m = N F_m$$

$$= -2NR\gamma_{lv}(\cos\theta_1 + \cos\theta_2) \int_{d-h_m}^{d} f\left(\frac{\Delta}{\Delta_c}\right) e^{\left\{\left[-\frac{1}{2}(h+\Delta)^2\right] - \frac{\rho^2}{2\beta^2}\right\}} \, d\rho\, d\Delta$$

$$(3.19)$$

One needs to obtain the value of h_m, which is the depth of the surface adsorbed film of fluid. If the surfaces are considered to be exposed to the normal environment, then a layer of water condenses on the rough surfaces, depth of which depends on the condensation activation time, which depends on the atmospheric pressure and that of saturation pressure. Riedo *et al.* (2002) show that the depth of water layer to be: $h_{mv} = \ln(t/t_a)[\ln(p_s/p_a)A_m\rho]^{-1}$. The same is not so easily obtained for thin, low, shear strength films of lubricant additives on rough engineering surfaces. Chapter 32 uses this relationship to obtain meniscus force when water condensation ingresses into MEMS gear contacts.

The above approach to determine the overall adhesion of two rough surfaces can equally be extended to the DMT model. Therefore, for a given

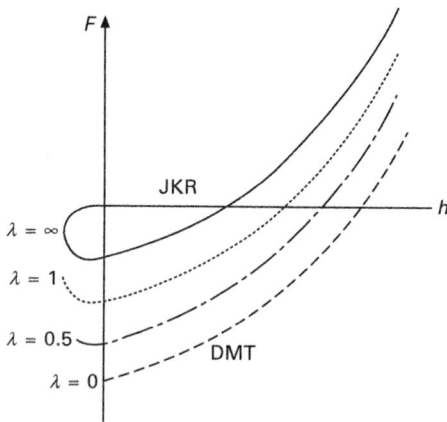

3.9 Force versus penetration depth for the DMT, JKR and Maugis models for different values.

pair of counterfaces it is necessary to decide which one of the two classical models (JKR or DMT) should be used. This distinction can be made by computing the Tabor parameter (Tabor, 1977) as: $\mu = (\beta^{1/3}\Delta\gamma^{2/3}/[E*/2]^{2/3}\varepsilon)$, where ε is the acting range for the surface forces (Johnson, 1997) and $\Delta\gamma = \gamma_1 + \gamma_2 + \gamma_{12}$. When the acting range is such that the surface attraction forces may be ignored in comparison with the Hertzian contact force between a pair of asperities, then the JKR model is used. For the JKR approach the Tabor parameter μ exceeds unity. On the other hand, the DMT takes into account the effect of surface forces of attraction and the Tabor parameter is below unity. The DMT model considers action of surface forces outside the contacting region between a pair of opposing asperities. An intermediate model was proposed by Maugis (1992) which takes into account the effect of surfaces force in a rim (ring zone) surrounding the Hertzian region of contact between a pair of hemispherical asperities. Thus, it can be used for any pair of contacting material type of high or low adhesion.

In the Maugis model, the extent of adhesion is determined by a parameter λ:

$$\lambda = 2\sigma_0 \left(\frac{R}{\pi\gamma E_n^2} \right)^{1/3} \tag{3.20}$$

where R is the equivalent radius of the hemispherical solid against a flat plane,

$$E_n = \frac{4}{3}\left(\frac{1-v_1^2}{E_1} + \frac{1-v_2^2}{E_2} \right)^{-1}$$

where v_1, v_2 are the sphere and flat-plane Poisson ratios; E_1, E_2 are the sphere and flat-plane Young moduli; σ_0 is a constant adhesion stress; γ is the work of adhesion.

The relationship between Tabor parameter μ and Maugis parameter λ is:

$$\lambda = 1.1570\,\mu \tag{3.21}$$

λ is often referred to as the *transition parameter*. If $\lambda > 5$, the JKR model applies, and if $\lambda < 0.1$, then the DMT model is used. Values between 0.1 and 5 correspond to the *transition regime* between JKR and DMT models. Figure 3.9 shows force–penetration depth variation for a pair of asperities for a range of values of **Maugis transition parameter** λ. Note that although Maugis's model applies to the full range of asperity compliance, those of JKR and DMT represent the extremes of the spectrum.

To compare the three adhesion models the normalised radius of the contact area $\overline{A}$, force $\overline{F}$ and penetration depth $\overline{h}$ are introduced as:

$$\bar{A} = \frac{A}{\left(\dfrac{\pi R^2 \gamma}{E_n}\right)^{1/3}} \tag{3.22}$$

$$\bar{F} = \frac{F}{\pi R \gamma} \tag{3.23}$$

$$\bar{h} = \frac{h}{\left(\dfrac{\pi^2 R \gamma^2}{E_n^2}\right)^{1/3}} \tag{3.24}$$

The quantitative summary of the DMT, JKR and Maugis models are shown in Table 3.4.

The Maugis correction to the Hertzian solution is expressed implicitly via parameter $n = a/c$; the ratio between the contact radius a and an outer radius c at which the gap between the surfaces constitutes diminution of adhesive stress.

3.7   Intermolecular interactions and near-surface effects

As mentioned at the beginning of Section 3.3, molecules of a fluid at its interface with another medium are subject to adhesive forces. These are in addition to the cohesive forces between themselves, which occur in the bulk of the fluid away from the interface. Therefore, the molecular organisation (packing) near another medium is unlike that in the bulk. The term adhesive force in this context is different from that dealt with in the previous section. It is a general term that describes a plethora of forces acting at close range at interfaces. The meniscus force mentioned above is a macroscopic explanation

Table 3.4 Quantitative comparison of DMT, JKR and Maugis models

Model	Normalised equations
DMT	$\bar{F} = \bar{A}^3 - 2; \; \bar{h} = \bar{A}^2$
JKR	$\bar{F} = \bar{A}^3 - \bar{A}\sqrt{6\bar{A}}; \; \bar{h} = \bar{A}^2 - \dfrac{2\sqrt{6\bar{A}}}{3}$
Maugis	$\dfrac{\lambda \bar{A}^2}{2} [\sqrt{n^2 - 1} + (n^2 - 2)\tan^{-1}\sqrt{n^2 - 1}] + \dfrac{4\lambda \bar{A}^2}{3} [1 - n + \sqrt{n^2 - 1}\tan^{-1}\sqrt{n^2 - 1}] = 1$
	$\bar{F} = \bar{A}^3 - \lambda \bar{A}^2[\sqrt{n^2 - 1} + n^2 \tan^{-1}\sqrt{n^2-1}]$
	$\bar{h} = \bar{A}^2 - \dfrac{4\lambda \bar{A}}{3}\sqrt{n^2 - 1}$

for formation of interfaces. There are intermolecular forces and a surface energy effect that act within this macrocosm in a scale of minutiae, usually in scale of a few to several molecular diameters of the fluid. The cohesive force also refers to all forms of interaction that occur between elements of the same medium, such as between molecules of a volume of a fluid or some times between these and atoms of an adjacent solid. The intermolecular forces of interest in nano-tribology are van der Waals interactions and electrostatic repulsion. Additionally, near very smooth barriers and at light loads (typically in nanotechnology) there is fluid density variation from that of the bulk fluid film. This is because of different packing order of fluid molecules near such barriers than that in the bulk of the fluid (recall the adhesive forces versus the cohesive forces). This density variation causes a oscillatory pressure variation, which leads to mainly repulsive action of fluid. The effect is known as solvation. Hydration is a form of solvation, also discussed briefly in this chapter. These forces are discussed in some detail by Israelachvili (1992) and Gohar and Rahnejat (2008). Here attention is paid to van der Waals forces which are always present and become quite significant at very close range. They can become important in microelectromechanical systems (MEMS), including tiny MEMS gears and micro-engines (see Chapter 32).

3.8   van der Waals forces

When two charges q^+ and q^- are separated by a small distance, r, they constitute an *electric dipole*, which is usually represented by a vector as shown in Fig. 3.10. The magnitude of this vector is the *dipole moment*; $\xi = qr$. Its direction is from the negative charge towards the positive one. The unit of ξ is C m (Coulomb metre), but the value of ξ is usually very small; of the order $10^{-30} - 10^{-29}$ C m, thus they are more conveniently expressed in the unit: debye ($1D = 3.336 \times 10^{-30}$ C m), after Peter Debye who was the pioneering scientist in the study of polarisability of molecules.

The electric properties of charges in the form of dipole moment extends to ions and molecules. When a pair of such molecules (*polar molecules*) are close to each other, their dipole moments interact. The nature of interactions of their dipole moments depends on their relative orientation. Since dipole moments are represented by vectors, it is possible to resolve their contributions into various components. Those contributions in a head-to-head direction are repulsive, while those in a head-to-tail orientation are attractive. The latter correspond to a lower energy level and on average more prevalent in a fluidic medium, giving rise to the van der Waals forces.

$$\xi = qr$$
$$q^- \xrightarrow{\ \ r\ \ } q^+$$

3.10 An electric dipole.

3.8.1   Polarisability

Although not all molecules are polar in nature, like atoms they are all polarisable. *Polarisability* of a molecule is due to the distortion of its electronic structure when subjected to an electric field, which is as the result of nearby polar molecules. The distortion in the electronic structure of a non-polar molecule arises when the field displaces its negatively charged electron cloud relative to its positively charged nucleus. Therefore, the induced dipole moment in a non-polar molecule is given as:

$$\xi_{ind} = \alpha_0 E_e \tag{3.25}$$

where E_e is the electric field and α_0 is defined as the polarisability of the molecule, and it can be shown that (Atkins, 1986):

$$\alpha_0 = 4\pi\varepsilon_0 R_a^3 \tag{3.26}$$

where R_a is the atomic radius.

For a polar molecule, there is the additional *orientational polarisability* due to its rotating dipole. It can be shown that:

$$\alpha_{orien} = \frac{\xi^2}{3KT} \tag{3.27}$$

Therefore, the total polarisability of a polar molecule is obtained as the addition of the above stated contributions, and is known as the *Debye–Langevin equation*.

Now in a given medium, with certain molecular structure and disposition there will be an assortment of molecular interactions due to the above described polarisation effects. The van der Waals forces arising from these can be obtained according to the types of interactions, described below. The method highlighted is simple to follow and illustrates the nature of interactions quite well. However, there are some major drawbacks, when applied to nano-tribology problems. The main shortcomings are that the dispersion forces (which dominate the interactions) assume that molecules have a single ionisation potential (i.e. one absorption frequency). Also, the method cannot handle the effect of medium/solvent in which the molecules reside. Nevertheless, an appropriate introduction to the various forms of interaction is a necessary prelude in order to proceed to the more appropriate and general case described later.

3.8.2   Various forms of intermolecular interactions

There are three distinct types of interaction, which contribute to the total long-range interaction of van der Waals force between molecules. These are highlighted here, but avid readers should refer to more specialist texts in physical chemistry, such as Atkins (1986) and Israelachvili (1992). These interactions are by: *induction*, *orientation* and *dispersion*. The interactions by *orientation*, often referred to as **Keesom interaction**, have already been described in the preceding section, where the average interaction energy between two polar molecules with dipole moments, ξ_1 and ξ_2, separated by a distance r is given as:

$$w = -\frac{2}{3KT}\left(\frac{\xi_1\xi_2}{4\pi\varepsilon_0}\right)^2 \frac{1}{r^6} \tag{3.28}$$

Note that the negative sign indicates the attractive nature of the interaction.

A polar molecule near another molecule, which may or may not be polar induces a dipole moment upon it. This form of interaction is referred to as **induction interaction**. This induced dipole interacts with the permanent dipole of the polar molecule itself, causing an attraction potential between them, where:

$$w = -\frac{\xi_1^2 \alpha_v}{4\pi\varepsilon_0} \frac{1}{r^6} \tag{3.29}$$

where ξ_1 is the permanent dipole moment of the polar molecule. α_v is defined as the polarisability volume of the non-polar molecule. It essentially arises from the rationalisation of the molecular polarisability α_0, which has the rather untidy unit of $J^{-1}c^{-2}m^2$, by defining $\alpha_v = \alpha_0/4\pi\varepsilon_0$. α_v has now the unit of volume.

Non-polar molecules do not possess permanent moments, but due to changing orbital disposition of their electrons one may regard them as possessing instantaneous dipoles, which alter in both magnitude and direction. A pair of such molecules in proximity of each other can, therefore, induce moments on each other merely by altering their electronic arrangement, leading to the development of instantaneous dipole moments. This induced-dipole–induced-dipole interaction is called the **dispersion interaction**. As a result the pair of molecules are attracted to each other, and changes in the instantaneous dipole of one induces an effect upon the other and vice versa. If the ionisation potentials of these molecules are denoted by I_1 and I_2, then the formula, obtained by London (1937) gives the interaction energy as:

$$w = -\frac{3}{2}\frac{\alpha_{v1}\alpha_{v2}}{(4\pi\varepsilon_0)^2 r^6}\frac{I_1 I_2}{(I_1 + I_2)} \tag{3.30}$$

where $I = \hbar\varpi$, with $\hbar$ being the **Planck constant** and ϖ the **absorption frequency**. The **ionisation potential** is the energy required to ionise an atom or a molecule.

It can be observed that like the *induction* and *orientation* interactions, the *dispersion* contribution also follows the inverse sixth power relationship with distance. Thus, the total attractive interaction energy between any pair of molecules can be regarded as the sum of all the three contributions.

While the above simple analysis serves the purpose of clarifying some basic principles behind the origin of van der Waals forces, it has two serious shortcomings. Firstly, it assumes that atoms and molecules have a single ionisation potential (or absorption frequency). This is of course not true. For example, ionisation of an atom takes place by electrons, with their ionisation potential (or the energy needed) being different according to their orbital position (i.e. the **orbiting frequency**, ϖ). This is also true of molecules, which have various ionisation potentials according to the state of their motion, which is clearly affected by the medium in which they reside and the interactions with other neighbouring molecules. Secondly, it is clear from the above formulae that an assumption of pair-wise additivity is inherent for interaction of a molecule with its neighbours. This, of course, ignores the effect of other molecules upon the polarisability of any interacting pair nearby. Because of these shortcomings Lifshitz (1955) proposed his theory of van der Waals forces, which does not depend on the structure of atoms and molecules. Instead the **Lifshitz theory** treats the problem as forces between bodies in a continuum, described by a quantum field. A detailed treatment of the theory is not intended here, only a note that the forces are obtained in terms of bulk properties of media such as refractive indices and dielectric constants, which are easy to determine and suit tribological studies. It is, however, important to remember that the relationships stated above for the interaction energies remain valid in the Lifshitz approach.

Using the Lifshitz theory the interaction energy between two surfaces separated by a very thin fluid film can be obtained, taking into account the interactions between the atoms of the two surfaces and these with the molecules of an intervening fluid film. In tribology one is interested in the contact pressure distribution arising from this interaction energy, which obviously depends on the geometry of the contact. Israelachvili (1992) provides analysis for van der Waals interaction energy per unit area for various contact geometries. Since the contact area is usually very small in nano-tribology, one can assume the interaction energy to act between two flat parallel planes, where:

$$w = \frac{-A}{12\pi h^2} \tag{3.31}$$

where A is the **Hamaker constant**, explained below, and h is the thickness of the fluid film. Then, the van der Waals pressure in the fluid gap between the flat planes is obtained as:

$$p_{vdw} = \frac{\partial w}{\partial h} = -\frac{A}{6\pi h^3} \qquad (3.32)$$

For other contact geometries one must resort to the **Derjaguin approximation**, which shows that the van der Waals force and pressure between two spheres of radii R_1 and R_2 can be obtained in terms of w as:

$$F \approx 2\pi \left(\frac{R_1 R_2}{R_1 + R_2} \right) w \qquad (3.33)$$

Note that the ratio in equation (3.33) is in fact the equivalent radius of the two contacting spheres, or the equivalent sphere/cylinder near a flat, semi-infinite plane (Chapter 4).

3.8.3   Calculation of Hamaker constant

The conventional Hamaker (1937) constant was originally set as:

$$A = C\pi^2 \rho_1 \rho_2 \qquad (3.34)$$

where C is the coefficient in atom–atom pair potential and ρ_1 and ρ_2 are the number of atoms per unit volume in the two contacting surfaces. Typical values for Hamaker constant of condensed phases in liquids or solids are $10^{-20} - 10^{-19}$ J.

It is clear that the previously mentioned assumption of simple pair-wise additivity is inherent in equation (3.34). This means that the effect of any neighbouring atoms to a pair of atoms: atom 1 and atom 2 are ignored in their interactions. However, if a third atom is considered, it will also be polarised by the instantaneous fields of atoms 1 and 2, and its induced dipole will also act upon them. Therefore, the field from atom 1 will not only directly affect atom 2, but also by *reflection* from atom 3.

Using Lifshitz theory, the Hamaker constant for interaction between two surfaces (denoted by subscripts 1 and 2), separated by a very thin fluid film (denoted by the subscript 3) can be obtained in terms of the continuum properties of the system:

$$A_{131} = -\frac{3}{2} KT \sum_{n=0}^{\infty} \int_{r_n}^{\infty} x [\ln(1 - \Delta_{13}^2 e^{-x}) + \ln(1 - \overline{\Delta}_{13}^2 e^{-x})] dx \qquad (3.35)$$

where:

$$\Delta_{jk} = \frac{\varepsilon_j S_k - \varepsilon_k S_j}{\varepsilon_j S_k + \varepsilon_k S_j}, \quad \overline{\Delta}_{jk} = \frac{S_k - S_j}{S_k + S_j}, \quad S_k^2 = x^2 + \left(\frac{2\omega_n \hbar}{c} \right)^2 (\varepsilon_k - \varepsilon_3),$$

$$r_n = \frac{2\hbar\omega_n \sqrt{\varepsilon_3}}{c}, \quad \omega_n = \frac{2\pi n KT}{\hbar}$$

and:

$$\varepsilon_k(i\omega_n) = \begin{cases} 1 + \dfrac{n_k^2 - 1}{1 + \omega_n^2/\varpi_k^2} & \text{(for } \omega_n > 0) \\[2ex] \varepsilon_{k0} & \text{(for } \omega_n = 0) \end{cases} \qquad (3.36)$$

where K is the **Boltzmann constant**, T is temperature in kelvin, c is the speed of light (300 000 km/s), n_k are the refractive indices of the various media, ϖ is the absorption frequency, ω_n is the **imaginary frequency** and $\hbar$ is Planck's constant. Once the Hamaker constant is obtained, equation (3.32) can be used to obtain the van der Waals pressure for a given film thickness, h. The value of h should be in the nano-scale for the van der Waals pressure to be significant. Thus, in conjunctions usually encountered in engine and drivetrain system of vehicles, van der Waals contribution is insignificant, but the same is not true in the scale of minutiae, such as in MEMS devices or indeed in many nano-biotribological systems such as the example of gecko feet or bio-inspired micro-fibres which are supposed to emulate them as sources of regenerative adhesion (see for example, Teodorescu *et al.*, 2008).

Returning to equation (3.32), and taking the Hamaker constant as $A = 10^{-20}$ J, one can observe that with a gap of 1 μm an attractive van der Waals pressure of −53 mPa results between two flat surfaces. If this was an elastohydrodynamic contact (recall that the two bodies flatten in the Hertzian region of the contact, see Chapters 4 and 6) then the contribution due to van der Waals interaction is negligible. Now consider a pair of flat surface with a separation of 1 nm. The van der Waals attractive pressure becomes −5305 kPa. This shows that van der Waals interactions are quite effective at small separations in very lightly loaded contact. Referring back to Fig. 3.6 the gecko toe pads have a density of 14 000 mm^{-2} setae, each with spatulae at their terminus, making for typically 14 million potential contact areas of 0.04 μm^2. Adhesion caused by van der Waals interaction can be found to be several times greater than the weight of the gecko. Therefore, many regard van der Waals attraction as the reason for geckos' stiction to surfaces in bizarre postures, seemingly in defiance of gravity. This also indicates that a pull-off force is required for the gecko to initiate motion. However, this explanation is unlikely to be strictly true except in vacuum! In the normal atmosphere other forces also operate between the rough surfaces of gecko spatulae and target solids. One such force is electrostatic repulsion. The balance between electrostatic and van der Waals forces is one of the success stories in nature, otherwise the cells or molecules of many substances would simply coagulate due to van der Waals attraction such as blood congealing and all internal organs forming a large coagulated mass! (See Leonard-Jones potential in Atkins, 1986, or Gohar and Rahnejat, 2008.)

3.9   Other near-surface effects

Forces such as van der Waals attraction, as shown above, are not of concern in traditional tribological contacts which are analysed in micro-scale, such as those in Chapters 5 and 6. They become important in very small-scale interactions near-surface asperities, which themselves are in sub-micrometre scale, often in nanometre range. Their importance with respect to the topic of this book is in either development of micro-engines, MEMS gears and mechanisms or in surface adsorbed films on very smooth surfaces. Chapter 32 provides some examples. When dealing with such problems, a growing trend with progressive miniaturisation of certain devices shows the behaviour of any intervening fluid itself deviates from its bulk properties such as its density. Therefore, the force laws near surfaces at close range deviate from bulk characteristics. One such force is solvation or structural force, described by Gohar and Rahnejat (2008), Matsuoka and Kato (1997), Al-Samieh and Rahnejat (2001) and Teodorescu *et al.* (2006).

Interaction pair potentials between pair of particles or molecules, given for example, by w in equations (3.28–3.30), are in free space. These are, in fact, often referred to as the free energy. However, when the interaction between, for example, two solute molecules in a solution is taken into account, then the effect is not just because of the pair of molecules themselves, but also by many of the solvent molecules nearby. For example, ionic crystals often dissolve in solvents, which consist of molecules that form an electrostatic association with them. This effect is called **solvation**. When the solvent is water, then the effect is called **hydration**. Therefore, two molecules in free space would not behave in the same manner as they would in a solution, irrespective of their polar or non-polar nature, because they are also affected by the presence of solvent molecules. In fact, the nature of their interactions may completely alter. The continuum of the solvent itself is also altered by the presence of solute or *guest* molecules, which are accommodated and an additional force comes into play called the solvation force. It is now clear that molecules of a fluid re-order when they meet a solid barrier depending upon molecular geometry, their packing formation and the local geometry of the surface (i.e. its topography). This means that the solvation force between the fluid molecules and the solid boundary dictates different ordering near its surface (several molecular diameter deep) than elsewhere within the bulk of the fluid. Therefore, density of fluid molecules near the surface is different. This effect becomes important in vanishingly small conjunctions of ultra-smooth solid boundaries. A force known as solvation or structural force is generated due to oscillatory variation in fluid density near the surfaces. This attractive–repulsive force results in drainage of fluid from the conjunction in discrete packets. Readers should refer to Chan and Horn (1984), Al-Samieh and Rahnejat (2001) and Matsuoka and Kato (1997).

Of particular interest is when such conditions occur in nano-scale conjunctions (close range) in the presence of moisture. Like the solvation effect near solid boundaries, the proposed oscillatory behaviour of hydration is noted only for fairly smooth surfaces. It has a hydrophobic potential, which accounts for rupture of water molecules from the surface. An everyday experience is the hydration effect of skin (beta-carotene), which acts against the stiction of objects. The exact nature of hydration is not well understood. The hydrophobic potential is thought to be an exponential decay of the form (Israelachvili, 1992):

$$W_H = 2\gamma_i\, e^{-h/\lambda_0} \tag{3.37}$$

where $\gamma_i = 10 - 50\,\text{mJ/m}^2$ and $\lambda_0 = 1.5\,\text{nm}$, which is six times that of a water molecule, $a_w = 0.25\,\text{nm}$.

A hydration pressure is then generated at very close range between a pair of surfaces as:

$$p_H = \frac{\partial W_H}{\partial h} = \frac{2\gamma_i}{\lambda_0}\, e^{-h/\lambda_0} \tag{3.38}$$

which indicates a monotonic rise with decreasing gap h. It is clear that this effect works at fairly close range as shown in the example used in Chapter 32.

3.10   Conclusion

This chapter has shown that a plethora of force laws operate in the scale of minutiae, below the usual hydrodynamic or elastohydrodynamic films that are common in today's engine and powertrain conjunctions. As miniaturisation of devices proceeds, the role of these largely empirical force laws should be understood in a more fundamental manner. Already a lack of in-depth understanding accounts for unreliability of key emerging technologies such as micro-engines, MEMS devices, high-storage, very smooth metallic devices with ultra-low flying head sliders and bio-inspired microfibres (Teodorescu *et al.*, 2008). This understanding will also be crucial in increasingly employed modified surfaces with refined etching processes and self-assembled monolayers (SAM).

3.11   References

Amontons, G., 'De la resistance causee dans les machines', *Memoires de l' Academie Royale A, Paris*, 1699

Al-Samieh, M.F. and Rahnejat, H., 'Ultra-thin lubricating films under transient conditions', *J. Phys., Part D: Appl. Phys.*, **32**, 2001, 2610–2621

Atkins, P.W., *Physical Chemistry*, Oxford University Press, New York, 3rd edition, 1986

Blake, T.D., *Wettability*, Marcel Dekker, New York, 1993

Bowden, F.P. and Tabor, D., *The Friction and Lubrication of Solids*, Clarendon Press, Oxford, 1950

Brandon, S. and Marmur, A., 'Simulation of contact angle hysteresis on chemically heterogeneous surfaces', *J. Colloid Interface Sci.*, **183**(2), 1996, 351–355.

Brillouin, L. and Brennan, S.J., *Tensors in Mechanics and Elasticity*, Academic Press, New York, London, 1964

Chan, D.Y.C. and Horn, R.G., 'The drainage of thin liquid films between solid surfaces', *J. Chem. Phys.*, **83**, 1984, 5311–5324

Christenson, H.K., 'Interactions between hydrocarbon surfaces in a nonpolar liquid: Effect of surface properties on solvation forces', *J. Phys. Chem.*, **90**, 1986, 4–6.

de Coulomb, C. A., 'Recherches theorique et experimentales sur la force de torsim et l'élasticité des fils de metal', *L' Académie Royal des Sciences, Paris*, 1784

Derjaguin, K.L., Muller, V.M. and Toporov, Y.P., 'Effect of contact deformation on the adhesion of particles', *J. Colloid Interface Sci.*, **53**(2), 1975, 314–326.

Elliott, G.E.P. and Riddiford, A.C., 'Dynamic contact angles: I. The effect of impressed motion', *J. Colloid Interface Sci.*, **23**(3), 1967, 389–398

Extrand, C.W., 'Contact angles and their hysteresis as a measure of liquid–solid adhesion', *Langmuir*, **20**, 2004, 4017–4021

Fadeev, A.Y. and McCarthy, T.J., 'Trialkylsilane monolayers covalently attached to silicon surfaces: wettability studies indicating that molecular topography contributes to contact angle hysteresis', *Langmuir*, **15**, 1999, 3759–3766

Fuller, K.N.G. and Tabor, D., 'The effect of surface roughness on the adhesion of elastic solids', *Proc. Roy. Soc., Series A*, **345–1642**, 1975, 327–342

Gohar, R. and Rahnejat, H., *Fundamentals of Tribology*, Imperial College Press, London, 2008

Good, R.J. and van Oss, C.J., *Modern Approaches to Wettability: Theory and Applications*, Plenum Press, New York, 1992

Hamaker, H.C., 'The London–van der Waals attraction between spherical particles', *Physica*, **4**(10), 1937, 1058–1072

Hennig, A., Eichhorn, K.J., Staudinger, U., Sahre, K., Rogalli, M., Stamm, M., Neumann, A.W. and Grundke, K., 'Contact angle hysteresis: study by dynamic cycling contact angle measurements and variable angle spectroscopic ellipsometry on polyimide', *Langmuir*, **20**, 2004, 6685–6691

Israelachvili, J.N., *Intermolecular and Surface Forces*, Academic Press, New York, 1992

Johnson, K.L., 'Adhesion and friction between a smooth elastic spherical asperity and a plane surface', *Proc. R. Soc., Series A*, **453**, 1997, 163–179

Johnson, K.L., Kendall, K. and Roberts, A.D., 'Surface energy and the contact of elastic solids', *Proc. Roy. Soc., Series A*, **324** (1558), 1971, 301–313

Johnson, R.E. Jr. and Dettre, R.H., 'Contact angle hysteresis, Part I. Study of an idealized rough surfaces', *Adv. Chem. Ser.*, **43**, 1964, 112–135

Johnson, R.E. Jr. and Dettre, R.H., 'Contact angle hysteresis, Part IV: Contact angle measurements on heterogeneous surfaces', *J. Phys. Chem.*, **69**, 1965, 1507–1514

Johnson, R.E. Jr., Dettre, R.H. and Brandreth, D.A., 'Dynamic contact angles and contact angle hysteresis', *J. Colloid Interface Sci.*, **62**, 1977, 205–212

Kepler, J., *Astronomia Nova*, Pragae, 1609

Lam, C.N.C., Kim, N., Hui, D., Kwok, D.Y., Hair, M.L. and Neumann, A.W., 'The effect of liquid properties to contact angle hysteresis', *Colloids Surf. A Physicochem. Eng. Asp.*, **189**, 2001, 265–278

Lam, C.N.C., Wu, R., Li, D., Hair, M.L. and Neumann, A.W., 'Study of the advancing and receding contact angles: liquid sorption as a cause of contact angle hysteresis', *Adv Colloid Interface Sci.*, **96**, 2002, 169–191

Lifshitz, E.M., 'The theory of molecular attractive forces between solids', *Soviet Phys., Zhur. Eksptl.' i Teoret. Fiz. JETP*, **29**, 1955, 94–110

Lin, F.Y.H., Li, D. and Neumann, A.W., 'Effect of surface roughness on the dependence of contact angles on drop size', *J. Colloid Interface Sci.*, **159**(1), 1993, 86–95

London, F., 'The general theory of molecular forces', *Trans. Faraday Soc.*, **33**, 1937, 8–26

Matsuoka, H. and Kato, T., 'An ultra-thin liquid film lubrication theory: calculation method of solvation pressure and its applications to EHL problem', *Trans. ASME, J. Trib.*, **119**, 1997, 217–226

Maugis, D., 'Adhesion of spheres: the JKR-DMT transition using a Dugdale model', *J. Colloid Interface Sci.*, **150**, 1992, 243–269

McHale, G., Shirtcliffe, N.J. and Newton, M.I., 'Contact-angle hysteresis on super-hydrophobic surfaces', *Langmuir*, **20**, 2004, 10146–10149

Middleman, S., *An Introduction to Fluid Dynamics: Principles of analysis and design*, John Wiley and Sons, New York, 1998

Neumann, A.W. and Good, R.J., 'Thermodynamics of contact angles. I. Heterogeneous solid surfaces', *J. Colloid Interface Sci.*, **38**(2), 1972, 341–358

Newton, I., *Philosophie Naturalis Principia Mathematica*, Royal Society, London, 1687

Pashley, R.M., McGuiggan, P.M., Ninham, B.W. and Evans, D.F., 'Attractive forces between uncharged hydrophobic surfaces: direct measurements in aqueous solution', *Science*, **229**, 1985, 1088–1089

Prokopovich, P. and Perni, S., 'An investigation of microbial adhesion to natural and synthetic polysaccharide-based films and its relationship with the surface energy components', *J. Materials Sci.: Materials in Medicine*, **20**, 2009, 195–202

Rahnejat, H., 'Physics of causality and continuum: questioning nature', *Proc. Instn. Mech. Engrs., Part K: J. Multi-body Dyn.*, 2008, 255–264

Rangwalla, H., Schwab, A.D., Yurdumakan, B., Yablon, D.G., Yeganeh, M.S. and Dhinojwala, A., 'Molecular structure of an alkyl-side-chain polymer–water interface: origins of contact angle hysteresis', *Langmuir*, **20**, 2004, 8625–8633

Riedo, E., Lévy, F. and Brune, H., 'Kinetics of capillary condensation in nanoscopic sliding friction', *Phys. Rev. Lett.*, **88**(18), 2002

Rusanov, A.I. and Prokhorov, V.A., In: Möbius D, Miller R (eds.), *Interfacial Tensiometry*, 3, Elsevier, Amsterdam, 1996.

Sedev, R.V., Petrov, J.G. and Neumann, A.W., 'Effect of swelling of a polymer surface on advancing and receding contact angles', *J Colloid Interface Sci.*, **180**(1), 1996, 36–42

Stribeck, R., 'Die wesentliechen ichen eigenschaften gleit und rollen lager or: Ball bearings for various loads', *Trans. ASME*, **29**, 1907, 420–463

Tabor, D., 'Surface forces and surface interactions', *J. Colloids Interface Sci.*, **58**, 1, 1977, 2–13

Teodorescu, M. and Rahnejat, H., 'Newtonian mechanics in scale of minutia', *Proc. Instn. Mech. Engrs., Part K: J. Multi-body Dyn.*, **222**, 2008a, 393–405

Teodorescu, M. and Rahnejat, H., 'Dry and wet nano-scale impact dynamics of rough surfaces with or without a self-assembled monolayer', *Proc. Instn. Mech. Engrs., Part N: J. Nanoengng. Nanosys.*, **221**, 2008b, 49–58

Teodorescu, M., Balakrishnan, S. and Rahnejat, H., 'Physics of ultra-thin surface films on molecularly smooth surfaces', *Proc. Instn. Mech. Engrs., Part N: J. Nanoengng. Nanosys.*, **220**(1), 2006, 7–19

Teodorescu, M., Majidi, C., Fearing, R.S. and Rahnejat, H., 'Effect of surface roughness on adhesion and friction of microfibers in side contact', *STLE/ASME Int. Joint Trib. Conf., Miami, Florida*, 20–22 October 2008

van Oss, C.J., Chaudhury, M.K. and Good, R.J., 'Interfacial Lifshitz–van der Waals and polar interactions in macroscopic systems', *Chem. Rev.*, **88**, 1988, 927–941

Wilhelmy, L., Über die Abhängigkeit der Capillaritäts-Constanten des Alkohols von Substanz und Gestalt des Benetzten Festen Körpers, *Annalen der Physik*, **119**, 1863, 177–179

Yasuda, T., Miyama, M. and Yasuda, H., 'Effect of water immersion on surface configuration of an ethylene–vinyl alcohol copolymer', *Langmuir*, **10**, 1994, 583–585

Zisman, W.A., 'Relation of the equilibrium contact angle to liquid and solid constitution', in: *Contact Angle: Wettability and Adhesion*, American Chemical Society Series 43, Washington, DC, 1–51, 1964

3.12   Nomenclature

a	Hertzian radius
A_{131}	Hamaker constant
A_e	Apparent contact area
A_m	Cross-sectional area of a meniscus
$E_{1,2}$	Young's modulus of elasticity of surfaces
E^*	Reduces elastic modulus
F	Force
F_e	Pull-off (break-off) force
F_m	Meniscus force per asperity contact
F_H	Hertzian force per asperity contact
F_a	Adhesive force per asperity contact
g	Gravitational acceleration
G	Universal gravitational constant
K	Boltzmann constant
h	Gap/film thickness or separation
h_{mw}	Molecular thickness of water
$\hbar$	Planck constant
I	Ionisation potential
m	Number of menisci bridges
m'	Mass of target body
M	Mass of source
N	Number of asperities per unit area
n'	Total number of asperity pair contacts

p	Pressure
p_a	Atmospheric pressure
p_H	Hydration pressure
p_s	Saturation pressure
r	Distance between bodies
$r_{1,2}$	Radii of a meniscus
R	Equivalent radius
R_a	Atomic radius
t	Time
t_a	Condensation time
T	Temperature in kelvin
w	Potential
w_H	Hydrophobic potential
W_{svl}	Work of adhesion
z	Mean asperity height
α_0	Polarisability of a molecule
β	Asperity tip radius
δ	Deflection
δ_a	Deformation due to adhesive force
δ_c	Maximum asperity stretching
ε_0	Vacuum permitivity
$\gamma_{1,2}$	Surface energies
γ_{12}	Interfacial surface energy
γ_{lv}	Liquid–vapour interfacial surface tension
γ_{sl}	Solid–liquid interfacial surface tension
γ_{sv}	Solid–vapour interfacialo surface tension
λ	Maugis parameter
λ_s	Stribeck's oil film parameter
μ	Tabor parameter
μ_f	Coefficient of friction
θ	Contact angles
ρ	Asperity tip curvature radius or density as described
σ	Surface roughness RMS
$\upsilon_{1,2}$	Poisson's ratio
ϖ	Absorption frequency
ω_n	Imaginary frequency
ξ	Dipole moment

4

Fundamentals of impact dynamics of semi-infinite and layered solids

M. TEODORESCU, Cranfield University, UK, V. VOTSIOS, Atos, Spain, P. M. JOHNS-RAHNEJAT, (formerly) Imperial College London, UK and H. RAHNEJAT, Loughborough University, UK

Abstract: This chapter deals with the fundamentals of contact mechanics and impact dynamics of ellipsoidal solids of revolution. It provides a brief review of Hertzian contact/impact dynamics with its underlying assumptions. It provides some examples of Hertzian contact/impact conditions for various contact geometries in rolling element bearings and cam-follower pairs. It also introduces the use of finite element analysis to study contact/impact problems in more detail. Other non-Hertzian type contact problems such as in non-conforming thin-layered solid pairs are studied by appropriate analytical approaches. A generic method decomposing pressure waves propagating into the bulk of solid contacting surfaces is also presented with generic results for both soft and hard-coated surfaces.

Key words: contact mechanics, impact dynamics, Hertzian, layered solids, propagating deformation waves.

4.1   Introduction

Often an important consideration in system dynamics is the effect of impulsive action and/or contact/ impact forces. Chapter 1 (Appendix) has already shown that behaviour of load-bearing conjunctions play an important role in the response of multi-bodies. Many noise, vibration and harshness (NVH) phenomena described in this book (e.g. rattle: Chapters 1, 25–29, clonk, Chapters 1, 21 and 30) are as a result of sudden load changes, which induce contact/impact reactions, the effect of which are transmitted through the system as a whole. These effects manifest themselves as various NVH concerns, which are named onomatopoeically, such as rattle, clatter and clonk. Therefore, as far as a system such as the drivetrain is concerned, the load change reaction in load-bearing conjunctions is often the *cause* of these phenomena and the *effect* is the manner in which they are perceived, such as vibration or noise or both. The severity of a load change reaction induces *deformation waves* which are either localised or propagate through the system. The latter may be regarded as waves of a global nature within the context of a bounded system. They may only induce rigid body inertial

105

motions (such as shuffle, see Chapters 21 and 23) or they may result in structural vibration, depending on system components' compliances. This chain of events does not only occur in engine and powertrain, but in all systems. In fact, it is underlying in nature itself.

The aforementioned waves are of deformation nature in the case of structural elements. If at the point of initiation the impact energy is regarded to be somewhat low, its effect is *localised* and in the case of impacting solids of revolution the classical theory, describing the ensuing effect is due to Hertz (1881). On the other hand, with sufficient energy and for compliant system elements their effect extends farther. The wave propagation is then regarded to be of a *global* nature, depending on the structural medium of propagation. The theories used to predict these structure-borne events were initiated by St Venant (1883). The waves are considered to move in structures according to their material of construction and shape, as well as wave parameters such as amplitude and frequency content. They are now referred to as mode shapes (see Chapters 1 and 30). The concept is not only fundamental, but intuitive to human perception. Therefore, when Newton (1687) declared his law of universal gravitation (see Chapter 1) in terms of **action-at-a-distance**, it infuriated the Cartesian school (followers of Decartes' philosophy). In Descartes' perspective the *quantity of motion* was conserved with the creation of universe and objects imparted motion onto each other through impulsions. The effect at a distance was due to a cause by *geometric extension* (Descartes, 1644). Action-at-a-distance without a medium of transmission as inferred by Newton would not do for D'Alembert (see Rahnejat, 2004), who was at the time the main Cartesian proponent. Newton, himself, was unhappy with the concept of action-at-a-distance. Cartesian *positivist* mechanical perspective required a matter-filled universe for the *impulsive* waves to travel through, rather similar to those of St Venant and the example of the clonk problem in this book. Newton, himself, was unhappy with the concept of action-at-a-distance without any intervening matter. Later, through **general relativity** an explanation was found as curvatures in space caused by the presence of large matter, upon which all other matter are compelled to reside (Einstein, 1916). Cartesian impulsions are, therefore, surmised to be gravitational waves on the fabric of space due to the pulsating bodies (rather similar to the mechanical perspective). These quite small amplitude waves are very difficult to measure as they would interact with electromagnetic phenomena (Rahnejat, 2005). However, some have measured background microwave radiations claimed to be the ongoing transient effect of the Big Bang (Penzias and Wilson, 1965).

The conclusion is that within the realm of mechanics, the concept of wave propagation is well understood and simpler to measure or predict, for example using harmonic analysis. To obtain the effect of global waves within an elastic solid, finite element analysis may be employed (see Chapter 1).

For localised waves the same approach may be employed, apart from the classical theories that can also be used under certain assumed conditions.

At the heart of the theory, describing *contact mechanics* and impact phenomena, resides continuum mechanics, which was established during the seventeenth to nineteenth centuries by a series of significant discoveries mainly by Newton, Poisson, Euler, St Venant and Hertz. For all the modifications and additions made to mechanics of contact to encompass the scheme of the very large and very small, Newtonian physics remains its cornerstone. St Venant's *global modal behaviour* and Hertz's *localised contact deformation* remain the classical solutions. Hertzian theory is based on small strains which remain local to the contact/impact domain, while St Venant considered propagation of deformation waves from the impact zone, due to their higher energy content.

Fundamentals of localised contact and impact phenomena have been studied for centuries. In engineering applications pioneering works are particularly those of Hertz (1881) and Boussinesq (1885), the latter through potential functions, representing elastic deformation. Their contributions cover most of the fundamentals of contact mechanics. In recent years, the use of coatings, engineered to reduce the untoward effects of friction, wear and fatigue of highly loaded contacts has led to observations that deviate from the classical theories and added an impetus for the better understanding of contact mechanics.

This chapter deals with the classical Hertzian contact/impact in an introductory manner. However, it is noted that in practice engineering surfaces are often rough and lubricated (see Chapters 3 and 7) and invariably protected by coatings or heat-treated surface layers. These are referred to as layered bonded solids and a treatment of this problem is provided here. Furthermore, in the scale of minutiae, impact dynamics' problems are often governed by forces other than those in Newtonian mechanics. These are described in Chapters 3 and 32 (also see Rahnejat *et al.*, 2009).

4.2   Basic aspects of contact mechanics for elastic solids

In all contact conjunctions, the solids of revolution flatten under sufficient applied loads and with increasing load the footprint area grows accordingly. This effect distributes the pressure over a larger area, decreasing the sub-surface maximum stresses. When small strains are considered, an analytic solution to this problem was found by Hertz (1881), which predicted the pressure distribution between a generic ellipsoidal rigid solid and a *semi-infinite elastic half-space* (a very large elastic flat plane compared with the dimensions of the deformed contact footprint). If the body is considered to be a perfectly smooth sphere, the initial contact is an idealised point which would grow to a *circular footprint* with an ellipsoidal pressure distribution acting upon it. If, on the other hand, the body is a perfectly smooth long right circular cylinder, then the pressure distribution is assumed to be parabolic in cross-section and uniform in its axial direction. The footprint shape would be a very narrow rectangle of infinite length ($L = 2a \gg 2b$) (very long line contact, see Fig. 4.1). Both cases are *counterformal contacts* of elastic solids of revolution. The dimensions of the contact are thus quite small compared with the principal radii of the bodies in contact (the semi-infinite assumption). Therefore, they are termed concentrated non-conforming (counterformal) contacts, unlike bodies that snug each other fully or partially, such as a ball-in-socket joint, a journal bearing and a piston ring-to-cylinder bore contact. These are considered as fully or partially conforming contacts with relatively large contact areas, where the contact dimensions are comparable with the radii of solids in contact. Thus, Hertzian theory does not apply well to these problems.

Hertz's mathematical solution is very elegant. However, as already mentioned it is applicable only to idealised contacts. For example, an *infinite line contact* does not exist and the theory can be used only for cases that the contact length is very large compared with its narrow width, for example in spur gears and only as an initial analysis. In roller bearings and cam-to-flat tappet contacts the footprint is of finite length. Such configurations are referred to as *finite line contact* (see Chapter 16 and Johns and Gohar, 1981). A characteristic of the contact footprint is its spreading at the edges (contact extremities), because of an abrupt change of profile there (Rahnejat and Gohar, 1979; Johns and Gohar, 1981). The footprint shape now becomes rather similar to a dumbbell (see Fig. 4.2), sometimes referred to as a *dog-bone footprint shape*. The edge stress discontinuities can cause localised fatigue spalls beneath the material surfaces, hence, the edges of the rollers are often

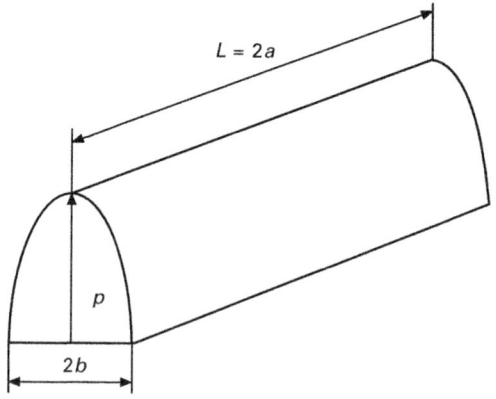

4.1 Long line contact.

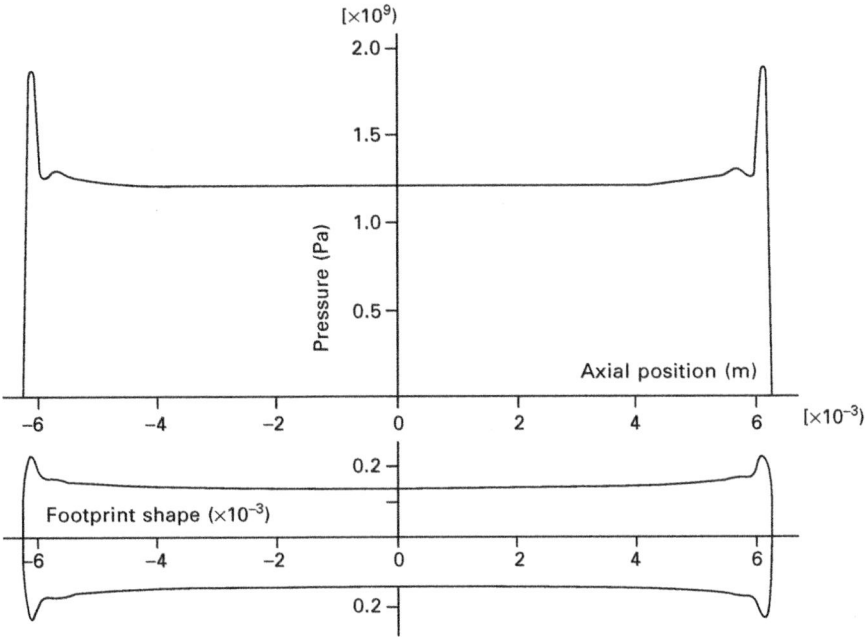

4.2 Central axial pressure distribution and footprint shape for an unblended roller against a race (Johns, 1978; Rahnejat *et al.*, 2009).

blended with relief radii known as ***dub-off*** or ***edge crown*** or indeed all the axial profile of the roller is crowned or barrelled, which has the additional benefit of self-alignment under applied loads. The effect is to reduce the edge pressures, also referred to as ***pressure spike*** or ***pressure pip*** (see Fig. 4.2). These edge stress discontinuities are not taken into account by the ***classical Hertzian theory***, because of its simplifying assumptions in order to arrive at an analytical solution. The results of a finite difference solution by Johns (1978) is shown in Fig. 4.2. Plate I (between pages 514 and 515) shows one half of typical contours of sub-surface stress beneath the surface due to localised deformation caused by pressures such as those shown in Fig. 4.2. Furthermore, engineering surfaces are rough (indeed all surfaces are rough, even those which are regarded as molecularly smooth, see Chapters 2 and 3). Therefore, the contacting bodies are subject to combined normal pressure distribution and tangential friction/traction (see Johnson, 1985). The effect is to skew the loading of rollers, for example, misalignment can result (see Johns and Gohar, 1981). In fact, rolling element bearings find instantaneous equilibrium conditions at a small misalignment as a result of contact reactions with inner and outer raceways as well as the contact of their dome ends with their retaining ribs (Rahnejat and Gohar, 1979).

For point contacts, a circular footprint shape must be considered as rather idealised. Such a condition arises from a perfectly round and smooth sphere on a perfectly flat smooth plane or two perfect right circular cylinders in contact with their axes orthogonal to each other. Under dynamic conditions no slippage is expected for a *circular point contact* condition to result. Ball bearings are retained in raceway grooves, which are only partially conforming to them, but with quite low conformance. This and other contact geometries, subject to any partial conformance and/or slipping in contact, result in an *elliptical footprint* (see Fig. 4.3). Some good examples of this are provided in Chapter 6. The same is also true of helical, hypoid and Novikov gears, where the contact ellipse is also subject to rotation because of varying principal contact radii of meshing pairs and also slide-roll ratio in the contact.

As already mentioned, all real surfaces are rough, thus the *real contact area* is often much smaller than that predicted by the Hertzian theory as the *apparent contact area*. Often the load is carried only by a few asperity tips of rough surfaces. The asperities on the contacting surfaces interact. They suffer high pressures which can cold weld the interacting pairs and generate friction. Also their oblique contact often means that harder asperities plough through the softer ones, further contributing to friction. These mechanisms are adequately described in Chapter 2. The asperities can undergo elastic deformation (implying that they recover after the load is removed, in some cases work hardened) or plastically deform and removed by various wear mechanisms (see Chapter 2). Therefore, within a rough contact quite complex interactions can occur. Additionally, the compliant asperities, when separating, can stretch to some extent. This means that some work is necessary to separate them, thus accounting for loss of energy. This effect is adhesion and is described in some detail in Chapter 3.

Finally, in most and some would say in all applications, a film of fluid or surface oxides intervenes between the contacting surfaces, which can act as a lubricant unless a good vacuum exists. This film, if fluidic, would act with

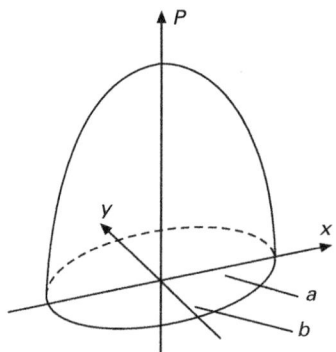

4.3 An elliptical footprint.

many facets, depending on its composition and the gap size (see Chapters 3, 5, 6 and 32). If the film is solid, such as an oxide, it would break down under tangential forces and the debris particles can act as ***third bodies*** in the contact. Depending upon their size and shape they can carry load in the conjunction as a form of ***boundary lubrication***. If their size is large relative to the gap and their shape irregular, inhibiting rolling motion, then they can act as sharp grits, causing gradual wear.

4.3   Hertzian theory

As already noted, ***Hertzian theory*** was derived for the case of the contact between an ellipsoidal solid of revolution and a semi-infinite elastic half-space. Often two bodies in contact have different contact radii (R_1 and R_2) and moduli of elasticity (E_1 and E_2). This means that the *Hertzian* equivalent is that of a rigid ellipsoidal solid of radius:

$$\frac{1}{R} = \frac{1}{R_1} + \frac{1}{R_2}$$

contacting a semi-infinite elastic half-space of reduced modulus:

$$\frac{1}{E'} = \frac{(1 - v_1^2)}{E_1} + \frac{(1 - v_2^2)}{E_2}$$

Both the ***equivalent radius*** and the ***effective modulus*** are referred to as *reduced*. The former, because its value is smaller than the radii of both the actual contacting solids and the latter, because its value is smaller than the sum of their individual moduli. If one of the bodies resides in the other, as in a ball in a raceway groove, with low degree of conformity (see Fig. 4.3), then one can use the same equivalent *Hertzian* model, where the radius of the concave surface is regarded as negative. In this case, the equivalent radius becomes:

$$\frac{1}{R} = \frac{1}{R_1} - \frac{1}{R_2}$$

which yields a value larger than both the contacting surfaces radii, thus referred to as the *increased* radius. Note that for closely conforming surfaces (such as journal bearings), this approach is not permitted as a semi-infinite elastic half-space cannot be assumed.

The parameters of interest in contact mechanics are the footprint dimensions, the pressure distribution and the elastic deflection. This varies according to contact geometry (discussed above). For an ***elastic line contact*** of length L, the Hertzian theory yields these as (see Fig. 4.1):

$$b = 2 \left(\frac{WR}{\pi L E'} \right)^{1/2} \tag{4.1}$$

$$p = p_0 \left(1 - \frac{x^2}{b^2} \right)^{1/2} \tag{4.2}$$

$$\delta = \frac{W}{\pi L E'} \left[\ln \left(\frac{\pi L^2 E'}{2RW} \right) + 1 \right] \tag{4.3}$$

where the maximum Hertzian pressure is:

$$p_0 = \left(\frac{WE'}{\pi L R} \right)^{1/2} \tag{4.4}$$

For a **circular point contact**:

$$a = \left(\frac{3WR}{4E'} \right)^{1/3} \tag{4.5}$$

$$p = p_0 \left(1 - \frac{r^2}{a^2} \right)^{1/2} \tag{4.6}$$

$$\delta = \frac{1}{2} \left(\frac{9W^2}{2E'^2 R} \right)^{1/3} \tag{4.7}$$

where the maximum Hertzian pressure is:

$$p_0 = \frac{1}{\pi} \left(\frac{6WE'^2}{R^2} \right)^{1/3} \tag{4.8}$$

For an **elliptical point contact** (see Fig. 4.3):

$$\sqrt{ab} = \left(\frac{3W\sqrt{R_x R_y}}{4E'} \right)^{1/3} \tag{4.9}$$

$$p = p_0 \left(1 - \frac{x^2}{a^2} - \frac{y^2}{b^2} \right)^{1/2} \tag{4.10}$$

$$\delta = \frac{1}{2} \left(\frac{9W^2}{2E'^2 \sqrt{R_x R_y}} \right)^{1/3} \tag{4.11}$$

where the maximum Hertzian pressure is:

$$p_0 = \frac{1}{\pi}\left(\frac{6WE'^2}{R_x R_y}\right)^{1/3} \tag{4.12}$$

To find a and b the semi-major and semi-minor contact half-widths of the elastostatic contact elliptical footprint, equation (4.9) can be used together with:

$$\frac{b}{a} \approx \left(\frac{R_y}{R_x}\right)^{2/3} \tag{4.13}$$

The equivalent radii R_x and R_y are those of the two ellipsoidal solids in the xz and yz planes respectively (for example, for a ball in a raceway groove in a ball bearing see Fig. 4.4). For a circular point contact, clearly: $R = R_x = R_y$. The ratio a/b is referred to as the **ellipticity ratio**, which has a value of unity for a circular point contact.

In dynamics analysis, Hertzian contacts are often represented in the form of their load–deflection characteristics. Such relationships are obtained by re-arranging equations (4.3), (4.7) or (4.11). For example, for a circular point contact, $W = \frac{4}{3}\sqrt{R}E'\delta^{3/2}$, which is often represented as:

$$W = K\delta^{3/2} \tag{4.14}$$

where $K = \frac{4}{3}\sqrt{R}E'$ is the constant of proportionality, referred to as **contact stiffness non-linearity**. This is because as the contact deflection is increased it becomes stiffer. One can observe that stiffness is defined as:

$$k = \frac{\partial W}{\partial \delta} = \frac{3}{2}K\delta^{1/2} = 2E'\sqrt{R\delta} \tag{4.15}$$

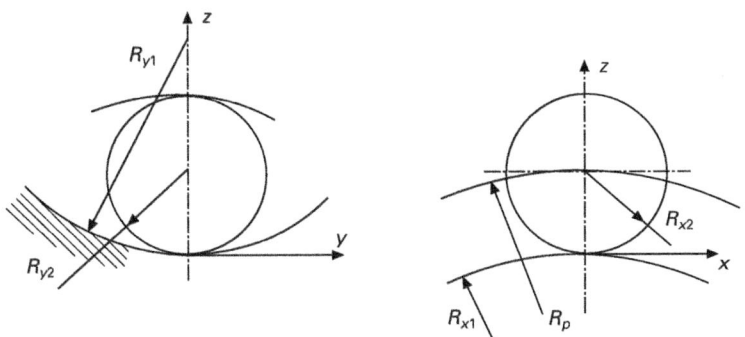

4.4 Principal radii of curvature of ball and raceway grooves in planes of contact.

For a linear spring k is a constant, while clearly for a ball bearing it is a function of its deflection and increases with it. Note from (4.5) and (4.7) that: $\delta = a/R^2$. This condition holds true for point contacts, where Hertzian conditions may be assumed (semi-infinite solid).

4.4   Analytical treatment of contact mechanics of layered solids

Without going into detailed stress analysis it transpires that contact mechanics behaviour of a coated surface deviates from the Hertzian theory, depending on the thickness of the coating. These surfaces are referred to as *layered bonded elastic solids*. Naghieh *et al.* (1998) have shown that when the ratio of contact radius to layer thickness; $a/d \geq 2$, then: $\delta = a^2/2R$ and the contact radius and pressure distribution for *compressible layers* (coatings); are obtained as:

$$a = \left(\frac{(1 - 2v)(1 + v)dRW}{\pi E(1 - v)} \right)^{1/4} \tag{4.16}$$

$$p = \frac{E(1 - v)}{2R(1 + v)(1 - 2v)} \frac{(a^2 - r^2)}{d} \tag{4.17}$$

These relationships apply for fairly thin layers, on top of an assumed rigid substrate. For quite thick layers, the contact mechanics behaviour begins to approach Hertzian conditions. In fact, this situation is found to be around: $a/d \leq 0.25$. For the intervening period (between very thick and very thin layers), $0.25 \leq a/d \leq 2$, finite element analysis is usually used. Later in Section 4.6 an alternative semi-analytical approach of generic nature is highlighted for all values of a/d.

It is clear from equation (4.16) that for incompressible materials ($v = 0.5$, a theoretical ideal where rubber-based material can tend to, with $v \leq 0.49$) the footprint radius diminishes and pressures reach infinity. This, of course, is not true and using the correct boundary conditions and constitutive equations, Naghieh *et al.* (1998) have shown that for *incompressible thin elastic layered solids*, $\delta = a^2/4R$:

$$a = 2 \left(\frac{3d^3 RW}{\pi E} \right)^{1/3} \tag{4.18}$$

$$p = \frac{3Ea^4}{64(1 + v)Rd^3} \left(1 - \frac{r^2}{a^2} \right)^2 \tag{4.19}$$

Finite element analysis carried out by Votsios (2003) for circular point contacts of coated surfaces has shown that the values obtained for small *a/b* ratios, typically below 0.25, agree with classical Hertzian theory. With given values for the coating thickness, *d*, the calculated contact radii *a* for various loads are in accord with the relationship, $W \propto a^3$. For semi-infinite solids in circular point contact, equations (4.14) and relationship $\delta = a^2/R$ yield the Hertzian relationship

$$W = K\left(\frac{a^2}{R}\right)^{3/2} = \frac{4E'}{3R}a^3$$

The finite element results also showed that for *a/d* > 2, the relationship was approximately $W \propto a^4$. This is in line with equation (4.16) for incompressible elastic layered bonded solids. Thus, detailed finite element analysis of Votsios (2003) confirmed the analytical approaches of Johnson (1985) and Naghieh *et al.* (1998) among others, based on the ratio *a/d*. In physical terms, with **hard wear-resistant coatings** high pressures are generated, which are well in excess of those predicted by the classical Hertzian theory (see Fig. 4.5 for TiN and the ceramic coating, being alumina). In fact hard coatings protect the substrate material (in this case steel) from the effect of sub-surface stresses. For soft coatings (see cases of silver and gold in the same figure), although

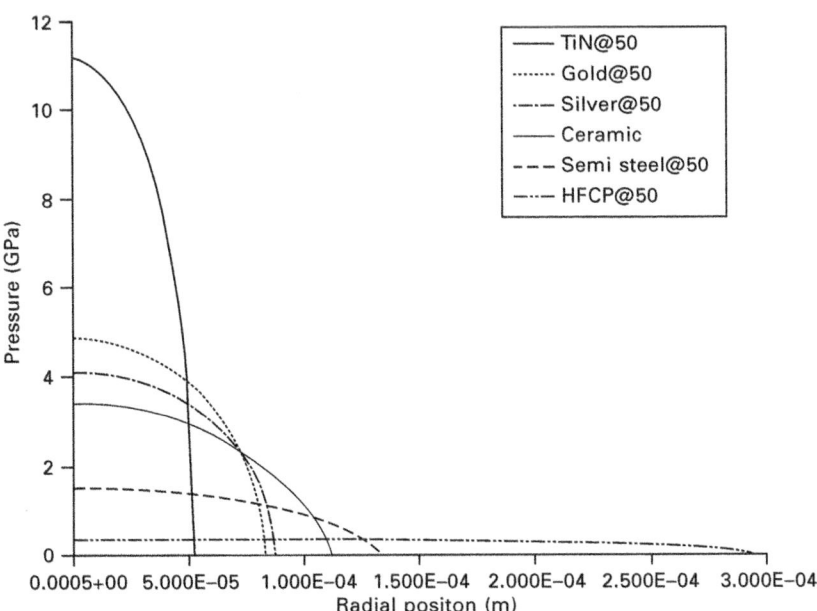

4.5 Bonded layered elastic behaviour in circular point contact predicted by finite element analysis and compared with the semi-infinite behaviour of the semi-infinite steel substrate.

the pressures are high the sub-surface stresses mainly occur in the substrate material. The *soft coatings*, such as those studied here, are often used to act as *solid lubricants*, particularly in space bearings. The relationship $W \propto a^n$, $3 < n < 4$ holds for these soft coatings, indicating an elastic response that suggests a behaviour between a compressible and an incompressible layer. In fact, this is party due to the relaxation behaviour of soft coatings and partly due to their elasto-plastic behaviour at higher contact loads. HFCP, in the same figure, refers to *highly filled carbon polymers*, which are used as a soft thin-bonded layer with applications in electrical systems, as protective layers and the emerging nanotechnologies, such as in nanotubes. They have quite low elastic moduli, and their behaviour is best described through viscoelasticity. In fact, one should note that most coating materials act in a viscoelastic manner, thus the relationships given here are for their instantaneous elastic response. Relaxation takes place thereafter, where the contact footprint spreads as the pressures fall (see Johnson, 1985; Naghieh *et al.*, 1998).

4.5   Impact dynamics

To a large extent contact mechanics behaviour may be regarded as a special case of impact dynamics. With the exception of propagation of light and electromagnetic phenomena *in vacuo*, all events are subject to a duration. This is the basis of modern physics, embodied in the *special theory of relativity* (Einstein, 1905). The theory is based on the declaration of constancy of speed of light *in vacuo*. This is a restatement of the Galilean proclamation that there is no distinction between *absolute motion* and *absolute rest*, both being the observation of propagation of light as an event which is the upper limit of its speed, which is omnipresent. All other events, therefore, have finite durations, including impact of solids. Contact, therefore, is a special case of impact with long enough duration in a *chosen space–time frame of reference* that one may regard as almost stationery. In Chapters 1 (Appendix) and 25–29, rattle as series of impacting gear teeth pairs is described. The impact durations depend on a host of physical parameters, such as geometry of teeth, mechanical properties of solids, rheology of lubricant and kinematics of contact. Irrespective of the mathematics of various models employed, the durations of the impacts are finite. The understanding has evolved since Newton and Poisson who assumed instantaneous impacts. Thus, pre- and post-dynamics were treated separately.

Works of Hertz and St Venant, described earlier, however, have shown that a host of events occur in impact of solids, so that decomposition of such events to their constituent parts show a host of responses with their distinct durations. In practice, for example in the case of impacting gear teeth pairs, the constituent phenomena include *localised deformation* of surfaces (formation of a dimple and subsequent elastic recovery) and **global structural deformation** (bending or rocking oscillations of teeth). In the case of the former, the deformation wave is confined to the surfaces, whereas in the latter case waves represent the modal behaviour of the solids and propagate farther. The localised impact can be discerned as an **accelerative noise** (mainly airborne), while in global deformation there is the additional ringing noise, when the **structure-borne vibration** can coincide with acoustic modal response of components (see the clonk phenomenon, Chapter 30). A good example is the church bell, when struck by its clapper. The initial impact has an almost instantaneous localised response, followed by **ringing noise**.

Hertzian impact theory, like his contact theory is based on small strains, and thus ignores modal behaviour of impacting solids. A Newton trolley comprises the impact of a number of spheres. Close examination of the transferred impact (impulse) through the system shows that the arc movement of the ultimate struck sphere is very similar to the initial arc of motion of the released striking sphere. This means that modal response of the spheres may be ignored, although present in an insignificant manner. Furthermore, losses due to friction are also ignored. Thus, the underlying assumptions in Hertzian impact becomes clear. The noise heard is accelerative, and the sound energy is quite low (small losses). If now one were to replace at least one of the solid spheres with a hollow ball of otherwise exact material and geometry, the motion of the ultimate struck sphere and that of the striking sphere deviate significantly enough to be noted by usual means of measurement. Furthermore, when a hollow ball is struck a ringing noise is heard in addition to the initial accelerative noise. Clearly, the hollow sphere undergoes some of its modal structural responses. The problem can no longer be treated as localised to the point of impact. With global deformation waves more energy is dissipated in the forms of sound, structural vibration as well as heat. These gradually dampen the effect of impact energy. However, these are not very ideal in drivetrains, owing to poor NVH refinement and in some cases structural integrity issues. They are, nevertheless, a constituent part of physics of nature itself, but do not enjoy the same privileges. For instance, the longer effect of impact of large space debris on Earth is somewhat mitigated by plastic deformation and intensive heat dissipation, and although some of these effects are catastrophic for the inhabitants of Earth, its own survival as an entity is assured. Smaller falling debris are dealt with by heat generation in the atmosphere (viscous friction). Several thousand debris items are dealt with in an almost daily basis.

In engineering one has to find solutions to reduce the effect of impact problems akin to nature, not contrary to its workings. Therefore, making hollow components, whilst increasing engine torques, although desired as an environmental and customer-driven concept is *flying in the face* of nature.

Given that structural deformation in most cases are limited by elastic limits, it is not surprising that hopes of attenuation are pinned, to a large extent, on the tribological issues, such as contact geometry, lubricant rheology, surface topography and composition. Therefore, contact mechanics is gaining in importance that it actually deserves.

This chapter deals with localised impact dynamics. The global effect of impacts is dealt with in other chapters (see Chapters 21, 25–30, for example). In these cases a palliative approach is usually used to reduce the effect of propagating deformation waves.

4.5.1   Hertzian impact

As already described **Hertzian impact** embodies the assumption of a localised nature, meaning small strains within the Hookean elastic limit and semi-infinite conditions (i.e. dimensions of the contact remain much smaller than the principal radii of bodies at the region of impact). Also, the depth of penetration is quite small (i.e. $\delta < a << R$). The impacting solid surfaces are considered to be perfectly smooth and frictionless. With these assumptions, it becomes quite clear that the deformation energy would be below that required to cause structural modal response and that the impacting solids would be of ellipsoidal form. Altogether, this scenario pertains to very idealised conditions, but the solution is analytical and quite elegant (Hertz, 1881). The theory also assumes no losses, as already described. Thus, the kinetic energy of impact is converted into potential energy in the form of stored strain energy, which upon rebound is converted back to kinetic energy again. Using Rayleigh's method, the maximum kinetic energy (at the instant of impact) equates the maximum stored strain energy at maximum contact deflection, if one converts any impacting pair to the Hertzian model of a rigid sphere of equivalent radius R impacting a semi-infinite elastic half space of effective modulus, E'. Thus:

$$\frac{1}{2}mv^2 = \frac{2}{5}K\delta_{max}^{5/2} \tag{4.20}$$

where m is the mass of the equivalent sphere. Also, note that according to Euler: $E = \int W d\delta$; the stored strain energy is the work done by the impact/contact force W in deforming the elastic half-space. Note that the contact force is given by equation (4.14), thus the stored energy at maximum penetration δ_{max} is the term on the right-hand side of equation (4.20).

At the instant of impact, $\dot{\delta} = v$, this being the impact velocity. If this is the impact of a sphere released, under the influence of gravity, from a height H upon a semi-infinite elastic half-space, then $v = \sqrt{2gH}$. The equation of motion is given as $\ddot{\delta} = -W/m$, where the impact force, W, is given by equation (4.14). The localised small penetration of the rigid sphere into the elastic half-space is decelerative ($\ddot{\delta} < 0$) due to the work done by the contact force to resist the motion. This resistance continues until the maximum penetration δ_{max}, where sufficient stored energy (given by the right-hand side of equation (4.20)) causes the rebound of the sphere. Thus, the motion of the penetrating sphere can be represented as:

$$\frac{1}{2}(\dot{\delta}^2 - v^2) = -\frac{2K}{5m}\delta^{5/2}$$

When $\dot{\delta} = 0$, the moment of maximum penetration is reached, $\delta = \delta_{max}$, yielding equation (4.20). The **impact time** is twice that from instant of impact to δ_{max}, obtained by solution of equation of motion (see Hertz, 1881; Gohar and Rahnejat, 2008):

$$t \simeq 2.94\frac{\delta_{max}}{v} \tag{4.21}$$

The same approach can be used for elastic line contacts (Hertz, 1881; Teodorescu et al., 2005):

$$t \simeq \frac{\pi\delta_{max}}{v} \tag{4.22}$$

Note that Hertzian impact applies to only a subset of impact phenomena. For coated surfaces the same procedure can be applied using the corresponding equation for contact force from equations provided in Section 4.4. Assuming small strains, the Hertzian impact theory also upholds the **principle of conservation of momentum**. Thus, using equations (4.21), (4.15) and (4.14) with the relationship: $Wt = mv$, the impact can be represented in terms of instantaneous stiffness k and, impacting mass m as:

$$Wt = \frac{2.94\delta_{max}}{v}\left(\frac{2}{3}k\delta_{max}^{-1/2}\delta_{max}^{3/2}\right) = mv \tag{4.23}$$

which leads to:

$$\frac{1.96\delta_{max}^2}{v^2} = \frac{m}{k} = \frac{1}{\omega_n^2}$$

at the instant of maximum penetration (for a point contact geometry), prior to rebound. Hence:

$$\omega_n \simeq \frac{0.7v}{\delta_{max}} \tag{4.24}$$

Clearly, v remains the same, as δ increases up to δ_{max} and reduces during rebound. This means that impact can be characterised as a march of natural contact frequency, with its minimum value at maximum penetration. This is valid only for small strains where no wave propagation occurs in the impacting solids. Thus, the impact time is very short-lived (typically a few tenths of a millisecond), The sound emanating from such an impact is regarded to be accelerative in nature and for most part airborne. If the material is of lower elastic modulus the depth of penetration becomes larger and the natural frequency decreases (larger δ_{max}, corresponding to a longer impact time). The airborne contact noise is thus reduced. For such solids and those of hollow structure, strains encountered exceed Hertzian conditions and their modal behaviour is noted by structure-borne noise.

In many applications contact deformation is encouraged in order to achieve elastohydrodynamic lubrication (EHL) conditions (see Chapter 6), promoting lubrication. However, larger deformation in impact can also lead to sound propagation. An example is the slapping noise from piston-cylinder bore or liner interactions. A lubricant film formed in the conjunction presents additional resistance to overcome. Thus, the principle of conservation of momentum is not upheld as some work is required to squeeze the entrapped lubricant film and increase its viscosity and density, so that it becomes an amorphous solid. This work translates to an increased impact time. Furthermore, a slightly greater value for δ_{max} is reached due to the resistance of the entrapped fluid (note that Hertzian theory implicitly assumes an impact *in vacuo*). This combination of events constitutes lower contact frequencies, thus reducing slapping noise. An analytical solution is difficult to obtain, but numerical predictions can be made. The section below provides a practical example.

4.5.2   Impacts zones in valve train systems

There are two potential impact zones in most **valve train systems** (see Chapter 17 and Teodorescu *et. al.*, 2005, 2006). One is between cam and the follower (see Fig. 4.6) due to **valve spring surge** or camshaft flexibility or both can result in contact separation. This occurs when the surfaces of cam and the follower separate such that a coherent film of lubricant may not be maintained. Upon rebound an impact occurs, which may be largely dry, thus it follows the Hertzian impact theory for an elastic line contact of finite length. However, as the lubricant entraining action takes place at the same time, a partially lubricated contact is most likely. A sharp rise rate in the cam lift profile and high rotational speed can also cause valve spring surge with an insufficient valve spring preload. This can result in impacts in the valve-to-valve seat contact or continuous **valve flutter** (see Fig. 4.6 (b)). An elastodynamic model of the valve train system, incorporating cam-follower tribology is necessary to study the effect of valve toss, rebound and these impacts (see Chapter 17, Teodorescu *et al.*, 2005, 2006; Gohar and Rahnejat, 2008).

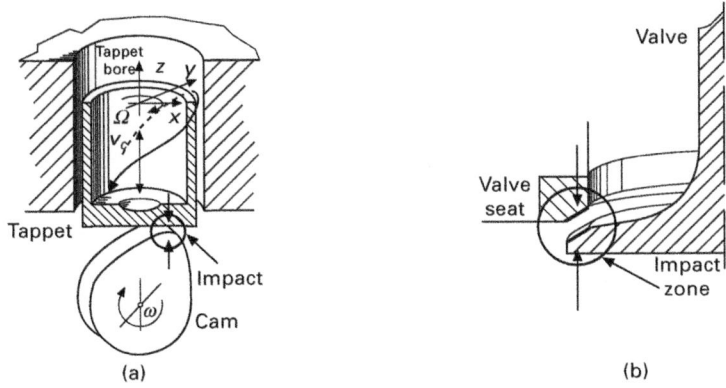

4.6 Impact zones in valve train systems: (a) cam–tappet, (b) valve–
valve seat.

To avoid the valve-to-valve seat impact, the cam profile is modified
to accommodate a slow rising ramp (see Chapter 17). However, to avoid
cam–tappet impacts, the entire dynamics of the valve train mechanism must
be considered. Therefore, during the design phase, for a valve train system
alternative configurations should be analysed, yielding operational limiting
speeds,

To illustrate the consequences of cam–tappet contact separation and
subsequent impact, a simplified example is considered here. Figure 4.7(a)
shows the geometrical lift, velocity and acceleration for an overhead valve
train system, while Fig. 4.7(b) provides the prediction of a dynamic model
for a camshaft speed of 1000 rpm. At this speed the contact between the
cam and the tappet is not lost. However, at the higher speed of 3000 rpm
loss of contact occurs, followed by sharp impacts. Figure 4.7(c) shows the
acceleration and Fig. 4.7(d) the variation of oil film thickness. In Figs 4.7
(b) and (c) experimentally obtained results by Teodorescu *et al.* (2005) are
also shown, validating the model used to study valve train dynamics.

The rate of change of film $\partial h/\partial t$ (squeeze velocity) is predicted by valve
train dynamics and the oil film thickness is obtained from an extrapolated
oil film thickness formula (see Chapter 17, Teodorescu and Taraza, 2004;
Gohar and Rahnejat, 2008). It can be seen that the rebounding valve does
tend to improve the film thickness marginally. This is due to an increase
in the squeeze film effect rather than the actual impact force, since in the
prevailing elastohydrodynamic regime of lubrication, the film thickness is
rather insensitive to load (see Chapter 6). However, the nature of squeeze
caving phenomenon under transient conditions is rather short-lived, owing
to the very short impact time and the fact that the lubricant film is very
stiff and diminishes rapidly as the bodies separate (i.e. the positive squeeze
film effect: separation). This explains the high-frequency oscillations of the

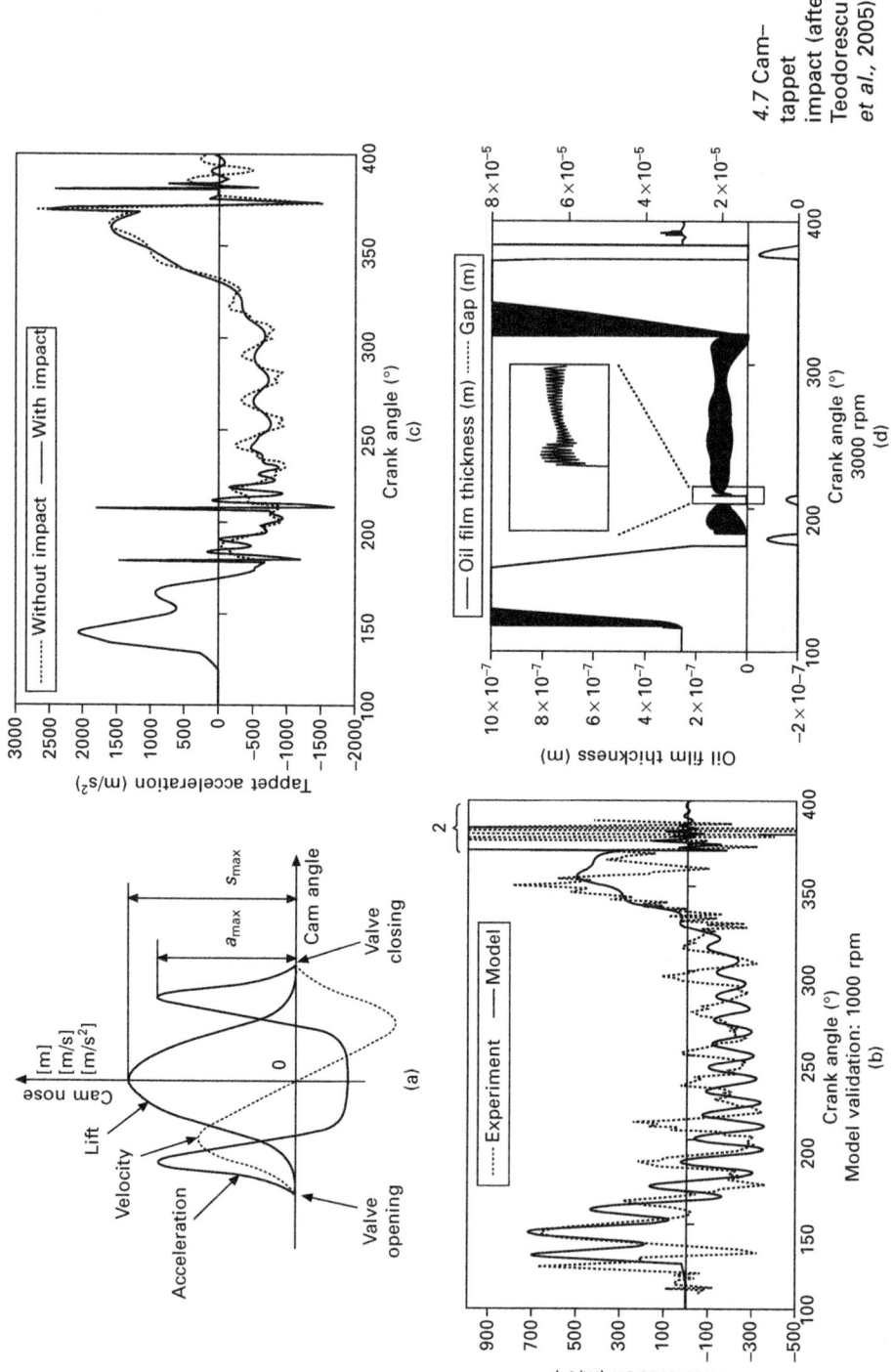

4.7 Cam–tappet impact (after Teodorescu et al., 2005).

lubricant film under impacting conditions, shown by the zoomed region in Fig. 4.7.

4.6   Contact mechanics based on action of deformation potential

The classical Hertzian theory embodies the idealised frictionless semi-infinite contact conditions. As already pointed out above, it does not apply to cases of *layered bonded elastic solids*, such as coatings or indeed treated surfaces such that a structure with variation of mechanical properties exist into the depth of the solid. For such cases one should resort to numerical analysis. Finite element and finite difference approaches are highlighted earlier in the chapter. Also, for special cases such as fairly thin surface layers closed form analytical solutions are provided, again for frictionless contacts and rigid substrates.

The need for prediction of behaviour of hard wear resistant coatings (e.g. for tools and highly loaded rolling element bearings) and soft lubricating layers (e.g. gold or MoS_2 coatings for space bearing applications) has led to the development of various simplified analytic solutions. The original work by Hannah (1951) has been extended, by Johnson (1985), Jaffar (1989), Barber (1990) and Naghieh et al. (1998) among others. These methods are fast, accurate and generally accepted by the scientific community. However, they are usually limited to certain layer thickness and elastic properties.

Predictions from the method described below are compared with Johnson's analytical expression for case of thin coatings. Johnson (1985) gives contact footprint and the pressure distribution as:

$$a = \sqrt[3]{\frac{3}{2} R l_1 \frac{1 - 2v_\phi}{1 - v_\phi} \frac{1 + v_\phi}{E_\phi} W} \qquad (4.25)$$

$$p(x) = \frac{1 - v_\phi}{1 - 2v_\phi} \frac{E_\phi}{1 + v_\phi} \frac{a^2}{2R l_1} (1 - x^2/a^2) \qquad (4.26)$$

The most complex situation is when the layer is neither thick nor thin, as often is the case in many engine applications, where the coating thickness has been arrived at in an experiential manner, rather than by any analytical rigour. When a predictive method is applied for these intermediate cases, the most widely used approach has been finite element method (Votsios, 2003; Konvopulos and Gong, 2003; Ovaert and Pan, 2002). However, accurate predictions require a very large number of nodes and can consequently lead to excessive computation. If the contact mechanics solution is to be included in a multi-physic approach, which accounts for a large number of physical phenomena within a convergence loop, the computation speed becomes a very important factor.

Recently, an additional family of contact mechanics solutions have been proposed as possible fast and accurate approaches. These methods are based on the original work of Sneddon (1951), later extended by Hooke (2000), Greenwood and Morales-Espejel (1993), Polansky and Keer (2000), Liu and Wang (2002) and Teodorescu *et al.* (2009), among other researchers. These methods predict that the response of the elastic materials to an applied pressure distribution can be computed firstly by decomposing the contact pressure into a harmonic series over a chosen domain. Then, the elastic response for each harmonic component of applied pressure is computed, and recomposed so that a final overall solution is obtained. The fundamental basis for such an approach lies in the Bernoulli's *principle of superposition*.

This section represents a brief description of this family of methods. For a detailed description, the interested reader should refer to Teodorescu *et al.* (2009). Figure 4.8 shows a generic contact between a cylindrical indenter ζ and a coated flat substrate ψ. The coating ϕ is applied to the flat substrate. A pressure distribution can be decomposed into a harmonic series as:

$$p(x) = \frac{1}{2} P_0 + \sum_{k=1}^{N \to \infty} P_k \cos(\alpha_k x - \varphi_k) \qquad (4.27)$$

where:

$$\alpha_k = 2\pi/\lambda_k \text{ and } \lambda_k = L_F/k.$$

By independently applying each component of the pressure distribution to the contact, the stress and strain fields can be computed.

Any coating alters the properties of the contact and, therefore, the sub-surface stress and strain fields, and contact deflection also alters accordingly. By independently applying each component of the pressure distribution, these can be obtained as Teodorescu *et al.* (2009):

$$\begin{cases} {}^q\sigma_{ij}^{\xi} = {}^q_0\sigma_{ij}^{\xi} + \sum_{k=1}^{N \to \infty} {}^q_k\sigma_{ij}^{\xi} \\ {}^q\varepsilon_i^{\xi} = {}^q_0\varepsilon_i^{\xi} + \sum_{k=1}^{N \to \infty} {}^q_k\varepsilon_i^{\xi} \quad \text{where: } \begin{cases} q \in \{C, T\} \\ \xi \in \{\phi, \psi, \zeta\} \\ i, j \in \{x, y\} \end{cases} \\ {}^qu_i^{\xi} = {}^q_0u_i^{\xi} + \sum_{k=1}^{N \to \infty} {}^q_ku_i^{\xi} \end{cases} \qquad (4.28)$$

Each line from equation (4.28) consists of two distinct terms. The first is due to the constant pressure component from equation (4.27) and the second term is the sum of the responses generated by the harmonic components of the pressure distribution. The predictive approach for each of these terms is fundamentally different (Teodorescu *et al.*, 2009).

The sub-surface stress field is dependent on the applied load, the elastic

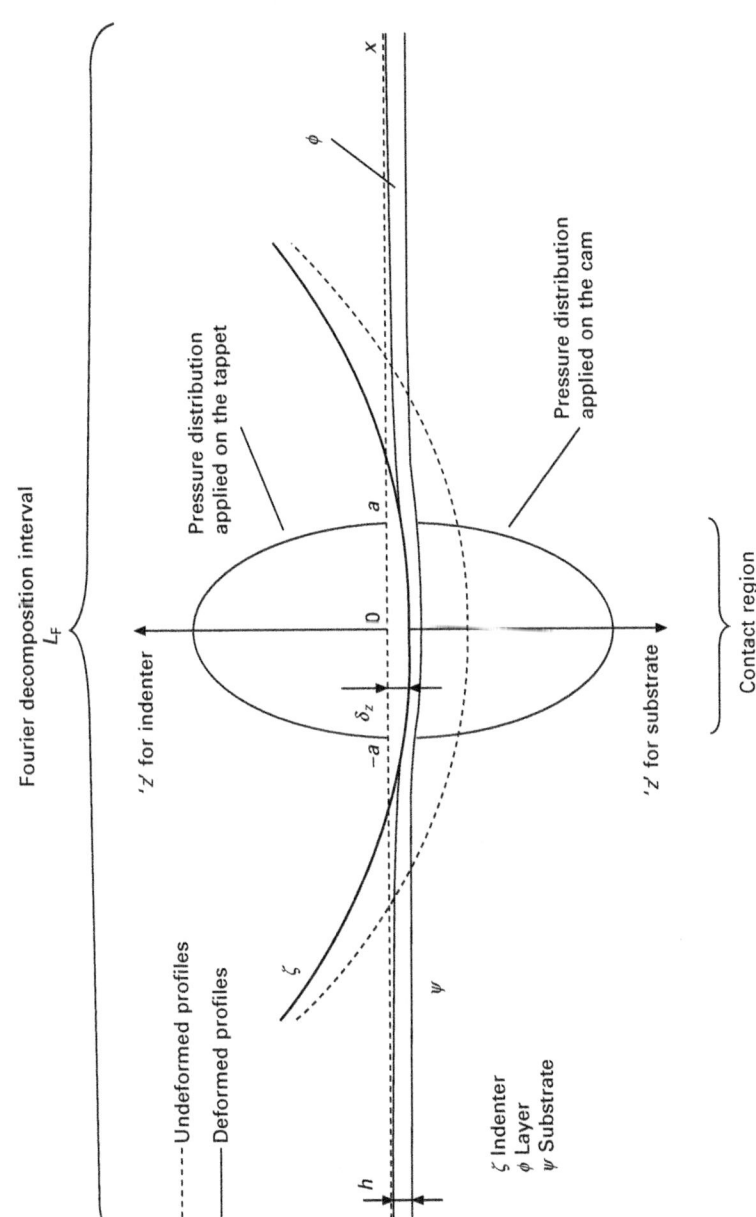

4.8 Loaded contact configuration.

properties of the solids and geometry. For each elastic media in the contact there is a unique scalar potential function, which can fully describe the field. Therefore, in the case of a single bonded elastic layer solid, two such functions are required; one for the layer and another for the substrate.

Using the notations from Fig. 4.8, each stress function should satisfy the following **biharmonic equation**:

$$\frac{\partial^4 \xi_k}{\partial x^4} + 2\frac{\partial^4 \xi_k}{\partial x^2 \partial y^2} + \frac{\partial^4 \xi_k}{\partial y^4} = 0$$

$$\xi_k = \{\phi_k, \psi_k, \zeta_k\}$$

(4.29)

Equation (4.29) shows a possible shape, which can fulfil the conditions imposed by equation (4.28) and preserves the physics of the problem. This approach is described in detail by Love (1927) and applied for a similar set of conditions by Teodorescu *et al.* (2009):

$$\xi_k = \cos(\alpha_k x - \varphi_k)(_kC_1^{\xi}\cosh \alpha_k y + _kC_2^{\xi}\sinh \alpha_k y$$

$$+ _kC_3^{\xi}y \cosh \alpha_k y + _kC_4^{\xi}\sinh \alpha_k x)$$

$$\xi_k = \{\phi_k, \psi_k, \zeta_k\}$$

(4.30)

For a harmonic wave applied along the *x*-axis in Fig. 4.8, this function preserves the harmonic order, as well as the phase angle, but it diminishes in amplitude into the depth of the elastic solid. However, the pressure wave completely decays as $y \to \infty$.

Teodorescu *et al.* (2009) showed that the resulting **sub-surface stress field**, as well as the surface deflection, can be expressed as:
For the substrate ($y \geq l_1$)

$$\begin{cases} _k\sigma_{xx}^{\psi}\,\text{sech}\,\alpha_k y = P_k \cos(\alpha_k x - \varphi_k)(1 - \tanh\alpha_k y)_k\Theta_2\{-\Theta\alpha_k l_1 + \overline{v}[\alpha_k(l_1-y)+1]\} \\ _k\sigma_{yy}^{\psi}\,\text{sech}\,\alpha_k y = P_k \cos(\alpha_k x - \varphi_k)(1 - \tanh\alpha_k y)_k\Theta_2\{\Theta\alpha_k l_1 + \overline{v}[\alpha_k(l_1-y)-1]\} \\ _k\sigma_{xy}^{\psi}\,\text{sech}\,\alpha_k y = P_k \cos(\alpha_k x - \varphi)(1 - \tanh\alpha_k y)_k\Theta_2[\Theta\alpha_k l_1 - \overline{v}\alpha_k(l_1-y)] \\ _k\sigma_{y}^{\psi}\,\text{sech}\,\alpha_k y = P_k \cos(\alpha_k x - \varphi_k)(1 - \tanh\alpha_k y)_k\Theta_2[_kH_2(\alpha_k y + 1) + _kH_3]/\alpha_k E^{\psi} \end{cases}$$

(4.31)

For the layer ($0 > y > l_1$)

$$\begin{cases} _k\sigma_{xx}^{\phi}\text{sech}\,\alpha_k y = P_k \cos(\alpha_k x - \varphi_k)[\tanh\alpha_k y(\alpha_k y - _k\Theta_1) + 1 - _k\Theta_1\alpha_k y] \\ _k\sigma_{yy}^{\phi}\text{sech}\,\alpha_k y = -P_k \cos(\alpha_k x - \varphi_k)[\tanh\alpha_k y(\alpha_k y + _k\Theta_1) - 1 - _k\Theta_1\alpha_k y] \\ _k\sigma_{xy}^{\phi}\text{sech}\,\alpha_k y = P_k \sin(\alpha_k x - \varphi_k)\alpha_k y[1 - _k\Theta_1\tanh \alpha_k y] \\ _k\delta_y^{\phi} = \cos(\alpha_k x - \varphi_k)P_k \,_k\Omega_y/E^{\psi} \end{cases}$$

(4.32)

where:

$$
{}_k\Omega_y = \frac{1}{\alpha_k}
\begin{pmatrix}
{}_k\Theta_2[H_2(\alpha_k l_1 +1) + H_3](\cosh\alpha_k l_1 - \sinh\alpha_k l_1 \\[4pt]
+ (v^\phi - 1)\dfrac{E^\psi}{E^\phi}\begin{bmatrix} \alpha_k l_1\cosh\alpha_k l_1 - \sinh\alpha_k l_1 + {}_k\Theta_1(\cosh\alpha_k l_1 - \alpha_k l_1\sinh\alpha_k l_1) \\[4pt] - \alpha_k y \cosh\alpha_k y + \sinh\alpha_k y - {}_k\Theta_1(\cosh\alpha_k y - \alpha_k y \sinh\alpha_k y) \end{bmatrix} \\[4pt]
- (v^\phi + 1)\dfrac{E^\psi}{E^\phi}\,[\cosh\alpha_k l_1 - {}_k\Theta_1\sinh\alpha_k l_1 - \cosh\alpha_k y + {}_k\Theta_1 \sinh\alpha_k y]
\end{pmatrix}
$$

$$(4.33)$$

As expected, the stress fields predicted for each individual harmonic component of the load have a harmonic form. The phase angle and the harmonic number of the stress or deflection wave is inherited from the applied load. However, the amplitude is attenuated into the depth of the solid.

The deflection at a specified depth in the layer is computed by adding the layer deflection to the deflection of the layer–substrate interface. This can be expressed as:

$$
{}_k\delta_y = {}_k\delta_{l_1} + \int_y^{l_1} {}_k\varepsilon_y^\phi \, dy
$$

For the extreme cases of an infinitely thick layer ($l_1 \rightarrow \infty$), and for an infinitely thin layer ($l_1 \rightarrow 0$), as well as for a bonded layer and the substrate of the same elastic properties, the sub-surface fields predicted reduce to that of semi-infinite solids. This is a further proof of the validity of the method developed.

Contacting surfaces finally converge into quasi-conforming deformed profiles within the elastic limit. This intrinsic property is used in the present study as one of the main convergence criteria (a form of imposed constraint function). The method assumes an initial pressure distribution and computes its harmonic components. Surface deflection due to the constant pressure (${}_0^q\delta_0$) is computed, using the method described by Johnson (1985) and the surface deflection (at $y = 0$) due to the harmonic components can also be computed as outlined above.

For Hertzian point contact condition the sub-surface absolute maximum shear stress value is $\tau_{max} \approx 0.31 p_0$, p_0 being the maximum Hertzian pressure. The depth at which the maximum shear stress occurs is $z = 0.47a$. For line contact condition $\tau_{max} \approx 0.3 p_0$ and $z = 0.78b$. Using **Tresca yield criterion**, the onset of yield is at $\tau_{max} = 1/2\,\sigma_Y$, where σ_Y is the yield stress of the elastic half-space. It can be shown that in both cases the maximum Hertzian pressure at the onset of local yielding is $p_{0|Y} \approx 0.6H$, where H is the hardness of the substrate. For most hard substrate (usually relatively brittle material) Tresca criterion is applied. For more ductile materials failure is often by von Mises yield criterion (see Johnson, 1985; Gohar and Rahnejat, 2008).

No solid acts as a rigid indenter (an idealistic assumption). Thus, the sub-surface absolute maximum shear stress is usually slightly higher than that stated by the Hertzian theory. This value increases further with hard coatings as the footprint area is reduced and correspondingly the pressures exceed those that are predicted by the Hertzian theory, as already described above. Figure 4.9 shows the maximum shear stress contours in both an elastic indenter and a substrate with a hard coating.

Note, firstly that the absolute maximum shear stress in the indenter exceeds its Hertzian value, here shown as: $\bar{\tau}_{max} = \tau_{max}/p_0$. Secondly, the hard coating (with an elastic modulus greater than that of the substrate) retains higher levels of shear stress. This is a characteristic sought for a hard coating, protecting the substrate material from effects of fatigue spalling and wear. However, the thickness of the layer should not be chosen arbitrarily as often is the case. The occurrence of high shear stress values at the interface between the layer and the substrate can lead to its exfoliation. The ratio of layer thickness to Hertzian radius or semi-half-width is quite important, as is the modulus ratio of the coating to that of the substrate. Readers should refer to Teodorescu *et al.* (2009) for a fundamental study and to Teodorescu and Rahnejat (2007) for the case of cam–tappet contact with wear-resistant coatings.

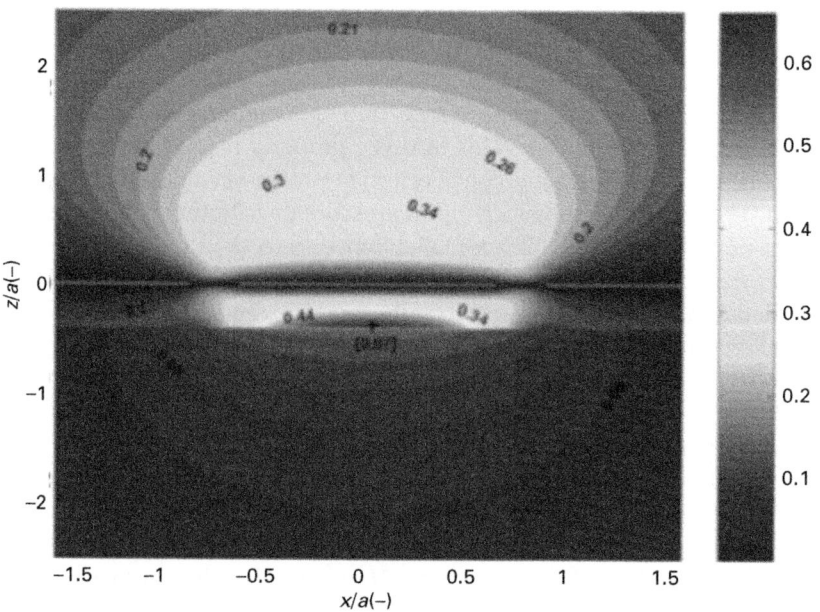

4.9 Sub-surface maximum shear stress field for an elastic indenter and a substrate with a hard coating.

4.7   References

Barber, J. R. (1990), 'Contact problems for the thin elastic layer', *Int. J. Mech. Sci.*, **32**, pp. 129–132.

Boussineq, J.V. (1885), *'Sur la resistance d'une sphere solide'*, L'Academie Des Science, Paris.

Descartes, R. (1644), *Principia Philosophiae*, Amsterdam.

Einstein, A. (1905), 'Ist die Trägheit eines Körpers von seinem Energiegehalt abhängig?', *Annalen der Physik*, **17**.

Einstein, A. (1916), 'Die Grundlage der allegmeinen relativitats-theorie', *Annalen der Physik*, **49**.

Gohar, R. and Rahnejat, H. (2008), *Fundamentals of Tribology*, Imperial College Press, London.

Greenwood, J.A. and Morales-Espejel, G.E. (1993), 'The behaviour of transverse roughness in EHL contacts', *Proc. Instn. Mech. Engrs, J. Engng. Trib.*, **208**, 121–132.

Hannah, M., (1951), 'Contact stress determination in a thin elastic layer', *Q. J. Mech. Appl. Math.*, **4**, 94–105.

Hertz, H. (1881), 'On the contact of elastic solids', *J. Reine Angew. Math.*

Hooke, C. J. (2000), 'The behaviour of low-amplitude surface roughness under line contacts: non-Newtonian fluids', *Proc. Instn. Mech. Engrs, Part J: J. Engng. Trib.*, **214**, 253–265.

Jaffar, M.J., (1989), 'Asymptotic behaviour of thin elastic layers bonded and unbonded to a rigid foundation', *Int. J. Mech. Sci.*, **31**, 229.

Johns, P.M. (1978), The design of cylindrical rollers for use in shaft and bearing systems, MSc Dissertation Imperial College of Science and Technology, University of London, London, UK.

Johns, P.M. and Gohar, R. (1981), 'Roller bearing under radial and eccentric loads', *Trib. Int.*, **13**, 131–136.

Johnson, K.L. (1985), *Contact Mechanics*, Cambridge University Press, Cambridge.

Konvopoulos, K. and Gong, Z.Q. (2003), 'Effect of surface patterning on contact deformation of elastic-plastic layered media', *Trans. ASME, J. Trib.*, **125**, 16–24.

Liu, S. and Wang, Q. (2002) 'Studying contact stress fields caused by surface tractions with a discrete convolution and fast fourier transform algorithm', *Trans. ASME, J. Trib.*, **124**, 36–45.

Love, A.E.H. (1927), *A Treatise on the Mathematical Theory of Elasticity*, Cambridge University Press, Cambridge.

Naghieh, G. R., Rahnejat, H. and Jin, Z.M. (1998), 'Characteristics of frictionless contact of bonded elastic and viscoelastic layered solids', *Wear*, **232**, 243–249.

Newton, I. (1687), *Philosophiae Naturalis Principia Mathematica*, Royal Society, London.

Ovaert, T. and Pan, J. (2002), 'Optimal design of layered structures under normal (frictionless) contact loading', *Trans. ASME, J. Trib.*, **124**, 438–442.

Penzias, A.A. and Wilson, R.W. (1965), 'A measurement of excess antenna temperature at 4080 Mc/s', *Astrophysical J.*, **142**, 419–421.

Polansky, I.A. and Keer, L.M. (2000), 'A fast and accurate method for numerical analysis of elastic layered contacts', *Trans ASME, J. Trib.*, **122**, 30–35.

Rahnejat, H. (2004), 'Foreword: a tribute to Jean D'Alembert and Albert Einstein, Geometrical interpretation of motion: An ironic legacy of apparently irreconcilable atomistic and continuum philosophies', in Rahnejat, H. and Rothberg, S.J. (eds.),

Multi-body Dynamics: Monitoring & Simulation Techniques, Professional Engineering Publishing (IMechE) Bury St. Edmunds, UK.

Rahnejat, H. (2005), 'Special relativity: interpretation and implications for space-time geometry', *Proc. Instn. Mech. Engrs., Part K: J. Multi–body Dyn.,* **219**(2), 133–146.

Rahnejat, H. and Gohar, R. (1979), 'Design of profiled taper roller bearings', *Trib. Int.,* **12**(6), 269–275.

Rahnejat, H., Johns-Rahnejat, P.M., Teodorescu, M., Votsios, V. and Kushwaha, M. (2009) 'A review of some tribo-dynamics phenomena from micro to nano-scale conjunctions', *Trib. Int.,* **42**(11), 1531–1541.

Sneddon, I.N. (1951), *Fourier Transforms,* McGraw-Hill Book Company, Inc., New York.

St Venant A.J.C.B. (1883), *Théorie de l'élasticité des corps solides,* L'Academie Des Science, Paris.

Teodorescu, M. and Rahnejat, H. (2007), 'Mathematical modelling of layered contact mechanics of cam–tappet conjunction', *Appl. Math. Modelling,* **31**, 2610–2627.

Teodorescu, M. and Taraza, D. (2004), 'Combined multi-body dynamics and experimental investigation for determination of the cam flat tapper contact condition', *Proc. Instn. Mech. Engrs., Part K:J. Multi Body Dyn.,* **218** (3), 133–142.

Teodorescu, M., Votsios, V. and Rahnejat, H. (2005), 'Multiphysics analysis for the determination of valvetrain characteristics', *Proc. Instn. Mech. Engrs., Part D: J. Automobile Engng.,* **219**(9), 1109–1117.

Teodorescu, M., Votsios, V., Rahnejat, H. and Taraza, D. (2006): 'Jounce and impact in cam–tappet conjunction induced by the elastodynamics of valve train system', *Meccanica,* **41**(2), 157–171.

Teodorescu, M. Rahnejat, H. Gohar, R. and Dowson, D. (2009) 'Harmonic decomposition analysis of contact mechanics of bonded layered elastic solids', *Appl. Math. Modelling,* **33**(1), 467–485.

Votsios, V. (2003), Contact mechanics and impact dynamics of non-conforming elastic and viscoelastic semi-infinite or thin bonded layered solids, PhD thesis, Loughborough University. Loughborough, UK.

4.8   Nomenclature

a	Major semi-half-width of the elastostatic contact ellipse
b	Minor semi-half-width of the elastostatic contact ellipse
d	Layer thickness
E'	Reduced modulus of elasticity
$E_{1,2}$	Modulus of elasticity
E^{ψ}	Modulus of elasticity of the substrate
$E^{\psi*}$	$= E^{\psi}/(1 - v^{\psi 2})$ plane strain modulus for the substrate
E^{ϕ}	Modulus of elasticity of the layer
$E^{\phi*}$	$= E^{\phi}/(1 - v^{\phi 2})$ plane strain modulus for the layer
H	Height
i, j	$\in \{x, y\}$
K	Contact stiffness non-linearity
K	Linearised stiffness (instantaneous)

k	Harmonic order (as a subscript)
l_1	Layer thickness
l_{max}	$\rightarrow \infty$ Maximum depth
L	Contact length
L_F	Fourier decomposition interval
m	Mass of the equivalent sphere
p	Contact pressure
p_0	Maximum Hertzian pressure
P_k	Amplitude of kth harmonic of the applied pressure $k = 1 \rightarrow N$
r	Radius
R	Reduced contact radius of counterformal contact
$R_{1,2}$	Contact radius
$R_{x,y}$	Reduced contact radii in x and y directions
t	Impact time
v	Impact velocity
W	Load
x, y, z	Cartesian co-ordinate set
Θ	$= [E^\phi(1 + v^\psi)]/[E^\psi(1 + v^\phi)]$ effective modulus ratio
$_k\Theta_1$	$= [\Theta + T_k]/[T_k\Theta + 1]$
$_k\Theta_2$	$= [T_k + 1]/[\Theta(T_k\Theta + 1)]$
$_k\Omega_y$	Generic deflection function
α_κ	$= 2k\pi/L$
$_k\sigma_{ij}^\psi$	kth harmonic of stress for the substrate $k = 0 \rightarrow N$
$_k\sigma_{ij}^\xi$	kth harmonic of stress for the layer $k = 0 \rightarrow N$
ε_i^ψ	Strain tensor in the substrate
ε_i^ξ	Strain tensor in the layer
$_k\varepsilon_i^\psi$	kth harmonic of strain in the substrate $k = 0 \rightarrow N$
$_k\varepsilon_i^\xi$	kth harmonic of strain in the layer $k = 0 \rightarrow N$
$v_{1,2}$	Poisson's ratio
v	Reduced Poisson's ratio
v^ψ	Poisson's ratio for the substrate
$v^{\psi*}$	$= v^\psi/(1 - v^\psi)$ plane strain Poisson function for the substrate
v^ϕ	Poisson's ratio for the layer
$v^{\phi*}$	$= v^\phi/(1 - v^\phi)$ plane strain Poisson function for the layer
ξ	$\in \{\phi, \psi\}$
λ_k	$= L/k$ wavelength for kth harmonic
φ_k	Phase angle for kth harmonic
δ	Local deflection
δ_{max}	Maximum local deflection
ω_n	Contact natural frequency

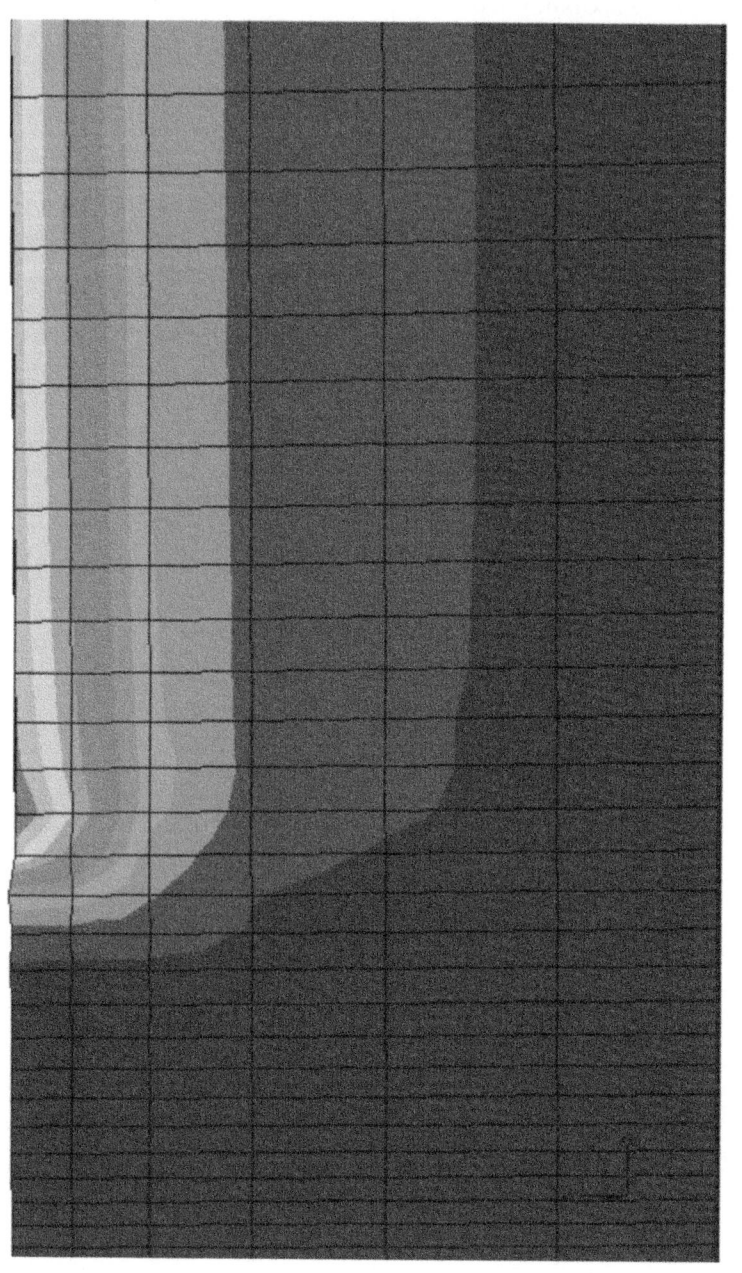

Plate I Contours of maximum sub-surface shear stress under a roller indenting an elastic half-space.

5

Fluid film lubrication

R. GOHAR, Imperial College London, UK and M. M. A. SAFA, Kingston University, UK

Abstract: This chapter discusses fluid film bearings, together with a proof of Reynolds equation and also the relevance of the Navier–Stokes equations for certain applications. Some examples of free surface oil entry and exit flow behaviour are given. For compressible flow, an example is given of a high-speed externally pressurised bearing with a squeeze film vibration damper in parallel with the rotor. There is also shown a simple thermal design procedure for hydrodynamic thrust bearings that uses isothermal numerical solution results as the basis of the method.

Key words: dynamic viscosity, piezoviscous, pressure viscosity index, Reynolds equation, non-Newtonian behaviour, Navier–Stokes equations, Reynolds number, convected heat, conducted heat, Peclet number, surface tension, EP gas bearings, oil thrust bearings.

5.1   Lubricant properties

5.1.1   Introduction

Friction and wear occur when the surfaces of solids are made to slide relative to each other (see Chapter 2). However, if in addition there is a viscous fluid film introduced between these surfaces, both friction and wear are considerably reduced. This is achieved, through the mechanism of *hydrodynamic lubrication*. This chapter will therefore concentrate mainly on liquid lubricant (oil) behaviour between rigid surfaces or subsequently as a free surface.

Before discussing hydrodynamic lubrication, the lubricant properties themselves must be analysed. The relative motion between two elements of a moving fluid is retarded by intermolecular interactions caused by *viscosity*, a real fluid property. This property is the most essential feature of fluid lubricants, as it enables them, given the right conditions, to separate the bodies when in relative motion.

5.1.2   Dynamic viscosity

Consider firstly an elastic element of solid of unit face width and thickness dz, with a shear stress, τ, put on its opposing faces, as in Fig. 5.1(a). The resulting movement of the top face relative to the base yields the relationship

132

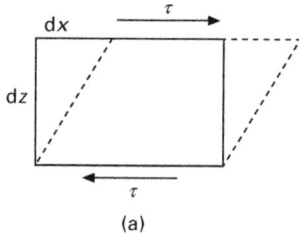 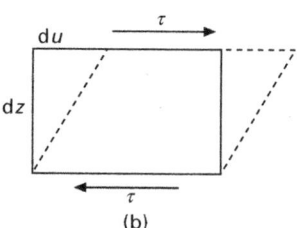

(a)   (b)

5.1 Dynamic viscosity.

$$G = \frac{\text{shear stress}}{\text{shear strain}} = \frac{\tau}{dx/dx}$$

G being the **shear modulus** of the material.

Considering now a thin layer of oil in one-dimensional flow. From Fig. 5.1(b), if du is the velocity difference between the top and bottom surfaces of the layer, then du/dz is the velocity gradient across it. This is called shear **strain rate** or strain rate. Furthermore, as $u = dx/dt$:

$$du/dz = \frac{d}{dz}\left(\frac{dx}{dt}\right) = \frac{d}{dt}\left(\frac{dx}{dz}\right) = \frac{d\,(\text{shear strain})}{dt} = \text{strain rate}$$

Thus, for fluids, shear strain is replaced by shear strain rate, du/dz, while G is replaced by the **dynamic viscosity**, η, of the fluid giving

$$\eta = \frac{\text{shear stress}}{\text{shear strain rate}}$$

or

$$\eta = \frac{\tau}{(du/dz)} \tag{5.1}$$

Equation (5.1) defines dynamic viscosity of a **Newtonian fluid**. An important property of a Newtonian fluid is that it must take the same velocity as any solid boundary it contacts. The units of dynamic viscosity in terms of (force, length, time) are

$$\frac{(F/L^2)L}{(L/T)} = \frac{FT}{L^2}$$

In the SI units system, dynamic viscosity is in: newtons per square metre seconds, or pascal seconds (Pa s).

In SI units in terms of mass, length and time, dynamic viscosity is $M/LT = kg/(m\ s)$. This linear relationship between stress and shear rate, given by equation (5.1), is analogous to Hooke's law in solid body mechanics. Thus Newtonian fluids are thus called after Isaac Newton who defined them.

Other frequently quoted units of dynamic viscosity follow the CGS system. To define the unit of viscosity in this system, we can state that:

If the shear stress is 1 dyne/cm^2, and the velocity gradient, du/dz, is s^{-1}, then the absolute unit of viscosity is 1 *poise* (P).

The conversion from CGS to SI units of dynamic viscosity is obtained from

$$1\,P = 0.1\,Pa\,s$$

Sometimes the *centipoise* (cP) is used,

$$1\,cP = 10^{-3}\,Pa\,s$$

Another frequently quoted unit of viscosity is the **kinematic viscosity**, v. If ρ is the lubricant density

$$v = \frac{\eta}{\rho} \tag{5.2}$$

ρ (dimensions of M/L^3), is usually about 850 kg/m^3 (0.85 g/cm^3). The dimensions of v are therefore: L^2/T. In SI units they are m^2/s but are rarely quoted as such in the commercial literature. In the CGS system, the unit of v is the *Stoke* (S) in (cm^2/s). More often, the *centistoke* (cSt) in (mm^2/s) is used where $1\,S = 100\,cSt$. Finally, converting v in cSt to the SI system:

$$\left(\frac{m^2}{s}\right) = cSt \times 10^{-6}$$

The value of v for a liquid lubricant (identified here as an oil) is determined experimentally with a **viscometer**. A discussion on some simple ways of measuring viscosity, and various applications involving it, is given by Middleman (1998).

5.1.3   Effect of temperature on viscosity

Unlike gases, where the viscosity increases with temperature, the viscosity of oil decreases strongly with temperature. The standard experiment to find the effect of temperature is to use a capillary tube viscometer under controlled temperatures, and then curve fit the results.

The simplest fit, called Reynolds viscosity equation, is

$$\eta = \eta_0 \exp(-\beta \Delta \theta) \tag{5.3}$$

Here, η_0 is the viscosity at some representative temperature, θ_i in °C, $\Delta\theta$ is the temperature rise from θ_i: $\theta = \theta_i + \Delta\theta$ (the temperature at a point in the oil film in °C) and β is the **viscosity–temperature coefficient**. Equation (5.2) is not accurate except over a very low temperature range. Another more accurate fit over a wider temperature range is by Vogel (1921) and further analysed by Cameron (1967):

$$\eta = a \, \exp\!\left(\frac{b}{\Theta - c}\right) \qquad (5.4)$$

To use equation (5.4), the constants a, b and c must be determined for each oil from three sets of data supplied by the manufacturer, and Θ being the required **absolute temperature** at a point in the film in relvin.

Another method of calculating viscosity is to use the American Society for Testing Materials (ASTM) chart that supplies a data sheet compiled by Walther and Sassenfeld (1954) and discussed by Vogel (1921). It is based on the expression

$$\log \log(v + 0.6) = d - e \, \log\Theta \qquad (5.5)$$

Equation (5.5) can therefore be conveniently plotted on special ASTM (log)–(log log) paper as a straight line. The solution is kinematic viscosity, v, expressed in cS, at supplied temperature Θ in K. In order to plot the line, its viscosity at two other temperatures must be known. These are generally supplied by the manufacturer. Another lubricant property is the **viscosity index** (VI) that determines the goodness of an oil, meaning how little viscosity varies with temperature (often a desirable feature).

5.1.4   Effect of pressure on viscosity

The behaviour of lubricants under pressure is most significant when there are concentrated contacts (a large normal force over a small, usually distorted, area), Barus (1893) showed how the lubricant viscosity varies with pressure at constant temperature (its **piezoviscous properties**):

$$\eta_s = \eta_0 \, \exp(\alpha p) \qquad (5.6)$$

where η_s is the oil viscosity at gauge pressure, p, η_0 is the viscosity at $p = 0$, and α is a constant, depending on the oil, called the pressure viscosity coefficient with units of m^2/N. Equation (5.6) can be quite inaccurate, one reason being because α itself can vary with both temperature and pressure. Equation (5.7) is, however, satisfactory for computing purposes where there is a rolling contact (for example, in a moderately loaded ball bearing). The *sliding* between the contacting surfaces is an important contribution to temperature rise (for example, between a pair of involute gear teeth). To give some idea of scale, the peak pressure in a ball bearing can reach 4 GPa where, at such pressures, the lubricant appears to have nearly solidified. In Chapter 6, the piezoviscous property of oil is shown to be important when dealing with **elastohydrodynamic lubrication**.

It was mentioned above that Barus's equation is inaccurate at very high pressures. This can be demonstrated by way of an example. Take a typical oil of dynamic viscosity of 0.03 Pa s at atmospheric pressure, and with a pressure viscosity coefficient of 10^{-8} Pa^{-1}. Using equation (5.6), $\bar{\eta} = \eta/\eta_0 = \exp(\alpha p)$. At $p = 1$ GPa, $\bar{\eta} = \exp(10) \approx 21\,365$. Thus, the viscosity of the lubricant appears to have increased by more than four orders of magnitude. At such high pressures the lubricant would become similar to an amorphous solid but, as shown below, predicting this level of increase in viscosity is erroneous.

A more accurate expression, found by Roelands (1966) and developed further by Houpert (1985), is often used in numerical solutions. It includes the effects of both temperature and pressure on the viscosity and is often expressed in the same form as equation (5.6) but with a modified pressure viscosity coefficient:

$$\eta_R = \eta_0 \exp(\alpha^* p) \tag{5.7}$$

where α^* is a function of both p and Θ:

$$\alpha^* = \frac{1}{p} \left[\ln(\eta_0) + 9.67\right] \left\{ -1 + \left(\frac{\Theta - 138}{\Theta_0 - 138}\right)^{-S_0} (1 + 5.1 \times 10^{-9} p)^z \right\} \tag{5.8}$$

Z and S_0 are constants, independent of temperature and pressure, defined below, p is in Pa, Θ and Θ_0 are in K ($\Theta_0 = \theta_0 + 273$ and $\Theta = \theta + 273$). The constants are obtained from:

$$Z = \frac{\alpha_0}{5.1 \times 10^{-9}[\ln(\eta_0) + 9.67]}, \quad S_0 = \frac{\beta_0(\Theta_0 - 138)}{\ln(\eta_0) + 9.67}$$

where β_0 and α_0 are at atmospheric temperature and pressure, supplied by the oil company. Comparing the viscosity based on equation (5.7) with that obtained from equation (5.8), in the above case $\eta_{Barus}/\eta_{Roelands} = 188$, a huge reduction to a more realistic value.

5.1.5   Lubricant density

Knowing how the **lubricant density** varies with pressure is important when the pressures are excessive, as will be seen later in the chapter on **elastohydrodynamics** (EHL) where numerical solutions to EHL problems are discussed. A formula that shows this variation is:

$$\rho = \rho_0 \left(1 + \frac{0.6 \times 10^{-9} p}{1 + 1.7 \times 10^{-9} p}\right) \tag{5.9}$$

where p is in N/m^2 and ρ_0 for oils is typically $870\,kg/m^2$ at $20\,^{\circ}C$. The variation of density with temperature is found to be negligible in most lubrication problems.

5.1.6   Effect of shear rate on viscosity

Fluid behaviour is described as **non-Newtonian** if the viscosity depends on the shear rate. Lubricants like grease suffer from **thixotropic behaviour** if there is a steady fall in viscosity with the duration of the shear action. The normal viscosity is slowly regained after the shear action ends.

Some oils can suffer from **shear thinning**, which makes their viscosities reduce at high shear rates ($>10^9\,s^{-1}$), their normal viscosities being regained as soon as the shear action ends. The shear rate action itself does not normally affect oils. In the case of EHL conditions, although the shear rate is usually below $10^9\,s^{-1}$, the shear stress within the lubricant can itself cause shear thinning. Shear thinning has a significant effect on the friction levels found in EHL contacts.

5.2   Reynolds equation

5.2.1   Introduction

Above, the properties of lubricants used in viscous film bearings were discussed. Below, will be stated, and sometimes derived, some of the equations needed to design these bearings. The approach will be as follows:

- Explain the principle of the hydrodynamic wedge, both by analogy and by dimensional analysis.
- Derive **Reynolds equation** from first principles.
- Define the **energy equation**, which accounts for thermal effects in bearings and discuss its significance when applied to bearings.
- Discuss some applications of the **Navier–Stokes** and Reynolds equation in some lubrication problems.

5.2.2   Physics of the Reynolds equation

Consider Fig. 5.2(a) where a Newtonian oil is entrained by the moving bottom surface into the thin parallel gap. The gap dimension in the y direction is very large compared with the other dimensions. The velocity distribution across the film must be triangular at all three stations shown because the oil adjacent to each surface must take its velocity there. Therefore, there is zero pressure and pressure gradient everywhere because of the unvarying geometry in the x direction.

In Fig. 5.2(b) a **wedge** has been formed. Therefore, in order to preserve

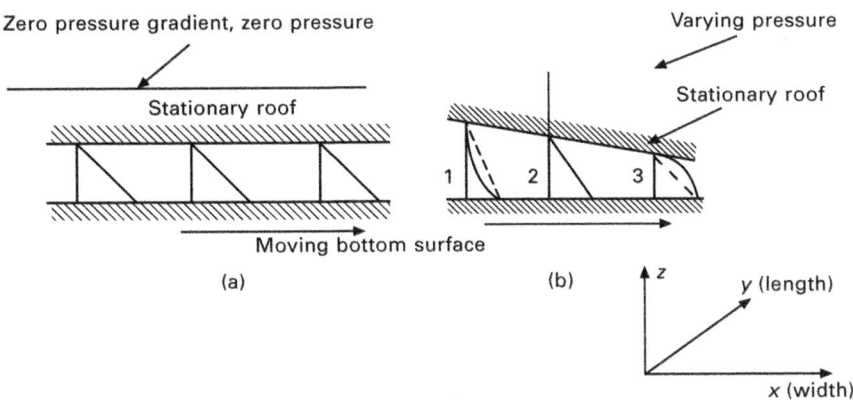

5.2 Principle of the wedge film.

continuity, the velocity distributions alter at each of the three stations by changing in the manner shown. Where the film height is greatest, in order to preserve the same area under each velocity distribution as previously, the local velocities decrease from their triangular shape distribution while at the thinnest third station, on the right, the local velocities increase for the same reason. Only at a certain intermediate station is the velocity distribution still triangular. Because of this behaviour, a pressure distribution is set up in the x direction, increasing to the left of this station and falling to the right of it, with zero gradient occurring at it. Thus, a bearing has been created by this arrangement. Being based on this concept, the remainder of the chapter will outline lubrication theory, together with some applications.

5.2.3   Derivation of Reynolds equation in three dimensions

Reynolds equation is the basis of lubrication theory. As was done in Section 5.2.2, some realistic assumptions are needed to derive it. Even though a hydrodynamic bearing has not yet been designed, firstly that the film is thin, so that in profile it has a large aspect ratio (plan width[1] : mean film thickness) that is typically 1000:1. As an example, for a thrust bearing, the minimum film thickness can be 25 μm with a wedge angle 0.075°. This implies that the film thickness dimensions, $h(x)$, are much less than those defining its top and bottom surfaces, B and L, as illustrated in Fig. 5.3.

Reynolds (1886) first obtained the differential equation governing pressure distribution in a Newtonian lubricant film. It may be derived from the full Navier–Stokes (NS) equations (see Cameron, 1967) by making simplifying

---

[1]Remember, *width* is measured here in the x direction.

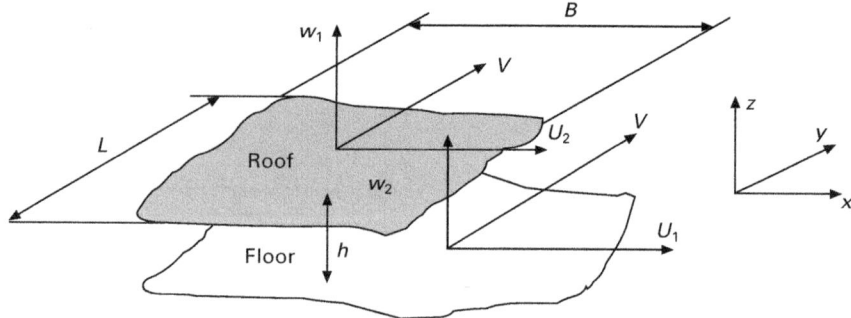

5.3 Two generalised surfaces, in relative motion, bounding an oil film.

assumptions at appropriate points in the analysis. The finished article is a perfectly generalised version of Reynolds equation. However, here the more direct engineering approach will be used: by making some simplifying assumptions at the start of the analysis. This approach offers more insight into the physics of the equation.

The following assumptions are made in the direct derivation of Reynolds equation:

1. the oil film, which is thin, has negligible mass (gravity forces neglected);
2. because it is thin, pressure is assumed to be constant across the film (z direction);
3. there is no slip at the boundaries (Newtonian fluid);
4. lubricant flow is laminar (low Reynolds numbers);
5. inertia and surface tension forces are negligible compared with viscous forces;
6. because it is thin, shear stresses and velocity gradients are only significant across the film (z direction);
7. the lubricant is Newtonian (high shear rates are not present);
8. the lubricant viscosity is constant across the film (z direction);
9. the boundary surfaces (roof and floor in Fig. 5.3) follow some designated geometry but are always at low angles to each other.

5.2.4   Equilibrium of forces on a lubricant element

Referring to Fig. 5.4, let the two bounding surfaces have perfectly general motion defined by their velocity vectors. Consider the local forces acting on a lubricant element, defined at x,y,z in a column of thickness, h. The element centre has velocity components u,v,w. Therefore, neglecting shear stress and velocity gradients in the x and y directions (assumption 6):

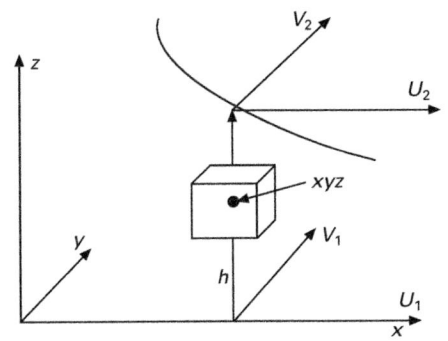

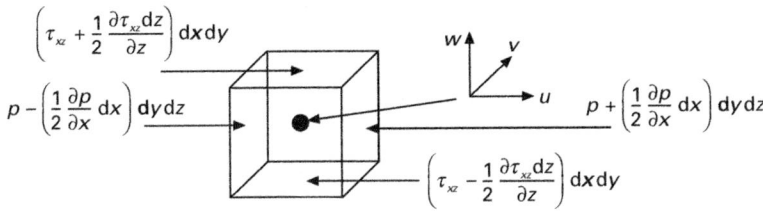

5.4 Forces on an element.

$$\Sigma F_x = 0$$

$$\therefore \frac{\partial \tau_{xz}}{\partial z} = \frac{\partial p}{\partial x} \qquad (5.10)$$

Here, τ_{xz} means shear stress in the x direction in a plane having z as its normal. Similarly

$$\Sigma F_y = 0$$

$$\therefore \frac{\partial \tau_{yz}}{\partial z} = \frac{\partial p}{\partial y} \qquad (5.11)$$

Also, from equation (5.1)

$$\tau_{xz} = \eta \frac{\partial u}{\partial z} \qquad (5.12)$$

$$\tau_{yz} = \eta \frac{\partial v}{\partial z} \qquad (5.13)$$

Combining equation (5.10) with (5.11) and equation (5.12) with (5.13):

$$\frac{\partial}{\partial z}\left[\eta \frac{\partial u}{\partial z}\right] = \frac{\partial p}{\partial x} \qquad (5.14)$$

$$\frac{\partial}{\partial z}\left[\eta \frac{\partial v}{\partial z}\right] = \frac{\partial p}{\partial y} \qquad (5.15)$$

5.2.5   Velocity distribution

Let η be invariable in the z direction (assumption 8). Also, let $\partial p/\partial x$ and $\partial p/\partial y$ not vary with z (assumption 2). For the x direction, integrate equation (5.14) twice with respect to z:

$$\eta u = \frac{\partial p}{\partial x} \frac{z^2}{2} + cz + d$$

Two boundary conditions are needed to find the constants c and d. These are: at $z = h$, $u = U_2$ and $z = 0$, $u = U_1$. Solving for the constants:

$$u = \frac{1}{2\eta} \frac{\partial p}{\partial x} (z^2 - zh) + \frac{z}{h} (U_2 - U_1) + U_1 \qquad (5.16)$$

Similarly, for the y direction, integrating equation (5.15) twice

$$v = \frac{1}{2\eta} \frac{\partial p}{\partial y} (z^2 - zh) + \frac{z}{h} (V_2 - V_1) + V_1 \qquad (5.17)$$

Just as was surmised from the wedge shape study above, equations (5.21) and (5.22) describe velocity distributions composed of two parts. There is a parabolic part due to the pressure gradient (**Poiseuille flow**) and a linear part due to the boundary surface velocities (**Couette flow**).

5.2.6   Mass continuity

To complete the full derivation of Reynolds equation, **mass continuity** must be invoked. This states that there is the same mass of fluid per second entering a column of oil height, h, as that leaving it. Here, it is not necessary to assume that the fluid is of constant density in the x and y directions. Referring to Fig. 5.5 let m_x and m_y be respectively the mass flows flow per unit width in the x and y directions through the column. The net mass flow out of the column in the z direction is $(\mathrm{d}/\mathrm{d}t) \rho h \mathrm{d}x \mathrm{d}y$ and the mass flows in the x and y directions are, by definition:

$$m_x = \rho q_x = \rho \int_0^h u \mathrm{d}z \qquad (5.18)$$

$$m_y = \rho q_y = \rho \int_0^h v \mathrm{d}z \qquad (5.19)$$

Also, from Fig. 5.5, equating the flows entering the column with those leaving it:

$$\frac{\partial m_x}{\partial x} + \frac{\partial m_y}{\partial y} + \frac{\partial}{\partial t} (\rho h) = 0 \qquad (5.20)$$

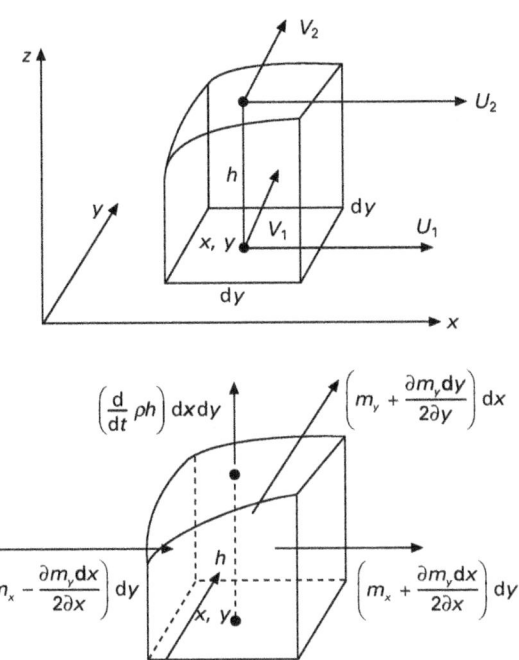

5.5 Flow through a column.

Additionally, from equations (5.16), (5.17), (5.18) and (5.19):

$$m_x = -\rho \int_0^h \left[\frac{1}{2}\eta \frac{\partial p}{\partial x}(z^2 - zh) + \frac{z}{h}(U_2 - U_1) + U_1 \right] dz \qquad (5.21)$$

Therefore

$$m_y = -\frac{\rho h^3}{12\eta}\left[\frac{\partial p}{\partial y}\right] + (V_1 + V_2)\left[\frac{\rho h}{2}\right] \qquad (5.22)$$

Similarly, in the y direction

$$m_y = -\frac{\rho h^3}{12\eta}\left[\frac{\partial p}{\partial y}\right] + (V_1 + V_2)\left[\frac{\rho h}{2}\right] \qquad (5.23)$$

Equations (5.22) and (5.23) give the mass flows of the lubricant per unit length in the x and y directions. Like equations (5.16) and (5.17) each is composed of a pressure-induced term (Poiseuille flow) and a boundary velocity-induced term (Couette flow). If there is only one surface velocity, say in the x direction, then $V_1 = V_2 = 0$. Equation (5.23), however, states that a mass flow, m_y, still occurs in that direction. This now due only to the pressure gradient, $\partial p/\partial y$, causing some lubricant to flow transversely from a high pressure to a lower pressure region in that direction. This action in a bearing is called *side leakage*.

Substitute equations (5.22) and (5.23) into equation (5.20) and rearrange the order so that the pressure-induced terms only are on the left-hand side. The result is the full Reynolds equation:

$$\frac{\partial}{\partial x}\left[\frac{\rho h^3}{\eta}\frac{\partial p}{\partial x}\right] + \frac{\partial}{\partial y}\left[\frac{\rho h^3}{\eta}\frac{\partial p}{\partial y}\right]$$

$$= 6\left\{\frac{\partial}{\partial x}[\rho h(U_1 + U_2)] + \frac{\partial}{\partial y}[\rho h(V_1 + V_2)] + 2\frac{d}{dt}(\rho h)\right\} \qquad (5.24)$$

Equation (5.24) is the fundamental equation of fluid film lubrication theory, with units of $kg/m^2\,s$. If the density, ρ, is constant, it cancels out, leaving terms of dimension m/s. It is the full Reynolds equation in three dimensions for compressible or incompressible flow of a Newtonian fluid. On the left-hand side are the Poiseuille pressure-induced terms. The right-hand side is composed of Couette terms that, from left to right, are divided into wedge and squeeze components. Equation (5.24) accounts for flow components in the x and y directions. Neither boundary surface need have a uniform velocity vector parallel to the xy plane because the components are respectively situated within the differentials on the right-hand side, $\partial/\partial x$ and $\partial/\partial y$. The same applies to the right hand side, where ρ and μ are again within the differential operators. Both may vary with x and y because of their sensitivities to pressure (demonstrated above that it rises and falls) and/or temperature. Finally, the squeeze term, $\partial/\partial t(\rho h)$ need not be uniform over the region of pressure. If the squeeze term is expanded, it can be written as:

$$\rho(w_1 - w_2) + h\frac{d\rho}{dt}$$

where $(w_1 - w_2) = dh/dt$ and w_1 and w_2 are respectively the roof and floor velocities, from whatever cause.

5.2.7   Simplifications of reynolds equation

Generally, simplified versions of equation (5.24) can be employed. For the two examples below, assume

- the oil density and viscosity are constant;
- the roof and floor of the film are non-porous and have no normal velocity components;
- the surface velocities are in the x direction only.

Long bearing

As a further simplification assume that the transverse length of the bearing, L (y direction), is effectively infinity. In practice it is called a **long bearing** (L much greater than the width, B, in the x direction). The long transverse length assumption makes the pressure distribution uniform in the y direction, except close to the edges, where the pressures must drop to zero there. The shape is illustrated in Fig. 5.6.

In this case, dp/dy can be neglected compared with dp/dx, so equation (5.29) becomes

$$\frac{d}{dx}\left(\frac{h^3}{\eta}\frac{dp}{dx}\right) = 6(U_1 + U_2)\frac{dh}{dx} \tag{5.25}$$

Equation (5.25) can be integrated with respect to x:

$$h^3\frac{dp}{dx} = 6(U_1 + U_2)\eta h + C$$

where C is an integration constant. Let $h = h_c$ (as yet unknown) at dp/dx = 0. Equation (5.25) then modifies to

$$\frac{dp}{dx} = 6(U_1 + U_2)\eta\left[\frac{h - h_c}{h^3}\right] \tag{5.26}$$

This equation is used frequently for approximate solutions. The long bearing assumption is reasonably accurate if $L/B > 3$. Below is an example.

Long bearing approximation for rigid cylinders

An important application of equation (5.26) is the lubrication of long rigid cylinders where additional boundary conditions are needed in Reynolds

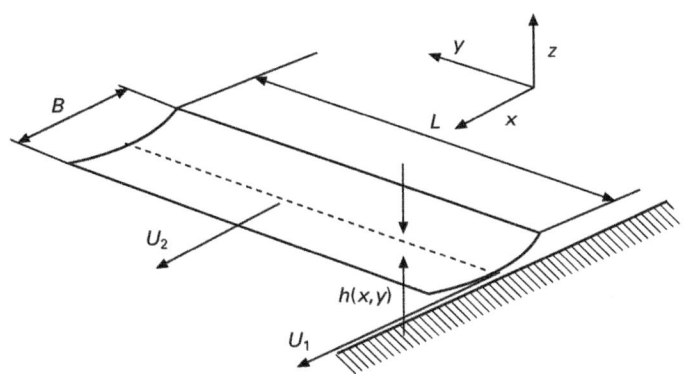

5.6 Long bearing $L \gg B$.

equation for a complete solution. The model below is a useful precursor to Chapter 6 on EHL where the bounding surfaces are assumed to be elastic, their shapes varying with the oil pressure.

Figure 5.7(a) shows two rigid cylinders, loaded on their circumferences, in rolling–sliding lubricated contact, producing a convergent–divergent wedge film. This arrangement is called a **hydrodynamic line contact**, even if the surfaces suffer elastic distortion. First find the pressure distribution created by the wedging action of the surfaces.

Line contact pressure distribution

The assumptions are:

1. the discs are considered long in the y direction, so that there is no transverse flow except near their ends;
2. $R \gg h$;
3. conditions are **isoviscous** throughout, so that $\eta = \eta_0 = $ constant.

Referring to Fig. 5.7(b) in order to simplify the co-ordinate system let the discs be replaced by an equivalent disc of **reduced radius**, R, in contact with a plane, where $1/R = 1/R_1 + 1/R_2$. In addition, in keeping with normal practice for lubricated concentrated contacts, let the mean velocity of the Couette flow component be $U = (U_1 + U_2)/2$ (called the entrainment velocity). Equation (5.26) therefore becomes

$$\frac{dp}{dx} = 12 U \eta_0 \frac{h - h_c}{h^3} \tag{5.27}$$

Also, from assumption 2 above and assuming the film shape is approximately a parabola in the vicinity of the contact, write

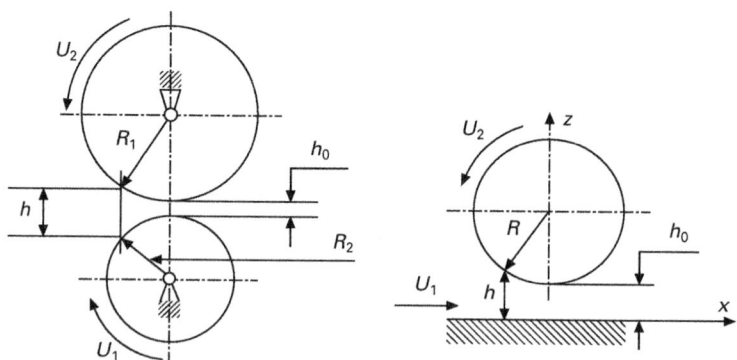

5.7 Line contact geometry: (a) two discs; (b) equivalent contact pressure distribution.

$$h \approx h_0 + \frac{x^2}{2R} \qquad (5.28)$$

It is usual practice to employ dimensionless groups, so let

$$\bar{x} = \frac{x}{\sqrt{2Rh_0}}, \bar{h} = \frac{h}{h_0}$$

Making this substitution, equation (5.28) becomes

$$\bar{h} = 1 + \bar{x}^2$$

Letting

$$\bar{p} = \frac{h_0^{3/2} p}{12 U \eta_0 \sqrt{2R}}$$

with $\bar{x}_c$ the $\bar{x}$ co-ordinate when $\bar{h} = \bar{h}_c$:

$$\frac{d\bar{p}}{d\bar{x}} = \frac{\bar{x}^2 - \bar{x}_c^2}{(1 + \bar{x}^2)^3} \qquad (5.29)$$

Integrating equation (5.29) with respect to $\bar{x}$

$$\bar{p} = \int \frac{\bar{x}^2 d\bar{x}}{(1 + \bar{x}^2)^3} - \bar{x}_c^2 \int \frac{d\bar{x}}{(1 + \bar{x}^2)^3} \qquad (5.30)$$

These are standard integrals that can be solved using Math CAD, or see Cameron (1967). The solution comes to

$$\bar{p} = \left[\frac{-\bar{x}}{4(1 + \bar{x}^2)^2} + \frac{\bar{x}}{8(1 + \bar{x}^2)} + \frac{1}{8} tg^{-1}(\bar{x}) \right]$$
$$- \bar{x}_c^2 \left[\frac{x}{4(1 + \bar{x}^2)^2} + \frac{3}{8} \frac{\bar{x}}{(1 + \bar{x}^2)} + \frac{3}{8} tg^{-1}\bar{x} \right] + C_1 \qquad (5.31)$$

There are still needed two additional boundary conditions needed to determine $\bar{x}_c$ and C_1. (Remember in deriving equation (5.27) boundary conditions were used at the top and bottom of the film (z direction). By this stage the boundary values are in the x direction). One pair, called the *full Sommerfeld condition*, by Sommerfeld (1904), and also discussed by Cameron (1967), is

$$p = 0 \text{ at } \bar{x} = \pm\infty$$

The condition at $\bar{x} = -\infty$ is called a *fully flooded* or *drowned inlet*.

Inserting the above two boundary conditions into equation (5.31), after some manipulation and reference to standard integrals:

$$\bar{p} = \frac{-\bar{x}}{3(1 + \bar{x}^2)^2}$$ (5.32)

Equation (5.32) has an antisymmetric shape producing zero load capacity because of the positive and pressure loops of equal area. One alternative model is to ignore the negative pressures. It is called the **half Sommerfeld** boundary condition: Sommerfeld (1904) and discussed further by Cameron (1967). This is curve (a) in Fig. 5.8. It is obtained from equation (5.32) with $p = 0$ when $\bar{x} < 0$ Although it gives reasonable approximate answers for the pressure distribution, the abrupt change of pressure gradient to zero at $x = 0$, cannot occur because flow continuity is contravened (the Poiseuille flow component has suddenly vanished, although the Couette flow component continues across the z axis). Steiber (1933) suggested that a more realistic boundary condition that does not contravene flow continuity is the **Reynolds exit boundary condition** (curve (b) in Fig. 5.8). It is often the one of choice in numerical solutions, especially when there are high loads, such as under EHL conditions (see Chapter 6). The Reynolds exit boundary condition states that the pressure exit boundary is where $p = dp/dx$ (a little beyond $x = 0$). Experiments show that where Reynolds condition *film rupture* occurs, the remaining Couette flow component has to expand into a widening gap, causing the flow to break up into oil carrying partitions (fingers) separated by air gaps (cavities). There is some discussion about these in Gohar (2001) and Gohar and Rahnejat (2008).

Line contact load

Using the half Sommerfeld boundary condition, integration of equation (5.32) between $x = -\infty$ and 0 produces a load per unit length (see Cameron, 1967) of

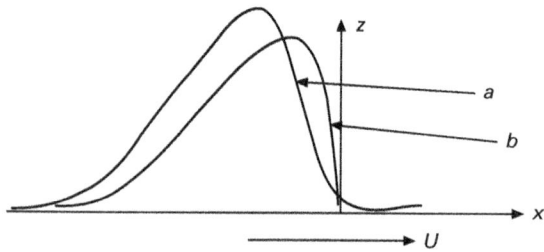

5.8 Illustration of line contact pressure distribution for a flooded inlet distance: (a) half Sommerfeld exit condition; (b) Reynolds exit condition.

$$\frac{W}{L} = \frac{4U\eta_0 R}{h_0} \tag{5.33}$$

Using the Reynolds exit boundary condition, the load per unit length is

$$\frac{W}{L} = \frac{4.9U\eta_0 R}{h_0} \tag{5.34}$$

Thus, for rigid line contacts with flooded inlets, the half Sommerfeld boundary condition produces a reasonable approximation to the Reynolds boundary condition load.

However, are these film thickness expressions realistic for this sort of geometry? To give some idea of scale, let $R = 0.0127\,\mathrm{m}$, $W = 5000\,\mathrm{N}$, $L = 0.025\,\mathrm{m}$, $U = 1\,\mathrm{m/s}$, $\eta_0 = 0.1\,\mathrm{N/m^2 s}$. Equation (5.34) gives $h_0 = 3 \times 10^{-8}\,\mathrm{m}$. This value is low in engineering terms because it can be below normal surface roughness height. What happens in practice is that the counterformally contacting surfaces distort elastically in addition to the oil viscosity increasing with pressure (Section 5.1.4), producing more a more realistic film thickness that is ten times this value. This behaviour is the basis of elastohydrodynamic lubrication (Chapter 6).

Squeeze film bearings

Equation (5.24) also applies to another type of bearing behaviour depending on the squeeze film effect. If a flat plate is placed on a uniform thin film of oil, it will sink down slowly. The more viscous the oil, the more slowly it will sink. This viscous resistance is governed by the ***squeeze film effect*** (see also Chapters 18 and 19). It is accommodated by the last term on the right-hand side of equation (5.24). Noting that at any instant h is constant for a flat plate, and assuming ρ is invariable, equation (5.24) becomes

$$\frac{\partial}{\partial x}\left[\frac{h^3}{\eta}\frac{\partial p}{\partial x}\right] + \frac{\partial}{\partial y}\left[\frac{h^3}{\eta}\frac{\partial p}{\partial y}\right] = 12\frac{dh}{dt} \tag{5.35}$$

For a circular flat plate or a journal bearing (see equation 5.46 below), Reynolds equation is better expressed in polar co-ordinates.

5.3   The energy equation

The other fundamental equation needed is the ***energy equation***, where a full proof is given in Cameron (1967). It is presented here two dimensionally, without proof, at a level suitable for demonstrating its principle.

$$vu\theta\frac{\partial p}{\partial x} \quad + \quad \eta\left(\frac{\partial u}{\partial z}\right)^2 \quad = \quad \rho u c_p \frac{\partial \theta}{\partial x} \quad - \quad k_t \frac{\partial^2 \theta}{\partial z^2} \qquad (5.36)$$

compressive heating   viscous heating^2 convection cooling   conduction cooling

where $\theta(x)$ is the temperature rise of the oil from the **wedge inlet**, v is its coefficient of thermal expansion, c_p its specific heat (at constant pressure), and k_t its thermal conductivity.

The following points should be noted in connection with equation (5.36):

- The compression heating term on the left-hand side, caused by the pressure distribution, is relatively insignificant and therefore will be ignored in the subsequent simple analysis. The viscous heating term comes from shearing in the x direction across the film.
- On the right-hand side of equation (5.36), the convection cooling term carries some of the heat away through the flow in the x direction, while the **conduction cooling** term carries some away in the z direction across the solid boundaries. Gohar (2001) discusses this term.
- Considering heat transfer in hydrodynamic bearings, the viscous heating and **convection cooling** terms are the most important because of the relatively thick films encountered.
- On the other hand, when EHL contacts are considered the viscous heating and conduction cooling terms are the most important because the oil films are much thinner than in hydrodynamic applications.
- Because the compression heating term is insignificant, equation (5.36) can be simplified considerably then it can be assumed that the bearing film is sensibly parallel and also is long in the y direction, with the dominant heating from Couette flow (caused only by the x direction velocities of the boundaries).

With the above assumptions equation (5.36) becomes

$$\eta\left(\frac{\partial u}{\partial z}\right)^2 = \rho u c_p \frac{\partial \theta}{\partial x} + k_t \frac{\partial^2 \theta}{\partial x^2} \qquad (5.37)$$

5.3.1   Significance of terms in the energy equation

To determine the relative significance of the right-hand side terms of equation (5.37) and using the above assumptions, Fig. 5.9 results.

---

[2]A more general expression for the viscous heating term is $q = \tau\,\partial u/\partial z$. Making $\tau = \eta\,\partial u/\partial z$ producing the form above for a Newtonian fluid.

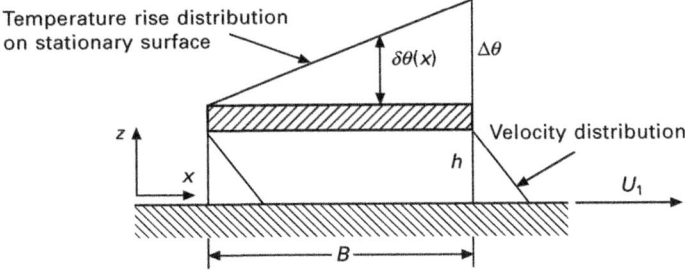

5.9 Thermal effects.

5.3.2   Convected heat only

Dealing with the convection cooling term only in equation (5.37), we can use that term to derive an order of magnitude expression for the heat convected away. Assume that in this case the bounding surfaces are completely insulated so that all the heat is carried away by convection along the film. Let the average fluid velocity be $U1/2$ and the maximum average temperature rise in the film $\Delta\theta/2$. Then, at position x, the convected heat flow across distance dx is

$$\rho c_p \frac{d\theta}{dx} \, dx \int_0^h u \, dz = \rho c_p \left(\frac{d\theta}{dx}\right)\left(\frac{U_1 h}{2}\right) dx$$

But at $x = B$, $d\theta/dx = \Delta\theta/2B$. Therefore, total convected heat flow out of the bearing end is

$$\frac{\rho c_p U_1 h}{2}\left(\frac{\Delta\theta}{2B}\right)\int_0^B dx = \frac{\rho c_p U_1 h \Delta\theta}{4} = \left(\frac{U_1 \rho h}{2}\right) \times c_p \times \frac{\Delta\theta}{2} \qquad (5.38)$$

The right-hand side of equation (5.38) is easy to remember if the convective heat flow rate is described as: mass flow per second × specific heat × temperature rise.

5.3.3   Conducted heat only

Now assume that all heat is removed by conduction. An order of magnitude solution assumes that the temperature gradient *across* the film varies linearly for each value of x, rising from zero on the moving surface (along $z = 0$) to $\delta\theta/h$ on the stationary top surface, giving a parabolic temperature distribution across the film. Therefore, integrating with respect to z, the heat flow rate into the top surface through a column of width dx and height, h, is

$$k_t dx \int_0^h (d^2\theta/dz^2) \, dz = k_t dx \, (\delta\theta/h) \qquad (5.39)$$

Substituting $\delta\theta = x\Delta\theta/B$ into equation (5.44), the total conducted heat flow through the whole top surface is:

$$\frac{k_t \Delta\theta}{Bh} \int_0^B x\mathrm{d}x = \frac{k_t\theta B}{2h} \tag{5.40}$$

Heat flow ratio

Therefore, from equations (5.38) and (5.40), the ratio of convected to conducted heat (defined here as the fluid **Peclet number**) is given by

$$\mathrm{Pe} = \frac{\left(\dfrac{U_1 h}{4} c_p \rho \Delta\theta\right)}{\left(\dfrac{\Delta\theta k_t B}{2h}\right)} = \frac{\dfrac{U_1 h^2}{2B}}{\left(\dfrac{k_t}{\rho c_p}\right)} \tag{5.41}$$

The bracketed group of material constants on the right hand side of equation (5.41) is called the **thermal diffusivity** (κ) of the fluid. For various fluids, Table 5.1 gives values of thermal diffusivity and Peclet numbers for a typical high-speed long journal bearing. Note that k_t has units of $kJ/m\,s\,k$ and has units of $kJ/kg\,k$. For the three fluids shown, oil is the best medium for carrying some of the heat away by convection, while air is best for transfer by conduction.

Had the model been two involute gear teeth meshing in oil operating under EHL conditions, typically of contact width 0.002 m, total surface speed 10 m/s and an average film thickness of 1 μm, then Pe = 0.0279, which is about 1/500 of that of the journal bearing above.

5.4   The Navier–Stokes equations

In Section 5.2, Reynolds equation was derived from first principles by employing various assumptions. It is interesting to examine further some of these assumptions because they explain a little more about the mechanics of fluid film lubrication when designing bearings. The *Navier–Stokes equations*, together with the continuity equation, are the fundamental equations of fluid flow. Reynolds equation can easily be derived from them by making the same assumptions as above and also by neglecting inertial flow. One simplified form of these equations for 2-dimensional hydrodynamic flow in

Table 5.1 Diffusivity and Peclet numbers

Fluid	Oil	Water	Air (STP)
Diffusivity (κ) (m²/s)	8.96×10^{-8}	1.4×10^{-7}	2.180×10^{-5}
Peclet number (Pe)	14.92	9.346	0.061

Cartesian co-ordinates for a long bearing and neglecting gravity terms is (Middleman, 1998):

$$\frac{\partial u}{\partial x} + \frac{\partial w}{\partial z} = 0 \tag{5.42}$$

$$\rho\left(\frac{\partial u}{\partial t} + u\frac{\partial u}{\partial x} + w\frac{\partial u}{\partial z}\right) = -\frac{\partial p}{\partial x} + \eta\left(\frac{\partial^2 u}{\partial x^2} + \frac{\partial^2 u}{\partial z^2}\right) \tag{5.43}$$

$$\rho\left(\frac{\partial w}{\partial t} + u\frac{\partial w}{\partial x} + w\frac{\partial w}{\partial z}\right) = -\frac{\partial p}{\partial x} + \eta\left(\frac{\partial^2 w}{\partial x^2} + \frac{\partial^2 w}{\partial z^2}\right) \tag{5.44}$$

At this stage it is appropriate to simplify these equations further by making them steady state and non-dimensional. The model in Fig. 5.10 is the *thrust-bearing* based on Fig. 5.4 where the entering lubricant enters at height H_1 and the pressure distribution commences there and ends at height H_0, the time-dependent terms having vanished.

Define: $\bar{x} = x/h_1$, $\bar{z} = z/h_1$, $\bar{u} = u/U_1$, $\bar{w} = w/U_1$ and $\bar{p} = ph_1/\eta U_1$. Putting these substitutions in equations (5.43) and (54.4) we get

$$\frac{\rho U_1 h_1}{\eta}\left(\bar{u}\frac{\partial\bar{u}}{\partial\bar{x}} + \bar{w}\frac{\partial\bar{u}}{\partial\bar{z}}\right) = -\frac{\partial\bar{p}}{\partial\bar{x}} + \left(\frac{\partial^2\bar{u}}{\partial\bar{x}^2} + \frac{\partial^2\bar{u}}{\partial\bar{z}^2}\right) \tag{5.45}$$

$$\frac{\rho U_1 h_1}{\eta}\left(\bar{u}\frac{\partial\bar{w}}{\partial\bar{x}} + \bar{w}\frac{\partial\bar{w}}{\partial\bar{z}}\right) = -\frac{\partial\bar{p}}{\partial\bar{z}} + \left(\frac{\partial^2\bar{w}}{\partial\bar{x}^2} + \frac{\partial^2\bar{w}}{\partial\bar{z}^2}\right) \tag{5.46}$$

The coefficient on the left-hand side of these equations is **Reynolds number (Re)** for thin film lubrication. Its magnitude defines the inertia effect in the film. The other terms on the left-hand sides are between zero and one, so that if the Reynolds number is sufficiently low, the right-hand sides of these equations can be neglected. For a typical thrust bearing pair: $\rho = 900\,\text{kg/m}^3$, $U_1 = 2\,\text{m/s}$, $_1 = 2 \times 10^{-5}\,\text{m}$, $\eta = 1\,\text{Pa s}$, making Re = 0.036 which is low enough to neglect the inertia terms as we did in Section 5.2.3. Another assumption made is

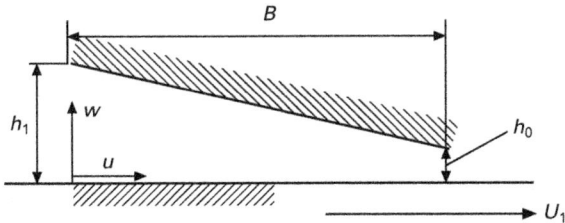

5.10 A single wedge pair.

$$\frac{\partial^2 \overline{u}}{\partial \overline{x}^2} \ll \frac{\partial^2 \overline{u}}{\partial \overline{z}^2}$$

This is not immediately obvious unless it is noted that the aspect ratio of a typical film (width/depth) is 2500 so the variation of $\overline{u}$ along the film width is much less than through its depth, where the lubricant dimensionless speed varies from zero to one. Middleman (1998) offers a justification of this assumption. With all the above assumptions equation (5.44) reduces to equation (5.14).

5.5   Free surface behaviour of lubricant films

An application of the full 3-dimensional Navier–Stokes equations is shown in Fig. 5.11. It is a computer model of a stationary rigid spherical surface above a rigid moving plane entraining an originally *free surface lubricant* into the variable geometry wedge. The picture (originally in colour) is on the vertical plane of symmetry. The full 3-dimensional equations are needed here because there is also side flow as well as free surfaces at both inlet and outlet, needing also an additional equation involving surface tension properties at the air/oil interfaces as well as *multiphase flow*.

Observe how the *free surface lubricant* has piled up at the inlet on the left and that the exiting layer is very thin. Practical concentrated contact problems often involve, thinner films, local elastic flattening of the surfaces and oil *piezoviscous properties* (EHL) (see also Chapter 6). These effects raise dramatically the load capacity of the arrangement. Normally, the boundary conditions used at the inlet assume a complete oil film across the gap that in this case would start further in to the contact where no multiphase flow exists. The shape of the approaching *free surface oil film* and the film wake are important features of the oil feed mechanism of rolling element bearings and gears (see Chan *et al.*, 2007). The transverse view of a wake that has become a free surface is shown by the computer solution in Fig. 5.12(a). The film thickness, is least in the centre, forming a track, with side ramps on either side. Figure 5.12(b) is a photo of a departing wake in a fully flooded rolling EHL contact.

If such is the shape of the wake in a mechanism like a ball bearing, the

5.11 Vertical section through a rigid sphere/plane oil film. Inlet on left; outlet on right; grey areas 2-phase flow; lower black area oil; upper black area air (internal report Imperial College, London).

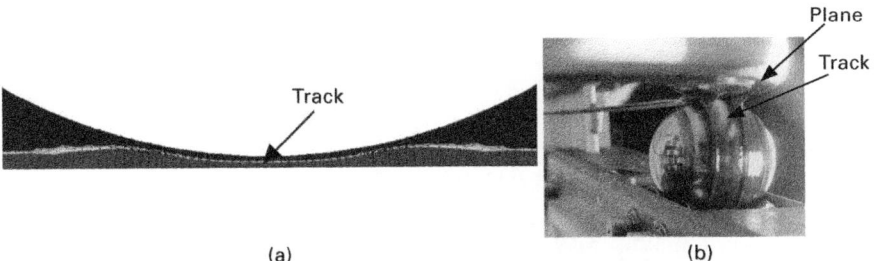

5.12 (a) Computer simulation of a transverse section of the emerging free surface wake from a rigid sphere/plane contact. Lower dark region is oil; upper dark region is air (internal report Imperial College, London) (b) Photo of a wake in a flooded ball/plane EHL contact.

rolling element following in the train might encounter the same shape as its inlet! Indeed, this is the case if there is no external oil feed onto the races, resulting in the wake not having time fully to recover. A computer solution uses the full Navier–Stokes equations, and the surface tension equation for the air/oil interface was solved by Yin *et al.* (1999). There is also some experimental evidence, from Nasser (1997), that confirms the **wake recovery** delay. He employed experimental two-beam interferometry contours to map the wake recovery, as demonstrated in Fig. 5.13, which shows a plan view of the emerging central track and one of its side ramps.

There is an initial recovery phase where the progressive narrowing of the marked lowest region (the track), from its flattened EHL shape, is a precursor to eventual complete film recovery. The reason why a recovery occurs is that the negative pressures, driven by **surface tension** (see Chapter 3) at the air/ oil interface, overcome the viscous forces in the wake, leading eventually to the undisturbed entry layer shape. Figure 5.14 shows experimentally successive transverse profiles of a recovering wake section that uses a laser-surf scanning microscope.

The lack of symmetry seen in Figs 5.13 and 5.14 is attributed to the track inner wall of the rotating glass plate (the plane) being shorter than the outer wall. Also note that the average height of the undisturbed film is 70 μm compared with a maximum horizontal distance seen of 13 000 μm, so the picture is really very flat. The track recovery seen in Figs 5.13 and 5.14 is extremely slow.

5.6   Externally pressurised (EP) gas journal bearings

The **EP gas bearing** is practically frictionless. The journal is separated from the bearing housing by a thin air film that is fed at high-pressure (typically

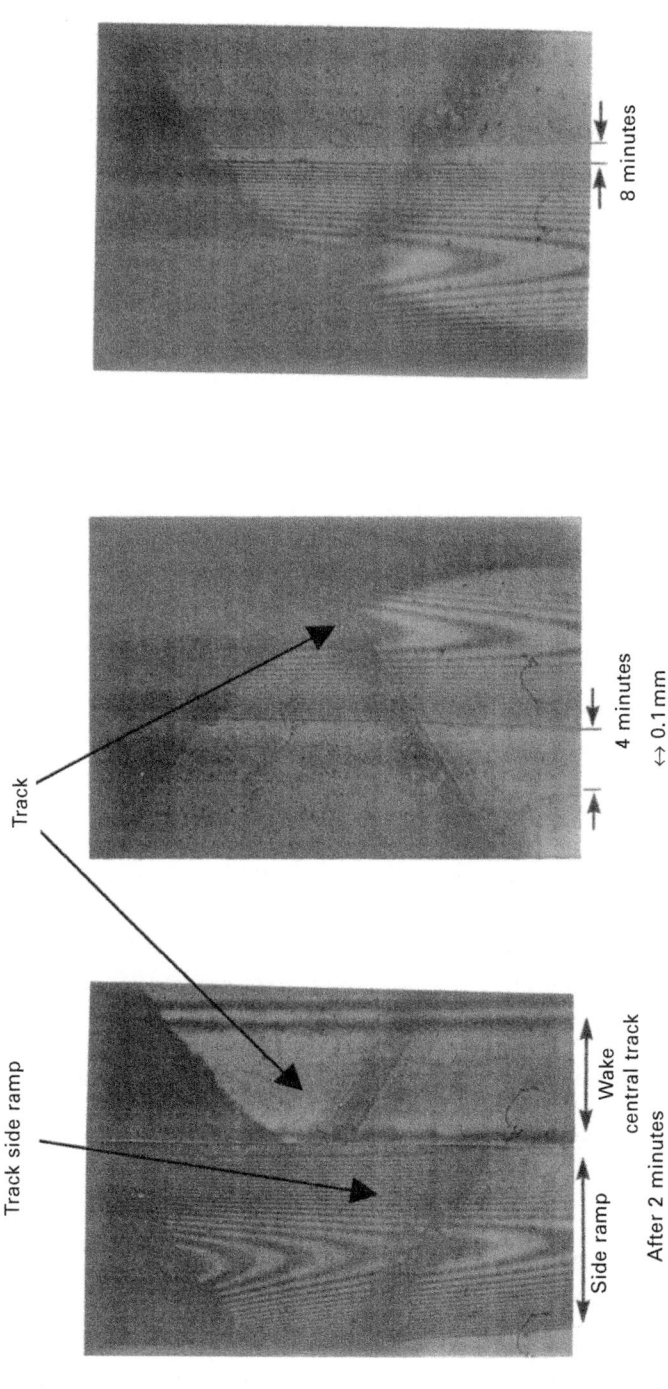

5.13 Contours showing the recovery with time of a wake, oil viscosity: 1.06 Pa. Height distance between adjacent dark fringes: 0.098 μm (after Nasser, 1997).

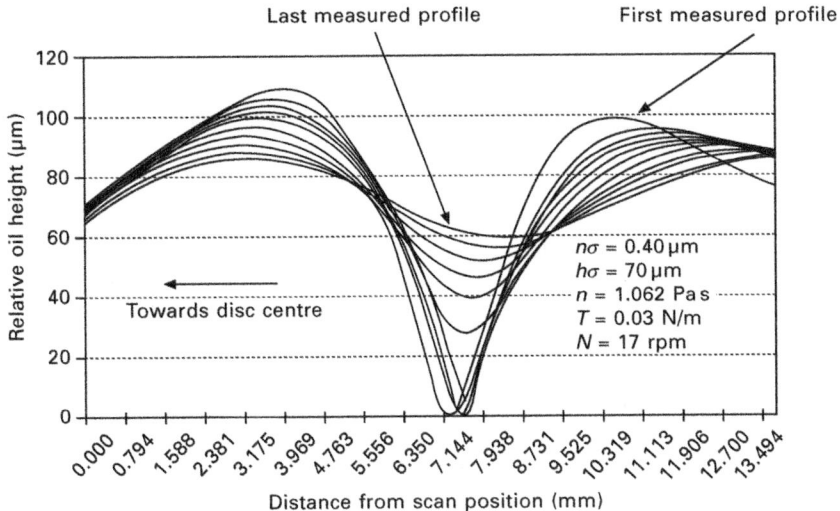

5.14 Successive profiles with time of a recovering wake (after Nasser, 1997).

2 atmospheres) from rows of equispaced discrete orifices drilled into the bearing housing that are themselves fed by high-pressure air from galleries within the housing (5 atmospheres). Because of the high pressure in the air film, unlike the pure hydrodynamic bearing, it can normally operate at nearly constant radial thickness throughout the bearing clearance. The air in the film then leaks out to atmospheric pressure at the bearing ends. In the case of a horizontal rotating EP journal bearing there can additionally be present a hydrodynamic effect that increases the load capacity, thus creating a *hybrid bearing*. Nevertheless, EP gas bearings have a much lower stiffness than equivalent oil hydrostatic bearings but can rotate in the air film at excessive speeds. A disadvantage is that they can become catastrophically unstable at a certain journal speed. The designs below outline how this critical speed can be increased considerably.

The bearing here is assumed to be *vertical* or under a negligible radial load and spinning about its geometric centre line at speed ω. Following a small radial perturbation the journal centre line itself is caused to rotate (*whirl*) about the stationary bearing housing centre line at its natural frequency, Ω (the *whirl onset speed*).

Reynolds equation (5.24) applied to gases in dimensionless terms is (Mori and Mori, 1985):

$$\frac{\partial}{\partial \theta_r}\left[H^3\frac{\partial P^2}{\partial \theta_r}\right] + \frac{\partial}{\partial \zeta}\left[H^3\frac{\partial P^2}{\partial \zeta}\right] = 2\,(\Lambda - \sigma)\frac{\partial (PH)}{\partial \theta_r} \qquad (5.47)$$

Here: $P = p/p_a$, $H = h/c$, $\Lambda = \dfrac{6\eta\omega}{p_a}\left(\dfrac{R}{c}\right)^2$ = **compressibility number**,

$\sigma = \dfrac{12\eta\Omega}{p_a}\left(\dfrac{R}{c}\right)^2$ = frequency number, $H = 1 \cos \theta_r$ = dimensionless film thickness, θ_r = angular co-ordinate from the line joining the journal centre to the fixed bearing centre, ω = spin speed of the journal, Ω = whirl speed of the journal, c = radial clearance, p_a = atmospheric pressure, η = gas viscosity, R = journal radius, $\zeta = y/R$ = dimensionless axial coordinate, y = axial co-ordinate. Another important group is the feeding factor (not shown) that describes the orifice characteristics in relation to the air properties.

From equation (5.47), if $\sigma = \Lambda$, the right-hand side becomes zero, demonstrating that such a disturbance will create an instability called **half-speed whirl** (Gohar and Rahnejat, 2008), causing the journal eventually contact the bearing surface. For this to occur, $\omega = 2\Omega$, where pressure, P, is found from the solution of Reynolds equation from which the gas film stiffness and damping coefficients are determined. Using additionally the equations of motion of the rotor, the bearing dynamic behaviour can be found.

In order to delay the whirl onset speed, a design based on Mori and Mori (1985) was employed. It used a **non-rotating sleeve** surrounding the journal. The sleeve itself contains a high-pressure air gallery supplied through flexible tubing, from which the air is fed from the orifices into two films, one between the housing and the sleeve (both non-rotating) and the other between the sleeve and the journal rotating at ωrad/s. The outer film acts as a damper for the inner film. One alternative configuration with a parallel damper, analysed by Cazan *et al.* (2001), is shown in Fig. 5.15 where. (a) is the undamped configuration, (b) is similar to Mori and Mori's design, while (c) is a mixed configuration having a parallel damper that has a lower maximum threshold whirl speed than (b) but is stiffer. Figure 5.16 is an example of a stability diagram for the mixed configuration. Note that, by altering the configuration geometry and/or sleeve mass, the device can be tuned to maximise the whirl onset speed.

For the mixed configuration (Fig. 5.15(c)) and a rotor diameter of 0.032 m, the maximum allowable value of Ω has increased from that of the single bearing arrangement (a) of 54 000 rev/min to 138 000 rev/min. A higher threshold speed is possible with arrangement (b) but on the downside, the bearing has a lower stiffness.

5.7   Approximate design of oil thrust bearings

As will be shown later, when designing **thrust bearings**, it is important to consider the effect of temperature on the viscosity of the lubricant. In the case of a multi-sector shaped pad arrangement, the mean lubricant viscosity

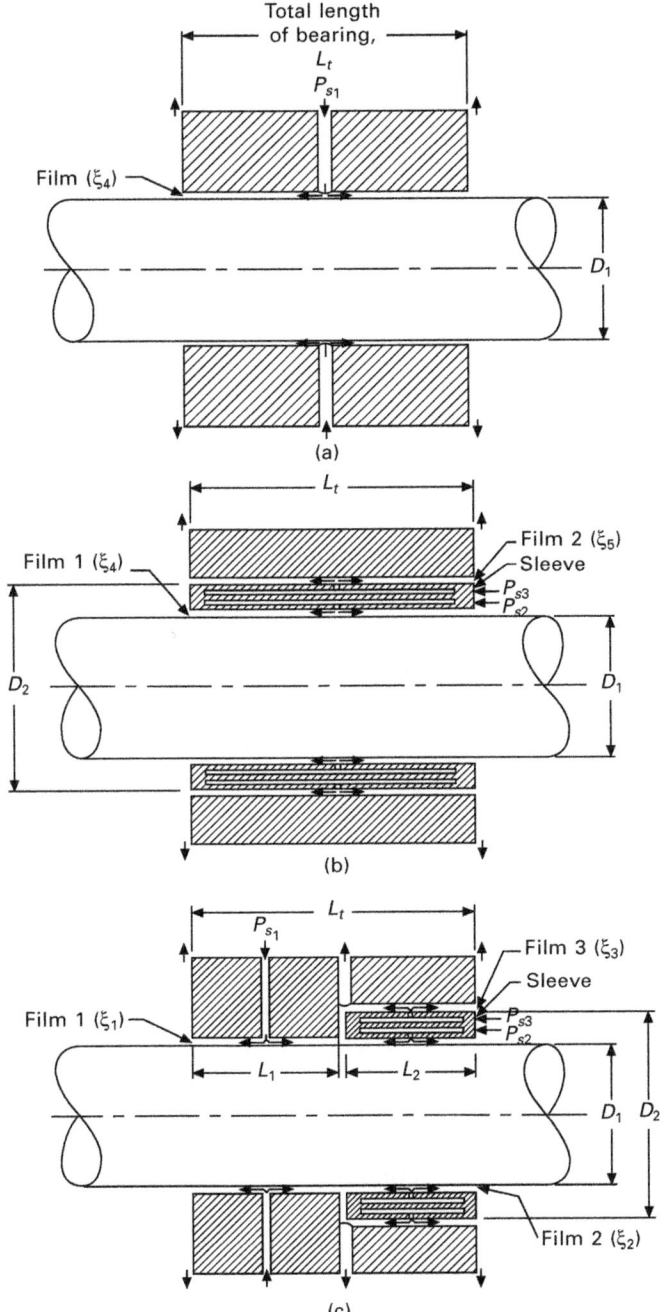

5.15 Designs of EP gas bearings with sleeved dampers (after Cazan *et al.*, 2001).

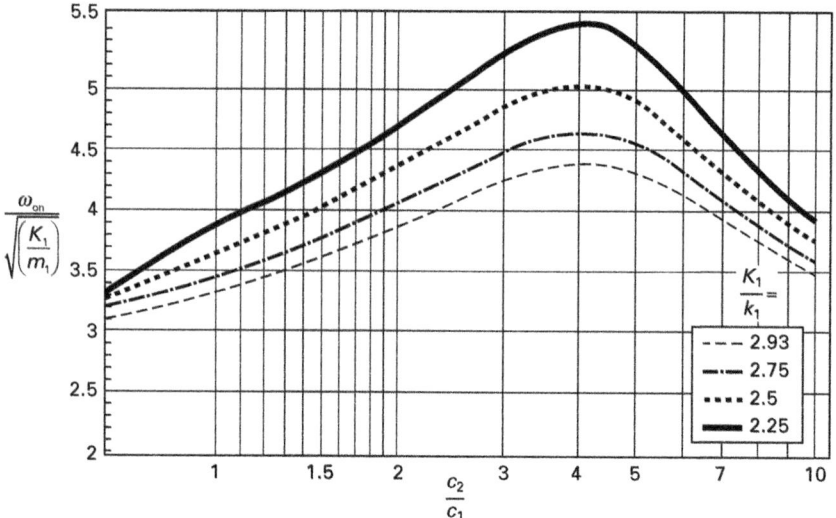

5.16 EP gas bearing with mixed configuration: stability diagram for Fig. 15.15(c). (after Cazan *et al.*, 2001).

can be 20–40% of the entry viscosity to the bearing housing before its distribution to the various wedge pairs. The full design procedure for a bearing with a plurality of such sector shaped pads, requires a complex numerical analysis involving the Reynolds and energy equations together with the lubricant temperature/viscosity equation. Also sometimes included in a complete solution, is pad distortion caused both by conductive heating as well as from the individual pressure distributions. There is insufficient space here to discuss a full numerical solution, but see Ettles and Cameron (1963). Instead, an approximate design procedure will be given for a thrust bearing having rigid plane finite length pads. The method estimates a single representative lubricant film temperature for a pair at a corresponding effective viscosity. The method is based on data from existing isothermal numerical solutions.

Referring to the energy equation (5.37) for a wedge pair, because the pressure distribution has little effect on the temperature (as assumed in Section 5.3), the film can be considered uniform when considering temperatures:

$$\eta \left(\frac{\partial u}{\partial z} \right)^2 = \rho u c_p \frac{\partial \theta}{\partial x} \tag{5.48}$$

The left-hand side of equation (5.48) is the viscous heating and the right-hand side is the convective cooling term. Integrating across the film of thickness h and assuming that η is constant *throughout* the film let it be called now the **effective viscosity**, η_e. Use this single value, η_e, to represent the normally variable viscosity:

$$\eta_e \int_0^h \left(\frac{\partial u}{\partial z}\right)^2 dz = \rho c_p \int_0^h u \frac{\partial \theta}{\partial z} dz \qquad (5.49)$$

Assuming, additionally a linear mean film temperature rise over width B of $\Delta\theta$, and ignoring pressure effects, if U_1 is the runner velocity

$$\frac{\partial u}{\partial z} = \frac{U_1}{h} \quad \frac{\partial \theta}{\partial z} = \frac{\Delta\theta}{B}$$

Now take both terms outside the integrals in equation (5.49). With these substitutions, after integrating and rearranging:

$$\frac{\Delta\theta}{\eta_e} = \frac{2BU_1}{\rho c_p h^2} \qquad (5.50)$$

Equation (5.50) could have been obtained directly by writing:

mass flow/s × specific heat × temperature rise

= work done per second (power)

that is

$$\frac{\rho U_1 h}{2} \times c_p \times \Delta\theta = \left(\frac{\eta_e U_1 B}{h}\right) U_1$$

Moreover, using this power balance, in general terms for a long bearing and *any* film geometry, including the pressure effects:

$$(Q\rho)c_p\Delta\theta = FU_1 \qquad (5.51)$$

where Q is the inlet flow into to the film and F is the **friction force**.

Using general expressions for F and Q in terms of their dimensionless groups defined in the appendix at the end of the chapter, equation (5.51) becomes

$$F^* \left(\frac{\eta_e U_1 B L}{h_0}\right) = Q^* (LU_1 h_0) \rho c_p \Delta\theta$$

Then letting $\Delta\theta^* = F^*/Q^*$:

$$\Delta\theta = k_1 \Delta\theta^* \left(\frac{\eta_e U_1 B}{\rho c_p h_0^2}\right) \qquad (5.52)$$

Note that equation (5.52) is perfectly general and applies to both long and *finite length bearings*. The empirical constant, k_1, accounts for the possibility that not all the heat is removed by convection from the bearing end, some escaping by conduction through the pad and runner. For a plane thrust bearing pair this is generally taken as $k_1 = 0.6$. All the above expressions for load, friction force, flow, temperature rise are summarised in the appendix at the end of the chapter.

5.7.1   Effective temperature of long bearings

Because here the pad is considered long in the y direction, only a bearing pair in isolation can be considered. Referring to Fig. 5.13, assuming the exit film thickness, h_0, the inlet temperature to the film, θ_1 and U_1 are all stipulated, then the effective viscosity, η_e, and total temperature rise *along the pad*, $\Delta\theta$, are unknowns in equation (5.52). An additional equation is therefore needed for a solution. The method is to seek a value of η_e that can represent the temperature-dependent variable viscosity throughout the film. Thus, η_e must have a corresponding *effective temperature*, θ_e, somewhere between the film inlet and exit temperatures. Thus,

$$\theta_e = (\theta_1 + k_2\Delta\theta) \tag{5.53}$$

Normally, for thrust bearings, θ_e is taken as the mean temperature of the pad, so that $k_2 = 0.5$. Equation (5.53) can also be used for the finite length bearing discussed below. It is also valid for journal bearings. A single viscosity, η_e, for the film at θ_e makes it possible to employ the isothermal expressions already derived above for determining the bearing performance characteristics.

5.7.2   Finite length plane thrust bearings

These bearings are the ones used in practice. Their pads are generally segmental in plan form with a number of them creating wedge pairs supporting the annular runner, as mentioned above. When analysing a single wedge pair, a *square pad* is a reasonable approximation to a segmental shape. The pressure distribution must fall to zero along the sides because each pad is now of finite length. As well as reducing the load capacity, these falling pressures cause *side leakage* of the oil, with the make up flow supplied through feedholes into the chambers between the wedges (Fig. 5.17). The steady state 2- dimensional Reynolds equation (5.27) is therefore required to analyse a finite length bearing. This is normally done numerically. Also needed is the energy equation (5.51) in addition to considering thermal and mechanical flexure of the pads if the load and speed are excessive. There is also a *thermal mixing* process occurring in the chambers. It is therefore important to understand such behaviour if a simple design procedure is to be used, as will be explained below.

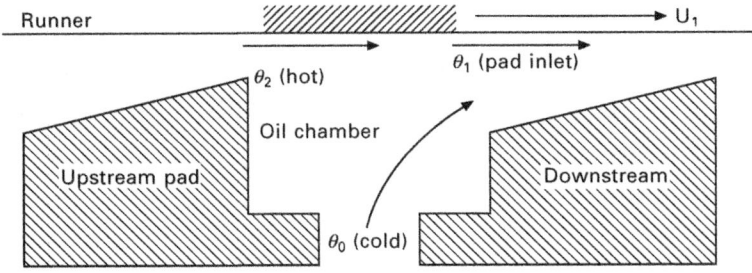

5.17 Lubricant mixing in a supply groove.

5.8   Thermal design of finite length bearings

It has been shown that, for a single *long pad bearing* pair, a convective flow analysis only requires the x direction flow entering and leaving the wedge film. In the case of a finite length-bearing pair, the situation is more complicated on account of the additional side flow. Pinkus (1990) discusses the resulting flow processes. The side flow causes some of the heat to be carried away by convection from the pad sides, in addition to the carry over flow from its end. Furthermore there is normally a plurality of these wedge pairs arranged circumferentially below an annular runner. Consider now Fig. 5.17. It depicts a supply chamber formed by the inlet and outlet profiles of two adjacent pads in the bearing, with the runner above moving at constant speed U_1, entraining the lubricant.

The finite lengths of the pads and their transverse pressure gradients (y direction) cause some of the lubricant to leak out from the film sides. This oil side flow then leaves the bearing and travels to a common sump from where it is pumped back and distributed via feed-holes to the oil chambers situated between the pads. When it re-enters the bearing films, the make up oil has cooled to bulk temperature θ_0. In each chamber, this cold entering oil mixes with the hot oil leaving the upstream pad film at temperature θ_2 and heats up to temperature θ_1, the entry temperature to the adjacent downstream film. The mixing of the hot and cold oil in the chamber is a complex process, both thermally and hydrodynamically. Nevertheless, if one of these bearing pairs is considered from a design perspective, the problem can be simplified.

5.8.1   Power balance for the effective viscosity

Figure 5.18 shows a pad film surrounded by a control volume. The power supplied to drive the runner is P, the externally pumped lubricant make up power is $c_p \rho Q_0 \theta_0$ and the power possessed by the side leakage lubricant is $c_p \rho Q_s \theta_s$.

Considering the external power balance, power going into the control volume equals power leaving it, that is:

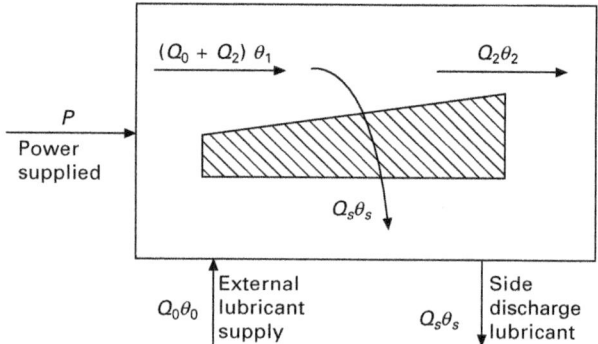

5.18 Control volume round a thrust bearing pair (ρc_p terms omitted).

$$P + \rho c_p Q_0 \theta_0 = \rho c_p Q_s \theta_s$$

Since $Q_0 = Q_s$

$$P = \rho c_p Q_s (\theta_s - \theta_0) \tag{5.54}$$

Here, $\theta_s - \theta_0$ represents the temperature rise from the inlet *supply chamber* to the equivalent temperature of the side flow. Consider now the power flow *within* the bearing film, assuming perfect thermal mixing in the supply chamber. Again write:

> Friction power inputted to the film = total power carried away from the film sides and its end

that is

$$P = \rho c_p [Q_s(\theta_s - \theta_1) + Q_2(\theta_2 - \theta_1)] \tag{5.55}$$

Equation (5.55) is not needed in this simple design procedure because the pad inlet and outlet temperatures are unknown. They can however be found if the empirical equation (5.53), describing the assumed position of the effective temperature along the film, is used in conjunction with equation (5.55). Also, equation (5.54) is adequate for the approximate solution because $\theta_s - \theta_0$ includes this local temperature rise. Another advantage of using equation (5.54) is that θ_0 can be chosen as input data, it being more easily measured experimentally. Therefore, the procedure discussed below attempts to find θ_s but cannot alone determine θ_1 or the maximum temperature, θ_2, at exit to the film.

5.8.2   Effective temperature design coefficients

From the definition of P and letting

$$\Delta\theta_s = \theta_s - \theta_0 \tag{5.56}$$

Because power = friction force times speed, equation (5.54) can be written as

$$F_0 U_1 = \rho c_p Q_s \Delta \theta_s \qquad (5.57)$$

It was also noted that equation (5.57) is perfectly general where the dimensionless coefficient $E^* = \dfrac{F_0 h_0}{LB\eta U_1}$ includes any chosen geometry for the bearing film. The same situation applies to a side flow factor defined as $Q_s^* = \dfrac{Q_s}{LU_1 h_0}$ and a load factor $W^* = \dfrac{\eta U_1 LB^2}{W h_0^2}$. Equation (5.4) is also needed as it defines the behaviour of the lubricant with temperature, the constants being calculated from data supplied by the manufacturer.

In addition, a proportion of the heat generated in the film, k_1, is assumed lost by conduction through the bounding solid surfaces. Putting these substitutions into equation (5.57) and letting $\Delta \theta_s^* = F^*/Q_s^*$:

$$\Delta \theta_s = \left(\frac{k_1}{\rho c_p} \right) \left(\frac{B\eta U_1}{h_0^2} \right) \Delta \theta_s^* \qquad (5.58)$$

The coefficients W^*, F^*, Q^*, Q_i^*, Q_s^* and $\Delta \theta_s^*$ are tabulated in Table A.1 in the appendix. There is now sufficient information for a design example to be tackled.

5.8.3 An example of thrust bearing thermal design

A fixed inclination plane thrust bearing has a square plan form with $B = L = 50$ mm, and a fixed taper, t, of 0.03 mm. The bearing is required to support a maximum load of 7 kN at a runner speed of 10 m/s. The behaviour of the mineral oil used is defined by the expression below and the supply temperature is 30 °C. Assuming that 60% of the heat flow in the film is carried away by convection, find the minimum film thickness and temperature rise under these conditions. The behaviour with temperature of the mineral oil used is obtained from equation (5.4) as

$$\ln(\eta) = -1.845 + \left[\frac{700.81}{\Theta_s - 203} \right]$$

η being in cp and Θ_s in K.

Solution

Referring to Fig. 5.13, the fixed inclination geometry of the pad is defined by

$$K = \frac{h_1 - h_0}{h_0} = \frac{t}{h_0} = \frac{3 \times 10^{-5}}{h_0} \tag{a}$$

so that this geometry factor K is not constant but will vary with the value of H_0. The effective viscosity, temperature rise and least film thickness are all unknown. The numerical data used for the solution are given in Table A.1, adapted from Jacobson and Floberg (1958), for a square pad. It supplies all the values of the dimensionless groups needed for the solution.

The method is to choose a series of different K and hence h_0 then put them in Table 5.2 together with the corresponding values of the relevant dimensionless groups found from Table A.1, defined again in the appendix. Assuming there is an adequate range for a solution, start the new solution (Table 5.2) reproducing the values of the dimensionless groups needed. Firstly, enter the chosen K and computed h_0 values, in columns 1 and 2, as shown in Table 5.2. The additional data needed can be found in the following way:

- In column 3, W^* from Table A.1 is required because a single load, W, is supplied.
- In column 4, $\Delta\theta_s^*$ is also needed because the problem involves a temperature rise. It can also be found from Table A.1.
- For column 5, use the appendix, for W, to obtain η for each value of h_0:

$$\eta = \frac{Wh_0^2}{W^* U_1 LB^2} = 5.6 \times 10^6 \left(\frac{h_0^2}{W^*}\right) \text{Pa s} \tag{b}$$

- For column 6, use the appendix, for $\Delta\theta_s$ where $k_1 = 0.6$. Thus

$$\Delta\theta_s = k_1\theta_s^*\left(\frac{\eta U_1 B}{\rho c_p h_0^2}\right) = \frac{0.60_s^*\eta_e \times 10 \times 0.05}{850 \times 1880 h_0^2} = \theta_s^* \times 1.877 \times 10^{-7}\left(\frac{\eta}{\eta_0^2}\right)$$

At this stage there are *two* temperature equations to be solved to complete Table 5.2.

Table 5.2 Tabulated results

1 K	2 $h_0 \times 10^{-5}$m	3 W^*	4 $\Delta\theta_s^*$	5 η (Pa s)	6 $\Delta\theta_s$(°C)	7 θ_{s1}(°C) = 30 + $\Delta\theta_s$	8 θ_{s2}(°C)
0.5	6	0.0575	6.76	0.501	123.44	153.4	20.96
1	3	0.0689	2.959	0.104	45.19	75.1	44.18
1.5	2	0.07	1.787	0.048	127.69	57.69	60.88
2	1.5	0.0671	1.242	0.027	19.43	49.35	76.68
3	1	0.0583	0.732	0.014	13.18	43.18	100.62
4	0.75	0.0503	0.461	0.00892	9.66	39.66	120.25

- Column 7 produces θ_{s1} by using column 6 and equation (5.56):

$$\theta_{s1} = \theta_0 + \Delta\theta_s$$

- Column 8 produces θ_{s2} by using η from column 5 in Vogel's equation (equation 5.4).

$$\ln(\eta) = -1.845 + \left(\frac{700.81}{\Theta_{s2} - 203} \right)$$

Note that in Vogel's equation η must be input in cP. Moreover, Θ_{s2} is output in K and must be converted to °C for Table 5.2.

For the correct solution, the values of θ_s from columns 7 and 8 must be the same at only one particular value of h_0, it being found by plotting both versions of θ_s against h_0. Where the two curves intersect, gives the solution. The graphs are shown in Fig. 5.19.

The answers read off are $h_1 = 2.1 \times 10^{-5}$ m and $\theta_s = 63\,°C$, $\eta_e = 38 \times 10^{-3}$ Pa s. The bearing is running fairly hot because of the relatively thin film. Knowing the correct values of θ_s, η_e and h_0, the other unknown design variables can then be obtained from the design coefficient expressions.

This type of problem is rapidly solved with a software package like *Mathcad* using its 'Vectorize' operator. It means, for example, that having entered K as a vector from the data table, a vector of h_0 can be found from equation (a) above. Having entered W^* from the data table, equation (b) above allows the vector η to be found, and so on building up Table 5.2. Finally, *Mathcad* allows Fig. 5.19 to be plotted with the intersection point determined, using its 'Trace' facility.

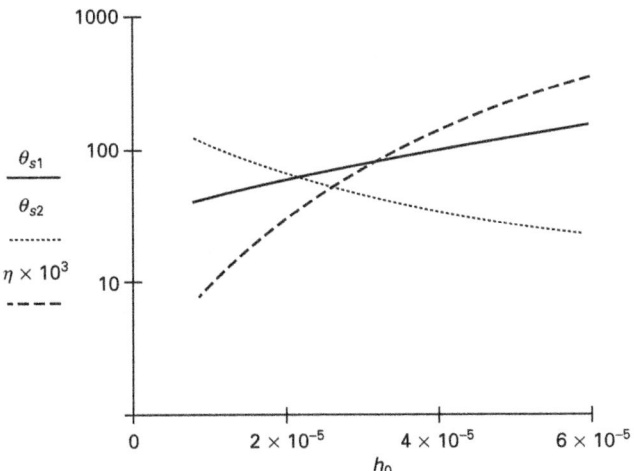

5.19 Solution curves for temperature and viscosity.

5.9   Review of some unusual and recent applications of fluid film lubrication

Because the hydrodynamic type of fluid film lubrication has been extensively researched and reviewed in the literature (see for example Khonsari, 1986), I propose to outline below a few more unusual designs of fluid film bearings.

5.9.1   Flexible support squeeze film gas bearings

The principles of hydrodynamic lubrication and externally pressurised gas bearings have been covered earlier in the chapter. However, there is another type of bearing that is seldom mentioned nowadays but could again become a research project topic.

The *squeeze film gas bearing* has similarities with the externally pressurised gas bearing (EP), an example of which was discussed in Section 5.6. Like the hydrodynamic bearing (or aerodynamic bearing for gases), all three types depend on viscous retardation within the film. The hydrodynamic bearing has wedge geometry to fulfil this function, while the EP bearing has an external source of highly pressurised air.

In the case of the squeeze film bearing, it can be composed of two parallel concentric steel discs, the top one carrying the load and the bottom one supported by, for example, a piezoelectric ceramic vibrator. In principle we have here a *squeeze film bearing*. If the stationary member of the bearing pair is vibrated sinusoidally, the way the gas behaves as a result of this process determines whether a fluid film bearing will result or not. When the forcing frequency is low, or there is present a large mean clearance, the air is forced out radially when the clearance decreases and is sucked in when it increases. Under these conditions the film behaves as a damper with viscous forces important. If the oscillation amplitude is increased, and/or the frequency is made very high and also the mean clearance is low, then compressibility effects are introduced, the gas alternately being compressed and expanded and following *Boyle's law*. In this case, the radial flow is then restricted by the local pumping action and will occur only in the narrow annulus of the outside edge region of the film. The bulk of the film, inboard, acts as a non-linear spring supporting the steady load. To increase the amplitude at the edge of the supporting member it can be made flexible so that its geometry alters according to the frequency.

In some applications the piezoelectric vibrator itself is one member of the pair. Alternative designs can be cylindrical, journal, spherical or conical bearing pairs. One application of these bearings has been gimbal support of two axis gyroscopes and other precision instruments. Some references that discuss squeeze film bearings are Salbu (1964), Pan (1964) and Beck and Strodman (1968). Lastly, Da Silva (1978) gives a detailed analysis of squeeze bearing behaviour.

5.9.2   Textured bearings

This design is an oil thrust bearing that, instead of having a uniform land or one that is stepped (see Cameron, 1967), has additionally cut into it a recess near the inlet. The idea is to create a sub-ambient pressure region within the recess, that will suck in more oil than would normally enter through entrainment due to shear.

Brajdic-Mitidieri *et al.* (2005) demonstrated this idea with models having a single recess or multiple recesses, the film itself being of the order of 1 μm. Both parallel and taper land long bearings with recesses were used, as well as a square recess in a finite length bearing. They showed that the presence of the recess does indeed increase the oil throughput as well as reducing the friction coefficient in comparison with non-recessed designs. In the parallel bearing Olver *et al.* (2006) find that the recess even enables it to function as a bearing. Without the recess it could not, assuming conditions remained isothermal. The reason the long recessed parallel bearing works is that there must subsequently be flow into the film downstream of the recess. They show that to cause this flow, there must be initially a positive pressure gradient there followed by a fall near exit in order for the oil to leave the bearing end.

The relatively deep recess employed in the models, means that Reynolds equation cannot strictly apply because inertia forces become important when the Reynolds number exceeds one for lubricant films (Section 5.4). Brajdic-Mitidieri *et al.* (2005) used a version of the Navier–Stokes equations for their solution that presumably included these inertia forces operating within the recess. If a commercially designed recessed thrust bearing with a film of the order of 10 μm is used as a model, then the Reynolds number everywhere becomes larger, making the NS equations even more appropriate.

5.10   References

Barus, C. (1893), 'Isothermals isopietics and isometrics in relation to viscosity', *American Journal of Science*, 3rd Series **45**, 87–96

Beck, J. V. and Strodman, C. L. (1968), 'Load support of squeeze film journal bearing of finite length', *Trans ASME J (F), J. Lub. Tech.*, **90**, 157–161

Brajdic-Mitidieri, P., Gosman, A. D., Ionnidis, E. and Spikes, H. A. (2005), 'CFD analysis of a low friction pocketed pad bearing', *Trans. ASME J. Tribology*, **127**, 803–812

Cameron, A. (1967), *Basic Lubrication Theory* London, Longman Ltd

Cazan, A., Gohar, R. and Safa, M. M. A. (2001), 'Externally pressurized gas bearings in a mixed configuration', *Proc. I. Mech. E. Part K*, **216**, 181–189

Chan, R. T. P., Martinez-Botas, R. F. and Gohar, R. (2007), 'Isoviscous flow past a rigid sphere partially immersed in a thin oil film', *Lubrication Science*, **19**, 117–212

Da Silva, F. P. (1978), 'Squeeze film air bearings with flexible supports', PhD thesis, Imperial College, London

Ettles, C. and Cameron, A. (1963), 'Thermal and elastic distortions in thrust bearings', paper I presented at the Institution of Mechanical Engineers Lubrication and Wear Convention, Bournemouth.

Gohar, R. (2001), *Elastohydrodynamics*, 2nd ed., Imperial College Press, London.

Gohar, R. and Rahnejat, H. (2008), *Fundamentals of Tribology*, Imperial College Press, London

Houpert, L.(1985), 'New results of traction force calculations in elastohydrodynamic contacts', *ASME J. Tribology*, **107**, 241–248

Jacobsen, B. and Floberg, L. (1958), *The Rectangular Plane Thrust Bearing'*, Trans Chalmers University Tech., Gothenburg, vol. **203**, Appendix 10, p. 42

Khonsari, M. M. (1986), 'A review of thermal effects in hydrodynamic bearings Part ll: journal bearings', *ASLE Transactions*, **30**, 26–33

Middleman, S. (1998), *An Introduction to Fluid Dynamics*, John Wiley and Sons Inc., New York

Mori, H. and Mori, A. (1985), 'Analysis of stability of floating bush journal bearings', *Bull. Jap. Soc. Mech. Engrs.*, **28**, 1001–1003

Nasser, H. F. (1997), 'Recovery of disturbed lubricant layers', MSc Thesis, Imperial College, London, (1997).

Olver, A.V., Fowell, M. T., Spikes, H. A. and Pegg, I. G. (2006), 'An inlet suction mechanism in non-convergent hydrodynamic bearings', *Proc. I. Mech. E, Part J, J. Engg Tribology*, **220**, 105–108

Pan, C. H. T. (1964), 'Analysis, design and prototype development of squeeze film bearings for the AB–5 Gyro', *MIT Report 65–TR–25*

Pinkus, O. (1990), *Thermal Aspects of Fluid film Lubrication*, ASME Press, New York

Reynolds, O. (1886), 'On the theory of lubrication and its applications to Mr Beauchamp Tower's experiments including an experimental determination of the viscosity of olive oil', *Phil. Trans.*, **177**, 157–234

Roelands, C. J. A. (1966), 'Correlation aspects of the viscosity–temperature–pressure relationships of lubricating oils', PhD thesis, Delft University of Technology, The Netherlands

Salbu, E. O. J. (1964),'Compressible squeeze films and squeeze bearings', *Trans ASME (D) Journal of Basic Engineering*, 355–364

Sommerfeld, A. (1904), Sur hydrodynamic Theorie der Schmierittelreibung', *Zeits. f. Math. u. Phys.*, **40**, 97–155

Steiber, W. (1933), *Das-Schwimnlager Krayn*, VDI, Berlin.

Vogel, H. (1921), *Physik*, **22**, 645

Walther, A. and Sassenfeld, H. (1954),'Erleitlager Berechnungen', VDI-*Forschungshaft*, **441**

Yin, J, Eissa, K., Gohar, R. and and Cann, P. M. E. (1999), 'Rebounding lubricant layers', *Lubrication Sci.*, **12**, 1–30.

5.11   Appendix: Design coefficients for plane thrust bearing pairs

$$W = W^* \left(\frac{U_1 \eta L B^2}{h_0^2} \right), \; F = F^* \left(\frac{L B \eta U_1}{h_0} \right), \; Q_s = Q_s^* (L U_1 h_0),$$

$$\Delta \theta_s^* = F^* / Q_s^*, \; \Delta \theta_s = k_1 \Delta \theta_s^* \left(\frac{\eta U_1 B}{\rho c_p h_0^2} \right)$$

Table A.1

K	0	0.5	1	1.5	2	3	4
W^*	0.0556	0.05575	0.0689	0.07	0.067	0.0584	0.0503
Q_s^*	0	0.122	0.246	0.371	0.496	0.75	1.01
F^*	1	0.825	0.728	0.663	0.616	0.549	0.466
$\Delta \theta_s^*$	∞	6.76	2.859	1.787	1.242	0.732	0.461

6

Elastohydrodynamic lubrication

F. SADEGHI, Purdue University, USA

Abstract: This chapter describes the lubrication mechanism for non-conformal contacts operating under heavy loads. The chapter provides an in-depth review of previous analytical and experimental work conducted in elastohydrodynamic lubrication of line and point contacts. The chapter also describes the lubricant pressure and film thickness distributions for line and point contacts as well as dry Hertz contact stresses between two general bodies, cylinders and spheres. Minimum film thickness equations for line and point contacts are presented and examples of lubricant film thickness and contact stress calculations for ball and cylindrical bearing are also discussed. An example for lubrication at the cam and follower contact is presented.

Key words: elastohydrodynamic lubrication, EHL, Hertz contact stresses, surface roughness, ball and roller element bearings.

6.1   Introduction

The formation of a thin lubricant film between the mating surfaces of rolling/sliding non-conformal machine elements is commonly referred to as *elastohydrodynamic lubrication*, which is abbreviated as EHL or EHD. The major factors influencing EHL are the elastic deformation of the contacting bodies due to the applied load, the hydrodynamic action entraining the lubricant between the contacting surfaces and the lubricant viscosity variation with pressure, hence the term elastohydrodynamic lubrication. The discovery and understanding of elastohydrodynamic lubrication represents a major milestone in the history of tribology and machine elements' design and performance. The thin film of lubricant separating the rolling/sliding non-conformal surfaces successfully predicted by the EHL theory provided the explanation for effective lubrication and satisfactory operation of many machine elements (e.g. ball and rolling element bearings, cams and gears) (see also Chapters 7, 14–17).

It was only a few decades ago that the EHL theory was developed. Previously, hydrodynamic lubrication of conformal surfaces was successfully studied using the Reynolds equation (Reynolds, 1886) of thin film lubrication or *creeping flow* (see Chapter 5). In the early 1920s, the hydrodynamic lubrication theory was used to predict film thickness for non-conformal gear contacts; however, the film thickness predicted was considerably

171

smaller than the surface roughness and thus could not explain the effective operation of gears. The breakthrough occurred as Ertel (1939) included the elastic deformation of the contacting solids and the viscosity variation of the lubricant with pressure in analysing the inlet zone of lubricated non-conformal contacts. His pioneering work showed significantly larger film thickness than those previously predicted by the hydrodynamic theory. Other notable pioneers from this era include Grubin and Vinogradova (1949) who performed a similar analysis, Petrusevich (1951) who provided the line contact numerical simulation, and Dowson and Higginson (1959) who presented an iterative procedure for the EHL solution and developed a minimum film thickness equation for line (rectangular) contacts. Crook (1958) experimentally confirmed the presence of film thickness in EHL line contacts, and later Crook (1961) and Gohar and Cameron (1963) experimentally verified the results obtained by Dowson and Higginson (1959). The basic governing equations of elastohydrodynamic lubrication are a set of integro-differential equations, namely the hydrodynamic Reynolds equation, the film thickness equation including elastic deformation of contacting surfaces, the viscosity pressure equation, and the force balance equation. The density variation with pressure is usually included to account for the compressibility of the lubricant.

Significant research and progress in elastohydrodynamic lubrication study continued in the following decades. Hamrock and Dowson (1976a,b) in a series of papers presented results for isothermal point (elliptical) contacts and developed formulae for minimum and central film thicknesses, which are widely used today. The investigators of the early years primarily concentrated on the basic EHL solution for isothermal lubrication with the *Newtonian fluid* model (see Chapter 5). With those solutions well understood, Cheng and Sternlicht (1965) and Dowson and Whitaker (1965) investigated thermal effects in elastohydrodynamic lubrication and others studied the effects of *non-Newtonian lubricant* models. Sadeghi and Sui (1990, 1991) developed a complete numerical model for Newtonian and non-Newtonian thermal elastohydrodynamic lubrication of line contacts. They used the Newton–Raphson technique to obtain a numerical solution and demonstrated that temperature effects in EHL contacts can be significant and in order to have a good estimate of velocity effects on friction, thermal effects need to be included in the model. The numerical simulations of EHL problems were improved in the late 1980s by the introduction of the multi-grid multi-level technique for faster convergence to solutions of isothermal Newtonian EHL of line and point contacts (Lubrecht, 1987; Venner 1991). Using this technique, Kim and Sadeghi (1991, 1992) further extended the modelling effort and developed numerical models for Newtonian and non-Newtonian thermal elastohydrodynamic lubrication of point contacts. The solution of time-dependent EHL problems then became a focus of numerous researchers.

Osborn and Sadeghi (1992) investigated the effects of a bump or dent on time-dependent EHL line contacts, and Venner and Lubrecht (1994a) studied the effects of a bump moving through the EHL circular contact.

More recently, another improvement of the numerical solution of EHL problems is in the calculation of the double integral in the film thickness equation for point contacts. By recognising this double integral as the convolution of the pressure and its elastic response, Stanley and Kato (1997) used fast Fourier transform (FFT) to evaluate the elastic deformation. This approach greatly reduced the computational efforts involved in obtaining a solution to elastic deformation from N^2 to $N \ln N$, about the same order of calculations as the multi-level integral method introduced by Brandt and Lubrecht (1990). These methods can be used to study the solid contact problems as well as EHL problems. Figure 6.1 depicts a brief history of various significant developments in EHL theoretical solutions in chronological order. Owing to the advance in computational methods and computer hardware, EHL investigators have been able to extend their efforts and develop models for *mixed EHL*. Mixed EHL is the lubrication condition where parts of the surfaces or surface asperities come in direct contact with each other during the lubrication process. An example of such a scenario resides in the start up condition of a normal EHL operation, where the contact starts as a direct solid-to-solid Hertzian contact, and then changes to a fully lubricated contact as the surfaces move toward the final operating velocity. Zhao and Sadeghi (2001) and Zhao et al. (2001) investigated the mixed EHL condition of smooth surfaces and demonstrated the transition from fully solid contact, to mixed solid–lubricated contact and finally to the fully lubricated contact. They also developed models to describe the surface temperature change during this process.

Parallel to the theoretical developments of EHL problems, there have been significant efforts in experimental investigation of EHL problems as well. Crook (1961) employed the capacitance technique to measure film thickness in EHL contacts. Sibley and Orcutt (1961) used the X-ray transmission technique to estimate lubricant film thickness. However, the most widely used experimental method for EHL studies today is optical interferometry. Gohar and Cameron (1963) provided the first clear interferometry picture of the EHL film. Koye and Winer (1981) used optical interferometry to successfully validate the formulae generated from the numerical analysis by Hamrock and Dowson (1976b). Kaneta et al. (1992) used optical interferometry to investigate the effects of a bump moving through EHL circular contact. Recently, the modified ultra-thin film interferometry was employed by Glovnea and Spikes (2001) to measure film thickness down to a few nanometres. They studied the behaviour of EHL contacts subjected to sudden reduction in speed as it happens during the shutdown stages of EHL operation. In addition to the above techniques to measure film thickness, thin film sensors, vacuum

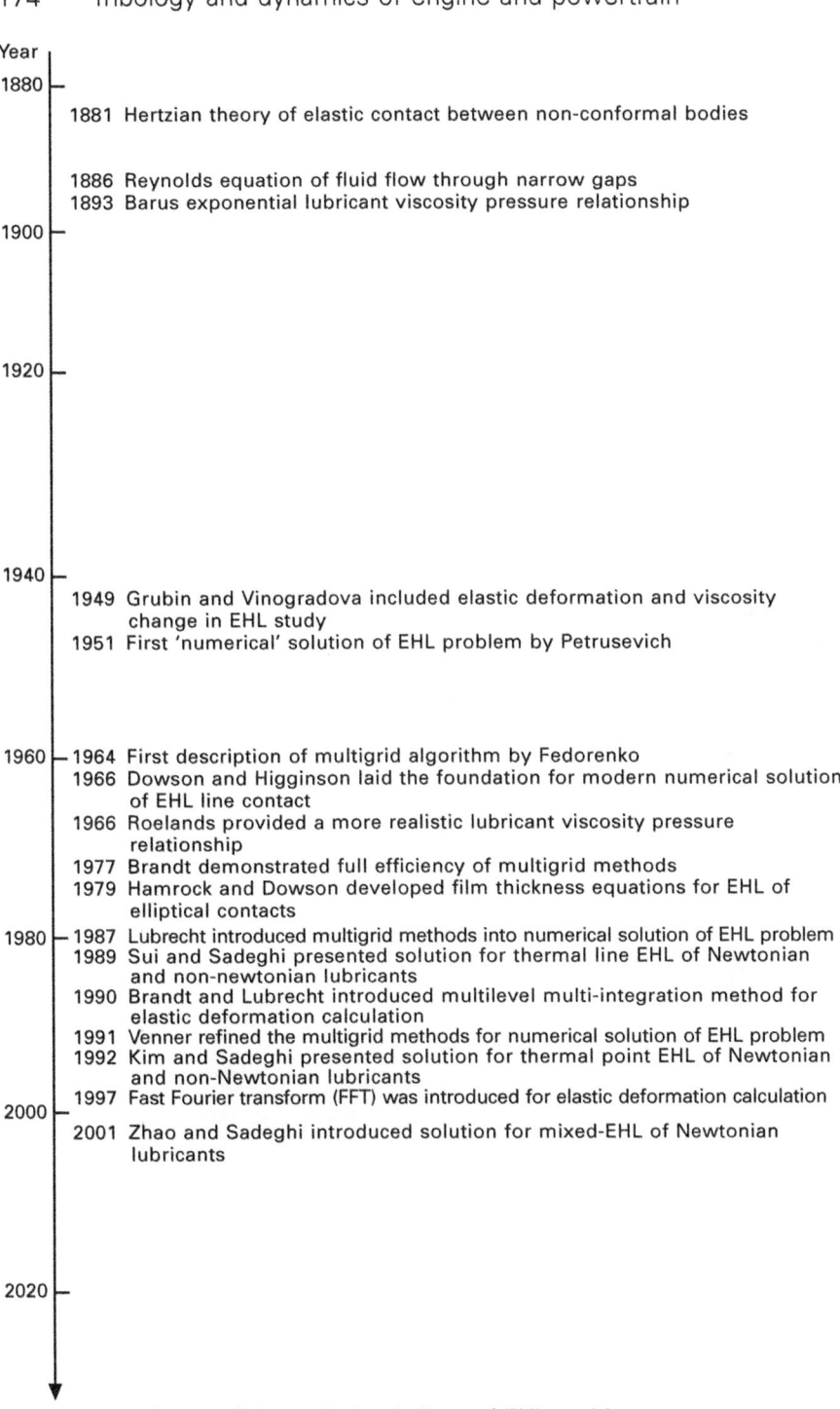

6.1 History of theoretical solutions of EHL problems.

deposited on one of the EHL contact surfaces, were used to measure pressure and temperature (Safa *et al.*, 1982; Johns-Rahnejat and Gohar, 1994; Nickel, 1999). Their pressure results confirmed the occurrence of the second local maxima in the pressure profile, as predicted by the EHL theories.

Previous paragraphs provide a brief history and introduction to EHL studies over the past few decades. More detailed discussion of various aspects of EHL problems is included in the following sections and additional references relevant to EHL are listed at the end of the chapter.

6.2   Conformal and non-conformal contacts

Bearing mating surfaces are generally divided into two categories: conformal and non-conformal contacts. Conformal contacts refer to bearing geometries that have a high degree of conformity (i.e. one surface fits relatively snugly in the other). The radial clearance between the bearing and the journal is about one-thousandth of the journal diameter. Conformal contacts support the applied load over a relatively large area. For example in a journal bearing the load is carried over the length of the bearing and approximately half of the circumference (see Chapters 18–20). The lubrication area for journal bearing is relatively large (2π times the radius of the journal times the length of the bearing) and the pressure generated during the bearing operation is usually less than 5 MPa. The minimum film thickness separating the surfaces is usually a few micrometres and the coefficient of friction is in order of 10^{-2}. Figure 6.2 depicts a journal bearing where the bearing is commonly referred to as a sleeve or bushing. Fatigue is of no concern in journal bearings in normal bearing operating conditions.

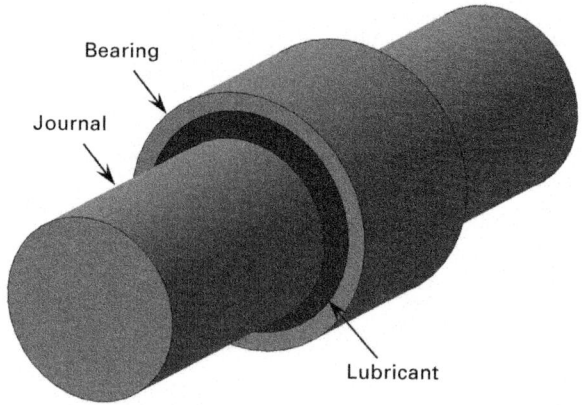

6.2 Journal bearing.

Non-conformal contacts refer to bearing geometries that do not conform to one another (e.g. ball and rolling element bearings, cams and gears). The minimum film thickness separating the surfaces is usually less than or equal to 1 micrometre. Figure 6.3 illustrates a non-conformal contact, here shown as a cylindrical roller bearing. The applied load is supported over a very small lubricated contact patch. This contact patch is usually a few hundred micrometres in length and/or width and increases with applied load. The pressure generated in this small contact patch can range from 0.5 to over 3 GPa with a friction coefficient in order of 10^{-2} to 10^{-1}. Note that a rolling element bearing can be thought of as a journal bearing, where several small journals are added to the original journal (shaft). This adds additional stiffness to the contact and converts the pure sliding motion to a nearly pure rolling motion. Fatigue is usually of concern; however, new theories indicate that if non-conformal contacts are maintained properly; fatigue can be of no concern.

6.3   Regimes of lubrication

A lubricant is a type of substance which can be brought into the contact of loaded rolling/sliding bodies to control friction and wear. Most lubricants are some type of fluid (e.g. mineral and synthetic oils). However, there are some solid lubricants (e.g. gold, silver, polymers). Liquid lubricants can be brought into a converging contact due to rotation and pressure generation between the bodies; they can lower the temperature of interacting surfaces and remove contaminants. Liquid lubricants can be mixed with other

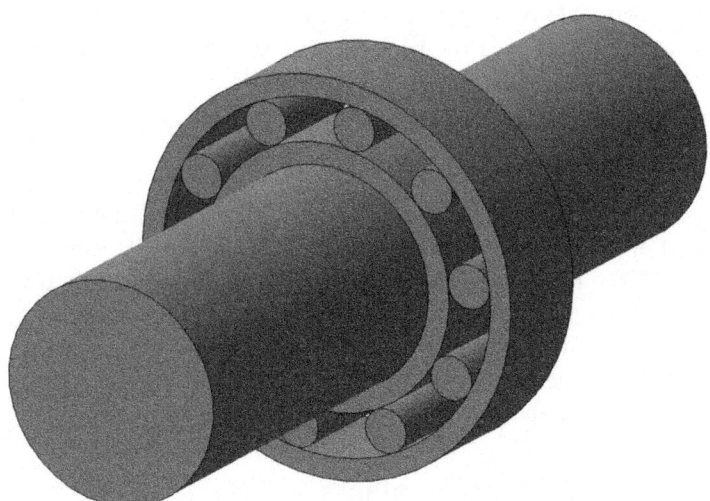

6.3 Rolling element bearing.

chemicals to provide additional properties (i.e. corrosion resistance, surface active layers, etc.).

Four different forms of lubrication can be identified for self-pressure generating lubricated contacts: hydrodynamic, elastohydrodynamic, partial or mixed boundary (see also Chapter 3).

Hydrodynamic or *full film lubrication* is the condition when the load-carrying surfaces are separated by a relatively thick film of lubricant. This is a stable regime of lubrication and metal-to-metal contact does not occur during the steady state operation of the bearing. The lubricant pressure is self-generated by the moving surfaces drawing the lubricant into the wedge formed by the bounding surfaces at a high enough velocity to generate the pressure to completely separate the surfaces and support the applied load (see Chapter 5).

Elastohydrodynamic lubrication is the condition that occurs when a lubricant is introduced between surfaces that are in rolling/sliding contact, such as ball and rolling element bearings, gears, etc. In this lubrication regime, the load is sufficiently high enough for the surfaces to elastically deform during the hydrodynamic action.

Partial or *mixed lubrication regime* deals with the condition when the speed is low, the load is high or the temperature is sufficiently large to significantly reduce lubricant viscosity – when any of these conditions occur, the tallest asperities of the bounding surfaces will protrude through the film and occasionally come into contact.

Boundary lubrication is the condition when the fluid films are negligible and there is considerable asperity contact. The physical and chemical properties of thin surface films are of significant importance while the properties of the bulk fluid lubricant are insignificant. Later in this chapter further description and modelling techniques for EHL and mixed lubrication is presented.

6.4   Elastohydrodynamic lubrication (EHL) minimum film thickness equations

The minimum film thickness separating rolling and sliding surfaces is of significant importance for tribological contacts, because its value is indicative of how well the lubricant is capable of supporting the load and preventing surface to surface interaction. The conditions that influence the minimum film thickness are: the material properties of the contacting bodies, the rheology of the lubricant, the force transmitted through the contact and the surface velocity of the contact.

6.4.1   Line and point contacts

As described in the previous section, ***line contact*** corresponds to conditions when two cylindrical bodies are in contact. In general, rolling element bearings, uncrowned spur and helical gears are considered as line contacts. *However, it is important to note that even cylindrical rolling element bearings usually have a crown; the crown is very large and therefore, the contact is usually considered as a line contact.* Ball bearings and most cam and followers have an elliptical contact and, therefore, their contact is commonly referred to as **point contact**. Note that the term point contact is usually used to describe circular contacts, but in this chapter the term point contact is used to refer to both elliptical and circular contacts.

The EHL line and point contact analysis as mentioned earlier has a long history. Dowson and Ehret (1999) provide a comprehensive review of EHL. In order to obtain a solution to EHL of line and point contacts, the Reynolds and elasticity equations are simultaneously solved subject to boundary conditions while allowing for viscosity and density variation with pressure.

- **Line EHL contact:** The steady state non-dimensionalised one-dimensional Reynolds equation with all appropriate assumptions is given as;

$$\frac{\partial}{\partial X}\left(\frac{\bar{\rho}H^3}{\bar{\eta}}\frac{\partial P}{\partial X}\right) - \lambda\frac{\partial}{\partial X}(\bar{\rho}H) = 0 \tag{6.1}$$

The one-dimensional film thickness equation including the elastic deformation is given as;

$$H(X) = H_0 + \frac{X^2}{2} - \frac{1}{\pi}\int_\Omega P(X)\ln|X - X'|\,dX' \tag{6.2}$$

The viscosity and density variation with pressure is:

$$\bar{\eta}(p) = \exp\{[\ln(\eta_0) + 9.67][(1 + 5.1\times 10^{-9}p)^z - 1]\} \tag{6.3}$$

$$\bar{\rho}(p) = \left(1 + \frac{0.6\times 10^{-9}p}{1 + 1.7\times 10^{-9}p}\right) \tag{6.4}$$

The boundary conditions for one-dimensional EHL are given by:

$$P = \begin{cases} 0, & \text{at } X = X_{\text{inlet}} \\ P = \dfrac{dP}{dX} = 0, & \text{at } X = X_{\text{exit}} \end{cases} \tag{6.5}$$

The load balance equation in non-dimensional form for one-dimensional EHL contact is:

$$\int_\Omega P(X)\,dX = \frac{\pi}{2} \tag{6.6}$$

- *Point EHL (elliptical and circular) contact:* The steady state non-dimensionalised two-dimensional Reynolds equation with appropriate assumptions is given as:

$$\frac{\partial}{\partial X}\left(\frac{\overline{\rho}H^3}{12\overline{\eta}}\frac{\partial P}{\partial X}\right) + \frac{1}{\kappa^2}\frac{\partial}{\partial Y}\left(\frac{\overline{\rho}H^3}{12\overline{\eta}}\frac{\partial P}{\partial X}\right) - \lambda\frac{\partial}{\partial X}(\overline{\rho}H) = 0 \qquad (6.7)$$

The two-dimensional film thickness equation including the elastic deformation is given as:

$$H(X, Y) = H_0 + \frac{X^2}{2} + \frac{\kappa^2 Y^2}{A_r\ 2} + \frac{2R_x P_H \kappa}{\pi E'b}\iint_\Omega \frac{P_l(X',Y')dX'dY'}{\sqrt{(X-X')^2 + (Y-Y')^2}}$$

$$(6.8)$$

The boundary conditions for the two-dimensional EHL are:

$$P = \begin{cases} 0, & \text{at } (X_{\text{inlet}}, Y),(X, Y_{\text{inlet}}),(X, Y_{\text{outlet}}) \\ P = \dfrac{dP}{dX} = 0, & \text{at } X = X_{\text{exit}} \end{cases} \qquad (6.9)$$

The load balance equation in non-dimensional form for two-dimensional EHL is given by:

$$\iint_\Omega P(X, Y)\ dX = W \qquad (6.10)$$

Figure 6.4 depicts the pressure and film thickness results for an EHL line contact. The results demonstrate that as the speed increases the well-known *pressure spike* moves toward the centre of the contact. Figure 6.4 also depicts the restriction in film thickness near the exit location. The pressure spike and/or film thickness restrictions are typical of EHL contacts (also see Chapter 16).

Figure 6.5 illustrates the film thickness, contour of the thickness and pressure for an EHL point (circular) contact. The results demonstrate the usual pressure spike near the end of the contact and the film reduction in the same areas. It is believed that this pressure spike contributes significantly to rolling contact fatigue.

Grubin and Vinogradova (1949) were the first to develop an elegant solution for the minimum film thickness for line contacts. Dowson and Higginson (1959) developed a similar minimum film thickness based on their numerical EHL solution. They computed minimum film thicknesses for a variety of loads, speeds and material parameters using their model and then curve fitted the results to obtain an equation for H_{min}.

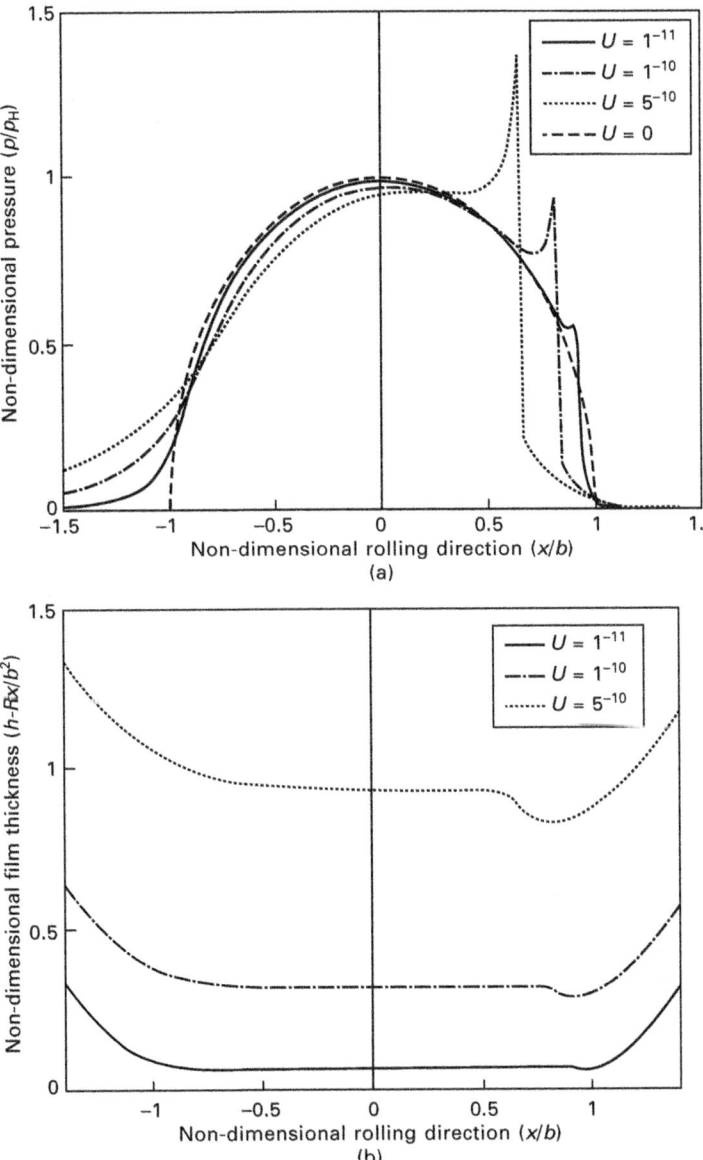

6.4 Pressure and film thickness results for EHL line contact at various speeds (W = 1e–4, G = 3742): (a) pressure; (b) film thickness.

Pan and Hamrock (1989) presented a curve-fit formula for line contact minimum and central film thickness as:

$$\frac{h_{min}}{R_x} = 1.714 \left(\frac{u_m \eta_0}{E' R_x}\right)^{0.694} (\alpha E')^{0.568} \left(\frac{w}{E' R_x}\right)^{-0.128} \qquad (6.11)$$

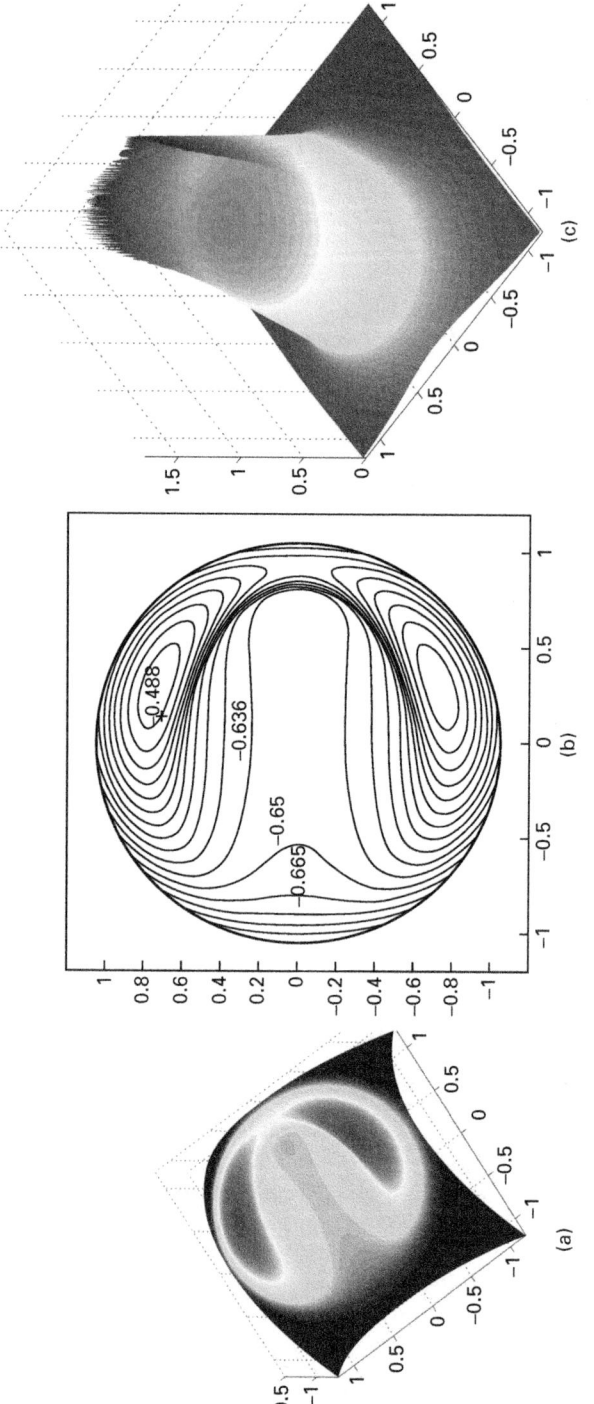

6.5 Analytical film thickness and pressure results at $W = 2.0e{-}5$, $U = 5.0e{-}9$, $G = 2800$: (a) analytical film thickness; (b) analytical film thickness contours; (c) analytical pressure distribution.

$$\frac{h_c}{R_x} = 2.922 \left(\frac{u_m \eta_0}{E'R_x}\right)^{0.692} (\alpha E')^{0.470} \left(\frac{w}{E'R_x}\right)^{-0.166} \qquad (6.12)$$

Hamrock and Dowson (HD) (1977a) developed the minimum and central film thickness for elliptical contacts:

$$\frac{h_{\min}}{R_x} = 3.63 \left(\frac{u_m \eta_0}{E'R_x}\right)^{0.68} (\alpha E')^{0.49} \left(\frac{w}{E'R_x^2}\right)^{-0.073} \left(1 - e^{-0.68\left(\frac{R_y}{R_x}\right)^{2/\pi}}\right) \qquad (6.13)$$

$$\frac{h_c}{R_x} = 2.69 \left(\frac{u_m \eta_0}{E'R_x}\right)^{0.67} (\alpha E')^{0.53} \left(\frac{w}{E'R_x^2}\right)^{-0.067} \left(1 - 0.61 e^{-0.73\left(\frac{R_y}{R_x}\right)^{2/\pi}}\right)$$

$$(6.14)$$

6.5   Experimental film thickness and corroboration with analytical results

The most widely used experimental method in investigating EHL contacts is *optical interferometry*, which reveals the details of EHL film shapes. Optical interferometry also has a rich history, starting with the work of Gohar and Cameron (1966). Koye and Winer (1981), Smeeth and Spikes (1997) and others all used optical interferometry to experimentally determine the minimum film thickness of point contacts and compare it with the results as predicted by the numerical analysis of Hamrock and Dowson (1977b). It was confirmed that the Hamrock and Dowson (1977b) equation predicts the minimum film thickness in EHL contact within the accuracy normally required in mechanical design calculations. To measure the film thickness in EHL contacts, a highly polished steel ball is brought in contact under load with a transparent disk (usually glass) coated in order to enhance and provide colour interference fringes. Plates II–IV (between pages 514 and 515) depict the experimental interference photo-micrograph and the corresponding analytical film thickness and pressure results for various loads, speeds and material parameter combinations. Plates III and IV depict that the pressure distribution exhibits the well-known *pressure spike* near the exit location of the contact and as the speed increases the pressure spike magnitude also increases.

Koye and Winer (1981) showed good correlation between measured and predicted minimum film thicknesses, except for higher loads, where the minimum film thickness measured falls below the predicted values. These results confirmed that Hamrock and Dowson (1977b) equation accurately predicts the lubrication condition in most EHL applications. Some recent studies indicate that at low speeds the Hamrock and Dowson (1977b) equation does not compare well with experimental results. Guangteng *et al.* (2000) illustrated that by applying various coatings to the transparent disk, film thickness down to a few nanometres can be measured.

6.6   Thermal effects in elastohydrodynamic lubrication (EHL) contacts

Newtonian isothermal EHL analyses have been well developed over the last few decades. These analyses predict film thicknesses that closely match the experimental results; however, they fail in predicting friction at the contact accurately. The need to include the thermal as well as ***non-Newtonian fluid*** effects or both has been the subject of research since the mid-1960s. In order to determine the thermal effects in EHL contacts, the thermal Reynolds (Newtonian and/or non-Newtonian), elasticity and energy equations with viscosity and density variation with pressure and temperature and appropriate boundary conditions need to be solved simultaneously. Cheng (1970) was among the first to investigate the effects of temperature on EHL contacts. Sadeghi and Sui (1990, 1991) investigated thermal effects on EHL line contacts using both a Newtonian and non-Newtonian (***Ree–Eyring***) fluid model. They demonstrated that the minimum film thickness in EHL contact is affected by the inclusion of thermal effects. Their results for higher loads and slide to roll ratios up to 30% indicated minimum film thickness reductions of up to 15%. They demonstrated that at elevated pressures and high slide to roll ratios the temperature within the lubricant film is significant and a better representation for friction variation with slide to roll ratio can be achieved when thermal effects are included in the analysis. Kim and Sadeghi (1991) followed the work of Sadeghi and Sui (1990, 1991) and presented results for thermal EHL of circular and elliptical contacts using both Newtonian and non-Newtonian (Ree–Eyring) fluid models. They demonstrated that for point contacts again, film thickness is slightly affected by temperature, although better corroboration between experimental friction measurements and analytical results can be achieved when thermal effects are included in the analysis. There have been a number of others who have developed models for thermal EHL contacts. Lee and Hamrock (1990) also developed a thermal EHL line contact model using a circular model. Gupta *et al.* (1991) presented a thermal correction factor formula for calculating the percentage of film thickness reduction due to inlet heating.

$$C_t = \frac{1 - 13.2\left(\frac{P_H}{E'}\right)(L^*)^{0.42}}{1 + 0.213(1 + 2.23S^{0.83})(L^*)^{0.64}}$$

(6.15)

where:

$$L^* = \left(-\frac{\partial \eta_0}{\partial t_m}\right)\frac{(u_m)^2}{K_f} \tag{6.16}$$

C_t is used to multiply the H_{min} equation and obtain the effects of temperature on film thickness. There have also been some attempts to measure the temperature in EHL contacts. Early measurements included using a thin film transducer deposited on the surface to measure and characterise the temperature variation as a function of load and slide to roll ratio. Using infrared thermometry Nagaraj et al. (1978) measured the surface as well as film temperatures in a circular contact. Measured temperatures were in agreement with those predicted from the Jaeger–Archard formula. Winer (1983) again employed infrared thermometry to measure the temperature between a steel ball sliding against a sapphire disk. He demonstrated a significant temperature rise in the contact. The measured temperature trends are similar to those reported by Sadeghi and Sui (1990, 1991) and Kim and Sadeghi (1992).

6.7   Non-newtonian fluid model

In general for EHL lubrication analysis, the lubricant is assumed to behave in a Newtonian manner. As mentioned earlier, previous investigations have shown that the Newtonian fluid model provides satisfactory solution in terms of fluid film thickness. However, the Newtonian fluid model overestimates the friction generated in EHL contacts. The discrepancy in friction measurements has been attributed to the Newtonian fluid model assumptions. For a Newtonian fluid shear stress varies linearly with shear strain. The **Ree–Eyring** fluid model has been used extensively to incorporate the effect of non-Newtonian fluids in EHL contacts. The Ree–Eyring fluid model is based on the activation energy concept and thus provides a physical understanding for decrease in viscosity with shear rate and change in viscosity with temperature. However, the Ree–Eyring fluid model cannot capture the fact that the shear stress is bounded by a critical value. Bair and Winer (1979) proposed a viscoelastic lubricant rheology model which limited the shear stress as the shear rate increased. Lee and Hamrock (1990) used a circular non-Newtonian fluid model to investigate thermal EHL of line contacts. Kim and Sadeghi (1991) used the Ree-Eyring fluid rheology model to investigate EHL of point contacts. Their results indicate that the pressure distribution maintains its main attribute as the Newtonian fluid model; however, the pressure spikes are reduced and the film thickness is essentially the same as that of the Newtonian fluid model.

6.8   Boundary lubrication

In boundary lubrication, the rolling and sliding bodies are separated only by a few layers of molecules and the film varies in thickness from 1 to a few tens of nanometres (see also Chapter 3). Boundary lubrication exists when a fluid film cannot be formed or sustained due to heavy loads, low running speeds, high surface roughness, and/or simply a lack of lubricant supply. The frictional characteristics of boundary lubrication depend on the rheological behaviour of the lubricant and the interaction (both physical and chemical) between the molecular surface film and the solids (see Chapter 3). The coefficient of friction for sliding boundary lubrication ranges from 0.05 to 0.15 with an average of 0.1, which is higher than that of elastohydrodynamic lubrication, although significantly smaller than that of the unlubricated contacts. It is also to be noted that rolling element bearings experience boundary lubrication; however, due to rolling condition, the coefficient of friction will most likely be less than the average of 0.1 experienced in sliding contacts. The wear rate for boundary lubrication is considerably higher than that of negligible values for hydrodynamic and elastohydrodynamic lubrications. However, it is much smaller than the wear rate for unlubricated contacts, for which scoring or seizure can quickly occur, leading to catastrophic failures. The different mechanisms for generating low-friction protective films on the contacting surfaces are discussed in detail by Stachowiak and Batchelor (2005). Typical applications of boundary lubrication can be found in low-cost, low-speed conditions such as door hinge contact, where as in other situations, the boundary lubrication provides the remedy if the normal lubrication (hydrodynamic or elastohydrodynamic) breaks down.

The analytical study of boundary lubrication has focused on the behaviour of thin surface film and numerical simulations to describe the phenomena using the molecular dynamics (MD) approach. Experimental studies of boundary lubrication involve using the surface or atomic force apparatus to measure the forces at the contact interface in controlled environments. A more detailed discussion of these topics can be found in Persson (2000) and Gohar and Rahnejat (2008).

6.9   Mixed elastohydrodynamic lubrication (EHL)

There are situations where both boundary lubrication and full film lubrication play an important role in the overall lubrication of the contacting bodies. The lubrication regime between boundary and elastohydrodynamic lubrication is termed as *partial EHL* or mixed EHL. Partial EHL deals with the simultaneously occurring solid to solid contact and elastohydrodynamic lubricated contact. It is generally believed that if the average film thickness is less than three times the composite surface roughness, the surface asperities will force direct solid to solid contact, resulting in mixed EHL. A dimensionless film parameter Λ is defined to distinguish the lubrication regimes possible for rough surfaces lubrication:

$$\Lambda = \frac{h_{\min}}{\sqrt{R_{q,a}^2 + R_{q,b}^2}} \tag{6.17}$$

where $h_{\min}$ is the minimum lubricant film thickness, $R_{q,a}$ and $R_{q,b}$ are the root mean square surface finish of contacting bodies respectively. The lubrication regimes are characterised as (see also Chapter 3):

$$\begin{cases} 5 \leq \Lambda < 100 & \text{hydrodynamic lubrication} \\ 3 \leq \Lambda < 10 & \text{elastohydrodynamic lubrication} \\ 1 \leq \Lambda < 5 & \text{partial or mixed lubrication} \\ \Lambda < 1 & \text{boundary lubrication} \end{cases}$$

The average film thickness in a partial lubrication is between 0.01 and 0.1 μm. Because of the partial load support by the bulk lubricant film, the coefficient of friction of mixed lubrication is usually less than 0.1.

Previous studies on rough surface EHL mainly focused on the contact with continuous EHL flow. Cheng and Dyson (1978) and Patir and Cheng (1978) investigated stochastically the effects of surface roughness in EHL contacts. Patir and Cheng (1978) modified the Reynolds equation with flow factor parameters along and across the rolling directions to handle the effects of surface roughness of any arbitrary surface pattern. Their approach served as the foundation for EHL of rough surfaces until more robust models for EHL were developed. Sadeghi (1991) investigated the effects of fluid models and surface roughness on EHL of rough surfaces. Greenwood and Morales-Espejel (1994) and Morales-Espejel et al. (1996) investigated the effects of transverse roughness and kinematics of roughness in EHL. Venner and Lubrecht (1994a) investigated the influence of a transverse ridge on the film thickness in a circular EHL contact under rolling/sliding conditions, assuming Newtonian, isothermal lubricant. Xu and Sadeghi (1996) investigated the EHL of circular contacts with measured surface roughness and thermal effects (see an example for big-end bearings in Chapter 20). These studies demonstrated that surface features generally affect the lubricant film thickness and pressure distributions as compared to the smooth surface results. Experimentally, optical interferometry has been used by Kaneta et al. (1992, and Kaneta (1993) to demonstrate the effects of bumps on EHL circular contacts. Under low surface velocity conditions, their results showed zero film thickness for a series of bumps, which suggested that direct solid to solid contact, occurs in EHL with non-smooth surfaces.

The developments of numerical simulation techniques together with the advancements of computer hardware enabled tribology researchers to investigate more complicated lubrication problems including mixed EHL. The attention therefore shifted from the steady state, smooth surface, isothermal and Newtonian EHL solutions to more realistic transient, rough surface, thermal and non-Newtonian EHL solutions. The challenge of dealing with the direct solid to solid contact in an elastohydrodynamic contact and its deterministic numerical solution was only met in the past decade. To study such an EHL problem when surface features protrude through the lubricant film, a mixed contact model is needed. Jiang *et al.* (1999) presented a deterministic mixed contact model and investigated solid to solid contact of surface asperities or features as they moved through an EHL contact region. In their models, the solid contact pressure was calculated using inverse FFT (fast Fourier transform) method. More recently, Wang *et al.* (2004) used a macro-micro model to study mixed EHL, which superimposed off-line asperity contact pressure calculation on to the elastohydrodynamic pressure determination from the average flow model developed by Patir and Cheng (1978). Hu *et al.* (1999) developed a deterministic mixed EHL model (see also Chapter 15).

Plate V (between pages 514 and 515) depicts the pressure and film thickness distribution for a *mixed EHL* contact. The pressure distribution along the centre of contact in the rolling direction is also shown. The results show that the pressure undergoes large fluctuations where there is *surface asperity contact*. The orthogonal shear stress along the centre line is also shown to demonstrate the effect of surface asperities on internal stresses.

For lower average film thickness ($h_a/R_q<1$) due to more severe surface asperity interaction, the solution for the asperity dry contact is suggested to provide the upper boundary for the pressure distribution. Recent developments in mixed EHL lubrication include the determination of flash temperatures and the effects of surface roughness on *fatigue life* predictions (Deolalikar and Sadeghi, 2007; Deolalikar *et al.* 2008). Their results, as shown in Fig. 6.6 for a pure sliding condition, indicated the life reduction due to the surface roughness. They show that when the equivalent rms of contacting surfaces is below 0.2 the relative life increases and therefore, surface roughness has little influence on fatigue of machine elements. In a more general sense, mixed EHL contact could also occur for smooth surfaces. Hu *et al.* (1999) and Hu and Zhu (2000) presented full numerical solutions to the mixed contact problems and studied smooth surface EHL with low rolling speeds, which enables lubricated contact and solid contact to happen at the same time. Mixed EHL also occurs during the start up condition of the normal EHL operations. During the start up condition, the contact starts as a direct solid-to-solid Hertzian contact as the surfaces are at rest under the applied load. As the surfaces start to move, the lubricant film in the inlet starts to

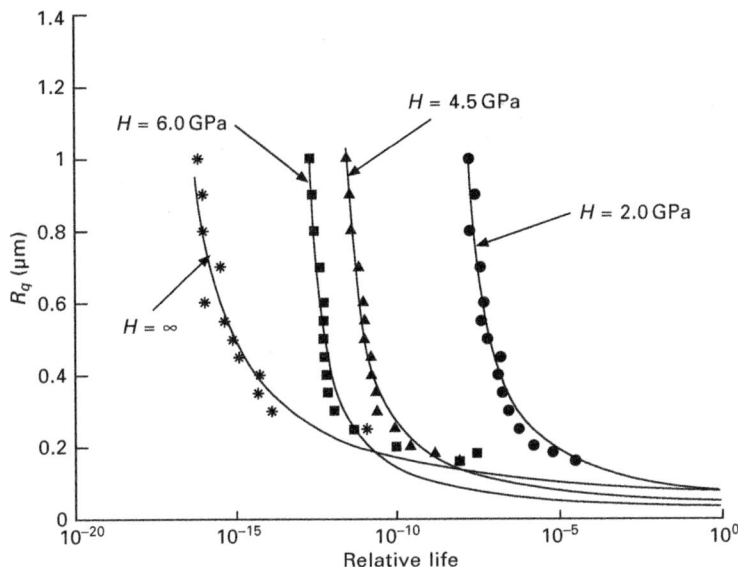

6.6 Variation of relative life for rough surfaces with different RMS values under pure sliding conditions.

build up through the contact area, breaking apart the two contacting surfaces. Finally a full EHL lubricant film completely separates the contacting surfaces. This phenomenon was investigated by Zhao *et al.* (2001) using a mixed lubrication contact model that uses the multi-grid multi-level method to solve the lubricated contact and a minimisation of complementary energy approach to solve the solid contact. The FFT method was incorporated to increase the efficiency of computing the film thickness. Holmes (2002) also used a numerical model using a new coupled differential deflection method to study the start-up condition. Glovnea and Spikes (2001) published their experimental results of steel ball on glass disc contact under start up conditions with controlled acceleration. The numerical and experimental results demonstrate that for *hard EHL* ($P_H > 0.5$ GPa), the lubricant film propagates through the contact area with the mean surface velocity. It is consistent with the analysis of the Reynolds equation that for heavily loaded contacts, the high pressure causes the pressure terms in the Reynolds equation to vanish and it simply becomes:

$$u_m \frac{\partial \rho h}{\partial x} + \frac{\partial \rho h}{\partial t} = 0 \qquad (6.18)$$

For most lubricants, the effect of compressibility is minimal and can be neglected. Thus the solution to the above equation is: $h \cong h(x - u_m t)$, which demonstrates that the film profile moves to the right with a velocity of u_m.

The overall film thickness propagation for the start up process is shown in Plate VI (between pages 514 and 515) for the experimental results (Glovenea and Spikes, 2001), together with the analytical and numerical results. Figure 6.7 depicts the film thickness and pressure distributions at the beginning, an intermediate and the final instants during the start up process. The solid and lubricated contact regions can be clearly visualised during the EHL start up process.

The start up time, defined as the time in which both the solid contact and the lubricated contact are simultaneously occurring, can simply be determined for a linear acceleration (constant acceleration rate) start up as:

$$t_s = \begin{cases} \sqrt{\dfrac{4b}{a}}, & \text{if } u_M \geq \sqrt{4ab} \\[2mm] \dfrac{2b}{u_M} + \dfrac{u_M}{2a}, & \text{if } u_M < \sqrt{4ab} \end{cases} \tag{6.19}$$

This parameter can be used in estimating how long the EHL contact is in

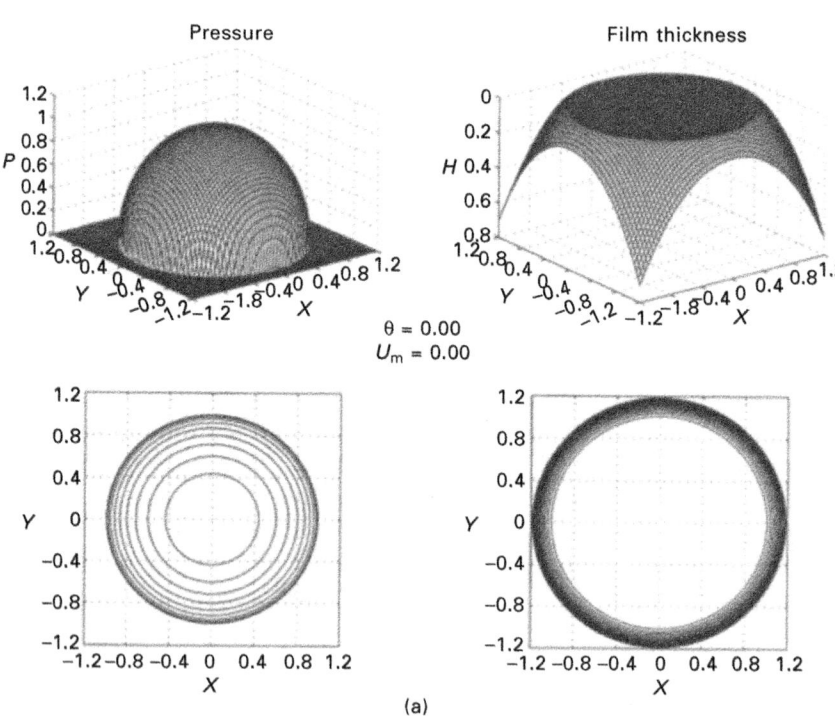

6.7 Lubricant film thickness and pressure distributions during start up process of EHL contact with maximum Hertzian pressure of 2.2 GPa and mean velocity of 11.54 m/s: (a) initial; (b) intermediate; (c) final.

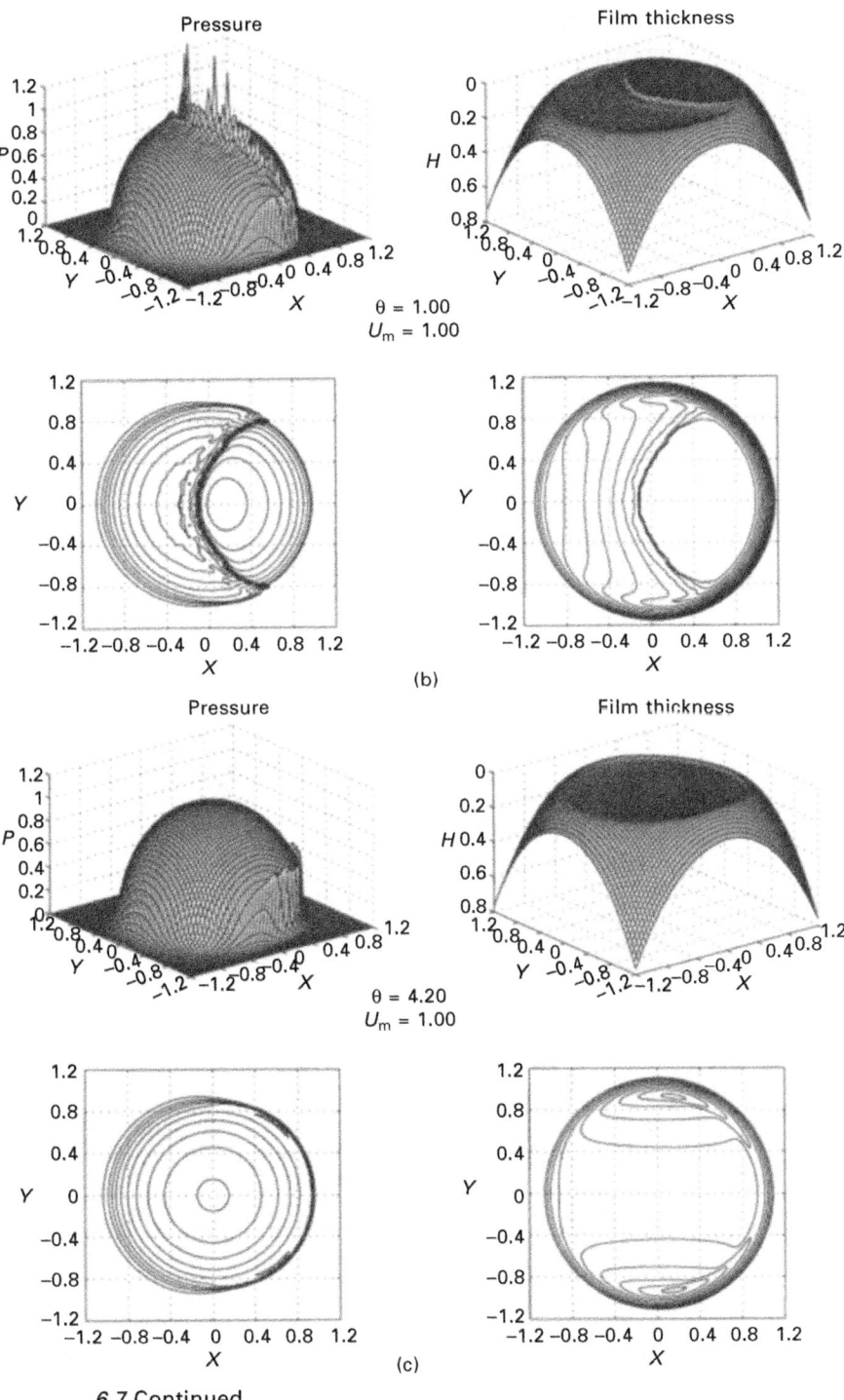

6.7 Continued

the critical condition as the friction in direct solid to solid contact is larger than the fully lubricated EHL contact. The *flash temperature* rise due to the friction during this start up process for dry contact surfaces (without a protective boundary film) was studied by Zhao and Sadegh: (2001). Their results indicated that the surfaces can experience a temperature rise of nearly 600 K for a slide to roll ratio of 0.2. As described in this section, mixed EHL study up to this point focused on developing models and generating solutions for individual EHL running conditions with specific surface roughness profiles. The results are more application specific, rather than being generally applicable to machine component design process. It is believed that, as more and more mixed EHL solutions are developed, general empirical relationships (similar to the ones developed by Hamrock and Dowson, 1976b, for isothermal smooth EHL) will be developed in the near future for the film thickness and pressure in mixed EHL as the functions of the running conditions and the characteristic parameters of the surface roughness.

6.10   Surface roughness

As described in the previous sections, surface roughness plays an important role in satisfactory operation of EHL of rolling/sliding contacts. Although machined surfaces appear smooth to the naked eye, they are quite rough at the microscopic levels. In the past few decades a number of different devices have been designed and developed in order to characterise the surface features of machine components. In the past decade, non-contacting (optical) surface *profilometry* has become the most widely used approach for characterising various surface features of machine components (see Chapter 2). Machine elements which have been prepared to operate in EHL regimes are usually quite smooth, with roughness being of the main concern. The following terms are usually used to describe a surface which has been prepared by machining, grinding or other processes. *Roughness* is fine irregularities that are produced during a machining process (grinding, polishing, etc.). *Waviness* is the result of unwanted vibration, runout, deflection, tool wear, misalignment, etc. during the manufacturing process which results in widely spaced surface profiles. *Error of form* is the result of errors in manufacturing process and results in deviations from the desired surface patterns.

A number of standard parameters are used to describe surface roughness of engineered surfaces. *Average roughness* (R_a) is the average of the individual heights (asperities) and depths from the arithmetic mean elevation of the profile.

$$R_a = \frac{1}{mn} \sum_{k=0}^{m-1} \sum_{l=0}^{n-1} |Z(x_k, y_l) - \mu|$$

(6.20)

where

$$\mu = \frac{1}{mn} \sum_{k=0}^{m-1} \sum_{l=0}^{n-1} Z(x_k, y_l) \tag{6.21}$$

Root mean square roughness (R_q) is the square root of the sum of the squares of the individual heights and depths from the mean line.

$$R_q = \left\{ \frac{1}{mn} \sum_{k=0}^{m-1} \sum_{l=0}^{n-1} [Z(x_k, y_l) - \mu]^2 \right\}^{0.5} \tag{6.22}$$

Skewness (R_{sk}) is a measure of the average of the first derivative of the surface (the departure of the surface from symmetry). A negative value of R_{sk} indicates that the surface is made up of valleys, whereas a surface with a positive skewness is said to contain mainly peaks and asperities. Therefore a negatively skewed surface is good for lubrication purposes.

$$R_{sk} = \frac{1}{mnR_q^3} \sum_{k=0}^{m-1} \sum_{l=0}^{n-1} [Z(x_k, y_l) - \mu]^3 \tag{6.23}$$

Kurtosis (R_{ku}) is a measure of sharpness of profile peaks.

$$R_{ku} = \frac{1}{mnR_q^4} \sum_{k=0}^{m-1} \sum_{l=0}^{n-1} [Z(x_k, y_l) - \mu]^4 \tag{6.24}$$

In order to determine that the lubrication condition has moved from the full film lubrication to mixed and boundary lubrication the minimum film thickness is divided by the square root of squares of R_q of the surfaces in contact.

Figure 6.8 illustrates an example of a ground surface profile obtained from an optical surface profilometer.

6.11   Contact and internal stresses

Load supported over a microscopic area of contact results in high stresses. This occurs when force is transmitted through bodies in contact. Two stationary ellipsoidal bodies in contact under a static load will generate an elliptical area of contact between them. Because there is no motion between the bodies, there is no shear force and therefore no shear stress is created and/or transmitted between them. From elementary mechanics of materials, planes which do not contain shear stress are planes of maximum normal stresses and, therefore, the stresses acting on these surfaces are principal stresses.

Figure 6.9 depicts the contact under a static load between two ellipsoids. The ellipsoids have two radii of curvature at the point of contact. R_x and R_y are the radii of curvature for the ellipsoid at the point of contact, or

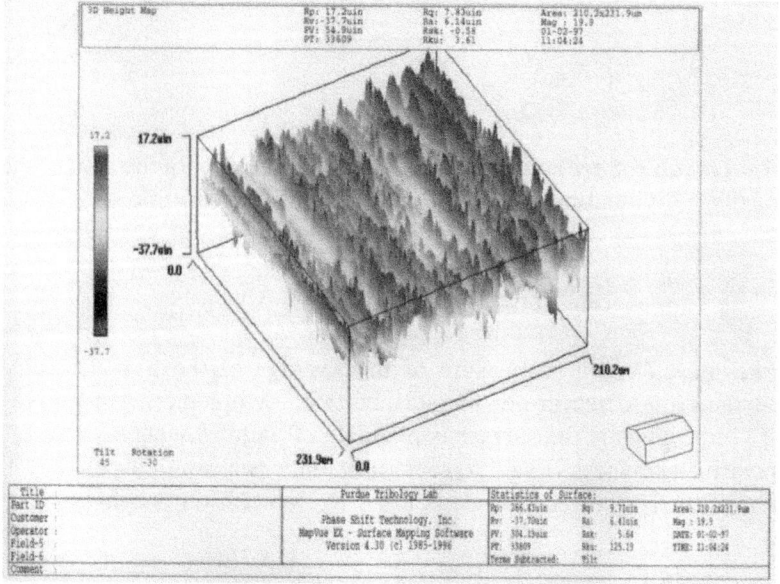

6.8 Surface profile of a ground surface.

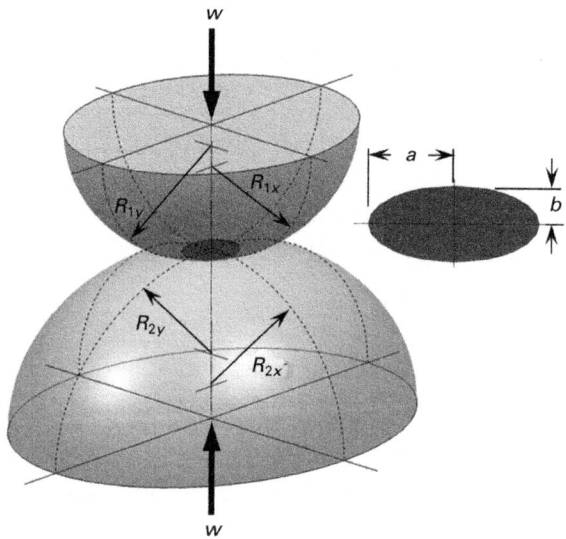

6.9 Two ellipsoids in contact.

the *equivalent radii of curvature* of two ellipsoids in contact (see also Chapter 3).

$$\frac{1}{R_x} = \frac{1}{R_{1x}} + \frac{1}{R_{2x}}$$ (6.25)

$$\frac{1}{R_y} = \frac{1}{R_{1y}} + \frac{1}{R_{2y}} \qquad (6.26)$$

Here, R_{1x}, R_{2x}, R_{1y} and R_{2y} are the radii of curvature of ellipsoids in the x and y directions in the vicinity of the contact for bodies 1 and 2, respectively. Please note that the convention to choose the co-ordinates x and y is such that $1/R_x \geq 1/R_y$. It is important to also note that the sign convention for the radii of curvature is such that it is positive if the corresponding centre of curvature is inside the body. If the centre of curvature is outside the body, the curvature is negative.

For static conditions the contacting bodies are only subjected to normal stresses and, therefore, the normal stresses generated between the bodies are principal stresses and the planes containing them are the principal planes. Thus, the planes of maximum shear stresses occur on planes which are 45° from the principal planes. Therefore, τ_{max} is given by:

$$t_{\text{max}} = \frac{\sigma_{\text{max}} - \sigma_{\text{min}}}{2} \qquad (6.27)$$

Under conditions when the load is sufficiently large the contact can experience plastic deformation. It is to be noted that τ_{max} magnitude varies within the contacting bodies and achieves its maximum value at some depth below the surface.

In general, however, as in the case of heavily loaded lubricated contacts the contact is subject to surface shear stresses due to friction, and therefore, a complete solution to elasticity problem is required in order to assess the state of internal stresses and maximum shear stresses.

In the next few paragraphs, the contact pressure and dimensions for **Hertzian contact** for various geometries (ellipsoidal, spherical and cylindrical) in contact are presented (see also Chapter 3).

General bodies in contact

When two bodies of general shape (Fig. 6.9) are brought together under static load, they elastically deform and generate a microscopic elliptical area of contact between them, where the load is supported.

The formulation and solution to this problem was obtained by Hertz (1882). The derivation of this theory is beyond the scope of this chapter and can be found in many texts. However, the equation describing the contact stresses and the microscopic area of contact generated between the bodies is critical for complete understanding of lubrication of heavily loaded lubricated contacts and therefore, they are repeated here. In this case, the contact area is an ellipse with semi-major and semi-minor axes of a and b, and the pressure over this area is given by:

$$P = P_{max}\left(1 - \frac{x^2}{a^2} - \frac{y^2}{b^2}\right)^{0.5} \tag{6.28}$$

$$P_{max} = \frac{3w}{2\pi ab} \tag{6.29}$$

where P_{max} is the maximum Hertzian pressure and w is the load applied to the contact. In order to determine the major and minor axes of the ellipse, let:

$$A + B = 0.5\left(\frac{1}{R_{1x}} + \frac{1}{R_{1y}} + \frac{1}{R_{2x}} + \frac{1}{R_{2y}}\right) \tag{6.30}$$

$$B - A = 0.5\left[\left(\frac{1}{R_{1x}} - \frac{1}{R_{1y}}\right)^2 + \left(\frac{1}{R_{2x}} - \frac{1}{R_{2y}}\right)^2\right.$$

$$\left. + 2\left(\frac{1}{R_{1x}} - \frac{1}{R_{1y}}\right)\left(\frac{1}{R_{2x}} - \frac{1}{R_{2y}}\right)\cos2\psi\right]^{0.5} \tag{6.31}$$

$$\frac{1}{E'} = \frac{1}{2}\cdot\left(\frac{1 - v_1^2}{E_1} + \frac{1 - v_2^2}{E_2}\right) \tag{6.32}$$

where ψ is the angle between the planes containing the minimum radii of curvature of body 1 and body 2. The major and minor axes of the ellipse are given by:

$$a = c_1\left[\frac{3w}{2E'(A + B)}\right]^{1/3}, b = c_2\left[\frac{3w}{2E'(A + B)}\right]^{1/3} \tag{6.33}$$

Using:

$$\cos\theta = \frac{B - A}{A + B} \tag{6.34}$$

values of c_1 and c_2 for various values of θ are given in Table 6.1. As the ratio of a/b increases, one would obtain narrower and narrower contact ellipses and at the limit, the condition of two cylinders in contact is reached.

Figure 6.10 depicts an ellipsoid inside another (ball in contact with outer race). In this case, the contact area is an ellipse. The equivalent radii of curvature are calculated as:

$$\frac{1}{R_x} = \frac{1}{R_{1x}} - \frac{1}{R_{2x}} \tag{6.35}$$

and

Table 6.1 Values of c_1 and c_2 used in equation (6.33)

θ(degrees)	c_1	c_2	θ(degrees)	c_1	c_2
0	∞	0	35	2.397	0.530
0.5	61.40	0.1018	40	2.136	0.567
1	36.80	0.1314	45	1.926	0.604
1.5	27.48	0.1522	50	1.754	0.641
2	22.26	0.1691	55	1.611	0.678
3	16.50	0.1964	60	1.486	0.717
4	13.31	0.2188	65	1.378	0.759
6	9.790	0.2552	70	1.284	0.802
8	7.860	0.2850	75	1.202	0.846
10	6.604	0.3112	80	1.128	0.893
20	3.778	0.4080	85	1.061	0.944
30	2.731	0.4930	90	1.000	1.000

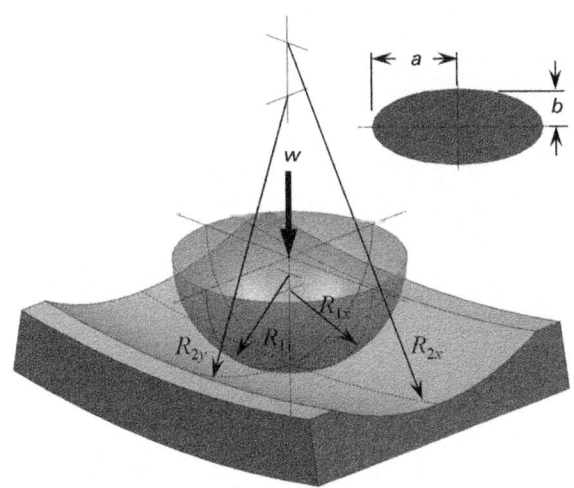

6.10 One ellipsoid inside another.

$$\frac{1}{R_y} = \frac{1}{R_{1y}} - \frac{1}{R_{2y}}$$

(6.36)

Spherical contact

When two spheres (Fig. 6.11) of radius R_1 and R_2 are brought in contact under load of w, a circular contact of radius a is obtained. The radius of contact is given by;

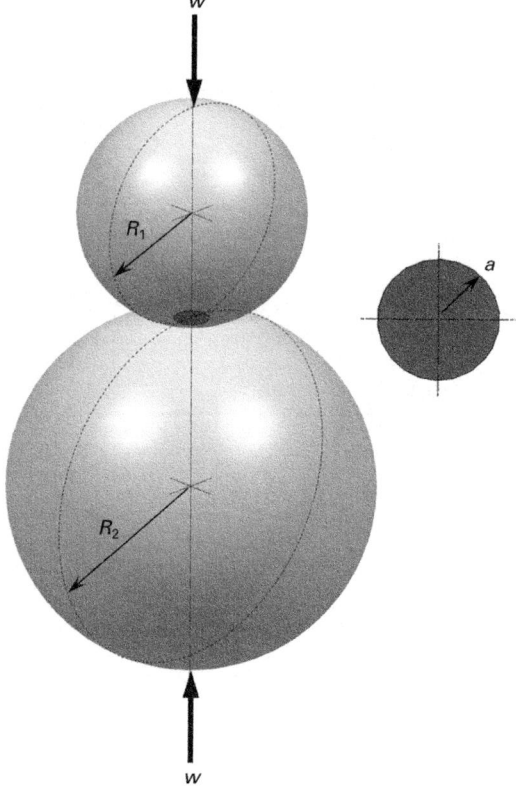

6.11 Two spheres in contact.

$$a = \left[\frac{3w}{2E'\left(\dfrac{1}{R_1} + \dfrac{1}{R_2} \right)} \right]^{1/3} \tag{6.37}$$

The pressure distribution within the contact is hemispherical and given as;

$$P = P_{\max}\left(1 - \frac{x^2}{a^2} - \frac{y^2}{a^2} \right)^{0.5} \tag{6.38}$$

where:

$$P_{\max} = \frac{3w}{2\pi a^2} \tag{6.39}$$

Equations (6.38) and (6.39) are general and note that when they are applied to the condition of a sphere on a plane (Fig. 6.12), the plane radius R is set

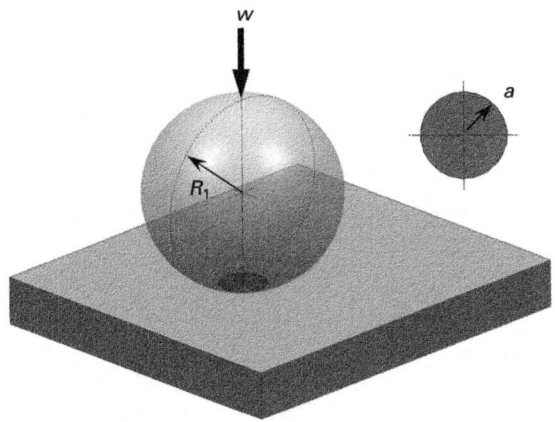

6.12 Sphere on a flat.

to infinity, or to the case of a sphere within another, the radius is expressed as a negative quantity.

The maximum internal stresses within the bodies occur along the z axis and these are the principal stresses given by:

$$\sigma_x = \sigma_y = -P_{max}\left[\left(1 - \left|\frac{z}{a}\right|\tan^{-1}\frac{1}{\left|\frac{z}{a}\right|}\right)(1+v) - \frac{1}{2\left(1+\frac{z^2}{a^2}\right)}\right] \qquad (6.40)$$

$$\sigma_z = \frac{-P_{max}}{\left(1+\frac{z^2}{a^2}\right)} \qquad (6.41)$$

Since there is no shear force, the state of stress is triaxial and therefore, σ_x, σ_y and σ_z are the principal stresses, σ_1, σ_2 and σ_3 respectively. Therefore, the **maximum shear stress** is given by:

$$\tau_{max} = \frac{\sigma_{max} - \sigma_{min}}{2} = \frac{\sigma_1 - \sigma_3}{2} \qquad (6.42)$$

Figure 6.13 depicts the internal stresses along the z axis for $v = 0.3$; τ_{max} achieves its maximum at $z = 0.48a$ below the surface and it is approximately equal to $0.31P_{max}$.

Cylindrical contact

When two cylinders of radius R_1 and R_2 and length l (Fig. 6.14) are brought in contact under a fixed load of w and with their axes parallel, a narrow

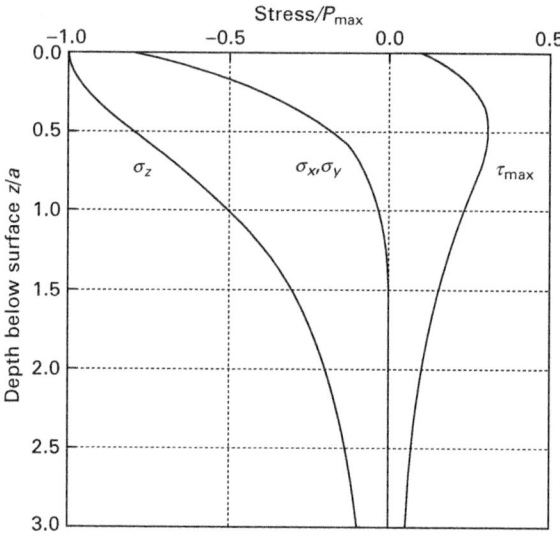

6.13 Internal stresses for two spheres in contact.

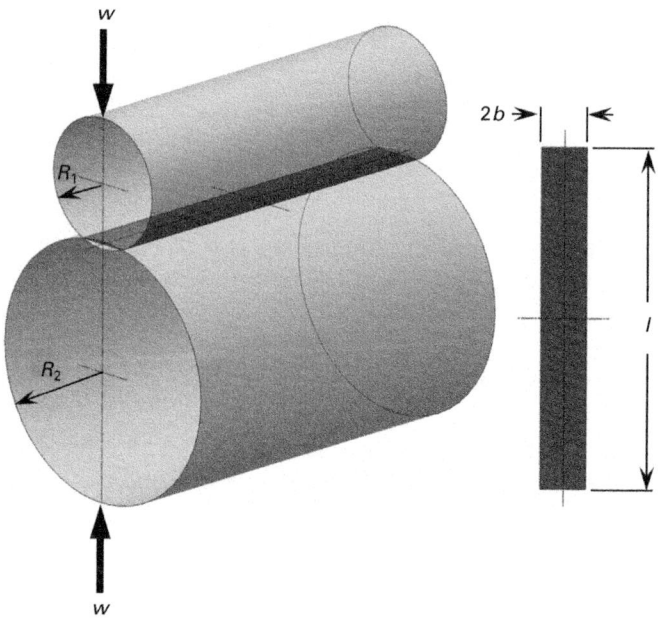

6.14 Two cylinders in contact.

rectangular area of contact with width $2b$ is obtained. The half-width of contact is given by:

$$b = \left[\frac{8w}{\pi l E'\left(\dfrac{1}{R_1} + \dfrac{1}{R_2}\right)} \right] \qquad (6.43)$$

The pressure distribution within the contact is elliptical and is given as:

$$P = P_{max}\left(1 - \frac{x^2}{b^2}\right)^{0.5} \qquad (6.44)$$

where:

$$P_{max} = \frac{2w}{\pi b l} \qquad (6.45)$$

Equations (6.44) and (6.45) are general and note that when they are applied to the condition of a cylinder on a plane, the plane radius R is set to infinity, or to the case of a cylinder within another, the radius is expressed as a negative quantity.

The maximum internal stresses within the bodies occur along the z axis and these are the ***principal stresses*** and given by:

$$\sigma_x = -2vP_{max}\left[\left(1 + \frac{z^2}{b^2}\right)^{0.5} - \left|\frac{z}{b}\right|\right] \qquad (6.46)$$

$$\sigma_y = -P_{max}\left[\frac{1 + 2\dfrac{z^2}{b^2}}{\left(1 + \dfrac{z^2}{b^2}\right)^{0.5}} - 2\left|\frac{z}{b}\right|\right] \qquad (6.47)$$

$$\sigma_z = \frac{-P_{max}}{\left(1 + \dfrac{z^2}{b^2}\right)^{0.5}} \qquad (6.48)$$

In this case due to the absence of shear force, the state of stress in triaxial and therefore, σ_x, σ_y and σ_z are principal stresses, σ_1, σ_2 and σ_3 respectively. Therefore, the maximum shear stress is given by:

$$\tau_{max} = \frac{\sigma_{max} - \sigma_{min}}{2} \qquad (6.49)$$

Figure 6.15 depicts the internal stresses along the z axis for $v = 0.3$; τ_{max}

achieves its maximum at $z = 0.78b$ below the surface and it is approximately equal to $0.3P_{max}$. For the region of $0 < z < 0.436b$, $\tau_{max} = (\sigma_x - \sigma_z)/2$ and for $z > 0.436b$, $\tau_{max} = (\sigma_y - \sigma_z)/2$. It is to be noted that when two cylinders are in contact as shown in Fig. 6.16, the area of contact is a circle, the analysis described for two cylinders does not apply and the general analysis for two ellipsoids in contact should be used.

6.12 Application of elastohydrodynamic lubrication (EHL) theory to machine components

As described earlier, bearings, gears, cam and followers, etc., are machine elements where the load is sufficiently high to elastically deform the mating surfaces during the hydrodynamic action. Contacts between such elements are classified as EHL contacts. The effectiveness of lubrication in these machine elements is governed by calculating the *oil film parameter* lambda (Λ) which is the ratio of the minimum film thickness to the composite root mean square roughness of the surfaces in contact. In this section, examples of film thickness calculation and lambda ratio for a ball and cylindrical roller bearing and a cam and follower are provided.

6.12.1 Ball bearings

Deep groove ball bearings are the most commonly bearings used in industry. In order to determine whether the bearing operates in EHL or mixed and/or

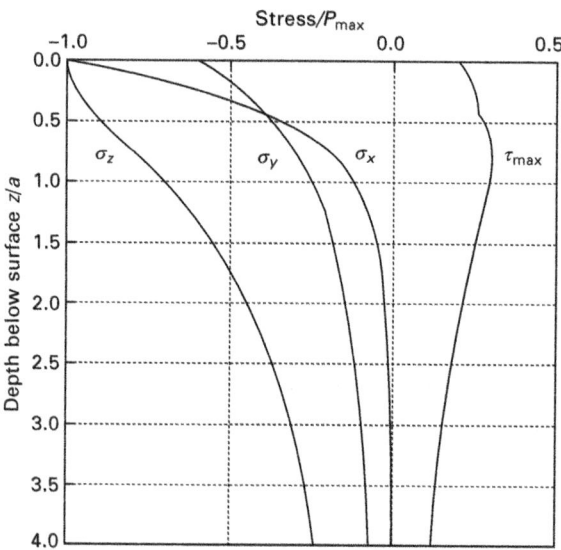

6.15 Internal stresses or two cylinders in contact.

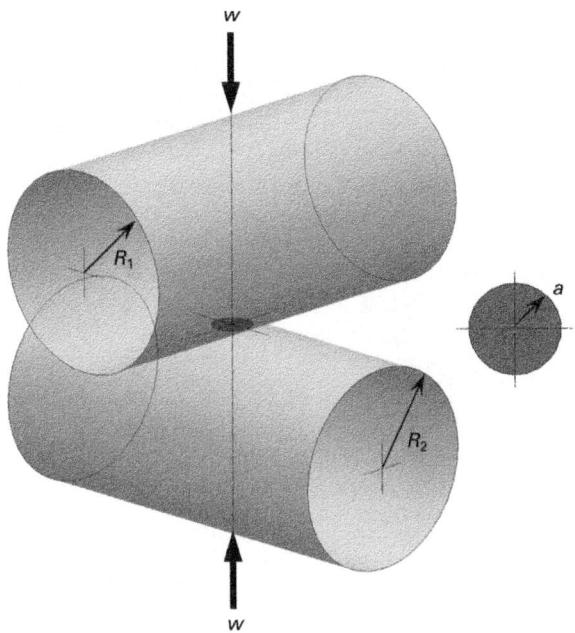

6.16 Two cylinders in contact.

boundary lubrication regimes, one needs to determine the minimum film thickness separating the surfaces (i.e. the ball with inner and outer races). However, it is also of importance to determine the Hertzian contact pressure. Here an approach commonly used is highlighted to determine the contact pressure and minimum film thickness for a deep groove ball bearing.

A radial load of 500 N is applied to a deep groove ball bearing which has 11 balls as shown in Fig. 6.17. The inner race is turning at a rate of 10 000 rpm and the outer race is fixed on the ground. The bearing is lubricated with an SAE 30 oil and operating at 100 °C. The bearing may be assumed to have the standard properties for steel. The root mean square surface roughness of the race and roller are 0.0447 μm and 0.01 μm respectively. From the given information several important parameters can be calculated.

Equivalent modulus of elasticity

$$\frac{1}{E'}\frac{1}{2}\left(\frac{1-v_1^2}{E_1}+\frac{1-v_2^2}{E_2}\right)=\frac{1}{2}\left(\frac{1-0.3^2}{210\,\text{GPa}}+\frac{1-0.3^2}{210\,\text{GPa}}\right)=\frac{1}{231\,\text{GPa}} \quad (6.50)$$

Pitch diameter:

$$d_e=\frac{1}{2}(d_{ir}+d_{or})=\frac{1}{2}(30.66+46.54)=38.6\,\text{mm} \quad (6.51)$$

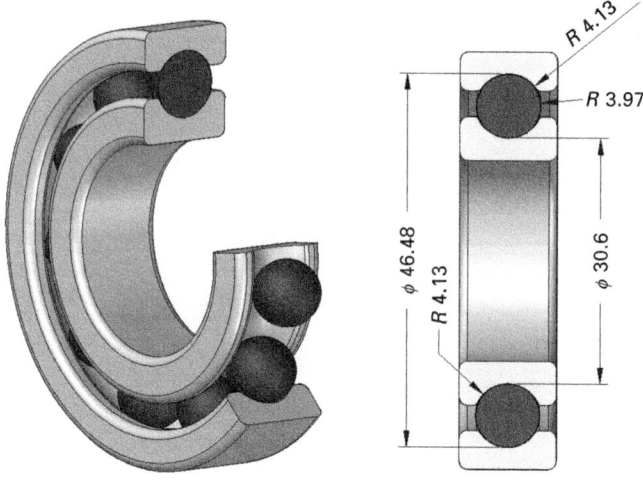

6.17 Dimensions for deep groove ball bearing (dimensions in mm).

Surface velocities assuming pure rolling conditions:

$$u = \frac{|\omega_o - \omega_i|(d_e^2 - d_b^2)}{4d_e} = \frac{1047\,\text{rad/s}\,(38.6^2 - 7.94^2)}{(4)(38.6)} = 9678\,\text{mm/s}$$

(6.52)

Equivalent radii in the rolling direction:

$$\frac{1}{R_{or-b,x}} = \frac{1}{R_{or,x}} + \frac{1}{R_{b,x}} = \frac{1}{-23.3} + \frac{1}{3.97} = 0.209\,\text{mm}^{-1}$$

(6.53)

$$\frac{1}{R_{or-b,y}} = \frac{1}{R_{or,y}} + \frac{1}{R_{b,y}} = \frac{1}{-4.13} + \frac{1}{3.97} = 0.00969\,\text{mm}^{-1}$$

(6.54)

$$\frac{1}{R_{ir-b,x}} = \frac{1}{R_{ir,x}} + \frac{1}{R_{b,x}} = \frac{1}{15.3} + \frac{1}{3.97} = 0.317\,\text{mm}^{-1}$$

(6.55)

$$\frac{1}{R_{ir-b,y}} = \frac{1}{R_{ir,y}} + \frac{1}{R_{b,y}} = \frac{1}{-4.13} + \frac{1}{3.97} = 0.00969\,\text{mm}^{-1}$$

(6.56)

Here, d and ω are the diameter and angular velocity while the subscripts *ir*, *or* and *b* denote inner race, outer race and ball. Using **Stribeck** approximation the force on the ball carrying the maximum load can be obtained:

$$w_{r,\text{max}} = \frac{5}{N_{\text{ball}}}\,w_r = \frac{5}{11}\,500\,\text{N} = 227\,\text{N}$$

(6.57)

In equation (6.57) $w_{r,max}$ is the force on the ball carrying the maximum load, N_{ball} is the number of balls in the bearing, and w_r is the radial load applied on the bearing.

Using equations described in Section 6.11, the maximum Hertzian contact pressure can be determined. Note β in equations below is the angle for angular contact ball bearings. For a deep groove ball bearing β is set to zero, thus:

$$A + B = \frac{1}{2}\left(\frac{1}{R_{or,x}} + \frac{1}{R_{or,y}} + \frac{1}{R_{b,x}\cos\beta} + \frac{1}{R_{b,y}}\right)$$

$$= \frac{1}{2}\left(\frac{1}{-23.3} + \frac{1}{-4.13} + \frac{1}{3.97\cos 0°} + \frac{1}{3.97}\right) = 0.109\,\text{mm}^{-1}$$

$$(6.58)$$

$$B - A = \frac{1}{2}\left[\left(\frac{1}{R_{or,x}} - \frac{1}{R_{or,y}}\right)^2 + \left(\frac{1}{R_{b,x}\cos\beta} - \frac{1}{R_{b,y}}\right)^2\right.$$

$$\left. + 2\left(\frac{1}{R_{or,x}} - \frac{1}{R_{or,y}}\right)\left(\frac{1}{R_{b,x}\cos\beta} - \frac{1}{R_{b,y}}\right)\cos 2\psi\right]^{1/2}$$

$$= \frac{1}{2}\left[\left(\frac{1}{-23.3} - \frac{1}{-4.13}\right)^2\right]^{1/2} = 0.0996\,\text{mm}^{-1} \qquad (6.59)$$

$$\theta = \cos^{-1}\left(\frac{B-A}{A+B}\right) = \cos^{-1}\frac{0.0996}{0.109} = 24.3° \qquad (6.60)$$

Ball-inner race contact:

$$A + B = \frac{1}{2}\left(\frac{1}{R_{ir,x}} + \frac{1}{R_{ir,y}} + \frac{1}{R_{b,x}\cos\beta} + \frac{1}{R_{b,y}}\right)$$

$$= \frac{1}{2}\left(\frac{1}{15.3} + \frac{1}{-4.13} + \frac{1}{3.97\cos 0°} + \frac{1}{3.97}\right) = 0.163\,\text{mm}^{-1} \quad (6.61)$$

$$B - A = \frac{1}{2}\left[\left(\frac{1}{R_{or,x}} - \frac{1}{R_{or,y}}\right)^2 + \left(\frac{1}{R_{b,x}\cos\beta} - \frac{1}{R_{b,y}}\right)^2\right.$$

$$\left. + 2\left(\frac{1}{R_{ir,x}} - \frac{1}{R_{ir,y}}\right)\left(\frac{1}{R_{b,x}\cos\beta} - \frac{1}{R_{b,y}}\right)\cos 2\psi\right]^{1/2}$$

$$= \frac{1}{2}\left[\left(\frac{1}{15.3} - \frac{1}{-4.13}\right)^2\right]^{1/2} = 0.153\,\text{mm}^{-1} \qquad (6.62)$$

$$\theta = \cos^{-1}\left(\frac{B-A}{A+B}\right) = \cos^{-1}\frac{0.153}{0.163} = 19.8° \tag{6.63}$$

Using values of θ from equations (6.60) and (6.63) the values of c_1 and c_2 can be obtained from Table 6.1 for both the inner race-ball and outer race-ball contacts as:

$$c_{1,or} = 3.328$$

$$c_{2,or} = 0.4446$$

$$c_{1,ir} = 3.778$$

$$c_{2,ir} = 0.4080$$

Now all parameters are available to calculate the Hertzian half-width and pressure.

Outer race-ball Hertzian contact:

$$a = c_1\left[\frac{3w}{2E'(A+B)}\right]^{1/3} = 3.328\left[\frac{(3)227\,\mathrm{N}}{(2)2.31 \times 10^{11}\mathrm{Pa} \times 109\,\mathrm{m}^{-1}}\right]^{1/3}$$
$$= 793\,\mu\mathrm{m}$$

$$\tag{6.64}$$

$$b = c_2\left[\frac{3w}{2E'(A+B)}\right]^{1/3} = 0.4446\left[\frac{(3)227\,\mathrm{N}}{(2)2.31 \times 10^{11}\mathrm{Pa} \times 109\,\mathrm{m}^{-1}}\right]^{1/3}$$
$$= 106\,\mu\mathrm{m}$$

$$\tag{6.65}$$

$$P_{\max} = \frac{3w}{2\pi ab} = \frac{3.227\,\mathrm{N}}{(2\pi)(793\,\mu\mathrm{m})(106\,\mu\mathrm{m})} = 1.29\,\mathrm{GPa} \tag{6.66}$$

Inner race-ball Hertzian contact:

$$a = c_1\left[\frac{3w}{2E'(A+B)}\right]^{1/3} = 3.778\left[\frac{(3)227\,\mathrm{N}}{(2)2.31 \times 10^{11}\mathrm{Pa} \times 163\,\mathrm{m}^{-1}}\right]^{1/3} \tag{6.67}$$
$$= 787\,\mu\mathrm{m}$$

$$b = c_2\left[\frac{3w}{2E'(A+B)}\right]^{1/3} = 0.4080\left[\frac{(3)227\,\mathrm{N}}{(2)2.31 \times 10^{11}\mathrm{Pa} \times 163\,\mathrm{m}^{-1}}\right]^{1/3}$$
$$= 85.0\,\mu\mathrm{m}$$

$$\tag{6.68}$$

$$P_{max} = \frac{3w}{2\pi ab} = \frac{3.227\,\text{N}}{(2\pi)787\,\mu\text{m} \times 85.0\,\mu\text{m}} = 1.62\,\text{GPa} \qquad (6.69)$$

The Hamrock–Dowson minimum film thickness equation can now be used to determine the minimum film thickness for the inner and outer race contacts.

Outer race ball minimum film thickness:

$$U = \frac{u\eta_o}{E'R_x} = \frac{9.68\,\text{m/s} \times 0.007\,\text{Pa s}}{2.31\,\text{GPa} \times 4.79\,\text{mm}} = 6.13 \times 10^{-11} \qquad (6.70)$$

$$G = \alpha E' = 1.54 \times 10^{-8}\,Pa^{-1}\,231\,\text{GPa} = 3554 \qquad (6.71)$$

$$W = \frac{w_{r,max}}{E'R_x^2} = \frac{227\,\text{N}}{231\,\text{GPa} \times (4.78\,\text{mm})^2} = 4.30 \times 10^{-5} \qquad (6.72)$$

$$k = \left(\frac{R_{or-b,y}}{R_{or-b,x}}\right)^{2/\pi} = \left(\frac{103}{4.78}\right)^{2/\pi} = 7.07 \qquad (6.73)$$

Note that the value of given in equation (6.73) provides an excellent approximation to the actual value of ellipticity ratio $\kappa = a/b = 793/106 = 7.48$.

$$H_{min} = 3.63U^{0.68}G^{0.49}W^{-0.073}(1 - e^{-0.68k}) = 4.68 \times 10^{-5} \qquad (6.74)$$

$$h_{min} = R_x H_{min} = 4.78\,\text{mm} \times 4.68 \times 10^{-5} = 0.224\,\mu\text{m} \qquad (6.75)$$

$$\Lambda = \frac{h_{min}}{\sqrt{Rq_{or}^2 + Rq_b^2}} = \frac{0.224\,\mu\text{m}}{\sqrt{(0.0447\,\mu\text{m})^2 + (0.01\mu\text{m})^2}} = 4.9 \qquad (6.76)$$

Inner race-ball minimum film thickness:

$$U = \frac{u\eta_o}{E'R_x} = \frac{9.68\,\text{m/s} \times 0.007\,\text{Pa·s}}{2.31\,\text{GPa} \times 3.15\,\text{mm}} = 9.31 \times 10^{-11} \qquad (6.77)$$

$$G = \alpha E' = 1.54 \times 10^{-8}\,\text{Pa}^{-1}\,231\,\text{GPa} = 3554 \qquad (6.78)$$

$$W = \frac{w_{r,max}}{E'R_x^2} = \frac{227\,\text{N}}{231\,\text{GPa}\,(3.15\,\text{mm})^2} = 9.91 \times 10^{-5} \qquad (6.79)$$

$$k = \left(\frac{R_{or-b,y}}{R_{or-b,x}}\right)^{2/\pi} = \left(\frac{103}{3.15}\right)^{2/\pi} = 9.22 \qquad (6.80)$$

$$H_{min} = 3.63U^{0.68}G^{0.49}W^{-0.073}(1 - e^{-0.68k}) = 5.89 \times 10^{-5} \qquad (6.81)$$

$$h_{min} = R_x H_{min} = 3.15\,\text{mm} \times 5.89 \times 10^{-5} = 0.186\,\mu\text{m} \qquad (6.82)$$

$$\Lambda = \frac{h_{min}}{\sqrt{Rq_{ir}^2 + Rq_b^2}} = \frac{0.200\,\mu\text{m}}{\sqrt{(0.0447\,\mu\text{m})^2 + (0.01\,\mu\text{m})^2}} \qquad (6.83)$$

In these equations η_o is the base viscosity, α is the Barus pressure viscosity coefficient, and R_q is the root mean square of surface roughness. Both the inner race and outer race contacts have a sufficient film thickness and are operating in the elastohydrodynamic regime.

6.12.2 Cylindrical bearing

The **cylindrical roller bearing** shown in Fig. 6.18 has 16 rollers and it is operating under a radial load of 1500 N at a speed of 10 000 rpm. It is lubricated with SAE 30 oil and operating at 100 °C. One would like to know the minimum film thickness and maximum contact pressure. The bearing may be assumed to have the standard properties for steel. The root mean square values of the races and roller are 0.15 and 0.06 μm respectively.

In order to achieve the objectives, a similar procedure as described for the ball bearing analysis is used in order to determine the maximum contact pressure and the minimum film thickness.

Equivalent modulus of elasticity:

$$\frac{1}{E'} = \frac{1}{2}\left(\frac{1-v_1^2}{E_1} + \frac{1-v_2^2}{E_2}\right) = \frac{1}{2}\left(\frac{1-0.3^2}{210\,\text{GPa}} + \frac{1-0.3^2}{210\,\text{GPa}}\right) = \frac{1}{231\,\text{GPa}} \qquad (6.84)$$

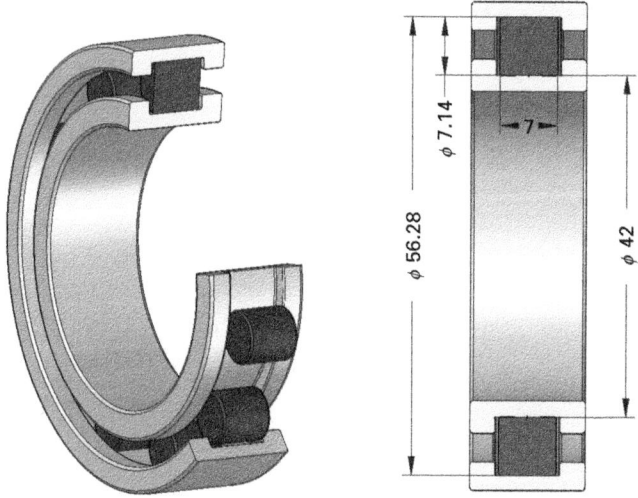

6.18 Dimensions for cylindrical roller bearing (dimensions in mm).

Pitch diameter:

$$d_e = \frac{1}{2}(d_{ir} + d_{or}) = \frac{1}{2}(42.00 + 56.28) = 49.14\,\text{mm} \qquad (6.85)$$

Surface velocities assuming pure rolling contacts:

$$u = \frac{|\omega_o - \omega_i|(d_e^2 - d_r^2)}{4d_e} = \frac{1047\,\text{rad/s}\,(49.14^2 - 7.14^2)}{(4)49.14} \qquad (6.86)$$

$$= 12.593\,\text{mm/s}$$

Equivalent radii in the rolling direction:

$$\frac{1}{R_{or-r,x}} = \frac{1}{R_{or,x}} + \frac{1}{R_{r,x}} = \frac{1}{-28.14} + \frac{1}{3.57} = 0.245\,\text{mm}^{-1} \qquad (6.87)$$

$$\frac{1}{R_{ir-r,x}} = \frac{1}{R_{ir,x}} + \frac{1}{R_{r,x}} = \frac{1}{21} + \frac{1}{3.57} = 0.328\,\text{mm}^{-1} \qquad (6.88)$$

Here, d and ω are the diameter and angular velocity while the subscripts ir, or, and r denote inner race, outer race and roller. In order to determine the maximum load on the roller, the diametral clearance must be calculated.

$$c_D = d_{or} - d_{ir} - 2d_r = 56.28 - 42 - (2)\,7.14 = 0.0\,\text{mm} \qquad (6.89)$$

Because the diametral clearance is zero, the force on the maximally loaded roller may be determined using the following equation:

$$w_{\text{rad,max}} = \frac{4}{N_{\text{roller}}}\,w_{\text{rad}} = \frac{4}{16}\,1500\,\text{N} = 375\,\text{N} \qquad (6.90)$$

$$w_{\text{rad,max-avg}} = \frac{w_{\text{rad,max}}}{l} = \frac{375\,\text{N}}{6.14\,\text{mm}} = 61.1\,\text{N/mm} \qquad (6.91)$$

Here, $w_{\text{rad,max}}$ is the force on the roller carrying the largest load, N_{roller} is the number of rollers in the bearing, and w_{rad} is the radial load on the bearing. Now the size of the Hertzian contacts at the race-roller contacts may be calculated.

Roller-outer race contact:

$$b = \left(\frac{8w_{\text{rad,max}}R_{or-r,x}}{\pi l E'}\right)^{1/2} = \left(\frac{8 \times 375\,\text{N} \times 4.09\,\text{mm}}{\pi \times 6.14\,\text{mm} \times 231\,\text{GPa}}\right)^{1/2} = 52.6\,\mu\text{m} \qquad (6.92)$$

$$P_{\text{max}} = \frac{2w_{\text{rad,max}}}{\pi b l} = \frac{2 \times 375\,\text{N}}{\pi \times 52.6\,\mu\text{m} \times 6.14\,\text{mm}} = 0.74\,\text{GPa} \qquad (6.93)$$

Roller-inner race contact

$$b = \left(\frac{8w_{\text{rad,max}} R_{ir-r,x}}{\pi l E'}\right)^{1/2} = \left(\frac{8 \times 375\,\text{N} \times 3.05\,\text{mm}}{\pi \times 6.14\,\text{mm} \times 231\,\text{GPa}}\right)^{1/2} = 45.3\,\mu\text{m}$$

(6.94)

$$P_{\text{max}} = \frac{2w_{\text{rad,max}}}{\pi b l} = \frac{2 \times 375\,\text{N}}{\pi \times 45.3\,\mu\text{m} \times 6.14\,\text{mm}} = 0.86\,\text{GPa} \qquad (6.95)$$

The HD minimum film thickness equation can be used to determine the minimum film thickness in the raceways.

Outer race-roller minimum film thickness:

$$U = \frac{u\eta_o}{E' R_x} = \frac{12.6\,\text{m/s} \times 0.007\,\text{Pa s}}{2.31\,\text{GPa} \times 4.09\,\text{mm}} = 9.34 \times 10^{-11} \qquad (6.96)$$

$$G = \alpha E' = 1.54 \times 10^{-8}\,\text{Pa}^{-1}\,231\,\text{GPa} = 3554 \qquad (6.97)$$

$$W = \frac{w_{r,\text{max}-\text{avg}}}{E' R_x} = \frac{61.1\,N/\text{mm}}{2.31\,\text{GPa} \times 4.09\,\text{mm}} = 6.47 \times 10^{-5} \qquad (6.98)$$

$$H_{\text{min}} = 1.714 U^{0.694} G^{0.568} W^{-0.128} = 6.71 \times 10^{-5} \qquad (6.99)$$

$$h_{\text{min}} = R_x H_{\text{min}} = 4.09\,\text{mm} \times 6.71 \times 10^{-5} = 0.274\,\mu\text{m} \qquad (6.100)$$

$$\Lambda = \frac{h_{\text{min}}}{\sqrt{Rq_{or}^2 + Rq_r^2}} = \frac{0.274\,\mu\text{m}}{\sqrt{(0.15\,\mu\text{m})^2 + (0.06\,\mu\text{m})^2}} = 1.7 \qquad (6.101)$$

Inner race-roller minimum film thickness:

$$U = \frac{u\eta_o}{E' R_x} = \frac{12.6\,\text{m/s} \times 0.007\,\text{Pa s}}{2.31\,\text{GPa} \times 3.05\,\text{mm}} = 1.25 \times 10^{-10} \qquad (6.102)$$

$$G = \alpha E' = 1.54 \times 10^{-8}\,\text{Pa}^{-1}\,231\,\text{GPa} = 3554 \qquad (6.103)$$

$$W = \frac{w_{r,\text{max}-\text{avg}}}{E' R_x} = \frac{61.1\,N/\text{mm}}{2.31\,\text{GPa} \times 3.05\,\text{mm}} = 8.67 \times 10^{-5} \qquad (6.104)$$

$$H_{\text{min}} = 1.714 U^{0.694} G^{0.568} W^{-0.128} = 7.91 \times 10^{-5} \qquad (6.105)$$

$$h_{\text{min}} = R_x H_{\text{min}} = 3.05\,\text{mm} \times 7.91 \times 10^{-5} = 0.241\,\mu\text{m} \qquad (6.106)$$

$$\Lambda = \frac{h_{\text{min}}}{\sqrt{Rq_{ir}^2 + Rq_r^2}} = \frac{0.241\,\mu\text{m}}{\sqrt{(0.15\,\mu\text{m})^2 + (0.06\,\mu\text{m})^2}} = 1.5 \qquad (6.107)$$

Both the inner race and outer race contacts do not have a sufficient film thickness to be operating in the elastohydrodynamic regime. The regime of lubrication is mixed or partial lubrication.

6.12.3 Cam and follower

Cam and follower mechanisms are used to convert rotational motion into translational motion (for more detailed analysis see Chapters 16 and 17). In contrast to a crank slider mechanism which also converts rotational motion into translational motion, cam and follower mechanism can undergo dwell (i.e. during a part of the cam rotation there is no translation of the follower). There are a number of different cam and follower configurations (cam and flat faced follower, cam and spherical faced follower, cam and roller follower, etc., see Chapter 16). Here the contact pressure, film thickness and lambda ratio calculations for a cam operating with a reciprocating roller follower is provided.

Figure 6.19 illustrates a cam and follower mechanism in contact. Diesel engine manufacturers commonly use the cam to operate the fuel injector systems. In order to atomise the fuel mixture, the cam and follower system is subject to heavy loads and therefore, stresses at the contact are of significant importance.

In order to determine the lubrication condition, the entraining velocity at the point of contact between the cam and follower is needed. In this analysis, the cam is assumed to be operating at 1000 rpm, the lubricant is SAE 30 oil, operating at 100 °C and the normal load at the point of contact is 2500 N. The entraining velocity for this configuration is given by (Matthews *et al.*, 1996):

$$u_E = \frac{u_c + u_f}{2} \tag{6.108}$$

where u_E is the entraining velocity, u_c and u_f are the cam and follower velocities respectively, which can be calculated given the cam profile, follower radius and cam rotational velocity (Mathews *et al.* 1996; Gohar and Rahnejat, 2008; also see Chapters 16 and 17). For the purposes of this example, assume the entraining velocity at the instant shown in Fig. 6.19 is 1.78 m/s.

The equivalent elastic modulus and equivalent radius in the rolling direction can be calculated as well.

$$\frac{1}{E'} = \frac{1}{2}\left(\frac{1-v_1^2}{E_1} + \frac{1-v_2^2}{E_2}\right) = \frac{1}{2}\left(\frac{1-0.3^2}{210\,\text{GPa}} + \frac{1-0.3^2}{210\,\text{GPa}}\right) = \frac{1}{231\,\text{GPa}}$$

$$\tag{6.109}$$

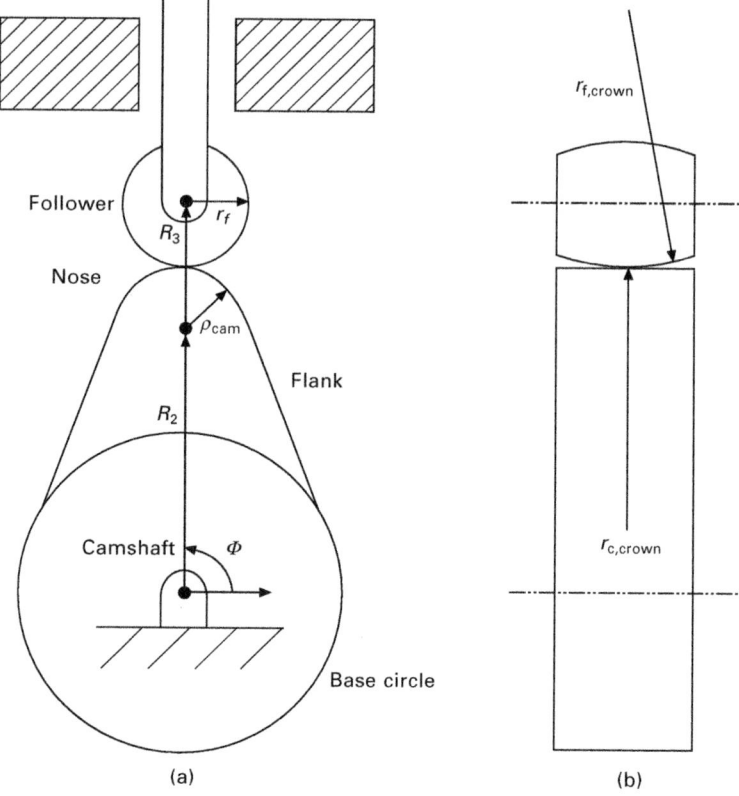

6.19 Cam and reciprocating follower mechanism: (a) front view; (b) side view.

$$\frac{1}{R_x} = \frac{1}{R_{c,x}} + \frac{1}{R_{f,x}} = \frac{1}{25} + \frac{1}{5} = 0.24\,\text{mm}^{-1} \tag{6.110}$$

$R_{c,x}$ and $R_{f,x}$ are the radius of curvature of the cam and follower at the point of contact. In this case the cam is flat in the y direction while the follower has a crown on $R_{f,y} = 400\,\text{mm}$.

The Hertzian pressure of the cam–follower contact can now be determined.

$$A + B = \frac{1}{2}\left(\frac{1}{R_{c,x}} + \frac{1}{R_{c,y}} + \frac{1}{R_{f,x}} + \frac{1}{R_{f,y}} \right)$$

$$= \frac{1}{2}\left(\frac{1}{25} + \frac{1}{\infty} + \frac{1}{5} + \frac{1}{400} \right) = 0.121\,\text{mm}^{-1} \tag{6.111}$$

$$B - A = \frac{1}{2}\left[\left(\frac{1}{R_{c,x}} - \frac{1}{R_{c,y}}\right)^2 + \left(\frac{1}{R_{f,x}} - \frac{1}{R_{f,y}}\right)^2\right.$$

$$\left. + 2\left(\frac{1}{R_{c,x}} - \frac{1}{R_{c,y}}\right)\left(\frac{1}{R_{f,x}} - \frac{1}{R_{f,y}}\right)\cos 2\psi\right]^{1/2}$$

$$= \frac{1}{2}\left[\left(\frac{1}{25} - \frac{1}{\infty}\right)^2 + \left(\frac{1}{5} - \frac{1}{400}\right)^2 + 2\left(\frac{1}{25} - \frac{1}{\infty}\right)\left(\frac{1}{5} - \frac{1}{400}\right)\cos 0\right]^{1/2}$$

$$= 0.119\,\text{mm}^{-1} \tag{6.112}$$

$$\theta = \cos^{-1}\left(\frac{B - A}{B + A}\right) = \cos^{-1}\frac{0.119}{0.121} = 11.7° \tag{6.113}$$

The values of c_1 and c_2 can be interpolated using the values in Table 6.1.

$c_1 = 6.136$

$c_2 = 0.3272$

$$a = c_1\left[\frac{3F}{2E'(B + A)}\right]^{1/3} = 6.136\left[\frac{(3)2500\,\text{N}}{(2)2.31 \times 10^{11}\text{Pa} \times 122\,\text{m}^{-1}}\right]^{1/3}$$

$$= 3140\,\mu\text{m} \tag{6.114}$$

$$b = c_2\left[\frac{3F}{2E'(A + B)}\right]^{1/3} = 0.3272\left[\frac{(3)2500\,\text{N}}{(2)2.31 \times 10^{11}\text{Pa} \times 122\,\text{m}^{-1}}\right]^{1/3}$$

$$= 167\,\mu\text{m} \tag{6.115}$$

$$P_{\text{max}} = \frac{3w}{2\pi ab} = \frac{(3)2500\,\text{N}}{(2\pi)3140\,\mu\text{m} \times 167\mu\text{m}} = 2.27\,\text{GPa} \tag{6.116}$$

Therefore, the minimum film thickness can be determined as:

$$U = \frac{u\eta_o}{E'R_x} = \frac{1.78\,\text{m/s} \times 0.007\,\text{Pa s}}{231\,\text{GPa} \times 4.17\,\text{m}} = 1.30 \times 10^{-10} \tag{6.117}$$

$$G = \alpha E' = 1.54 \times 10^{-8}\text{Pa}^{-1}\ 231\,\text{GPa} = 3554 \tag{6.118}$$

$$W = \frac{w}{E'R_x^2} = \frac{2500\,\text{N}}{2131\,\text{GPa} \times (4.17\,\text{mm})^2} = 6.24 \times 10^{-4} \tag{6.119}$$

$$k = \frac{a}{b} = \frac{3140}{167} = 18.8 \tag{6.120}$$

$$H_{min} = 3.63 U^{0.68} G^{0.49} W^{-0.073} (1 - e^{-0.68k}) = 6.46 \times 10^{-5} \tag{6.121}$$

$$h_{min} = R_x H_{min} = 4.17\,\text{mm} \times 6.45 \times 10^{-4} = 0.269\,\mu\text{m} \tag{6.122}$$

The R_q for a typical cam and follower system are 0.0202 and 0.0067 mm respectively, thus the lambda ratio is:

$$\Lambda = \frac{h_{min}}{\sqrt{Rq_c^2 + Rq_f^2}} = \frac{0.269\,\mu\text{m}}{\sqrt{(0.12\,\mu\text{m})^2 + (0.11\,\mu\text{m})^2}} = 1.7 \tag{6.123}$$

Thus, the cam and follower contact is operating in the mixed or partial lubrication regime. Finally, note that the analysis in Section 6.12 provides steady state solution for pure rolling contact condition. In practice transient conditions lead to squeeze film motion which enhances the load carrying capacity (somewhat larger film thickness). For more detailed analysis see Chapter 17.

6.13   References and further reading

Ai, X. and Cheng, H. S., The influence of moving dent on point EHL contact, *STLE Tribology Trans.*, 1994, **37**, 323–335.

Bair, S. and Winer, W. O., A rheological model for elastohydrodynamic contacts based on primary laboratory data, *Trans. ASME F, J. Lubric. Technol.*, 1979, **101**, 258–265.

Bair, S. and Winer, W. O., A rheological basis for concentrated contact friction. In *Proceedings of 20th Leeds–Lyon Symposium on Tribology*, 1994, 37–44.

Barus, C., Isotherms, isopiestics, and isometrics relative to viscosity, *Am. J Sci.*, 1893, **45**, 87–96.

Barwell, F. T., *The Founder of Modern Tribology* (Eds D. M. McDowell and J. D. Jackson), 1970, Manchester University Press, Manchester.

Bedewi, M. A., Dowson, D. and Taylor, C. M., Elastohydrodynamic lubrication of line contacts subjected to time dependent loading with particular reference to roller bearings and cams and follower. In *Mechanisms and Surface Distress, Proceedings of 12th Leeds–Lyon Symposium on Tribology*, 1986, 289–304, Butterworths, London.

Blok, H., Theoretical study of temperature rise at surfaces of actual contact under oiliness conditions, *Proc. Inst. Mech. Engi. Gen. Discussion Lubric.*, 1937, **2**, 222–235.

Brandt, A., Multi–level adaptive solutions to boundary-value problems. *Mathematics of Computation*, 1977, **31**(138), 333–389.

Brandt, A. and Lubrecht, A. A., Multilevel matrix multiplication and fast solution of integral equations, *J. Computational Phys.*, 1990, **90**, 348–370.

Cameron, A. and Gohar, R., Theoretical and experimental studies of the oil in lubricated point contact, *Proc. R. Soc. London*, 1966, **A291**, 520–536.

Castle, P. and Dowson, D., A theoretical analysis of the starved elastohydrodynamic lubrication problem for cylinders in line contact. In *Proceedings of 2nd IMechE Symposium on Elastohydrodynamic Lubrication*, 1972, 131–137.

Chang, L. and Zhai, X., A transient thermal model for mixed film contacts, *Tribology Transactions*, 2000, **43**, 427–434.

Chang, L., Cusano, C. and Conry, T. F., Effects of lubricant rheology and kinematic conditions on micro-elastohydrodynamic lubrication. *Trans. ASME F, J. Tribology*, 1989, **111**, 344–351.

Cheng, H. S., *Calculation of Elastohydrodynamic Film Thickness in High-Speed Rolling and Sliding Contacts*, Rep. No. MTI-67TR24, Mechanical Technology Inc., Latham, New York, May 1967.

Cheng, H. S., A numerical solution of the elastohydrodynamic film thickness in an elliptical contact. *Trans. ASME F, J. Lubric. Technol.*, 1970, **92**(1), 155–162.

Cheng, H. S., *Application of Elastohydrodynamics of Rolling Element Bearings,* ASME Paper 74-DE-32, American Society of Mechanical Engineers, New York, 1974.

Cheng, H. S. and Dyson, A., Elastohydrodynamic lubrication of circumferentially ground disks, *ASLE Trans.*, 1978, **21**(1), 25–40.

Cheng, H. S. and Sternlicht, B., A numerical solution for pressure, temperature and film thickness between two infinitely long rolling and sliding cylinders under heavy load, *J. Basic Eng., Trans. ASME*, 1965, **87**(3), 695.

Chevalier, F., Lubrecht, A. A., Cann, P. M. E., Colin, F. and Dalmaz, G., Film thickness in starved EHL point contact, *Trans. ASME F, J. Tribology*, 1998, **120**(1), 126–133.

Chittenden, R. J., Dowson, D., Dunn, J. and Taylor, C. M., A theoretical analysis of isothermal EHL concentrated contacts: Parts I and II, *Proc. R. Soc. Lond.*, 1985, **A97**(12), 245–294.

Christensen, H., Elastohydrodynamic theory of spherical bodies in normal approach, *Trans. ASME, J. Lubric. Technol.*, 1970, **92**, 145–154.

Clyens, S., Evans, C. R. and Johnson, K. L., Measurement of viscosity of supercooled liquids at high shear rates with a Hopkinson torsion bar, *Proc. R. Soc. Lond.*, 1985, **A381**, 195–214.

Crook, A. W., The lubrication of rollers, I. *Phil. Trans. Roy. Soc.* London, 1958, **A250**, 387–409.

Crook, A. W., The lubrication of rollers, II – film thickness with relation to viscosity and speed, *Phil. Trans.*, 1961, **A254**, 223–236.

Dawson, P. H., Effect of metallic contact on the pitting of lubricated rolling surface, *Proc. Instn Mech. Engrs, J. Mech. Eng. Sci.*, 1962, **4**(1), 16–21.

Deolalikar, N. and Sadeghi, F., Fatigue life reduction in mixed lubricated elliptical contacts. *Tribology Letters*, 2007, **27**, 2, 197–209.

Deolalikar, N., Sadeghi, F, and Marble, S., Numerical modeling of mixed lubrication and flash temperature in EHL elliptical contacts. *ASME, J. Tribology*, 20078 130(1),011004

Dowson, D., *Elastohydrodynamic Lubrication, Interdisciplinary Approach to the Lubrication of Concentrated Contacts*, Spec. Publ. No. NASA SP–237, National Aeronautics and Space Administration, Washington, DC, 1970, 34.

Dowson, D., and Ehret, P., Past, present and future studies in elastohydrodynamics, *Proc. Inst. of Mech. Eng.*, 1999, **213**, 317–333.

Dowson, D. and Higginson, G. R., A numerical solution to the elastohydrodynamic problem, *J. Mech. Eng. Sci.*, 1959, **1**(1), 6–15.

Dowson, D. and Higginson, G. R., *Elastohydrodynamic Lubrication*, Pergamon Press, Oxford, 1966.

Dowson, U. and Higginson, G. R., *Elastohydrodynamic Lubrication*, Pergamon Press, Oxford, 1977.

Dowson, D. and Wang, D., An analysis of the normal bouncing of a solid elastic ball on an oily plate, *Wear*, 1994, **179**, 29–37.

Dowson, D. and Whittaker, B. A., A numerical procedure for the solution of the elastohydrodynamic problem of rolling and sliding contacts lubricated by a Newtonian fluid, *Proc. Inst. Mech. Eng.*, 1965, **180**(3B), 57.

Dowson, D., Higginson, G. R., and Whitaker, A. V., Elastohydrodynamic lubrication – a survey of isothermal solutions, *J. Mech. Eng. Sci.*, 1962, **4**(2), 121.

Dowson, D., Higginson, G. R. and Whitaker, A. V., Stress distribution in lubricated rolling contacts. In *Proceedings of IMechE Symposium on Fatigue in Rolling Contact*, paper 6.66, 1963, 66–75.

Dowson, D., Harrison, P. and Taylor, C. M., The lubrication of automotive cams and followers. In *Mechanisms and Surface Distress, Proceedings of 11th Leeds–Lyon Symposium on Tribology*, 1986, 305–322, Butterworths, London.

Dowson, D., Taylor, C. M. and Zhu, G., A transient elastohydrodynamic lubrication analysis of a cam and follower, *J. Phys. D, Appl. Phys., Frontiers Tribology*, 1992, **25**(1a), A313–A320.

Ehret, P., Dowson, D., Taylor, C. M. and Wang, D., Analysis of isothermal elastohydrodynamic point contacts lubricated by Newtonian fluids using multigrid methods, *Proc. Inst. Mech. Eng., Part C, J. Mech. Eng. Sci.*, 1997, **211**(C7), 493–508.

Ehret, P., Dowson, B. and Taylor, C. M., On lubricant transport conditions in elastohydrodynamic conjunctions, *Proc. R. Soc. Lond.*, 1998, **A454**(1971), 763–787.

Ertel, A. M., Hydrodynamic lubrication based on new principles, *Akad. Nauk SSSR. Prikadnaya Mathematica i Mekhanika*, 1939, **3**(2), 41–52.

Evans, H. P. and Snidle, R. W., Inverse solution of Reynolds equation of lubrication under point contact elastohydrodynamic conditions, *Trans. ASME F, J. Lubric. Technol.*, 1981, **103**(4), 539–546.

Evans, H. P. and Snidle, R. W., The elastohydrodynamic lubrication of point contacts at heavy loads, *Proc. R. Soc.*, 1982, **A382**, 183–199.

Foord, C. A. *et al.*, Optical elastohydrodynamics, *Proc. Inst. Mech. Eng.*, 1969, **184**(1), 487.

Fowles, P. E., The application of elastohydrodynamic theory to individual asperity–asperity collisions, *J. Lubr. Tech., Trans. ASME*, 1969, **91**, 464.

Gao, J., Lee, S. C., Ai, X. and Nixon, H., 'An FFT-based transient flash temperature model for general three-dimensional rough surface contacts,' *J. Tribology*, 2000, **122**, 519–523.

Glovnea, R. P. and Spikes, H. A., Elastohydrodynamic film collapse during rapid deceleration – Part I: Experimental results, ASME *J. Tribology*, 2001, **123**, 254–261.

Goglia, P. R., Conry, T. F. and Cusano, C., The effect of surface irregularities on the elastohydrodynamic lubrication sliding line contacts. Part I – Single irregularities. *Trans. ASME F, J. Lubric. Technol.*, 1981, **106**, 104–112.

Goglia, P. R., Conry, T. F. and Cusano, C., The effect of surface irregularities on the elastohydrodynamic lubrication sliding line contacts. Part II – Wavy surfaces. *Trans. ASME F, J. Lubric. Technol.*, 1981, **106**, 113–119.

Gohar, R. and Cameron, A., Optical measurement of oil film thickness under EHD lubrication, Nature, 1963, **200**, 458–459.

Gohar, R., and Cameron, A., Theoretical and experimental studies of oil film in lubricated point contact, *Proc. Roy. Soc., A*, 1966, **291**, 520.

Gohar, R. and Cameron, A., The mapping of EHD contacts, *ASLE Trans.*, 1967, **10**, 214.

Gohar, R. and Rahnejat, H., *Fundamentals of Tribology*, Imperial College Press, London, 2008

Greenwood, J. and Kanzlarich, J., Inlet shear heating in elastohydrodynamic lubrication, *J. Lubr. Technol. Trans. ASME*, 1973, **95**(4), 417.

Greenwood, J. A. and Morales-Espejel, G. E., The behaviour of transverse roughness in EHL contacts. *Proc. Inst. Mech. Eng., Part J, J. Eng. Tribology*, 1994, **208**(J2), 121–132.

Grubin, A. N., *Contact Stresses in Toothed Gears and Worm Gears*, Central Scientific Research Institute for Technology and Mechanical Engineering, 1949, Book No. 30, Moscow, English Translation No. 337.

Grubin, A. N. and Vinogradova, I. E., *Fundamentals of the Hydrodynamic Theory of Lubrication of Heavily Loaded Cylindrical Surfaces*, Central Scientific Research Institute for Technology and Mechanical Engineering, Book No. 30. Moscow, Translation No. 337 into English by the Department of Science and Industrial Research, UK, 1949.

Guangteng, G., *et al.*, An experimental study of film thickness between rough surfaces in EHD contacts. *Tribology International*, 2000, **33**, 3–4, 183–189.

Gupta, P. K., *et al.*, Visco-elastic effect in Mil-L-7808 type lubricant, Part I: Analytical Formulation, *STLE Tribol. Trans.*, 1991, **34**, 4, 608–617.

Hamrock, B. J. and Dowson, D., Isothermal elastohydrodynamic lubrication of point contact, Part I – Theoretical formulation, *J. Lubrication Technol.*, 1976a, **98**, 223–229.

Hamrock, B. J. and Dowson, D., Isothermal elastohydrodynamic lubrication of point contacts. Part II – Ellipticity parameter results, *Trans. ASME F, J. Tribology*, 1976b, **98**(3), 375–383.

Hamrock, B. J. and Dowson, D., Isothermal elastohydrodynamic lubrication of point contacts. Part III – Fully flooded results, *Trans. ASME F, J. Tribology*, 1977a, **99**(2), 264–276.

Hamrock, B. J. and Dowson, D., Isothermal elastohydrodynamic lubrication of point contacts. Part IV – Starvation results, *Trans. ASME F, J1 Tribology*, 1977b, **99**(1), 15–23.

Hamrock, B. J. and Dowson, D., *Ball Bearing Lubrication. The Elastohydrodynamics of Elliptical Contacts*, John Wiley, New York, 1981.

Hamrock, B. J., Schmid, S. R. and Jacobson, B. O., *Fundamentals of Fluid Film Lubrication*, McGraw-Hill, Inc., New York, 2nd edition, 2004.

Hertz, H., On the contact of elastic solid, *J. reine und angewandte Mathematik*, 1882, **92**, 156–171

Hoglund, E. and Jacobson, B., Experimental investigation of the shear strength of lubricant subjected to high pressure, *Trans. ASME, J. Lubric. Technol.*, 1986, **108**, 571–578.

Hoglund, E. and Larsson, R., Modelling non-steady EHL with focus on lubricant density. In *Elastohydrodynamics '96, Fundamentals and Application in Lubrication and Traction, Proceedings of 23rd Leeds–Lyon Symposium*, Leeds, 1997, 511–521.

Holmes, M. J. A. Transient analysis of the point contact elastohydrodynamic lubrication problem using coupled solution methods. PhD thesis, Cardiff University, 2002.

Hu, Y. and Zhu, D., A full numerical solution to mixed lubrication in point contacts, *J. Tribology*, 2000, **122**, 1–9.

Hu, Y. et al., Numerical analysis for the elastic contact of real rough surfaces, *Tribology Trans.*, 1999, **42**, 3, 443–452.

Hua, D. V. and Khonsari, M. M., Application of transient elastohydrodynamic lubrication analysis for gear transmissions, *STLE, Tribology Trans.*, 1995, **38**(4), 905–913.

Hua, D. Y., Zhang, H. H. and Zhan, S. L., Transient elastohydrodynamic lubrication of involute spur gear. In *Proceedings of Japan International Tribology Conference, Nagoya, Japan*, 1990, 1641–1646.

Ioannides, E. and Harris, T. A., A new fatigue life model for rolling bearings, *J. Tribology*, 1985, **107**, 367–378.

Ioannides, E., Bergling, G. and Gabelli, A., an analytical formulation for the life of rolling bearings, *Acta Polytechnica Scandinavia, Mech. Eng. Series* 137, 1999.

Jacobson, B., On the lubrication of heavily loaded spherical surfaces considering surface deformations and solidification of the lubricant, *Acta Polytechn. Scand. Mech. Eng. Ser.*, 1970, 54.

Jacobson, B. O., A high pressure-short time shear strength analyser for lubricants, *Trans. ASME F, J. Tribology*, 1985, **107**, 220–223.

Jaeger, J. C., Moving sources of heat and the temperature at sliding contacts, *Roy. Soc. New South Wales – J. Proc.*, 1943, **76**, 203–224.

Jiang, X., Hua, D. Y., Cheng, H. S., Ai, X. and Lee, S. C., A mixed elastohydrodynamic lubrication model with asperity contact, *J. Tribology*, 1999, **121**, 481–491.

Johnson, K. L., Correlation of theory and experiment in research on fatigue in rolling contact. In *Proceedings of IMechE Symposium on Fatigue in Rolling Contact*, 1963, 155–159.

Johnson, K. L., *Contact Mechanics*, Cambridge University Press, Cambridge, 1985.

Johnson, K. L. and Cameron, R., Shear behavior of elastohydrodynamic oil film at high rolling contact pressures, *Proc. Inst. Mech. Eng.*, 1967, **182**, 307.

Johnson, K. L. and Greenwood, J. A., Thermal analysis of an Eyring fluid in elastohydrodynamic traction, *Wear*, 1980, **61**, 353–374.

Johnson, K. L., Greenwood, J. A. and Poon, S. V., A simple theory of asperity contact in elastohydrodynamic lubrication, *Wear*, 1972, **19**, 91–108.

Johns-Rahnejat, P.M. and Gohar, R., Measuring contact pressure distribution under elastohydrodynamic point contacts, *Tribo-test*, 1994, **1**, 33–53.

Kaneta, M., Necessity of reconstruction of EHL theory. *Jap. J. Tribology*, 1993, **38**, 860–868.

Kaneta, M., Nishikawa, H., Kameishi, K., Sakal, T. and Ohno, N., Effects of elastic moduli of contact surfaces in elastohydrodynamic lubrication, *Trans. ASME F, J. Tribology*, 1992, **114**, 75–80.

Kannel, J. W., Measurements of pressures in rolling contact. *Proc. Inst. Mech. Eng., 1965–1966, Elastohydrodynamic Lubrication*, 1966, **180**(3B), 135–142.

Kannel, J. W., Zugaro, F. F. and Dow, T. A., A method for measuring surface temperature between rolling/sliding steel cylinders, *J. Lubr. Technol. Trans. ASME*, 1978, **100**(1), 100.

Kim, K. and Sadeghi, F., Non-Newtonian elastohydrodynamic lubrication of point contact, *ASME J. Tribology*, 1991, **113**, 703–711.

Kim, K. and Sadeghi, F., Three dimensional temperature distribution in EHD lubrication, *ASME J. Tribology*, 1992, **114**, 32–41.

Koye, K. A. and Winer, W. O., An experimental evaluation of the Hamrock and Dowson minimum film thickness equation for fully flooded EHD point contacts, *J. Lubrication Technology*, 1981, **103**, 284–294.

Kumar, A., Sadeghi, F. and Krousgrill C. M., 'Effects of surface roughness on normal contact compression response,' *Proc. Inst. Mech. Eng., Part J, J. Eng. Tribology*, 2006, **220**, 65–77.

Kweh, C. C., Evans, H. P. and Snidle, R. W., Microelastohydrodynamic lubrication of elliptical contact with transverse and three-dimensional roughness, *Trans. ASME F, J. Tribology*, 1989, **111**, 577–584.

Lee, R.-T. and Hamrock, B. J., A circular non-Newtonian model: Part I – Used in elastohydrodynamic lubrication, *Trans. ASME F, J. Tribology*, 1990, **112**, 386–496.

Lubrecht, A. A., Numerical solution of the EHL line and point contact problem using multigrid techniques. PhD thesis, Twente University, The Netherlands, 1987.

Lubrecht, A. A., ten Napel, W. E. and Bosma, R., Multigrid, an alternative method of solution for two-dimension elastohydrodynamically lubricated point contact calculations. *Trans. ASME F, J. Tribology*, 1986, **108**(3), 551–556.

Lundberg, G. and Palmgren, A., Dynamic capacity of roller bearings, *Acta Polytechnica – Mech. Eng. Series*, 1952, **2**, 1–32.

Majumbar, B. C. and Hamrock, B. J., Effect of surface roughness on elastohydrodynamics line contact, *Trans. ASME, J. Lubric. Technol.*, 1982, **104**(3), 401–409.

Martin, H. M., Lubrication of gear teeth, *Engineering, Lond.*, 1916, **102**, 199.

Matthews, J. A. and Sadeghi, F., 'Kinematics and lubrication of camshaft roller follower mechanisms,' *Soc. Tribologists Lubri. Eng., Tribology Trans.*, 1996, **39**, 2, 425–433.

Matthews, J. A., Sadeghi, F. and Cipra, R. J., 'Radius of curvature and entraining velocity of cam follower mechanisms,' *Soc. Tribologists Lubric. Eng., Tribology Transa.*, **39**, 4, 899–907.

Moes, H. and Bosma, R., Film thickness and traction in EHL at point contact. In *Proceedings of 1972 IMechE Symposium on Elastohydrodynamic Lubrication,* 1972, 149–152.

Morales-Espejel, G. E., Greenwood, J. and Melgar, J. L., Kinematics of roughness in EHL. In *Proceedings of 22nd Leeds–Lyon Symposium on Tribology*, 1996, pp. 501–513.

Mostofi, A. and Gohar, R., Oil film thickness and pressure distribution in elastohydrodynamic point contacts, *J. Mech. Engng Sci.*, 1982, **24**(4), 173–182.

Murch, L. E. and Wilson, W. R. B., A thermal elastohydrodynamic inlet zone analysis, *J. Lubric. Technol., Trans. ASME*, 1975, **97**(2), 212.

Nagaraj, H. S., Sanborn, D. M. and Winer, W. O., Direct surface temperature measurement by infrared radiation in elastohydrodynamic contacts and the correlation with the Block temperature theory, *Wear*, 1978, **49**(1), 43–59.

Nickel, D. A., Experimental methods for the assessment of tribological hard contacts at the device, component, and surface scales. PhD Thesis, Purdue University, West Lafayette, USA, 1999.

Okamura, H., A contribution to the numerical analysis of isothermal elastohydrodynamic lubrication. In *Proceedings of 9th Leeds–Lyon Symposium on Tribology*, 1982, 313–320.

Osborn, K. F. and Sadeghi F., Time dependent line EHD lubrication using the multigrid/multilevel technique, *J. Tribology*, 1992, **114**(1), 68–74.

Pan, P. and Hamrock, B. J., Simple formulas for performance parameters used in elastohydrodynamically lubricated line contacts, *J. Tribology*, 1989, **111**(2), 246–251.

Patir, N. and Cheng, H. S., An average model for determining effects of three dimensional roughness on partial hydrodynamic lubrication, *Trans. ASME, J. Lubric. Technol.*, 1978, **100**, 12–17.

Persson, B. N. J., *Sliding Friction : Physical Principles and Applications*, Springer-Verlag: Berlin Heidelberg, 2000.

Petrusevich, A. I., Fundamental conclusions from the contact-hydrodynamic theory of lubrication, *Izv. Akad. Nauk. SSSR(OTN)*, 1951, **2**, 209–223.

Ramesh, K. T., On the rheology of a traction fluid, *Trans. ASME F, J. Tribology*, 1989, **111**, 614–619.

Reynolds, O., On the theory of lubrication and its application to Mr. Beauchamp Tower's experiments, including an experimental determination of viscosity of olive oil, *Philos. Trans. R. Soc. London Ser. A*, 1886, **177**, 157–234.

Rodkiewicz, C. M. and Srinivanasan, V., EHD lubrication in rolling and sliding contacts, *J. Lubr. Technol. Trans. ASME*, 1972, **94**(4), 324.

Roelands, C. J. A., Vlugter, J. and Waterman, H., The viscosity–temperature–pressure relationship of lubricating oils and its correlation with chemical constitution, *J. Basic Eng.*, 1963, 601–610.

Sadeghi, F., A comparison of the fluid models effect on the internal stresses of rough surfaces, *Trans. ASME F J. Tribology*, 1991, **113**, 142–149.

Sadeghi, F. and Sui, P. C., Thermal elastohydrodynamic lubrication of rolling/sliding contacts, *Journal of Tribology*, 1990, **112**(2), 189–195.

Sadeghi, F. and Sui, P. C., Non-Newtonian thermal elastohydrodynamic lubrication, *Journal of Tribology*, 1991, **113**(2), 390–397.

Safa, M. M. A., Anderson, J. C. and Leather, J. A., Transducers for pressure, temperature and oil film thickness measurement in bearings, Sensors and Actuators, 1982, **3**, 119–128.

Sibley, L. B. and Orcutt, F. K., Elastohydrodynamic lubrication of rolling contact surfaces, *Am. Soc. Lubr. Eng. Trans.*, 1961, **4**(2), 234.

Smeeth, M. and Spikes, H. A., Central and minimum elastohydrodynamic film thickness at high contact pressure. *ASME*, 1997, **119**(2), 291–296.

Spikes, H. A., Mixed lubrication – an overview. In *Proceedings of Symposium on Tribology-Solving Friction and Wear Problems*, Esslingen, 1996, 1713–1735.

Stachowiak, G.W. and Batchelor A.W., *Engineering Tribology*, Elsevier Butterworth-Heinemann, London, 2005.

Stanley, H. M. and Kato, T., An FFT-based method for rough surface contact, *J. Tribology*, 1997, **119**, 481–485.

Tian, X. and Kennedy Jr., F. E., Maximum and average flash temperatures in sliding contacts, *J. Tribology*, 1994, **116**, 167–174.

Venner, C. H., Multilevel solution of the EHL line and point contact problems. PhD thesis, Twente University, The Netherlands, 1991.

Venner, C. H. and Lubrecht, A. A., Numerical simulation of a transverse ridge in a circular EHL contact under rolling/sliding, *Trans. ASME F, J. Tribology*, 1994a, **116**, 751–761.

Venner, C. H. and Lubrecht, A. A., Numerical simulation of waviness in a circular EHL contact, under rolling/sliding. In *Lubricant and Lubrication, Proceedings of 21st Leeds–Lyon Symposium on Tribology*, Leeds, 1994b, 259–272.

Wang, Q. J., Zhu, D., Cheng, H. S., Yu, T., Jiang, X. and Liu, S., Mixed lubrication analysis by macro-micro approach and a full-scale mixed EHL model, *J. Tribology*, 2004, **126**, 81–91.

Wedevan, L. D., Optical measurements in EHD rolling-contact bearings, PhD thesis, University of London, March 1970.

Wedeven, L. D., Evans, D. and Cameron, A., Optical analysis of ball bearing starvation. *Trans. ASME, J Lubric. Technol.*, 1971, **97**, 321–363.

Weibull, W., *The Phenomenon of Rupture in Solids*, Ingeniors Vetenskaps Akademien Handlingar, 1939, 153.

Whitehouse, D. J. and Archard, J. F., The properties of random surfaces of significance in their contact, *Proc. R. Soc. London.*, 1970, **A316**, 97.

Winer, W. O., Temperature effects in elastohydrodynamically lubricated contacts. In *Tribology in the 80's*. NASA CP-2300, 1983, Vol. II, 533–543.

Xu, G and Sadeghi, F., Thermal EHL analysis of circular contacts with measured surface roughness, *Trans. ASME, J. Tribology*, 1996, **118**, 473–483.

Zhao, J. and Sadeghi, F., Analyis of EHL circular contact start up: Part II – Surface temperature rise model and results, *J. Tribology*, 2001, **123**, 75–82.

Zhao, J. and Sadeghi, F., The effects of a stationary surface pocket on ehl line contact start up, *J. Tribology*, 2002, **126**, 672–680.

Zhao, J., Sadeghi, F. and Hoeprich, M. H., Analysis of EHL circular contact start up: Part I – Mixed contact model with pressure and film thickness results, *J. Tribology*, 2001 a, **123**, 67–74.

6.14   Nomenclature

a	half-width of Hertzian contact across rolling direction, m
b	half-width of Hertzian contact along rolling direction, m
E'	equivalent modulus of elasticity, $2((1 - v_1^2)/E_1 + (1 - v_2^2)/E_2)^{-1}$, Pa
E_1, E_2	modulus of elasticity of contacting surfaces 1 and 2, Pa
f_c	coefficient of friction
G	dimensionless material parameter, $\alpha E'$
H	dimensionless film thickness, hR_x/b^2
H_{min}	dimensionless minimum film thickness
H_0	dimensionless film thickness constant
h	film thickness, m
h_c	film thickness at centre of contact, m
h_{min}	minimum film thickness, m
K_f	thermal conductivity of lubricant, W/(m K)
P	dimensionless pressure, p/P_H
P_H	maximum Hertzian pressure, Pa
P_1	dimensionless lubricant pressure
p	pressure, Pa
R	radius ratio, R_y/R_x
R_a	average roughness, m
R_{ku}	kurtosis
R_q	root mean square roughness, m
R_{sk}	skewness
R_x	equivalent radius of curvature in the x-direction, m
R_y	equivalent radius of curvature in the y-direction, m
S	slide to roll ratio, u_s/u_m
t_m	lubricant temperature, °C
U	dimensionless speed parameter, $u_m\eta_0/E'R_x$
u_1, u_2	velocities of contacting surfaces 1 and 2 in rolling direction, m/s
u_m	average surface velocity, $(u_1 + u_2)/2$, m/s
u_m^0	initial steady state mean velocity of contacting surfaces in rolling direction, m/s

u_M	maximum (desired) mean velocity of contacting surfaces in rolling direction, m/s
u_s	sliding velocity, $u_1 - u_2$, m/s
U_1, U_2	dimensionless velocities of contacting surfaces 1 and 2 in rolling direction, u_1/u_m, u_2/u_m
U_m	dimensionless mean velocity of contacting surfaces in rolling direction, u_m/u_M
W	dimensionless load parameter, $w/E'R_x^2$
w	external normal load, N
X	dimensionless location along rolling direction, x/b
x	co-ordinate in rolling direction, m
Y	dimensionless location across rolling direction, y/b
y	co-ordinate across rolling direction, m
z	Roelands pressure viscosity index of lubricant
α	Barus pressure viscosity coefficient, Pa^{-1}
α_s	thermal diffusivity of solid, m^2/s
η	viscosity of lubricant, Pa s
η_0	absolute viscosity of the lubricant at $p = 0$ and constant temperature, Pa s
$\bar{\eta}$	dimensionless absolute viscosity, η/η_0
κ	ellipticity ratio, a/b
λ	dimensionless speed parameter, $\eta_0 u_m R_x^2/b^3 P_H$
v_1, v_2	Poisson's ratio for contacting surfaces 1 and 2
ρ	density of lubricant, kg/m^3
ρ_0	ambient density of lubricant, kg/m^3
$\bar{\rho}$	dimensionless density, ρ/ρ_0
μ_l	coefficient of friction for lubricated contact
μ_s	coefficient of friction for solid contact

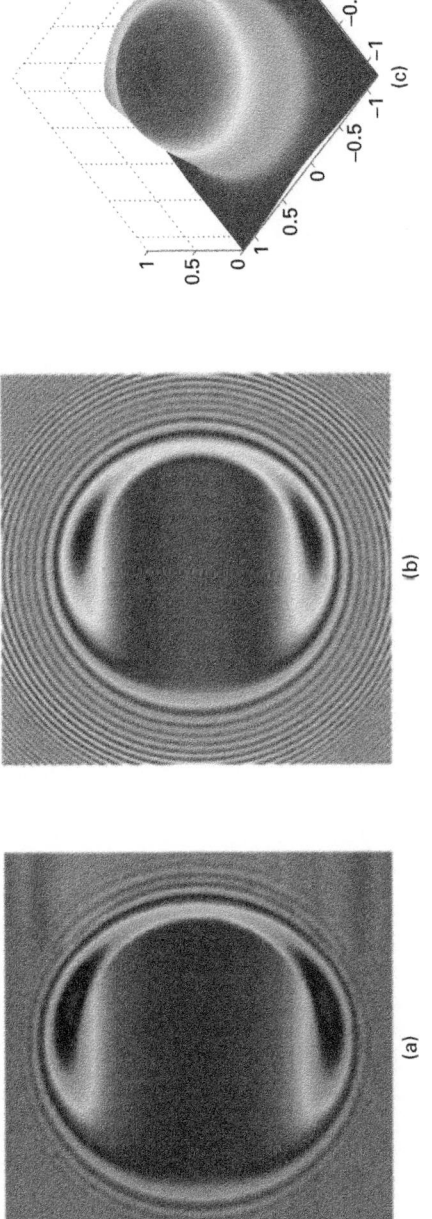

Plate II Experimental interference photomicrograph and corresponding analytical film thickness and pressure results at $W = 3.4e{-}6$, $U = 5.8e{-}11$, $G = 2400$: (a) Experimental interference photomicrograph; (b) analytical results; (c) analytical pressure distribution.

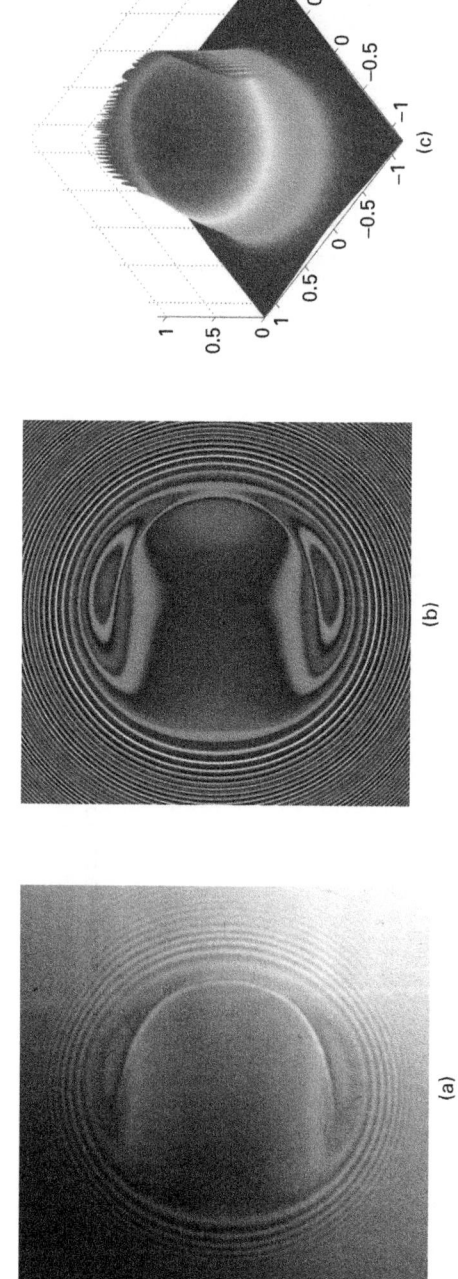

(a)

(b)

(c)

Plate III Experimental interference photomicrograph and corresponding analytical film thickness and pressure results at $W = 4e{-}6$, $U = 9.7e{-}11$, $G = 4127$: (a) experimental interference photomicrograph; (b) analytical results; (c) analytical pressure distribution.

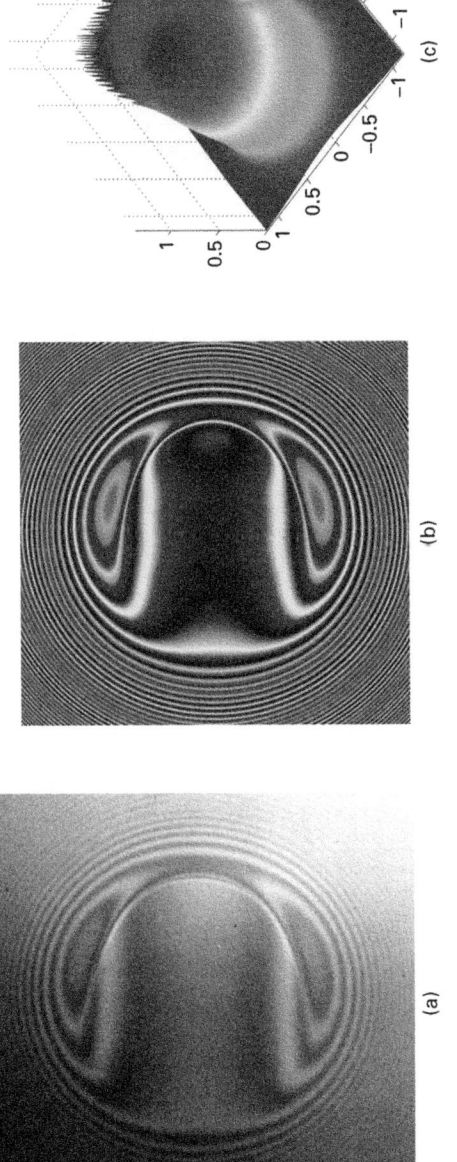

Plate IV Experimental interference photomicrograph and corresponding analytical film thickness and pressure results at $W = 4e{-}6$, $U = 1.94e{-}10$, $G = 4127$: (a) experimental interference photomicrograph; (b) analytical results; (c) analytical pressure distribution.

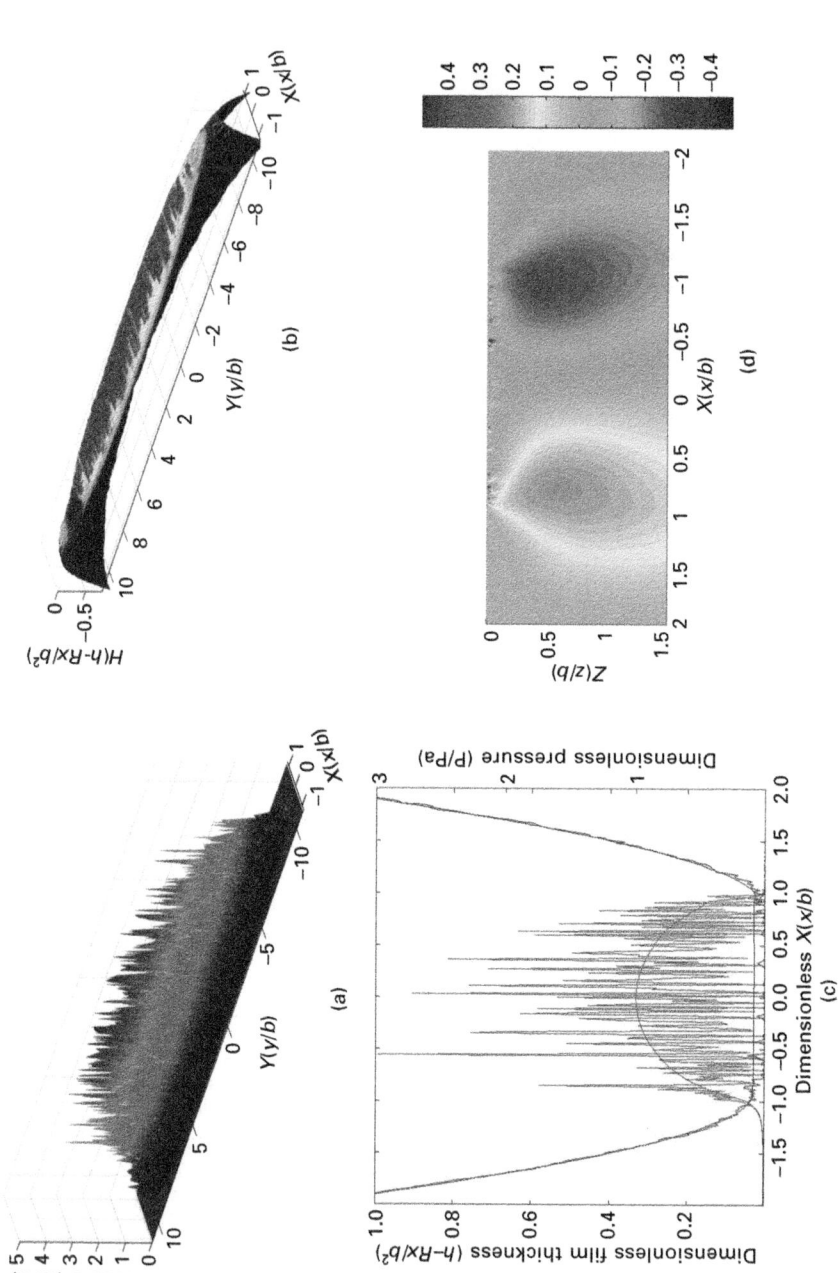

Plate V Mixed-EHL contacts ($W = 7.85\text{e}{-}5$, $U = 1.36\text{e}{-}11$, $G = 2787$): (a) pressure profile; (b) film thickness profile; (c) centre line pressure and film thickness; (d) centre line subsurface shear stress.

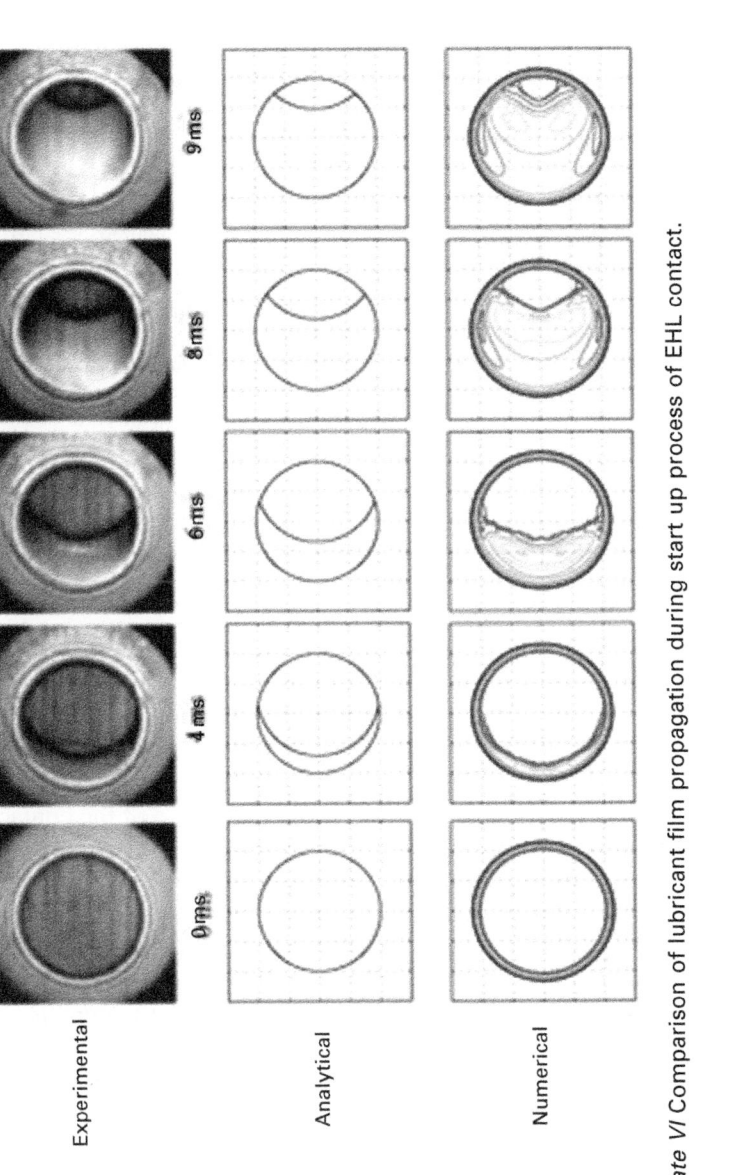

Plate VI Comparison of lubricant film propagation during start up process of EHL contact.

7

Measurement of contact pressure under elastohydrodynamic lubrication conditions

R. GOHAR, Imperial College London, UK and
M. M. A. SAFA, Kingston University, UK

Abstract: This chapter discusses some experimental methods used to measure pressure distribution in elastohydrodynamic lubrication (EHL) contacts. Most of these concentrate on results obtained from deposited manganin pressure gauges on one of the film bounding surfaces. A recent and very promising technique, which relies directly on the EHL film properties (Raman) to determine the pressure distribution, is discussed. Another alternative is mentioned, where an indirect measurement of the wanted pressure distribution is found from a number of discrete points that describe the film geometry.

Key words: manometry, counterformal geometry, EHL, alumna, manganin pressure transducer, piezo-resistive properties, high vacuum, evaporated films, sputtering, masks, flash evaporation, foil mask, Wheatstone bridge, Raman scattering, photoelasticity.

7.1   Introduction

This chapter discusses some direct and indirect methods of measuring the pressure distribution in *elastohydrodynamic lubrication* (EHL) *contacts*. In hydrodynamic lubrication, where the pressures are much lower but the area of pressure is large because of the conformal nature of the contact, one mechanical method uses *manometry*, where the pressure encountered at a point in the film is detected by a hole drilled into one of the bodies at that point. This hole is connected to one arm of a U-tube that is counterbalanced by a static column of the oil in the other arm. For higher pressures in the film the counterbalance can be a pressure gauge. When manometry is applied to EHL, the area of pressure is far smaller on account of the counterformal geometry of the contacting distorted body surfaces. To prevent the measuring instrument itself interfering with the actual pressure distribution, both the contacting bodies must have a large contact reduced radius of curvature and/or a relatively small diameter hole as a pressure tapping. Thus, Dowson and Longfield (1963) employed manometry in a sliding contact between a stationary brass shoe and a steel disc, the contact having a reduced radius of $1.98\,\mathrm{m}$ with a $200\,\mu\mathrm{m}$ diameter pressure tapping in the shoe.

Now EHL contacts can have an average film thickness typically of $1\,\mu\mathrm{m}$.

222

For example, ball bearing rolling element contacts can have a plan-form of minor axis 2000 μm and major axis 500 μm. At a typical rolling speed of the bearing, a fluid element entering the pressure zone takes about 200 μs to traverse its width. If both surfaces are moving, then, because of its slow response time, a manometer-type pressure gauge must be ruled out for such applications. Instead, a vacuum-type miniature transducer is ideal both for measuring pressure and temperature. If located on one of the rolling bodies, it responds immediately to the high pressures or temperatures encountered while traversing the conjunction.

7.2   Gauge manufacturing process

One design of pressure transducer assembly by Safa (1982) comprises a non-metal undercoat layer of alumina insulating the gauge from its steel substrate. The transducer, of manganin, is then deposited on the undercoat through a mask that shapes the active part of the gauge as well as its immediate leads. Sometimes a second insulation overcoat layer of alumina is deposited above the gauge to protect it from the other disc. There is always a problem that the assembly will be so thick as to influence the oil film being measured; so the thinner the assembly the better. Illustrations of a *manganin pressure tranducer* assembly, deposited on a disc surface, are shown in Fig. 7.1. The

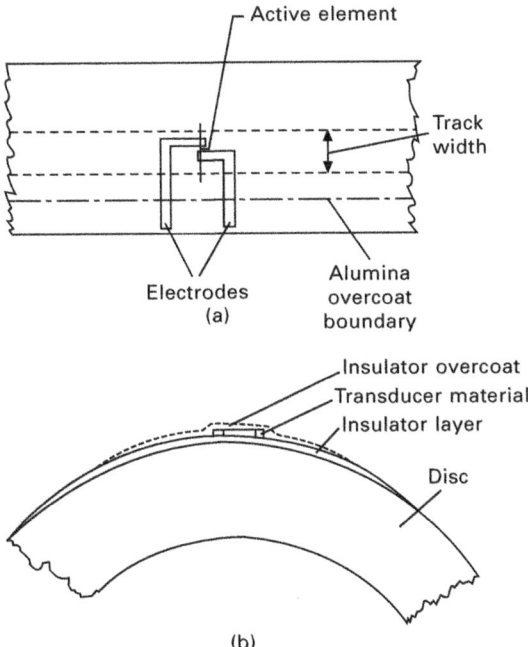

7.1 Plan and side view illustrations of a thin film pressure transducer.

undercoat and overcoat layers are made to build up and down gradually over about one-third of the circumference of one of the discs used in the experiment. The total thickness of the composite layer in this case is about 1.5 μm deep at its apex, while the active element seldom exceeds 0.1 μm thickness. The width of the active element must be sufficiently fine as to detect discreet pressure widths, an example being the EHL contact trailing edge narrow pressure peak predicted in numerical solutions (see Chapter 6). Another property is that it must have a short response time to pressure signals because the measured pressure records on a time base. As an example, a transducer on a stationary surface must react accurately and immediately to a ball rolling over it at high speed. All these properties are found in a manganin gauge that also is sufficiently robust to withstand the repeated normal and tangential stresses encountered in EHL contacts. Furthermore, it has excellent *piezo-resistive properties*, especially when it is very thin. It has a linear change of electrical resistance over a wide pressure range and its resistance coefficient changes little with temperature. Figure 7.2 is an example of a disc machine application. The deposited manganin gauge has produced a raw pressure trace on its time base (Safa, 1982).

7.2.1 Vacuum evaporation

There are two main methods of depositing thin films, both under high vacuum: by evaporation or by *sputtering*. Evaporated films result from heating a material in a vacuum enclosure to a temperature high enough to cause large

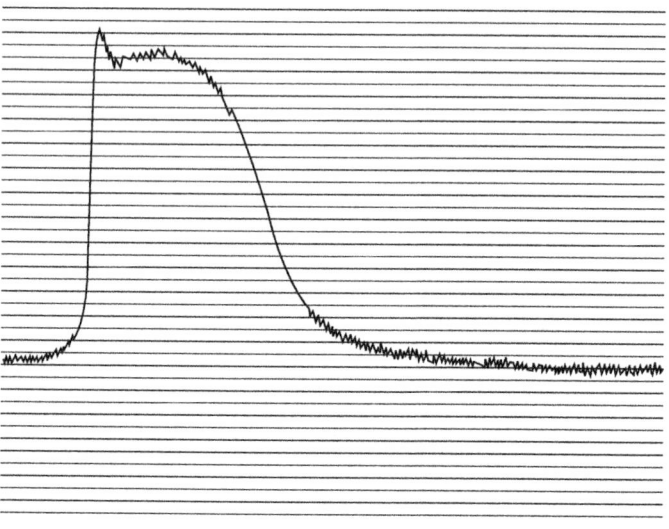

7.2 Raw time base–pressure trace.

numbers of molecules to leave the surface of the material and deposit on the substrate (the surface where the film is to be). A simple *evaporation apparatus* is shown in Fig. 7.3. This arrangement is adequate for evaporating most elements, but if an alloy, like manganin is to be evaporated, what is known as *fractionation* may occur owing to the different vapour pressures of its constituent elements. The effect is to cause the constituent elements to be evaporated individually at different times, resulting in the deposition of three distinct layers of, in this case, manganese, copper and nickel. To control the evaporation technique and the alloy composition, *flash evaporation* is used where tiny particles of the alloy are dropped onto a surface that is so hot that they evaporate instantaneously. Flash evaporation has, however, a problem called spitting, associated with the large gas content of the finely divided particles. The resultant rapid release of this gas produces high background pressures that cause the molten solid to leave the *source and strike the substrate,* an untoward effect. But if a slow and steady feeding of the manganin pieces to the hot boat is employed, the result is on the whole a homogeneous deposit.

7.2.2   Sputtering

This method of thin film deposition relies on the bombardment of a target, of the material to be deposited, by high-energy gas molecules in the form of a plasma of an inert gas, such as argon, in the vacuum chamber. Because

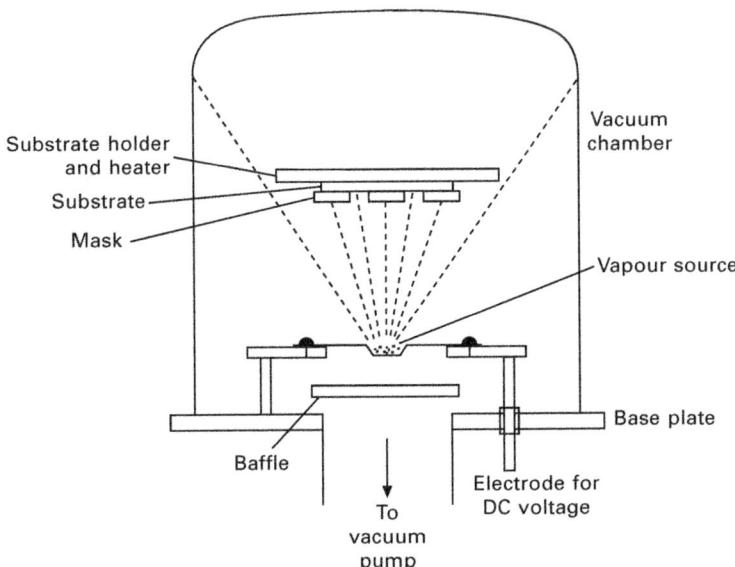

7.3 Principle of the evaporation apparatus.

the energy of the molecules ejected from the target are much higher than those of evaporated molecules, thin films deposited by **sputtering** show a better adhesion than films deposited by evaporation. To obtain a pure film the chamber has to be initially evacuated to a pressure of less than 10^{-6} torr. Then the chamber must be back filled with the pure inert gas to a pressure of around 10^{-3} torr. Figure 7.4 illustrates a simple **sputtering apparatus** where the target (the cathode) is surrounded by a shield. Electrical power is connected to the target, producing a negative voltage between it and the earthed substrate assembly. It is this voltage that causes the ionisation of the low-pressure argon in the chamber, thus setting up the required plasma where it is confined to the central region by means of a magnetic field produced by a circular electromagnet arrangement.

Radio frequency (RF) sputtering allows for insulators also to be sputtered and is the method used in some of the experiments described below. The process overcomes the problem of charge build up when *the insulating material is the target.*

7.2.3   Fabrication of the thin film pressure transducers

To deposit transducers on the periphery of a steel disc of a disc machine, a special purpose vacuum system has to be developed to handle the disc assembly including rotating the disc and fabricating the transducers on it

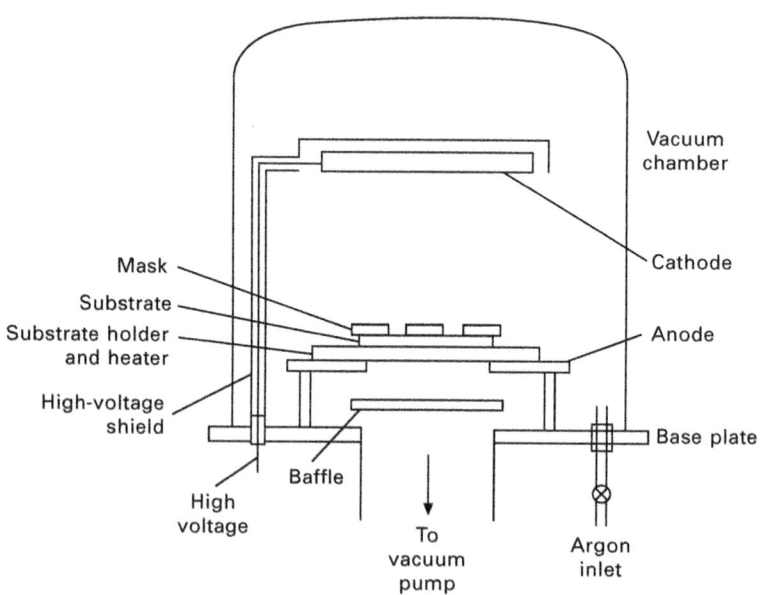

Mask

Substrate

Substrate holder and heater

High-voltage shield

High voltage

Baffle

To vacuum pump

Argon inlet

Vacuum chamber

Cathode

Anode

Base plate

7.4 Sputtering apparatus.

with as few vacuum breaks as possible. The vacuum-generating equipment composes the chamber to be evacuated placed on a base plate mounted above a pump. The chamber is a glass cylinder onto the base plate through which is fed the various electrical and mechanical feeding devices. A stand is used for the substrate holder and the masks. The substrate is situated directly above the boat. With the final arrangement it must be possible to sputter-etch the substrate, sputter-deposit the first insulator layer, deposit all the sensors and finally overcoat the assembly with another layer of ceramic material, all the time keeping air out of the system. The transducers are fabricated through masks made of molybdenum foil. To cut the required pattern from the foil, a computer-controlled x–y platform carries it underneath a ruby laser. This can be focused down to $1.25\,\mu m$. In this way an active element of $10 \times 30\,\mu m^3$ is achieved in the transducer, with the $10\,\mu m$ dimension oriented in the direction of rolling.

7.2.4   Calibration of the transducers

The linearity of the thin film manganin as a pressure transducer is checked with the help of a hydrostatic pressure vessel specifically designed for the purpose. Electrical contacts are made by soldering copper wires to the electrodes part of the pressure transducers patterned out of thin films deposited on glass slides. These copper wires are then used to form electrical connections to the measuring circuits outside the pressure vessel. Figure 7.5 shows the percentage change in resistance due to the application of pressure. Maximum pressure of about $500\,MN/m^2$ could be achieved in the pressure vessel. Pressure coefficient of resistivity is found to vary between deposits, there being some slight changes of linearity. The 25.4 mm diameter cylindrical pressure chamber hydrostatic pressure vessel was too small to calibrate the pressure transducers which were fabricated on a 76.2 mm diameter steel ring. The calculated Hertzian contact pressure was equated to the centre of the pressure profile. This value was recorded by a sensor passing through the contact region under moderate load and very low rolling speed.

7.3   Applications using pressure gauges

7.3.1   Disc machine experiments

Laboratory disc machines are ideal for studying the behaviour of EHL contacts. Below are examples of some experimental results obtained by Safa (1982). Using the relevant fabrication method outlined above, the test disc was first sputter coated with a thin alumina film to provide insulation. Then a manganin pressure gauge was deposited by flash evaporation through a

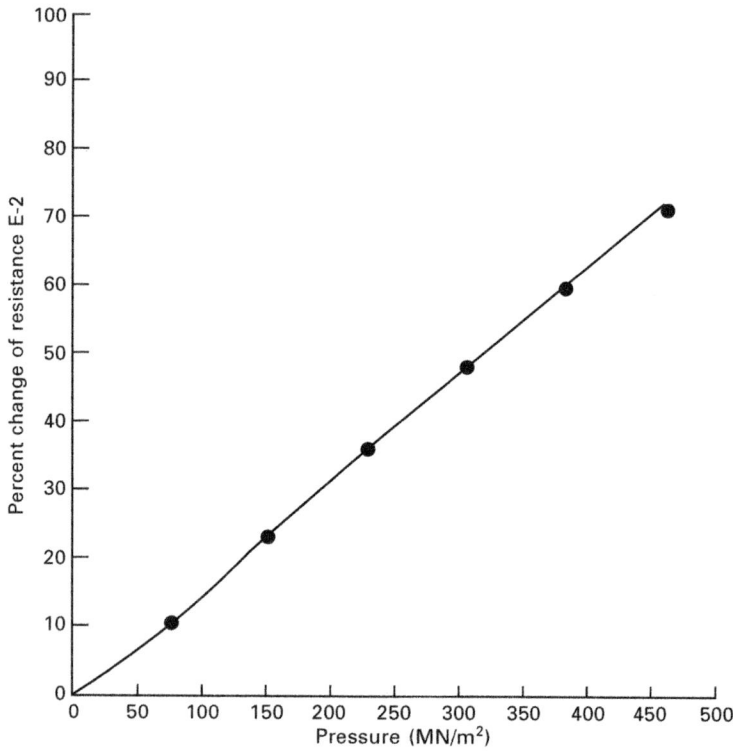

7.5 Gauge calibration in a pressure vessel.

molybdenum foil mask cut by a laser to define the pattern, as shown in Fig. 7.6(a). The active element was usually $10 \times 30 \, \mu m^2$ with the $10 \, \mu m$ width dimension in the rolling direction. The compound gauges seen in (b) and (c) in Fig. 7.6, respectively measured temperature and film thickness at the same time as pressure.

Figure 7.7 shows the pressure distributions obtained for various entrainment velocities and one load. In all cases the inlet pressure sweep follows approximately the Hertzian pressure distribution (dotted) and see also Chapter 6 on EHL, while at outlet there is a ***pressure spike*** appearing prior to exit. At the lowest speed (a), the spike is hardly seen because its width becomes too narrow for the gauge to register. Such a narrow width agrees with numerical predictions by Dowson and Higginson (1977). To confirm this, a numerical solution, taken from Dowson and Higginson, was used in Fig. 7.8. The experiment gauge widths were then varied in a series of tests to see ascertain their effect on the pressure spike heights. Observe that the narrowest gauge has produced the greatest height relative to the numerical solution.

Although this chapter concentrates on measuring EHL pressure, some

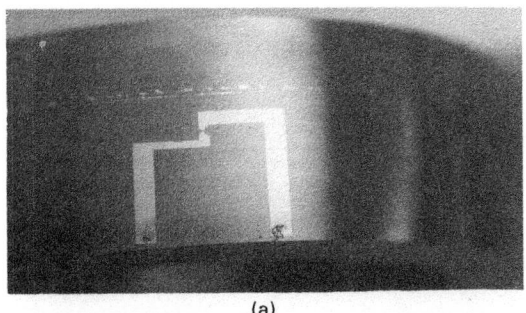

(a)

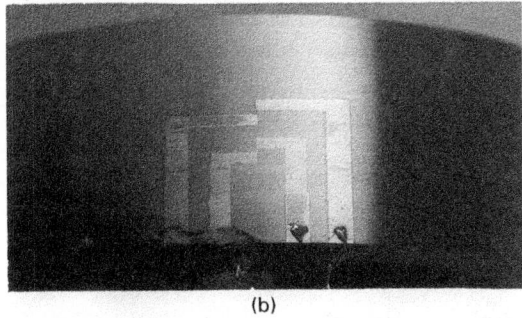

(b)

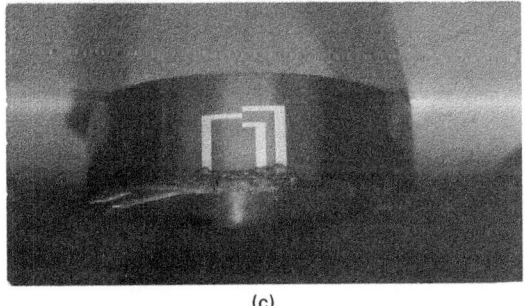

(c)

7.6 (a) Pressure transducer; (b) combined pressure and temperature transducers; (c) combined pressure and film thickness transducers.

surface temperature distribution results are relevant. To do this a ***resistive transducer*** consisting of a thin strip of titanium was used, being deposited through a foil mask aligned with the pressure. These results also helped in the calibration of the ***temperature transducer*** because its temperature coefficient varies with pressure. An assembly of the two such gauges is shown above in Fig. 7.6(b). The actual raw traces can be seen in Safa (1982). A series of temperature profiles for varying slide roll ratios (sliding speed/constant entrainment speed) are shown in Fig. 7.9. Curve (a), for pure rolling, shows the temperature effect of ***inlet compressive heating*** followed by some cooling within the parallel part of the conjunction ($x < 0.4$ mm). At higher sliding

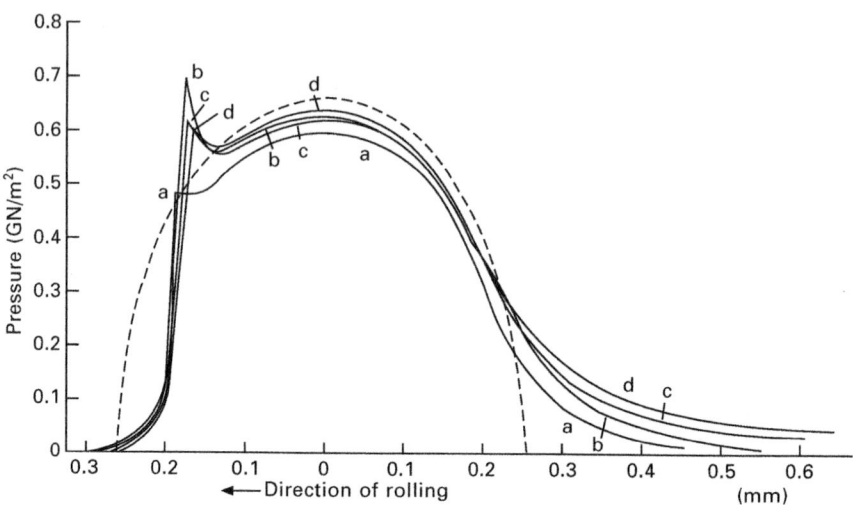

7.7 Variation of pressure distribution with entrainment velocity.

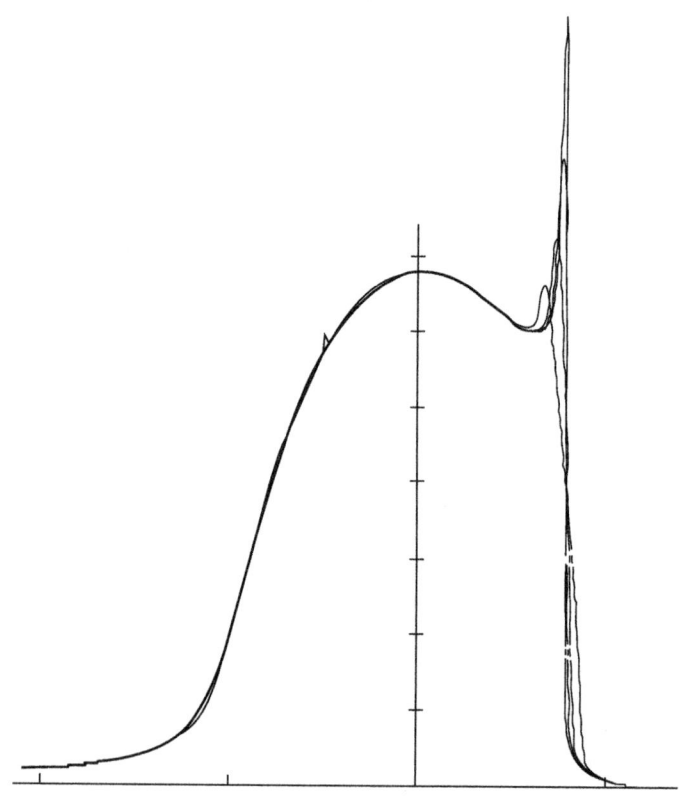

7.8 Effect of the gauge width on the EHL line contact pressure spike; comparison with a numerical solution (Dowson and Higginson, 1977).

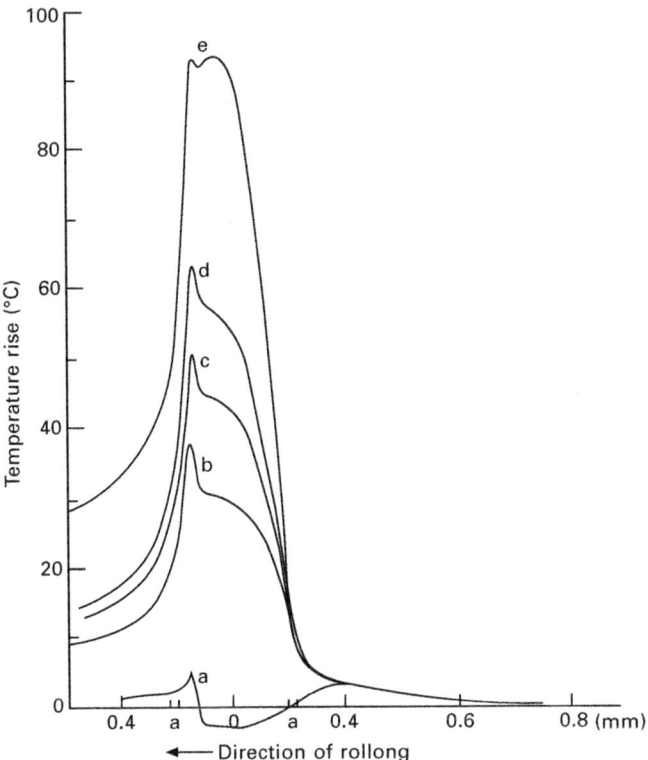

7.9 Profiles showing temperature variation.

speeds the greater heat generated masks that of the rolling speed occurring only within the conjunction and not in the inlet region, where there is always the same temperature rise there ($x > 0.4$ mm). Note also that the surface temperature curves follow in general the shapes of the pressure distributions seen above.

7.3.2   Roller bearings

In this example, both the film thickness and pressure distribution were measured by Safa *et al.* (1982/83) *in situ* in a roller bearing. The bearing was one member of a roller bearing pair supporting a central load applied hydraulically through a ball bearing. To do this, synchronised pressure and film thickness gauges were fabricated on the bottom of the inside surface of the stationary outer race of the test bearing which was 42 mm outside diameter with the 14 rollers each 5.5 mm diameter. The direct laser milled pressure transducer had to be reduced to 5 μm in the rolling direction and 10 μm long. For the film thickness measurement an electrical contact had to be made to the test roller that was rolling between two oil films between the inner and outer races. To do this, a small *axial* hole was spark eroded through the roller centre into which a fine needle rigidly mounted. The tip of this needle brushed over a copper contact *on the outer race* that spanned the region covered by the transducer. Thus, the capacity between the test roller and the transducer electrode could be measured. For a full description of the film thickness measurement technique, see Safa (1982). Figure 7.10 shows some results of pressure distribution for various entrainment speeds. The pressure distribution lacks the detail shown in Fig. 7.7, especially the spike at the exit. This is because of the necessary minaturisation of the transducer for this application. In the next section some experiments will describe measurement of EHL pressure distribution under impact loads.

7.3.3   Impact experiments

In EHL, there are sometimes conditions where the distorted surfaces, separated by the oil film, are subjected to a squeeze load, which when of short duration becomes an impact. A practical application is in ball bearings subject to environmental shock loads. The measurement of pressure by the technique discussed above is made quite simple if, for example, the model

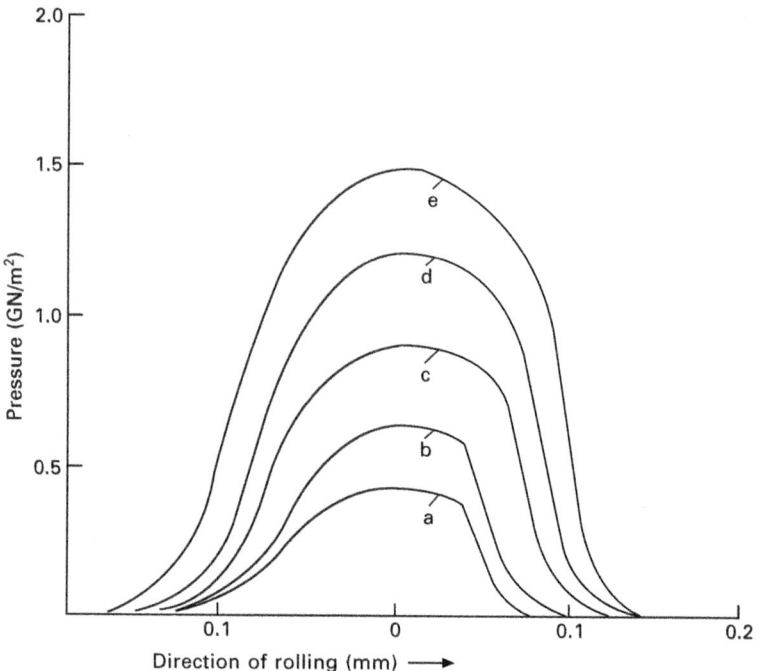

7.10 Pressure distribution in a roller bearing contact.

is a steel ball that is dropped on a stationary plane surface having on it a manganin transducer. Safa and Gohar (1985) used such an arrangement to obtain the ***impact pressure distribution***.

An illustration of a rig that fulfils this purpose is shown in Fig. 7.11. A chrome steel sphere, guided by a Perspex tube, was allowed to fall onto a glass plate on which was deposited a ***manganin pressure transducer*** 200 Å thick. The plate itself was glued to another thicker glass plate to give rigidity to the assembly. A droplet of oil was deposited over the manganin gauge.

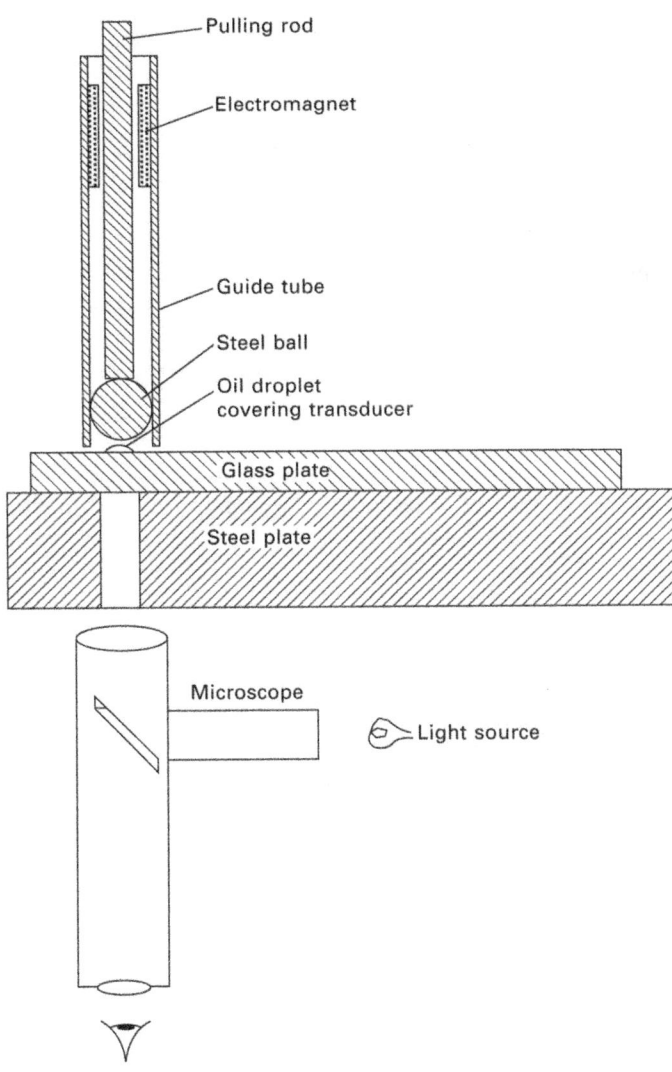

7.11 Impact apparatus.

The guide tube was attached to a three-axis micrometer microscope rack in order to locate accurately the drop coordinates and also measure the fall height. To ensure the correct location of the gauge with respect to the area of pressure caused by the impact event, another microscope with a long focal length objective could view the impact area glass plate from below. This aided the location of the gauge co-ordinates so that the same gauge could map the complete pressure distribution, because it was axisymmetric. Measurements needed only to be taken along a single radius, the location accuracy of which being 1 μm.

The ball was held by magnetised pulling rod that could raise it to a known height above the plate and release it at the appropriate instant. The glass plate itself was firmly attached to a 0.0127 m thick backing steel plate below it, with the glass just protruding beyond its edge, thus forming a short stiff glass cantilever. The exact location of the double pressure gauge was obtained from the microscope below the plate assembly using a sodium lamp to illuminate the area of impact pressure, as illustrated in Fig. 7.12. The weak *Newton's rings* thus formed enabled the precise position of the transducer to be found relative to the impact centre. The hatched region is the calculated Hertzian contact circle diameter of 146 μm for a load of 2.28 N. The transducer and its electrical feed leads are about 200 Å thick, which is approximately 20% of the least EHL film thickness detectable by *two-beam interferometry*.

Signal monitoring and trigger facility

The pressure coefficient of resistance of manganin as a thin film is such that for a pressure change of 0.7 GPa, there is a change in resistance of the transducer of around 1%. Therefore low noise and sensitive resistance monitors are required. Basically, the transducer is one arm of a commercial

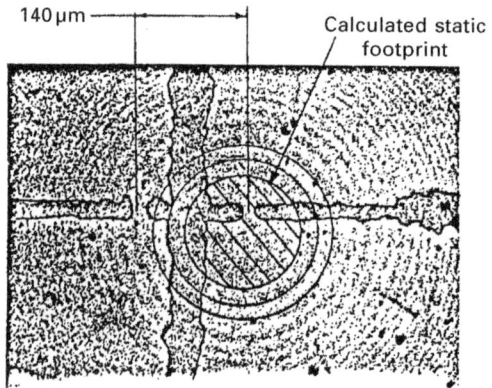

7.12 Double pressure gauge with Newton's rings.

Wheatstone bridge that is balanced at its resistance value under atmospheric pressure. The unbalanced voltage produced by the bridge is reasonably linear relative to small changes in transducer resistance produced. The output from the bridge is amplified and fed to a digital storage oscilloscope, typically with a maximum sampling rate of 1 μs and a total capacity of 1024 samples. For these experiments, the start of the pressure signal commencement triggered the digital oscilloscope.

Fabrication

To fabricate the transducer on a glass slide (an insulator), the method of direct milling of a patch of manganin deposited on the glass slide was used. The gauge assembly was sufficiently thin for it to be semi-transparent at 200 Å. Thus, when Newton's rings were observed between the ball and plate surfaces when the gap was less than 3 μm, the gauge assembly was clearly visible, as Fig. 7.12 has shown. (*In this example, there are two gauges in series, the maximum pressure registered at the first one acting as a trigger for the second one.*)

Calibration

Each transducer, after deposition on a glass slide, was calibrated in a hydrostatic pressure vessel. An alternative and more direct approach was sometimes used, which applied a very slow rolling motion load to the oil-covered glass plate, thus producing Newton's rings in the gap between it, with the oil providing an insulating overcoat, and the steel ball surface. As the load is increased, the almost Hertzian contact circle enlarges. Knowing its radius for a given load and assuming an elliptical pressure distribution, the actual pressure at the gauge can be found and related to the electrical signal height (without time base) registered on an oscilloscope.

Experimental method

The glass slide with its transducer was firmly attached to the glass plate with clear glue. A drop of the test oil was deposited over the transducer and the surrounding area. The tube and ball assembly was then moved with the micrometer microscope as to be directly over the pressure gauge at a known radial distance (r) from the impact centre and from a given ball dropping height. After some adjustments a series of results of the time-based pressure distributions was obtained, at different radii from the impact centre. These experiments were repeated for various drop heights. Because the pressure distribution is axisymmetric, a complete picture of the impact pressure distributions were obtained pressure with time at the impact centre for a given drop height. Figure 7.13 shows the pressure variation with time at various radii during the impact event. Note that these time-based pressure distribution shapes are similar to the steady the steady state ones produced on a distance base (Dowson and Longfield, 1963; see also Chapter 6 on EHL).

The last term on the right-hand side of equation (5.24) is

$$2\frac{\mathrm{d}}{\mathrm{d}t}(\rho h)$$

describing the **squeeze film behaviour**. Alternatively, assuming ρ is constant, it can be written as

$$2\frac{\mathrm{d}h}{\mathrm{d}x}\frac{\mathrm{d}x}{\mathrm{d}t}$$

which is $2\mathrm{d}h/\mathrm{d}x \times$ radial velocity. The first term on the right side of equation (5.24) is

$$\frac{\partial}{\partial x}[h(U_1 + U_2)]$$

so the two terms are of the same form in Reynolds equation, both producing similar pressure distribution shapes, one on a time base and the other on a distance base.

Figure 7.14 shows one-half of the spatial pressure variation at different time instants during the event. Note that the maximum pressure achieved exceeds 3 GPa after 257 µs while surrounding pressures have diminished more compared with those of earlier times. The total time exceeds the time

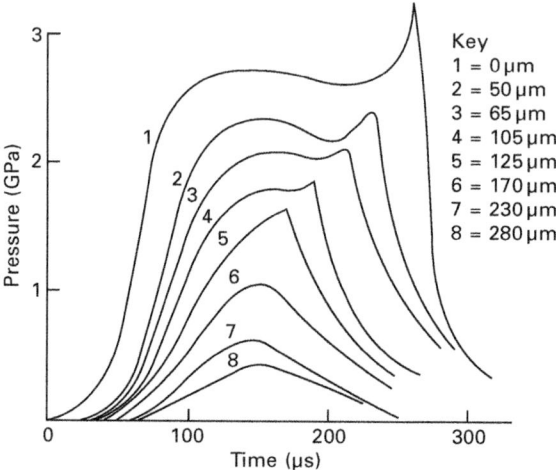

7.13 Pressure against time during impact.

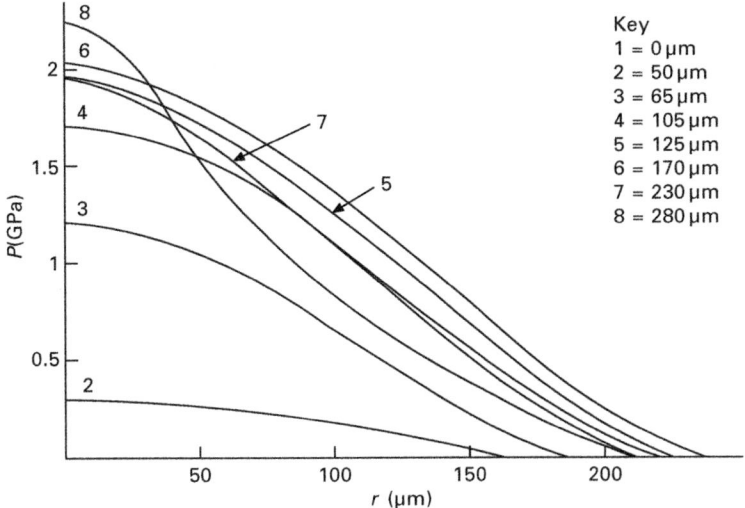

7.14 Impact pressure distributions at each instant of time.

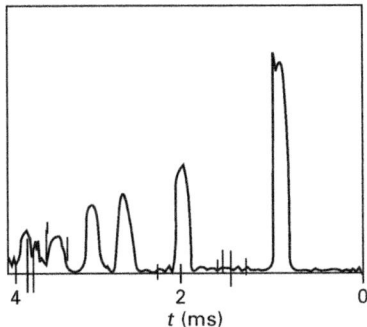

7.15 Impact sequence of a bouncing ball.

had the fall been without any oil present. This is because, with oil present, energy is absorbed over a longer period (Safa and Gohar, 1985). Figure 7.15 shows the decaying pressure distributions produced from a ball bouncing on the glass oily plate, with the time between bounces diminishing, as would be expected.

As for the distorted elastic geometry of the contact, Fig. 7.16 shows, for comparison, a numerical solution of the pressure distribution and film shape when a ball impacts an oily elastic surface was given by Al-Samieh and Rahnejat (2002). Observe the similarity with the experimental pressure shapes seen in Fig. 7.14. The sequence shown does not show the rebound phase. The complete event appears in Al-Samieh and Rahnejat (2002).

In Fig. 7.17, an impact pressure trace, taken, from one of the experimental

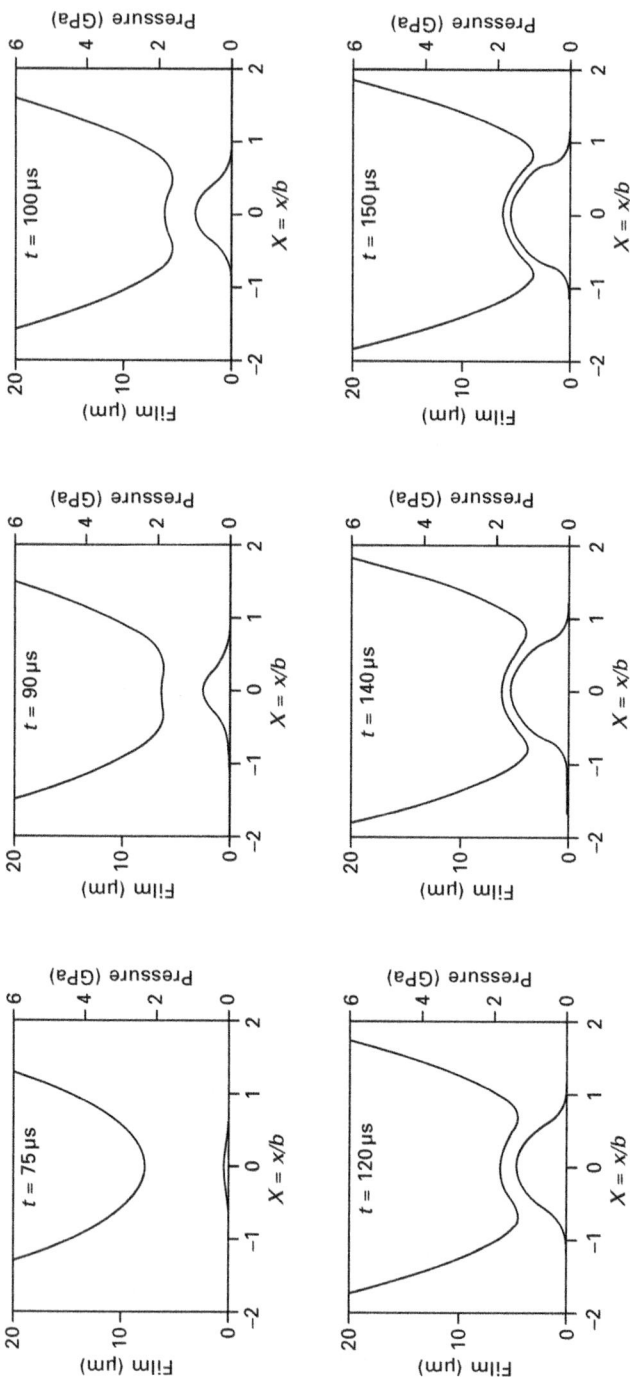

7.16 Numerical solution for a bouncing ball (from Al-Samieh and Rahnejat, 2002).

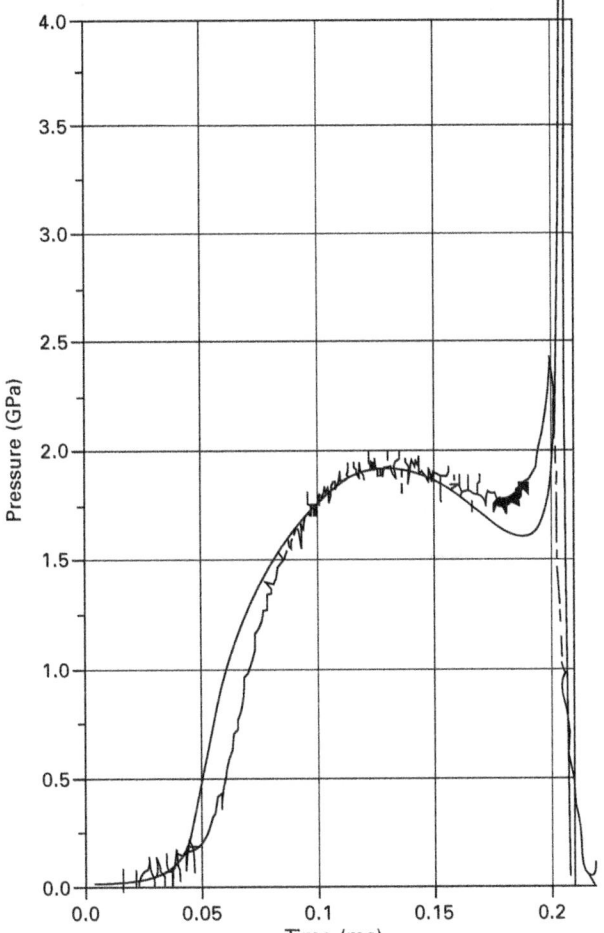

7.17 Impact pressure distribution compared with a numerical solution (Nishkawa *et al.*, 1995).

results, described above, is compared with a numerical solution by Dowson and Wang (1994). For the reasons discussed above, the experimental exit spike registers a lower value than the numerical solution.

Another experimental example of the respective distorted elastic shapes of a ball impacting an oily plane disc producing pure squeeze action, was shown in Fig. 7.13 of a paper by Nishikawa *et al.* (1995). Using a ball and disc machine, they subjected the pair to vibrational axial squeeze motion thus producing a series of cyclic impacts similar to the bouncing ball example in Fig. 7.15 . The measurement of the composite film shape was by the widely used method of *optical interferometry*, see for example, Gohar (2001).

Knowing the respective Young's moduli of the ball and disc, the deformation dimples in both surfaces at one stage in the impacts were obtained and are clearly seen in that paper.

7.3.4   Pressure distribution in a rolling point contact

A further application of manganin pressure gauges was by Johns-Rahnejat and Gohar (1994). They designed a machine that had nearly pure rolling point contact. It had a stationary glass plate, with the manganin gauge flash deposited on it. The glass plate was supported by three equispaced rolling balls in a cage, these being driven from below by a lower steel plate. The cage assembly speed was therefore about half the lower plate rotational speed.

To position the ball and cage assembly with respect to the gauge on the stationary floating upper plate, three adjustable screws, in a slightly larger diameter fixed collar surrounding it, could locate it radially. Adjusting the screws up to 0.1 μm enabled the position of the gauge within the contact area to be varied.

The plate was lapped to a finish of 0.1 μm centre line average (CLA) while the balls had a surface finish of 0.05 μm CLA and sphericity of 0.1 μm. The axial load on the top race was applied through an externally pressurised air thrust bearing and a lever arm weighted hanger assembly. The transducer and disc arrangement allows the position of the transducer relative to the test ball to be adjusted so that the whole anticipated contact area was covered. One pass of the ball, under the gauge at a chosen location, resulted in a pressure–time trace, from zero at the local inlet, back to zero at the local outlet. The gauge position was then changed for further traces until the whole anticipated contact area could be covered.

Figure 7.18 shows point contact **EHL pressure distributions** through various vertical sections in the entrainment direction for a given entrainment speed. The distributions follow the usual EHL point contact shapes seen in Gohar (2001). The inlet sweep commencements are very close to the start of the equivalent Hertzian pressure distribution, suggesting some starvation there, This is quite a common occurrence in ball bearings that are not continuously flooded with oil from an external source. The oil, initially deposited on the plate (or race), is pushed out transversely by the rolling elements to leave a track at exit with ramped sides. There is a full discussion of lubricated concentrated contact exit shape by Yin *et al.* (1999) and Chan *et al.* (2007).

Isobars covering a complete contact area are shown in Fig. 7.19(a). The inset is an interference fringe contour map of a typical EHL point contact. Note the similar exit boundaries of the **cavitation wake**. Figure 7.19(b) shows a similar experimental EHL isobar map found by employing **Raman microspectroscopy**, to be discussed in the next section. The lowest recorded pressure isobar in (b),

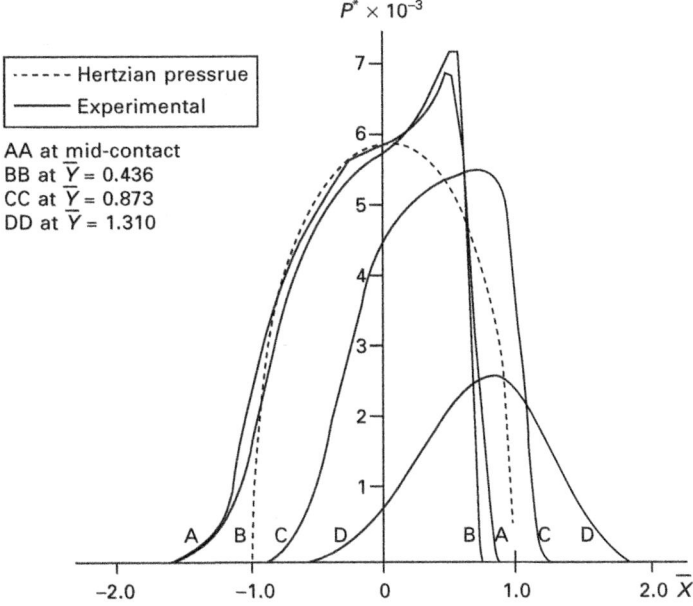

7.18 Rolling point contact pressure distributions.

seen in the inlet region, is similar to that of a rigid isoviscous solution found by Chan *et al.* (2007). There is a small bulge on the principal axis that may be associated with a ramped film shape of the approaching oil-free surface. In that small area probably some reverse flow is occurring. Finally, Fig. 7.19(c) shows isobars from a flooded inlet numerical solution for a much higher load (Gohar, 2001). Note the crescent shape of the isobars near the exit, with local peaks appearing on them in all three examples.

Disadvantages of employing manganin pressure gauges (and alternative materials for film shape and temperature measurement gauges) are the complexities of depositing them and the subsequent wear that occurs during frequent rollovers of the contiguous body. Below, we discuss an alternative direct method that measures the pressure distribution from the changing behaviour with pressure of oil film seen over the contact. Some indirect methods of solution re also discussed.

7.4   Alternative methods of measuring contact pressure

7.4.1   Direct methods: Raman spectroscopy

If a monochromatic light beam is focused on a sample, such as the thin oil film in an EHL rolling contact, for example, a ball rolling over a sapphire plate, part of light is absorbed and part is scattered. The scattered light has

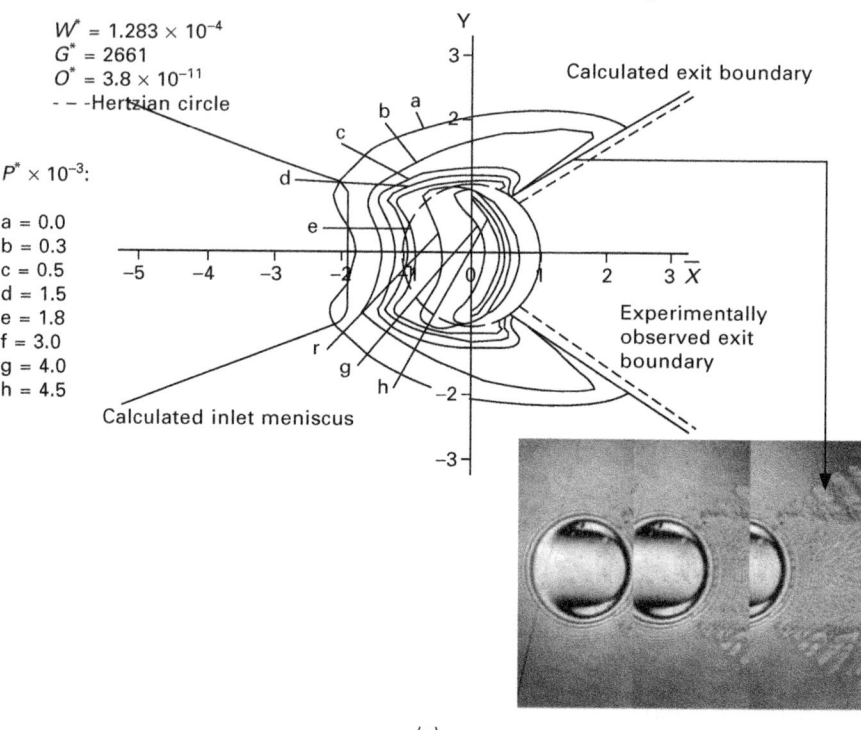

Pressure isobars for $U^* = 3.8 \times 10^{-11}$ and $W^* = 1.283 \times 10^{-4}$ using Shell Vitrea 220 oil

$W^* = 1.283 \times 10^{-4}$
$G^* = 2661$
$O^* = 3.8 \times 10^{-11}$
– – –Hertzian circle

$P^* \times 10^{-3}$:

a = 0.0
b = 0.3
c = 0.5
d = 1.5
e = 1.8
f = 3.0
g = 4.0
h = 4.5

Calculated exit boundary

Experimentally observed exit boundary

Calculated inlet meniscus

(a)

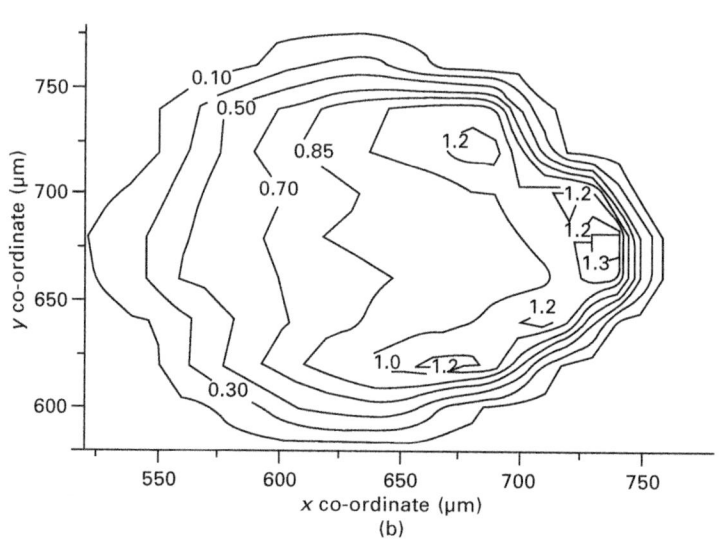

(b)

7.19 (a) Pressure isobars using a manganin pressure gauge, (b) using Raman microspectroscopy (c) a numerical solution.

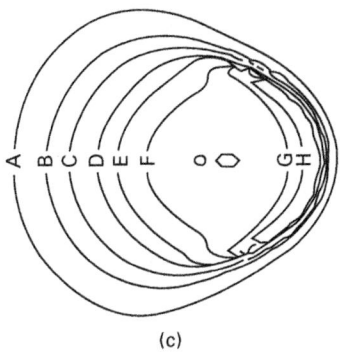

(c)

7.19 Continued

an elastic component (***Raleigh scattering***) that suffers no frequency change. However, there is also a much smaller proportion that has inelastic scattering. The latter is called ***Raman scattering*** after the discovery of its existence by Raman for which he received the Nobel Prize in 1930. For this component, the gain or loss of the energy quanta of photons in a sample is related to the vibrational energies of the molecules present. Under the effect of changing the applied pressure, the equilibrium distances between neighbouring atoms of these molecules, decreases. This causes their frequencies of vibration to alter and therefore there is a shift of the corresponding ***Raman lines*** in the Raman spectrum seen. The changing behaviour of these lines within the spectrum can be related to the pressure there. The technique was first employed in an EHL entrapment by Gardiner *et al.* (1983).

For calibration, the shifts in vibration frequency of the lubricant molecules in relation to pressure, were exploited by Jubault and Mansot (2002). They employed a diamond anvil cell that exploits the known pressure dependence of the fluorescence spectrum of ruby. An example of their experimental results is an isobar map of an EHL rolling point contact, using Raman spectroscopy. It is reproduced above in Fig. 7.19(b).

Later, Jubault *et al.* (2003) performed similar experiments to measure the pressure distribution but this time in a rolling-sliding point contact. Coloulon *et al.* (2004) have also employed the Raman technique to measure point contact EHL pressure profiles where one of the surfaces had a dent of known shape inserted in it.

7.4.2   Indirect methods for determining pressure distribution

By photoelasticity

Klemze *et al.* (1971) used photoelasticity to find the lines of principal stress difference in the glass member of a lubricated disc machine, the other member being of steel. Knowing the relationship between these principal stress lines and the direct stresses, it was possible to find the corresponding pressure distribution. Because of the glass member and the necessary disc thickness, pressures were relatively low and so the expected end spike was not seen. The EHL film shape was therefore determined from its relationship with pressure distribution and its thickness from the film geometry equation. Its shape coincided with what was expected at low loads with some **end closure** present. The clarity of the stress lines in this finite line contact was quite poor owing to the stress discontinuity at the line contact end (part of the exit pressure spike seen in an infinite line contact or in a **finite length contact** far from the ends).

Using an experimental film thickness contour map

In the indirect method, the experimentally determined EHL film thickness distribution is used as data input. To measure the film thickness, interferometry is employed because it has now become very accurate right down to fractions of a micrometer. Assuming such an experimental film thickness contour map is available, the corresponding pressure distribution and apparent viscosities can be computed from the elastic deformation and load equations. Various researchers have used this approach. A problem is that the estimated 'apparent' viscosity in the EHL conjunction, which can be very high because of its exponential variation with pressure, suffers large fluctuations from small pressure variations, themselves caused by modest changes in film thickness. A more recent approach, which removes these pressure fluctuations, employs an inverse model. It assumes that the boundary conditions of the region of pressure are initially unknown, these being represented as a series containing a large number of unknown coefficients that need to be found. Another variation of the inverse method is by Lee *et al.* (2003). Their method required only a few measuring points of film thickness data to estimate the wanted pressure distribution. They claimed it was quite accurate when compared with a numerical solution.

7.5 Conclusions

This chapter has discussed the various experimental methods used to measure pressure distribution in EHL contacts. Most of these have concentrated on results obtained from deposited manganin pressure gauges on one of the film bounding surfaces. A recent and very promising technique relies directly on the EHL film properties (Raman) to determine the pressure distribution. Alternatively, an indirect measurement of the pressure distribution employs a number of discrete points describing the film geometry to solve for the pressure distribution.

7.6   References

Al-Samieh, M. F. and Rahnejat, H. (2002), 'Physics of lubricated impact of a sphere on a plate in a narrow continuum to gaps of molecular dimensions', *J. Phys D: Appl. Phys.*, **35**(18), 2311–2326

Chan, R. T. P., Martinez-Botas, R. F. and Gohar, R. (2007), 'Isoviscous flow past a rigid sphere partially immersed in a thin oil film', *Lubrication Sci.*, **19**, 117–212

Coloulon, S., Jubault, J., Lubrecht, A. A., Ville, F. and Vergne, P. (2004), 'Pressure profiles measured within lubricated contacts in presence of dented surfaces', *Tribology Int.*, **37**, 111–117

Dowson, D. and Higginson, G. R. (1977), *Elastohydrodynamic Lubrication*, Pergamon, Oxford

Dowson, D. and Longfield , M. D. (1963), 'An elastohydrodynamic lubrication experiment', *Nature*, **197**, 586

Dowson, D. and Wang, D. (1994), 'An analysis of the normal bouncing of a solid elastic ball on an oily plate', *Wear*, **179**, 29–32

Gardiner, D. J., Baird, E., Gorvin, A. C., Marshall, W. E. and Dale-Edwards, M. P. (1983), 'Raman spectra of lubricants in elastohydrodynamic entrapments', *Wear*, **91**, 111–114

Gohar, R. (2001), *Elastohydrodynamics*, 2nd Ed., Imperial College Press, London

Johns-Rahnejat, P. M. and Gohar, R. (1994), 'Measuring contact pressure distributions under elastohydrodynamic point contacts', *Tribotest J.*, **1**, 33–53

Jubault, I. and Mansot, J. L. (2002), '*In situ* pressure measurements using Raman microspectroscopy in a rolling elastohydrodynamic contact', *Trans ASME J. Tribology*, **124**, 114–120

Jubault, I., Molimard, J., Lubrechet, A. A., Mansot, J. L. and Vergne, P. (2003), '*In situ* pressure measurements and film thickness in rolling/sliding lubricated point contacts', *Tribology Lett.*, **15**, 421–429

Klemze, B. L., Gohar, R. and Cameron, A. (1971), 'Photoelastic studies of lubricated line contacts', *Inst. Mech. Engrs. Tribology Convention*, London, 18–30

Lee, R.-T., Chu, H.-M. and Chiou, Y.-C. (2003), 'Inverse approach for estimating pressure and viscosity in the elastohydrodynamic lubrication of circular contacts', *Proc. Inst. Mech. E, Part J, J. Eng. Tribology*, **217**(4), 277–288

Nishikawa, H., Handa, K. and Kaneta, M. (1995), 'Behaviour of EHL films in reciprocating motion', *JSME Int. J. Ser. C*, **38**(3), 558–567

Safa, M. M. A. (1982), 'Elastohydrodynamic studies using thin film transducers', PhD thesis, Imperial College, University of London

Safa, M. M. A. and Gohar, R. (1985), 'Pressure distribution under a ball impacting a thin lubricant layer', *Trans ASME J. Lub. Tech.*, **108**, 372–376

Safa, M. M. A., Anderson, J. C. and Leather, J. A. (1982/83), 'Transducers for pressure, temperature and oil film thickness measurement in bearings,' *Sensors and Actuators*, **3**, 119–128

Yin, J., Eissa, K., Gohar, R. and Cann, P. M. E. (1999), 'Rebounding lubricant layers', *Lubrication Sci.*, **12**, 1–30

Part II

Engine and powertrain technologies and applications

Section II.I

Overview

8

Tribological considerations in internal combustion engines

D. R. ADAMS, Ford Dagenham Development Centre, UK

Abstract: This chapter presents a practitioner's view of engine tribology in an internal combustion engine. The tribological performance of an engine in large measure still relies on simplified models, empirical knowledge, experience and the results of rig and engine testing. Advances in the mathematical treatment of component tribology, although not yet sufficiently advanced to predict some of the tribological phenomena, nevertheless provide useful guidance for optimisation of tribological performance.

Key words: IC engines, tribology, pistons, rings, bearings.

8.1 Introduction

Tribology is defined as the study of interacting surfaces in relative motion and the practices related thereto (Jost Report, 1966). Tribology is fundamentally about materials, surfaces and lubrication and the interplay that determines the wear and frictional properties that are so sought by the designer (see Chapters 2 and 5). The function of an internal combustion (IC) engine is dependent on upwards of 150–200 moving parts operating reliably for anywhere between 5000 and 50 000 hours in applications ranging from a passenger car to a large commercial vehicle. Without an understanding of tribology and the application of its principles the chances of an engine working reliably would be very small.

This chapter is primarily about the tribology of an IC engine where the lubricant is engine oil. The tribology of other systems, for example the exhaust valve seat, the *exhaust gas recirculation* (EGR) valve, the turbocharger waste gate are no less important but for reasons of space have not been included. The main engine systems where fluid lubrication dominates the tribology are the crank train, the valve train and the turbocharger (Figs 8.1 and 8.2). In this chapter the discussion will be confined to the so-called power conversion system, i.e. the piston, connecting rod and bearing assembly.

The treatment of engine tribology in this chapter has avoided a mathematical approach, partly because an up-to-date description of numerical analysis for many engine systems is covered elsewhere in this volume (see Chapters 10–20). However, it also seemed more appropriate to present a practitioner's view of engine tribology. Despite the many advances in computer modelling

251

8.1 Crank and valve train systems.

8.2 Turbo charger.

and technical understanding the tribological performance of an engine in large measure still relies on simplified models, empirical knowledge, experience and the results of rig and engine testing. While computer-aided engineering (CAE) treatments are well developed for structural analysis, in general the mathematical treatment of component tribology, at least from the perspective of a design tool, remains rather immature, hampered by a still-evolving understanding of the complex physics of wear and dynamics in some engine systems and the development of validated damage models. Further comments on the role of testing and CAE in the design of engine systems will be made in later sections.

8.2   Issues of cost, competition and reliability in internal combustion (IC) engine tribology

As a topic, *cost* rarely, if ever, appears in a textbook on tribology. However, in an industry that is as cost conscious as the automotive industry it deserves a mention. Automotive design is about value engineering. In respect of an IC engine the aims of the designer of any moving part are about meeting the desired functionality (life, oil consumption, noise, vibration and harshness (NVH), weight, etc.), while minimising cost. Consideration of value for money underpins the designer's decisions. Higher-cost parts must deliver proportionately improved performance or solve a real problem in order to be adopted. The 'best' product is rarely affordable. Although cost is a big constraint it adds a dimension that can make automotive engineering a much more interesting task – achieving value for money creates as many opportunities for creativity as pure research. No further mention of cost will be made explicitly in this chapter but the reader needs to be aware of this underlying preoccupation as much as the physical principles employed shapes the actions and decisions of the automotive design engineer.

In the early decades of engine design, tribological failures were not unusual. Intervals between services were short as were engine lives. In some respects the consumer was probably more tolerant of problems – after all engines were complex systems and design and manufacturing methods were relatively unsophisticated. Today, however, the modern consumer's expectations have risen by orders of magnitude. Partly led by the kind of quality ethos championed by the Japanese original equipment manufacturers (OEMs) in the 1970s and 80s (Womach *et al.*, 1990) changes in the industry have transformed personal mobility from a convenient mode of transport to fashion statement. Poor reliability has no place in the modern automotive industry. With the advent of reliability surveys such as JD Power, *What Car?* and others the consumer has immediate access to information that will influence the buying decision. Continued poor reliability will gradually erode consumers' perception of a brand, a process that can take years to reverse. Thus, reliability is an additional preoccupation of the engine designer.

8.3   Drivers for tribological design and innovation

Ever since the first oil crisis of 1971, when the oil price effectively trebled overnight, the price of crude oil has steadily risen and with it concerns about the environmental impact of the IC engine.

Today, CO_2 emission from engines is becoming an increasingly important aspect of engine design as designers strive to improve fuel economy. Together with the requirement to meet ever-tightening legislative targets on controlled gaseous emissions these two factors are driving enormous changes in engine design. For example:

- Downsizing of engines and increased use of turbocharging resulting in higher power densities. Although lower friction losses are a consequence of downsizing this is offset by the higher demands on components operating at higher power densities.
- More fuel-efficient combustion technologies such as *gasoline direct injection* (GDI) introduce more 'diesel-like' loads in gasoline engines. Cylinder pressures in a conventional port fuelled injection (PFI) engine are typically 80–90 bar. In a diesel engine pressures range from 130 to 200 bar. In a GDI engine cylinder pressures range from 100 to 130 bar.
- The use of particulate filters in exhaust systems on diesel engines is mandatory for EU5 engines. To prevent filters becoming clogged the particulates are regularly burnt-off in a process known as regeneration using additional fuel injected after the combustion process. This post-injection of fuel in diesel engines impinges on the bore wall, causing local depletion of lubricant. Some of this fuel inevitably ends up in the sump where dilution levels of >10% can be found. This has the effect of reducing oil film thicknesses throughout the engine.
- Biofuels are being introduced to supplement fossil fuels. The use of alternative fuels, particularly when they find their way into the lube oil as with post-injection, can have a significant effect on oil properties and hence component tribology.
- To improve vehicle drivability there is a demand for higher engine torque at low speeds. This generates very low oil film thickness in lubricated parts, particularly for crankshaft and connecting rod bearings of turbocharged diesel engines.
- Maximum cylinder pressures have risen significantly over the past 10 years driven by the requirement to achieve emission pollutant regulation. This pressure has risen from 130 bars to over 200 bars in turbocharged

diesel engines. Again, the pressure increase affects the behaviour of the crankshaft bearings, the connecting rod bearings and the piston/ring/bore interfaces.

- There has been a trend to lower lubricant viscosities as one means of reducing engine friction and fuel consumption. From using 15W40 10 years ago, European engines are now filled with oil 5W30. This results in a reduction of approximately 50% in the nominal dynamic oil viscosity. This reduction affects all the engine components operating in hydrodynamic lubrication.
- The effect of changes in combustion on the lubricant has to be taken in consideration. Today, the high soot oil content due to EGR, the oil consumption reduction currently controlled by ring loadings and the fuel dilution in the oil, cause significantly increased wear of rings and other components.
- The average temperature in engine use has increased in direct relation to the specific power increase. Today the highest rated passenger car turbo diesel engine in production today has a *specific power* of 75 kW/L up from 35 kW/L 10 years ago. Such specific power allows car manufacturers to develop smaller engines, but more loaded, in order to use them in more efficient condition to save energy.
- Increased service intervals.

The effect of all of these changes is that oil films are thinner and components are spending a greater proportion of time running at or close to a mixed lubrication regime. In short the tribological environment is becoming much more challenging. Simultaneously, new targets for engine reliability must be achieved by car manufacturers. Commonly an engine is considered to be reliable when it operates for a certain mileage without any problem on the main components. If one problem occurs, it must happen with a very low probability. This presents significant challenges for engine testing, an issue that will be returned to in a later section of this chapter.

Owing to the increased time spent under *mixed lubrication* conditions (hydrodynamics and asperity interactions) the ability to predict realistic component damage is necessary. Damage modelling is well developed for structural performance but remains relatively in its infancy for wear and other derived tribological factors (see Chapter 2).

8.4 A systems view of the piston/ring/cylinder bore interface

Before going into details about each component in the *power conversion system* (PCS) it is appropriate to stand back and consider the function of the assembly from a systems perspective. In this way it will be easier to appreciate some of the interactions that determine the performance of the assembly. Figure 8.3 shows a tribological overview of the PCS. From a functional perspective, the purpose of the PCS is the efficient conversion of fuel energy to mechanical energy at the crankshaft. Undesirable 'outputs' or 'error states' of the piston assembly are friction, oil consumption and noise. These error states are the result of interactions between several tribological systems. The aim of the designer is to minimise these undesirable error states without compromising the energy transfer. A brief overview of factors involved in generating these error states will be given in this section. This is followed by more detailed descriptions of the factors involved in design optimisation (also see Chapters 10–20).

8.4.1   Friction

Estimates of the break-down in **engine friction** from several sources (Parker and Adams, 1982; Bishop, 1964; Ku and Patterson, 1988) suggest that the mechanical losses in an engine are equivalent to about 40% of engine output: approximately 40% of these losses of the engine originate from the PCS. Thus, the frictional losses in the PCS could, if reduced to zero, improve fuel economy by around 16%. For this reason the PCS plays an important role in the drive to reduce CO_2 emissions.

In general a strategy for friction reduction involves developing designs that:

- have reduced area of contact enabling lower oil film thickness and hence operation near the bottom of the **Stribeck curve** (see Rhodes and Parker, 1984);

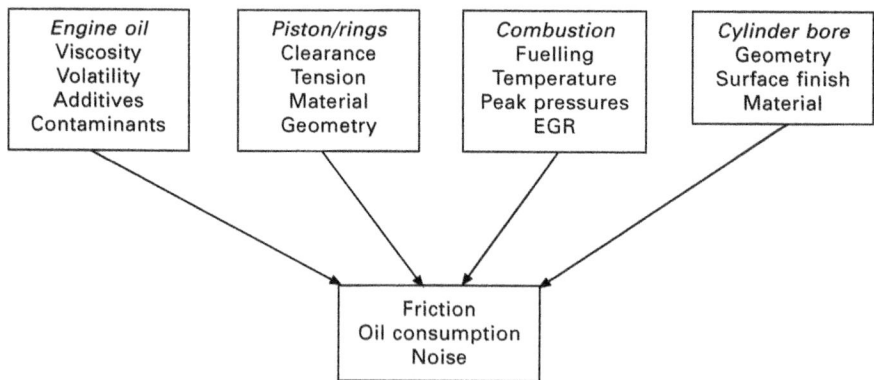

8.3 Overview of interactions in the power conversion system (published with permission of Nagel GmbH).

- are more conformable and therefore tolerant of shape errors in the contact area;
- have reduced contact loads;
- have low friction surfaces/coatings (see Chapter 7).

Enablers for these designs include:

- reducing surface roughness, enabling hydrodynamic lubrication to be maintained at thinner oil films (see Chapters 3 and 5);
- achieving desired surface form with fewer shape errors, e.g. better roundness of crank journals and cylinder bores (see Chapters 18–20);
- cleaner engine builds to avoid contamination;
- better oil filtration to remove wear particles and other contamination.

8.4.2   Oil consumption

Several engine systems are involved in the control of *oil consumption*:

- The piston ring/bore interface controls approximately 60–80% of total oil consumption. The main mechanisms of oil loss are oil evaporation from the surfaces of the cylinder wall and piston, oil throw off due to inertia and oil carry over in the gas flows from the rings to the combustion chamber. There are many excellent publications explaining these mechanisms in detail (McGeehan, 1979; De Petris *et al.*, 1996; Yilmaz *et al.*, 2004; Hill, 2001)
- Oil loss through the turbo-charger bearings and valve stem seals accounts for around 5–10% of total oil consumption. Developments in valve stem seals enable virtually zero oil loss.
- Oil loss through the *positive crankcase ventilation* (PCV) system accounts for the balance of oil consumption. The PCV ensures that blow-by gases are purged from the sump and drawn back into the intake manifold. Inevitably a small amount of oil is picked up by the blow-by gases and this contributes to overall oil consumption. The oil is separated by passing the gasses though a baffle, filter or mesh but a badly designed separator can be responsible for poor oil consumption performance. In the worst cases the PCV system can be responsible for 20–30% of total engine oil consumption.

Control of oil consumption has become much more important in recent years since the quantity of oil passing into the exhaust system has a direct influence on the performance of the after-treatment system and hence emissions compliance. Longer service intervals and the expectation of fewer oil top-ups is also a factor in driving lower levels of oil consumption. Subsequent chapters will expand on the way lower oil consumption is being addressed.

8.4.3   Noise

Although noise is not an obvious characteristic of engine tribology the optimisation of **NVH** involves many of the design features relevant in component tribology.

Sources of noise from the PCS include the following:

* Impact between the piston and cylinder bore, known as **piston slap**. This can be controlled by careful attention to piston clearance, control of **piston secondary motion** (see Chapters 14 and 15) – mostly by optimising the offset of the pin centre relative to the piston axis – and the shape and stiffness of the piston skirt. There are several types of piston impact known by the nature of their noise signature:
  - **piston rattle**, where the top land impacts on anti-thrust side just before the top dead centre (**TDC**) on compression stroke;
  - **croaking**, where the top land impacts the thrust side just after TDC firing.
* Bearing noise which relates to the interaction between the pin/journal and bearing. One of the most common noises originates from the interaction between the gudgeon pin and connecting rod small end and is referred to as **pin tick**. Excessive and/or non-circular clearance is responsible for this type of noise (Moshrefi *et al.*, 2007)

8.5   The development process in internal combustion (IC) engines

8.5.1   Overview

It is impossible in confines of a single chapter to describe the complete design process. However, the steps that will be outlined below should give a basic understanding of the processes that govern the tribological performance of the PCS and provide a framework for outlining some of the main tribological considerations of the system.

A sketch of the PCS (Fig. 8.4) illustrates the main regions where tribological factors concern the designer:

* Loading at each of the joints in the linkage – main bearings, big-end bearings, small-end bush and piston boss. In common with all bearings the criteria for a successful design in each case is a low friction, low wearing joint that can carry the loads for the life of the engine.
* Ring to groove lubrication: the most critical interaction is at the top ring due to the high temperatures and exposure to unburnt fuel and combustion products.
* Ring to bore contact: the ring to bore interaction determines the sealing efficiency of the ring pack, specifically blow-by and oil consumption.

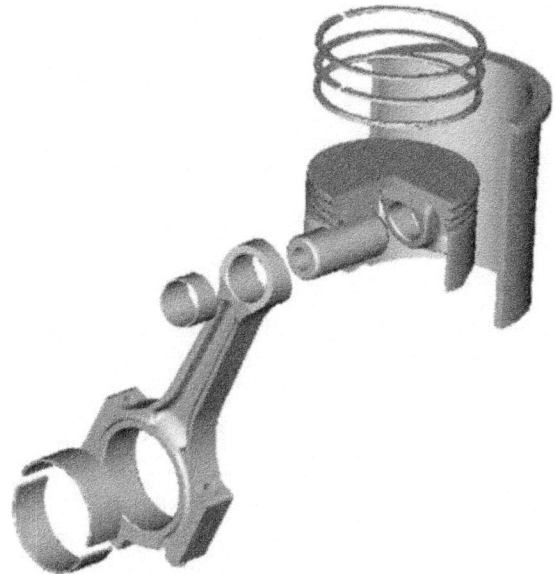

8.4 Power conversion system.

The performance of the **ring pack** (see Chapters 10 and 15) is influenced not only by the shape of the bore surface but also the surface finish.

• Piston to bore contact. The piston supports the axial loads arising from gas pressure and inertia effects.

Each of these areas will now be expanded in more detail.

8.5.2   Main and big-end bearings: background and general considerations

A typical construction of an **engine bearing** is shown is Fig. 8.5. Two important requirements for a bearing material are (a) that it is strong enough to support the loads and (b) that the wear and seizure resistance is consistent with bearing oil film thickness. For the most highly loaded components – the lower main and upper **big-end rod bearing** (see Chapters 19 and 20) – the bearing is conventionally designed as a three-layer or tri-metal system: the steel backing, a bearing substrate and a surface layer or overlay. Simplistically the load capability of the bearing is governed by the substrate material, typically a bronze-based material, and the wear properties are controlled by the overlay. In practice the overlay must exhibit both good wear and fatigue strength. For more lightly loaded bearings a cheaper **bi-metal construction** involving the steel backing and an aluminium backing is invariably used.

The trend towards higher power densities due to downsizing, higher cylinder pressures, higher torque requirements at low speeds and higher temperatures

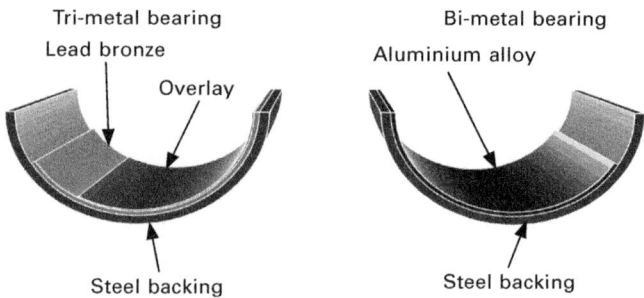

8.5 Construction of a half-shell bearing (published with permission of Mahle Engine Systems UK Ltd).

has placed extra demands on bearing materials to support increased loads and thinner oil films. In an ideal world the bearing designer would be given the freedom to define the complete geometry of the bearing system – diameter, width and shape of the mating surfaces. In practice the engine designers' need to achieve an overall package size for the engine inevitably constrains the available space for the bearing. Thus, bearing suppliers have had to invest heavily in bearing technology – materials, coatings, CAE – to meet the demands of modern engines (see Chapter 18). In the late 1990s one of the most significant changes in bearing design came about with the development of *sputter overlays*, a process in which the bearing overlay is deposited in a vacuum. This created a stronger, more wear-resistant surface that significantly increased the load-carrying capacity of the big-end bearing and was one of the enablers for the introduction of modern high-speed passenger car diesel engines.

A further challenge in the last 10 years has been the impact of the ***End-of-Life Vehicle (ELV) Directive***. This Directive aims to reduce the amount of waste from vehicles (cars and vans) when they are finally scrapped, which is forcing the elimination of hazardous elements to enable recycling of vehicle parts. The Directive demands the elimination of lead in bearings by 2008. This has been particularly challenging for bearings in diesel engines which, because of their aggressive environment, have relied on tri-metal bronze bearings for successful operation. The bearings industry has responded in a number of ways – the substitution of lead by bismuth has enabled many of the original properties of the bronze bearing to be retained without loss of functionality. In addition the effect of the ELV Directive has been to drive the development of stronger aluminium bi-metals, thus avoiding the need for the original bronze alloy while at the same time reducing cost. These developments have been successful and in the case of Ford have enabled all new diesel engines in the last 3 years to be launched with lead-free bearings. However, one of functions of an overlay is the ability to absorb foreign particles and conform to minor geometrical imperfections in the bearing system. This property of embedability has not been so easy to replicate in aluminium bearings and in at least one case has prevented the introduction of a bi-metal (see Chapter 18).

The behaviour of journal bearings is critical to an engine's operation from the point of view of performance and durability. Hence, the development of techniques for engine bearing analysis has received extensive attention over the years. The methodology for designing a bearing commonly starts with simplified models that can be used to rapidly assess a range of potential design solutions. These so-called rapid methods employ simplified bearing analysis techniques such as Mobility (see Chapter 19) to get quick estimates of bearing performance. Later on, further improvements in computational power led to the development of more rigorous analyses allowing additional features such as grooving, oil properties, non-circular film shapes and oil supply to be incorporated (see Chapter 20). Graphical aids could also be constructed by running many cases using such rigorous methods, and then using regression and curve fitting to generate simple to use charting methods (see Chapter 18).

However, with trends towards higher power output and lighter weight components it is apparent that as the bearing environment becomes more aggressive so the assumptions about some of the boundary conditions, particularly the rigid bearing housing have become increasingly important as determinants of actual bearing performance. To address this *elastohydrodynamic lubrication* (EHL) techniques have been developed for bearings (see Chapters 6, 14–17 and 20). A good example of the application of EHL is the effect of crankshaft journal shape (see Chapter 20). With minimum oil films typically less than 1 µm it is not too difficult to imagine that in respect of the running surfaces small deviations in the journal shape would disrupt the oil film and significantly reduce the load-carrying capacity of an engine. Assumed to be perfectly cylindrical, it is often in practice barrelled and may contain high-order deviations in roundness known as lobing. The work of Wang and Parker (2003) using an elastohydrodynamic model of a 2 L diesel engine found that deviations from roundness of the order of as little as 1 µm can have a profound effect on the lubrication of engine bearings (Fig. 8.6).

The future development of the bearing system requires as much emphasis on improving manufacturing processes, for example improving the fidelity of mating surfaces, as it does on materials and design development. Other factors of importance in achieving reliable bearing performance include engine cleanliness and filtration systems. Thus, further advances in engine bearing performance will need to focus on the whole bearing system, not just the bearing itself. A quote from one of Glacier Metal's bearing experts frequently echoes in my mind: 'Bearings themselves never fail; it is only bearing systems that fail'.

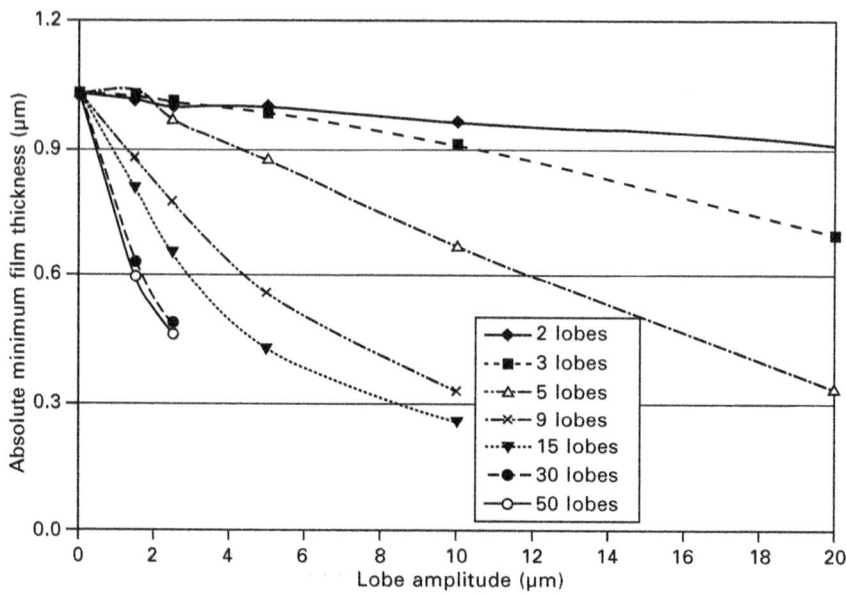

8.6 Effect of crankshaft lobing on bearing oil film thickness (published with permission of Mahle Engine Systems UK Ltd).

8.5.3   Small-end and piston boss

As with the crank bearings the main concerns are to achieve an adequate oil film and strength. Unlike the crank bearings the tools available for the designer are largely empirical. In many ways the piston-to-pin joint (***small-end bearing*** or ***wrist pin bearing***) is far from an ideal bearing design inasmuch as the pin forms a cantilever to support the reciprocating motion and axial loading of the piston causing considerable deformations of the pin to take place. A typical pin shape under loading is shown in Fig. 8.7. In early designs of pistons and rods, up to 15–20 years ago, the constraints of manufacturing technology would have limited the shapes of the connecting small end and ***piston boss*** to be cylindrical. Thus, the interaction between the pin and rod/piston created load concentrations at the edges of the rod and piston that limited the load-carrying capacity of the bearings. As engine speeds and loads have increased so the demand was for an ability to machine profile forms onto the piston boss and rod small-end that would spread the loads more uniformly and increase the load capacity of these joints.

The range of profile forms in common use in the pin hole of a direct injection (***DI***) ***diesel piston*** is shown in Fig. 8.8. There is a trade-off between the pin hole stresses and crown stresses and in a given application the optimum relies on balancing the different requirements of the two regions.

Lubrication features are similarly employed in the ***connecting rod small-end***

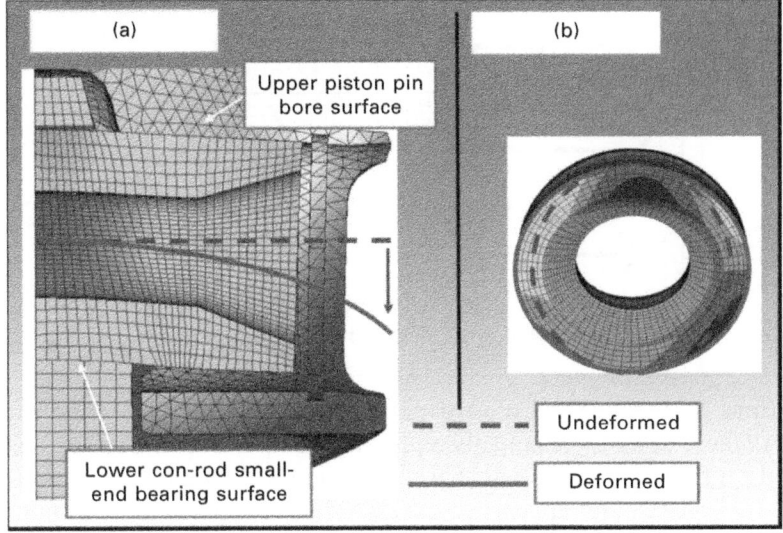

8.7 Typical pin shape under load: (a) pin bending; (b) pin ovalising (published with permission of Federal-Mogul Corporation).

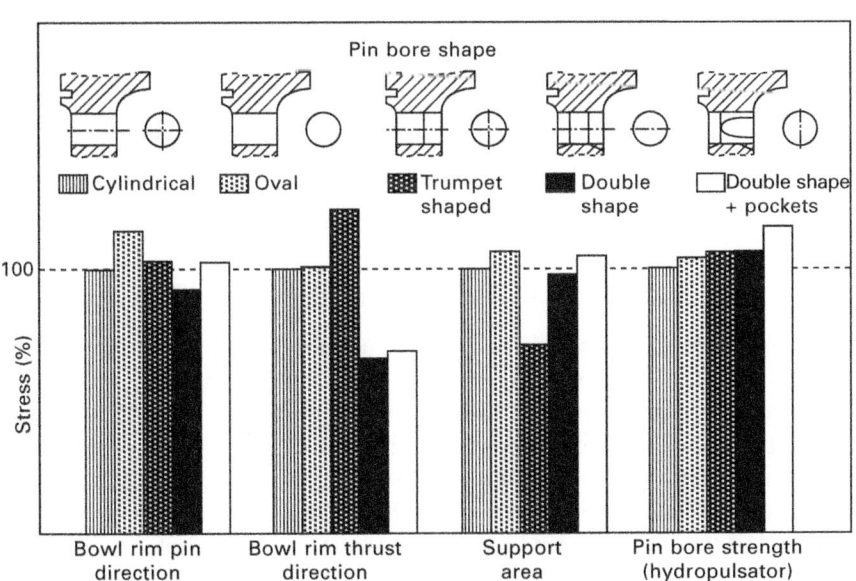

8.8 Influence of the pin hole shape on piston stresses (published with permission of Mahle International GmbH).

bore. In this case, the lack of any underlying understanding of the lubrication conditions is evident in the range of lubricant features employed in different engines. Figure 8.9 shows in diagrammatic form the unwrapped features, in

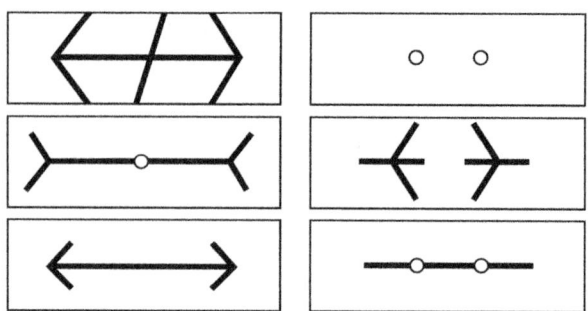

8.9 Alternative connecting rod small-end features.

terms of drillings and grooves, in a range of diesel connecting rod bushes. All these designs presumably represent attempts to ensure an adequate supply of oil to the interface between the pin and rod. In most cases it is assumed that these features are carried over from earlier applications without any real attempt at optimisation. The current trend in the most recently introduced engines is to employ axial profiling and oval features of the type already described for the piston pin hole, thereby eliminating the need for grooves or drillings.

In terms of designing the pin contact conditions simple analytical procedures and chart-based rules of thumb indicate the acceptable bearing specific loads for various types of profile feature. An example is shown in Fig. 8.10. Increasingly these are being supplemented by more detailed contact analysis which includes the effect of thermal and mechanical distortions of the components on the distribution of contact loads (Fig. 8.11).

8.6   The piston in internal combustion (IC) engines

The function of the *piston* is to provide a stable platform for the piston rings, to transmit the combustion loads to the crankshaft and to provide a heat transfer path from the combustion process to the cylinder wall and lubricant. The piston plays a key role in the efficiency, reliability and life of the engine. *Frictional losses* of the piston account for approximately 15% of the total mechanical losses in an engine with the majority of the friction being generated at the interface between the piston skirt and cylinder bore.

Pistons are conventionally manufactured by gravity die casting in a eutectic aluminium–silicon alloy. The silicon confers strength, relatively low expansion and good wear resistance. Steel pistons are used in some heavy duty commercial vehicles but the vast majority of pistons are still made in the light alloy.

Despite the presence of silicon the piston alloy the *thermal expansion* coefficient is about 70% higher than the block material. This relative difference

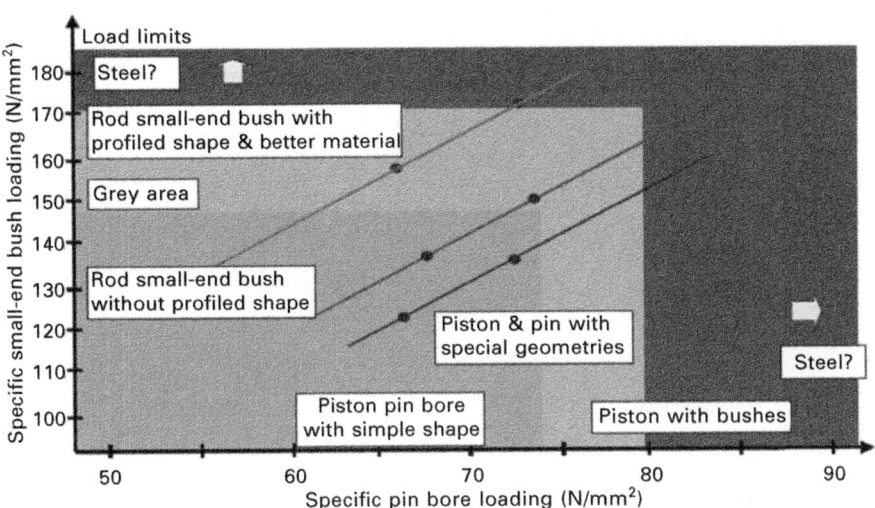

8.10 Design guide for connecting rod small end and piston pin bore features (published with permission of Federal-Mogul Corporation).

in expansion necessitates that the piston is fitted with a clearance at ambient conditions – typically 50–70 μm in diameter. In operation, because of the differences in temperatures and expansion of the piston and cylinder, the piston will almost certainly operate at a theoretical interference. In practice elastic distortion of the piston under side load ensures some clearance exists under most conditions.

During operation piston dynamic motion consists of a primary axial movement due to the kinematics of the crankshaft and a *secondary transverse motion* due to the moment of the axial force about the pin and crank axes. Although the magnitude of secondary motion is very small it influences the distribution of contact loads over the skirt and generates impulsive loads – *piston slap* – that contribute to overall engine noise levels.

The main tribological failure mode of the piston skirt is *scuffing* (Fig. 8.12), caused by a breakdown in the oil film, local welding between asperities and subsequent transfer of material from the skirt to the cylinder bore. It may lead to seizure and total engine failure.

The design clearance is essentially a trade-off between the risk of seizure if the clearance is too low and the risk of audible piston slap if the clearance is too large. The factor that is key to achieving an almost, scuff-free design is the skirt shape. A typical skirt profile combines an axial form (Fig. 8.13), which is designed to generate a convergent oil film on both up and down

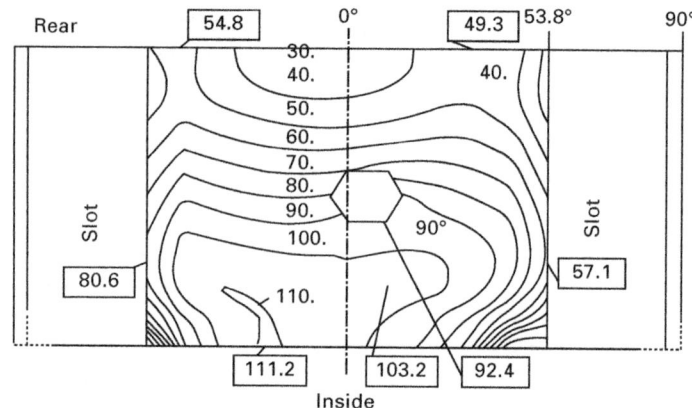

Contact pressure in MPa at pin bore by max. gas force

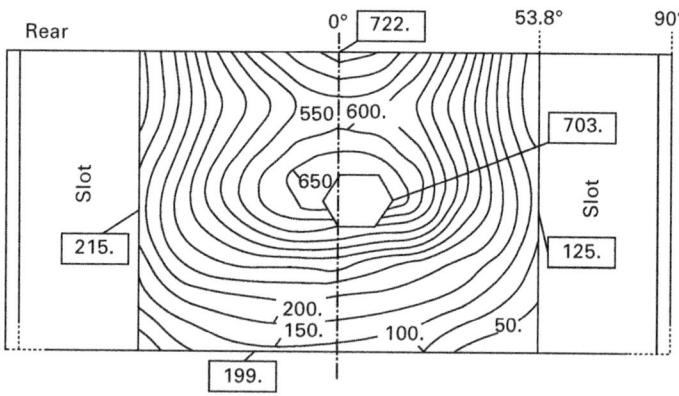

8.11 Predicted contact pressures between piston and pin (published with permission of KS Kolbenschmidt GmbH).

strokes, with *circumferential ovality* (Fig. 8.14) to optimise the extent of the loaded area to around 90° of the bore circumference.

Originally the skirt form would be cam turned, often reducing the scope and accuracy of controlling the piston shape. Modern computer numerical control (CNC) manufacturing processes have greatly contributed to the process of optimising the shape of the skirt and enabled complex skirt shapes to be customised more precisely for the demands of the application. Examples include the *AEconoguide skirt* (Fig. 8.15) form developed by AE in the 1980s and the *plateau profile* (Fig. 8.16) introduced more recently by Kolbenschmidt. Both types of design have demonstrated up to 2% brahe specific fuel consumption (BSFC) improvement. Both of these designs are exploiting the Stribeck relationship between coefficient of friction and oil

8.12 Example of piston scuffing.

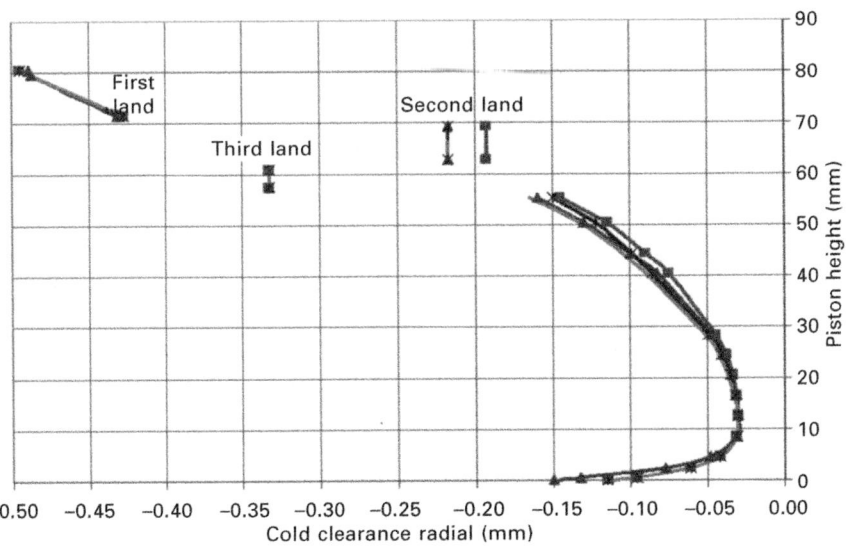

8.13 Typical piston skirt profile (published with permission of Federal-Mogul Corporation).

film thickness by reducing the area of contact to generate thinner oil films. A more direct way of achieving the same result is to simply reduce the area of contact of a piston skirt. This approach is embodied in the latest designs from most piston suppliers.

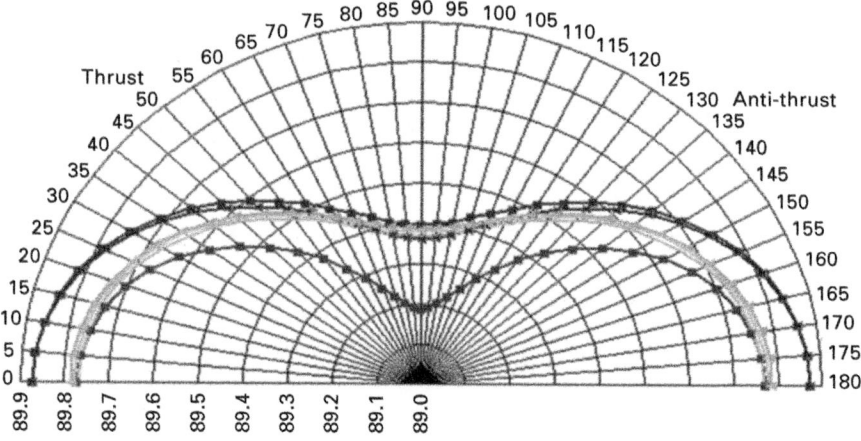

8.14 Typical circumferential skirt profiles (published with permission of Federal-Mogul Corporation).

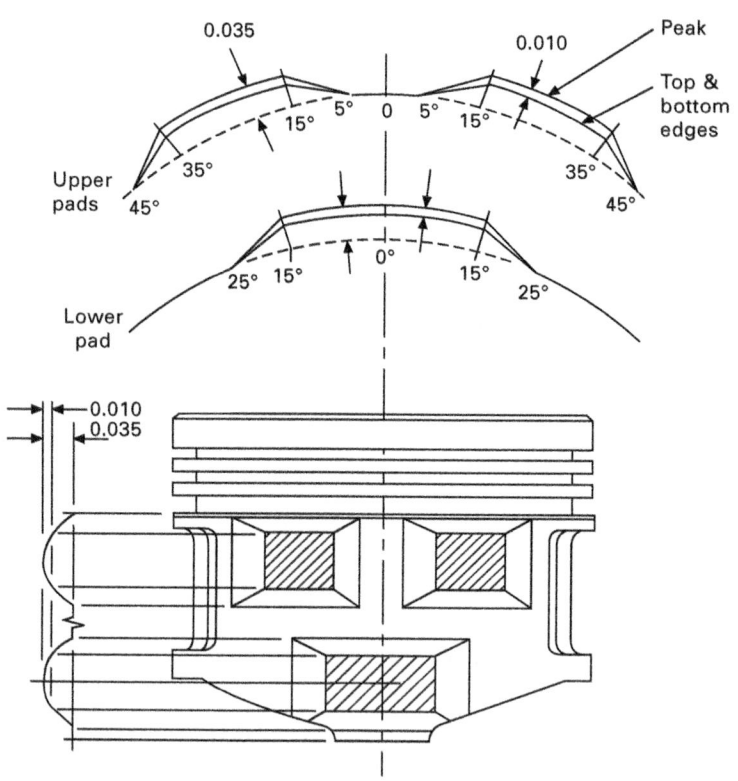

8.15 Raised skirt pad features on AEconoguide low friction piston (published with permission of Federal-Mogul Corporation).

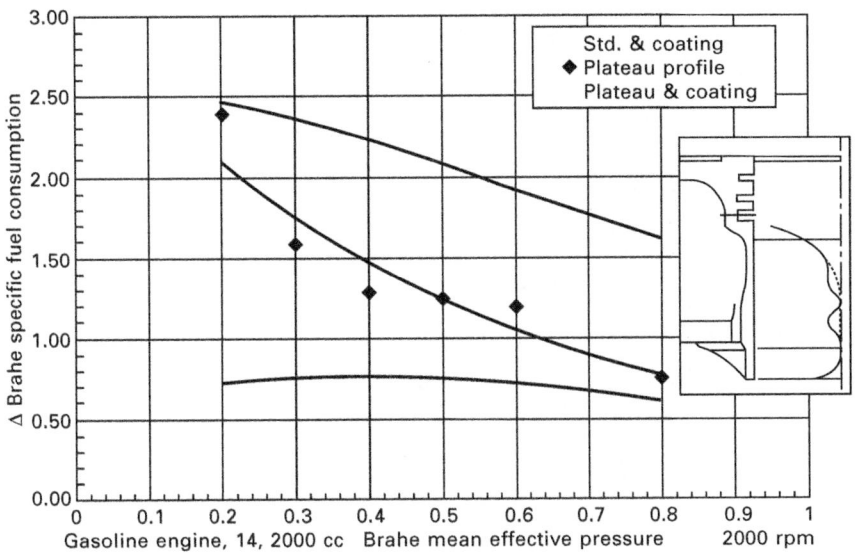

8.16 Friction reduction effect of KS plateau skirt profile (published with permission of KS Kolbenschmidt GmbH).

Similar claims for friction reduction have been made for the use of low-friction skirt coatings which are in use in many applications. However, the piston skirt is hydrodynamically lubricated throughout most of the engine cycle and therefore the properties of the coating would not be expected to influence the frictional losses. The mechanism of friction reduction with a coating is more probably a consequence of the modification of the surface finish (making it smoother) resulting from the application of the coating.

Designing a robust skirt form must take account not only of the skirt profile but also the influence of the skirt and cylinder bore stiffness and the thermal expansion. In the early days of piston design the skirt profile would have been developed iteratively through a combination of experience and empirical optimisation. However, CAE techniques have developed to the point where piston temperatures can now be accurately predicted and the hydrodynamics of the skirt modelled, taking into account both elastic effects and component dynamics. Figure 8.17 shows the predicted film pressures between the skirt and cylinder bore in a modern V6 diesel engine. Such tools are becoming essential for the development of robust designs since they can also be used to determine sensitivities to tolerance conditions, conduct design of experiments, etc., much more easily than relying on empirical methods.

Mention has already been made of the contact conditions between the pin and piston. The other tribologically important area of the piston is the contact between the top ring and its groove. The combination of inertia, friction and gas forces together with the transverse secondary motion of the

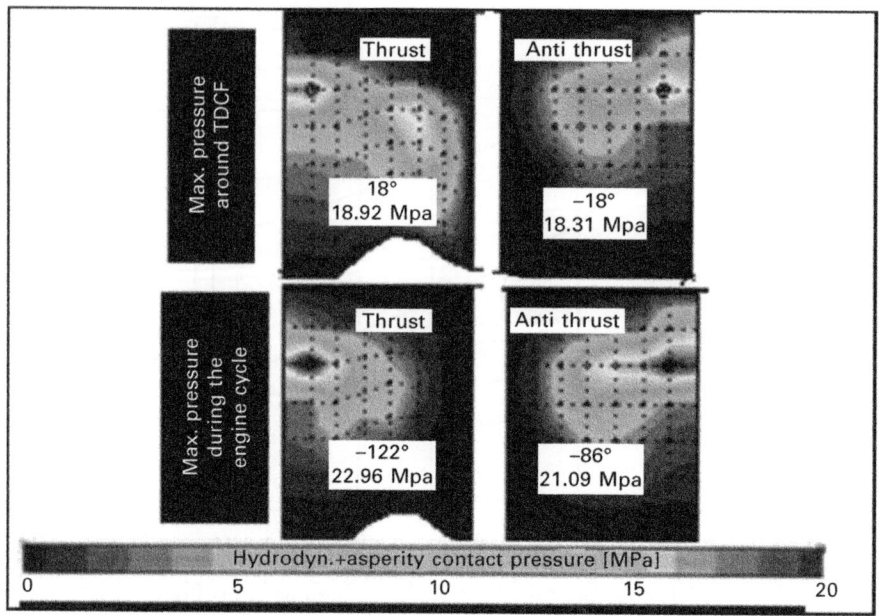

8.17 Predicted skirt contact pressures (published with permission of Federal-Mogul Corporation).

piston introduces both sliding and impact conditions. These conditions are most severe for the top ring in a gasoline engine, leading to a phenomenon known as *microwelding* – essentially transfer of groove material to the side face of the ring Fig. 8.18. The consequent roughening of the interface leads to loss of oil and blow-by control and often necessitates some form of groove reinforcement of surface treatment. In diesel applications the traditional solution is to cast in an austenitic iron insert during manufacture of the piston. In conventional gasoline PFI applications, hard anodising is commonly used for increasing the durability of the groove.

8.7   Piston rings in internal combustion (IC) engines

The function of the *piston ring assembly* (Fig. 8.19) is to seal the combustion chamber to prevent gas flow and combustion products from entering the rest of the engine, control the consumption of oil and to act as a path for heat flow from the piston to the cylinder wall. The piston ring assembly plays a key role in the efficiency, reliability and performance of the engine. The rings account for approximately 25% of the total mechanical losses in the engine, are responsible for between 50 and 80% of the engine oil consumption and play an important role in cooling the piston with around 30% of heat flowing into the piston being transferred to the cylinder via the

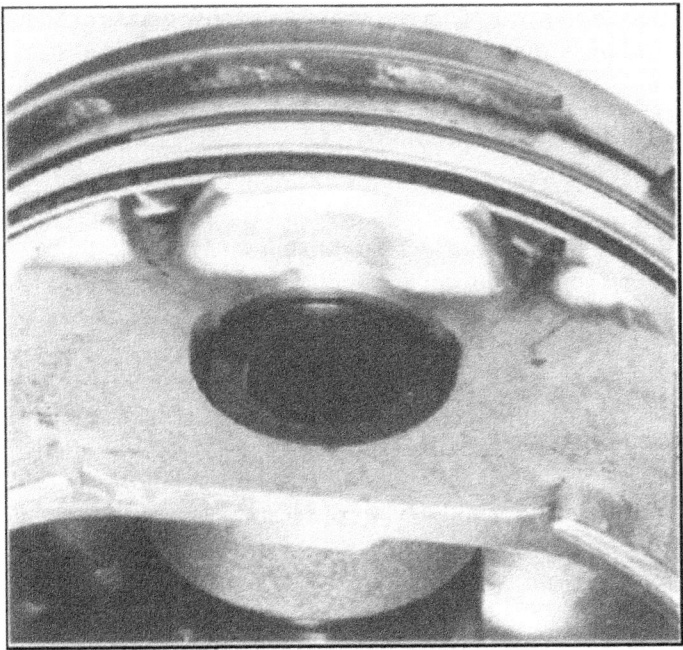

8.18 Example of top ring welding (published with permission of Federal-Mogul Corporation).

8.19 Typical gasoline piston ring pack (published with permission of Federal-Mogul Corporation).

rings. Conventionally, there are three rings in the pack – simplistically the function of the top ring is biased towards gas sealing, the third ring meters the oil flow to the top two rings and the second ring both controls the oil and acts as a further gas seal. However, the rings, in conjunction with the piston and cylinder bore operate as an interactive dynamic system and their functions cannot be considered independently.

Following extensive work carried out by many workers over the last 30–40 years there is now a very good basic understanding of the tribology of piston rings. As a consequence the requirements of ring designs capable of achieving satisfactory performance and maintaining that performance over the life of the engine – 150 000 miles (250 000 km) for passenger cars and over a million miles (1 600 000 km) for commercial vehicles – are now well established. As each ring is subject to cyclically varying inertia, friction and gas loading an important principle of ring assembly design is to achieve dynamically stable behaviour avoiding conditions such as *ring flutter*, a condition of rapid axial or radial motion of the ring relative to the ring groove. It is therefore important that the gas forces within the rings are managed through careful control of the inter-ring volumes and ring gaps.

The basic design principles of a ring pack are as follows. The first/*top compression ring* is generally a rectangular ring with a barrelled running face to promote hydrodynamic action. The majority of gasoline top rings in use in Europe are nitrided steel, while the majority of passenger car diesel top rings are manufactured from a high-strength ductile iron with a composite chromium running coating which provides good wear and scuffing performance. The base material has to be chosen for its fatigue strength and for its wear and impact resistance against the ring groove.

The *second compression ring* is taper-faced to introduce a biased downward scraping action for oil control. The ring relies on its own inherent tension plus gas pressure leaking past the top ring to generate a force against the bore wall. The ring is usually cast iron which may be untreated or phosphated.

The *oil control ring* is spring loaded to generate a high force which is supported by two thin rails against the cylinder bore. The high contact load at the rails scrapes off the majority of the oil from the cylinder bore to leave a 1–2 μm film of oil to lubricate the upper compression rings. The spring load ensures that the ring conforms to the circumferential shape of the cylinder bore. The rail material is either cast iron or steel. If cast iron the running face is invariably chrome plated for wear resistance. If steel, the surface is usually nitrided all over for wear resistance.

The trend over recent years is to axially thinner ring designs which offer fuel economy benefits, associated with lower friction. Today a typical gasoline engine ring pack would comprise a 1.2 mm top ring, 1.5 mm second ring and 1.5 mm oil control ring. Thinner rings have also been introduced on diesel engines, but the high cylinder pressure constrains the thickness of the top ring to around 1.75–2 mm to maintain adequate torsional stiffness. Ring suppliers have responded by developing radially narrower rings that conform more readily to bore out-of-roundness.

Successful ring design is about attention to detail. The following outlines the relevance of a number of key geometrical ring features that are employed to optimise the tribological performance of the assembly (Fig. 8.20):

- The sharpness of the bottom edge radius on the top ring has a significant effect on the *oil scraping efficiency* of the ring, but there is a trade-off with the cost of manufacture.
- Tight control of the closed ring gap is essential for control of blow-by and ring dynamics. In operation the top ring gap should be as small as possible. The second ring gap should be optimised to prevent gross gas flow to the sump but should not be so small as to cause excessive pressures to build up below the top ring that can cause it to lose contact with the groove and become unstable.
- Internal chamfers are used to bias the rotation of the ring cross-section when inserted in the engine. The rotation or twist of the ring plays a role in both oil consumption and blow-by control.
- Control of ring circumferential shape ensures uniform contact with the cylinder bore.
- The running profile controls the hydrodynamics and dynamics of the ring during operation.
- Side face surface finish influences gas and oil sealing.

Over the past decade or so there has been a convergence towards generic ring pack designs for gasoline and for diesel engines (Fig. 8.21) in passenger car engines. Thus for any new engine the starting point for a design is dictated by this experience. The design process beyond that stage can best be described as optimisation taking account of the specific idiosyncratic nature of the engine – dictated by the uniqueness of the engine in terms of speeds,

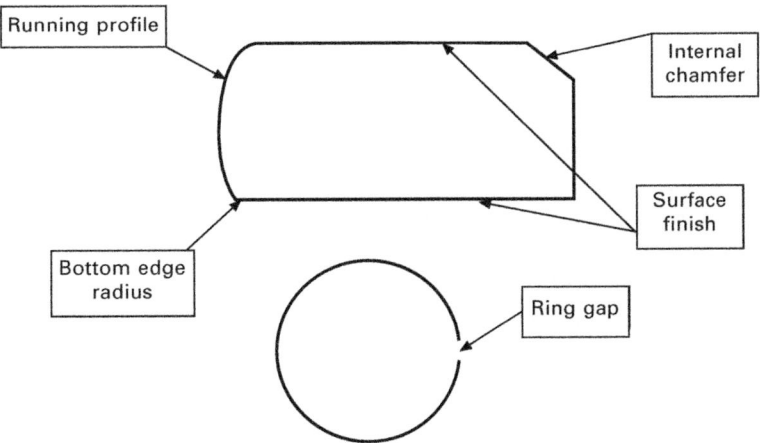

8.20 Feature detail on top ring.

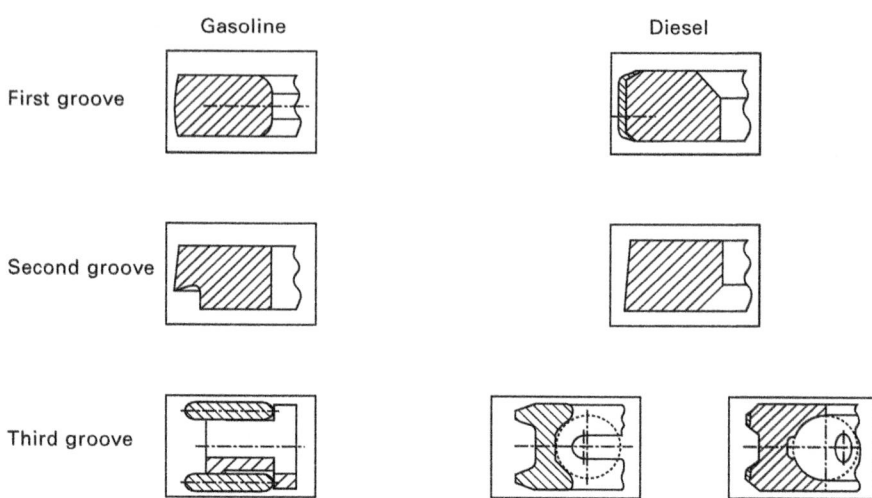

8.21 Typical ring packs for gasoline and diesel engines (published with permission of Federal-Mogul Corporation).

cylinder pressure, temperature, bore shape, performance targets, etc. There is a wealth of published work covering the design factors that influence ring performance, some of it going back to the 1970s (Tian *et al.*, 1997; Ruddy *et al.*, 1979; Jakobs and Westbrooke, 1990; Priebsch and Herbst, 1999). This body of knowledge is a fundamental source of information for applications engineers when fine tuning the design of a ring pack.

Little has been said about the use of CAE to support ring modelling. From a practical point of view, the writer's experience is that in engine development today CAE plays a relatively small role in ring pack optimisation compared with the application of experience. That is not to belittle the excellent work devoted to understanding the physics of piston ring behaviour at centres of excellence both in the UK and elsewhere. However, this effort has yet to result in practical design tools for piston rings. That is not to say that there are no useful models within the automotive community both commercially and in-house. Indeed a number of publications in recent years have demonstrated an ability to simulate various features of ring behaviour very successfully, specifically gas leakage (blow-by) and ring dynamics. However, the holy grail in ring performance prediction is oil consumption and to date such an all embracing model does not yet exist. The possible reasons for this are expanded upon in Section 8.8.

8.8   The cylinder bore surface

The main functional requirement of the bore surface is that it should run-in quickly and subsequently provide a stable and uniform surface for the remainder of the life of the engine. Two attributes of the cylinder bore that are key determinants of piston assembly function are shape and surface finish.

8.8.1   Bore shape

Cylinder liner bore distortion (Fig. 8.22) is one of today's most important topics in crankcase structure development. Poor ring sealing can be a major cause of high oil consumption and blow-by in an engine. Pachernegg (1971) showed that a ring passing over a depression in the bore surface of no more than 4 μm would suffer excessive oil consumption. Low bore distortion clearly opens up the potential for optimising the piston group. As the piston rings achieve better sealing characteristics in a low-deformation cylinder liner, oil consumption and blow-by are reduced. Without affecting oil consumption and blow-by, engine friction and hence fuel consumption can be reduced by decreasing the pre-tension of the piston rings. Additionally, from the acoustical point of view one factor in the attenuation of piston-slap noise is that of minimising bore distortion.

   To enable good ring conformability and sealing performance the ideal shape of the cylinder bore is clearly cylindrical. In practice the effects of manufacturing variability, temperature gradients, stress relaxation and assembly stresses introduce significant deviations from this ideal condition. Ring conformability becomes more demanding at higher orders of bore distortion with fourth order usually being the most critical. Using simple design methods it is possible to calculate the ring conformability at different orders of bore distortion. This can highlight the need for either change in block design to reduce distortion levels or improvement in *ring conformability* (see Chapter 20). Bore distortions at seven different levels of height for a six cylinder diesel engine illustrate the typical range seen in a modern well-designed application (Fig. 8.23). The horizontal bars indicate the ring conformability limits, in this case showing no anticipated issue with ring sealing.

   There is good understanding of the factors that cause bore distortion and the measures that can be used for improvement. These include optimisation of the block design – improved coolant flow around the bore, increased section

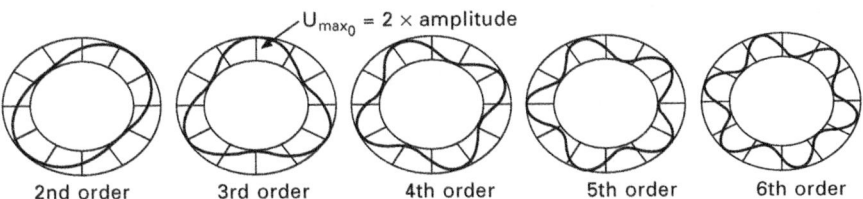

U_{max_0} = 2 × amplitude

2nd order   3rd order   4th order   5th order   6th order

8.22 Type of bore distortion (published with permission of Federal-Mogul Corporation).

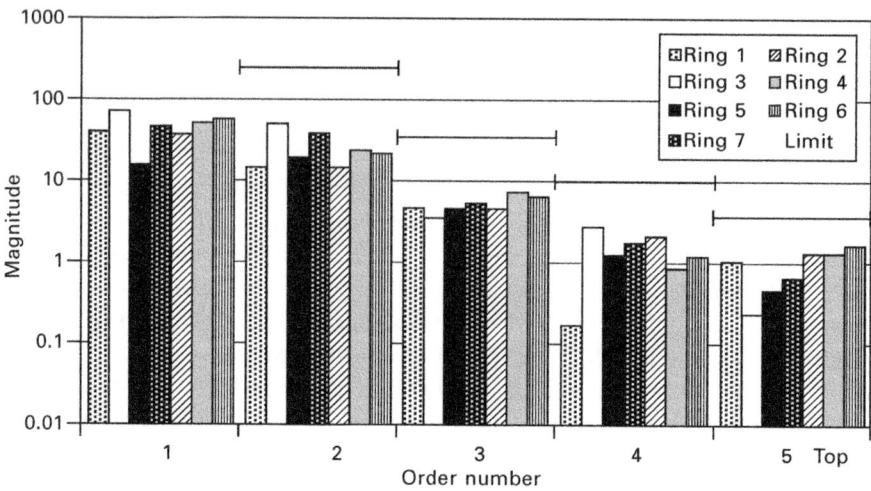

8.23 Evaluation of ring conformability at different order bore distortions.

moduli, reduced bolt loads, open deck design – deck plate honing and low distortion head gasket design (Loenne and Ziemba, 1988). Invariably bore distortions are worse in engines with no inter-bore cooling (siamesed) and engines with light alloy blocks. The technology also exists to hone the bore to a non-cylindrical shape that compensates for the assembled bore distortion, resulting in a more or less cylindrical shape under running conditions (Flores *et al.*, 2008).

8.8.2   Bore surface finish

The introduction of stricter emissions legislation on gasoline and diesel engines demands lower oil consumption with particular attention to running-in behaviour and scuffing resistance. Additionally extended service intervals beyond 20 000 km necessitate a reduction in oil consumption to avoid the need for oil top-ups between services. Maintaining compliance with vehicle emissions requirements also requires low oil consumption to ensure optimum performance of the after-treatment system throughout its life. Modern rings with coatings, optimised geometry and surface finish together with more accurate bore shapes and improved engine oils allow the thickness of the oil film on the cylinder bore to be reduced. Thinner films demand smoother cylinder bore surfaces to take full advantage of the latest ring and lubricant technology. In the last 5 years new *honing* developments have resulted in the introduction of significantly smoother bore surfaces. Previously a typical bore surface would have had a finish of $Ra = 0.8 - 1.0\,\mu m$. Today values of Ra of the order $0.15–0.4\,\mu m$ are not uncommon.

Ra is not a good parameter to use to characterise surface finish from a functional perspective as it does not discriminate well between different types of surface. Use of the **Abbott–Firestone parameters** (Rp_k, R_k, R_{vk}, M_{r1}, M_{r2}), based on DIN 4776, are becoming a common measure of cylinder bore surfaces. These 2D parameters basically define the bearing area curve as a function of depth from the surface (see Fig. 8.24). The sixth parameter, V_o, also shown in the chart represents the oil reservoir within the valleys on the surface. This is sometimes referred to as the oil void volume, calculated using the equation:

$$V_o = 100-M_{r2})*R_{vk}/200$$

(Note M_{r1} represents the fraction of the surface which consists of small peaks above the plateau and M_{r2}, the fraction of the surface which will carry load during the practical lifetime of the part. The construction of these parameters is shown in Fig. 8.24.)

The ideal surface would be perfectly smooth with a consistent series of grooves deep enough to provide lubrication reservoirs for the life of the engine, as shown in Fig. 8.25. However, honing is not an isolated manufacturing process. Inevitably the optimisation of the surface finish must be carried out taking account of other factors such as cycle time constraints, tool life and size tolerances. Nevertheless modern bore finishes are approaching this ideal as honing technology advances. Typical values of the above parameters in use today are: R_{pk} = <0.2; R_k = 0.3 – 0.8; R_{vk} = 1.0 – 1.5. Of course these parameters do not describe the complete surface. The functional performance of the finished surface is also influenced by the quality of the resulting surface, i.e. it must be free of folded metal, avoid significant cold working and exhibit uniformly cut honing grooves. Figure 8.26 shows a good quality honed bore surface.

There is a growing body of published empirical and analytical work aimed at improving the understanding of the surface parameters that determine its

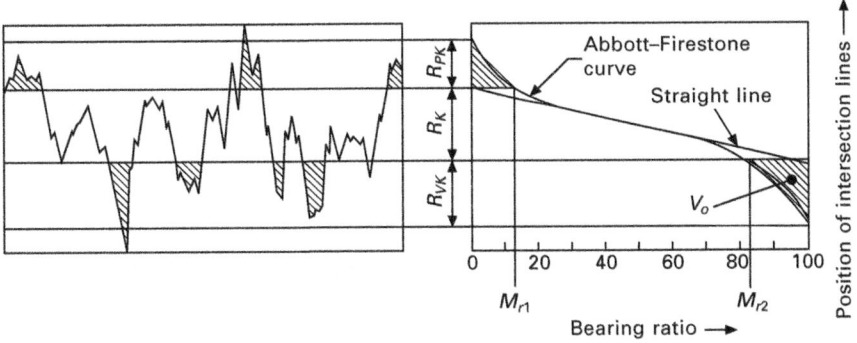

8.24 Definition of Abbott–Firestone surface finish parameters.

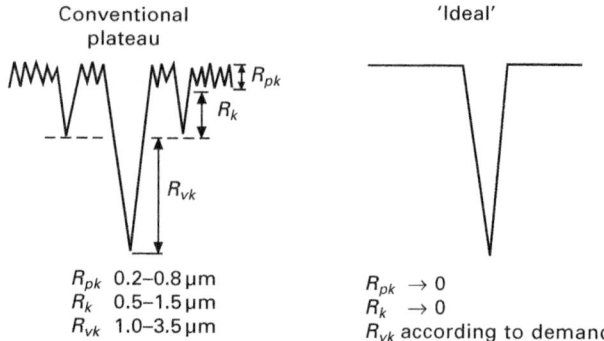

Conventional plateau   'Ideal'

R_{pk} 0.2–0.8 µm $R_{pk} \rightarrow 0$
R_k 0.5–1.5 µm $R_k \rightarrow 0$
R_{vk} 1.0–3.5 µm R_{vk} according to demand

8.25 The ideal cylinder bore surface (published with permission of Nagel GmbH).

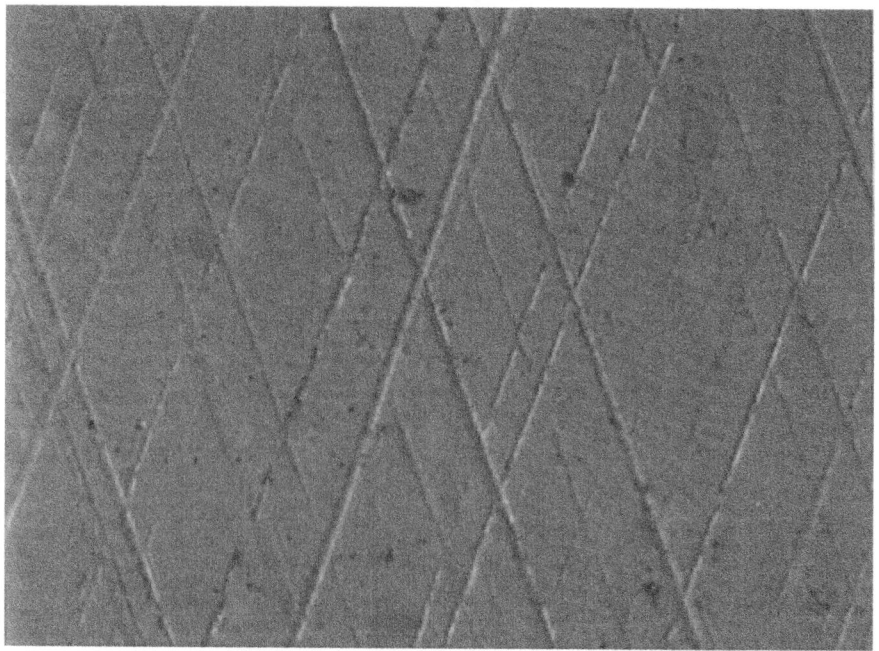

8.26 Typical good quality honed bore (published with permission of Nagel GmbH).

functional performance. This includes the use of 3D surface metrology which will enable more sophisticated analysis of the character of a surface (Dong *et al.*, 1995; Sayles, 2001). However, it would require a complete chapter in this volume to do this topic justice. One example taken from the work of Hill (2001) confirms that the trend towards the ideal surface referred to above is consistent with reduced oil consumption (Fig. 8.27).

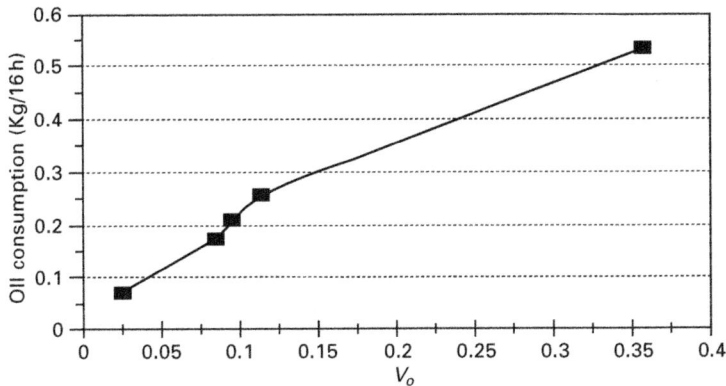

8.27 Influence of crevice volume on oil consumption (published with permission of Mahle Engine Components USA Inc.).

8.9   Design validation of internal combustion (IC) engines

The typical design life for a passenger car engine is 10 years/150 000 miles. Throughout its life the engine is expected to perform reliably with only minor degradation in performance, fuel economy and emissions. Commonly an engine is considered to be reliable when it operates for a certain mileage without any problem on the main components. If one problem occurs, it must happen with a very low probability. Failure rates are typically monitored in terms of numbers of repairs per thousand vehicles ($R/1000$). Should a failure occur in the tribological system of an engine there is a high probability that it will result in total engine failure and a costly repair or engine replacement. *Engine reliability* is based on the cumulative probability of failure of each of the many subsystems. To achieve a satisfactorily low overall failure rate means that each sub-system needs to be designed to achieve a very high degree of reliability – typically the target described in statistical terms is 3.4 ppm (referred to as *6 sigma*). Engine testing is used by all OEMs to validate the performance and reliability of an engine. The challenge is to demonstrate this level of reliability in a relatively short engine test with relatively few engines. In a typical engine programme for a new engine typically fewer than 500 engines would be built and run before engine launch. The question is how reliabilities of around 1 to 2 $R/1000$ can be statistically proven with such a relatively small sample.

The general approach taken to prove a design of a system is to assess its robustness to every known failure mode of that system. The techniques for doing this are well established in the industry and involve the use of failure mode and effects analysis (FMEA) to assess the risk that a given design will succumb to a particular failure mode. This allows the high-risk failure modes to be anticipated and steps taken to reduce or eliminate the risk at the design stage. It is also important to recognise the presence of variability due to, for example, dimensional tolerances, the effect of wear over time and external factors such as customer usage, climate, etc. An effective programme of testing verifies that the system is robust to the combined effect of each of these 'noises' by ensuring that test engines are built and tested to represent worst case conditions. If this procedure is followed rigorously then it will be possible to have a high level of confidence that the design is robust to all known failure modes under any reasonable set of conditions.

So what is the role of computer simulation in designing an engine since, with a good enough computer model, it should in principle be possible to dispense with testing – in other words 'right first time design'.

Although major advances have been made in structural analysis through the application of the finite element method, allowing reasonably accurate predictions of fatigue life and other mechanisms of structural failure to be made, prediction of system performance at a tribological level remains relatively immature. In some fields, for example friction modelling, there exist design tools that can be used to assist in the initial stages of engine optimisation (Lee *et al.*, 1999; Sandoval and Heywood, 2003).These tools include a mixture of analytical and empirical component models that are immensely useful when conducting assessments of alternative concepts prior to defining an engine's architecture.

It was also noted in earlier sections that both simple models and more sophisticated *elastohydrodynamic* modelling can be applied in a practical way to assist in the design of skirt shapes, pin contact conditions and bearing design. Whilst none of these models are sufficiently accurate to dispense with engine testing entirely their use greatly reduces the number of iterations necessary to achieve an optimised design.

However, at the other extreme the prediction of oil consumption remains extremely challenging. Today there are few if any design aids routinely applied in engine design to model ring pack performance. There are several reasons for this:

- The quantities of oil consumed per engine cycle are extremely small, thus small errors in calculated quantities will have a disproportionately large influence on prediction. Typically in a four cylinder 2.0 l diesel engine running at 4000 rpm, WOT (wide open throttle) total oil consumption is 10 g/h. This equates to 10 µg/cylinder/rev.
- The scale of the problem. There are many phenomena and interfaces involved in oil consumption behaviour – ring-groove interface, ring to bore interface, piston to bore, ring structure response, ring dynamics, oil evaporation, gas dynamics, ring & bore surface finish, component wear, deposit formation, oil chemistry, PCV performance – where the

physics needs to be understood and translated into robust and validated sub-models.

- The experimental validation of the individual sub-models requires very sophisticated measurement technologies.

Given the scale of the task it seems doubtful that a reliable model of oil consumption will emerge within the next 5 years, i.e. one that demonstrates the observed sensitivities. Nevertheless even without a complete model of ring behaviour, the degree of knowledge gained from the many years of empirical observation by the ring and piston community has led to highly refined ring designs and a deep understanding of cause and effect. Within this scenario the use of even approximate analysis techniques, though lacking in absolute accuracy, often predict the correct trends and therefore provide valuable insights that can be exploited to support optimisation and problem solving.

In summary, design validation requires the on-going use of CAE in combination with testing that includes robustness factors to ensure that engines will operate reliably at the extremes of usage.

8.10   Future trends

Despite the early hype about fuel cell technology IC engines are likely to be around for at least the next 15–20 years. I am confident that the next volume of engine tribology will still be addressing the same kind of topics that are included in this.

Given the lead time to develop and introduce a new engine, typically 3–4 years, the future is already known. The technologies for engines that will be introduced into the market in 2011/2012 are already known and must have been proven for mass production.

The drivers for change were listed in Section 8.3. How these drivers will shape future engines is a matter of conjecture. However, some trends seem clear:

- The continued pressure on cost is likely to drive component integration where feasible – for example integrated bearings by metal spray technologies may be employed to eliminate separate shell bearings and bushings in the rod and crankshaft. Such technologies are already in development.
- With the wider introduction of direct injection of gasoline (*GDI*) there will be a gradual convergence between component designs for diesel and gasoline engines. Cylinder pressures in GDI engines will increase and there will be demands for the type of groove reinforcement and cooling requirements commonly seen in diesel pistons.
- The use of biofuels will accelerate. Their poorer lubricity will require

improvements in materials and coatings in the hot parts of the engine.

- Enhanced coatings, e.g. DLC (***diamond-like coating***), will find increasing use to reduce friction and increase durability so that emissions constraints can be met at longer engine lives.
- Downsizing of engines will continue in order to achieve the kind of CO_2 reductions needed by the industry. This in turn will drive the need for new ring and piston materials to meet the increased power densities.
- Friction reduction will continue to play a key role in CO_2 reduction. The trade-off with cost may actually support increasing feature costs to achieve what may ultimately become mandatory CO_2 targets.
- Hybrid technology will grow. Micro-hybrids employing integrated starter generators (ISG) will be found on delivery vehicles where frequent start–stop operation will benefit. This will impact on main bearing technologies due to additional side loads applied to the crankshaft to drive the ISG.
- Improvements in manufacturing technology will be needed to achieve more idealised surfaces in terms of shape control and reduced geometric variability.
- Design improvements to reduce cylinder bore distortion will be a significant enabler for the use of lower load and hence lower friction ring packs.
- CAE techniques will continue to advance and, if not displace engine testing, at least eliminate a large proportion of iterations.

In summary, I fully expect that engine tribology is going to provide future generations of engineers with the same kind of technical challenges that we are enjoying today.

8.11   References

Bishop, I.N. (1964), Effect of design variables on friction and economy, SAE Paper 812A.

De Petris, C., Giglio, V. and Police, G. (1996), Some insights on mechanisms of oil consumption, SAE International Spring Fuels and Lubricants Meeting. SAE 961216.

Dong, W.P., Davis, E.J., Butler, D.L. and Stout, K.J. (1995), Topographic features of cylinder liners – an application of three-dimensional characterization techniques, *Tribology International*, **28**(7).

Flores, G., Klink, U. and Abelyn, T. (2008) Honen von Funktions formen in Zylinderkorbelgehausen. UDI-Bericht Nr. 1994.

Hill, S.H. (2001), Cylinder bore finishes and their effect on oil consumption, International Fall Fuels and Lubricants Meeting, SAE 2001-01-3550.

Jakobs, R.J. and Westbrooke, K. (1990), Aspects of influencing oil consumption in diesel engines for low emissions, SAE Paper 900587.

Jost Report (1966), HMSO, Lubrication, Tribology, Education and Research, DES Report, London, UK.

Ku, Y. and Patterson, D. (1988), Piston and ring friction by the fixed sleeve method, SAE 880571.

Lee, So-duk, Shannon, B.A., Mikulec, A. and Vrsek, G. (1999), Applications of friction algorothms for rapid engine concept assessments, SAE 1999-01-0558.

Loenne, K. and Ziemba, R. (1988), The Goetze cylinder distortion measurement system and the possibilities of reducing cylinder distortions, SAE 880142.

McGeehan, J. (1979) A survey of the mechanical design factors affecting engine oil consumption, SAE 790864.

Moshrefi, N., Mazzella, G., Yeager, D. and Homco, S. (2007), Gasoline engine piston pin tick noise, SAE Noise and Vibration Conference and Exhibition, Paper No. 2007-01-2290.

Pachernegg, S.J. (1971), The hydraulics of oil scraping, SAE 710816.

Parker, D.A. and Adams, D.R. (1982), Friction losses in the reciprocating internal combustion engine, I. Mech E Conference, Paper no. C5/82.

Priebsch, H.H. and Herbst, M.H. (1999), Simulation of effects of piston ring parameters on ring movement, friction, blow-by and LOC, MTZ 60.

Rhodes, M.L.P. and Parker, D.A. (1984), A Econoguide – the low friction piston, SAE 840181.

Ruddy, B.L., Dowson, D., Economou, P.N. and Baker, A.J.S. (1979), Piston ring lubrication – Part 3, the influence of ring dynamics and ring twist energy conservation through fluid film lubrication technology, ASME Winter Meeting, pp 191–215.

Sandoval, D. and Heywood, J.B. (2003), An improved friction model for spark-ignition engines, SAE 2003-01-0725.

Sayles, R.S. (2001), How two- and three-dimensional surface metrology data are being used to improve the tribological performance and life of some common machine elements, Tribology International, 34, 299–305.

Tian, T., Rabute, R., Wong, V.W. and Heywood, J.B. (1997), Effects of piston ring dynamics on ring/groove wear and oil consumption in a diesel engine, Development of New Diesel Engines and Components Design, SP-1245, Paper No. 970835.

Wang, D. and Parker, D.D. (2003), Effects of crank-pin surface circumferential waviness on the EHL of a big-end bearing in a diesel engine, SAE 03FFL-235.

Womach, J.P., Jones, D.T. and Roos, D. (1990), The Machine That Changed the World: The story of lean production, Harper Perennial.

Yilmaz, E., Tian, T., Wong, V.W. and Heywood, J.B. (2004), The contribution of different oil consumption sources to total oil consumption in a spark ignition engine, SAE 2004-01-2909.

9

Predictive methods for tribological performance in internal combustion engines

I. MCLUCKIE, AIES Ltd, UK

Abstract: This chapter demonstrates user-friendly computer-aided engineering (CAE) methods in engine tribology and dynamics: firstly with a study of a race engine big-end bearing with multi-body dynamics (MBD), elastohydrodynamics (EHD) and thermoelastohydrodynamic (TEHD) methods; secondly with a piston and liner study citing the methodology using MBD, rigid hydrodynamics (RHD) and incorporating thermal and manufacturing effects; thirdly by a study of turbocharger full floating bearings, using MBD, RHD and rotor dynamics demonstrating a dynamic phenomenon only seen previously in an independent experimental study; and finally engine system friction demonstrates overall system and component interactions applicable with multi-fidelity CAE like short bearing approximation (SBA), short and long bearing approximation (SALBA), RHD, EHD, TEHD, MBD, finite element analysis (FEA) and finite difference (FD).

Key words: RHD, EHD, TEHD, MBD, bearings, piston, liner, turbocharger bearing, sub-synchronous whirl, IKBS.

9.1   Introduction

Engine tribology and dynamics, generally known as tribo-dynamics, is a fascinating subject as it encompasses many, if not all, aspects of physics together in a single simulation and experimentation field. In this branch of engineering the simulation of crank pin bearing, piston, ring and liner tribology is one of the most exacting and rewarding areas. The performance of the crank pin, piston and liner involve the multiple interactions of thermal, structural, dynamic, and fluid structure interactions, plus the effects of manufacturing, assembly and loading imposed by the vehicle duty cycle and terrain.

Modelling a real engine system involves systematic and sequential simulations in fuel injection, combustion simulation, cycle simulation, heat transfer and cooling, thermal stress analysis, lubrication and friction, structural dynamics and noise, vibration and harshness (NVH), durability (fatigue) and wear. Therefore accurate modelling of engine friction requires the prediction of all the engine's tribological systems and components such as the valve train, power cylinder system, crank train, auxiliary drives, turbocharger and

284

the lubricating system and pump. See details of powertrain system analyses in McLuckie (2003).

To model such a complex system, the tribo-dynamicist needs the cooperation and input of engineers from a number of disciplines, in order that the tribological analyses accurately predict the engine performance of the test bed and the engines service conditions.

The formulation of the models used to solve these complex engine inter-relationships is by a form of Reynolds equation solved simultaneously with the equations of motion (***multi-body dynamics***) of the engine system. This chapter considers their application to a number of engine tribology or multi-physics problems, via coupled fluid structural interactions, in the form of crank pin lubrication, piston and liner lubrication, turbocharger bearing performance and overall engine system friction predictions. It also shows how improved accuracy needs detailed understanding and more detailed numerical formulations.

The generalised nature of the methods described here means that the methods can be applied to any scale of engine size. At one end of the extreme there is the large ship's slow speed diesel engine (see Ma *et al.*, 2001, for analysis of a large ship's diesel engine big-end bearing), and at the other end of the scale you have a high-speed (20 000 rpm) race engine (see McLuckie and Barrett, 2005a) or a small ecofriendly gasoline/diesel hybrid system. All are applicable to the methods described in this chapter and this book.

9.2   Integrated knowledge-based tribology systems

Modelling engine tribology systems involves multi-physics and multi-fidelity analyses and this requires a suitable analysis framework and simulation approach. The integrated knowledge base system (IKBS) allows the building of such systems with integrated physics such as FD (finite difference), FEA (finite element analysis), FV (finite volume) and BEA (boundary element analysis) and analytical models which support Reynolds equation solutions and the equations of motion for multi-body systems. See Barrett, *et al.* (2002) for BEA and FD coupling applied to elastohydrodynamics (EHD).

Currently most computer-aided design/engineering (CAD/CAE) integration methods limit the structural models with tetrahedral elements, while the oil film mesh is generally depicted by quadrilaterals in finite difference form. Some work arounds have been developed but on the whole the approach is unsatisfactory both in being time consuming and compromising solution accuracy. Figure 9.1 shows the comparison of different methods for a crank pin EHD simulation. A detailed description of the methodology can be seen in McLuckie and Barrett (2005b).

In order to improve productivity and solution accuracy we introduce the IKBS concept of model creation and preparation which enables rapid

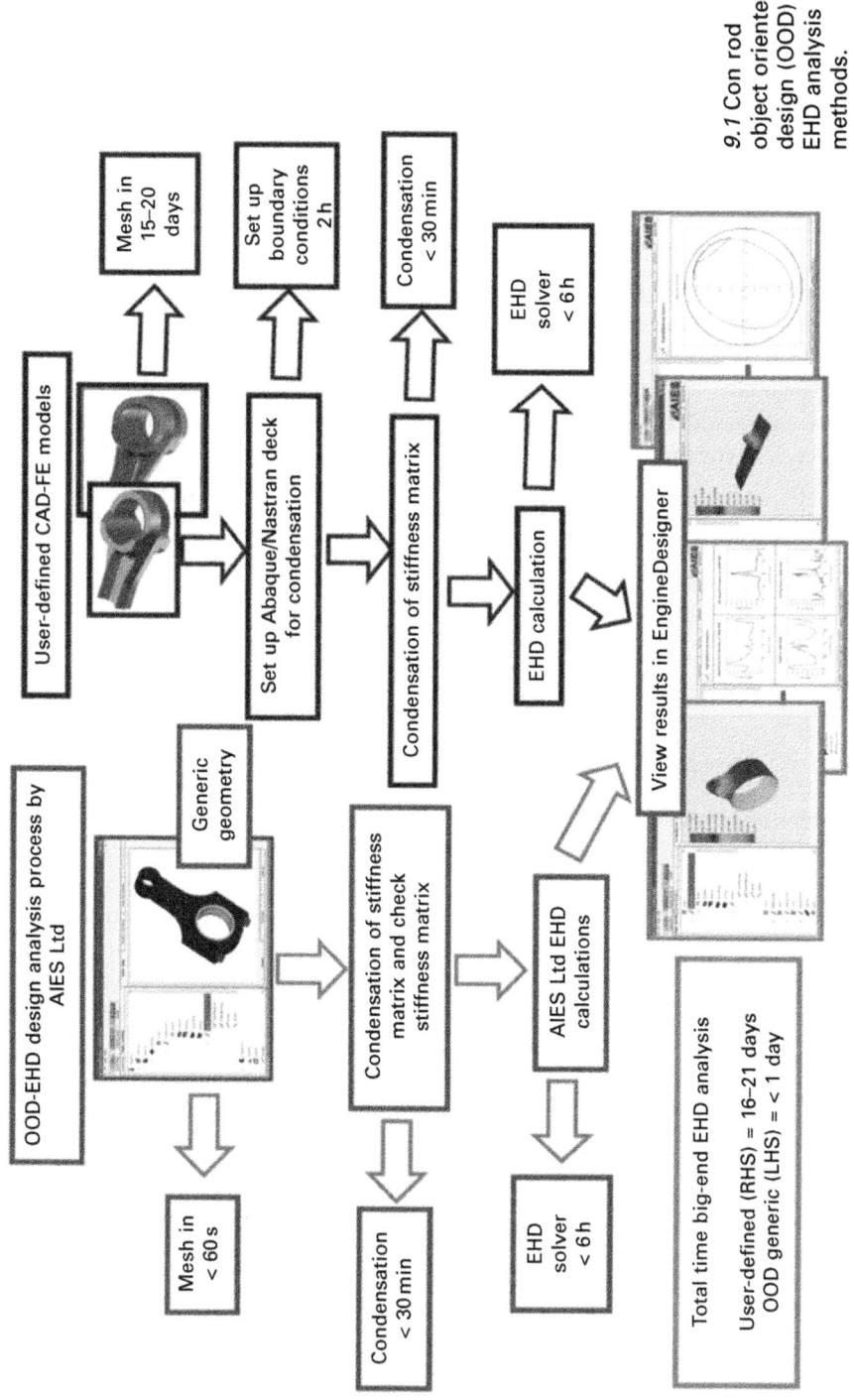

9.1 Con rod object oriented design (OOD) EHD analysis methods.

hexahedral meshes to be coupled to finite difference quadrilateral meshes. This is a much more reliable and suitable approach to solving tribo-dynamic system problems. Detailed discussions and methodology are given in McLuckie (2007a).

Figure 9.2 shows a crankshaft built from finite objects generated by a newly patented approach that integrates the hexahedral mesh with the solid model. The tribological objects of the crankshaft are the main bearing journals and crank pin journal, which contain the geometrical and surface finish data and the positions and size of oil holes.

Figure 9.3 shows the valve train sub-assembly built of finite objects which has camshaft bearings, cam/tappet tribology objects and tappet to

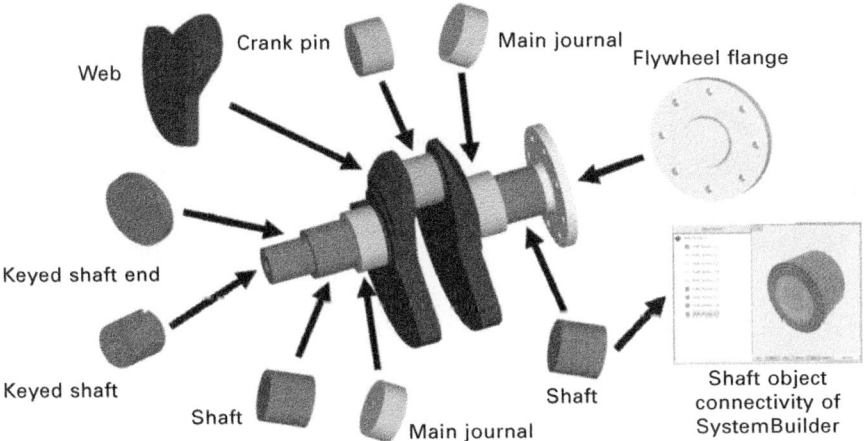

9.2 Crankshaft constructed of reusable objects.

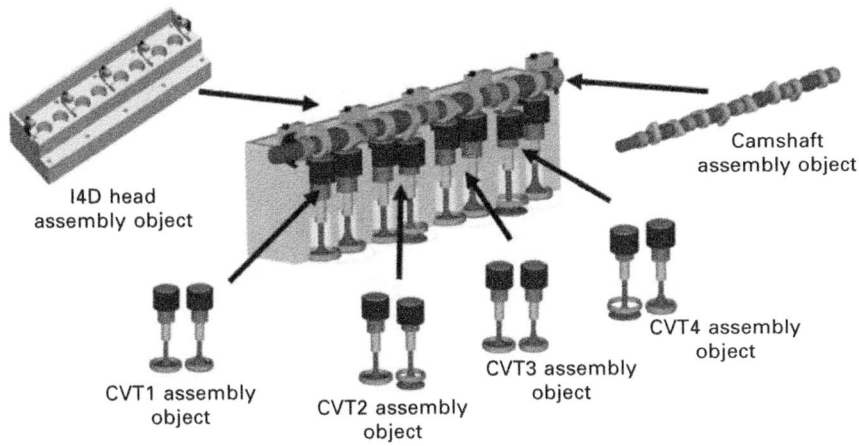

9.3 Valve train subsystem built with objects.

head tribology objects. See Barrett, *et al.* (2000) for a detailed comparative study of three engine valve trains.

Figure 9.4 shows the crank train subsystem with crankshaft, conrod, piston, and liner tribology objects (models). Finally, Fig. 9.5 shows the engine system with crank train and valve train systems. Thus, the engine system comprises solid and thin film tribological objects (interface objects) which are generic and reusable around the engine system. Finite objects build

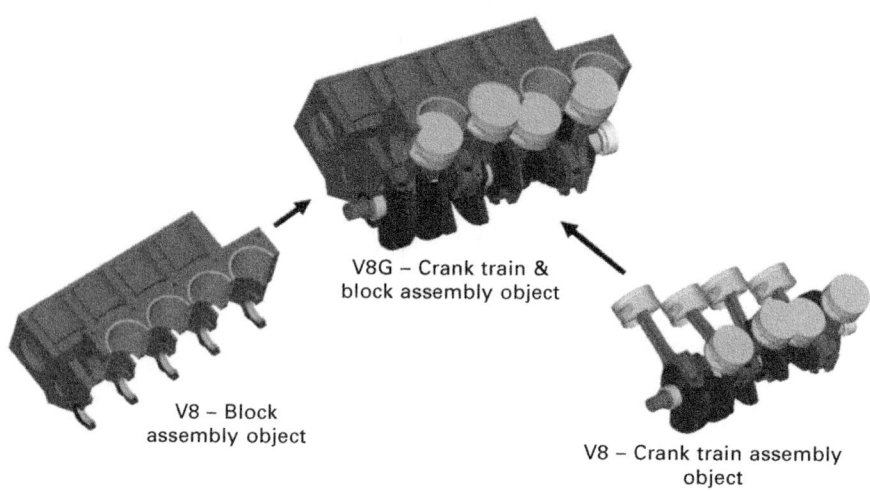

9.4 Crank train subsystem built with objects.

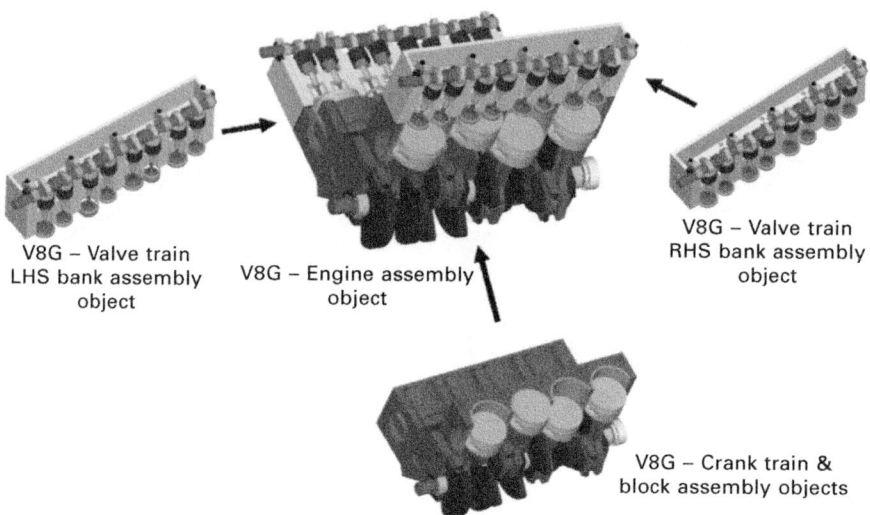

9.5 Engine system built with reusable objects.

engine objects, such as tappets, cams, camshafts, pistons, crankshafts, con rods and bearings. See McLuckie (2007a,b) for a more detailed description of the benefits of IKBS methods.

The interface objects build tribology objects from library of solution methods, for example a bearing may be represented as a short bearing approximation (SBA) developed by Ocvirk and Dubois (1953), or as a short and long bearing approximation (SALBA) initially developed by Reason and Narang (1982) but now modified to include squeeze film effects, rigid hydrodynamics (RHD) (see Cameron, 1983), elastohydrodynamics (EHD) (see Ma *et al.*, 2001), thermoelastohydrodynamics (TEHD) (see Pinkus, 1990) or an elastohydrodynamic lubrication (EHL) model (see Gohar, 1988).

All models have a mesh to display the pressure, film thickness and oil flow, so they are interchangeable to enable rapid comparisons in solution method and allow trade-offs in solution time and accuracy to be made. For example an engineer could examine a particular difficult design with EHD while the other bearings in the system were modelled with RHD, SBA and SALBA to reduce overall system computation time.

As the solvers are generic they can be applied to a number of problems namely: cam bearing oil films, big-end bearings, main bearings, piston liner oil films, cam tappet oil films, ring and liner oil films and turbocharger bearings.

These objects/solvers are used around the engine to model system behaviour. Figure 9.38 in Section 9.6 shows a single cylinder engine example with the tribological objects used around the engine.

9.3   Application of integrated knowledge-based systems (IKBS) and elastohydrodynamics (EHD) to a race engine crank pin

Tribological thin film fluid structure coupling within the engine consists of thin film hydrodynamics, in the form of Reynolds equation coupled with the equations of motion and the equations governing heat transfer (see Chapter 5). Reynolds equation is formulated to consider the effects of cavitation, oil film history and effects of surface finish (asperity contact) by flow factors (see Chapter 15). Reynolds equation is solved simultaneously with the equations of motion plus the coupled equations.

Reynolds equation formulated in generalised form below is:

$$\frac{\partial}{\partial x}\left(\frac{1}{12\eta}\,\theta h^3\,\frac{\partial p}{\partial x}\right) + \frac{\partial}{\partial z}\left(\frac{1}{12\eta}\,\theta h^3\,\frac{\partial p}{\partial z}\right) = (U_1 + U_2)\,\frac{\partial(\theta h)}{\partial x} + \frac{\partial(\theta h)}{\partial t}$$

(9.1)

where p = pressure, h = film thickness, x = circumferential position, z = axial position, η = dynamic viscosity, and U = circumferential velocity. Note that U is not constant around the bearing surface.

The equations of motion for the journal below are:

$$M_j\ddot{x}_j = F_{HD}(x_j, \dot{x}_j) + F_{external}(t) - D\dot{x}_j - Kx_j \tag{9.2}$$

and are coupled to the system of equations of motion for the bearing housing node points through:

$$M_n\ddot{x}_n = F_{HD}(x_n, \dot{x}_n) - D\dot{x}_n - Kx_n \tag{9.3}$$

The equations of motion now form a large system of equations which include the journal motion and the motion of all the active nodes in the bearing housing.

The assembled stiffness matrix is now partitioned into active nodes (to be retained), free nodes (no constraint) and constrained nodes. As **multi-body dynamic** (MBD) calculations are being carried out, the mass matrix needs to be included. See Rahnejat (1998), Geradin and Cardona (2001) and Rahnejat and Rothberg (2004) for a detailed account of MBD methods.

The mass and stiffness matrices are condensed using a suitable reduction routine such as the Guyan reduction or through a commercial program such as Abaqus or Nastran. Having obtained the reduced matrices the following equation of motion is solved, which is similar to equation 9.3 but contains many more nodes:

$$M_n\ddot{x}_n = F_{HD}(x_n, \dot{x}_n) - D\dot{x}_n - Kx_n \tag{9.4}$$

Figure 9.6 shows the race engine con rod geometry analysed as in this example. It shows the model is generated by a number of components which are used to determine the stiffness and mass (dynamic) characteristics needed to carry out the EHD simulation.

Figure 9.7 shows the model used to study the affect of assembling the bearing shell into the **big-end**, and bolting of the end cap into place. This involves a non-linear contact FE analysis of the sub-assembly and, owing to axisymmetry, this can be carried out with a ¼ finite element model.

The deformations of the shell are then added to the initial bearing shell shape or if the bearing shell is machined after assembly, the final machined shape replaces the initial bearing shape (plus residual stresses).

The effects of heat generation due to viscous shearing of the oil film can be included via the coupled heat equations or with an estimate of the heat generated in the oil film which can be determined by assuming some percentage (80% in this case) of the heat leaves with the oil and the remaining percentage (20% in this case) is conducted into the **bearing shell** and the journal.

An approximate method of determining the heat dissipated in the shell

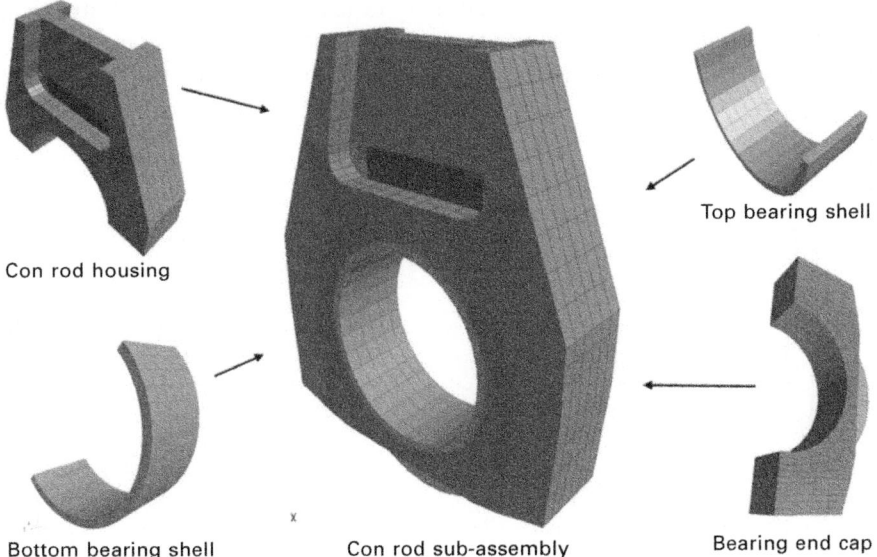

Con rod housing

Top bearing shell

Bottom bearing shell   Con rod sub-assembly   Bearing end cap

9.6 Con rod object with bearing shells, end cap and rod body.

is to assume that the heat flux is generated by viscous shear proportional to the pressure field around the bearing surface. The temperature rise is approximated thus:

$$\delta T = \frac{P}{\dot{Q}\rho C} \tag{9.5}$$

where Q = the fill ratio.

This temperature rise can then be prescribed on to the FE model of the con rod big-end, which includes the convective heat transfer effects cooling the big-end. The convective heat transfer being a function of the mass flow of oil washing over it. See Pinkus (1990) for a detailed account of TEHD methods.

Figure 9.8 shows the FE model of the racing engine big-end with the temperature distribution and the deformations of the shell due to thermal expansion. The thermal deformations tend to compensate for the pressure deformations and are added to the initial bearing shell shape further modifying its clearance map.

The following example shows a tribo-dynamic study of a race engine at maximum power and operating at around 16 000 rpm, illustrating the use of tribological numerical methods to better understand the phenomena seen in tests and in the engine.

Figure 9.9 shows the comparison of an RHD solution with an EHD solution for the ***big-end bearing***. Immediately it can be seen that the EHD solution generates orbits with eccentricity ratios greater than 1.0. With RHD solutions eccentricity can be used as a means of determining the minimum film thickness, as the journal and shell are assumed rigid. With EHD the

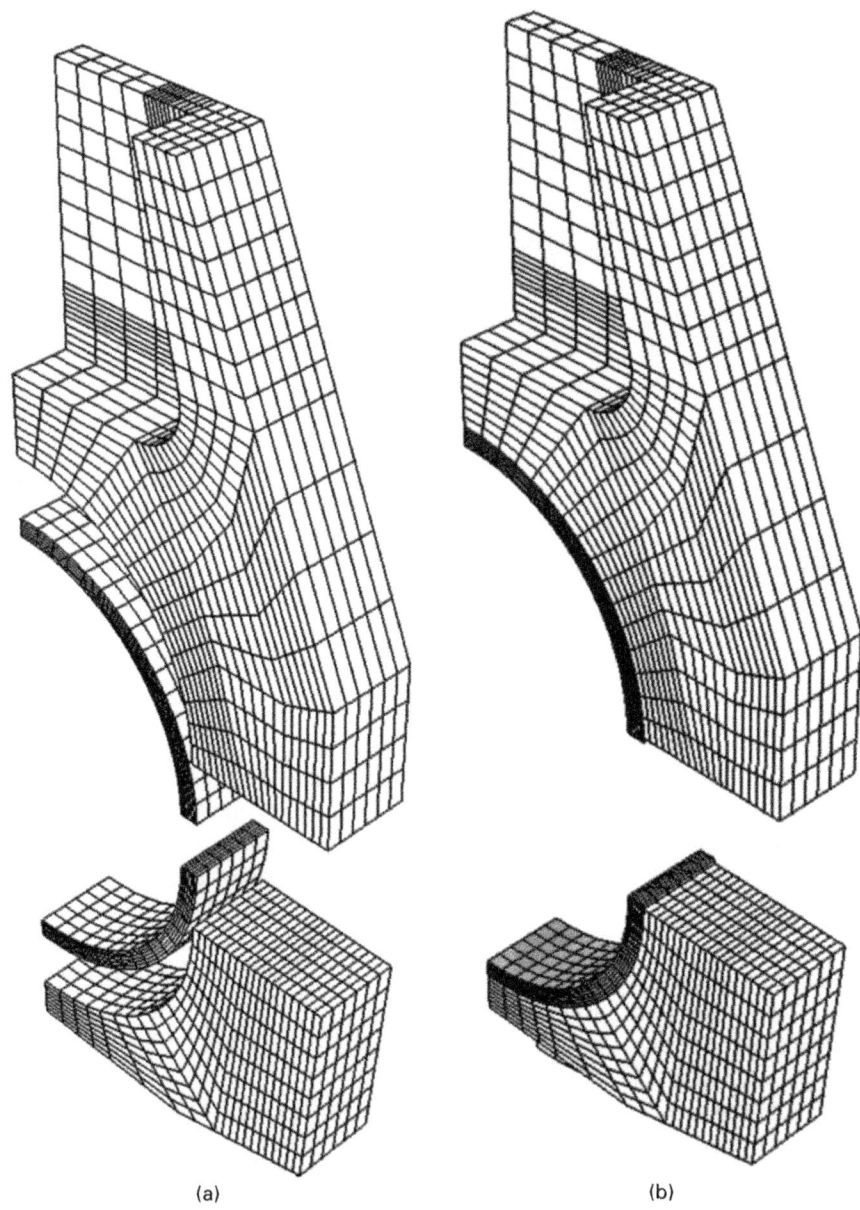

(a)   (b)

9.7 Assembly modelling and bolt preloading of the bearing shells fitted into the con rod big-end.

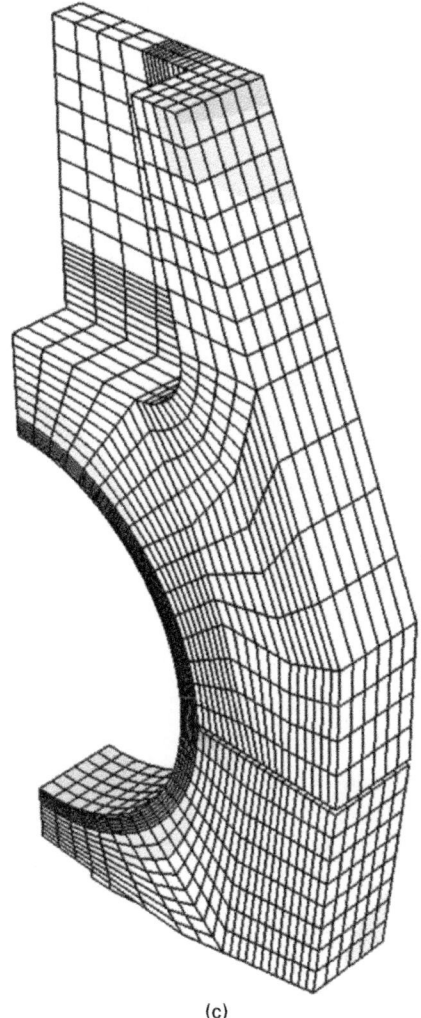

(c)

9.7 Continued

eccentricity ratio is many times greater than 1.0 so the minimum oil film thickness (MOFT) must be used as an assessment of performance and clearance.

Figure 9.10 shows the EHD results of the deformed shape of the big-end shell as it deforms and wraps itself around the journal, at the instance of top dead centre (TDC) non-firing. Figure 9.11 shows the comparison of RHD and EHD maximum oil film pressures. The EHD solution has lower pressures than for RHD due to the shell wrapping itself elastically around the journal. Figure 9.12 shows the MOFT comparisons of RHD with EHD and it can be seen that EHD gives much larger oil films than for RHD.

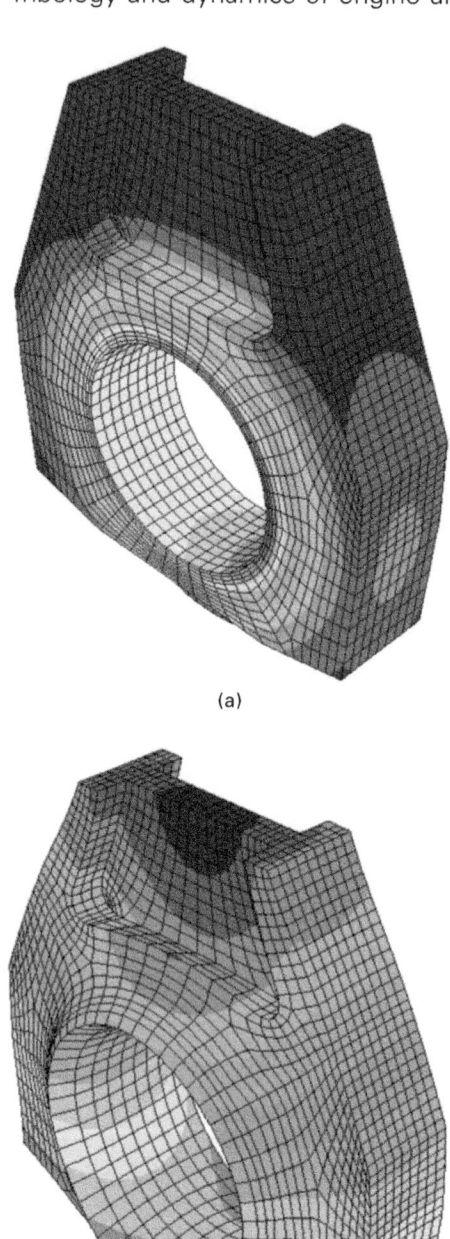

(a)

(b)

9.8 Con rod big-end heat flux distribution and thermal distortions.

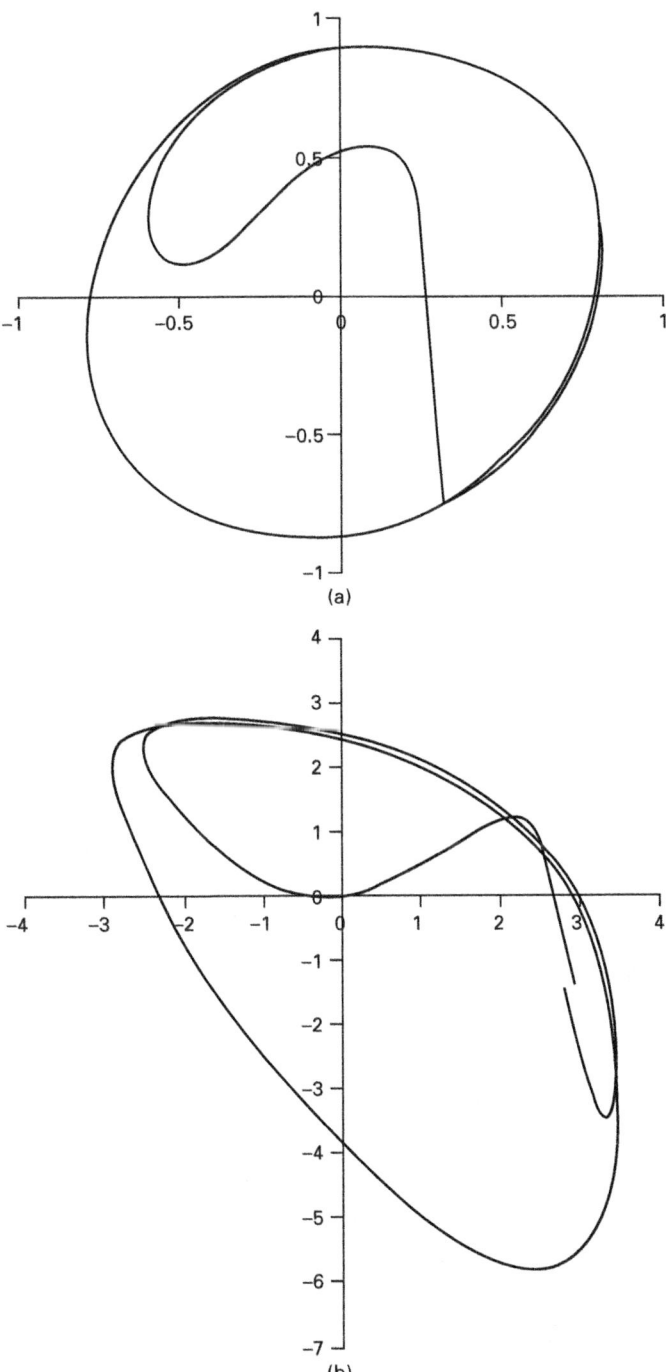

9.9 Comparison of RHD (a) and EHD (b) journal orbits.

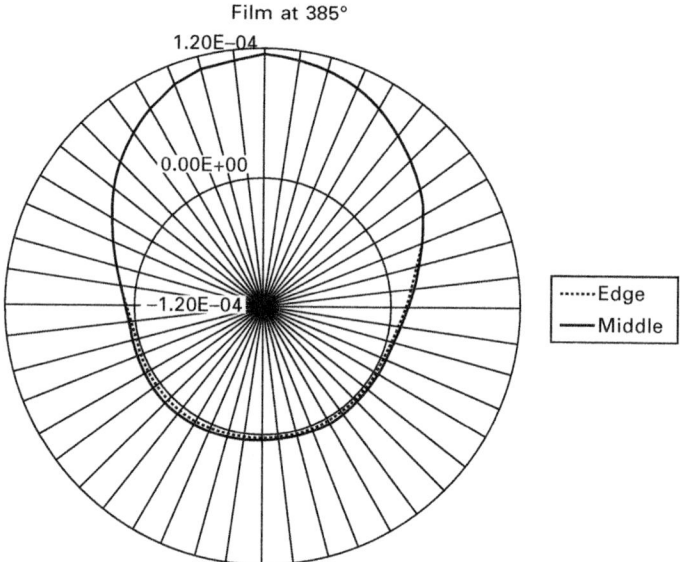

9.10 Deformed shape near to top dead centre non-firing.

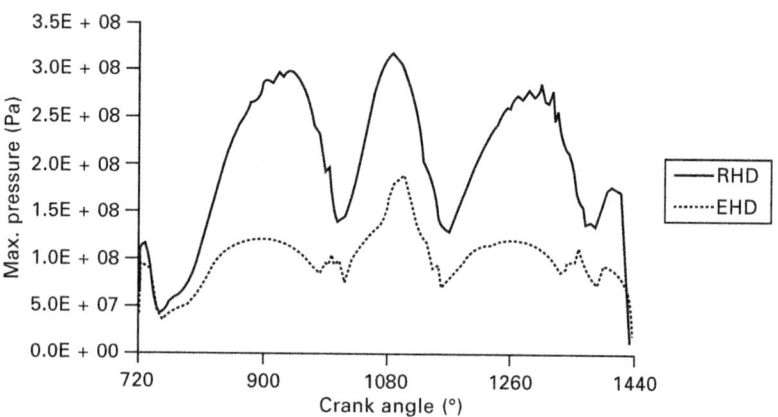

9.11 Comparison RHD and EHD maximum oil film pressure.

Figure 9.13 shows the comparison of RHD and EHD power losses. It can be seen that EHD has the greatest power loss as more oil film is present as the big-end wraps itself further around the crank pin journal.

Positioning of an *oil supply hole* is critical in preserving the operation of the big-end under extreme operation of load and speed as continuous oil supply to the bearing oil film is essential for such arduous bearing applications.

The position of the oil hole for this big-end example was varied from 30°, 0° and 60° TDC. The generally accepted position is 60° but it has been

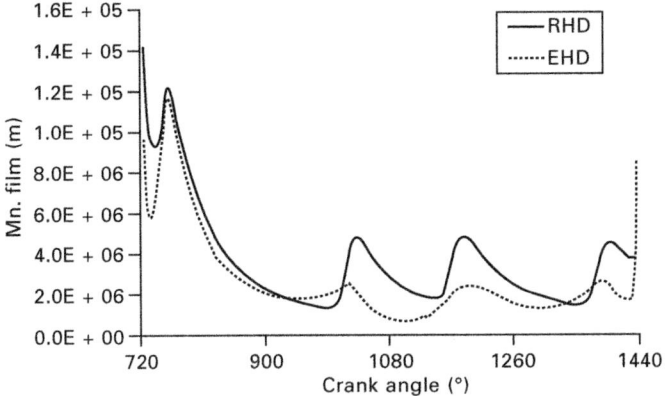

9.12 Comparison RHD and EHD minimum oil film thickness.

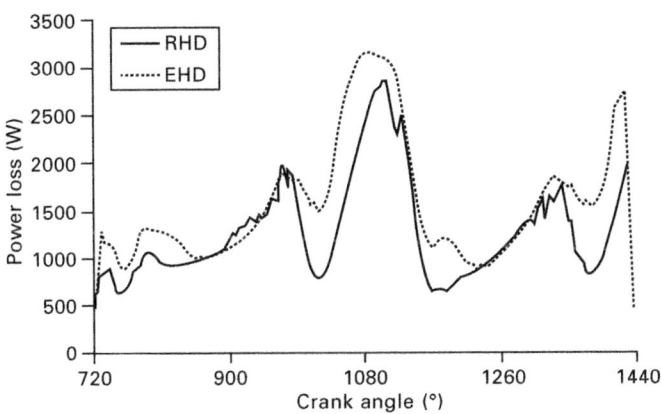

9.13 Comparison RHD and EHD power loss.

found that 30° gives a better bearing life. It was also found from testing that a wrongly placed oil hole at 60° induced cavitation erosion as the oil hole was far too remote to supply effective flow into the oil film.

Figure 9.14 shows the crankshaft big-end journal (crank pin) depicting the oil hole position. Figure 9.15 shows the comparison of EHD calculations with oil holes drilled at 30°, 0° and 60°. Although the maximum pressure, minimum film thickness and power loss are not affected much, the next figure shows what happens to the oil film pressure distribution.

For 30° it can be seen in Figure 9.16 that the oil hole is catching the trailing edge of the pressure film, at 0° the oil hole seriously encroaches into the pressure field, but for 60° the oil hole misses the pressure field entirely, being ineffective. Thus 30° should be a problem with tests; however, in reality it was found not to be the case. So EHD alone was not sufficient.

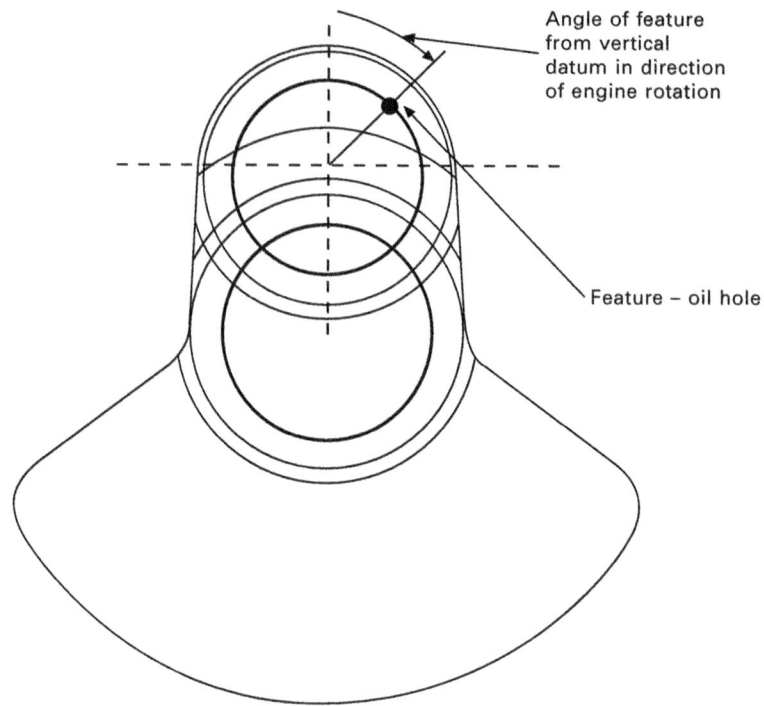

9.14 Definition of oil hole position in the big-end journal.

Therefore the effects of thermal deformation was considered next, where the above method was used, and the thermal deformations of the big-end shell were added to the EHD deformations and the initial shape of the bearing.

Figure 9.17 shows the orbit of the thermally deformed bearing which has increased in pressure from 200 MPa with EHD to 250 MPa for the thermally deformed bearing (TEHD). Figure 9.18 shows that for 30° the oil hole almost misses the pressure field, but again at 0° the pressure field is again seriously disrupted, generating a much higher ***pressure spike*** than the EHD solution indicated.

Figure 9.19 shows the comparison of the EHD and thermal deformed bearing solution (TEHD) it can be seen that the thermally deformed bearing shell has less wrap around, and from Fig. 9.18 it can be seen that the pressure at 10° crank angle has gone up from 100 MPa with EHD to 120 MPa for the TEHD analysis.

The effect of having a ***non-circular bearing*** shape (also see Chapter 20) to account for crush relief was also investigated, but the analysis was carried out with the cold shape only. Figure 9.20 shows the assembled initial bearing shape and curve fit, and Fig. 9.21 shows the bearing wrap around to be less than for the EHD analysis and that the film shape diverges more rapidly than the EHD solution.

Figure 9.22 shows the orbit of the non-circular bearing to have increased considerably in the horizontal axis due to the local crush relief and in addition due to lower wrap around of the bearing shell this leads to much higher maximum pressures of 500–600 MPa from the reduction in squeeze retention of the bearing shell. This compares to a maximum pressure of 200 MPa for EHD and 250 MPa for the thermal (TEHD) case. The power

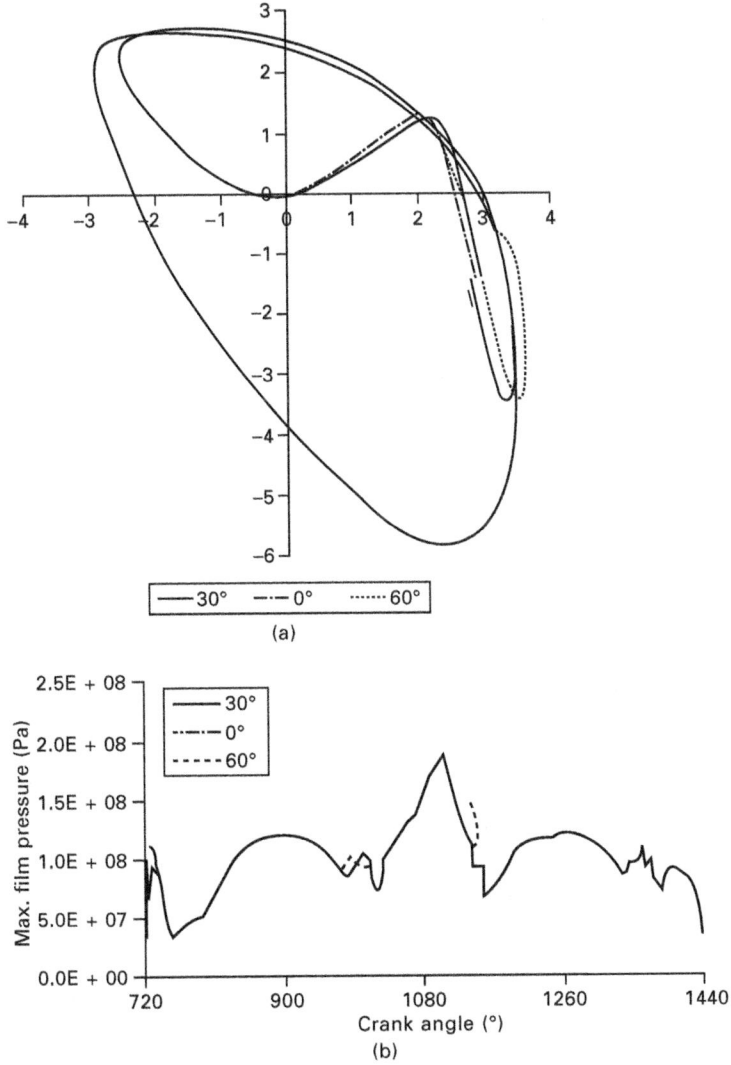

9.15 Effect of oil hole position on EHD models.

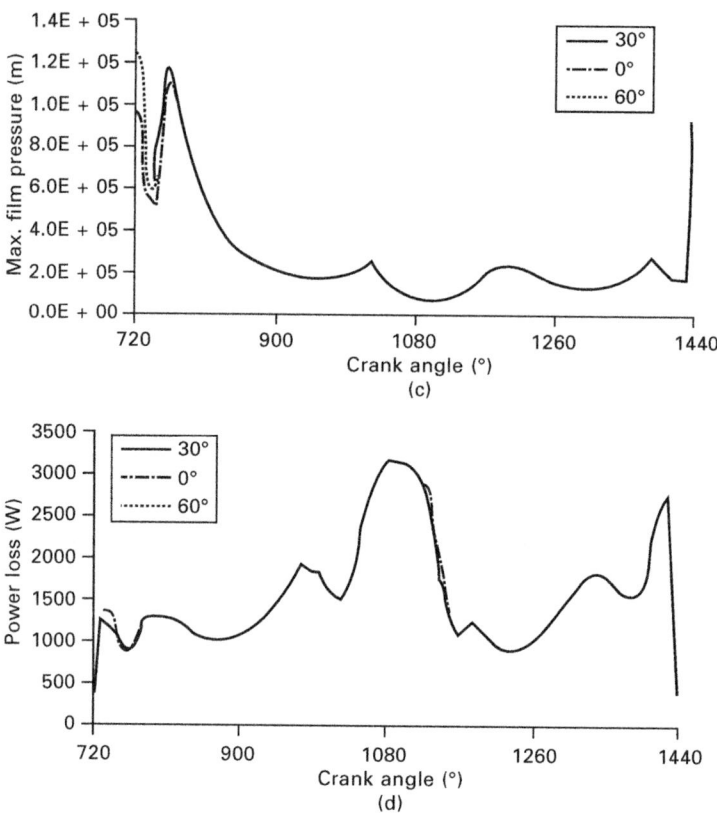

9.15 Continued

loss is the lowest, however, at 1800 W. These increased pressures would necessitate better bearing shell materials than normally used, for example the use of sputtered bearings.

Figure 9.23 shows the oil hole's effect on the pressure distributions at 10° crank angle. It can be seen that the oil hole performs best for this application as at 10° crank angle, the oil hole at 30° completely misses the pressure field, but at 0° the oil generates much higher twin peaks and again at 60° misses the pressure field and is ineffective. Thus, this study shows that the deformed shape of the bearing bush has a significant influence over the maximum pressure seen in the cycle, plus the power loss and the minimum film thickness, and hence bearing life and scuffing resistance of the bush.

Although non-circular shells are used to ensure clearance at the split line, the practice of machining bush diameters after assembly can be shown to be advantageous over non-circular forms when it comes to overall bearing performance. For very large bearings it may not be practicable to assemble

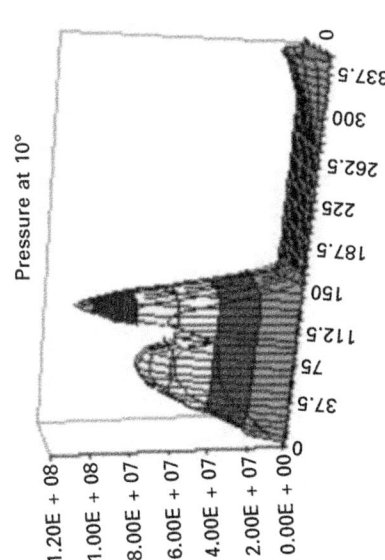

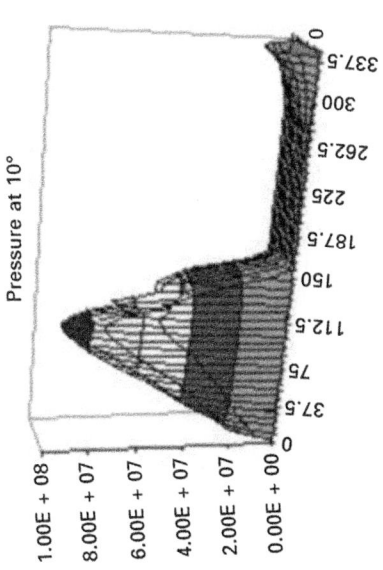

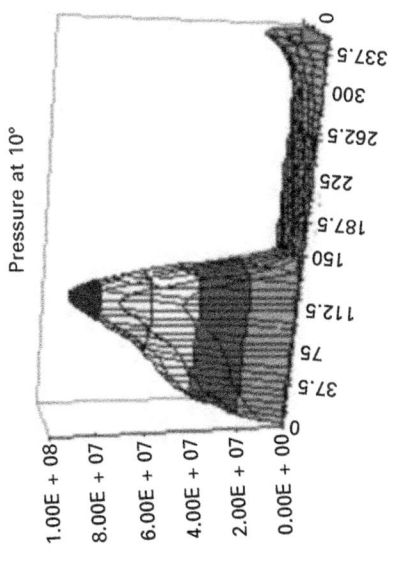

9.16 Pressure distribution for different oil hole positions at 10° crank angle – EHD (oil hole positions from top left – 30°, 0°, 60°).

and machine, so a shell with *crush relief* or non-circular circular profile may be the only option.

9.4   Application of integrated knowledge-based systems (IKBS) and rigid hydrodynamics (RHD) to piston and liner

The extension of Reynolds equation into the *piston* and liner interface oil film requires no modification of Reynolds equation, as the formulation remains

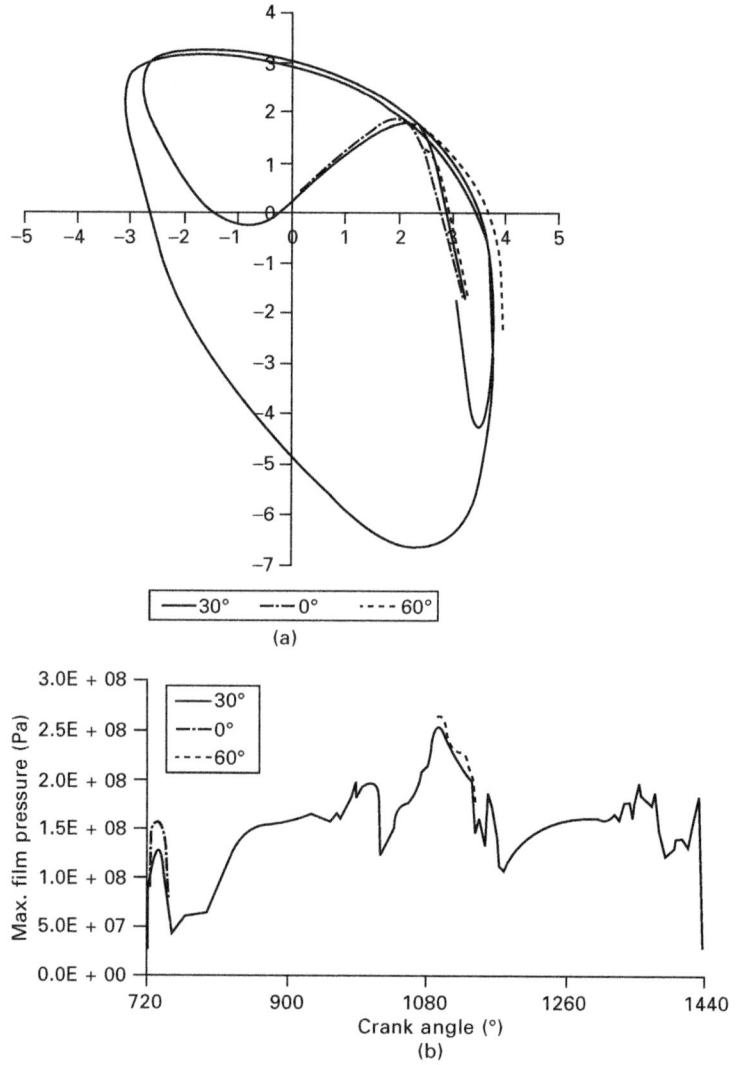

9.17 Effect of oil hole position – thermally deformed bearing.

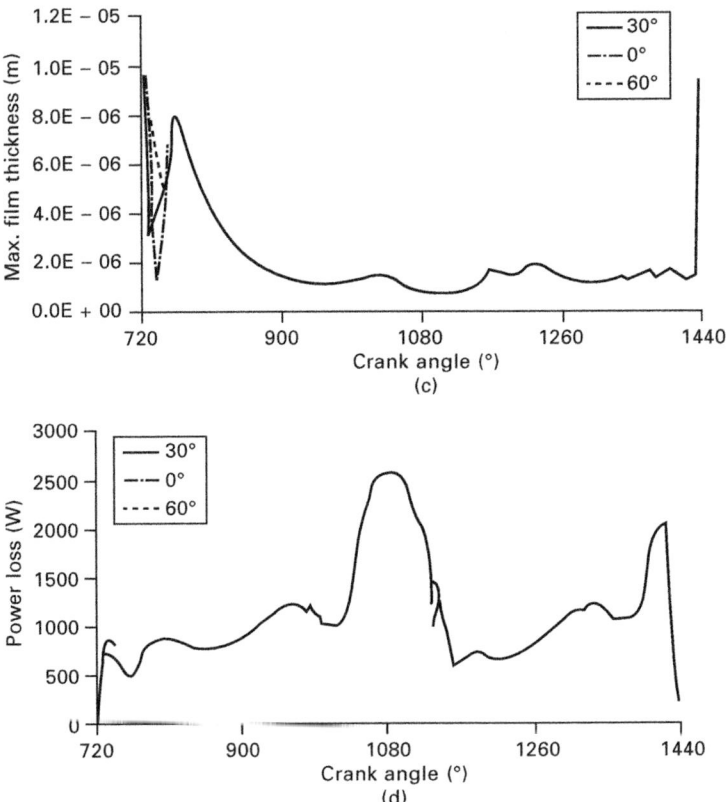

9.17 Continued

the same as for a journal bearing but the calculation of position and velocity is different since the sliding velocity is now in line with the cylinder axis. The equations of motion are also a little different because any axial offset between the piston centre of pressure and the piston pin centre line causes the piston to tilt. The equation of motion is as follows:

$$M\ddot{x} = F_{HD}(x, \dot{x}, \theta, \dot{\theta}) + F_{\text{external}}(t) - D\dot{x} - Kx \tag{9.6}$$

with the rotary equation of motion defined as:

$$I\ddot{\theta} = M_{HD}(x_j, \dot{x}_j, \theta, \dot{\theta}) + M_{\text{external}}(t) - D\dot{\theta} - K\theta \tag{9.7}$$

where the external forces and moments are calculated from the dynamics of the crank mechanism. The oil film moment is produced by the hydrodynamic forces and the difference in axial position between the **piston pin** and centre line.

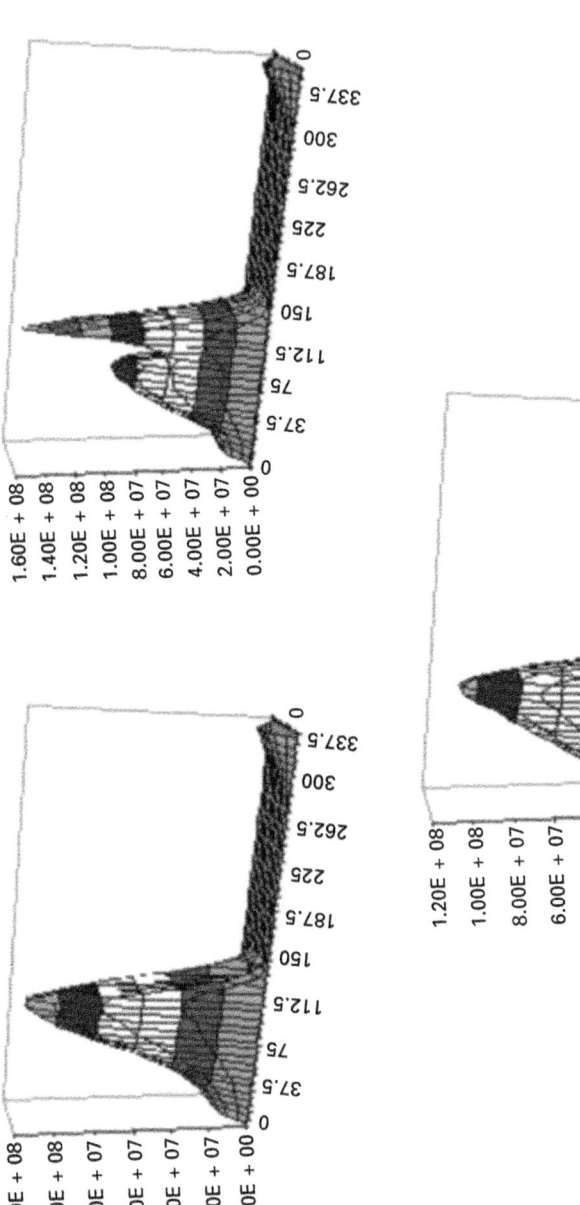

9.18 Pressure distribution at 10° crank angle – thermally deformed bearing (oil hole positions from top left – 30°, 0°, 60°).

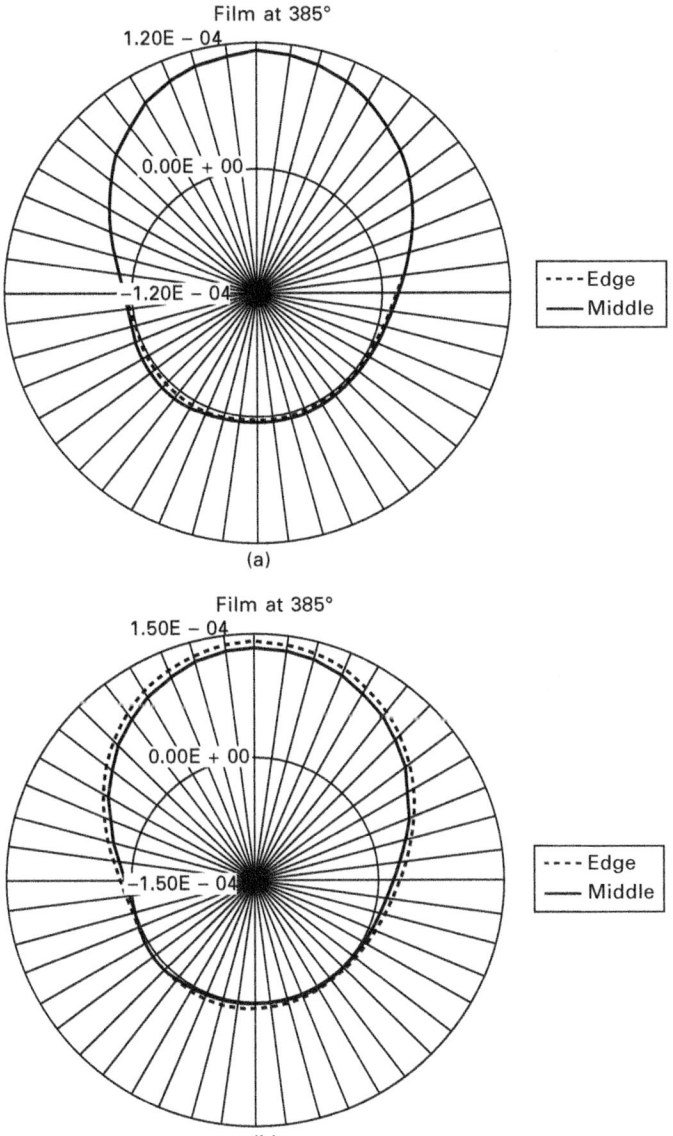

9.19 Typical bearing deformed shapes near to TDC non-firing – EHD circular bearing (a) versus thermally deformed bearing (b).

As was pointed out in the previous section the calculation of oil film performance is not only the effects of the oil film. Piston and liner tribology requires the use of cold and thermal distortions, in addition to elastic and dynamic deformations. How these are determined for a piston and liner

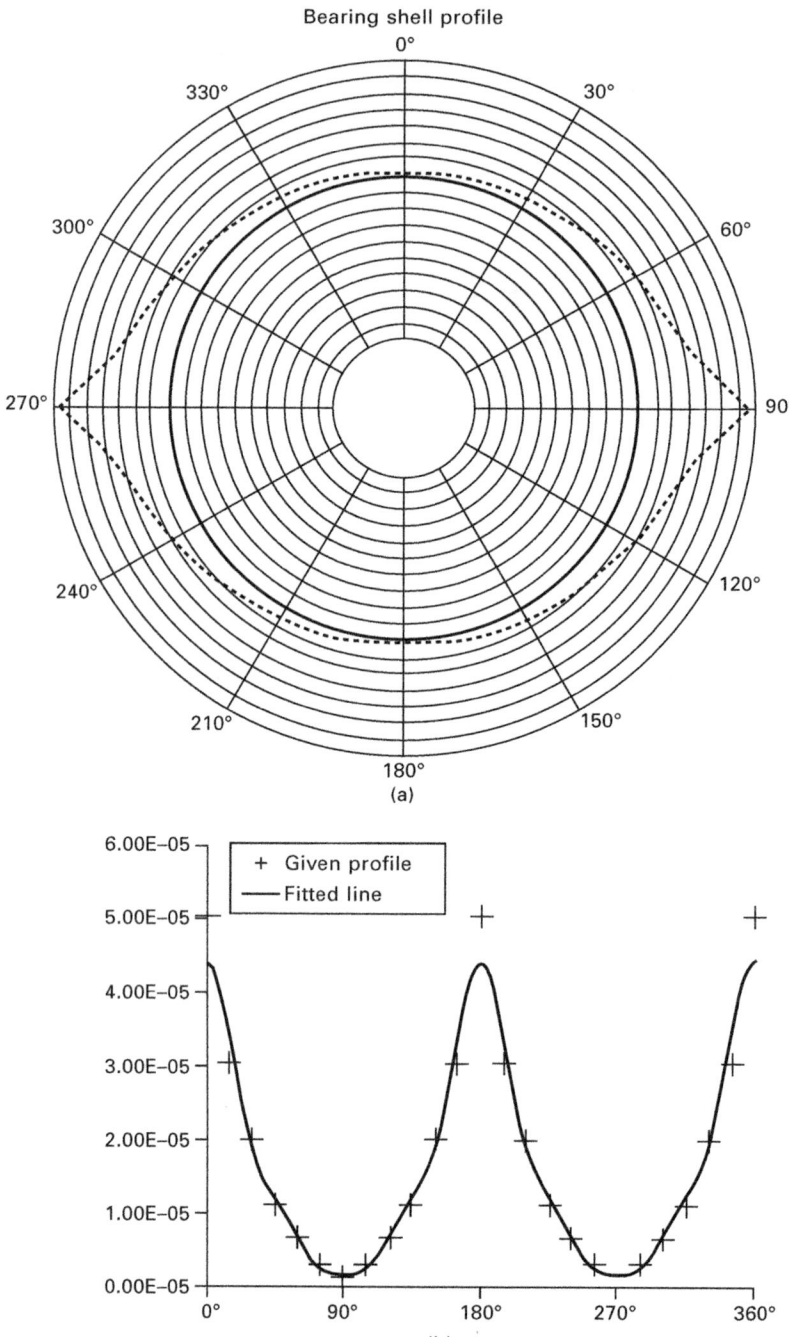

9.20 Measured assembled bearing profile and curve fit used in analysis.

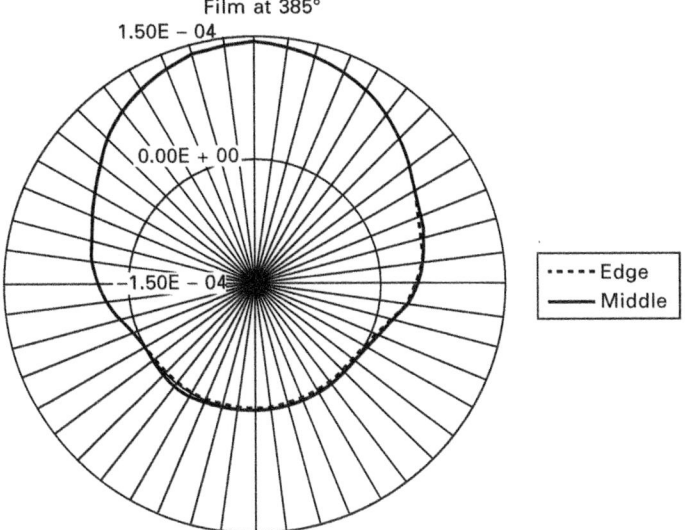

9.21 Bearing shape near TDC non-firing – EHD – non-circular bearing.

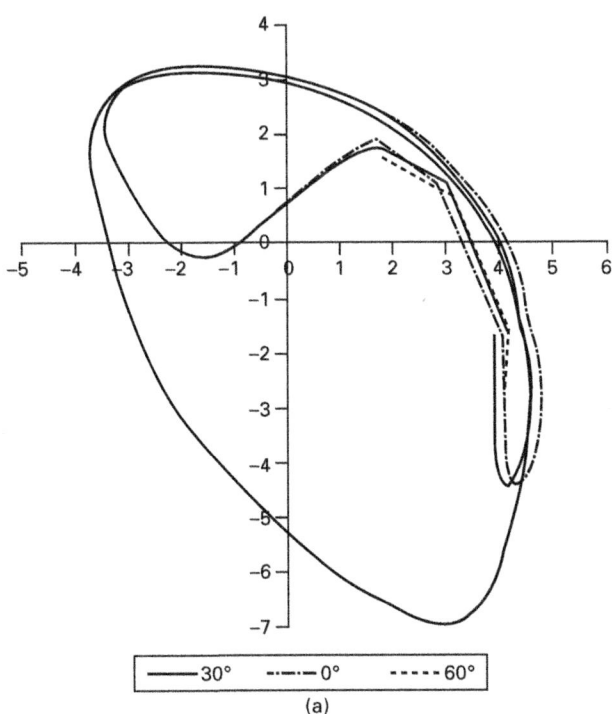

9.22 Non-circular bearing comparison of effects of oil hole position.

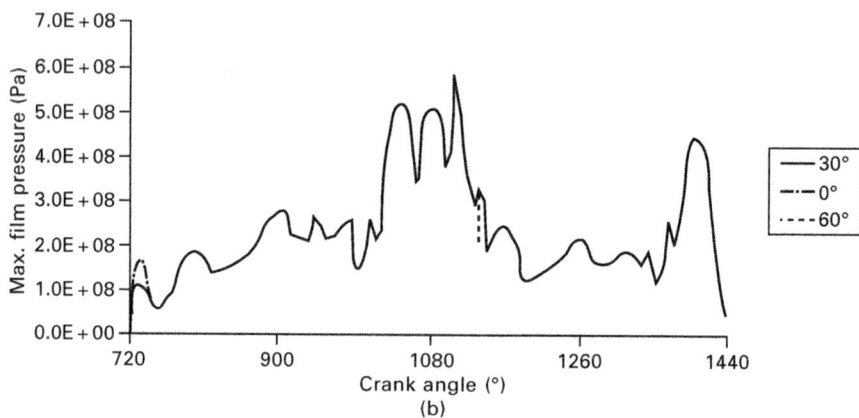

(b)

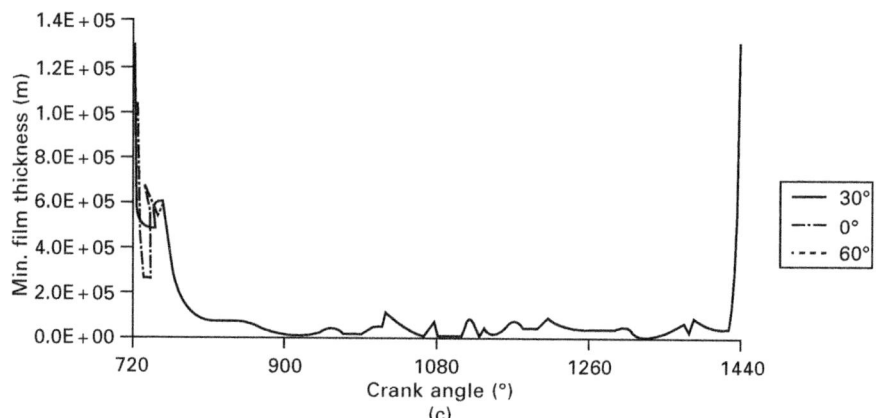

(c)

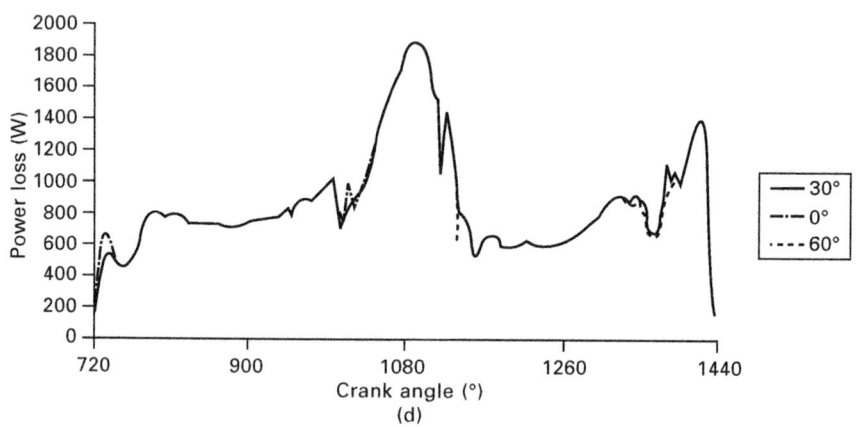

(d)

9.22 Continued

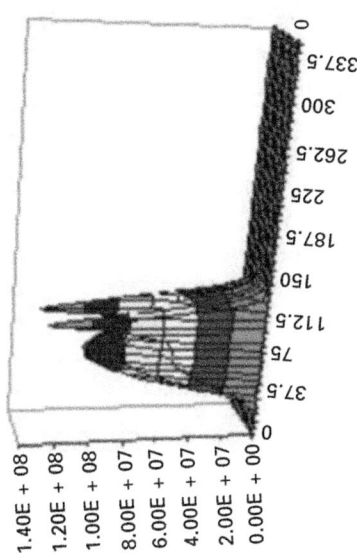

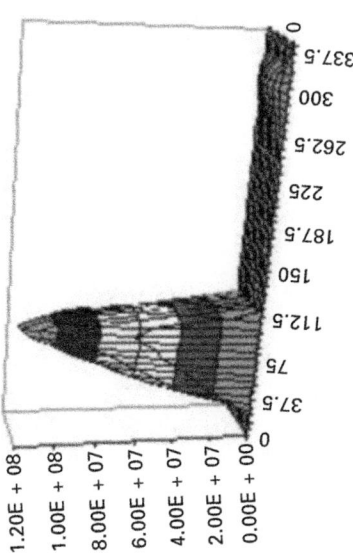

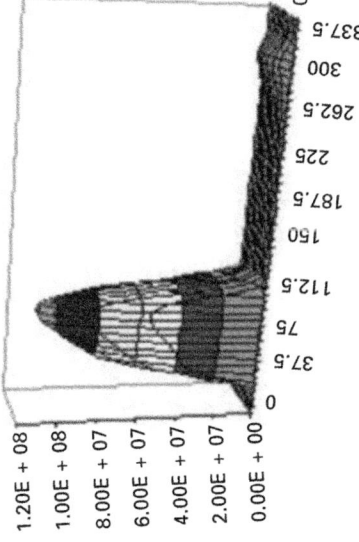

9.23 Non-circular bearing – effect of oil hole on pressure distribution at 10° shaft angle for oil hole at 30°, 0°, 60°.

assembly is much more involved than for a big-end bearing, although the principles are the same. The process is as follows:

1. Firstly the determination of the **liner** profile. This can be measured cold and is done so as far up the liner as is possible with a CMM (**co-ordinate measuring machine**). The data are then generated into a series of polar plots at various heights along the liner length. Figure 9.24 shows a typical liner cold bore distortion.

2. Next determination of the piston skirt cold profile. This is normally done by measuring or defining a profile as a set of x and y coordinates, plus an ovality profile in order to give clearance to the non-contacting sides of the piston. Figure 9.25 shows a typical cold **piston skirt** profile. The profile data for the skirt need to be a continuous curve fit for best results to ensure scuffing does not occur and secondly to ensure stability in the numerical calculations.

3. Next temperatures of the piston and liner need to be obtained. In the case of the piston the heat flux calculated by a thermodynamic cycle simulation program is used over the crown and wetted surfaces of the

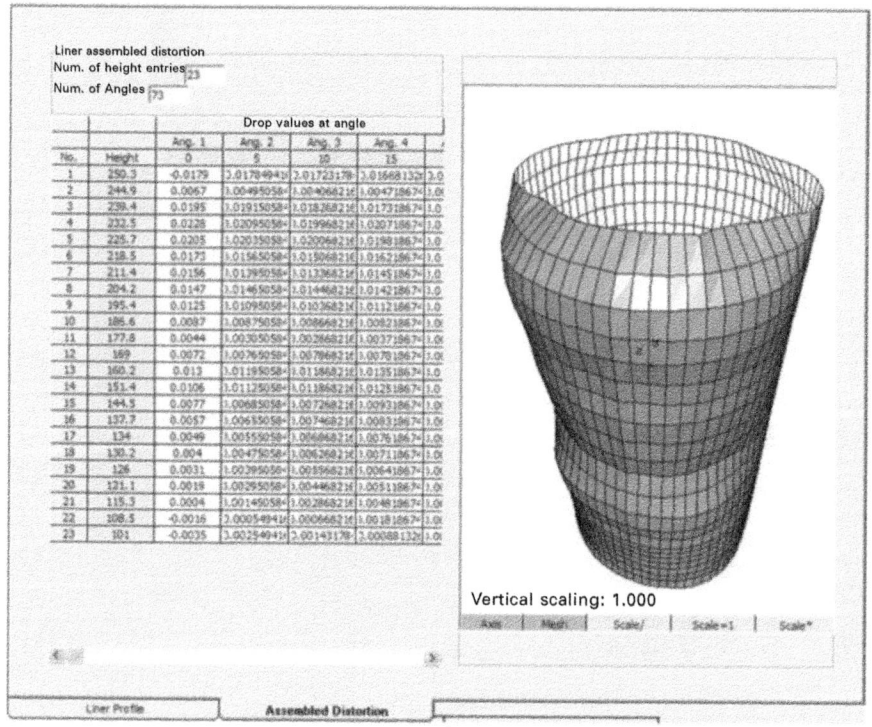

9.24 Liner cold assembly deformations.

piston for the full engine cycle. Figure 9.25 also shows a typical piston temperature distribution and a thermally distorted piston profile for the TF (thrust face) and the ATF (anti-thrust face).

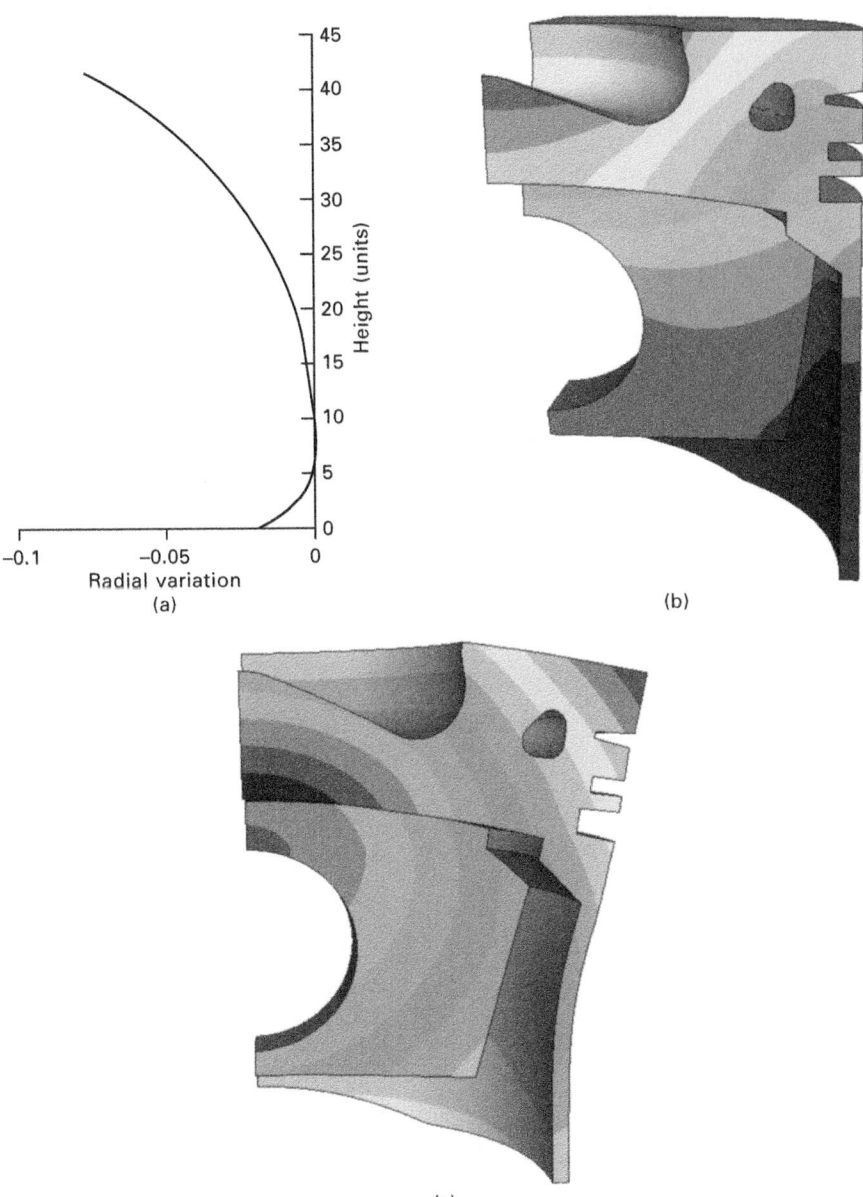

9.25 (a) Diesel piston skirt cold crown profile on thrust face (TF) and anti thrust face (ATF); (b) piston temperature profile and (c) piston thermally distorted shape.

For the liner, which is part of the head and block compound, the thermal distortions are normally derived from a conjugate heat transfer analysis of the head and block compound which includes the support constraints at the positions of the engine mounts.

Figure 9.26 shows a typical liner thermal bore distortions at an operating speed of around 5000 rpm. The deformed shape includes the affect of

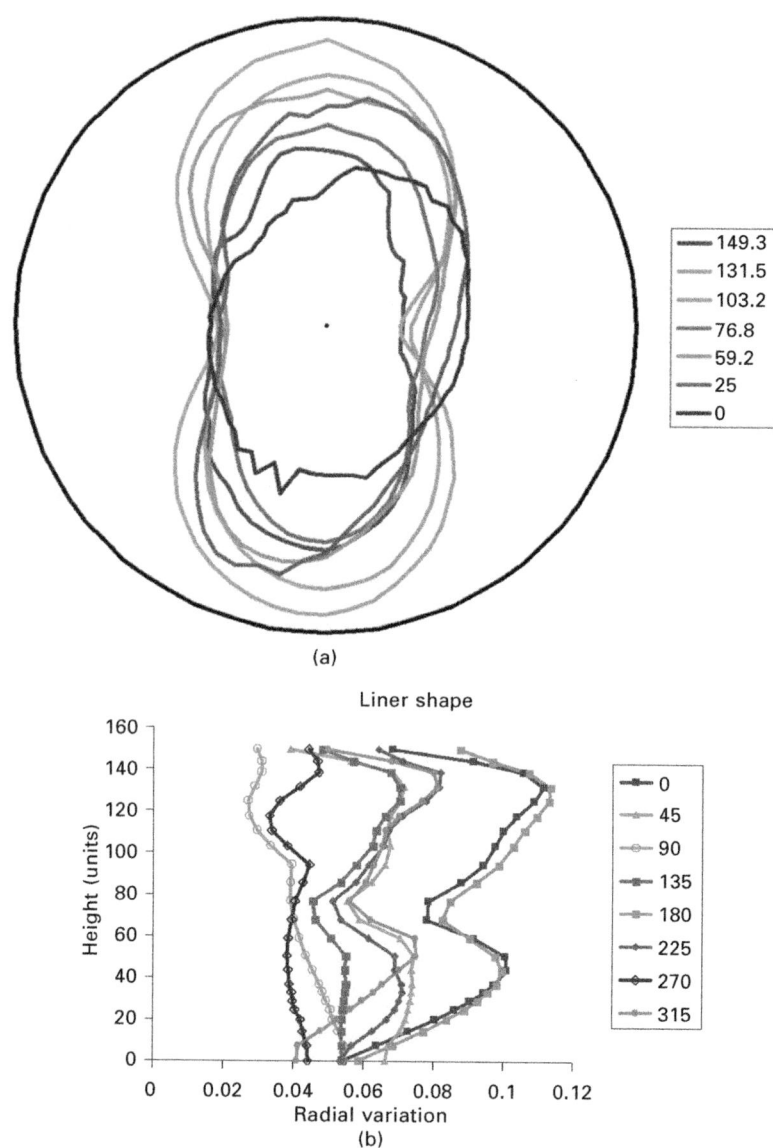

(a)

(b)

9.26 Liner thermal bore distortions at 5000 rpm.

cylinder head bolt clamping, and one can see the liner is crimped in four places, where the bolts are located. The liner is also seen to be grossly out of round, thus the necessity for an oval piston shape.

It is worth saying although all the pistons may be subjected to similar thermal distortions, this is not true for the liners, which will see different temperatures at each cylinder and will see transient temperatures that are out of phase by the firing order of the engine.

Additionally the distortions of each cylinder will be different due to the asymmetric structural stiffness and the thermal response of the head and block, plus the effect of the engine's mounting constraints (locations). Bending of the head and block compound, under gravity, causes the liners to squash into an oval or elliptical shape. Thermal arching of the head and block compound further aggravates the ovalisation of the cylinders and liners.

4. Lastly the cold and thermal deformations over the cycle are collected together at the required speeds of interest (as the thermal response of the piston and liner are assumed constant over the cycle, but vary with speed).

Although the thermal variations are assumed to be fairly steady and vary only with speed, the multi-body dynamic response of the oil film varies over the cycle and with speed.

Figure 9.27 shows a typical example of a piston trajectory (orbit) and its tilt at a speed of 5000 rpm. The thermally distorted liner gives greater excursions in the piston liner orbit than the straight liner. The choice of cold **piston skirt profile** has to compensate for the thermally distorted profiles and has to be acceptable for the full speed and load range of the engine and each cylinder liner. The wrong choice of cold profile will lead to high peak pressures and the likely hood of scuffing at the highest peak pressure over the cycle and speed.

Figure 9.28 shows the MOFT, max pressure, FMEP and oil flow at a speed of 5000 rpm. Figure 9.29 shows the 3D piston pressure at 3000 rpm with crowning and ovality and the 3D liner pressures at 1000 rpm. Figure 9.30 shows an example of **piston secondary motion** with exaggerated manufactured and thermal distortions.

9.5   Application of integrated knowledge-based systems (IKBS) and rigid hydrodynamics (RHD) to turbocharger bearings

The next example to demonstrate the flexibility of tribo-dynamics is the application of a **turbocharger bearing**, in particular the FFB or **full floating bearing**, which is used in most turbocharger systems. It utilises the engine oil

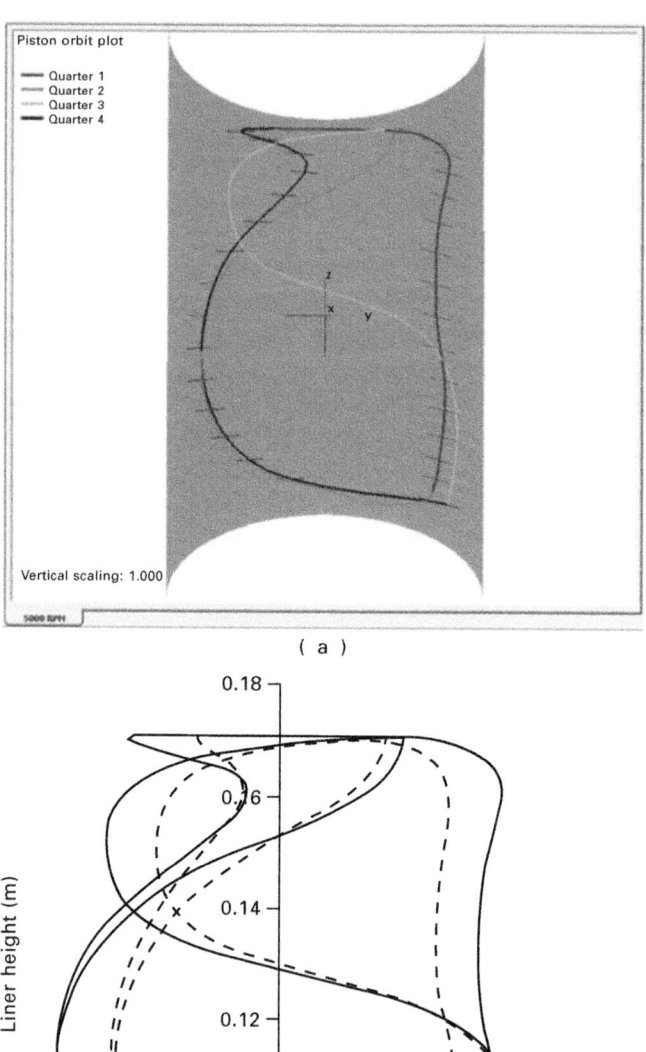

9.27 (a) Piston orbit and tilt in ¼ cycles at 5000 rpm and (b) piston orbits with straight and thermally distorted liner.

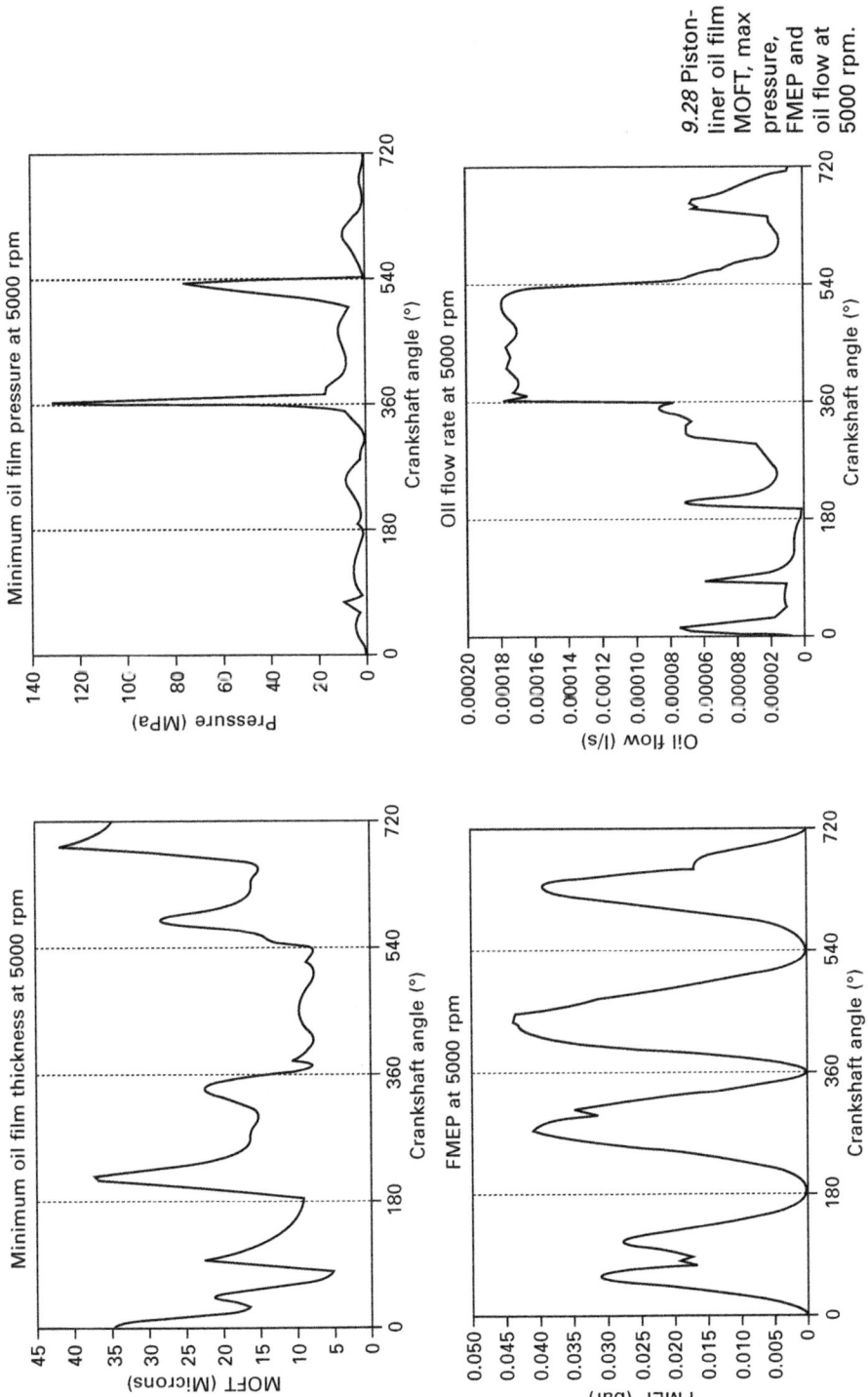

9.28 Piston-liner oil film MOFT, max pressure, FMEP and oil flow at 5000 rpm.

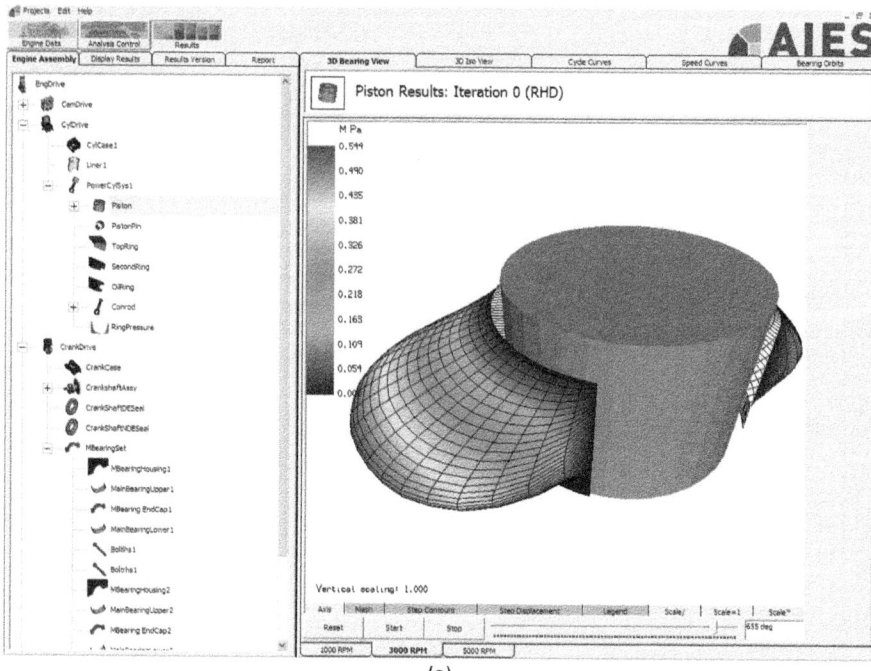

(a)

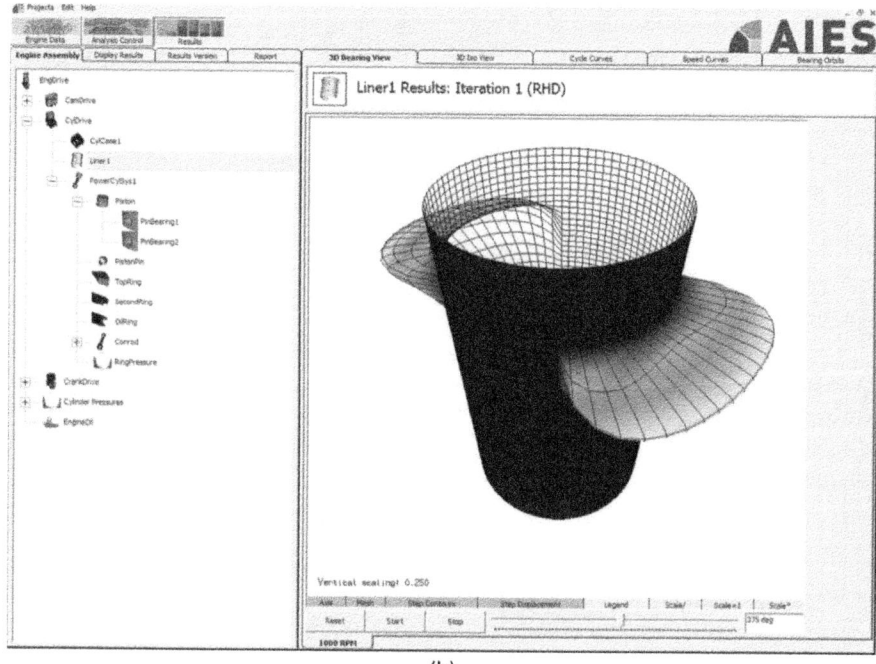

(b)

9.29 (a) Piston pressure at 3000 rpm, with crowning and ovality and (b) liner pressures at 1000 rpm.

9.30 Example of piston secondary motion at 1000 rpm, with exaggerated piston manufactured and thermal distortions.

used by all the other tribology objects of the engine and runs up to speeds of 100 000 rpm and beyond. Most FFB designs are developed by a test program carried out on a gas stand, which is isolated from external excitation to the bearing housings. External excitation sources are primarily the reason some systems fail to work in engines, as engine excitation generates vibrations in the housing which render the FFB bearings to become unstable. For the detailed study see McLuckie, *et al.* (2006).

Thus, in order to design a stable bearing system for a turbocharger the use of Reynolds equation and the equations of motion are combined in a multi-body solution scheme. Here the FFB has been studied to demonstrate the usefulness of the methodology. Reynolds equation is formulated in the following form

$$\frac{\partial}{\partial x}\left(h^3 \frac{\partial p}{\partial x}\right) + \frac{\partial}{\partial z}\left(h^3 \frac{\partial p}{\partial z}\right) = 6\eta(U_1 + U_2)\frac{\partial h}{\partial x} + 12\eta(V_2 - V_1) \qquad (9.8)$$

where: p = pressure, h = film thickness, x = circumferential position, z = axial position, η = absolute viscosity, U = circumferential velocity and V = radial velocity. Note that U and V are not constant around the bearing surface.

The oil film forces are calculated, and for this example the housing is stationary; however, the journal and ring move. The equations of motion are in vector form and include the rotational acceleration terms for the floating ring.

For the journal:

$$M_j \ddot{x}_j = F_{RHD}(x_j, \dot{x}_j) + F_{external}(t) - D\dot{x}_j - Kx_j \tag{9.9}$$

For the ring:

$$M_r \ddot{x}_r = F_{RHD}(x_r, \dot{x}_r) - F_{RHD}(x_j, \dot{x}_j) - D\dot{x}_r - Kx_r \tag{9.10}$$

The equations are solved here by the Bulirsch–Stoer numerical integration method to obtain velocity and position at the next step from acceleration and velocity at the current step exactly like the other bearing examples shown previously.

The full scope of this study can be seen in McLuckie *et al.* (2006), and has been compared to the work of San Andres and Kerth (2004) and Naranjo *et al.* (2001), but this rotor is assumed to be rigid, that is, operating below its first bending mode of vibration. A dynamic transient forced response of the turbocharger turbine end bearing is shown here from speeds near standstill up to 100 000 rpm.

The loads applied to the rotor system were gravity and unbalance, but aerodynamic forcing can be easily accommodated, if it is known how it varies over the cycle (revolution) of the rotor. The unbalance load was varied from 0.023 g mm to 0.23 g mm and 2.3 g mm. The lowest unbalance figure is based on BS (6861-1) grade G1. The others were G10 and G100. Note that most rigid rotor turbochargers are balanced to G2.5, so the unbalance range encompasses common practice.

The FFB ring had an inner bore of 8 mm and a length of 3.5 mm with a 1 mm wide circumferential groove, the outside diameter was 14 mm and 6.5 mm long with a 1 mm wide circumferential groove. For the FFB the inner clearance was 2.5 μm and outer clearance was 7 μm. Supply pressure to the bearings was varied as 4 bar, 2 bar and 0 bar. The oil for the bearing was assumed to be 10W40 with an inlet temperature of 100 °C.

Figure 9.31 shows the turbocharger configuration and the FFB. FFB bearings are commonly used as they are believed to have better stability performance threshold. The FFB calculations have been carried out with a constant ring to rotor speed ratio of 0.25 (25 000 rpm to 100 000 rpm).

Figure 9.32 shows the FFB whirl orbits for an unbalance of 0.23 g mm, 2.5 μm inner clearance and 7 μm outer clearance and a supply pressure of 4 bar. The first orbit shows **synchronous whirl** at 1000 rpm, the next three plots show **sub-synchronous whirl** at 3000 rpm, 30 000 rpm and 100 000 rpm respectively. This is four times the normal balance standard of grade G2.5 yet the inner film is unstable.

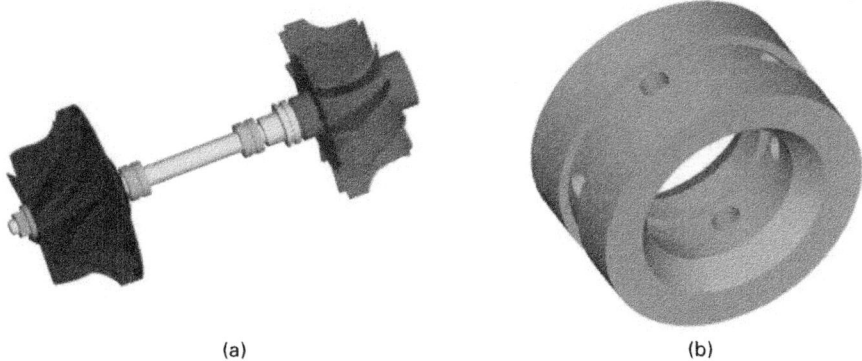

(a)   (b)

9.31 (a) Turbocharger rotor system with FFB bearing (b) typical full floating bearing shell.

Figure 9.33 shows a typical set of results for the turbine FFB inner oil film response with a diametral clearance of 2.5 μm, a supply pressure of 4 bar and gross unbalance of 2.3 g mm. The inner response is stable right up to 100 000 rpm.

Figure 9.34 shows a set of results for the turbine FFB outer oil film response running with an outer diametral clearance of 7 μm and a supply pressure of 4 bar and unbalance of 0.23 g mm. The first plot shows a 1/8 sub-synchronous whirl at 6000 rpm. The second plot shows a 1/8 sub-synchronous orbit at 14 000 rpm. The third plot shows a synchronous orbit onset at about 40 000 rpm and the fourth plot shows a sub-synchronous orbit at 100 000 rpm.

Figure 9.35 shows the FFB outer film response running with diametral clearance of 7 μm and a supply pressure of 4 bar and an unbalance of 2.3 g mm. The first curve shows sub-synchronous whirl at 6000 rpm and the second curve shows sub-synchronous whirl at 14 000 rpm. The third plot shows sub-synchronous whirl at 40 000 rpm and the fourth plot shows synchronous whirl at 100 000 rpm taking up 15% of the clearance space.

Figure 9.36(a) shows the FFB inner oil film response for eccentricity versus speed under gravity and unbalance forcing; with unbalances of 0.023 g mm and 0.23 g mm the inner film becomes unstable at around 60 000 rpm. But with an unbalance of 2.3 g mm the inner film is stable up to 100 000 rpm. Figure 9.36(b) shows the outer bearing oil film response for eccentricity versus speed under gravity and unbalance forcing for all unbalance values the outer film is stable up to 100 000 rpm, but only the 2.3 g mm value is permissible as the inner film is unstable with 0.023 g mm and 0.23 g mm unbalance.

Figure 9.37(a) shows the waterfall frequency spectra for the FFB, with 0.23 g mm unbalance and 2.5/7 μm clearance ratio with 4 bar supply pressure. The plot shows mainly 1/8 shaft order which is ½ of the ring. It also shows

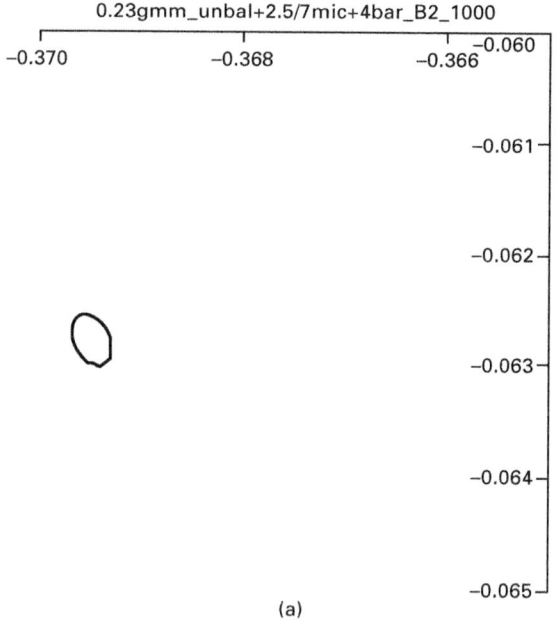

(a)

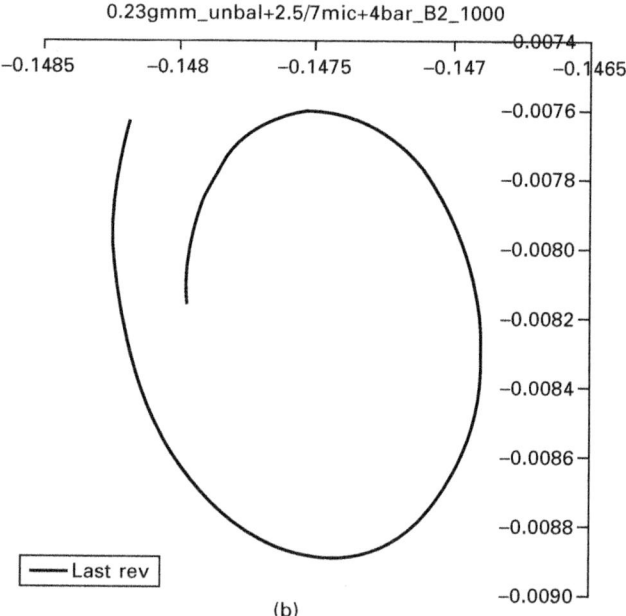

(b)

9.32 Turbine bearing FFB2 (L-R) (a) synchronous whirl at 1000 rpm
(b) onset of instability at 3000 rpm (c) sub-synchronous whirl 30 000
rpm (d) sub-synchronous whirl at 100 000 rpm (0.23 g mm, 2.5 μm
inner/7 μm outer clearance, and supply pressure of 4 bar).

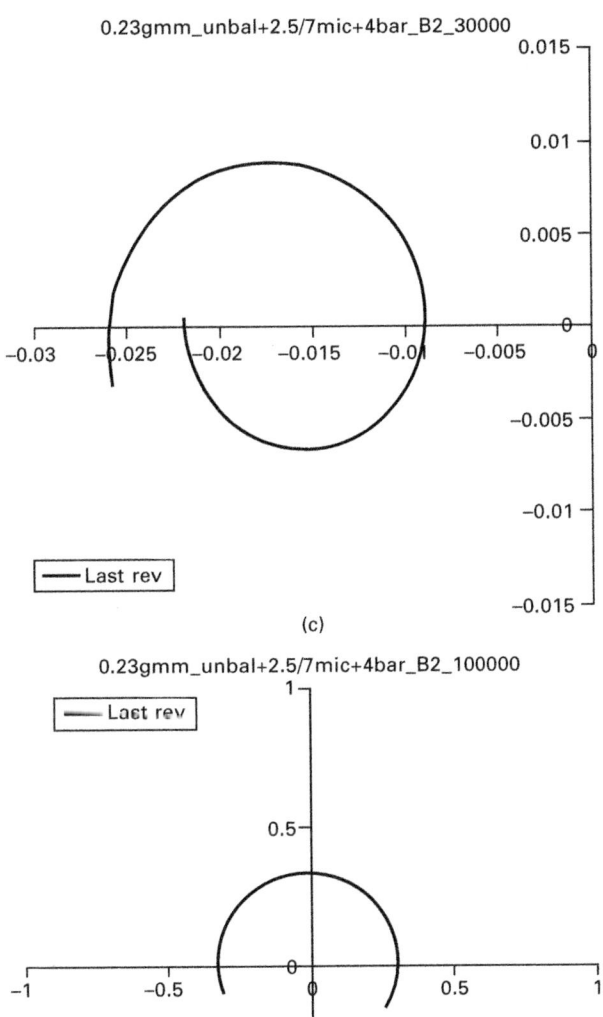

0.23gmm_unbal+2.5/7mic+4bar_B2_30000

(c)

0.23gmm_unbal+2.5/7mic+4bar_B2_100000

(d)

9.32 Continued

5/8 shaft order which is ½ shaft order of the rotor + ring. It also shows a synchronous 1x shaft order component.

Figure 9.37(b) shows the waterfall frequency spectra for the FFB, with

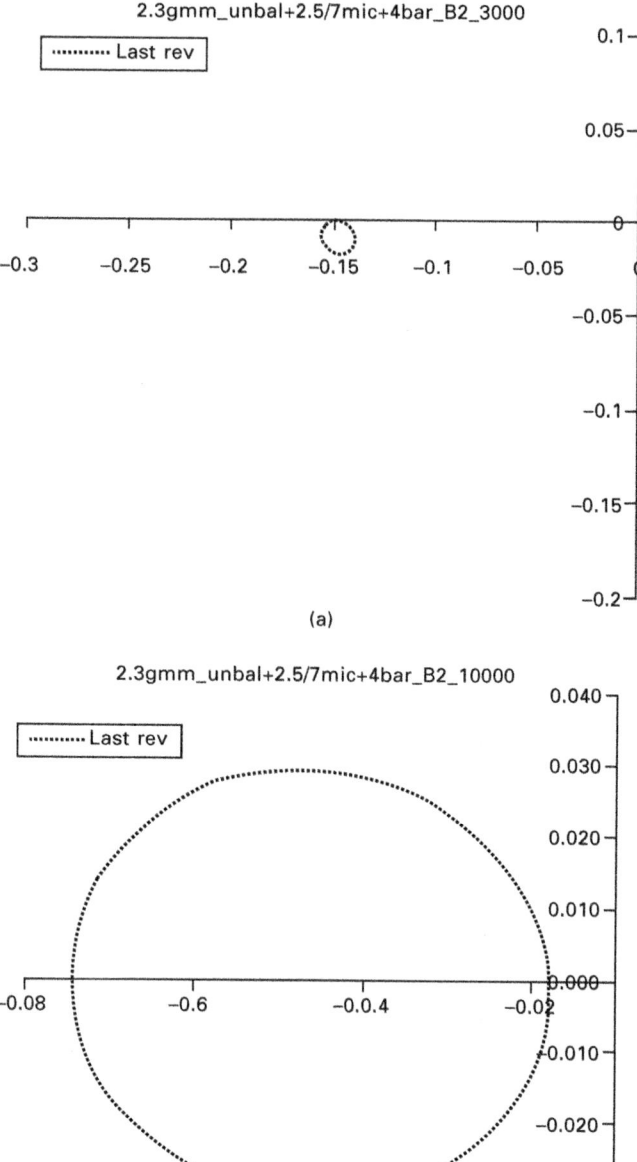

9.33 Turbine bearing FFB2 (L-R) (a) synchronous whirl at 3000 rpm (b) synchronous whirl at 10 000 rpm (c) synchronous whirl at 12 000 rpm (d) synchronous whirl at 100 000 rpm (2.3 g mm, 2.5 μm inner/7 μm outer clearance, 4 bar).

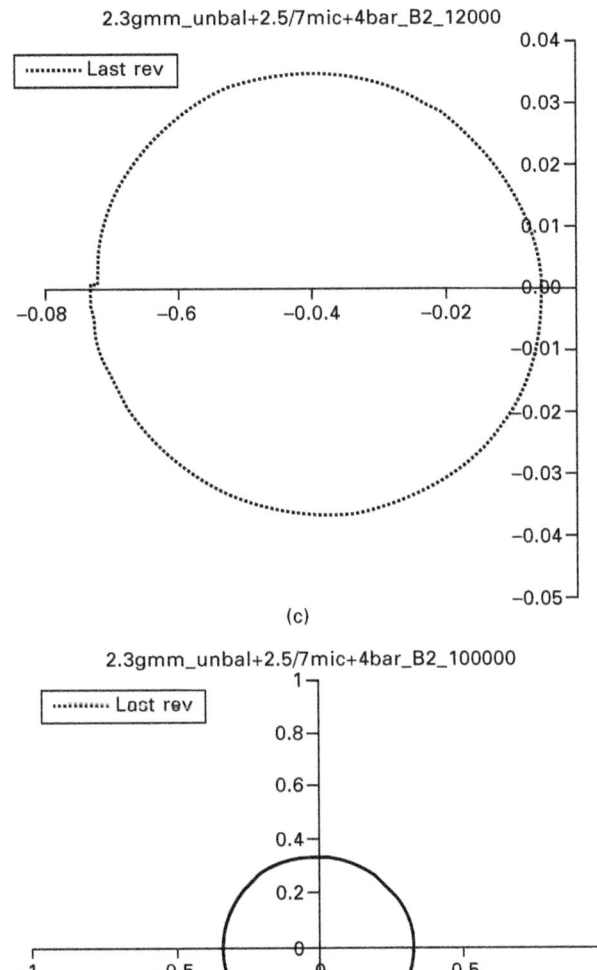

(c)

(d)

9.33 Continued

an unbalance of 2.3 g mm and a clearance ratio of 2.5/7 µm with a supply pressure of 4 bar. The plot shows a 1/8 shaft order (sub-synchronous) up to about 40 000 to 45 000, where upon the response jumps to a 1x shaft order

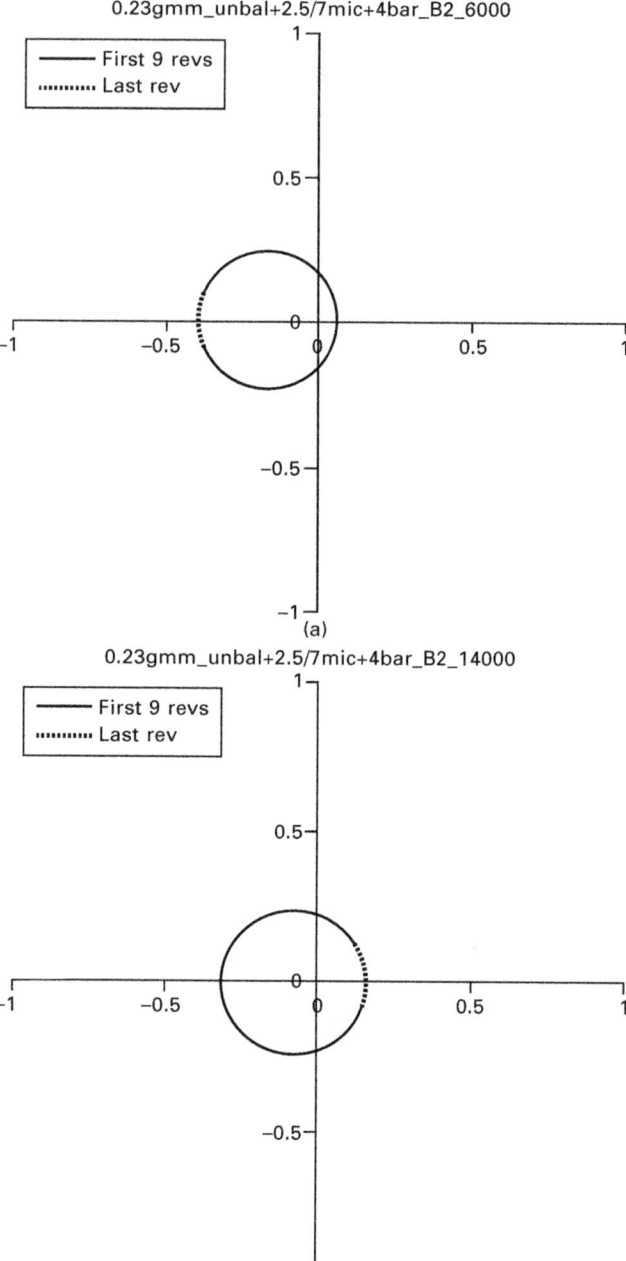

9.34 Turbine bearing FFB2 (L-R) (a) ring sub-synchronous whirl at 6000 rpm (b) ring sub-synchronous whirl at 14 000 rpm (c) ring sub-synchronous whirl at 30 000 rpm (d) ring sub-synchronous whirl at 100 000 rpm (0.23 g mm, 2.5 μm inner/7 μm outer clearance, 4 bar).

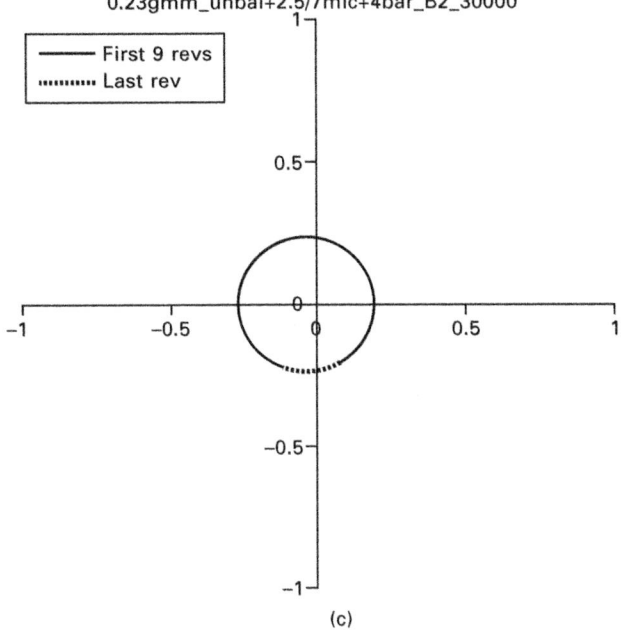

(c)

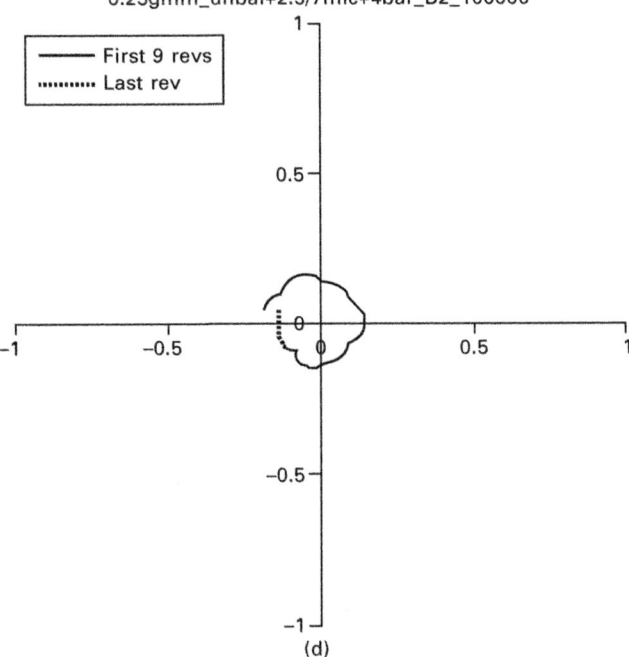

(d)

9.34 Continued

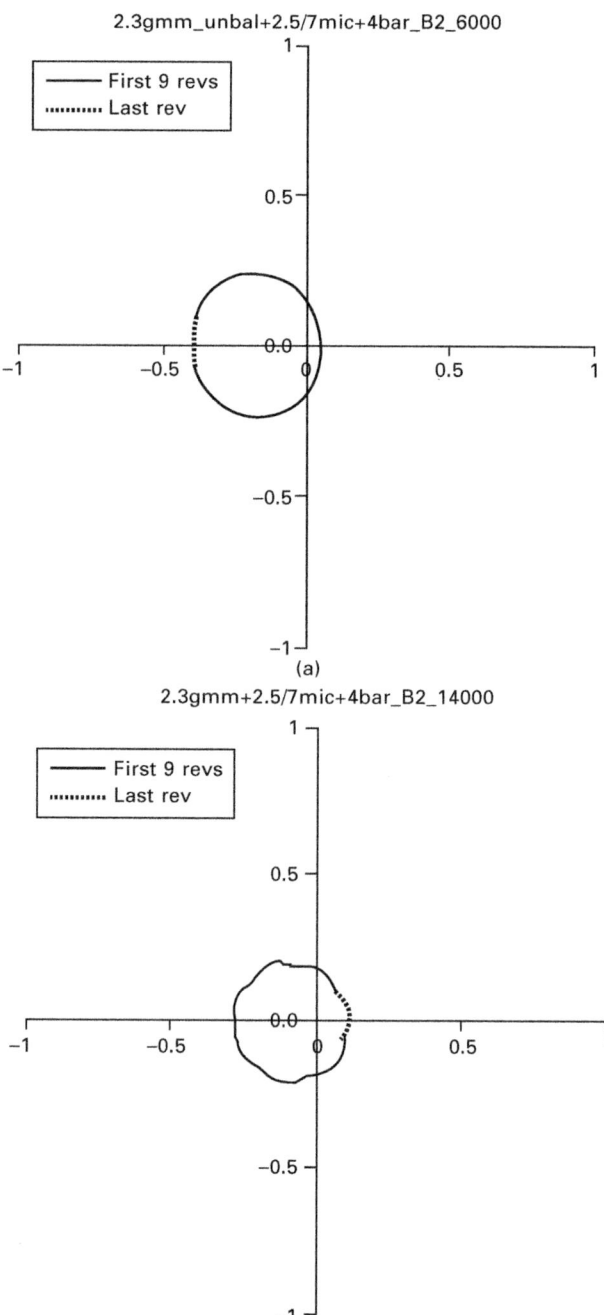

9.35 Turbine bearing FFB2 (a) ring sub-synchronous whirl at 6000 rpm (b) ring sub-synchronous whirl at 14000 rpm (c) ring synchronous whirl onset at 40000 rpm (d) ring synchronous whirl at 100000 rpm (2.3 g mm, 2.5 μm inner/7 μm outer clearance, 4 bar).

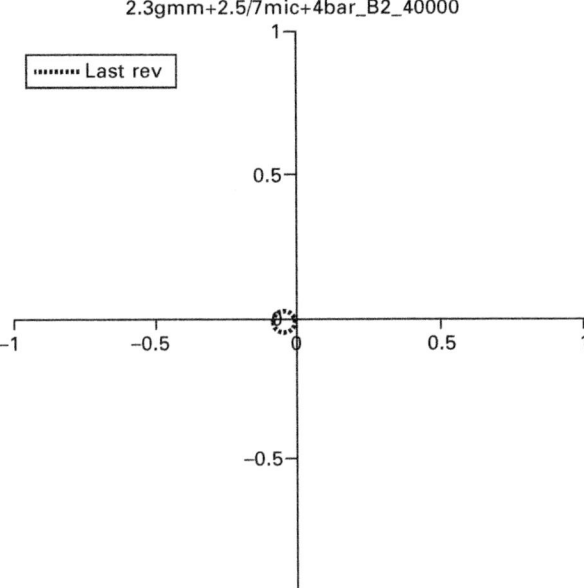

(c)

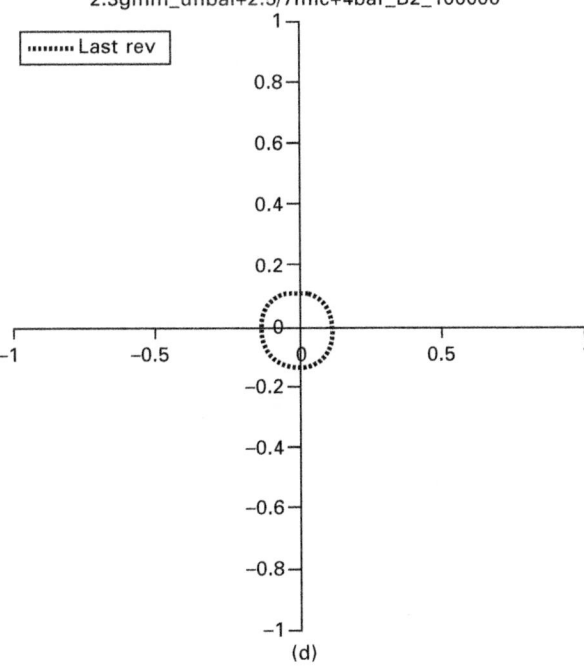

(d)

9.35 Continued

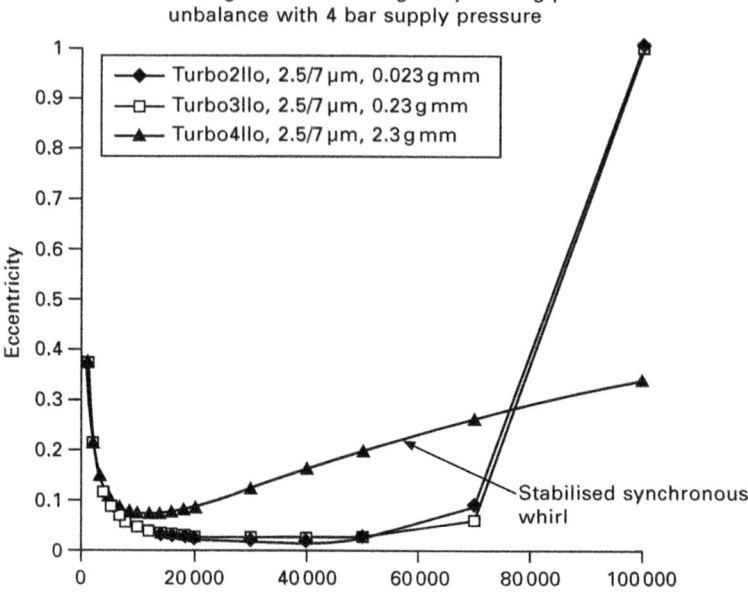

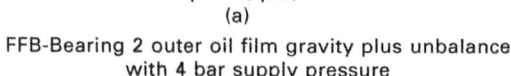

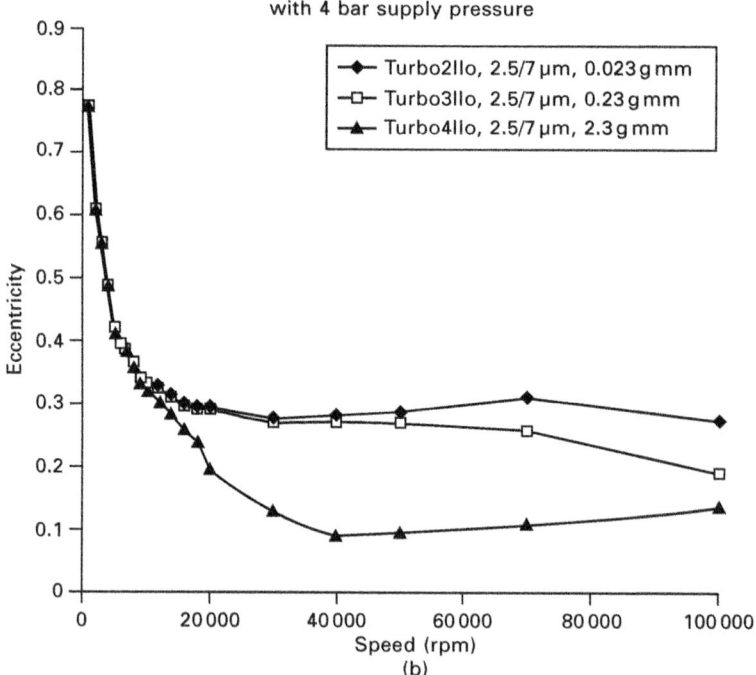

9.36 Turbine bearing FFB2 (L-R) (a) inner bearing eccentricity vs. speed (b) outer bearing eccentricity v speed (2.5/7 μm clearance (0.023 – 2.3 g mm unbalance, 4 bar).

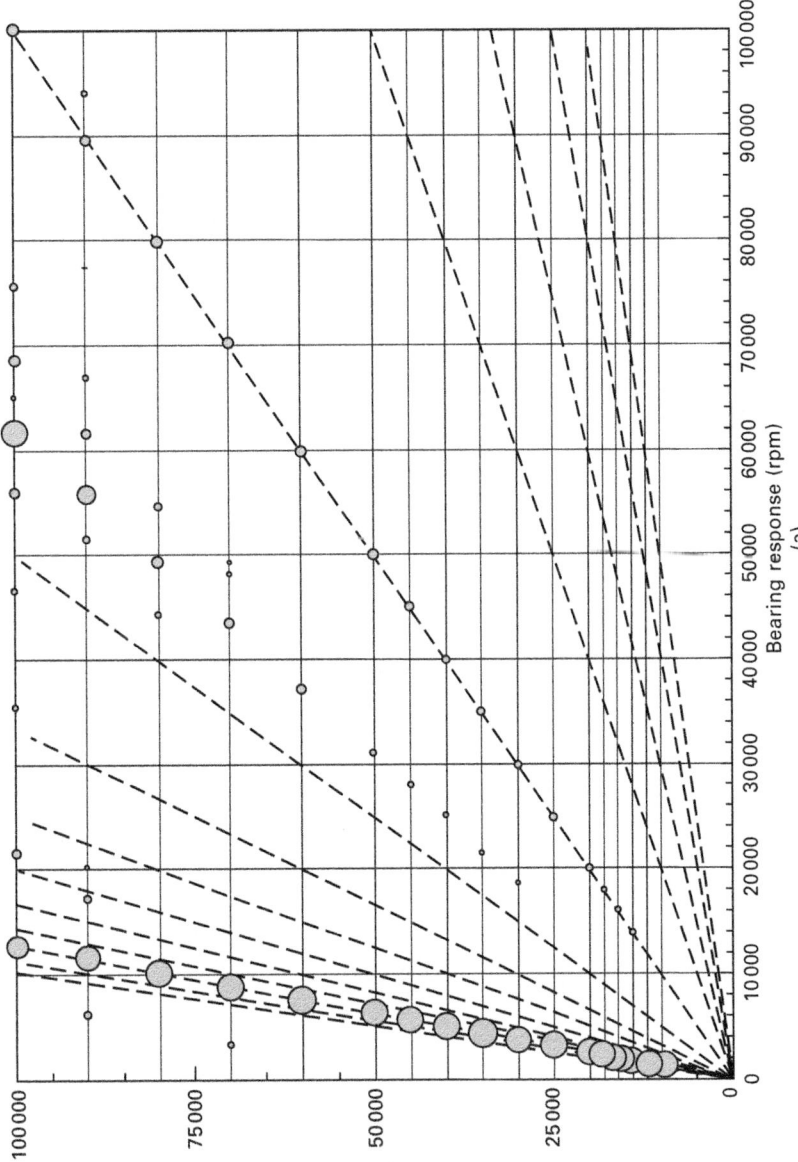

9.37 A waterfall bubble plot of turbine FFB (a) frequency spectra for 0.23 g mm unbalance, 2.5/7 μm inner/outer clearance, 4 bar supply pressure (b) frequency spectra for 2.3 g mm gross unbalance, 2.5/7 μm inner/outer clearance, 4 bar supply pressure.

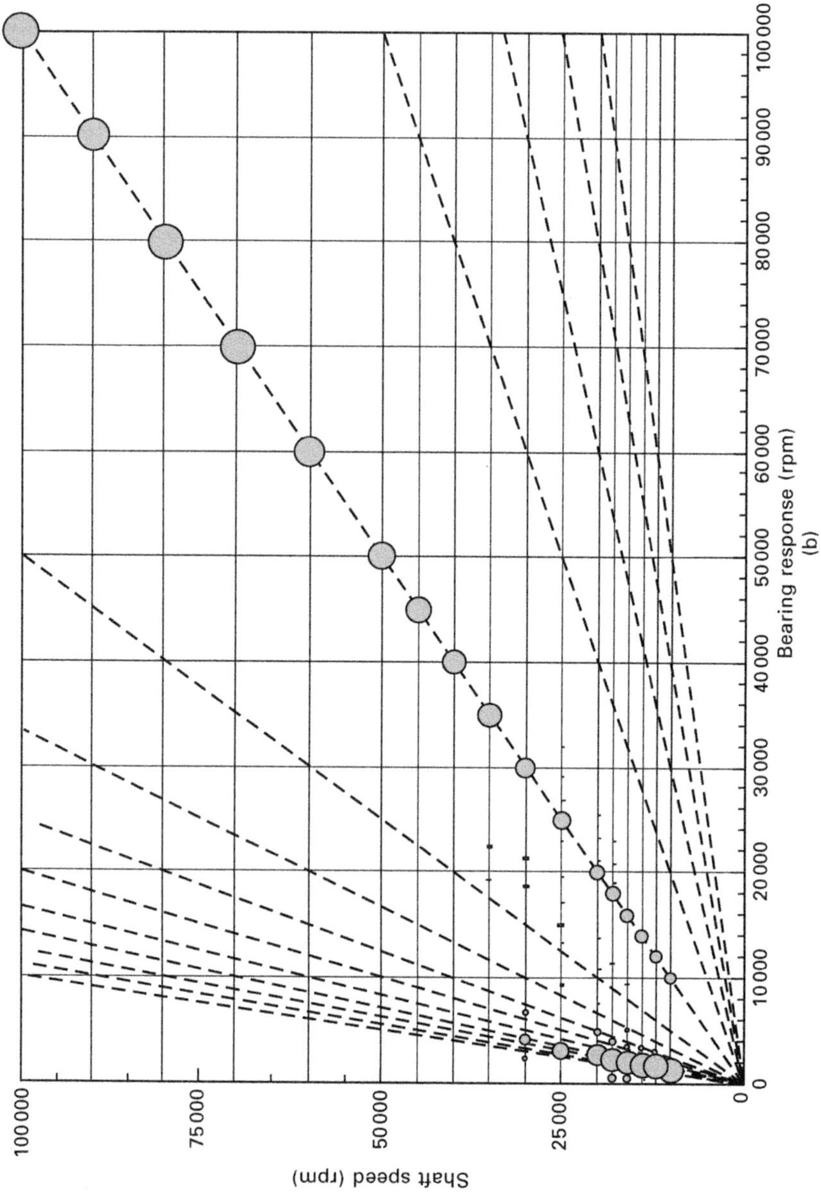

9.37 Continued

which is then stable up to 100 000 rpm. There is also a small 5/8 shaft order which jumps to 1x shaft order at 40 000 to 45 000 rpm. This last plot shows a phenomenon that has been observed experimentally by San Andres and Kerth (2004) and Naranjo *et al.* (2001) where the results of a turbocharger system shows that bearings can be seen to operate sub-synchronously and then suddenly jump to synchronous operation at some particular speed.

The control of stability by gross unbalance has been observed and recorded by many in industry and here it has been modelled using tribo-dynamic methods by models that can be applied to many other bearing applications in order to better understand the physical behaviour of engine systems.

9.6   Engine friction: building a better understanding

Being able to reduce systems to collections of individual components allows a better understanding of a component as it interacts with others in a subsystem or system environment.

Basic tribo-dynamic simulations with low-fidelity models such as the SBA, SALBA and Booker Mobility methods (see Chapter 19) enable designers to obtain a basic understanding of the component or the systems. However, when the basic models fail, or the basic design limits are exceeded, or subsequent engine tests fail to agree with the basic calculations, a more detailed and complex system model is required with a more rigorous use of detailed CAE methods, such as FEA, FD and CFD.

Although more detailed models such as RHD, EHD and TEHD generate a better understanding of the system, the designer and tribo-dynamicist need a greater knowledge of the engine's physics, such as cooling systems, cycle simulation, combustion, structures, dynamics and of course tribology to make best use of these detailed models. The data to populate these models may not be fully available in the early stages of design.

Thus systems which allow these complex interactions to be rapidly modelled, studied and understood prior to manufacture, are indeed the way forward, whereby a clear design and analysis process is mapped out for the engine design team. Engine friction simulation is a design process that needs clear mapping to enable integration of the analyses of various tribological components to fit together as one coherent system calculation. Engine system friction determines power losses from various subsystems such as the valve train (see Chapters 16 and 17), power cylinder subsystem, crank train and turbocharger. These subsystems make up the engine system containing the principal tribology bearing systems of the engine. The lubrication network can also be included, but this is outside the scope of this chapter.

The principal bearings are termed here as tribological interfaces or objects which can be seen in Fig. 9.38 for a single cylinder engine. Each of the

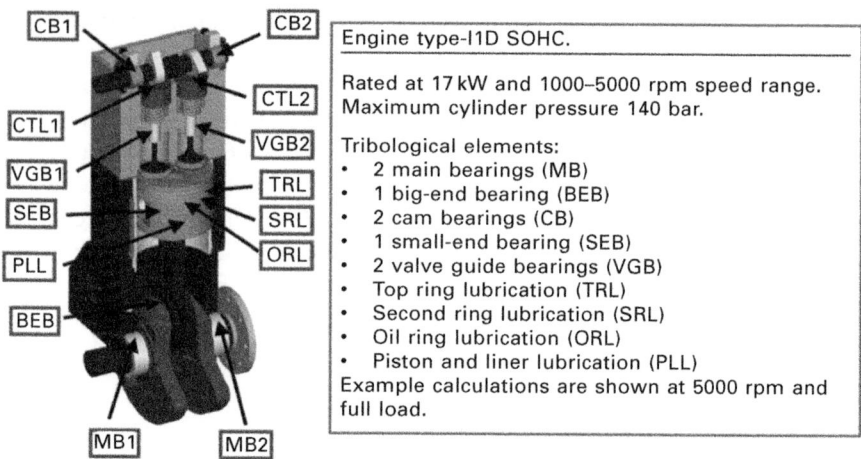

Engine type-I1D SOHC.

Rated at 17 kW and 1000–5000 rpm speed range.
Maximum cylinder pressure 140 bar.

Tribological elements:
- 2 main bearings (MB)
- 1 big-end bearing (BEB)
- 2 cam bearings (CB)
- 1 small-end bearing (SEB)
- 2 valve guide bearings (VGB)
- Top ring lubrication (TRL)
- Second ring lubrication (SRL)
- Oil ring lubrication (ORL)
- Piston and liner lubrication (PLL)

Example calculations are shown at 5000 rpm and full load.

9.38 Tribology objects used within engine lubrication simulations.

tribological components are summed together, having been subjected to loads and operating conditions to produce component, subsystem and system frictional behaviour in terms of *frictional mean effect pressure* (FMEP), which is a measure of the parasitic pressure loss averaged over the cycle. The overall FMEP can be compared with engine tests and in the early stages of design compared against FMEP design targets.

The following figures show the results of a frictional study of a typical four cylinder diesel engine, highlighting how components and subsystems make up the overall engine frictional losses. Figure 9.39 shows a typical *cam bearing* FMEP and frictional behaviour for a cam bearing showing reducing friction resistance as speed increases, due to high preloading from the belts and valve springs (see Chapter 17). Figure 9.40 shows the cam and tappet frictional resistance whereby the losses can be seen to reduce with speed showing how the *cam* and *tappet* operate mainly in the boundary/mixed lubrication regime. Figure 9.41 shows the summation of all the valve train components to generate an overall frictional resistance. This again shows that the valve train operates in the boundary/mixed regime, never truly being hydrodynamic even though it contains cam bearings. The major loss contributor can be seen as the cam and tappet.

Figure 9.42 shows the power cylinder system losses showing mainly the piston pins and small-ends, and the *regime of lubrication* can be seen to be boundary at low speed and mixed over a large proportion of the speed range barely becoming hydrodynamic at high speed, thus illustrating the difficulty of modelling and designing piston pin bearings and small ends, as they operate mainly through squeeze action and inertial oil recirculation. The piston and rings are not included in this particular piece of work (see Chapters 10–15).

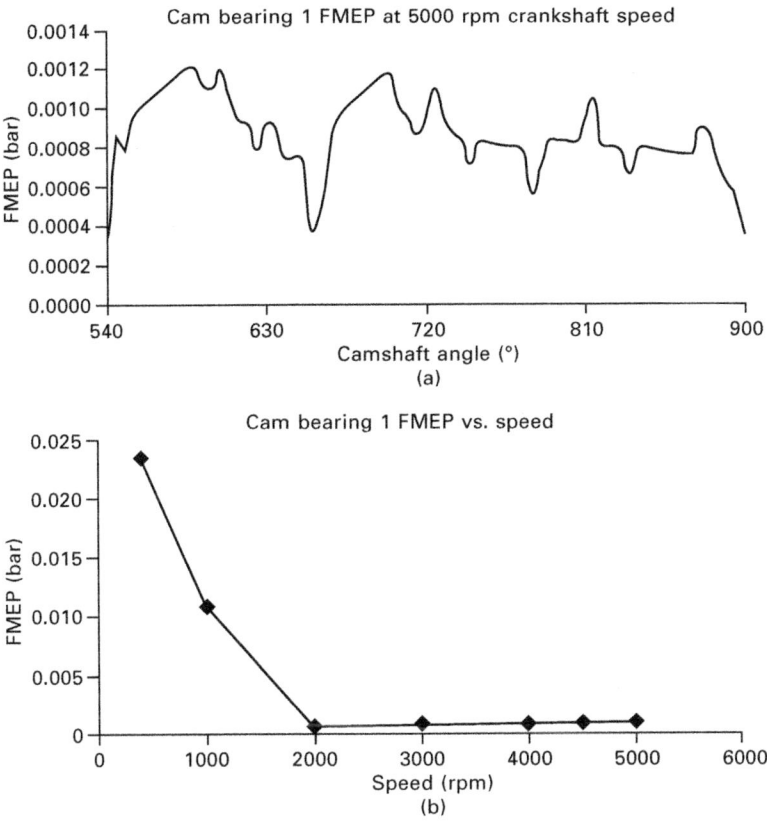

9.39 Cam bearing FMEP over the cycle with varying load and speed.

Figure 9.43 shows the crankshaft system friction losses, the lubrication regime can be seen to be mixed at low speed and becomes hydrodynamic after 1000 rpm. The big-end can be seen to be boundary/mixed at low speed, becoming hydrodynamic after 1000 rpm; these bearings operate both through squeeze and sliding action. The main bearings can be seen to mainly operate in the hydrodynamic lubrication regime or through sliding action.

Figure 9.44 shows the frictional losses of the whole system which includes the crank train, valve train, and power cylinder system, but no pistons or rings. In this instance it shows the crank train to have the highest losses, for this particular study (as the pistons and rings were absent) and is purely illustrative of system integration methods.

Figure 9.45 shows the comparison of various bearing oil film solutions for a gasoline engine running at 8000 rpm, assuming a rigid journal and housing. It can be seen that SBA under predicts orbit excursions where as SALBA and RHD give very similar orbit excursion results. It also shows

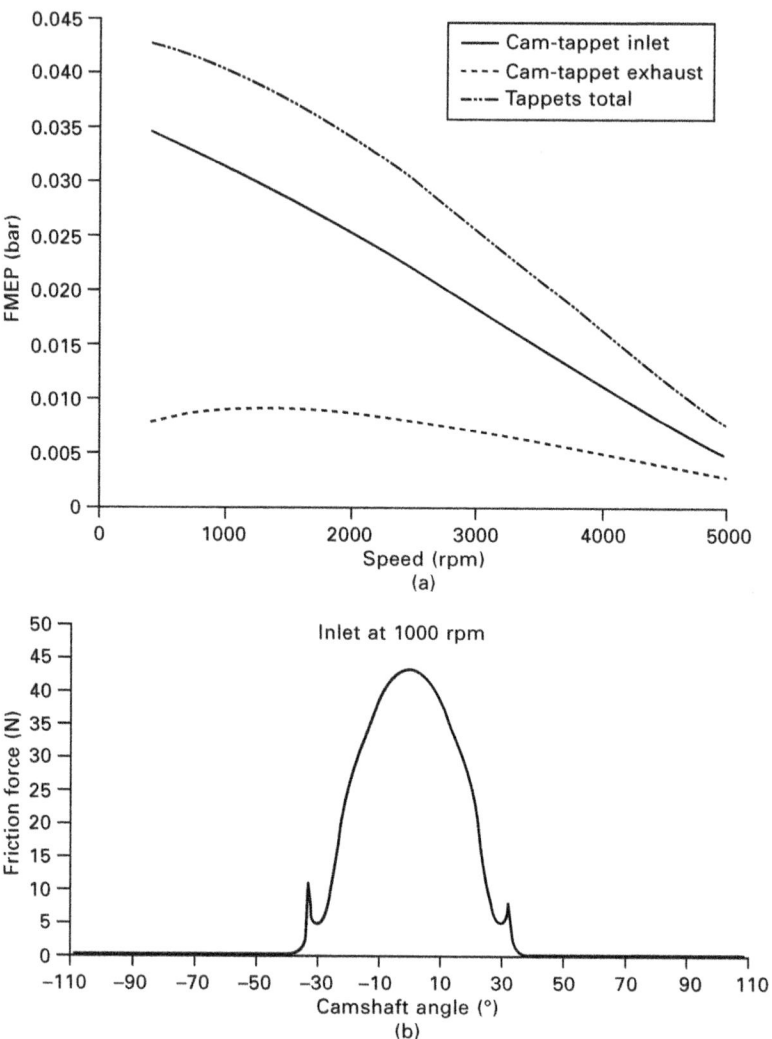

9.40 Cam and tappet friction and total cam and tappet FMEP.

that solution accuracy of Bulirsch-Stoer and the Newmark-Beta method are similar.

To obtain a closer realisation of system understanding the effects of elastic distortions and dynamic response need to be considered in order to predict engine performance closer to test results prior to manufacture.

In the initial design stages SBA, SALBA, RHD and Booker Mobility may be valid if the engine structure is close to being rigid (or stiffness is not known), which is generally not the case, as most engine structures are becoming more elastic due to reductions in weight and the use of aluminium,

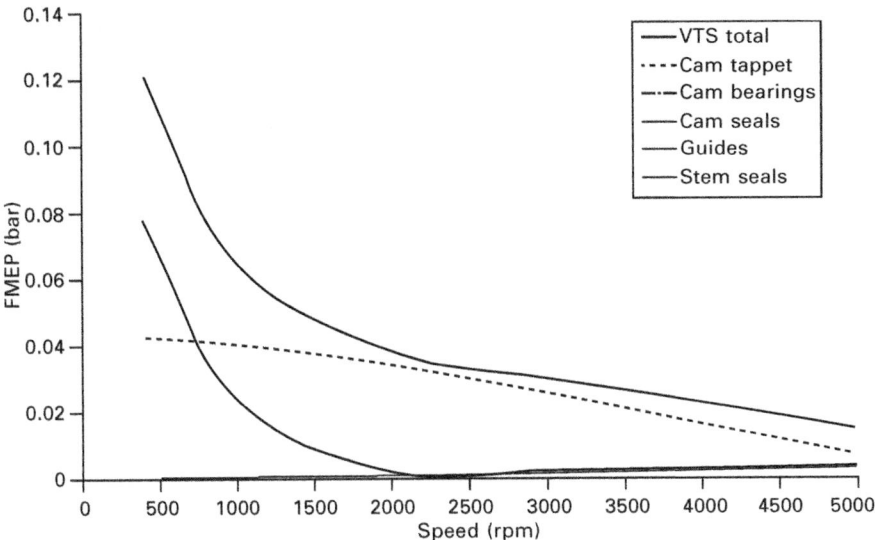

9.41 Valve train subsystem FMEP.

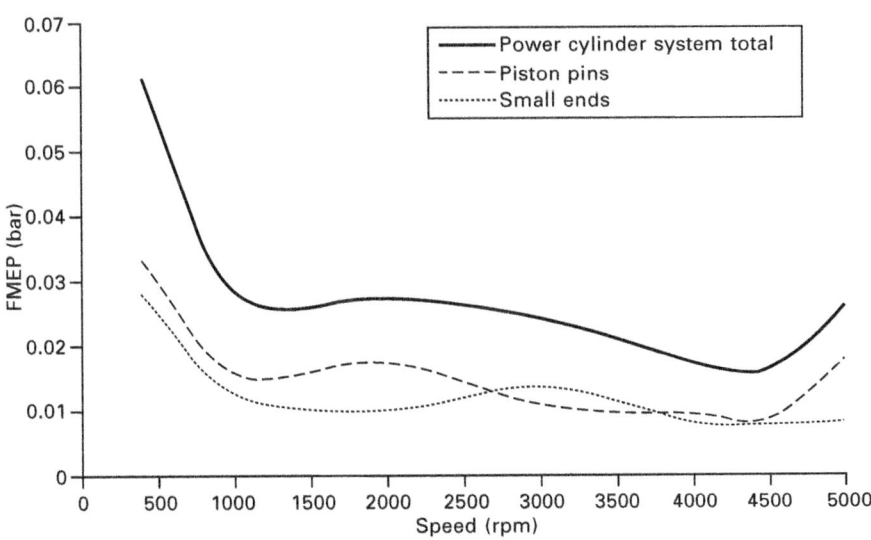

9.42 Power cylinder system (PCS) FMEP.

and the use of titanium for con rods. If the elastic deformations are small then a rigid analysis may be accurate enough, for example a very robust engine that is very lightly loaded.

Compare RHD against EHD calculations in Fig. 9.46 where the maximum eccentricity for EHD is 6.5 times that of RHD. Also the maximum pressures

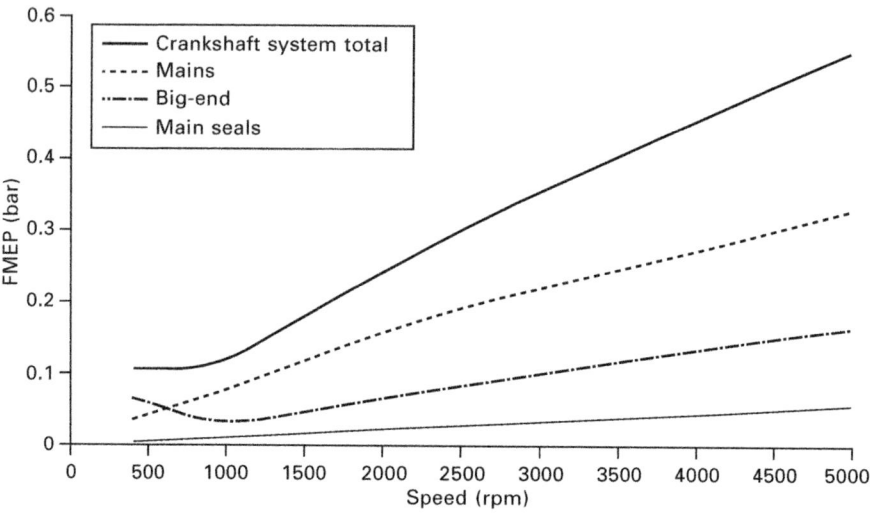

9.43 Crankshaft subsystem (CSS) FMEP.

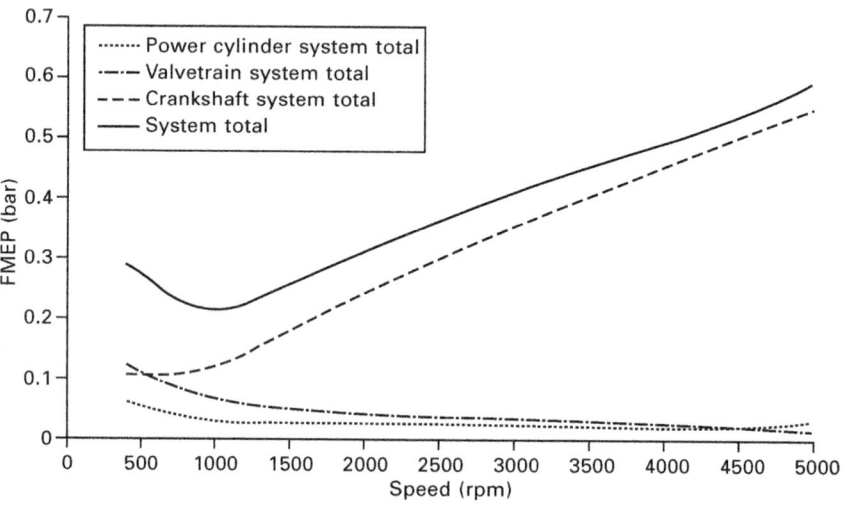

9.44 Total engine friction (FMEP).

for RHD will be at least twice that for an EHD solution as seen in Fig. 9.11 and RHD will underestimate power losses as RHD cannot model the effect of bearing wrap around or elasticity.

In order to enable EHD to be used more readily in the design process it needs to be made much more user friendly, that is, as user friendly as SBA, SALBA and the Booker Mobility method are now. This means improving its

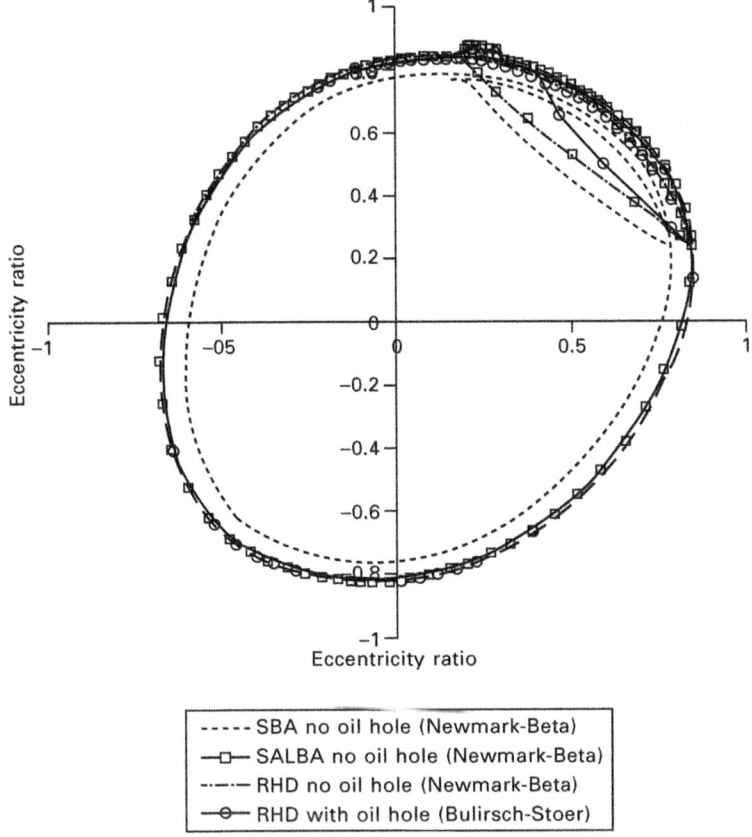

9.45 SBA, SALBA and RHD comparisons at 8000 rpm.

methodology and thus allowing engine designers the possibility of assessing bearings with EHD tools.

Figure 9.1 in Section 9.2 showed a novel approach to EHD where the time-consuming effort in generating realistic EHD models is preparing the model, as the solution time is down to about 6 hours. Preparing a good mesh is essential, otherwise solution time can be 6–7 days. This type of user-friendly approach is essential and thus the way forward if EHD is to appear more readily in the early stages of an engines design.

9.7   Conclusions

This chapter has highlighted some of the issues that need resolving in order that detailed analysis methods, such as EHD and MBD, are seen up front in the engine design and development process. Great leaps forward have been made in engine technology and reliability, but if the 12 month engine and

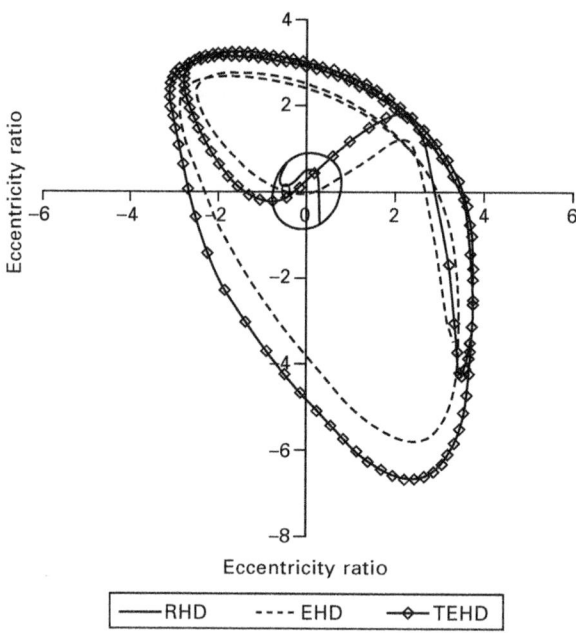

9.46 RHD, EHD and TEHD comparisons at 16 000 rpm.

vehicle development timescale is to become commonplace then advanced methods such as EHD and MBD need to be included in the early stages of the design and development process in order to reduce the time spent in the test house.

Where we are currently with technology allows at best an incremental change in production timescales and engine reliability. However, improvements in the engine model preparation time and process, as discussed here, will allow more detailed calculations to be taken up earlier, resulting in improved performance and reliability.

The hiatus in the design and development process will then become the time it takes for the current solvers to run EHD and TEHD calculations, not their preparation times or complicated processes.

When the current solver times have been reduced considerably (say to an hour or less), engine CAE design models can then evolve to include sub-system and system models as outlined here. The modelling of real world effects in component and system interactions needs to be solved in parallel, which will eventually be possible when CPU speeds have improved considerably.

Thus tribology system calculations will then more readily involve the influence of analyses such as: combustion, heat transfer (HT) and cooling, thermal stress, distortions and lubrication (plus the effects of pump oil flow).

System analysis methods described here will allow step changes in performance and reliability and the methods allow critical areas of the design to be identified in the early stages, rather than being found out on the test stand or even discovered as a reliability and warranty issue in service.

Rigorous numerical methods as described in this chapter and book need to displace the approximate methods used for design engine systems. This will only happen if it is made easier for design and development engineers to use these tools, and thus show that they can indeed obtain better correlation with them, rather than the approximate methods they use now. The use of approximate analysis methods have their place but produce approximate answers and thus uncertainty and risk.

The global issues of low emissions and hence low friction pave the way for more rigorous methods to be adopted, in order to explain why troublesome engines arise and how it can best be avoided early on in the design stage, rather than using expensive palliatives at the development stage when the hardware is already fully committed.

Scuffing is a subject that has often been found to manifest itself on the test stand, and this is generally because approximate methods have been used in the design process, and whereby expensive fixes have been applied to the piston and liner in order to fix the engine. However, scuffing is a system effect and if a systematic EHD system methodology is used, as pointed out in this chapter, it can be avoided at an early stage of the design by simple changes to the geometry of interacting components, and the proper choice of material and manufacturing process.

Once EHD and TEHD methods are improved, so the methods can be adopted early on in the design process, I have no doubt that this will result in more reliable and cost-effective engines being built in much shorter timescales. This will then truly promote the benefits of engine CAE as a virtual design and test system for the design, development and manufacture of engine systems.

9.8   Acknowledgements

I would like to thank all the team at Advanced Integrated Engineering Solutions Ltd for the effort that has been put into the engine and system design CAE software.

9.9   References

Barrett, D.J.S., McLuckie, I.R.W. and Ma, M.-T. (2000), 'A Valve Train Design Analysis Comparative Study of Three Current Diesel Engines'. *Der Virtuelle Motor Conference*. Munich, Germany. Haus Der Technik, October 2000.
Barrett, D.J.S., El-Zafrany, A. and McLuckie, I.R.W. (2002), 'Elasto-Hydrodynamic

Analysis of Bearings by Boundary Element and Finite Difference Methods', *Boundary Elements XXIV*, Sintra, 2002.

Cameron, A. (1983), *Basic Lubrication Theory*, Ellis Horwood.

Geradin, M. and Cardona, A. (2001), *Flexible Multi-Body Dynamics. A Finite Element Approach*, John Wiley & Sons.

Gohar, R. (1988), *Elastohydrodynamics*, Ellis Horwood.

Ma, M.-T., McLuckie, I.R.W., Poynton, A. and Loibnegger, B. (2001), 'An EHD Study of a Connecting Rod Big End Bearing Including Elasticity and Inertia Effects of the Bearing Structure', *World Tribology Conference 2001*, Vienna, Austria.

McLuckie, I. (2003), 'From Virtual Engine to Virtual Vehicle and Beyond', *NAFEMS World Congress 2003*, May 27–31, Orlando, Florida.

McLuckie, I. (2007a), 'System Integration by Faster CAE Solutions', *2007 Automotive VPD Conference*, 21–22 March, Munich, Germany.

McLuckie, I. (2007b), 'System Integration – CAE with a New Science', *NAFEMS World Congress 2007*, 22–25 May, Toronto, Canada.

McLuckie, I. and Barrett, S. (2005a), 'V8 Engine Bearing Dynamics – High Performance with Minimum Friction', *Proceedings of The ASME IDETC/CIE Conference*, September 24–28, 2005, Long Beach, California.

McLuckie, I. and Barrett, S. (2005b), 'Objective Faster Real World Solutions', *NAFEMS World Congress 2005*, 17–21 May, Malta.

McLuckie, I., Barrett, S. and Teo, B.-K. (2006), 'Plain & Full Floating Bearing Simulations with Rigid Shaft Dynamics', *The 8th International Conference On Turbochargers And Turbo Charging*, 17–18 May 2006, London.

Naranjo, J., Holt, C. and San Andres, L. (2001), 'Dynamic Response Of A Rotor Supported in a Floating Ring Bearing', *1st International Conference on Rotordynamics of Machinery*, ISCRMA1, Paper 2005, Lake Tahoe, Nevada, 2001.

Ocvirk, F.W. and Dubois, G.B. (1953), *NASA Tech. Report No. 1157*.

Pinkus, O. (1990), *Thermal Aspects of Fluid Film Tribology*, ASME Press.

Rahnejat, H. (1998), *Multi-Body Dynamics. Vehicles, Machines and Mechanisms*, Professional Engineering Publishing.

Rahnejat, H. and Rothberg, S. (2004), *Multi-Body Dynamics. Monitoring and Simulation Techniques – III*, Professional Engineering Publishing.

Reason, B.R. and Narang, I.P. (1982), 'Rapid Design and Performance Evaluation of Steady State Journal Bearings', *ASLE Transactions*, Vol. 25, pp. 429–444.

San Andres, L. and Kerth, J. (2004), 'Thermal Effects on the performance of Floating Ring Bearings for Turbochargers' *Proceedings of the IMechE*, Vol. 218, Part J, 2004.

Section II.II
Tribology of piston systems

10

Fundamentals of lubrication and friction of piston ring contact

V. D'AGOSTINO and A. SENATORE,
University of Salerno, Italy

Abstracts: The piston ring-pack exhibits a complex dynamic behaviour, which includes gas and oil flows, twisting motion of each ring and its influence on ring-liner and ring-groove lubrication and contact, as well as unsteady oil supply. In modern automotive engines, the dynamics result in a significant share of the total friction power loss and plays a crucial role in the piston assembly response in terms of blowby, wear and oil consumption. This chapter illustrates the physics of piston ring lubrication and friction, the main mathematical and computer models and the fundamentals of the ring-pack role. The chapter is completed by future trends and opportunities of research into piston ring modelling and improvements to the frictional behaviour.

Key words: piston ring, lubrication models, friction losses, piston assembly dynamics, asperity contacts, engine mechanical efficiency.

10.1   Introduction

This chapter aims to illustrate the physics of piston ring lubrication and friction characteristics involving the fundamentals of their role in the complex engine system and the main mathematical and computer models about the theoretical reported investigations. Sections 10.2 and 10.3 focus on the ring classifications underlining the different shapes and functions, with a brief reference to the engine friction loss share due to the piston ring pack.

Sections 10.4–10.7 introduce a short review of some published work on piston ring lubrication and then provide the reader with an analytical approach for solving the lubrication equation in its simpler formulation for two kinds of ring shape. A study on the boundary conditions of ring hydrodynamic lubrication regime is presented, while an investigation about the mixed-lubrication regime is introduced.

Sections 10.8–10.12 discuss the effect of oil temperature and its effect on viscosity, coupling the energy equation with the already mentioned lubrication equation. It also addresses the influence of piston secondary motion and ring flexibility on the non-symmetric ring lubrication and ring/liner asperity interactions. Also studied are ring flutter and collapse phenomena and the effect of cylinder bore distortion on ring motion. Other important issues

343

are oil consumption, ring contact mechanical efficiency and improvements due to laser textured surfaces and the high level of losses in the warm-up engine period. The chapter is completed by a look into the future trends and opportunities of research on piston ring modelling and improvements to its frictional behaviour.

10.2   Piston ring: history and basics

At the outset, piston rings were used for the purpose of sealing the combustion chamber gases, thus preventing their passage through the piston/wall clearance into the crankcase. The ring for a steam engine designed by Ramsbottom in 1852 (Ramsbottom 1855) was a single-piece metallic ring; its free diameter was 10% larger than the diameter of the cylinder bore. When fitted into a groove in a piston, the ring was pressed against the cylinder bore by its own elasticity. The modification introduced by Miller in 1862 (Priest and Taylor 2000) consisted of allowing the steam pressure to act on the back of the ring (the ring-groove side), hence providing a higher sealing force.

This new solution enabled the use of more flexible rings, achieving the function of closely conforming to their grooves in the piston (designed for thermal expansion, with an optimised skirt-liner gap). The sealing ability of the ring pack depends on a number of factors, such as ring and liner conformability, radial tension and gas force distribution on the ring faces. The forces acting on the ring in the radial direction are discussed in Section 10.7.

Since a part of the combustion chamber heat energy is transferred through the piston to the piston boundaries, i.e. the piston skirt and rings, the secondary role of the piston ring is to transfer the generated heat away from the piston to the cylinder wall, and then into the engine cooling system. As their third task, the piston rings avoid excess lubrication oil from moving into the combustion chamber by scraping the oil from the liner wall during the piston downstroke, limiting the amount of oil flow transported from the crankcase to the combustion chamber. This oil flow is probably the largest share of engine *oil consumption* and leads to increased harmful exhaust emissions as the oil mixes and reacts with the other contents of the combustion chamber.

The piston rings sustain the piston in the radial plane, reducing the lateral motion of the piston and attenuate the slap noise effect (Cho *et al.*, 2002), particularly during cold starts, where the clearance is greater than in the steady running conditions. The rings are generally open at one location, at the ring gap, hence easily assembled on the piston.

In order to fulfil all the above-mentioned tasks, the dynamic behaviour of the rings is strongly influenced by the engine operating conditions and by all the design parameters of piston and cylinder (see Fig. 10.1). For this reason, the piston ring pack exhibits a complex dynamic behaviour, which includes

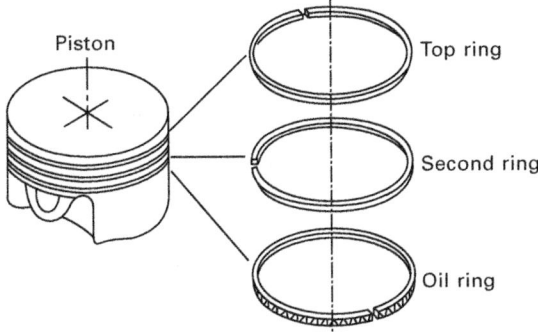

Piston

Top ring

Second ring

Oil ring

10.1 Piston and ring names and positions.

the dynamics of the rings and related gas flow and oil flows, twisting of the piston rings and its influence on ring-liner and ring-groove lubrication and contact, and unsteady oil supply to upper piston rings and its effects on ring-liner lubrication. These dynamic phenomena play a crucial role in the piston assembly response such as *blowby*, friction, wear and oil consumption and it is of both theoretical and practical interest to understand and model them.

In fact, a significant share of the total power loss in a modern automotive engine is due to the ring pack/cylinder wall friction. On this basis, the lubrication of the piston ring has been an important research matter for many years, because it is generally accepted that the interaction at the ring–cylinder wall interface provides substantial effects on friction, wear, oil consumption and power loss in internal combustion engines (ICEs). The analysis of McGeehan (1978) underlined that piston assembly friction could account for 58–75% of the total mechanical friction of an ICE. Recent literature suggests that the total friction due to pistons, rings and connecting rod contributes for 40–55% in a modern spark-ignited (SI) or diesel engine. The pie charts in Fig. 10.2 depict the friction loss share due to the piston skirt friction and the ring pack. If one analyses the friction loss shares as shown in the right-hand pie chart, engine friction loss including piston skirt friction, piston rings and bearings account for 66% of the total friction loss; then, the valve train, crankshaft, transmission and gears are approximately 34%. Concerning powertrain friction loss only, sliding of the piston rings and piston skirt against the cylinder wall is undoubtedly the largest contribution to friction in a powertrain system.

The plots in Fig. 10.3 introduce the friction work in a whole thermodynamic cycle, expressed in the conventional terms of *friction mean effective pressure* (*FMEP*), i.e. the friction work divide by the cylinder displacement; in the first figure the shares of each ring in a three-set is shown with the part of

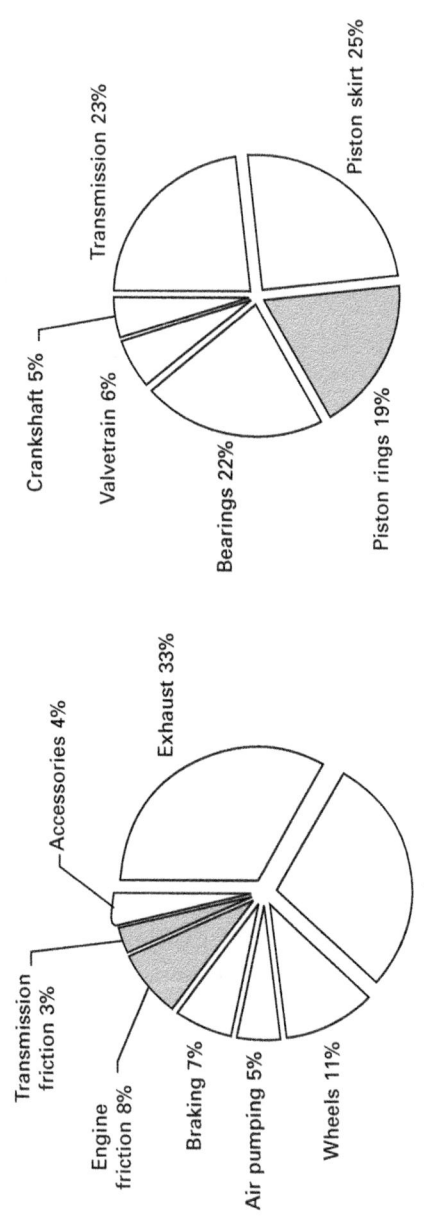

10.2 Distribution in a light-duty vehicle of (a) powertrain losses (b) engine/transmission friction losses (after Tung and McMillan, 2004).

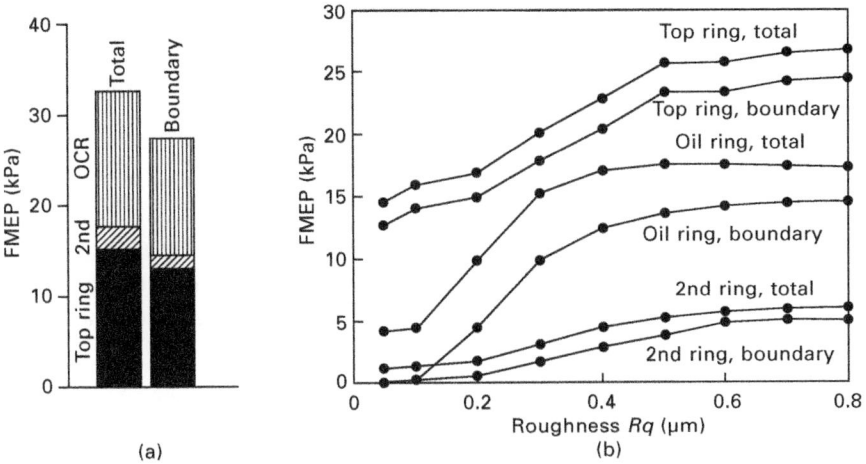

10.3 FMEP from different rings and lubrication mechanisms (after Tian, 2002b).

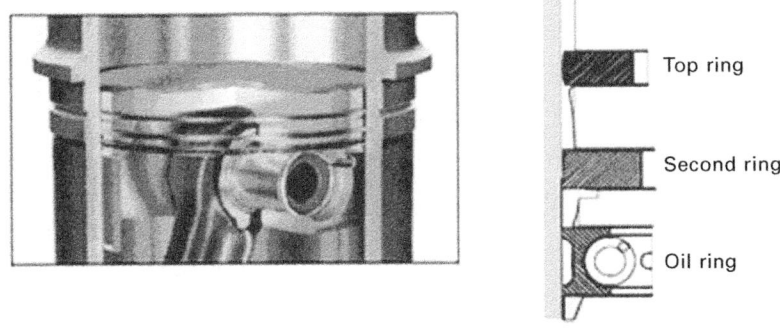

10.4 Typical piston ring pack.

total loss due to the boundary component. The second one gives a picture of the influence of surface roughness of the rings on the FMEP trend.

10.3   Piston rings classification

The piston rings form the so-called 'ring pack' (Fig. 10.4), which typically includes up to five rings, with at least one *compression ring* (the 'top ring'). The number of rings depends on the engine type, but usually comprises 2–4 compression rings and 0–3 *oil control rings* or simply *oil ring* (Andersson *et al.*, 2002).

For example, fast speed four-stroke diesel engines have two or three compression rings and a single oil control ring; the latter used in diesel engines are two-piece assemblies (Fig. 10.5) and SI engine oil control rings may be three-piece assemblies as well. In addition to the general compression rings and oil control rings there are **scraper rings**, which have the tasks of both sealing and scraping off the oil from the liner wall.

10.3.1 The role of the compression rings

The **compression ring** acts as a gas seal between the piston and the liner wall, preventing the combustion gases' passage from the combustion chamber to the crankcase (also see Chapter 15). These rings have a given mounting

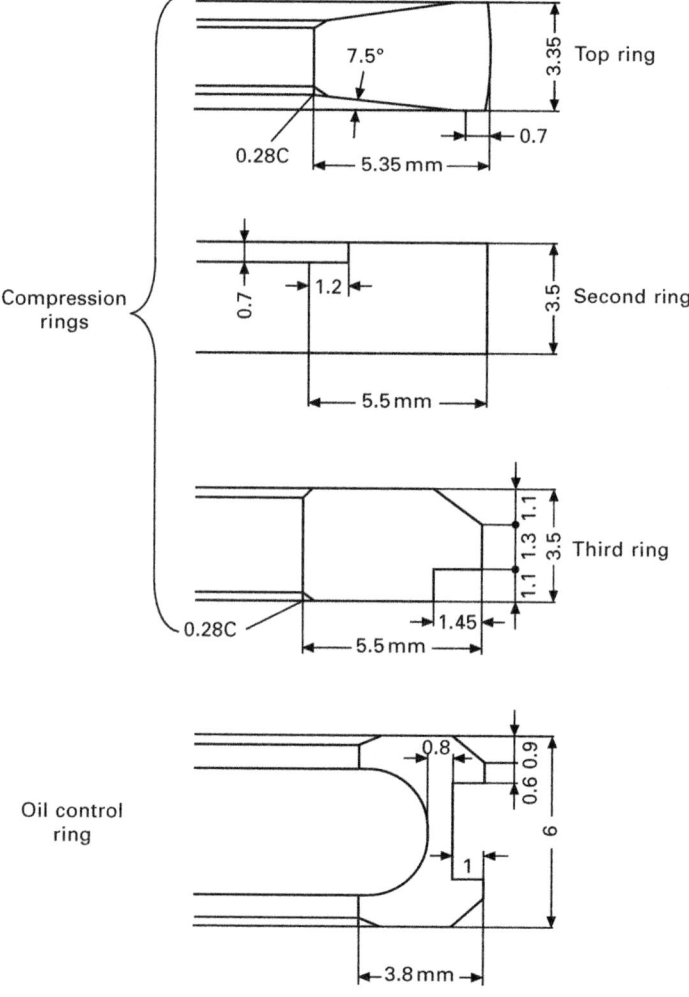

10.5 Ring pack composed by four rings.

pretension; in fact, they have a free diameter larger than the cylinder bore, which allows the ring to match the liner size: the radial tension ($F_{R,t}$) has influence in ring modelling.

The combustion chamber pressure acts on the back-side of the ring, especially on the top ring, pressing it against the liner. The ring force distribution depends on the face form. With a rectangular face profile the force is higher than with a barrel-shaped face (the top ring of Fig. 10.5), as the compression pressure is able to act on the face-side of the barrel-shaped ring against the ring pre-tension action. Plain compression rings, with a rectangular cross-section (the second ring in the same figure), meet satisfactorily the sealing demands of ordinary running conditions and this type of compression ring is the most common one (Andersson *et al.*, 2002).

The ring may have a conical shape in order to shorten the running-in period; the tapered profile enables the compression gas pressure to act on the face-side as well and thus relieve the pressure against the liner wall, which reduces the wear rate during running-in. This kind of profile produces good scraping performance and the ring can be used both as oil-scraping ring or as compression ring.

10.3.2 The role of the oil rings

In addition to the task of the compression rings to seal off the combustion chamber from the crankcase, there is the need to distribute the lubricant on the cylinder liner surface. The number of *oil control rings* in a ring pack varies between 0 and 2. Normally a single oil control ring is sufficient but some technical solutions host a second ring. The appearance of the oil control ring differs from that of the compression ring; see Figs 10.4 and 10.5.

The oil control ring has slots in the peripheral region (Fig. 10.6), which provides a way for the excess oil to leave the ring pack area. The scraped oil is collected in the oil control ring groove and transported through the piston down to the crankcase. The scraped oil may also run through the possible gap between the liner wall and the piston skirt and then to be forced in front of the oil control ring.

The oil control rings usually have a reinforced coil spring, as the pre-tension of the ring is not sufficient in all instances. The additional force on the oil control rings causes them to have the most extreme lubrication conditions (boundary, mixed lubrication), even with no significant gas pressure acting on their back-side. Oil control rings are not always necessary, contrary to the compression rings. Two-stroke engines, for example, have the lubricant mixed in the fuel, and therefore need no oil ring (Andersson *et al.*, 2002).

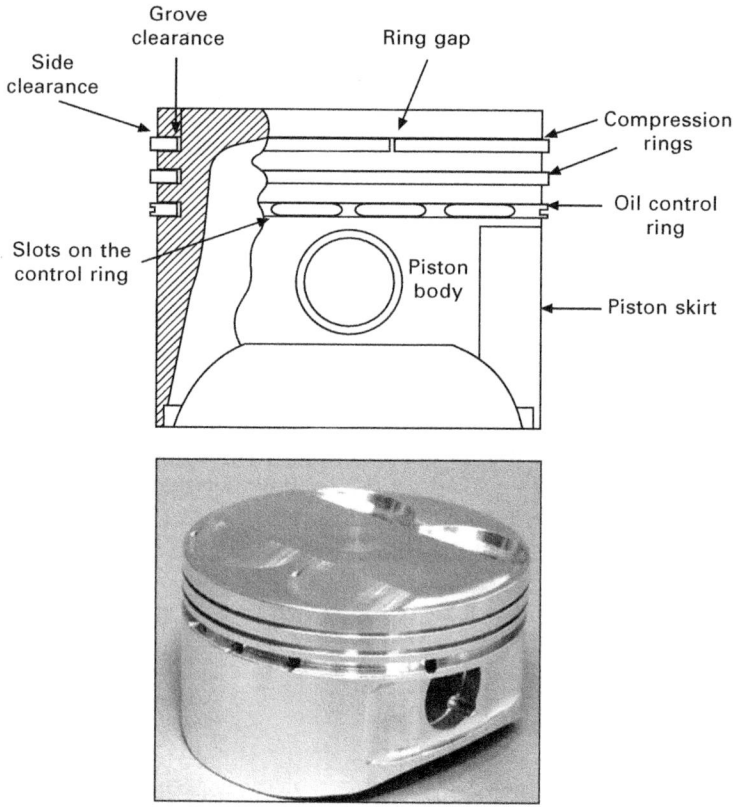

10.6 Piston and ring geometrical features, SI engine.

10.4   Lubrication models

The piston ring is the most complicated tribological component in the internal combustion engine to analyse, because of large variations of load, speed, temperature and lubricant availability. In a single stroke of the piston, the piston ring interface with the cylinder wall may experience boundary, mixed and full fluid film lubrication (Ruddy *et al.*, 1982).

There are many theoretical models on piston ring lubrication; common to almost all of these models is that they are based on the Reynolds equation which includes parameters of the geometry, viscosity, pressure and surface velocities. The equation can be solved in order to predict pressure distribution, load capacity, friction force and oil flow.

Oil film thickness calculations consider various lubrication mechanisms: full hydrodynamic (HDL), mixed (ML) and boundary lubrication (BL). Full HDL occurs when there is no surface asperity contact and a continuous oil film is interposed between the ring face and the cylinder wall. This is possible in the mid-stroke area, where the relative surface velocity is at its highest; this lubrication mechanism requires the ring area to be flooded, i.e. there is always a sufficient amount of oil available at the inlet side of the ring in order to prevent direct surface contact (see also Section 10.5).

Where the oil film pressure is particularly high, the surfaces start to elastically deform and the surfaces approach each other more than the clearance would allow. In this case, an elastohydrodynamic lubrication regime (EHL) model has to be included in the ring frictional simulation. According to this theory, the ring and liner surfaces are still separated, but the contact is much more concentrated, the films are thinner and other physical phenomena as elastic distortion of the surfaces and the effect of pressure on dynamic viscosity are influential (Taylor, 1998). The elasticity effect of the surfaces has been considered as an extension of pure hydrodynamic lubrication by Dowson *et al.* (1983) and Qingmin *et al.* (1996).

Ring pack simulation models include a variety of phenomena. Some physical phenomena are opportunely neglected from certain models, as they are not considered important for the purpose for which the model has been developed. In fact, second order effects often involve long computation times with no significant improvements of the simulation/experiments agreement. Almost every hydrodynamic lubrication simulation algorithm includes an iterative scheme on the oil film pressure in order to find an equilibrium of the force acting on the ring in radial plane.

The ring/liner gap is usually modelled by taking into account the axis-symmetry hypothesis (one-dimensional Reynolds equation). This is not the case in actual engine operating conditions, as the piston experiences variable forces in the radial plane. As a consequence, the ring-liner gaps are different on the *thrust side* and *anti-thrust side*, also as an effect of local contact and ring flexibility (Fig. 10.7). For this purpose the axis-symmetry hypothesis will be removed in Sections 10.6 and 10.8.2.

Experimental work (Richardson and Borman, 1992) has shown that the film thickness of the oil control ring differs significantly from a theoretical value at the beginning of the downstroke. The measured oil film thickness is greater than that calculated. The reason is presumed to be additional oil transported from the piston skirt and piston approaching motion to cylinder wall (a form of squeeze film action).

The oil film thickness at the oil control ring increases when the ring tangential tension is reduced; the film thickness decreases when the ring width is reduced (Seki *et al.*, 2000). The oil supply to the ring/liner wedge is modelled in different way in literature. The oil film in front of the ring (inlet meniscus) can be considered as being of constant thickness. A model which considers the oil film thickness trailing the previous ring as the input oil film thickness for the following ring is more realistic (Fig. 10.9). However, even this approach does not fully represent reality, because of the inter-ring

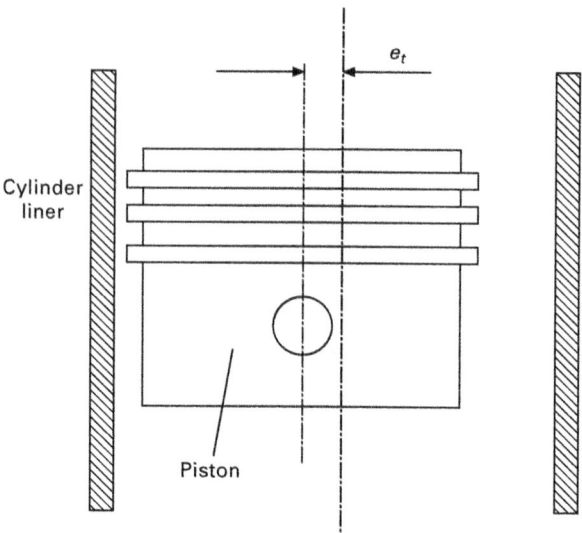

10.7 Piston top displacement, e_t.

region has a non-negligible inference on the ring lubrication (Andersson *et al.*, 2002; Lotz Felter, 2008).

In the oil film thickness calculation the choice of a set of hypotheses allows the model complexity to be kept within reasonable limits. The lubricant is usually considered Newtonian, while some applications include shear stress state effect on lubricant viscosity.

The oil pressure in hydrodynamic regime is solved through lubrication equations, but the pressure description under boundary conditions remains an open problem. It is well known that the lubricant additives have so far been demonstrated to have greatest benefit in the boundary lubrication regime, when the film thickness of the lubricating base is no longer sufficient to maintain separation of the opposing metal surfaces.

A study by Gao *et al.* (1998) has described how molecular dynamics simulations might reveal the nature of dynamical states and structural characteristics of confined sheared molecular films, as is typical of lubrication. The study of fluid transport near/or in contact with walls is a possible option as shown by molecular simulation works. However, applying these and other modelling techniques to lubricants also incorporating innovative nanoparticle-based additives presents several challenges. Normally, the oil pressure at inlet and outlet flow section is assumed to be equal to the gas pressure on the respective side of the ring. On the front side of the ring, the remaining oil on the cylinder wall either passes under the ring through the gap or accumulates in front of it. If the oil amount is not adequate to fill the ring-wall gap, the ring lubrication is called starved. In this case, the oil film exhibits less thrust (radial) load than a fully flooded ring. There are numerous approaches for analysing this boundary condition in the lubrication models (Section 10.5).

10.4.1 Friction analysis and simulation

The friction models that have been included in the computer simulations of the researchers vary depending the main focus of the specific investigation. Wakuri *et al.* (1995) note that since perfectly hydrodynamic lubrication in the ring pack cannot be ensured, the theoretical estimation of the friction should always include a mixed lubrication model with asperity contacts (Section 10.7; see also Chapter 15).

The interaction between the ring and liner wall becomes more and more important as the oil film thickness decreases. In a hydrodynamic regime, a decreasing oil film thickness allows lower friction loss. It is evident that direct surface contact would occur with thin oil films. The surface contact increases the total friction losses; thus small friction loss levels require compromises between hydrodynamic and sliding direct contact friction. The oil viscosity decreases for increasing temperature, while the viscosity decreases when the shear stress increases. The latter effect is called *shear thinning* and allows improved agreement of simulation results with experimental data (Tian *et al.*, 1996).

The top ring exhibits a higher frictional loss due to direct surface contact rather than hydrodynamic action, since the high-pressure combustion gas acts on the groove side of the ring and presses it against the cylinder liner (Section 10.7). Thus the sliding frictional power at the top ring surface is increased at increasing load. The sealing action of the first compression ring yields to low values of gas pressure on the back-side of the other piston rings whose frictional power is almost not affected by load variations (Yun *et al.*, 1995).

Studies of two-ring pistons in SI engines have been made by Takiguchi *et al.* (1996) in order to explain the interdependence between oil film thickness, ring friction and oil consumption. They underlined that a two-ring pack provides a greater blow by than a three-ring one; the oil film for two-ring pistons in turn becomes thinner, which leads to a very small raise in oil consumption. Moreover, the authors conclude that the ring friction loss can be reduced regardless of the number of piston rings by reducing their pre-tension.

10.4.2 Lubrication equation

The first approach to the one-dimensional Reynolds equation (see Chapter 5) for a piston ring will be analysed under the following assumptions (Taraza *et al.*, 1997):

- the flow is laminar;
- the surfaces are perfectly smooth and rigid;
- the ring does not tilt or rotate during operation;
- lubricant viscosity throughout the film is constant at any crank angle by using a mean value;
- cavitation is not considered and a half-Sommerfeld condition is assumed, i.e. the load-carrying capacity being ensured only by the leading edge of the ring profile with respect to the direction of motion.

Under these hypotheses, the Reynolds equation can be written as follows:

$$\frac{d}{dx}\left(h^3\frac{dp}{dx}\right) = 6\mu U\frac{dh}{dx} + 12\mu\frac{dh}{dt} \tag{10.1}$$

By integrating two times in the axial direction:

$$\frac{dp}{dx} = 6\mu U\frac{1}{h^2} + 12\mu\frac{x}{h^3}\frac{dh}{dt} + 12\mu\frac{C_1}{h^3} \tag{10.2}$$

$$p(x) = 6\mu U\int\frac{dx}{h^2} + 12\mu\frac{dh}{dt}\int\frac{x}{h^3}dx + 12\mu\int\frac{C_1}{h^3}dx + C_2 \tag{10.3}$$

If the ring profile is described by a parabolic law (barrel-shape profile), the oil film thickness is given by:

$$h(x) = h_0 + c\left(\frac{x}{a}\right)^2 = h_0 + Ax^2 \tag{10.4}$$

The above-mentioned indefinite integrals are given by:

$$I_0 = \int\frac{dx}{h} = \int\frac{dx}{x_0 + Ax^2} = \frac{1}{h_0}\int\frac{dx}{1 + \left(\frac{A}{h_0}\right)x^2}$$

$$= \frac{1}{\sqrt{Ah_0}}\,\text{arctg}\sqrt{\frac{A}{h_0}}\,x \tag{10.5}$$

$$I_1 = \int\frac{dx}{h^2} = \frac{1}{2h_0}\left(\frac{x}{h_0 + Ax^2} + I_0\right) \tag{10.6}$$

$$I_2 = \int\frac{dx}{h^3} = \frac{1}{4h_0}\left[\frac{x}{(h_0 + Ax^2)^2} + 3I_1\right] \tag{10.7}$$

$$I_3 = \int\frac{x}{h^3}dx = -\frac{1}{4A}\frac{1}{(h_0 + Ax^2)^2} \tag{10.8}$$

Considering an upstroke piston motion, i.e. $U > 0$ (Fig. 10.8):

$$x = 0 \Rightarrow p = 0 \tag{10.9}$$

$$x = -L \Rightarrow p = p_1$$

$$C_1 = \frac{1}{I_2(-L)} \left\{ -\frac{U}{2} I_1(-L) + \frac{dh}{dt} [I_3(0) - I_3(L)] + \frac{p_1}{12\mu} \right\} \tag{10.10}$$

$$C_2 = -\frac{dh}{dt} I_3(0) \tag{10.11}$$

The radial equilibrium condition is given by:

$$\frac{1}{2L} \int_{x=-L}^{x=0} p(x)\, dx = p_g + p_t \tag{10.12}$$

where p_g is the gas pressure acting on the ring back-side; p_t is the pressure due to the ring radial pre-tension given by:

$$p_t = \frac{F_{R,t}}{2\pi BL} \tag{10.13}$$

where B is the piston bore and $2L$ is the ring length.
 Considering a downstroke piston motion ($U < 0$):

$$x = 0 \Rightarrow p = 0 \tag{10.14}$$

$$x = L \Rightarrow p = p_2$$

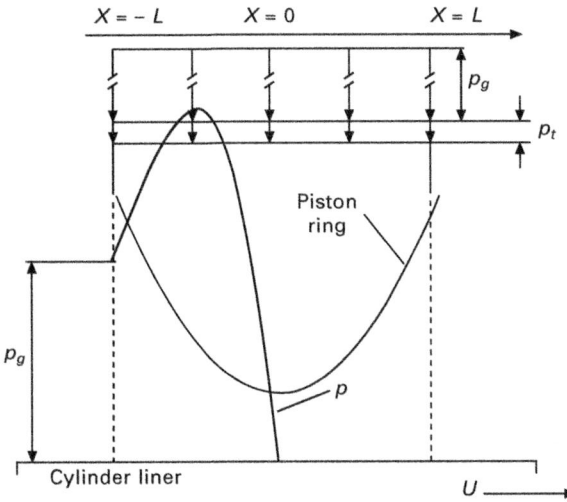

10.8 Balancing of back force with fluid film force (upstroke motion).

$$C_2 = -\frac{dh}{dt} I_3(0) \tag{10.15}$$

$$C_1 = \frac{1}{I_2(L)} \left\{ -\frac{U}{2} I_1(L) + \frac{dh}{dt} [I_3(0) - I_3(L)] + \frac{p_2}{12\mu} \right\} \tag{10.16}$$

$$\frac{1}{2L} \int_{x=0}^{x=L} p(x) dx = p_g + p_t \tag{10.17}$$

For a linear profile (tapered face profile):

$$h(x) = h_0 + A x \tag{10.18}$$

$$I_1 = \int \frac{dx}{h^2} = -\frac{1}{A(h_0 + Ax)} \tag{10.19}$$

$$I_2 = \int \frac{dx}{h^3} = -\frac{1}{2 A(h_0 + Ax)^2} \tag{10.20}$$

$$I_3 = \int \frac{x}{h^3} dx = -\frac{h_0 + 2Ax}{2 A^2 (h_0 + Ax)^2} \tag{10.21}$$

The viscous friction force is calculated by integrating the shear stress:

$$\tau_H(x) = -\mu \frac{U}{h(x)} - \frac{h(x)}{2} \frac{dp}{dx} \tag{10.22}$$

over the HDL effective ring domain.

10.5   A brief analysis of the main assumptions on the boundary conditions

Various boundary conditions in the lubrication simulation are used; some models have been criticised for causing inaccurate results owing to what boundary conditions have been assumed. The *Sommerfeld condition* allows both positive and negative pressure values. The *half-Sommerfeld condition* sets all negative pressure values to zero, and this is designated as the *cavitation zone*. Furthermore, mass-conserving algorithms are used in order to take into account the effect of cavitation zones on the oil availability. The oil flow pattern in the ring pack during a reciprocating motion is described in Fig. 10.9.

Experimental investigations have shown that the ring face is not fully lubricated, i.e. the ring is partially or totally starved. Han and Lee (1998) have developed a new model, where the inlet region is in a starved condition and the outlet region has an open-end assumption. These assumptions cause

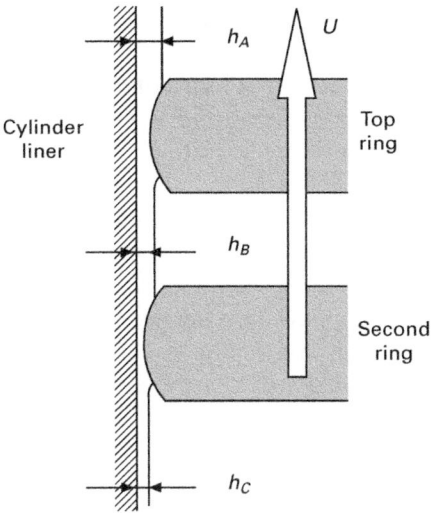

10.9 Oil flow pattern during an upstroke piston motion.

the effective width of the ring face to be 20–30% of the whole ring width. Still, using this model, the ring face works under flooded conditions at the vicinity of the dead centres.

Ma and Smith (1996) have compared two different oil availability models, namely a fully flooded model and a flow-continuity model. The fully flooded model comprises a model, in which the piston rings are considered to have a sufficient amount of oil, while the flow-continuity model comprises a model in which the oil film thickness of the preceding ring is considered as the available oil film thickness for the trailing ring. The authors conclude that only approximately 10–40% of the ring face is covered by an oil film. Therefore a flow-continuity model should be used, rather than a fully flooded model (Ma *et al.*, 1997).

When the ring pack moves along the cylinder axis, the oil on the wall fills the space between the ring and the wall and this generates a hydrodynamic pressure. The portion of the oil left behind by the preceding ring maintains its volume as long as there is no external supply of the oil and it is delivered to the next ring. If the amount of the oil is insufficient for filling the ring-wall gap, the ring is said to be operating under oil *starvation*.

In contrast, if the amount of oil is more than sufficient to fill the gap between the ring surface and the wall, the excess oil accumulates in front of the ring. Because in a multi-ring pack an oil scraper ring prevents the inflow of excess oil, most multi-ring packs are operating under oil starvation. Similarly, when the amount of oil seeping through the gap between the ring and the wall is not enough to fill the open space between the rear side of the ring and the wall, the posterior portion of the ring surface is operating without lubricating oil.

These are not only important phenomena in terms of flow continuity, but also direct causes for decrease in the effective width of the ring surface. The status of oil pressure in the gap under these conditions is described in Fig. 10.10(a), according to the ***Reynolds cavitation condition***.

Figure 10.10(b) describes the ***open cavitation condition*** (Ma *et al.*, 1997). As shown in Fig. 10.10(b), the pressure in the outlet region does not drop steeply. The status of oil pressure under the open-end assumption is shown in Fig. 10.10(c). Unlike the previous two figures, the pressure within the computational area does go down drastically and at times reaches the saturation pressure. The pressure profile will change abruptly at the points where hydrodynamic pressure is generated or stopped.

The pressure within the effective width is assumed to be lower than that of the boundary pressure (the inter-ring gas pressure) and the minimum pressure is assumed to be higher than the saturation pressure. There will be no disruption of the oil film due to ***Poiseuille flow*** (see Chapter 5), and the pressure within the effective width will eventually rise to the level of the boundary pressure. With this assumption and with preset boundary pressure conditions, an analysis on lubrication is performed to estimate the length of the posterior portion of the effective width. This permits flow continuity and makes it possible to apply the Reynolds equation properly and the only other condition necessary to set is the ***saturation pressure*** level. The boundary condition applied at the outlet is shown as follows:

$$p_w = p(w_e) = p_2 \tag{10.23}$$

This is basically the ***open-end assumption***. In the analysis described by Han and Lee (1998), the saturation pressure is set at atmospheric pressure.

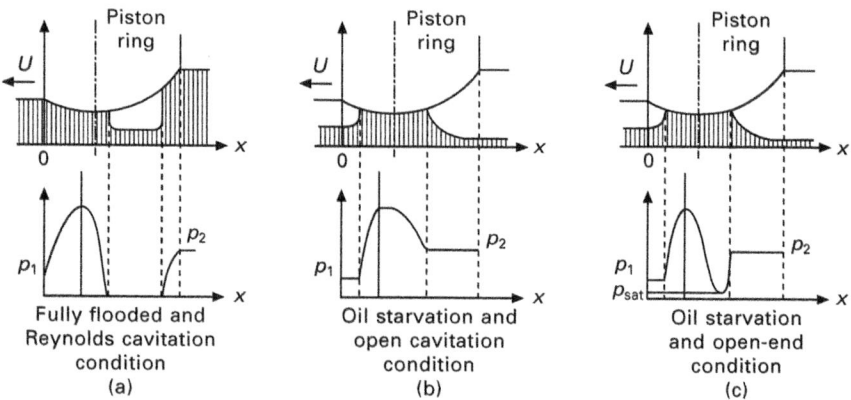

10.10 Boundary conditions assumption for the piston compression rings (after Han and Lee, 1998).

Numerical routines must be involved for analysing the solution of the open-end lubrication problem, but also approximate closed-form approaches have been proposed (D'Agostino *et al.*, 2005). A result in terms of hydrodynamic pressure profile and ring effective length for different boundary conditions is depicted in Fig. 10.11.

Recent approaches to the numerical simulation of piston ring lubrication has aimed to analyse also the free surface of the oil film outside the piston rings: an analysis based on the Navier–Stokes equations has been introduced by Lotz Felter (2008).

10.6   Simplified two-dimensional Reynolds equation for oil ring

In this section a simplified approach to the ring lubrication phenomena in presence of non-axial symmetry oil film thickness is described. This kind of effort may be seen as a tool to take into account the effect of piston lateral displacement (***piston secondary motions***) with the consequent drag action on the ring-pack lubrication behaviour (see Fig. 10.7).

Assuming the following film thickness expression:

$$h(x, \alpha) = h_0 + cx - e_t \cos \alpha \qquad (10.24)$$

where e_t is the piston lateral displacement at the ring pack top and α the circumferential co-ordinate ($\alpha = 0$ at the piston-cylinder approach side), the mathematical investigation about the hydrodynamic pressure passes through the choice of a solution of a *limiting case* of the two-dimensional Reynolds

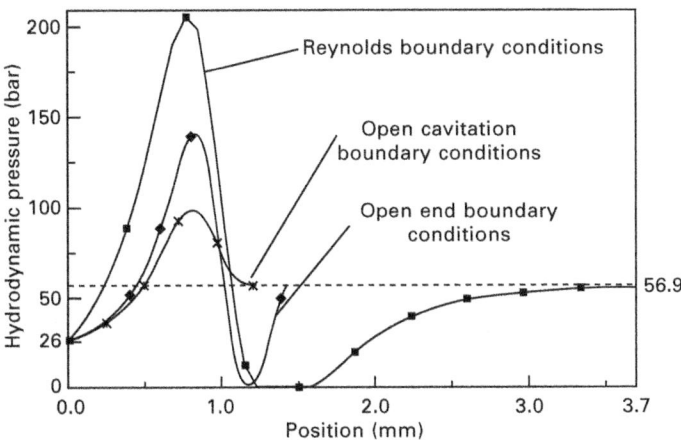

10.11 Ring HDL effective length for different boundary conditions (after Han and Lee, 1998).

equation. In fact, by assuming a very high ratio cylinder bore/ring length ($\pi D/L \gg 1$), one finds the same outcome as that of Ockvirk and Dubois (1953) short-bearing solution of Reynolds equation, where the oil film thickness relationship provides a 2D pressure profile even from the one-dimensional Reynolds equation.

Considering the simple condition of no-squeeze effect, the oil pressure solution is:

$$p(x,\alpha) = \frac{6\mu Uc\; x\;(x-L)}{h^2(x,\alpha)\,(2h_0 + cL - 2e_t\cos\alpha)} \tag{10.25}$$

Now, this limiting case solution can be rewritten with a product function in order to force along the circumferential direction the *shape* of the oil pressure, by allowing the differential problem to provide the pressure along the x-axis (*integral method*); a possible choice is given by:

$$p_i(x,\alpha) = a(x)b(\alpha) = a(x)\,\frac{1}{h^2(x = L/2,\, \alpha)\,(2h_0 + cL - 2e_t\cos\alpha)} \tag{10.26}$$

The following steps are the substitution of the product function in the two-dimensional Reynolds equation and the integration of the whole equation in the α co-ordinate on the hydrodynamic effective range: $[-\pi, \pi]$ or a part of it. This integration aims to fulfil the Reynolds equation, not locally but over a circumferential stripe.

At this stage, the α-integrated equation is an ordinary differential equation in the $a(x)$ unknown function. In the case of the oil ring the boundary conditions on the relative pressure give:

$$p(x = 0,\, \alpha) = 0 \Rightarrow a(x = 0) = 0$$
$$p(x = L,\, \alpha) = 0 \Rightarrow a(x = L) = 0 \tag{10.27}$$

According to the ring profile other approximations on the ordinary equation may be necessary in order to have a closed form solution. However, by considering the linear shape of the ring profile (10.24) all the solutions of this method are very close to the $p(x, \alpha)$ in (10.25). The graphical representation in terms of oil pressure and shear stress:

$$\tau_H(x, \alpha) = -\mu\frac{U}{h(x,\alpha)} - \frac{h(x,\alpha)}{2}\frac{\mathrm{d}}{\mathrm{d}x}p(x,\alpha) \tag{10.28}$$

is given in Fig. 10.12.

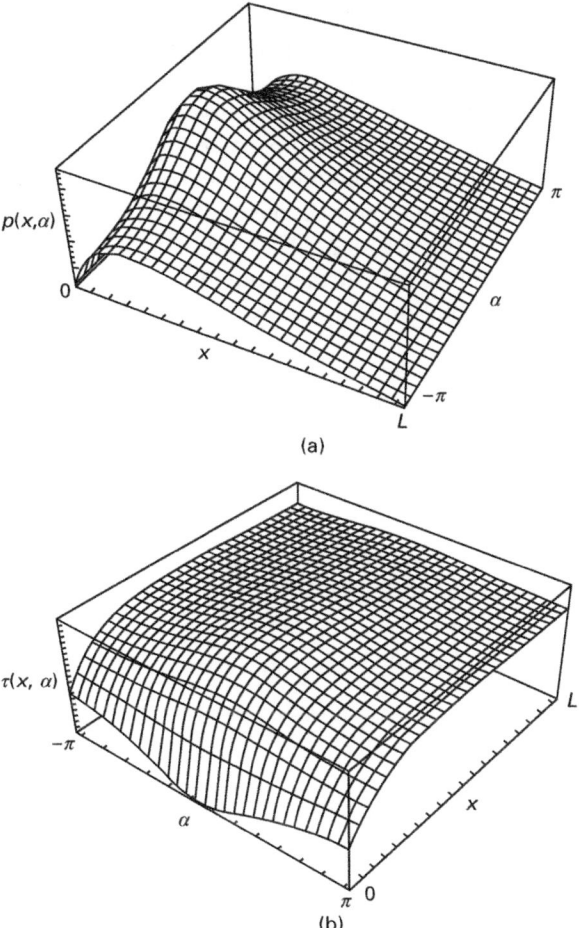

$p(x,\alpha)$

π

α

$-\pi$

x

L

(a)

$\tau(x, \alpha)$

L

$-\pi$

α

x

π 0

(b)

10.12 Two-dimensional solution for oil film pressure (a) and shear stress (b) HDL regime.

10.7   The contact between the asperities: mixed-lubrication regime

The first consequence of piston/liner relative velocity decreasing is a reducing of oil film thickness. The asperity contact can occur if the oil film becomes thin enough. In order to numerically describe these effects, the computer simulations of ring performance generally include *mixed regime of lubrication* models.

The oil film boundary value for the mixed lubrication model can be determined in many ways, but it always depends on surface roughness. Boundary lubrication occurs when the surface contact becomes continuous.

The oil film thickness has decreased to such a low level that the oil film only provides lubrication between the asperities: the load is carried by the surface peaks and not by the oil film.

Mixed lubrication of piston rings can be investigated by applying the simplified average Reynolds equation presented in the work of Wu and Zheng (1989). The mathematical model is based on the Patir and Cheng (1979) average flow model, which was reconstructed from original Reynolds equation by introducing flow factors ϕ_x, ϕ_y, ϕ_s and contact factor ϕ_c in order to consider the effect of lubricated surface roughness (also see Chapter 15). Their average Reynolds equation is as follows:

$$\frac{\partial}{\partial x}\left(\phi_x \frac{h^3}{12\mu}\frac{\partial p}{\partial x}\right) + \frac{\partial}{\partial y}\left(\phi_y \frac{h^3}{12\mu}\frac{\partial p}{\partial y}\right) = \phi_c\left(\frac{U_1 + U_2}{2}\frac{\partial h}{\partial x} + \frac{\partial h}{\partial t}\right)$$

$$+ \frac{U_1 - U_2}{2}\sigma\frac{\partial \phi_s}{\partial x} \tag{10.29}$$

where the average gap $\overline{h}_T$ that considers the asperity contact between lubricated surfaces is defined as follows:

$$\overline{h}_T = \int_{-h}^{\infty}(h + \delta)f(\delta)\,\mathrm{d}\delta \tag{10.30}$$

and

$$\phi_c = \frac{\partial \overline{h}_T}{\partial h} = \int_{-h}^{\infty}\varphi(s)\,\mathrm{d}s \tag{10.31}$$

according to the Patir and Cheng (1979) theory.

For the application of (10.29) into the lubrication analysis of piston ring assembly, the oil film thickness between piston ring and cylinder wall is assumed as an infinite flat bearing, and it is also assumed that piston ring is fixed and cylinder wall moves with a speed U. Then, (10.29) can be reduced as follows (Yun *et al.*, 1995):

$$\frac{\partial}{\partial x}\left(\phi_x \frac{h^3}{12\mu}\frac{\partial p}{\partial x}\right) = \phi_c\left(\frac{U}{2}\frac{\partial h}{\partial x} + \frac{\partial h}{\partial t}\right) + \frac{U}{2}\sigma\frac{\partial \phi_s}{\partial x} \tag{10.32}$$

with the boundary conditions:

$$p(0, t) = p_1(t)$$

$$p(L, t) = p_2(t) \tag{10.33}$$

The oil film thickness can be expressed in terms of ring-face profile and a time-variant component as shown in Fig. 10.13.

Introducing the following dimensionless variables:

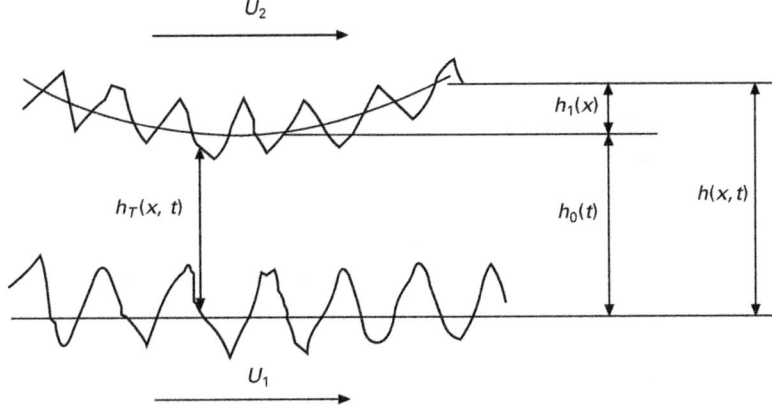

10.13 Film thickness function with ring and liner rough surfaces.

$$\bar{x} = \frac{x}{L} \quad \bar{H} = \frac{h}{\sigma} \quad \bar{t} = \omega\, t \quad \bar{h}_0 = \frac{h_0}{\sigma}$$

$$\bar{p} = \frac{p}{p_0} \quad \bar{h}_t = \frac{h_t}{\sigma} \quad \bar{\mu} = \frac{\mu}{\mu_0} \quad \bar{U} = \frac{U}{U_0} \tag{10.34}$$

After two *x*-integrations of Reynolds equation (10.32), the following result is obtained:

$$\bar{p}(\bar{x},\bar{t}) = \bar{p}_1(\bar{t}) + [\bar{p}_2(\bar{t}) - \bar{p}_1(\bar{t})]\frac{I_3(\bar{x},\bar{t})}{I_3(1,\bar{t})}$$

$$+ \bar{\mu}\,\bar{U}\left[I_2(\bar{x},\bar{t}) - I_2(1,\bar{t})\frac{I_3(\bar{x},\bar{t})}{I_3(1,\bar{t})}\right]$$

$$+ 2\frac{L}{R_c}\bar{\mu}\frac{d\bar{H}}{d\bar{t}}\left[I_1(\bar{x},\bar{t}) - I_1(1,\bar{t})\frac{I_3(\bar{x},\bar{t})}{I_3(1,\bar{t})}\right] \tag{10.35}$$

where:

$$I_1(\bar{x},\bar{t}) = \int_0^{\bar{x}} \frac{\phi_c(H)\,\xi}{\phi_x(H,\gamma)H^3(\xi,\bar{t})}\, d\xi \tag{10.36}$$

$$I_2(\bar{x},\bar{t}) = \int_0^{\bar{x}} \frac{1}{\phi_x(H,\gamma)H^3(\xi,\gamma)}\left[\int_0^{H(\xi,\bar{t})} \phi_c(H)\, dH + \phi_s(H,\gamma)\right] d\xi$$

$$\tag{10.37}$$

$$I_3(\bar{x}, \bar{t}) = \int_0^{\bar{x}} \frac{d\xi}{\phi_x(H, \gamma)H^3(\xi, \bar{t})}$$

(10.38)

The force equilibrium condition is applied in the radial direction of a ring to find a possible constraint (Fig. 10.14):

$$F_{tot} = F_{c,oil} + F_{c,asp} - F_{R,t} - F_{R,g} = 0$$

(10.39)

The $F_{c,oil}$ term is the oil thrust load that can be calculated by applying half-Sommerfeld boundary condition (Section 10.5); the normal load about the asperities ($F_{c,asp}$) can be obtained by using the Greenwood and Williamson (1966) equation. This approach allows deriving the solid-to-solid contact force per unit area and the contact area too; under the hypothesis of two rough surfaces with Gaussian distributions:

$$F_{c,asp} = 2\pi R \int_0^B P_{AC}(h) \, dx$$

(10.40)

$$P_{AC}(h) = \frac{6\sqrt{2}}{15} \pi (\eta\beta\sigma)^2 E \sqrt{\frac{\sigma}{\beta}} F_{5/2}\left(\frac{h}{\sigma}\right)$$

(10.41)

$$A_{AC}(h) = \pi^2 (\eta\beta\sigma)^2 F_2\left(\frac{h}{\sigma}\right)$$

(10.42)

where:

$$F_n\left(\frac{h}{\sigma}\right) = \frac{1}{\sqrt{2\pi}} \int_{\frac{h}{\sigma}}^{\infty} \left(S - \frac{h}{\sigma}\right)^n e^{\frac{S^2}{2}} \, dS$$

(10.43)

By neglecting the force caused by the boundary pressure on the ring regions

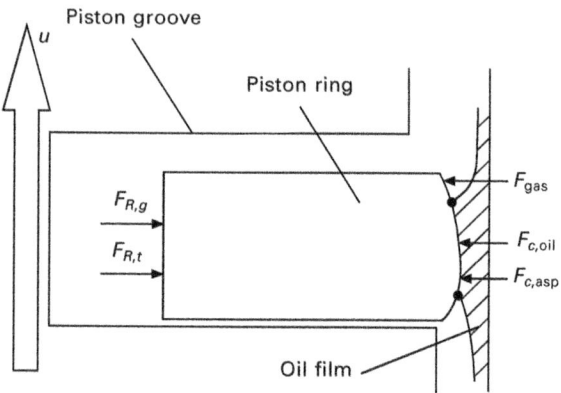

10.14 Forces acting on the ring in the radial direction.

exposed to the gas, the force due to the gas pressure acting on the ring groove side ($F_{R,g}$) can be written as follows:

$$F_{R,g} = 2\pi BL \, p_g \qquad (10.44)$$

The inter-ring gas pressures, as well as the pressure on the back side of each ring, are normally computed from the experimental cylinder pressure data (p_1) and assumptions on the blowby and ring's *sealing effect*. The $F_{R,t}$ term in the ring radial equilibrium equation is given by the elastic ring tension:

$$F_{R,t} = K_r (D_{fp} - B - 2h_0) \qquad (10.45)$$

where D_{fp} identifies the *free diameter* of the ring, i.e. without radial tension.

Implementing an iterative procedure imposing a first check on the convergence of F_{tot} and a second one on the oil film thickness the following results can be introduced about the top ring friction forces (Fig. 10.15), the

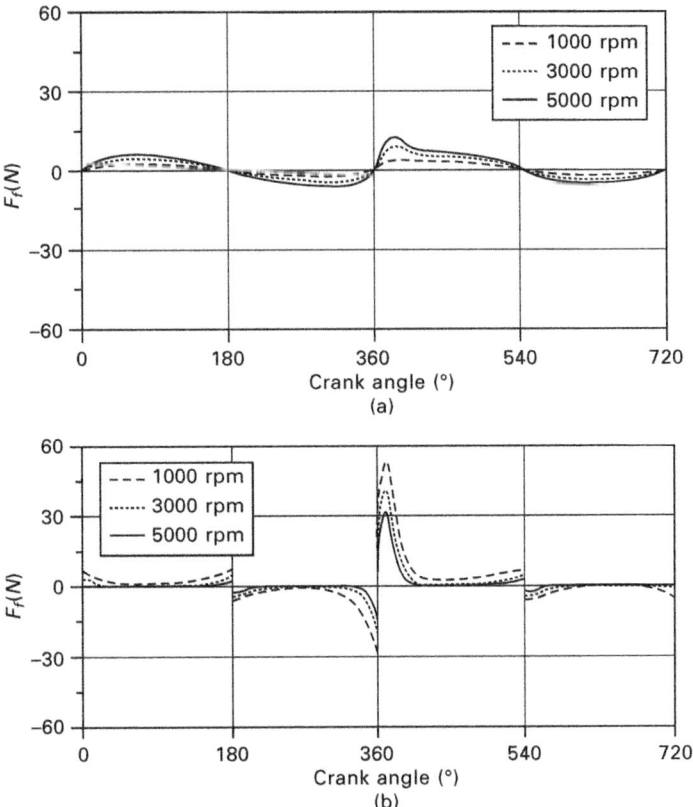

10.15 Top ring barrel-shaped viscous (a) and asperities (b) friction forces (after Yun *et al.*, 1995).

oil ring ones (Fig. 10.16) and the oil film thickness for both ring types (Fig. 10.17).

These outcomes can be synthetically delineated:

- The hydrodynamic regime of lubrication is dominant in the middle of the stroke, while a mixed regime characteristic prevails near the dead centres.
- The wedge between the oil ring and cylinder wall is lowest, because it has a relatively large radial tension and narrow lubricated surfaces, if compared with the others.
- The top ring experiences the minimum film thickness immediately after the top dead centre piston position (expansion) due to the gas pressure on the ring back-side.
- The film thickness is significantly increased with the engine speed, due to the higher hydrodynamic thrust load.

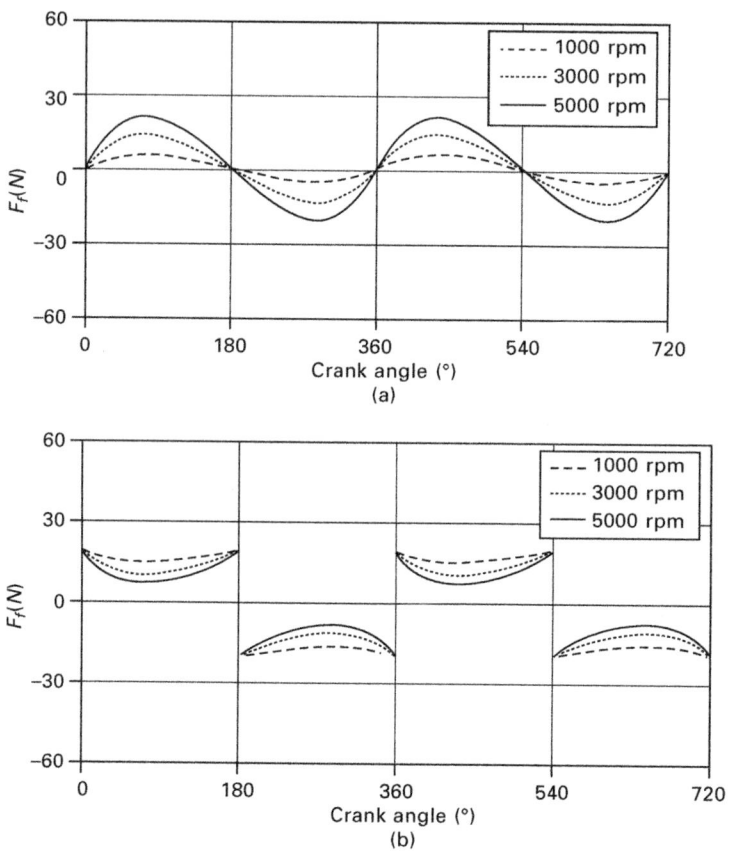

10.16 Oil ring 3-piece type viscous (a) and asperities (b) friction forces (after Yun *et al.*, 1995).

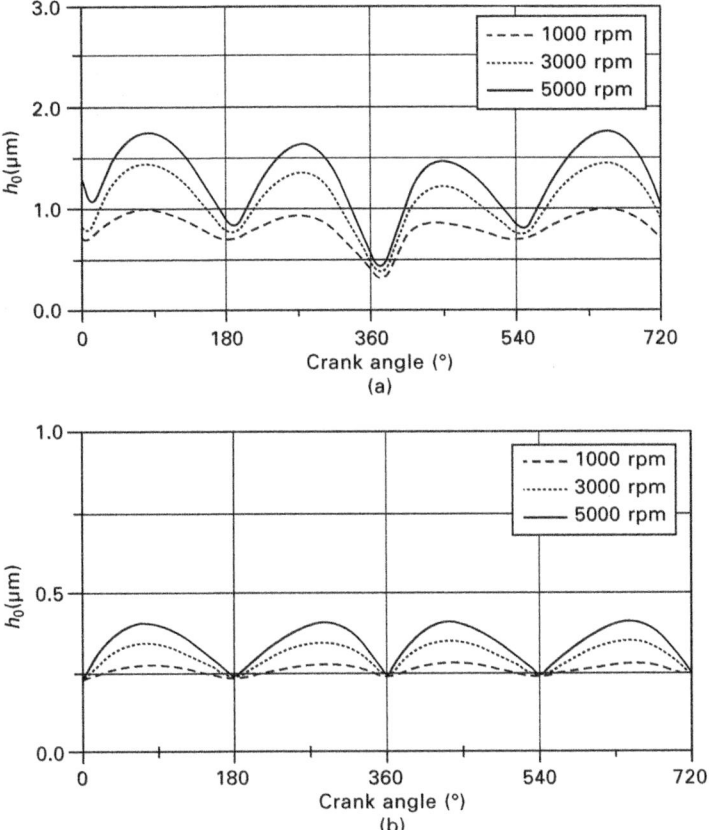

10.17 Minimum film thickness: (a) top ring and (b) oil ring (after Yun *et al.*, 1995).

- The friction force presents an increase with the engine speed near the middle of the stroke (hydrodynamic regime) while a decrease with the engine speed can be observed near the dead centres (mixed lubrication regime is predominant).
- The mixed-lubrication regime appears through the whole piston stroke for the oil ring.

An axis-symmetric, hydrodynamic, mixed-lubrication model using the averaged Reynolds equation and asperity contact approach for simulating frictional performance of rings in highly loaded diesel engine has been introduced by Akalin and Newaz (2001a).

The friction force between ring and cylinder bore is predicted by taking into account rupture location, surface flow factors, surface roughness and metal-to-metal contact loading. A fully flooded inlet boundary condition and Reynolds boundary conditions for cavitation outlet zone are assumed.

Reynolds boundary conditions have been modified for non-cavitation zones. The pressure distribution along the ring thickness and the lubricant film thickness are determined for each crank angle degree. The analysis has been carried out for the top compression ring by assuming cast-iron cylinder bore and chromium plated ring as mating materials, 700 rpm as engine speed and 70 °C average oil temperature. Results show that the hydrodynamic lubrication regime occurs in most part of the stroke while significant increase in the friction coefficient near the top dead centre (TDC) and bottom dead centre (BDC) is due to the mixed lubrication as a result of the high cylinder pressure and low speed in these locations.

The authors have compared the results obtained by simulation (Fig. 10.18) to those obtained on a test bench and found that the friction results correlate well (Akalin and Newaz, 2001b). The results show that the temperature, surface roughness and running speed are the most important parameters, as they affect the lubrication regime the most. The effect of the normal load on the friction coefficient during mixed lubrication was low.

An efficient time-saving numerical algorithm can be based on the sharing of the total load acting in the ring radial direction between the solid-to-solid interactions along the asperities of the surfaces and the hydrodynamic force by considering, as a first approximation, ring/liner smooth surfaces in the lubrication equation. Since the one-dimensional Reynolds equation has a closed form solution (see Section 10.4.2) and the radial contact force can

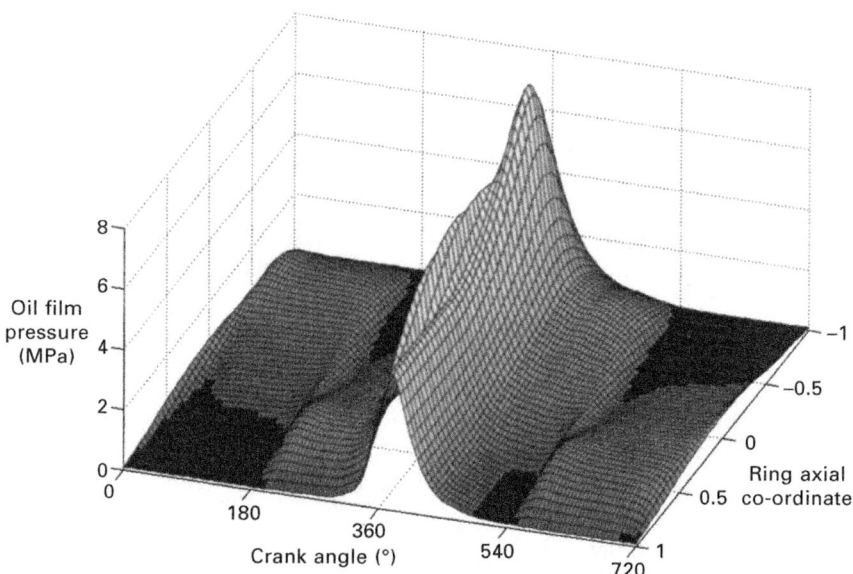

10.18 Oil film pressure; highly loaded diesel engine (700 rpm, 70 °C average oil temperature) (after Akalin and Newaz, 2001a).

be successfully replaced by a well-fitted spline, for a certain parameter set the relationship radial force vs. film thickness can be inverted (Senatore and Ciortan, 2004).

This fast procedure, described by Fig. 10.19 for a piston top ring, yields

$$F_{c,asp}(h_0) + F_{c,oil}(h_0, U, p_1, p_2) = F_{R,t} + F_{R,g}(p_g) \tag{10.46}$$

$$F_{c,asp}(h_0) + F_{c,oil}(h_0, U, p_1, p_2) - F_{R,t} - F_{R,g}(p_g) = 0 \Rightarrow h_0 \tag{10.47}$$

In (10.46) and (10.47), the pressure values p_1, p_g and p_2 are the pressure in the combustion chamber, the pressure acting on the ring back-side and the inter-ring gas pressure, i.e. above the second ring, respectively.

Once the previous equation is solved and the minimum ring distance from the cylinder wall h_0 is determined, the friction forces and the friction coefficient can be calculated by using (Gelinck and Schipper, 2000):

$$F_f = F_{f,asp} + F_{f,oil} = \sum_{i=1}^{N_a} \iint_{AC_i} \tau_{C_i} \, \mathrm{d}A_{C_i} + \iint_{A_H} \tau_H \, \mathrm{d}A_H \tag{10.48}$$

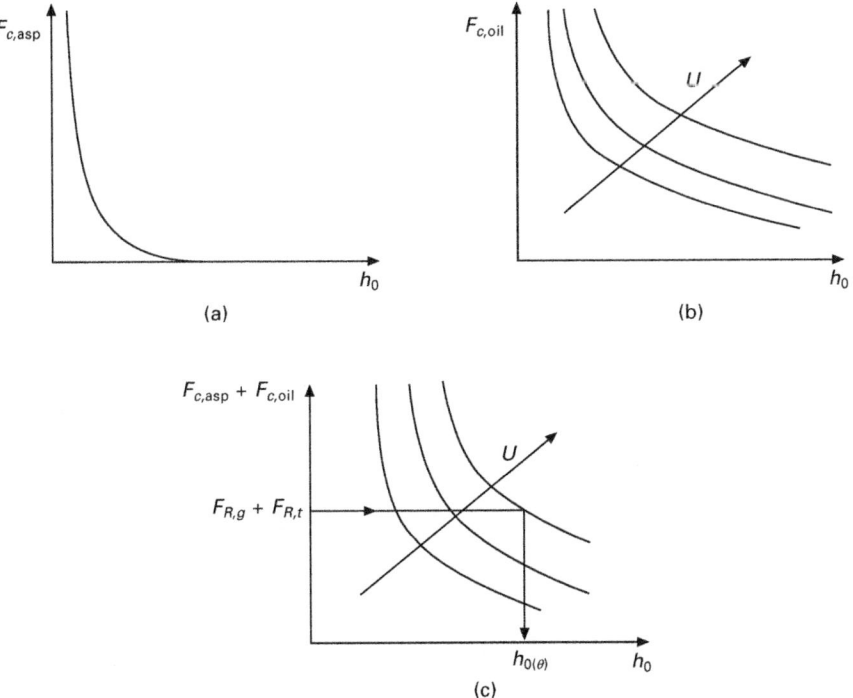

10.19 (a) Radial asperities ring force vs. minimum ring/liner distance (b) HDL thrust load (c) total radial ring load, cylinder liner side.

where the Greenwood and Tripp equation is involved for the friction force due to the contacts and the viscous shear stress is described in analytical way by:

$$\tau_H(x) = -\mu \frac{U}{h(x)} - \frac{h(x)}{2} \frac{dp}{dx}$$

(10.49)

Thus, the friction coefficient is:

$$f = \frac{F_f}{F_R} = \frac{F_{f,\text{asp}} + F_{f,\text{oil}}}{F_{c,\text{asp}} + F_{c,\text{oil}}} = \frac{\sum_{i=1}^{N_a} \iint_{A_{Ci}} \tau_{C_i} \, dA_{C_i} + \iint_{A_H} \tau_H \, dA_H}{F_{c,\text{asp}} + F_{c,\text{oil}}}$$

(10.50)

The plots (Fig. 10.20) of the F_f in (10.48) using the simulation input parameters of Table 10.1 – four-inline cylinder SI engine, 311 cm^3 unit displacement – illustrate the effect of oil viscosity on total top ring friction force and lubrication mechanisms transitions (from mixed to hydrodynamic and back). The analysis takes into account lubricant viscosity in the range 5.0–25.0 mPa s, typical values for the commercial defined 0W30 or 5W30 along the temperature range 70–100 °C.

For higher viscosities, the lubrication regime is hydrodynamic over most of the stroke, also in presence of the high radial load during the expansion stroke, as shown in Fig. 10.20; analysing for example the intake stroke, $\theta \in$ [0, 180°], the boundary friction force declines to zero with the delay of the

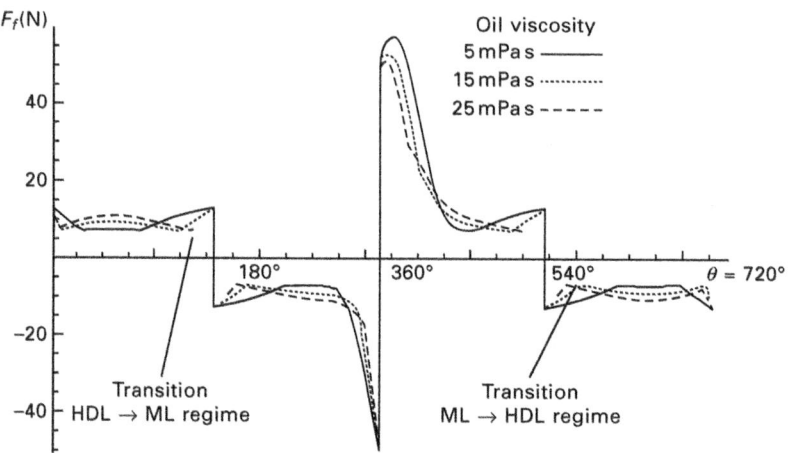

10.20 Total top ring friction force; 5000 rpm, $p_{\text{cyl,max}}$ = 10 bar; SI engine.

Table 10.1 Input parameters for ML simulation about piston top ring

Parameter	Symbol	Value	Unit
Density of asperities	n	10^{11}	m^{-2}
Average radius of asperities	β	10	mm
Standard deviation of the asperities	σ_s	1.00	mm
Reduced modulus of elasticity	E'	231	GPa
Length of the top ring	L_1	1.50	mm
Cylinder bore	B	71.0	mm
Oil viscosity	μ	5.0–15.0–25.0	mPa s
Connecting rod length	L_{CR}	129	mm
Characteristic ratio of the slider-crank mechanism	L_{CR}/R_C	3.30	–

transition ML → HDL regime for the lower viscosity oil up to 80 crankshaft degrees.

In the same figure, the lower viscosity oil confirms a better economy only in the pure HDL regime. Simulations on a broad range of engine speed/load show that the oil with low viscosity exhibits lower friction force for high speed and low radial load, while high viscosity has better effect on the mechanical efficiency on the opposite side.

10.8   The multi-physics approach to ring friction

10.8.1   Effect of oil film temperature

The analysis of the oil film thickness can be performed by using the Reynolds equation coupled with the two-dimensional energy equation (see Chapter 5), in order to take into account the oil temperature increasing due to the heat generated from the viscous dissipation (Harigaya *et al.*, 2003). In fact, using the energy equation, temperature distributions in the oil film as well as the average value can be calculated; then, the oil viscosity is estimated by using the mean oil film temperature.

Since the oil film temperature between the ring and the liner varies from the liner to the ring as well as from the inlet to outlet, it is necessary to be cautious about the prediction of oil film thickness using the original Reynolds equation in which viscosity is assumed to be a constant in the vertical direction of oil film.

Under the assumption of incompressible oil film with Newtonian behaviour, uniform viscosity along the whole ring surface, laminar flow, uniform properties as specific heat, heat conductivity, in a engine cycle while the oil viscosity is a function of temperature, fully flooded inlet and Reynolds boundary conditions (the oil starvation is not considered), the one-dimensional unsteady Reynolds equation can be coupled with the *energy equation* for two-dimensional flow:

$$\rho C \left(\frac{\partial T}{\partial t} + u \frac{\partial T}{\partial x} + v \frac{\partial T}{\partial y} \right) = \kappa \left(\frac{\partial^2 T}{\partial x^2} + \frac{\partial^2 T}{\partial y^2} \right) + \phi \qquad (10.51)$$

with:

$$\phi = 2\mu \left[\left(\frac{\partial u}{\partial x} \right)^2 + \left(\frac{\partial v}{\partial y} \right)^2 + \frac{1}{2} \left(\frac{\partial u}{\partial y} + \frac{\partial v}{\partial x} \right)^2 \right]$$

$$(10.52)$$

in which ρ is the oil density, μ the local oil dynamic viscosity, κ the oil film thermal conductivity, while the last equation represents the viscous dissipation in the oil film. The boundary conditions on the temperature field are:

$T = T_{Liner}$	at the inflow on inlet side	(10.53)
$dT/dx = 0$	at the outflow on inlet side	(10.54)
$dT/dx = 0$	at the downstream side where the oil flow breaks down	(10.55)
$T = T_{Liner}$	at the liner surface	(10.56)
$T = T_{Ring}$	at the ring surface	(10.57)

The temperature distribution along the cylinder liner has a higher slope near the top dead centre and a lower slope near the bottom dead centre, as observed through experimental analysis. An interesting approach to the liner temperature is to use the measured temperatures at the **TDC**, mid-stroke point and **BDC**, and to adopt an approximate expression to evaluate the liner temperature distribution.

By means of the numerical algorithm steps:

1. the oil film thickness h and the viscosity μ of the oil film at arbitrary crank angles is assumed;
2. the Reynolds equation is solved numerically by finite difference method, and the pressure, velocity and temperature distributions are calculated;
3. the lubricant viscosity is calculated, using the average oil temperature, and the squeeze term is calculated;
4. The calculations of h, T_m and μ are repeated until the solution converges to a certain condition.

Temperature and velocity distributions in oil film between piston ring and cylinder liners can be described through the plots in Figs 10.21 and 10.22. The mean oil film temperature and viscosity are depicted in Fig 10.23(a) and (b), respectively.

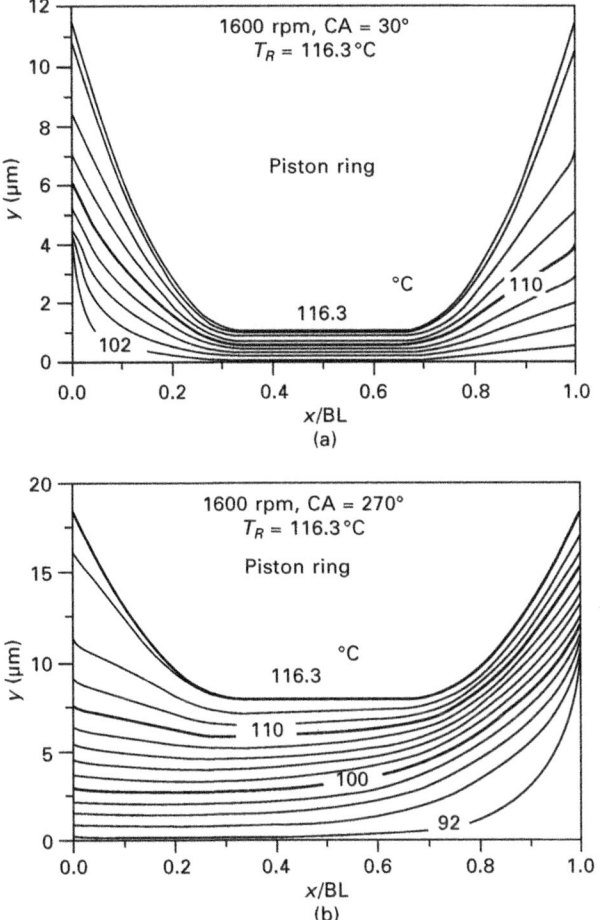

10.21 Temperature distribution in the oil film (ring/liner wedge); 1600 rpm, no load (a) crankshaft angle $\theta = 30°$ (b) $\theta = 270°$ (after Harigaya *et al.*, 2003).

10.8.2 The effects of the piston lateral dynamics and ring flexibility

This section introduces the results of simulations about the interaction of a SI engine piston ring taking into account the mixed lubrication regime, ring flexibility and piston motion in the radial plane. The reference frame in this plane is introduced in Fig. 10.24.

The model assumes a mixed regime of lubrication: the total load acting in the radial direction on the ring is shared between the hydrodynamic force and the solid-to-solid surface interactions between the asperities of the surfaces, ring and cylinder liner (see also Section 10.7).

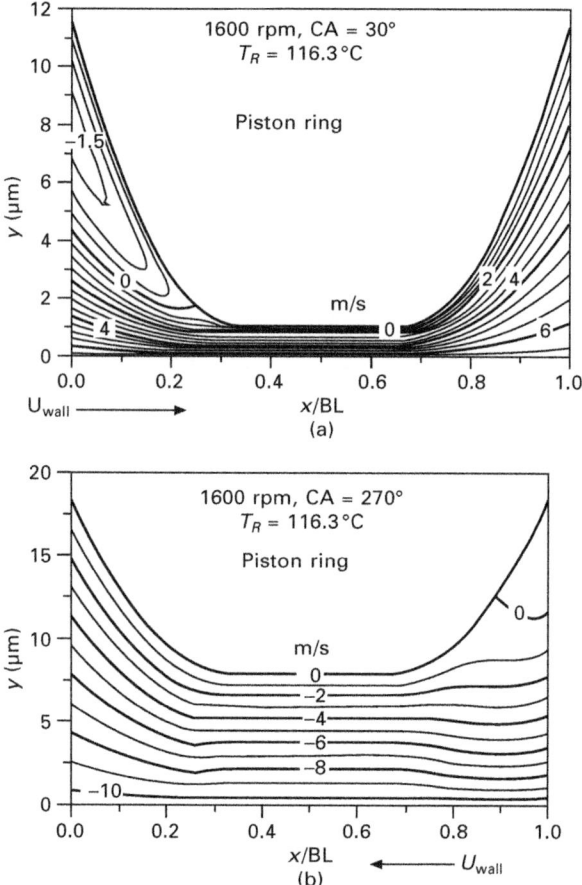

10.22 Velocity distribution in the oil film (ring/liner wedge); 1600 rpm, no load (a) crankshaft angle $\theta = 30°$ (b) $\theta = 270°$. (after Harigaya *et al.*, 2008)

Ring flexibility

The piston ring flexibility has been taken into account with an FEM model of a tapered steel ring with the geometry in Fig. 10.24(c). The actual ring shape during simulation is calculated considering the mixed-lubrication regime on the cylinder-liner side (the external one) and the contact interaction with a rigid piston along the groove surfaces. The figure shows the reference system about the ring circumferential α co-ordinate.

Oil film pressure and radial hydrodynamic load on the ring, as well as radial load due to the asperity contacts can be computed at each step of numerical algorithm, updating the ring geometry by means of the FEM ring bending model; the piston–ring contact forces are computed using lumped spring-damper parameters (also see the analytical model in Chapter 15).

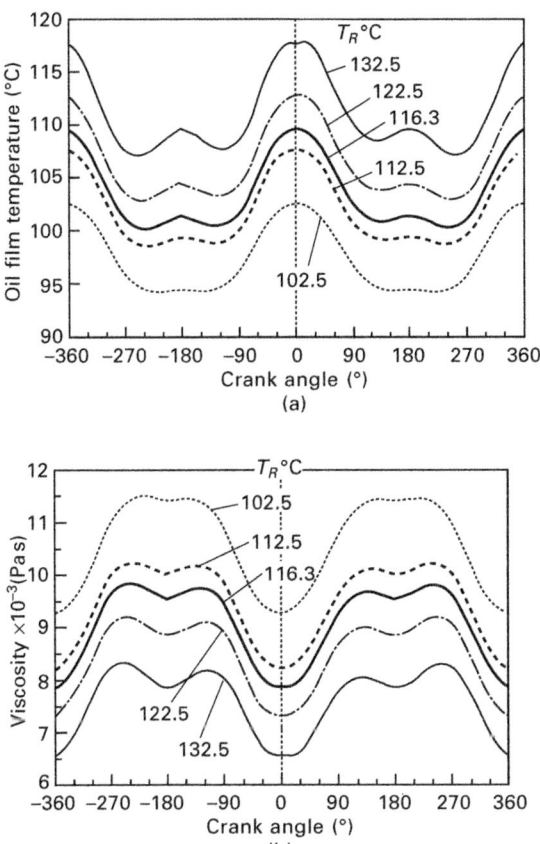

10.23 Mean oil film temperature (a) and viscosity (b) vs. crankshaft angle; 1600 rpm, no load (after Harigaya *et al.*, 2003).

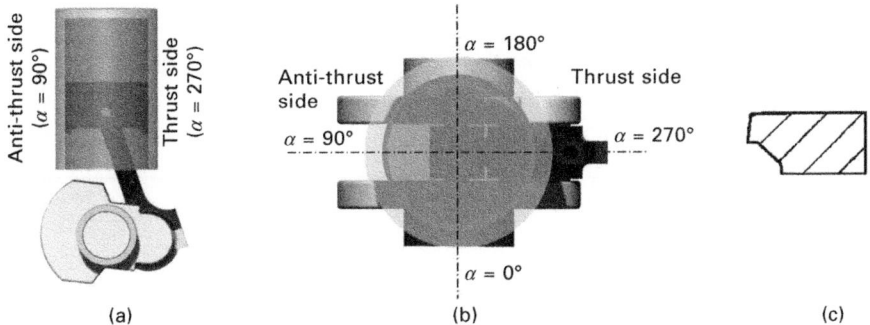

10.24 Piston/cylinder system and circumferential co-ordinate (a, b) ring profile (c).

Piston lateral dynamics

It is well known that ***piston secondary motions*** and the impacts with the cylinder wall represent one of the main sources of noise and vibration in the ICEs (Cho *et al.*, 2002). The radial forces developed during an impact are a function of the inertia force system acting on the piston and the connecting rod, the lubricating fluid film force between piston skirt and cylinder liner, as well as of the gas force due to the thermodynamic cycle (Perera *et al.*, 2007). This analysis is mainly addressed to the effects of this characteristic piston behaviour on the friction force due to the piston ring in mixed lubrication. This analysis assumes lateral drag motion of the entire ring profile due to the approaching motion of the piston to the cylinder bore, while no feedback on the piston dynamics is considered, i.e. the action of the piston ring radial forces on the piston dynamics is neglected.

Figure 10.25 shows the results of Dursunkaya *et al.* (1994) on the piston lateral dynamics: for this study the piston is assumed to be a rigid body with two degrees of freedom which drags the ring along the axial and radial cylinder directions, since the tilting motion is not considered. In the same figure, during a piston motion towards the thrust side (for example, expansion phase), $e_t/C < 0$ and the piston approaches the cylinder liner at $\alpha = 270°$.

In the following descriptions, the crank angle 0° is set at the TDC before expansion according to this plot. The simulation results about the flexible ring are presented in terms of oil film thickness at four circumferential ring locations, α-step = 90°, in order to highlight the effect of flexibility and piston lateral drag motion and show the substantial impact of whole piston assembly motion on the interaction of the single ring with cylinder liner and the piston (Fig. 10.26).

The figure underlines the effect of piston approach to the cylinder liner

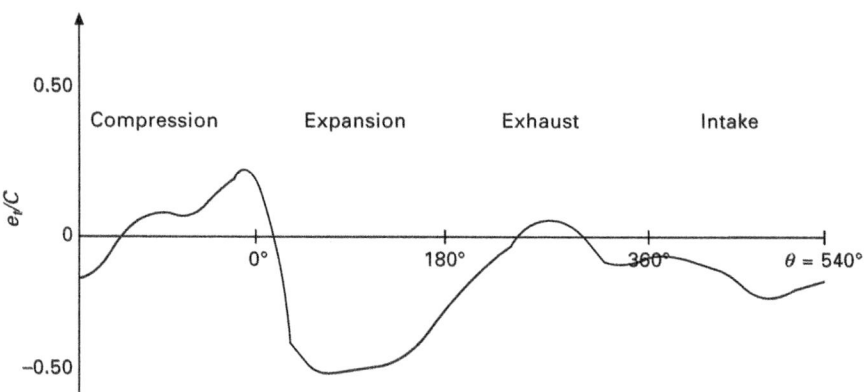

10.25 Rigid piston lateral motion: dimensionless position e_t/C vs. crank angle.

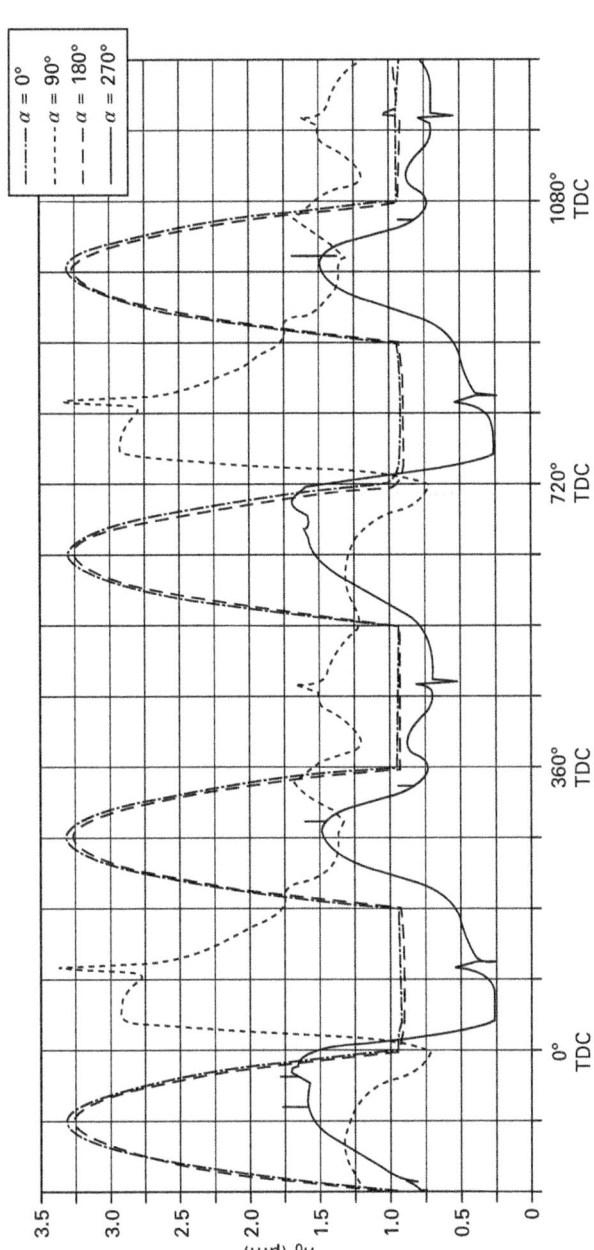

10.26 Oil film thickness at four circumferential positions (α-step = 90°) during two 4-stroke. Engine cycles, $\theta \in [-180°,$ 1260°] (after D'Agostino and Senatore 2005).

during expansion [0°, 180°] and intake strokes [360°, 540°] with evident differences on the oil thickness: 0.25 μm at $\alpha = 270°$, 3.3 μm at $\alpha = 90°$ around the mid-expansion stroke ($\theta \approx 100°$). Furthermore, the piston lateral motion is the cause of different transition crank angle between ML and full developed HDL regimes in the upward strokes with a bigger mixed-lubrication portion of stroke for the ring portions on the *thrust side–anti-thrust side* plane.

In the same plot a weak influence of the piston lateral dynamics may be observed on the ring locations in the plane perpendicular to the slider-crank one, where the oil ring behaves as a slider bearing in transition between ML and full developed HDL regime and vice versa.

10.9   Ring flutter and collapse

The position of a ring inside the ring groove is determined by gas pressures, inertial force, frictional force and the pressure distribution on the flanks of the ring as well as on the running surface of the ring; a piston ring moves up and down inside the ring groove several times in a cycle due to the variation in the driving forces.

Ring flutter is defined as the phenomenon that the ring oscillates inside the groove several times within a portion of a cycle. Ring flutter could occur either with or without strong gas flow through the ring groove and gas pressure fluctuation. The second case takes place when the land pressures are almost equal everywhere and the reason for the oscillation is simply that it is difficult for the ring to find the balanced position after the ring makes much localised contact with the groove. The discussion will thus focus on the first case.

Ring flutter, which is mainly driven by the competing of the gas pressure loading and the inertial force, has significant effects on gas flow, as well as the oil flow. This type of ring flutter can occur for all the three rings in a piston ring pack when conditions are right, and has been widely described in the work of Tian (2002a,b).

The ring flutter occurs with the following steps: firstly, the ring moves to the other side of the ring groove due to a change in the direction of the net force on the ring; then, the ring motion gradually leaves a large enough clearance between the ring and the groove. This introduces strong gas flow through the groove and fast land pressure variation. As a result of the land pressure variation, the net force reverses the direction and the ring moves back. The top-ring flutter needs to be controlled because it creates great *blowby* and subsequent oil consumption problems through the positive crankcase ventilation system; in order to control this phenomenon, it is necessary to create a positive top ring/groove relative angle and to find ways to reduce ring dynamic twist during dynamic situations. Unlike top ring flutter, the second ring flutter has a much smaller effect on blowby. Any increase in blowby due to second ring flutter is probably because second ring flutter reduces the amount of reverse gas flow through the top ring by quickly releasing the second land gas to the lower land and reducing the second land pressure. Therefore, second ring flutter may be beneficial to oil consumption reduction if it indeed results in higher blowby. Once a second ring is designed with a negative static twist, it flutters in a wide range of operating conditions.

Radial *ring collapse* mainly occurs when a ring completely seals the gas flow path through the upper side of ring-groove clearance by making contact with the outer diameter location of the ring groove contact corner on the upper side of the groove. By having such contact, the ring can stay on the top with maximum stability under gas pressure of the upper land as illustrated in Fig. 10.27. Generally, ring collapse requires a positive relative angle between the upper flanks of the ring and the groove. As an effect, ring radial collapse is more likely to occur for the second ring with a positive static twist (Napier or simple taper face), because the second land pressure rises relatively slowly and thus the inertial force has the opportunity to bring the second ring up at a certain point of the late part of the compression stroke.

Minimum net downward force
from gas pressure when ring is up

Minimum net downward force from
gas pressure when ring is down

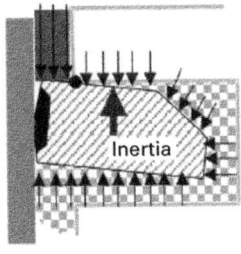
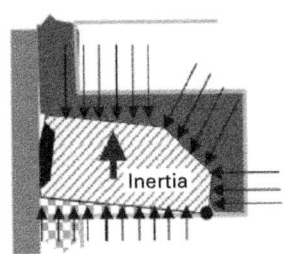

Positive relative
angle

Stable

Minimum net downward force
from gas pressure when ring is up

Minimum net downward force from
gas pressure when ring is down

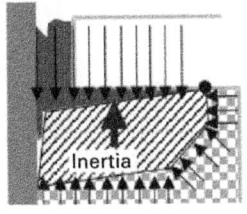
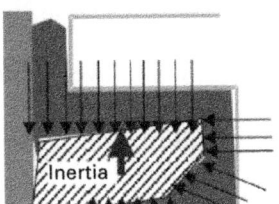

Negative relative
angle

Unstable

10.27 Ring stability (after Tian, 2002a).

10.10   Bore distortion in lubrication models

Cylinder bore geometry plays a significant role in the conformability of the piston ring. The bore distortion directly affects the piston/cylinder liner blowby, which in turn is critical for oil consumption, emissions and a sufficient lubricant supply. Therefore, it is essential to include this geometrical feature in the simulation computer codes.

The effect of the bore axial distortion on ring radial dynamics has been accounted in the ring pack phenomenological models and numerically investigated (Piao and Gulwadi, 2003). The predictions from simulations show good qualitative comparisons with experimental results on a similar engine. Some conclusions can be drawn from the mentioned study:

- Radial ring motion is significantly influenced by axial bore profiles. An analytical expression has been developed to represent the radial inertia of a ring due to an axial bore profile, which provides an insight to the phenomena. There can be situations of 'ring separation' from the bore depending on (a) bore profile geometry, (b) engine speed, (c) ring tension, and (d) gas loads (some are discussed in Chapter 15).
- The ring separation distance can be substantial so as to allow oil to flow past the ring towards the chamber, potentially leading to higher oil consumption. Large predicted values of ring minimum oil film thickness during ring separation are accompanied by substantial oil flow rates below the ring-face during this event.

10.11   Laser-textured surfaces

The influence of laser-textured cylinder bore and piston ring surfaces on the friction has been studied (Ronen *et al.*, 2001; Kligerman *et al.*, 2005) (also see Chapters 13 and 14). In particular, theoretical models have been described in these papers to analyse the potential of reducing the friction force between a partially surface textured piston ring and a cylinder liner through the collective hydrodynamic effect of micro-dimples. The authors achieved these results here summarised:

- the average friction force is not affected by the dimple diameter and it decreases monotonically with increasing dimple area density (Chapter 13);
- the minimum average friction force for the optimum partially textured piston ring is considerably lower than that for the corresponding optimum fully textured ring. The difference varies from about 30% reduction for narrow rings to about 55% reduction in wide rings.

10.12   Warm-up effect

A model for the prediction of the engine friction has been presented by Taylor (1997). The results of his work include simulations for fully warmed-up conditions and cold-start conditions, where the total engine friction is investigated. According to the results, the total engine friction immediately after a cold-start is four to five times higher than at fully warmed up conditions.

Froelund *et al.* (2001) approached the ring/liner oil film development theoretically during the warm-up phase for a SI engine; the main outcomes of the simulations are as follows:

- For all temperatures between 20 °C and the warm condition (100 °C), all rings change lubrication regime near the TDC to the mixed lubrication. Thereby, asperity contact occurs during the whole warm-up phase.
- The minimum oil film thickness between the oil ring and cylinder liner scales fairly well with the square root of the viscosity. The thickness between other rings in the ring pack does not scale with viscosity in any simple manner. While the thickness of the oil left on liner is coupled with the minimum oil film thickness of the compression ring, no simple scaling is found;
- The cycle-averaged ring pack *friction mean effective pressure* (FMEP) increases four to five times at cold conditions (20 °C) compared with the warm condition (100 °C) for the baseline SAE 10W30 oil. By averaging the above-mentioned values of FMEP over the entire warm-up phase, the average 'warm-up FMEP' (range 20 → 100 °C) is twice the warm engine FMEP at 100 °C.

10.13   Future trends

The engine operating characteristics in modern automotive applications bring new challenges to piston ring design in controlling engine blowby, friction, wear and oil consumption. Under the modern high engine speeds the ring instability in both axial and radial directions is a marked phenomenon; flutter and ring radial collapse generate more direct and stronger gas flow all around the circumference of the piston than otherwise only through ring gaps under low speed conditions. Thus, the main route for blowby and oil consumption are different from those low-under speed conditions and design strategies to control blowby and oil consumption under high-speed conditions have to be adjusted. Furthermore, at high engine load conditions, high bore expansion in engines with aluminium blocks gives significant increase to ring gaps when ring travels at top liner locations (Rabuté and Tian, 2001).

Increases of piston temperature and adaptation of low-friction oils in modern SI engines bring more and more problems in micro-welding between the top ring and the groove as well as scuffing between the top ring and the liner. While changes in materials, coatings and surface treatment may ultimately be needed, magnitude of asperity contact pressure between the top ring and its mating parts can be reduced by making simple modifications in the geometry of the piston and top ring. In this scenario, in which the range of factors that can influence the tribological behaviour of the piston assembly is alarmingly extensive, the following list describes possible future analyses and design challenges (Taylor, 1998):

- enhanced ring and lubrication model linked to wear model and engine life history predictions;
- improved integration of piston ring and piston analyses (secondary motions, mutual forces, flexibility);
- improved ring materials/coating considerations linked to durability and failure;
- enhanced understanding of lubricant transport mechanisms and three-dimensional analysis improvements, including non-circular bore/piston considerations;
- lubricant degradation action and its influence, including also the behaviour of innovative nanoparticle-based additives;
- lubricant chemistry/reaction film studies.

10.14   References and further reading

Akalin, O., Newaz, G. M., Piston ring-cylinder bore friction modelling in mixed lubrication regime: Part I – Analytical Results, *ASME Journal of Tribology*, vol. 123, pp. 211–218, 2001a.

Akalin, O., Newaz, G. M., Piston ring-cylinder bore friction modelling in mixed lubrication regime: Part II – Correlation with bench test data, *ASME J. of Tribology*, vol. 123, pp. 219–223, 2001b.

Andersson, P., Tamminen, J., Sandström, C. E., *Piston Ring Tribology: A Literature Survey*, VTT Research Notes 2178, Espoo, Finland, 2002.

Cho, S. H., Ahn, S. T., Kim, Y. H., A simple model to estimate the impact force induced by piston slap, *Journal of Sound and Vibrations*, vol. 255, pp. 229–242, 2002.

D'Agostino, V., Senatore, A., Piston ring behaviour simulation considering mixed-lubrication and flexibility, *Proceedings of IDETC/CIE 2005, ASME International Design Engineering Technical Conferences & Computers and Information in Engineering Conference*, Long Beach, California, 24–28, September 2005.

D'Agostino, V., della Valle, S., Ruggiero, A., Senatore, A., On the effective length of the top piston ring involved in hydrodynamic lubrication, *Lubrication Science, Leaf Coppin*, vol. 17–3, pp. 309–318, 2005.

Dowson, D., Ruddy, B. L., Economou, P. N., The elastohydrodynamic lubrication of piston rings, *Proceedings of The Royal Society of London, Series A, Mathematical and Physical Sciences*, vol. 386, issue 1791, pp. 409–430, 1983.

Dursunkaya, Z., Keribar, R., Ganapathy, V., A model of piston secondary motion and elastohydrodynamic skirt lubrication, *ASME, Journal of Tribology*, vol. 116, pp. 777–785, 1994.

Froelund, K., Schramm, J., Tian, T., Wong, V., Hochgreb, S., Analysis of the piston ring/liner oil film development during warm-up for an SI-engine, *Journal of Engineering for Gas Turbines and Power, ASME*, vol. 123, pp. 109–116, 2001.

Gao, J., Luedtke, W. D., Landman, U., Friction control in thin-film lubrication, *Journal of Physical Chemistry B*, vol. 102 (26), pp. 5033–5037, 1998.

Gelinck, E. R. M., Schipper, D. J., Calculation of Stribeck curves for line contacts, *Tribology International*, vol. 33, pp. 175–181, 2000.

Greenwood, J. A., Williamson, J. B. P., Contact of nominally flat surfaces, *Philosophical Transactions of the Royal Society London, Series A*, vol. 19, pp. 295–300, 1966.

Han, D. C., Lee, J. S., Analysis of the piston ring lubrication with a new boundary condition, *Tribology International*, vol. 31, No. 12, pp. 753–760, 1998.

Harigaya, Y., Suzuki, M., Takiguchi, M., Analysis of oil film thickness on a piston ring of diesel engine: effect of oil film temperature, *Journal of Engineering for Gas Turbines and Power, ASME*, vol. 125, pp. 596–604, 2003.

Kligerman, Y., Etsion, I., Shinkarenko, A., Improving tribological performance of piston rings by partial surface texturing, *Journal of Tribology, ASME*, vol. 127, pp. 632–639, 2005.

Lotz Felter, C., Numerical simulation of piston ring lubrication, *Tribology International*, vol. 41, No. 9–10, pp. 914–919, 2008.

Ma, M. T., Smith, E. H., Implementation of an algorithm to model the starved lubrication of a piston ring in distorted bores: prediction of oil flow and onset of gas blow-by. *Proceedings of the Institution of Mechanical Engineers, Part J*, vol. 210, pp. 29–44, 1996.

Ma, M. T., Smith, E. H., Sherrington, I., Analysis of lubrication and friction for a complete piston-ring pack with an improved oil availability model – Part 2: Circumferentially variable film. *Proceedings of the Institution of Mechanical Engineers, Part J, Journal of Engineering Tribology*, vol. 211, pp. 17–27, 1997.

McGeehan, J. A., *Literature Review of the Effects of Piston Ring Friction and Lubricating Oil Viscosity on Fuel Economy*, SAE Paper 780673, 1978.

Nakayama, T., Seki, M., Someya, T. T., Furuhama, S. *Effect of Oil Ring Geometry on Oil Film Thickness in the Circumferential Direction of the Cylinder*, SAE Technical Paper Series 982578, 1998.

Ockvirk, F. W., Dubois, G. B., *Analytical Derivation and Experimental Evaluation of Short Bearing Approximation for Full Journal Bearings*, NACA rep. no. 1157, 1953.

Patir, N., Cheng, H. S., Application of average flow model to lubrication between rough sliding surfaces, *ASME Journal of Lubrication Technology*, vol. 101, pp. 220–230, 1979.

Perera, M. S. M., Theodossiades, S., Rahnejat, H., A multi-physics multi-scale approach in engine design analysis, *Proceedings of the Institution of Mechanical Engineers Part K, Journal of Multi-body Dynamics*, vol. 221, no. 3, pp. 335–348, 2007.

Piao, Y., Gulwadi, S. D., Numerical investigation of the effects of axial cylinder bore profiles on piston ring radial dynamics, *J. Engineering for Gas Turbines and Power*, vol. 125, pp. 1081–1089, 2003.

Priest, M., Taylor, M., Automobile engine tribology – approaching the surface, *Wear*, vol. 241(2), pp. 193–203, 2000.

Qingmin Y., Keith T. G., An elastohydrodynamic cavitation algorithm for piston ring lubrication, *International J. Multiphase Flow*, vol 22, supplement 1, pp. 150–159, 1996.

Rabuté, R., Tian, T., Challenges involved in piston top ring designs for modern SI engines, *J. Engineering for Gas Turbines and Power*, vol. 123, pp. 448–459, 2001.

Ramsbotton, J., On the construction of packed rings for pistons, *Proc. Inst. Mech. Engrs.*, vol. 6, pp. 206–208, 1855.

Richardson, D. E., Borman, G. L., *Theoretical and Experimental Investigation of Oil Films for Application to Piston Ring Lubrication*, Society of Automotive Engineers, Inc., SAE Paper 922341, 1992.

Ronen, A., Kligerman, Y., Etsion, I., Different approaches for analysis of friction in surface textured reciprocating components, *Proceedings of the 2nd World Tribology Congress, WTC 2001*, Vienna, 2001.

Ruddy, B. L., Dowson, D., Economou, P. N., Baker, A. J. S. Piston ring lubrication. Part III: the influence of ring dynamics and ring twist, in *Energy Conservation, Through Fluid Film Lubrication Technology: Frontiers in Research and Design*, Winter Annual Meeting of ASME, pp. 191, 215, 1979.

Ruddy, B. L., Dowson, D., Economou, P. N., A review of studies of piston ring lubrication. *Proceedings of 9th Leeds–Lyon Symposium on Tribology: Tribology of Reciprocating Engines*, Paper V(i), pp. 109–121, 1982.

Seki, T., Nakayama, K., Yamada, T., Yoshida, A., Takiguchi, M., A study on variation in oil film thickness of a piston ring package: variation of oil film thickness in piston sliding direction, *JSAE Review*, vol. 21(3), pp. 315–320, 2000.

Senatore, A., Ciortan, S., Oil ring friction loss simulation considering the mixed lubrication regime, *Proceedings of the 12th Conference on Elastohydrodynamic Lubrication and Traction, VAREHD 12*, Suceava (RO), 8–9 October 2004.

Spearot, J. A., Friction, wear, health, and environmental impacts – tribology in the new millennium, keynote lecture at the *STLE Annual Meeting*, Nashville, Tennessee, May 2000.

Taraza, D., Heinen, N. A., Stabley, R., Teodorescu, M., Friction in multicylinder engines, *Arc Annual Meeting*, University of Michigan, 3–4 June 1997.

Takiguchi, M., Ando, H., Takimoto, T., Uratsuka, A., Characteristics of friction and lubrication of two-ring piston, *JSAE Review*, vol. 17, pp. 11–16, 1996.

Taylor, C. M., Automobile engine tribology – design considerations for efficiency and durability, *Wear*, vol. 221(1), pp. 1–8, 1998.

Taylor, R. I., Engine friction: the influence of lubricant rheology. *Proceedings of the Institution of Mechanical Engineers Part J, Journal of Engineering Tribology*, vol. 211(3), pp. 235–246, 1997.

Tian, T., Wong, V. W., Heywood, J. B., *A Piston Ring-Pack Film Thickness and Friction Model for Multigrade Oils and Rough Surfaces*, SAE Technical Paper Series 962032, 1996.

Tian, T., Dynamic behaviours of piston rings and their practical impact. Part 1: ring flutter and ring collapse and their effects on gas flow and oil transport, *Proceedings of the Institution of Mechanical Engineers, Part J: J Engineering Tribology*, vol. 216, pp. 209–228, 2002a.

Tian, T., Dynamic behaviours of piston rings and their practical impact. Part 2: oil transport, friction and wear of ring/ liner interface and the effects of piston and ring dynamics, *Proceedings of the Institution of Mechanical Engineers, Part J: J Engineering Tribology*, vol. 216, pp. 229–248, 2002b.

Tung, S. C., McMillan, M. L., Automotive tribology overview of current advances and challenges for the future, *Tribology International*, vol. 37, 517–536, 2004.

Wakuri, Y., Soejima, M., Ejima, Y., Hamatake, T., Kitahara, T., *Studies on Friction Characteristics of Reciprocating Engines.*, Society of Automotive Engineers, Inc., SAE Technical Paper Series 952471, 1995.

Wu, C., Zheng, L., An average Reynolds equation for partial film lubrication with a contact factor, *Journal of Tribology, ASME*, vol. 111, pp. 188–191, 1989.

Yun, J. E., Chung, Y., Chun, S. M., Lee, K. Y., *An Application of Simplified Average Reynolds Equation for Mixed Lubrication Analysis of Piston Ring Assembly in an Internal Combustion Engine*, SAE paper no. 952562, SAE, 1995.

10.15 Notation

A_{Ci}	asperity contact area
A_H	effective area for the hydrodynamic film
B	piston bore
C	oil specific heat
D_{fp}	ring free diameter
E	Young modulus
e_t	piston top displacement
$F_{c,\mathrm{asp}}$	radial force due to asperity contact
$F_{c,\mathrm{oil}}$	radial force due to hydrodynamic pressure
F_f	friction force
$F_{f,\mathrm{asp}}$	friction force due to asperity contact
$F_{f,\mathrm{oil}}$	friction force due to oil film
F_n	asperity contact function
$F_{R,g}$	radial force due to gas pressure (groove side)
$F_{R,t}$	radial ring pre-tension
h	oil film thickness
h_{fp}	ring position with no radial tension
h_0	minimum oil film thickness
$\bar{h}_T$	average ring-liner gap
K_r	ring stiffness constant
L	ring half-length
N_a	number of contact asperity
p_g	gas pressure acting on the ring back-side
p_t	pressure due to the ring radial pre-tension
p	oil pressure
p_1	pressure at ring inlet side
p_2	pressure at ring outlet side
P_{AC}	asperity contact force per unit area
p_w	pressure at $x = w_e$
T	temperature
t	time
u	axial direction oil velocity
U	piston speed
v	radial direction oil velocity
x	axial co-ordinate

α	circumferential co-ordinate
β	mean asperity curvature
δ	combined roughness
φ	standard probability density function
κ	oil film thermal conductivity
ϕ	viscous dissipation in the oil film
ϕ_c	contact factor
ϕ_x, ϕ_y	pressure flow factors
ϕ_s	shear stress factor
η	asperity density
μ	oil dynamic viscosity
ρ	oil density
w_e	effective length of the ring
σ	average RMS roughness
τ_{Ci}	asperity contact shear stress
τ_H	viscous shear stress

11
Measurement techniques for piston-ring tribology

I. SHERRINGTON, University of Central Lancashire, UK

Abstract: This chapter reviews experimental methods that have been used to study various aspects of the tribology of piston rings in operating internal combustion engines. The discussion covers methods used to determine the thickness of the lubricating film between the ring and the cylinder wall, the axial and angular position of the ring within the ring groove, ring wear, piston assembly friction, ring zone temperatures, lubricant flow through the ring pack, the condition of the lubricant around the ring groove and oil consumption through mechanisms associated with ring pack lubrication.

Key words: piston-ring tribology, piston-ring lubrication, piston-ring wear, oil film thickness measurement, piston-ring friction, ring zone temperatures, ring pack lubricant flow, ring zone lubricant condition, oil consumption.

11.1 Introduction

11.1.1 Chapter structure

This chapter reviews state of the art of methods applied to elucidate the phenomena related to tribology which arise in the operation of piston rings in internal combustion (IC) engines. The author recognises that many investigations on this topic have been conducted through the testing of engine components in bench apparatus and through the use of motored engines. However, these studies are for the most part not considered by this chapter; the focus of the text is firmly upon the measurement of relevant parameters in firing engines.

The chapter includes descriptions of methods used to determine the precise position of the ring, both in relation to the cylinder wall, as measurements of the lubricating film thickness, and as axial and angular positions within the piston-ring groove. Additionally, it also considers methods that have been applied to determine ring wear, piston assembly friction, ring zone temperatures, lubricant flow through the ring pack and the condition of the lubricant around the ring groove. Methods to evaluate oil consumption through routes associated with ring pack lubrication are also considered.

Each of these sections is accompanied by a table of selected references to published work which give details of specific experimental techniques as well as investigations on engines where they have been employed. A final short

387

section presents a list of more general sources where further information on these topics can be obtained.

11.1.2 Functions of piston rings

The dynamic sliding components which form the seal between the piston and cylinder wall of a reciprocating IC engine have a number of important functions and play a critical role in the efficiency and performance of engines. Their principal purpose is to prevent the transfer of combustion chamber gases into the crankcase so that expansion in the combustion process is able to deliver maximum work to the piston. However, piston rings also perform a number of other roles. These include cooling the piston crown by conducting heat away from the piston to the cylinder wall, where it can be transferred out of the engine by the cooling system, and controlling gaseous emission and oil consumption by scraping the oil from the cylinder wall back into the sump.

Normally a piston will have several rings, which are referred to as a *ring pack*. Typically, a modern engine for a passenger or commercial vehicle will use three rings as illustrated in Fig. 11.1. Each ring in the pack is

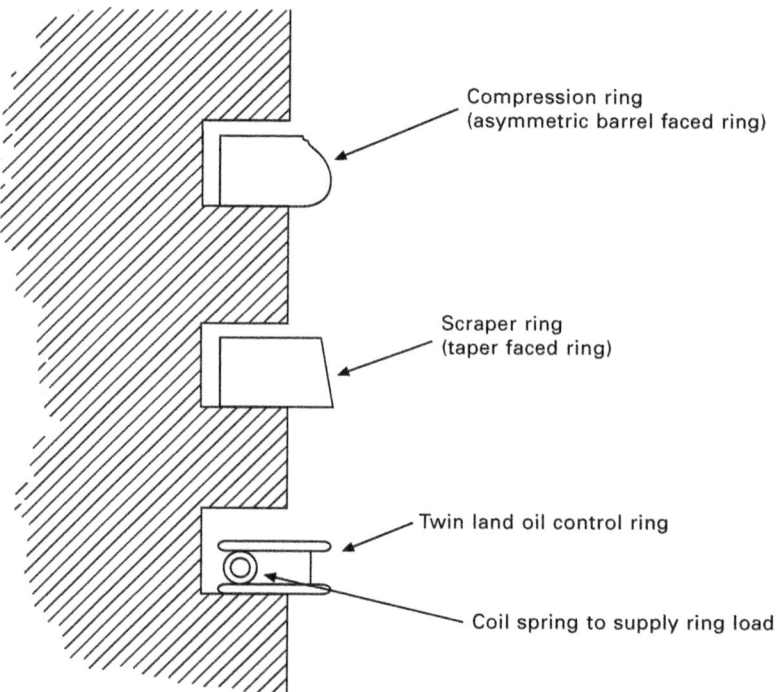

Compression ring
(asymmetric barrel faced ring)

Scraper ring
(taper faced ring)

Twin land oil control ring

Coil spring to supply ring load

11.1 Detail of a typical ring pack for a passenger or commercial vehicle.

designed to perform a preferential function. The main function of the top ring (***compression ring***) is to seal gas pressure. The second ring is sometimes referred to as the second compression ring, but it is also sometimes called the ***scraper ring***. It also functions as a gas seal, but generally additionally performs a significant role in the control of oil flow. The third ring is called the ***oil control ring***. It does not have any gas sealing function, but serves to regulate the flow of oil to the other rings in the ring pack.

Each ring has design elements characteristic of its function. Compression rings are barrel faced and are typically coated with ***wear-resistant coatings*** such as chromium or plasma-sprayed molybdenum. Scraper rings are generally taper faced or stepped and may also be coated with the same materials as compression rings. Oil control rings have two narrow contact rails which are loaded against the cylinder wall by means of a spring which is incorporated into the back of the ring.

Piston rings are located in ring grooves on the piston circumference where they are free to rotate, twist and undergo limited axial movement. The first and second rings are loaded against the cylinder wall by gas pressure acting at the back face of the ring. For the largest part of the stroke this pressure is balanced by hydrodynamic pressure which develops in a lubricant film generated by lubricant entering a convergent gap between the ring face and the cylinder wall as illustrated in Fig. 11.2 (see also Chapter 10). Around the top dead centre (TDC) and bottom dead centre (BDC) of the stroke, where the velocity of the piston is reduced and momentarily zero, this hydrodynamic action is reduced and the lubricating film is thinner, leading to mixed and boundary lubrication regimes in these regions. The variation of the thickness of the lubricant film along the stroke roughly follows the velocity profile of the piston, but has a magnitude which varies with the ring design. Typical variations of oil film thickness in a small diesel engine, as a function of crank angle, are illustrated in Fig. 11.3. This film thickness is of great interest as it is this factor which governs such factors as power loss, lubricant consumption and exhaust emission levels.

11.1.3 Influences on ring pack design

Ring pack designs must meet a number of criteria which in some regard are mutually exclusive. The most obvious of these are that the ring pack must be:

- effective in sealing, minimising blowby effects;
- wear resistant, offering long service;
- efficient in operation, minimising contact friction and lubricant shear effects;
- effective in controlling the flow of lubricant to minimise oil consumption.

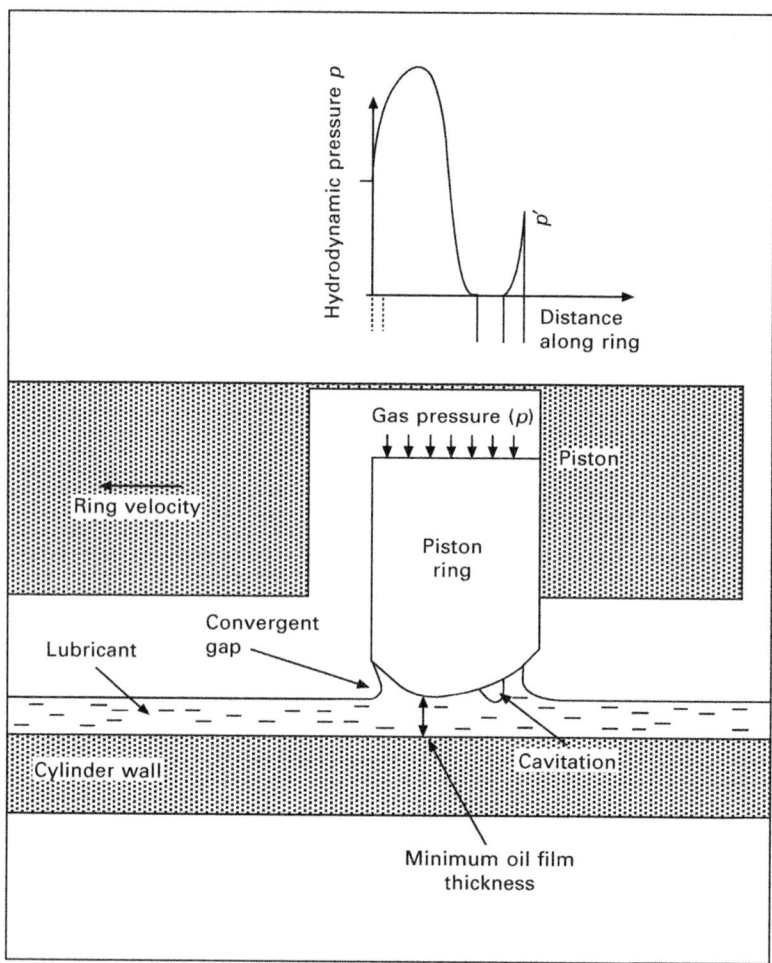

11.2 Hydrodynamic lubrication of a piston ring.

In addition, ring pack design must also respond to the demands of customers and society. To achieve this means they must constantly become longer lasting, more efficient and environmentally acceptable. The technical consequences of this are probably that compression ring designs of the future must operate under conditions of increasingly higher temperatures and pressure with thinner hydrodynamic films. Additionally, oil control rings must regulate a lower delivery rate of lubricant. All this must happen while tribologically useful, but environmentally damaging materials such as chromium and phosphate, are removed from the manufacturing process and lubricating mechanism.

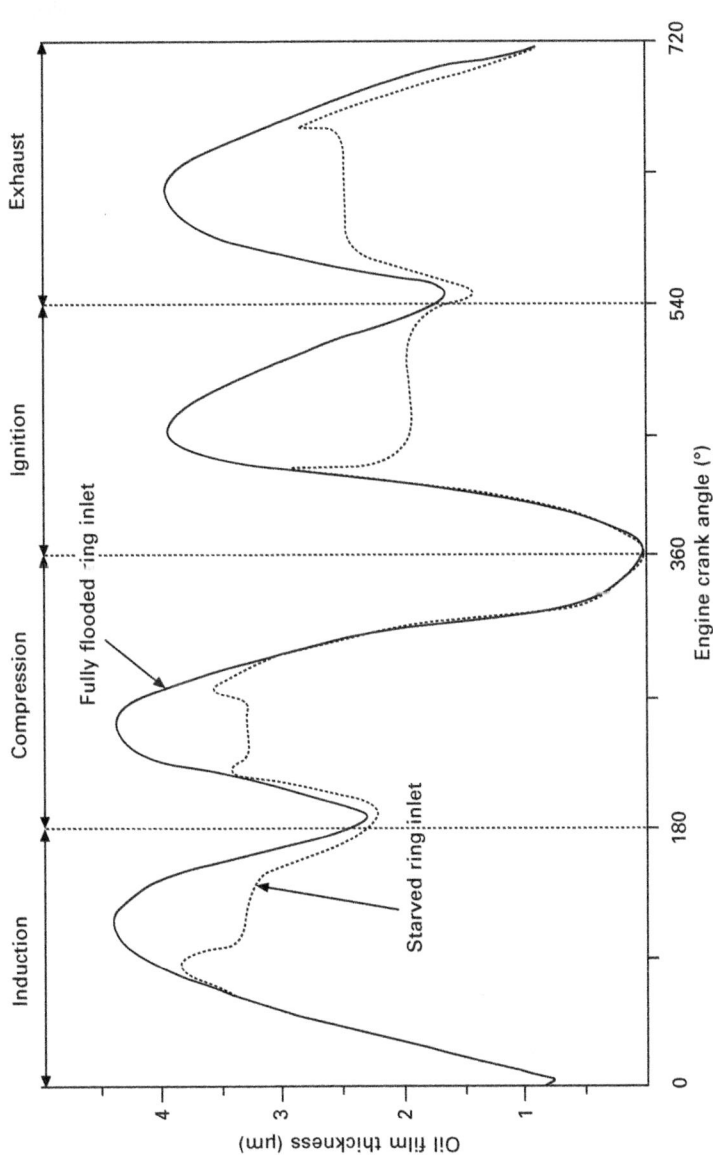

11.3 Schematic representation of typical variations of oil film thickness in a small diesel engine (graph based on theoretical analyses).

11.1.4 Verification of predictive software

In the 1960s investigators began to develop computer-based models of *piston-ring lubrication* based on the assumption that piston rings operated in the hydrodynamic regime over most of the stroke and that the behaviour of the lubricant could be described by the Reynolds equation and a series of appropriate boundary conditions. Over time these computer models have become relatively sophisticated, incorporating a wide range of physical influences, such as ring twist, axial cylinder temperature profile, piston-ring profile, gas pressure flow in the ring pack, cylinder wall surface topography and lubricant flow interactions with the piston skirt. The use of these models has led to significant advances in the understanding of the operation of piston rings and subsequently to improvements in the design of rings packs. However, sophisticated as they are, these models remain incomplete as many important engine-related factors are difficult to incorporate in the models. Consequently, as further developments in computer simulations emerge, it is crucial to compare their predictions with measurements from operating engines. Oil film thickness data provide a convenient means for comparing the reliability of computer models with engine performance. This arises partly because oil film thickness is relatively straightforward to measure and partly because this parameter is a primary influence on other aspects of engine performance such as oil consumption and gaseous emission.

A range of techniques have become well established as methods for measuring the thickness of piston-ring lubricant films. Principally, they fall into three groups: electrical methods, optical methods and acoustic methods. Electrical methods are based on resistance, inductance and capacitance, optical methods are mainly based on the use of laser-induced fluorescence (LIF) and evolving acoustic methods are based on ultrasonic technology.

11.2   Measurement of lubricating film thickness

11.2.1 Capacitance methods

The basic principle of oil film thickness measurement using the *capacitance technique* involves the formation of the equivalent of a miniature parallel plate capacitor between the sensor and the ring, or a ring mounted sensor and the cylinder wall. The sensor normally consists of a small central electrode, typically up to 0.25 mm in diameter which is enclosed by an insulating material and hypodermic tube, which acts as a screen.

Placing the sensor in a piston ring allows a continuous record of oil film thickness to be obtained over the entire engine cycle. However, it also adds considerable complication as the approach requires the ring to be fixed in position by a pin to prevent rotation and a linkage to pass the signal cable from the rapidly moving big-end to the crank case or sump so that it can be routed out of the engine. Additionally, film thickness is only measured on one point on the ring and this can be prone to misinterpretation if appropriate steps are not taken to account for ring tilt. Mounting the sensor in the cylinder wall circumvents all of these problems, but only permits measurements of film thickness at specific crank angles.

Unless measurements are to be made on very large engines, such as marine engines or stationary generators, it is normally necessary for investigators to manufacture custom-made sensors as commercially available devices, with active electrodes in the millimetre range, tend to be too large. To a first approximation the capacitance, C, of an electrode in a capacitance sensor is given by:

$$C = \frac{\varepsilon_0 \varepsilon_r A}{d}$$

where A is the area of the electrode, ε_0 is the permittivity of free space, ε_r is the relative permittivity of the lubricant (typically in the range 1.8 to 2.1), and d is the separation between the faces of the capacitor. For the size of transducers normally used in internal combustion engines for road-going applications, capacitance values are normally in the range of 10^{-15} F. Such small experimental values mean that the use of specialist charge amplifiers and careful attention to the screening of stray capacitance with low noise cable is required in order to obtain reliable film thickness data.

Capacitance sensors offer a relatively non-invasive means of measuring oil film thickness. The sensors themselves are comparatively easy to install and robust enough to withstand the conditions of high temperature and pressure found in the engine cylinder for extended periods, allowing long duration experiments. Additionally, manipulation of the electrical output is convenient and can be readily digitised for real time processing or post-processing in a PC.

Limitations

Limitations arise when using capacitance sensors, but they are, in general, considered to be less constraining than those presented by some other methods of oil film thickness measurement. The most common issues are discussed below.

Periodic electrical contact, i.e. short-circuits, are sometimes observed. When lubricating film thickness is of the same order as the surface roughness, contact can arise between the sensor electrode and the opposing component. This effect is normally most prevalent during the early use of newly installed sensors and tends to reduce with running time as the surface of the devices is 'run-in'. It is generally accepted that this effect can be reduced by coating the face of the electrodes with an insulting coating such as aluminium oxide.

Cavitation is commonly observed at the outlet of piston rings. Where this occurs there will be change in the relative permittivity of the gap medium leading to a change in the apparent shape of the piston-ring profile. Most investigators do not perceive this effect as a problem as the effect is readily identified and, additionally, can even be regarded as helpful as it gives information about lubricant film extent around the ring.

Capacitance-based systems have an inherent response time which is governed by the amplifier electronics. Care needs to be taken to select amplifier electronics which have response times which are considerably shorter than the passing time of the piston ring at the point of measurement if cylinder mounted sensors are adopted. Frequency modulated (FM) amplifier systems have much shorter response times than amplitude modulated (AM) systems. Typically, the minimum response time for an FM system is in the order of about 10 microseconds.

It is often not straightforward to calibrate capacitance systems *in situ* so a representative sample sensor would normally be used for this purpose. Clearly, it is unlikely that the sample sensor will be identical to the other devices it represents. For example, there may be differences in the face area of the active electrode, so errors may arise. Additionally, while it is relatively routine to calibrate the AM-based systems using a device to create accurate static micrometre scale displacements, the calibration of FM systems is more demanding. FM systems only respond to changes in displacement, so it is necessary to generate oscillating gaps on a micrometre scale, this can be achieved through the use of feedback linked piezo-translation devices.

Applications

Capacitance sensors have been applied in engines to measure piston-ring oil film thickness by a number of investigators. Table 11.1 gives details of examples of some investigations that have been reported in the literature.

11.2.2 Inductance methods

The operation of **inductance sensors** depends on the fact that the inductance of a circuit element, typically a wire coil, is dependent on its proximity to magnetic material. Inductance sensors are used with an oscillating voltage supply, by measuring the lag of the current behind voltage signal, the proximity of a target can be determined.

Inductance sensors for oil film thickness (OFT) measurement are normally miniature coils, typically containing only a few windings. Experimental devices may be of self-inductance, mutual inductance or eddy current type. These systems work on similar principles, as described below.

In self-inductance systems a single coil energised by a high frequency

Table 11.1 Examples of the use of capacitance sensors in engines

Authors	Details
Hamilton and Moore (1974)	Employed cylinder-based transducers Multiple ring passes were required to form ring profile
Parker *et al.* (1975)	Used a ring-based transducer
Brown and Hamilton (1977)	Employed cylinder-based transducers Investigated oil starvation
Moore and Hamilton (1980)	Investigated the behaviour of the oil film close to TDC
Furuhama *et al.* (1982)	Adopted ring-based transducer
Shin *et al.* (1983)	Adopted transducer located in ring
Grice *et al.* (1990)	Described the operation of a system to record a piston-ring profile in a single pass
Grice and Sherrington (1993)	Used cylinder-based transducers Investigated cavitation effects
Mattson (1995)	Used cylinder-based transducers Investigated the effect of engine operating conditions
Takaguchi *et al.* (2000)	Adopted a transducer located in ring

signal is employed. The rise and fall of flux through the coil generates a magnetic field which induces a current which opposes the current that excites the coil. The presence of magnetic material close by will modify the flux passing through the coil and change the current in the circuit. In self-inductance coils the self-inductance L, for a coil with N turns carrying a current I with flux linkage Φ is defined by:

$$L = \frac{N\Phi}{I}$$

If the coil is excited by an alternating signal E_t, the self-induced voltage is given by:

$$E_t = -L\frac{dI}{dt}$$

Consequently, if the electromotive force (emf) in the coil is calibrated as the displacement of a nearby magnetic target is adjusted, the system can be used as a proximity sensor because the target will modify the flux linkage as its position is adjusted.

In mutual inductance systems two coils are employed. The rise and collapse of the magnetic field in the primary coil generate an induced field in the secondary coil. If magnetic material is placed in the vicinity of the magnetic field generated by the primary coil the flux linkage is modified changing the magnitude of the induced signal. The mutual inductance, M, of the system is defined by:

$$M = \frac{N_s \Phi_s}{I_p}$$

where I_p is the current in the primary coil, N_s is the number of turns on the secondary coil and Φ_s is the flux linking the secondary coil. The induced emf E_t, given by:

$$E_t = -M \frac{\mathrm{d}I_p}{\mathrm{d}t}$$

can then be used to obtain the displacement of a local target when the change in E as a function of distance is known from calibration.

In eddy current systems a permanent magnet or electromagnet is used to generate a static field which becomes disturbed by the magnetic field generated by eddy currents which develop in a moving magnetic target such as a piston ring. These field perturbations are in turn detected by the multi-turn coil which forms the sensor. Again since the induced voltage E_t is given by:

$$E_t = N \frac{\mathrm{d}\Phi}{\mathrm{d}t}$$

Calibration of the induced signal as a function of the displacement of a target can be used as a sensitive means to determine displacement of the target in an experimental arrangement.

In order to convey a satisfactorily small lateral resolution to observe the passage of a ring if the transducer is mounted in the cylinder wall, *inductance sensors* of small dimensions are required and, like capacitance sensors, they are normally custom made by investigators. Transducers of this type are also sometimes fitted into piston rings, which are generally pegged to prevent rotation and breakage of the transducer connections, to monitor oil film thickness over a complete engine cycle. Miniature devices are also essential for this type of application to allow installation in the ring. In recent years the advent of 'off the shelf' modelling software for electromagnetic systems has greatly aided the design of these highly miniaturised systems, eliminating the need for large numbers of trial and error designs to achieve satisfactory performance.

Limitations

The output of signal amplifiers when used in association with inductive devices is non-linear. Conveniently, as the greatest sensitivity occurs at the lowest film thicknesses, they can be valuable in some instances for measuring small changes when films are thin. Unfortunately, inductive devices have found limited use in oil film thickness measurement applications as they exhibit significant sensitivity both to temperature variations and to the material type of the target. Both of these effects need to be considered when making measurements. Clearly, significant temperature variations occur in IC engines and the need to obtain an assessment of the temperature influence for a given sensor and then apply appropriate compensation to measurement data is a disadvantage. Additionally, the performance of the sensors is strongly influenced by the magnetic properties of the target and, to some extent, neighbouring material. This means that calibration should ideally be done with the actual target. However, the challenge of accounting for the neighbouring material is more difficult to address.

Applications

Inductive sensors have been applied in engines to measure piston-ring oil film thickness by a number of investigators. Table 11.2 gives details of some investigations that have been reported in the literature.

11.2.3 Resistive methods

Measuring the ***resistance of an oil film*** as its thickness changes over the stroke appears to be an obvious and convenient approach to assessing oil film thickness. It is not surprising, therefore, that it was the first technique to be applied to attempt to measure this parameter in the 1940s.

The general approach involves electrically insulating a piston ring, or a ring segment and connecting a small DC voltage and a simple resistance bridge circuit to detect small changes in resistance in large overall resistance values. Owing to requirements related to supply voltage, voltage changes for complete oil films are normally in the milli-volt range. Calibration data are required to infer film thickness. Use of a complete piston ring or ring segment

Table 11.2 Examples of the use of inductive sensors in engines

Authors	Details
Wing and Saunders (1972)	Used two self-inductance transducers placed at opposite sides of the piston, behind the piston ring, to infer film thickness
Lewis (1974)	Developed a mutual inductance transducer with improved thermal stability, but did not deploy it in an engine
Dow *et al.* (1983)	Mounted an inductance transducer in the end face of a piston ring and used variations in gap size to infer average film thickness
Tamminen *et al.* (2006)	Used inductance transducers mounted in the piston, behind the rings, study the impact of engine operating conditions

gives allows the measurement of average film thickness values. However, a recent trend to develop **resistance transducers** with small electrodes has meant that they can also be developed as devices with relatively small lateral resolution for spot film thickness measurements.

Limitations

Early approaches to measuring oil film thickness using resistance techniques tended to be limited in success, because the output from such systems was dominated by electrical breakdown, probably arising as a result of many intermittent contacts between asperities on the opposing components. Electrical measurements are also influenced by the formation of deposits on the surfaces of the ring or cylinder. Deposits may arise due to beneficial **tribo-chemical action**, e.g. the formation of anti-wear layers, or the formation of soot layers arising as a consequence of the combustion process.

Applications

The use of resistance techniques to measure piston-ring oil film thickness has only been reported by a small number of investigators, reflecting the limitation that contact breakdown imposes on the ability of this technique to yield useful information. The use of resistance sensors may change this in the future as they seem to offer a route forward to reducing breakdown by offering a smaller and position controllable area than a ring or ring segment. Table 11.3 gives details of examples of some resistance measurement investigations on engines that have been reported in the literature.

11.2.4 Ultrasound methods

Acoustic methods (**ultrasonic sensors**) are the most recent technology to be developed as a tool for measuring oil film thickness. To date they have not been applied to measure this parameter in firing IC engines. Fundamentally, the

Table 11.3 Examples of the use of resistance methods in engines

Authors	Details
Courtney-Pratt and Tudor (1946)	Adopted insulated ring, noted multiple short circuits throughout the stroke
Furuhama and Sumi (1961)	Employed a piston ring pinned at one end with the free end connected to the anode of a valve to measure average film thickness
Saad *et al.* (2007)	Measured changes in voltage between the piston ring and two electrically insulated electrodes located in the cylinder wall

process involves determining the reflection coefficient for an ultrasonic pulse from a lubricating layer in a manner similar to that used in non-destructive testing. Several types of pulse mode can be used to obtain data about the OFT. Additionally, this technique can also be applied to measure the viscosity characteristics of the film within the contact, offering the promise of being able to obtain more accurate data about the lubricating film directly.

Limitations

Ultrasound measurements offer a minimally invasive approach to the measurement of the OFT which is of use in many applications. However, in investigations conducted to date the rate at which film thickness measurements have been made has been slow, i.e. the time between collecting one data point and the next has been relatively long. Consequently, one of the challenges currently being addressed is to increase the repetition rate for measurement so that sufficient data points can be collected across the face of a piston ring to allow point to point interpolation to form a reasonable record of the face profile and an accurate minimum film thickness assessment.

Applications

The **ultrasound technique** has been applied to measure the thickness of the oil film between a piston skirt and a cylinder wall. However, to date there have been no published reports of its use to measure the OFT between piston rings and cylinder walls.

11.2.5   LIF

In the mid-1970s it was found that the fluorescence of oils in ultra-violet (UV) light could be used to estimate the thickness of lubricant films on open surfaces. Illuminating a thin film of lubricant with light causes molecules to absorb energy. To return to their original state the molecules will re-emit this energy as light of a different, specific wavelength. It can be shown that the intensity of the re-emitted light I_f, is dependent on the thickness h, of the fluorescent layer according to:

$$I_f = I_o \Phi_q (1 - 10^{-\varepsilon \chi h})$$

where I_o is the intensity of the incident light, ε is the **molar absorptivity** and χ is the molar concentration of the fluorescing molecules. In the 1980s optical arrangements were devised to exploit this principle in bench equipment and transparent engines to measure oil film thickness on sliders and ring packs. By the 1990s methods had been developed using various types of combustion cylinder window or optical fibres to gain optical access to the sliding ring/cylinder wall interface, so that the method could be applied to firing engines. Additionally, by this time, it had been established that UV light was not essential for the fluorescence process and lasers, for example helium–cadmium lasers emitting blue light, had become standard as a more convenient replacement for UV light sources.

LIF is considered by some investigators to be an awkward method to apply to measure OFT (see Limitations below). However, it does have considerable advantages as a tool for aiding the visualisation of lubricant around the rings and in other parts of the piston assembly. The technique readily adapts to applications which involve mapping of the distribution of lubricant around the ring pack and is able to estimate the thickness of lubricant films outside as well as inside contacts.

Limitations

LIF suffers from several drawbacks. It requires a significant amount of specialist equipment and operator knowledge and although it may be possible to simplify the experimental arrangement, it normally consists of a moderately complex optical arrangement involving a laser. Some investigators also advocate the use of a dye in the lubricant to enhance the fluorescent properties of the oil.

The output of these systems is non-linear and somewhat temperature dependent, so temperature compensation is essential. In addition, the target oil/ dyes suffer from an effect known as ***photo-bleaching*** in which the fluorescent effect gradually becomes less sensitive to the illumination. This is a particular problem during static gap calibration processes as it makes it necessary to ensure a constant flow of fresh oil in the calibration gap during the process.

The *fluorescent response* of the oil also depends on the condition of the oil and changes as the oil deteriorates through oxidation and the incorporation of combustion products, etc. To get around these and other problems, investigators sometimes resort to various types of *in situ* calibration, for example the use of shallow grooves of known depth somewhere close to the target on the piston skirt. Measurement of the OFT by a more established secondary transducer, such as a capacitance transducer, has also been employed.

Finally, the method is also sensitive to the reflectivity of the target. This is a difficult problem to overcome and some investigators try to resolve it simply by making the reflectivity of the calibration target as similar to that of the real target as possible.

Applications

LIF was first reported as a technique which had potential for measurement of the OFT around piston rings in 1980 when it was used in association with a transparent cylinder engine at Ford Research. Since then it has been widely applied in piston-ring lubrication studies over the last decade and a half as can be seen from the examples listed in Table 11.4. Although it is difficult to calibrate to obtain reliable measurements of the OFT, it has the advantage that it is able to give significant information about oil film extent and the thickness of partial oil layers in an interface.

11.3   Measurement of piston-ring friction

The piston assembly contributes the biggest proportion of the frictional losses in an IC engine (also see Chapters 10 and 15). It is, therefore, a focus of attention for many investigations and development projects that seek to reduce frictional losses in engines.

Frictional losses in an engine are presented in several differing contexts and this can cause confusion about perceptions of their magnitude and subsequent impact on engine performance. Table 11.5 lists two examples of piston assembly friction losses presented in different terms. Friction losses

Table 11.4 Examples of the use of the LIF technique in engines

Authors	Details
Hoult *et al.* (1988)	Described a calibration method for LIF
Richardson and Borman (1991)	Used a fibre optic based method to covey light to the engine
Shaw *et al.* (1992)	Demonstrated improved signal to noise ratio (SNR) and vibration resistance
Phen *et al.* (1993)	Employed multi-mode optical fibres
Sanda *et al.* (1993)	Developed scanning LIF and applied it in a firing engine to measure OFT throughout the engine stroke
Stiyer and Ghandi (1997)	Described the use of LIF in an engine running on propane
Yoshida *et al.* (2000)	Investigated oil control ring effect on oil consumption
Weimar and Spicher (2003)	Investigated HC emissions and OFT simultaneously
Przesmitzki and Tian (2007)	Studied oil transport through the ring pack

Table 11.5 Piston-ring and skirt losses

Investigator	Type of evaluation	Parameter assessed	Piston and piston skirt losses
Pinkus and Wilcock (1977)	Environmental protection agency (EPA) driving cycle	% of energy input	3%
Ryk *et al.* (2005)	Result of informal review of published results	% of friction losses	Approximately 50%

may also be expressed in the context of percentage of output power and percentage of mechanical output.

To develop a fundamental understanding of the mechanism of piston assembly friction, it is necessary to measure instantaneous friction over the complete engine cycle. A schematic representation of a typical instantaneous piston assembly friction curve is given in Fig. 11.4. It can be seen that the highest friction forces occur around the dead centres where piston velocity is at its lowest and mixed or boundary friction dominates. Elsewhere in the cycle, friction forces are dominated by shear in the hydrodynamic film, but do not necessarily remain proportional to piston velocity due to starvation effects. It should also be noted that total power loss is little affected by the large friction spikes around the dead centres as the piston velocity in this region is at, or close to, zero. Piston assembly power loss is, therefore, largely a consequence of lubricant shear.

Measuring instantaneous piston assembly friction tends to require more invasive engine modification that is required for measurements of lubricant film thickness. Essentially, there are only two forms of engine modification that have been used for formal motored and firing tests on engines. These are the *floating liner* **method** and the ***instantaneous mean effective pressure (IMEP) method***.

11.3.1 Floating liner method

The *floating liner* **method** of measuring piston assembly friction involves separating the cylinder liner from an engine by machining small clearances between the cylinder and crankcase and supporting the cylinder on elastic or low friction constraints so that it can move freely in the axial direction, but is constrained in the radial direction. This allows the friction force due to the piston skirt and ring pack to act through the cylinder and be measured by *force transducers*. The force transducers are typically located at the base of the liner to avoid effects due to the high-temperature area around the combustion chamber. To allow free motion the cylinder cannot be in contact with the cylinder head in the normal way in an arrangement of this type. Consequently, either the upper open surface of the cylinder needs to be isolated from the combustion pressure without impeding the liner motion, or the combustion chamber should be sealed by a device termed a 'force balancing seal'.

Isolation from combustion pressure

One approach to isolating combustion the pressure which has been adopted by a number of investigators is to incorporate two hydrostatic bearings to support the cylinder. In this arrangement, illustrated in Fig. 11.5, one bearing

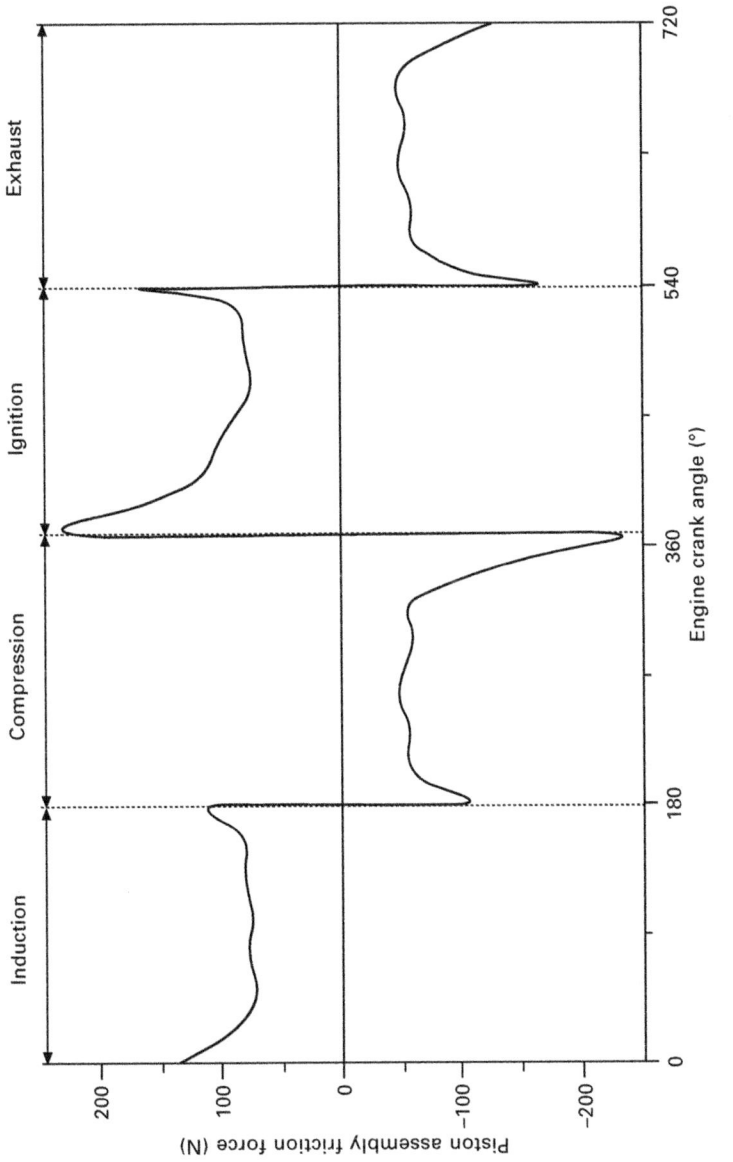

11.4 Schematic representation of instantaneous piston assembly friction (graph based on experimental data).

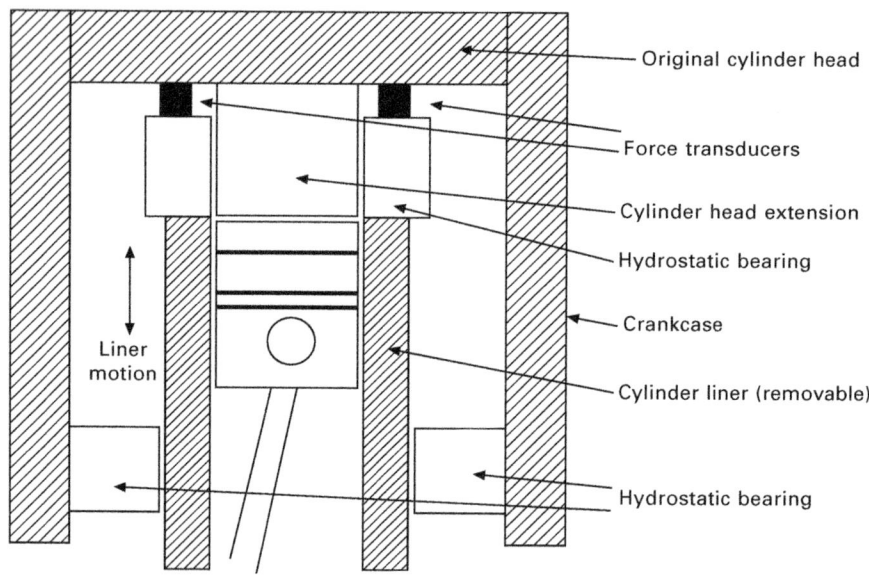

11.5 Floating liner arrangement with hydrostatic bearings.

is located at the base on the outside of the liner. The other is located at the top of the cylinder and supports a liner extension from the inside. This arrangement allows axial motion, provides radial rigidity and forms a seal which prevents gas pressure from acting on the top face of the cylinder. However, the drawbacks to this arrangement are that extensive engine modification is required and, during operation, oil flows from the upper hydrostatic bearing into the combustion chamber modifying the normal lubrication conditions.

Force balancing seal

An alternative to the use of a hydrostatic bearing is to use a force balancing seal, as illustrated in Fig. 11.6. In principle, a force balancing seal allows the gas force acting on the top of liner to be balanced by gas forces acting on other parts of the arrangement and the small forces due to any deflection. It also provides radial rigidity at the top end of the cylinder and flexes to allow axial motion to transfers friction forces to transducers located at the base of the cylinder. A number of variants of this type of approach have been used. However, they divide into two types, those based on the use of O-ring seals and those based on the use of customised annular rings.

The use of O-ring seals tends to lead to a fairly complex arrangement requiring separate components to control force balancing and the radial displacement of the liner. Annular rings are able to provide both functions

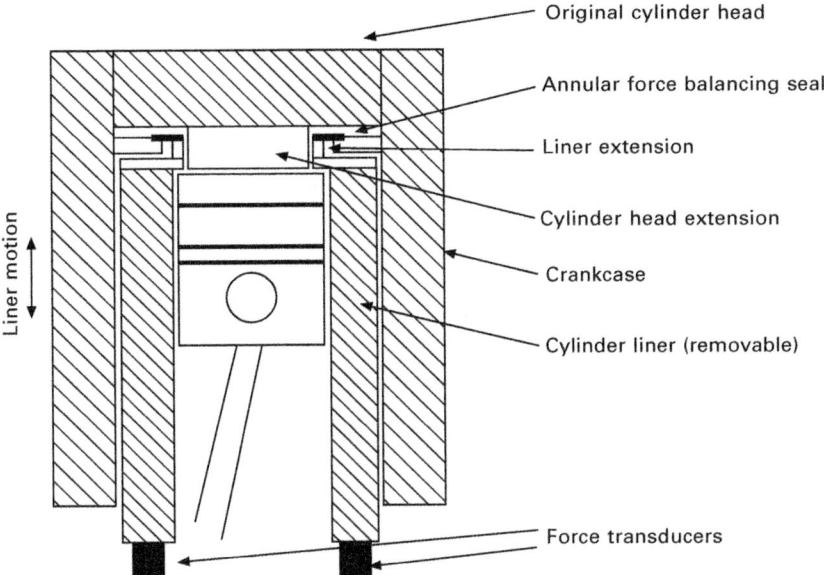

Liner motion

Original cylinder head

Annular force balancing seal

Liner extension

Cylinder head extension

Crankcase

Cylinder liner (removable)

Force transducers

11.6 Floating liner arrangement with force balancing seal.

in a single component. The design process for a seal of this type is described by Clarke *et al.* (1990).

Limitations

Floating liner measurement systems are not straightforward. They all require fairly extensive engine modifications and most appear to suffer from vibration instability when operating at medium to high engine speeds. It is also not possible to separate the contributions to friction force from the different piston components of ring pack and skirt, only overall friction forces can be measured. This is a frustrating feature for those who try to assess the impact of changes in the design of individual rings in a ring pack.

Applications

A number of investigators have conducted investigations into piston-assembly friction using floating liners. Examples of the use of the three main types of floating liner arrangement are described in the publications listed in Table 11.6.

11.3.2 IMEP method

The *IMEP* method of assessing piston friction is an indirect technique. It involves making very accurate measurements of instantaneous connecting rod

Table 11.6 Examples of the use of floating liner arrangements

Investigators	Details
Furuhama and Takiguchi (1979)	Employed an O-ring seal system
Mollenhauer and Bruchner (1980)	Adopted a hydrostatic bearing system
Clarke *et al.* (1990)	Adopted a steel annulus and described the design process for the force balancing arrangement

Table 11.7 Examples of the use of the IMEP method

Investigators	Comments
Uras and Patterson (1983)	Described a pioneering implementation of the IMEP method
Mufti and Priest (2005)	Applied the IMEP method to study the impact of engine operating conditions and lubricant formulation

forces, cylinder pressure, engine speed and crank position. This information, along with details of the piston acceleration, allows the theoretical forces expected in the connecting rod to be calculated and compared with the measured force. The difference in these values is ascribed to friction.

Limitations

The principal drawback with this problem is that it involves subtracting two large forces to obtain a small difference which is the friction force. Small errors in the measured or calculated connecting rod forces can lead to significant errors in the friction estimate. However, although the system requires a linkage to route signal cables away from the connecting rod, the engine itself remains in a condition which is highly representative or the lubricating conditions of an unmodified engine.

Applications

Relatively few investigators have applied this technique to investigate piston assembly friction. However, two examples are described in the papers listed in Table 11.7.

11.3.3 Other methods

Several approaches are used to assess piston assembly friction in a comparative way for engine acceptance and development testing. These include the **Morse test**, the **Wilans line test**, dynamometer testing and the deceleration test for example as described by Greene and Lucas (1969). However, these approaches are not generally used for detailed assessment of piston-ring and piston assembly friction for fundamental research purposes.

11.4   Measurement of piston-ring movement

Piston rings are subjected to a range of forces during their operation. These forces generate a range of ring motion including ring twist, axial motion and rotation. These motions influence the performance of the ring. ***Ring twist*** is a particularly strong influence on the wear of both the ring face profile and the ring groove. Wear of this type is critical to the operation of the engine as it changes the characteristics of the oil film-forming capability of the ring which affects liner wear as well as changing the oil flow characteristics of the engine incrementally during its life. Ring motion is influenced by engine operating conditions and by detrimental influences on the engine such as the formation of soot deposits in the rings grooves. ***Ring motion*** is, therefore, of considerable interest to tribologists and engine designers. A range of methods have been applied to investigate ring motion, but electrical methods and radioactive tracer-based methods are the most widely used.

11.4.1 Electrical methods

The same methods which are employed to investigate piston-ring oil film thickness can, in principle, also be used to investigate piston-ring motion as both measurements are fundamentally similar. They both involve the measurement of displacements, and capacitance, inductive and resistive methods can all be applied. Electrical methods tend to be used to investigate the axial motion of rings, rather than ring rotation. Typical arrangements involve the placement of sensors at several points above and/or below the top and bottom of the ring groove to detect the location of the upper or lower face of the ring. Several sensors are required because it is typical for rings to be differentially located on the upper and lower parts of the grooves due to piston tilt and other influences.

Limitations

Sensors located in the piston all require an arrangement to transfer data signals from the reciprocating elements of the engine out through the crankcase. This is most commonly achieved through either a mechanical linkage or a radio transmitter system. The design of mechanical linkages is considered a specialist skill. They must be durable to allow their extended use and provide an appropriate routing for data transfer cables. The most common form of failure for these systems is fatigue in the data transfer cables. This problem can be reduced by placing the cables into torsion, rather than adopting plain bending configurations. Transfer of data can also be achieved through use of a radio transmitter system located in the piston/connecting rod arrangement.

In addition to these methods, approaches involving the use of on-board storage of data in memory devices within the piston are now feasible. Following experiments collected data can be downloaded to a laptop or PC.

Both of the latter approaches negate the need for a transmitted or a linkage, but are more difficult to apply with sensors that require special forms of signal processing, such as capacitance sensors, as commercial amplifier systems tend to be large with inbuilt 240 V mains power supplies as they are designed mainly for bench-top use.

Applications

Investigations of piston-ring motion using electrical systems have been conducted by many investigators. Often studies of this type also involve the measurement of other parameters, such as piston-ring OFT, in order to investigate related effects. Some examples are listed in Table 11.8.

11.4.2 Radioactive tracers

Relatively few studies have been conducted to measure ring rotation. In contrast to the measurement of axial motion, which tends to favour the use of electrical sensing, a method involving the use of *radioactive tracers* has been successfully adopted to measure piston-ring rotation. The method involves incorporating two different radioactive tracers into the ring to be investigated. The ring angular position of the piston ring can then be determined by measuring the emission rates of the tracer isotopes using a gamma ray detector mounted outside the engine during operation of the engine.

Limitations

Radioactive tracer experiments provide an accurate and *non-invasive* method for measuring the rotation of piston rings. However, in line with the handling

Table 11.8 Examples of the electrical measurements of piston-ring motion

Investigator	Comments
Furuhama and Hiruma (1972)	Used capacitance-based sensors. Also measured combustion chamber pressure and inter-ring gas pressure
Furuhama *et al.* (1979)	Used capacitance-based sensors. Also considered piston-ring OFT and lubricant consumption
Taylor and Evans (2004)	Used eddy current sensors to measure ring lift. Also measured piston tilt and ring OFT

of all radioactive materials special precautions are required in order to handle, prepare and dispose of test specimens. Other associated engine parts used in the test, lubricants and gaseous emissions from the engine also need to be monitored and managed to maintain health and safety requirements.

Applications

Only a few detailed studies of methods for measuring instantaneous ring rotation have been published. Some examples are listed in Table 11.9.

11.4.3 Other methods

Engines with quartz windows to permit optical access are widely used in combustion studies. In principle, it would be possible to arrange quartz windows in engines to permit the study of piston-ring movement by optical means. However, the author has not been able to trace published reports of studies of this type.

11.5   Measurement of piston-ring wear

The characteristics of an IC engine change over time due to wear of engine components. This is particularly true for the piston rings: wear of the ring profile will influence the frictional power loss of the ring pack as well as the emission characteristics of the engine since changes in the ring profile will affect the generation of hydrodynamic films and the oil transport characteristics of the pack. Piston rings are subjected to a range of lubrication regimes during a single stroke of operation. However, the majority of wear of the ring profile occurs at the top and bottom dead centres where the hydrodynamic film tends to break down, leading to mixed and boundary lubrication.

11.5.1 Laboratory methods

Piston-ring wear is most commonly studied through a process of intermittent measurement of the ring profile shape. Normally, this involves measurement of

Table 11.9 Examples of the radioactive measurements of piston-ring motion

Investigator	Comment
Schneider and Blossfeld (1990)	First published description of tracer method
Schneider *et al.* (1993)	Investigated the effect of out of roundness on ring rotation and oil consumption
Kim *et al.* (2000)	Investigated the impact of relative positions of ring gaps on oil consumption

the initial (new) ring profile followed by a series of measurements conducted during disassembly of the engine after significant periods of operation. Measurements of the ring profiles can be made routinely using either optical or stylus-based surface roughness measurement instruments while the ring is held vertically in a suitable clamp. In doing this it is valuable to record or mark the measurement positions so that subsequent measurements can be made at the same peripheral location on the ring and compared.

Limitations

Intermittent measurement of the wear of ring profiles requires regular disassembly and rebuilding of the test engine, which is a time-consuming process and needs to be done often to form a detailed picture of the changes in the ring profile over a period of operation. The method also requires long engine test times, which can be inconvenient and costly. However, the process is technically straightforward and falls within the capability of most university and research departments.

Applications

Laboratory measurements of piston-ring profiles are relatively commonplace and can be found in many publications. Some examples of work of this type are described in the publications listed in Table 11.10.

11.5.2 On-line methods

Methods based on the use of *radioactive tracers* have been developed for non-invasive instantaneous measurement of *piston-ring wear*. (The method can also be used to simultaneously measure cylinder wall wear.) The technique is highly sensitive and permits rapid measurement of wear in specific engine and material arrangements. The principle involves irradiating piston rings using a neutron source to generate short-lived radio-isotopes in thin surface layers prior to the installation of the rings in the engine. During engine operation the oil lubricating the rings accumulates radioactive wear debris. By routing the oil circuit so that it passes an X-ray detector levels of radiation

Table 11.10 Examples of laboratory measurement of piston-ring profiles

Investigators	Comments
Schuster *et al.* (1999)	Compared wear rates of piston rings with a range of surface coatings
Priest *et al.* (1999)	Measured piston-ring profiles in order to evaluate a predictive model for piston-ring wear

can be detected and decay rate corrected intensity values can be related to the volume of wear debris present in the lubricant.

Limitations

As with all experiments involving the use of radioactive materials, special precautions are required in order to handle, prepare and dispose of the test specimens. Other engine parts used in the test, as well as lubricants and gaseous emissions from the engine also need to be monitored and properly managed to maintain effective health and safety arrangements.

Applications

Thin layer activation is broadly speaking a well-established experimental technique. However, it remains relatively little used in studies of wear in IC engines. Some investigations relating to the use of this method in assessing the wear of piston rings are detailed in Table 11.11.

11.6   Measurement of ring zone temperature

Measurements of *ring zone temperature* are rarely made as a specific investigation. However, it is relatively common for experimentalists to measure piston temperature distributions as a whole, as thermal effects make important contributions to the stresses and deformations of this component. The piston rings form a vital route for the conduction of heat from the piston to the cylinder wall where it can be removed from the engine by the cooling system (see also Chapter 5). Consequently, considerable heat transfer takes place across the hydrodynamic film which lubricates the ring for the large majority of the stroke. Knowledge of this parameter is, therefore, likely to make a significant contribution to the accuracy to which the viscosity of the lubricating film is known. (It is interesting in this respect that the majority of piston-ring lubrication models consider only the cylinder wall temperature in estimating the instantaneous viscosity of the lubricating film.)

Table 11.11 Examples of the use of thin layer activation to measure piston-ring wear

Investigators	Comments
Schneider and Blossfeld (2003)	Demonstrated the high sensitivity of the thin layer activation method for piston-ring wear measurement
Schneider and Blossfeld (2004)	Measured piston-ring running-in and steady state wear under a range of operating conditions

Commonly, piston temperature measurements are made using a ***thermistor*** or thermocouples which are connected to an amplifier and power supply embedded within the piston. It is also quite common for temperature data to be transferred from the piston to the data recording system outside the engine by telemetric means rather than by cables and a linkage as this is a relatively straightforward approach for this type of measurement. (Methods of data transfer involving electromagnetic induction and contact with a data transfer connector, located at the base of the stroke, are also employed.) Other methods of measuring piston temperature have also been employed less frequently. These include the use of ***thermographic phosphors***. These materials can be stimulated to emit light by a short duration external light source; the decay rate of the emitted light is a function of the temperature of the phosphor. Phosphor coatings can be placed on components or, more conveniently, embedded in optical fibres. In engines, coatings can be placed on the piston which is then illuminated and viewed through an optical window in the engine. Other, more conventional, optical approaches involving the use of optical windows to allow the observation of the piston in the infra-red range to infer surface temperature have also been employed.

Limitations

Electrical systems such as thermocouples and thermistors are highly convenient and well-established technologies for temperature measurement. They can be relatively easily incorporated into pistons and are able to measure internal and surface temperatures. However, they normally record average temperature and are not normally able to respond to the rapid and potentially wide-ranging changes in temperature which occur at the surface of the piston along its stroke and as a consequence of combustion gas flow effects.

Optical systems are largely valuable for measuring the surface temperature of engine components. The exception to this is the use of ***thermographic phosphor doped optical fibres*** which can also be inserted into components for localised measurements. Optical methods also offer the possibility to respond to rapid transient changes in surface temperature which is potentially of considerable value in lubrication problems.

Applications

Many experimental studies of piston temperature have been conducted. Table 11.12 gives details of a few of these studies. (Examples have been selected which illustrate the methods used in preference to presenting studies which give information specifically about ***ring zone temperatures***.)

Table 11.12 Examples of the use of various techniques to measure piston temperature

Investigators	Comments
Assanis and Friedmann (1991)	Used a novel linkage system with fast thermocouples
Chang *et al.* (1993)	Adopted the use of an infra-red system and sapphire window to measure transient changes (includes ring and ring-land data)
Huisberg (2005)	Conducted temperature measurements on a piston using thermographic phosphors

11.7   Observation and measurement of lubricant movement and consumption

The flow of lubricant around a ring pack is a subject of considerable interest. Oil lubricating the piston rings may flow around the ring pack and eventually return to the sump or, alternatively, leave the engine via the exhaust gases. In either case the journey is detrimental in some way. Oil which is returned to the sump from the ring pack is degraded, both thermally and through exhaustion of additive content. It may also transport back undesirable materials, for example abrasive particles, corrosive combustion products or small amounts of fuel, which dilutes the lubricant. Oil which leaves the engine with the exhaust gases, in burnt or un-burnt form, is an undesirable emission, reducing the sump volume and contributing to pollution. Degradation and loss contribute to the need to monitor the lubricant quality and quantity and lead to the necessity for replacement and refilling at regular intervals. *Oil consumption* rates are relatively easily measured. However, developing a knowledge of oil flow patterns in piston-ring packs presents a more serious challenge. Over the years engineers and technologists have sought to improve understanding of oil flow in IC engine ring packs through experimental investigation. Many of these studies have involved the use engines with transparent cylinders operating under motored conditions. Conducting studies of oil flow in firing engines is more challenging; however, scientists and engineers are now finding ways to address data collection in this situation.

11.7.1 Measurement of oil consumption

Techniques for the measurement of *macroscopic oil consumption* in passenger and commercial vehicle engines are outlined in a range of standards at national, international and motor industry level.

Conventionally, measurement of oil consumption involves measurement of the mass or volume of oil before and after an engine test. The approach is extremely simple, but prone to error as the process commonly involves the subtraction of two very similar large values. This can be further complicated by the fact that fuel and/or water may become included in the oil over the period of the test. Due to the time required to consume a measurable volume of oil, the approach also generally precludes any opportunity to investigate the impact of transient effects on instantaneous rates of oil consumption.

Several techniques have been developed to allow oil consumption rates to be measured accurately over short timescales. The most commonly used methods are based on the analysis exhaust gas using radioactive materials or sulphur levels as tracers to track oil loss.

Several slightly different approaches based on the use of radioactive tracer elements in sump soils are in use. In these approaches radioactive isotopes are used to label oil fractions which can then be detected in exhaust emissions. Levels of radioactivity in exhaust emissions can be related to the lubricant consumption rate. Details of the tracers used and techniques for calibrating the radioactive measurements vary.

Using sulphur as a tracer material is also a common approach and circumvents the need for the extra safety precautions required when handling radioactive materials. The method involves using low sulphur fuel along with sulphated lubricants. By measuring the mass flow rates of fuel and air along with the sulphur concentration in the exhaust gases during engine operation, lubricant consumption can be calculated. Various means can be adopted to detect sulphur in the exhaust, but generally the equipment to perform this task must be relatively sophisticated, suitable instruments include: mass spectrometers, Fourier transform infra-red spectrometers and UV differential optical absorption spectrometers.

Limitations

Measurement of near instantaneous oil consumption requires a sophisticated experimental arrangement, careful calibration and an array of secondary measurements, e.g. air and fuel flow rates, on the engine. In some cases, where radioactive tracer materials are employed, special precautions for handling and disposing of radioactive materials are also needed. Consequently, although the science of the techniques themselves is well established, such measurements cannot be considered routine in the broadest sense as they are generally only conducted in specialist engine laboratories.

Applications

Many investigations have been conducted to develop and improve established tracer methods for measuring near instantaneous oil consumption. Some example of consumption measurements and measurement techniques are listed in Table 11.13.

Table 11.13 Examples of the use of various techniques to measure oil consumption

Investigators	Comment
Wong and Hoult (1991)	Employed a radioactive technique to measure oil consumption whilst also assessing operating oil film thickness
Froelund *et al.* (2001)	Adopted SO_2 tracer technique. Compared engine and ring pack variability with respect to oil consumption
Yilmaz *et al.* (2002)	Presents experimental measurements of oil consumption made using the SO_2 tracer technique with theoretical studies and LIF measurements of oil distribution
Pisiano *et al.* (2003)	Described a UV-based spectrometer for measuring SO_2 emissions
Zellbeck *et al.* (2006)	Described a radioactive method and studied oil consumption in relation to lubricant chemistry

11.7.2 Top ring zone oil sampling

Collection of the lubricant at the back of the piston-ring grooves, or at the face of the ring on the cylinder wall near the TDC and BDC, is now a well-established technique. It is achieved by connecting small diameter sampling pipes to the ring groove or cylinder wall to convey a small flow of lubricant into a sample collection bottle and is generally used to study changes in the chemistry of lubricants in the contact zone between the ring and liner. Such *oil sampling* is done because lubricant in the ring grooves and between the ring and liner suffers considerable thermal degradation and physical stress and component exhaustion in this local operating environment. This results in both physical and chemical changes which modify the lubrication mechanism for this contact. These changes include effects on viscosity, total acid number (TAN) and total base number (TBN), water content, fuel dilution, exhaustion of anti-oxidants, exhaustion of anti-wear additives, breakdown of viscosity index improver, etc.

Top ring zone sampling can also be used to obtain information about lubricant flow rates and residence times in these ring pack area. One approach to evaluating residence times involves comparing the rate of build-up of oxidation products in the engine sump with the prevailing concentration of the same products in the ring-groove samples. These measurements can also be used to evaluate useful secondary information such as the number of times the sump volume has passed through the ring pack.

Limitations

Ring zone sampling is a relatively straightforward method to apply. However, deriving useful information from the samples obtained requires significant expertise in chemical analysis and access to sophisticated equipment for chemical analysis. Additionally, for the most part, the technique does not lead to information about the condition of the engine or the lubricant in real time.

Applications

The use of **ring zone sampling** is not a particularly widely applied technique. However, it is able to give exceptionally valuable information about the real state of the lubricant which sustains the operation of the ring pack in firing engines. Some examples of the use of the method are detailed in Table 11.14.

11.7.3 Ring pack lubricant flow studies

Ring zone sampling is able to supply information about flow rates and residence times, but it is not generally able to elucidate the mechanisms by which lubricant is propelled through the ring pack. It is well known that there are many mechanisms of oil flow in ring packs and the problem for experimental investigators is to establish the comparative impact of the separate mechanisms at work. The most obvious approach to this issue is to view the flow of lubricant using optical access in firing engines. Several investigators have adopted this approach, mainly in association with the use of LIF to improve the visibility of the flow and in some cases to supply quantitative information about the thickness of lubricant films inside and outside contacts. Observational studies of this type have also proven valuable in identifying how fuel dilution of the lubricant layer on the cylinder wall can develop.

Limitations

While there have been several studies of lubricant flow in engines with complete transparent cylinders and motored pistons, most optical systems in firing engines are only able to access a circumferential section of an engine over a restricted range of crank angles preventing visualisation of the entire process.

Table 11.14 Examples of the use of ring zone sampling

Investigators	Comments
Picken *et al.* (1991)	Presents details of the ring zone sampling technique and lubricant degradation data
Fox *et al.* (1997)	Discusses trends in mass flow rates and lubrication regime at TDC
Stark *et al.* (2005)	Investigated degradation of lubricants and lubricant flow rates

Applications

Table 11.15 gives details of investigations which have adopted the use of optical access in order to study the movement or physical condition of lubricant films around the piston or ring pack.

11.8 Future trends

This chapter has reviewed a range of techniques and sensors for studying different aspects of the lubrication of ring packs in firing IC engines. Owing to the impact that piston assemblies have on the overall performance of IC engines, research on piston-ring lubrication is intense, driven by legislative requirements on the motor industry through governments and transnational organisations such as the EU Parliament. The need for further, more detailed information and imminent short-term technological developments mean that rapid developments, such as those listed in Table 11.16, are quite likely to be seen in the next five years or so.

All of the sensor systems described in Table 11.16 will also potentially benefit from developments in manufacturing technology including thin film technologies (which could, for example, be used to deposit sensors onto surfaces such as the piston-ring and cylinder liner) and micro-machining / micro-mechanical technologies (MEMS) which could be implemented as sensors or components in signal processing/transmission systems.

Of course, much engineering data collection today is dependent on the massive leaps in the development that computing technology has made in respect of speed and memory capacity and this will continue to be the case. This point is worth a special note here because the opportunity for collecting more data on isolated piston mounted systems for downloading after experiments, or through transmission by proprietary means, such as Bluetooth, has increased over the last few years. Fast compact memory systems and wireless transmission systems are now routinely and inexpensively available off the shelf.

Many of the methods discussed in this chapter are not routine, and require

Table 11.15 Examples of studies of oil movement and condition by optical access

Investigators	Comments
Sanda and Konomi (1993)	Used scanning LIF to study oil movement in ring packs and along the piston skirt
Thirouard *et al.* (1998)	Used LIF to study oil accumulation on the crown land/second land as well as inertial and gas flow driven oil movement
Parks *et al.* (1998)	Employed LIF to study fuel accumulation in the oil layer on a cylinder wall

Table 11.16 Possible short-term developments in technology to investigate piston assembly performance

Topic	Development
Oil film thickness measurement	
Capacitance methods	Coated sensors for very thin OFT measurement
Inductance methods	Thermal compensation built into sensors
Resistive methods	Greater use of resistance sensors to reduce breakdown
Ultrasound methods	Increased data repetition rates /sensor with better lateral resolution
LIF	More compact 'off the shelf' components
Piston assembly friction measurement	
Floating liner method	Better control of response at high engine speeds
IMEP method	Improved linkage arrangements for sensor wires
Other methods	Use of 'friction sensors'
Piston-ring movement	
Electrical methods	
Radioactive tracers	Development of detectors with improved resolution
Other methods	Use of ultrasonic sensors for ring gap detection
Measurement of ring wear	
Radioactive tracer methods	Development of more sensitive detectors/reduced need for radioactive methods due to improvements in other chemical and electrical approaches
Other methods	Improvements in the sensitivity of chemical sensor technology leading to the use of chemical tracers in preference to radioactive tracers
	Greater use of electrical systems with modified ring designs to enhance changes in magnetic effects during ring wear
Measurement of ring zone temperature	
	Further use of optical systems to map oil film temperature
	Use of ultrasonic systems to measure ring gap viscosity to infer oil film temperature
Measurement of lubricant movement and consumption	
	More extensive use of ring zone sampling, and sampling in other areas around the piston
	Use of real time chemical analysis with dedicated sensors to determine oil flow details and/or the chemical behaviour of new lubricant formulations

specialist knowledge and extensive equipment in order to apply them. Often they exploit practice and technologies from other discipline areas and the further development of methods to investigate piston assembly tribology clearly innovation in these areas will continue to be have dependence on 'importing' new techniques.

Conversely, some of the sensor technologies used to investigate piston assembly performance are relatively mature and have the potential to be incorporated into production engines to provide real-time feedback on the lubrication at the ring cylinder wall contact for purposes of engine performance optimisation. This is likely to be the most significant development in piston assembly related measurement in the longer term. Initially, developments are most likely to occur in the low volume high cost engines, typically marine engines, large engines for stationary generators, engines for railway carriages, etc.

One notable example of such an approach, involving the use of OFT sensors to provide feedback to regulate the delivery of lubricant to the ring pack, has already been described in patents (Sherrington, 2002). The system is illustrated in Fig. 11.7. Such systems potentially have significant value in a number of applications, but the initial focus is on large two-stroke marine diesel engines where there is considerable annual cost for lubricants and a significant loss of lubricant combustion products into the atmosphere. These engines have a total loss arrangement to lubricate the piston assembly in which lubricant is sprayed through holes called 'quills' in the cylinder liner to lubricate the rings and piston skirt. If the volume of lubricant supplied is controlled to the bare minimum amount required to maintain mid-stroke hydrodynamic lubrication, this will reduce the consumption of lubricant, lower emissions and increase engine efficiency by allowing the rings to operate at the lowest possible point on the Stribeck curve. As well as being able to control the supply of lubricant, such systems also offer the opportunity to control the supply rate of anti-corrosion additives to neutralise the effect of sulphuric acid generated in combustion and to incorporate engine and lubricant condition monitoring functions.

Similarly, some of the technologies described above are linked to the chemical analysis of the lubricant. As these methods become developed, miniaturised and less expensive, it may become feasible to include selected elements of these systems into lubricant condition monitoring systems on passenger and commercial vehicles. For example, to monitor the levels of various lubricant additives to advise the driver or service engineer when the oil was no longer fit for continued service. They could also potentially be to monitor the chemical composition of exhaust gas to provide real-time wear assessment of piston assemblies or other additional fault diagnosis for the engine.

Finally, it is worth noting that developments in IC engine technology and lubricant formulation have, hitherto, been largely incremental and continuous. Research has contributed to a more or less single line of development of petrol and diesel-based engines. This is not likely to continue in the long-

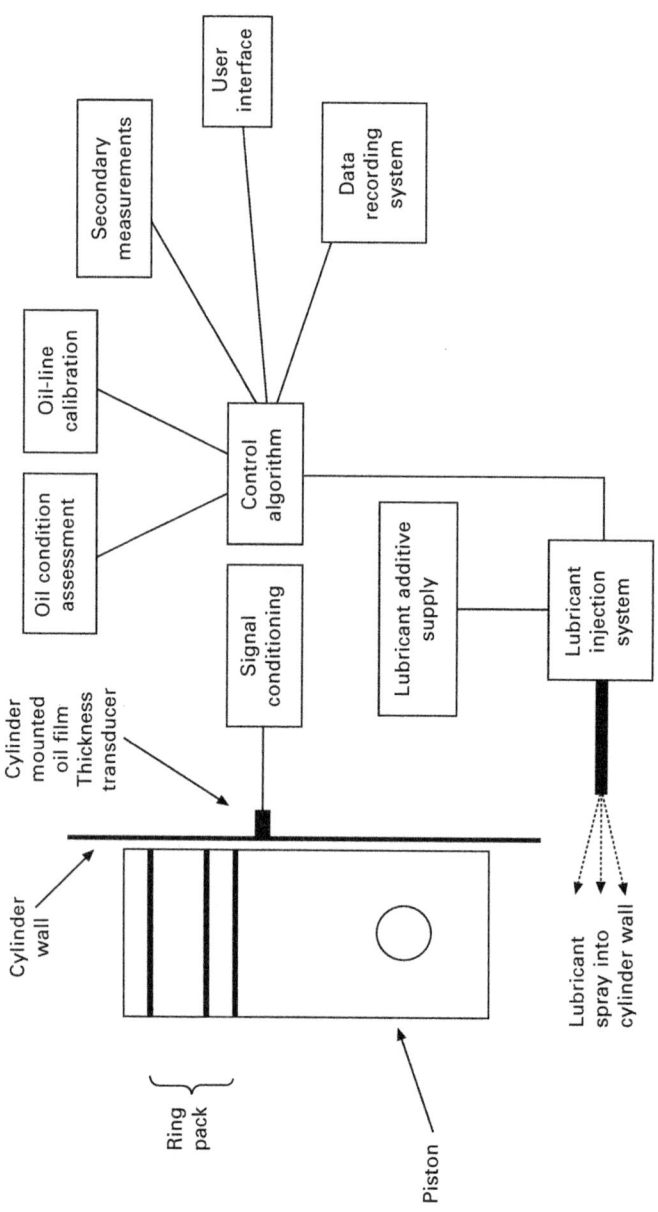

11.7 System for feedback regulation of ring pack lubrication.

term future. Technologies for hybrid systems, hydrogen powered vehicles and electric systems will be developed in parallel to liquid fuel technologies and it is likely future engineers will need to evaluate completely new types of bearing and sealing systems, materials, lubricants and additives for the engines of the future.

11.9   Sources of further information

A wide range of information is available about the design, performance and testing of piston ring assemblies and their lubricants. The list below gives a few examples of sources where readers can obtain further information on this topic as well as general information on tribology. Many conferences and congresses, not listed below, also discuss topics related to piston assembly lubrication.

11.9.1 Websites

For publications on experimental and theoretical investigations on engines:

- SAE International – www.sae.org
- Society of Tribologists and Lubrication Engineers (STLE) – www.stle.org
- Thesis collection for MIT – www.dspace@MIT.edu
- General tribology booklist – www.tribology.group.shef.ac.uk/teaching/books.html
- General tribology data and web calculators – www.tribology-abc.com
- European project details (COST 532 – Triboscience and tribotechnology – Superior friction and wear control in engines and transmissions) – Information on various sites.

11.9.2 Standards

- ASTM – www.astm.org
- ISO – www.iso.org

11.9.3 Books

- Taylor, C. M. (Ed.) (1993) *Engine Tribology*, Elsevier, ISBN-13: 978-0444897558
- Heywood, J. B. (1988) *Internal Combustion Engine Fundamentals*, McGraw-Hill, ISBN-13: 978-0070286375

11.9.4 Journals

- ASME – *Journal of Tribology*
- Emerald Insight – *Industrial lubrication and tribology*
- STLE – *Tribology transactions*
- STLE – *Tribology and lubrication technology*
- Springer – *Tribology letters*
- PEP – *Proceedings of the IMechE Part J (Engineering tribology)*
- PEP – *Proceedings of the IMechE Part D (Automoblie engineering)*
- Elsevier – *Tribology International*
- Vieweg Publishing – *MTZ Motortechnische Zeitschrift*
- Wiley Interscience – *Tribotest*

11.10   References

Assanis, D. N. and Friedmann, F. (1991) 'A telemetry linkage system for piston temperature measurements in a diesel engine' SAE paper 910299.

Brown, S. R. and Hamilton, G. L. (1977) 'The partially lubricated piston-ring' *J. Mech. Eng. Sci.* **19**(2) pp 81–89.

Chang, H. S., Wayte, R. and Spikes, H. A. (1993) 'Measurement of piston-ring and land temperature in a firing diesel engine' *Tribology Trans.* **26**(1) pp 104–112.

Clarke, D. G., Sherrington, I. and Smith, E. H. (1990) 'The floating liner method applied to measure instantaneous piston assembly friction in a motored engine' Proc. of Nordtrib '90', 4th Nordic Symposium on Tribology, Lubrication, Friction and Wear, Hirtshals, Denmark, June 1990.

Courtney-Pratt, B. E. and Tudor, G. K. (1946) 'An analysis of the lubrication between the piston-rings and cylinder wall of a running engine' *Proc. IMechE.* **155** pp 293–299.

Dow, T. A., Schiele, C. A. and Stockwell, R. D. (1983) 'Technique for experimental evaluation of piston-ring/cylinder film thickness' *J. Lubr. Technol.* **105** pp 353–360.

Fox, M. F. Jones, C. J., Picken, J. and Stow, C. G. (1997) 'The "limits of lubrication" concept applied to piston ring zone lubrication of modern engines' *Trib. Lett.* **3** pp 99–106.

Froelund, K., Menezes, L. A., Johnson, H. R. and Wolfgang, R. O. (2001) 'Real-time transient and steady state measurement of oil consumption for several production SI-engines' SAE paper 2001-01-1902.

Furuhama, S. and Hiruma, M. (1972) 'Axial movement of piston rings in the groove' *Tribology Trans.* **15**(4) pp 278–287.

Furuhama, S. and Sumi, T. A. (1961) 'A dynamic theory of piston ring lubrication, 3rd report: measurement of oil film thickness' *Bull JSME* **4**(16) pp 744–752.

Furuhama, S. and Takiguchi, M. (1979) 'Measurement of piston fictional force in actual operating engine' SAE paper 790855

Furuhama, S., Hiruma, M. and Tzuzita, M. (1979) 'Piston ring motion and its influence on engine tribology' SAE paper 790860.

Furuhama, S., Asahi, C. and Hiruma, M. (1982) 'Measurement of piston ring oil film thickness in an operating engine' *Trans ASLE* **26** pp 325–332.

Greene, A. B. and Lucas, G. G. (1969) *The Testing of Internal Combustion Engines.* English Universities Press, London.

Grice, N. and Sherrington, I. (1993) 'An experimental investigation into the lubrication of piston-rings in an internal combustion engine – oil film thickness trends, film stability and cavitation' SAE International Congress and Exposition (March, 1993) Detroit, USA. SAE paper 930688.

Grice, N., Sherrington, I., Smith, E. H., O' Donnell, S. G. and Stringfellow, J. F. (1990) 'A capacitance based system for high resolution measurement of lubricant film thickness' Proc. of 'Nordtrib '90'. 4th Nordic Symposium on Tribology, Lubrication, Friction and Wear, Hirtshals, Denmark 10–13 June 1990.

Hamilton, G. M. and Moore, S. L. (1974) 'Measurement of the oil film thickness between the piston-rings and liner of a small diesel engine' Proc. IMechE 188 pp 253–261.

Hoult, D.P, Lux, J. P., Wong, V. W. and Billan, S. A. (1988) 'Calibration of laser fluorescence measurements of lubricant film thickness in engines' SAE paper 881587.

Husiberg, T. (2005) 'Piston temperature measurement by use of thermographic phosphors in a heavy duty diesel engine run under partly premixed conditions' SAE paper 2005-01-1646.

Kim, J. S., Min, B. S., Oh, D. Y. and Choi, J. K. (2000) 'Effects of piston ring gap positions on the oil consumption rate and thermal load of piston' Proc. JSAE Annual Congress pp 9–12.

Lewis, M. G. (1974) 'A miniature mutual inductive proximity transducer' J. Phys. E. 7 pp 269–271.

Mattson, C. (1995) 'Measurement of the oil film thickness between the cylinder liner and the piston rings in a heavy duty directly injected diesel engine' SAE paper 952469

Mollenhauer, K. and Bruchner, K. (1980) 'Contribution to the dtermination pf the influence of cylinder pressure and engine speed on engine friction' MTZ Motortechnische Zeitschrift 41 pp 265–268 (in German).

Moore, S. L. and Hamilton, G. M. (1980) 'The piston-ring at top dead centre' Proc IMechE 194(36) pp 373–381.

Mufti, R. A. and Priest, M. (2005) 'Experimental evaluation of piston-assembly friction under motored and fired conditions in a gasoline engine' ASME J. Tribology 127(4) pp 826–836.

Parker, D. A., Stafford, J. V., Kenrick, M. and Graham, N. A. (1975) 'Experimental measurements of the quantities necessary to predict piston/cylinder bore oil-film thickness, and the oil film thickness itself in two particular engines' Proc. IMechE paper number C71/75, 79.

Parks, J. E., Armfield, J. S., Barber, T. E., Storey, J. M. E. and Wachter, E. A. (1998) 'In situ measurement of fuel in the cylinder wall oil film of a combustion engine by LIF spectroscopy'. Appl. Sectrosc. 52 pp 112–118.

Phen, R. V., Richardson, D. and Borman, G. (1993) 'Measurements of cylinder liner oil film thickness in a motored diesel engine' SAE Paper 932789.

Picken, D. J., Fox, M. F. and Preston, W. H. (1991) 'Lubricating oil – extraction and chemical analysis' Proc. of 'Experimental methods and engine research and development '91', MEP, pp 115–120.

Pinkus, O. and Wilcock, D. F. (Ed.) (1977) Strategy for Energy Conservation Through Tribology ASME, New York.

Pisano, J. T., Sauer, C. G., Robbins, J., Miller, J. W., Gamble, H. and Durbin, T. D. (2003) 'A UV differential optical absorption spectrometer for the measurement of sulfur dioxide emissions from vehicles' Meas. Sci. Technol. 14 pp 2089–2095.

Priest, M., Dowson, D. and Taylor, C. M. (1999) 'Predictive wear modelling of lubricated piston rings in a diesel engine' Wear 231 pp 89–101.

Przesmitzki, S. and Tian, T. (2007) 'Oil transport inside the power cylinder during transient load changes' SAE paper 2007-01-1054.

Richardson, D. E. and Borman, G. L. (1991) 'Using fibre optics and laser fluorescence for measuring thin oil films with application to engines' SAE paper 912399.

Ryk, G., Kligerman, Y., Etsion, I. and Shinkarenko, A. (2005) 'Experimental investigation of partial laser surface texturing for piston-ring friction reduction' *Tribology Trans* **48**(4) pp 583–588.

Saad, P., Kamo, L., Mekari, M., Bryzik, W., Wong, V., Dmitrichenko, N. and Mnatsakanov, R. (2007) 'Modelling and measurement of tribological parameters between piston rings and liner in a turobocharged diesel engine' SAE paper 2007-01-1440.

Sanda, S. and Konomi, T. (1993) 'Development of a scanning laser-induced-fluorescence method for analysing piston oil film behaviour' Paper number C465/014, IMechE.

Sanda, S., Saito, A., Konomi, T. and Nohira, H. (1993) 'Development of a scanning laser induced fluorescence method for analysing piston oil film behaviour' IMechE paper number C465/014.

Schneider, E. W. and Blossfeld, D. H. (1990) 'Method for measurement of piston ring rotation in an operating engine' SAE paper 900224.

Schneider, E. W. and Blossfeld, D. H. (2003) 'Radiotracer method for measuring real-time piston-ring and cyclinder bore wear in spark ignition engines' *Nuclear Instruments and Methods in Physics Research (Section A: Accelerators, spectrometers and associated equipment)* **505** (1–2) pp 559–563.

Schneider, E. W. and Blossfeld, D. H. (2004) 'Effect of break-in and operating conditions on piston ring and cylinder bore wear in spark ignition engines' *SAE Trans.* **113**(3) pp 1357–1369.

Schneider, E. W., Blossfeld, D. H., Lechman, D. C., Hill, R. F., Reising, R. F. and Brevick, J. E. (1993) 'Effect of cylinder bore out-of-roundness on piston ring rotation and engine oil consumption' SAE paper 930796.

Schuster, M., Maler, F. and Crysler, D. (1999) 'Metallurgical and metrological examinations of the cylinder liner-piston ring surfaces after heavy duty diesel engine testing' *Tribology Trans* **42**(1) pp 116–125.

Shaw, B. T., Hoult, D. P. and Wong, V. W. (1992) 'Development of engine lubricant film diagnostics using fibre optics and laser fluorescence' SAE paper 920651.

Shin, K., Tateishi, Y. and Furuhama, S. (1983) 'Measurement of oil film thickness between piston-ring and cylinder' SAE paper 830068.

Sherrington, I. (2002) 'Lubrication Control System' http://v3.espacenet.com/ textdoc?DB=EPODOC&IDX=EP1240455

Stark, M. S., Wilkinson, J. J., Lee, P. M., Lindsay Smith, J. R., Priest, M., Taylor, R. I. and Chung, S. (2005) 'The degradation of lubricants in gasoline engines: Lubricant flow and degradation in the piston assembly' *Proc. Leeds–Lyon Symposium*, Leeds, 2004, Elsevier, pp 779–786.

Stiyer, M. J. and Ghandi, J. B. (1997) 'Direct calibration of LIF measurements of the oil film thickness using capacitance technique' SAE paper 972859.

Takaguchi, M., Sasaki, R., Takahashi, I., Ishibashi, F., Furuhama, S., Kai, R. and Sato, M. (2000) 'Oil film thickness measurement and analysis of a three ring pack in an operating diesel engine' SAE paper 2000-01-1787.

Tamminen, J., Sandstrom, C-E. and Andersson, P. (2006) 'Influence of load on the tribological conditions in piston ring and cylinder liner contacts in a medium-speed diesel engine' *Trib. Int.* **39** pp 1643–1652.

Taylor, R. I. and Evans, P. G. (2004) '*In-situ* piston measurements' *Proc. IMechE Part J (Engineering Tribology)* **281** pp 185–200.

Thirouard, B. P., Tian, T. and Hart, D. P. (1998) 'Investigation of oil transport mechanisms in the piston ring pack of a single cylinder diesel engine, using two dimensional laser induced fluorescence' SAE paper 982658.

Uras, H. M. and Patterson, D. J. (1983) 'Measurement of piston and piston ring assembly friction instantaneous IMEP method' SAE paper 830416.

Weimar, H. J. and Spicher, U. (2003) 'Crank angle resolved oil film thickness measurement between piston-ring and cylinder liner in a spark ignition engine' ASME ICES 2003-544.

Wing, R. D. and Saunders, O. (1972) 'Oil film temperature and thickness measurements on the piston-rings of a diesel engine' *Proc. IMechE* **186** pp 1–9.

Wong, V. W. and Hoult, D. P. (1991) 'Experimental survey of lubricant-film characteristics and oil consumption in a small diesel engine' SAE paper 910741.

Yilmaz, E., Tian, T., Wong, V. W. and Heywood, J. B. (2002) 'An experimental and theoretical study of the contribution of oil evaporation to oil consumption' SAE paper 2002-01-2684.

Yoshida, H., Kobyashi, H., Yamada, T., Takiguchi, M. and Kuwada, K. (2000) 'Effects of narrow width, low tangential tension, 3 piece oil ring on oil consumption' *JSAE Review* **21** pp 21–27.

Zellbeck, H. Bergmann, M., Röthig, J., Seibold, J. and Zeuner, A. (2006) 'A method of measuring oil consumption by labelling with radioactive bromine' *Tribotest* **6**(3) pp 251–265.

12

An ultrasonic approach for the measurement of oil films in the piston zone

R. S. DWYER-JOYCE, University of Sheffield, UK

Abstract: Control of the lubrication of the piston reciprocating in the liner is a key part of the efficient operation of an internal combustion engine. A thick lubricant film leads to excessive oil in the combustion zone, while an inadequate film leads to bore polishing. In this work a method for the non-invasive measurement of the oil film thickness using reflected ultrasound is described. The method is explained in relation to two cases studied; the first on the contact between the piston skirt and liner in a fired engine, the second on the ring liner contacts in a bench test apparatus.

Key words: piston ring – liner oil film, ultrasound, oil film thickness measurement.

12.1   Introduction

The *piston ring pack* plays an important part in the operation of an internal combustion (IC) engine. The seal it forms to prevent escape of high-pressure gases during combustion is integral to engine efficiency. Ring pack lubrication is essential for smooth running and long life. Sufficient separation of piston skirt and ring pack from the cylinder wall reduces the *frictional losses* in an IC engine. It is reported that this accounts for nearly 40–50% of all such losses (Andersson, 1991; Taylor, 1993). Clearly, a separating lubricant film keeps this friction to a minimum. However, excessive lubrication exposes quantities of oil on the cylinder walls during combustion. This leads to elevated emissions and the possibility of the combustion product build-up on the mating surfaces which may lead to ring stick.

Clearly, it is important to optimise the lubricant film formation to minimise friction and yet limit emissions. Therefore, design of piston skirt and rings to encourage lubricant entrainment into the contact are important criteria. Their axial profiles include chamfered edges or relief radii to create the necessary wedge effect (see Chapter 5) for entraining of the lubricant with relative motion of surfaces. Cylinder bores or liners may be treated to provide lubricant retaining features (see Chapters 13 and 14). This promotes film by entrapment with cessation of entraining motion at dead centres, where there is no relative motion of surfaces.

There are a number of modelling techniques for the prediction of film

426

formation for the piston rings and skirt based on solutions to the Reynolds equation (see for example Lloyd, 1969; Hamilton and Moore, 1974; Dowson *et al.*, 1983; Ma *et al.*, 1995; Balakrishnan and Rahnejat, 2005; also Chapters 10, 14, 15). The models by their very nature are complex and can incorporate a number of different characteristics such as 3D geometrical variations, elastic deformation of the components, viscosity variation, thermal effects and cavitation. It is important to ascertain the validity of these predictions through measurement of film thickness.

Measurement of film thickness in the piston zone has proven to be difficult, owing to the inaccessible nature of the contact and the hostile environment. The oil films are thin and so can generally not be measured by the movement or position of the piston components. In-cylinder oil film thickness measurement approaches can broadly be divided into three categories: electrical, optical and acoustic (also see Chapter 11).

Early optical approaches to oil film measurement were conducted using a complete transparent cylinder. Shaw and Nussdorfer (1946) used a glass liner to explore the operation of the ring pack and showed the importance of side forces. They also used ultraviolet light imaging with an oil containing a fluorescent dye. Greene (1969) used a similar apparatus to produce stroboscopic photographic images of oil film formation and the presence or absence of oil in regions around the piston. A recent study (Todsen and Niethus, 2006) used a transparent liner to view the formation of an oil layer around the piston skirt. The work showed images of intermittent film formation and were able to identify how this contributed to regions of reduced film in piston ring contacts.

If the oil constituents naturally fluoresce, or if a fluorescent compound is added, then this can be used to give a quantitative measure of the volume of oil present. Several workers have used this approach of *laser-induced fluorescence* in piston ring studies (Ting, 1980; Richardson and Borman, 1991; Shaw *et al.*, 1992; Sanda *et al.*, 1997; Baba *et al.*, 2006). A small transparent window, either quartz or silica, is inserted into the cylinder. A laser is channelled through a fibre optic to the window. The amount of light reflected from the fluorescent film depends on the volume of the film. Technical challenges include getting the window flush with the liner bore, achieving sufficient laser light to the oil film, and the removal of stray reflections and illumination from combustion. Further, an independent calibration is required to determine the oil volume for fluorescence signals; which is a non-linear relationship. Much has been obtained from such optical measurements on test engines. However, the nature of the experiment requires considerable modification to an engine and some careful challenging experimental work.

Electrical methods potentially give an easier approach to quantifying the oil film (see Chapter 11). Resistance, capacitance, and inductance methods have all been used (see Sherrington and Smith, 1985, for a review and Chapter 11).

Oil acts as an insulator so if the piston ring is isolated from the piston then the potential difference can be measured across the oil film (Courtney-Pratt and Tudor 1945; Furuhama and Sumi, 1961). When the oil film collapses, at the top dead centre *(TDC)* and bottom dead centre *(BDC)*, then the resistance falls to zero. The resistance is clearly measured around the whole circumference of the piston ring; this can be related to a volume of oil but not to how it is spatially distributed. Data from these kinds of experiments tend to be full of peaks where instantaneous asperity contact occurs and the resistance falls to almost zero.

Inductance has also been used in cylinder measurements. Wing and Saunders (1972) mounted inductance coils behind a piston ring either side of the piston and monitored the gap during running, hence deducing oil film thickness. However, the coils in such a sensor change resistance with temperature, so the output is highly sensitive to temperature changes during operation.

Alternative electrical method is to measure the capacitance of the oil film, which also depends on the film thickness. Hamilton and Moore (1974) designed a small diameter capacitance probe within a protective screen. Probes were calibrated outside the engine using a known film thickness. The probes were inserted at several locations through the liner wall of a small fired diesel engine. Typically results were averaged over many cylinder passes. Measurements of film thickness were obtained, as well as the shape of the ring during the stroke. Furuhama *et al.* (1982) used a similar concept transducer mounted inside a piston ring. The use of such localised measurement probes mean that quantitative film thickness data can be obtained at a single location in the piston ring cycle. Some recent developments (Sherrington *et al.*, 2002) have seen the development of probes that combine capacitance and induction probes in a single sensor.

The last class of methods for the measurement of oil film thickness reviewed here use ultrasound as their basis. An ultrasonic wave can be propagated non-invasively through a material and be reflected from an oil film. The proportion of the wave reflected depends on, among other things, the thickness of the oil film. This method has been widely used for oil film measurements in hydrodynamic and elastohydrodynamic bearings but has seen little application in piston–cylinder applications.

In this chapter the ultrasonic method for measuring oil film thickness is described and two case studies are described. The first is on the measurement of oil film formation between the piston skirt and liner. This can be achieved using a simple contact type transducer. The second case study uses a bench-top piston ring apparatus, where a special focusing transducer has been used to provide the resolution necessary to measure the oil film between the ring and liner.

12.2   Ultrasonic measurement of oil film thickness

12.2.1 Reflection of ultrasound from a boundary

When **ultrasound** is incident on a boundary between two different media, some of the energy is reflected and some transmitted. The reflection and transmission behaviour at the boundary is dependent on the acoustic properties of the two media. The proportion of the amplitude of an incident pulse that is reflected is known as the reflection coefficient, R. There is a simple relationship between the **reflection coefficient** and the acoustic properties of the materials either side of the boundary:

$$R = \frac{z_1 - z_2}{z_1 + z_2} \tag{12.1}$$

where z is the **acoustic impedance** of the media (given by the product of density and speed of sound) and the subscripts refer to the two media. The proportion of the incident signal transmitted (or transmission coefficient, T) can then be calculated as:

$$T = 1 - R \tag{12.2}$$

If ultrasound is incident on a multi-layered system, then the signal transmitted or reflected is the superposition of the result of the application of equations (12.1) and (12.2) at each boundary. An oil film between metals surfaces such as a piston and liner represents such a multi-layered system.

Figure 12.1 schematically shows an ultrasonic beam incident on a **metal–lubricant–metal layered system**. The metal bodies are large when

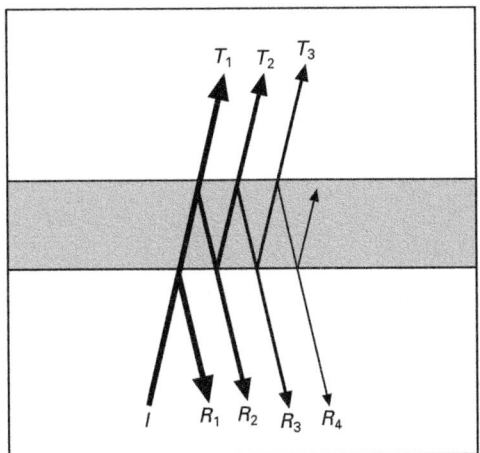

12.1 Schematic of an ultrasonic beam incident on a lubricated contact. The incident wave undergoes multiple internal reflections at each interface.

compared with the wavelength of ultrasound and so they can be modelled as semi-infinite. If the lubricant layer is sufficiently thick or the ultrasonic wave packet small, then the reflections from either side of the oil film are discrete in time. This means that if the speed of sound in the lubricant is known, so the thickness of the lubricant film can be determined by measuring the *time-of-flight* (ToF) between the two reflections. The amplitude of these reflected pulses can also be calculated from equations (12.1) and (12.2), assuming that the losses in the lubricant are small. This ToF approach is commonly used for thickness gauging of structural components such as pressure vessels or pipelines. Typically liquid layers or solid bodies thicker than around 100 μm can be measured in this way.

Oil films, however, are considerably thinner than this. For these very thin layers, the reflected pulses overlap and it becomes impossible to distinguish the discrete reflections. A time domain-based method is not longer possible and an alternative approach is required.

12.2.2 Reflection of ultrasound from a thin oil film

In the case where the lubricant film is thin the amplitude of the reflected signal can be measured and interpreted using a *spring interface model*. This approach is now discussed in detail.

When the layer is thin compared with the wavelength of ultrasound then a quasi-static modelling approach can be used (Tattersall, 1973). Equations for the stress and displacement at each boundary are written and then compatibility and equilibrium conditions are applied. For a full derivation see Reddyhoff *et al.* (2005); this paper also evaluates the effects of mass and damping of the oil film layer and shows this to have a negligible contribution to the response. The result, commonly known as the **spring model of reflection** shows that the reflection coefficient depends on the stiffness of the thin layer, K, according to:

$$R = \frac{z_1 - z_2 + i\omega(z_1 z_2 / K)}{z_1 + z_2 + i\omega(z_1 z_2 / K)}$$

(12.3)

where ω is the angular frequency of the ultrasonic wave ($\omega = 2\pi f$). The reflection coefficient is a complex parameter. In most practical cases the amplitude of the incident and reflected signals are recorded so the modulus of reflection coefficient is used:

$$|R| = \sqrt{\frac{(\omega z_1 z_2)^2 + K^2(z_1 - z_2)^2}{(\omega z_1 z_2)^2 + K^2(z_1 + z_2)^2}}$$

(12.4)

If the materials either side of the interface are identical, then this reduces to a simple relationship between reflection coefficient and stiffness:

$$|R| = \frac{1}{\sqrt{1 + (2K/\omega z)^2}} \tag{12.5}$$

12.2.3 The stiffness of an oil film layer

If the oil film of thickness, h, is subjected to a pressure of p, then its stiffness is given by:

$$K = -\frac{\mathrm{d}p}{\mathrm{d}h} \tag{12.6}$$

It is assumed that the ultrasonic wave has a large wavelength compared with the film thickness; so applying a pressure causes a negligible volume of oil to be squeezed out of the sides of the layer. Then, the deflection of the layer is found from the definition of **bulk modulus**, B:

$$B = -\frac{\mathrm{d}p}{\mathrm{d}V/V} = \frac{\mathrm{d}p}{\mathrm{d}h/h} \tag{12.7}$$

where V is the volume of oil subjected to the pressure ($V = Ah$, where A is the area over which the pressure acts and remains constant). The **film stiffness** then becomes:

$$K = \frac{B}{h} \tag{12.8}$$

Thus, stiffness is inversely proportional to the thickness of the oil film. The bulk modulus can be expressed in terms of the acoustic properties of the liquid by noting that the speed of sound through a liquid, c, is related to the density, ρ, and bulk modulus by:

$$c = \sqrt{\frac{B}{\rho}} \tag{12.9}$$

Combining (12.6) and (12.7) gives the stiffness of a liquid layer in terms of its acoustic properties:

$$K = \frac{\rho c^2}{h} \tag{12.10}$$

It is worth noting that equation (12.6) holds for any interface between two materials where the wavelength is large compared with the interface dimensions. Therefore, it can also be applied to dry surfaces in *incomplete* contact where the interface consists of an array of asperity contact regions and air gaps. The stiffness (also obtained from equation (12.5)) is then a function of the number of **asperity contacts**, their size, and distribution.

This phenomenon has been used to study imperfect adhesive bonds (Nagy, 1992), aspects of the dry rough surface contact (Kendall and Tabor, 1971; Krolikowski and Szczepek, 1993; Dwyer-Joyce *et al.*, 2001; Baltazar *et al.*, 2002), and mixed lubrication (Gonzalez-Valadez *et al.*, 2004).

12.2.4 Oil film thickness from ultrasonic reflection

Combining equations (12.4) and (12.10) gives a convenient relationship for the *oil film thickness* in terms of the ultrasonic reflection coefficient and materials properties:

$$h = \frac{\rho c^2}{\omega z_1 z_2} \sqrt{\frac{R^2 (z_1 + z_2)^2 - (z_1 - z_2)^2}{1 - R^2}} \tag{12.11}$$

Or for identical materials either side of the interface:

$$h = \frac{2\rho c^2}{\omega z} \sqrt{\frac{R^2}{1 - R^2}} \tag{12.12}$$

Figure 12.2 shows a plot of the *reflection coefficient*, predicted by equation (12.4), against frequency as the thickness of the oil film is varied. The plot

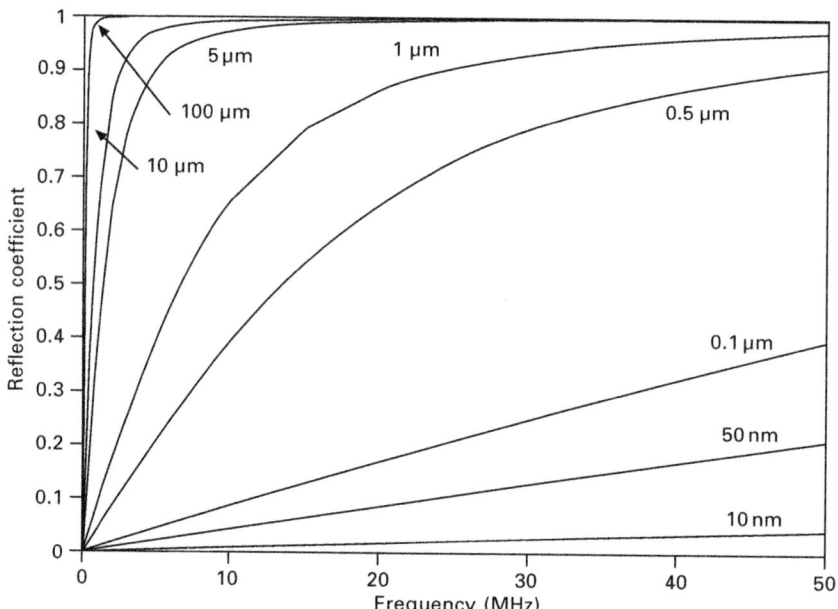

12.2 The response of an interface to an ultrasonic wave (determined for a steel–oil–steel contact) according to the spring model equation (12.4). As the interface becomes stiffer less of the wave is reflected.

shows what reflection coefficient would be expected for a particular oil film thickness and ***ultrasonic wave frequency***. It is best to have a reflection coefficient that shows sensitivity to film thickness at a given frequency. Thus, small film thicknesses require a high frequency, and large thicknesses a low frequency.

It is this spring model equation that forms the basis for the measurement procedure. If the reflection from an oil film can be measured then this can readily be turned into an oil film thickness if the properties (i.e. bulk modulus) of the oil are known.

Dwyer-Joyce *et al.* (2002) demonstrated that this relationship holds for thin oil films such as those found in ***hydrodynamic*** and ***elastohydrodynamic*** contacts. This approach has subsequently been used for the measurement of oil film thickness in power station thrust bearings (Dwyer-Joyce *et al.*, 2006); face seals (Reddyhoff *et al.*, 2008), ball bearings (Zhang *et al.*, 2006) and prosthetic hip joints (Brockett *et al.*, 2008).

12.3   Ultrasonic measurement equipment

12.3.1 Generation and capture of signals

An ***ultrasonic pulser/receiver*** (UPR) is used to send a high-frequency top-hat voltage pulse (typically 10 to 100 V) to the transducer. The width of the pulse is tuned, in order to excite the transducer at its resonant frequency to give the maximum power output. Typically the pulse rise time (the square-ness of the top-hat function) is less than 9 ns. The pulse width is typically 100 ns.

A typical schematic layout of the apparatus is shown in Fig. 12.3. In a ***pulse-echo arrangement*** (most convenient for oil film measurement) the pulse is transmitted to the probe along a cable that is also used to receive

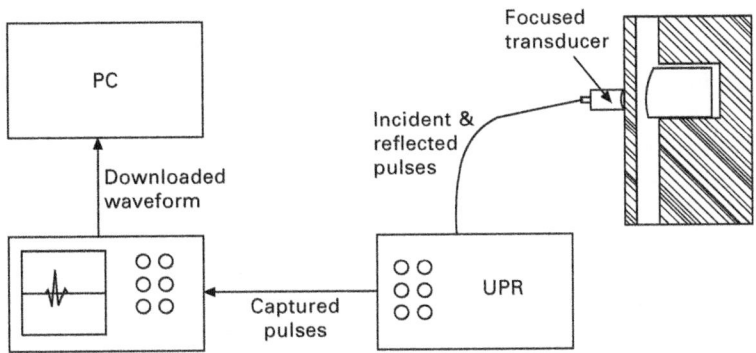

12.3 Schematic diagram of the equipment used for generating, capturing and storing ultrasonic signals.

the reflected signal (the transducer acts as both an emitter and receiver). The receiver section of the UPR receives the reflected signal from the transducer, typically 30 mV. This signal is then amplified using a low noise amplifier and sent to the digitiser. The digitiser, in this case a high-frequency 500 MHz oscilloscope, is used on the signal from the UPR and passes the data to a computer for processing.

12.3.2 Ultrasonic transducers: coupling and focusing

The ultrasonic transducers consist of a piezoelectric material with electrodes bonded to each side of the element. As the ultrasonic signal would be heavily attenuated in air the transducer must be coupled in some way to the outer surface. There are two generic types of transducer that can be used for these kinds of measurements; contact type or an immersion type (as shown in Fig. 12.4).

The contact transducer is coupled directly to the surface with a water-based gel (Fig. 12.4(a)). A simpler and cheaper approach is to use a piezo-element directly and permanently bonded to the test piece (Fig. 12.4(b)). This transducer consists of a thin slice of piezoelectric material; each side of which has been coated with an electrode. The figure shows a wrap-around electrode type, where the lower electrode has been wrapped over onto the top surfaces. In this way both wires can be conveniently soldered directly onto the top face. The drawback of using a contact type transducer is the spatial resolution. The ultrasonic beam will spread out in the test piece and by the time it reaches the oil film the spatial resolution of the measurement is low. Typically the transducer will be measuring over an area of 5–10 mm. This is satisfactory for piston skirt film formation measurements but not suitable for piston ring contacts.

An alternative design is the immersion type transducer (Fig. 12.4(c)). This is used in a water bath to couple the ultrasound to the test piece. This type of transducer can be modified to achieve a focusing of the ultrasonic wave. This is achieved by means of a concave lens bonded to the piezo-element.

Figure 12.5 schematically shows the focusing method. The focal length in water is defined by the lens diameter. The wave will be refracted at the water–solid boundary. The true focal length is thus slightly less than the apparent focal length. Snell's law of refraction is used to determine the true focal length:

$$\frac{\sin \theta_1}{\sin \theta_2} = \frac{c_1}{c_2} \qquad (12.13)$$

where θ_1 and θ_2 are the incident and refracted angles as shown on Fig. 12.5, and c_1 and c_2 are the speed of sound in the water and test material respectively.

(a)

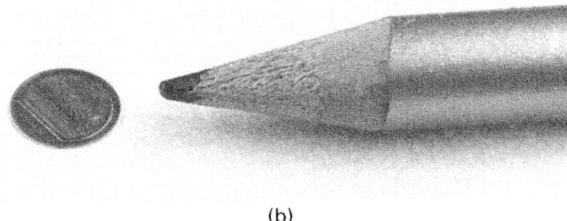

(b)

(c)

12.4 Ultrasonic transducers used in oil film measuring (a) conventional contact transducer (b) piezo-electric element (c) conventional immersion transducer.

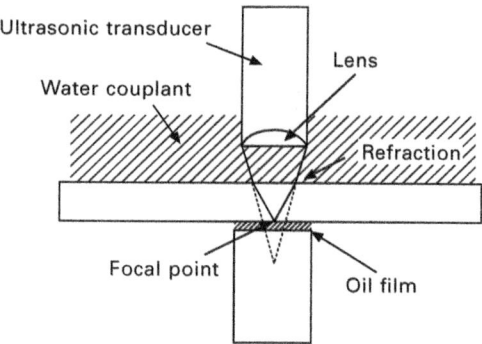

12.5 Schematic diagram of the operation of a focusing ultrasonic transducer.

The spot size of the focused wave varies with the frequency of the ultrasound according to the empirical relationship (Silk, 1984):

$$d_{f_{(-6\mathrm{dB})}} = \frac{1.028Fc}{fD} \tag{12.14}$$

where d_f is the diameter of the spot size (where the signal has reduced to −6 dB of its peak value), F is the transducer focusing length, f is the wave frequency, c is the wave speed in water and D is the diameter of the element.

A typical transducer of element size 7.5 mm and focal length 25 mm would have spot sizes of 1 mm, 0.5 mm and 100 μm if driven at 5, 10 and 50 MHz respectively. Clearly the higher frequency transducer will provide the best resolution and if the target is moving will maximise the number of measurement points that can be achieved. The relationship between the piston speed, ring thickness, transducer spot size, and ultrasonic pulsing rate is explored in a later section.

In the case studies that follow in this chapter, contact transducers have been used to measure the piston skirt oil film and a focused transducer used to measure a piston ring oil film.

12.3.3 Signal processing

The signal processing procedure is similar regardless of the type of transducer that is used. Initially a pulse is reflected back from the liner inside wall when there is no piston ring or skirt oil film in position in the measurement zone. The pulse is thus reflecting from a steel–air contact and, according to equation (12.1) one would expect complete reflection. This means that the reflected pulse is equal to the incident pulse. The time domain pulse (amplitude vs. time) is passed through a fast Fourier transform (***FFT***) to give the amplitude spectrum (amplitude vs. frequency). This is used as a reference signal.

Reflections are then recorded from the target oil film. Each reflected pulse is passed through an FFT and the resulting reflection spectra are divided by the reference signal. This gives the reflection coefficient spectra. This is then put into equation (12.12) to produce the thickness of the oil film. The output is then the oil film thickness measured at all the frequencies present in the ultrasonic pulse. The film thickness should be independent of frequency; in practice beyond the limits of the pulse bandwidth (where the energy of the emitted frequencies is low) the measurement becomes erratic and unreliable. Only data from the centre of the pulse bandwidth (with 6 dB) are used. The signal processing procedure is summarised in Fig. 12.6.

12.3.4 Pulsing rate and data capture

The piston is a fast-moving component and the sensor is stationary with respect to the liner. This means that a high pulsing rate is required to capture the transient oil film as it forms in the measurement zone.

The ultrasonic transducer pulses repeatedly. Typically a pulse repetition rate of 20 kHz is achieved (i.e. a pulse is generated every 50 μs) with conventional ultrasonic pulsing equipment. It is not possible to digitise and process the data at these sorts of speeds. Instead the data are stored on the digitiser as a series of pulse segments (the segment is the required oil film reflection extracted from the whole received signal). The array of segments is then downloaded to the PC and each segment extracted and processed using the route described above. The number of pulse segments that can be stored depends on the onboard memory. For example an oscilloscope limited to 250 k points per channel would be able to store 250 pulse segments each with 1000 points. Typically 250 to 500 pulse segments can be stored in this way, giving the same number of fluid film thickness readings recorded every 50 μs.

The number of signals that can be recorded as the piston ring sweeps past depends on this pulsing frequency and both the speed of the ring pack and the thickness of the individual ring. The relationship between the piston speed and the number of measurements that can be taken as it sweeps past is plotted as Fig. 12.7. This has been calculated for a 2 mm wide piston ring. Clearly as the pulsing frequency increases, more measurements can be taken. For realistic piston speeds it is apparent that pulsing frequencies of 50 kHz are desirable to obtain a reasonable resolution.

12.4   Case study: measurement from a piston skirt

The reduction in friction between the piston skirt and the liner is an important part of reducing in-cylinder losses. This is governed by operating conditions

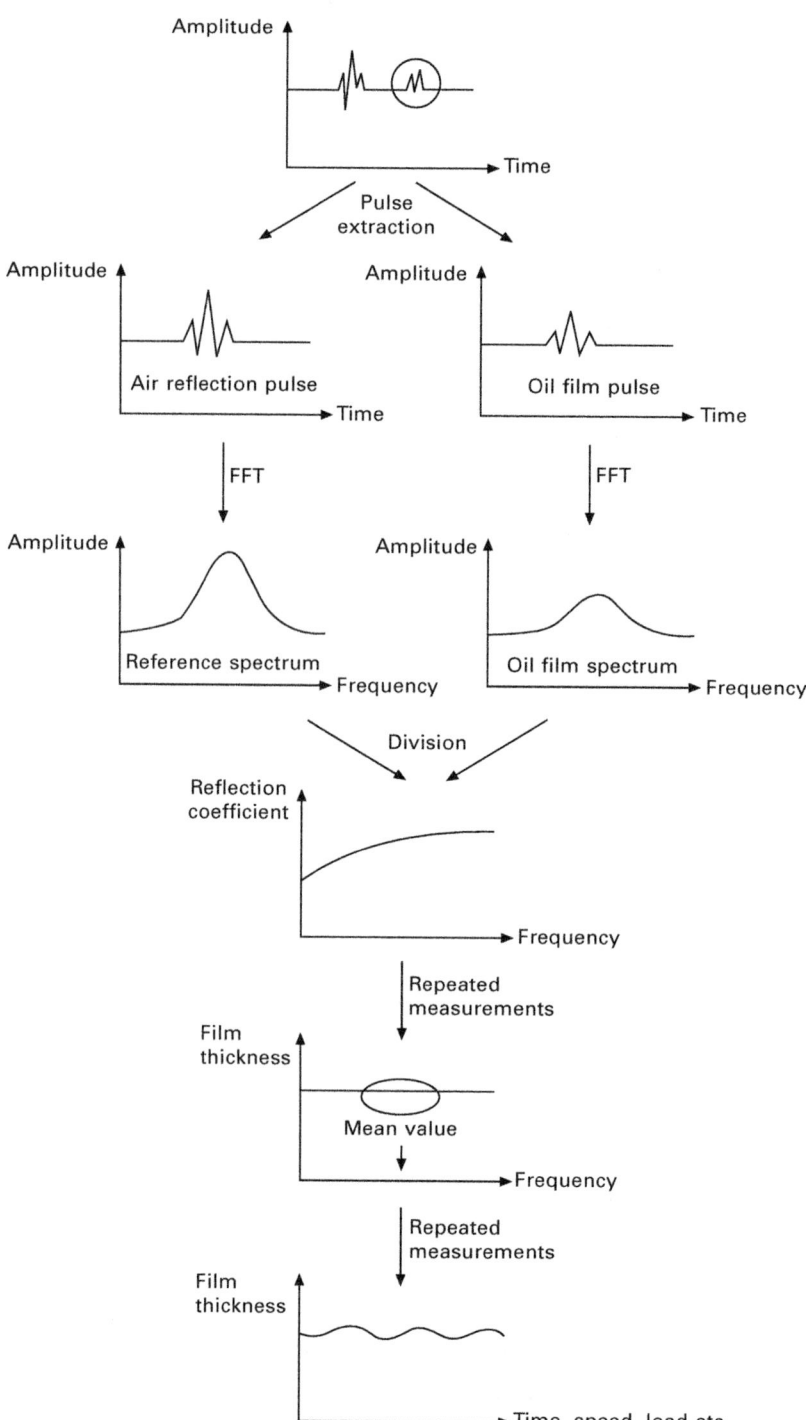

12.6 Flow chart describing the signal processing methods.

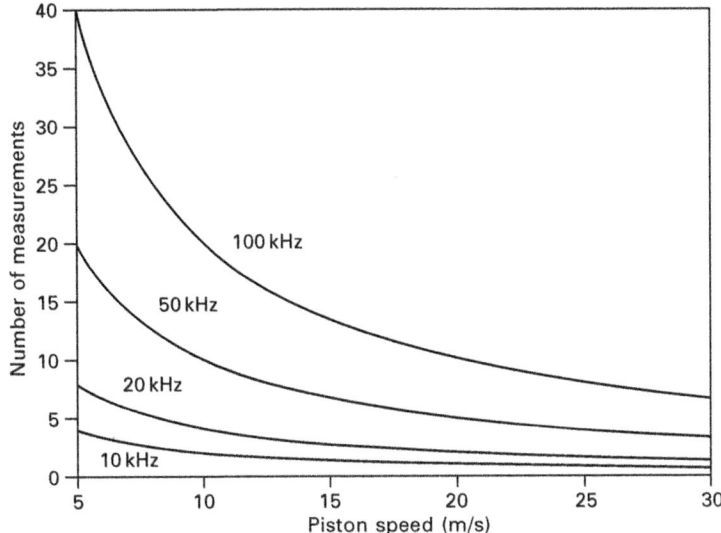

12.7 Relationship between pulse repetition rate, piston speed and the number of measurments recorded per piston ring sweep (calculated for a 2 mm wide piston ring).

and rheological properties of the lubricant. It is important to optimise the lubricant film formation to minimise friction and yet limit emissions. Therefore, the design of piston skirt and rings to encourage lubricant entrainment into the contact is important. Their axial profiles include chamfered edges or relief radii to create the necessary wedge effect for entraining of the lubricant with relative motion of surfaces. The material composition of the contacting surfaces is also important, particularly encouraging localised deformation under load, enhancing the separation and inducing formation of a thicker coherent film. Finally, cylinder bores or liners may be treated to provide lubricant retaining features. This promotes film by entrapment with cessation of entraining motion at dead centres, where there is no relative motion of surfaces. There are numerical models to predict such behaviour (for example, Balakrishnan and Rahnejat, 2005) but these need validating against experimental results.

The oil film that forms between the piston skirt and the liner occurs over a relatively large area (principally hydrodynamic film formation). This means that it is not essential to focus the transducer into a small contact region. This makes the use of simple glued-on piezo-electric transducers ideal.

12.4.1 Engine and dynamometer test bed

Figure 12.8 shows a 10 MHz *piezoelectric sensor* that was adhesively bonded to the wet side of the cylinder liner. The location of the sensor was such that

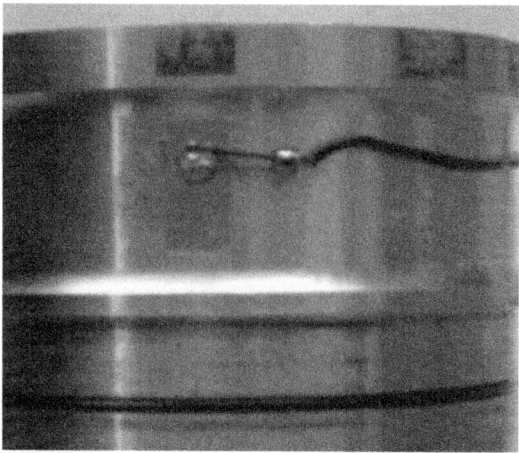

12.8 Photograph of the cylinder liner with the ultrasonic transducer bonded to the wet side.

12.9 Photograph of the piston; the area where average film measured on the piston skirt is indicated by the circle.

it was coincident with the piston skirt while the piston is at the TDC as shown in Fig. 12.3. The transducer was a thin element of piezo-electric material of diameter 7 mm and thickness 0.2 mm with 'wrap-around' electrodes as described in Fig. 12.4(b). The wires were fed out of the liner through the water jacket to UPR and digitising apparatus as described in Section 12.3.1.

Figure 12.9 shows the piston. The approximate dimensions of the measurement zone are shown on the figure. This zone will move up and down with respect to the piston as the cycle progresses. Clearly the resolution of this kind of transducer is low and it would not be possible to pick out the effect of the rings (the transducer would be recording form a region consisting of the rings and inter-ring gaps).

The test engine is a liquid-cooled, $449\,cm^3$ four-stroke four-valve single overhead-camshaft single cylinder engine. It produces $41\,kW$ at 9000 rpm, and is resisted by a transient A/C dynamometer (see Fig. 12.10). The stroke is 64 mm and the bore diameter is 96 mm. The nominal skirt clearance is $150\,\mu m$. The barrel of the engine is adapted to accept wet liners. A *TDC pulse* obtained from a digital rotary encoder is used to trigger the data acquisition system.

Results

A pulse is recorded from a reflection from the liner front face when the piston is remote from the sensor location. This pulse is reflected from a liner–air interface. This reflected pulse is then equal to the incident signal (since a wave is almost completely reflected at a solid–air interface) provides the reference signal (as described in Section 12.3.3). Figure 12.11 shows a pulse reflected back from the liner–air and liner–oil film–piston skirt interfaces. The later pulse is reduced in amplitude because part of the wave has been transmitted through the oil film to the piston.

The next step is to pass the reference signal and reflected pulses through a fast Fourier transform (FFT) to give an amplitude spectrum. Figure 12.12 shows the data of Fig. 12.10 as amplitude spectra. The data demonstrate that

12.10 Single cylinder test engine.

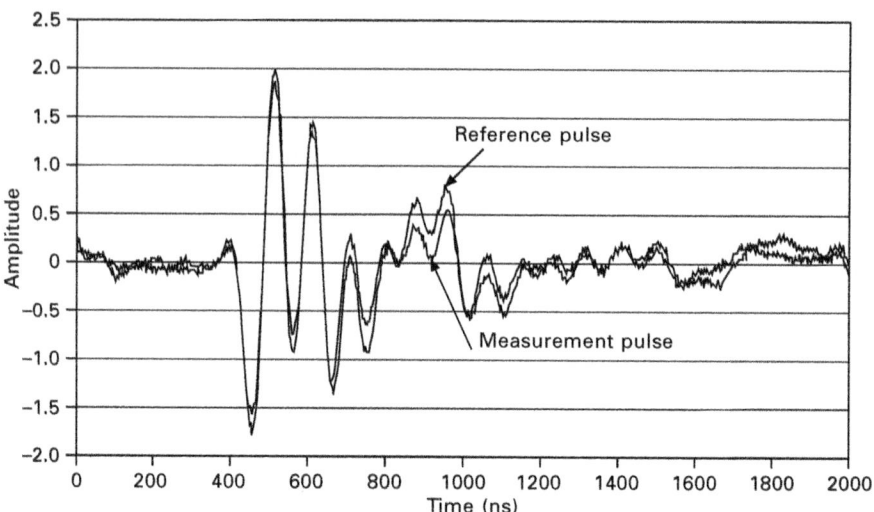

12.11 Sample pulses reflected from a liner–air interface and a liner–oil film–piston skirt interface.

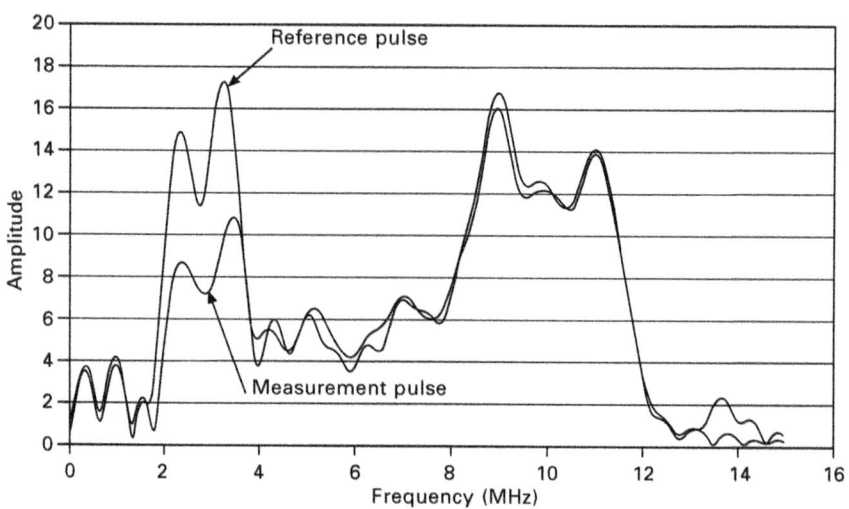

12.12 Amplitude spectra for the pulses of Fig. 12.4 obtained by performing a fast Fourier transform (FFT).

the centre frequency of the probe was 3 MHz. The figure also shows that the transducer has useful energy in the range 2 to 4 MHz. Dividing the oil film pulse by the reference pulse gives the reflection coefficient, as shown in Fig. 12.13 for a series of oil films.

In principle the reflection coefficient at any of the frequencies shown

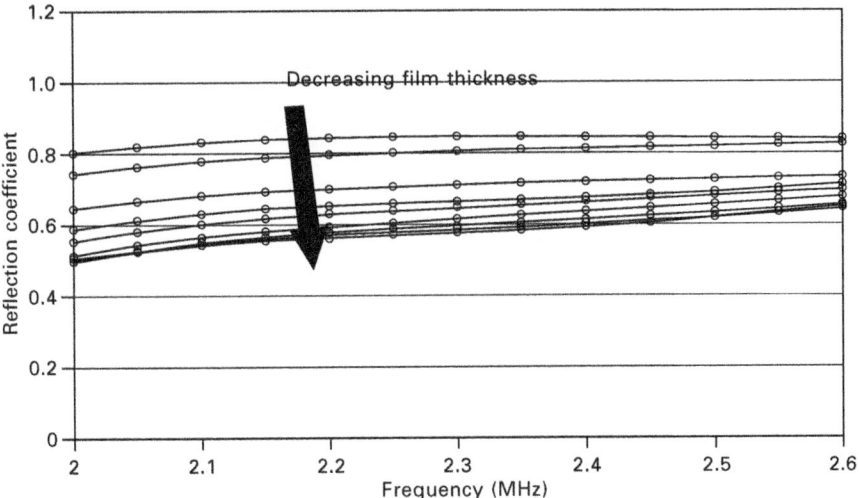

12.13 Reflection coefficient spectra obtained by dividing the oil film FFT by the reference FFT. The different curves represent different engine conditions and hence film thicknesses.

in Fig. 12.6 can be used to determine the film thickness using equation (12.11), provided it is within the bandwidth of the transducer. Typically the film thickness is determined over a range of frequencies and a mean presented.

12.4.2 Motored tests

To demonstrate how the ultrasonic transducer resolves the lubricant film thickness, a series of motored tests were carried out initially. Figure 12.14 shows a typical measurement of the *reflection coefficient* recorded as the piston passes the sensor location (recorded at a motor speed of 850 rpm).

Zone A represents the piston wall area above the ring zone. Zone B corresponds to the passage of the piston rings, where no measurement of oil film was recorded. Zone C represents the piston skirt passing the transducer.

Minimum film thickness of 9.4 μm was recorded when at the TDC. During the subsequent down-stroke the film thickness increases to 16.5 μm as the residual oil on the cylinder walls is entrained into the contact. Motored tests were performed at 850, 1800 and 6000 rpm. Figure 12.15 displays the increasing oil film thicknesses with increasing engine speed.

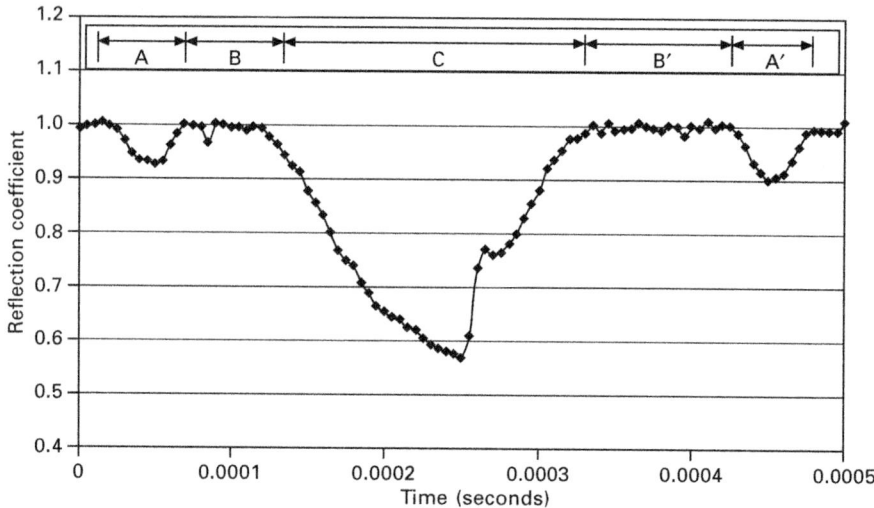

12.14 Plot of reflection coefficient recorded as the piston passes over the sensor location at an engine speed of 850 rpm.

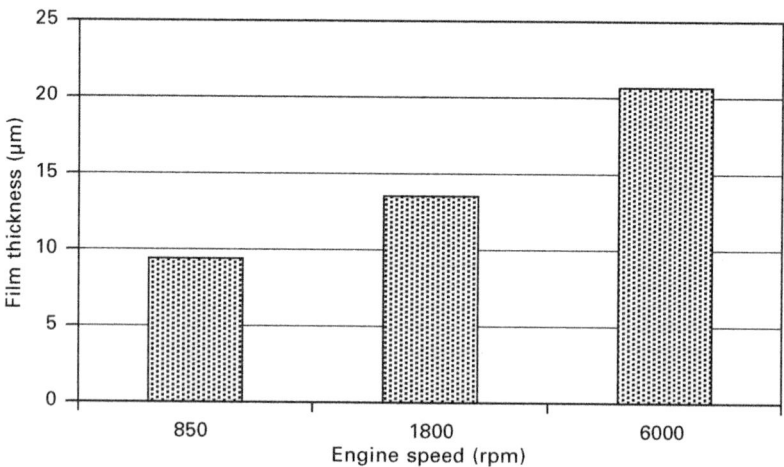

12.15 Measured liner–piston skirt oil film thickness for three motored engine tests.

12.4.3 Fired tests

The engine was run at different speeds at wide-open throttle condition, fully resisted by the dynamometer. For each test, repeatable conditions were ensured by monitoring air–fuel ratio, test cell pressure, humidity, coolant, and bulk oil temperature.

Figure 12.16 shows the film thickness measured at the idle speed of

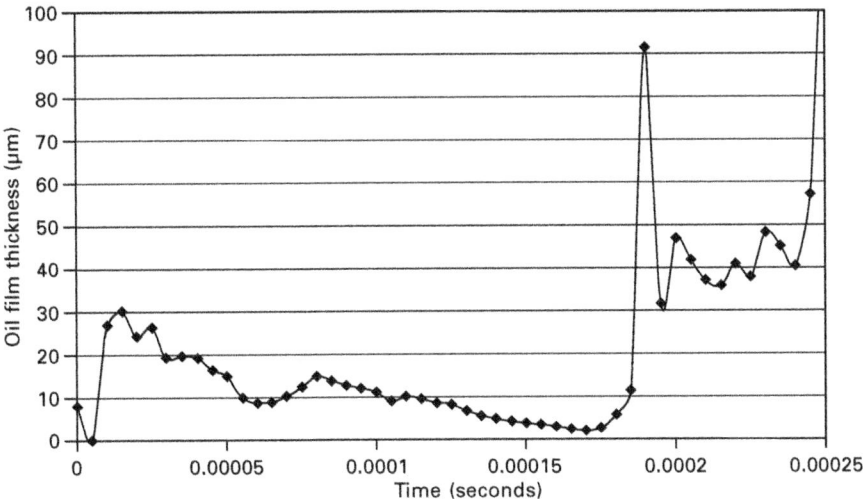

12.16 Plot of measured oil film thickness for a fired test performed at an engine speed of 1800 rpm.

1800 rpm. The measurement is recorded as the piston skirt sweeps passed the sensor (from ~0.5 to 1 μs) and then back again (from ~1 to 1.5 μs). At this stage it is not possible to exactly match the film measurement with the piston skirt geometry, as the piston location is not exactly defined with respect to the sensor (the TDC is approximately at 1 μs on the plot). A minimum film of approximately 2 μm is observed towards the end of the return stroke.

Prior numerical results, based on transient analysis of piston–cylinder liner contact, reported by Balakrishnan and Rahnejat (2002), predict a piston tilt of 0.085° with a side force of 4800 N. Under this condition the analysis indicates a minimum film thickness of 1.94 μm (see Fig. 12.17). Good agreement is observed between numerical prediction and measurement. Further validation could be obtained by measurements and modelling of the film at other locations on the piston stroke.

12.5   Case study: piston rings in a test bench

This study demonstrates the possibility of measuring the film from the contact between the liner and the rings themselves. It is a significantly more difficult experiment to perform because the small size of the lubricated region means that the ultrasonic pulse must be focused. This in turn requires the use of a coupling liquid that must be positioned between the transducer and the cylinder.

In this case study a cylinder with a double acting piston has been isolated on a test bench. This provides a convenient easy access platform to experiment

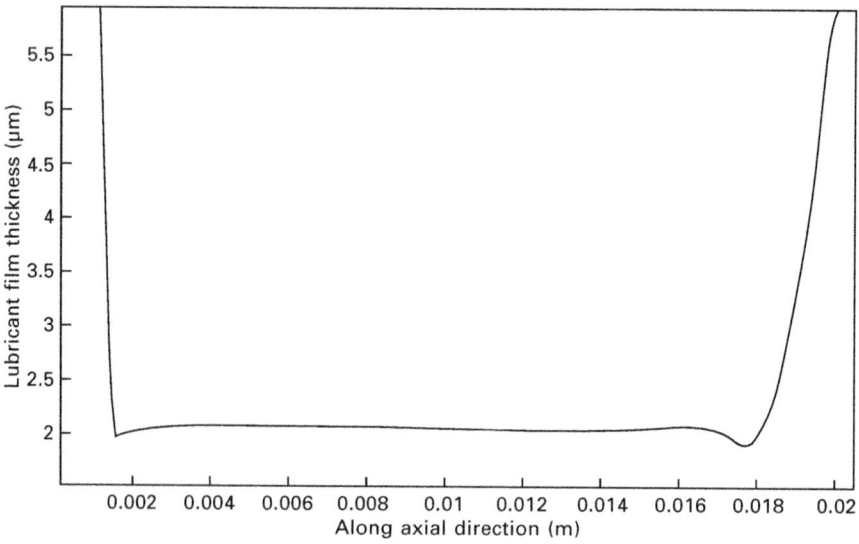

12.17 Axial lubricant film variation along the piston skirt at TDC from the numerical mode.

with the approach. In this instance the piston and cylinder have been modelled on those from a hydraulic motor; but the concept is the same as that of an automotive engine.

Several practical aspects were investigated such as attenuation in the cylinder material, response time and transducer resolution. While this study demonstrated that film thickness measurement is feasible, there are a number of practical considerations that require further work, principally the focusing and coupling of the ultrasonic transducer and the response time.

12.5.1   Piston test bench

The bench test apparatus used was originally developed to study the performance of different piston ring and cylinder bore designs by Sjödin and Olofsson (2003) during sliding motion. In this work a transducer was mounted on one of a pair of test cylinders. Figure 12.18 shows a schematic of the test bench. Two cylinders are mounted on the bench as shown. Both pistons are driven by means of a connecting rod attached to a crank on a motor. Each piston has two piston rings; one at each end. High-pressure oil (Shell Tellus T32) is fed into the cavity between the two rings (as shown in Fig. 12.19) and passes over the piston rings to the low pressure oil outlets. During the tests the pressure was varied from 10 to 30 MPa.

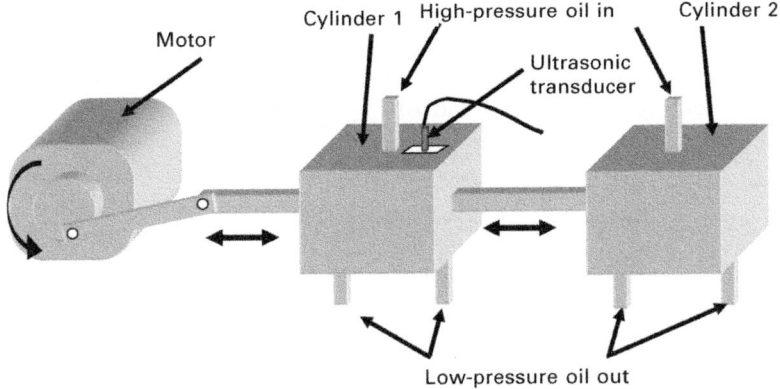

12.18 Schematic representation of the hydraulic motor piston ring test bench.

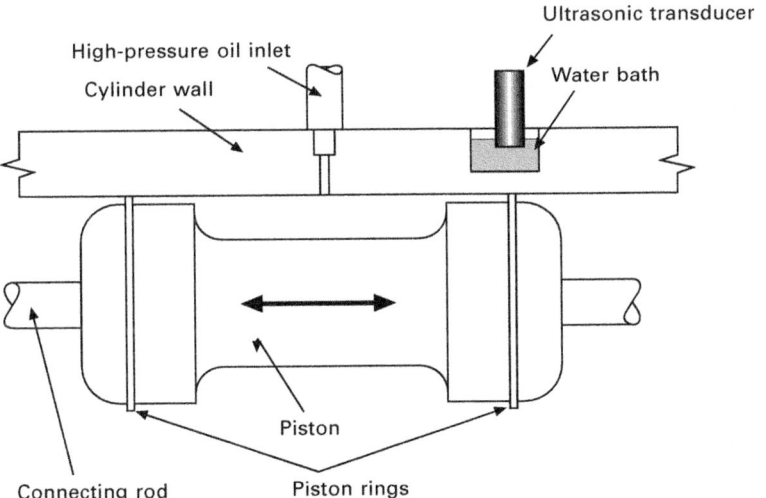

12.19 Schematic section showing the piston, rings, cylinder, high-pressure oil inlet and transducer position.

12.5.2 Piston ring and cylinder

In this study a split ring type with asymmetric outer crowning were used. The piston rings were of diameter 75 mm, thickness 2 mm and made of hardened bearing steel, while the piston and cylinder bore were made of grey cast iron and nodular cast iron respectively. The material was found to attenuate ultrasound to such an extent that the signal could not be adequately propagated through the full cylinder thickness. Some of the thickness was machined away to provide a shorter path for the ultrasound (and also to act as a water bath).

The ultrasonic measurement system was turned on and then the piston rig was started. The ultrasonic measurement system continuously measured the reflected signal while the piston oscillated past the measurement point. The reflected signal was recorded for post-processing to maximise the capture speed.

12.5.3 Ultrasonic instrumentation

In this work it is necessary to focus the wave onto the oil film and a transducer similar to that shown in Fig. 12.4(c) was used. The transducer was a nominal 10 MHz (the centre frequency was at 8.8 MHz) and 90% bandwidth. Focusing is achieved by means of a concave lens bonded to the piezo-element (as shown in Fig. 12.4(c)). The transducer has a focal distance of 75 mm in water and a corresponding focal area of diameter ~520 μm. The piston ring has a thickness of 2 mm with a slight crowning. The ultrasonic spot size falls within a region of the ring face that is virtually flat. Harper *et al.* (2005) describe in more detail the focusing of ultrasonic waves and the resulting spatial resolution.

The transducer has a focal distance of 75 mm in water and a corresponding focal area of diameter ~520 μm. The piston ring has a thickness of 2 mm with a slight crowning. The ultrasonic spot size falls within a region of the ring face that is virtually flat. Dwyer-Joyce *et al.* (2003) describe in more detail the focusing of ultrasonic waves and the resulting spatial resolution.

Figure 12.20 shows a schematic of the transducer location and focusing system. The transducer was mounted in a small water bath above the piston ring rig. A positioning fixture allows the accurate location of the transducer

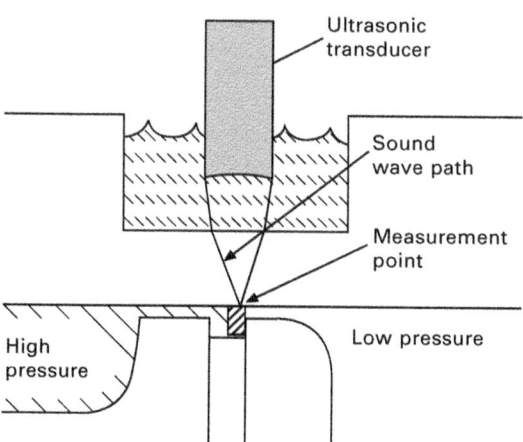

12.20 Schematic diagram of the transducer location and focusing through a water bath and the cylinder wall.

over the piston ring contact. The instrumentation described in Section 12.3.1 was used. In this test series, however, high-frequency data capture was not used (the software and systems were put in place after this work was completed). So results were limited to real time processing with relatively low pulsing frequencies.

12.5.4 Results

The transducer was set to pulse continuously at a repetition rate of 0.1 kHz. However, the rate at which each pulse could be captured and stored was 0.125 s. This proved to be a factor limiting the speed of operation of the piston. Reflected signals were recorded for all positions of the piston with respect to the transducer. The reflected pulse was divided by the reference pulse (as described in Section 12.3.3). Figure 12.21 shows the *reflection coefficient* variation as the piston ring passes under the transducer location on both its forward and backward stroke.

Two troughs are clearly seen when the piston passes beneath the measurement location. When an oil film is present more of the sound wave is transmitted and so the reflected amplitude drops. Ideally the reflection coefficient when recording away from this location should be unity; since the reflection is from a liner–air interface (i.e. equal to the reference signal) or liner–oil interface which would be very close to complete reflection. There is some variation as can be seen; however, the observed peaks and troughs are symmetrical, suggesting that it is a geometry effect.

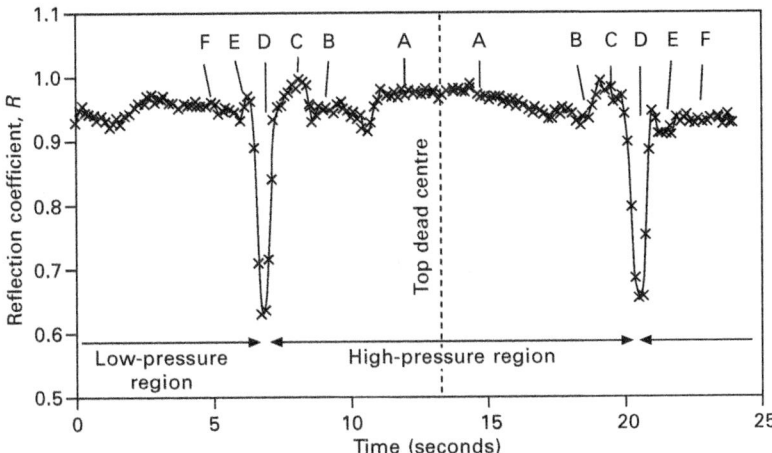

12.21 Plot of the recorded reflection coefficient as the piston moves from bottom dead centre (BDC) to dead centre (TDC) and back. Also shown are measurement points A, B, C, D, E, F.

Figure 12.22 shows a close-up of the piston ring and surrounding land. The measurement locations are approximately correlated with the features on the amplitude plot. The additional peaks are thought to be caused by either some residual oil left on the inside of the liner (in the low-pressure region only), or the extra thick *film* between the piston profile and the cylinder wall. However, at location D there are two distinct measurement points which give the same reading. These were taken as the minimum amplitude signal and were recorded for each passage of the piston.

Figure 12.23 shows a series of pulses recorded as the film reduces in size as the piston measurement location moves from C to D. The Fourier transform of a series of pulses is shown in Fig. 12.24. The reference pulse is marked. Each of the reflected pulses is divided by the reference pulse to give the reflection coefficient (shown in Fig. 12.25).

Finally equation (12.11) is used to obtain film thickness. This is plotted as Fig. 12.26. The measured film thickness data are, as expected, largely independent of frequency. The film thickness will be an aggregate of that which occurs over the spatial resolution of the transducer (520 µm), thus the crowning of the piston ring will tend to increase the measured thickness value. However, the level of crowning on the piston ring is small (over the distance of the spot size) that the effect on the film thickness measured is likely to be small.

Measurements were taken for three ring sliding speeds, 6, 9 and 12 mm/s and a range of contact pressures (from 10 to 30 MPa). The sliding speed of the piston was limited to 12 mm/s due to the maximum capture speed of the equipment. The film thickness remained at similar levels throughout most of the testing. Both increasing speed and reducing pressure caused a thicker

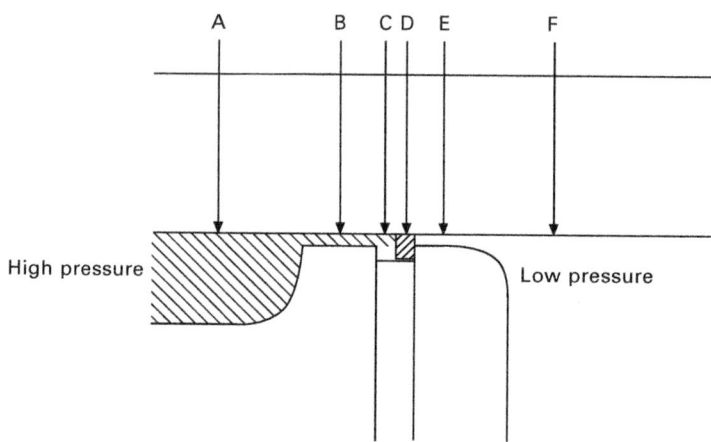

12.22 Close-up view of the piston profile either side of the piston ring showing six measurement points, A, B, C, D, E, F.

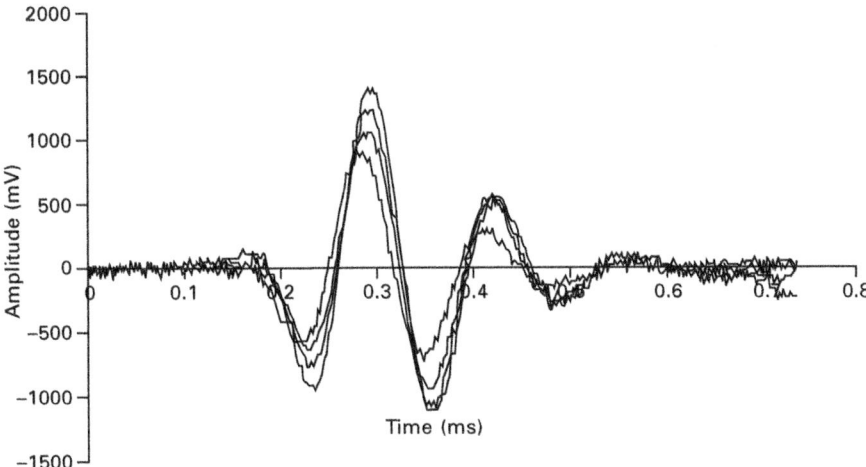

12.23 Pulses reflected from four oil films recorded as the ring moves from C to D (as the oil film reduces in thickness the amplitude decreases).

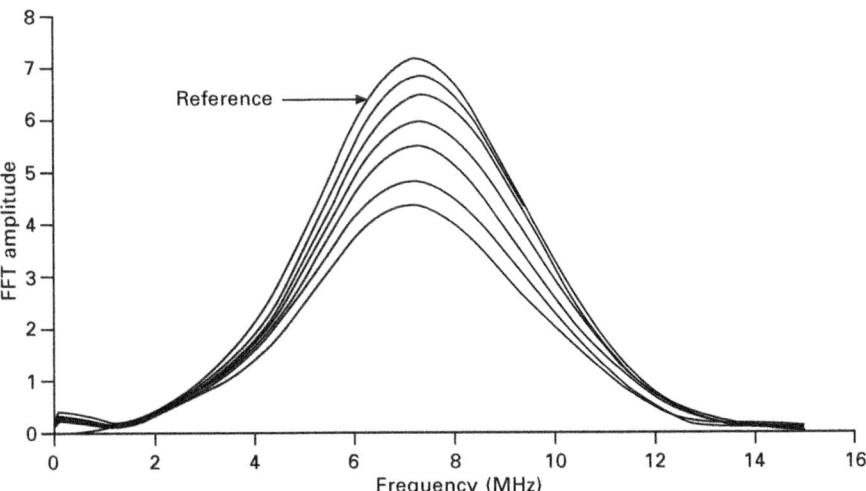

12.24 Fourier transform of the pulses shown in Fig. 12.23 (with two additional measurements added) recorded as the ring moves from C to D. The reference pulse is also shown.

oil film; but a great deal of scatter is observed. Figure 12.27 shows a series of film thickness measurements with changing speed and pressure.

It was believed that the spring loading caused by the piston ring was a more significant parameter than sliding speed or applied pressure. It should be noted that these tests were not intended as a comprehensive study of film

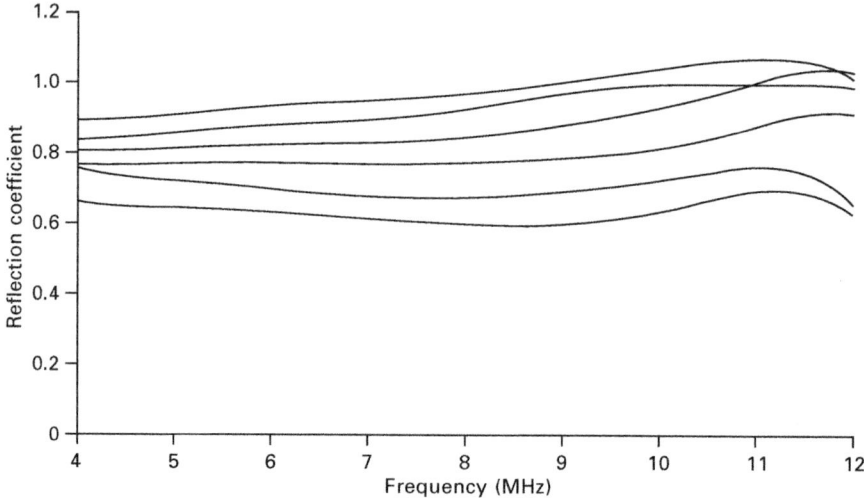

12.25 Reflection coefficient spectra obtained for the pulses of Fig. 12.23 recorded as the ring moves from C to D.

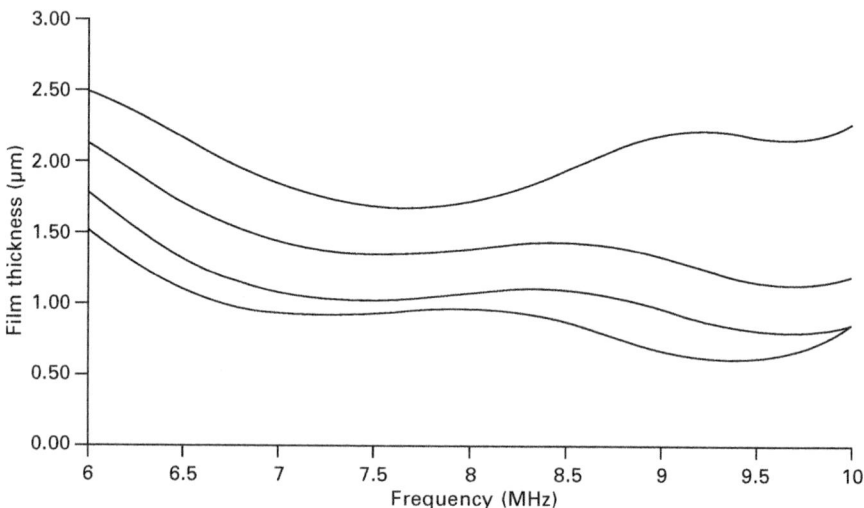

12.26 Film thickness determined using equation (12.11) for each frequency from the pulses of Fig. 12.23 recorded at the ring moves from C to D.

thickness effects, but rather as an indication of the possibility of obtaining measurements.

12.6   Overview

Measurement of the oil film formation between a piston and liner is always going to be a difficult task. The oil film is thin, rapidly moving and occurs

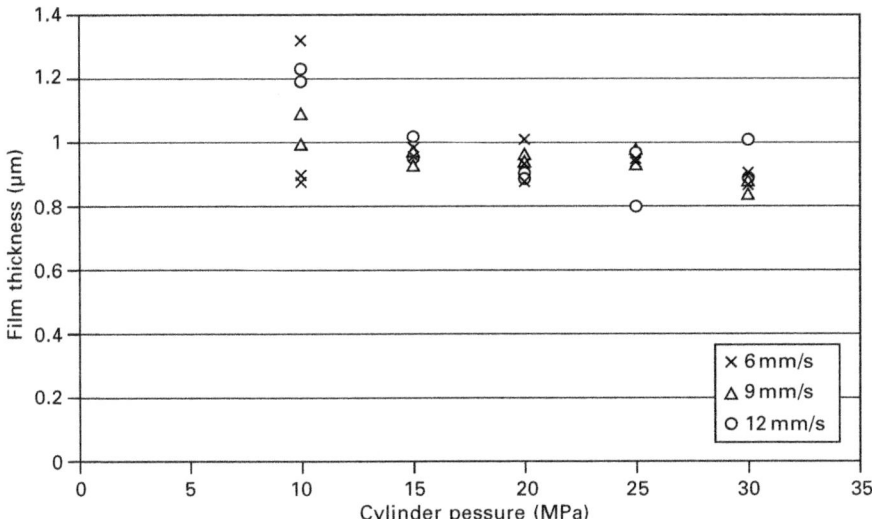

12.27 Oil film thickness between the piston ring and cylinder versus pressure for three different speeds.

over a small area. This is coupled with an engine environment that is subject to vibration, electrical noise and thermal gradients. Laser fluorescence, capacitance and ultrasonic reflection have all been demonstrated to be viable methods for measuring the film. However, the experimental details are tricky and this has limited measurements to laboratory-based simulators or engines under dynamometer testing. The prospect of installation of these kinds of sensors in a fleet car as part of an online condition monitoring system seems a long way off.

Nevertheless, the piston ring oil film is a critical part of the function of the engine. With the drive for emission reduction, and reduced fuel consumption this is an absolutely key parameter, and one that would be extremely useful to measure reliably and easily. If there were some viable method for its measurement the thickness could be controlled much more closely and the possibility of altering the film thickness (by means of restricting or enhancing oil flow, for example) for different regimes of engine operation arises.

This chapter has described the ultrasonic reflection approach. The most serious limitation of this method is the difficulty in measuring the small size of the oil film. The low film thickness does not pose a problem – measurements down to the sub-micron region are quite feasible – rather it is the fact that the rings are thin and so the target is small. This means that some method of focusing the ultrasonic wave is needed. This requires a lens in front of the transducer and hence the need for a liquid coupling bath.

Installing a transducer like this in an engine is possible (the water jacket provides a suitable couplant bath) but requires some considerable modification

(Green, 2008, has attempted this). One method by which this problem of focusing may be overcome is by the use of a different kind of ultrasonic sensor. There are some new sputter-coated aluminium nitride films that exhibit piezo-electric properties. These have two significant advantages; firstly they are very high frequencies (up to around 200–300 MHz) and also they can be activated by a very small electrode patterned onto the surface of the film. Drinkwater *et al.* (2008) describes a study where these kinds of transducer have been used to measure the oil film in a ball bearing as the ball passes rapidly over the transducer location. This technique could certainly be used with the piezo-coatings applied either to the liner or to the ring.

Another important parameter in cylinder operation is the residual film that remains on the liner wall. This strongly affects the formation of films under the rings and also how much oil ends up being burnt in the combustion chamber. At this stage it is not possible to ensure thin 'free-surface' oil films like this using an ultrasonic approach. The methods described in this chapter work best for a layered system where the layers either side of the liquid are solid (and of similar acoustic properties). So although metal–oil–metal is an ideal system, metal–oil–air is not; and equation (12.11) predicts a reflection coefficient very close to one for all but very thick oil layers. Free surface oil film measurement is much better approached using imbedded capacitance probes (see for example Sherrington *et al.*, 2002).

12.7   Conclusions

Oil film formation in the piston zone is a critical part of engine function. Measurement of the presence of the oil film and its thickness is difficult both because of the transient nature of the film and its geometrical dimensions. This chapter has described the use of ultrasonic reflection to measure oil films between sliding solid surfaces. The proportion of an ultrasonic wave that reflects from an oil film can be recorded and transformed to determine the oil film thickness, if the properties of both the oil and the solid surfaces are known. A key requirement of the method is to get the sound wave incident on the oil film and to capture and record at high speeds.

Two case studies have been presented. The first was a measurement of the film that forms between the piston skirt and the cylinder. This film forms hydrodynamically over a large region and so simple glued-on piezo-electric disks were suitable for supplying the ultrasonic wave. The oil film was measured in motor and fired engines and results agreed well with published numerical models.

The second case study looked at the film formation between the rings and the liner in a laboratory bench piston and cylinder apparatus. Focusing transducers were used to achieve measurements. The possibility was demonstrated, albeit at much lower operation speeds that would be found in the real engine.

These two studies demonstrate that it is possible to measure both skirt and ring film in an engine, and this should be possible to achieve in a fired engine on a dynamometer. However, the complexity of the experimental requirements means that application for condition monitoring in a fleet vehicle is probably some way off.

12.8   Acknowledgements

The author would like to thank PhD students and research assistants in the tribology group at Sheffield, especially David Green, Phil Harper, Yusuf Avan, Tom Reddyhoff, and Salmiah Kasolang.

Much of this work has arisen out of valuable collaborations with Prof Homer Rahnejat at Loughborough University, and Prof Ulf Olofsson at Royal Institute of Technology, Stockholm, Perfect Bore Ltd (now Capricorn Automotive) (Performance Motorsport Inc.), and Hagglunds, Sweden. The author would like to express his gratitude to all parties.

12.9   References and further reading

Andersson, B.S., (1991), Company's perspective in vehicle tribology. *Leeds–Lyon Symposium on Tribology*. Elsevier, pp. 503–506.

Baba, Y., Taneichi, Y., Ishima, T. and Obokata, T., (2006), Fundamental study on lubricant oil film behaviour by LIF and PIV, *Trans. JSME*, **72**, pp. 1001–1006.

Balakrishnan, S. and Rahnejat, H., (2002), Combined secondary piston dynamics and transient elastohydrodynamic analysis of piston skirt to cylinder contact conjunctions, AIMETA Symposium, Salerno, Italy.

Balakrishnan, S. and Rahnejat, H., (2005), Isothermal transient analysis of piston skirt-to-cylinder wall contacts under combined axial, lateral and tilting motion, *J. Phys. D: Appl. Phys.*, **38** pp. 787–799.

Baltazar, A., Rokhlin, S. and Pecorari, C., (2002), On the relationship between ultrasonic and micromechanical properties of contacting rough surfaces, *J. Mech. Phys. Solids*, **50**, pp. 1397–1416.

Brockett, C., Harper, P., Williams, S., Isaac, G., Dwyer-Joyce, R., Jin, Z. and Fisher, J., (2008), The influence of clearance on friction, lubrication and squeaking in large diameter metal-on-metal hip replacements, *J. Mater. Sci.: Mater. Med.*, **19**, pp. 1575–1579.

Brown, M.A., McCann, H. and Thompson, D.M., (1993), Characterization of the oil film behaviour between the liner and piston of a heavy duty diesel engine. *Tribological Insights and Performance Characteristics of Modern Engine Lubricants*, SAE/SP-93/996/932784

Courtney-Pratt, B.E. and Tudor, G.K., (1945), An analysis of the lubrication between the piston rings and cylinder wall of a running engine, *Proc. IMech.E.*, **152**, p. 384.

Dowson, D., Ruddy, B.L. and Economou, P.N., (1983), The elastohydrodynamic lubrication of piston-rings, *Proc. Roy. Soc.*, **386**, pp. 409–430.

Drinkwater, B.W., Zhang, K., Kirk, K.J., Elgoyhen, J. and Dwyer-Joyce, R.S., (2008), Measurement of rolling bearing lubrication using piezoelectric thin films, *ASME J. Tribology*, **131**(1), 011502.DOI:10.1115/13002324.

Dwyer-Joyce, R.S., Drinkwater, B.W. and Quinn, A.M., (2001), The use of ultrasound in the investigation of rough surface interfaces, *ASME J. of Tribology*, **123**, pp. 8–16.

Dwyer-Joyce, R.S., Drinkwater, B.W. and Donohoe, C.J., (2002), The measurement of lubricant film thickness using ultrasound, *Proc. Roy. Soc.*, **459A**, pp 957–976.

Dwyer-Joyce, R.S., Harper, P. and Drinkwater, B., (2003), A method for the measurement of hydrodynamic oil films using ultrasonic reflection, *Tribology Lett.*, **17**, pp. 337–348.

Dwyer-Joyce, R.S., Reddyhoff, T. and Drinkwater, B., (2004), Operating limits for acoustic measurement of rolling bearing oil film thickness, *STLE Trib. Trans.*, **47**, pp. 366–375.

Dwyer-Joyce, R.S., Harper, P. and Drinkwater, B., (2006), Oil film measurement in PTFE faced thrust pad bearings for hydrogenerator applications, *Proc. I.Mech.E. part A, J. Power & Energy*, **220**, pp. 619–628.

Furuhama, S. and Sumi, T.A., (1961), A dynamic theory of piston-ring lubrication: measurement of oil film thickness, *Bull. JSME*, **4**, pp. 744–752.

Furuhama, S., Asahi, C. and Hiruam, M., (1982), Measurement of piston-ring oil film thickness in an operating engine, *ASLE Trans*, **26**, pp. 325–332.

Gonzalez-Valadez, M., Dwyer-Joyce, R.S. and Lewis, R., (2004), Ultrasonic reflection from mixed liquid–solid contacts and the determination of interface stiffness, *Proceedings of the 31th Leeds–Lyon Symposium on Tribology*, Elsevier Tribology Series No. 48, pp. 313–322.

Graddage, M.J., Czysz, F.J. and Killinger, A., (1993), Field testing to validate models used in explaining a piston problem in a large diesel engine, *Trans ASME J. Eng. Gas Turb. Power*, **115**, pp. 721–727.

Green, D., (2008), Department of Mechanical Engineering, University of Sheffield, PhD Thesis.

Greene, A.B., (1969), Initial visual studies of piston–cylinder dynamic oil film behaviour, *Wear*, **13**, pp. 345–369.

Grice, N. and Sherrington, I., (1993), An experimental investigation into the lubrication of piston-rings in an internal combustion engine – oil film thickness trends, film stability and cavitation, SAE Paper 920651.

Hamilton, G.M. and Moore, S.L., (1974), Measurement of oil film thickness between the piston rings and liner of a small diesel engine, *Proc. I. Mech. E.*, **188**, pp. 253–261.

Harper, P., Dwyer-Joyce, R.S., Sjödin, U. and Olofsson, U., (2005), Evaluation of an ultrasonic method for measurement of oil film thickness in a hydraulic motor piston ring, *Proceedings of the 31th Leeds–Lyon Symposium on Tribology*, Elsevier Tribology Series No. 48, pp. 305–312.

Kendall, K. and Tabor, D., (1971), An ultrasonic study of the area of contact between stationary and sliding surfaces, *Proc. R. Soc. Lond. A*, **323**, pp. 321–340.

Krolikowski, J. and Szczepek, J., (1991), Prediction of contact parameters using ultrasonic method, *Wear*, **148**, pp. 181–195.

Krolikowski, J. and Szczepek, J., (1993), Assessment of tangential and normal stiffness of contact between rough surfaces using ultrasonic method, *Wear*, **160**, pp. 253–258.

Lloyd, T., (1969), The hydrodynamic lubrication of piston rings, *Proc. I. Mech. E.*, **183**, pp. 28–34.

Ma, M.-T., Smith, E.H. and Sherrington, I., (1995), A three dimensional analysis of piston ring lubrication, part I Modelling, *Proc. Inst. Mech. Eng.*, **209**, pp. 1–15.

Nagy, P.B., (1992), Ultrasonic classification of imperfect interfaces, *J. Non-destructive Evaluation*, **11**, pp. 127–139.

Pialucha, T. and Cawley, P., (1994), The detection of thin embedded layers using normal incidence ultrasound, *Ultrasonics*, **32**, pp. 431–440.

Reddyhoff, T., Kasolang, S., Dwyer-Joyce, R.S. and Drinkwater, B., (2005), The phase shift of an ultrasonic pulse at an oil layer and determination of film thickness and damping, *Proc. I. Mech. E. part J: J. Eng. Tribology*, **219**, pp. 387–400.

Reddyhoff, T., Dwyer-Joyce, R.S. and Harper, P., (2008), A new approach for the measurement of film thickness in liquid face seals, *Tribology Trans*, **51**, No. 2, pp. 140–149.

Richardson, D.A. and Borman, G.L., (1991), Using fibre optics and laser fluorescence for measuring thin oil films with applications to engines, SAE Tech. Paper Ser., No. 912388, Society of Automotive Engineers, Warrendale, Pennsylvania.

Sanda, S., Murakami, M., Noda, T. and Konomi, T., (1997), Analysis of lubrication of a piston ring package, *Trans. JSME*, **40**, pp. 478–486.

Seki, T., Nakayama, K., Yamada, T., Yoshida, A. and Takiguchi, M., (2000), A study on variation in oil film thickness of a piston ring package: variation of oil film thickness in piston sliding direction, *Jap. Soc. Auto. Eng. Review*, **21**, pp. 315–320.

Shaw, B.T., Hoult, D.P. and Wong, V.W., (1992), Development of engine lubricant film thickness diagnostics using fiber optics and laser fluorescence, SAE Tech. Paper Ser., No. 920651, Society of Automotive Engineers, Warrendale, Pennsylvania.

Shaw, M.C. and Nussdorfer, T.J., (1946), Camera techniques expose oil–film behaviour, *SAE J.*, **54**, 41.

Sherrington, I. and Smith, E.H., (1985), Experimental methods for measuring the oil film thickness between the piston rings and cylinder wall of internal combustion engines, *Trib. Int.*, **18**, 315–320.

Sherrington, I., Freeman, S.A., Grice, N. and Smith, E.H., (2002), Simultaneous measurement of oil availability and minimum operating film thickness in a piston-ring simulator, *Proceedings of 10th Nordic Symposium on Tribology*, Nordtrib 2002.

Silk, M.G., (1984), *Ultrasonic Transducers for Nondestructive Testing*, Hilger, Bristol.

Sjödin, U.I. and Olofsson, U.L.-O., (2003), Initial sliding wear on piston rings in a hydraulic motor, *Wear*, **254**, 1208–1215.

Tattersall, A.G., (1973), The ultrasonic pulse-echo technique as applied to adhesion testing, *J. Phys. D: Appl. Phys*, **6**, 819–832.

Taylor, C.M., (1993), Lubrication regimes and the internal combustion engine, *Engine Tribology*, Elsevier, pp. 75–87.

Ting, L.L., (1980), Development of a laser fluorescence technique for measuring piston ring oil film thickness, *ASME Transactions*, **102**, pp. 165–170.

Todsen, U. and Niethus, K.U., (2006), Optical ways to improve the tribological system piston–ring–liner, SAE paper 2006-01-0527.

Wing, R.D. and Saunders, O., (1972), Oil film temperature and thickness measurements on the piston rings of a diesel engine, *Proc. I.Mech.E.*, **186**, p. 1.

Zhang, J., Drinkwater, B.W. and Dwyer-Joyce, R.S., (2006), Monitoring of lubricant film failure in a ball bearing using ultrasound, *ASME J. Tribology*, **128**, pp. 612–618.

13

Surface texturing for in-cylinder friction reduction

I. ETSION, Technion, Israel

Abstract: This chapter reviews the current effort being made worldwide on surface texturing in general and on laser surface texturing (LST) for in-cylinder friction reduction in particular. The chapter reviews briefly the work that was done on LST for in-cylinder application prior to 2005 followed by the new developments since 2005. LST may be successfully applied to piston rings and cylinder liners resulting in up to 4.5% reduction in fuel consumption or engine torque.

Key words: surface texturing, friction reduction, fuel efficiency, energy conservation.

13.1 Introduction

Surface texturing has emerged in the last decade as a viable option of surface engineering, resulting in significant improvement in load capacity, wear resistance, friction coefficient, etc., of tribological mechanical components. This chapter reviews the current effort being made worldwide on surface texturing in general and on laser surface texturing for in-cylinder friction reduction in particular.

Perhaps the most familiar and earliest commercial application of surface texturing is that of cylinder liner honing (Jeng, 1996; Willis, 1986). Today surfaces of modern magnetic storage devices are commonly textured and surface texturing is also considered as a means for overcoming adhesion and stiction in *micro-electro mechanical systems (MEMS)*. Fundamental research work on various forms and shapes of surface texturing for tribological applications is carried out by several research groups worldwide and various texturing techniques are employed in these studies including machining, ion beam texturing, etching techniques and laser texturing. Interestingly almost all these fundamental works are experimental in nature and most of them are motivated by the idea that the surface texturing provides micro-reservoirs to enhance lubricant retention or micro-traps to capture wear debris. Usually, optimisation of the texturing dimensions is done by a trial and error approach.

A more analytical approach to surface texturing started some 40 years ago. Hamilton *et al.* (1966) presented surface texturing in the form of micro-

458

asperities that act as micro-hydrodynamic bearings. This idea was promoted mainly for parallel sliding, as is the case in mechanical seals (Anno *et al.*, 1968, 1969). An etching technique was used for the texturing, and both theoretical and experimental work was performed in an attempt to optimise the texturing dimensions. Some 30 years later, Etsion and Burstein (1996) presented a model for mechanical seals with regular micro-surface structure, showing a substantial improvement in seal performance when evenly distributed hemispherical *micro-dimples* are present on one of the mating seal faces. The work in Etsion and Burstein (1996) was followed by an experimental study (Etsion *et al.*, 1997) in which laser-textured seal rings were tested in oil showing that the spherical dimple shape can be optimised and that an optimum dimple depth over dimple diameter ratio exists that maximises the film stiffness and the pressure velocity (PV) factor at seizure inception.

Various forms and techniques of surface texturing were developed over the years for enhancing tribological performance. The *vibro-rolling method* was developed by Schneider (1984). It consists of producing shallow grooves by plastic deformation using a hard indenter on metallic parts. Extensive work has been done on vibro-rolling in Eastern Europe (Bulatov *et al.*, 1997) that somehow went unnoticed in the western world. At about the same time Suh and co-workers (Saka *et al.*, 1984) in the US presented the idea of modulated surface for removing oxide wear debris from the interface of electrical contacts. They initially used an etching technique which was later replaced by abrasive machining to form grooves (Saka *et al.*, 1989; Tian *et al.*, 1989, Suh *et al.*, 1994; Mosleh *et al.*, 1999) that they termed *undulated surfaces*. Like Saka *et al.* (1984) the function of the undulations is to act as traps for wear debris, thereby reducing the ploughing and deformation components of friction and wear.

Reactive ion etching (RIE) was employed by a group lead by Kato in Japan (Wang *et al.*, 2002, 2003; Wang and Kato, 2003) to study the effect of surface texturing, in the form of micro-dimples, on parallel sliding faces of SiC in water. Other techniques include *abrasive jet machining* (Wakuda *et al.*, 2003), Lithographie, Galvanoformung, Abformung (LIGA) (Stephens *et al.*, 2004), and *lithography* and *anisotropic etching* (Pettersson and Jacobson, 2003). Table 13.1 summarises these various techniques and shows the global spread of interest in surface texturing.

As can be seen various techniques can be employed for surface texturing but *laser surface texturing* (LST) is probably the most advanced so far. LST produces a very large number of micro-dimples on the surface (see Fig. 13.1) and each of these micro-dimples can serve either as a micro-hydrodynamic bearing in cases of full or mixed lubrication, a micro-reservoir for lubricant in cases of starved lubrication conditions, or a micro-trap for wear debris in either lubricated or dry sliding. A good review of the state of the art of LST covering this subject to 2005 can be found in Etsion (2005).

Table 13.1 Various techniques utilised for surface texturing other than LST

Technique	Institute	References	Comments
Vibro-rolling	Inst. of Mechanical Engineering Problems, Petersburg, Russia	Schneider (1984), Bulatov *et al.* (1997)	Various mechanical components
Undulated surfaces	MIT, Cambridge, USA	Saka *et al.* (1984, 1989), Tian *et al.* (1989), Suh *et al.* (1994), Mosleh *et al.* (1999)	Wear particle trapping sites
Reactive ion etching (RIE)	Tohoku Univ., Sendai, Japan	Wang *et al.* (2002, 2003), Wang and Kato, (2003)	Lab. tests and limited theoretical modelling
Abrasive jet machining & eximer laser	FCRA & AIST, Nagoya, Japan	Wakuda *et al.* (2003)	Pin on disk tests
LIGA	Univ. of Kentucky, Lexington, USA	Stephens *et al.* (2004)	Lab. tests on thrust rings. Limited modelling
Lithography & anisotropic etching	Uppsala Univ., Uppsala, Sweden	Pettersson and Jacobson (2003)	Reciprocating test rig

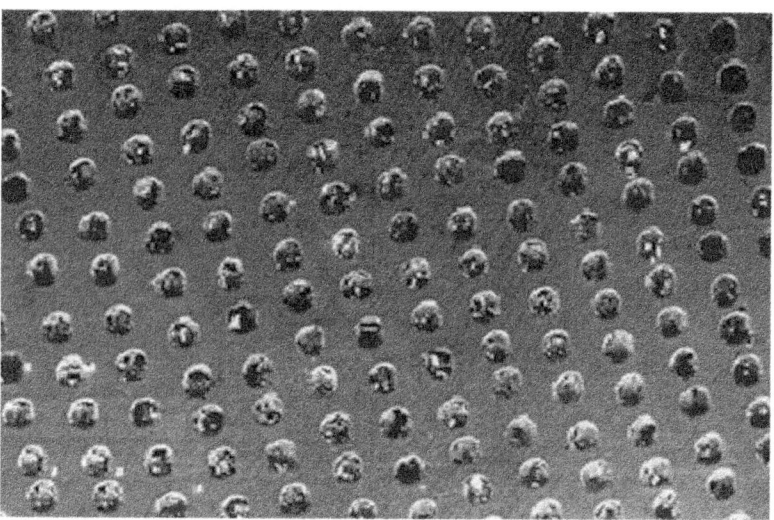

13.1 LST regular micro-surface structure in the form of micro-dimples.

In the next section the work that was done on LST for in-cylinder application prior to 2005 will be described briefly followed by the new developments since 2005.

13.2   Laser surface texturing (LST) for friction reduction in engines

Most of the work on laser surface texturing for in-cylinder friction reduction was devoted to piston rings. The potential benefits of applying LST to piston rings has been demonstrated theoretically by Ronen *et al.* (2001) who solved simultaneously the Reynolds equation and a dynamic equation of the ring radial motion. Optimum texturing parameters for minimum friction force were found in Ronen *et al.* (2001), showing a potential reduction of about 30% compared with non-textured rings under full lubrication conditions. Good agreement was found with laboratory test results in the experimental work performed by Ryk *et al.* (2002). In addition it was found that optimum LST is beneficial under starvation as well, where the dimples serve as micro-reservoirs for lubricant.

In 2004 the use of laser texturing in the form of micro-grooves on cylinder liners of internal combustion engines was presented at the 14th International Colloquium Tribology in Esslingen (see for example Golloch *et al.*, 2004) showing lower fuel consumption and wear. This technique, called *laser honing*, is now commercially available from the Gehring Company in Germany (http://www. Gehring.de).

The early work on LST piston rings (Ronen *et al.*, 2001; Ryk *et al.*, 2002) considered full width texturing but very soon it was realised that partial texturing may be more beneficial. The difference between these two types of LST is demonstrated in Fig. 13.2 and the rational for the better performance of partial LST is fully described in Brizmer *et al.* (2003). Basically, in the full width LST the area density of the dimples is relatively small and each

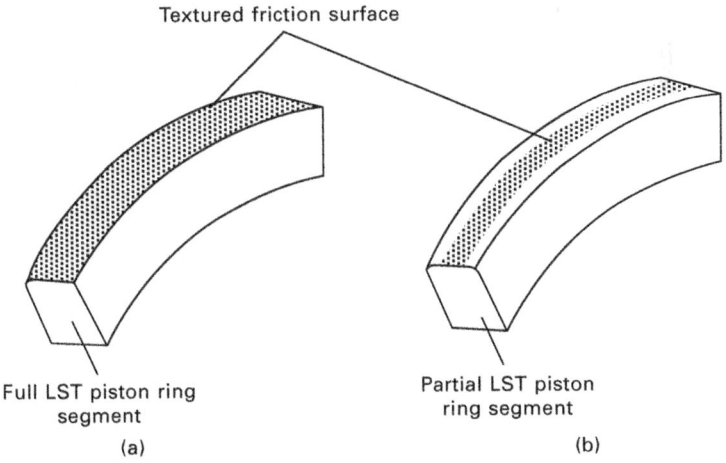

Textured friction surface

Full LST piston ring segment

(a)

Partial LST piston ring segment

(b)

13.2 Segments of piston rings: (a) fully textured; (b) partially textured.

dimple acts individually as a micro-hydrodynamic bearing with negligible interaction between neighbouring dimples. In the partial LST case the dimple area density is higher and the dimples act collectively to form an equivalent step bearing with higher load-carrying capacity and much better performance under high-pressure differential. Both theoretical modelling (Kligerman *et al.*, 2005) and experimental verification (Ryk *et al.*, 2005) of the concept were done on relatively simple flat face *piston ring* specimens. A schematic of a test for such flat and parallel rubbing surfaces is presented in Fig. 13.3, taken from Ryk *et al.* (2005). A steel specimen (1) with two chrome coated flat surfaces, each having a width of 3 mm and length of 11 mm, is fixed in a special holder. The reciprocating planar plate (2) is made of cast-iron. The operating normal load F_e is applied to the specimen's holder by means of accurate weights. Fully formulated engine oil Ultra-40 (equivalent of SAE 40) with a viscosity index 95 was used and the test rig was provided with a heat source for heating the friction zone and maintaining a selected ambient temperature to simulate as close as possible the lubricant viscosity conditions in an internal combustion engine. The LST parameters were: dimple diameter of about 80 μm, dimple depth of about 8 μm, and area densities of 10% for full LST and 50% for partial LST. It was shown in Kligerman *et al.* (2005) that in partial LST an optimum textured portion (the ratio between the width of the textured portion to the total ring width) of 0.6 holds for a wide range of LST parameters and operating conditions regardless of the position of this textured portion. Hence, a textured portion of 0.6 was applied to the partial LST specimens symmetrically at their ends (see Fig. 13.3).

Typical results of the average friction force versus crank angular velocity are shown in Fig. 13.4 for the reference untextured case and for the two modes of full and partial LST cases. As can be seen the average friction increases with speed in all three cases as would be expected. Clearly the LST has a substantial effect on friction reduction compared with the untextured reference case. The average friction obtained with the full LST is about 40 to 45% lower than in the reference case at low speeds around 500 rpm,

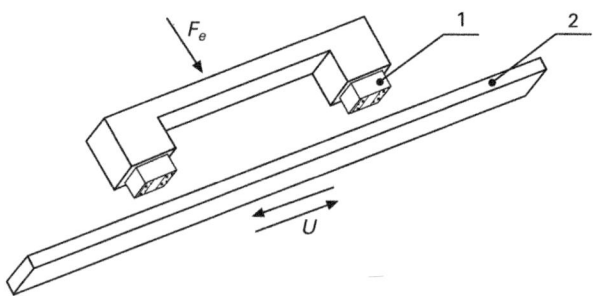

13.3 Schematic of the test (Ryk *et al.*, 2005).

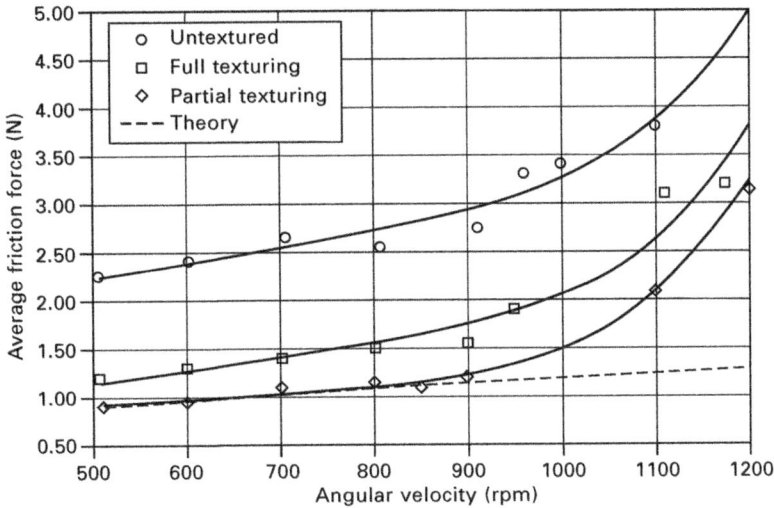

13.4 Time average friction force vs. crank angular velocity for external normal pressure 0.3 MPa (Ryk *et al.*, 2005).

and 23 to 35% lower at higher speeds around 1200 rpm. These percentage differences between the average friction in the untextured and full LST cases were almost independent of the external normal load, and slightly decrease with increasing angular velocity. The results in Fig. 13.4 clearly show the additional reduction in friction that can be obtained with partial LST over that of the full LST case as was predicted in Kligerman *et al.* (2005). This additional reduction varies from 12 to 29% depending on the load and speed. In the tests described in Ryk *et al.* (2005) the maximum benefit of the partial LST was obtained with the combination of lowest speed and highest load.

Some preliminary real firing engine tests that were performed with LST barrel shape rings showed very little friction reduction compared with the same untextured rings at low speeds below 2000 rev/min. Above 2000 rev/min this little benefit of the LST vanished completely. It seems that the barrel shape, which presumably was arrived at by trial and error experience over many years (Taylor, 1998), is not a good candidate for LST. The crowning of the ring face by itself provides strong hydrodynamic effect that masks the weaker hydrodynamic effect of the surface texturing especially at high speeds. Indeed, a more appropriate comparison between the performance of non-textured barrel shape and optimum partial LST cylindrical shape rings, which was performed on a laboratory reciprocating test rig (Ryk and Etsion, 2005), showed that a friction reduction of up to about 25% can be obtained with partial LST cylindrical face rings.

A four-cylinder Ford Transit naturally aspirated 2500 cm³ diesel engine was used to test the effect of the LST as applied to the upper set of rings

(Etsion and Sher, 2009). The engine was mounted on a Hofmann eddy-current dynamometer test bench and a control system controlled the fuel metering of the engine to keep a constant prescribed engine speed and a constant prescribed level of engine partial load. Half keystone top piston rings were obtained for testing. The rings outer diameter was 93.7 mm and their nominal width was 2.5 mm. The peripheral faces of the rings are coated with a chrome base coating that forms the ring profile in contact with the cylinder liner. Figure 13.5(a) shows a cross-section of a ring with a cylindrical face profile to which partial laser texturing was applied. Figure 13.5(b) shows a ring with a barrel face profile that is a series production ring and was used as the baseline without texturing. In addition, cylindrical face rings identical to these shown in Fig. 13.5 but without the chrome coating were also obtained for texturing. The *laser texturing* was applied at both axial ends of the cylindrical face rings as shown schematically in Fig. 13.6 with a textured portion $B/W = 0.6$.

Plate VII (between pages 514 and 515) shows a 3D optical profilometer scan of the partial LST Cr coated cylindrical face ring (series 2 in the following). The dimples are located symmetrically along the circumference of the ring on both ends of its width, leaving the central portion of the ring width untextured. The softer uncoated ring (series 3) resulted in somewhat larger and deeper dimples with higher area density compared with the Cr coated ring. Note also from Plate VII that the laser texturing results in bulges of raised material around the rim of the dimples. The height of these bulges was about 2 μm for the Cr coated rings and about 4 μm for the uncoated rings. From previous test rig tests it was found that these bulges are easily

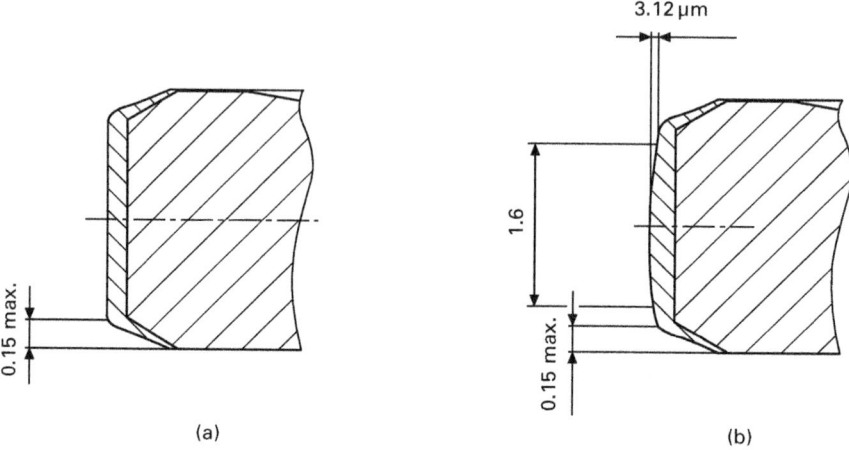

13.5 Cross-sections of cylindrical (a) and barrel shape (b) Cr coated piston rings.

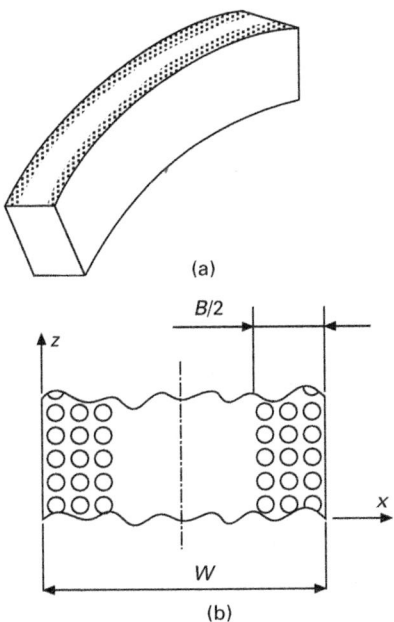

(a)

(b)

13.6 Partial LST cylindrical face piston ring (a) schematic of a partial LST ring segment (b) top view of the two symmetrically located LST zones of width *B/2* each at both axial ends of the piston ring having a face width *W*.

removed during the first few reciprocating cycles and hence, no special post LST process is needed to remove them prior to testing.

A comparison between the performance of a reference untextured conventional barrel shape rings with Cr coating (series 1) and optimum partial LST cylindrical shape rings (series 2 Cr coated, and series 3 uncoated), is depicted in Fig. 13.7. The laser-treated Cr coated rings (series 2) seem to perform somewhat better at low engine speeds while the laser treated without the Cr coating rings (series 3) seem to be in favour at high engine speeds. This result is probably due to the somewhat deeper dimples of the series 3 rings and is in agreement with the model prediction in Kligerman *et al.* (2005) where deeper dimples gave lower friction as the speed increases. The laser-treated rings, however, are superior to the baseline reference rings (series 1) over the entire range of engine speed. It is clearly shown in Fig. 13.7 that the partial LST piston rings exhibited up to 4% lower fuel consumption. This level of fuel economy was obtained at 1800 rpm, which corresponds to the maximum torque of the engine.

Other in-cylinder components that were studied recently for the effect of surface texturing are the cylinder liner (Rahnejat *et al.*, 2006) and the piston pin (Etsion *et al.*, 2006). The aim of the research in Rahnejat *et al.*

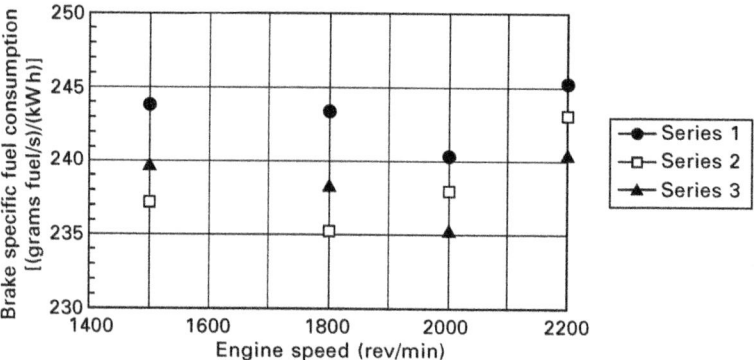

13.7 Engine specific fuel consumption vs. engine speed. Series 1: Barrel, chrome coated, baseline, Series 2: Flat, chrome coated, laser treated, Series 3: Flat, no chrome, laser treated (Etsion and Sher, 2009).

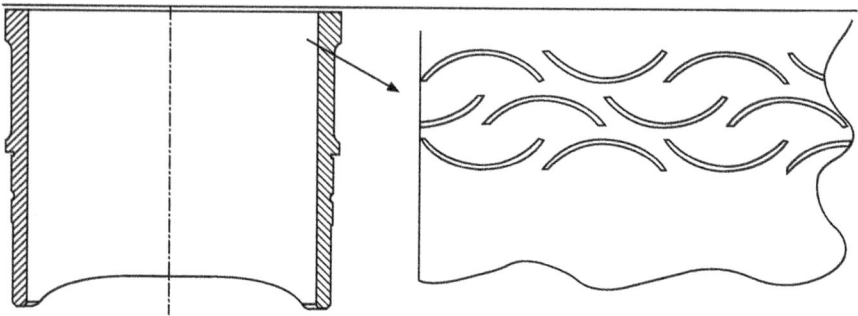

13.8 Laser-etched cylinder liner (Rahnejat *et al.*, 2006).

(2006) was to undertake a comparative study between standard cross-hatched (honed) super-finish cylinder liners for high-performance engines and those of identical material construction but with surface laser-etched profiles (see Fig. 13.8) to retain a lubricant film through entrapment. The hypothesis is that, contrary to the belief that smooth surfaces promote reduced friction, the inclusion of laser-etched features actually improves lubricant film, which can reduce frictional losses. The groove patterns and their interspacing and depth were optimised through a number of numerical simulation studies in order to maximise film thickness at reversal positions. Following this theoretical study to optimise the geometry of the laser-etched features, a 449 cm³ four-stroke, single-cylinder engine was used to test the concept. Interestingly a 4.5% gain in torque was obtained with the laser etched pattern cylinder in Rahnejat *et al.* (2006), which is very similar to the percentage gain in fuel consumption reported in Etsion and Sher (2009). Here too the maximum benefit was obtained at the pick torque.

Scuffing resistance of piston pin provided by laser surface texturing in comparison with CrN and *diamond-like carbon* (*DLC*) coatings and a base-line standard piston pin was studied in Etsion *et al.* (2006). Scuffing inception could be obtained with low-viscosity base oil. In this case, all the treated pins performed better than the standard one, with the laser surface texturing offering the best performance.

The search for better texturing in reciprocating sliding suitable for in-cylinder application is still going on as can be seen for example in Weidner *et al.* (2006), which proposes a 3D roughness evaluation method based on morphological algorithms to describe surfaces with laser honing and laser texturing, and in Costa and Hutchings (2007), where the influence of surface topography on lubricant film thickness has been investigated for reciprocating sliding of patterned plane steel surfaces against cylindrical counter-bodies under conditions of hydrodynamic lubrication.

13.3   Summary

Surface texturing in general and laser surface texturing in particular has emerged in recent years as a viable means of enhancing tribological performance. A great deal of fundamental research work is still going on worldwide, utilising various texturing techniques, to explore the benefits of surface texturing and to optimise the texturing forms and dimensions under various operating conditions. Of all the practical micro-surface patterning methods it seems that laser surface texturing (LST) is the most promising concept. This is because the laser is extremely fast, clean to the environment and provides excellent control of the shape and size of the micro-dimples, which allows realisation of optimum designs. LST is starting to gain more and more attention in the tribology community as is evident from the growing number of publications on this subject.

LST may be successfully applied to piston rings and cylinder liners resulting in up to 4.5% reduction in fuel consumption or engine torque. This success is attributed to the theoretical modelling of LST under full fluid film conditions, which gave good agreement with laboratory tests and permitted optimisation of the LST parameters.

13.4   References

Anno, J.N., Walowit, J.A. and Allen, C.M., 1968, 'Microasperity Lubrication,' *J. of Lubrication Technology Trans. ASME*, **90**(2), pp. 351–355.

Anno, J.N., Walowit, J.A. and Allen, C.M., 1969, 'Load support and leakage from microasperity-lubricated face seals,' *J. of Lubrication Technology Trans. ASME*, **91**(4), pp. 726–731.

Brizmer, V., Kligerman, Y. and Etsion, I., 2003, 'A laser surface textured parallel thrust bearing,' *Tribology Transactions*, **46**(3), pp. 397–403.

Bulatov, V.P., Krasny, V.A. and Schneider, Y.G., 1997, 'Basics of machining methods to yield wear and fretting resistive surfaces, having regular roughness patterns,' *Wear*, **208**, pp. 132–137.

Costa, H.L. and Hutchings, I.M., 2007, 'Hydrodynamic lubrication of textured steel surfaces under reciprocating sliding conditions,' *Tribology International*, **40**, pp. 1227–1238.

Etsion, I., 2005, 'State of the art in laser surface texturing', *J. of Tribology Trans. ASME*, **127**(1), pp. 248–253.

Etsion, I. and Burstein, L., 1996, 'A model for mechanical seals with regular microsurface structure,' *Tribology Transactions*, **39**(3), pp. 677–683.

Etsion, I. and Sher, E., 2009, 'Improving fuel efficiency with laser surface textured piston rings,' *Tribology International*, **42**, pp. 542–547.

Etsion, I., Halperin, G. and Greenberg, Y., 1997, 'Increasing mechanical seal life with laser–textured seal faces,' 15th Int. Conf. On Fluid Sealing BHR Group, Maastricht, pp. 3–11.

Etsion, I., Halperin, G. and Becker, E., 2006, 'The effect of various surface treatments on piston pin scuffing resistance,' *Wear*, **261**, pp. 785–791.

Golloch, R., Merker, G.P., Kessen, U. and Brinkmann, S., 2004, 'Benefits of laser-structured cylinder liners for internal combustion engines,' Proc. 14th International Colloquium Tribology, 13–15, Jan Esslingen, pp. 321–328.

Hamilton, D.B., Walowit, J.A. and Allen, C.M., 1966, 'A theory of lubrication by microasperities,' *J. of Basic Engineering Trans. ASME*, **88**(1), pp. 177–185.

Jeng, Y.R., 1996, 'Impact of plateaued surfaces on tribological performance,' *Tribology Transactions*, **39**(2), pp. 354–361.

Kligerman, Y., Etsion, I. and Shinkarenko, A., 2005, 'Improving tribological performance of piston rings by partial surface texturing,' *J. of Tribology, Trans. ASME*, **127**(3) pp. 632–638.

Mosleh, M., Laube, S.J.P. and Suh, N.P., 1999, 'Friction of undulated surfaces coated with MoS_2 by pulsed laser deposition,' *Tribology Transactions*, **42**(3), pp. 495–502.

Pettersson, U. and Jacobson, S., 2003, 'Influence of surface texture on boundary lubricated sliding contacts,' *Tribology International*, **36**(11), pp. 857–864.

Rahnejat, H., Balakrishnan, S., King, P.D. and Howell-Smith, S., 2006, 'In-cylinder friction reduction using a surface finish optimization technique,' *Proc. I. Mech. E. Part D: J. Automobile Engineering*, **220**, pp. 1309–1318.

Ronen, A., Etsion, I. and Kligerman, Y., 2001, 'Friction-reducing surface texturing in reciprocating automotive components,' *Tribology Transactions*, **44**(3), pp. 359–366.

Ryk, G. and Etsion, I., 2005, 'Testing piston rings with partial laser surface texturing for friction reduction,' *Tribology Transactions*, **48**(4), pp. 583–588.

Ryk, G., Kligerman, Y. and Etsion, I., 2002, 'Experimental investigation of laser surface texturing for reciprocating automotive components,' *Tribology Transactions*, **45**(4), pp. 444–449.

Ryk, G., Kligerman, Y., Etsion, I. and Shinkarenko, A., 2005, 'Experimental investigation of partial laser surface texturing for piston rings friction reduction,' *Tribology Transactions*, **48**(4), pp. 583–588.

Saka, A., Liou, M.J. and Suh, N.P., 1984, 'The role of tribology in electrical contact phenomena,' *Wear*, **100**, pp. 77–105.

Saka, N., Tian, H. and Suh, N.P., 1989, 'Boundary lubrication of undulated metal surfaces at elevated temperatures,' *Tribology Transactions*, **32**(3), pp. 389–385.

Schneider, Y.G., 1984, 'Formation of surfaces with uniform micropatterns on precision machine and instrument parts,' *Precision Engineering*, **6**, pp. 219–225.

Stephens, L.S., Siripuram, R., Hyden, M. and McCartt, B., 2004, 'Deterministic micro asperities on bearings and seals using a modified LIGA process,' *J. Eng. Gas Turb. Power*, **126**(1), pp. 147–154.

Suh, N.P., Mosleh, M. and Howard, P.S., 1994, 'Control of friction,' *Wear*, **175**, pp. 151–158.

Taylor, C.M., 1998, 'Automobile engine tribology – design considerations for efficiency and durability,' *Wear*, **221**, pp. 1–8.

Tian, H., Saka, N. and Suh, N.P., 1989, 'Boundary lubrication studies on undulated titanium surfaces,' *Tribology Transactions*, **32**(3), pp. 289–296.

Wakuda, M., Yamauchi, Y., Kanzaki, S. and Yasuda, Y., 2003, 'Effect of surface texturing on friction reduction between ceramic and steel materials under lubricated sliding contact,' *Wear*, **254**, pp. 356–363.

Wang, X. and Kato, K., 2003, 'Improving the anti-seizure ability of SiC seal in water with RIE texturing,' *Tribology Letters*, **14**(4), pp. 275–280.

Wang, X., Kato, K. and Adachi, K., 2002, 'The lubrication effect of micro-pits on parallel sliding faces of SiC in Water,' *Tribology Transactions*, **45** (3), pp. 294–301.

Wang, X., Kato, K., Adachi, K. and Aizawa, K., 2003, 'Loads carrying capacity map for the surface texture design of SiC thrust bearing sliding in water,' *Tribology International*, **36**(3), pp. 189–197.

Weidner, A., Seewig, J. and Reithmeier, E., 2006, '3D roughness evaluation of cylinder liner surfaces based on structure–oriented parameters,' *Meas. Sci. Technol.*, **17**(3), pp. 477–482.

Willis, E., 1986, 'Surface finish in relation to cylinder liners,' *Wear*, **109**, pp. 351–366.

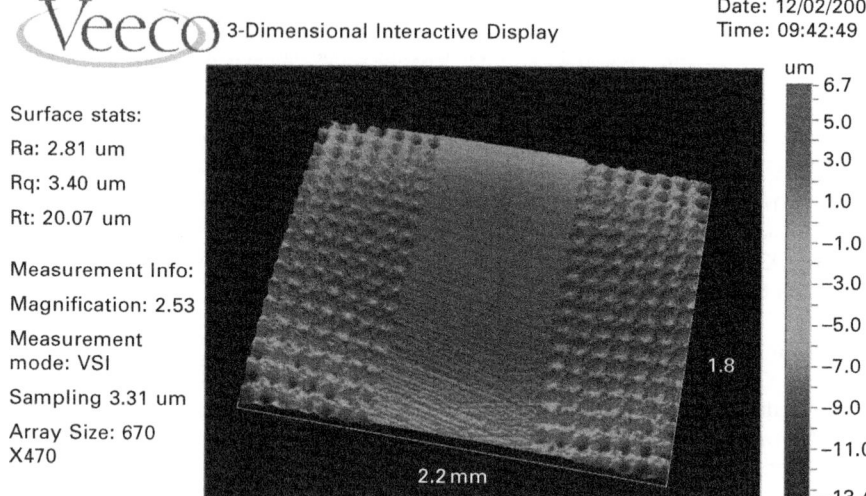

Surface stats:
Ra: 2.81 um
Rq: 3.40 um
Rt: 20.07 um

Measurement Info:
Magnification: 2.53
Measurement
mode: VSI
Sampling 3.31 um
Array Size: 670
X470

Title: Subregion
Note: X offset: 37 Y offset: 0

Plate VII Partial LST Cr coated cylindrical face piston ring (Etsion and Sher, 2009).

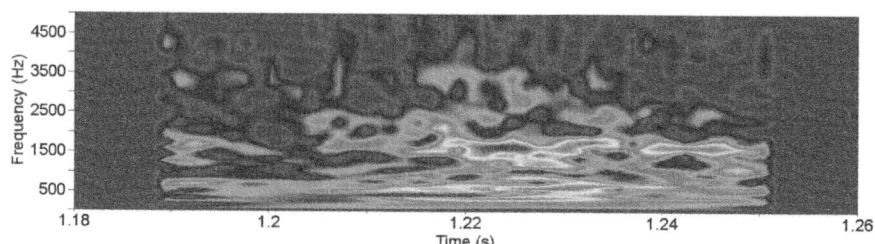

Plate VIII A wavelet spectrum of noise monitored from the first tube.

14

Optimised textured surfaces with application in piston ring/cylinder liner contact

R. RAHMANI, Loughborough University, UK and
A. SHIRVANI and H. SHIRVANI, Anglia Ruskin
University, UK

Abstract: The application of textured surfaces in tribology has recently gained a huge momentum. In this chapter, a systematic approach to investigate the maximum outcomes from employing such surfaces is introduced with an insight into their application in internal combustion engines. A combination of various affecting parameters on the tribological performance of such surfaces is studied and the optimum results are introduced. The effect of employing such optimised textures in enhancing the lubrication condition in piston ring/cylinder liner contact is also studied.

Key words: surface texturing, piston ring/cylinder liner contact, optimisation, slider bearings.

14.1   Introduction

In general, friction is inherent to and is produced between bodies in contact with relative motion. Friction tends to oppose this relative motion between the bodies through loss of energy mostly in the form of heat, mechanical vibration and noise. Therefore, in most cases, friction is an unfavourable phenomenon. In some cases introduction of friction is desired, for example in order to slow down a moving car, the brakes are used to reduce the existing kinetic energy in the wheels by means of producing frictional losses. However, in most of the cases, facilitating this relative motion between bodies is of concern and as a result, one needs to overcome friction so that the relative motion pursues with a minimal loss of energy.

Excessive friction can cause damage to the surfaces in contact in several ways and as a result make them wear. Since the surfaces are damaged due to wear, the rate of energy dissipation due to friction increases and consequently amplifies the rate of wear itself (see Chapter 2).

The reason for the existence of friction is mainly that, in reality, there is no perfect smooth surface. In fact, surfaces have a degree of roughness in the form of small 'hills' and 'valleys' no matter how well they are prepared. When two surfaces, which are in contact with each other, are put into relative motion, the asperities on the opposing surfaces become locked hence inducing

470

friction as initially proposed by Amontons for onset of motion and later under kinetic conditions by Coulomb (see Chapter 2). Wear can ensue and as a result, the surface landscape changes and new 'hills' and 'valleys' are produced, while each time some of the material from the contacting surfaces are removed and the surfaces become worn. Some of these broken pieces can still remain in contact and produce even more resistance against the motion. As a result a form of wear, which is called *abrasive wear*, will occur.

Therefore, the inherent roughness of surfaces can be blamed for existence of friction and occurrence of wear. However, this is not the entire story. It has been observed that, although friction can be reduced by polishing of the contacting surfaces to a very fine degree, after some point, both the friction and wear will start to increase again. The reason for this is that as the surfaces become smoother, the electromagnetic forces between molecules and/or atoms of the opposing surfaces become significant and as a result the surfaces may 'stick' to each other so that again an extra energy is required to overcome these attractive forces (see Chapter 3). However, the given energy would not necessarily break these forces evenly and, as a result, some of the material from each surface will stick to the other surface. This form of wear is usually referenced as *adhesive wear*, which is the primary cause of scuffing. Even a combination of pieces from each surface, which are attached to each other by electromagnetic forces, can be released into the contact and cause some degree of abrasive wear as well.

In practice, both of these phenomena can subsist in any contact. Here it should be noted that although there are other forms of wear that may happen in various circumstances, the two mentioned wear mechanisms are considered as the major wear causes. For example, the surfaces may enter into a chemical reaction with each other or with environment (such as air) and, as a result, *corrosive wear* may occur.

Since long ago, it has been discovered that introducing a third element in the contact can reduce friction and wear. These elements can keep the asperities apart from direct contact, become an insulator against intermolecular/atomic forces and protect the surfaces from chemical reactions. Such an element, usually of low shear strength, is generally termed a *lubricant*.

A lubricant can be solid, liquid or gas. Liquid lubricants, usually in the form of oils, are the most common. Designing this type of lubricant involves good knowledge of rheology (see Chapters 3 and 5). The lubricating oils that are nowadays used in industry evolved in the last century and can be mineral, organic or synthetic. In addition, depending on application, a combination of various additives can be introduced in the base oil to provide a suitable lubricant.

Along with designing appropriate lubricants, improvements to tribological characteristics of contacting surfaces are also achieved through surface engineering. In general, surface engineering can be considered in two categories: surface coating (see Chapter 4) and surface modification (treatment) (also see Chapter 13). In surface coating technology a layer (or layers) of a material (or materials) are implanted over the main contacting surfaces in order to enhance their tribological characteristics such as wear resistance. The coating process can also be achieved by altering the chemical composition of the surface itself. Methods in which the surface is modified without changing its composition such as polishing, honing, heating or employing local melting–solidification are called the *surface modification* (or treatment) methods.

In recent years, efforts for better controlling of friction and wear have focused on the modification of surface topography, including parameters such as surface roughness (see Chapter 2) and skewness, in a controlled manner. Surface finishing processes, such as turning, shaving, grinding, honing and polishing generate surfaces with textures of specific topography. It is believed that creating surfaces with controlled micro-geometry can be an effective approach in improving their tribological performance.

In addition to these enhancement processes on the inherent topography of the surfaces, the idea of introducing artificial and deterministic surface micro-structures in the form of microscopic surface features, having a regular geometry and a repeatable pattern have been of increasing interest. It is believed that somehow these artificial features can be designed so that they can provide a desired tribological performance.

14.2   Surface texturing

Several mechanisms may contribute to hydrodynamic lift in bearings with parallel flat and smooth surfaces that according to the conventional theory of lubrication would produce no hydrodynamic lift due to a lack of surface gradient (see Chapters 5 and 6). These mechanisms are lubricant density change, thermal and viscosity wedge, non-Newtonian effects, squeeze film action, eccentric rotation, wobble and bounce as well as surface roughness or waviness. Of these mechanisms, the influence of surface roughness/waviness is of interest in this chapter. It is also observed that by implementing macro/ micro-surface structure(s) or irregularities in the form of asperities or cavities on one of the sliding surfaces, an additional load carrying capacity within a coherent lubricant film may be developed. These results were achieved in the mid-1960s, when Hamilton *et al.* (1966) at BMI, USA, investigated the effect of surface irregularities on lubrication and promoting self-sustaining films in parallel rotary-shaft mechanical face seals. They examined both asperities and cavities. Their work was continued by Anno *et al.* (1968, 1969). Since then, it has been recognised that micro-surface patterns, such as radial tapers, hydro-pads, lobes and grooves can introduce beneficial effects on lubrication of mechanical seals and enhance their axial stiffness.

This concept was developed further from mid and late 1990s, mainly thanks to the works conducted by Etsion and Burstein (1996) and Etsion *et al.* (1999). They employed laser technology to produce surface textures in the form of shallow dimples or pores.

Different terms in literature have been used to describe these artificially implanted surface features such as ***deterministically controlled textures*** (Wong *et al.*, 2006) and ***intentionally created surface undulations*** (Blatter *et al.*, 1999). They may also be called (macro- or micro-) surface features, structures, geometries or topographies, as well as textures. In general, these surface textures can be of two types: they may be produced by *engraving* the surface, whether in the form of an individual *spot* or a narrow *channel*. In the former case, they are usually named ***dimples***, depressions, pits, pockets, pores, indents or dents, recesses, holes or cavities in the literature. These terms are also often accompanied with adjectives such as shallow or hollow. In the latter case, they are commonly called ***grooves*** or sometimes valleys, treads, troughs or corrugations. In addition, they might be produced as protruding objects on surfaces, whether in a continuously projected line form or as an individually projected object or *spot*, which in this case are usually called protrudes or protrusions, peaks, bumps, posts, (micro-) islands or more commonly asperities. It is also common to use the terms ***negative textures*** or ***positive textures*** to distinguish the engraved form from the protruding type textures. It is noted that in the tribology texts, surface asperity is sometimes used with the same meaning as surface roughness and, hence, these deterministic artificially created asperities should not be misinterpreted as inherent roughness or random asperities of the surface.

Early textures were limited to grooves or troughs, while new techniques have allowed complex patterns of different shapes, including circular, triangular and other geometric shapes, to be used. Asperity shape, geometry, depth, area ratio (the ratio of asperity to the flat area) and orientation can all influence the tribological effectiveness of a sliding surface.

Depending on the requirements for a specific application, the depth of these dimples can range from a few to several micrometres (normally 4–10 μm) and their diameter can be several tens of micrometres (normally 70–100 μm) (Erdemir *et al.*, 2004b, 2005).

14.3   Application of surface texturing in tribology

Surface texturing is receiving ever-increasing attention in tribological design as the previous studies have shown its potential to improve lubrication and reduce friction and wear. Nowadays, the concept has become common practice in some industrial applications. As an example, very small textures are employed in the *landing* sections of magnetic storage disc surfaces to prevent adhesion when the recording heads contact the surface (Etsion,

2005; Wong *et al.*, 2006). Surface texturing is also considered as a means for overcoming adhesion and stiction in the MEMS devices (Etsion, 2005).

Micro-texturing of functional tool surfaces can help reduce friction and wear during critical forming processes such as cold forging (Wagner *et al.*, 2006, 2008), sheet rolling (Ike *et al.*, 2002) and deep drawing (Vermeulen and Scheers, 2001). This increases tool life and reduces the amount of consumed lubricants.

In recent years, great strides have also been made in the texturing of various tribological surfaces including mechanical seals (Etsion and Halperin, 2002; Yu *et al.*, 2002; Yagi *et al.*, 2008; Yi and Dang-Sheng, 2008) and dry gas seals (McNickle and Etsion, 2004), thrust bearings (Etsion *et al.*, 2004) or journal bearings (Lu and Khonsary, 2007), ring/liner assembly of internal combustion (IC) engines (Ryk *et al.*, 2002; Rahnejat *et al.*, 2006; Ryk and Etsion, 2006; Etsion and Sher, 2009), roller/piston contact in hydraulic motors (Pettersson and Jacobson, 2007). The textures produced on these surfaces can improve the hydrodynamic efficiency by altering the regime of lubrication or performance of sliding surfaces. Surface texturing is a cheap technology for applications in mechanical face seals to produce load-carrying capacity, when compared with other methods such as spiral grooves, waviness, etc. This technology has also been successfully applied to mechanical seals used in operating pumps in the field (Etsion *et al.*, 1999).

A well-known example of tribological improvement through surface texturing is in the automotive industry where the honing of the cylinder liners for internal combustion engines can aid support the oil retention and re-supply. Surface texturing has also been proposed to be employed in engine powertrain and drivetrain components and wrist pin bearings, where the supply of lubricant is usually rather poor, etc. (Ryk *et al.*, 2002; Etsion *et al.*, 2006; Lisowsky, 2006; Borghi *et al.*, 2008).

14.4   Surface texturing methods

Several methods are available for creating the surface micro-topographies. Mechanical techniques such as ***vibro-rolling*** and ***abrasive machining*** can be used to create grooves (Wong *et al.*, 2006). Micro-dimpling has also been achieved with abrasive jet machining (AJM) by Wakuda *et al.* (2003), while using mechanical scribing through utilising a Rockwell indenter. Hsu *et al.* (2003) and Krupka and Hartl (2007) have produced grooves and dimples on the metal surface. Examples of textured surfaces produced by indentation using a hardness tester can be found in Pettersson (2005), for instance.

Embossing offers the possibility to transfer a textured pattern on all plastically deformable materials including steel (Pettersson, 2005; Pettersson and Jacobson, 2006).

Methods including deep reactive ion etching (RIE) (Wang *et al.*, 2003), different forms of etching and *UV lithography* can produce a variety of shapes in both metals and ceramics (Alberdi *et al.*, 2004; Wong *et al.*, 2006). Early studies have used *photo-etching* for micro-asperity fabrication (Hsu *et al.*, 2003; Siripuram, 2003). The electrical discharge etching has also been used by Ito *et al.* (2000). Using combined techniques of *physical vapour deposition* (PVD) and UV photo-lithography has been reported in the literature as well (Alberdi *et al.*, 2004). Costa and Hutchings (2007) have used photochemical etching, which comprises masking by photolithography followed by chemical etching. The lithography, electroplating, moulding (LIGA) process has been widely used in the pattern generation in silicon wafer processing, but less so in robust applications such as surface texturing for tribological applications (Kortikar, 2004).

Each method has advantages and disadvantages, and some may be more appropriate for use in a given application than the other methods. For instance, although etching is flexible in producing different shapes, the process is time-consuming and the profiles of the features are determined by the chemical erosion process and cannot be controlled (e.g. round-shape textures may not be possible). Although, methods such as photo-etching are old traditional methods, they are limited to surfaces made of, for example, copper, while erosion is an effective technique for hard materials (Wang *et al.*, 2003). Regular and abrasive-jet machining also have some limitations on profile shape and may not be appropriate for cylinder liner texturing if machining heads are too large to access the inner liner surface (Wong *et al.*, 2006). According to Wang *et al.* (2003), micro-blasting is a suitable technique to produce surface texture on ceramics. The LIGA process can be used for fabricating positive asperities as well and offers an advantage of achieving higher aspect ratios. The modified photo-lithography process introduced by Kortikar (2004) is an alternative to fabricate both deterministic micro-asperities and micro-dimples with high/low aspect ratios on surface of any arbitrary cross-section and orientation.

In addition to the above-mentioned methods, laser ablation or *laser texturing* is a new energy beam technique to achieve controlled surface textures (Andersson *et al.*, 2007). Using laser material processing, topographical features on almost any material such as metal, crystalline structures and glass, ceramics and polymers can be produced (Haefke *et al.*, 2000; Etsion, 2005; Neves *et al.*, 2006). The method is the most efficient and convenient approach for metals (Wang *et al.*, 2003). Laser surface texturing (LST) is a flexible and high-speed texturing method that can provide well-controlled surface characteristics for a variety of materials. Examples of textured surface produced using LST technology can be found in Pettersson (2005) and Etsion (2005). By controlling the laser beam parameters, it is possible to control the diameter, depth and area density of the micro-dimples accurately.

Recent progress in laser surface texturing using focused UV-lasers has extended the possibilities for tribological surface topography modification. Excimer laser beam machining (LBM) is used by Wakuda *et al.* (2003) can induce nano to micro-scale patterns on hard coatings such as TiN, TiCN and diamond-like carbon (DLC) films, utilising lasers with pico- to femtosecond pulsation capability.

In general, laser-texturing technology is currently believed to be a very promising technique based on its flexibility and speed (Etsion, 2005). It is clean, environmentally friendly and provides an excellent control on the size of the dimples. However, it has a possible drawback that the laser technique may create *burrs* or *bulges* of melted and re-deposited material around the edges (rims) of dimpled areas with disadvantageous tribological effects and the surfaces may require a subsequent polishing step (Wong *et al.*, 2006). One of the other limitations of the laser texturing is that it can only produce dimples and of only spherical and conical shapes. In addition, using this method, the average pore depth can be produced is around $1-30\,\mu m$ and the pore diameter is limited to $95-230\,\mu m$ and at lowest aspect ratios of $0.004-0.3$ (Kortikar, 2004). The concerns about the heat affected zone and resulting cracks on the surface are other issues, which should be considered.

14.5   The mechanisms behind tribological improvements through surface texturing

Both the need to reduce friction and increase the load-carrying capacity in bearings requires an effective lubrication strategy for sliding surfaces. Surface texturing can provide such a benefit in a number of ways. In this regard, several physical mechanisms are proposed to describe the potential effect of the micro-features.

The micro-structures in the form of intentionally created undulations or minute cavities, grooves or valleys in the surface can prohibit wear debris by entrapping (accommodating) them, whether the contact is dry or lubricated. They suppress abrasion and ploughing friction and third-body wear, resulting in improved fretting fatigue resistance, longer durability and hence reliability. This is very common idea amongst different researchers such as Blatter *et al.* (1999), Fenske *et al.* (2003), Alberdi *et al.* (2004), Erdemir *et al.* (2004a,b), Kovalchenko *et al.* (2005), Etsion (2005), Wong *et al.* (2006), Andersson *et al.* (2007), Costa and Hutchings (2007) and Krupka and Hartl (2007).

In addition, these surface micro-structures can act as lubricant micro-reservoirs or the so-called lubricant (micro-) pockets or *lubricant capacitors* in order to improve localised lubrication by retaining the lubricant in desired locations (Blatter *et al.*, 1999; Etsion, 2005; Wong *et al.*, 2006; Krupka and Hartl, 2007). These features can retain lubricant to release when needed, thus increasing the quantity of the lubricant locally even under high pressures (Blatter *et al.*, 1999; Fenske *et al.*, 2003; Alberdi *et al.*, 2004; Uehara *et al.*,

2004; Mourier *et al.*, 2006; Andersson *et al.*, 2007). The liquid trapped in the low valleys of the textured surfaces can be considered as a secondary source, which is then drawn by the relative movement of the surfaces to seep into the surrounding areas by capillarity action, and therefore, reduce friction and reduce the chance of galling (Wang *et al.*, 2001; 2003). They can also transport lubricant to a desired location and subsequently expel it into the contact, resulting in much longer life for lubrication (Haefke *et al.*, 2000; Kovalchenko *et al.*, 2005).

These mechanisms are thought to be important under boundary or starved lubrication conditions (Hupp, 2004; Etsion, 2005). A normal ground surface roughness does not effectively entrap lubricant or wear debris since there is no physical mechanism by which the lubricant can be *held* in place. Therefore, an *ideal* surface texture should consist of a nearly uniform array of microscopic depressions each having geometrical dimensions similar to those of its neighbours (Haefke *et al.*, 2000).

In instances where there are frequent start/stop operations, it is thought that the lubricant remaining in the pores can avoid an abnormal temperature rise caused by dry running conditions (Wakuda *et al.*, 2003). Experimental results also indicate that surface texture helps the running-in progress to smoothen the contact surfaces and reduce friction (Wang *et al.*, 2006).

At sufficient density, the surface irregularities can improve the wetting of the surface by lubricant, and thereby support the formation of a lubricating film. At higher sliding speeds and with a sufficient supply of lubricant, surface cavities can take action as hydrodynamic pressure sources to produce hydrodynamic fluid lift and reduce friction, unless the benefit is counteracted by turbulence or cavitation at the cavities (Kovalchenko *et al.*, 2005; Andersson *et al.*, 2007).

It is suggested that during the initial stages of sliding, the leading edge of each asperity becomes slightly worn so that it acts as an inclined surface that promotes hydrodynamic lift (Hupp, 2004). In this regard, each micro-dimple acts as a micro-hydrodynamic bearing to enhance lubrication even in the partial lubrication condition (Etsion, 2005; Wang and Zhu, 2005). This leading edge hypothesis is also supported by experimental data that indicate the initial friction at the start-up is much higher than the average values (Hupp, 2004). Another hypothesis on the working mechanism of the micro-features is based on occurrence of cavitation. According to this hypothesis, an asymmetric hydrodynamic pressure distribution over each dimple owing to the *local cavitation* in the diverging clearance of the dimple provides a significant amount of load-carrying capacity. Generally, the pressure increases in the converging film regions, while it decreases in the diverging film regions for incompressible fluids. The cavity generated in the diverging film regions is, theoretically, an isobaric region. The pressure in this region cannot be lower than the cavity pressure, which is the fluid vapour pressure or the pressure at which the lubricant is saturated with the gas dissolved in it. As a result, the pressure rise in the converging film regions can be much larger than the pressure drop in the diverging film regions and so additional load capacity is generated (Wang *et al.*, 2001, 2006). This type of enhancement in the load support by a hydrodynamic effect is often called ***micro-hydrodynamics*** of surface features (Mourier *et al.*, 2006; Wong *et al.*, 2006).

Surface texturing can have a significant beneficial effect by increasing the boundaries of hydrodynamic lubrication. In this case, the performance improvement is thought to result from the increased hydrodynamic efficiency of such a sliding surface under the boundary-lubricated sliding condition (Erdemir, 2005). In addition to these, ***micro-plasto-hydrostatic*** lubrication effects (Hsu *et al.*, 2005) and reduction of stiction owing to the smaller *real* contact area (Wang *et al.*, 2006) are also enumerated as beneficial outcomes from the action of textured surfaces. The former is related to application of laser texturing in metal forming in which the micro-features act as ***micro-pools*** or a micro-plastic hydrodynamic tool (Etsion, 2005) and the latter is associated with hard disk memory devices.

14.6   Debates surrounding surface texturing

Despite the considerable emphasis in some literature put on the tribological benefits and advantages associated with surface texturing, there are some considerations and debates as well. In fact, despite a reported decrease in friction and wear between two sliding surfaces through use of textured surfaces, some experimental results also show negative effects on contact lubrication and durability (Mourier *et al.*, 2006; Wong *et al.*, 2006). Andersson *et al.* (2007) point out a drawback of the surface irregularities is abrasive wear of a soft counter-surface at high contact pressures. In addition, experimental work by Erdemir *et al.* (2004a) has verified that dimpled surfaces work extremely well under mixed or hydrodynamic sliding regimes of lubrication. However, under boundary lubricated sliding conditions (i.e. high load and low speed conditions) they only provide marginal improvements in friction, but may cause increased wear losses. Hsu (2004a) also confirms that the ability of micro-texturing to reduce friction does not seem to be effective under boundary regime of lubrication. Under these conditions, the edge stresses around the dimples increase friction rather than decrease it.

The behaviour of textured surfaces can be quite variable. Some studies show a substantial reduction in coefficient of friction due to surface texturing (e.g. Aldajah *et al.*, 2005), while others report virtually no difference between textured and untextured surfaces (e.g. Pettersson and Jacobson, 2003). Sometimes texturing is not optimised for a given case, in others there is no optimal case and any kind of texturing may be worse than a nominally smooth surface (Wong *et al.*, 2006). For example, some studies show that surface texturing can reduce friction only if the pattern features are smaller than the contact width (e.g. Wang *et al.*, 2001; Pettersson, 2005), while others show that friction increases under these conditions (e.g. Wakuda *et al.*, 2003). The extent of the reported effect has also been variable (Costa and Hutchings, 2007). In some tests, adding texturing can increase friction or have no effect on the friction coefficient, but the mechanism is not clear (Hsu *et al.*, 2003; Blau and Qu, 2004). Although some studies have shown that texturing removes wear particles from a sliding interface, a good understanding of the chemical interaction of the materials involved in sliding is required to predict the possible effects on friction. For instance, Kovalchenko *et al.* (2005) have observed that, when the dimples are relatively deep or when the oil viscosity is relatively high, LST might be detrimental to tribological performance under starved lubrication conditions.

Nevertheless, despite the above-mentioned facts, the use of surface texturing has opened a new avenue to explore friction reduction especially in engine applications. Research and analysis presented to date demonstrates both the potential to improve tribological properties via surface texturing, and the need to understand materials, lubricants and running conditions before a surface texture is applied (Wong *et al.*, 2006). According to Hsu *et al.* (2005), there is lack of theory and understanding in designing of these features and the mechanism(s) and set of conditions in which these features can influence friction and durability. Detailed mechanisms of friction reduction are not fully understood and the lubrication phenomena induced by the micro-geometry distribution remain widely unknown (Mourier *et al.*, 2006). Moreover, the influence of size and shape of the surface texture on their tribological performance is not well known either (Alberdi *et al.*, 2004). In fact, the maximum benefits that can be accrued from sophisticated texturing are unclear and this area is relatively unexplored.

According to Hsu (2004a,b), the effect of dimples on friction reduction appears to depend on textures' size, their pattern and density on the contacting surfaces. The dimensions and area ratio of the textures are considered as important parameters related to the generation of hydrodynamic pressures (Wang *et al.*, 2006). Surface texture shape and orientation may also have a significant effect on friction and especially wear (Hsu, 2004a,b). On the other hand, Ronen *et al.* (2001) and Ryk *et al.* (2002) point out that the main geometrical parameters of laser texturing, which can affect friction, are the ratio of dimple depth to its diameter and the area density of the dimples.

As Wong *et al.* (2006) mention that the outcome of research works to date indicates that optimal surface texturing parameters depend on the running conditions and on the dominant regime of lubrication. Each optimisation process requires an accurate correlation of the micro-geometry dimensions and density with its induced lubrication influence for the relevant range of operating conditions, which is still difficult to obtain as noted by Mourier *et al.*

(2006). Hence, the fundamental understanding of the lubrication phenomena induced by a single micro-feature, as a function of its dimensions and the operating conditions is an important step.

A review of the literature indicates that there is no general agreement on the related issues such as what types of textures should be used and under what conditions. In addition, there is no general agreement on what texturing parameters are the most important in friction reduction or load support and even what the effects of various texture parameters may be. For instance, the results from Kovalchenko *et al.* (2005) show a better performance for textured surfaces in boundary and mixed lubrication condition, while almost no improvement in hydrodynamic lubrication conditions ensue. On the other hand, the results from Etsion *et al.* (2004) show a 50% reduction in friction by applying micro-textures in hydrodynamic regime of lubrication.

As another example, Hsu *et al.* (2005) state that in high-speed, low-load conditions, surface feature shape and size has great influence on friction reduction, while Etsion *et al.* (1999), under almost identical conditions, reveal that the actual shape of the micro-dimple is insignificant. Pettersson and Jacobson (2003) observed no improvement by laser texturing in dry and lubricated cases and Blau and Qu (2004) using solid lubricants, while according to some other literature it is believed that even in these conditions, the textured surfaces can be beneficial by trapping wear debris and reducing friction as a result (e.g. Erdemir *et al.*, 2004a,b). In addition, the results of Aldajah *et al.* (2005) show lower coefficients of friction for textured surfaces when compared with untextured ones for a whole range of boundary to mixed and early hydrodynamic regimes of lubrication. According to Hupp (2004) and Petterson (2005), the surfaces with a higher percentage of surface area taken up by micro-pores have a lower coefficient of friction under the hydrodynamic and mixed regimes of lubrication regimes, while Wang *et al.* (2001, 2003) propose a very small area ratio for pores as the optimum case.

Through the published experimental research, one cannot reach a specific conclusion. For example, in the work by Blatter *et al.* (1999) although the distance between grooves is kept constant, it is not known that the better performance of the *finer* grooves is due to their smaller width or due to their shallower depth. Furthermore, study of the literature shows a lack of systematic approach in the study of the textured surfaces and the corresponding optimisation procedures. As Etsion (2005) also confirms, a trial and error approach is adopted whenever optimisation of the texturing dimensions is attempted. Consequently, a relatively large range of optimum parameters has been attained in different studies.

Of course, this trial and error approach in some cases is the only option in the study of the performance of textured surfaces such as dry or boundary lubrication conditions, where there is a lack of basic theoretical modelling approach. However, in the rest of the cases such as under hydrodynamics, there is a considerable theoretical basis. Nevertheless, even in these cases, there remains a lack of a general and systematic optimisation approach and a few efforts have been reported, largely through numerical trial and error (e.g. Brizmer *et al.*, 2003). As Wong *et al.* (2006) suggest, much work yet remains in this field before a good understanding of the effects of surface texturing is achieved.

14.7   Surface texturing technology and internal combustion (IC) engines

The frictional losses in IC engines are an important factor in influencing fuel economy and performance of vehicles. Estimates show that nearly 40–50% of the total frictional losses of an IC engine are due to the piston/cylinder system of which 70–80% are attributed to the piston rings (Ronen *et al.*, 2001; Ryk *et al.*, 2002; Ryk and Etsion, 2006). Reducing these frictional losses is a key factor in reducing fuel consumption and protecting the environment. In addition, the internal combustion engine contributes to environmental pollution through particulates, NO_x and HC and to the greenhouse effect via CO_2 emissions (Taylor and Coy, 2000; Tung and McMillan, 2004).

Improving fuel efficiency of IC engines while still enabling them to meet environmental requirements presents both design and material challenges. Reducing parasitic frictional losses in engines can be achieved by a combination of strategies such as those noted by Blau and Qu (2004):

- redesigning the engine components;
- reformulating the lubricants;
- improving methods of lubricant filtration and supply;
- reducing churning losses in fluids;
- changing the operating conditions of the engine;
- substituting more durable, low-friction materials;
- altering the finish or micro-scale geometry of the bearing surfaces.

Among these factors, proper lubrication and surface roughness are key issues in reducing friction in the piston/cylinder system. Improvement to surface finish is one of the most reasonable methods for friction reduction. This has therefore promoted an increase in the number of automotive components whose surfaces are finished with grinding with subsequent additional processes such as lapping or super-finishing. However, the attainable surface roughness is ultimately limited by the material and machining considerations. The earliest and well-known industrial application of effective surface treatment in the automotive industry is of course cylinder liner honing.

On the other hand, surface texturing can be regarded as another striking approach for elements subjected to sliding contact. This area of research is widely unexplored and needs extensive research (Hsu, 2004b). Although the textured surfaces have shown the potential to reduce friction under reciprocating sliding conditions, there is a lack of agreement over the underlying mechanism(s) involved and the optimum condition with which maximum benefits can be accrued (Wong *et al.*, 2006).

Overall, it is expected that by producing optimised textures on various engine and drivetrain components, a great deal of improved fuel economy would be achieved owing to subsequent reductions in friction, gaining longer durability and hence reliability through suppression of wear.

14.8   The basic equations of tribology

14.8.1 The Reynolds hydrodynamic lubrication equation

As is common in tribology of bearings in general, the Reynolds hydrodynamic lubrication equation is used for the analysis the textured surfaces as well.

The Reynolds equation can be derived either by writing the equilibrium equations for an infinitesimal element inside the lubricant film and considering the mass continuity equation (e.g. Stachowiack and Batchelor, 2001) or by simplifying the Navier–Stokes (N-S) equations considering a series of assumptions (e.g. Pinkus and Strenlicht, 1961). It can also be obtained rather simply through dimensional analysis of fluid flow (Gohar and Rahnejat, 2008). The derivation of Reynolds equation is carried out in Chapter 5.

Consider Fig. 14.1 for an arbitrary bearing geometry, a general form of Reynolds hydrodynamic equation in 2D/3D form for an arbitrary reference point in space in the Cartesian co-ordinate system may be stated as follows:

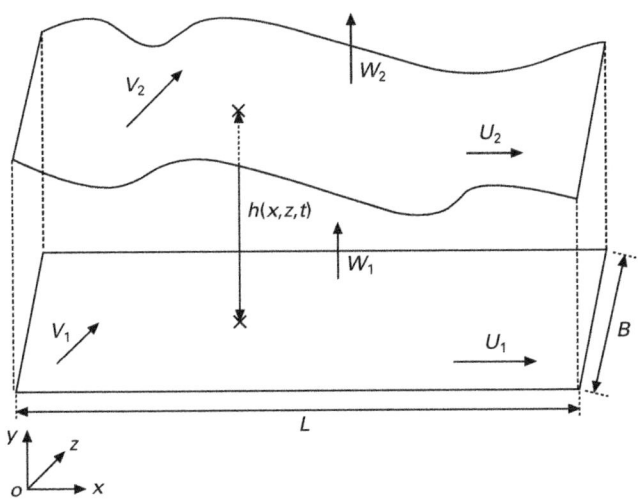

14.1 An arbitrary bearing geometry.

$$\frac{\partial}{\partial x}\left[\frac{\rho}{\eta_0}\left(\frac{\partial p}{\partial x}\right)(h_2 - h_1)^3\right] + \frac{\partial}{\partial z}\left[\frac{\rho}{\eta_0}\left(\frac{\partial p}{\partial z}\right)(h_2 - h_1)^3\right] = 12\rho\,(W_2 - W_1)$$

$$+\, 6\left\{\rho(U_1 + U_2)\frac{\partial}{\partial x}(h_2 + h_1) + (h_2 - h_1)\frac{\partial}{\partial x}[\rho(U_1 + U_2)]\right\} \qquad (14.1)$$

$$+\, 6\left\{\rho(V_1 + V_2)\frac{\partial}{\partial z}(h_2 + h_1) + (h_2 - h_1)\frac{\partial}{\partial z}[\rho(V_1 + V_2)]\right\}$$

In this equation, the second and the third terms on the right-hand side are called the wedge terms (or Couette flow terms), while the first term on the right-hand side is called the squeeze film term. Wedge terms correspond to the flow due to surface velocities, while the squeeze film term accounts for flow into the contact due to lubricant entrapment by mutual approach of surfaces. The pressure build-up is due to changing of film thickness with time (Lubbinge, 1999). In the latter case, changes in the film thickness can be induced indirectly because of variations of the sliding velocity (wedge term) or it can be caused as a direct consequence of variations in the normal velocity of the bearing surface(s).

It is noted that since in the presented equation above the pressure distribution in the y-direction is neglected; therefore it can be considered as a 2D equation. On the other hand, since this equation is capable of taking into account the variations of surface geometry and clearance between the surfaces in contact, which are stated in the y-direction, therefore it can be considered as a 3D equation. Thus, it is more convenient to use the terms 'quasi-3D' or '2D/3D' for the given equation.

The Reynolds equation in general form stated in equation (14.1). It is often somewhat simplified to represent a specific problem. By considering the co-ordinate system attached to the bottom surface, then $h_1 = 0$ and $h_2 = h$. The squeeze film term is also usually specified based on the variations of the distance between bearing surfaces with time, i.e. $W_2 - W_1 = \partial h/\partial t$. In addition, the transversal sliding components can also be neglected in the most of engineering problems, i.e. $V_1 = V_2 = 0$. Furthermore, it can be assumed that the bearing surfaces are inelastic in the x-direction (Pinkus and Strenlicht, 1961); i.e. $\partial(U_1 + U_2)/\partial x = 0$. In addition, if the lubricant is considered an incompressible and an iso-viscous fluid, which results in $\partial\rho/\partial x = \partial\rho/\partial z = 0$ and $\partial\mu/\partial x = \partial\mu/\partial z = 0$, respectively, the Reynolds equation can be downsized as follows:

$$\frac{\partial}{\partial x}\left[h^3\left(\frac{\partial p}{\partial x}\right)\right] + \frac{\partial}{\partial z}\left[h^3\left(\frac{\partial p}{\partial z}\right)\right] = 6\eta_0\,(U_1 + U_2)\frac{\partial h}{\partial x} + 12\eta_0\frac{\partial h}{\partial t} \qquad (14.2)$$

It is noted that the general Reynolds equation in 2D/3D form is an elliptic second-order partial differential equation (PDE) with respect to pressure. In the 1D/2D though, the Reynolds equation turns into a non-complete second-order linear non-homogenous variable coefficient ordinary differential equation (ODE) with respect to pressure.

The load-carrying capacity of a bearing can be calculated through integrating the obtained pressure distribution over the bearing profile. In addition, the total (viscous) friction force for each surface is obtained from calculating the (viscous) shear stress based on the velocity profile inside the film. Hence, the friction coefficient can be obtained based on the well-known Amonton or Coulomb's law as follows:

$$\eta = \frac{F_v}{W_H} = \frac{\int_0^B \int_0^L \left[\pm \left(\frac{\partial p}{\partial x}\right) \frac{(h_2 - h_1)}{2} + \frac{\eta_0 (U_2 - U_1)}{h_2 - h_1} \right] dx dz}{\int_0^B \int_0^L p \, dx dz} \qquad (14.3)$$

where the negative sign is used for the lower surface and positive sign for the upper.

Lubricant flow rate inside the bearing can be another interesting parameter in order to analyse the bearings under hydrodynamic regime of lubrication. It is defined as (see Chapter 5):

$$Q_x = \int_0^B \left[-\frac{1}{12\eta_0} \left(\frac{\partial p}{\partial x}\right) (h_2 - h_1)^3 + \frac{1}{2} (U_2 + U_1)(h_2 - h_1) \right] dz \qquad (14.4)$$

A similar type of equation can also be obtained for the lubricant flow rate in the transversal direction.

14.8.2 Bearing dynamics equation

When analysing the bearings, one may often come across situation(s) where the bearing surfaces are under external loads as well. These loads can be constant or variable with time and both types may exist at the same time originating from various sources. A good example of this is the case of piston ring/cylinder liner contact, where the ring is under constant and variable loads. In addition, the generated hydrodynamic load capacity by the ring is variable too due to its reciprocating motion. The dynamics of the ring in its retaining groove can be modelled by considering the forces acting on a section of the ring.

By neglecting ring twist, Fig. 14.2 illustrates the forces exerted on a piston ring in the top ring groove of a gasoline engine (see Barrell *et al.*, 2000). Applying Newton's second law of motion in the axial and radial directions gives:

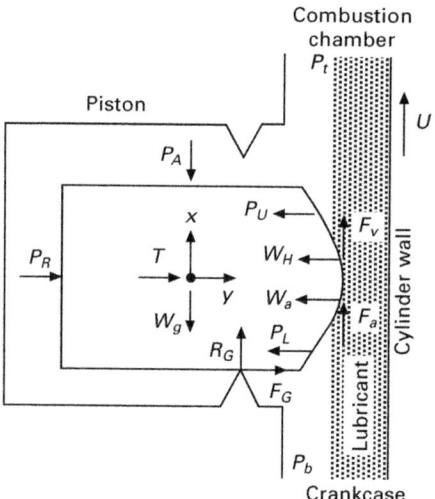

14.2 Forces diagram on a piston ring in an IC engine (Barrell *et al.*, 2000; Priest and Taylor, 2000).

$$\begin{cases} m\dfrac{\mathrm{d}^2 x}{\mathrm{d}t^2} = R_G + F_a + F_v - P_A - W_g \\[2mm] m\dfrac{\mathrm{d}^2 y}{\mathrm{d}t^2} = P_R + T + F_G - P_U - P_L - W_H - W_a \end{cases} \qquad (14.5)$$

Since the variations of oil film thickness with time, i.e. $\partial h/\partial t$, is of concern; therefore, the dynamics of the ring in the radial direction is more of interest. If,

$$F = P_R + T + F_G - P_U - P_L - W_a \qquad (14.6)$$

and allowing the co-ordinate system to be on the cylinder liner with y-axis facing the ring and replacing y with the minimum clearance or film thickness, h_m, the second equation in (14.5) can be written as follows:

$$m\dfrac{\mathrm{d}^2 h_m}{\mathrm{d}t^2} = W_H - F \qquad (14.7)$$

14.9   Modelling of the textured surfaces

14.9.1 Terminology of the textured surfaces

In Fig. 14.3, the following dimensionless parameters may be defined in order to describe the geometry of the textured surface:

• Leading edge length-to-bearing length ratio (leading length ratio) for both x- and z-directions:

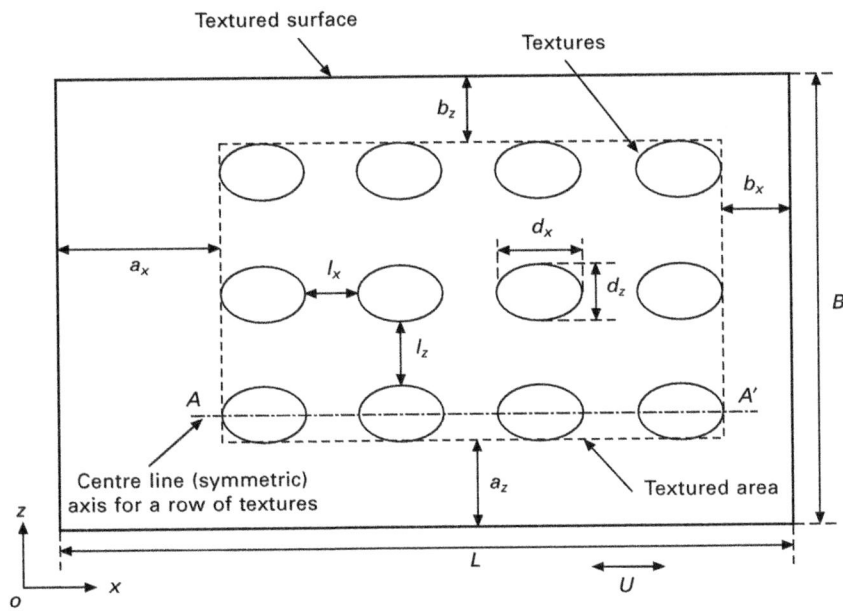

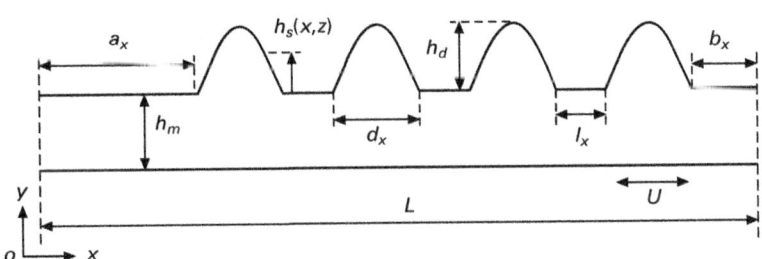

14.3 Schematic of a textured bearing surface and the corresponding geometrical parameters.

$$\alpha_x = \frac{a_x}{L} \quad \text{and} \quad \alpha_z = \frac{a_z}{B} \tag{14.8}$$

- Trailing edge length-to-bearing length ratio (trailing length ratio) for both x- and z-directions:

$$\beta_x = \frac{b_x}{L} \quad \text{and} \quad \beta_z = \frac{b_z}{B} \tag{14.9}$$

- The area density of textures in x- and z-directions:

$$Sp_x = \frac{d_x}{l_x} \quad \text{and} \quad Sp_z = \frac{d_z}{l_z} \tag{14.10}$$

- The (general) area density of textures:

$$Sp = \frac{A_d}{l_x \cdot l_z}$$
(14.11)

in which A_d denotes the surface area of the base of an individual texture and the denominator is the surface area of an imaginary rectangular cell around the texture as defined in Brizmer *et al.* (2003).

- The **texture height ratio**:

$$\xi = 1 + \frac{h_d}{h_m}$$
(14.12)

- *Textured area ratio* (textured portion):

$$\kappa = \frac{[L - (a_x + b_x)] \cdot [B - (a_z + b_z)]}{L \cdot B}$$

$$= \frac{[N_x d_x + (N_x - 1)\,l_x] \cdot [N_z d_z + (N_z - 1)l_z]}{L \cdot B}$$
(14.13)

- The aspect ratio for the base of texture feature:

$$\phi_d = \frac{d_x}{d_z}$$
(14.14)

14.9.2 The associated dimensionless groups and the dimensionless form of the equations

Non-dimensionalising of the time-independent Reynolds equation

If the bearing length, L, minimum clearance between the bearing surfaces, h_m, and the ambient pressure, P_a, are considered as the reference length, height and pressure, respectively, the dimensionless length, width, height and pressure can be defined as (see Fig. 14.3):

$$x^* = \frac{x}{L}, \quad z^* = \frac{z}{B}, \quad h^* = \frac{h}{h_m}, \quad p^* = \Lambda\left(\frac{p - P_a}{P_a}\right)$$
(14.15)

in which Λ is the *bearing number* defined as:

$$\Lambda = \frac{P_a h_m^2}{\eta_0 U L}$$
(14.16)

By defining the *bearing aspect ratio* in the xz-plane as $\varphi = B/L$, the dimensionless steady Reynolds equation becomes:

$$\frac{\partial}{\partial x^*}\left(h^{*3}\frac{\partial p^*}{\partial x^*}\right)+\left(\frac{1}{\phi}\right)^2\frac{\partial}{\partial z^*}\left(h^{*3}\frac{\partial p^*}{\partial z^*}\right)=6\frac{\partial h^*}{\partial x^*} \qquad (14.17)$$

In addition, the dimensionless load capacity, friction force and lubricant flow rate in the x-direction can be written as:

$$W_H^*=\frac{(W_H-P_aLB)h_m^2}{\eta_0UL^2B},\ F_v^*=\frac{F_vh_m}{\eta_0ULB},\ Q_x^*=\frac{Q_x}{UBh_m} \qquad (14.18)$$

Note that the term η_0ULB/h_m is in fact the friction force for a bearing with parallel and smooth flat surfaces in the absence of any pressure difference between bearing ends. In addition, the lubricant flow for such a case would be $UBh_m/2$. Therefore, the absolute dimensionless friction force and lubricant flow for such a bearing in the absence of pressure difference between bearing ends would be 1 and 1/2, respectively (Rahmani, 2009).

The *modified* friction coefficient based on the dimensionless load capacity and friction force can also be defined as:

$$\eta'=\frac{F_v^*}{W_H^*} \qquad (14.19)$$

Non-dimensionalising of the time-dependent Reynolds equation

The transient Reynolds equation is expressed in dimensionless form using some different (forms of) non-dimensionalising parameters than were used before in the steady state case. The main reason is the presence of the new parameter, time, and having a variable minimum film thickness and sliding velocities, which can take different values including zero (especially for the latter case) and, hence, cannot appear in (the denominators of) the non-dimensional groups defined in the steady state analysis. The dimensionless time and pressure are defined as:

$$t^*=t\frac{P_a}{\eta_0},\ p^*=\left(\frac{p-P_a}{P_a}\right) \qquad (14.20)$$

In the transient analysis, the film thickness in the general form $h=h(x,z,t)$ can be decomposed into two spatial and temporal components as follows:

$$h(x,z,t)=h_m(t)+h_s(x,z) \qquad (14.21)$$

Since in the transient analysis, the temporal term is variable, therefore, it cannot be chosen for non-dimensionalisation. For this reason, an arbitrary parameter such as the bearing length can be considered as a suitable parameter for this purpose:

$$h^* = \frac{h}{L}$$ (14.22)

To obtain a relation for the sliding velocity, consider the case of a piston–cylinder mechanism in the IC engines as schematically shown in Fig. 14.4. If the crank angle is assumed to be zero when the piston is at TDC, the sliding velocity of piston at each crank angle can be expressed as follows:

$$v_p = \dot{s} = -r\omega \sin\theta \left(1 + \lambda \frac{\cos\theta}{\sqrt{1 - \lambda^2 \sin^2\theta}}\right)$$ (14.23)

Considering that $\theta = \omega \cdot t$, and by defining the **transient bearing number** as

$$\Lambda_\omega = \frac{\eta_0 \omega}{P_a}$$ (14.24)

the dimensionless form of the time-dependent Reynolds equation can be written as:

$$\frac{\partial}{\partial x^*}\left(h^{*3}\frac{\partial p^*}{\partial x^*}\right) + \left(\frac{1}{\varphi}\right)^2 \frac{\partial}{\partial z^*}\left(h^{*3}\frac{\partial p^*}{\partial z^*}\right) = 6\Lambda_\omega r^* \Xi(t^*)\frac{\partial h^*}{\partial x^*} + 12\frac{\partial h^*}{\partial t^*}$$

(14.25)

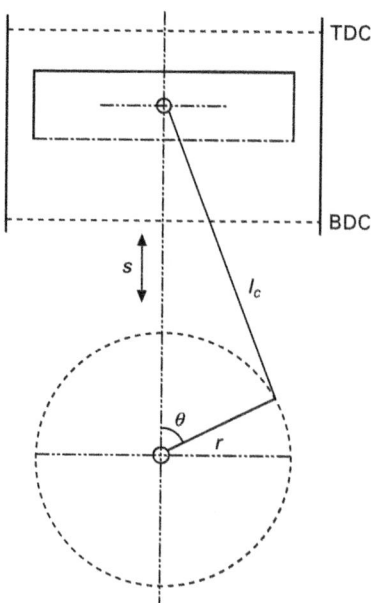

14.4 Schematic of a piston–cylinder mechanism.

in which:

$$r^* = \frac{r}{L} \tag{14.26}$$

and

$$\Xi\,(t^*) = -\sin\,(\Lambda_\omega t^*)\left[1 + \lambda\frac{\cos\,(\Lambda_\omega t^*)}{\sqrt{1 - \lambda^2\sin^2\,(\Lambda_\omega t^*)}}\right] \tag{14.27}$$

Non-dimensionalising of the ring dynamic equation

The dynamic equation for a typical piston ring given in equation (14.7) can also be expressed in the dimensionless form as:

$$\Lambda_M\frac{d^2 h_m^*}{dt^{*2}} = W_H^* - F^* \tag{14.28}$$

in which:

$$W_H^* = \frac{W_H}{P_a BL} \quad \text{and} \quad F^* = \frac{F}{P_a BL} \tag{14.29}$$

and Λ_M is called the dimensionless inertia parameter, defined as follows:

$$\Lambda_M = \frac{mP_a}{\eta_0^2 B} \tag{14.30}$$

Finally, the dimensionless *squeeze velocity* may also be defined as follows:

$$W^* = \frac{\eta_0 W}{P_a L} \tag{14.31}$$

14.10   Solution methods

14.10.1 Analytical approach

Knowing the profile of the bearing surfaces, the steady state Reynolds equation can be solved analytically by employing full-Sommerfeld or *Gumbel boundary conditions*. However, the obtained results are of no practical importance when cavitation occurs. For example, an analytical solution for a barrel-shaped piston ring profile obtained using full-Sommerfeld boundary conditions shows the existence of no net hydrodynamic load support, while the experiments have shown that such a surface profile can develop significant load-carrying capacity. Therefore, to obtain valuable results, the cavitation boundary conditions must be employed. However, since in the given cavitation models, the rupture point is not already known, implementing these models in an analytical solution would not be a very easy task, especially if the surface profile has a complex geometry. Despite this, an example of an analytical solution for a parabolic bearing surface profile, considering the Reynolds–Swift–Stieber (RSS) cavitation boundary condition, can be found in Rahmani (2009).

One of the main interests for developing the analytical solution methods for the textured slider bearings is the computation times' considerations associated with the numerical approaches, especially when the number of texture features grows. In the case of textured surfaces, the numerical mesh inside each single texture should be sufficiently dense to capture the surface profile gradient appropriately. In addition, an appropriate meshing scheme should also be employed for the *edge areas* of features, where the textured profile meets the bearing surface. Of course, for a bearing with a large number of textures this will cause the numerical approach to be very time-consuming. This problem may escalate when the optimisation of textured surfaces is of concern as the common optimisation approaches may need considerable number of the cases to be evaluated.

Another main concern with the numerical approach is that special means should be considered in order to tackle the cases, where the texture profile has a sharp edge, such as those, which may appear in a rectangular shaped texture, implanted on a slider bearing with parallel flat surfaces. This may also increase the time penalty of the numerical approach.

For surfaces with geometrical (profile) discontinuities such as those that comprise a number of regular textured patterns, an analytical approach has been introduced by Rahmani *et al.* (2007). The method has been developed in Rahmani *et al.* (2010) to include any given texture profile either in negative or positive form. The given analytical relations in Rahmani *et al.* (2010) for some texture profiles can be used for the cases where the texturing pattern and/or pressure boundary conditions are such that cavitation is not a dominant phenomenon. It is also shown that the given approach can be employed for the surfaces with asymmetric texturing patterns with satisfactory results. Pascovici *et al.* (2004, 2009) have also attempted to produce analytical solutions for the problem of partial textured surfaces.

14.10.2 Numerical approach

The finite difference method is a common approach in order to discretise the governing Reynolds equation. The obtained system of linear algebraic equations can be solved by direct decomposition methods such as Gauss, Choleski and LUD methods (Booker, 1988; see also Chapter 19). However, indirect iteration methods are the most common. In the iterative approach, starting from the initial condition given for pressure (e.g. ambient pressure), in each iteration step, the pressure at each node is calculated using the pressures at its neighbourhood found from the previous iteration. This is called the *Jacobi iteration method*. However, the convergence rate can increase by 100% if the current values of the dependent variable are used to compute the neighbouring points as soon as they are available (Hoffmann and Chiang, 1993). This method is acknowledged as the *point Gauss–Seidel iteration* method. In addition, the solution can be accelerated using a method called *point successive over-relaxation* (PSOR) or simply the *successive over-relaxation* (SOR) technique (see Chapters 6, 17 and 20).

Implementation of the Reynolds cavitation boundary condition into the numerical computations is very simple and straightforward. For iterative solution methods, the Christopherson method is employed (Christopherson, 1941). In this method the linear equation set obtained by discretising of the Reynolds lubrication equation is solved by employing the desired numerical iterative method (see next section), in which case after each iteration step, any computed pressures lower than the cavitation pressure are set equal to the cavitation pressure and the process is repeated until satisfactory convergence is achieved.

In the 2D/3D analysis, determining of the boundary conditions at the bearings laterals are also of demand. In this case, determining the boundary conditions is completely problem dependent. In general, two assumptions can be made. In the first assumption, it is supposed that the bearing has a finite width. Therefore, the leakage from bearing laterals should be considered. In this case, the boundary conditions are of Dirichlet type, where the pressures at the leading and trailing edges in the transversal direction (i.e. z-direction) (or lateral edges) are set to the desired pressure values. It should also be noted that in this case, the textures on the bearing surface (if it is textured) could be considered with either finite width (i.e. pores or asperities) or infinite width (i.e. grooves in either positive or negative form). In the second assumption, it is supposed that the bearing has infinite width. Therefore, if the bearing is either textured with parallel positive or negative grooves or not textured, the problem can be simplified to a 1D/2D model. However, if the bearing has finite width textures such as pores or asperities, then the problem is still in 2D/3D form. In this case, the conditions at the transversal (lateral) leading and trailing edges are considered to be periodic spatially, where the pressure gradient is set to zero at these boundaries. Hence, the boundary conditions are the specific form of the Neuman-type boundary conditions.

Both of these assumptions can be of interest in the study of the textured surfaces. For example, if the numerical analyses attempt to model the behaviour of the textured surfaces being tested experimentally on a tribo-testrig, where two finite size flat surfaces are sliding against each other, then the application of the first assumption seems to be more reasonable. However, if the numerical analyses are intended to simulate the conditions (e.g. between piston ring and cylinder liner in an engine), then owing to the symmetry of the ring profile in the circumferential direction, the second assumption for the lateral boundary conditions may be more realistic.

14.11   Optimisation of textured surfaces

14.11.1 Parameters involved in the optimisation of textured surfaces

In a broad spectrum, there are relatively considerable amounts of parameters that may alter the performance of a textured bearing in practice. Therefore, it is very important to address these parameters before any attempt is made for their optimisation. A list of parameters that in any modelling approach are given as follows (Rahmani, 2009):

- The profile of the surface(s) that will host the textures. This is one of the most important parameters. So far, most of the numerical analyses have focused on the textured bearings with parallel flat surfaces. Figure 14.5(a) demonstrates a series of textures implanted on a flat surface, while the case for a curved surface such as a barrel-shaped piston top ring is shown in Fig. 14.5(b).
- 1D/2D or 2D/3D bearing profile assumption. In the former, the textures are considered infinite in width such as grooves and in the latter; the textures may be considered with finite width such as pores. Figure 14.6(a) represents a schematic of textured bearing with infinite width textures, while in Fig. 14.6(b) a series of textures with finite width are shown. Note that the finite width textures can still be considered on an infinite width bearing. In this case, a symmetrical boundary condition approach may be used in the lateral direction.
- Arrangement of textures on the bearing surface or the textured pattern.

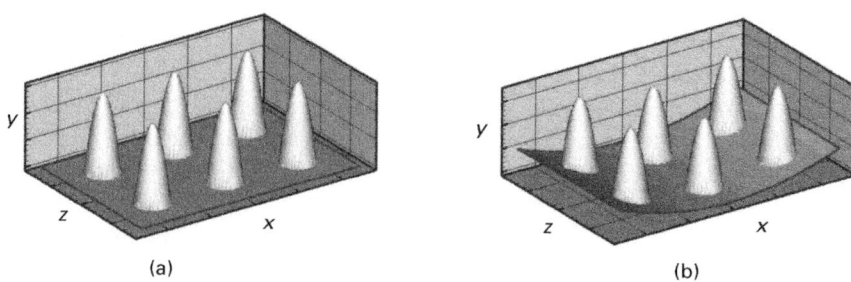

(a)   (b)

14.5 A 3D representation of textured (a) flat and (b) parabolic bearing surface profiles.

In general, three main arrangements of textures on a bearing surface can be considered including asymmetric partially textured, symmetric partially textured and fully textured patterns, which are shown in Fig. 14.7 for 1D/2D configurations. The same arrangement may also exist in the z-direction for finite width textures. In the **asymmetric partially textured pattern**, the textures are concentrated in only one side (left or right-hand side) of the bearing. In this case, depending to the direction of sliding, the outcomes from the bearing can differ considerably. In the favoured sliding direction, the hydrodynamic load support can become significant, while in the opposite sliding direction; the textures may not contribute to the load-carrying capacity. In the **symmetric partially textured pattern**, the textures are concentrated in the central part of the surface so that the lengths of leading and trailing edges are higher than the edge-to-edge distance between the textures. In the **fully textured pattern**, the textures are spread along the bearing and the lengths of leading and trailing edges are less than the edge-to-edge distance between the textures.

- The type of texturing can be in the form of cavities (negative) or asperities (positive). Figure 14.6(a) above is a demonstration of positive

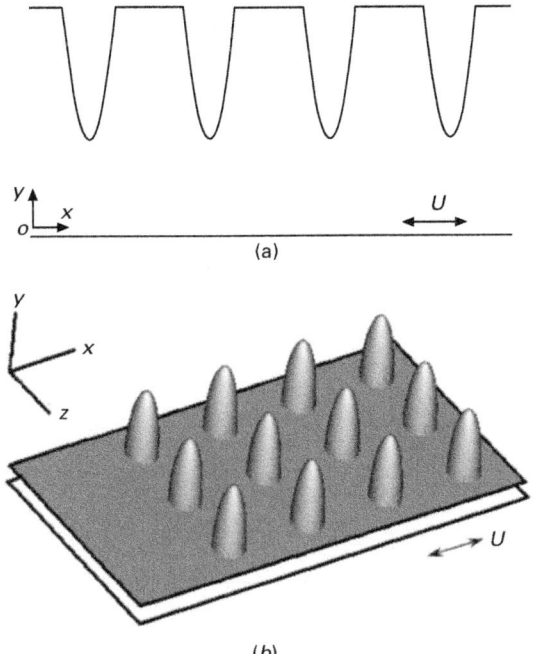

(a)

(b)

14.6 Schematic of (a) a 1D/2D model of a textured bearing with infinite width textures and (b) a 2D/3D model of textured bearing surface with finite with textures.

type infinite width textures, while those in Fig. 14.6(b) represent finite width negative type textures. In addition, the textured surfaces shown in Fig. 14.7 above are all negative infinite width textures.

- The shape of textures: In general, various texture profiles can be imagined. Different texture profiles for finite width textures are shown in Fig. 14.8. Equivalent type of texture profiles may also be defined in the case of infinite width textures such as rectangular, triangular, parabolic, elliptic, etc. In addition, in the case of finite width textures, the shape of the base of the textures can be of importance as well, which one may need to include as another effective parameter in the associated computations.

- The geometrical parameters specifying the dimension of the textures. The parameters such as textures' maximum height (or depth), length (and width) of the textures, the pitch or distance between textures, the number of textures in different directions, etc., are other important parameters, particularly from the optimisation point of view. A suitable form of these parameters was defined in dimensionless forms above.

- Variations of the geometry or geometrical parameters of textures alongside the bearing. Depending on application, some geometrical parameters of textures may vary alongside the bearing's axial and/or transverse directions. For example, Fig. 14.9 represents a textured bearing with constant height textures compared with the same bearing, but with uniformly variable texture heights. Such cases have been investigated in Rahmani (2009).

Some other parameters also exist that may be categorised as *operational parameters* and can affect the performance of a bearing in general and the

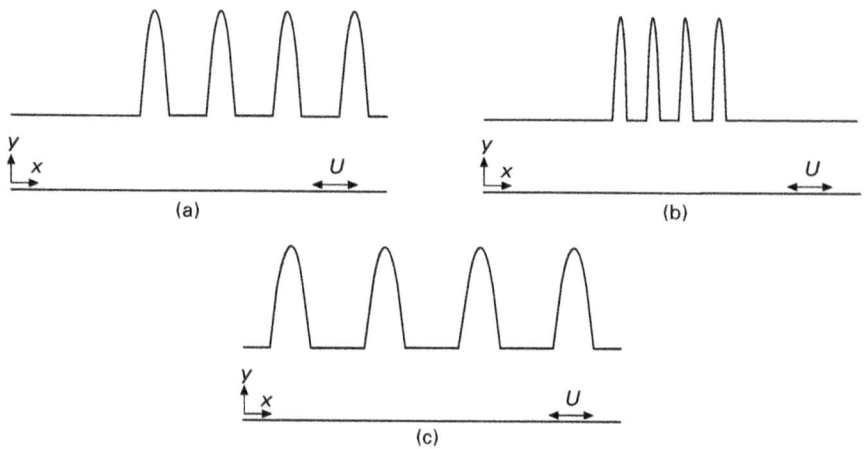

14.7 Various texturing patterns (a) asymmetrically partial (b) symmetrically partial and (c) fully textured surfaces.

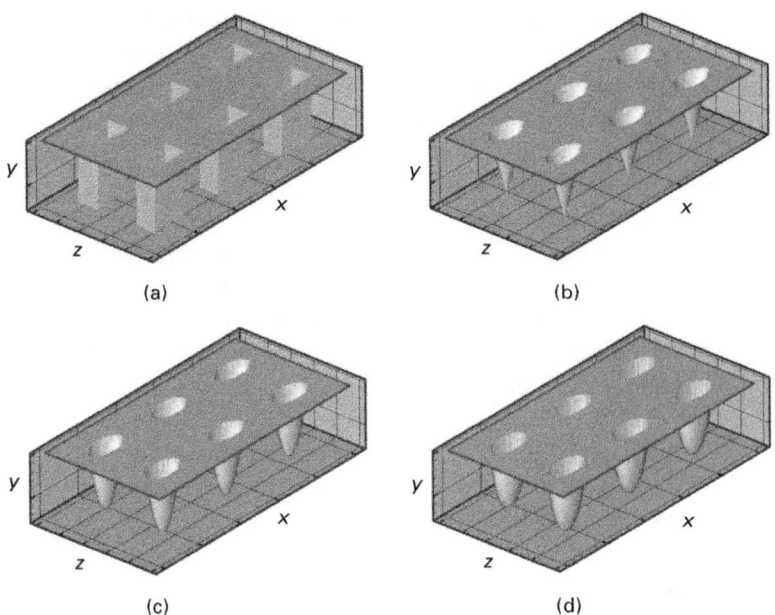

14.8 Various texture profiles in 3D with rectangular-shape base for (a) rectangular prism with the longer lateral in the *x*-direction and an elliptical base for other texture profiles including (b) conic (c) paraboloid and (d) ellipsoid with the major axis in the *x*-direction.

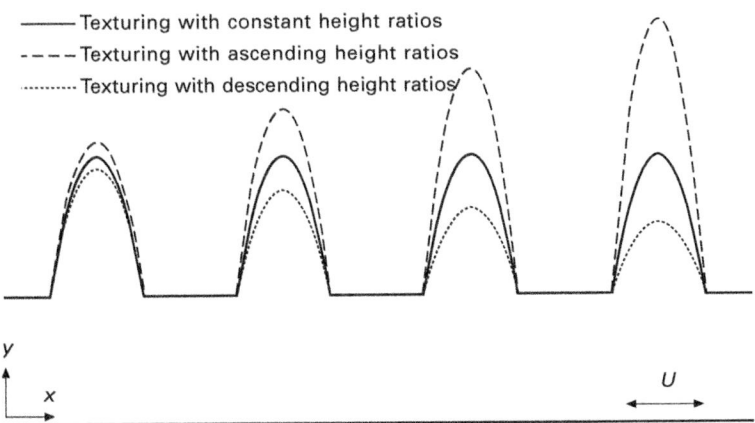

14.9 Schematic of textured bearing with constant, ascending and descending height ratios along the bearing length.

textured bearing in particular. The working conditions including parameters such as variations of pressure at the boundaries or the effect of surface roughness, rheological properties of lubricant, etc., are of this type. In addition, the lubrication regime in which these micro-features work, e.g. hydrodynamic, elastohydrodynamic, mixed or boundary, is another main parameter, which needs to be addressed. The performance of textures and the basic philosophy of employing these macro-/micro-features can be fundamentally different under different regimes of lubrication.

Another important parameter for optimisation is defining or determining an *objective function*. Two main objective functions may be defined in the analysis of the bearings in general: load capacity of the bearing and coefficient of friction. In the former case, the aim is to maximise the load capacity of the bearing, while in the latter minimisation of the function is of interest. Other parameters such as load capacity-to-lubricant flow rate ratio, which accounts for lubricant consumption to produce a desired hydrodynamic lift have also been suggested (Rahmani *et al.*, 2010). Using an appropriate optimisation technique, a weighted objective function, which considers a trade-off between various performance parameters of a bearing may also be defined and employed.

14.11.2 Optimisation results

Typical results from an optimisation process for a symmetrically textured infinite width bearing are demonstrated in Fig. 14.10. The results are obtained for a fixed bearing length and the objective function was the total coefficient of friction of the bearing. Five texture profiles, including rectangular, trapezoidal, isosceles triangular, parabolic and elliptic in both negative and positive forms, were considered. To overcome the negative effect of sharp edges in the convergency of numerical computations related to the rectangular shaped textures, the edges were considered to be slightly inclined. The trapezoidal shaped textures were also considered and isosceles with its upper lateral equal to the one-third of the lower lateral. The lengths of leading and trailing edges were set as a function of the edge-to-edge distance between the textures. Therefore, they can vary with the area density of the textures. For each texture type, profile and with different number of textures, the textures' height ratios as well as area density of textures were set as the optimisation parameters.

The range of variations of the textures' height ratio was between 1 and 5, although for the area density of textures the range of variations is ideally from 0 to 1, while in practice this range (especially the upper limit) may become restricted owing to the computational penalty in refining the numerical mesh.

Figure 14.10(a) represents the obtained optimum texture height ratios in each case. As it can be seen, the variations of optimum texture height ratio with the number of textures are not very significant for higher numbers of textures. In addition, apart from negative rectangular textures, the rest of

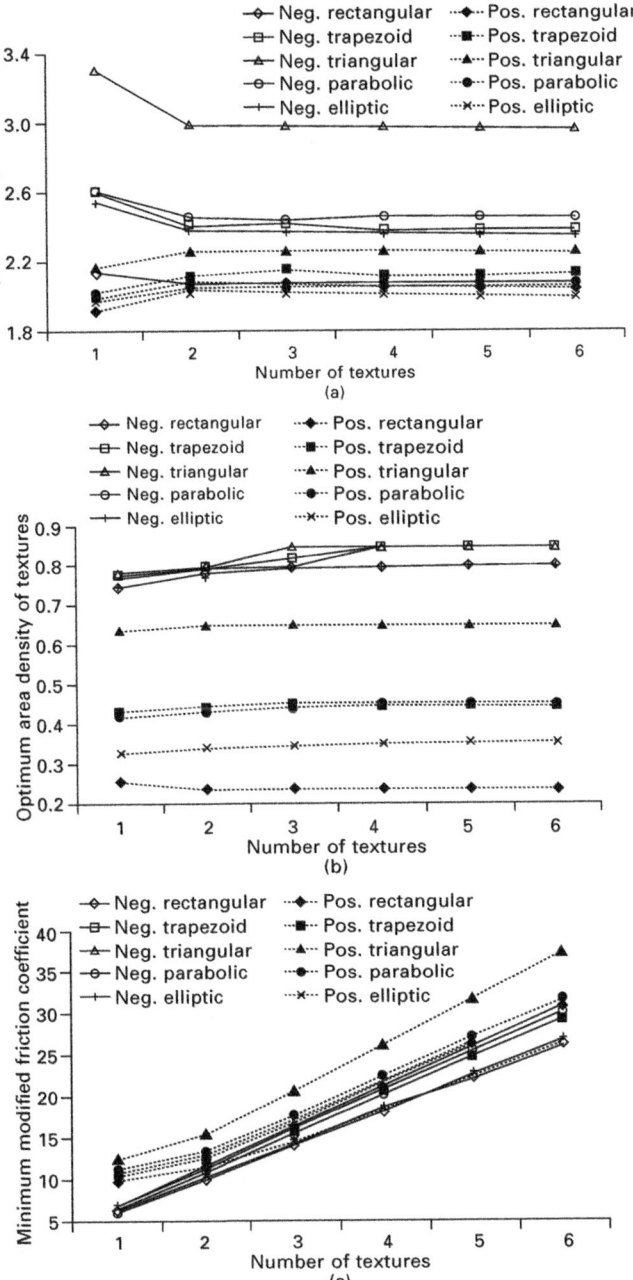

14.10 Optimum (a) height ratio and (b) area density of textures as well as (c) resulting minimum modified friction coefficients for a symmetrically textured infinite width bearing with various texture types, numbers and profiles based on minimisation of modified friction coefficients.

negative type textures have higher optimum height ratios than the positive ones. For the negative textures, the triangular and rectangular textures have the highest and lowest optimum height ratios, while the rest have very close optimum values. For the positive textures, the optimum values for different profiles are closer to each other in general and a similar pattern can also be observed except for the positive elliptic textures, which have the lowest optimum values at higher numbers of textures.

The corresponding optimum area densities demonstrated in Fig. 14.10(b) show distinctively the higher, but very close values for negative textures, which slightly increases at the lower numbers of textures. For this type of textures, it seems that the shape of textures has weak influence on the optimum area density of the textures. On the other hand, for the positive textures, unlike their negative counterparts, the textures' profiles have significant influence on the optimum area density values.

Comparing the performance of the studied textures, based on the coefficient of friction, shown in Fig. 14.10(c), indicates the lowest coefficient of friction results for the negative rectangular and elliptic features, as well as positive rectangular (at higher number of textures) textures, while the positive triangular textures prove to be the least favoured of all. It can also be seen that the coefficient of friction increases as the number of textures increases. In addition, the distinction between the performances of the different textured profiles slightly increases as the number of textures rises.

Figure 14.11(a) and (b) compare the obtained optimum configuration for the studied textured surfaces with six textures for negative and positive types respectively. In addition, the corresponding dimensionless pressure distribution for each case is shown in Fig. 14.11(c).

A similar approach can be employed for the optimisation of textured surfaces, comprising finite width textures. However, it is noted that some extra parameters may have to be considered in the case of finite width textures such as the aspect ratio of the textures' base, the lengths of the lateral edges, distance between textures, etc.

A comparison between the performance of infinite and finite width textures in their optimum configuration is given in Fig. 14.12 for various numbers of textures, types and profiles. To obtain these results, the load capacity was set as the objective function and a fully textured pattern was considered in both cases, while the base aspect ratio for the finite width textures was set to unity. In addition, the area density for infinite width and the axial area density for finite width textures were equal. Therefore, textures' height ratio was the only optimisation parameter. In order to reduce CPU times, only half a row of textures in the axial direction was considered (corresponding to the A–A' line in Fig. 14.3) and the results obtained for the load capacity were then doubled. Furthermore, the bearing edge in the lateral direction was also set equal to the bearing leading (and therefore, trailing) edges in

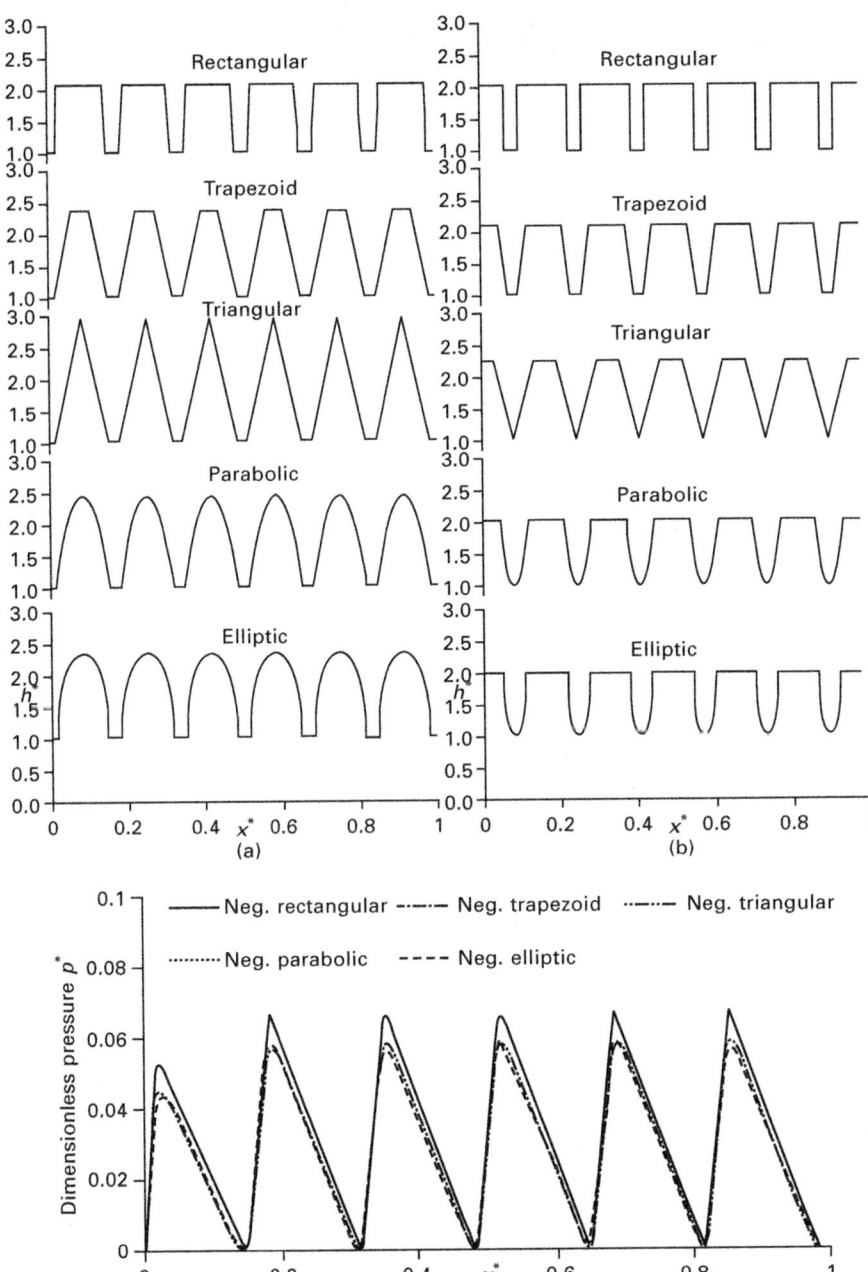

14.11 The optimised textured bearings with (a) negative and (b) positive textures and correspondent pressure distributions for (c) negative and (d) positive textures at texture number of six for partially texturing pattern.

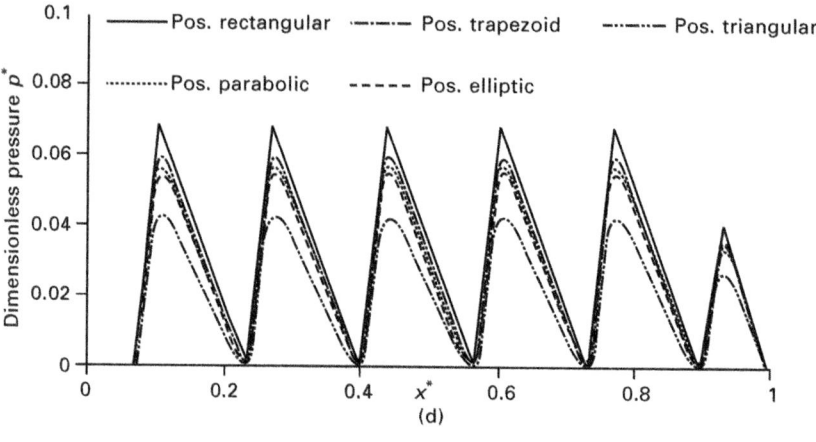

14.11 Continued

the axial direction. Thus, the final results for the load capacity of the finite width textures was in fact for a single row of textures in the axial direction including the effect of lateral edges.

A cross-sectional view of the optimum configurations for the studies of 3D textures along the axis of symmetry are presented in Fig. 14.12(a) and (b) for negative and positive type textures respectively, with six textures. As can be seen, the optimum height ratios for negative textures vary more with the profile of the texture, when compared with those obtained for positive textures. As a result, it is seen that for a given texturing pattern, the optimum height ratio for the positive textures remains almost independent of their profile. In addition, in the case of negative textures, although the conic textures demand the highest optimum height ratios, they still seem to have the lowest load-carrying capacity as is shown in Fig. 14.12(c) which shows the corresponding dimensionless pressure distribution for each case along the axis of symmetry. As this figure indicates, although the profile of the textures does not have any considerable effect on the optimum height ratios for the positive textures, its effect on the performance of the bearing is relatively more significant than that of negative textures.

Figure 14.13(a) compares the performances (in terms of maximum attainable load capacities) of various infinite and finite width negative texture profiles at their corresponding optimum height ratios and with different numbers of textures. The results for the infinite width textures were obtained in the same manner as discussed above for their finite width counterparts. As can be seen, for both cases the performance reduces as the number of textures increases. For a lower number of textures the infinite width textures show higher load-carrying capacity although as the number of textures increases, the difference in the results becomes less pronounced. It is also noted that

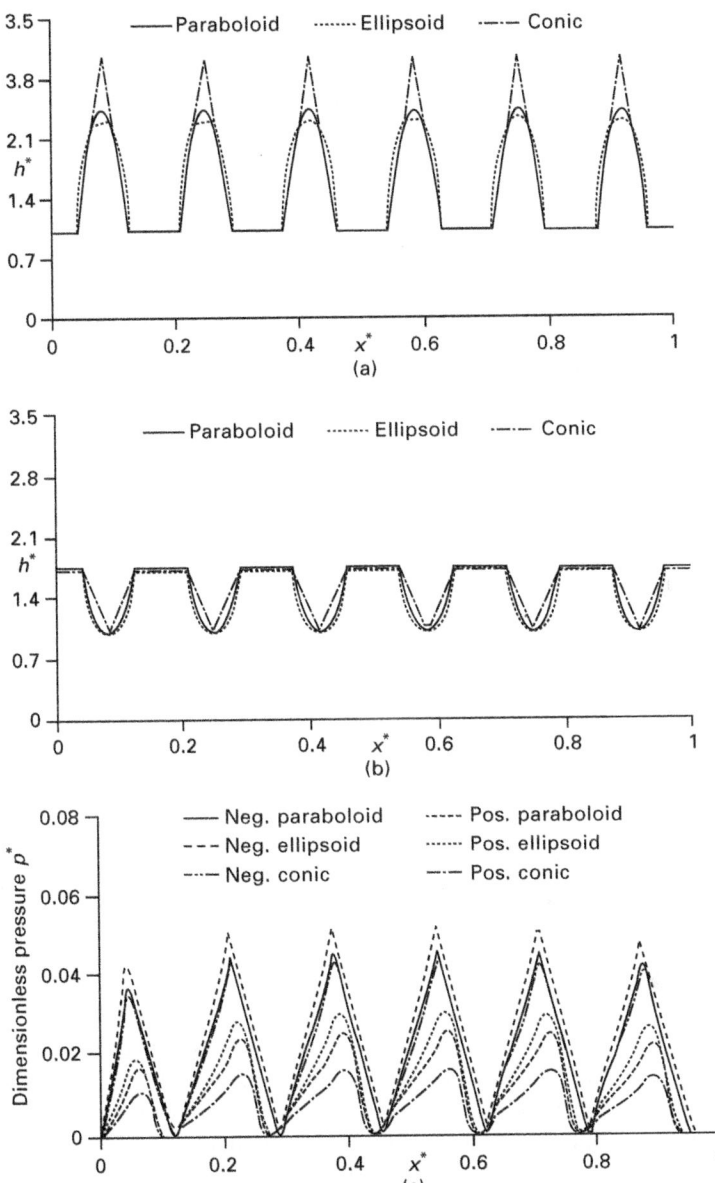

14.12 The optimised textured bearings with (a) negative and (b) positive dimples and (c) corresponding pressure distributions at texture number of six for fully texturing pattern.

for the finite width textures, the ellipsoidal textures have the highest load carrying capacities, while the conic ones are the poorest.

For the positive type textures, the gap between the maximum load capacities,

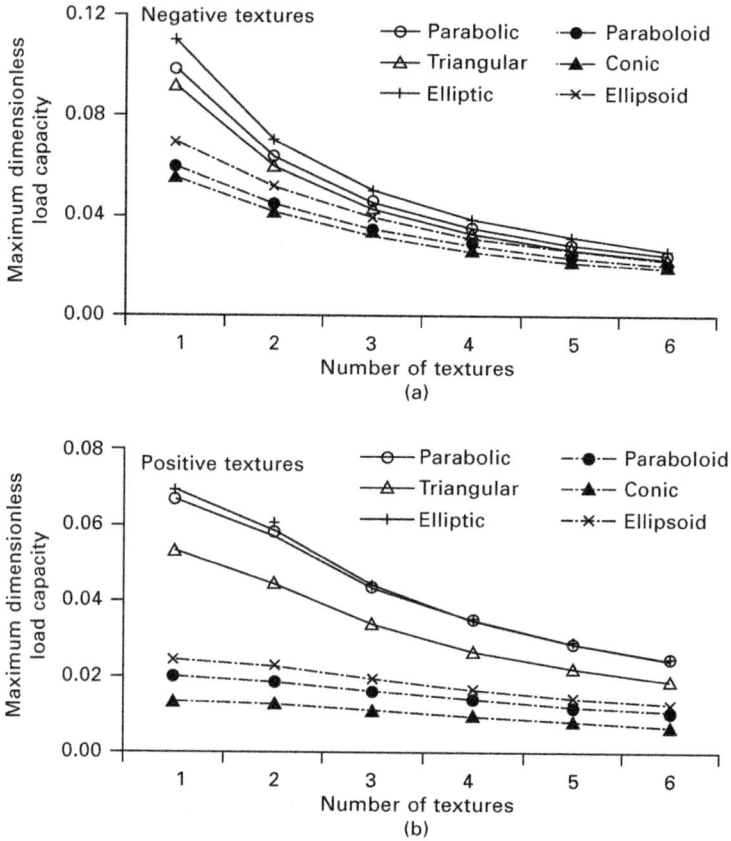

14.13 Comparison of maximum dimensionless load capacities for infinite and finite width textures in fully texturing pattern.

in general, for infinite width textures and those for finite width ones are higher as is shown in Fig. 14.13(b). However, similar to the negative textures, this difference reduces as the number of textures grows. In addition, the infinite width textures seem to be more sensitive to the variations in the number of textures. A similar trend for the performances of the infinite width textures in terms of their shapes can be observed as in the case of negative types

Considering the given results in Fig. 14.13, the advantages of employing the infinite width textures over finite width textures is obvious. The reason for this is simply that in the case of finite width textures, since the lateral distance between textures is a flat surface, it will not therefore contribute to the production of the hydrodynamic lift, when the bearing's surfaces slide relative to each other. In fact, any build-up of hydrodynamic pressure in that region is as a consequence of the existence of textures.

14.12   Application of the optimum results in the piston ring/cylinder liner contact

The lubrication of the piston ring/cylinder contact is of main interest in the study of the tribology of IC engines. In this regard, the top ring has attracted much the attention (see Chapter 15). This is because the contribution of top ring to the frictional losses in the piston and piston ring/cylinder liner contact is the highest. Investigations show that although the top ring works in the hydrodynamic regime of lubrication at mid-stroke, this usually turns to mixed (or partial) lubrication regime (and even boundary lubrication in the case of heavy-duty diesel engines) near the top dead centre (TDC) and bottom dead centre (BDC) with the consequent increase in friction and wear. This is because in the mixed regime of lubrication, the asperities of ring and liner surfaces are in contact (see Chapters 2 and 3). Considering the conventional Stribeck diagram shown in Fig. 14.14, the *ideal* regime of lubrication for the piston ring is the boundary between hydrodynamic and mixed lubrication regimes. Therefore, an ideal working condition for a piston ring would be the case in which the oil film thickness in the mid-stroke is reduced in order to minimise lubricant consumption and the hydrodynamic friction force, while it is sufficiently thick near the TDC and BDC, so that asperity contact is avoided as much as possible.

In this section, a comparison between the performances of conventional untextured barrel-shaped top ring profile and the optimised textured flat surfaces is made.

The general family of convex (converging–diverging) bearing surface profiles are defined as follows:

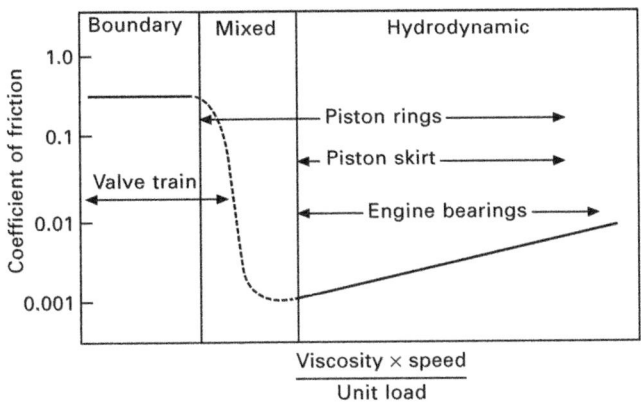

14.14 The Stribeck diagram based on Sommerfeld parameter (Tung and McMillan, 2004).

$$h(x) = h_m + h_R \left(2\frac{|x|}{L} \right)^n$$ (14.32)

in which h_R is the height of curvature and n may be considered as the *bearing shape power*. A parameter called *crown height ratio* for this kind of bearings can also be defined as follows:

$$\zeta = 1 + \frac{h_R}{h_m}$$ (14.33)

Considering $n = 2$, equation (14.32) would result in a parabolic profile, which is usually employed to approximate the piston top ring profile with sufficient accuracy (Jeng, 1992; Baek *et al.*, 2005; Mishra *et al.*, 2009) and is also called the barrel-shaped profile. In this case, the bearing surface profile may be written as follows (see Fig. 14.15):

$$h(x) = h_m + qx^2$$ (14.34)

where $q = 4h_R/L^2$.

An investigation into the optimisation of convex-shaped bearing profiles shows that in the absence of a pressure difference between the bearing ends, $n = 2$ is the optimum shape power for convex bearing families as described above. In this case, the optimum crown height ratio of $\zeta_{opt} = 2.2$ would provide the maximum dimensionless load capacity of $(W_H^*)_{max} \approx 0.0637$. In addition, by considering $\zeta_{opt} = 2.6$, the minimum modified friction coefficient of $\eta'_{min} \approx 12.53$ can be achieved (Rahmani, 2009).

Since the working conditions are time-dependent (transient), one needs to solve the transient Reynolds equation in conjunction with the dynamic force balance equation as introduced earlier. For the barrel-shaped ring profile, $\zeta_{opt} = 2.2$, which provides the highest load capacity. This is the case considered here. In addition, two texture types and profiles of negative parabolic and

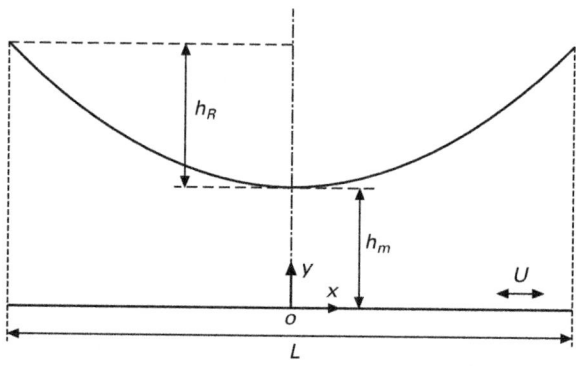

14.15 Schematic of a slider bearing with parabolic surface profile.

positive triangular were employed for comparison purposes. In both cases, a fully textured pattern was utilised. The optimum results for this texturing pattern have already been presented above. In order to reduce the CPU time, a 1D/2D model with only three textures was investigated. The values of the parameters that were used for the study were $r^* = 14.87$, $\omega = 1145.0$ rpm, $\delta_0 = 3.72 \times 10^{-4}$, $F^* = 1.24 \times 10^3$, $\Lambda_\omega = 3.33 \times 10^{-4}$, $\Lambda_M = 1.25 \times 10^8$, and $\iota = 0.001$.

A fourth-order Runge–Kutta approach is employed for this purpose (see e.g. Gerald and Wheatley, 1992; Burden and Faires, 1997). Although the results for a full engine cycle, i.e. $\theta = 0$ to $720°$, are of interest, some extra crank angles should also be considered in order to achieve a fully periodic solution. For the current simulations, the termination condition was achieved after around 50 iterations starting from the second period.

In addition, since the optimum height ratio for the textured surfaces and also the optimum crown height ratio for the barrel-shaped ring profile were introduced, based on the minimum film thickness, the solution approach described above needs to be repeated until the desired minimum clearance is achieved. Thus, at the start of the analysis, an initial guess for the minimum film thickness is made and the optimum height for the textures and the crown height for the ring are calculated. Once the solution is completed, the predicted minimum film thickness is used to recalculate the optimum heights for the ring and the textures. This is repeated until the difference in the obtained minimum clearances between two successive solutions meets the stated criterion. In the studied cases, the final solution was achieved usually after four iterations, as beyond that there was no significant change in the attained minimum clearance. The variations of minimum dimensionless clearance for five solution stages in the case of barrel-shaped ring profile are shown in Fig. 14.16. It should be noted that in this figure and in the following figures

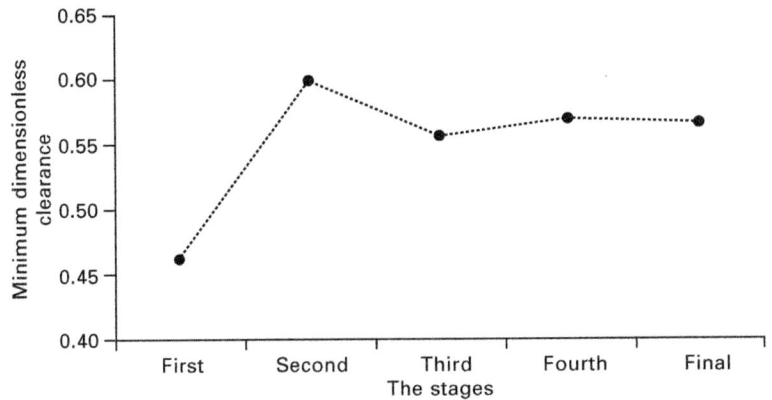

14.16 The minimum dimensionless clearance at each solution stage.

the film thickness is non-dimensionalised, based on an initial guess for the minimum clearance.

Figure 14.17 shows the obtained values for the minimum clearances in the studied optimum cases in the final stage of iteration for a full four-stroke cycle of the engine. As it can be seen, the optimum positive triangular textures produce lower clearances at the mid-stroke, while the trade-off is the lower film clearance at and near the TDC and BDC, compared with the barrel-shaped ring profile. On the other hand, the optimum negative type parabolic-shaped textures have the lower film thickness in the mid-stroke, compared with the optimised barrel-shaped piston ring profile, while producing a higher film thickness at and near the TDC and BDC. The results in this figure show both the importance of type and shape of textures, as well as the effectiveness of textured surfaces, compared with the usual barrel-shaped piston rings.

Here an important question may arise about the effect of squeeze velocity on the obtained optimum values since the optimisation was conducted based on the steady state form of Reynolds equation, neglecting the squeeze film effect. In Fig. 14.18 the variations of piston axial velocity, minimum film thickness and the squeeze velocity with the crank angle for a barrel-shaped ring profile for the full four-stroke of an IC engine is shown. In this figure, the results for sliding and squeeze velocities are scaled in the range of [−1,1] and for minimum film thickness in the range of [0,1] for the sake of demonstration.

As can be seen in Fig. 14.18, the squeeze velocity has its absolute minimum value at the mid-strokes as well as at the TDCs and BDCs. Therefore, one can observe that the squeeze velocity has a small influence in these regions and hence, the obtained optimum values from steady state analyses are a good representation. This is in line with the fact that the original aim of

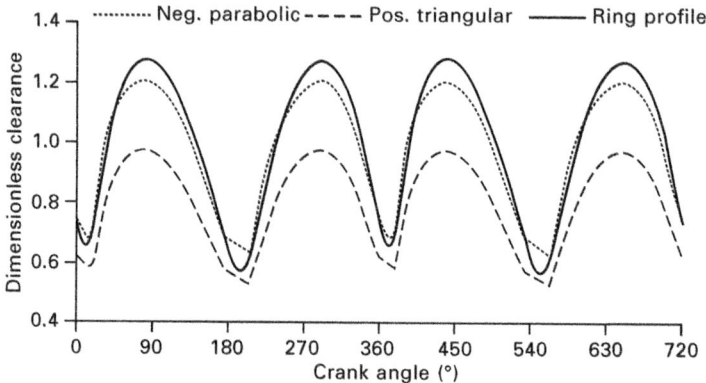

14.17 The minimum dimensionless clearance at the final stage for the studied optimum textured and untextured ring profiles.

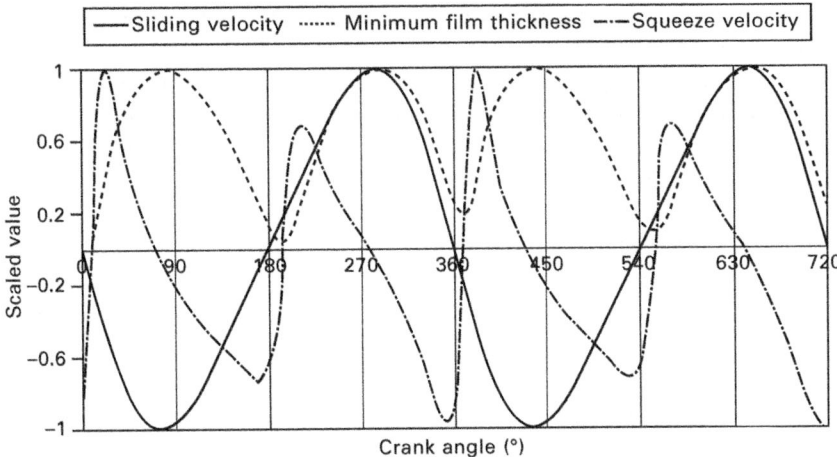

14.18 The variations of dimensionless squeeze velocity with the crank angle for barrel-shaped piston ring profile.

optimisation was to gain the highest possible performance at the crucial points of an IC engine cycle (i.e. TDCs and BDCs).

14.13   Conclusions

From hydrodynamic lubrication point of view, the performance of the textured surfaces reduces by increasing the number of textures in the axial direction. Therefore, for a slider bearing working under full fluid film lubrication, the number of textures should be as low as possible. For bearings, which may work in other lubricating regime(s) in all or part of their sliding course or encountering lubricant starvation, an optimum number of textures that makes a trade-off between their hydrodynamic performance and their role as the oil reservoirs and debris traps should be sought with further investigations.

For a lower number of textures, the obtained optimum values for texture height ratio, area density and/or optimum textured area ratio can vary in different ways, dependent on the parameters such as texture types and/or profiles. Nevertheless, as the number of textures rises, the trend(s) in the variations of the optimum parameters with the number of textures for a given texture type and profile can be predicted since the variations become uniform and steady.

For infinite width negative type textures, the rectangular (and, with a minor difference, the elliptical) shape grooves are the best while with the positive side, the rectangular ones are the best by far. In both negative and positive type textures, the triangular shape grooves are the worst. In addition, the negative infinite width textures seem to have better performance than the positive ones.

For finite width textures, those with ellipsoidal and conical profiles were found to perform the best and worst respectively, in both positive and negative forms. However, based on the obtained results for the infinite width textures, it is expected that the rectangular prism shape textures in their optimum configuration perform even better than the ellipsoidal ones. In addition, the optimised negative form of the finite width textures provided better performances, compared with their positive counterparts.

It was observed that for textures in positive form, the infinite width variety has certain advantages over the finite width ones in general. Although the same is true for the negative type textures, the difference is less pronounced than in the case of positive textures and it reduces further as the number of textures increases.

The obtained optimum results based on steady state form of Reynolds equation can be applied to enhance the lubrication conditions at the critical points near TDC and BDC reversals and/or control the lubricant flow through mid-strokes in IC engines, as the squeeze velocity becomes zero at such instances. In fact, employing the appropriate *optimally designed* textures, based on the critical conditions in the vicinity of TDC and BDC, not only one can enhance the lubrication condition by increasing the minimum clearance, it can also control the lubricant flow through the mid-strokes by minimising the minimum clearance there. Therefore, the appropriate and optimised textures in the piston-ring/cylinder liner contact could be more beneficial than employing the common barrel-shaped piston rings from tribological point of view.

14.14 **References**

Alberdi A., Merino S., Barriga J., Aranzabe A. (2004) Microstructured surfaces for tribological applications, Tribology and Lubricant Engineering, 14th International Colloquium Tribology; Esslingen, 13–15 January, Germany

Aldajah S., Ajayi O.O., Fenske G.R. (2005) Investigation of scuffing resistance and tribological performance of laser modified surfaces, The 6th Annual U.A.E. University Research Conference, College of Engineering, April, UAE, pp. Eng. 224–232

Andersson P., Koskinen J., Varjus S., Gerbig Y., Haefke H., Georgiou S., Zhmud B., Buss W. (2007) Microlubrication effect by laser-textured steel surfaces, *Wear*, Vol. 262, pp. 369–379

Anno J.N., Walowit J.A., Allen C.M. (1968) Microasperity lubrication, *Journal of Lubrication Technology, Transactions of the ASME*, Vol. 90, pp. 351–355

Anno J.N., Walowit J.A., Allen C.M. (1969) Load support and leakage from microasperity-lubricated face seals, *Journal of Lubrication Technology, Transactions of the ASME*, Vol. 91, pp. 726–731

Baek J.S., Groll E.A., Lawless P.B. (2005) Piston–cylinder work producing expansion device in a transcritical carbon dioxide cycle: Part II: Theoretical Model, *International Journal of Refrigeration*, Vol. 28, pp. 152–164

Barrell D.J.W., Priest M., Taylor C.M. (2000) The axial motion of the piston ring in the top ring groove of a gasoline engine, Proceedings of the 27th Leeds–Lyon Symposium on Tribology, Lyon, France

Blatter A., Maillat M., Pimenov S.M., Shafeev G.A., Simakin A.V., Loubnin E.N. (1999) Lubricated sliding performance of laser-patterned sapphire, *Wear*, Vol. 232, pp. 226–230

Blau P.J., Qu J. (2004) Laser surface texturing of lubricated ceramic parts, FY 2004 Progress Report on Heavy Vehicle Propulsion Materials, pp. 123–128, URL: http://www1.eere.energy.gov/vehiclesandfuels/pdfs/hv_propulsion_04/4k_blau-laser.pdf (accessed 4 December 2009)

Booker J.F. (1988) *Classic Cavitation Models for Finite Element Analysis*, NASA Technical Memorandum No. TM-103184, pp. 39–40

Borghi A., Gualtieri E., Marchetto D., Moretti L., Valeri S. (2008) Tribological effects of surface texturing on nitriding steel for high-performance engine applications, *Wear*, Vol. 265, pp. 1046–1051

Brizmer V., Kligerman Y., Etsion I. (2003) A laser surface textured parallel thrust bearing, *Tribology Transactions*, Vol. 46, No. 3, pp. 397–403

Burden R.L., Faires J.D. (1997) *Numerical Analysis*, 6th Edition, Brooks/Cole Publishing Co., US

Christopherson D.G. (1941) A new mathematical model for the solution of film lubrication problems, *Proceedings of the Institution of Mechanical Engineers London*, Vol. 146, Part 3, pp. 126–135

Costa H.L., Hutchings I.M. (2007) Hydrodynamic lubrication of textured steel surfaces under reciprocating sliding conditions, *Tribology International*, Vol. 40, pp. 1227–1238

Erdemir A. (2005) Review of engineered tribological interfaces for improved boundary lubrication, *Tribology International*, Vol. 38, pp. 249–256

Erdemir A., Ajayi L., Eryilmaz O., Kazmanli K., Etsion I. (2004a) Superhard nanocrystalline coatings for wear and friction reduction, *FY 2004 Annual Report on Heavy Vehicle Systems Optimization Program*, pp. 121–126, URL: http://www1.eere.energy.gov/vehiclesandfuels/pdfs/program/2004_hv_optimization.pdf (accessed 4 December 2009)

Erdemir A., Ajayi O., Kovalchenko A., Kazmanli K., Erck R., Eryilmaz O., Fenske G. (2004b) Laser texturing of materials, *FY 2004 Progress Report on High Strength Weight Reduction Materials*, pp. 71–76, URL: http://www1.eere.energy.gov/vehiclesandfuels/pdfs/hswr_2004/fy04_hswr_4c.pdf (accessed 4 December 2009)

Erdemir A., Ajayi O., Kovalchenko A., Kazmanli K., Erck R., Eryilmaz O., Fenske G. (2005) Laser texturing of propulsion materials, *FY 2005 Progress Report on High Strength Weight Reduction Materials*, pp. 95–102, URL: http://www1.eere.energy.gov/vehiclesandfuels/pdfs/hswr_2005/fy05_hswr_4f.pdf (accessed 4 December 2009)

Etsion I. (2005) State of the art in laser surface texturing, *ASME Journal of Tribology*, Vol. 127, pp. 248–253

Etsion I., Burstein L. (1996) A model for mechanical seals with regular microsurface structure, *Tribology Transactions*, Vol. 39, No. 3, pp. 677–683

Etsion I., Halperin G. (2002) A laser surface textured hydrostatic mechanical seal, *Tribology Transactions*, Vol. 45, No. 3, pp. 430–434.

Etsion I., Sher E. (2009) Improving fuel efficiency with laser surface textured piston rings, *Tribology International*, Vol. 42, pp. 542–547

Etsion I., Kligerman Y., Halperin G. (1999) Analytical and experimental investigation of laser-textured mechanical seal faces, *Tribology Transactions*, Vol. 42, No. 3, pp. 511–516

Etsion I., Halperin G., Brizmer V., Kligerman Y. (2004) Experimental investigation of laser surface textured parallel thrust bearings, *Tribology Letters*, Vol. 17, No. 2, pp. 295–300

Etsion I., Halperin G., Becker E. (2006) The effect of various surface treatments on piston pin scuffing resistance, *Wear*, Vol. 261, pp. 785–791

Fenske G., Erdemir A., Ajayi L., Kovalchenko A. (2003) Parasitic engine loss models, *FY 2003 Annual Report on Heavy Vehicle Systems Optimization Program*, pp. 101–106, URL: http://www1.eere.energy.gov/vehiclesandfuels/pdfs/program/2003_hv_optimization.pdf (accessed 4 December 2009)

Gerald C.F., Wheatley P.O. (1992) *Applied Numerical Analysis*, 4th Edition, Addison-Wesley Publishing Co., US

Gohar R., Rahnejat H. (2008) *Fundamentals of Tribology*, Imperial College Press, Singapore

Haefke H., Gerbig Y., Dumitru G., Romano V. (2000) Microtexturing of functional surfaces for improving their tribological performance, Proceedings of the International Tribology Conference, Nagasaki, Japan, pp. 217–221

Hamilton D.B., Walowit J.A., Allen C.M. (1966) A theory of lubrication by micro-irregularities, *Journal of Basic Engineering, Transaction of the ASME*, Vol. 88, pp. 177–185

Hoffmann K.A., Chiang S.T. (1993) *Computational Fluid Dynamics for Engineers*, Engineering Education System, Wichita, Kansas, US

Hsu S., Wang X., Chae Y., Ives L.K. (2003) Modification of engineering materials for heavy-vehicle applications, *FY 2003 Progress Report on Heavy Vehicle Propulsion Materials Program*, URL: http://www1.eere.energy.gov/vehiclesandfuels/pdfs/hv_propulsion/5_test_and_materials_standards.pdf (accessed 4 December 2009)

Hsu S. (2004a) Surface modification of engineering materials for heavy vehicle applications, *FY 2004 Quarterly Progress Report on Heavy Vehicle Propulsion Materials*, URL: http://www.ornl.gov/sci/propulsionmaterials/pdfs/HV_04_1.pdf (accessed 4 December 2009)

Hsu S. (2004b) Surface modification of engineering materials for heavy vehicle applications, *FY 2004 Progress Report on Heavy Vehicle Propulsion Materials*, pp. 205–209, URL: http://www1.eere.energy.gov/vehiclesandfuels/pdfs/hv_propulsion_04/5_test_and_materials_standards.pdf (accessed 4 December 2009)

Hsu S., Wang X., Ives L.K., Zhang H., Liang Y., Ying C. (2005) An integrated surface modification of engineering materials for heavy vehicle applications, *FY 2005 Progress Report on Heavy Vehicle Propulsion Materials*, pp. 231–235, URL: http://www1.eere.energy.gov/vehiclesandfuels/pdfs/hv_propulsion_05/5d_hsu.pdf (accessed 4 December 2009)

Hupp S.J. (2004) A tribological study of the interaction between surface micro texturing and viscoelastic lubricants, Master Thesis, Department of Mechanical Engineering, Massachusetts Institute of Technology (MIT), US

Ike H., Tsuji K., Takase M. (2002) *In situ* observation of a rolling interface and modelling of the surface texturing of rolled sheets, *Wear*, Vol. 252, pp. 48–62

Ito H., Kaneda K., Yuhta T., Nishimura I., Yasuda K., Matsuno T. (2000) Reduction of polyethylene wear by concave dimples on the frictional surface in artificial hip joints, *The Journal of Arthroplasty*, Vol. 15, No. 3, pp. 332–338

Jeng Y.-R. (1992) Theoretical analysis of piston-ring lubrication, Part I – fully flooded lubrication, *ASME Tribology Transactions*, Vol. 35, No. 4, pp. 696–706

Kortikar S.N. (2004) Fabrication and characterization of deterministic microasperities on thrust surfaces, Master Thesis, College of Engineering, University of Kentucky, Lexington, Kentucky, US

Kovalchenko A., Ajayi O., Erdemir A., Fenske G., Etsion I. (2005) The effect of laser surface texturing on transitions in lubrication regimes during unidirectional sliding contact, *Tribology International*, Vol. 38, pp. 219–225

Krupka I., Hartl M. (2007) The effect of surface texturing on thin EHD lubrication films, *Tribology International*, Vol. 40, pp. 1100–1110

Lisowsky B. (2006) Efficiency improvement through reduction of friction and wear in powertrain systems, *FY 2006 Progress Report on Heavy Vehicle Systems Optimization Program*, pp. 1013–114, URL: http://www1.eere.energy.gov/vehiclesandfuels/pdfs/program/2006_hvsop_report.pdf (accessed 4 December 2009)

Lu X., Khonsary M. (2007) An experimental investigation of dimple effect on the Stribeck curve of journal bearings, *Tribology Letters*, Vol. 27, pp. 169–176

Lubbinge H. (1999) On the lubrication of mechanical face seals, PhD Thesis, University of Twente, Enschede, the Netherlands

McNickle A.D., Etsion I. (2004) Near-contact laser surface textured dry gas seals, *Journal of Tribology, Transactions of the ASME*, Vol. 126, No. 4, pp. 788–794

Mishra P.C., Rahnejat H., King P.D. (2009) Tribology of the ring–bore conjunction subject to a mixed regime of lubrication, *Proceedings of the Institution of Mechanical Engineers, Part C: Journal of Mechanical Engineering Science*, Vol. 223, No. 4, pp. 987–998

Mourier L., Mazuyer D., Lubrecht A.A., Donnet C. (2006) Transient increase of film thickness in micro-textured EHL contacts, *Tribology International*, Vol. 39, pp. 1745–1756

Neves D., Diniz A.E., de Lima M.S.F. (2006) Efficiency of the laser texturing on the adhesion of the coated twist drills, *Journal of Materials Processing Technology*, Vol. 179, pp. 139–145

Pascovici M., Marian V., Gaman, D. (2004) Analytical and numerical approach of load carrying capacity for partially textured slider, Proceedings of International Nanotribology Conference, Nano Sikkim II: Friction and Biotribology, 8–12 November, Peeling, Sikkim, India

Pascovici M., Cicone T., Fillon, M., Dobrica, M.B. (2009) Analytical investigation of a partially textured parallel slider, *Proceedings of IMechE, Part J: Journal of Engineering Tribology*, Vol. 223, pp. 151–158

Pettersson U. (2005) Surfaces designed for high and low friction, PhD Thesis, Department of Engineering Sciences, Uppsala University, Uppsala, Sweden

Pettersson U., Jacobson S. (2003) Influence of surface texture on boundary lubricated sliding contacts, *Tribology International*, Vol. 36, pp. 857–864

Pettersson U., Jacobson S. (2006) Tribological texturing of steel surfaces with a novel diamond embossing tool technique, *Tribology International*, Vol. 39, pp. 695–700

Pettersson U., Jacobson S. (2007) Textured surfaces for improved lubrication at high pressure and low sliding speed of roller/piston in hydraulic motors, *Tribology International*, Vol. 40, pp. 355–359

Pinkus O., Strenlicht B. (1961) *Theory of Hydrodynamic Lubrication*, McGraw-Hill Inc., US

Priest M., Taylor C.M. (2000) Automobile engine tribology – approaching the surface, *Wear*, Vol. 241, pp. 193–203

Rahmani R. (2009) An investigation into analysis and optimisation of textured slider bearings with application in piston-ring/cylinder liner contact, PhD Thesis, Anglia Ruskin University, UK

Rahmani R., Shirvani A., Shirvani H. (2007) optimization of partially textured parallel thrust bearings with square-shaped micro-dimples, *STLE Tribology Transactions*, Vol. 50, No. 3, pp. 401–406

Rahmani R., Mirzaee I., Shirvani A., Shirvani H. (2010) An analytical approach for analysis and optimisation of slider bearings with infinite width parallel textures, *Tribology International*, Vol. 43, No. 8, pp. 1551–1565.

Rahnejat H., Balakrishnan S., King P.D., Howell-Smith S. (2006) in-cylinder friction reduction using a surface finish optimization technique, *Proceedings of IMechE, Part D: Journal of Automobile Engineering*, Vol. 220, pp. 1309–1318

Ronen A., Etsion I., Kligerman Y. (2001) Friction-reducing surface-texturing in reciprocating automotive components, *Tribology Transactions*, Vol. 44, No. 3, pp. 359–366

Ryk G., Etsion I. (2006) Testing piston rings with partial laser surface texturing for friction reduction, *Wear*, Vol. 261, pp. 792–796

Ryk G., Kligerman Y., Etsion I. (2002) experimental investigation of laser surface texturing for reciprocating automotive components, *Tribology Transactions*, Vol. 45, No. 4, pp. 444–449

Siripuram R.B. (2003) Analysis of hydrodynamic effects of microasperity shapes on thrust bearing surfaces, Master Thesis, University of Kentucky, Kentucky, US

Stachowiack G.W., Batchelor A.W. (2001) *Engineering Tribology*, 2nd Edition, Butterworth-Heinemann, US

Taylor R.I., Coy R.C. (2000) Improved fuel efficiency by lubricant design: a review, *Proceedings of the IMechE, Part J Journal of Engineering Tribology*, Vol. 214, No. 1, pp. 1–15

Tung S.C., McMillan M.L. (2004) Automotive tribology overview of current advances and challenges for the future, *Tribology International*, Vol. 37, pp. 517–536

Uehara Y., Wakuda M., Yamauchi Y., Kanzaki S., Sakaguchi S. (2004) Tribological properties of dimpled silicon nitride under oil lubrication, *Journal of the European Ceramic Society*, Vol. 24, pp. 369–373

Vermeulen M., Scheers J. (2001) Micro-hydrodynamic effects in EBT textured steel sheet, *International Journal of Machine Tools & Manufacture*, Vol. 41, pp. 1941–1951

Wagner K., Volkl R., Engel U. (2006) Tool life enhancement in cold forging by locally optimized surfaces, *Journal of Materials Processing Technology*, Vol. 177, pp. 206–209

Wagner K., Putz A., Engel U. (2008) Improvement of tool life in cold forging by locally optimized surfaces, *Journal of Materials Processing Technology*, Vol. 201, pp. 2–8

Wakuda M., Yamauchi Y., Kanzaki S., Yasuda Y. (2003) Effect of surface texturing on friction reduction between ceramic and steel materials under lubricated sliding contact, *Wear*, Vol. 254, pp. 356–363

Wang Q.-J., Zhu D. (2005) Virtual texturing: modelling the performance of lubricated contacts of engineered surfaces, *Journal of Tribology, Transactions of the ASME*, Vol. 127, No. 4, pp. 722–728

Wang X., Kato K., Adachi K., Aizawa K. (2001) The effect of laser texturing of sic surface on the critical load for the transition of water lubrication mode from hydrodynamic to mixed, *Tribology International*, Vol. 34, pp. 703–711

Wang X., Kato K., Adachi K., Aizawa K. (2003) Loads carrying capacity map for the surface texture design of SIC thrust bearing sliding in water, *Tribology International*, Vol. 36, pp. 189–197

Wang X., Adachi K., Otsuka K., Kato K. (2006) Optimization of the surface texture for silicon carbide sliding in water, *Applied Surface Science*, Vol. 253, pp. 1282–1286

Wong V., Tian T., Moughon L., Takata R., Jocsak J., Stanglmaier R., Bestor T., Evans K., Quillen K. (2006) Low-engine-friction technology for advanced natural-gas reciprocating engines, *Annual Technical Progress Report for April 2005 to May 2006*, Massachusetts Institute of Technology (MIT), submitted to Department of Energy of National Energy Technology Laboratory, US

Yagi K., Takedomi W., Tanaka H., Sugimura J. (2008) Improvement of lubrication performance by micro pit surfaces, *Tribology Online*, Vol. 3, No. 5, pp. 285–288

Yi W., Dang-Sheng X. (2008) The effect of laser surface texturing on frictional performance of face seal, *Journal of Materials Processing Technology*, Vol. 197, pp. 96–100

Yu X.Q., He S., Cai R.L. (2002) Friction characteristics of mechanical seals with a laser-textured seal face, *Journal of Materials Processing Technology*, Vol. 129, pp. 463–466

14.15 Nomenclature

A_d	surface area of the base of each individual texture
a	leading edge length
B	bearing width
b	trailing edge length
d	length (diameter) of the texture
F	sum of forces in the radial direction excluding the hydrodynamic force
F_a	friction force due to asperity contact
F_G	radial friction force between groove and ring
F_v	viscous friction force
h	normal distance between two surfaces at any location and time
h_1	normal distance of the lower surface from xz-plane at any location
h_2	normal distance of the upper surface from xz-plane at any location
h_d	maximum height (depth) of the texture measured from texture's base
h_m	minimum distance between two bearing surfaces at any time
h_s	bearing surface profile without considering the minimum clearance
L	bearing length
l	edge-to-edge distance between two successive textures
l_c	connecting rod (conrod) length
m	(piston ring) mass
N	number of textures
n	shape power for convex (converging–diverging) bearing profiles
P_A	axial gas pressure force
P_a	ambient (reference) pressure
P_b	gas pressure at crankcase

P_L	radial gas pressure force exerted at the unwetted lower edge of the ring
P_R	radial component of blowby gas pressure force
P_t	gas pressure in the combustion chamber
P_U	radial gas pressure force on the unwetted upper edge of the ring
p	pressure
Q	volume flow rate of lubricant
q	shape factor for barrel-shape bearing surface profile
R_G	axial reaction force from groove onto the ring
r	crank radius
Sp	the area density of the textures
s	instantaneous centre-to-centre distance between piston and crankshaft
T	radial force due to elastic tension
t	time
U	sliding velocity
U_1	tangential component of velocity vector of lower surface
U_2	tangential component of velocity vector of upper surface
v_p	axial velocity of piston
V_1	transverse component of velocity vector of lower surface
V_2	transverse component of velocity vector of upper surface
W_1	normal component of velocity vector of lower surface
W_2	normal component of velocity vector of upper surface
W_a	radial force due to asperity contact
W_g	weight (force)
W_H	hydrodynamic lift force (load carrying capacity)
x,y,z	Cartesian (spatial) x-, y- and z-co-ordinates

Greek symbols

α	leading edge length to bearing length ratio (leading length ratio)
β	trailing edge length to bearing length ratio (trailing length ratio)
δ_0	the initial minimum clearance to bearing length ratio
ζ	crown height ratio for barrel-shape ring profile
η	friction coefficient
η'	modified friction coefficient
θ	crank angle
ι	termination condition for transient analysis
κ	textured area ratio (textured portion)
Λ	bearing number
Λ_M	dimensionless inertia parameter
Λ_ω	transient bearing number
λ	connecting rod (conrod) ratio

η_0 dynamic viscosity
Ξ dimensionless trigonometric function for piston velocity
ξ textures height ratio
ρ density
φ the bearing aspect ratio in the xz-plane
φ_d the base aspect ratio of textures
ω angular velocity of the crank

Mathematical notations

$\approx$ approximately equal
$\in$ membership
$[a, b]$ closed interval (a and $b \in [a,b]$)
$^\circ$ degree
∂ operator for derivation for multivariable function
$\int$ operator for continuous summation (integration)
$\%$ per cent
cos   trigonometric cosine function
sin   trigonometric sine function

Superscripts

$*$ dimensionless form

Subscripts

1   related to the lower surface of the bearing
2   related to the upper surface of the bearing
min   minimum
max   maximum
opt   optimum
x in the x-direction
y in the y-direction
z in the z-direction

Units

μ micro-

Abbreviations

1D/2D   one/two-dimensional (quasi-two-dimensional)
2D/3D   two/three-dimensional (quasi-three-dimensional)

AJM   abrasive jet machining
BDC   bottom dead centre
BMI   Battelle Memorial Institute
CO_2 carbon dioxide
CPU   central processing unit
DLC   diamond-like carbon
EHD   elastohydrodynamic (lubrication)
EHL   elastohydrodynamic lubrication
FDM   finite difference method
HC   hydrocarbon emissions
IC   internal combustion
LBM   laser beam machining
LIGA   *Lithiographie, Galvanoformung, Abformung* (lithography, electroplating, moulding)
LST   laser surface texturing (textured)
LUD   lower and upper decomposition
MEMS   micro-electromechanical systems
Neg   negative
NO_x nitrogen oxides
N-S   Navier–Stokes (equations)
ODE   ordinary differential equation
PDE   partial differential equation
Pos   positive
PSOR   point successive over-relaxation method
PVD   physical vapour deposition
RIE   reactive ion etching
RSS   Reynolds–Swift–Stieber cavitation model
SiC   silicon carbide
SOR   successive over-relaxation method
TDC   top dead centre
TiCN   titanium carbonitride
TiN   titanium nitride
US   United States
UV   ultra-violet

15
Transient thermo-elastohydrodynamics of rough piston ring conjunction

P.C. MISHRA, H. RAHNEJAT and P. KING,
Loughborough University, UK

Abstract: One of the key conjunctions in internal combustion engines is the piston compression ring to cylinder bore or liner. Its main functions are to act as a seal against the combustion gases as well as lubricant ingression into the combustion chamber. Another key function is to conduct the heat away to the piston and the liner. Both these requirements call for minimisation of the gap between the ring and the bore. However, this has the main drawback of poor lubrication, thus increased frictional losses and reduced fuel efficiency. This conjunction, in fact, accounts for 70–80% of piston system losses, which itself is responsible for some 45% of all frictional losses in an engine. The tribological conditions are subject to transient conditions on account of the reciprocating action of the piston, thus changing contact kinematics and loading due to combustion cycle. Prediction of tribological performance of this conjunction is, therefore, quite important in terms of emission, fuel efficiency, as well as its reliability. The analysis should therefore be holistic and necessarily detailed. This chapter provides a comprehensive thermo-elastohydrodynamic analysis of compression ring conjunction, including the effect of surface topography of mating surfaces.

Key words: piston, compression ring, ring dynamics, thermo-elastohydrodynamics of rough surfaces, frictional losses.

15.1 Introduction

Pistons are furnished with a number of rings, each with important functions to perform. These have been described in the previous chapters. The *compression ring*, subject of analysis in this chapter, seals the combustion chamber, guarding against ingression of lubricant into it. This function also mitigates entry of combustion gases and products into the bottom end of the engine. Furthermore, the ring conducts the heat from the piston to the cylinder wall or the liner. It is, therefore, subjected to the combustion pressure on its top and back faces, hence the highest side forces. It also has the lowest film thickness among the ring pack and thus account for the highest frictional losses. This is approximately 70% of all the piston system *parasitic losses*, which itself accounts for nearly 45% of engine mechanical losses (see Fig. 15.1). Therefore, tribological study of piston compression ring is quite important in order to improve upon engine efficiency. There

518

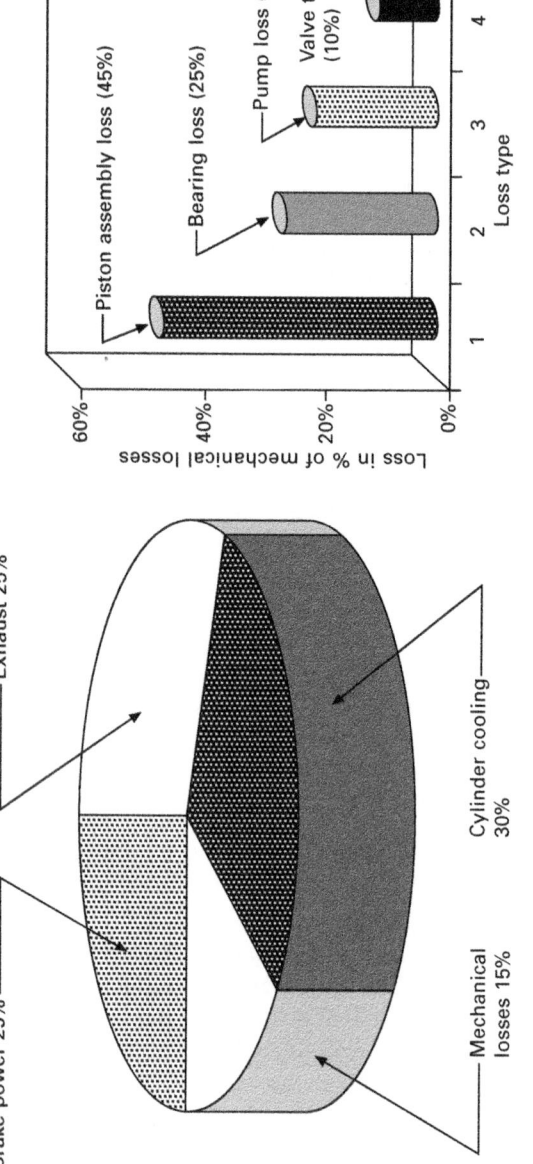

15.1 Dissipated fuel energy (left) and mechanical losses (right).

has been a plethora of investigations in this regard. However, very detailed investigations require a multi-physics approach involving many interactions. These include relatively complex dynamics; inertial rigid body motions and modal behaviour. The former accounts for the axial oscillations of the ring relative to the piston within the confine of its retaining groove, known as *ring flutter*. It also includes small rotations of the ring within the groove. Most importantly the ring is subject to elastic modal deformation within its radial plane (in-plane deformation) and bending and twist (out-of-plane deformation). These are modal behaviour of the ring (elastodynamics, see Chapter 1) and are regarded as global deformation as opposed to any localised deformation in its contact with the cylinder wall.

Aside from the rather complex dynamics, the thin lubricant film, usually of the order of a few tenths to a couple of micrometres is subject to mixed regime of lubrication (see Chapter 3). Thus, the mechanism of friction generation is due to a combination of viscous shear of the lubricant and asperity-pair tip interactions. In the case of the former, thinner films and higher pressures can result in the piezo-viscous behaviour of the lubricant, which can occur in the engine cycle at locations where maximum combustion pressure occurs (usually $10°–15°$) past the *top dead centre* (TDC). Localised deformation of contiguous solids may occur, leading to elastohydrodynamic regime of lubrication (EHL; see Chapter 6). The piezo-viscous conditions is not often noted due to relatively large area of contact (as opposed to small Hertzian contact of ball and rolling element bearings and cam-follower pairs, see Chapters 6, 16 and 17). If it takes place it is usually at the maximum combustion pressure in high-performance engines, such as in the niche original equipment manufacturer (OEM) or motorsport with significant side forces, thus high load intensity (load per unit area). The film thickness increases at piston mid-span and hydrodynamic regime of lubrication becomes dominant, and so viscous shear is the main mechanism of friction. High-performance engines such as motocross motorsport and, superbikes have short-stroke high-piston speed engines, thus high shear rates are experienced, which can induce lubricant thixotropic action. Therefore, non-Newtonian behaviour of lubricant should be considered, which at high enough shear rates can promote viscoelastic behaviour of the lubricant (Evans and Johnson 1986; Gohar and Rahnejat, 2008).

Since real surfaces are rough, any realistic detailed analysis should take surface topography into account. Most analyses ignore the influence of surface roughness of contiguous solids in contact. However, rough surfaces can promote a host of effects, which are described in some detail in Chapters 2 and 3. The most important in the case of compression ring, at least in the micro-scale, are adhesive friction and entrapment of a film of fluid in the valleys of rough surfaces. The former also accounts for the boundary contribution to the mixed regime of lubrication, while the latter is quite important to promote squeeze film action at the dead centres (top and bottom) to provide a degree of load-carrying capacity, with momentary cessation of entraining motion and subsequent inlet boundary reversal (Balakrishnan and Rahnejat, 2005). Other surface effects are thought to be not significant, certainly in micro-scale analyses. However, it is clear that roughness interactions are a problem of scale (see Chapter 3) and the topography is also hierarchical. Thus, with an insufficient load intensity and surface topography, one can envisage conditions that asperity pair adhesion and other surface phenomena can become important.

The tribology of compression ring conjunction was referred to above as a multi-physics problem, because it involves rigid body inertial dynamics, modal behaviour, tribology and surface engineering. In fact, surface modifications are used to retain tiny reservoirs of lubricant where idealised nominally smooth surfaces indicate direct interactions (see Chapters 13 and 14).

Shear of the lubricant and interaction of asperity pairs cause heat generation, which in turn affects the lubricant viscosity (shear thinning). This reduces the lubricant load-carrying capacity. Therefore, a realistic analysis should take into account thermal balance in any conjunction (Gohar and Rahnejat, 2008). Therefore, the problem occurs at various physical scales; sub-micro-scale asperity interactions, micrometre-scale films, sub-millimetre modal behaviour/vibrations and, relative to these, large rigid body motions (thus a multi-scale problem). Hence, the phrase *multi-physics multi-scale* is now firmly associated with such multi-interaction problems.

15.2   A brief review

A brief, and not in any way an exhaustive, review may be instructive. The compression ring is an incomplete circular ring developed and used ever since the detection of leakage in the James Watt engine. The basic shape has remained almost the same as those early developments, but modifications/ evolutions include the bevel shape in the axial ring profile (to cause a hydrodynamic wedge, see Chapter 5), material choice to guard against mismatch in thermal expansion of ring and liner and modern physical vapour deposition (PVD) wear-resistant coatings (see Chapter 4).

As already noted above, a piston compression ring is designed for the main purpose of preventing the leakage of combustion gases. It reduces blow by and helps to maintain a healthy power. As noted, it is an incomplete ring, which carries a free end-gap, typically 7–10 mm. It is compressed, fitted into the piston groove and installed in the cylinder bore. The ring tension strives to return it to its original shape, thus conform to the bore geometry. The ring gap then reduces typically to 0.1 mm. This elastically deformed shape retains an outward elastic pressure acting on the bore surface. The end-gap is further reduced to less than a micrometre with application of pressure, thus ideally completely seals the combustion chamber. The gas force acts at the back of the ring, thus increasing the pressure at its contacting face with the cylinder bore. When a film of lubricant is entrained into the contact, it should withstand this total applied pressure.

The minimum film thickness and the tribological performance of the ring–bore conjunction are controlled through the transient changes in the combustion pressure. As a result, the regime of lubrication alters in a transient manner depending on the contact load and kinematics. In most cases, as already noted in the introduction, the regime of lubrication is mixed: boundary interactions with hydrodynamic or elastohydrodynamic conditions (see Fig. 15.2). This differs from the highly loaded concentrated contact of cam-follower, which often yields mixed or boundary regime of lubrication in the loaded part of the cycle between the inlet reversal points through the cam nose contact (see Chapters 16 and 17). On the base circle and on the cam flank contact, a hydrodynamic regime of lubrication is prevalent. In the case of engine bearings (see Chapter 18 for example), the regime of lubrication is often hydrodynamic, because of the conforming nature of the contacting surfaces, unless a soft overlay is used and the bearing operates at high eccentricity ratios. Figure 15.2 underpins these differences between the various engine conjunctions. Some representative literature articles are cited below.

Love (1944) through his treatise in mathematical theory of elasticity worked out the global deformation of an incomplete circular ring in terms of ring tension, flexural rigidity and shear force. The incomplete circular ring was then considered with different types of force application such as slightly bent by a couple applied at its ends in its radial plane (the central-line remaining undeformed), the ends subject to opposing tensile forces or

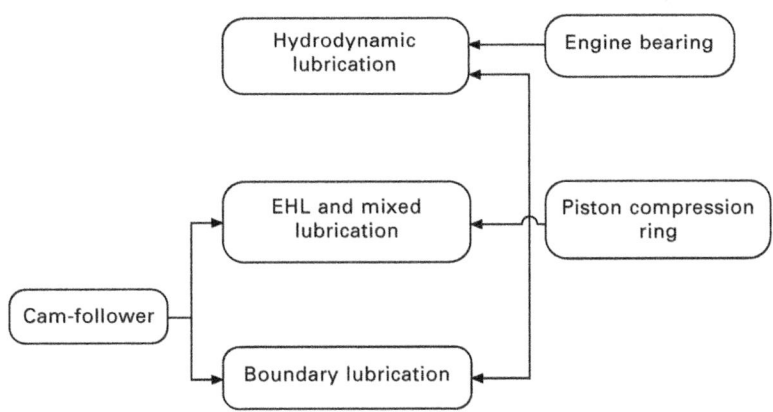

15.2 Regime of lubrication in various engine load-bearing conjunctions.

the case where a couple is applied at its ends perpendicular to the ring radial plane.

Dowson *et al.* (1982) investigated the gas pressure within the ring-pack with the objective of highlighting the technical problems inherent in sealing a moving piston. It was observed that the leakage path for gas could be the ring-face gap and that between the flank of the ring and the ring groove. It was also assumed that a ring fitted to the top groove would experience pressures approximately equal to the combustion pressure and that during periods of high-pressure loading it would press into the lower groove flank. Under such conditions there would also be some leakage. However, it was concluded that gas flow within the ring pack, caused by the passage of gas around the piston, would be significant if the flow rate exceeded 10^{-2} kg s^{-1}. They considered both hydrodynamic and elastohydrodynamic lubrication of the piston ring.

Knoll and Peeken (1982) modelled hydrodynamic lubrication of piston skirt and cylinder liner conjunction through an iterative method, using open end boundary conditions to estimate the generated pressures. They included the piston secondary motions.

Ma *et al.* (1997) analysed lubrication and friction for a complete piston ring pack with an improved oil availability model. Piston ring lubrication and friction, capable of analysing non-asymmetric conditions was developed in the model, describing relative rings' locations, oil accumulation and mixed and boundary lubrication in a ring pack.

Okamoto and Sakai (2001) estimated contact pressure distribution for a piston ring by implementing the governing elasticity equation for a curved beam, first measuring the undeformed piston ring shape and then calculating the contact pressure distribution on the ring circumference.

Akalin and Newaz (2001a) presented an axisymmetric hydrodynamic mixed lubrication model using average Reynolds equation with asperity contact in order to simulate frictional performance of piston ring and cylinder liner contact. Friction between piston ring and cylinder liner contact was predicted, considering the film rupture location, surface flow factors, surface roughness and metal-to-metal contact loading. Furthermore, Akalin and Newaz (2001b) developed a friction bench test for piston ring and cylinder liner contact, which had a large stroke and a large contact width to verify their analytical mixed lubrication model in Akalin and Newaz (2001a). This test bench enabled control of sliding speed, temperature and lubricant flow rate.

Taylor and Evans (2003) reviewed the fluid film lubrication of internal combustion engines for engine bearings, valve trains and piston assembly in which the non-Newtonian behaviour, influence of grooving and thermal analysis were considered to be the most important issues for future development and estimation of exact film thickness for all these cases.

Bolander *et al.* (2005) analysed the transitions in the regime of lubrication for the piston ring–cylinder liner contact through numerical analysis and experimental monitoring and determined lubrication conditions and frictional losses. Reynolds equation and film thickness equation subjected to suitable boundary conditions were solved simultaneously. Boundary and mixed regimes of lubrication were considered using the Greenwood and Tripp (1971) stochastic model.

Balakrishnan and Rahnejat (2005) studied the transient conditions in the contact of piston skirt and ring pack against cylinder liners during piston reversal. Their study showed changes in the regime of lubrication during reversal at or near the top dead centre. They also showed that fluid film lubrication can be encouraged by introduction of lubricant retaining surface features. Their model only took into account local deformation of the contact caused by generated elastohydrodynamic pressures. The ring model was extended to include roughness and surface modified features on the ring by Teodorescu *et al.* (2004), who showed an enhanced lubricant film thickness for rough contacts. Experimental validation of the predictions were undertaken by measurement of film in a fired engine using an ultrasonic means, reported by Dwyer-Joyce *et al.* (2006). This showed remarkable agreement with the results of Rahnejat *et al.* (2006), but only for the piston-skirt-to-cylinder liner contact as the sensing head was larger than that required to resolve necessary measurements from the compression ring conjunction.

Mishra *et al.* (2008a) carried out tribological analysis of compression ring and cylinder bore contact at reversal. Later Mishra *et al.* (2008b) analysed a racing car engine's compression ring and cylinder conjunction at high sliding velocities with mixed lubrication and predicted the minimum film, friction force and friction power. They also showed remarkable agreement with experimental findings of Furuhama and Sasaki (1983) under the same conditions.

15.3   Compression ring cylinder liner conformability

While formulating a lubrication model for ring-liner conjuction, the conformability of an incomplete ring fitted to an out-of-round cylinder is important as described in Section 15.1. The initial step for this is to estimate the ring radial deformation, which plays a crucial role in estimation of ring-bore conformance, thus calculation of the gap. This problem is addressed by Okamoto and Sakai (2001). Mishra *et al.* (2008b) obtained the ring radial deformation with a free end gap of 7 mm when fitted *in situ* in a single cylinder motocross high-performance engine (see Fig. 15.3).

Mishra *et al.* (2008b) also validated the numerically computed ring deformation with the experimental measurements, using a ***co-ordinate measuring machine*** (CMM, see Fig. 15.4) and found good agreement

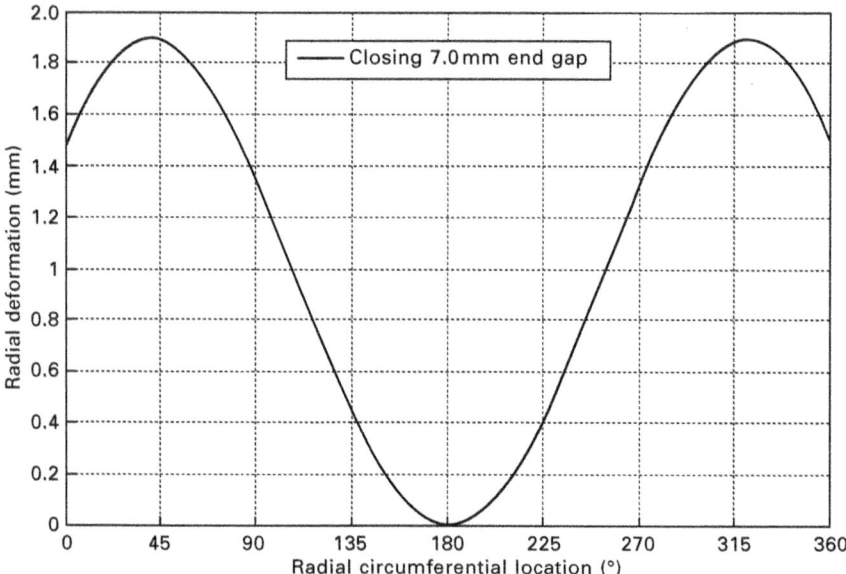

15.3 A computed deformed ring shape.

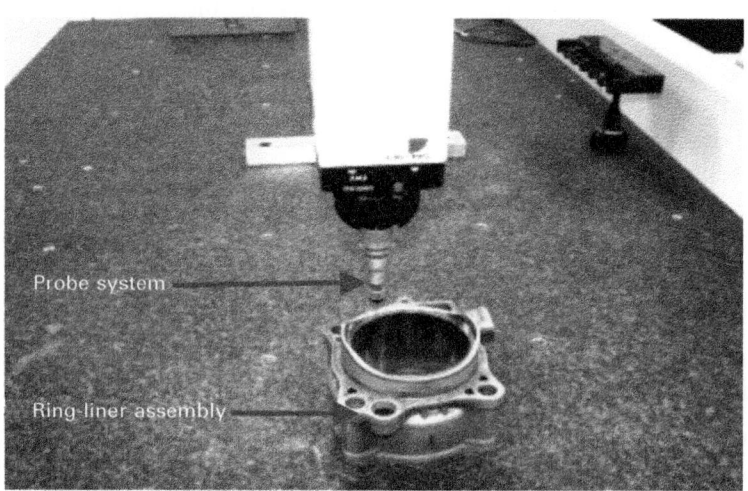

15.4 Measuring a fitted ring within a liner.

between predictions and measurements (see Fig. 15.5). The accuracy of CMM measurements was ±1 µm. CMM has sufficient speed control to capture large number of data points. It uses a contact type probing device, which uses either a ball or needle-type stylus.

The compression ring, when fitted into the piston groove and installed in an engine, exerts an outward elastic pressure due to its flexibility (as noted in Section 15.1). This elastic pressure depends on the ring geometry

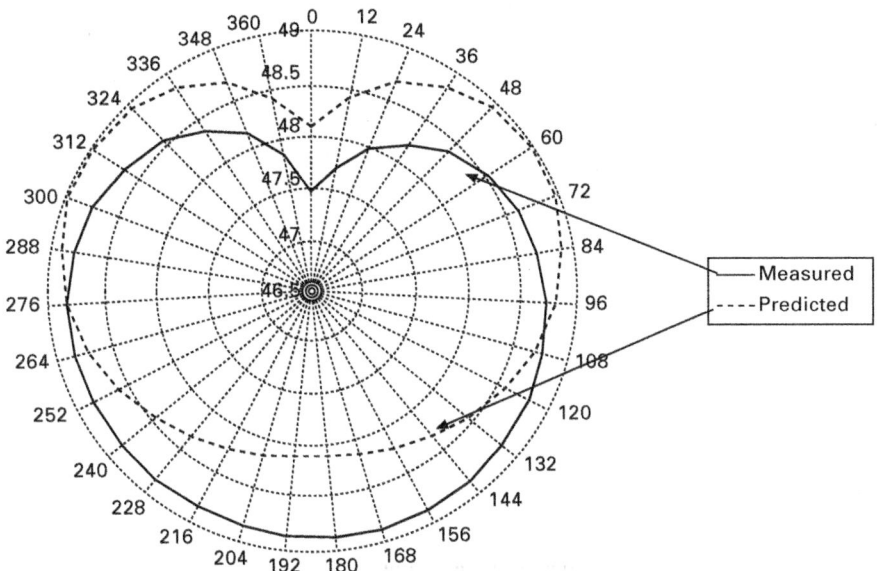

15.5 Measured and predicted ring shapes.

and material. Also the free end gap determines the elastic pressure, which enables the ring to conform to the *bore shape*. In the running state of the engine, gas pressure along with the elastic pressure make the ring-liner more conformable. The degree of conformability also depends on the order of the bore as described by Mishra *et al.* (2008b). The *cylinder bore order* is the polynomial order that best fits the bore's out-of-round shape. In the case of the engine studied by Mishra *et al.* (2008b) a polynomial of 8th order shows reasonably good ring-liner conformance (see Fig. 15.6). The minimum film thickness is normally the variable gap between the ring and the liner. Such a gap is variable during an engine cycle due to the variation in the applied gas pressure (Fig. 15.7). Hence the gross film shape (the *elastic film shape*) is the sum of the initial ring-liner gap, ring profile, global deformed ring shape and the local ring deformation (that due to the generated contact pressures):

$$h_{ij} = h_0 + s_{ij} + \delta_{ij} + \Delta_{ij} \qquad (15.1)$$

The above individual contributions are described in Mishra *et al.* (2008b).

15.4   Tribology of ring–bore conjunction

15.4.1 Fluid film viscous action

The motion of the ring results in the entrainment of lubricant into the ring–bore conjunction (into the gap described by equation (15.1)). The

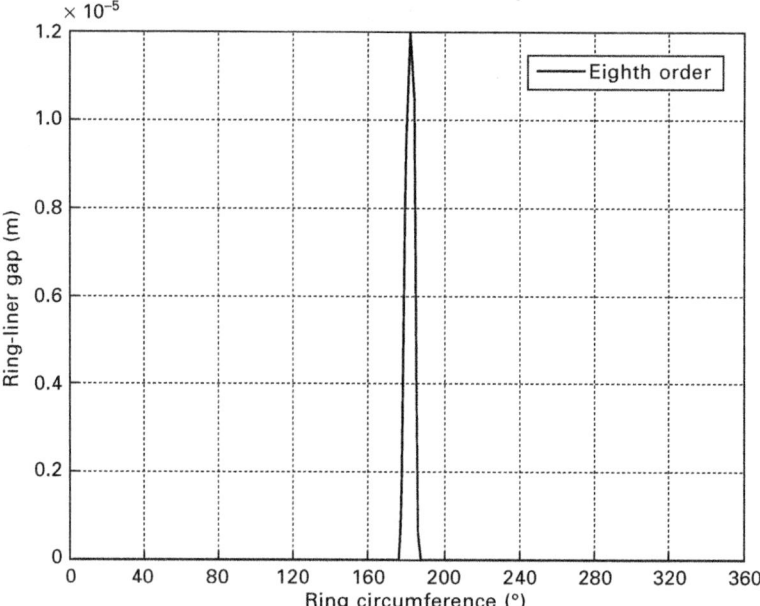

15.6 Predicted order of the bore.

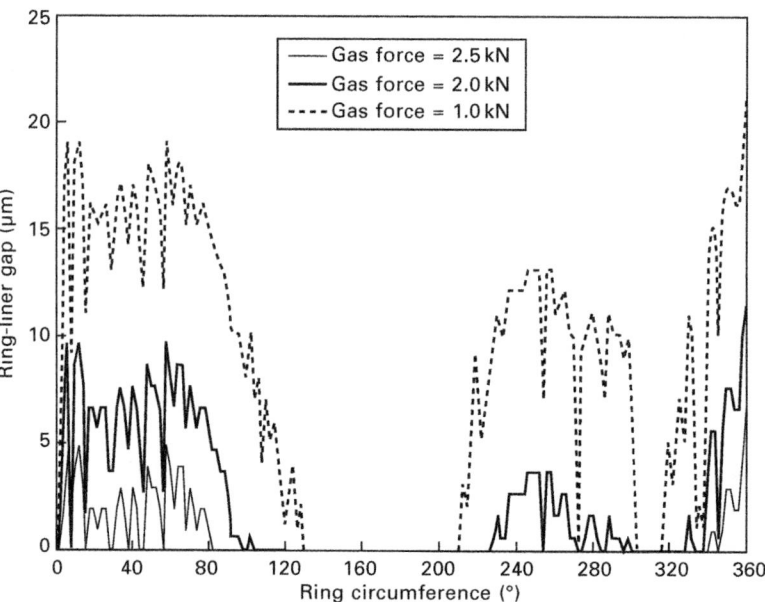

15.7 Effect of gas pressure variation on ring-liner conformance.

generated pressures are obtained from **Reynolds equation** (see Chapters 5 and 6):

$$\frac{\partial}{\partial x}\left(\frac{\rho h^3}{\eta}\frac{\partial p_h}{\partial x}\right) + \frac{\partial}{\partial y}\left(\frac{\rho h^3}{\eta}\frac{\partial p_h}{\partial y}\right) = 12\left[U\frac{\partial}{\partial x}(\rho h) + V\frac{\partial}{\partial y}(\rho h) + \frac{\partial}{\partial t}(\rho h)\right]$$

(15.2)

where the instantaneous contact kinematics determines $U = \frac{1}{2}\dot{x}$ ($\dot{x}$ being the sliding velocity of the ring), $V = 0$ (no side-leakage of lubricant in the circumferential contact direction) and the **squeeze film action** is obtained as:

$$\frac{\partial h_{ij}}{\partial t} \approx \frac{\Delta h_{ij}}{\Delta t} = \frac{h_{ij}^k - h_{ij}^{k-1}}{\Delta t}$$

(first order approximation and $\partial \rho/\partial t = 0$). Note that for the approaching contiguous surfaces in contact: $\partial h_{ij}/\partial t < 0$.

Simultaneous solution of equations (15.1) and (15.2) requires the determination of the bulk rheological state of the lubricant in the contact. For an isothermal analysis, these are given as follows.

For lubricant viscosity's pressure-dependence (Roelands, 1966):

$$\bar{\eta} = \frac{\eta}{\eta_0} = \exp\left[(\ln \eta_0 + 9.67)(-1 + (1 + 5.1 \times 10^{-9} p_h))\right]$$

(15.3)

For lubricant density's pressure-dependence (Dowson and Higginson, 1966):

$$\bar{\rho} = \frac{\rho}{\rho_0} = 1 + \frac{0.6 p_h}{1 + 1.7 p_h}$$

(15.4)

When the generated pressures are sufficiently large, localised deformation of the contiguous surfaces may also take place (δ_{ij} in equation (15.1)). This is obtained through use of generalised contact elasticity integral as:

$$\delta_{i,j} = \frac{p_{hij}}{\pi E'} D^*$$

(15.5)

where the influence coefficient matrix D^* is given by Balakrishnan and Rahnejat (2005) and:

$$\frac{1}{E'} = \frac{1}{2}\left[(1 - v_r^2)/E_r + (1 - v_b^2)/E_b\right]$$

Boundary conditions used here are for an assumed fully flooded inlet: $p_h = 0$ at $x = -\infty$, and **Reynolds exit boundary condition** $p_h = dp_h/dx = 0$ at $x = x_c$. Two main shortcomings remain.

Firstly, observations show that in piston reversals at the dead centres due to momentary cessation of entraining motion ($U = 0$) and insignificant squeeze film motion ($\partial h/\partial t = 0$), formation of a coherent lubricant film is not assured (note the right-hand side of Reynolds equation $\rightarrow 0$). Thus, direct surface-to-surface contact through asperity interactions can occur. This leads to a mixed regime of lubrication (an interrupted fluid film).

Secondly, lubricant viscosity is reduced with rising contact temperatures at high shear rates, particularly at high sliding velocities, typical of high-performance engines. Owing to thin films in the compression ring–bore conjunction, heat removal takes place mainly by conduction through the contacting bodies, which in the case of the ring can cause further global thermoelastic distortion.

15.4.2 Asperity interactions

With an insufficient film of lubricant, *asperity interactions* occur between any pair of rough surfaces in close contiguity. Greenwood and Tripp (1971) proposed a model to obtain the pressure distribution between two rough surfaces with normally distributed asperity heights:

$$p_a = K^* E' F_{2.5}(\lambda) \tag{15.6}$$

The surface roughness of both the ring and the bore or liner are assumed to be isotropic.

The function $F_{2.5}(\lambda)$ relates to the probability distribution of asperity heights. For a Gaussian distribution of asperities, $F_{2.5}(\lambda)$ has the following form (Hu *et al.*, 1994):

$$F_{2.5}(\lambda) = \frac{1}{\sqrt{2\pi}} \int_\lambda^\infty (s - \lambda)^{5/2}\, e^{\frac{s^2}{2}}\, ds \tag{15.7}$$

A curve-fit of the function is more suited to numerical analysis. For typical ring–bore contact Hu *et al.* (1994) state that:

$$F_{2.5}(\lambda) = \begin{cases} A(\beta - \lambda)^z & \lambda \leq \beta \\ 0 & \lambda > \beta \end{cases} \tag{15.8}$$

where: $\beta = 4$, $A = 4.4068 \times 10^{-5}$, $z = 6.804$ and $\lambda = h/\sigma_{\mathrm{rms}}$ (Stribeck's oil film parameter, see Chapter 3).

K^* in equation (15.6) is a function of surface roughness as: $K^* = 5.318748 \times 10^{10}\ \sigma_{\mathrm{rms}}^{5/2}$.

The generated contact pressures are, therefore, owing to the viscous action of the fluid film (hydrodynamic/elastohydrodynamic) and the asperity contact pressures. At any instant of time in the engine cycle, the applied force acting on the ring–bore conjunction is obtained as:

$$W = \int \int (p_h + p_a) \, dx \, dy \tag{15.9}$$

15.4.3 Conjunctional friction

An isothermal analysis can be carried out with the approach expounded thus far. However, friction and heat generated as the result of friction play key roles in the tribology of compression ring–cylinder conjunction. A salient point has already been alluded to; two key mechanisms of friction are at play; viscous shear of the lubricant film and asperity interactions. Friction due to asperity interactions is contributed by adhesive as well as deformation friction, both of which may be within the elastic limit or accompanied by plastic deformation. Some basic friction models are adequately described in Chapter 2. More complex models relating the applied normal force to mechanism of friction generation, and relating these to average asperity geometry within assumed Gaussian distribution of these can be found in Arnell *et al.* (1991), Bhushan (1999) and Gohar and Rahnejat (2008). A simple approach for asperity friction (boundary contribution) was proposed by Bolander *et al.* (2005), where:

$$f_a = \mu_a \int \int p_a \, dx \, dy \tag{15.10}$$

where an asperity coefficient of friction in the range 0.1–0.15 is suggested. This approach must be considered as rather simplified, because the link between normal pressure distribution and friction is quite complex, and still the subject of some debate and research.

Viscous friction is due to generated shear stress arising from the entraining motion of the lubricant as well as the pressure gradient in a converging–diverging wedge:

$$\tau_{ij} = \frac{h_{ij}}{2} \frac{\partial p_h}{\partial x} + \frac{\eta_{ij} U}{h_{ij}} \tag{15.11}$$

where it is assumed that: $\partial p_h / \partial x \gg \partial p_h / \partial y$. In fact, the pressure-induced shear in partially conforming contacts such as the ring–bore conjunction is much smaller than that due to the entraining motion of the fluid. However, both effects are included in the current analysis, except for the cavitation region, where the first term in equation (15.11) is ignored. Therefore, the total friction force is: $f = f_v + f_a$.

15.4.4 Thermal effects

Friction in the conjunction causes generated heat, which in turn affects the effective viscosity of the lubricant. In Chapter 20, the procedure for heat generation in the contact is described. However, unlike journal bearings, the variation in film thickness in the compression ring–cylinder bore/liner means that the mechanism that is responsible for carrying the heat away from the contact varies. In Chapter 20, for big-end bearings, it is noted that heat is carried away by the lubricant film through convection. However, as shown by Gohar and Rahnejat (2008), simple analytical solutions based on the Peclet number indicates the dominance of convection or conduction. Their approach is based upon determination of the Peclet number by assuming cooling through convection or conduction only. Then the ratio of this is the Peclet number. For large Peclet numbers (much greater than unity), one can assume convection cooling alone, whereas for small Peclet numbers (typically less than unity) one may assume conduction alone. Certainly low contact pressures for partially conforming contacts such as the conjunction under investigation, compressive heating may be neglected (see energy equation in Chapter 5), when compared with heating caused by friction. Then, an average temperature rise in the contact is obtained by assuming viscous heating and convection or conduction cooling only. This approach is detailed in Gohar and Rahnejat (2008) for a rectangular band contact, which one may assume for the conjunction of the ring with the liner as a simplification. This is reported in Chapter 31, and is thus not detailed here again.

The Peclet number is given as:

$$\mathrm{Pe} = \frac{\rho c_p U h^2}{2 b k_t}$$

It can be seen that as the speed of entraining motion increases, and thus the film thickness, convection cooling becomes more dominant. Thus, in parts of the engine cycle with thicker films, under dominant hydrodynamic conditions, the lubricant film carries the heat away. In other parts of the cycle, with thin films, cooling is mainly by conduction, this being one of the key functions of the compression ring.

Owing to shear heating of the lubricant, its effective viscosity is reduced. This is not taken into account by equation (15.3). Thus, one can either use a modified version of Roelands equation (see Chapter 31) or simply adjust the value of η_0 as η_θ in (15.3) using: $\eta_\theta = \eta_0 \, e^{-\beta \Delta \theta}$, where β is the gradient of viscosity variation with temperature rise $\Delta \theta$ in a log-linear plot, and is often in the range 0.005–0.01. Then, η in equation (15.3) is now known as η_e; the lubricant effective viscosity.

15.5   Results and discussion

Simulations are carried out for the entire engine cycle (i.e. 720° crank angle rotation) in the case of a single cylinder high-performance motocross bike engine, with maximum power of 50 bhp and maximum torque of 200 N m at 6500 rpm. The simulations presented here are for the top engine speed of 13 000 rpm.

Figure 15.8 shows the input to the model; the combustion curve, with maximum gas pressure of 115 bar at the crank angle of 373° (i.e. 13° past the TDC in the power stroke of this four-stroke engine). The entraining velocity in the compression ring–cylinder liner conjunction is also shown in the figure. Note that 0°, 360° and 720° correspond to the position TDC. In these positions and those at the BDC, reversals take place (i.e. the velocity of entraining changes direction). At these positions the piston orientation alters as the loaded region of the ring–liner contacts goes through a transition from thrust to anti-thrust side or *vice versa*. This causes a squeeze film effect as shown in Fig. 15.9, which enhances the load-carrying capacity of the lubricant film, and is particularly important as the main mechanism of lubrication (the entraining motion of the lubricant) is diminished. A negative

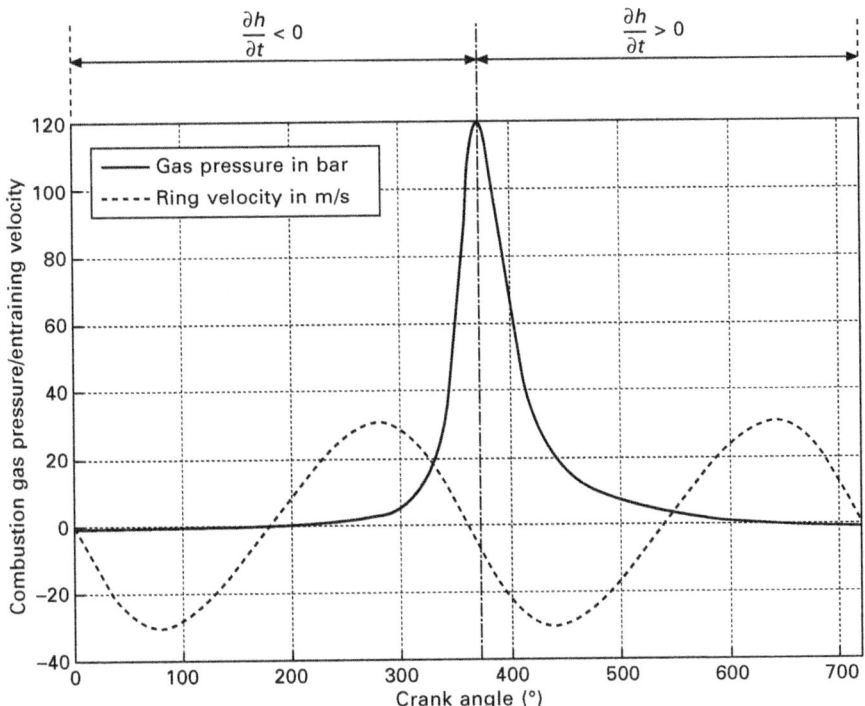

15.8 Combustion gas pressure and speed of entraining motion.

squeeze film velocity indicates convergence of bounding contacting surfaces, while a positive value corresponds to their separation. With the latter, load-carrying capacity is reduced. Squeeze film action has the positive effect of entrapping a film of lubricant which otherwise would cease to exist in nominally smooth surfaces with the cessation of entraining motion at dead centres. This reduces boundary friction in real rough surfaces. Figure 15.10(a) shows the contribution to friction due to asperity interactions (boundary friction), whereas the contribution due to viscous shear of the lubricant film is shown in Fig. 15.10(b). In both cases friction diminishes at the dead centres (cessation of motion), but increases sharply prior to and immediately after the reversals. The overall friction is the addition of these two contributions. After reversals the speed of entraining motion of the lubricant into the contact increases, resulting in thicker films and a rapid reduction in friction. The sudden rise in the power stroke around the crank angle 373° is because of an increase in the contact force due to the maximum combustion pressure.

Friction causes parasitic power loss as $P = 2FU$, where $2U$ is the sliding velocity of the ring relative to the liner. This is shown in Fig. 15.11. It is clear that, owing to thinner films in the power stroke, the losses are higher than the other strokes. The total loss per engine cycle is the sum of all these losses, any reduction of which would improve the engine fuel efficiency, as described in the introduction to this chapter. The effect of friction is generated heat in the conjunction, which is carried away either by the lubricant film or by conduction through the bounding solids. The compression ring plays a key role in this, conducting the heat to the piston (which is cooled by impinging jets of lubricant) or to the cylinder liner/bore, which is cooled externally by a water jacket. The temperature rise due to friction is shown in Fig. 15.12.

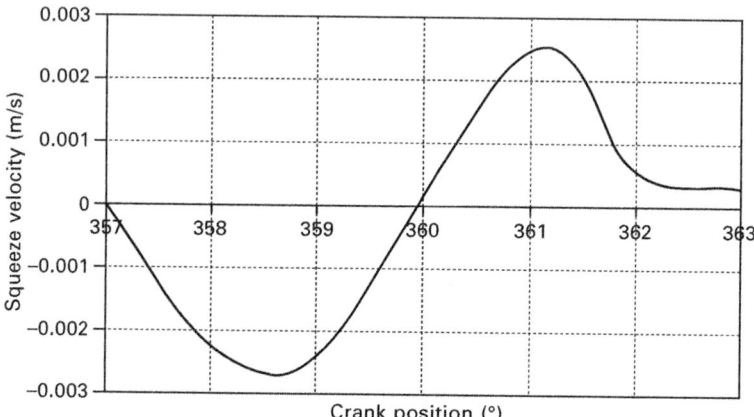

15.9 Squeeze velocity variation during reversal at the TDC in transition from compression to power stroke.

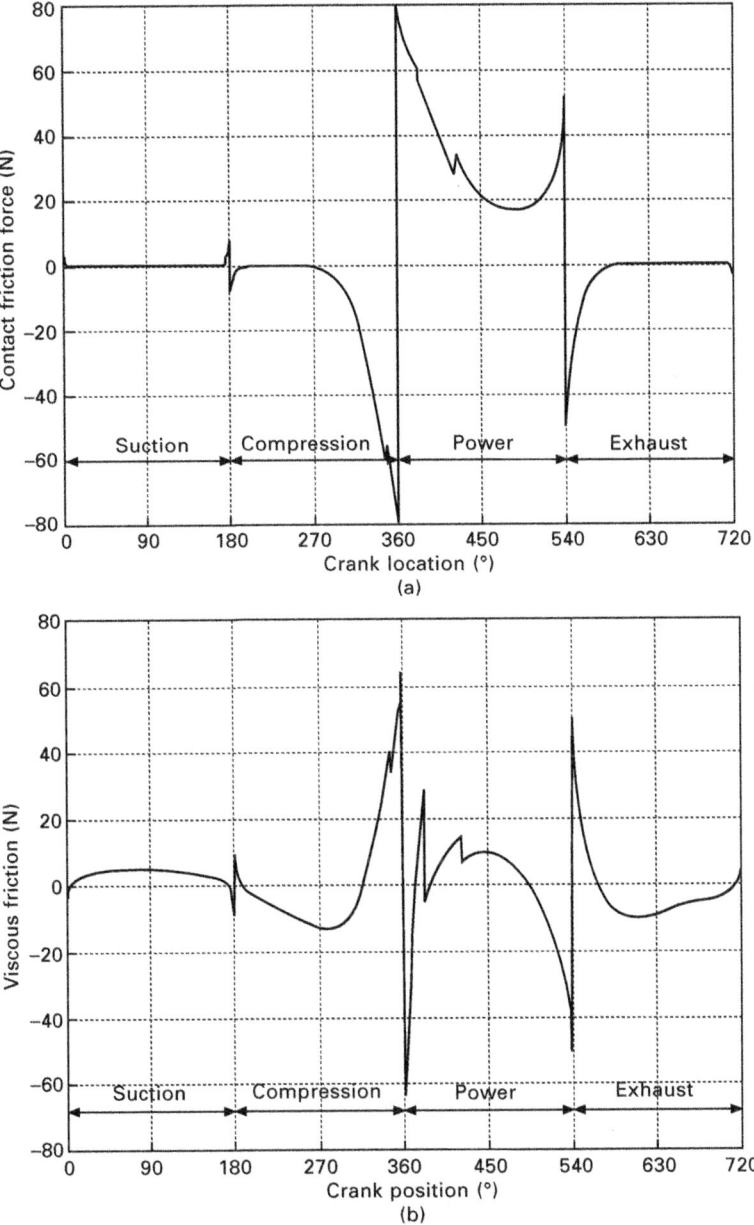

15.10 Contributory mechanisms to the overall friction: (a) asperity friction contribution (b) viscous friction contribution.

In the suction and exhaust strokes temperature rises as the result of viscous heating of the lubricant film, reaches a plateau and falls with decreasing shear rate. The same occurs in the compression stroke, except that as the

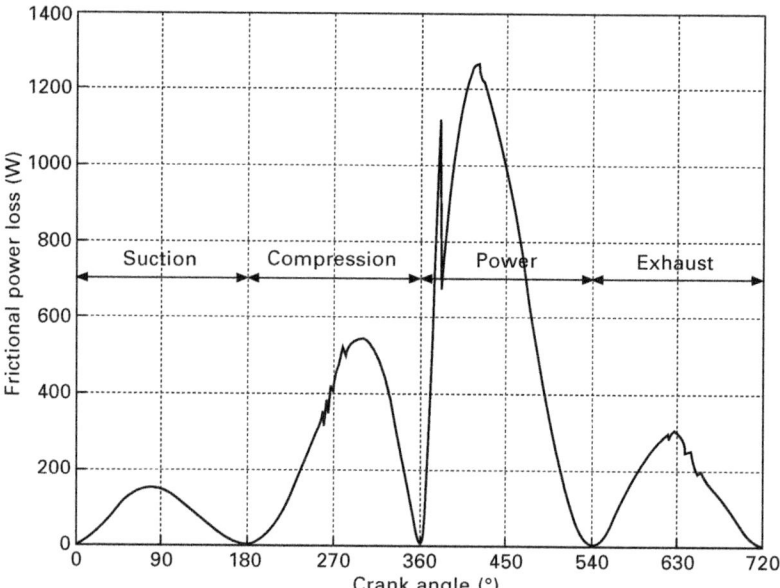

15.11 Frictional power loss.

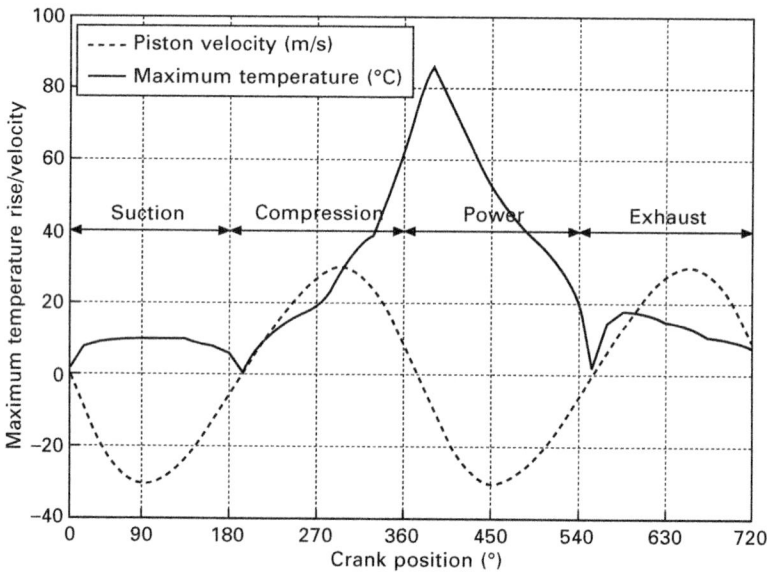

15.12 Temperature rise in the conjunction.

piston advances towards the TDC, the film has thinned sufficiently that the correspondence of temperature variation with sliding velocity is lost. This is indicative of mixed regime of lubrication, where asperity interactions in the form of adhesive friction is independent of sliding speed (see Chapter 3). The temperature rise continues in the power stroke up to the position of maximum combustion pressure as the combined result of asperity interactions and to a lesser extent compressive heating of lubricant. Thereafter, the temperature rise is reduced.

As already noted the generated heat is carried away either by convection or through conduction. Figure 15.13 shows the ratio of proportion of convected to conducted heat. This ratio is known as the **Peclet number**. Large values of Peclet number correspond to thicker hydrodynamic films, which exist in the suction and exhaust strokes, except for the dead centres at the beginning and end of each stroke (beginning and end of each line graph in the figure). The situation is quite different in the compression and power strokes as the film thickness is considerably reduced, particularly in the case of the latter. Thus, for most of these strokes the generated heat is removed mainly by conduction. Figure 15.14 shows a zoomed-in version of the variations in the power stroke shown in Fig. 15.13.

An example of the diminished film thickness during the power stroke is shown in Fig. 15.15 at 373° crank angle (maximum combustion pressure). It is also shown that the film thickness is reduced by 20–25% when shear thinning, giving rise to contact temperature and thus reduced effective viscosity of the lubricant is taken into account. Such thin films promote mixed regime

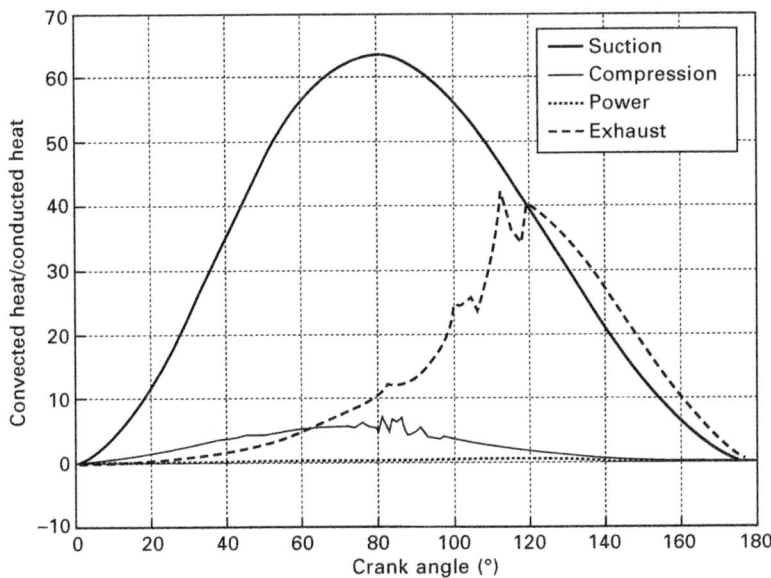

15.13 Variation of heat transfer ratio (Peclet number) during various strokes.

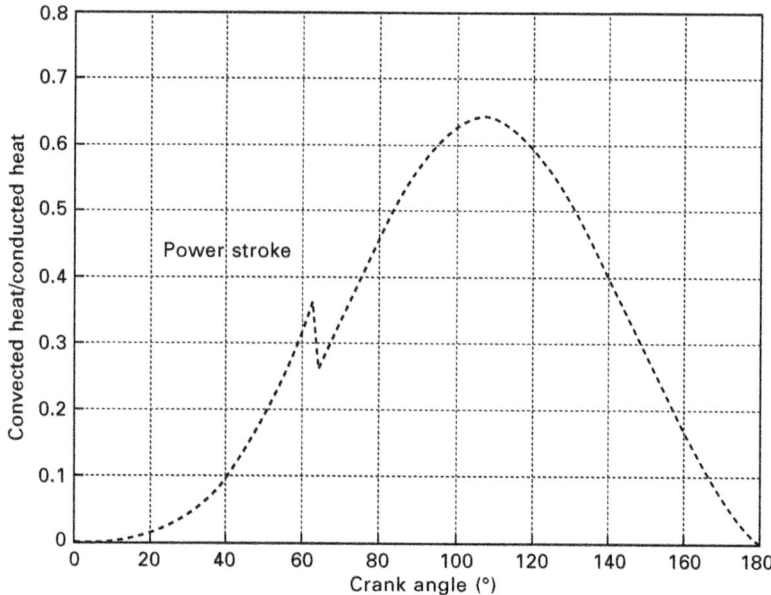

15.14 Variation of Peclet number during the power stroke.

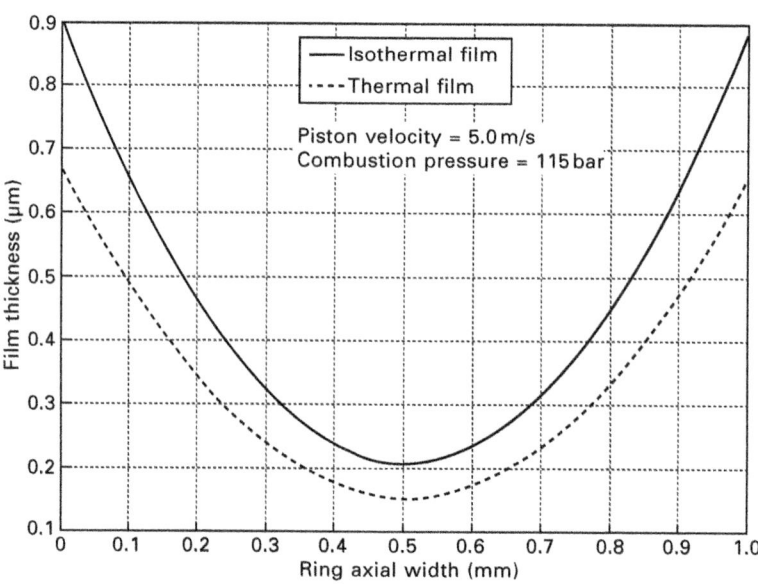

15.15 Film thickness in ring–liner conjunction.

of lubrication and heat transfer to the bounding contacting surfaces through conduction.

The corresponding pressures are generated due to a combination of hydrodynamic action of the lubricant film and asperity interactions. This is shown in Fig. 15.16. Note that in the case of thermal contact the share of load carried by asperity contact is increased due to thinning of the lubricant film. In both cases a hydrodynamic film only supports the load for part of the contact, some parts of the leading and trailing edges of which are subject to boundary regime of lubrication (note that the axes for each pressure component are different). This is often the reason for evidence of wear scars in the high-pressure zone of the compression-power stroke transition at the TDC.

15.6   Future trends

The current analysis shows the complex nature of interactions in compression ring–bore/liner conjunction. There are a host of salient practical issues to be considered. This includes the ring–bore conformance and global elastodynamic behaviour of the ring (only in-plane radial deformation is considered here). In fact, motions of the ring are quite complex, comprising rigid body flutter (axial motion relative to the piston within its retaining groove) and flexural motion (out-of-plane bending and twist modes). Furthermore, in practice, both the ring and the liner/bore are subject to thermal distortion. Also, the radial out-of-roundness of the bore is only taken into account here, whereas in reality there is also axial variation of the same. Hence, inclusion of other salient features will change the analysis outcome.

The analysis presented here, however, shows that the ring–bore conjunction is a major source of frictional losses and that the transient nature of regime

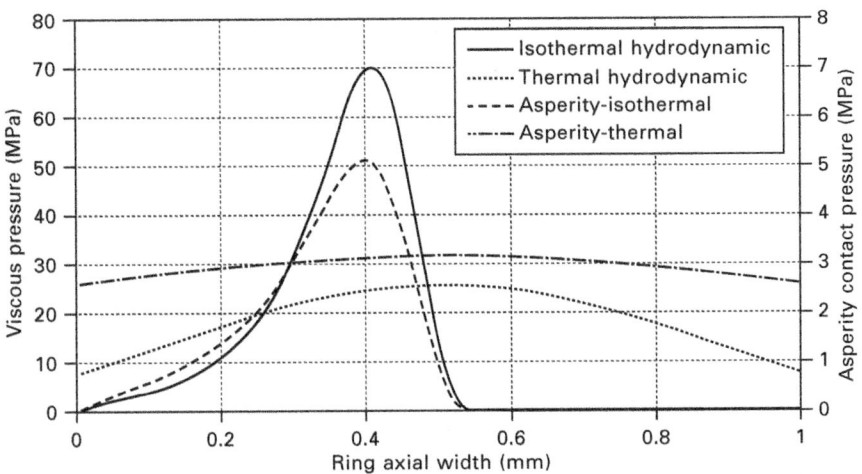

15.16 Share of contact load carried by viscous and asperity pressures.

of lubrication is quite complex. The film thickness is particularly poor in the power stroke even at fairly high speeds of entraining motion. Thus, wear-resistant coatings and use of friction modifiers are and have been common methods to mitigate frictional losses. Other recent developments are surface modifications to entrap/retain lubricant on contiguous surfaces. Optimisation of such features as well as engineering of lubricant rheology to encourage formation of better surface films of low shear strength will constitute further future developments.

15.7   Acknowledgements

This work is carried out under the EPSRC funded Encyclopaedic Program Grant. The authors would like to express their gratitude to the EPSRC (Engineering and Physical Sciences Research Council) and all the partner organisations in the project, including Aston Martin Lagonda, Capricorn Automotive, BP Castrol, ProDrive, Ricardo Consulting Engineers, ES Technology, Cranfield and Sheffield Universities (www.encyclopaedic.org)

15.8   References

Akalin, O. and Newaz, G.M., 'Piston ring-cylinder bore friction modelling in mixed lubrication regime: Part I – Analytical results', *Trans. ASME, J. Trib..* **123**, 2001a, pp. 211–218.

Akalin, O. and Newaz, G.M., 'Piston ring-cylinder bore friction modelling in mixed lubrication regime: Part II – Correlation between bench test data', *Trans. ASME, J. Trib.*, **123**, 2001b, pp. 219–223.

Arnell, R.D., Davies, P.B., Halling, J. and Whomes, T.W., *Tribology Principles and Design*, Macmillan, 1991.

Balakrishnan, S. and Rahnejat, H., 'Isothermal transient analysis of piston skirt-to-cylinder wall contacts under combined axial-lateral-tilting motion', *J. Phys., Part D: Appl. Phys.*, **38**, 2005, pp. 787–799.

Bolander, N.W., Steenwyk, B.D., Sadeghi, F. and Gerber, G.R., 'Lubrication regime transitions at the piston ring–cylinder liner interface', *Proc. Instn. Mech Engrs., Part J: J. Engng. Trib.*, **129**, 2005, pp. 19–31.

Bhushan, B., *Principles and Applications of Tribology*, John Wiley & Sons, 1999.

Dowson, D. and Higginson, G.R., *Elastohydrodynamic Lubrication*, Pergamon Press Ltd, 1966.

Dowson, D., Ruddy, B.L. and Economou, P.N., 'Elastohydrodynamic lubrication of piston rings', *Proc. Roy. Soc.*, **386**(A), 1982, pp. 409–430.

Dwyer-Joyce, R.S., Green, D.A., Balakrishnan, S., Harper, P., Lewis, R., Howell- Smith, S., King, P.D. and Rahnejat, H., 'The measurement of liner-piston skirt oil film thickness by an ultrasonic means', *SAE Trans., J. Engines*, Pap. 2006-01-0648, 2006.

Evans, C.R. and Johnson, K.L., 'The rheological properties of EHD lubricants', *Proc. Instn. Mech. Engrs., Part C*, **200**, 1986, pp. 303–312.

Furuhama, S. and Sasaki, S., 'New device for the measurement of piston frictional forces in small engines', *Trans. ASME*, SAE technical paper 831284, 1983.

Gohar, R. and Rahnejat, H., *Fundamentals of Tribology*, Imperial College Press, 2008.

Greenwood, J.A. and Tripp, J.H., 'The contact of two nominally flat rough surfaces', *Proc. Instn. Mech. Engrs*, 185, 1971, pp. 625–633.

Hu, Y., Cheng, H.S., Arai, T., Kobayashi, Y. and Ayoma, S., 'Numerical simulation of piston ring in Mixed lubrication – A non-axi-symmetrical analysis', *Trans. ASME, J. Trib.*, **116**, 1994, pp. 470–478.

Knoll, G.D. and Peeken, H.J., 'Hydrodynamic lubrication of piston skirt' *Trans. ASME, J. Lubn. Tech.*, **104**, 1982, pp. 505–509.

Love, A.E.H., *A Treatise on Mathematical Theory of Elasticity*, Dover, 1944.

Ma, M.-T., Sherrington, I. and Smith, E.H., 'Analysis of lubrication and friction for a complete piston-ring pack with an improved oil availability model: part 1: circumferentially uniform film', *Proc. Instn. Mech. Engrs., Part C: J. Mech. Engng. Sci.*, **211**, 1997, pp. 1–15.

Mishra, P.C., Rahnejat, H. and Balakrishnan, S., 'Tribology of piston compression ring and cylinder contact at reversal', *Proc. Instn, Mech. Engrs., Part J: J. Engng. Trib.*, 222, 2008a pp. 815–826.

Mishra, P.C., Rahnejat, H. and King, P.D., 'Tribology of ring-bore conjunction subjected to mixed regime of lubrication', *Proc. Instn. Mech. Engrs., Part.C: J. Mech. Engng. Sci.*, **223**, 2008b pp. 987–998.

Okamoto, M. and Sakai, I., 'Contact pressure distribution of piston rings – calculation based on piston ring contour', SAE Technical Paper 2001-01-0571, 2001, pp. 1–7.

Rahnejat, H., Balakrishnan, S., King, P.D. and Howell-Smith, S., 'In-cylinder friction reduction using a surface finish optimization technique', *Proc. Instn. Mech. Engrs., Part D: J. Auto. Engng.*, **220**(D9), 2006, pp. 1309–1318.

Roelands, C.J.A., *Correlation Aspects of the Viscosity–Temperature–Pressure Relationships of Lubricating Oils*, Druk VRB Kleine der A3-4, Groningen, 1966.

Taylor, R.I. and Evans, P.G., '*In-situ* piston measurements', *Proc. Instn. Mech. Engrs., Part. J: J. Engng. Trib.*, **218**, 2003, pp. 185–200.

Teodorescu, M., Balakrishnan, S., Rahnejat, H., Howell-Smith, S. and Dowson, D., 'Tribological analysis within a multi-physics framework', *Proc. 31st Leeds–Lyon Symposium*, Leeds, UK, September 2004, Elsevier.

15.9   Nomenclature

b	Ring face-width	m
c_p	Specific heat at constant pressure of lubricant	$J\,kg^{-1}\,K^{-1}$
E_b	Modulus of elasticity of the bore/liner	$N\,m^{-2}$
E_r	Modulus of elasticity of the ring	$N\,m^{-2}$
E'	Reduced modulus of elasticity of the contacting pair	$N\,m^{-2}$
f_a	Adhesive (boundary) friction	N
f_v	Viscous friction	N
h_0, h_{ij}	Gap, film thickness	m
k_t	Thermal conductivity of lubricant	$J\,m^{-1}\,s^{-1}\,K^{-1}$
p_a	Asperity pressure	$N\,m^{-2}$
p_{ij}	Pressure at any location ij	$N\,m^{-2}$
p_h	Hydrodynamic pressure	$N\,m^{-2}$

s_{ij}	Undeformed axial ring profile	m
t	Time	s
U	Speed of entraining motion	$\mathrm{m\,s^{-1}}$
V	Speed of side-leakage	$\mathrm{m\,s^{-1}}$
x, y	Co-ordinate directions	
x_c	Cavitation boundary (film rupture location)	m
$\dot{x}$	Sliding velocity of the ring	$\mathrm{m\,s^{-1}}$
δ_{ij}	Local elastic deformation of the ring	m
Δ_{ij}	Global elastic in-plane deformation of the ring	m
Δ_{ij}	Time step of analysis	s
η	Lubricant dynamic viscosity at pressure p_{ij}	Pa s
η_0	Lubricant dynamic viscosity at atmospheric pressure	Pa s
λ	Stribeck's oil film parameter	–
ρ	Lubricant density at pressure p_{ij}	$\mathrm{kg\,m^{-3}}$
ρ_0	Lubricant density at atmospheric pressure	
σ_{rms}	Root mean square roughness of counterfaces	m
τ	Shear stress	$\mathrm{N\,m^{-2}}$
υ_b	Poisson's ratio of bore/liner material	–
υ_r	Poisson's ratio of ring material	–

Section II.III

Valve train systems

16

Tribological issues in cam–tappet contacts

M. KUSHWAHU, Ford Motor Company, UK

Abstract: This chapter deals with the general case of finite line contact under elastohydrodynamic regime of lubrication. This condition occurs in many concentrated non-conforming contacting pairs, such as roller-to-raceways' contacts in rolling element bearings, meshing gear teeth pairs for spur and some helical gears as well as for cam-tappets. The chapter deals specifically with the case of cam-tappet contact in automotive valve trains. It shows that the regime of lubrication is elastohydrodynamic in the contacting pair through the event angle; from valve opening to its closure. The finite length nature of the very narrow width contact gives rise to high pressure peaks at contact edges, which inhibit flow of lubricant, thus lead to very thin films. This can lead to wear in practice. Additionally, the high pressure spikes can promote fatigue spalling.

Key words: valve train system, elastohydrodynamics, finite line contact, pressure spike.

16.1   Introduction

As mentioned in Chapter 15, progressive demand for high-performance engines with better fuel economy and emissions underpins the continuous challenge for today's automotive original equipment manufactures (OEMs). Sustaining such operating environments require faster acting, lighter, but durable mechanical components.

The operational life and efficiency of such engineering mechanisms rely totally on the structural integrity of its constituent inertial elements. For these machines to function ideally, the load or torque from one component should transfer to another through the load-bearing surfaces in relative motion and with minimum friction. This relative motion of the contiguous surfaces, which can take the form of either rolling, sliding (rubbing) or spinning, leads to the generation of friction. Onset of direct contact of surfaces can result in wear or seizure, thereby, affecting the operation of the entire system.

Introduction of a lubricant film between these mating members simply guards against the direct interaction of the two moving surfaces, when they are constantly separated by a protective film, which is ideally sheared with minimum effort (Dowson, 1995). It is, therefore, necessary to arrange, wherever possible, an everlasting presence of a lubricant film of sufficient thickness to ensure that no metallic contact occurs between the opposing

545

surfaces, hence, considerably minimising the undesired effects of friction and wear. In most engineering applications, the form of lubricant used is usually mineral or synthetic oil.

This chapter looks at the contact conjunction between the *cam* and the *tappet* in the valve train system, which is the most loaded 'high-pair' contact in an internal combustion (IC) engine. The same dynamic principles apply to a modified cycloidal cam employed in a direct over-head cam (DOHC) to tappet arrangement as in Chapter 17. This chapter covers transient events in the cam–tappet contact during a steady state cycle.

16.2   The cam geometric profile

One of the most recognised ways of increasing the power output of the IC engine is by improved cylinder breathing, which is solely dependent upon the precise timings for opening and closing of the inlet and exhaust valves (Morel *et al.*, 1990). This particular phenomenon is controlled by the 'high-pair' contact between the cam and the tappet. Under ideal conditions, the cam should remain in contact with the tappet at all times, otherwise the loss of lubricant may ensue. However, owing to the dynamics of the system involved (discussed briefly in the lubrication section), this may not generally be the case. Hence, it becomes important to understand the kinematics of the cam geometrical profile and tailor it to optimise the *valve timing* and dynamics of the system, and consequently improve the breathing capability of the cylinders (Morel *et al.*, 1990).

The basic geometry of a *cam profile* consists of three arcs that outline the dwell–rise–return–dwell (D-R-R-D) periods in a cam rotation cycle (Reeve, 1997). These arcs are generally defined as the base circle, the flank region and the nose. The angle that defines the time period between where the valve commences opening to its complete closure, is known as the *cam event angle*. This is an important parameter as it governs the entire translational activity of the valve.

To generate a cam profile, the designer has to first establish the desired amplitude of valve displacement and the corresponding time interval for which this displacement has to be maintained. After this, the cam profile has to be carefully determined, bearing in mind the effect that this action may have on the dynamic and the noise vibration and harshness (NVH) output of the entire system (Reeve, 1997). For example, in terms of inertial dynamics, it is imperative to have a profile which gives a continuous acceleration response since any abrupt changes in acceleration would introduce a *jerk* input into the system. Referring to the force–acceleration relationship in Newton's second law of motion, if there is any sudden sharp change in acceleration, then a similar rise in inertial force will ensue. These sudden changes in force cause noticeable inertial imbalances within the system which, with their cyclic nature can affect the structural integrity of the system (Jeon *et al.*, 1988; Yan *et al.*, 1996; Reeve, 1997).

16.2.1 Kinematic characteristics of a modified cycloidal cam

Many cam profile designs are available, as each one has been tailored for a particular use in various machines. In the case of automotive cams, the modified cycloidal cam constitutes the basis for the development of new generation of cam profiles, although, it is still widely used across many engineering applications which range from knitting to food processing machinery.

The **modified cycloidal cam** profile is also a three-phase D-R-R-D cam that has a distinguishable characteristic, which makes it ideal for use in high-speed machinery (Reeve, 1997). Owing to the broad nature of the cam's geometry (i.e. broad cam nose), the transition between the three phases, for an object sliding on the cam's surface is very smooth. Because of this reason, the acceleration characteristics of this cam are continuous, thus yielding low noise, vibration and wear responses. Figure 16.1(a) shows a typical modified cycloidal cam profile which can be generated from the following relation:

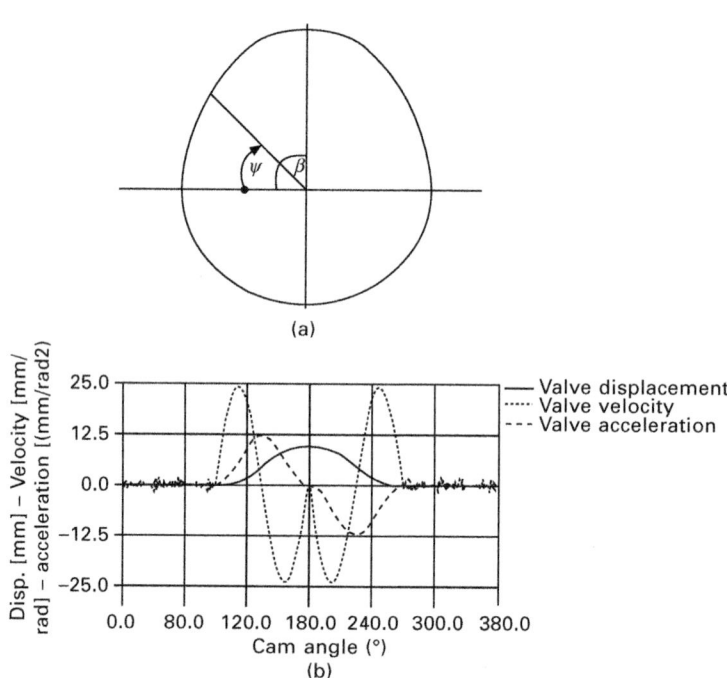

16.1 The modified cycloidal cam profile with its kinematic characteristics.

$$s = s_{\max} \left\{ \frac{\psi}{\beta} - \left[\frac{1}{2\pi} \sin \left(\frac{2\pi\psi}{\beta} \right) \right] \right\} \tag{16.1}$$

where: s is the valve lift, $s_{\max}$ denotes the maximum allowable **valve lift**, ψ is the angular rotation of the cam measured from the beginning of the event angle and β is the actual duration of the **cam event angle**.

Figure 16.1(b) illustrates the displacement (lift), velocity and acceleration characteristics generated by the modified cycloidal cam. As can be seen, the acceleration curve follows a smooth sinusoidal profile, where the nose region denotes the point of acceleration reversal. This transition is quite smooth, because of the flat period at the nose, which allows a gradual change in velocity as depicted by the point of inflexion in the lift velocity profile. The perturbations shown over the base circle region of the cam are merely due to numerical errors.

16.3   Lubrication analysis of the cam and tappet conjunction

A number of governing parameters are needed to formulate a representative solution of the lubricated contact between the cam and the flat tappet. All these factors form individual components of the four main equations, which are then required to be solved simultaneously to obtain the lubricant pressure distribution and film thickness within the contact conjunction. The formulation of these equations is defined in the following sections.

16.3.1 The contact footprint

To analyse the lubrication mechanism experienced by the contact conjunction between the cam and the tappet, it is important to first understand the type of **contact footprint** formed between the two contiguous surfaces under elastostatic conditions. In this case, the geometry of the contiguous surfaces is considered as that of a roller on a flat plane, although, it is known that the cam radius at the point of contact changes for the duration of its event angle.

The width of the cam on a flat tappet forms a finite line of contact, thus the contact condition is referred to as a *finite-line contact*. As the two bodies are pressed together under the application of normal load, they deform and this line expands in width to resemble a *dog-bone* shape (Mostofi, 1981; Rahnejat, 1984). This is shown in Fig. 16.2. It should be noted that the tappet surface is considered to be much larger than the width of the cam, and also, the contact footprint is far smaller than the cam width itself. This satisfies one of the main Hertzian assumptions of the contact being formed on a *semi-infinite solid*.

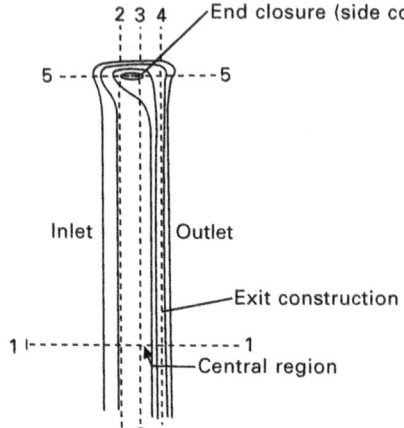

2 3 4 End closure (side construction)

5 ---   ---5

Inlet   Outlet

Exit construction

1 |-------   ------1
Central region

2 3 4

Section key:

- Section 1-1: Centre line through the contact in the entraining direction

- Section 2-2: Line through the contact along the cam width just prior to the contact centre

- Section 3-3: Centre line through the contact along the cam width

- Section 4-4: Line through the contact along the cam width just after to the contact centre

- Section 5-5: Line along the side construction of the contact along the entraining direction

16.2 The dog-bone contact footprint.

For finite-line contacts, the length and width of the footprint are given by the following set of equations which are in-line with the Hertzian assumptions. Hertz defines the contact patch in terms of semi-major and semi-minor half-widths which are basically half of the contact footprint. Hence:

- The semi-major width a of the footprint is simply half the width of the cam:

$$a = \frac{l}{2} \tag{16.2}$$

where: l is the total width of the cam.
- The semi-minor width b is defined as:

$$b = \sqrt{\frac{4WR_e}{\pi E'l}} \tag{16.3}$$

where: W is the applied load, R_e is the equivalent radius of the mating members at the point of contact and E' is the reduced modulus of elasticity which depends on the Young's modulus and the Poisson's ratio of the materials employed.

16.3.2 Contact load and pressure distribution

The applied load W results from the inertial dynamics of the valve train assembly and the valve spring (see Chapter 15). The load can be defined as follows:

$$W = k_{sp}\, s + F_p \tag{16.4}$$

where: k_{sp} is the valve spring stiffness, s is the valve lift from equation (16.1) and F_p is the valve spring pre-load. It is obvious from the above equation that the contact load will vary according to the cam lift profile which occurs for the duration of the event angle.

This load induces a pressure distribution over the contact area and is approximated by a parabolic shape across the width of the contact. This is known as the **elastostatic Hertzian pressure distribution** over the contact and is considered as maximum along the centre line of the contact. Figure 16.3 clearly indicates this distribution.

Based on Fig. 16.3, the parabolic pressure distribution can be expressed as:

$$P_{(x)} = P_h \sqrt{1 - \frac{x^2}{b^2}} \tag{16.5}$$

where: P_h is the maximum Hertzian pressure at the centre of the contact and is defined by:

$$P_h = \frac{2W}{\pi b l} \tag{16.6}$$

In some cases, an elliptical pressure distribution is assumed along the length of the contact when either the applied load is concentrated at the centre point of the contact length, or, when the contacting member is barrel shaped along its entire length.

Contact deflection

The contact deflection is obtained by employing the formulation outlined by Timoshenko and Goodier (1951) and Dowson and Higginson (1966), and is given as:

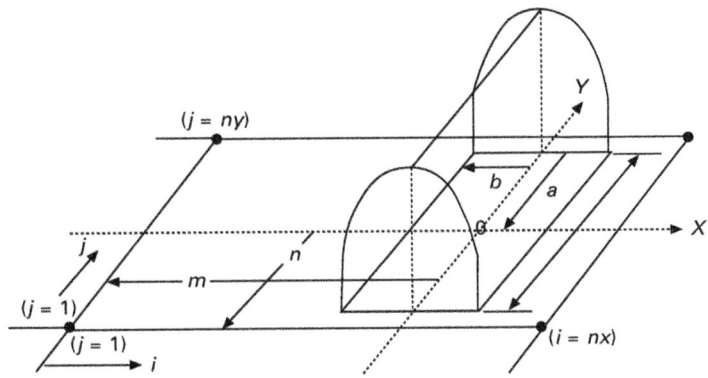

16.3 Hertzian pressure distribution.

$$\delta(x, y) = \frac{2p}{\pi E'} \int_{-b}^{b} \int_{-a}^{a} \frac{dx_r dy_r}{\sqrt{(y - y_r)^2 + (x - x_r)^2}} \tag{16.7}$$

where p is the pressure distribution given by equation (16.5), d and b are given by equations (16.2) and (16.3) respectively, x and y define the current position of the centre of the element where the deflection is being calculated and x_r and y_r give the position of the an element which is at some radial distance from the current element.

16.3.3 The entraining velocity

The entraining velocity quantifies the speed with which the lubricant is drawn into the contact of the bodies in relative motion. In order to calculate this parameter, the instantaneous radius of curvature ρ at the point of contact between the cam and the tappet needs to be first established.

Instantaneous radius of curvature

The instantaneous radius of curvature is dependent on the cam profile and the contact kinematics with the flat tappet. Owing to the varying nature of the cam profile, this radius constantly changes with the cam rotation angle during its event cycle. Over the base circle, it remains constant and is equal to the base circle radius of the cam. ρ can be derived from analysing the contact kinematics between the cam and the tappet (Gohar and Rahnejat, 2008). Figure 16.4 illustrates this point.

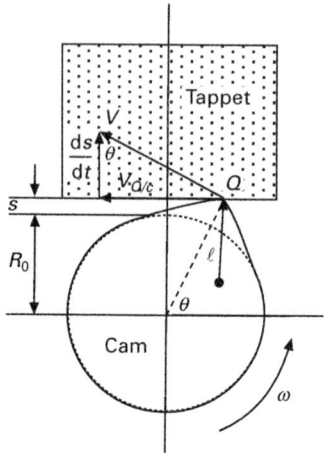

16.4 Cam to tappet contact kinematics.

If the rotation of the tappet is ignored (i.e. tappet spin), the velocity of the instantaneous contact point Q (with respect to the tappet) as it slides along the flat surface of the tappet can be obtained in terms of the camshaft speed ω as:

$$V_{Q/t} = -\omega j_\theta \qquad (16.8)$$

where j_θ is referred to as the **geometric acceleration** of the tappet caused by the change in the cam profile. This is defined as:

$$j_\theta = \frac{d^2 s}{d\theta^2} \qquad (16.9)$$

If the camshaft is considered to be rotating with a constant angular velocity (i.e. ω is constant), then $\theta = \omega t$. Therefore, substituting for θ in the above equation yields:

$$j_\theta = \frac{d^2 s}{d(\omega t)^2} = \frac{1}{\omega^2} \frac{d^2 s}{dt^2} \qquad (16.10)$$

The velocity of the contact point Q can also be represented in terms of the instantaneous radius of curvature of the cam profile at the point of contact. This can be given as:

$$V_{Q/c} = V \sin \theta - \omega \rho \qquad (16.11)$$

where:

$$V \sin \theta = \omega (R_o + s) \qquad (16.12)$$

Since no slip condition is assumed between the cam and the tappet, equation (16.11) equals equation (16.8). Hence:

$$\omega (R_o + s) - \omega \rho = -\omega j_\theta \qquad (16.13)$$

Therefore, the instantaneous radius of curvature at the point of contact is obtained as:

$$\rho = R_o + s + j_\theta \qquad (16.14)$$

where: R_o is the base circle radius.

Equivalent radius

The **equivalent radius**, as mentioned above, represents the replacement of radii of the two curved bodies/surfaces in contact by a radius of an equivalent curved body contacting a flat surface. This can be obtained from the simple relationship below which is applicable to counterformal contacts only:

$$\frac{1}{R_e} = \frac{1}{\rho} + \frac{1}{R_t} \qquad (16.15)$$

where R_t is the radius of the tappet which is 'flat'. Hence, $R_t = \infty$. Substituting this in the above equation, the equivalent radius simply becomes equal to the instantaneous radius of curvature:

$$R_e = \rho \qquad (16.16)$$

The tangential surface velocity of the contact point along the horizontal with respect to the cam is equal to $\omega\rho$. Therefore, the **entraining velocity** u with which the lubricant enters the contact is given as the average of the surface velocities of the tappet and the cam at the point of contact. Hence:

$$u = \frac{1}{2} (\omega j_\theta + \omega\rho) \qquad (16.17)$$

Replacing ρ into equation (16.14) gives:

$$u = \frac{1}{2} [\omega j_\theta + \omega (R_o + s + j_\theta)] \qquad (16.18)$$

$$= \frac{1}{2} \omega (R_o + s + 2j_\theta) \qquad (16.19)$$

The variation of the instantaneous radius of curvature at the point of contact between the cam and the flat tappet is shown in Fig. 16.5(a). It replicates the acceleration curve of the cam illustrated in Fig. 16.1(b). The radius of curvature remains constant over the base circle, but increases on the flank which occurs at 90° and signifies the opening of the valve. In the vicinity and prior to the cam nose, the curvature radius decreases, but increases again when the cam reaches the nose at (i.e. the maximum lift) at which the valve is fully open. The increase in the radius at the nose is due to the cam's flatness in this area. The radius of curvature then reverses as the cam commences the valve closing cycle.

Figure 16.5(b) is the corresponding generated entraining velocities and shows a qualitative comparison to the shape of the radius of curvature variation. In short, the greater the radius of curvature at the point of contact, the greater is the entraining velocity of the lubricant for a constant angular velocity of the camshaft. As the camshaft speed is increased, so does the relative surface velocity between the two mating surfaces.

One of the most important observations is that the entraining velocity for a **modified cycloidal cam** undergoes reversals in four places (in the proximity of the cam nose), i.e. it changes direction. Figure 16.5(b) shows this phenomenon at different engine speeds. In the normal automotive cams, this inlet reversal occurs only twice, because the automotive cam profile is

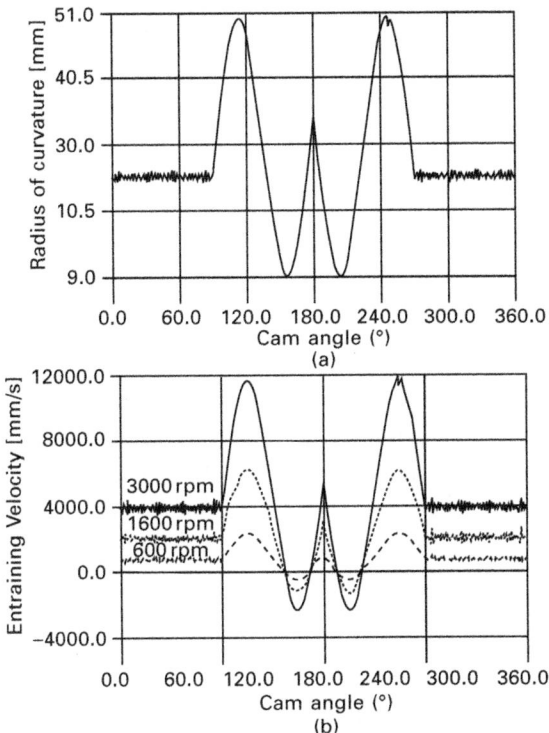

16.5 Equivalent radius of curvature and entraining velocity at point of contact.

based on a polynomial law. It is important to remember this characteristic since it *implies* that at the point where the velocity is zero, there is no lubricant film present within the contact. This would mean that the two metallic bodies can actually come into contact under ***boundary regime of lubrication*** (metal-to-metal contact). However, there is still an absence of wear of the bodies! Section 16.6 explains why this can be the case.

16.3.4 The elastic film shape

The ***elastic film shape*** of the lubricant in the contact conjunction is the same as the deformed profile of the two surfaces between which it is trapped. Under loaded conditions, the deformation of the solids results in a change of their surface geometrical profiles. Since an equivalent system for the cam and tappet is considered as that of a roller in contact with a flat elastic half-space at any small step of time, only the deformation of the elastic surface is considered. The geometrical profile of the cam h_{pr} is thus assumed to be parabolic in its undeformed state in the X-direction. This is given as:

$$h_{pr} = \frac{x^2}{2R_e}$$ (16.20)

In the Y-direction, the profile is assumed to be flat. Therefore, by making use of equations (16.7) and (16.20) the lubricant film shape in an elastic contact can be obtained as:

$$h = h_o + h_{pr} + \delta$$ (16.21)

where h_o is the initial gap separating the two bodies. Figure 16.6 depicts the different terms in the above equation.

16.3.5 Density and viscosity

The density and viscosity of the lubricant are also required to understand its behaviour in the contact since they explain the rheological variations with respect to pressure and temperature. The effect of temperature is ignored in this chapter which can be significant. The density–pressure dependence ρ is defined by Dowson and Higginson (1966) as:

$$\rho = 1 + \frac{\alpha p}{1 + \gamma p}$$ (16.22)

where α and γ are constants dependent upon the properties of the lubricant.

The viscosity η for highly loaded regions given by Roelands (1966) is:

$$\ln \eta + 1.2 = (\ln \eta_o + 1.2)\left(1 + \frac{p}{2000}\right)^{\kappa}$$ (16.23)

where $\kappa = 0.67$ is the viscosity–pressure index and η_o is the viscosity measured at atmospheric pressure and temperature.

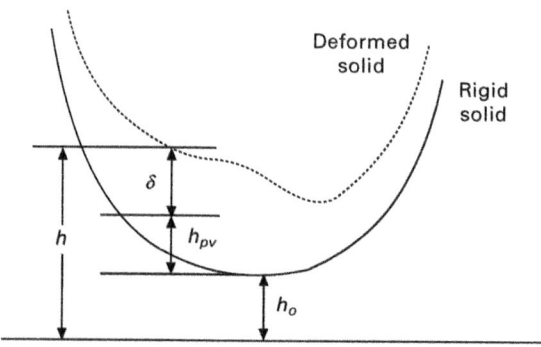

16.6 Components of the elastic film shape.

16.3.6 Reynolds equation

The **Reynolds equation** is:

$$\frac{\partial}{\partial x}\left(\frac{\rho h^3}{\eta}\frac{\partial p}{\partial x}\right) + \frac{\partial}{\partial y}\left(\frac{\rho h^3}{\eta}\frac{\partial p}{\partial y}\right) = 12\left[u\,\frac{\partial(\rho h)}{\partial x} + v\,\frac{\partial(\rho h)}{\partial y} + \frac{\partial(\rho h)}{\partial t}\right]$$

$$(16.24)$$

The explanation for this equation has already been given in Chapter 5. Here the side leakage velocity $v = 0$, as lubricant flow in the Y-direction is ignored. This simplifies the above equation to:

$$\frac{\partial}{\partial x}\left(\frac{\rho h^3}{\eta}\frac{\partial p}{\partial x}\right) + \frac{\partial}{\partial y}\left(\frac{\rho h^3}{\eta}\frac{\partial p}{\partial y}\right) = 12\left[u\,\frac{\partial(\rho h)}{\partial x} + \frac{\partial(\rho h)}{\partial t}\right] \qquad (16.25)$$

16.4   The solution procedure

To resolve for the fluid film thickness and pressures within the contact, equations (16.21)–(16.23) and (16.25) need to be solved simultaneously with the necessary boundary, initial and convergence conditions. The numerical method used to carry out this computation is based on the modified low relaxation Newton–Raphson method (Kushwaha and Rahnejat, 2003). This is carried out over a meshed computation zone indicated in Fig. 16.3. The size of this zone is broken down into 64 elements in the X-direction and 162 elements in the Y-direction.

16.4.1 Boundary conditions

There are two boundary conditions employed to limit the solution domain:

- The pressures are considered to be zero at the edges of the computational zone.
- To avoid tensile pressures in the the fluid film (which are negative) at the cavitation boundary, the pressure gradients are set to zero, i.e.:

$$\frac{\partial p}{\partial x} = \frac{\partial p}{\partial y} = 0$$

16.4.2 Initial conditions

The pressure distribution inside the elastostatic footprint for a particular load is assumed to be Hertzian (as indicated by equation (16.5)), while outside this area, the pressure equates to zero.

16.4.3 The convergence criterion

For the numerical method to reach an approximate solution for the non-linear equations stated above, an error tolerance must be specified. If the solution obtained is within the limits of the required tolerance, then the numerical procedure is deemed to have converged and a solution found.

There are two convergence criteria, required in order to attain the overall solution to the coupled non-linear equations. The first is for the pressure which in turn is fed back into the numerical procedure to evaluate the load. The second convergence criterion is for the evaluated load (obtained from the converged pressures) which should approximately equal the initial input load. This is achieved by adjusting the elastic film shape.

Therefore, for the pressures to converge, the following criterion is employed:

$$\frac{\sum\limits_{i=1}^{nx} \sum\limits_{j=1}^{ny} \left| p_{i,j}^{new} - p_{i,j}^{old} \right|}{\sum\limits_{i=1}^{nx} \sum\limits_{j=1}^{ny} p_{i,j}^{new}} \leq Err_p \tag{16.26}$$

where nx and ny are the total number of elements in the X and Y-directions respectively, and the error tolerance for the pressure Err_p is set equal to 0.0001. If the convergence is not achieved, the input pressures into the next iteration loop are updated, using the following expression:

$$p_{i,j}^{new} = p_{i,j}^{new} + \varepsilon \Delta p_{i,j} \tag{16.27}$$

where ε is the under-relaxation factor which ranges from 0.1 to 1.0, depending on the type of elastohydrodynamic lubrication problem at hand and $\Delta p_{i,j}$ is the difference between the new and the old calculated pressures.

Similarly, for the load to reach a converged solution, the criterion used is:

$$\left| W^{new} - W^{old} \right| \leq Err_w | \tag{16.28}$$

Here, the error tolerance for load convergence Err_w is set equal to 0.0001. If the load has not converged, then the central film thickness is adjusted according to the evaluated unbalanced load, using the following formulation:

$$h^{new} = h_o^{old} + \lambda \left(W^{old} - W^{new} \right) \tag{16.29}$$

where λ is the damping coefficient used to dampen the sudden changes in the central oil film thickness, and lies between 0.001 and 0.05. Hence, if the evaluated load W^{new} is less than W^{old}, then the central film thickness is further reduced and vice versa.

Since the solution is transient in nature, the deflection, the instantaneous radius of curvature, the pressures and the film thickness become functions of time. As the solution progresses in the time domain, these variables are updated accordingly, with reference to their values at the previous time step.

16.5   Simulation conditions

The values of the required input parameters (i.e. R_e, u, ∂t and W) for the *transient elastohydrodynamic* model described above, are obtained from a dynamic multi-body model for one steady state cycle of the cam rotation. This is gathered at the camshaft angular velocity of 3000 rpm (since the multi-body model exhibits *valve spring surge* at this speed) which converts to 0.02 s for one cam revolution. The analysis covers the period of the event angle only (i.e. for $2\beta = 180°$), since the base circle of the cam maintains the same results for the remaining 180° due to the constant values of the input parameters mentioned above. For the purpose of smooth numerical integration, the cycle time is covered over 500 time steps. Hence, the time interval per step is: $\partial t = \Delta t = 0.02/500 = 40\,\mu s$.

The transient EHL model for the above simulation conditions was run with a mesh density of 64×162 elements and took between 60 and 70 days on an Intel Pentium II 400 MHz (512 RAM) machine. However, with increasing processor capability, the analysis time is continually decreasing.

16.6   Results

Traditionally, cam lift curves are obtained under kinematic considerations described in Section 16.2. In such analyses, the valve spring rate is considered to be constant with a predefined preload value. Furthermore, the dynamic effects that lead to contact separation have not usually been taken into account in the lubrication study of cam to flat followers. A significant difference between the dynamic contact loads and those obtained by the imposition of kinematic curve–curve adherence constraints is the loss of contact that can ensue during valve acceleration. This effect has been ignored in the present analysis, although loss of contact would constitute depleted lubrication and a subsequent impact loading in the contact region can lead to generation of significant pressures. However, the non-linear dynamic behaviour of the valve spring coils (having different pitch) has been incorporated in the present analysis.

The surge effect caused by the variable stiffness of the valve spring coils is responsible for the momentary loss of contact and the subsequent high rate of approach in the contacting region. Therefore, the surge effect through elastic deformation of the solids has been included in the model, although contact separation is not allowed due to the presence of the kinematic constraints in the analysis described here.

A number of investigators have employed the predicted contact loads and entraining speeds under kinematic valve train motions in order to undertake numerical solutions of the lubricated contact between the cam and a flat follower (see Chapter 15). Hitherto, all such solutions that have been carried out under non-steady conditions have assumed infinite line contact conditions between the cam and the follower. These studies include the works reported by Ai and Yu (1988, 1989), Dowson *et al.* (1992) and Mei and Xie (1996). As the speed of entraining motion diminishes at four symmetric locations for the modified cycloidal cam on the either sides of the cam nose to the flat–follower contact, the inlet lubrication boundary reverses in direction as the speed of entraining motion increases to a finite value. These locations, therefore, are significant in the study of cam to follower lubrication, because the retention of a coherent film would be largely dependent only on the squeeze film motion and the entrapment of a volume of lubricant. Furthermore, as it has been shown by all the aforementioned authors, the transit time through these regions is very fast indeed, resulting in rapid lubricant replenishment.

A comprehensive study of the modified cycloidal cam has been carried out by Fessler and Ham (1990). They employed the Dowson and Higginson (1961) regression formula for the central oil film thickness to predict the lubricant film height at various locations in a cam cycle, superimposing, the effect of pure squeeze action in an infinite line contact solution. The authors showed that under such quasi-static conditions, the zero films obtained by the latter, increased marginally to a finite value. They also showed that for the modified cycloidal cams, four values of central film thickness minima were obtained, which validates the observations made earlier (where the extrapolated central oil film thickness equation, includes the effect of elastic body squeeze). From their calculations, under practical conditions and at idle engine speeds, the minimum central oil film thickness in the vicinity of the inlet reversals were of the order of 50 nm at high temperatures with modern low-viscosity oils. They measured the mean surface roughness of the contiguous bodies to be in the region 200 nm, indicating that for much of the contact between the symmetrical inlet reversal positions, a ***mixed regime of lubrication*** may have been prevalent which, in extreme cases, may tend to boundary lubrication conditions. Clearly, with an increased engine speed, fluid film lubrication becomes progressively dominant. This effect has been shown in Fig. 16.5(b) under quasi-static conditions from a low speed of 600 rpm to an engine speed of 6000 rpm (camshaft speed of 3000 rpm).

The flat cam profile in the transverse direction (into the depth of the cam) makes a narrow and long contact with a flat follower. Because of this, the aforementioned analyses of the lubricated contact have all assumed line contact conditions. However, the length of contact in the transverse direction of the cam is finite, which approaches the full width of the cam, and thus, an infinite line contact consideration is not very practical. Furthermore, owing to the unblended nature of the cam's lateral profile, edge stresses are generated at the extremities of its contact with the flat follower. Under lubricated conditions, these edge stresses cause high-pressure spikes which can be considerably larger in magnitude than the elastostatic Hertzian pressures and constitute the determining factors in the fatigue spalling of the loaded mating members in concentrated counterformal contacts. The presence of these pressure spikes inhibit the flow of lubricant through the edges of the contact. This phenomenon results in lubricant film thickness, being at its absolute minimum value in these regions, and referred to as the *side exit constriction* (Fig. 16.2). These minimum film thickness regions are more susceptible to wear, and the most likely zones, where full fluid lubrication may cease to exist. Therefore, the one-dimensional analyses of cam to follower contact, hitherto reported by other authors disregards the limiting cases that determine both fatigue and wear behaviour of the contact.

Figure 16.7(a) shows the transient lubricant film thickness at the central and the minimum central exit positions during a *cam event cycle*. Note that the same figure cannot be easily determined for the absolute minimum lubricant

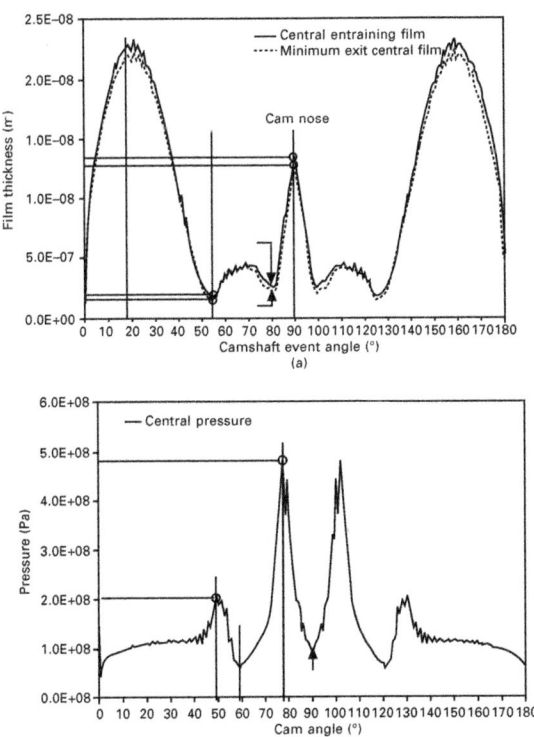

16.7 Central and minimum film thickness and pressures along section 1-1.

film thickness at the side exit constriction as this point moves instantaneously in space-time. Referring back to Fig. 16.7(a), for this cycloidal cam, four positions of inlet boundary reversal, corresponding to momentary zero entraining motion can be observed. These are at ±36° (point B on the figure) and ±12° to the cam nose position. The minimum lubricant film thickness in both places occurs in these regions, where the lubricant film is sustained primarily by the squeeze film action and by rapid replenishment due to the commencement of entraining action. These findings for the cycloidal cam are qualitatively in line with the findings of Fessler and Ham (1990). The film thickness is in the region of 0.15–0.2 μm in regions described by the minima rising to 1.2–1.4 μm at the nose, with the flank region having a higher film thickness up to 2.25 μm

For the segments of the cam in which both the entraining speed and the radius of curvature approach maximum values, the lubricant film thickness also approaches its maximum value. These regions occur along the cam flank. In these regions, the lubrication of cam and follower lies in the hydrodynamic regime of lubrication . With the modified cycloidal cam, the radius of curvature at the nose is larger than those for a polynomial automotive cam and the speed of entraining motion is also larger, resulting in a significant rise in the lubricant film thickness from the corresponding value obtained at the positions of minima. These observations are not in accord with the results obtained for an automotive cam, which has a small radius of curvature at the cam tip (see for example Dowson *et al.*, 1992, or Mei and Xie, 1996). The results obtained here, however, are in line with the findings of Fessler and Ham (1990), who have employed a similar (but not identical) modified cycloidal cam. Sections through the side constriction of the contact in the entraining and transverse directions have been discussed at locations 'A' (on the flank) and at the cam nose in this chapter. In the positions of zero entraining motion (point 'B'), the *squeeze film action* tends to squeeze the lubricant out of the contact, but since the interval is short lived, the entraining flow is replenished (Kushwaha and Rahnejat, 2003; see also Chapter 17).

It must be noted that an insignificant lubricant film may exist on the base circle of the cam due to the clearance (also referred to as lash and is usually in the order of 0.1 to 0.5 mm) in the conjunction, but as the cam comes into contact with the follower after taking up the lash (i.e. on the flank), the valve opening event commences and a lubricant film starts to form due to the increasing contact load from the valve spring. A clearer picture of valve train dynamics under lubricated conditions in a cam-to-follower contact emerges, when the corresponding transit pressure peak values to those of the oil film thickness in Fig. 16.7(a) are plotted (see Fig. 16.8(a) and (b)). The maximum pressure time history at the centre of the contact is shown in Fig. 16.7(b). This pressure is lower than those at the edges of the contact.

In the above figures, the variation of lubricant pressure is small due to smaller

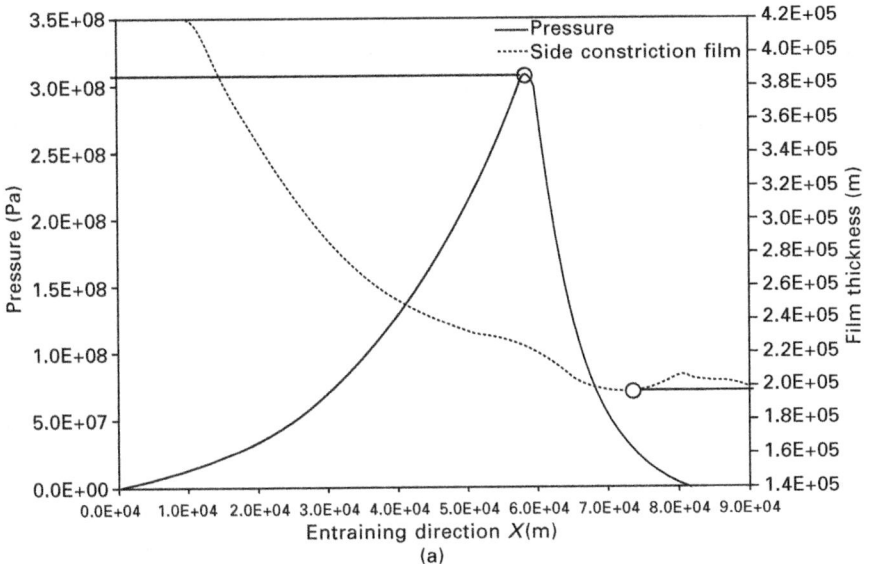

(a)

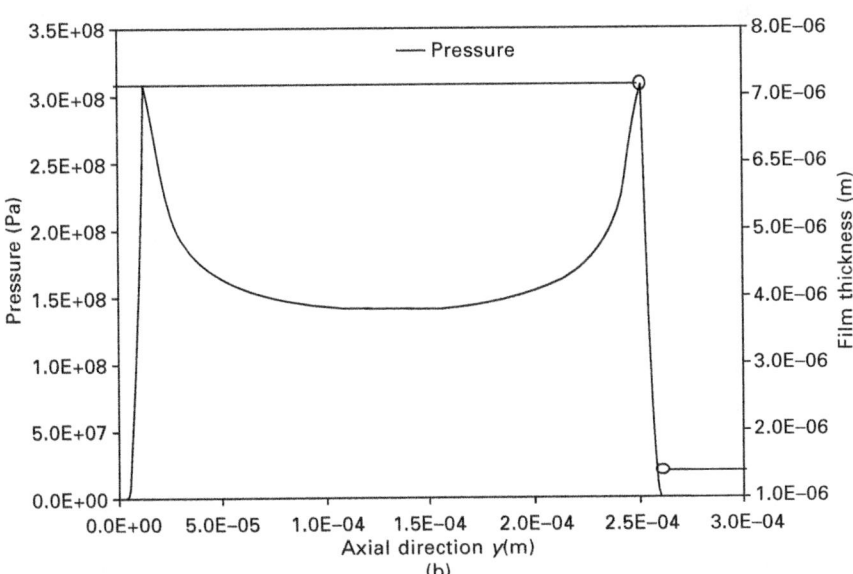

(b)

16.8 Entraining and transverse film thickness and pressures along section 5-5 and 4-4 on the flank.

elastic deformation of the contiguous bodies since the load on the flank is still relatively low. In fact, the dominant regime of lubrication is hydrodynamic. The primary pressure peak at the contact extremities is approximately 1.02 times its equivalent Hertzian value with a magnitude of 307 MPa (see Fig.

16.8(a)). In the same figure, the minimum exit film, along the same cross-section is found to be 1.97 μm. The absolute minimum side constriction film thickness is in an axial cut through the contact and in the vicinity of this maximum pressure, with a magnitude of 1.45 μm (see Fig. 16.8(b)). Note that the pressures in the central region and elsewhere do not reach the Hertzian equivalent, another confirmation for the prevailing hydrodynamic conditions (the position indicated by point A in Fig. 16.7(a)).

With contact transitions from the flank towards the nose, the film thickness decreases dramatically due to a reduction in the entraining velocity. At the positions of boundary reversal when the entraining velocity momentarily ceases, the lubricant in the contact is sustained primarily by the squeeze film effect. The film shape exhibits significant elastic deformation and the pressure reaches a maximum value (in the order of 1.6 GPa (Kushwaha and Rahnejat, 2003). The dominant conditions are due to elastohydrodynamic lubrication (EHL). Thereafter (after the position of *inlet reversal*), the speed of entraining motion increases, resulting in an increase in the film thickness. A corresponding reduction in load and an increase in the radius of curvature constitute reduced elastic deformation of the contiguous solids. This yields decreasing pressures. This trend has also been observed by Mei and Xie (1996) but the difference is that the current analysis is for a cycloidal cam with two inlet reversals in each half cycle of the cam. Nevertheless, the physics of the problem is the same here as that also noted by Mei and Xie (1996).

The other position of interest to investigate under transient contact dynamic conditions is the cam nose-to-flat follower contact. Figure 16.9(a) shows the generated pressure and the corresponding film shape in the direction of entraining motion through the side exit constriction. The figure shows that the maximum pressure occurs in the side exit constriction region, reaching a maximum value of approximately 1.1 times the maximum Hertzian pressure. However, the central pressures are approximately 65–70% of the Hertzian elastostatic values. The maximum pressure at the edges of the contact are 223 MPa. This can also be seen in Fig. 16.9(b), showing the film and pressure variations in the axial lateral direction. In this figure the absolute minimum exit film is 0.725 μm.

It can be seen from the above results that the pressures generated on the cam nose at the side constriction are not as high as generally thought. In fact, as already pointed out, the worst pressures occur during boundary reversals and shown in Fig. 16.10 (as reported in Kushwaha and Rahnejat, 2003).

16.7   Conclusion

A number of conclusions can be arrived at based on the discussions carried out in this chapter:

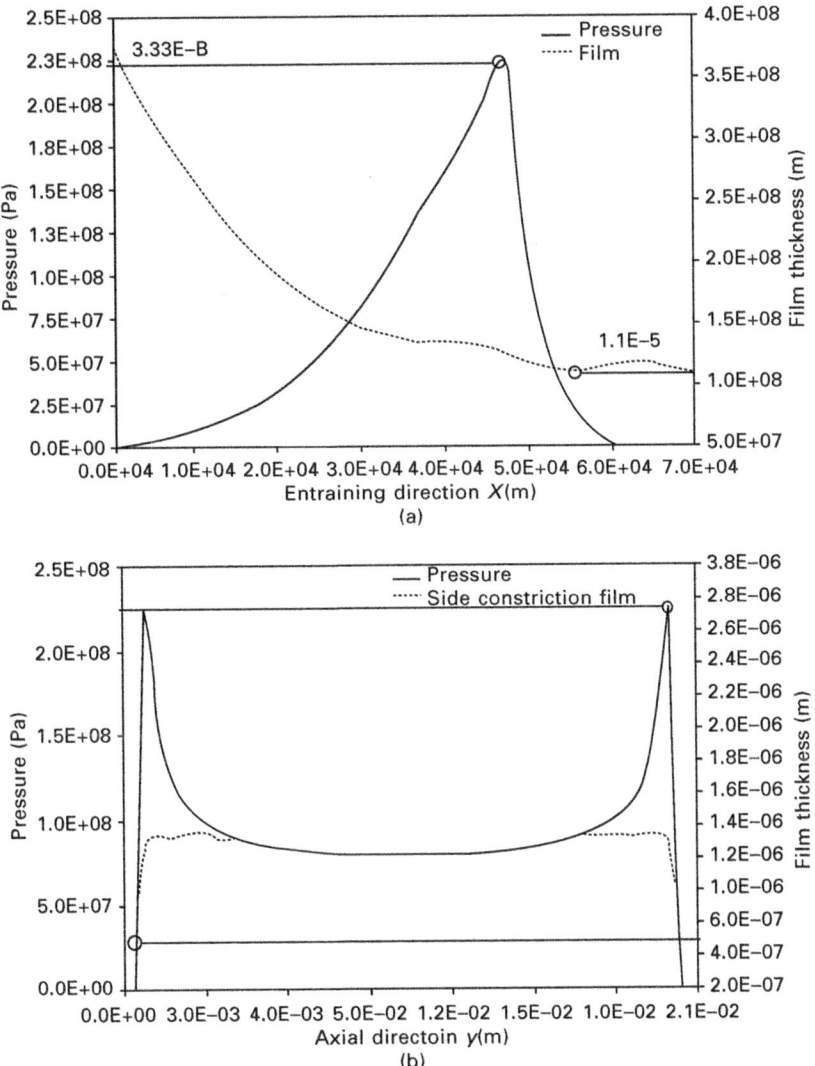

16.9 Entraining and transverse film thickness and pressures along section 5-5 and 4-4 on the nose.

- Valve train tribology and dynamics are inter-related and the contact conditions are therefore, transient in nature.
- The most important issue is to be able to accurately predict film thickness and pressure distribution in the cam-follower contact and realise that these conditions *cannot* be judged based the elastostatic (dry) Hertzian contact stress theory (note the lubricant pressure spikes at the contact extremities).

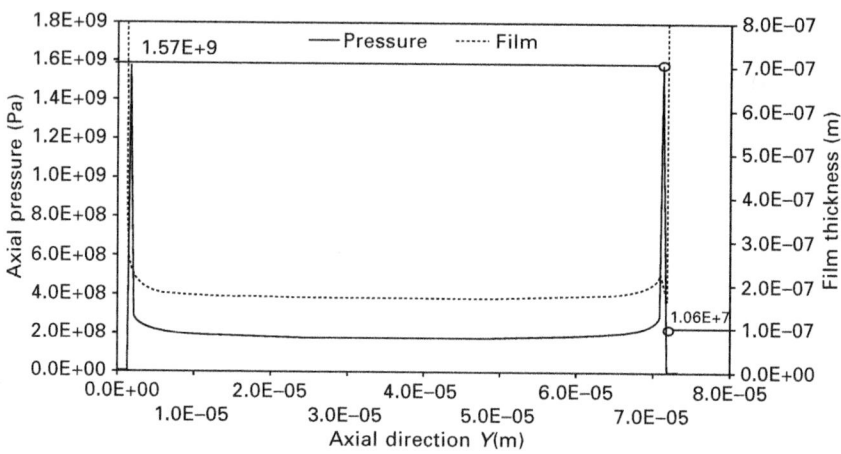

16.10 Entraining and transverse film thickness and pressures along section 5-5 and 4-4 at point B.

- This analysis, as stated in the latter point above, is necessary in order to predict the onset of wear due to the diminution of the film thickness and contact fatigue spalling resulting form excessive pressures.
- Although this chapter has addressed some of these issues, it has not looked at friction and thermal effects in the contact. These play an important role, because 10–15% of all mechanical and frictional losses in an IC engine are attributed directly to the valve train system, 70% of which originate from the cam–tappet contact.

16.8   References

Ai H and Yu H, 'A full numerical solution for general transient elastohydrodynamic line contacts and its application', *Wear*, 1988, Vol. 121, pp. 143–159.

Ai H and Yu H, 'A numerical analysis for the transient ehl process of a cam–tappet pair in I. C. Engine', *Trans. ASME. J. Tribology*, July 1989, Vol. 111, pp. 413–417.

Dowson D, 'Elastohydrodynamic and micro-elastohydrodynamic lubrication', *Wear*, 1995, Vol. 190, pp. 125–138.

Dowson D and Higginson G R, 'New roller bearing lubrication formula', *Engineering*, 1961, Vol. 192, pp. 158–159.

Dowson D and Higginson G R, *Elastohydrodynamic Lubrication*, Pergamon Press, New York, 1966.

Dowson D, Taylor C M and Zhu G, 'A transient elastohydrodynamic lubrication analysis of a cam and follower', *J. Phys. D: Appl. Phys.*, 1992, Vol. 25, pp. A313–320.

Fessler H and Ham R, 'Lubrication and stress analysis as a basis for camshaft optimisation', *Proc. XXIII FISITA Congress*, Torino (Italy), May 1990, pp. 565–579.

Gohar R and Rahnejat H, *Fundamentals of Tribology*, Imperial College Press, London, 2008.

Jeon H S, Park K J and Park Y S, 'An optimal cam profile design considering dynamic characteristics of a cam-valve system', *SEM Spring Conference on Experimental Mechanics*, Portland, OR, June 5-10, 1988, pp. 357–363.

Kushwaha M and Rahnejat H, 'Transient elastohydrodynamic lubrication of finite line conjunction of cam to follower concentrated contact', *J. Phys. D: Appl. Phys.*, 2003, Vol. 35, pp. 2872–2890.

Mei X and Xie Y, 'A numerical analysis of the nonsteady EHL process in high-speed rotating engine cam/tappet pairs', *Trans. ASME. Journal of Tribology*, July 1996, Vol. 118, pp. 637–643.

Morel T, Flemming M F and Buuck B A, 'Evaluation of variable camshaft effects on performance of a high output, 4-valve SI engine', SAE Pap. No. 905173, 1990, pp. 397–433.

Mostofi A, 'Oil film thickness and pressure distribution in elastohydrodynamic elliptical contacts', PhD Thesis, Imperial College of Science and Technology, University of London, 1981.

Rahnejat H, 'Influence of vibration on oil film in concentrated contacts', PhD Thesis, Imperial College of Science and Technology, University of London, 1984.

Reeve J., *Cams for Industry*, Mechanical Engineering Publications, Bury St Edmunds, 1997.

Roelands C J A, 'Correlation aspects of viscosity–temperature–pressure relationship of lubricating oils, PhD Thesis, Delft University of Technology, The Netherlands, 1966.

Timoshenko S P and Goodier J N, *Theory of Elasticity*, McGraw Hill, New York, 1951.

Yan H S, Tsai M C and Hsu M H, 'An experimental study of the effects of cam speeds on cam-follower systems', *Mech. Mach. Theory*, 1996, Vol. 31, No. 4, pp. 397–412.

17
A multi-scale approach to analysis of valve train systems

M. TEODORESCU, Cranfield University, UK

Abstract: The valve train system synchronises the engine timing (inlet and exhaust) with the thermodynamics/combustion process. Thus, its optimal operation is vital for the health of the entire engine. The interplay between various forms of dynamics (e.g inertial, structural and impact/contact) and reactions due to loaded tribological conjunctions leads to a complex problem. The solution requires simplifications and assumptions based on careful analysis. The current chapter describes a modelling approach which integrates the fundamental aspects of valve train behaviour in a fast converging methodology.

Key words: valve train systems, cam-tappet contact, NVH, impact dynamics, friction prediction, tribology of rough sourfaces.

17.1 Background

Long before combustion engines were built or even thought of, humans saw fire spreading quicker when wind blew. They did not understand the underlying principles of oxidation or combustion. However, they understood that a healthy fire needed a plentiful supply of air. This observation of nature became more useful after it was noted that at high temperatures certain rocks melted down into metal. In some cases it was possible to use an open fire, but the process was faster and the results were more consistent if a furnace with air steadily pumped through it was used. Consequently, right from the beginning, a series of complicated air-pumping devices became the heart of all furnaces.

In modern internal combustion (IC) engines a steady supply of air continues to be one of the key elements of engine performance. However, due to the cyclic nature of IC engine operation, the intake of fresh air and the exhaust of burnt gases also follow a cyclic pattern. The basic principle is very simple and is inherited from those earlier findings, chiefly from the steam engines. A set of valves open progressively during the intake part of the cycle, allowing air into the cylinder, and firmly closes during the remaining part of the cycle. Another set of valves open during the exhaust part of the cycle and close during the remaining part of the cycle. For every mechanism, the opening and closing events are carefully designed to synchronise their mechanical motions

567

with the combustion cycle, thus achieving maximum efficiency. Although the main design characteristics have evolved considerably since the early days, the basic principle remains unchanged. The differences arise from the significant increase in engine demand, cylinder pressures and operational speeds. At the time of early steam engines the operational expectancy was very long and the over-design was usual. Therefore, valve train components failed mostly due to progressive corrosion of the unloaded conjunctions rather than any mechanical failure of the loaded components. With later steam engines and early IC engines, the focus had changed towards smaller, faster and more efficient mechanisms. This brought additional challenges, including the need to reduce mechanical and frictional losses. New generations of valve trains became significantly lighter, faster and subjected to increasingly higher loads. These have inevitably led to increased research in valve trains, furthering understanding of the subject.

A typical *valve train* system comprises a significant number of contacting parts, with interactions governed by many coupled phenomena. Consequently, a prerequisite for a pertinent prediction is an integrated model of the inertial dynamics (usually expressed in Lagrangian or Newton–Euler formulation), tribology (governed by Reynolds equation), contact mechanics, surface characteristics and physical chemistry of the lubricant. Such an approach is commonly referred to as *multi-physics*. Historically (from a tribological point of view), two different classes of solutions have emerged. The first one proposes solution of Reynolds equation, ideally integrated with an accurate prediction of the dynamic behaviour of the mechanism (Teodorescu *et al.*, 2007). Although this approach (if applied correctly) has proven to be very accurate, any simulation would be quite time consuming and require advanced computer programming skills. Moreover, the timeframe required for such simulations can be long, and therefore, possibly daunting for the uninitiated or for industrial applications. For a detailed description of the assumptions and methodology necessary for a full solution of Reynolds equation in the elastohydrodynamics (EHD) conjunction, the reader should refer to Chapter 16, and for additional information to the works of Dowson and Higginson (1966), Hamrock (1994) and Jalali-Vahid *et al.* (2001). The second approach predicts the behaviour of the tribological conjunctions, using *extrapolated oil film thickness formulae*. These are equations obtained through regression of numerical studies, which correlate the applied load (lubricant reaction), entrainment velocity, lubricant viscosity and mechanical properties of contacting solids for given types of contact geometry with a representative value for the oil film thickness. Although the predictions, using such formulae, have certain limitations (due to various simplifying assumptions), they are relatively easy to use and provide fairly accurate and quick results.

17.2   Aspects of valve train geometry and construction

The purpose of a mathematical model is to help the designer predict friction and wear and highlight means to reduce them. Often separate computation modules are developed for system dynamics, lubrication and frictional behaviour of individual components of a valve train system. These are then integrated into an overall model to predict *power loss* in the system. In most cases increasing the accuracy of a theoretical simulation increases the extent of computation. Here a simplified model is used to describe the physical phenomena within an acceptable computation time. To increase the overall efficiency, simpler models are used in the first instance and integrated into a larger and more comprehensive model for the dynamics and friction of the entire valve train system. In these cases, the computation speed of each individual model is crucial for the feasibility of the integrated package. In practice, more elaborate models are used at an ultimate stage, after a fast model singles out specific problems, which may require further detailed analysis.

The *valve train* is a highly loaded mechanism characterised by an oscillatory cyclic motion, where each component is described by its mass, stiffness and damping. The fast cyclic motion of the cam can induce a significant level of vibration, which has to be taken into consideration. Figure 17.1(a) shows a classical geometry for the valve train system, comprising the *cam*, *tappet*, pushrod, rocker-arm, valve and spring. This configuration is often used for larger engines (e.g. heavy-duty trucks or naval engines), but it is rarely applicable for lighter configurations, due to high levels of vibration. Figure 17.1(b) is a schematic of several alternatives used for lighter applications (e.g. automobiles). These are usually less loaded and much faster mechanisms, where the reduced inertia represents a significant advantage.

The opening and closing strategy for the valve train must be correlated with the engine cycle. Therefore, there are distinctive valve train configurations for four-stroke and two-stroke engines. In a *four-stroke engine*, the combustion cycle spans over two full crankshaft rotations (720°). During the cycle each valve should fully open and close. This can be achieved if the camshaft speed is half the crankshaft speed. In a *two-stroke engine*, the combustion cycle spans over one full rotation of the crankshaft (360°). Therefore, the camshaft and the crankshaft have the same speed. It is sensible for either configuration to link the camshaft and crankshaft rotation, using an appropriate mechanism. Traditionally, there are two solutions used to power the camshaft using the energy provided by the crankshaft. Figure 17.2 shows a schematic view of these possible solutions. The advantage of using a *timing belt* (left view) is that the location of the camshaft is not constrained by the engine geometry. This solution allows the camshaft to be mounted *overhead*, very close to

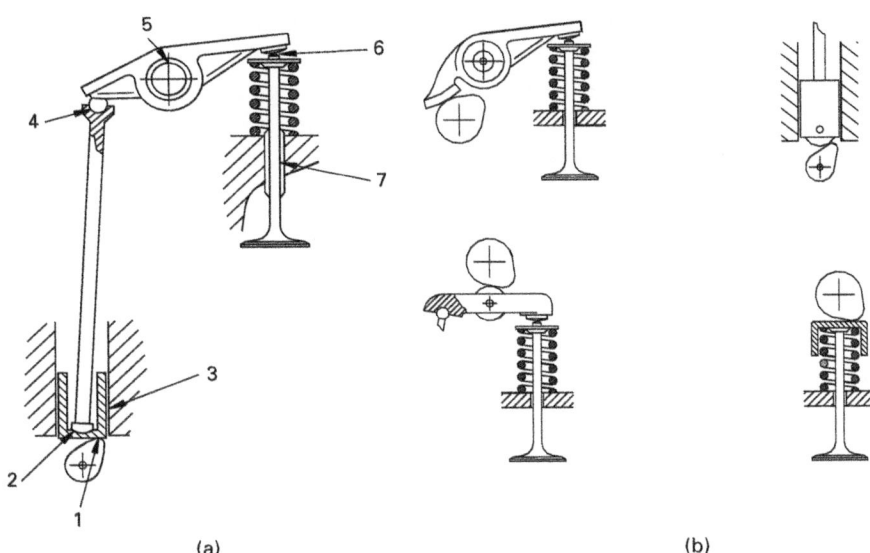

17.1 Valve train structure. The main frictional losses in the valve train occur between the cam and the tappet (1), between the tappet and its bore (3), in the rocker arm bearing (5), between the valve stem and the valve guide (7) and in the camshaft bearings. The friction forces at the two ends of the push rod (2, 4) and the friction between the valve stem top and the rocker arm end (6) (a) Schematic of the push rod valve train and the major components of the friction losses (b) different valve trains.

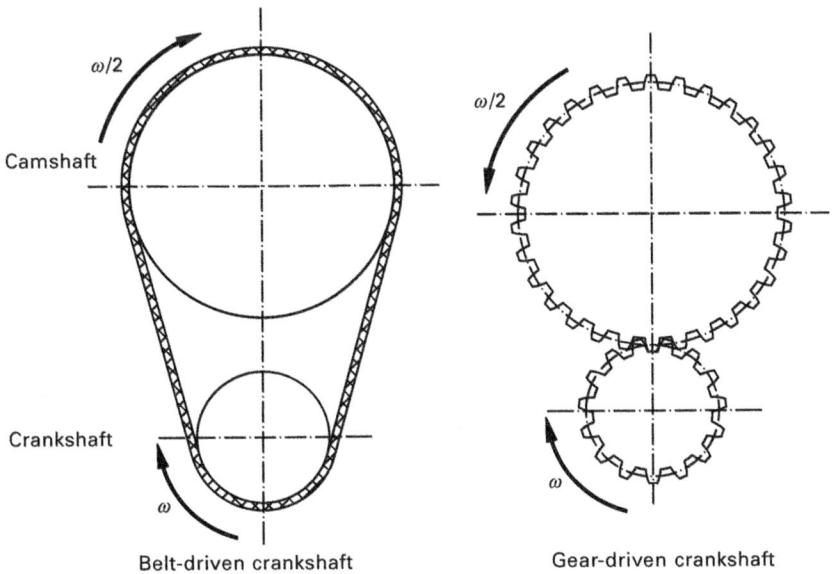

17.2 Overhead cam operation for a four-stroke engine.

the valve itself (see Fig. 17.1(b) – except for the top right option), which leads to a compact engine configuration. This is advantageous if the engine compartment space is regarded as an important design parameter. Thus, this represents the solution of choice for most modern vehicles. However, the life of the mechanism is limited by the timing belt itself and, therefore, this solution must be considered with care if the valve train load and service time are important issues. Consequently, for heavy duty, long-life applications (e.g. naval engines) this option may not be suitable, and the use of a *pushrod* gear driven valve train is preferred.

17.3   Valve train as an integrated problem

Engine friction is responsible for 10–20% of the engine fuel consumption (Straton and Willermet, 1983; Uras and Patterson, 1983). Although this lost energy has always been recognised as worrisome, it has become a particularly important issue in recent times, because of the growing importance attached to fuel economy, as well as environmental impacts. There are three major sources of frictional losses in an IC engine: piston–liner interactions, valve train system and engine bearings. It is generally accepted that the piston–liner interface is responsible for most of the engine frictional inefficiency. However, the valve train accounts for a significant portion as well. Even though the exact proportion depends on many parameters, it was found that under certain operational conditions the *valve train losses* could be as high as 5–10% of the total engine friction (Wakuri *et al.*, 1995).

To underline the importance of an integrated tribo-dynamics model for the valve train, it is significant to show how different components contribute to the total valve train losses. The valve train is a complex mechanism and its components undergo different motions, resulting in various frictional characteristics. The total valve train energy loss is the sum of the friction energy consumed by all its individual components. For the valve train shown in Fig. 17.1(a), the main losses occur between the cam and the tappet (1), between the tappet and its bore (3), in the *rocker arm bearing* (5), between the valve stem and the valve guide (7) and in the camshaft bearings. The friction forces at the two ends of the push rod (2, 4) and the friction between the valve stem top and the rocker arm end (6) are much lower than in the other components. They appear as a residual term in the overall balance of frictional losses.

The amplitude of valve train vibration increases with engine speed. Therefore, applied forces on the load bearing conjunctions can increase significantly (Norton *et al.*, 1999; Teodorescu *et al.*, 2002). Figure 17.3 (Teodorescu *et al.*, 2002) shows a break-down of the energy lost between the different valve train components during engine operation. The total energy dissipated by friction decreases with an increase in engine speed. This is

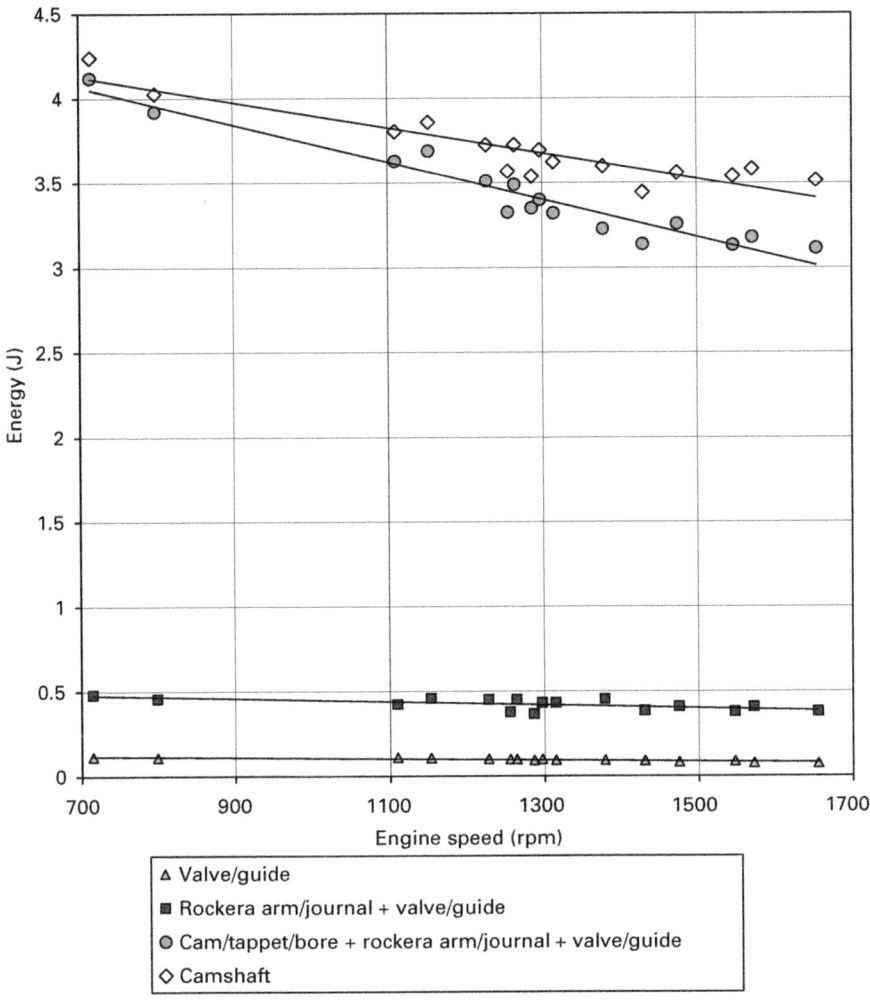

17.3 The energy dissipated by friction during the operation of the exhaust valve (after Teodorescu *et al.*, 2002).

because of the *elastohydrodynamic* lubricated contact in the cam–tappet conjunction, which promotes a thicker oil film at higher speeds of entraining motion, thus lowering friction. Up to 85% of the energy dissipated in the valve train is lost in the cam–tappet conjunction. There is little influence from speed on the dissipated friction energy in the valve/guide assembly and in the rocker arm bearing. Friction between the valve stem and its guide is in fact quite low, representing only about 2–3% of the total energy lost in the valve train, while the friction in the rocker arm bearing represents 10–11% of the of the total. The difference between the total *valve train frictional losses* measured on the camshaft and that obtained by the cumulating sum of cam/tappet/bore, rocker arm bearing and the valve/guide friction increases from ~1–2% at 750 rpm to about 12% at 1680 rpm. This increase is mainly attributed to the friction in the camshaft bearing, which was included in the camshaft torque measurement. Consequently, for most valve trains (and especially for the overhead cam systems) it is imperative to focus the modelling work on the lubrication/friction in the cam–tappet conjunction.

17.4   Valve train kinematics: cam-to-flat follower contact

The starting point for modelling the cam–tappet conjunction is a through understanding of the *kinematics* of their contact. Figure 17.4 shows a schematic view of a cam–tappet contact. Although in most practical applications a roller tappet is used (see Fig. 17.1(b)), the difference is qualitative and does not extend to the physical meaning of the tribological phenomena analysed here. Interested readers are referred to the work of Jensen (1965), Koster (1975) or Taylor (1993) for a full description of contact kinematics for a large range of cam–roller configurations. Once the adjustment is made for the practical case, the rest of the analysis can be easily updated. In Fig. 17.4, the global co-ordinate system XY is fixed at the centre of the cam and the contact is made at a typical point P on the cam flank. The valve lift is denoted by s and the angular velocity of the camshaft is ω.

The *speed of entraining motion* in the contact is obtained as:

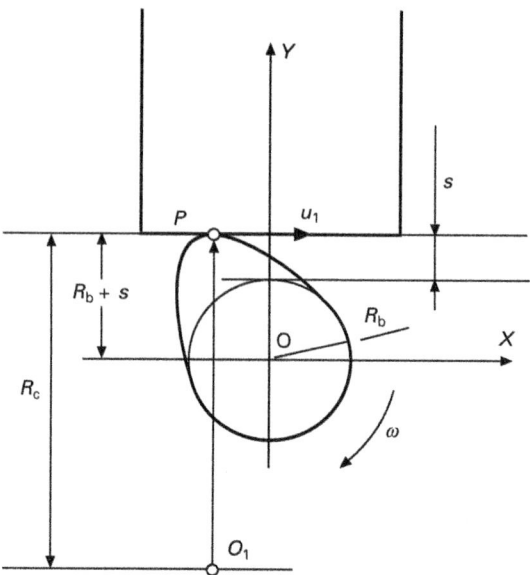

17.4 The geometry of the cam–tappet contact.

$$u = (u_1 + u_2)/2 \qquad (17.1)$$

This represents the speed by which the lubricant is drawn into the contact due to the relative motion of the contiguous surfaces to form a film. Together with the contact deflection, this accounts for the instantaneous gap formed between the load-bearing surfaces. For simplicity the surface speed of the flat follower is considered to be zero. It is advantageous to explain the concept using a *flat* tappet. For a roller, u_2 cannot be neglected. For a flat tappet the speed of lubricant entrainment is simply half that of the instantaneous speed of the cam surface at the point of contact, thus:

$$u_1 = (R_b + s) \cdot \omega \qquad (17.2.)$$

The instantaneous radius of curvature at the point of contact should be computed taking into account the geometrical acceleration, caused by the cam lift, which causes the tappet to accelerate in an upward direction, hence:

$$R_c = R_b + s + \ddot{s}/\omega^2 \qquad (17.3)$$

Using equations (17.1)–(17.3) as a starting point, the mathematical approach proposed by Taylor (1993) yields the speed of entraining motion as:

$$u = \omega [R_c + \ddot{s}/\omega^2]/2 \qquad (17.4)$$

17.5   Valve train dynamics

Several attempts to develop a suitable model for *valve train dynamics* behaviour have been made (Kurisu *et al.*, 1991; Roß and Arnold, 1993; Schamel *et al.*, 1993; Norton *et al.*, 1999; Teodorescu *et al.*, 2005). The ideal approach for the valve train problem requires the solutions for all the physical interacting phenomena, which contribute to the valve train operation. These should be integrated into a unique framework, which is then solved numerically. Such an approach begins with the multi-body formulation (such as Lagrangian dynamics, see Chapter 1) for rigid body motions and the Reynolds equation for lubricated conjunctions (see Chapter 5). Both these require solution of differential equations, taking into account the assembly constraint functions and elasticity equation for all the intervening contacts. This set of (differential and algebraic) equations must be solved simultaneously, in time and space domains (see Chapter 1). Although very elegant, the main drawbacks of this approach are the size of the system of equations and the relatively advanced numerical expertise required to solve them. Alternatively, the valve train system can be approached in a modular fashion, where the independent mathematical modules (each viewed as a black box) describe the individual physical phenomenon. The *Newton–Euler formulation* for multi-body dynamics is used as the backbone required to link these individual modules. Each module has a specified number of inputs and outputs, which provide the link with the other interacting modules. The output from one module becomes the input to the next and so forth. This is a versatile approach where individual modules can be altered as needed without changing the entire model. In an industrial environment, the designer can access several alternative modules for each physical phenomenon. These can embody an increasing degree of complexity and accuracy. At the initial stages, a fast (but simplified) model would suffice. However, in later stages of the design process, more advanced (but slower in computation terms) modules could be used for specific objectives.

Figure 17.5 shows an example, where the valve train (Fig. 17.5(a)) is considered, comprising lumped mass elements connected to each other by mass-less elements with stiffness and damping (Fig. 17.5(b)). The method described by Teodorescu *et al.* (2007) is adopted. Individual valve trains are discretised either by a single or a two degrees of freedom system and the link between them is accounted for by the elasticity of the connecting section of the crankshaft. One of the masses represents the motion of the tappet, the push-rod and the translational proportion of the **rocker arm** (on the push-rod side of the mechanism). The other mass represents the valve, the effective mass of the valve spring (one-third of its actual mass) and the remaining part of the rocker arm inertial contribution.

For a mass m_i (which can be either a linear as well as torsional element) the two equations of motion are written in the Newton–Euler formulation as:

$$L_i \ddot{\xi}_i + (c_{e_i} + c_{e_{i+1}}) \dot{\xi}_i - c_{e_i} \dot{\xi}_{i-1} - c_{e_{i+1}} \dot{\xi}_{i+1}$$

$$+ (k_{e_i} + k_{e_{i+1}}) \xi_i - k_{e_i} \xi_{i-1} - k_{e_{i+1}} \xi_{i+1} + \Phi_i = 0 \tag{17.5}$$

$$\xi_i \in \{x_i, \varphi_i\}, \ L_i \in \{m_i, J_i\} \quad \text{and} \quad \Phi_i \in \{F_i, M_i\};$$

$$k_{e_i} = k_i k_{i+1} / (k_i + k_{i+1}); \ c_{e_i} = c_i c_{i+1} / (c_i + c_{i+1}); \ i = 1, 2$$

where l_i is the displacement, F_i^m the applied force, k_{ei} the equivalent elasticity and c_{ei} the equivalent damping ratio.

Solution to these equations requires the use of a time marching integration method, based on backward difference formulae with first order approximation in the case of first derivatives, and trapezoidal rule for the second derivatives with respect to time. Thus:

$$\dot{l}_i = (l_i - l_{i-1})/\Delta t; \ \ddot{l}_i = (l_{i+1} - 2l_i + l_{i-1})/\Delta t \tag{17.6}$$

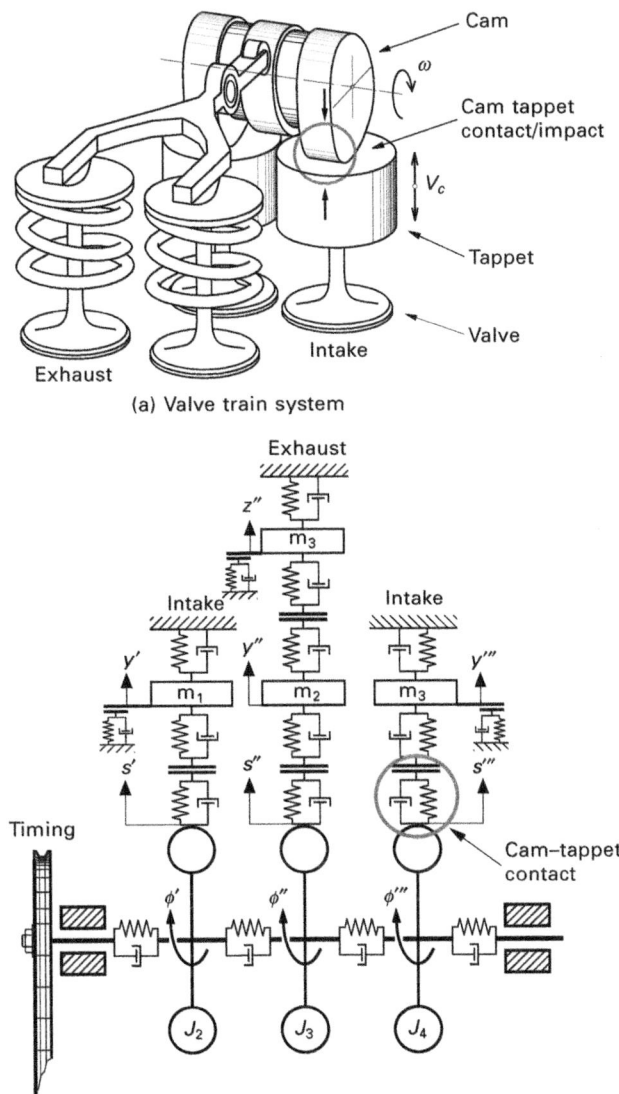

(a) Valve train system

(b) Equivalent system

17.5 Dynamic equivalent model for a valve train system.

17.6   Valve lift and cam profile

To choose a suitable *cam lift*, several possible approaches are commonly used. Figure 17.6 shows some of these possible approaches.

Although practical applications could require specific cam lift profiles, the

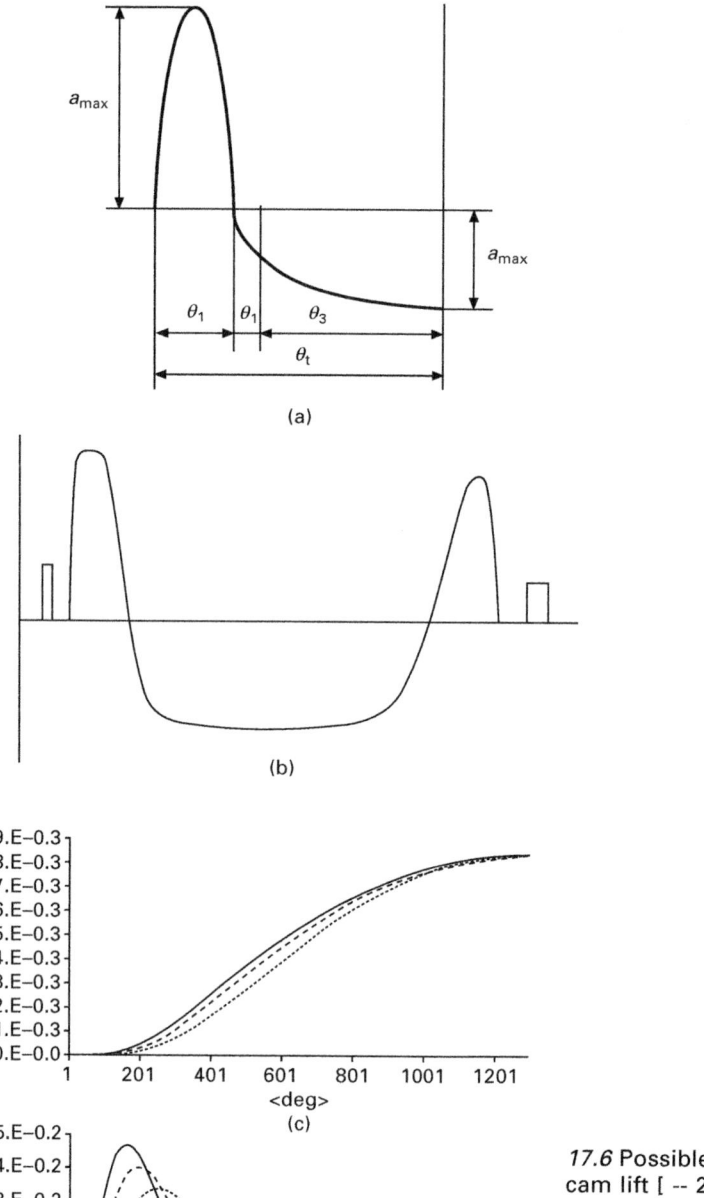

17.6 Possible types of cam lift [-- 2-8-14-20-26; -- 2-10-18-26-34; --- 2-12-22-32-42]; (a) continuous sinusoidal acceleration (b) cubic spline acceleration (c) valve list for three polynomial laws (d) acceleration for three polynomial laws.

current analysis considers a **polynomial cam** profile following the classical approach described by Jensen (1965) and Koster (1975). This was used as a starting point and carefully adopted. The valve lift is considered symmetrical with both events (opening and closing), fully described by a polynomial function of the following form:

$$x = x_0 \left(1 + \sum_i^k C_i \cdot \alpha^{r_i} \right) \quad C_i = \prod_{j=1;\, j \neq i}^{k} r_j \Big/ \prod_{j=1;\, j \neq i}^{k} (r_j - r_i) \qquad (17.7)$$

A large number of possible polynomials have been successively tested and the closest agreement between the proposed valve lift and the experimentally measured one was reached for the following conditions:

$$k = 5;\ r_1 = 2;\ r_2 = 8;\ r_3 = 14;\ r_4 = 20;\ r_5 = 26 \qquad (17.8)$$

A successful valve train mechanism ensures a full closing of the valve during engine operation and avoids impacts due to sharp closing and opening of the valves. Therefore, it is common practice to use a *ramp* to blend in the lift profile into the base circle. Figure 17.7 shows several ramp profiles.

17.7   Cam–tappet tribology: lubricant reaction

At various loads and speeds the cam–tappet regime of lubrication changes from hydrodynamic (on the base circle) to elastohydrodynamic (in the vicinity of the cam nose). In the latter case, the **cam–tappet friction** force has two main components: (i) **boundary friction**, given by the contact between asperities tips, and (ii) **viscous friction**, given by the non-Newtonian behaviour of the oil trapped in the elastohydrodynamic contact. To predict the behaviour of the oil film in the heavily loaded conjunction, the Reynolds equation must be solved. A full solution to this problem is provided in Chapter 16. This chapter considers a set of simplifying assumptions, which increases the speed of the predictions, facilitating its integration into a full dynamic solution for the entire valve train.

The pressure gradient along the length of an equivalent roller is much smaller than the one in the direction of lubricant entrainment: $\partial P/\partial y \ll \partial P/\partial x$. Therefore, the cam–tappet contact can be represented as an infinitely long line contact between an equivalent roller and a semi-infinite elastic half-space at any instant of time. This approach must be considered with care, since it neglects the stress concentrations at the edges of the cam–tappet contact (in the direction of the cam-width), predicting the minimum film thickness in the vicinity of the exit boundary (instead of the sides of the contact as would be expected). Additionally, this approach neglects the complex tappet dynamics and side leakage of the lubricant. Thus, the one-dimensional Reynolds equation becomes:

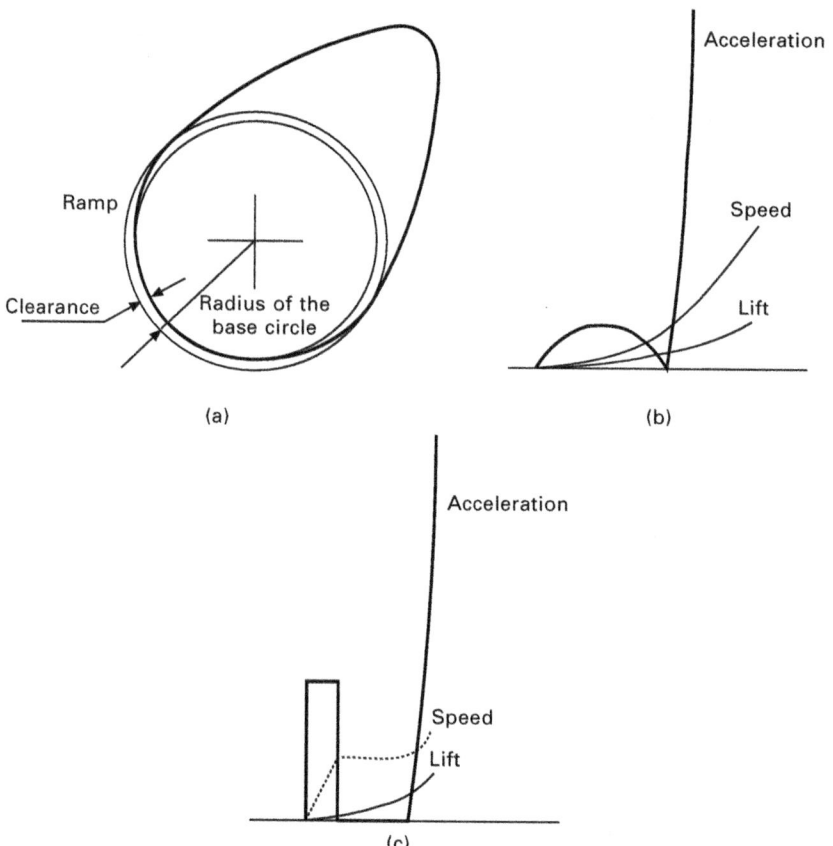

Acceleration

Ramp

Speed

Clearance

Radius of the
base circle

Lift

(a)

(b)

Acceleration

Speed

Lift

(c)

17.7 Possible cam ramps; (a) blending the base cycle into the active
cam profile (b) sinusoidal acceleration (c) constant acceleration.

$$\frac{\partial}{\partial x}\left(\frac{\rho h^3}{\eta}\frac{\partial P}{\partial x}\right) = 12\left[u\frac{\partial(\rho h)}{\partial x} + \frac{\partial(\rho h)}{\partial t}\right] \qquad (17.9)$$

The solution to this one-dimensional equation is fast compared with the
two-dimensional case. However, some practical applications require an even
faster approach. Therefore, there are two possible approaches, both of which
will be described in the following sections. The first one outlines full solution
of equation (17.9), while the second one uses an extrapolated oil film thickness
formula to obtain an approximate value for the contact parameters.

17.7.1 Full solution

The **elastic film shape** is given by:

$$h = h_0 + \delta \qquad (17.10)$$

where h_0 is the instantaneous rigid separation, including the profile of the cam at any location (in the x-direction) and δ is the instantaneous local deflection at any location in the contact. The latter is obtained by solution of the contact elasticity problem as:

$$\delta(x) = \frac{1}{\pi E'} \int_{x^* = x_{inlet}}^{x^* = x_{outlet}} P(x^*) D(x, x^*) \, dx^* \tag{17.11}$$

where $E' = 2[(1 - v_1^2)/E_1 + (1 - v_2^2)/E_2]^{-1}$ is the **reduced Young's modulus of elasticity** of contacting pairs and D is the influence coefficient matrix (see Hamrock, 1994).

The pressure distribution from inlet to outlet must be determined. Therefore, an outlet boundary condition is required. Reynolds boundary condition is implemented as:

$$P|_{x_{inlet}} = 0 \text{ at } x = -\infty \text{(fully flooded), and } P|_{x_{outlet}} = \left.\frac{\partial P}{\partial x}\right|_{x_{outlet}} = 0 \tag{17.12}$$

To solve the Reynolds equation, the rheological state of the lubricant (η, ρ) is required (see also Chapter 5). Viscosity variation with pressure could be taken into account using Roelands equation (Hamrock, 1994):

$$\eta = \eta_0, \exp\left[\ln \eta_0 + 9.67\right]\left[(1 + 5.1 \times 10^{-9} P_h)^z - 1\right]$$

where: $Z = \alpha/(5.1 \times 10^{-9} [\ln \eta_0 + 9.67])$. Density variation with pressure is given by Dowson and Higginson (1966)

$$\rho = \rho_0 \left(\frac{1 + 0.6 P}{1 + 1.7 P}\right)$$

Finally, the lubricant reaction is found as: $F_i = \int_{inlet}^{outlet} P \, dx$.

Kinematics of the contact $(u, \partial h/\partial t)$ are essential for the solution of the Reynolds equation. The speed of entraining motion of the lubricant is the average velocity of the contacting surfaces in the x-direction. Disregarding tappet spin (see Zhu and Taylor, 2001; Teodorescu $et\ al.$, 2003) this becomes:

$$u = [R_c + \omega^{-2}(\partial^2 s/\partial t^2)]\omega/2 \tag{17.13}$$

The elastic squeeze film velocity is: $\partial h/\partial t = \partial h_0/\partial t + \partial \delta/\partial t$ (see equation (17.4)). This shows that the approach/separation of the contacting surfaces takes place by a combination of rigid body motion (the first term, when negative indicates approach of surfaces) and their local rate of deformation (when positive indicates an emerging gap by local deformation of the surfaces). This is easily computed under transient conditions by order approximation as: $\partial h/\partial t = (h^j - h^{j-1})/\Delta t$, where j denotes the time step, and the analysis time step size. Therefore, in the transient analysis $\partial h/\partial t$ represents historical variation in local contact conditions at any location x, and can be termed the *film memory*.

17.7.2 Extrapolated solution

Ideally, the transient lubricated contact condition would be investigated through simultaneous solution of the Reynolds equation with the inclusion of the *elastic squeeze film* effect and the elastic film shape for this finite line concentrated contact conjunction. However, from a practical viewpoint, a realistic solution should be fast. Thus, a simplified, yet sensible and representative, method is required. Use is made of extrapolated oil film formulae for the right contact geometry, in this case for a finite line configuration, in which the width of the cam accounts for the finite length of the contact, and a typical geometry is best described by a 'dog-bone' or a 'dumbbell' shape (see Chapter 16). Equations for such conjunctions have been obtained by Mostofi and Gohar (1983) and Rahnejat (1984) for the study of lubricated gear teeth meshing dynamics and found very good agreement with the experimental findings of Dareing and Johnson (1975).

The extrapolated equation for the central oil film thickness for such a conjunction was derived for combined entraining and squeeze film actions by Mostofi and Gohar (1983) and is given as:

$$h_0^* = 1.67 G^{*0.421} U^{*0.541} W^{*0.059} e^{-96.775 w_s^*} \tag{17.14}$$

The dimensionless parameters on the right-hand side of the above equation relate to *materials parameter*, G^*, *rolling viscosity parameter*, U^*, and *loading parameter*, W^*. The *squeeze–roll ratio* is given by the exponent $w_s^* = (dh_0/dt)/u_e$.

The contact load is the integrated contact pressure distribution, which is the same as the force responsible for the local deformation of the contact. The contact deflection can be approximated by $\delta_0 = W/k$.

17.8   Tribology of rough surfaces

The narrow conjunction of an EHD contact is meshed with a rectangle grid along x–y. Therefore, for an infinitesimal element $dA = dxdy$ the *asperity contact area* and the load carried by the asperities is computed, based upon the model proposed by Greenwood and Tripp (1970–71) as:

$$dA_a = \pi^2 (\zeta \beta \sigma)^2 F_2(\lambda) \, dA$$

and $$dF_a = \frac{8\sqrt{2}}{15} \pi (\zeta \beta \sigma)^2 \sqrt{\frac{\sigma}{\beta}} E^* F_{5/2}(\lambda) \, dA \tag{17.15}$$

The two statistical functions F_2 and $F_{5/2}$ are obtained from:

$$F_n(\lambda) = \frac{1}{\sqrt{2\pi}} \int_\lambda^\infty (s - \lambda)^n e^{-s^{2/2}} ds$$

In order to speed up the numerical predictions, the method proposed by Teodorescu *et al.* (2003) and extended by Teodorescu *et al.* (2004) is adopted. For this, a fifth degree polynomial is fitted to each corresponding statistical function. Figure 17.8 shows the prediction of the fifth degree polynomial versus the values computed, using equation (17.15).

The **boundary friction force** results from the shearing of a very thin film (down to several layers of molecules), which prevails in the contact between the asperity tips during their contact. The friction force is computed as $dF_b = \tau \, dA_a$.

The **non-Newtonian shear stress** is given as a function of the normal load component as (see Evans and Johnson, 1986): $\tau = \tau_0 + m(dF_a/dA_a)$. Therefore, the boundary friction force can be expressed as:

$$dF_b = dA(\zeta\beta\sigma)^2 \left[\tau_0 \pi^2 F_2(\lambda) + m\pi \frac{8\sqrt{2}}{15} \sqrt{\frac{\sigma}{\beta}} E^* F_{5/2}(\lambda) \right]$$

An interesting observation is that the boundary friction force can be expressed as a generic function of λ multiplied with the area dA, thus:

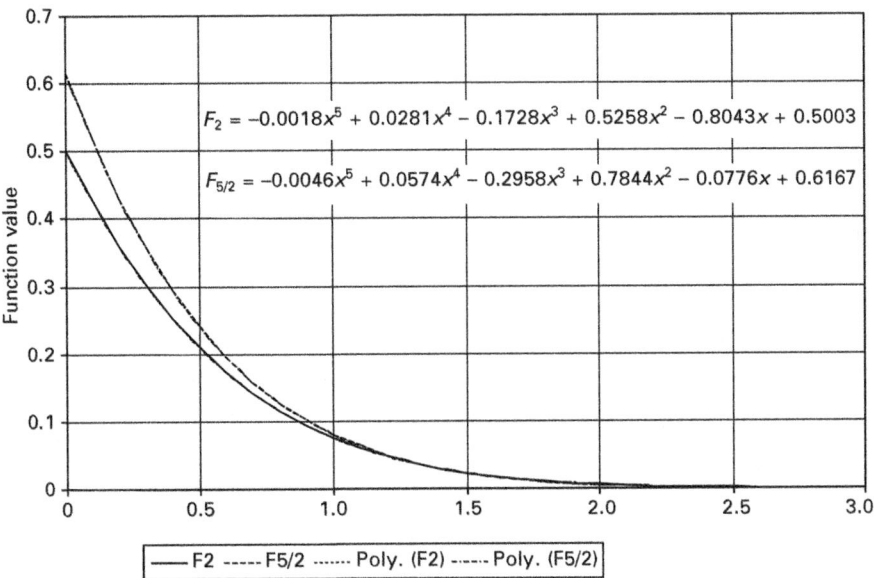

17.8 Characteristics of the statistical functions (after Teodorescu *et al.*, 2004).

$$dF_b = dA \cdot f(\lambda) \tag{17.16}$$

The general form of viscous friction force only takes into account the shear of the lubricant film, which is trapped between the sliding surfaces outside the asperity contact areas as $dF_v = \tau(dA - dA_a)$.

The oil shear stress is computed, considering the combined Newtonian and non-Newtonian behaviours of the lubricant, thus:

$$\tau = \begin{cases} \dfrac{\eta V}{h} \to & \dfrac{\eta V}{h} \leq \tau_0 \approx 2\,\text{MPa} \\[3mm] \tau_0 + \gamma \dfrac{dF - dF_a}{dA} \to & \dfrac{\eta V}{h} \geq \tau_0 \approx 2\,\text{MPa} \end{cases} \tag{17.17}$$

The total friction force, acting on an element dA is computed as the sum of the viscous and boundary friction components as:

$$dF_f = dF_b + dF_v, \; F_f = \int \int dF_f \, dx \, dy \tag{17.18}$$

17.9   Applications

Teodorescu *et al.* (2007) proposed an integrated analysis to investigate the tribology of the cam–tappet interface as well as the friction forces and the subsurface stress field (see Fig. 17.9). Their analysis considers the transient behaviour of the oil film in the most failure-prompt conditions: inlet reversal during wind-up. Under these conditions, the entraining velocity in the conjunction reverses. Two major problems can be noted. Firstly, the inlet reversal means that for a very short period of time there is cessation of entraining motion of the lubricant. Therefore, the first term on the right-hand side in equation (17.9) becomes zero and the only mechanism maintaining a film of fluid is the squeeze effect. Secondly, flow reversal means that the pressure spike (near the exit constriction) suddenly appears at the new inlet position, where the pressure is usually much lower. This is an unstable condition and the pressure spike traverses through the contact. This action creates a transitory pressure wave. Although in the first instance this development might look damaging to the film of lubricant, it in fact has the beneficial effect inducing a film squeeze cavity or *dimple* (see Kushwaha and Rahnejat, 2002). To understand the full effect that the pressure wave propagation has on the contact, the resulting subsurface stress field should be investigated. However, this also requires prediction of the contact friction force. This is obtained with a model similar to that described in Section 17.8. The final subsurface stress field appears quite *distorted*. The travelling waves in the narrow conjunction generates additional islands of stress (isoclines), which can contribute to local fatigue spalls, eventually leading to pits. It was noted

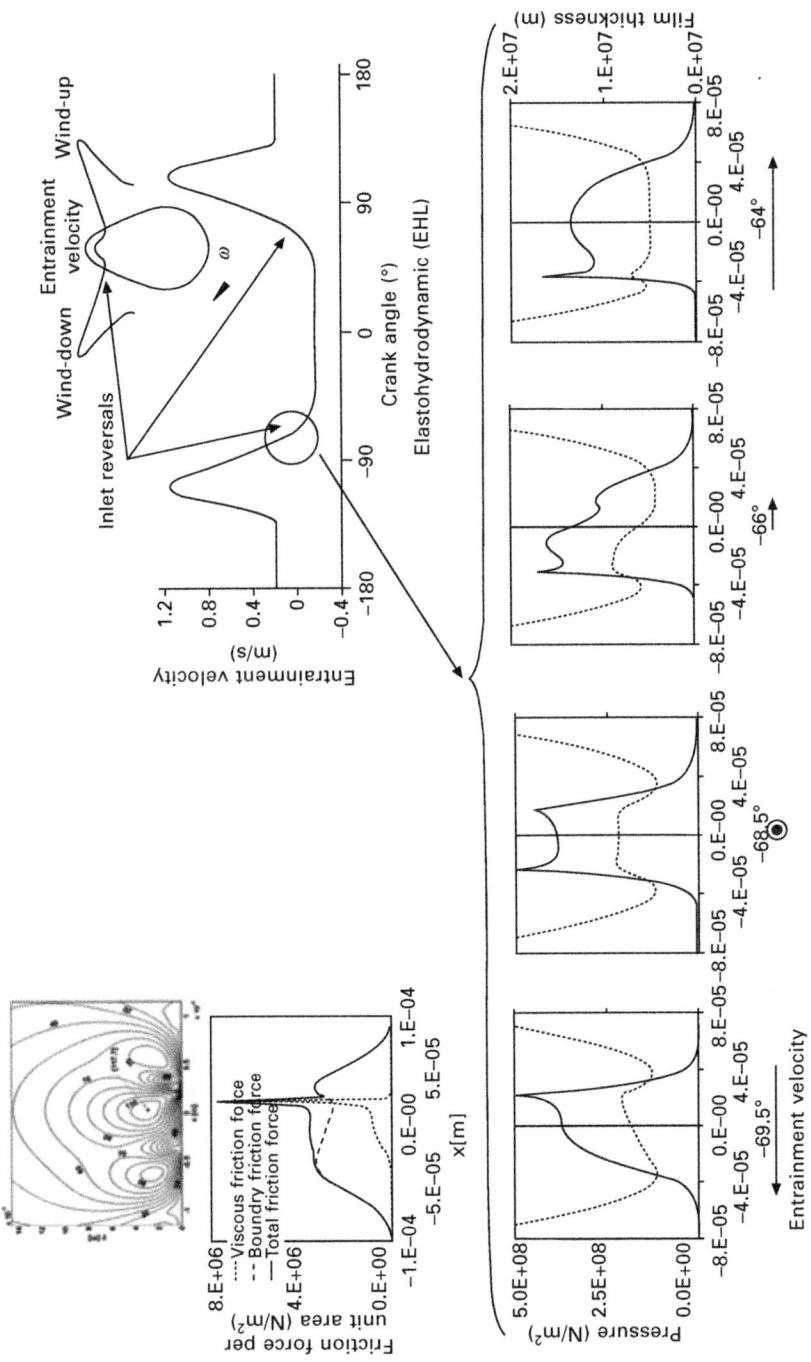

17.9 Transient behaviour of the oil film at inlet reversal during wind-up.

that the worst wear is not encountered on the cam nose (where the curvature radius is the smallest), but in the inlet reversal points. This drawback can be tentatively improved by encouraging **tappet spin**, which can reduce friction (inhibiting chance of localized adhesion, which is not described here).

17.10 References

Dareing, D. W. and Johnson, K. L. (1975) 'Fluid film damping of rolling contact vibrations', *J. Mech. Eng. Sci.*, **17**(4), pp. 214–218.

Dowson, D. and Higginson, G. R. (1966) *Elastohydrodynamic Lubrication: The Fundamentals of Roller and Gear Lubrication*, Pergamon, Oxford.

Evans, C. R. and Johnson, K. L. (1986) 'The rheological properties of elastohydrodynamic lubricants', *Proc. Instn. Mech. Engrs*, **200**(C5), 303–312.

Greenwood, J. A. and Tripp, J. H. (1970–71) 'The Contact of Two Nominally Flat Rough Surfaces.' *Proc. Instn. Mech. Engrs*, **185**(48/71), 625–633.

Hamrock, B. J. (1994) *Fundamentals of Fluid Film Lubrication*, McGraw-Hill International Editions, New York, London, New Mexico.

Jalali-Vahid, D., Rahnejat, H., Jin, Z. M. and Dowson, D. (2001) 'Transient analysis of isothermal elastohydrodynamic circular point contacts', *Proc. IMechE – Part C: J. Mech. Engrs.*, **215**, 1159–1173.

Jensen, P. W. (1965) *Cam Design and Manufacturing*, The Industrial Press, New York.

Koster, M. P. (1975) 'Effect of flexibility of driving shaft on the dynamic behaviour of a cam mechanism', *Trans. ASME, J. Engng for Industry*, pp. 595–602.

Kurisu, T., Hatamura, K. and Omoti, H. (1991) 'A study of jump and bounce in valve train', SAE Technical Paper Series, 910426.

Kushwaha, M. and Rahnejat, H. (2002) 'Transient elastohydrodynamic lubrication of finite line conjunction of cam to follower concentrated contact', *J. Phys. D: Appl. Phys.*, **35**, pp. 2872–2890.

Mostofi, A. and Gohar, R. (1983) 'Elastohydrodynamic lubrication of finite line contacts', *Trans. ASME. J. Lub. Tech.*, **105**, pp. 598–604.

Norton, R. L., Eovaldi, D., Westbrook, J. I. and Stene, R. L. (1999) 'Effect of valve-cam ramps on valve train dynamics., SAE Technical Paper Series, 1999-01-0801.

Rahnejat, H. (1984) 'Influence of vibration on oil film in concentrated contacts', PhD Thesis Imperial College of Science and Technology, University of London.

Roß, J. and Arnold, M. (1993) 'Analysis of dynamic interactions in valve train systems of IC-engines by using a simulation model', SAE Technical Paper Series, 930616.

Schamel, A. R., Hammacher, J. and Utsch, D. (1993) 'Modeling and measurement techniques for valve spring dynamics in high revving internal combustion engines', SAE Technical Paper Series, 930615.

Straton, J. T. and Willermet, P. A. (1983) 'An analysis of valve train friction in terms of lubrication principles', SAE Technical Paper Series, 830165.

Taylor, C. M. (1993) *Engine Tribology*, Elsevier Science Publisher B.V. Amsterdam: 15–51.

Teodorescu, M., Taraza, D., Henein, N. A. and Bryzik, W. (2002) 'Experimental analysis of dynamics and friction in valve train systems', *SAE Transactions – Journal of Engines*, 2002-01-0484, pp. 1027–1037.

Teodorescu, M., Taraza, D., Henein, N. A. and Bryzik, W. (2003) 'Simplified elastohydrodynamic friction model of the cam–tappet contact', *SAE Transactions – Journal of Engines*, 2003-01-0985, pp. 1271–1282.

Teodorescu, M., Balakrishnan, S. and Rahnejat, H. (2004) 'Integrated Tribological Analysis within a Multi- physics Approach to System Dynamics', 31st Leeds–Lyon Symposium – Life cycle Tribological analysis, Leeds, UK.

Teodorescu, M., Kushwaha, M., Rahnejat, H. and Taraza, D. (2005) 'Elastodynamic transient analysis of a four-cylinder valvetrain system with camshaft flexibility', *Proc. IMechE – Part K: J. Multi-Body Dynamics*, **219** (1), pp. 13–25.

Teodorescu, M., Kushwaha, M., Rahnejat, H. and Rothberg, S. (2007) 'Multi-physics analysis of valve train systems: from system level to microscale interactions', *Proc. IMechE – Part K: J. Multi-Body Dynamics*, **221** (3), pp. 349–361.

Uras, H. and Patterson, D. (1983) *Measurement of piston ring assembly friction force instantaneous IMEP method*, SAE Paper 830416 (1983).

Wakuri, Y., Soejima, M., Ejima, Y., Hamatake, T. and Kitahara, T. (1995) 'Studies on friction characteristics of reciprocating engines', SAE Technical Paper Series, 952471.

Zhu, G. and Taylor, C. M. (2001) *Tribological Analysis and Design of Modern Automobile Cam and Follower*, Professional Engineering Publishing Limited, London and Bury St. Edmunds, UK.

17.11   Nomenclature

A	Hertzian contact area (m^2)
A_a	actual asperity contact area (m^2)
E_1, E_2	elastic moduli for cam and follower materials (Pa)
E'	$\dfrac{1}{E'} = \dfrac{1}{2}\left\{\dfrac{1-v^2}{E_1} + \dfrac{1-v^2}{E_2}\right\}$
F_b	boundary friction force [N]
F_f	friction force between cam and tappet [N]
F_v	viscous friction force [N]
$F_n(\lambda)$	$= \dfrac{1}{\sqrt{2\pi}}\displaystyle\int_{\lambda}^{\infty}(s-\lambda)^n e^{-s^{2/2}}\,ds$ ($n = 2$ or $5/2$) statistical functions [–]
G^*	$= \alpha E'$ non-dimensional material parameter [–]
I_R	rocker arm moment of inertia (kg m^2)
$K.E.$	rocker ram kinetic energy (J)
L	cam width (m)
R_b	base circle radius (m)
R_c	instantaneous curvature radius (m)
U^*	$= (\eta_0 u_e)/(E' R_c)$ non-dimensional speed (–)
W	contact load (N)
W^*	$= W/(E' R_c L)$ non-dimensional load (–)
$\overline{W}$	oil film pressure (N/m^2)
W_a	load carried by asperities in the contact area (N)
X, Y	global frame of reference

c_i	damping coefficient (N s/m)
c_{ei}	equivalent damping coefficient (N s/m)
f_i	friction forces associated with mass 'i' (N)
h_0	central oil film thickness (m)
i	order number or compliance identity
k	contact stiffness (N/m)
k_{ei}	equivalent stiffness (N/m)
k_i	stiffness (N/m)
m_1	valve assembly equivalent mass (kg)
m_2	pushrod and tappet equivalent mass (kg)
m	pressure coeff. of the boundary shear strength (–)
r	distance between rocker arm bearing and rocker arm – pushrod contact (m)
s	floor movement (m)
u	speed of entraining motion (m/s)
u_1	instantaneous cam surface speed (m/s)
u_2	surface speed of the flat follower (m/s)
w^*_s	squeeze–roll ratio (–)
y	mass m_1 movement (m)
z	mass m_2 movement (m)
α	piezo-viscosity index (m^2/N)
β	asperity radius of curvature (m)
δ_0	contact deflection (m)
Δt	time increment (s)
γ	rate of change of the shear stress with pressure (–) or phase vector of cylinder firing in cam-star
η_0	atmospheric dynamic viscosity of the oil (Pa s)
λ	$= h_0/\sigma$ film thickness parameter (–)
v	Poisson's ratio (–)
σ	combined surface roughness (m)
τ	shear strength (Pa)
τ_L	limiting shear stress (Pa)
τ_0	Eyring stress of the oil (Pa)
ζ	surface density of asperity peaks (1/m^2)
θ_R	rocker arm oscillation angle (rad)
ω	camshaft angular velocity (rad/s)

Section II.IV

Engine bearings

18

Fundamentals of hydrodynamic journal bearings: an analytical approach

S. BALAKRISHNAN, Mercedes Benz High Performance Engines, UK; C. MCMINN, Ford Motor Company, UK and C. E. BAKER and H. RAHNEJAT, Loughborough University, UK

Abstract: This chapter provides the fundamental aspects of journal bearings. These include hydrodynamic lubrication, friction and heat generation. An analytical approach is highlighted to calculate pressure distribution, lubricant film thickness, viscous friction and generated heat. The method is extended to include the soft elastohydrodynamic (iso-viscous elastic) conditions in overlay/thin shell bearings, using simple numerical analysis. In particular, the case of crankshaft main support bearings is taken into account, including the procedure for determination of bearing loading. A six cylinder engine with four main support bearings is used as an example.

Key words: journal bearings, crankshaft main support bearings, thermo-hydrodynamics, thin shell/overlay bearings, bearing materials, multi-cylinder engines.

18.1 Introduction

Journal bearings are one of the most common types of hydrodynamic bearings. Their primary purpose is to support a rotating shaft. They are used in various subsystems in engines and power trains, for example for support of both crankshaft and camshaft. They are also used in the rocker shaft of rocker-arm valve train systems. The piston pin-bore bearing also acts as a form of journal bearing, so do the big-end bearings, covered in Chapters 19 and 20. Journal bearings usually operate in a hydrodynamic regime of lubrication as the generated pressures are low compared with those experienced by ball and rolling element bearings, gears and cam–follower pairs. Unlike these counterforming pairs, where the area of contact is very small, the journal conforms reasonably well to the bearing bushing (but not completely), allowing a wedge shape to form a film of lubricant, drawn into the contact. Thus, a large area of contact results in generated lubricant pressures which are typically at least an order of magnitude less than those in concentrated counterforming contacts. These are usually from a few to tens of MPa, which are normally insufficient to cause localised deformation

591

of surfaces in contact, unlike elastohydrodynamic conditions in the rolling element bearings and cam–follower pairs (see Chapters 6 and 16). The film thickness is also of the order of a few to several micrometres unlike a few tenths to a couple of micrometres in ball and rolling element bearings.

18.2   Bearing geometry

Figure 18.1 shows a journal of radius R_j within a bush of radius R_b. Under the influence of an applied load and the rotation of the journal relative to the stationary bush, a *hydrodynamic wedge* is formed (see Chapter 5). The lubricant is then drawn into this wedge (as the result of surface velocities) and is pressurised to form a film to carry the applied load. Therefore, the centre of the journal O is at an eccentric position (distance e) from the centre of the bush C. Lubricant pressure distribution forms typically over half of the conjunction between the journal and the bush, but its extent can be a small fraction of this in many cases.

The magnitude of pressures generated depends on the gap between the contiguous surfaces or in other words the film thickness. This is a function of the initial clearance c, the *journal eccentricity*, e, which is a function of

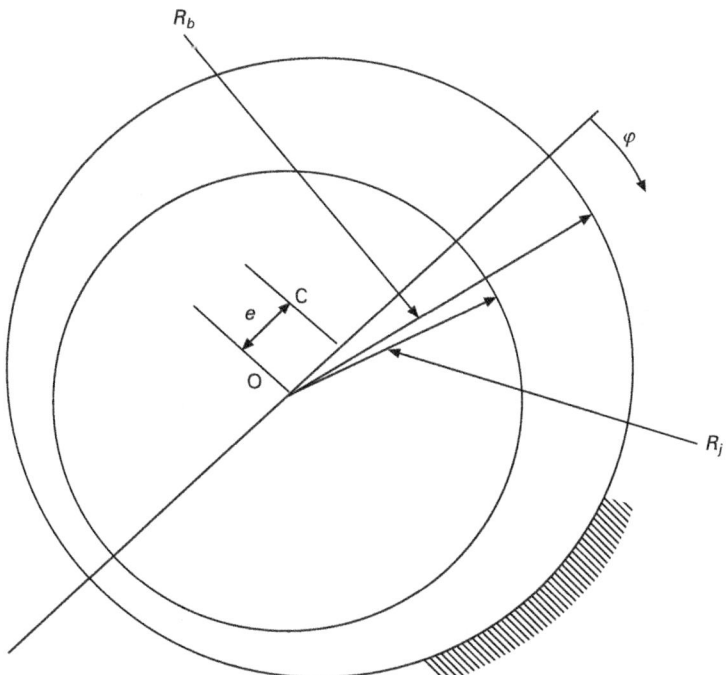

18.1 Geometry of journal bearing.

applied load, and the extent of the film domain. Gohar and Rahnejat (2008) show that:

$$h = c + e \cos \varphi \tag{18.1}$$

where h is the film thickness at any position φ, an angle measured from the line of centres (see Fig. 18.1). Therefore, for the case shown in the figure the minimum film is at $\varphi = \pi$, thus: $h_{min} = c - e$ and the maximum at $\varphi = 0$, $h_{max} = c + e$.

It is, therefore, clear that a sufficient clearance must exist such that: $c > e$. Since e is a function of system dynamics (bearing loading, due to combustion and gravity loads) one can obtain the minimum film thickness. It is also important that c be fairly small to generate high enough pressures to counteract the applied loading. These points are discussed later, but in general manufacture of very closely fitting bearings has cost implications. One important parameter is the c/R_j ratio, which has traditionally been in the range 1/250 to 1/2000.

Another important dimensionless parameter is the *eccentricity ratio*, defined as $\varepsilon = e/c$. Therefore, equation (18.1) is usually given as:

$$h = c \, (1 + \varepsilon \cos \varphi) \tag{18.2}$$

The eccentricity ratio defines the film shape.

18.3   Simple analytical solutions for journal bearings

There are two possible ways of predicting pressure distribution and corresponding film for journal bearings; analytical and numerical. Clearly, one would favour an analytical solution as this is fast and suitable for industrial use. However, this is only reasonable when certain simplifying assumptions can be made. Cameron (1964) and Gohar and Rahnejat (2008) among others show that for certain diameter-to-length ratio of journal bearings D/L ($D = 2R_j$) such analytic solutions are permissible as originally expounded by Sommerfeld. The length of the bearing, L, referred to by some as bearing width is that into the plane of paper in Fig. 18.1.

A *long bearing* is when $L/D \approx 2$ and a *short bearing* when $L/D \approx 0.5$ or 0.33 according to some. The analytical solutions are made as a result of a one-dimensional solution of the Reynolds equation, where for a long bearing $\partial p/\partial Y \approx 0$ and for a short bearing: $\partial p/\partial X \approx 0$ ($X = R_j \varphi$ denotes the direction of lubricant entrainment, if the journal bearing were visualised as *unwrapped*). Thus, for a long bearing, Reynolds equation becomes:

$$\frac{\mathrm{d}}{\mathrm{d}X} \left(h^3 \frac{\mathrm{d}p}{\mathrm{d}X} \right) = 6 \, u \eta_0 \frac{\mathrm{d}h}{\mathrm{d}X} \tag{18.3}$$

where *iso-viscous* condition (no change of viscosity with pressure) has been assumed, and u is the speed of entraining motion: $u = \frac{1}{2} \omega R_j$ (the average speed of contiguous surfaces, where the bush is usually stationary). Furthermore, it is assumed that no side leakage takes place along the bearing length, $v = 0$ in the generalised Reynolds equation and the condition is regarded as steady state, when $\partial h/\partial t = 0$ (see Chapters 5 and 6).

For a short bearing, the Reynolds equation becomes:

$$\frac{d}{dY}\left(h^3 \frac{dp}{dY}\right) = 6 u \eta_0 \frac{dh}{dX} \tag{18.4}$$

with the same assumptions regarding flow, iso-viscous behaviour and steady state conditions.

Another key assumption made thus far is isothermal behaviour, which affects the results quite significantly, as can be seen later. Therefore, analytical solutions can be very approximate as they are usually isothermal, iso-viscous and steady state. In practice, engine journal bearings are subjected to transient conditions due to combustion loading in an engine cycle, as well as cycle-to-cycle variations. Also, heat is generated in bearings due to friction, which cannot be ignored.

Here a solution for short bearings is given to familiarise readers as well as providing an approximate solution which may be used by industrialists in a host of applications that a *short bearing approximation* is deemed sensible, such as for most camshaft journal bearings, most turbocharger bearings, and the centre-bearing in long wheel-base rear wheel driveline systems.

Integrating equation (18.4), noting that $h \neq f(Y)$ and using the boundary conditions: $p = 0$ at $\pm L/2$ (either ends of the bearing length or width):

$$p = 6 u \eta_0 \frac{dh/dX}{h^3} (Y^2 - L^2/4) \tag{18.5}$$

Now letting $X = R_j \varphi$, thus $dX = R_j d\varphi$ and from equation (18.2): $dh/d\varphi = -c\varepsilon R_j \sin \varphi$, hence:

$$p = \frac{6 u \eta_0 c \varepsilon \sin \varphi}{R_j c^3 (1 + \varepsilon \cos \varphi)^3}\left(\frac{L^2}{4} - Y^2\right) \tag{18.6}$$

The maximum pressure and minimum film would be of interest in a design exercise. The minimum film is discussed above. The maximum pressure is when: $dp/d\varphi = 0$ at some circumferential position $\varphi = \varphi_m$. The maximum pressure is:

$$p_m = \frac{6 u \eta_0 \varepsilon}{4 R_j}\left(\frac{L}{c}\right)^2 \frac{\sin \varphi_m}{(1 + \varepsilon \cos \varphi_m)^3} \tag{18.7}$$

where:

$$\varphi_m = \cos^{-1}\left(\frac{1 - \sqrt{1 + 24\varepsilon^2}}{4\varepsilon}\right) \tag{18.8}$$

Note that the position of the maximum pressure is not the same as that of the minimum film thickness.

Remember the corresponding film is given by equation (18.2). Next, one can use these few relationships in the simplest of design procedures to select a suitable journal bearing. This is highlighted below.

18.4   Simple bearing selection

The very few simple relationships described thus far can be used to select a suitable bearing for a given application, if the shaft speed and bearing load or eccentricity are known. For a **crankshaft support bearing** the applied load is due to combustion and inertial force. The crank nominal speed is also a specified parameter (this is subject to fluctuations due to combustion and inertial dynamics, see Rahnejat, 1998). The procedure uses a number of design charts and is usually iterative.

Firstly, one should note that the load carried by the lubricant film is (see Gohar and Rahnejat, 2008):

$$W = \frac{u\eta_0 L^3}{c^2}\frac{\pi\varepsilon}{4(1-\varepsilon^2)^2}\left[\left(\frac{16}{\pi^2}-1\right)\varepsilon^2 + 1\right]^{1/2} \tag{18.9}$$

This is a relationship between load and eccentricity. A more common form is to express this equation in terms of the **Sommerfeld number** S. Since $u = \pi D N$, N being rev/s, then:

$$S = \frac{W}{NDL\eta_0}\left(\frac{c}{R}\right)^2 \tag{18.10}$$

Replacing for W from (18.9) yields:

$$S = \left(\frac{L}{D}\right)^2\frac{\pi^2\varepsilon}{(1-\varepsilon^2)^2}(0.62\,\varepsilon^2 + 1)^{1/2} \tag{18.11}$$

The relationship between Sommerfeld number and eccentricity ratio for finite length bearings which are much more common as in the case of the crankshaft deviates from the above. Moreover, the eccentricity ratio changes dynamically. Fortunately, a large number of full solutions of Reynolds equation for various L/D ratios have resulted in charts of the form shown in Fig. 18.2. In the figure the Sommerfeld number is the original definition

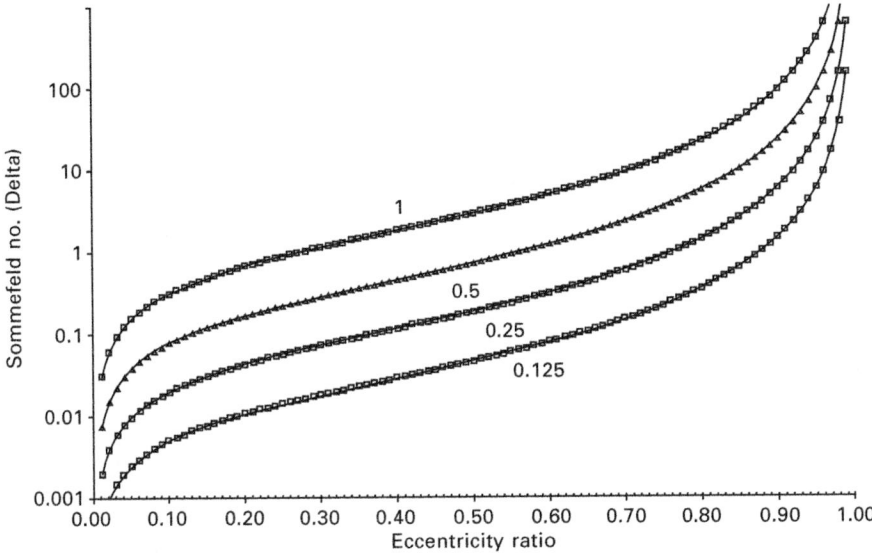

18.2 Sommerfeld chart for various *L/D* ratios.

Δ in terms of u, the speed of lubricant entrainment into the contact, being half the surface speed of the journal for a stationary bushing:

$$\Delta = \frac{W/L}{u\eta_0}\left(\frac{c}{R_j}\right)^2 \qquad (18.12)$$

Now for a chosen L/D ratio and for a given maximum inertial and combustion loads (with an appropriate measure of safety) a particular oil (viscosity) and crankshaft speed, the value of ε can be obtained from the chart. Maximum pressure and minimum film thickness can also be obtained from equations (18.7), using (18.8) and (18.2) (let $\varphi = \pi$) respectively.

Two further issues should be taken into account for this simple isothermal analysis. Firstly, the viscosity of lubricant increases with pressure, thus it may be necessary to adjust for this, using: $\eta = \eta_0 e^{\alpha p}$ (this equation is due to Barus, 1893, and is applicable for pressures up to several tens of MPa). This means that as the pressure is found the whole procedure should be repeated until the change in pressure is within a specified limit. However, for most journal bearing applications pressures are a few MPa, while pressure–viscosity coefficient $\alpha' \approx 10^{-8}\,\mathrm{Pa}^{-1}$, thus: $\alpha'\, p$ is usually quite small and one may discount the effect of pressure on viscosity in the first instance. Secondly, once the final solution by adjusting for viscosity is obtained, it is necessary to check that no direct metal-to-metal contact occurs. To ascertain this, one should have at least the most basic idea of surface topography. If the root mean square roughness of the journal and the bush is: $\sigma_{rms} = (Ra_j^2 + Ra_b^2)^{1/2}$, then the **Stribeck oil film parameter** is defined as: $\lambda = h_{min}/\sigma_{rms}$.

Stribeck (1907) noted that if $\lambda < 3$, then metal-to-metal contact is likely and the conditions are particularly alarming if this ratio is less than unity (see Chapter 3). This may mean that the c/R_j ratio chosen is inappropriate and should be revised. Providing that the value of R_j is fixed, then a number of alternatives exist. Firstly, one can increase the clearance c. However, this can lead to loss of pressure and hydrodynamic wedge (which can lead to a number of problems such as oil whirl, see Chapter 20). Secondly, one can use a bush with a lower modulus of elasticity or with relatively soft coating/overlay or a thinner shell wall thickness that would elastically deflect under load, creating an additional gap for the lubricant to occupy (increasing the film thickness). This has been the trend in the past couple of decades. The resulting journal bearings are known as **thin shell bearings**.

Referring to equation (18.12) it is clear that the applied load W is required to obtain the Sommerfeld number. This load is as the result of combustion loading as well as inertial forces acting, which alter on the various crankshaft supporting bearings during the engine cycle. Chapter 20 provides these forces, applied to the big-end bearing. The procedure to obtain the applied loads on support bearings is briefly highlighted below.

18.5   Determination of bearing loads

The procedure to obtain bearing loads, critical in predicting lubricant film thickness and thus the tribological conditions, is quite tedious in real engines, because of a host of issues. These include engine firing order, variations in combustion pressure within a cylinder and between the different cylinders, interactions between inertial loading and combustion force and bearing direct loading as well as moment contributions. A convenient point to start this rather lengthy procedure is to consider bearing loading as a result of a single cylinder. The reciprocating inertial translational force acting at the piston pin is given as (Rahnejat, 1998) (see Fig. 18.3):

$$F_{y'} = m_{y'}\omega^2 r \left[\cos\theta + \frac{r}{l}\cos 2\theta \right] \tag{18.13}$$

where y' denotes the translational motion along the axis of the cylinder (see Fig. 18.3), $m_{y'}$ is the effective mass in translation; that of the piston, piston pin and the proportion of the connecting rod in translational motion. Calculation of this may be found in any standard text on engine dynamics. Equation (18.13) only takes into account the primary and the first harmonic of inertial imbalance (i.e. ω and 2ω), since $\theta = \omega t$. In fact many higher order harmonics also exist. However, their effects may be neglected as their multiplying term is the ascending orders of the ratio r/l which diminish as this ratio is usually a fraction of unity.

The lateral inertial force is:

$$F_{x'} = m_{x'} \omega^2 r \qquad (18.14)$$

where subscript x' denotes the lateral direction at the piston pin perpendicular to the cylinder axis. The mass $m_{x'}$ includes the proportion of the mass of the connecting rod in pure rotation.

The total loading at any instant of time is that of inertial force and the gas force on the piston pin:

$$F_g = \pi p_g (D_b/2)^2 \qquad (18.15)$$

where D_b is the nominal bore diameter (note that the bore is actually out-of-round).

Figure 18.3 shows three co-ordinate frames, (x', y', z') related to the cylinder, (x, y, z) attached to the connecting rod at the piston pin and (x'', y'', z'') attached to the crank-pin. The z, z' and z'' axes are all co-directed. The support bearing loads are required for the co-ordinate frame (x'', y'', z''). These forces are initiated by the combustion gas force and the inertial imbalance acting on the piston pin and transmitted by the connecting rod to the crank-pin. The components of inertial and gas forces acting on the connecting rod bearing are those given above multiplied by the secant of its articulation angle Φ. This takes into account the connecting rod obliquity. The angle Φ is that between the connecting rod centre line and the axis of the cylinder (i.e. the y' direction) (Rahnejat, 1998):

$$\Phi = \sin^{-1}\left(\frac{r}{l}\sin\theta\right) \qquad (18.16)$$

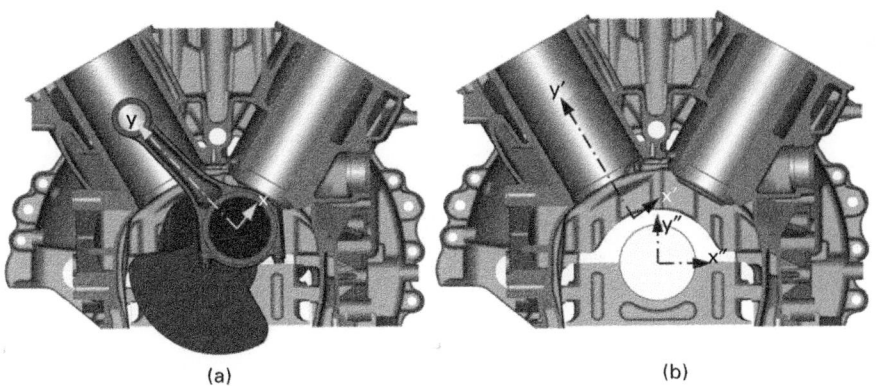

(a)   (b)

18.3 Co-ordinate directions for calculation of nearing loads (a) for big-end bearings (b) for piston pin and main bearings.

Additionally, the lateral inertial force (the rotational imbalance) in equation (18.14) also contributes to the total loading, thus:

$$F_y = (F_{y'} + F_g) \sec \Phi + F_{x'} \csc \Phi \tag{18.17}$$

Also in the x-direction: $F_x = F_{x'} \sin \theta$. The relationships between the above-mentioned frames of reference are shown in Fig. 18.4(a) and (b).

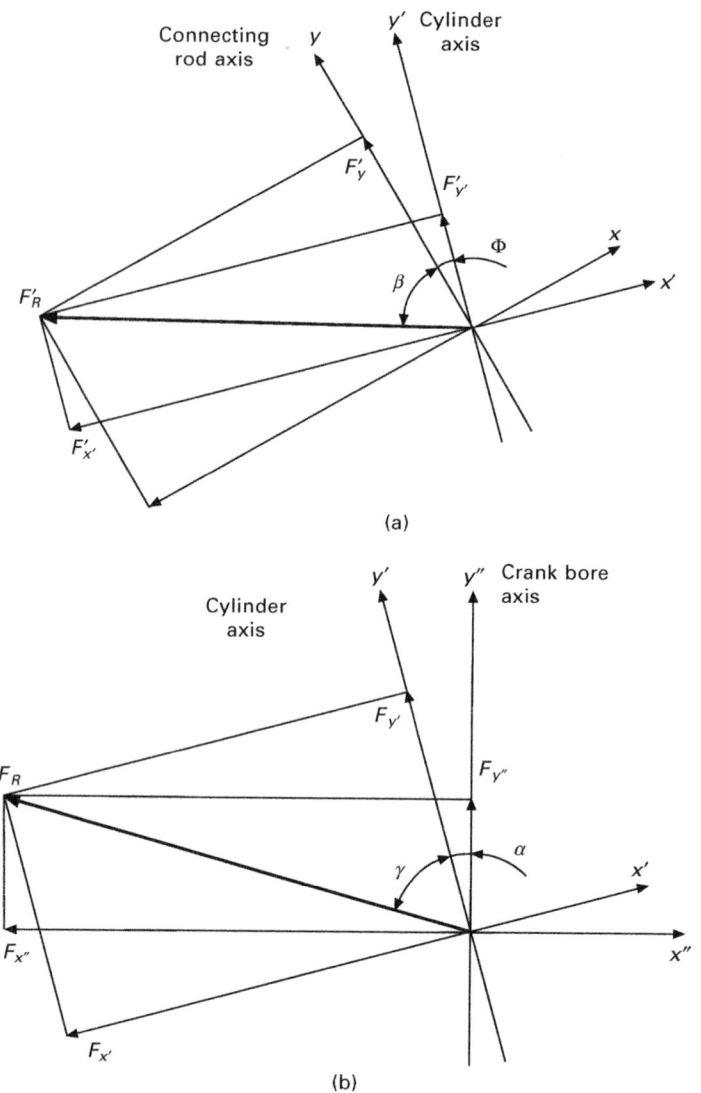

(a)

(b)

18.4 Relationships between the co-ordinate sets attached to piston, piston pin and the crank-pin.

Now the connecting rod bearing loads should be transformed to the crank bore axis. Initially, undertaking this transformation for the translational inertial force $F_{y'}$ only, the contributions to applied main support journal bearings are obtained, also taking into account the engine V-block angle; α (± sign representing left and right banks of the V-block). Thus:

$$F_{x''} = -\frac{F_{y'}\sin(\gamma \pm \alpha)}{\cos\gamma} \tag{18.18}$$

where $\gamma = -\tan^{-1}(F_x/F_y)$ is the crank pin load angle with respect to the cylinder axis. In the case studied here: $\alpha = \pi/6$. Also:

$$F_{y''} = \frac{F_{y'}\cos(\gamma \pm \alpha)}{\cos\gamma} \tag{18.19}$$

Similarly, one can state:

$$F_{x'} = -\frac{F_y\sin(\beta + \Phi)}{\cos\beta} \tag{18.20}$$

and:

$$F_{y'} = \frac{F_y\cos(\beta + \Phi)}{\cos\beta} \tag{18.21}$$

where: $\beta = -\tan^{-1}(F_x/F_y)$ (see Fig. 18.4(a)).

Using the above procedure the main bearing reaction forces with respect to the crank bore axes; x'', y'', resulting directly from the connecting rod bearing loads can be determined for the left and right-hand cylinder banks. Figure 18.5 shows these main support bearing reactions (opposing the calculated loading). Of course, it is clear that this loading diagram is rather simplistic, because it does not take into account the engine firing order and moment loading contributions. Nevertheless, the simple procedure used thus far indicates that bearing loading is not as simple as the approaches often used in bearing lubrication analyses, which are only applicable to the connecting rod (big-end) bearings (see Chapter 20). The results shown in Fig. 18.5 correspond to a V6 engine (LH and RH banks), where $\alpha = \pi/6$. Note the anti-symmetry in the bearing loading in V-engines.

To obtain the correct bearing loading a number of actions are required. Firstly, refer back to equation (18.15) to note that the combustion pressure p is not a constant. It varies with time. It also varies with engine rpm. Thus, the measured combustion pressure is required against crank angle for all the cylinders. Thus, the instantaneous load vector per cylinder can be determined. Secondly, it is also necessary to implement the engine firing order (see typical firing orders in Rahnejat, 1998). This means that the combustion loading

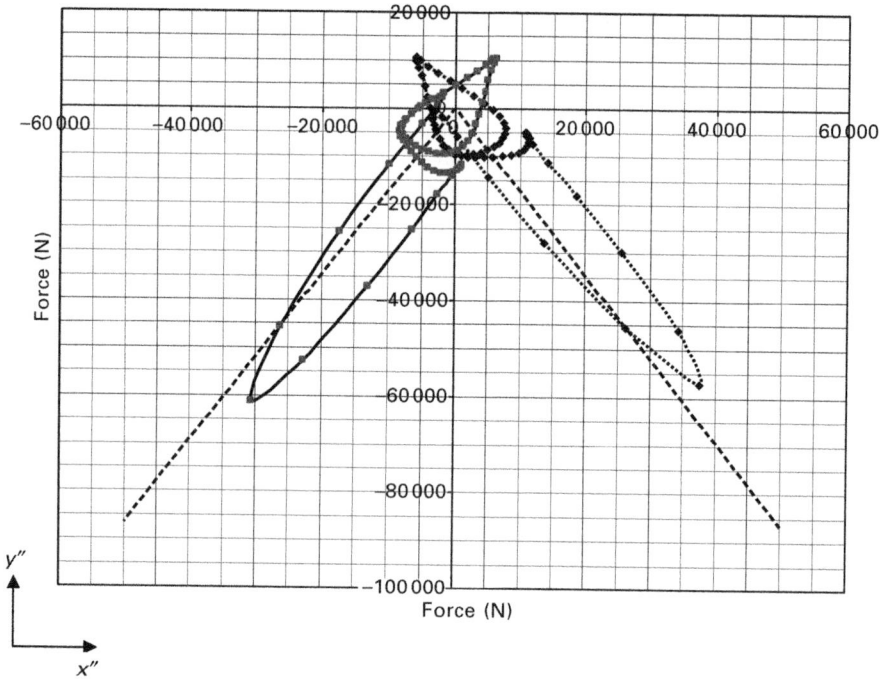

18.5 Main bearing reactions.

contributions for all the cylinders at any instant of time are known, as well
as the inertial forces, where $\theta = \omega t + \zeta_j$ for the cylinder j in all the equations
stated above. For a six cylinder engine, as an example,

$$\{\zeta_j\}_{j=1 \rightarrow 6} = \{0,\ 2\pi/3,\ 4\pi/3,\ 2\pi,\ 8\pi/3,\ 10\pi/3\}^{\mathrm{T}} \tag{18.22}$$

with a firing order 1-4-2-5-3-6. Thirdly, the engine is modelled as a series of
equivalent masses; those in pure translation or rotation (as described above)
and the crankshaft as a series of pins, webs and counter-balance masses.
These elements contribute to bearing loading $F_{x''}$ or $F_{y''}$ or both. Finally, it is
important to add the effect of moment loading on the main support bearings,
remote from the cylinders' locations (Fig. 18.6 shows an example of a six-
cylinder engine with four main support bearings). The results in Fig. 18.5
correspond to this engine configuration. Taking into account all the above
noted considerations for the engine depicted in Fig. 18.6, the loading for the
four main crankshaft support bearings are shown in Fig. 18.7.

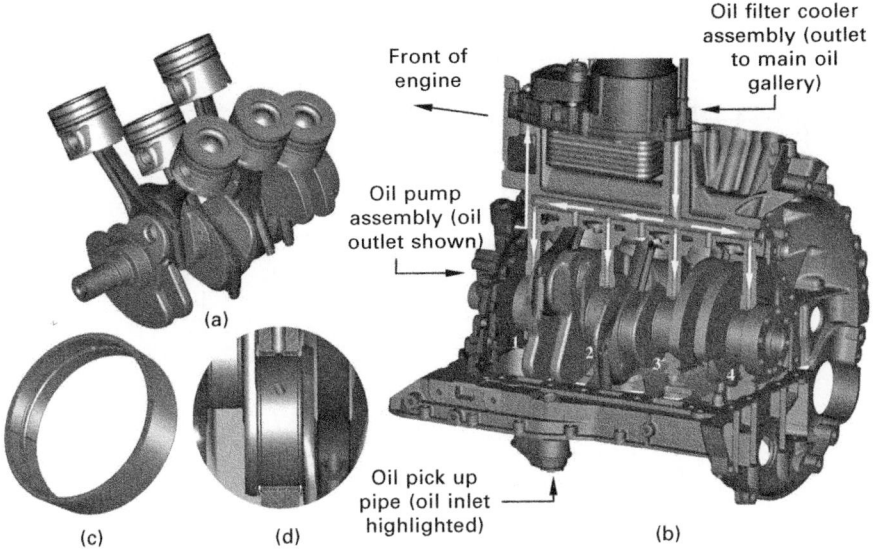

Oil filter cooler
assembly (outlet
to main oil
gallery)

Front of
engine

Oil pump
assembly (oil
outlet shown)

(a)

Oil pick up
pipe (oil inlet
highlighted)

(c)   (d)   (b)

18.6 A typical V6 engine.

The characteristic shapes of the main polar load diagrams can be explained with respect to the engine geometry. The main bearing #1 load increases sharply as cylinder #1 fires at 0° crank angle. The centre of pin journal #1 is closest to main journal #1 and hence the significant proportion of moment loading is supported by journal #1. At 120° crank angle the pin #4 is at the top dead centre (TDC), but a much lower proportion of the firing force is supported at journal #1 because of small moment loading contribution there. In the same manner the load share of all journals are related to their respective peak cylinder pressures at a given instant of time. The cylinder arrangement and the v-angle of the engine is evident with load vectors tending towards the $-x''$, $-y''$ quadrant corresponding to the left-hand bank of cylinders 4, 5 and 6 with the load vectors tending towards the x'', $-y''$ quadrant, corresponding to the right-hand bank of cylinders 1, 2 and 3. Note that the magnitudes of the main bearing loads are not just as the result of the combustion loading, but also the inertial loads through the piston-connecting rod subsystems and the centripetal forces due to the masses of the crankshaft elements, as described above. Finally, for each main crankshaft bearing the resultant load is calculated from the polar loading plots at any given engine speed.

18.6   Thin shell or soft overlay bearings

So far hydrodynamic journal bearings have been dealt with here. Large journal bearings in power plant applications have thick-walled bushing. Relatively low hydrodynamic pressures are insufficient to cause contact deformation,

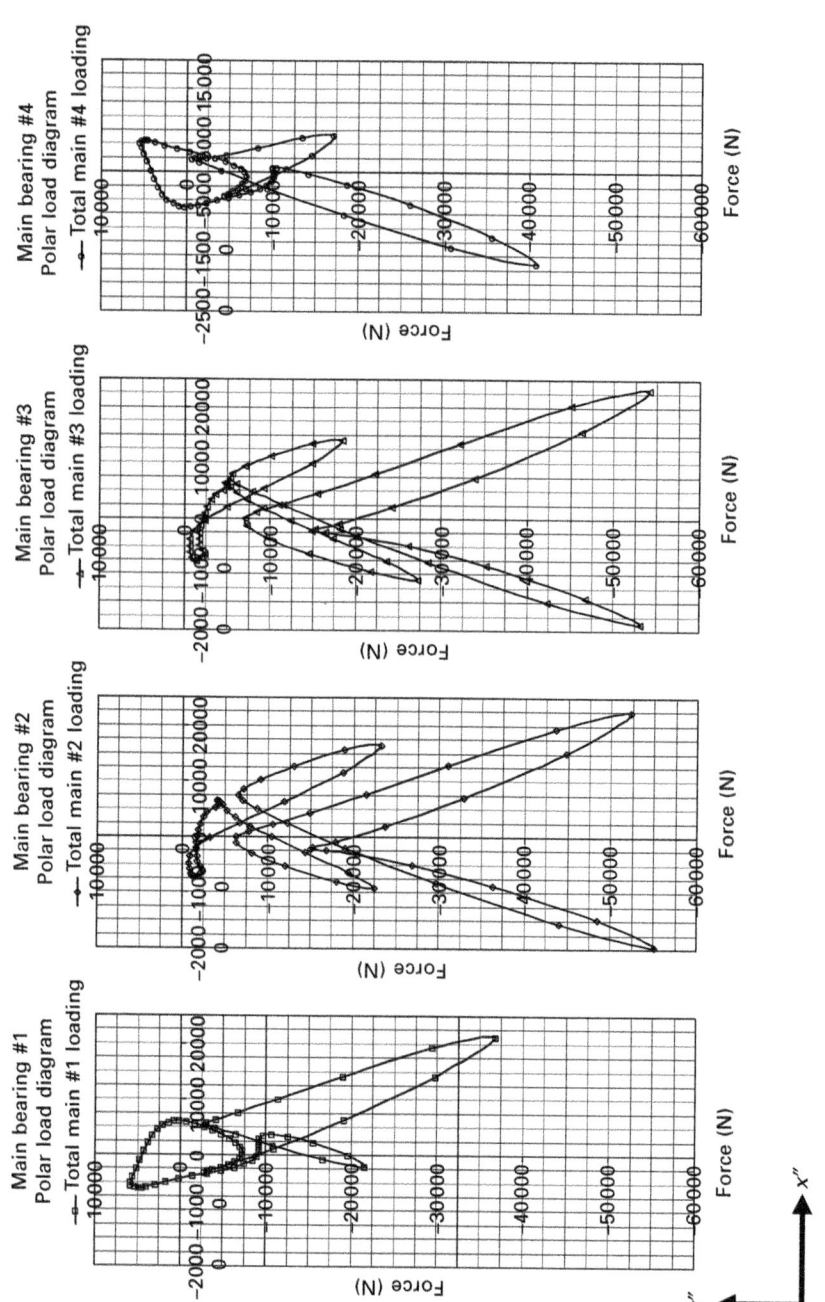

18.7 Polar load diagrams for the crankshaft support bearings with respect to crank bore axis at 4000 rpm with full load (maximum).

which can enhance the film thickness. Thus, in automobile engine bearings an overlay of materials of lower elastic modulus have been used, laid upon the steel substrate (see Fig. 18.8). Throughout the years such materials for *soft bearing overlay* have included aluminium alloys, silver, brass, lead or tin-based alloys such as Babbitt or a combination of such layers. The overall layer has a typical thickness of several micrometres to tenths of a millimetre. The generated pressures may now be sufficiently high to cause deformation of the bearing bush layer or in general as casted layer on a steel backing.

Hamrock *et al.* (2004), Bhushan (1991) and Booser and Wilcock (1957) state that no single material is available that can satisfy the requirements of the perfect journal bearing. The conflicting macroscopic as well as the intrinsic material properties of the bearing material are dependent upon the intended application for a bearing. These macroscopic bearing material properties are defined as follows.

18.6.1 Compatibility

Although a hydrodynamic bearing is separated by a lubricant film, there will inevitably be occasions when the shaft and the bearing bushing surfaces come into direct contact. This occurs typically under engine stop–start conditions or due to geometrical deviations in the journal profile, asperity conditions during the running in of an engine and when there is a reduction in lubricant viscosity (predominantly due to temperature, built-up of soot and/or fuel dilution). Under these conditions the breakdown of the oil film leads to localised heating at the asperities, which when significant enough causes the two materials to cold weld. Wear of the bearing is a function of the *compatibility* of the mating materials and their resistance to cold weld. Messler (1999) states that *cold welding* is a function of both surface materials' mutual solid solubility which is determined at least partially for unlimited solubility by the Hume–Rothery rules (Askeland and Phulé, 2006):

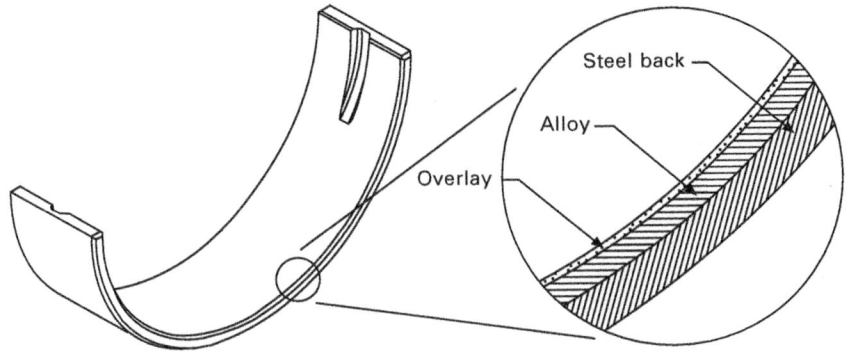

18.8 Typical engine bearing bushing construction.

- The atoms or ions of each material must be of similar size, with no more than a 15% difference in atomic radius in order to minimise lattice strain.
- The materials must have the same crystalline structure.
- The atoms or ions must have the same valency; otherwise any valent electron difference encourages the formation of compounds rather than solutions.
- The atoms must have approximately the same electronegativity, otherwise again compounds can be formed.

Surface energy (see Chapter 3) of the two mating materials is a further atomic/molecular property which leads to materials' ability to bond together. The higher the surface energy, the stronger the bonds and a greater potential for bonding between the two surfaces. Pure iron has a **surface energy** of 1700 mN/m and aluminium 900 mN/m, which are two of the highest values for the materials considered. Lead, in comparison, has a surface energy of only 450 mN/m and has the added benefit of being completely insoluble in iron. Surface energy can be reduced by lubrication, because that of most liquids is typically two orders of magnitude lower than that for solids (see Chapter 3). Non-polar lubricants have even lower surface energies than the polar lubricants, although this may negatively affect the formation of boundary films from the lubricant additives (Hamrock *et al.*, 2004).

18.6.2 Embeddability

Contamination in the bearing clearance can be caused by foreign particles, either intrinsically found in the engine from the manufacturing process, or from assembly operations or the wear debris transported by the lubricant. The bearing material is required to accommodate these contaminants or at least be tolerant of the scoring damage that they may cause. Failure of the bearing would occur due to oil film breakdown and/or localised asperity contact around the contamination or at the ploughed surfaces, caused by the contaminant (the so-called third body). **Embeddability** is the bearing's ability to absorb contamination and has been shown, together with conformability to have a direct relationship with the hardness and Young's modulus of a material (Booser and Wilcock, 1957).

18.6.3 Conformability

Conformability is a measure of the bearing's ability to conform to relatively macroscopic deviations from the perfect journal geometry. The contributing factors are the misalignment of the shaft relative to the bearing bushing, out-of-roundness of crankshaft journal dimensions, irregularities in the engine block geometry and the crank journal profile. Failure of a bearing to conform to the bushing may lead to localised contact stresses and high levels of seizure or wear. Shear strength of a material is a measure of its response to shearing strains and describes the materials resistance to wear during asperity contact. Shear modulus and bulk modulus have a direct relationship with the conformability of a material.

18.6.4 Corrosion resistance

The bearing material must be able to withstand and resist chemical attack by the lubricant oxidation by-products generated when the lubricant begins to break down or from any other contaminants in the oil, caused by the combustion process (i.e. soot or fuel dilution). Oxidation inhibitors in the oil need to be reviewed to ensure compatibility with the bearing material. The effects of oil under high temperature and pressure must also be considered although this is usually only achieved through engine testing. Typical corrosive products are alcohols, aliphatic and aromatic hydrocarbons, ethers, and weak acids and alkalis.

18.6.5 Fatigue resistance

Fatigue resistance is a measure of the bearing's strength especially with regard to combustion engines, where the loading is cyclic in direction as well as its intensity. Fatigue failure of the bearing, referred to as *fatigue spalling* is usually initiated in the depth of the material by subsurface stress field, quite close to the contacting surface (Gohar and Rahnejat, 2008). It can grow (propagate) to the surface in the form of a micro-spall or crack or a pit. Localised inelastic subsurface deformations such as cracks can be arrested by high contact pressures, acting in compression. However, the generated pressures, being rather low to medium in journal bearings, are usually insufficient and thus small cracks in the bearing surface can readily propagate throughout the bearing material. Where cracks interact and the crack geometry encompasses a now isolated piece of bearing material, it is likely that the bearing material would break up, disrupt the oil film in the region of highest load and lowest oil film thickness. This would cause bearing seizure due to the mechanisms described under contamination (see above). Furthermore, the detached piece of material is more likely to bond to the bearing or shaft under pressure.

18.6.6 Dimensional and thermal stability

Thermal characteristics of the bearing material concern heat dissipation and thermal distortion under the relatively high temperature operating conditions of the engine and under occurrence of intermittent boundary lubrication regimes. The bearing material must have high enough thermal conductivity to ensure maximum heat dissipation under unfavourable running conditions. Geometrical stability is required at high temperatures so as not to diminish the bearing clearance, which can exacerbate asperity interactions. Hence, the coefficient of thermal expansion should be sufficiently low for all the bearing materials. Specific heat capacity is the measure of the heat required to increase the temperature of a body through certain temperature rise. Specific heat capacity for bearing materials is required to be high to avoid localised heating and degradation of the material, leading to seizure. Melting point temperatures are important due to the relatively high expected operating temperatures.

Specific material mechanical properties which have been identified, relating to the macroscopic bearing property requirements stated above, are tensile strength, Young's modulus, shear modulus, bulk modulus and hardness, as well as affiliated properties such as Poisson's ratio and yield/proof stress. The conflicting requirements are for the materials to have low values of these mechanical properties for conformability and embeddability characteristics, whereas high values of mechanical properties are required for compatibility (wear), fatigue resistance (strength) and dimensional stability.

The overlay (or plate) material is mostly trimetal to add a soft deformable layer to the otherwise hard bimetal. The overlay is thin enough so that its strength is not compromised. The load is supported by the substrate alloy which in turn is supported by the steel backing material. Soft thin coatings transmit the high contact stresses to the higher elastic modulus substrate as shown by Teodorescu et al. (2009). These overlays also provide for better contact surface deformation which enhances the thickness of the lubricant film as shown by Gohar and Rahnejat (2008). The enhanced film thickness, referred to as the elastic film becomes:

$$h = c(1 + \varepsilon \cos \varphi) + \delta \qquad (18.23)$$

where $\delta = f(\varphi)$ within the contact and should be determined. At any location within the contact δ is function of pressures generated throughout the loaded zone. This is explained in Chapter 6 for elastohydrodynamics. However, for softer materials the effect of pressure directly above the location where δ is to be evaluated dominates. Thus, a simplified method is used to calculate deflection due to the contact pressure element there. This method is referred to as the **column method**. Rahnejat (2000) and Gohar and Rahnejat (2008) describe the method, which results in:

$$\delta = \frac{(1 - 2\upsilon)(1 + \upsilon)d}{E(1 - \upsilon)} p \qquad (18.24)$$

Now equations (18.23) and (18.24) can be solved simultaneously with

Reynolds equation. Assuming no side-leakage, but taking into account the history of film thickness variation (i.e. a transient solution), for a finite width bearing:

$$\frac{\partial}{\partial X}\left(h^3 \frac{\partial p}{\partial X}\right) + \frac{\partial}{\partial Y}\left(h^3 \frac{\partial p}{\partial Y}\right) = 12\eta_0 \left(u \frac{\partial h}{\partial X} + \frac{\partial h}{\partial t}\right) \qquad (18.25)$$

This form of the Reynolds equation is for the iso-viscous condition.

Note that as first order approximation:

$$\frac{\partial h}{\partial t} = \frac{h_i - h_{i-1}}{\Delta t}$$

It is clear that the solution using this approach is suitable for *iso-viscous elastic* condition, which is often referred to as *soft elastohydrodynamics*. It is also clear that simultaneous solution of equations (18.23)–(18.25), for which the unknowns are h, δ and p would be necessarily numerical. However, Gohar and Rahnejat (2008) provide a simpler semi-analytical solution for a short-width bearing under steady state condition (i.e. $\partial h/\partial t = 0$), using equations (18.23), (18.24), (18.7) and (18.8) for values of ε obtained by a dynamics analysis. This approach can be regarded as *quasi-static* (as dynamics and tribology are not integrated and simultaneous).

Before proceeding to provide some representative results of the transient analysis, involving simultaneous solution of equations (18.23)–(18.25), it is appropriate to note that thermal effects play a very significant role.

18.7   Thermo-hydrodynamics

The approaches described thus far do not take into account the effect of temperature on lubricant viscosity. This reduces the film thickness and generates higher pressures. One advantage is increased eccentricity ratio and enhanced load-carrying capacity. On the other hand, the reduced film thickness can result in a mixed regime of lubrication, thus the need for application of soft overlay as described above becomes quite important.

Here heat is assumed to be generated by viscous shear of the lubricant in the contact conjunction only. In practice, quite thin films can be commonplace, thus viscous friction is supplemented by asperity friction, caused by their adhesion or ploughing action. This is the reason for care to be taken in the choice of overlay bearing materials, as discussed above. With thin films and increased friction, the top layer is traditionally made of lead, but similar soft alloys or polymers are now being used. These undergo plastic deformation as a soft layer. This effect also contributes to the generation of heat.

In the analysis presented here only viscous shear is taken into account. Viscous friction is caused by the pressure gradient as well as the main contributory source, being due to contact velocities (the Couette flow). The derivation of viscous shear is provided by Gohar and Rahnejat (2008), yielding:

$$F_f = \pm \frac{c\varepsilon W}{2R_j} \sin\psi + \frac{2\pi\eta u R_j L}{c(1-\varepsilon^2)^{1/2}}$$

(18.26)

where:

$$\psi = \tan^{-1}\left\{\frac{\pi\sqrt{1-\varepsilon^2}}{4\varepsilon}\right\}$$

is known as the attitude angle, being the angle that the resultant bearing reaction makes with the line of centres in Fig. 18.1.

The coefficient of friction is obtained as:

$$\mu = \frac{F_f}{W} = \frac{c\varepsilon}{2R_j} \sin\psi + \frac{2\pi\eta u R_j L}{Wc(1-\varepsilon^2)^{1/2}}$$

(18.27)

Using the Sommerfeld number, S and replacing for W:

$$\mu\frac{R_j}{c} = \mu^* = \frac{\varepsilon}{2} \sin\psi + \frac{2\pi^2}{(1-\varepsilon^2)^{1/2}}\left(\frac{1}{S}\right)$$

(18.28)

To obtain the **effective viscosity** of the lubricant due to heat generated, one may assume that the generated heat is mainly convected away from the conjunction by the flow of the lubricant. This assumption renders an analytical solution and is reasonable in most applications of journal bearings where the film thickness is several micrometres thick. Since in many engine bearings film thickness can become quite thin, at least at some times during the engine operation, then a factor k, less than unity, may be used to recognise that some heat will be lost by conduction through the bounding surfaces. Thus, the frictional power due to the viscous action of the fluid is:

$$P = kF_f u$$

(18.29)

k is usually taken to be in the range 0.75–0.9. This generated heat, is carried away by the side flow in a finite width bearing as $Q_s \rho c_p \Delta\Theta$. The side leakage flow Q_s is obtained as the net flow over the pressurised region of the bearing (i.e. between the maximum and minimum films, $\varphi = 0$ and $\varphi = \pi$ respectively, thus (Cameron, 1964; Gohar and Rahnejat, 2008):

$$Q_s = 2\pi R_j NLc\varepsilon$$

(18.30)

Note that Poiseuille flow is neglected in the case of short width bearings. Now a side flow factor is defined as:

$$Q_s^* = \frac{Q_s}{\pi R_j N L c}$$

thus it is clear that for a short-width bearing:

$$Q_s^* = 2\varepsilon \tag{18.31}$$

Equating the generated and convected, $kFu = Q_s \rho c_p \Delta\Theta$.
   Letting: $K = k/(\rho c_p)$, and using equations (18.27), (18.28) and (18.31):

$$\Delta\Theta = \left(\frac{2KW}{R_j L}\right)\frac{\mu^*}{Q_s^*} \tag{18.32}$$

With the above procedure the temperature rise is obtained. This is added to the inlet supply temperature to obtain the lubricant temperature in the conjunction. Then, **Vogel's equation** is used to obtain the effective viscosity as:

$$\ln \eta_e = -1.845 + \left(\frac{700.81}{\Theta_e - 203}\right) \tag{18.33}$$

where $\Theta_e = \Theta_i + \Delta\Theta + 273$.
   It is clear that as the temperature rises, the effective viscosity is reduced. This results in a corresponding decrease in the Sommerfeld number (see equation (18.10)). There is a corresponding change in the eccentricity ratio ε (see equation (18.11) or the Sommerfeld chart of Fig. 18.2 for any *L/D* ratio). Thus, for any crank angle position an iterative procedure is used to compute the values of bearing load, friction, lubricant temperature and eccentricity. This approach can be applied to both the *rigid* bearings or to the *elastic* overlay bearings.

18.8   Tribological conditions

Thus far the bearing loads at any given engine speed have been determined for the engine shown in Fig. 18.6. As an example, the cases considered below correspond to bearing loads in Fig. 18.7 and for bearing #1 at the engine speed of 4000 rpm. An iterative procedure described in the preceding paragraph is used. The iteration process ceases, when the value of either the eccentricity ratio or the bearing reaction converges to within a specified limit. Four types of analyses have been described in this chapter with an increasing degree of complexity:

1. An isothermal rigid bearing (this being the standard **Sommerfeld short-width bearing** analysis).
2. A thermal rigid bearing (the standard Sommerfeld short-width bearing with effective lubricant viscosity).
3. An isothermal elastic bearing (applicable to thin shell or **overlay bearing**).

4. A thermal elastic bearing (for overlay bearings with effective lubricant viscosity).

For the latter two, a thin soft overlay of 100 μm thick with a modulus of elasticity of 60 GPa is assumed. Note that the analyses for these two cases is numerical, based on the full solution of the Reynolds equation, together with the column method, elastic film shape equation and Vogel's equation (in the case of thermal elastic analysis). The first two *rigid* bearing cases are analytical.

Figure 18.9 shows the film thickness variation with the crank-angle. With the rigid isothermal analysis, it is clear that the film thickness in bearing #1 reduces quite dramatically at zero crank-angle and every 720° as the nearest cylinder to its position (cylinder 1) fires. The film thickness increases and reduces in an oscillatory fashion as cylinders remote or close by to the bearing #1 fire. In fact, the analysis suggests that the film thickness increases or reduces by nearly three-fold. Correspondingly the eccentricity ratio alters from 0.55 to 0.93, which is rather unexpected. Low values of eccentricity ratio and relatively large film thickness reduce the load-carrying capacity, indicating a tendency towards whirl instability. However, in practice friction and heat generated reduce the film thickness and increase the pressures in the converging wedge. An analysis taking this into account is the case for

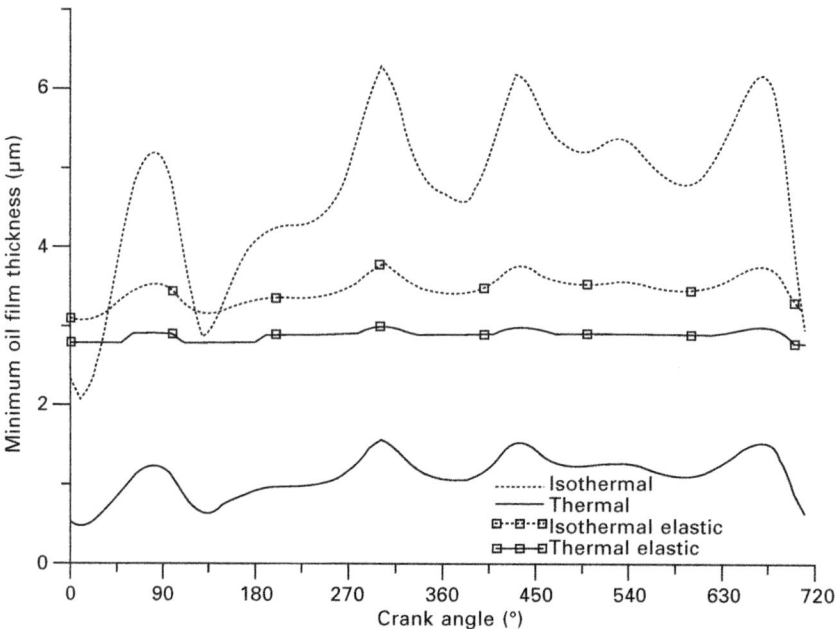

18.9 Minimum film thickness variation predicted with different analyses.

thermal rigid film in the figure. Two things are noted. Firstly, the oscillatory nature of the film thickness is reduced significantly. Secondly, much thinner films result; less than a quarter of that for the isothermal rigid case. However, the corresponding eccentricity ratio is in the range 0.9–0.98, which indicates direct surface interactions. This is the reason why overlay bearings are used, where deflection of the soft layers increases the gap by deformation. This is the isothermal elastic case in the figure. Note that the film thickness has increased by three-fold from the rigid thermal case, with an almost steady minimum film thickness and a corresponding eccentricity ratio of 0.9. This seems to be an ideal solution, except that friction and generated heat should also be taken into account, which means the thermal elastic case. The film thickness is now reduced because of reduced effective lubricant viscosity, thus the load-carrying capacity. The film thickness is still twice or more that of the thermal rigid case, with eccentricity ratio in the range 0.9–0.94.

There are two measures usually taken to mitigate the adverse effects of direct surface interactions, which are not included in these analyses. Firstly, the friction modifiers in the base lubricant form low shear strength boundary films that reduce friction and thus temperature, somewhat guarding against reductions in effective lubricant viscosity predicted here (note that the bulk rheological model used in this analysis does not include the action of lubricant additives). Secondly, the top layer in a multi-layered overlay is usually quite soft and in some cases acts elasto-plastically, breaking down with pressure and shear and reform again (acting as a solid lubricant, similar to soft coatings in rolling element bearing for space applications, see Chapter 4).

18.9   Concluding remarks

In conclusion it is clear that simple analytical solutions such as the standard Sommerfeld short-width bearing suffice for rigid bearings under light to moderate loads and at low shear rates. They also provide a reasonable first estimate of tribological conditions for other cases, but not of sufficient salient details. With smaller bearings increasingly used in progressively more compact engines, undergoing reasonably high direct and shear loads, it has been necessary to introduced soft overlays to encourage iso-viscous elastic (soft elastohydrodynamic lubrication) conditions to guard against boundary and mixed regimes of lubrication. This means that deformation of layers should be taken into account, which thus necessitates the more complex numerical solutions. Thermal solutions are essential as heat affects lubricant viscosity dramatically. This further complicates the analysis. The methods presented here should be improved to take into account friction and heat generation due to asperity interactions, potential mixing of plastically deformed top layers with the lubricant and its additives, forming deformable gel-type surface films with different flow behaviour according to shear rates.

There is significant research to be carried out in these areas as the operating loads and speeds are on the increase in modern engines.

18.10 References

Askeland, D.R. and Phulé, P.P., *The Science and Engineering of Materials*, 5th Edition. Nelson. Toronto, Canada, 2006

Barus, C., 'Isothermals isopietics and isometrics in relation to viscosity', *American J. Sci.*, 3rd Series, **45**, 1893, pp. 87–96

Bhushan, B., *Handbook of Tribology: Materials, Coatings and Surface Treatments*, Krieger Publishing Company. Florida, USA, 1991

Booser, E.R. and Wilcock, D.F., *Bearing Design and Application*, McGraw Hill, New York, USA, 1957

Cameron, A., *Principles of Lubrication*, Longmans, London, 1964

Gohar, R. and Rahnejat, H., *Fundamentals of Tribology*, Imperial College Press, London, 2008

Hamrock, B.J., Schmid, S.R. and Jacobson, B.O., *Fundamentals of Fluid Film Lubrication*, 2nd Edition. Marcel Dekker, New York, USA, 2004

Messler, E.W., *Principles of Welding. Processes, Physics, Chemistry, and Metallurgy*, John Wiley & Sons. New York, USA, 1999

Rahnejat, H., *Multi-body Dynamics: Vehicles, Machines and Mechanisms*, Professional Engineering Publishing (UK) and Society of Automotive Engineers (USA), 1998

Rahnejat, H., 'Multi-body dynamics: historical evolution and application', *Proc. Instn. Mech. Engrs. Part C, J. Mech. Engng. Sci.*, **214C1**, 2000, pp 149–173

Stribeck, R., 'Die wesentliechen ichen eigenschaften gleit und rollen lager' or 'Ball bearings for various loads', *Trans. ASME*, **29**, 1907, pp. 420–463

Teodorescu, M., Rahnejat, H., Gohar, R. and Dowson, D., 'Harmonic decomposition analysis of contact mechanics of bonded layered elastic solids', *Appl. Math. Modelling*, **32**, 2009, pp. 467–485

18.11 Nomenclature

c	Clearance
c_p	Lubricant specific heat at constant pressure
d	Thickness of overlay
D	Journal diameter
D_b	Cylinder bore diameter
e	Eccentricity
E	Modulus of elasticity
F_f	Friction
F_g	Combustion gas force
h	Film thickness
h_{max}	Maximum film thickness
h_{min}	Minimum film thickness
l	Connecting rod length
L	Bearing length (width)

m	Mass
N	Angular speed of journal in rev/s
p	Hydrodynamic pressure
p_g	Combustion gas pressure
p_m	Maximum hydrodynamic pressure
P	Frictional power loss
Q_s	Side leakage flow
r	Crank-pin radius
R_b	Radius of bearing bushing or shell
R_j	Radius of journal
S	Sommerfeld number as specified by the SAE
t	Time
u	Speed of entraining motion
v	Speed of side leakage flow
W	Bearing reaction
X	Direction of entraining motion
Y	Direction of side leakage
(x, y, z)	Cylinder co-ordinate system
(x', y', z')	Connecting rod co-ordinate system
(x'', y'', z'')	Crank-bore co-ordinate system
α	Engine V-block angle
α'	Lubricant pressure-viscosity coefficient
δ	Deflection
Δ	The original form of Sommerfeld number
Δt	Interval of time
θ	Crank angle
Θ_e	Effective temperature in K
Θ_i	Inlet temperature
$\Delta\Theta$	Rise in lubricant temperature
ε	Eccentricity ratio
η	Lubricant dynamic viscosity
η_0	Lubricant dynamic viscosity at ambient pressure
η_e	Effective dynamic viscosity at lubricant contact temperature
φ	Angle measured from line of centres
φ_m	Position of maximum contact pressure
Φ	Connecting rod angle (obliquity)
λ	Stribeck's oil film parameter
μ	Coefficient of friction
Θ	Lubricant contact temperature
ρ	Lubricant density
σ_{rms}	Composite root mean square surface roughness
υ	Poisson's ratio
ω	Crankshaft angular velocity in rad/s
ψ	Attitude angle

19

Practical tribological issues in big-end bearings

S. BOEDO, Rochester Institute of Technology, USA

Abstract: This chapter focuses upon practical design guidelines for big-end connecting rod journal bearings which allow for rapid prediction of three key tribological performance measures: cyclic minimum film thickness, cyclic average oil flow and cyclic average power loss. Design information, provided in both graphical and equation form, apply specifically to a grooveless big-end bearing pressure-lubricated by a single crank-pin feed hole. In addition, this chapter reviews the current state of tribological research aimed at improving design capabilities for big-end bearings and suggests related topics for further study.

Key words: connecting rod, journal bearings, oil flow, power loss, film thickness.

19.1   Introduction

The design of big-end connecting rod bearings for reciprocating machinery often poses a challenge to the non-specialist due to the complex nature of load variation combined with non-steady journal and sleeve rotation. Moreover, the designer often simply needs a first-order assessment of design trends as part of an overall engine design, usually combined with a tight timeframe for delivery.

In this chapter, practical design guidelines are discussed for big-end connecting rod bearings found in four-stroke automotive and diesel engines which provides rapid prediction of the three key tribological performance measures of cyclic minimum film thickness, cyclic average oil flow and cyclic average power loss.

Design charts are provided for a ***grooveless big-end bearing*** pressure-lubricated by a single feed hole drilled through the crank-pin. Included are new correlations of cyclic minimum film thickness for finite-length bearings applicable for four-stroke engines, while performance charts for oil flow and power loss have been compiled and assembled here from pertinent published sources (also see Chapter 18).

19.2   Bearing duty for big-end bearings

Computation of *bearing duty* (loads and kinematics) for a big-end connecting rod bearing is based on consideration of a composite crankshaft–connecting rod–piston rigid body mechanism with crank radius r and conrod length L connected with and supported by zero-clearance pin-jointed bearings as shown in Fig. 19.1. Such decoupling of load and bearing motion has been shown to be sufficiently accurate (Martin, 1983; Boedo, 1986) for bearing clearances typically found in practice. The X_1, Y_1, Z_1 inertial block co-ordinate frame (with unit vectors i_1, j_1, k_1) is fixed to the rigid engine block and the frame origin is situated at the crank main bearing. The cylinder axis lies along the X_1 axis, and all motion takes place in the $Z_1 = 0$ plane. The crankshaft rotates at a constant angular velocity $\omega = d\theta/dt$ about the Z_1 axis. Also shown are X_2, Y_2, Z_2 and X_3, Y_3, Z_3 frames (with unit vectors i_2, j_2, k_2 and i_3, j_3, k_3, respectively) which are attached to the moving crank and rod, respectively.

Connecting rod angular position, angular velocity, and angular acceleration in terms of crankangle and crank angular velocity follow from the specified kinematic constraints and are given by:

$$\sin \varphi = -\frac{r \sin \theta}{L} \tag{19.1}$$

$$\cos \varphi = (1 - \sin^2 \varphi)^{1/2} \tag{19.2}$$

$$d\varphi/dt = -\frac{r\omega \cos\theta}{L \cos\varphi} \tag{19.3}$$

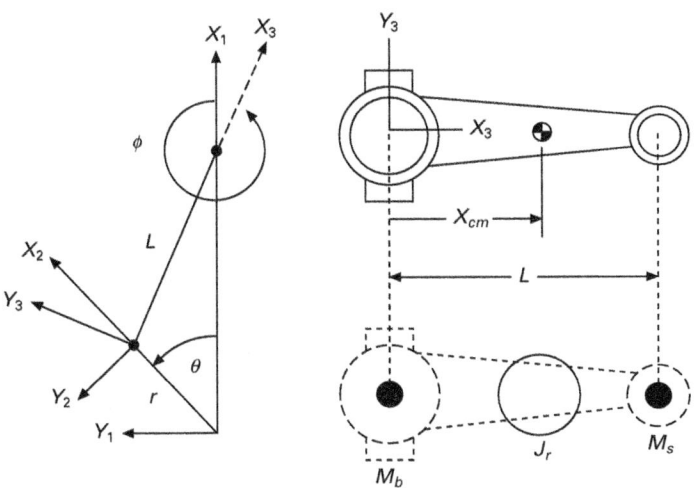

19.1 Connecting rod geometry.

$$d^2\varphi/dt^2 = \frac{r\omega^2 \sin\theta + L(d\varphi/dt)^2 \sin\varphi}{L\cos\varphi} \qquad (19.4)$$

For convenience, the actual connecting rod with centre-to-centre length L, mass M, centre of mass x_{cm}, and polar moment of inertia J (about the X_3, Y_3, Z_3 origin) is replaced by a dynamically equivalent model comprising small and big end point masses M_s, and M_b, respectively, and a massless residual inertia J_r (Paul, 1979) as shown in Fig. 19.1. In order that the actual and equivalent models have identical centres of mass, total mass, and polar moment of inertia, it follows that (Boedo and Booker, 2001):

$$M_s = Mx_{cm}/L \qquad (19.5)$$

$$M_b = M(1 - x_{cm}/L) \qquad (19.6)$$

$$J_r = J - Mx_{cm}L \qquad (19.7)$$

Further assuming that the piston and small end masses are constrained to reciprocate along the X_1 axis, and ignoring effects of gravity and cylinder offset, the external load vector $F(\theta)$ transmitted from the journal to the **big-end sleeve** has block-frame components (Boedo and Booker, 2001):

$$F^{X_1}(\theta) = P_{cyl}(\theta) A_{cyl} - M_{rot} r \omega^2 \cos\theta$$
$$- M_{rec} [r \omega^2 \cos\theta + L (d\varphi/dt)^2 \cos\varphi + L (d^2\varphi/dt^2) \sin\varphi] \qquad (19.8)$$

$$F^{Y_1}(\theta) = - M_{rot} r\omega^2 \sin\theta + (M_{rot} r\omega^2 \cos\theta + F^{X_1}) \tan\varphi$$
$$- (J_r\, d^2\varphi/dt^2)/(L\cos\varphi) \qquad (19.9)$$

in terms of reciprocating and rotating masses:

$$M_{rec} \equiv M_s + M_{pist}$$
$$M_{rot} \equiv M_b$$

with piston mass M_{pist} and cylinder pressure history $P_{cyl}(\theta)$. Load relations incorporating gravity have been formulated (Boedo and Booker, 1989); however, gravitational effects on bearing load are generally found to be important only for large-scale, low-speed engine designs typically found in large ships and powerplants. For four-stroke engines, cylinder pressure history and thus load components are periodic over 720° of crank rotation. Load components in the crank and rod reference frames can be found through the co-ordinate transformations:

$$F^{X_2} = +F^{X_1}\cos\theta + F^{Y_1}\sin\theta \qquad (19.10)$$

$$F^{Y_2} = -F^{X_1}\sin\theta + F^{Y_1}\cos\theta \qquad (19.11)$$

$$F^{X_3} = +F^{X_1}\cos\varphi + F^{Y_1}\sin\varphi \qquad\qquad (19.12)$$

$$F^{Y_3} = -F^{X_1}\sin\varphi + F^{Y_1}\cos\varphi \qquad\qquad (19.13)$$

Journal and sleeve angular velocity components in the block, crank and rod reference frames are given by:

$$\omega_j^{Z_1} = \omega \qquad\qquad \omega_s^{Z_1} = d\varphi/dt$$

$$\omega_j^{Z_2} = 0 \qquad\qquad \omega_s^{Z_2} = d\varphi/dt - \omega$$

$$\omega_j^{Z_3} = \omega - d\varphi/dt \quad \omega_s^{Z_3} = 0$$

Normalising the bearing load by rotating load parameter $F_{rot} \equiv M_{rot}\, r\omega^2$ defines dimensionless load vector $\boldsymbol{f}(\theta) \equiv \boldsymbol{F}(\theta)/F_{rot}$ which can be shown to depend upon the following dimensionless load parameters:

$$P_{cyl}{}^* A_{cyl}/F_{rot} \qquad M_{rec}/M_{rot}$$

$$r/L \qquad\qquad\quad J_r/(M_{rot}L^2)$$

as well as upon normalised cylinder pressure history (Fig. 19.2):

$$p_{cyl}(\theta) \equiv P_{cyl}(\theta)/P_{cyl}{}^*$$

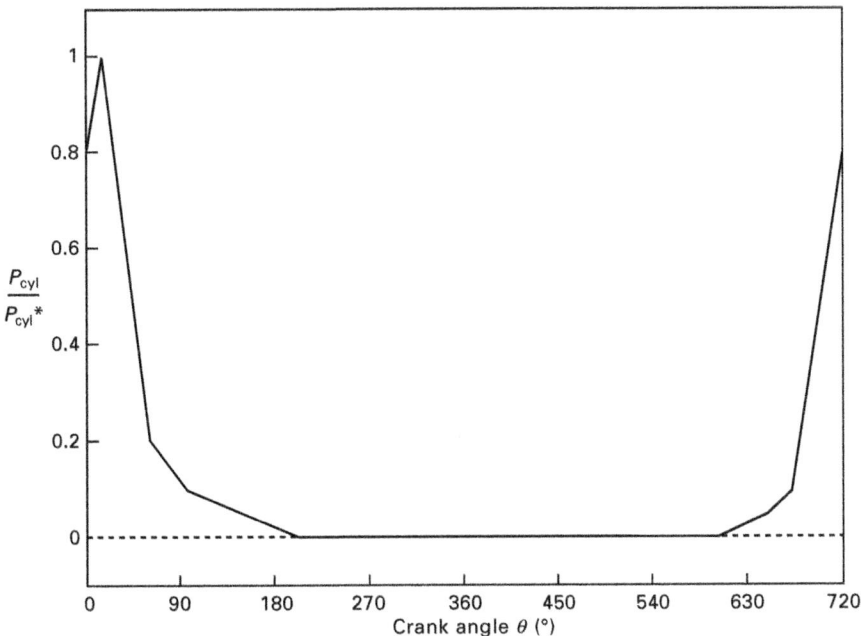

19.2 Normalised cylinder gas pressure history.

where $P_{cyl}*$ is the peak cylinder pressure obtained over the engine cycle. The normalised residual inertia term $J_r/(M_{rot}L^2)$ is generally found to be quite small and can be usually neglected for most connecting rod designs found in four-stroke automotive and diesel applications.

Figure 19.3 shows rod-frame components of periodic normalised load history f plotted in polar form for values of $P_{cyl}*A_{cyl}/F_{rot}$ and M_{rec}/M_{rot} typically found in practice, using the **normalised cylinder pressure history curve** of Fig. 19.2. The polar load plots are essentially unchanged for r/L ratios in the range 0.25–0.4. Practical deviations in the shape of the normalised cylinder pressure history curve also do not significantly affect the shape of the polar load diagrams. Additional load diagram shapes over a wider range of parameters are provided by Martin *et al.* (1987).

19.3   Geometry of big-end bearings

One common **big-end connecting rod bearing** design configuration found in practice employs an ungrooved sleeve with length B and diameter D as shown in Fig. 19.4. A feed passage drilled through the crank-pin journal supplies oil through a single feed hole of diameter d_h located on the bearing midplane and oriented at an angle β relative to the X_2, Y_2, Z_2 crank reference frame. The crank frame has its origin at the journal centre and the X_2–Y_2 plane passes through the bearing midplane. Diametrically opposed cross-drilled feed holes and conrod bearings employing partial sleeve groove geometry are also found in many production engines, but these feed arrangements will not be discussed here.

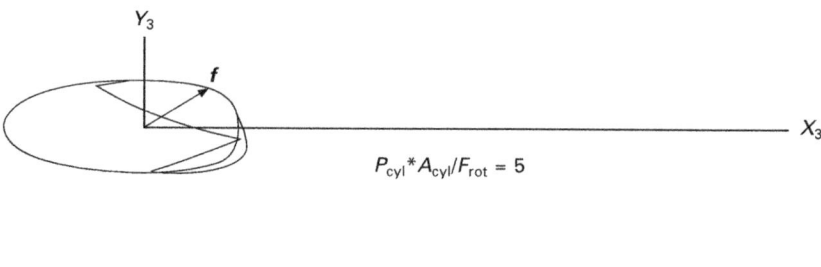

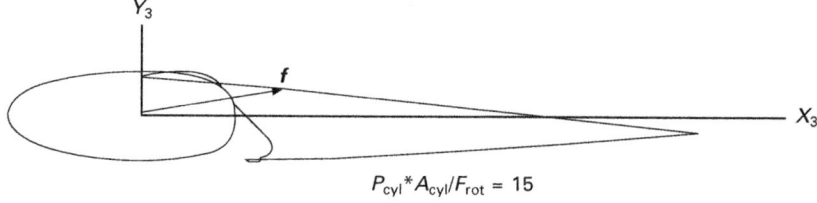

19.3 Bearing load shape diagrams, $r/L = 0.3$, $M_{rec}/M_{rot} = 1.5$.

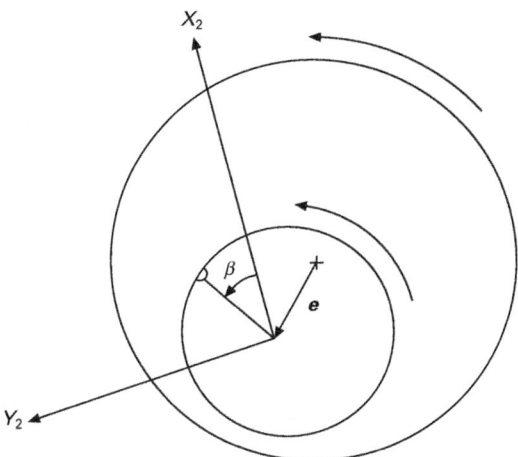

19.4 Bearing geometry: big-end connecting rod.

19.4   Cyclic minimum film thickness

The **mobility method** of solution (Booker, 1965, 1971) is a rapid means to predict the time history of journal centre motion, or orbit, of an **ungrooved journal bearing** within its sleeve when subjected to dynamically varying loads and kinematics. Mobility methods assume Newtonian rheology (see Chapter 5), spatially uniform clearance, circumferential symmetry, bearing rigidity and isoviscosity. Circumferential symmetry in particular ignores feed hole effects on journal motion; this assumption is generally good for feed hole locations typically found in practice. Mobility methods also employ a quasi-static cavitation model with zero (ambient) cavitation threshold pressure (LaBouff and Booker, 1985) which does not properly account for film density in cavitated regions and is thus not a reliable predictor of oil flow.

Journal eccentricity vector e (with components e^{X_2} and e^{Y_2} in the X_2–Y_2 plane) measures the position of the journal centre relative to the sleeve centre, as shown in Fig. 19.4. At any instant, the **minimum film thickness** h_{min} is given by (also see Chapters 18 and 20):

$$h_{min} = C - |e| \tag{19.14}$$

where C is (uniform) bearing radial clearance. Periodic time histories of bearing duty give rise to periodic time histories of journal centre motion, or periodic journal orbits, and the important parameter to the designer is the so-called cyclic minimum film thickness $min(h_{min})$ obtained over the duty cycle.

Treating the components of journal eccentricity as state variables, journal centre motion is represented by the vector state rate equation:

$$de/dt = \{[|F| \, (C/R)^2]/(\mu BD/C)\} \, M(e/C, \, B/D) + <\omega> \times e \tag{19.15}$$

where R is bearing radius, μ is fluid **dynamic viscosity** (see Chapter 5) and $<\omega>$ is the average angular velocity of journal and sleeve given by:

$$<\omega> = (1/2)(\omega_j^{Z_2} + \omega_s^{Z_2})k_2 \qquad (19.16)$$

relative to the X_2, Y_2, Z_2 crank reference frame of Fig. 19.4. Mobility vector **M** is a function of eccentricity ratio vector e/C and **bearing aspect ratio** B/D with vector components which are initially computed in a load frame of reference and subsequently transformed into the X_2-Y_2 plane (Booker, 1971). Mobility vector **M** generated from **short-bearing** (see Chapters 5 and 18) approximations of the Reynolds equation is independent of B/D (Booker, 1965).

Introducing (dimensionless) journal eccentricity ratio $\varepsilon = e/C$, the periodic (dimensionless) time history of journal eccentricity ratio $\varepsilon(\theta)$ (through numerical integration of equation (19.15) as an initial value problem) depends upon not only dimensionless load parameters and normalised cylinder pressure history defined in Section 19.2, but also upon two additional dimensionless bearing parameters (Martin and Booker, 1967):

$$[F_{rot}(C/R)^2]/(\mu \omega BD) \quad \text{and} \quad B/D$$

Figures 19.5–19.9 provide design charts to obtain the cyclic maximum **eccentricity ratio** ε_{max} for big-end connecting rod bearings constructed using

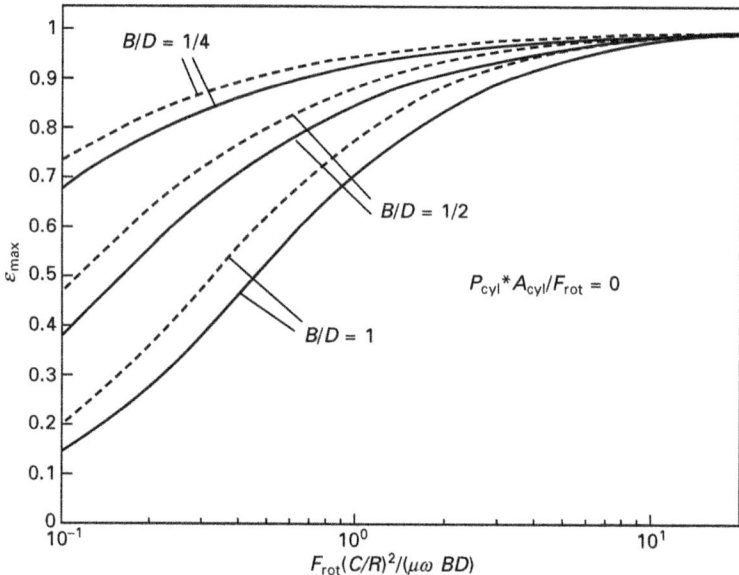

19.5 Cyclic-maximum eccentricity ratio, $P_{cyl}*A_{cyl}/F_{rot} = 0$, ——— $M_{rec}/M_{rot} = 1$, - - - - - - - $M_{rec}/M_{rot} = 2$.

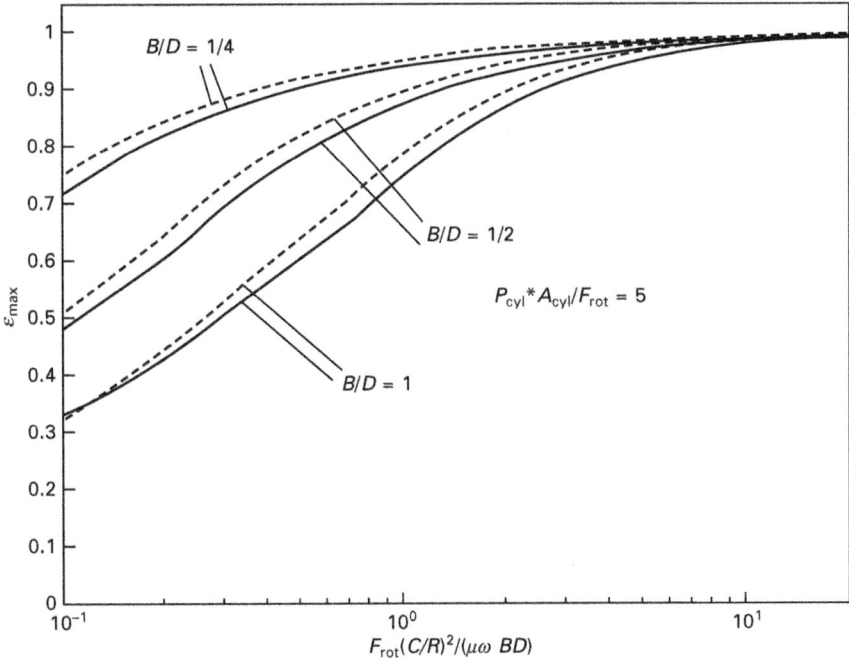

19.6 Cyclic-maximum eccentricity ratio, $P_{cyl}*A_{cyl}/F_{rot} = 5$,
———— $M_{rec}/M_{rot} = 1$, - - - - - - $M_{rec}/M_{rot} = 2$.

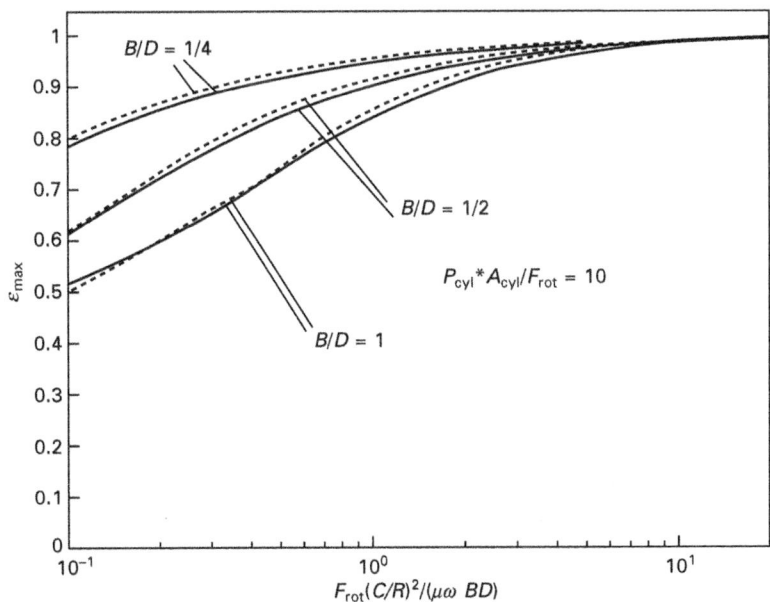

19.7 Cyclic-maximum eccentricity ratio, $P_{cyl}*A_{cyl}/F_{rot} = 10$,
———— $M_{rec}/M_{rot} = 1$, - - - - - - $M_{rec}/M_{rot} = 2$.

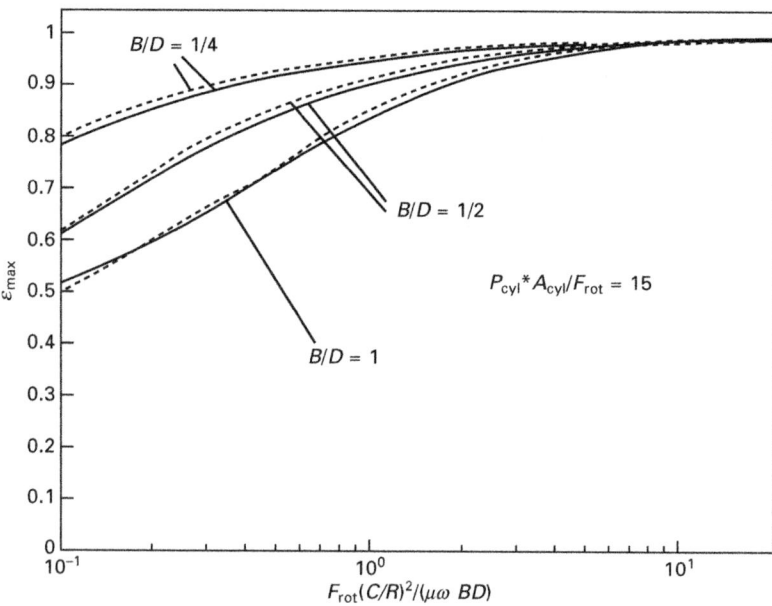

19.8 Cyclic-maximum eccentricity ratio, $P_{cyl}*A_{cyl}/F_{rot} = 15$,
——— $M_{rec}/M_{rot} = 1$, - - - - - - $M_{rec}/M_{rot} = 2$.

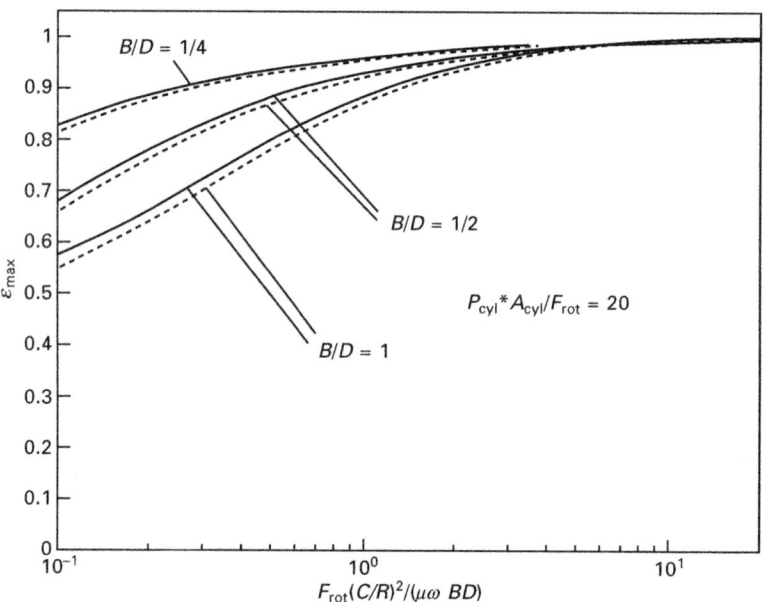

19.9 Cyclic-maximum eccentricity ratio, $P_{cyl}*A_{cyl}/F_{rot} = 20$,
——— $M_{rec}/M_{rot} = 1$, - - - - - - $M_{rec}/M_{rot} = 2$.

a *finite element mobility map* for *finite length journal bearings* (Goenka, 1984) (also see Chapter 5). The cyclic minimum film thickness and the cyclic maximum eccentricity ratio are related by:

$$\min(h_{\min}) = C(1 - \varepsilon_{\max}) \qquad (19.17)$$

These charts were constructed by running hundreds of bearing orbit simulations over a practical range of dimensionless load and bearing parameters (see Chapters 5 and 18). The charts are strictly valid for $r/L = 0.3$; however, journal motion was found to be insensitive to practical variation of r/L, $J_r/(M_{\rm rot}L^2)$ and $p_{\rm cyl}(\theta)$. Unlike similar design charts based on short bearing theory (Booker, 1984), cyclic maximum eccentricity ratio obtained using a finite bearing mobility map is dependent upon B/D ratio and is observed to be more sensitive to load parameter $P_{\rm cyl}*A_{\rm cyl}/F_{\rm rot}$. Short bearing-based charts assume that cyclic maximum eccentricity ratio is essentially independent of $P_{\rm cyl}*A_{\rm cyl}/F_{\rm rot}$ when peak gas force $P_{\rm cyl}*A_{\rm cyl}$ is no more than about seven times the rotating inertia load $F_{\rm rot}$ (Booker, 1984).

19.5   Oil flow in big-end bearings

For any lubricated bearing subjected to a periodic duty cycle, the cycle averaged oil flow entering through an inlet feed arrangement must equal the cycle averaged oil flow leaving the bearing ends. Instantaneous inlet and outlet flows generally do not balance, with the exception of steadily loaded journal bearings under steady rotation.

Computation of oil flow requires solution of the complete Reynolds equation coupled with proper temporal and spatial tracking of lubricant density (or mass volume fraction) within the cavitated region of the oil film. Established methods include an oil film history model (Jones, 1983), a finite element mass-conserving cavitation algorithm (Kumar and Booker, 1991), and a finite difference method involving switching functions (Elrod and Adams, 1975; Brewe, 1986).

Both oil flow and journal orbits can be affected by grooving arrangements (Martin, 1983) and parameters associated with the cavitation algorithm (Talmage *et al.*, 2002). However, qualitative estimation of journal orbit and oil flow as a means to establish design trends can be generally viewed as a decoupled process for the feed bearing geometry described here. The journal orbit is first computed using the mobility method as described in Section 19.3, which excludes oil feed hole geometry. The resulting oil flow is computed subsequently from the resulting journal motion.

For ungrooved big-end bearings lubricated by a single feed hole, estimates of cycle-averaged end flow $<Q>$ can be found from the empirical relation (Martin and Xu, 1993):

$$<Q> = [<Q_h> + <Q_p> - 0.3 (<Q_h> <Q_p>)^{1/2}]^S <Q_p>^{1-S} \qquad (19.18)$$

where $<Q_p>$ and $<Q_h>$ are so-called **cycle-averaged feed pressure** and hydrodynamic flows, respectively, and S is an empirical factor given by:

$$S = (d_h/B)^n$$

where:

$$n = 0.19 \, [G_1 \, (P_{cyl}*A_{cyl}/F_{rot})^{-1.4} + 1]$$

with:

$$G_1 = [2\pi \, F_{rot}(C/R)^2]/(\mu\omega BD) \quad \text{for} \; [2\pi \, F_{rot}(C/R)^2]/(\mu\omega BD) < 12$$

$$\quad = 12 \qquad\qquad\qquad\qquad\quad \text{for} \; [2\pi \, F_{rot}(C/R)^2]/(\mu\omega BD) \geq 12$$

Feed pressure flow is that computed by consideration of circular **hole feed geometry**, supply pressure, and journal position alone, ignoring effects of journal rotation. Hydrodynamic flow is that computed from consideration of a rotating hole-free grooveless plain bearing employing quasi-static cavitation. It should be noted that feed pressure and hydrodynamic flow computations for hole-fed big-end connecting rod bearings under consideration are each fictitious component approximations to the actual flow – neither satisfies the Reynolds equation using actual bearing boundary conditions and cavitation requirements.

Relations for dimensional instantaneous **feed pressure flow** Q_p and **hydrodynamic flow** Q_h in the X_2, Y_2, Z_2 reference frame of Fig. 19.4 have the form (Booker, 1979; Martin and Lee, 1983):

$$Q_p = (0.675/\mu) \, C^3 P_f \, [1 - (e^{X2}/C) \cos \beta - (e^{Y2}/C) \sin \beta]^3 \times$$

$$(d_h/B + 0.4)^{1.75} \tag{19.19}$$

$$Q_h = [R|F| \, (C/R)^3/\mu] \, |M| \tag{19.20}$$

where P_f is feed supply pressure. It can be shown that corresponding dimensionless feed pressure and hydrodynamic flows $Q_p \, \mu/(C^3 P_f)$ and $Q_h \, \mu/[RF_{rot}(C/R)^3]$ depend on (dimensionless) load and bearing parameters (Sections 19.2 and 19.4) as well as on two additional (dimensionless) feed hole parameters d_h/B and β.

Cycle-averaged feed pressure and hydrodynamic flows $<Q_p>$ and $<Q_h>$ computed from hundreds of numerical simulations over a wide range of operating parameters are given empirically as (Martin and Stanojevic, 1991):

$$<Q_p> = 21.5 \, C^3 \, P_f G_2/\mu \tag{19.21}$$

$$<Q_h> = (1.25/\pi) \, BD\omega C \, G_3 \tag{19.22}$$

where G_2 and G_3 are empirical (dimensionless) factors given by:

$$G_2 = (d_h/B)^{0.8} (M_{rec}/M_{rot})^{0.2} \{[2\pi \, F_{rot}(C/R)^2]/(\mu\omega BD)\}^{0.12}$$
$$\times (P_{cyl}*A_{cyl}/F_{rot})^{0.06} (180\beta/\pi)^{-0.75} (B/D)^{-0.28}$$
$$G_3 = (M_{rec}/M_{rot})^{0.2} \{[2\pi \, F_{rot}(C/R)^2]/(\mu\omega BD)\}^{0.07}$$
$$\times (P_{cyl}*A_{cyl}/F_{rot})^{-0.12} (B/D)^{-0.18}$$

Alternative empirical formulae for calculating $<Q_h>$ and $<Q_p>$ are available (Boedo and Booker, 1991) which cover a different range of operating conditions than that provided here.

19.6   Power loss in big-end bearings

The mechanical power H generated by bearing loads and kinematics is dissipated by heat conduction through the bearing and by heat convection via oil flow through the bearing (see also Chapter 5). Estimations of instantaneous and *cycle-averaged power loss* thus serve to estimate bearing temperature, friction and operating viscosity.

A conservative estimate of instantaneous power loss employs a grooveless, *hole-free plain journal* used in consideration of hydrodynamic flow calculations of Section 19.4. Mechanical power can be written as (Booker *et al.*, 1982; Booker, 1989):

$$H = H_{shear} + H_{squeeze} \geq 0 \tag{19.23}$$

where H_{shear} and $H_{squeeze}$ are power losses generated by shear and *squeeze bearing* actions, respectively. In vector forms (Martin *et al.*, 1987):

$$H_{shear} = T_{shear} \bullet (\omega_j - \omega_s) \geq 0 \tag{19.24}$$

$$H_{squeeze} = F \bullet v_{squeeze} \geq 0 \tag{19.25}$$

where, in the X_2, Y_2, Z_2 crank reference frame of Fig. 19.4, $\omega_j = \omega_j^{Z2} k_2$ and $\omega_s = \omega_s^{Z2} k_2$. Bearing shear torque vector T_{shear} and journal *squeeze velocity* vector $v_{squeeze}$ are found from:

$$T_{shear} = 2\pi\mu R^3 B <1/h> (\omega_j - \omega_s) \tag{19.26}$$

$$v_{squeeze} = de/dt - <\omega> \times e \tag{19.27}$$

with average inverse film thickness:

$$<1/h> = (C^2 - |e|^2)^{-1/2} \tag{19.28}$$

based on a cavitation-free oil film (Martin *et al.*, 1987). Since the shear component of power loss usually dominates for connecting rod bearings, a cavitation-free assumption provides a conservative estimate of actual power loss in the bearing.

Introducing dimensionless parameters defined above, it can be shown that dimensionless instantaneous *total power loss*:

$$H \, (C/R)/(\mu\omega^2 D^2 B)$$

depends only upon dimensionless load and bearing parameters defined in Sections 19.2 and 19.4, respectively (Martin *et al.*, 1987). However, empirical relations for cycle-averaged power loss <*H*> using bearing orbits computed a *finite-length mobility map* can be condensed to compact chart form as shown in Fig. 19.10, based on hundreds of numerical simulations performed over a wide range of operating parameters (Martin *et al.*, 1987). Power loss estimations based on short-bearing theory are also provided by Martin *et al.* (1987) but are not presented here.

The power loss data in Fig. 19.10 is valid for cyclic maximum eccentricity ratios in the range $0.7 \leq \varepsilon_{max} \leq 0.99$ and for bearing aspect ratios in the range $0.2 \leq B/D \leq 0.6$. The power loss data are strictly valid for $r/L = 0.3$, but it is believed accurate for practical variations in the range $0.25 \leq r/L \leq 0.4$.

19.7   Sample application of design charts: four-stroke automotive engine

The following example illustrates the computation of cyclic values of minimum film thickness, average oil flow and average power loss using direct numerical

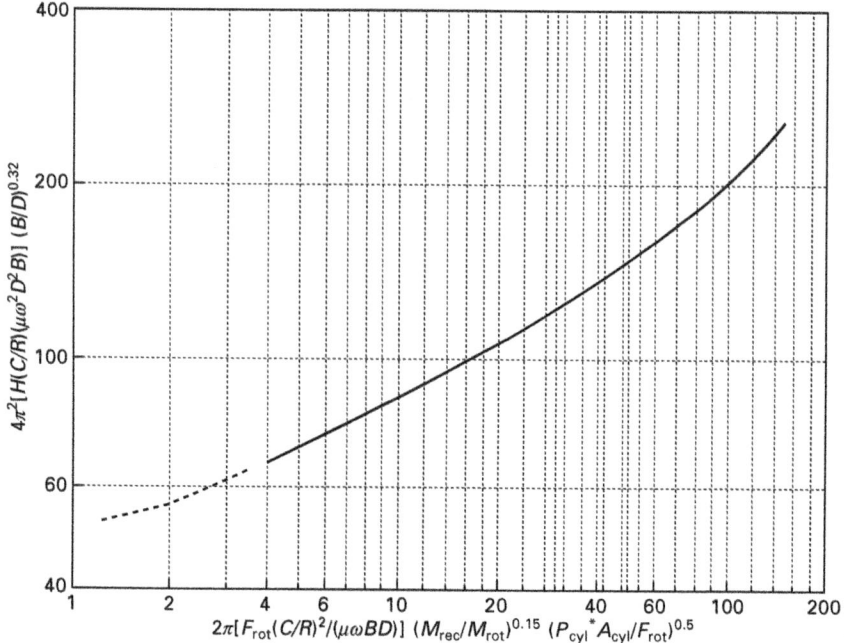

19.10 Total power loss (Martin *et al.*, 1987), $r/L = 0.3$, $0.2 \leq B/D \leq 0.6$.

simulation and using rapid design charts for a sample connecting rod bearing. Table 19.1 provides dimensional engine and bearing specifications typical of those found in four-stroke automotive engines.

Figure 19.11 shows that crank-frame periodic journal orbits computed using the mobility method are in good qualitative agreement with that obtained using a rigorous mass-conserving finite element method (Kumar and Booker, 1991). Quantitative comparison of cyclic extrema and averages provided in Table 19.2 indicate that the rapid assessment tools are reasonably accurate predictors of bearing performance.

19.8   Experimental results pertaining to big-end bearings

There is surprisingly little published experimental data that is directly applicable to the performance measures and bearing geometry described here. Several papers describe methodologies for measuring conrod bearing film thickness as applied to single and multi-cylinder engines (Bates *et al.*, 1990; Tseregounis *et al.*, 1998; Goodwin *et al.*, 2000; Paranjpe *et al.*, 2000; Moreau *et al.*, 2002). For big-end connecting rod bearings lubricated by a fully-circumferential feed groove along the bearing midplane, the mobility method provides a good qualitative estimation of the shape of the *periodic journal orbit* when compared with experimental measurements (Martin, 1983). The corresponding cyclic minimum film thickness obtained using the mobility method is quantitatively similar to that determined experimentally, provided that sleeve flexibility is not large (Martin, 1983; Aitken and

Table 19.1 Engine and bearing data (sample application)

Engine specifications

M_{rot}	= 0.32	kg
M_{rec}	= 0.48	kg
r	= 36	mm
L	= 120	mm
P_{cyl}^*	= 7.0	MN/m^2
A_{cyl}	= 4300	mm^2
ω	= $4000(2\pi/60)$	rad/s
J_r	= 0	$kg\,m^2$

Bearing specifications

D	= 42	mm
B	= 16.8	mm
C	= 20	μm
μ	= 4	mPa s
P_f	= 0.4	MN/m^2
d_h	= 6	mm
β	= $30(\pi/180)$	rad

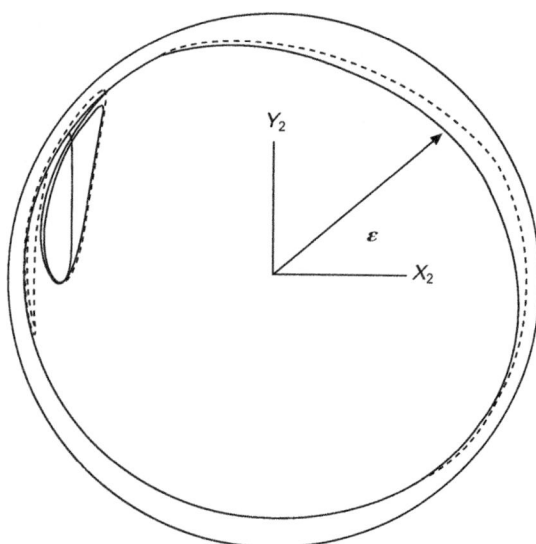

19.11 Bearing orbit comparison: ——— mobility with Goenka finite-bearing mobility map, - - - - - rigorous solution (mass-conserving, finite element method, p_{cav} = −101 kPa).

Table 19.2 Bearing performance comparison (sample application)

	Cyclic h_{min} (μm)	$<Q>$ (cm³/s)	$<H>$ (W)
Design charts[6]	1.2	2.687	94.2
Mobility	1.024[1]	2.595[1]	94.43[1]
		2.53[2]	93.9[3]
Finite element mass-conserving cavitation model	0.915[4]	2.281[4]	89.39[4]
	0.931[5]	2.005[5]	88.91[5]
Oil film history model	–	2.43[2]	–

[1]Finite element mobility map (Goenka, 1984), $<Q>$ computed using equation (19.18) with $<Q_p>$ and $<Q_h>$ found from journal orbit and equations (19.19) and (19.20).
[2]Martin and Stanojevic (1991).
[3]Martin *et al.* (1987).
[4]Cavitation threshold pressure p_{cav} = 0.
[5]Cavitation threshold pressure p_{cav} = −101.3 kPa.
[6]Figures 19.5–19.10, equations (19.18), (19.21) and (19.22).

McCallion, 1991). Choi *et al.* (1993) employed an experimental setup using an oil 'spurt' hole of unknown diameter, unknown supply pressure, and an unknown location. However, their measured cyclic minimum film thickness trends with engine speed variation agreed qualitatively with simulations using a short-bearing mobility formulation. It would be expected in this case that better quantitative results would be obtained using the finite element-based results provided here, but incomplete test specifications do not allow the possibility for comparison. *Cycle-averaged oil flow* in a connecting rod bearing fed through a single feed hole has been experimentally determined on a simulator test rig (Martin, 1983). The measured flow results agree quite well with those predicted using an oil film history model (Jones, 1983), again for bearings where structural rigidity can be assumed. To the author's best knowledge, experimental evaluation of power loss directly associated with connecting rod bearings is absent from the literature. Frictional losses in a piston assembly have been recently measured in a single cylinder engine (Mufti and Priest, 2005), which might allow for estimation of power loss in individual components.

19.9   Future trends

Since the mid-1980s, enhanced computational resources have encouraged the development of sophisticated bearing analysis predictive tools that are applicable to connecting rod bearings. Computational methods that take into account realistic connecting rod features such as structural flexibility, lubricant cavitation, structural body forces, surface roughness, heat generation and heat convection/conduction have been recently developed (Oh and Goenka, 1985; McIvor and Fenner, 1989; Kumar *et al.*, 1990; Bonneau *et al.*, 1995; Knoll *et al.*, 1996; Boedo and Booker, 1997a,b; Piffeteau *et al.*, 2000; Boedo and Booker, 2001; Kim and Kim, 2001; Stefani and Rebora, 2002; Fatu *et al.*, 2006). However, the relative importance of each of these realistic features for connecting rods over the applicable range of engine and bearing parameters is currently not well understood; clearly this is an item for future development. One suggestion is to build upon the work of Aitken and McCallion (1992), who provided an early attempt to develop design charts to ascertain the effect of conrod flexibility on bearing performance. Reliable experimental data for connecting rods remains a difficult task, particularly when such data are taken from instrumented production-level engines at realistic engine speeds. Bench test rigs offer the best means to validate recently developed software and to assist in establishing design trends for connecting rod bearings.

The effect of peak oil film pressure on bearing performance has not been covered in this chapter. Although mobility-based rapid methods for both short and finite-length journal bearings exist to calculate maximum film pressure and its location on the bearing surface (Booker, 1969; Goenka, 1984), design charts applicable to big-end connecting rod bearings are not found in the literature. Yahraus (1987) points out the need to characterise bearing materials using peak film pressure instead of conventional methods employing unit loading. Mihara and Someya (2002) report on new methods to measure and locate peak oil film pressure in dynamically loaded journal bearings using small piezo-resistive thin-film sensors, as well as provide a good literature review on the subject.

19.10 References

Aitken M B, McCallion H (1991), 'Elastohydrodynamic lubrication of big-end bearings part 2: ratification, *Proc Inst Mech Eng*, **205**, 107–119.

Aitken M B, McCallion H (1992), 'Parametric minimum film thickness performance of an elastic big-end bearing under inertial load', *Proc Inst Mech Eng*, **206**, 3–12.

Bates T W, Fantino B, Launay L, Frene J (1990), 'Oil film thickness in an elastic connecting-rod bearing: comparison between theory and experiment', *Trib Trans*, **33**, 254–266.

Boedo S (1986), 'Dynamics of engine bearing systems: rigid body analysis', MS thesis, Cornell University.

Boedo S, Booker J F (1989), 'Transient dynamics of engine bearing systems', in Dowson D, Taylor C M, Berthe D, *Tribological Design of Machine Elements*, Elsevier, 323–332.

Boedo S, Booker J F (1991), 'Feed pressure flow in connecting rod bearings', in Dowson D, Taylor C M, Godet M, *Vehicle Tribology*, Elsevier, 55–62, 509.

Boedo S, Booker J F (1997a), 'Surface roughness and structural inertia in a mode-based mass-conserving elastohydrodynamic lubrication model', *J Trib*, **119**, 449–455.

Boedo S, Booker J F (1997b), 'Mode stiffness variation in elastohydrodynamic bearing design', in Dowson D, *Elastohydrodynamics '96*, Elsevier, 685–697, 740–741.

Boedo S, Booker J F (2001), 'Finite element analysis of elastic engine bearing lubrication: application', *Revue Européenne des Éléments Finis*, **10**, 725–740.

Bonneau D, Guines D, Frene J, Toplosky J (1995), 'EHD analysis, including structural inertia effects and a mass conserving cavitation model', *J Trib*, **117**, 540–547.

Booker J F (1965), 'Dynamically loaded journal bearings: mobility method of solution', *J Basic Eng*, **87**, 537.

Booker J F (1969), 'Dynamically loaded journal bearings: maximum film pressure', *J Lub Tech*, **91**, 534.

Booker J F (1971), 'Dynamically loaded journal bearings: numerical application of the mobility method', *J Lubr Tech*, **93**, 168–176, 315.

Booker J F (1979), 'Design of dynamically loaded journal bearings', in Rhode S M, Maday C J, Allaire, P E, *Fundamentals of the Design of Fluid Film Bearings*, ASME, 31.

Booker, J F (1984), 'Squeeze films and bearing dynamics', in *CRC Handbook of Lubrication*, CRC Press.

Booker J F (1989), 'Basic equations for fluid films with variable properties', *J Trib*, **111**, 475–483.

Booker J F, Goenka P K, van Leeuwen H J (1982), 'Dynamic analysis of rocking journal bearings with multiple offset segments', *J Lub Tech*, **104**, 478–490, Addendum: **105**, 220.

Brewe D E (1986), 'Theoretical modeling of vapor cavitation in dynamically loaded journal bearings', *J Trib*, **108**, 628–638.

Choi J-K, Hur K, Han D-C (1993), 'Oil film thickness in engine connecting-rod bearing:

comparison between calculation and experiment', in *Tribology of Engines and Engine Oils (SP-959)*, SAE paper 930694.

Elrod H G, Adams M L (1975), 'A computer program for cavitation and starvation problems', in Dowson D, Godet M, Taylor C M, *Cavitation and Related Phenomena in Lubrication*, Mech Eng Pub Ltd, 37–41.

Fatu A, Hajjam M, Bonneau D (2006), 'A new model of thermoelastohydrodynamic lubrication in dynamically loaded journal bearings', *J Trib*, **128**, 85–95.

Goenka P K (1984), 'Analytical curve fits for solution parameters of dynamically loaded journal bearings', *J Trib*, **106**, 421–428.

Goodwin M J, Groves C, Nikolajsen J, Ogrodnik P J (2000), 'Experimental measurement of big-end bearing journal orbits', *Proc Inst Mech Eng Part J*, **214**, 219–228.

Jones G J (1983), 'Crankshaft bearings: oil film history', in Dowson D, Taylor C M, Godet M, Berthe D, *Tribology of Reciprocating Engines*, Butterworths, 83–88.

Kim B-J, Kim K-W (2001), 'Thermo-elastohydrodynamic analysis of connecting rod bearing in internal combustion engine', *J Trib*, **123**, 444–454.

Knoll, G, Lang J, Reinacker A (1996), 'Transient EHD connecting rod analysis: full dynamic versus quasi-static deformation', *J Trib*, **118**, 349–355.

Kumar A, Booker J F (1991), 'A finite element cavitation algorithm', *J Trib*, **113**, 276–286.

Kumar A, Goenka P K, Booker J F (1990), 'Modal analysis of elastohydrodynamic lubrication: a connecting rod application', *J Trib*, **112**, 524–534.

LaBouff G A, Booker J F (1985), 'Dynamically loaded journal bearings: a finite element treatment for rigid and elastic surfaces', *J Trib*, **107**, 505–515.

Martin F A (1983), 'Developments in engine bearing design', *Tribology International*, **16**, 147–164.

Martin F A, Booker J F (1967), 'Influence of engine inertia forces on minimum film thickness in connecting-rod big-end bearings', *Proc Inst Mech Eng*, **181**, 749–764.

Martin F A, Lee C S (1983), 'Feed-pressure flow in plain journal bearings', *ASLE*, **26**, 381–392.

Martin F A, Stanojevic M (1991), 'Oil flow in connecting rod bearings', in Dowson D, Taylor C M, Godet M, *Vehicle Tribology*, Elsevier, 69–80.

Martin F A, Xu H (1993), 'Improved oil flow prediction method for connecting rod bearings', in *Tribology of Engines and Engine Oils (SP-959)*, SAE paper 930791.

Martin F A, Lo P M, Booker J F (1987), 'Power loss in connecting rod bearings', in *Tribology, Friction, Lubrication, and Wear, Fifty Years On*, IMechE, Vol II, 701–708.

McIvor J D C, Fenner D N (1989), 'Finite element analysis of dynamically loaded flexible journal bearings: a fast Newton–Raphson method', *J Trib*, **111**, 597–604.

Mihara Y, Someya T (2002), 'Measurement of oil-film pressure in engine bearings using a thin film sensor', *Trib Trans*, **45**, 11–20.

Moreau H, Maspeyrot P, Bonneau D, Frene J (2002), 'Comparison between experimental film thickness measurements and elastohydrodynamic analysis in a connecting-rod bearing', *Proc Inst Mech Eng Part J*, **216**, 195–208.

Mufti R A, Priest M (2005), 'Experimental evaluation of piston assembly friction under motored and fired conditions in a gasoline engine', *J Trib*, **127**, 826–836.

Oh K P, Goenka P K (1985), 'The elastohydrodynamic solution of a journal bearing under dynamic loading', *J Trib*, **107**, 389–395.

Paranjpe R S, Tseregounis S I, Viola M B (2000), 'Comparison between theoretical calculations and oil film thickness measurements using the total capacitance method for crankshaft bearings in a firing engine', *Trib Trans*, **43**, 345–356.

Paul B (1979), *Kinematics and Dynamics of Planar Machinery*, Prentice Hall.

Piffeteau S, Souchet D, Bonneau D (2000), 'Influence of thermal and elastic deformations on connecting rod big end bearing lubrication under dynamic loading', *J Trib*, **122**, 181–191.

Stefani F A, Rebora A U (2002), 'Finite element analysis of dynamically loaded journal bearings: influence of bolt preload', *J Trib*, **124**, 486–493.

Talmage G, Carpino M, Sneck H (2002), 'Performance of a plain journal bearing with flooded ends', *Trib Trans*, **45**, 310–317.

Tseregounis S I, Viola M B, Paranjpe R S (1998), 'Determination of bearing oil film thickness (BOFT) for various engine oils in an automotive gasoline engine using capacitive measurements and analytical predictions', in *Advances in Powertrain Tribology (SP-1390)*, SAE paper 982661.

Yahraus W A (1987), 'Rating sleeve bearing material fatigue life in terms of peak oil film pressure', SAE paper 871685.

19.11   Principal nomenclature

r	crank radius	(m)
x_{cm}	conrod centre of mass	(m)
A_{cyl}	cylinder area	(m^2)
F	bearing load	(N)
F_{rot}	rotating inertia load	(N)
J	conrod polar moment of inertia	(kg m^2)
J_r	residual inertia	(kg m^2)
L	rod length	(m)
M	conrod total mass	(kg)
M_{rec}	reciprocating mass	(kg)
M_{rot}	rotating mass	(kg)
M_b	big end point mass	(kg)
M_s	small end point mass	(kg)
M_{pist}	piston mass	(kg)
P_{cyl}	cylinder gas pressure	(N/m^2)
$P_{cyl}{}^*$	peak cylinder gas pressure	(N/m^2)
θ	crank angle	(rad)
ω	crank angular velocity	(rad/s)
d_h	feed hole diameter	(m)
e	journal eccentricity	(m)
h_{min}	minimum film thickness	(m)
B	bearing length	(m)
C	radial clearance	(m)
D	bearing diameter	(m)
H	total power loss	(N m/s)
H_{shear}	power loss due to shear action	(N m/s)
$H_{squeeze}$	power loss due to squeeze action	(N m/s)

M	mobility	$(-)$
P_f	feed pressure	(N/m^2)
Q	bearing endflow	(m^3/s)
Q_h	hydrodynamic flow	(m^3/s)
Q_p	feed pressure flow	(m^3/s)
R	bearing radius	(m)
β	feed angle	(rad)
ε	eccentricity ratio	$(-)$
μ	fluid viscosity	$(N\,s/m^2)$
ω_j	journal angular velocity	(rad/s)
ω_s	sleeve angular velocity	(rad/s)
$<\ >$	average	
d/dt	time derivative	(s^{-1})
t	time	(s)

20
Tribology of big-end bearings

P.C. MISHRA and H. RAHNEJAT, Loughborough
University, UK

Abstract: Big-end or connecting rod bearings are subjected to large
variations in loads and kinematic conditions during the engine cycle. Owing
to these variations the tribological conditions are transient. Generated heat,
large transmitted forces and, in many instances, low speeds of entraining
motion can lead to mixed regime of lubrication and reduced load-carrying
capacity of thin films. As a result these bearings are usually made in
elliptic form in order to create multiple wedge shapes, thus improving their
load capacity and reduce chance of whirl instability. Therefore, accurate
prediction of transient tribological conditions is quite important, involving
complex numerical analysis. The chapter provides comprehensive solution
method, involving the Reynolds equation for transient analysis, together with
energy, lubricant rheological and film thickness equations.

Key words: big-end bearing, elliptic bore bearings, mixed
thermohydrodynamics, asperity interactions.

20.1   Introduction

The *parasitic frictional losses* in an engine are estimated to account for
20% of the total engine losses. These include those due to piston assembly
(45%), bearings (25%), pumping action (20%) and the valve train system
(10%). Thus, a quarter of all the parasitic losses are due to engine bearings,
which include the main crankshaft support bearings (see Chapter 18), big-
end bearing (see also Chapter 19), and other camshaft and any rocker arm
bearings. Though all these bearings are essentially hydrodynamic journal
bearings, the nature of load application, construction and design is specific
to each case. The main bearings have stationary bushings (see also Chapter
18), while the bore of the big-end bearing is oscillating and subjected to
cyclic fluctuating loads from repetitive combustion pressure.

The role of the *big-end bearing* is to sustain the transmitted forces through
the connecting rod, which are due to combustion pressure and inertial
imbalance, as well as transmitting the torque to drive the crankshaft. If it
malfunctions for any reason, then this would lead to catastrophic engine
failure. The understanding of journal bearing concept is essential for detailed
analysis of big-end bearings (see Chapter 18).

635

20.2   Brief review of literature

Hydrodynamic bearings are widely used in machines, particularly in engines and power plants. The bearing consists of a rotating journal inside the bore or a sleeve. The pressure developed in the variable-shaped *converging wedge* filled by a lubricant film supports the load. The load-carrying capacity at the desired speed is the prime objective in journal bearing design. The thickness of the lubricant film should also be sufficient to guard against interaction of mating surfaces. This requires the roughness of the surfaces to be included in any study. Furthermore, the bearing bushing or sleeve would in practice be out-of-round, either in manufacture or through wear or erosion caused at high shear rates of the film of lubricant. Finally, friction caused by viscous shear of the lubricant film or any asperity interaction of contiguous surfaces generates heat which in turn affects the lubricant viscosity, thus its load-carrying capacity.

In practice the design process involves a combination of simplified analytical methods and charts based on the *Sommerfeld number*, eccentricity ratio, bearing geometry (such as length-to-diameter ratio and nominal clearance-to-radius ratio). These parameters, together with lubricant viscosity variation with pressure and temperature (see Chapter 5) and operating conditions; load and journal speed go a long way in the design process (see Chapters 18 and 19). However, the simplified analytical methods are based on assumptions that disregard certain detail to yield solutions within often very limited industrial timescales and also take into account the commercial viability of design and development as pointed out in Chapters 8 and 9. Reason and Narang (1981) have presented a simple technique amenable to hand calculation for rapid design of steady state journal bearing behaviour. For cases that simplified solutions are deemed to yield inaccurate predictions, design charts are made from previous numerical undertakings, available in the literature as pointed out in Chapter 18 and by Gohar and Rahnejat (2008). For example, Pinkus (1960) has developed a complete set of numerical solutions in the form of design charts for load vector in any arbitrary position with respect to two oil grooves of the journal bearing. Values of *eccentricity ratio*, power loss and oil flow are evaluated as functions of *load angle* (sometimes referred to as the *attitude angle*). The optimal position of the load angle is obtained, at which the load capacity and the lubricant flow are maximised for least power loss. Other approaches have also been used such as the mobility method, described in Chapter 19.

Journal bearing behaviour, however, is quite complex, owing to its imperfect geometry as already mentioned and the fact that ideal smooth contiguous surfaces cannot be assumed (see also Chapters 2 and 3). Furthermore, bearings are subject to stop–start motions or acceleration–deceleration in use as well as cyclic loading (for example due to combustion loading or inertial imbalance or both). Although approximate solutions for some of these effects are found, such as the mobility method (see Chapter 19), a more comprehensive numerical analysis cannot be avoided if detailed data (at least at level of research work) is required. There have been many such solutions. For example, Hasimoto (1992, 1997) provided dynamic behaviour of an elliptical journal bearing with turbulent inertial flow for various ellipticity ratios. He provided the transient journal centre motion and the corresponding pressure distribution in a graphic form. Crosby (1992) also investigated the performance of a journal bearing with a slightly irregular bore, assumed to be elliptical.

Other salient features of bearing surfaces also play crucial roles in journal bearings; for example, under stop–start or cold start conditions with large loads, depletion of the lubricant film can lead to asperity interactions. Therefore, mechanical and topographical properties of surfaces play important roles in tribological performance (see Chapters 2 and 3). For bearings subject to high loads and thin films often a soft overlay on bearing shell/bushing is used which deforms under sufficient generated lubricant pressures and enhance the gap between the contiguous solids (see also Chapters 4, 8 and 18). The condition is a form of soft elastohydrodynamic lubrication (EHL), sometimes referred to as *iso-viscous elastic*, because often the pressures are insufficient to alter the lubricant viscosity in an appreciable manner (see Gohar and Rahnejat, 2008). These soft overlays also act as a form of filter where wear debris floating in engine oil are embedded into them (see also Chapter 8). Often very few of the contact asperities themselves carry the not inappreciable loads and deform plastically (see Chapter 2). There have been various in-depth studies of these conditions, such as by Elrod (1973) who investigated thin film lubrication of journal bearings for Newtonian fluids (see Chapter 5) with surfaces possessing striated roughness or grooving. He proposed a modified version of the Reynolds equation for ultra-thin gas films, estimating their load-carrying capacity.

Majumdar (1999) studied the influence of roughness parameter and pattern (longitudinal, transverse and isotropic) on steady state and dynamic characteristics of hydrodynamic journal bearings. He showed that transverse roughness tended to increase the load-carrying capacity and journal stability (see Chapters 18 and 19), while isotropic roughness decreased these. Longitudinal roughness showed a small reduction in load capacity. This approach is useful in control of manufacturing processes used to prepare the surfaces of both the journal and its bushing.

Two other aspects, highlighted above have also been subjected to investigations, namely lubricant shear characteristics and thermal effects (see Chapters 5 and 6). Majumdar (1997) provided a thermo-hydrodynamic analysis of submerged oil journal bearing, taking into account rough contacting surfaces. He took into account the total load-carrying capacity due to combined thermo-hydrodynamics and asperity contact pressures. A Gaussian distribution of the surface asperity heights was considered (see Chapters 2 and 3). The behaviour of very thin films in bearings subject to high shear rates is viscoelastic as noted by Evans and Johnson (1986), and simplified procedures can be used to account for this as shown by Gohar and Rahnejat (2008), who refer to the works of Evans and Johnson (1986). For finite-width journal bearings Jang and Chang (1988) have presented solutions for non-Newtonian lubricant behaviour (see also Chapter 6), using a power law model for *lubricant thixotropy*.

This rather brief review points to the rather complex behaviour of journal bearings. Most of the phenomena highlighted occur in the case of big-end bearings. Thus, except for a need for quick analyses and in the case of rather less complex camshaft and crankshaft main support bearings, a more realistic yet complex analysis is usually justified. For the case of big-end bearings a realistic analysis should include geometrical irregularities such as bore ellipticity, as well as surface roughness and thixotropy of lubricant. This is the approach highlighted in the rest of this chapter.

20.3   Bearing geometry

For the conventional *assumed* circular bushing the film shape is that described in Chapter 18. The simple geometrical proof is provided by Gohar and Rahnejat (2008). Thus:

$$h = c(1 + \varepsilon \cos\varphi) \tag{20.1}$$

where $c = r_b - r_j$ is the nominal designed clearance, and $\varepsilon = e/c$ is the *eccentricity ratio*, with e the eccentricity between the centre of the journal from that of the bushing. φ is the angle measured from the line of centres (see Fig. 20.1).

However, in practice, due to manufacturing irregularities or as a result of operational effects (thermoelastic stresses in the bore), an *elliptic bore* shape results. This is particularly the case in big-end bearings, some of which are referred to as *lemon shape bearings*. The ellipticity of the bore alters the film profile and subsequently the tribological performance of the bearing. In such case, the film shape is corrected according to the non-circularity (out-of-roundness) of the bore. For an elliptic bore:

$$h = c (1 + G \cos^2 \varphi + \varepsilon \cos \varphi) \tag{20.2}$$

where

$$G = \left(\frac{R_{maj} - R_{min}}{c} \right) \tag{20.3}$$

where R_{min} and R_{maj} are shown in Fig. 20.2.

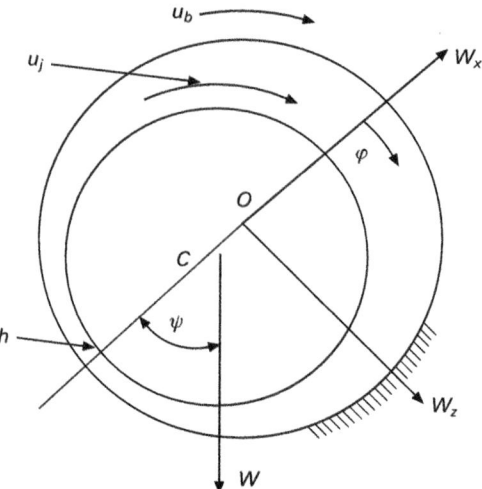

20.1 Geometry, kinematics and forces in a journal bearing.

The non-circularity, G, is usually in the range $(0 \leq G \leq 1.0)$. Figure 20.3 shows the effect of bore ellipticity on film thickness for a given ellipicity ratio and nominal clearance. As the figure shows the film geometry for a circular bushing/sleeve or bore yields a defined conjunction (***wedge***), where the lubricant film is formed. This forms the basis of most analytical models, such as that in Chapters 5 and 18. However, any deviation from circularity creates secondary conjunctions due to the ***multi-lobe effect***. This effect, in fact helps journal stability, a reason for deliberate introduction of so-called *lemon-shaped bearings*.

20.4   Lubricant rheology

Rheology is the study of lubricant and its properties. Chapter 5 deals with lubricant ***viscosity*** and how this is affected by pressure and temperature. Density of the lubricant also alters when subjected to pressure, such as in EHL (see Chapters 5 and 6). Other properties of lubricants include surface tension, aiding wetting of surfaces (see Gohar and Rahnejat, 2008, and Chapters 3 and 33), as well as its thermal conductivity to carry away the generated heat in conjunctions by convection cooling (see Chapter 5). Additives are often added to lubricants to enhance these rheological properties or enable them to form thin, low-shear strength surface films which act as an anti-wear layer, particularly at start up (see Chapters 5 and 8), when there is insufficient entraining motion to form a hydrodynamic film. The low-shear strength surface films do not follow the bulk rheological characteristics of the lubricant. In particular their behaviour is molecular near surface asperities.

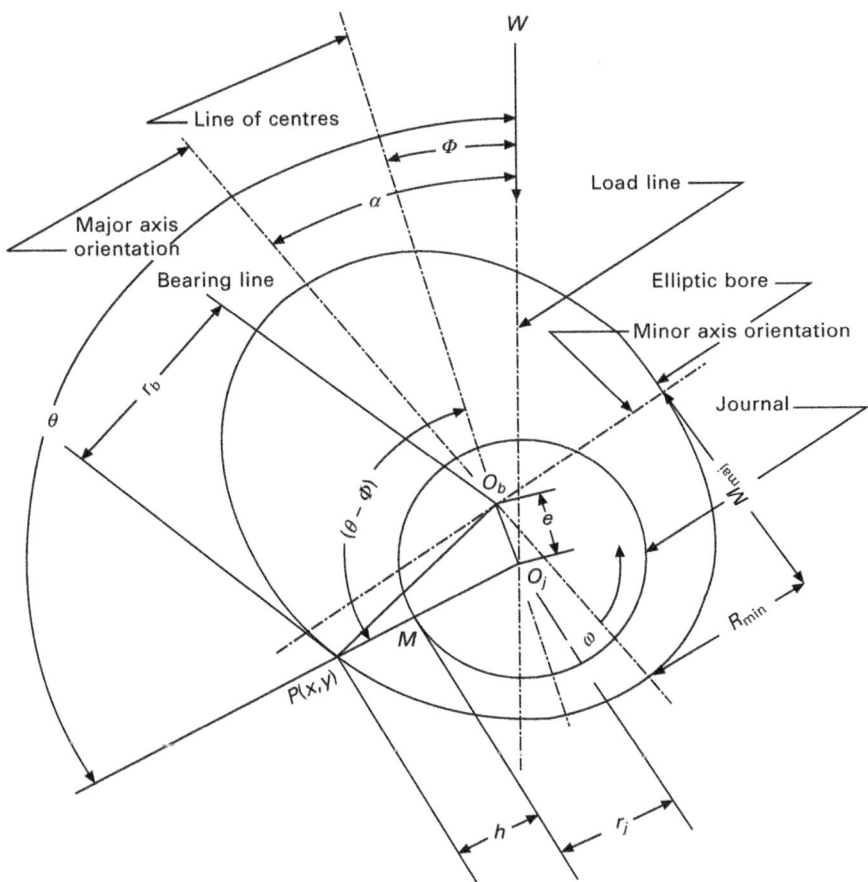

20.2 An elliptic bore journal bearing.

Some aspects of this behaviour at low loads are described in chapter 3. Thus, in the model in this chapter the two basic lubricant properties are density and viscosity.

20.4.1 Lubricant density

Pressure influences *lubricant density* only at significant values. Therefore, under hydrodynamic conditions, typical of most *engine bearings* due to the conforming nature of contact (relatively large contact area), density hardly alters. Under EHL conditions, the lubricant density increases after a *solidification pressure* is reached, making it an amorphous solid. Such conditions occur regularly in concentrated counterforming contacts, such as in cam–follower pairs (see Chapters 6 and 16). However, they can also occur in big-end bearings when subjected to quite high combustion forces,

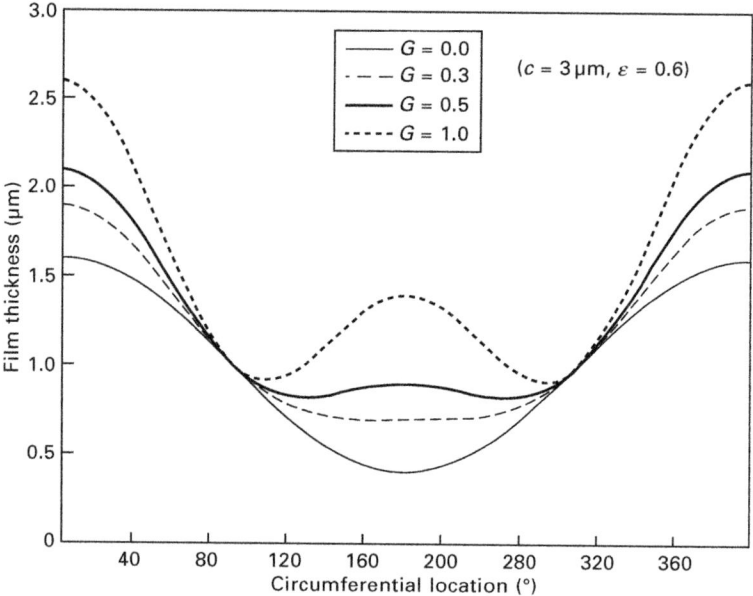

20.3 Effect of bore ellipticity on film thickness.

typical of diesel engines. Hence, change in density due to pressure should be included using a relationship, such as that given by Dowson and Higginson (1959):

$$\bar{\rho} = \frac{\rho}{\rho_0} = 1 + \frac{0.6p}{1 + 1.7p} \tag{20.4}$$

where p is pressure at which the density of lubricant is ρ, and ρ_0 is the density at atmospheric pressure.

20.4.2 Lubricant viscosity

Lubricant viscosity is the internal resistance of a fluid to flow (see Chapter 5). Dynamic *viscosity* is expressed in terms of velocity gradient and shear stress. It varies with both pressure and temperature.

Temperature dependency of viscosity can be obtained as (Cameron, 1970; Gohar and Rahnejat, 2008):

$$\eta = \eta_0 e^{-\beta_{thermal}\Delta\theta} \tag{20.5}$$

where $\Delta\theta$ is the temperature rise in the conjunction relative to that of the bulk oil temperature, where the viscosity η_s is for the supplied flow to the bearing.

The pressure dependence of viscosity is given by the **Barus law** (Barus, 1893) (also see Chapter 5) as:

$$\eta = \eta_0 e^{\alpha p} \tag{20.6}$$

When the pressures are high, such that $\alpha p \gg 1$ (typical of **EHL**), then predictions by Barus law become significantly erroneous and the **Roelands equation** (Roelands, 1966) is used instead (see Chapter 5).

For journal bearings one can combine (20.5) and (20.6) to yield:

$$\eta = \eta_0 e^{\alpha p - \beta_{\text{thermal}} \Delta \theta} \tag{20.7}$$

For higher pressures, combined with temperature effect a modified form of viscosity equation is described in Chapter 5 (also see Gohar and Rahnejat, 2008).

20.4.3 Pressure distribution: the Reynolds equation

Pressures in any conjunction are obtained by solution of the **Reynolds equation**, which describes the flow. This is a fundamental equation in lubricated contacts and is derived and described adequately in Chapter 5:

$$\frac{1}{r_j^2} \frac{\partial}{\partial \varphi}\left(\frac{\rho h^3}{\eta} \frac{\partial p}{\partial \varphi} \right) + \frac{\partial}{\partial y}\left(\frac{\rho h^3}{\eta} \frac{\partial p}{\partial y} \right) = 12\left(u_{\text{av}} \frac{1}{r_j} \frac{\partial \rho h}{\partial \varphi} + v_{\text{av}} \frac{\partial \rho h}{\partial y} + \frac{\partial h}{\partial t} \right) \tag{20.8}$$

where the **speed of entraining motion** in the direction of rotation of line of centres (note: $x = r_j \varphi$) is $u_{av} = \frac{1}{2}(u_j + u_b) = \frac{1}{2}\omega_j r_j$, when the bush is assumed to be stationary and that of **side-leakage** along the y direction, $v_{\text{av}} = \frac{1}{2}(v_j + v_b)$.

To include the effect of roughness of contiguous surfaces (journal and bushing) in the analysis, the approach highlighted by Christensen and Tonder (1973) is used. In this approach the fluid film geometry is split up into two parts; a nominal film thickness which measures large-scale variations (long wavelength disturbances) and roughness measured from the nominal level. Let h_n be the nominal film thickness and h_s the random or stochastic film thickness, then:

$$h = h_n + h_s \tag{20.9}$$

Therefore, the Reynolds equation is modified to (assuming no side-leakage):

$$\frac{1}{r_j^2} \frac{\partial}{\partial \phi}\left[\frac{\rho E(h^3)}{\eta} \frac{\partial p}{\partial \phi} \right] + \frac{\partial}{\partial y}\left[\frac{\rho E(h^3)}{\eta} \frac{\partial p}{\partial y} \right] = 12\left[\frac{1}{r_j} u_{av} \frac{\partial E(h)}{\partial \phi} + \frac{\partial E(h)}{\partial t} \right] \tag{20.10}$$

where $E(h)$ denotes the expectancy of the film thickness h. This form of the Reynolds equation can be generalised as (in this instance with no squeeze film action, i.e. steady state entraining):

$$\frac{1}{r^2}\frac{\partial}{\partial\phi}\left[\frac{\partial P}{\partial\phi}\Psi_1(h)\right] + \frac{\partial}{\partial y}\left[\frac{\partial P}{\partial y}\Psi_2(h)\right] = \frac{12}{r}\eta u\frac{\partial}{\partial\phi}\Psi_3(h) \qquad (20.11)$$

where $\Psi_1(h)$, $\Psi_2(h)$ and $\Psi_3(h)$ for different roughness are given in Table 20.1.

The expectancy of film as a random process depends on the surface roughness and orientation. For isotropic roughness (uniform orientation):

$$E(h) = h_n \qquad (20.12)$$

and

$$E(h^3) = \frac{1}{c^3}\left[h_n^3 + \frac{h_n K_r^2}{3}\right]$$

where:

$$K_r = \frac{\sigma_{rms}}{c}$$

The value of film expectancy for other roughness types can be obtained from Christensen and Tonder (1973). Though there is a chance of asperity interaction in boundary regime of lubrication (see Chapters 3 and 6), a rough surface would usually offer better tribological performance. Chapter 3 describes the reasons for this in some detail.

20.4.4 Thermal effects: energy equation

Rapid shearing of the lubricant occurs in many tribological conjunctions. This leads to a rise in temperature, which affects the lubricant viscosity. This constitutes the ***thixotropic behaviour*** of the lubricant and accounts for the ***shear thinning*** of the film. Thermal contact analysis commenced in the early 1950s with Crook (1961). It remains a major topic in both tribology and rheology. Cheng (1983) carried out an adiabatic analysis of finite-width journal bearing using a power law model. Gohar and Rahnejat (2008) describe

Table 20.1 Film function interpretation

Film function	Longitudinal	Transverse	Isotropic/Uniform
$\Psi_1(h)$	$E(h^3)$	$[1/E(1/h^3]$	$E(h^3)$
$\Psi_2(h)$	$[1/E(1/h^3]$	$E(h^3)$	$E(h^3)$
$\Psi_3(h)$	$[1/E(1/h)]$	$4E(1/h)$	$E(h)$

the **Peclet number** (see also Chapter 5) to determine the dominant mode of heat transfer (convection or conduction) from a lubricated contact. Mishra (2007) improved the thermal model of Jang and Chang (1988) for an elliptic bore journal bearing. All these analyses suggest a combined solution of the Reynolds equation and energy equation would be necessary to fully describe the contact conditions (film thickness, pressure and temperature).

Gohar and Rahnejat (2008) show that the thicker films formed in conforming contacts, such as in journal bearings, carry away most of the generated heat. Thus, the assumption of convection cooling can be made and temperature variation across the oil film may be neglected. The **energy equation** in this case becomes:

$$\rho C_p \left(\frac{u}{r} \frac{\partial \theta}{\partial \varphi} + w \frac{\partial \theta}{\partial y} \right) = \eta \left[\left(\frac{\partial u}{\partial z} \right)^2 + \left(\frac{\partial w}{\partial z} \right)^2 \right] \tag{20.13}$$

This can be re-written in the form proposed by Jang and Chang (1988) as:

$$A_{\text{energy}} \frac{\partial \theta^*}{\partial \varphi^*} + B_{\text{energy}} \frac{\partial \theta^*}{\partial y^*} = \alpha_{\text{dissipation}} \frac{\eta^*}{h^{*2}} (E_{\text{energy}} + F_{\text{energy}}) \tag{20.14}$$

where:

$$A_{\text{energy}} = \frac{1}{2} - \frac{1}{12} \left(\frac{h^{*2}}{\eta^*} \frac{\partial p^*}{\partial \varphi^*} \right), \quad B_{\text{energy}} = \frac{1}{12} \left(\frac{h^{*2}}{\eta^*} \frac{\partial p^*}{\partial y^*} \right), \quad E_{\text{energy}}$$

$$= 1 + \frac{1}{12} \left(\frac{h^{*2}}{\eta^*} \right)^2 \left(\frac{\partial p^*}{\partial \varphi^*} \right)^2 \tag{20.15}$$

and

$$F_{\text{energy}} = \frac{1}{12} \left(\frac{h^*}{\eta^*} \right)^2 \left(\frac{\partial p^*}{\partial y^*} \right)^2$$

Equations (20.14) and (20.15) are for a non-Newtonian lubricant. Mishra (2007) simplified the same and derived an expression for **contact temperature** in the case of a Newtonian fluid as:

$$\theta^* = \frac{\left[\frac{1}{2} \frac{\partial \theta^*}{\partial \varphi^*} - \left\{ \begin{array}{c} \frac{\alpha_{\text{dissipation}}}{12} h^{*2} \left(\frac{\partial p^*}{\partial \varphi^*} \right)^2 + \\ \frac{\alpha_{\text{dissipation}}}{12} h^{*2} \left(\frac{\partial p^*}{\partial y^*} \right)^2 \end{array} \right\} \right] \left[\left\{ \begin{array}{c} \frac{1}{12} h^{*2} \frac{\partial p^*}{\partial \varphi^*} \frac{\partial \theta^*}{\partial \varphi^*} \\ + \frac{1}{12} h^{*2} \frac{\partial p^*}{\partial y^*} \frac{\partial \theta^*}{\partial y^*} \end{array} \right\} + \frac{\alpha_{\text{dissipation}}}{h^{*2}} \right]}{\left[\left\{ \frac{1}{12} h^{*2} \frac{\partial p^*}{\partial \varphi^*} \frac{\partial \theta^*}{\partial \varphi^*} + \frac{1}{12} h^{*2} \frac{\partial p^*}{\partial y^*} \frac{\partial \theta^*}{\partial y^*} \right\} - \frac{2\alpha_{\text{dissipation}}}{h^{*2}} \right]} \tag{20.16}$$

where the **heat dissipation factor** is:

$$\alpha_{\text{dissipation}} = \frac{(2\pi N)\eta_0 \, \beta_{\text{thermal}}}{\rho C_p}\left(\frac{r_j}{c}\right)^2$$
(20.17)

20.5   Bearing load

To undertake a realistic analysis of big-end bearings the applied load should be obtained at any instant of time. This requires a simultaneous solution for crank-connecting rod–piston system with instantaneous applied gas pressure and the rough thermo-hydrodynamic/elastohydrodynamic of the journal–bush conjunction. This constitutes a full transient analysis. However, the simulation time would be quite long and instructive representative information can equally well be obtained by undertaking steady state solutions at the most severe conditions (e.g. high loads and speeds).

At any instant of time, the applied load to a big-end bearing can be regarded as a sum of the forces due to the combustion gas force and inertial imbalances due to rotational and reciprocating masses. The former is the mass of the crank and proportion of the mass of the connecting rod in rotation, while the latter is the total mass in translation; that of piston assembly and proportion of mass of the connecting rod in translation. Rahnejat (1998) shows how these can be obtained. Three other simplifications can also be made to render a quicker solution. Firstly, the net bearing load is assumed to be due to the firing cylinder directly acting upon it. This means that the analysis is valid for a single cylinder (such as that reported below) and only approximate for a multi-cylinder engine. Secondly, the effect of *engine order* fluctuations due to combustion signature and inertial imbalances can be confined to a few harmonics only. Rahnejat (1998) shows how these variations affect engine dynamics, while Kushwaha *et al.* (2002) take the effect of *engine roughness* due to structural flexibility into account (see also Chapter 1). The third simplifying assumption here is to assume no component flexibility. In reality the inertial imbalance force transmitted through the connecting rod has been increasing in magnitude in modern engines (due to an ever-increasing desired output power). Additionally, there has been a trend in the past two decades to reduce *inertial imbalance* by reducing the mass of moving components. This has had the effect of structural deformation (within elastic limit) of components subject to load. The connecting rod is no exception, and the above trends have led to its extension and deflection, with significant amplitudes for high performance engines.

With these assumptions and observations taken into account the bearing load is obtained as a combination of the following.

20.5.1 Applied combustion force

This is the **gas force** due to instantaneous combustion pressure acting on the piston crown surface area as:

$$F_{gas} = \frac{p_{gas}A}{\left[1 - \left(\frac{R}{L}\right)^2 \sin^2 \psi\right]^{1/2}} \tag{20.18}$$

where $\psi = \omega_j t$ is the crank angle.

20.5.2 Induced inertial imbalance forces

The motion of the piston causes a **translational imbalance** (induced inertial force), which also has a contribution due to mass of the connecting rod. For low values of the ratio R/L higher order engine harmonics may be neglected, thus:

$$F_{reciprocating} = m_{reciprocation} r_j \frac{\omega_j^2}{\left[1 - \left(\frac{R}{L}\right)^2 \sin^2 \psi\right]^{1/2}} \left(\cos\psi + \frac{R}{L}\cos 2\psi\right)$$

$$\tag{20.19}$$

The **rotational imbalance** is also obtained as:

$$F_{rotating} = m_{rotating} R\omega_j^2 \tag{20.20}$$

Thus, the bearing load is:

$$F = F_{gas} + F_{reciprocating} + F_{rotating} \tag{20.21}$$

20.6   Method of solution

A simultaneous solution to equations (20.2)–(20.4), (20.7), (20.11)–(20.12), (20.16)–(20.17) and (20.21) is required. The iterative process comprises two convergence steps: pressure convergence and load balance.

Boundary conditions are needed for the solution of the Reynolds equation (20.11). A convenient assumption for inlet to the converging wedge of the journal–bushing conjunction is to assume **fully flooded condition**: $p = 0$ at $\varphi \approx 0$. This is what is strived for in practice, but in reality it is hardly ever achieved, as the lubricant is mostly gravity fed into the bearing. Thus, starved conditions are most likely and a detailed analysis would have to take this into account (see Cameron, 1970). With higher speeds, film **starvation** becomes more pronounced as an insufficient reservoir of lubricant exists for the entraining action to draw upon. At the outlet of the film the diverging gap induces a fall in pressure, which can cause **cavitation**, forming finger-shaped air bubbles in the lubricant. This phenomenon is similar to the frothy water in the wake of a speed boat. Bursting bubbles of air can contribute to the erosion of the bushing surface. Reynolds or Swift–Steiber boundary condition may be used at the film outlet to specify the film rupture point, beyond which **cavitation** occurs: $p = dp/d\varphi = 0$ at $\varphi \approx \pi$ (in computation this translates to discarding all the negative pressures, thus demarcating the region of pressure).

The pressure convergence uses an under-relaxation factor Ω within each iteration step K for all elemental pressures that make-up the lubricant pressure distribution, $p_{i,j}$ (i, j being the grid point in the finite difference *computational molecule*) (see Gohar, 2001). Thus:

$$p_{i,j}^K = p_{i,j}^K + \Omega(p_{i,j}^K - p_{i,j}^{K-1}) \tag{20.22}$$

The convergence criterion for pressure is:

$$\text{Error} = \sum_i^n \sum_j^m \left| \frac{p_{i,j}^K - p_{i,j}^{K-1}}{p_{i,j}^K} \right| \leq 0.01 \tag{20.23}$$

Once the above convergence is met, the integrated pressure distribution (lubricant reaction) is obtained as: $W = r_j d\varphi \, dy \sum_i \sum_j p_{i,j}$ for a given journal *eccentricity ratio* ε. However, the true eccentricity is obtained by a balance of load between the applied load (equation (20.21)) and the calculated lubricant reaction F (the integrated pressure distribution). Thus, the eccentricity ratio should be varied until a load balance is achieved as:

$$\left| \frac{F - W}{F} \right| \leq 0.01 \tag{20.24}$$

If the above condition is not satisfied the eccentricity ratio is altered as follows, and the entire solution method is repeated:

$$\varepsilon^K = \varepsilon^{K-1} - \left(\xi \left| \frac{F - W}{F} \right| \right) \tag{20.25}$$

where ξ is referred to as the damping factor in the eccentricity relaxation method.

In a transient analysis the above procedure is repeated at small increments of the crank angle ψ (i.e. at small steps of time). The gas pressure then alters according to the combustion curve as $p_{\text{gas}} = f(\psi)$. Note also that in transient analysis the squeeze film term is:

$$\frac{\partial h}{\partial t} \approx \frac{h_{i,j}^{K} - h_{i,j}^{K-1}}{\Delta t}$$

(first order approximation), where Δt is a small step of time in the simulation process (typically in micro-second range for tribological problems).

20.7   A case study

To carry out an analysis one needs to know a number of input parameters. Those necessary are given in Table 20.2.

Figure 20.4 shows the variation of applied load components (described in Section 20.5). Note that for a certain engine speed ω_j (here at $N = 13\,000$ rpm), the rotational imbalance force remains constant throughout the loading cycle, while the translational (reciprocating) imbalance is a function of the speed, as well as the crank angle ψ. Its value attains a maximum value at the top dead centre (TDC) ($\psi = 360°$). The combustion gas force has its maximum value at $\psi = 373°$ (13° past the TDC). This is often the case in order to lessen the effect of *piston slapping* action. Therefore, total applied force on the big-end bearing is usually the largest at this position, marked on the figure by a vertical line. The result of an example analysis is shown at this location in Fig. 20.4.

Figure 20.5 shows the three-dimensional pressure distribution for a bearing with an eccentricity ratio of $\varepsilon = 0.6$, radial clearance of $c = 3\,\mu m$ and ellipticity parameter $G = 0.3$, at $\psi = 373°$. Note that the region of pressure is approximately: $0 \le \varphi \le \pi$ circumferentially and axially along the bearing width. The film thickness for this condition in the circumferential direction is

Table 20.2 Data for analysis

Attributes	Value
Rotating mass ($m_{rotation}$)	0.32 kg
Reciprocating mass ($m_{reciprocating}$)	0.48 kg
Crank throw (R)	0.036 m
Connecting rod length (L)	0.12 m
Piston crown surface area (A)	0.0043 m^2
Engine speed (N)	13000 rpm
Lubricant density (ρ)	881.46 kg m^{-3}
Specific heat (C_p)	1840 J kg^{-1} °C^{-1}
Reference viscosity (η_0)	4.06 × 10^{-3} Pa s
Coefficient of thermal expansion ($\beta_{thermal}$)	0.0315 °C^{-1}
Coefficient of piezo viscosity (α)	10^{-8} m^2 N^{-1}
Journal radius (r_j)	29.28 mm
Damping coefficient (Ω)	(10^{-8} – 10^{-10})
Ambient temperature (θ_0)	38 °C
Bore minor radius (R_{min})	29.283 mm
Radial clearance (c)	3 μm

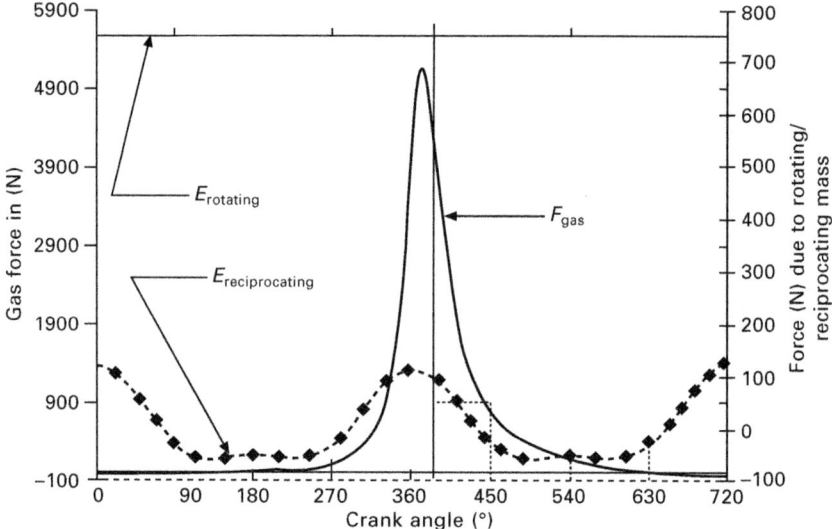

20.4 Force applied to the crankshaft bearing in an engine cycle.

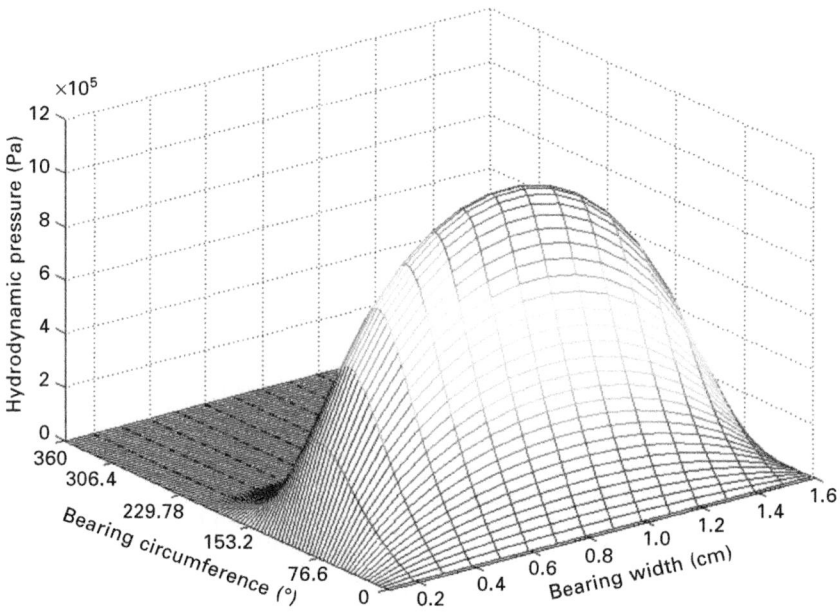

20.5 Hydrodynamic pressure distribution of an elliptic bore bearing (at $\psi = 373°$).

shown in Fig. 20.3. A minimum film thickness of 0.7 μm is predicted for the assumed isotropic roughness. Note that as the ellipticity parameter increases a number of conjunctions emerge (instead of the usual single conjunction for an assumed idealised circular bearing $G = 0$). This enhances the load-carrying capacity and stability of the bearing.

The other important design parameter, required from a detailed analysis, is the temperature distribution. For the example analysis shown here, the temperature distribution is shown in Fig. 20.6. For a given heat dissipation factor, given by (20.17), the temperature distribution is a function of the film thickness and pressure gradients in the direction of entraining motion and due to side leakage (equation (20.16)). Therefore, maximum temperatures are encountered in the converging wedge and within the conjunction.

With predicted film thickness, pressure and temperature distributions all the necessary basic design calculations can be carried out as described in Chapter 18.

20.8   Effect of surface roughness and pattern

Lubricant film shear causes viscous friction. Sometimes due to heavy or fluctuating loads the journal becomes more eccentric, encouraging severe asperity interactions. Hence, inclusion of rough surface in big-end bearing analysis becomes important.

The different roughness patterns are shown in Fig. 20.7. These are transverse, longitudinal and isotropic.

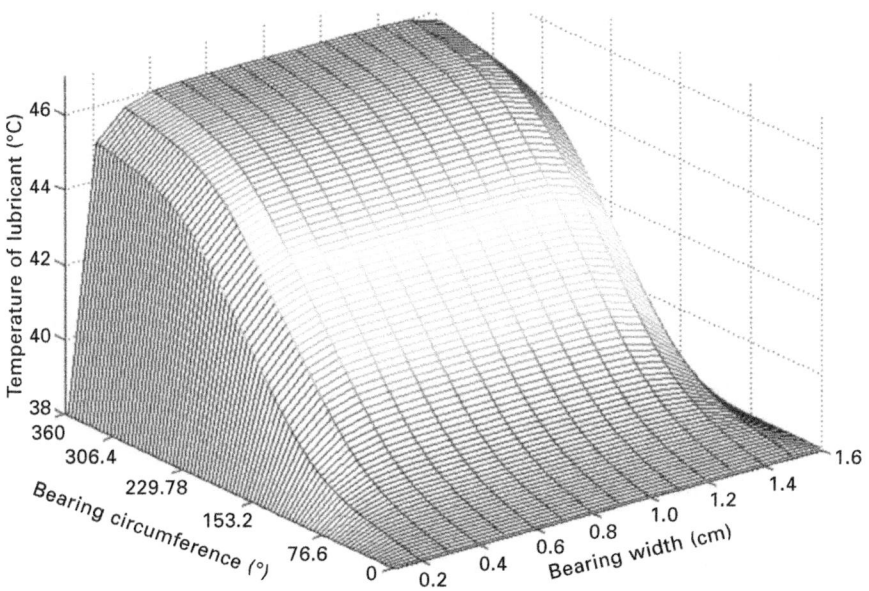

20.6 Three-dimensional temperature profile (at ψ 373°).

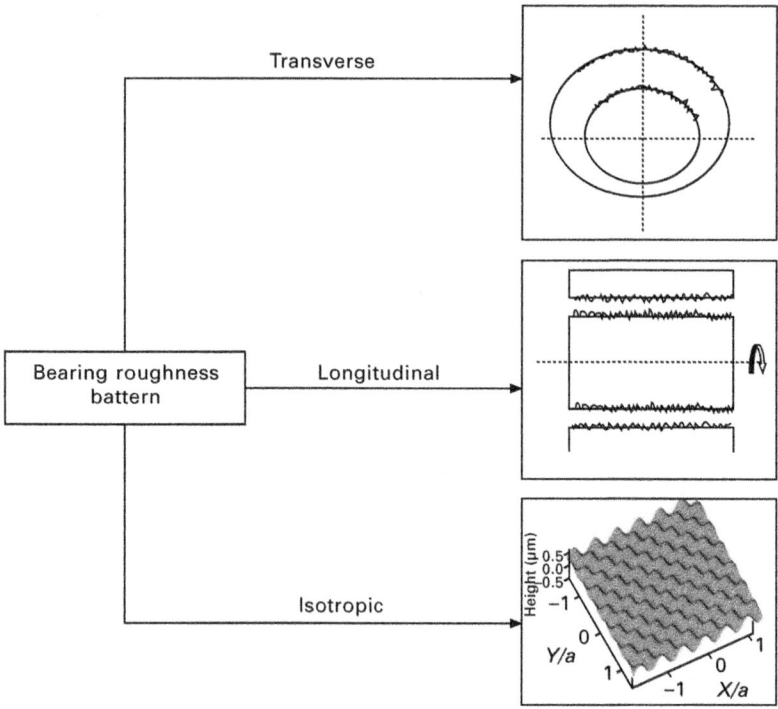

20.7 Roughness pattern.

20.8.1 Transverse roughness pattern

In the case of ***transverse roughness***, the valleys and ridges of the roughness pattern run along the bearing circumference. Such a pattern is able to develop significant hydrodynamic pressures and the flow rate of the lubricant is reduced. Some problems associated with this roughness pattern are heat generation and potential oil degradation.

20.8.2 Longitudinal roughness pattern

The valleys and ridges of the roughness pattern run along the axial width of the bearing in the case of ***longitudinal roughness***. This can cause comparatively higher lubricant side leakage. It can, therefore, aid convection cooling, but reduce hydrodynamic pressures and thus decrease the load-carrying capacity. Loss of lubricant can also lead to increased metal-to-metal contact and increased boundary friction.

20.8.3 Isotropic roughness

Isotropic roughness refers to the pattern of roughness with valleys and ridges running in both the axial and circumferential directions. Therefore, it aids both cooling and generation of higher hydrodynamic pressures, thus is most suitable for journal bearing surfaces. A rough surface has better oil retention properties, developing some micro-conjunctions. The principle is used in surface texturing (see also Chapters 13 and 14, and Gohar and Rahnejat, 2008).

Figure 20.8 shows a comparison in generated hydrodynamic pressures for the case of smooth and rough surfaces, where $K_r = \sigma_{rms}/c$. For the cases shown, the net hydrodynamic pressures in the conjunction are higher (higher load-carrying capacity) for a rough bearing. The analysis is based on the stochastic approach. However, the implementation of roughness effect on the Reynolds equation needs careful evaluation of the 3-D roughness topography and should ideally be deterministic in nature. Stochastic or statistical representation of roughness can significantly deviate from the actual real surfaces.

Figure 20.9 shows the ratio of generated friction in an elliptic bore bearing to that of a circular one with thermal effect taken into account in both cases. It is clear that non-circularity causes multiple wedge effects, thus enhancing film thickness and consequently reducing friction. Hence, the values of the ratio are all below unity and decrease further with greater non-circularity. As eccentricity ratio is increased friction also increases as thinner films are formed with decreasing number of acting effective wedges.

20.9   Tribological problems in big-end bearing

Big-end bearing instability can often result due to the blockage of the oil supply hole or an insufficient oil pressure or due to vibration. The instability

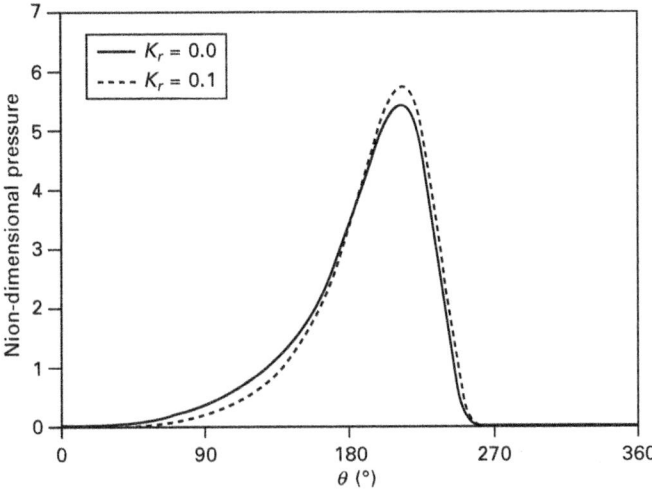

20.8 Comparison of smooth and rough bearing pressure (at 373°).

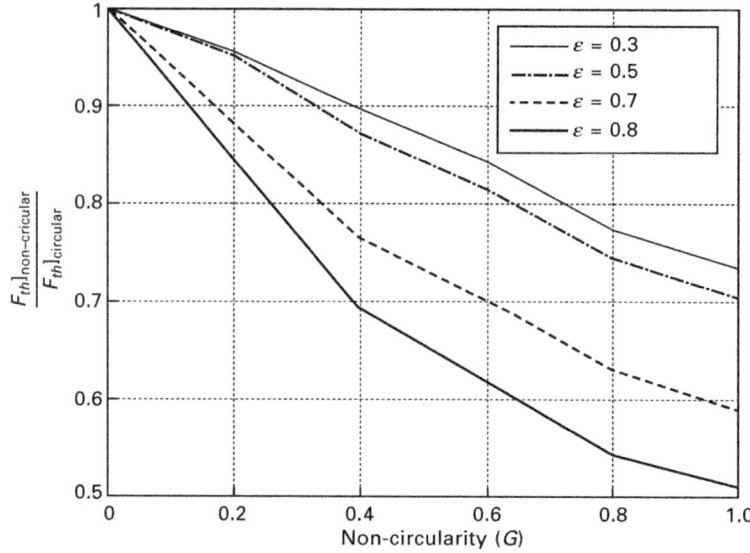

20.9 Variation of friction force ratio with bore non-circularity (at 373°).

leads to oil whirl or oil whip. This is a main source of primary component failure in engines. Chapter 18 provides some description of this (also see Gohar and Rahnejat, 2008).

20.9.1 Oil whirl

Oil whirl is probably the most common cause of sub-synchronous instability in hydrodynamic journal bearings. Typically, the oil film itself flows around the journal to lubricate and cool the bearing. This develops an average speed slightly less than 50% of the journal surface speed. Normally, the shaft rides on the crest of an oil pressure gradient, rising slightly up the side of the bearing somewhat off vertical at a given stable attitude angle and eccentricity. The amount of rise depends on the rotor speed, rotor weight and oil pressure. With the shaft operating eccentrically relative to the bearing centre, the lubricant is drawn into a wedge to produce a pressurised load-carrying film.

If the shaft receives a disturbing force such as a sudden shock, it can momentarily increase the eccentricity from its equilibrium position. When this occurs, additional oil is immediately pumped into the space vacated by the shaft. This results in an increased pressure in the load-carrying film, creating an additional force between the oil film and the shaft. In this case, the oil film can actually drive the shaft ahead of it in a forward circular motion and into a whirling path around the bearing within the bearing clearance. If there is sufficient damping within the system, the shaft can be returned to its normal position and stability. Otherwise, the shaft would continue in its whirling motion, which may become rather violent depending on several parameters.

20.9.2 Oil whirl instability

Oil whirl demonstrates the following characteristics:

- It can be induced by several conditions including:
  - light dynamic and preload forces,
  - excessive bearing wear or clearance,
  - a change in oil properties (primarily shear viscosity),
  - an increase or decrease in oil pressure or oil temperature; improper bearing design (sometimes an over design for the actual shaft loading),
  - fluid leakage from the shaft labyrinth seals (so-called the **Alford force** or the *aerodynamic force*),
  - change in internal damping (hysteretic, or material damping, or dry (coulomb) friction),
  - gyroscopic effects, especially on overhung rotors with excessive overhang.
  Any of these conditions can induce oil whirl after a disturbing force induces an initial rotor deflection.
- Sometimes machines exhibit oil whirl intermittently due to externally applied vibratory forces, which are transmitted to the bearing. In these cases, the vibratory forces can have the same frequency as the oil whirl frequency and cause resonant conditions. This would depend on the magnitude of the vibratory forces.
- Oil whirl is easily recognised by its unusual vibration frequency which is generally 40–48% of shaft angular velocity. This is uncomfortably close to the fundamental firing frequency of the four-stroke process. However, sufficient component rigidity and firing phases in multi-cylinder engines guard against this problem. However, half-engine order is quite pronounced in single cylinder engines and with progressively reduced component rigidity (the trend in use of lighter components) there is an increased change of vibration-induced whirl.
- As a rough guide, oil whirl can become significant when the vibration amplitudes reach 40–50% of the nominal bearing clearance.
- Various corrective measures can be sought to guard against whirl conditions. Chapter 18 and Gohar and Rahnejat (2008) point to increased mass to achieve higher eccentricity ratios and thus higher pressures, or increase speed at lower eccentricity values to operate in stable regions of bearing performance. Other temporary corrective measures include changing the temperature of the oil (thus, its viscosity), purposely

introducing a slight unbalance or misalignment to increase the loading, temporarily shifting the alignment by heating or cooling support legs or in some cases grooving the bearing surface to disrupt the lubricant wedge.

- Permanent corrective steps to resolve the oil whirl problem include installing a new bearing shell with proper clearances, preloading the bearing by an internal oil pressure dam, or completely changing the bearing type to oil film bearings that are less susceptible to oil whirl (including axial-groove bearings, lobed bearings or tilting pad bearings). The tilting pad bearing is a good choice because each segment or pad develops a pressurised oil wedge tending to centre the shaft in the bearing, thereby increasing the system damping and overall stability.

20.9.3 Oil whip instability

Oil whip occurs on those machines which are subject to oil whirl when the oil whirl frequency coincides with and becomes locked into a system's natural frequency (often a rotor balance or critical speed frequency). Oil whip is a lateral forward precessional sub-harmonic vibration of the rotor, notably 50% of the speed of the shaft. When a shaft goes into oil whip, its dominant dynamic factors become mass and stiffness in particular; and its amplitude is limited only by the bearing clearance. Left uncorrected, oil whip may cause destructive vibration, resulting in catastrophic failure, often in a relatively short period of time.

20.9.4 Dry whip

Dry whip occurs in journal bearings, subjected to either lack of lubrication or the use of a wrong lubricant. When this occurs, excessive friction is generated between the stationary bearing bushing and the rotating journal. This friction can cause self-excited vibration in the bearing. Dry whip can also be caused by journal bearings with excessive nominal clearance, as well as those with insufficient clearance.

The dry whip condition is similar to rubbing a moistened finger over a dry pane of glass. It generates a frequency specifically dependent upon the shaft and construction materials, geometries and lubricant properties. Normally, this frequency emanates as a high *squealing noise* similar to that generated by dry rolling element bearings or a phenomenon observed in vehicle brake systems (brake squeal). The frequency content itself is usually not an integer multiple of the operating speed. When dry whip is suspected, it must be taken care of quickly in order to prevent a potential catastrophic failure.

Some other tribological problems in big-end bearing are listed in Table 20.3.

Table 20.3 Some potential tribological problems in big-end bearings

Sl. no	Failure mode	Cause of failure	Effects	Remedial measures
1	Cavitation/aeration	Reduced oil supply Pressure	Pre-starvation Shock Metal erosion	Regulate pump pressure to avoid pressure drop
2	Loss of lubrication/starvation	Oil whirl/oil whip Oil supply disruption Blockage in the oil supply hole due to severe aeration	Metal-to-metal contact, wear and tear of bearing surfaces	Oil impurity check, continuous oil supply, higher oil supply pressure
3	Metal erosion	Shock load Cavitation Starvation Overload	Reduced bearing life Catastrophic failure	Avoid oil degradation Limiting load/speed Coating/white metalling
4	Mechanical shocking	Oil whirl Oil whip Dry whip	Strong vibration Bearing instability	Proper bearing design Temperature control Intentional misalignment Bearing grooving
5	Overheating	Load/speed over limit for prolonged times Surface contact	Lubrication failure due to oil degradation Asperity contact due to starvation	Running-in speed/load control Vibration reduction
6	System component failure	Metal erosion Vibrating at natural frequency Oil whirl/ oil whip	Connecting rod throw Engine failure Subsystem failure	Monitor and control load/speed

20.10   References

Barus, C. (1893), 'Isothermals, isopietics and isometrics in relation to viscosity', *American J. Sci.*, 3rd Series, **45**, pp. 87–96

Cameron, A. (1970), *Basic Lubrication Theory*, Longman Ltd

Cheng. H.S. (1983), 'Tribology in the 80s', (NASA CP-2300) pp. 161–70

Christensen, H. and Tonder, K. (1973), 'The hydrodynamic lubrication of rough journal bearings', *Trans. ASME, J. Lubn. Tech.*, **72**, pp. 166–172

Crook, A.W. (1961), 'The lubrication of rollers: part III: A theoretical discussion of friction and the temperatures in the oil film', *Phil. Trans. A*, **254**, 237–258

Crosby, W.A. (1992), 'An investigation of performance of journal bearings with slightly irregular bore', *Trib. Int.*, **25**(3), pp. 199–204

Dowson, D. and Higginson, G.R. (1959), 'A numerical solution to the elastohydrodynamic problem', *Proc. Instn., Mech. Engrs., Part C: J. Mech. Engng. Sci.*, **1**, pp. 6–15.

Elrod, H.G. (1973), 'Thin-film Lubrication Theory for Newtonian Fluids with surfaces possessing striated roughness or grooving', *Trans. ASME, J. Lubn. Tech.*, **73**, pp. 484–489

Evans, C.R. and Johnson, K.L. (1986), 'The rheological properties of EHD lubricants', *Proc. Instn. Mech. Engrs., Part C: J. Mech. Engng. Sci.*, **200**, pp. 313–324

Gohar, R. (2001), *Elastohydrodynamics*, Imperial College, London

Gohar, R. and Rahnejat, H. (2008), *Fundamental of Tribology*, Imperial College Press, London

Hasimoto, H. (1992), 'Dynamic characteristics analysis of short elliptical journal bearings in turbulent inertial flow regime', *STLE Trib. Trans.*, **35**(4), pp. 619–626

Hasimoto, H. (1997), 'Optimum design of high speed short journal bearing by mathematical programming', *STLE Trib. Trans.*, **40**(2), pp. 283–293

Jang, J.Y. and Chang, C.C. (1988), 'Adiabatic analysis of finite width journal bearing with non-Newtonian lubricants', *Wear*, **112**, pp. 63–75

Kushwaha, M., Gupta, S., Kelly, P. and Rahnejat, H. (2002), 'Elasto-multi-body dynamics of a multi-cylinder internal combustion engine', *Proc. Instn. Mech. Engrs., Part K: J. Multi-body Dyn.*, **216**, pp. 281–293

Majumdar, B.C. (1997), 'Thermodynamic analysis of submerged oil journal bearings considering surface roughness effects', *Trans. ASME, J. Lubn. Tech.* **119**, pp. 100–106

Majumdar, B.C. (1999), 'The effect of roughness parameter on the performance of hydrodynamic journal bearings with rough surfaces', *Trib. Int.*, **32**, pp. 231–236

Mishra, P.C. (2007), 'Thermal analysis of elliptic bore journal bearing', *Tribology Transaction: STLE*, **50**, pp 137–144

Pinkus, O. (1960), 'Solution of Reynolds equation for arbitrarily loaded Journal bearing', ASME No-**60**-Lub-3, pp. 145–152

Rahnejat, H. (1998), *Multi-body Dynamics: Vehicles, Machines and Mechanisms*, Professional Engineering Publishing/SAE.

Reason, B.R. and Narang, I.P. (1981), 'Rapid design and performance evaluation of steady state journal bearings – a technique amendable to programmable hand calculator', *Trans. ASLE*, **25**(4), pp. 429–444

Roelands, C.J.A. (1966), 'Correlation aspects of the viscosity-temperature-pressure relationships of lubricating oils', Druk VRB Kleine der A3-4 Groningen

20.11 Nomenclature

A	Bore area	m^2
A_{energy}	Term used in the energy equation	
B_{energy}	Term used in the energy equation	
c	Radial clearance	m
C_p	Specific heat at constant pressure	
e	Eccentricity of the journal	
E_{energy}	Term used in the energy equation	
$E\ ()$	Film expectancy function	
F	Applied force	N
F_{gas}	Gas force	N
F_{energy}	Term used in the energy equation	N
$F_{reciprocating}$	Inertial force due to reciprocating mass	N
$F_{rotating}$	Inertial force due to rotating mass	N
G	Non-circularity/out-of-roundness	
$h_{0,h}$	Minimum film thickness, film thickness	μm
h^*	Non-dimensional film thickness	μm
h_s	Stochastic film	μm
h_n	Nominal film	
K_r	Roughness parameter	
L	Width of the bearing	m
$m_{reciprocating}$	Mass in reciprocation	kg
$m_{rotation}$	Mass in rotation	kg
N	Journal speed	rps
p	Hydrodynamic pressure	$N\,m^{-2}$
P_{gas}	Gas pressure in the combustion chamber	$N\,m^{-2}$
r_b	Radius of bore	m
r_j	Radius of journal	m
R_{min}	Elliptic bore minor radius	m
R_{min}	Elliptic bore major radius	m
R	Crank radius	m
t	Time	s
u_{av}	Average velocity in the direction of rotation	$m\,s^{-1}$
u_j	Velocity of journal in the direction of rotation	$m\,s^{-1}$
u_b	Velocity of bore in the direction of rotation	$m\,s^{-1}$
v_b	Velocity of bore in direction side leakage	$m\,s^{-1}$
v_j	Velocity of journal in direction side leakage	$m\,s^{-1}$
W	Load applied	N
x, y	Co-ordinate directions	
x^*	Non-dimensional axial location	
α	Piezo-viscosity index	
$\alpha_{dissipation}$	Heat dissipation constant	

θ^*	Non-dimensional circumferential location	
θ_0	Ambient temperature	°C
$\Delta\theta$	Temperature difference	°C
φ	Angular location of bearing	radian
ε	Eccentricity ratio	
ξ	Damping factor	
$\rho/\rho_0/\rho^*$	Lubricant density/reference density/non-dimensional density	
$\eta/\eta_0/\eta^*$	Lubricant viscosity/reference density/non-dimensional density	
β_{thermal}	Coefficient of thermal expansion	
σ_{rms}	rms value of surface roughness	μm
Ψ_1, Ψ_2, Ψ_3,	Expectancy operators	

Non-dimensional parameters

Non dimensional parameter	Dimensional interpretation
p^*	$\dfrac{p - p_0}{2\pi N (R/c)^2}$
φ	(x/R)
y^*	$(2y/L)$
ρ^*	(ρ/ρ_0)
η^*	(η/η_0)
W^*	$\dfrac{Wc^2}{(2\pi N)\eta_0 R^3 L}$
F^*	$\dfrac{cF}{(2\pi N)\eta_0 R^2 L}$
ρ	(e/c)

Section II.V

Product Portfolio

Section II.V

Drivetrain systems

21

An introduction to noise and vibration issues in the automotive drivetrain and the role of tribology

M. MENDAY, Loughborough University, UK

Abstract: Nearly all drivetrain noise, vibration and harshness (NVH) issues appear to suggest a component or system malfunction, but in fact they are invariably error states of the complete vehicle dynamics system. Some drivetrain NVH issues are related to the conditions of lubrication in impact zones, some to engine irregularities (combustion or inertial dynamics), some to the driveline torsional eigen-modes, and others to resonant behaviour in the transfer path from source to body. All are highly complex non-linear system problems which are not easily or inexpensively palliated. Inevitably there is a need to fundamentally understand the noise generation mechanism, which can only partly be achieved through experimentation. There is a continuing strong case for the use of accurate CAE to describe and to predict how the initial perturbation can develop into an NVH issue. The modelling tool may then be used to identify design strategies for robust functionality.

Key words: noise, vibration, harshness (NVH), shuffle, clonk, whine, boom, rattle.

21.1 Introduction to drivetrain noise, vibration, harshness (NVH)

Recent improvements and refinements in fuel consumption and powertrain efficiency, and lower exhaust emissions, have ironically led to the emergence of a number of related drivetrain *NVH* (noise, vibration and harshness) concerns, some of which are described and discussed in this chapter. They include judder and shuffle (vibration) and clonk (also see Chapters 1 and 30), rattle (also see Chapters 25–29), whine and boom (noise). These phenomena are largely the result of an ongoing drive for lightweight and fuel-efficient drivetrains, and as a result the industry has been presented with a set of urgent challenges for critical examination. The growing significance of NVH is due to a number of contributory reasons, some of which are the following:

- Vehicle performance is now more responsive, with higher engine torque rise rates.
- Lubrication fluids are generally less viscous and are subjected to higher temperature and pressure levels.

663

- The vehicle power-to-weight ratio has increased over time.
- Construction materials are lightweight and have much lower intrinsic damping levels.
- Background sound levels in the vehicle are lower, giving greater emphasis to unwanted noises.
- Customer awareness and expectation is higher than ever.

The net result has been a general increase in drivetrain NVH sensitivity to engine-based excitation.

21.1.1 Nature of the problem

The above-mentioned phenomena become noticeable under different driving conditions; they all have different frequency characteristics and they all have a different subjective impact on the driver. However, they all share and respond to a similar generic trait, namely:

- they are subjected to complex sources of excitation;
- they are part of a complex and non-linear torsional spring mass and damper systems;
- they contribute to different forms of energy dissipation; airborne and structure-borne;
- the vibration paths from source to the driver are complex.

Each phenomenon requires careful resolution treatment, with a prevailing understanding that the resolution of one must not adversely affect another. The strategy is to reduce the excitation or desensitise the system or interrupt the vibration or noise path, or any combination of these, in order that the system behaves robustly as intended without error.

21.2 The application of multi-body dynamics (MBD) analysis

NVH problems can be notoriously difficult to investigate in a complaint vehicle. The problem is that the test, environmental and driving conditions have to be the same every time in order to excite the NVH problem in a repeatable way, and especially if a resolution is being tested. Often the problem does not manifest even between identical vehicles. There is always a risk that erroneous conclusions can be drawn with respect to both the problem and the solution, and especially if a palliative fix gives only a marginal improvement.

The preferred strategy is to simulate the vehicle conditions with a model to identify the causal mechanism, to then identify a solution, and to apply the results to an actual vehicle for objective and subjective confirmation (for an example, see Chapter 1 and Ambrosi and Orofino, 1992). Much will, therefore, depend on the accuracy of the engineering assumptions made when the model is assembled, especially with regard to conditions at impact and the vibration paths to the driving compartment.

21.3   Noise, vibration, harshness (NVH) characteristics

The common feature with all of the noise issues – clonk, rattle, whine and boom – is that they all share a similar mechanism, namely, lubrication or combustion impact(s), followed by energy dissipation into a nearby resonating elastic structure, with the resultant **airborne noise** following a path to the driver's ear. The NVH phenomenon occurs when the frequency of excitation matches the natural frequency of the elastic system (see Chapter 30). However, the necessary frequency match is not a difficult event to achieve. For example, the Fourier spectrum for a pure impulse is a continuous function of frequency, so any single impact will readily excite *all* the neighbouring modes of an elastic structure. In fact, one of the accepted methods of reducing the level of excitation is to ensure that the impact is *not* a pure impulse/Dirac function, a process which is referred to as **pulse conditioning**. Excitation may then be minimised if the impact is engineered to have a pulse shape with a frequency content which is *not* exactly matched to that of the structure being excited. A half sine pulse in the time domain, for example, has a rapidly decaying frequency component.

While the above is true for single event impacts such as clonk, the analysis is more difficult for multiple collisions which occur in rattle and whine. These excitation signals are complex, the frequency content is also complex over a very wide range. Nevertheless, one method of reducing structural excitation for single or multiple impacts is to exercise more tribological control in the lubricated contact zone and to effect better cushioning of the impact.

There are several lubrication regimes in the drivetrain where impacts occur. The transmission and axle gear teeth make lash contact in an oil bath splash environment. The axle half-shafts and transmission input shaft splines (in the presence of lash) and universal joints make contact through a greased connection. The clutch provides a lash generated impulse to the drivetrain through a dry contact in the clutch plate.

By comparison, interior boom is not initiated by lubricated impacts but is caused by second order engine firing forces typically entering the body through suspension mounts, engine and transmission mounts, exhaust mounts and driveshaft mounts.

21.4   Summary of tribological contacts

There are four lubrication modes which may be conveniently defined by the well-known oil film to surface roughness ratio λ (also see Chapter 3), where λ = thickness of oil film/Σ surface roughness.

21.4.1 Hydrodynamic lubrication (full fluid film) $5 \leq \lambda$

The fluid film is sufficiently thick to prevent any metallic contact (see Chapter 5). The oil property is therefore relatively unimportant. With this form of hydrodynamic lubricated contact, there is an insignificant elastic deformation and it is small compared with the oil film thickness. The lubricant viscosity is also not significantly affected during the application of load. Idle transmission rattle in neutral gear (loose gear rattle) would be expected under these conditions. Little or no wave propagation would result. Other examples are thrust bearings and lightly loaded ball bearings.

21.4.2 Elastohydrodynamic lubrication (EHL) $3 \leq \lambda \leq 5$

Here, the elasticity of the solids in Hertzian contact in the conjunction becomes important, and the lubricant piezo-viscous action occurs. Elastohydrodynamic lubrication (EHL) conditions can apply in normal driveline impacts, and can overlap into the mixed and boundary lubrication regimes; for example, pairs of teeth in meshing gears, rolling elements and their races in ball bearings, seals, cams, and splined connections.

Piezo is derived from the Greek *piezein*, which means to squeeze or press. With this form of lubricated contact, there will be significant local elastic deformation and also significant changes to the lubricant viscosity under pressure and temperature in the contact zone as the load varies with time. This is the EHL regime (see Chapter 6). The elastic deflection is comparable to the fluid film thickness under this type of contact. The pressure in the contact is also sufficiently high to cause an exponential increase in fluid viscosity. The contact stiffness is large enough to be comparable to bending stiffness. Over-run or coast gear rattle, and driveline clonk, are typical NVH conditions resulting from this form of contact condition.

The prediction for the elasto-piezoviscous oil film thickness was initially made by Dowson and Higginson (1959). A form of EHL is the iso-viscous elastic (Gohar and Rahnejat, 2008). Under this form of lubricated contact, there will be significant local elastic deformation; however, the lubricant viscosity will not significantly change. This situation is often found when materials of low elastic modulus are brought into contact under low load applications. Viscosity effects are not significant because either the lubricant has a low viscosity (it may be water) or the generated interface pressures are low such as in crankshaft support bearings with thin layers of soft overlay (see Chapter 18).

The contact stiffness is fairly low, and is much lower than the bending stiffness (when gear teeth or spline teeth loading are being considered). Examples may be seals, human joints and elastomers. So this is not relevant to contact in the driveline torque path. Initial predictions for the elastic iso-viscous oil film thickness were derived by Herrebrugh (1968).

21.4.3 Mixed lubrication $\lambda \leq 3$

Both the surface and oil film effects are important lubricant factors: for example, gear sets, bearings and seals.

21.4.4 Boundary lubrication $\lambda \leq 1$

Here $\lambda \leq 1$ so the oil viscosity has little influence on the lubricated contact. There is a direct surface to surface interaction.

21.5   Airborne and structure-borne noise

Drivetrain **NVH** phenomena generally have both *airborne* and *structure-borne* content. *Airborne noise* radiates from a noise source and travels directly through the air to the ear. *Structure-borne noise* is that which is propagated through structures as a vibration and is subsequently radiated as noise. Of course, all structure-borne sound must eventually become airborne sound in order for it to be heard by the driver. It is true to say that clonk, rattle, whine and boom are airborne noises, radiating from elastic structures which have been excited, sometimes into resonance, by an impact or impacts.

21.6   Noise and vibration paths from the driveline into the body

Extreme care must be exercised when investigating NVH problems. The driving condition (gear, engine speed, laden weight, incline, etc.) must be clearly defined. Precise microphone positions are important to ensure repeatability of measurement; and a *binaural head* is often a preferred choice. When investigating body boom a number of microphones are required to capture the three-dimensional nature of the standing acoustic waves. Similarly, several or many accelerometers placed on the body and suspension may be required to measure and confirm using correlation analysis, the vibration paths from the point of disturbance to the driver's ear.

The normal investigative procedure is to obtain microphone and accelerometer time histories along with the engine speed, for the driving conditions which cause a given concern. A Fourier analysis of the results may then be prepared in the form of a 'waterfall' plot, which is useful in showing whether a captured signal is speed-dependent. A speed-dependent signal will be shown as a radiating line on the waterfall plot, and the angle of the line will indicate the multiple, or order, of the engine speed. A second order signal from an I4 engine, for example, will be firing frequency which is the major torsional exciting force in the driveline.

A vertical line on the waterfall plot (with frequency on the x-axis) which is independent of engine speed will indicate that there is a resonance in the system. One should always be interested when a radiating **second order engine excitation** crosses a standing system resonance and to check if the match corresponds to an NVH issue in the vehicle.

21.7   Signal analysis

In the case of short transient signals arising from a single impact having rapid non-periodic fluctuations, **wavelet analysis** is particularly suitable and this is preferable to a Fourier-based system. Wavelet analysis (see Chapter 30) is able to extract a transient signal from a noisy background, and is particularly suitable for the study of transient structural and acoustic wave propagation that arise from impact forces. Wavelet analysis transfers 2D frequency and time data from a measured time history, and the time component may be used to study the propagation in some detail. For example, this has been successfully used to identify the time sequence of structural collapse and occupants from a vehicle crash test time history (Onsay, 1995; Cheng, 2002).

Further work in the analysis of short duration transients has been conducted by Vafaei *et al.* (2001).

21.8   Examples of high-energy impacts in the drivetrain

21.8.1   Boom

Interior **boom** is a low-frequency audible noise. It is usually related to the third torsional mode at 40–80 Hz, the frequency being dependent on the rear wheel drive configuration. Engine disturbances are fed into the driveline, then through the engine and transmission mounts, the exhaust mounts, the driveshaft mounts and suspension mounts, to excite the vehicle body. The body panels may then be excited into resonance. Boom occurs when the body panel modes coincide with the acoustic modes of the body cavity.

Idle boom occurs in drive at engine idle speed and although of short duration, is particularly noticeable against a low noise background. Mid and high-speed booms are of longer duration, but are less distinct against higher engine, road and wind noises.

The fundamental low-frequency acoustic cavity modes are almost inaudible, giving rise to a pressure undulation discomfort in the ears. The higher modes are driven by panel vibration. A characteristic of boom is that very little energy is required to excite the boom, and the cavity modes are well spaced throughout the vehicle speed range. They are hard to avoid. Changing gear or engine speed, or turning on the radio, are the only palliatives that can be taken while driving.

Acoustic pressure variation in the body cavity is a function of frequency and boundary excitation and boundary shape. The low-frequency modes are placed from end to end where the body dimensions are at a maximum, or from corner to corner. The antinodes are at the ends of the modes and are driven by the panels.

A commercial van has high panel energies and the surfaces are reflecting and undamped. The acoustic pressure variation may or may not be at a maximum at the driver/passenger head location. The acoustic pressure field may have high gradients especially as the frequencies increase, so spatial information is very important. The higher frequencies are more annoyingly audible and are driven from side to side or from floor to roof. Since the side to side dimensions and the floor to roof dimensions are similar, there can be a heavy modal coupling and a complex situation can emerge.

A useful way of breaking up the booms in a commercial van is to fit a bulkhead between the driver compartment and load compartment. This can be made to act as a Helmholtz resonator. More useful, is to use the clutch or DMF (see Chapter 28) to detune the driveline from the cavity.

It is a fairly simple task to generate a 3D finite element model of the cavity to determine all the cavity mode frequencies and their shapes using a passive boundary, especially those modes in the speed range of the vehicle. Since these are excited at firing frequency (two times engine speed for an I4 engine) the frequency range of interest is 30 to 150 Hz. If a frequency-dependent active boundary is fed into the model and used to drive the cavity acoustically, a useful exercise can be conducted to determine which parts of the boundary are most responsible for energising the standing waves. That can lead to an examination of the most sensitive parts of the body and panels, and the most likely mount entry points to palliate.

21.8.2 Clonk

Driveline clonk (also see Chapters 1 and 30) is an unacceptable audible and tactile driveline response which may occur during several possible driving conditions:

- *Tip-in clonk*, when the throttle is rapidly applied from coast.

- *Tip-out clonk*, when the throttle is rapidly released from drive.
- *Engagement clonk*, when the clutch is engaged rapidly following gear selection.
- *Clutch clonk*, when the drive is taken up during a low-speed creep manoeuvre.
- *Shift clonk*, when a gear up-shift has been made.
- *Over-run clonk*, when the road input accelerates through a driveline lash.

In all of the above driving conditions, the resultant torsional impulse to the driveline gives rise to a short duration vehicle jerk and an accompanying metallic clonk or *thud* noise (Krenz, 1985).

Clonk is always associated with lash in the driveline, for which there are many locations. When the entire lash has been taken up, a high-energy impact pulse is delivered to the driveline which excites the several thin elastic shells (transmission casing, clutch bell housing, drive shafts, etc.) into resonance. Elastic wave propagation and airborne noise follow (Menday *et al.*, 1999).

As has been described earlier, impacts in the many driveline lash zones have different characteristics. The severity of each impact will depend on the lubricant, the masses of the bodies in impact, the degree of lash damping, the velocities of the parts in contact, the restraining forces, and other factors. Tribology plays a key part in the definition and cushioning of the impulse characteristic and the ultimate severity of the clonk. There is a similar parallel to the pulse conditioning that is advocated for the solution to shuffle – see later. The key difference is that there is frequency matching at play in clonk, and the challenge is to remove the high frequency components of the impact signal which would otherwise excite neighbouring structures into noise emitters.

Clonk and *shuffle* (see Section 21.9) are related in much the same way as *judder* (Section 21.8.4) and shuffle – see later. Following a throttle tip-in, a shuffle response may be excited which is the first torsion mode of the driveline. The driveline is lightly damped and so several shuffle cycles may occur. Each cycle of shuffle may give rise to an audible clonk as the driveline lash is fully traversed.

The coupled *shuffle* and clonk phenomena may be readily simulated with a multi-body dynamic model of the driveline (Arrundale *et al.*, 1998). The results of this analysis showed that following a Dirac-type impulse to the system, a short duration high-frequency signal (clonk) was coupled to a low-frequency carrier wave (shuffle) pertaining to the rear driveshaft tube of a rear wheeled commercial van. The analysis of the high-frequency clonk signal showed clusters of resonances between 1.4 and 4.4 kHz. A further examination of the driveshaft tube acoustic cavity showed frequency coincidence with the tube structural resonances.

Vehicle tests for clonk have been conducted on a chassis dynamometer (Menday, 2003) with a front wheel drive vehicle in a series of controlled tests to investigate the contribution from potential clonk causal factors.

A design of experiments (DOE) was conducted to investigate the system robustness to five factors. The factors were chosen which would also have minimum disruptive effect to the testing. The factor with the most influence on the clonk was the engine entry speed. The clutch was rapidly engaged at either 700 or 900 rpm (below and above the normal idle speed of 825 rpm). In all the 16 test runs, the lower engine speed produced unacceptable clonk, and the higher engine speed produced acceptable/inaudible clonk. This factor had more effect on clonk than transmission oil viscosity; driveshaft construction; the type of clutch disc; clutch pedal engagement rates; or any combination of these.

A single clonk event from the testing was analysed using a Wavelet diagram, which showed:

- The clonk frequency range was 1500 to 5000 Hz, resulting from a hard impact.
- The actual impact time was circa 2 ms.
- The total clonk duration was 100 ms, including the undamped structural ringing.
- Three successive impact responses were recorded and measured at 6.3 Hz following the initial impact, and this was assumed to be the shuffle frequency.

21.8.3 Axle whine

Rear *axle whine* is a continuous, steady state, high-frequency tonal noise which is emitted from the meshing differential unit gears and manifests itself at the gear meshing frequency and its harmonics. It is torque-induced and can be caused by engine torque fluctuations and compounded by transmission error (i.e. the deviation from perfect motion transfer of gears) during the gear meshing cycle. Typically, it is engine speed-dependent over a limited speed range, but is independent of gear. Axle gear whine frequency content is 400–800 Hz.

Whine is a complex noise problem, which is related to gear set quality, vehicle sensitivity, vibration path(s), wear and driveline dynamics. There is very little opportunity for diagnosis once a complaint vehicle has been identified in service. The service action is maybe limited to an axle replacement, and the prospect of this action being effective is small. It is a costly action and bound to further aggravate customer relations. Whine, like most drivetrain NVH issues, especially requires preventative, not corrective, action.

Experimental measurements have revealed significant correlation between noise in the passenger compartment and vibration at the differential housings. A significant noise path is from the axle assembly into the body as structure-borne noise via suspension mounts, shock absorber mounts and driveshaft mounts. Once the body is active there is a further tendency for the body acoustic modes to be excited. The installation of a costly steel bulkhead usually modulates the problem. Another noise path is the transmission of vibration from the gear shafts and bearings to the differential housing, which radiates noise.

Although the resulting whine noise in the cabin is relatively low in comparison to road noise levels, it is very apparent due to its higher frequency content, with often dominant amplitude and frequency modulation effects due to mounting eccentricities, teeth contact stiffness variations and manufacturing errors.

Dynamic interactions between the differential unit and driveshafts can occur and this often generates excessive tonal noises, which are the result of coupled bending and torsional component resonances. These resonances may have a magnification effect on the source itself, by exciting the gear shafts and distorting the alignment of the gear sets.

The standard approach used to reduce whine noise has been to minimise the gear transmission error, which is induced either by manufacturing errors of the tooth surface or by gear tooth deflections caused by the transmitted load. However, even when gear quality is high – which is a high-cost route – noise levels may still exceed acceptable limits, because the drivetrain and vehicle body may be highly sensitive even to minimal values of manufacturing error. Hence there is a need for a complete vehicle simulation to address the system dynamics of whine, and noise generation into the cabin, including tribological effects at the gear set and the acoustic cavity couplings.

Several whine palliatives have been used to achieve more acceptable subjective ratings:

- Driveline tuning by addition of an inertia disk and/or a mass near rear axle.
- Vibration decoupling using appropriately selected rubber couplings.
- Use of internal cardboard liners in driveshafts to increase damping.
- Structural stiffening of the differential housing to shift natural frequencies.
- Modification to mounting attachment points to the body.
- Internal body modifications including bulkhead.

21.8.4 Judder

This is fully covered in Chapter 22.

21.8.5 Rattle

Gear rattle is a hard metallic noise which radiates from the transmission and driveshafts, but the analysis of rattle must necessarily include the whole torsion system (Hagiwara, 1982; also see Chapters 25–29). Although rattle between gears with lash separation occurs when the torsional vibration at the gear mesh location exceeds gear inertia and oil drag, the whole driveline system dynamics must be analysed. For a rear wheel drive vehicle the second or third torsional mode is normally at 40–80 Hz, dependent on configuration, with a large anti-node located at the transmission. This is the mode responsible for rattle. Gear rattle severity depends on the relative velocity between meshing gears and oil viscosity/oil temperature. Rattle occurs between unloaded gears, and so gear sets transmitting torque do not usually rattle.

One can consider the possibility that wave propagation would result from gear or spline teeth impact, and to understand how the resultant impact energy could be coupled to the attached dynamic structures. One would apply the criterion that if contact stiffness is of the same magnitude as tooth bending stiffness then a wave will emanate from the impact site. However, if the contact stiffness is of a lower order of magnitude than the bending stiffness, then it may be assumed that a Hertz-type contact has occurred and the impact will be *locally* confined to the local elastic area of contact, and so no wave motion from the site would result. In other words, under normal gear tooth impact conditions either the impact energy will be absorbed locally at the point of contact and no resulting wave motion occurs, or local absorption will be insufficient and gear tooth vibration/wave propagation would result and the gear set will transmit impulsive energy to the driveline.

Gear rattle is a relatively recent phenomenon, which is due largely to low-viscosity oil, lightweight driveline, increased road traffic density, thin-walled aluminium transmission and clutch bell housing, higher combustion pressures with diesel engines, drivetrain torsion modes, increased transmission shaft centre distances, lower background noise levels, and increased customer perception and awareness.

Rattle is both airborne and structure-borne. In both cases, the impact energy from meshing gears first enters the transmission casing. The main rattle airborne path is radiated from the transmission housing and reflects off the road and enters the driving compartment through an open window. This airborne noise is experienced by the driver and by any outside observer. The structure-borne rattle path enters the body from the transmission casing via the engine and transmission mounts and gear selector mechanism. Once the structure-borne energy is body side of the transmission, the noise is radiated into the driving compartment.

We now consider the various gear rattle modes which can occur under different driving conditions.

Idle/neutral transmission rattle

Idle rattle will occur when the transmission is hot and in neutral gear, and the engine is at a steady idling speed, as in a normal stationary traffic condition. Idle rattle is particularly audible and annoying, and dominates over engine and road noises. It is eliminated if the clutch pedal is depressed, thus decoupling the engine from the transmission. It occurs because the engine speed irregularities are not sufficiently damped by the flywheel and they enter the transmission coincident with the third torsion mode. Fortunately, idle rattle is readily eliminated even with the conventional clutch and flywheel. It is necessary to ensure that the first stage of the **clutch torsional damper** is wider than the speed oscillations, and that the first stage torsion rate is soft enough (less than 1 N m/deg) to lower the natural frequency of the system to below the engine speed.

Care has to be taken to obtain the first stage angle wide enough, or torsion impulses will occur when the idle clutch torque capacity is exceeded and the engine torque will impact against the main clutch damper stage. The compromise is that as the first stage angle becomes wider, the non-linearity is worsened and there is also a potential repercussion that other NVH concerns such as clonk and shuffle will be adversely affected due to the effective increase in driveline lash.

There are two other theoretical remedies for idle rattle. One is to increase the flywheel inertia, but this will have negative effects on economy and responsiveness/performance. The other is to slightly increase the idle speed, which can be very effective, but reduces fuel economy.

Transmission idle rattle is aggravated when ancillaries are switched on. Air conditioning is a particularly good example. The need for air conditioning normally coincides with high ambient temperatures, which cause a lower transmission oil viscosity and a lower gear drag torque. Another ancillary is that of vehicle headlights. Both these examples have the effect of increasing angular acceleration at the flywheel and worsening idle rattle.

Drive rattle

Drive rattle occurs in drive mode with a throttle opening. It is particularly disagreeable in high gear and at engine speeds below 2000 rpm, and when the transmission oil is hot (low oil viscosity at gear contact with harder impacts). It is relatively common for modern engines to allow the vehicle to labour in high gear with low engine speeds and to induce rattle. The same problem resolution could be applied as described for idle rattle (a 1 N m/deg damper rate), but this is not feasible given the restricted spring travel in a typical clutch disc. The best compromise is to minimise the damper rate and maximise the travel, and also increase the clutch hysteresis, to best control the rattle mode. The minimum damper rate is set by the physical limitations in the clutch damper plate and the stop pins. Otherwise, there is a danger of engine torque exceeding clutch torque capacity and bottoming out.

Even if this strategy improves rattle it will probably worsen tip-in and tip-out. However, a fully effective isolation cannot be achieved across the drive range with a two stage conventional clutch. So the addition of damper stages can help to achieve acceptable rattle, but only in lower gears and with a light throttle. The use of a higher-viscosity oil to increase drag torque and reduce rattle will, however, have a negative effect on transmission shiftability.

Coast rattle

Coast rattle occurs when the throttle is closed and the vehicle is coasting on level ground or on a decline with the wheels providing the driving energy. Torque excitation is low, rattle frequencies are higher, and it is possible to find a damper solution on the coast side of the clutch disc damper.

Dual mass flywheel (DMF)

The limitations of the conventional flywheel and clutch arrangement as a resolution for gear rattle have already been described. Now the twin/dual mass flywheel DMF with a simple clutch plate, almost completely eliminates rattle (also see Chapter 28). This is achieved because the system resonance with a **DMF** is reduced to 10–30 Hz, which is below normal driving conditions.

It must be said, however, that with every ignition start the driveline must pass through the sub-idle speed resonance, causing internal damage to the DMF due to the resultant extremely high internal torques. Similar damage can occur when the clutch is engaged very quickly, causing the two flywheel inertias to collide at high speed and again cause very high peak torques.

Another driving mode which can cause internal damage to the DMF is when the vehicle is allowed to 'lug' in high gear with the engine speed falling below the idle speed. Finally, the internal soft arc springs operate at a greater radius than the conventional clutch and these are exposed to high centrifugal forces.

The DMF replaces a conventional flywheel, with one half of the inertia attached to the engine crank and the other half attached to the transmission, both inertias are coupled with a low rate torsion damper. The smaller primary engine flywheel is bolted to the engine crankshaft and provides the ring gear drive to the starter motor. The lower inertia actually increases torsional irregularities, but the DMF secondary inertia eliminates these from the drivetrain. The added secondary transmission inertia is independent of the primary inertia, and this allows the secondary inertia to be decoupled during gear shifting and the synchroniser cones to operate without the burden of an additional inertia.

21.9   Shuffle

Shuffle can occur following a rapid tip-in or tip-out, and is characterised by a fore and aft low-frequency undamped vehicle vibration and which may continue for several cycles. The oscillations may be accompanied by audible clonks as the torque traverses the driveline lash during each cycle (Krenz, 1985; Arrundale *et al.*, 1998). The shuffle oscillations occur at the fundamental eigen-frequency of the driveline.

Provided that the throttle tip-in torque profile is sufficiently long and progressive, then a good shuffle response could possibly be achieved. However, this would not be acceptable in terms of 'fun to drive'.

Alternatively, shuffle response may be reduced or eliminated by increasing the driveline damping, or by modifying the excitation torque. A cancellation effect can be obtained by modifying and optimising the input torque profile to the driveline. The technique is to apply engine control strategies and calibration to achieve the desired torque profile, although the beneficial effects can be jeopardised if the throttle tip-in is applied from a closed throttle. Under these conditions the driveline lash is fully traversed and the full cancellation effects are not achieved. It then becomes a necessity to reduce the driveline lash, especially in the clutch plate, if this plate is fitted with a wide angle predamper.

Several shuffle palliatives have been used to achieve more acceptable subjective ratings:

- Engine management to achieve optimised torque profile.
- Delete clutch predamper/minimise driveline lash.
- Increase flywheel inertia.
- Soften torque ramp/rise rate.
- Electronic clutch engagement slip control.
- DMF.

21.10   Whoop (clutch in-cycle vibration)

Whoop is a disagreeable noise and vibration felt by the driver when disengaging and engaging the clutch (Rahnejat *et al.*, 1997, Kelly and Biermann, 2004). The noise, which originates from the driver footwell area, actually resembles a growl and is accompanied by a rough tactile clutch pedal response. The frequency range of interest is 150–500 Hz. Whoop is related to cable control systems (not hydraulic) and is evident only during pedal actuation, particularly the disengagement process, and not when the pedal is stationary. The problem is worse with diesel engines, particularly the pedal vibration. If the pedal box is removed from the vehicle and then operated outside the vehicle, the problem is not evident, hence the evidence that the problem derives from structure-borne activity.

Whoop has been investigated using a simulation and actual vehicle tests (Kelly and Rahnejat, 1997; Kelly and Biermann, 2004). Owing to the large number of potential causal factors, and the possibility of factor interaction, a series of DOE tests were performed on a vehicle. The complaint was readily excited and was sufficiently repeatable for test purposes.

A costly palliative is used to disguise and dampen the phenomenon. Therefore, no account is taken of the mechanism of the problem, the body sensitivity, the root cause or the vibration path. The resolution difficulties will increase as the auto industry moves towards direct injection and higher levels of excitation to the clutch pedal.

Background experimental research had established several plausible explanations for whoop:

- Flywheel nodding frequency of *ca* 250 Hz in line with crankshaft bending.
- Pedal vibration was mostly affected by the fourth cylinder (nearest the flywheel).
- Half order fourth cylinder impulse firing made the most contribution to crankshaft bending.
- Binaural head noise frequency band of interest 100–250 Hz.
- Pedal vibration band of interest 100–400 Hz.
- Third and fourth cylinder firing was the excitation source to the flywheel; crankshaft bending then led to flywheel nodding as the source of whoop vibration.
- The vibration path from the flywheel led directly through the actuation mechanism to the clutch pedal.

Even though the flywheel was nodding continuously, the whoop was evident only when the clutch pedal was moving. The explanation given for this was a higher level of friction control in the cable when it was not being used.

A controlled set of DOE tests were conducted using a Taguchi array to determine the factors having most influence on whoop (Kelly and Biermann, 2004), to determine whether there was factor interaction, and to determine whether the clutch actuation system could be made functionally robust against noise factors. The procedure also allowed the system performance to be predicted. Seven possible control factors were identified as well as two noise factors (which by definition are not controllable) and these were submitted for the test program. As a result, several factors and interacting factors later emerged as being significant contributors to whoop.

As mentioned earlier, the palliation for whoop is an expensive and space-consuming mass damper fixed to the release lever in the bell housing.

21.11   Summary of drivetrain noise, vibration, harshness (NVH) issues

Nearly all drivetrain *NVH* issues may superficially suggest a component malfunction, but they are invariably error states of the complete vehicle dynamics system. Some NVH issues are related to the conditions of lubrication at impact, some to engine irregularities, some to the driveline torsion eigen-modes, and others to resonant behaviour in the transfer path from source to body. All are highly complex non-linear system problems which are not easily or cheaply palliated. Inevitably there is a need to fundamentally understand the noise generation mechanism, which can only partly be achieved by experimentation. There is a continuing strong case for the use of accurate computer-aided engineering (CAE) to predict and to describe how the initial perturbation can develop into an NVH issue. The tool can then be profitably used to identify design strategies for robust functionality.

21.12   Future trends

Given the general trend towards lightweight structures, explosive power units, legislation, increasingly discerning customers, shorter program development times, lower product costs, and competition within the automotive industry, there will be an ongoing need for sophisticated and accurate prediction tools to be used in the search for robust drivetrain NVH.

There should be further advances towards a more objective – and less subjective – vehicle sign off, and a philosophy of prediction and prevention, rather than reaction and correction. This means that where appropriate, mathematical simulations must be used to augment the vehicle development program as pragmatically as possible. One would expect that a robust design initiative would be taken from the mathematical model and then fitted to a test vehicle for verification and sign off.

This necessarily means that modelling must continue to stretch its ability to fully represent the engineering system, and the modellers must also continue to improve and exercise their abilities to represent the engineering systems. We have seen in this chapter how complex the drivetrain and its vibration paths can be. Numerical predictions can only be as good as the accuracy and assumptions made in the modelling.

21.13   References

Ambrosi G, Orofino L, Driveline vibration simulation in a four wheel drive vehicle, *Proc. Insn.t Mech. Engrs*. C389/154, pp. 105–115, 1992.
Arrundale D P, Rahnejat H, Menday M, *Multi Body Dynamics of Automobile Drivelines: An Investigation into the Interaction between Shuffle and Clonk*, ISATA 99SF023 1998.

Cheng Z, Analysis of Automobile Crash Response Using Wavelets, SAE Paper 2002-01 0183, 2002.

Dowson D, Higginson G R, A numerical solution to the elastohydrodynamic problem, *J. Mech. Eng. Sci.* Vol. **1**, No 1, pp. 6–51, 1959.

Gohar R, Rahnejat H, *Fundamentals of Tribology*, Imperial College Press, 2008.

Hagiwara B, Analysis of Non-Linear Vibration of Drive Train for Heavy Duty Vehicle, SAE Paper 82067, 1982.

Herrebrugh K, Solving the incompressible and isothermal problem in elastohydrodynamic lubrication through an integral equation, *Trans. Am. Soc. Mech. Engrs.* Vol. **90**, Series F, pp. 262–270, 1968.

Kelly P, Biermann J W, Using Taguchi methods to aid understanding of a multi-body clutch pedal noise and vibration phenomenon, *Multi-Body Dynamics: Monitoring and Simulation Techniques*, Mechanical Engineering Publications Ltd, 2004.

Kelly P, Rahnejat H, Clutch pedal dynamic noise and vibration investigation, *Multi-Body Dynamics: Monitoring and Simulation Techniques*, Mechanical Engineering Publications Ltd, 1997.

Krenz R A, Vehicle Response to Throttle Tip-In/Tip-Out, SAE Paper 850967, 1985

Menday M, Multi-Body Dynamics Analysis and Experimental Investigations fo the Determination of the Physics of Drivetrain Vibro-Impact Induced Elasto-Acoustic Coupling, PhD Thesis, Loughborough University, 2003.

Menday M, Rahnejat H, Ebrahimi M, Clonk, An Onomatopoeic Response in Torsional Impact of Automotive Drivelines, *Proc. Inst. Mech. Engrs. Part D: J. Automobile Engineering* Vol. **213** No. 4 pp. 349–357, 1999.

Onsay T, The Use of Wavelet Transform and Frames in NVH Applications, SAE Paper 951364, 1995

Rahnejat H, Centea D and Kelly P, Non-linear multi-body dynamic analysis for the study of in-cycle vibrations (whoop) of cable operated clutch systems, Proc. 30th ISATA, Florence, Italy, June 1997, pp. 245–252.

Vafaei S, Menday M, Rahnejat H, Transient high frequency elasto-acoustic response of vehicular drivetrain to sudden throttle demand, *Proc. Instn. Mech. Engrs. Part K*, Vol. **215** pp. 35–52, 2001.

22

Friction lining characteristics and the clutch take-up judder phenomenon with manual transmission

M. MENDAY and H. RAHNEJAT, Loughborough University, UK

Abstract: Unlike many contacting pairs in engines and powertrains, where the emphasis is on reduction of friction, in the clutch systems maintenance of friction in the clutch disc interfaces with the flywheel and the pressure plate is the key requirement for any lining. Variations of coefficient of friction with slip speed during clutch actuation is one of the main root causes of a number of noise, vibration and harshness (NVH) phenomena, not least the clutch take-up judder. These characteristics can cause stick–slip perturbation during clutch actuation and are affected by lining composition, surface topography, contact pressures (due to clamp load) and generated, often localised, temperatures. This chapter deals with the problem of clutch take-up judder.

Key words: multi-body dynamics, NVH, clutch take-up judder, stick-slip friction, clutch lining interface.

22.1 Introduction

This chapter considers the clutch take-up judder phenomenon in rear wheel drive (RWD) automotive drivelines, fitted with manual transmissions and dry friction clutches. It discusses how and why the phenomenon occurs, the underlying system dynamics, the tribological contributions to the problem, the dominant influential factors, and the prospects for a design solution which would negate the need for costly palliation.

22.2 Background

Clutch take-up judder is a back and forth vibration of a vehicle in the frequency range 5–20 Hz, caused by the torsional vibrations of the driveline, which occur during the clutch engagement process. Judder is a significant concern in the automotive industry. The problem is often sensitive to the demands of specific work patterns, such as stop–start movement in heavy traffic or laden pull aways up an incline. However, it may also be an intermittent

680

problem, failing to repeat itself when expected, even though environmental factors may not have changed.

Sensitivity is also a real issue, since judder can occur when least expected and also fail to occur when most expected. Judder does not seem to behave in a predictable way. Two test vehicles with identical specification, but exposed to slightly different work regimes, may often not exhibit the same proneness to judder. This may suggest that the sensitivity is temperature related, and indeed there is an ongoing design tendency for smaller clutches to transmit higher torques at higher speeds with less opportunity for the heat to escape from the rubbing faces.

This annoying lack of predictable behaviour is partly explained by the non-linearity of the driveline system, and partly by the complexity of component parts in the driveline, and the variability of many potential causal factors. Here, *driveline* is defined as the assembly of all parts between the engine crankshaft to the rear road wheels (as in Rahnejat, 1998). Several researchers have concluded that judder is not only caused by clutch friction linings, but also by mechanical excitation (misalignment effects), when the frequency of excitation is matched to the natural frequency of the driveline system.

Judder may be easily confused with a similar problem known as *shuffle* (see Chapter 21). Judder and shuffle are excited differently and have different palliation strategies. This means that care must always be taken with every reported case of judder in the field, and to ensure the correct palliation methodology is applied.

As discussed later in more detail, judder is excited during clutch engagement, and when certain rubbing friction conditions are satisfied. In the field, the minimum palliation correction would be a replacement clutch. Often it will include the flywheel, if heat damage is found. Indeed, even these actions may not necessarily yield the required improvement. Hence, in order to avoid a costly and embarrassing recovery programme, it is essential to understand the phenomenon and the potential root cause solution. The objective is to design a vehicle driveline which can perform the required functions in a robust manner without being sensitive to external factors.

22.3   Description of clutch take-up judder

Clutch take-up judder is so-called because vehicle judder may occur when the clutch is being actuated during pull away. It is a low-frequency tactile vibration which may occur when the pull away gear has been selected, and the clutch pedal has been operated to bring the clutch into progressive slip engagement. Take-up judder is a low-frequency (5–20 Hz) longitudinal vibration felt by the driver through the seat during pull away, and while slipping the clutch. In fact, the vehicle may be almost stationary when judder is felt in the vehicle. It more readily occurs in reverse gear, on an incline pull away, or when the vehicle is fully loaded, or when the vehicle steering system is at full lock. Judder most often occurs when the clutch friction linings are hot, although it may occur less frequently when cold or when the linings are damp or moist. These conditions represent the critical role that tribology plays in the initiation of judder.

Take-up judder can be an intermittent problem. It is often difficult to replicate the condition even with a complaint vehicle! This alone necessitates the need for an analytical approach to the problem.

The role of the *dry clutch* in the take off manoeuvre is to synchronise the rotational speed of the engine crankshaft with the transmission input shaft and the driveline. When there is no relative angular velocity, the clutch is said to be fully clamped and the clutch pedal may be released, and the vehicle is then propelled forward under throttle control. However, during the clutch slip transition phase, with relative angular velocity low-frequency judder can occur.

Clutch engagement excites the first torsional eigen-mode of the driveline and a sensitive vehicle may then respond to judder. This torsional driveline mode is lightly damped and is readily excited. For a rear wheel drive vehicle, this mode has the maximum amplitude response at the flywheel and has an in-phase response of minimum amplitude at the rear axle. The mode is reacted at the tyre-to-road contact patch, and as a result a low-frequency *longitudinal* oscillation of the vehicle occurs. This is generally felt in the driver compartment and particularly at the driver's seat.

Similar modal behaviour has been noted for a rear wheel driveline with automatic transmission and a wet clutch (Hwang *et al.*, 1998) and similar conclusions were drawn with regards to the excitation of the negative wet clutch friction characteristics (as described later).

Newcomb and Spurr (1972) argued that judder was a mechanical resonance phenomenon and that all causes of judder, whether friction or mechanical, were consistent with the single mechanism, that the frequency of the exciting force matched the natural frequency of the (driveline) system.

The first mode may be readily confirmed by the use of a digital stroboscope, pointing at the engine support mounts, while the clutch is slowly slipped. This exercise is best conducted on a rolling road to allow the rear wheels the freedom to rotate while the stroboscope measures the judder frequency.

The other lower-order driveline torsional modes and their mode shapes may be readily computed with a simple three degrees of freedom lumped mass model of the driveline. For example, the second driveline torsional mode for a rear wheel drive vehicle has the engine and flywheel motion out-of-phase with the clutch and transmission. If gear rattle is a problem with this mode, then the engine input to the transmission may be modulated by eliminating the first stage characteristics of the clutch disc (to palliate idle rattle, see Chapters 21, 25–29) or to soften the second stage (to palliate drive rattle, see Chapter 1). The third torsional mode at around 30–40 Hz is often the driving force for internal acoustic boom problems (see Chapter 21).

The judder longitudinal vibration is felt by the driver at the driver's seat. There is a strong correlation between the seat track longitudinal acceleration and the transmission input shaft speed variation (the torsional mode), during the clutch slip phase. The transmission input shaft speed signal is obviously more difficult to access than the driver's seat. It is, therefore, normal practice to measure instead the seat track acceleration and the corresponding driver subjective reaction, as suitable and reliable alternative indicators of judder.

The many factors which may contribute to judder and how it is related to the frictional interface characteristics in the clutch are considered in this chapter. The tribology of interfacial rubbing action in the clutch friction lining is described later. It will be seen that the take-up judder phenomenon is a fairly complex, non-linear, dynamic problem, having a central excitation source at the clutch rubbing interfaces, but which also has many other contributory factors, many of which are not controllable.

22.4   Judder and shuffle

It should be noted that following clutch slip, when the clutch is fully engaged (fully clamped) and when the transmission is in low gear, the first torsional mode may also be excited by a *throttle tip-in* action, especially when the torque ramp rise time is faster than 300 ms (Hawthorn, 1990). Throttle tip-in is a rapid opening of the throttle. This throttle actuation may then cause a *shuffle* response (see Chapter 21 and Rahnejat, 1998), and is referred to as *tip-in shuffle*. It may also be accompanied by an audible *clonk* (see Chapters 1 and 30). Krenz (1985) has also written an authoritative paper on this subject. Shuffle can also occur as a result of *throttle tip-out* (rapid closing of the throttle) normally during a vehicle descent.

A shuffle response gives the same longitudinal vibration as judder, except it always occurs with the vehicle in motion. This longitudinal motion is sometimes referred to as *shunt*. Often the only way of removing this driving condition is to disengage the clutch and then re-introduce the engine torque at a slower rate with a more progressive clutch slip. It is important to recognise that shuffle is a product of the first torsional mode, and has the same tactile characteristics as judder. However, it is unrelated to clutch judder since the clutch is fully clamped when the throttle tip-in is applied and the clutch friction characteristics, therefore, have no bearing on the excitation of shuffle.

22.5   Multi-body dynamics analysis of judder

A multi-body system analysis approach may be undertaken to simulate the judder phenomenon. The model may also be used to understand the contributions from the clutch friction disc characteristics, and their effects on judder. Although take-up judder is excited at the clutch, it requires a whole vehicle simulation analysis from the clutch pedal and from the engine, to the rear road wheels in ground contact.

Centea *et al.* (1999, 2001) reported a ***multi-body dynamics*** model of a cable operated ***clutch*** system (see Fig. 22.1). The model was built in the commercial multi-body code; ADAMS (an acronym for Automatic Dynamic Analysis of Mechanical Systems, a trademark of MSC Software).

22.5.1   Inertial elements

The main parts of the clutch are shown in Fig. 22.1:

- The crankshaft (1).
- The flywheel (2).
- The input shaft (10).
- The pressure plate (4).
- The clutch cover (7).
- The straps (5) keep the plate and the cover rotating with the same speed and also, through their elasticity in the longitudinal direction of the clutch, permit an axial displacement of the pressure plate against the cover.

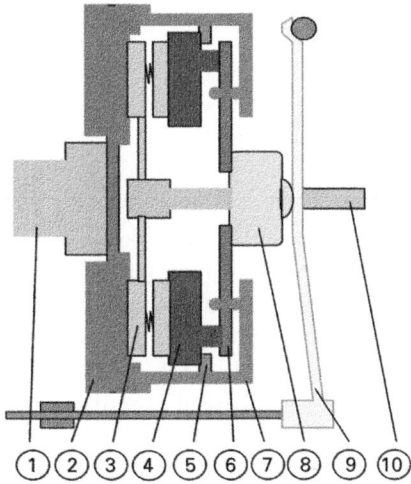

22.1 Main parts of the clutch used for studying the judder phenomena.

- The friction disc (3) is free to float between the flywheel and the pressure plate through a hub, splined to the input shaft of the gearbox. The friction disc is pressed between the pressure plate and the flywheel by the clamp force F_n, provided by the diaphragm spring.
- The diaphragm spring (6).

The ***clutch engagement*** is obtained through the application of the clamp force, provided by the diaphragm spring when the clutch is mounted on the flywheel.

In the ***clutch disengagement*** process, the force applied by the driver to the pedal is transmitted through the pedal quadrant to one end of the cable. The other end of the cable is mounted through a spherical type joint to the release lever.

- The release lever (9).
- The motion of the cable is transferred to the release lever, which rotates and pushes the release bearing (8) against the diaphragm spring fingers.

The diaphragm spring pivots on a fulcrum ring which is riveted onto the cover and the clamp force is subsequently reduced. The cushion spring and the straps pull back the pressure plate from the friction disc. The reducing friction torque permits a progressive braking of the torque transmitted by the engine through the flywheel to the driveline. The engagement process is similar to the disengagement process and occurs when the driver releases the pedal.

22.5.2 Constraints

As explained in Chapter 1, a multi-body mechanical dynamics model comprises a number of *parts*, assembled together by a series of constraint functions; in the case of a mechanical system, represented by joints. Certain *parts* in the assembly are regarded as compliant for the loading condition that affect the phenomenon investigated. Thus, their stiffness characteristics are included in the model. In the case of judder, these include the various springs in the clutch system. The system is then subjected to forces and motions which the system is subjected to in the course of its normal intended operation, such as the clamp load, engine torque and the input shaft angular velocity. The *parts* in the multi-body clutch model are listed in Table 22.1. Of note is the inclusion of vehicle inertia, which as described above plays a significant role in judder oscillations.

Note that the model includes the inertial properties of the transmission and rear axle *reduced* to the position of the clutch as their inertias, as already pointed out, affect the torsional oscillations during the take-up judder. The

Table 22.1 The clutch judder model

No.	Part name	Mass (kg)	Inertia (kg m^2)	Ratio
1	Crankshaft	(~10)	1	1
2	Flywheel	(14.5)	0.25	1
3	Clutch cover	(1.6)	0.03	1
4	Pressure_plate	(4.15)	0.04	1
5	Friction_disc	(1.45)	0.065	1
6	Hub	(1)	0.00001	1
7	Input shaft	(1.5)	0.0025	1
8	Gearbox	(~20)	0.002	1
9	Differential	(~20)	0.045	3.89
10	Wheels	(~10)	2	3.89 * 4.56
11	Vehicle	(2900)	210	3.89 * 4.56
12	Housing	(~10)	–	–
13	Sleeve	(0.5)	–	–
14	Bearing	0.2	–	–
15	Release lever	1.5	–	–
16	Cable_lvr	~0.2	–	–
17	Cable_guide	~0.1	–	–
18	Cable_qua	~0.2	–	–
19	Quadrant	~0.2	–	–
20	Pedal	~1.5	–	–

gear ratios given are those of the first gear at 3.89 and the differential (final drive) at 4.56. To ensure correct functionality of the clutch system, as in many other mechanical multi-bodies, some components of the system are often represented by a combination of *parts*. For example, the clutch pedal cable is defined by two separate parts; one is considered to be connected to the release lever, and called cable_lvr, and the other is connected to the clutch quadrant and is called cable_qua. The motions of the two ends of the cable are then coupled by a 1:1 ratio coupler, introducing a scalar constraint function. The joints used to assemble the clutch judder multi-body model are given in Table 22.2.

In the table, the constraints are specified between neighbouring *parts*, denoted by Part I and Part J respectively. The constraints also include other algebraic functions than combinations which represent mechanical joints such as revolute, translational, spherical, cylindrical and in-plane, some of which are discussed in Chapter 1 and details of all of which can be found in Rahnejat (1998). Note that fixed joints remove all the degrees of freedom. A curve–curve constraint function describes their adherence. The other constraints are couplers, specified motions and kinematic gearing functions such as that of a rack-pinion arrangement, relating a rotational function to a translational motion. Now using Gruebler–Kutzbach expression (see Chapter 1):

Table 22.2 Constraints in the clutch judder model

No.	Part I	Part J	Constraint type	No. of constraints
1	Housing	Ground	Fixed	6
2	Flywheel	Housing	Revolute	5
3	Friction_disc	Flywheel	Inplane	1
4	Hub	friction_disc	Revolute	5
5	Hub	Shaft	Translational	5
6	Shaft	Housing	Revolute	5
7	Gearbox	Shaft	Revolute	5
8	Differential	Gearbox	Revolute	5
9	Wheels	Differential	Revolute	5
10	Vehicle	Wheels	Revolute	5
11	Cover	Flywheel	Fixed	6
12	Pressure-plate	Cover	Translational	5
13	Pressure-plate	Bearing	Coupler	1
14	Bearing	Sleeve	Translational	5
15	Sleeve	Housing	Fixed	6
16	Bearing	Lever	Curve–curve	2
17	Lever	Housing	Cylindrical	4
18	Cable_lvr	Lever	Cylindrical	4
19	Cable_lvr	Cable_guide	Translational	5
20	Cable-guide	Housing	Spherical	3
21	Cable_lvr	Cable_pdl	Coupler	1
22	Cable-pdl	Ground	Translational	5
23	Cable_pdl	Quadrant	Rack-pin	1
24	Quadrant	Pedal	Fixed	6
25	Pedal	Ground	Revolute	5
26	Crankshaft	Flywheel	Fixed	6
27	Motion	Pedal	Function	1
	Σ constraints			113

$$\text{DOF} = 6(\text{No. of parts} - 1 - \Sigma \text{ constraints} = 6(21-1) - 113 = 7$$

$$(22.1)$$

Once a multi-body model is made, it is important to identify the remaining degrees of freedom. The seven degrees of freedom are the angular displacements of the flywheel, friction disc, hub, gearbox, differential, rear road wheels and the back and forth motion of the vehicle itself. The crankshaft, clutch cover and pressure plate have the same displacements as the flywheel.

The next step is to include the system compliances and applied forces.

22.5.3 Compliances and forces

The mechanical behaviour of a dry clutch can be characterised by the clamp load variation, pressure plate lift to contact the friction disc surface and the release bearing travel as shown in Fig. 22.2. These *clutch characteristics* include the effects of various compliant members in clamping and releasing

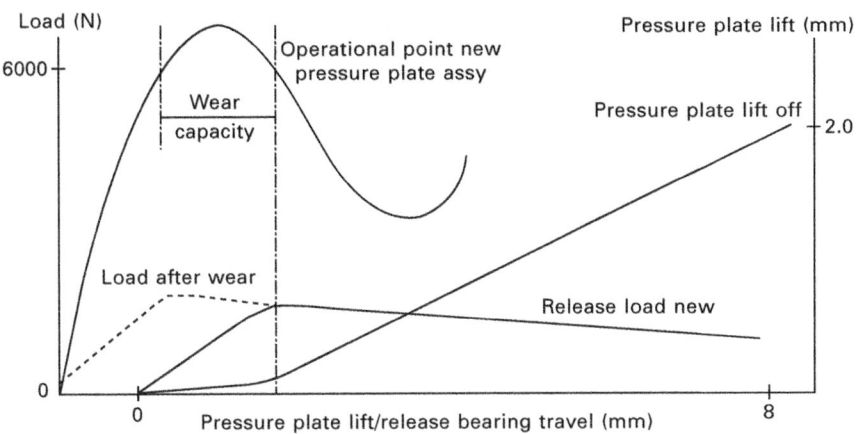

22.2 Clutch characteristics.

the clutch, such as the cushion springs (these are fitted between each friction lining to soften the engagement forces), the friction disc springs and the clutch cover diaphragm springs. For the sake of brevity, these individual characteristics are not included here. Interested readers should refer to Rahnejat (1998).

Returning to the characteristics in Fig. 22.2, the clamp load varies with pressure plate lift. The lower end of the graph represents point of separation of the pressure plate from the clutch cover diaphragm springs, thus no clamp load. With progressive movement (lift) of the pressure plate, the diaphragm springs are compressed between it and the cover, and the clamp load increases to a maximum value, followed by a decrease until the *mounting position* of a new clutch to the flywheel, a position referred to as the *operating point*. Gradual wear of the friction lining pads (facings) on the friction disc, causes the clamp load to occur with a larger pressure plate lift, with an initial increase in the load, which eventually returns to its original value after the wear limit is reached. Similar behaviour in release load characteristics with wear is also noted in the figure. The release load characteristics show the variation of the load on the release bearing with its sliding motion along the shaft. The pressure plate lift characteristics represent the travel of the pressure plate against the release bearing. The gradient of the first part of the characteristics is determined by the clutch cover stiffness, while the sharper characteristic is due to diaphragm springs.

Another source of compliance, key to torsional behaviour of the clutch is the characteristics of the hub springs mounted in the clutch disc assembly (Fig. 22.3). These show two slopes (two values of stiffness). In order to represent these, the model includes a frictional torque, *M* which is dependent on the relative angle θ between the friction disc and the hub upon which it is mounted (Centea *et al.*, 1999, 2001):

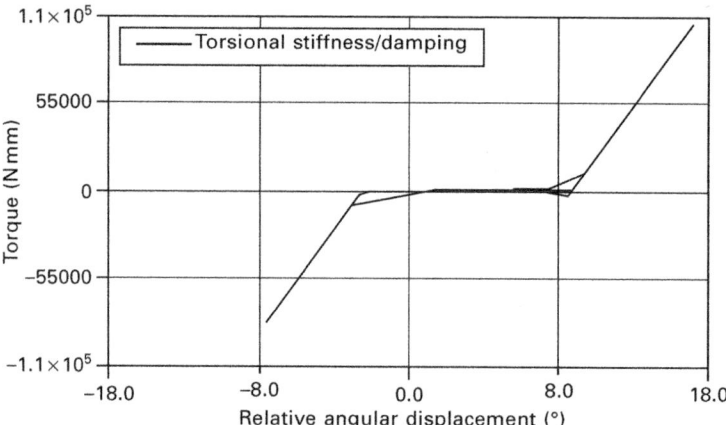

22.3 Characteristics of torsional clutch springs.

$$M = -k_{1-}\,\theta_{-} - k_{2-}\,\theta \qquad \text{if } \theta < -\theta_{-}$$
$$M = -k_{1-}\,\theta \qquad \text{if } -\theta_{-} < \theta < 0 \qquad\qquad (22.2)$$
$$M = k_{1+}\,\theta \qquad \text{if } 0 < \theta < \theta_{+}$$
$$M = k_{1+}\,\theta_{+} + k_{1+}\,\theta \qquad \text{if } \theta > \theta_{+}$$

k_{1-} is the stiffness of the torsional springs (situated between the friction disc and the hub) on the negative side of the characteristic curve, when the angle varies between zero and θ_{-}. k_{2-} is the torsional stiffness on the negative side of the characteristics, when the angle is less than zero. k_{1+} and k_{2+} are the corresponding values on the positive side of the characteristic curve. θ_{-} and θ_{+} are the angles where the characteristics alter (change of stiffness).

The negative side of the clutch characteristics represents the *coast* condition, when the throttle is closed and the road wheels drive the clutch. The positive side of the clutch characteristics represents the *drive* condition, when the throttle is open and the engine drives the clutch.

The cushion springs and straps between the clutch cover and the clutch housing also present stiffness characteristics which are included in the clutch model (see Centea *et al.*, 1999, 2001). Not shown is the longitudinal stiffness of tyres resisting the fore and aft motion of the vehicle associated with judder. These characteristics change with vehicle speed, best observed as the longitudinal rolling resistance. Tyre pressure can also affect judder response. At take-off, under judder condition, the longitudinal tyre forces are tractive at very low speed. When the clutch is fully actuated, after any judder, the longitudinal tyre force alters with road speed, and these characteristics affect vehicle shuffle with any throttle tip in or back out (see Chapter 23).

22.5.4   Frictional behaviour of clutch linings

By far the most important characteristics influencing take-up judder is the frictional behaviour of the interface between the driver (the flywheel) and the friction disc and the latter and the pressure plate (the driven members) as the clamping action occurs. The clamping action is not an instantaneous process, during the finite duration of which friction is developed at these interfaces. Therefore, friction is a function of slip speed.

Friction is generated between the **friction lining** on the surface of the friction disc and the mating surfaces on its either sides (flywheel and pressure plate surfaces). Friction is as a result of asperities of the harder surfaces of flywheel and pressure plate attempting to plough through the rough elastomeric surface of the friction pads. This is the main mechanism of friction generation at micro-scale, with contributions in the form of adhesive friction, fusing the asperities under the application of clamp load. These mechanisms of friction are discussed in Chapters 2 and 3. The mechanisms at work here are somewhat more complex as the friction lining behaviour is viscoelastic and also temperature dependent. This can be one of the reasons for changes of frictional behaviour with not only slip speed, but also with clamp load variation, effectively altering the contact temperature. Therefore, variation of friction with slip speed or changes in clamp load alter the generated friction torque, and thus the behaviour of the system. A simple representation of the friction torque M is:

$$M = 2FR = 2\mu WR \tag{22.3}$$

where the multiplier 2 recognises the contact of friction disc with both the flywheel and the pressure plate, μ is the kinetic coefficient of friction, W is the clamp load (normal force acting upon the friction surfaces) and R the mean radius of the friction surface defined as:

$$R = \frac{2}{3} \cdot \frac{R_o^3 - R_i^3}{R_o^2 - R_i^2} \tag{22.4}$$

where R_o is the outer radius of friction annulus and R_i is the inner radius of friction annulus.

As the engagement occurs in a gradual manner, the friction torque alters as the *driver* and the *driven* speeds equilibrate. This gradual change is represented as:

$$\frac{dM}{d\omega} = \frac{d(2\mu WR)}{d\omega} = 2WR\frac{d\mu}{d\omega} \tag{22.5}$$

And noting that:

$$\frac{d\mu}{d\omega} = \frac{d\mu}{dv} \cdot \frac{dv}{d\omega} = \frac{d\mu}{dv} \cdot R$$

thus:

$$\frac{dM}{d\omega} = 2WR^2 \frac{d\mu}{dv} \tag{22.6}$$

where v is the slip speed, ω is the angular velocity of the driver (engine) relative to the driven (the driveline). For a given clamp load W and mean radius of contact R, the variation should be a function of slip speed only. This supposition is correct as **ploughing friction** in asperity contact is a function of velocity (see Bhushan, 1999; Gohar and Rahnejat, 2008). This also indicates that with the assumed behaviour the adhesive friction, not being velocity dependent, would play a less significant role than the asperity ploughing action.

Kani *et al.* (1992) and Centea *et al.* (1999) represent the equation of motion for the driveline system during clutch slipping in the form:

$$m\ddot{x} + \left(c + W\,\frac{d\mu}{dv}\right)\dot{x} + kx = 0$$

where c and k are the effective stiffness and damping constant of the driveline system. For quasi-harmonic response it is clear that the solution is of the form: $x(t) = C \cdot e^{s \cdot t}$, where

$$s = -\frac{c + W\dfrac{d\mu}{dv}}{2m} + \sqrt{\left(\frac{c + W\dfrac{d\mu}{dv}}{2m}\right)^2 - \frac{k}{m}}$$

Substituting this into the equation of motion and after some manipulation, Kani *et al.* (1992) show, not surprisingly, that the behaviour of the system is contingent upon its damping characteristic: $c + W\,d\mu/dv$.

In general the damping constant c of the driveline system is quite low (a characteristic of drivelines, and the reason for many noise, vibration and harshness (NVH) issues, such as clonk; see Chapters 21 and 30). The effect of c is even less pronounced during clutch engagement, with the main sources of damping coming from $W\,d\mu/dv$. Thus, it is clear that both the clamp load and the friction lining characteristics are the important governing parameters.

Any loss of clamp load W reduces the product shown above. This will result in reduced damping and hence lead to oscillatory behaviour during engagement, which may have diverging harmonics. Such loss of clamp load can occur through hurried driver behaviour such as side-slipping off the clutch pedal. Note that rapid clutch actuation can also lead to other NVH concerns by an impulsive input to the driveline system (thump and clonk, Chapters 21 and 30). Loss of clamp load can also take place with misaligned contact during clutch actuation (this is observed in practice as noted in the introduction) (see Maucher, 1990). One telltale sign is often the presence of hotspots on the contact surfaces of the flywheel and the pressure plate and/ or uneven wear of friction disc pads.

Kinetic coefficient of friction alters with slip speed between any pair of surfaces. This is explained in a fundamental manner by a host of interacting phenomena, such as rate of ploughing of asperities on the softer counterface by opposing asperities on the harder mating surface. Also, a film of oxide is usually present on any surface, exposed to the environment (except completely clean surfaces *in vacuo*). These thin oxide films have lower shear strength than the bulk solid and thus aid **boundary lubrication**. As the slip speed increases, heat is generated by friction, which in turn aids the oxidation process, thus reduce friction. Heat generated can also soften asperities which are ploughed easier. Increased pressure also causes local yielding of asperities which plough easier with plastic deformation. Thus, pressure and temperature also affect the coefficient of friction–slip speed characteristics. This plethora of interacting phenomena underlies the noted variation of coefficient of friction with slip speed from its static value specific to a pair of surfaces of given topography (first expounded by Amontons, see Chapter 2) to its reduced kinetic value (first reported by Coulomb, but possibly noted earlier by Cavendish). Karnopp (1985) and Haessig and Friedland (1991) show various models and characteristics for $\mu - v$ characteristics.

In practice, it is necessary to measure the $\mu - v$ characteristics for a pair of counterfaces at different applied pressures (clamp load), while measuring the contact temperature. Thus, a piece of friction lining pad material is loaded against a sliding disc, representing the pressure plate for example. The contact temperature is measured by a rubber thermocouple carefully inserted through the pad and placed very close to the contacting surfaces. A series of measurements are obtained for different constant contact pressures, temperature and measured friction torque. These are sorted into a series of curves at constant pressure and temperature. Many such tests results in a carpet plot. A regression analysis can be carried out to obtain a relationship of the form $\mu = nv + C$. When $v = 0$, the static coefficient of friction is obtained. The kinetic coefficient of friction is dependent on slip speed and by the slope $d\mu/dv = n$, its value is reduced with a negative slope, when $n < 0$.

22.6   Results and discussion of numerical findings of multi-body dynamics analysis of judder

Centea *et al.* (1999) use $C = 0.43$ for a given range of friction linings. They used a series of values for n, both positive and negative. The value of n in any case is quite small in the region $0.002 \le |n| \le 0.02$. One would normally expect negative values for *n* in accord with the previous discussions. This means that the tendency to judder is inherent in frictional sliding contacts during engagement, at least in the form of stick–slip motions. However, with low negative slopes the oscillation amplitudes are restricted and possibly not discerned. As the negative slope is increased the rapidly falling coefficient of friction at higher slip speeds is transmitted through the system, through to the tyre contact patches as noted by Centea *et al.* (1999), thus easily discerned at the driver's seat. Figure 22.4 shows the driveline angular velocity variation for two values of *n* = –0.004 s/m and *n* = –0.016 s/m. Note that the engine speed drops slightly during the engagement coupling as that of the driveline increases towards eventual coincidence. The oscillation prior to the speed unification are as the result of take-up judder. Clearly, the ***friction lining***

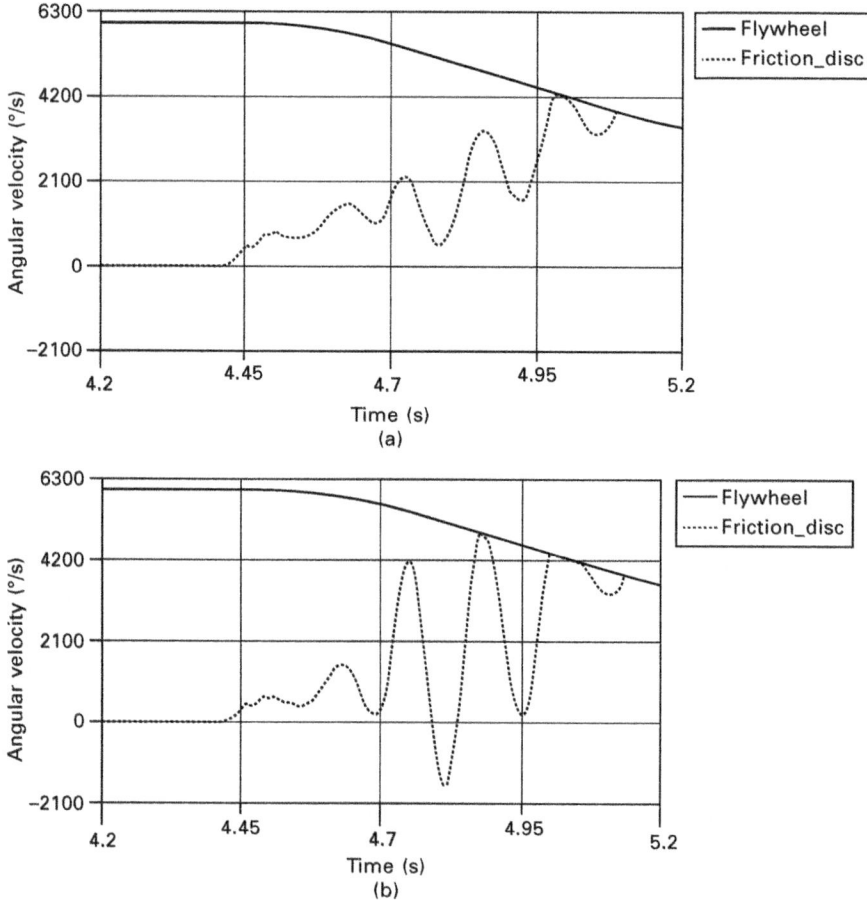

22.4 Clutch take-up judder with different friction lining characteristics (after Centea *et al.*, 1999) (a) *n* = –0.004 s/m (b) *n* = –0.016 s/m.

with a larger negative slope demonstrates significantly larger amplitudes of oscillation, indicating lower damping. The frequency of oscillation remains the same, being function of system stiffness and not significantly affected by low damping ratios.

Therefore, it is clear that a negative value for n increases the propensity to judder and this may be discerned by the amplitude of oscillations with a given threshold of detection. In industry NVH phenomena such as judder are noted subjectively by a rating system referred to as VER (vehicle evaluation rating). This acknowledges the perceptive resolving differences of individuals. Thus, judder to one may not be noted by another.

A point not addressed by Maucher (1990) and Centea *et al.* (1999), among others, is any underlying cause which would promote a positive value for n = $d\mu/dv$, which indicates a rise in the kinetic coefficient of friction in excess of its static value, an explanation which would be contrary to the generally accepted norm ever since Coulomb, but nevertheless was noted in experiments of the type explained above. The same explanation provided for reduced *ploughing friction* as a function of sliding velocity given above and noted as a function of the form: $\propto (1 - e^{\Phi v})$ by Bhushan (1999) and Gohar and Rahnejat (2008) is clearly not appropriate as Φ is essentially a function of topography and material properties which largely remains the same for any counterface pair at a given temperature. However, two other mechanisms are at play. One is the viscoelastic behaviour of the elastomeric-based lining. This means that during engagement stress relaxation of asperities occur (reduced pressures) as the area of contact increases accordingly (see, for example, Chapter 4; Johnson, 1985; Naghieh *et al.*, 1998). This means that the μ–v behaviour does not follow a specific curve, but across the carpet plot. This can occur when the relaxation time of the lining material is shorter than the stick–slip behaviour of judder. Such behaviour is noted in the case of tyres or rubber seals, in the case of the former affecting vehicle shuffle (see Chapter 23). More asperities in the contact gives higher friction which is perceived by an increase in the value of n.

By no means is this the only potential explanation. Unlike lubricated contacts that a rise in temperature reduces viscosity, thus the shear stress τ = $v\eta/h$, hence the viscous friction, the coefficient of friction in solid surfaces is directly proportional to the contact temperature. Nominally flat surfaces of the friction discs pads, flywheel and pressure plate do not form a flat contact. In fact, planar contact of mating surfaces is an idealised assumption made by Pascal, which only holds for fairly small localities (Hertzian contact is an example). The contact of the aforementioned nominally flat surfaces takes place over a few regions, often observed as hotspots (mentioned above). Note also that the flywheel *nods* during the engagement process which further promotes localised contacts (see Kushwaha *et al.*, 2002, on description of clutch whoop problem). The observed hotspots are circular footprints and the direct correspondence of coefficient of friction with **contact temperature** is obtained by Archard (1953) as:

$$\mu \approx 3.226 \frac{\Theta}{F_n} \sqrt{\frac{\rho c_p k_t a^3}{v}}$$
(22.7)

where Θ is the average contact temperature (in this case within a hotspot, see Fig. 22.5), F_n is a proportion of the clamp load W, depending on the number of hotspots, a the radius of the assumed Hertzian contact (the hotspot radius, comprising many asperity pairs), ρ the density of the lining material, c_p specific heat at constant pressure for the material and k_t its thermal conductivity. The calculation of Hertzian radius based on F_n and mechanical properties of the lining can be made using the relationships in Chapter 4.

Metallurgical examination of these hotspots on the flywheel and pressure plate has revealed that the intense localised friction heat causes an irreversible transformation to a Martensitic needle-like structure having increased hardness. Associated with this structure change is a significant increase in localised volume, causing the hotspot to rise above the nearby surrounding plateau. Thus, the cycle of localised heating is maintained.

The key point is that the coefficient of friction may be viewed as: $\mu \propto \theta/(p\sqrt{v})$, $p \propto F_n$ (p being the Hertzian pressure in a hotspot) for any given lining material (ρ, c_p, k_t). Now, it is clear that for a given contact pressure, slip speed reduces the coefficient of friction (as usually expected), but any rise in the contact temperature has the opposite effect. For the isothermal analysis undertaken by Centea *et al.* (1999, 2001) and Maucher (1990) a rise in the value of n is noted as a by-product of contact temperature. Thus, $n > 0$ values are due to thermal effect of solid sliding surfaces. It is clear that this effect was not noted by Coulomb in 1785. Lining materials with poor thermal conductivity would clearly promote rising contact temperatures. Additional work must be done to overcome the increased friction, which is a form of energy sink, thus enhanced damping. Figure 22.6 shows the reduced oscillations for the case of a positive $\mu - v$ characteristics.

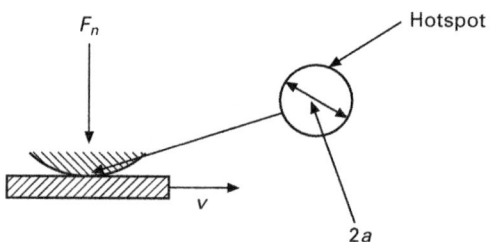

22.5 A hotspot localised contact.

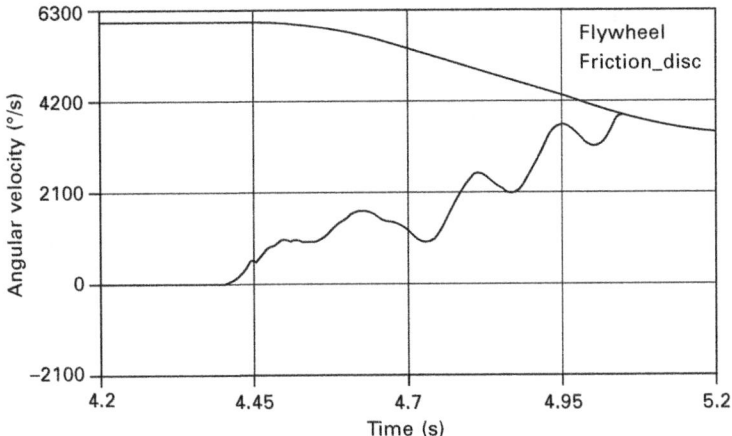

22.6 Clutch take-up judder with different friction lining characteristics (after Centea *et al.*, 1999) for n = 0.008 s/m.

The findings of the numerical analysis above need to be corroborated by field studies on vehicles or ideally on rigs correlated with vehicle conditions.

22.7   Vehicle studies and design of experiments for multi-body dynamics analysis of judder

A simulation model may be used to evaluate the contribution of potential causal factors, and to indicate which palliative action would be most cost effective. Given the potentially large number of active causal factors in the driveline system, a simulation may be effectively used to discover the ranking of these factors, or a combination of them in terms of their remedial cost effectiveness. The technique is referred to as the Taguchi design of experiments (DOE). It may be applied in a driveline mathematical simulation, or experimentally with a complaint vehicle. The use of DOE is particularly useful when there is a potential interaction between known factors, so the experimental approach is then preferred. However, it is absolutely essential that the DOE test program is completely repeatable for every factor combination that is tested. This is a significant task. The main benefit of using experimental DOE in a judder study is that the localised temperatures and their distortional effects within the clutch are fully considered under actual working conditions. Two such DOE studies were undertaken by Mike Menday with regard to judder.

The first was conducted on a complete vehicle at MIRA (Motor Industry Research Association). Five factors were investigated – friction material, engine mounts, misalignment, clutch clamp force and payload (as a noise factor). A total of 16 test runs were conducted to evaluate factor and factor interaction influences on judder. Several pickup signals were also monitored to best ensure repeatability. The outcome of the testing (unsurprisingly, perhaps) was that friction material had the single most influence on judder, corroborating the in-depth analysis highlighted in Section 22.6.

The second test program was conducted with a friction lining supplier. Different lining types were evaluated on a test rig, in conjunction with other system factors, to determine the best lining type to minimise judder excitation.

22.8   Overall conclusions of multi-body dynamics analysis of judder

The conclusion of the numerical analysis and the test programs can be summarised as follows.

22.8.1 Excitation of judder

In the case of a vehicle which is sensitive to take-up judder, the first order torsional mode may be excited in two ways:

- friction lining excitation;
- periodic clamp force excitation.

In both cases, the normally low level of vehicle driveline damping is insufficient to prevent an energetic normal mode response.

Friction lining excitation

It was shown in Section 22.6 and observed in the DOE test programs that judder may be excited when the clutch friction gradient $n = d\mu/dv$ becomes negative. When the friction gradient is zero or positive, the likelihood of judder is greatly reduced. Linings with positive gradients provide improved damping and little or no self-excitation.

Empirical data have indicated that judder can occur when both a significant vehicle mileage and a high working temperature have been reached. At lower working temperatures and with lower vehicle mileages, the lining friction behaviour is more predictable. Static friction is at a maximum before slip, but as slip speed increases the kinetic friction usually decreases, it being a function of interface pressure, local temperature, slip velocity and the number of engagements. The drop in kinetic friction leads to a further increase in surface temperatures. When local temperatures exceed 200 °C the lining permanently degrades and friction performance is irreversibly damaged. The temperature at the slip face is particularly difficult to predict due to the high-temperature gradients in the lining, pressure plate and flywheel. Clutch size is unfortunately not often determined by consideration of heat sink, but rather by the amount of space available.

Friction discs are radially grooved to allow lining debris to be readily expelled from the slip surfaces and to keep the working surfaces clean. But without exception, the grooves in worn linings from judder complaint vehicles were found to be either fully worn or were blocked by lining debris. This meant that the normal friction conditions would not apply. In addition, as wear progressed there was a greater propensity for the disc to stick to the flywheel and to the pressure plate, and for fade to occur.

Periodic clamp force variation

Judder may occur not only in the presence of negative clutch friction gradients, but also when other forcing functions are active. This includes clamp force variation as a function of clutch speed.

Heat is a major factor in these forcing functions and it is generated during clutch slip as localised heat, particularly if the first gear ratio is unsuited for the vehicle and greater slip energy is required for pull away. Additionally, one would be concerned with heat migration from the engine and transmission as global heat. Significant heat will also be generated during vehicle braking if the clutch is used as part of the braking effort.

The effects of clutch heat have been accumulated from laboratory evidence and from actual clutch complaint parts. They include the following empirical evidence:

- **Hot scorch markings:** These markings are found at the inner and outer radii of the friction disc on clutches from judder complaint vehicles. The scorch marks indicate a breakdown of the normal surface carbon composition, which greatly affects the friction behaviour.

  The pressure plate is manufactured with a small positive cone angle, the design intent being that this angle reduces to zero, and thereby achieves a full clamp contact, when under normal working temperature conditions. The cone angle is the angle between the normal and the actual rubbing surface. However, the pressure plate may flip from a positive to a negative cone angle especially when/if exposed to high thermal gradients. As a result the friction clamp is then concentrated at the inner or outer radii only, causing rapid heat build up and annular scorching.

- **Equi-spaced hotspots around the friction lining:** Again, these markings are also found only on clutches from judder complaint vehicles. When carefully measured, the adjacent pressure plate shows matching circumferential high spots. The high spots occur as a result of permanent metallurgical changes in the pressure plate due to localised heating of the plate. These measurements are taken when the plate is cold; it will be found that similar but smaller topological contours are present on a new unused pressure plate. Of course, the high spots are even more significant when hot.

- **Cushion segment springs:** There are several cushion segment springs fitted between the disc plate and the friction linings, and some or all of these may be heat attacked. Consequently, there will be impairment to the smoothness of torque take-up through the clutch.
- **Diaphragm spring fingers:** Softening of some or all of the fingers due to exposure to elevated temperatures, will cause finger distortion and lead to clamp force variation.

There are other *non-heat-related* causes of clamp force variation and judder. These include the following:

- **Clutch to flywheel assembly:** During assembly, the clutch is aligned and assembled to the flywheel by the use of dowel pins, to axially locate the clutch. Hence, it is there will be a necessarily tight assembly condition when the fixing screws are applied, so an unknown proportion of the applied fixing screw torque will be lost in overcoming this tight condition. It is important to make sure that every screw fixing location is equally clamped with same friction torque loss, to ensure minimum clamp force variation in the clutch.
- **Hub spline cleanliness:** Fretting corrosion or damage to the internal hub splines, or contamination of the spline grease with lining debris, or grease degradation due to heat, all prevent smooth axial movement of the friction disc to and from the flywheel rubbing face. This has been witnessed on judder complaint vehicles. There is no external force within the clutch assembly to mandate unrestricted movement of the disc away from the pressure plate along the splines.
- **Driveline lash:** There are several lash locations in the driveline, and free play will accumulate and aggravate the non-linearity of the system. This also includes the low-rate take-up spring in the clutch disc hub.
- **Clutch engagement rate:** In conjunction with the throttle demand, the clutch engagement rate is controlled by the driver and is a significant judder causal factor, both directly and indirectly. Driver behaviour in loss of clamp load has been simulated and reported by Centea *et al.* (1999, 2001).
- **Pull away gear ratio:** A low numeric pull away gear ratio will increase clutch slip times and heat generation during take-off, especially when the vehicle is highly loaded or negotiating a pull away on an incline. This is another major judder causal factor.
- **Alignment error:** The transmission input shaft and the engine crankshaft may become misaligned due to the adverse tolerance build-up through several component parts.

- **Lining suction:** As the linings and grooves wear and conform more closely to the pressure plate, a vacuum is established which increases the bond between disc and pressure plate and affects the normal slip friction condition.
- **Engine torque irregularities:** Judder occurs at low engine speeds when periodic combustion variation from cylinder to cylinder, random variation from cycle to cycle, and uncompensated forces in the reciprocating components, are important excitors.

22.9   Considerations for judder elimination/resolution

22.9.1   Friction material characteristics

Simulations and experimentation have shown that the potential for judder is greatly reduced if the clutch lining friction coefficient gradient is held to be positive or at least zero for its useful service life. This is clearly a demanding requirement given the abuse and misuse that a clutch may be subjected to.

The friction gradient is, however, only one of several friction material properties that must be considered for a clutch. Most other lining characteristics demand that a compromise is made when making a choice. Other considerations apart from friction characteristics include wear resistance, thermal stability, burst strength, fade resistance, inertia, dimensional stability and drag resistance.

22.9.2   Increased driveline damping

This is not a realistic solution since the general trend is a reduction in damping.

22.9.3   Controlled clutch engagements

Fast engagements increase the likelihood of judder, and numerical and experimental studies have additionally identified how fast the clutch engagement rate can be before the risk of engine stall (Rabieh and Crolla, 1996).

22.9.4   Clutch tuning

Judder is not resolved by clutch disc tuning, although clutch lash and soft torque transition can be helpful. It should be noted that soft lash is introduced to the clutch disc mainly as a means of controlling transmission gear idle rattle.

22.10 Future trends

22.10.1 Specification control

Suitable design specifications for the friction lining characteristics need to be established for worn/hot/abused/sensitive vehicle/other conditions. It is not sufficient to develop linings with positive friction gradient characteristics. The objective is to ensure that judder is prevented even when the lining is under mechanical and heat stress.

22.10.2 Test procedure

A suitable clutch lining test procedure needs to be established which meets the above design criteria. This will include the necessary means of accelerating the linings and clutch assembly to a realistic wear and heat-distressed state.

22.10.3 Lining design and development

The clutch is an important serviceable item in a vehicle. In the effort to extend the clutch lifetime, linings have become more likely to cause judder. What is needed are linings with positive gradients for all service conditions, and which achieve objective and extended lifetimes.

22.11 References

Archard, J.F., 'Contact and rubbing of flat surfaces', *J. Appl. Phys.*, **24**, 1953, pp. 981–988.

Bhushan, B., *Handbook of Micro/Nanotribology*, 2nd Edition, CRC Press, Boca Raton, Florida, 1999.

Centea, D., Rahnejat, H. and Menday, M., 'The influence of interface coefficient of friction upon propensity to judder in automotive clutches', *Proc Inst Mech Engrs, Part D: J. Automobile Engng.*, **213**, 1999, pp. 245–258.

Centea, D., Rahnejat, H. and Menday, M.T., 'Non-linear multi-body dynamic analysis for the study of clutch torsional vibrations (judder)', *Appl. Math. Modelling*, **25**, 2001, pp. 177–192.

Gohar, R. and Rahnejat, H., *Fundamentals of Tribology*, Imperial College Press, London, 2008.

Haessig, D.A. and Friedland, B., 'On the modelling and simulation of friction', *Trans. ASME, J. of Dynamic Systems, Measurement and Control*, **113**, 1991, pp. 354–362.

Hawthorn, J., 'A mathematical investigation of driveability', Inst. Mech. Engrs., Pap. C420//003, 1990.

Hwang, S.-J., Stout, J.L. and Ling, C.-C., 'Modelling and analysis of powertrain torsional response', SAE Paper 980276, 1998.

Johnson, K.L., *Contact Mechanics*, Cambridge University Press, Cambridge, 1985.

Kani, H., Miyake, J. and Ninomiya, T., 'Analysis of the friction surface on clutch judder', *JSAE Review Technical Notes*, **13**(1), 1992, pp. 82–84.

Karnopp, D., 'Computer simulation of the stick-slip friction in mechanical dynamical systems', *Trans. ASME, J. Dynamic Systems, Measurement and Control*, **107**, 1985, pp. 100–103.

Krenz, R.A., 'Vehicle response to throttle tip-in/tip-out', SAE Paper 850967, 1985.

Kushwaha, M., Gupta, S., Kelly, P. and Rahnejat, H., 'Elasto-multi-body dynamics of a multi-cylinder internal combustion engine', *Proc. Inst. Mech. Engrs., Part K: J. Multi-body Dynamics*, **216**, 2002, pp 281–293.

Maucher, P., 'Clutch chatter', *Proc 4th Int. Sympos. on Torsional Vibrations in the Drive Trains*, Baden Baden, Germany, April 1990, pp. 109–124.

Naghieh, G.R., Jin, Z.M. and Rahnejat, H., 'Contact characteristics of viscoelastic bonded layers', *Appl. Math. Modelling*, **22**, 1998, pp. 569–581.

Newcomb, T.P. and Spurr, R.T., 'Clutch judder', *XIV International Automobile Technical Congress of FISITA*, 1972.

Rabieh, E.M.A. and Crolla, D.A., 'Intelligent control of clutch judder and shunt phenomena in vehicle drivelines', *Int. J. Veh. Des.*, **17**, 1996, pp 318–332.

Rahnejat, H., *Multi-body Dynamics: Vehicles, Machines and Mechanisms*, Professional Engineering Publishing/SAE Bury St Edmunds, UK/Warrandale, PA, 1998.

22.12   Nomenclature

a	Hertzian contact radius per hot spot
c_p	Specific heat at constant pressure
DOF	Degrees of freedom
F	Friction
F_n	Share of clamp load per hot spot
h	Film thickness
k_t	Thermal conductivity
K_1, k_2	Stiffness stages of clutch torsional spring
m	Mass
M	Friction torque
p	Hertzian pressure
R	Mean radius of friction lining surface
R_t	Inner radius of friction annulus
R_o	Outer radius of friction annulus
v	Slip speed
W	Clamp load
η	Dynamic viscosity
μ	Kinetic coefficient of friction
θ	Angle of twist
Θ	Average contact temperature
ρ	Density
τ	Shear stress
ω	Angular velocity

23

Contact mechanics of tyre–road interactions and its role in vehicle shuffle

G. MAVROS, Loughborough University, UK

Abstract: The close relationship between tyre contact mechanics and low-frequency driveline dynamics is examined in this chapter. First, a dynamic model of a typical rear wheel drive driveline is developed. The model is then used to demonstrate the shuffle error state. Focus is subsequently moved to tyre–road interaction. A number of frequently used tyre modelling approaches are presented, initially for steady state operation and then for the investigation of low bandwidth transient responses. Such responses are treated by integrating a simple steady state tyre model with the relaxation length concept. Finally, the influence of tyre dynamics on shuffle is investigated by implementing a linear transient tyre model into the driveline model. Simulation results demonstrate the strong dependency of the rate of attenuation of shuffle oscillations on forward speed. It is shown that capturing such effects depends greatly on the tyre model of choice.

Key words: shuffle, driveline model, steady state tyre model, transient tyre model, friction, simulation.

23.1   Introduction

The *pneumatic tyre* represents one of the most important components of a road vehicle. It is the medium by which all control forces are generated and, in combination with the suspension reacts to the vehicle's weight in a manner that provides enhanced passenger comfort.

In the context of drivetrain dynamics the tyre represents the final stage between a vehicular driveline and the road, acting as a motion transformer that converts the rotational motion of the wheels into translational motion of the vehicle. When viewed as a driveline component, the tyre is inherently compliant compared with other driveline components, such as gears, half-shafts and propshafts. This is a direct result of the multi-dimensional role of the tyre which is also required to act as a means of filtering-out short-wavelength road irregularities.

Maintaining a driveline perspective, it is initially convenient to view the tyre as a compliant torsional spring connecting the wheel-rim to the road surface. Such a representation of the tyre immediately implies that the torsional response of the driveline might be influenced by tyre stiffness. In turn, this torsional response is closely related to *shuffle*, which manifests

703

itself as a low-frequency oscillation excited by sudden application or release of throttle, especially at low travelling speeds (throttle *tip-in* and *tip-out*). What is not immediately apparent from this initial representation of the tyre as a torsional spring is the importance of factors such as the rolling motion of the tyre, or the frictional behaviour of the contact patch. To address these issues a systematic modelling approach is required which provides an insight into the mechanics of the pneumatic tyre in relation to *drivetrain shuffle*.

In the following sections a detailed description of shuffle is provided, including the formulation of the relevant equations. A number of tyre modelling approaches are presented and their suitability for the study of shuffle is addressed. While some of the latest trends in tyre modelling are included, a simple yet effective approach that can be readily implemented by the reader is also provided. To this effect, a simulation case study provides some guidance and forms the basis for some interesting observations.

23.2   Shuffle as a drivetrain error state

Shuffle is a well-established drivetrain error state (Krenz 1985; Rahnejat, 1998; Farshidianfar, *et al.*, 2002; Stewart and Fleming, 2004; Schulz, 2005), resulting from the coupling of the first torsional mode of the driveline with the *fore–aft motion* of the vehicle body, during engine torque variations. It is a low-frequency phenomenon, typically in the region between 2 and 8 Hz and can be sensed by the passengers as a fore–aft oscillation of the vehicle body following throttle tip-in or tip-out commands (also see Chapter 30). This oscillation might become severe, especially at low travelling speeds, leading to the degradation of the drivability of the vehicle. Broadly speaking, drivability is a measure of how faithfully a vehicle follows its driver's commands and is adversely affected by an excessively oscillatory transient behaviour or an excessively lagging response prior to achieving steady state. It should be noted that *shuffle* usually refers to a vehicle operating in-gear with the clutch fully engaged. Torque variations are also experienced when shifting gear and subsequently re-engaging the clutch. Under these circumstances a shuffle response might occur; however, attention is usually moved to the take-up of various lash-zones along the *driveline* and the resulting impact-related phenomenon referred to as the clonk error state (Biermann and Hagerodt 1999) (also see Chapter 30).

An experimentally obtained shuffle oscillation is depicted in Fig. 23.1. The figure shows the forward acceleration of a vehicle, as measured during a throttle tip-in command while rolling at low forward speed. The low-frequency oscillatory behaviour is clearly evident in the acceleration trace, while a clearer picture of the frequencies involved is provided by the rms spectrum shown in Fig. 23.2.

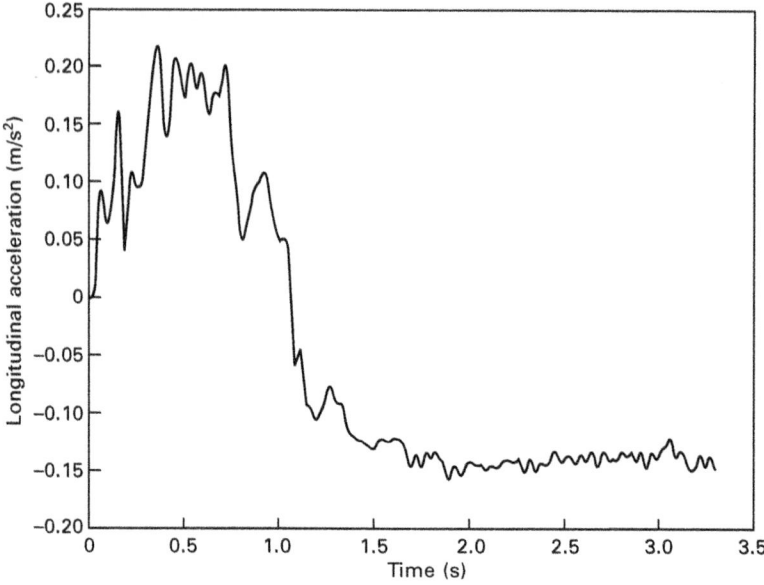

23.1 Experimentally measured shuffle oscillation in the time domain.

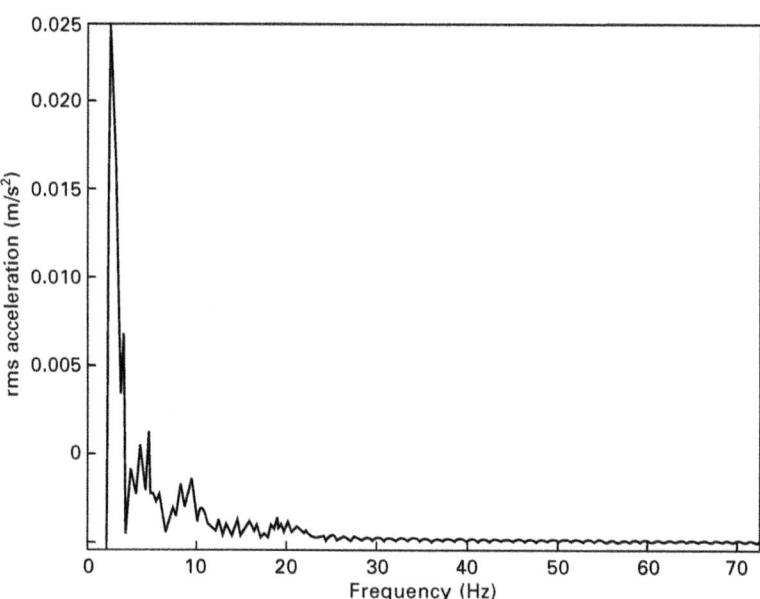

23.2 Frequency content of experimentally measured shuffle oscillation.

23.3   Basic model of vehicular driveline

Insight into the dynamics of shuffle is gained by developing a basic model of a vehicular driveline. For the purpose of shuffle investigations the torque input point is usually taken at the flywheel (Farshidianfar *et al.*, 2002; Fredriksson *et al.*, 2002) and all driveline components are considered, moving from the flywheel towards the wheels and tyres. A fundamental difference between candidate driveline models relates to whether the tyres are considered firmly connected to the ground (Farshidianfar *et al.*, 2002), or are allowed to roll, thus permitting the coupling of driveline torsional motion with the fore–aft motion of the vehicle (Schulz, 2005). In this section the equations of motion are derived for the simple case of a rear wheel driven vehicle. Initially the tyres are treated as rolling torsional springs. This is a rather rough approximation of real tyre behaviour that imposes limitations on the validity of the approach. Later, this representation of the tyres is replaced by a more accurate one which allows the transmission of motion to the vehicle body in a more realistic manner. Another important modelling point relates to the linearity of shuffle. Considering a fully engaged clutch and the lash-zones fully taken-up, shuffle is a primary candidate for linear analysis, since all dead-zones and *stick–slip*/impact phenomena are eliminated.

A typical driveline layout is illustrated in Fig. 23.3. Engine torque is applied to the flywheel which is connected to the clutch and subsequently to the transmission using compliant shafts. The transmission is represented by the meshing of two sets of gears. The influence of the intermediate shaft within the transmission can be safely neglected, since the high stiffness and relatively low inertia of this component would primarily contribute to higher-order oscillations which are irrelevant to shuffle. The output of

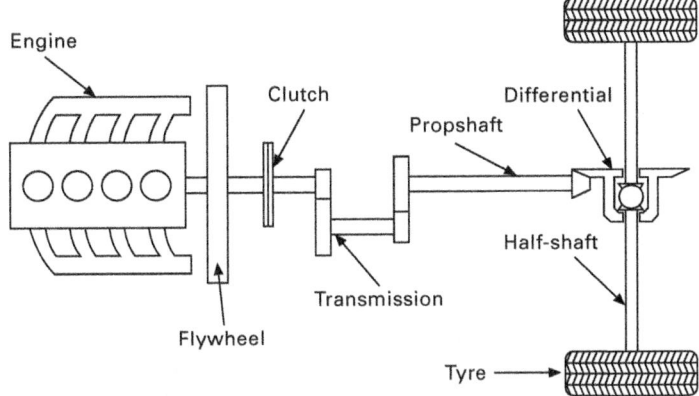

23.3 Frequency content of experimentally measured shuffle oscillation.

the transmission is connected to the propeller shaft which, in turn, passes the motion to the pinion gear of the differential. Subsequently, through the reduction ratio of the differential, motion is transferred to the wheel-rims via the flexible half-shafts. Finally, the wheel rims are connected to the ground through tyre compliance. The tyres are represented by torsional springs which have both the ability to wind-up and to roll without slipping. The rolling action accounts for the coupling of the rotational motion of the wheels with the fore–aft motion of the vehicle.

Owing to the focus of the analysis being directed to the first torsional mode of the driveline, a *lumped parameter approach* is often the choice (Farshidianfar *et al.*, 2002; Fredriksson *et al.*, 2002; Schulz, 2005). In particular, all inertia is assumed to lie within the primary inertia contributors such as the flywheel, gears and wheel-rims, while all stiffness is concentrated in the primary compliance contributors, such as the half-shafts, propeller shaft and tyres. For example, the clutch together with its input and output shafts are combined into a single equivalent shaft with its compliance containing the influence of all three individual elements. Also, the two stages of the transmission are reduced to one stage (one set of gears) with the same reduction ratio and with the inertia and stiffness of the intermediate shaft neglected. Based on these considerations/simplifications, an equivalent dynamic model of the driveline is depicted in Fig. 23.4. The schematic shows the degrees of freedom considered and all driveline components with their lumped properties. For the sake of simplicity it is assumed that there is no differential rotation between the wheels, so the two half-shafts and wheels/tyres are combined together and are represented by a single shaft and wheel with double inertia, stiffness and damping, while the torsional stiffness of the tyre is also increased by a factor of two. The differential equations of

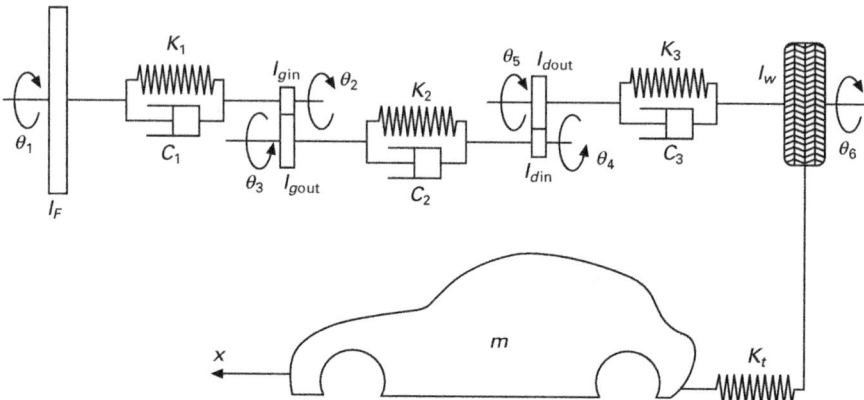

23.4 Reduced dynamic model of the driveline using the lumped parameter approach.

motion can be derived for each one of the inertias involved. The resulting system of equations can then be simplified/reduced, taking into account the kinematic relations between various degrees of freedom, as dictated by the gear ratios.

Referring to Fig. 23.4, the equation of motion of the flywheel becomes:

$$I_F \frac{d^2\theta_1}{dt^2} + C_1\left(\frac{d\theta_1}{dt} - \frac{d\theta_2}{dt}\right) + K_1(\theta_1 - \theta_2) - T_1 = 0 \tag{23.1}$$

where I_F is the moment of inertia of the flywheel, θ_1, θ_2 are the angular displacements of the flywheel and the transmission input gear, respectively, C_1 is the damping coefficient of the coupling between the flywheel and the transmission, K_1 is the corresponding torsional stiffness and T_1 is the torque applied to the flywheel by the engine.

In a similar manner, the equations for all angular degrees of freedom and the longitudinal motion of the vehicle body can be derived as follows:

$$I_{gin} \frac{d^2\theta_2}{dt^2} + C_1\left(\frac{d\theta_2}{dt} - \frac{d\theta_1}{dt}\right) + K_1(\theta_2 - \theta_1) + T_2 = 0 \tag{23.2}$$

$$I_{gout} \frac{d^2\theta_3}{dt^2} + C_2\left(\frac{d\theta_3}{dt} - \frac{d\theta_4}{dt}\right) + K_2(\theta_3 - \theta_4) - T_3 = 0 \tag{23.3}$$

$$I_{din} \frac{d^2\theta_4}{dt^2} + C_2\left(\frac{d\theta_4}{dt} - \frac{d\theta_3}{dt}\right) + K_2(\theta_4 - \theta_3) + T_4 = 0 \tag{23.4}$$

$$I_{dout} \frac{d^2\theta_5}{dt^2} + C_3\left(\frac{d\theta_5}{dt} - \frac{d\theta_6}{dt}\right) + K_3(\theta_5 - \theta_6) - T_5 = 0 \tag{23.5}$$

$$I_w \frac{d^2\theta_6}{dt^2} + C_3\left(\frac{d\theta_6}{dt} - \frac{d\theta_5}{dt}\right) + K_3(\theta_6 - \theta_5) + K_t\left(\theta_6 - \frac{x}{R}\right) = 0 \tag{23.6}$$

$$m \frac{d^2x}{dt^2} + \frac{K_t}{R}\left(\frac{x}{R} - \theta_6\right) = 0 \tag{23.7}$$

The moments of inertia I_F, I_{gin}, I_{gout}, I_{din}, I_{dout} and I_w involved in the above equations are shown in Fig. 23.4. In addition, m represents the total mass of the vehicle and x is the longitudinal displacement of the vehicle. The torques $T_2 - T_5$ are the reaction torques applied to the corresponding gear as a result of its meshing with the next gear. Finally, K_t is the torsional tyre stiffness and R represents the tyre radius, providing a kinematic relationship between the rotational motion of the wheel and the translational motion of the vehicle.

The transmission gears are considered rigid, so the following kinematic relationship applies between the rotations θ_2 and θ_3:

$$\theta_3 = \frac{\theta_2}{n_t} \tag{23.8}$$

where n_t is the transmission gear ratio expressed in terms of the corresponding gear radii as follows:

$$n_t = \frac{R_3}{R_2} \tag{23.9}$$

where R_2, R_3 are the radii of the input and output transmission gears, respectively.

The above gear ratio also provides a relationship between the torques T_2 and T_3:

$$T_3 = T_2 n_t \tag{23.10}$$

Using equations (23.9) and (23.10), equations (23.2) and (23.3) can be combined into a single equation with respect to the angular degree of freedom, θ_2. Similarly, the differential gear ratio, $n_d = R_5/R_4$ can be used in order to eliminate θ_5, thus combining equations (23.4) and (23.5) into a single equation with respect to the angular degree of freedom, θ_4. The resulting system of equations taking into account the algebraic relations imposed by the transmission and differential gear ratios is provided below:

$$I_F \frac{d^2\theta_1}{dt^2} + C_1\left(\frac{d\theta_1}{dt} - \frac{d\theta_2}{dt}\right) + K_1(\theta_1 - \theta_2) - T_1 = 0 \tag{23.11}$$

$$\left(I_{gin} + \frac{I_{gout}}{n_t^2}\right)\frac{d^2\theta_2}{dt^2} + \frac{C_2}{n_t}\left(\frac{1}{n_t}\frac{d\theta_2}{dt} - \frac{d\theta_4}{dt}\right) + C_1\left(\frac{d\theta_2}{dt} - \frac{d\theta_1}{dt}\right)$$

$$+ \frac{K_2}{n_t}\left(\frac{\theta_2}{n_t} - \theta_4\right) + K_1(\theta_2 - \theta_1) = 0 \tag{23.12}$$

$$\left(I_{din} + \frac{I_{dout}}{n_d^2}\right)\frac{d^2\theta_4}{dt^2} + \frac{C_3}{n_d}\left(\frac{1}{n_d}\frac{d\theta_4}{dt} - \frac{d\theta_6}{dt}\right) + C_2\left(\frac{d\theta_4}{dt} - \frac{1}{n_t}\frac{d\theta_2}{dt}\right)$$

$$+ \frac{K_3}{n_d}\left(\frac{\theta_4}{n_d} - \theta_6\right) + K_2\left(\theta_4 - \frac{\theta_2}{n_t}\right) = 0 \tag{23.13}$$

$$I_w \frac{d^2\theta_6}{dt^2} + C_3\left(\frac{d\theta_6}{dt} - \frac{1}{n_d}\frac{d\theta_4}{dt}\right) + K_3\left(\theta_6 - \frac{\theta_4}{n_d}\right) + K_t\left(\theta_6 - \frac{x}{R}\right) = 0$$

$$\tag{23.14}$$

$$m\frac{d^2x}{dt^2} + \frac{K_t}{R}\left(\frac{x}{R} - \theta_6\right) = 0 \tag{23.15}$$

The system of five second-order differential equations of motion is finally augmented by the following equations:

$$\frac{d\theta_1}{dt} = \omega_1 \tag{23.16}$$

$$\frac{d\theta_2}{dt} = \omega_2 \tag{23.17}$$

$$\frac{d\theta_4}{dt} = \omega_4 \tag{23.18}$$

$$\frac{d\theta_6}{dt} = \omega_6 \tag{23.19}$$

$$\frac{dx}{dt} = U \tag{23.20}$$

where $\omega_1 - \omega_6$ denote angular velocities and U is the forward speed of the vehicle.

The system of ten equations ((23.11)–(23.20)) is readily available for a **state-space representation** using the following state vector:

$$x = [\omega_1 \quad \omega_2 \quad \omega_4 \quad \omega_6 \quad U \quad \theta_1 \quad \theta_2 \quad \theta_4 \quad \theta_6 \quad x]^T \tag{23.21}$$

The equations of motion are written as:

$$A\dot{x} = Bx + T \tag{23.22}$$

Matrix A reads:

$$
A =
\begin{bmatrix}
\begin{bmatrix}
I_F & 0 & 0 & 0 & 0 \\
0 & I_{c1} & 0 & 0 & 0 \\
0 & 0 & I_{c2} & 0 & 0 \\
0 & 0 & 0 & I_w & 0 \\
0 & 0 & 0 & 0 & m
\end{bmatrix}
&
\begin{bmatrix}
C_1 & -C_1 & 0 & 0 & 0 \\
-C_1 & C_{c1} & \frac{-C_2}{n_t} & 0 & 0 \\
0 & \frac{-C_2}{n_t} & C_{c2} & \frac{-C_3}{n_d} & 0 \\
0 & 0 & \frac{-C_3}{n_d} & C_3 & 0 \\
0 & 0 & 0 & 0 & 0
\end{bmatrix}
\\[2em]
[\mathbf{0}]
&
\begin{bmatrix}
1 & 0 & 0 & 0 & 0 \\
0 & 1 & 0 & 0 & 0 \\
0 & 0 & 1 & 0 & 0 \\
0 & 0 & 0 & 1 & 0 \\
0 & 0 & 0 & 0 & 1
\end{bmatrix}
\end{bmatrix}
\tag{23.23}
$$

with:

$$I_{c1} = I_{gin} + \frac{I_{gout}}{n_t^2}, \ I_{c2} = I_{din} + \frac{I_{dout}}{n_d^2}, \ C_{c1} = C_1 + \frac{C_2}{n_t^2}, \ C_{c2} = C_2 + \frac{C_3}{n_d^2}$$

(23.24)

Matrix **B** reads:

$$B = \left[\begin{array}{cc} [0] & \begin{bmatrix} -K_1 & K_1 & 0 & 0 & 0 \\ K_1 & K_{c1} & \frac{K_2}{n_t} & 0 & 0 \\ 0 & \frac{K_2}{n_t} & K_{c2} & \frac{K_3}{n_d} & 0 \\ 0 & 0 & \frac{K_3}{n_d} & K_{c3} & \frac{K_t}{R} \\ 0 & 0 & 0 & \frac{K_t}{R} & -\frac{K_t}{R^2} \end{bmatrix} \\ \begin{bmatrix} 1 & 0 & 0 & 0 & 0 \\ 0 & 1 & 0 & 0 & 0 \\ 0 & 0 & 1 & 0 & 0 \\ 0 & 0 & 0 & 1 & 0 \\ 0 & 0 & 0 & 0 & 1 \end{bmatrix} & [0] \end{array} \right]$$

(23.25)

with:

$$K_{c1} = -K_1 - \frac{K_2}{n_t^2}, \ K_{c2} = -K_2 - \frac{K_3}{n_d^2}, \ K_{c3} = -K_t - K_3$$

(23.26)

The input vector, **T**, reads:

$$T = [T_1 \ 0 \ 0 \ 0 \ 0 \ 0 \ 0 \ 0 \ 0]^T$$

(23.27)

Finally, equation (23.22) is written in **state-space form** as:

$$\dot{x} = A^{-1} Bx + A^{-1} T \ \Rightarrow \ \dot{x} = Fx + GT$$

(23.28)

23.4   Results of driveline simulation

The state-space form of the equations of motion (eq. (23.28)) can be solved numerically, provided a known torque input, **T**. Prior to proceeding with this approach it is beneficial to carry out an *eigen-frequency analysis*, in order to determine the frequencies observed along the **driveline**. The eigen-values, λ, of the system are calculated by the following equation:

$$\det(F - \lambda I) = 0 \tag{23.29}$$

where I is a 10×10 unity matrix.

Table 23.1 provides a list of typical driveline parameter values. Solving equation (23.29) for the parameters contained in the Table 23.1 yields the **eigen-values** listed in Table 23.2.

The analytical results contained in Table 23.2 can be confirmed by simulation. Figure 23.5 illustrates the **fore–aft vehicle acceleration** as a result of a step-increase in the engine torque equal to 40 N m. The primary **shuffle** oscillation at 2.4 Hz is evident both in the time domain and in the power spectrum of the longitudinal acceleration signal, shown in Fig. 23.6. The time-response shows an oscillation which attenuates at a rather slow rate. This is due to the lightly damped driveline and the lack of torsional tyre damping. It should be emphasised that tyre damping is deliberately omitted here, in order to reveal the influence of systematic tyre modelling,

Table 23.1 Typical driveline parameters corresponding to the driveline model shown in Fig. 23.4

$I_F (\text{kg m}^2)$	$I_{gin} (\text{kg m}^2)$	$I_{gout} (\text{kg m}^2)$	$I_{din} (\text{kg m}^2)$	$I_{dout} (\text{kg m}^2)$	$I_w (\text{kg m}^2)$
0.29	0.003	0.004	0.001	0.006	2×0.4
$K_1 \left(\dfrac{\text{N m}}{\text{rad}}\right)$	$K_2 \left(\dfrac{\text{N m}}{\text{rad}}\right)$	$K_3 \left(\dfrac{\text{N m}}{\text{rad}}\right)$	$C_1 \left(\dfrac{\text{N m s}}{\text{rad}}\right)$	$C_2 \left(\dfrac{\text{N m s}}{\text{rad}}\right)$	$C_3 \left(\dfrac{\text{N m s}}{\text{rad}}\right)$
600	11 500	$2 \times 12\,000$	0.26	0.14	0.24
$R (\text{m})$	n_t	n_d	$m (\text{kg})$	$K_t \left(\dfrac{\text{N m}}{\text{rad}}\right)$	
0.32	3	4	1600	2×6827	

Table 23.2 Results of the eigen-frequency analysis carried out on the driveline model

Roots of char. eq. (23.29)	Eigen-values ($\times 10^3$)	Damped frequencies (Hz)
$\lambda_{1,2} =$	$-0.0581 + 3.1289i$ $-0.0581 - 3.1289i$	498.0
$\lambda_{3,4} =$	$-0.0364 + 0.4650i$ $-0.0364 - 0.4650i$	74.0
$\lambda_{5,6} =$	$-0.0014 + 0.1939i$ $-0.0014 - 0.1939i$	30.9
$\lambda_{7,8} =$	$-0.0000 + 0.0150i$ $-0.0000 - 0.0150i$	2.4
$\lambda_{9,10} =$	$-0.0000 + 0.0000i$ $-0.0000 + 0.0000i$	0

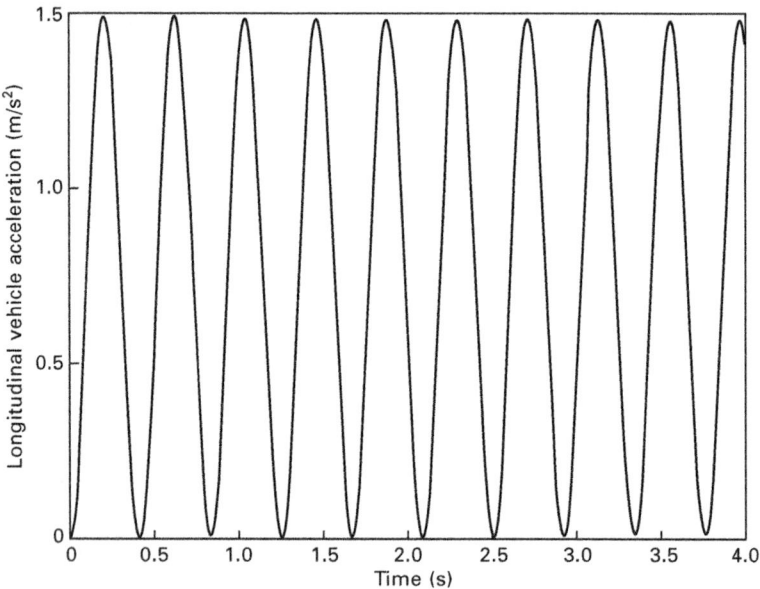

23.5 Shuffle response of the driveline model to a step-torque input.

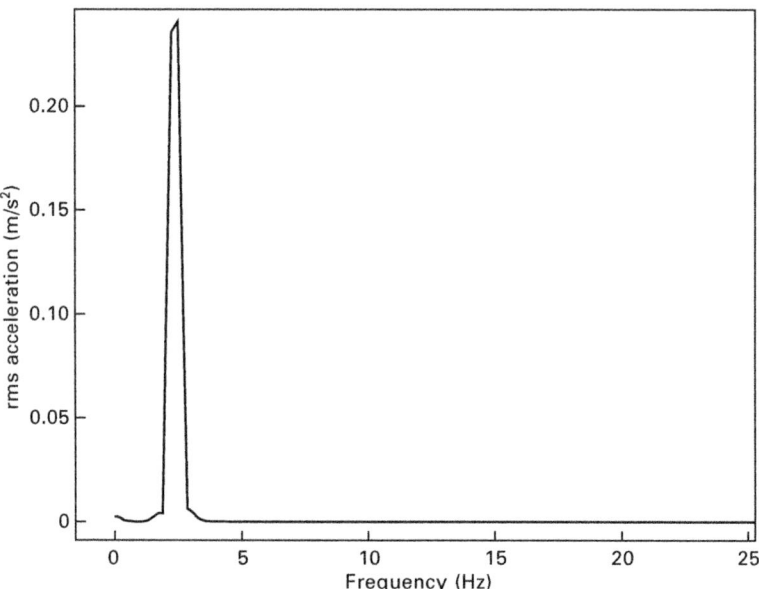

23.6 Frequency content of the shuffle response of the driveline model.

in the case study presented in Section 23.11. The two last eigen-values in Table 23.2 correspond to a rigid mode, which practically reflects the forward travel of the vehicle without any *driveline wind-up*.

As expected, the narrowband step input excites only the lower frequencies of the system. Moreover, eigen-vector analysis reveals that the frequencies listed in Table 23.2 are not spread evenly – in terms of their amplitude – along the degrees of freedom of the driveline. In order to excite higher frequencies a broadband random torque signal can be used as an input. As an example, Fig. 23.7 illustrates a white noise torque input applied at the flywheel. The resulting rms spectrum of the acceleration response of the pinion gear of the differential, $\dot{\omega}_4$, is provided in Fig. 23.8, which depicts significant contributions at approximately 2.5, 30.2, 69.5 and 499.3 Hz.

23.5   Tyre modelling for shuffle analysis

The treatment of the tyre as a torsional spring, detailed in the previous section, is appropriate for the determination of the modes of *shuffle* oscillation, assuming that tyre stiffness is known. This approach fails to capture the interactions at contact patch level and leads to rather unrealistic time domain responses. In turn, time domain characteristics such as the amount of acceleration overshoot and the rate of attenuation play a more important role in determining the severity of shuffle, than the exact frequency content of the phenomenon. In order to realistically predict the time response, some insight is required into both steady state and transient tyre force generation theory.

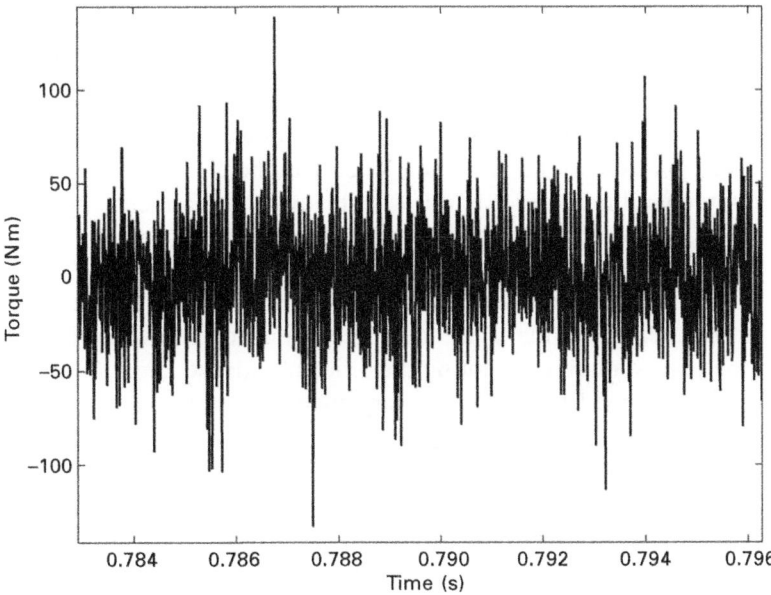

23.7 Broadband random-torque input used to excite all modes of the system.

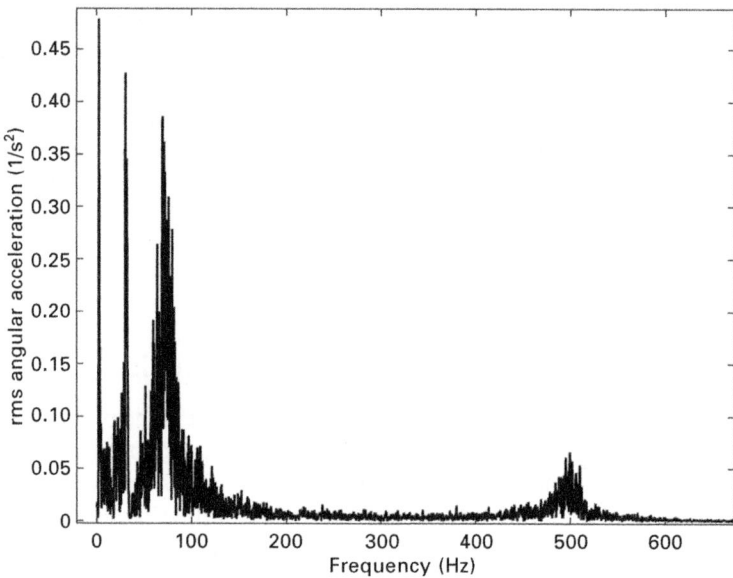

23.8 Frequency content of pinion-gear acceleration capturing the whole frequency spectrum predicted by eigen-frequency analysis.

23.6   Steady state tyre modelling

A detailed account of the fundamental approaches to steady state *tyre force* generation is provided in Pacejka and Sharp (1991). Steady state tyre models aim to calculate the shear forces, and, possibly, the resulting moments developed at the *contact patch* as a result of lateral and/or *longitudinal slip*, *camber* angle and the vertical load applied on the wheel. It is needless to mention that all the aforementioned parameters are assumed constant or slowly charging in order to allow a steadystate treatment. Steady state tyre models can be divided into two main subcategories, namely empirical/semi-empirical and physical models.

Empirical models rely on a large number of steady state tyre-force measurements and the subsequent generation of maps that link the forces to the corresponding operating conditions, with the lateral/longitudinal slip and vertical load being the primary independent variables. An alternative to tyre-force mapping is the generation of mathematical functions that are capable of accurately fitting experimental measurements within a wide range of operating conditions. Owing to the highly non-linear nature of tyre forces and their complex dependency on the independent variables, various attempts based on exponential and polynomial functions have seen limited applicability. The most successful such model is the well-known *magic formula tyre model* which appears in a number of variations (Bakker *et al.*, 1987, 1989; Pacejka and Bakker, 1993; Pacejka and Besselink, 1997; Pacejka, 2002). The magic formula is based on the implementation of a saturating sinusoidal function that captures the dependency of tyre shear forces, primarily with respect to the *lateral/longitudinal slip* and the vertical load. The parameters used in the formula can be identified by experimentally measuring the tyre forces throughout a wide range of steady state slip and loading conditions. Because some of the parameters involved in the magic formula have direct physical significance, the magic formula is often referred to as a semi-empirical tyre model.

Contrary to all the empirical/semi-empirical models, physical tyre models aim to mathematically describe the physical processes involved in tyre force generation. The level of available friction, the compliance of the tread and the gradual build-up of tread deformation as a result of tyre slip are all considered within a mechanical model for the calculation of shear forces. A large number of physical steady state models (Pacejka and Sharp, 1991) are based on the representation of the tyre as a rigid disk with a row of flexible bristles evenly distributed along the disk's circumference. The adoption of the *bristle concept* has resulted in the characterisation of such models as brush models. These models offer invaluable insight into the physical force-generation process. For this reason a simple *brush model* will be considered initially for the calculation of the longitudinal force generated by a slipping tyre under the influence of *traction*. This case is clearly relevant to *driveline shuffle*.

23.7   A brush-type model for shuffle analysis

The approach described herein is typical of a number of brush-type models and further details regarding its application can be found in Pacejka (2002). The tyre is initially represented by a disk with a single row of flexible bristles along its circumference, as shown in Fig. 23.9. Tyre forces are observed with respect to the *SAE frame of reference*, which is shown in the same figure. The x-axis of the SAE frame results from the intersection of the wheel plane and the road plane. The positive direction of the x-axis coincides with the forward rolling direction of the tyre. The y-axis results from the projection of the wheel spin axis to the ground and the positive direction is towards the right-hand side of the x-axis. Finally, the z-axis passes from the intersection of the x- and y-axes and is directed downwards, so as to generate a right-hand-side system. The SAE frame is a moving frame of reference, following the tyre along its path with the same forward speed.

Initially, a free-rolling tyre is assumed, so that the forward speed, V_x, at the centre of the wheel equals the linear speed of rolling, V_r, as shown in equation (23.30):

$$V_x = V_r = \Omega R \qquad (23.30)$$

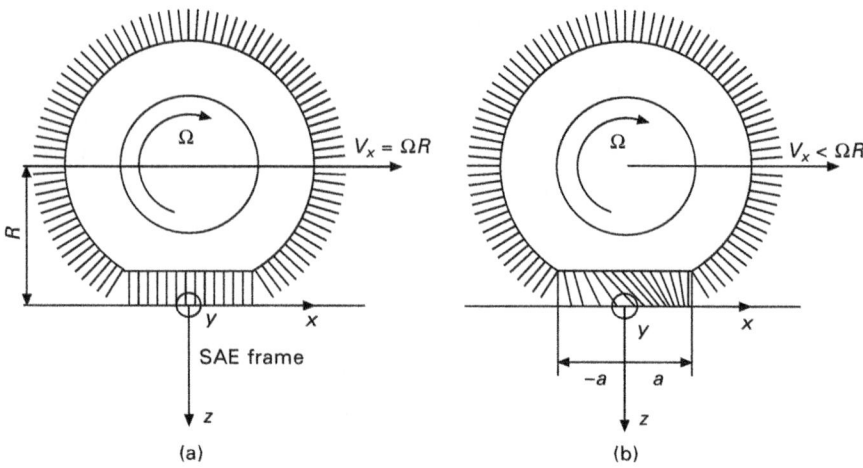

23.9 (a) Free rolling brush model (b) deformed brush model as a result of traction.

where Ω is the angular speed about the wheel's spin axis and R is the effective radius of the tyre.

For a wheel rolling forward, i.e. towards the positive x-axis, the angular speed, Ω points towards the negative y-axis, therefore the (–) sign should be used in equation (23.30) to maintain consistency with the forward speed, V_x. However, in driveline dynamics the wheel rotation that results in a forward motion of the vehicle is usually considered positive, so equation (23.30) is written without the (–) sign.

The wheel is rolling steadily, so by following the motion of a single bristle it is possible to obtain the motion of all bristles, as each bristle may be assumed to follow exactly the same path. This is a fundamental observation that allows the treatment of steady state force generation using the **brush concept**. A single representative bristle enters the contact patch at the leading edge. The bristle travels towards the negative x-axis of the SAE frame with speed ΩR until it reaches the trailing edge of the contact and finally exits. Because the SAE frame travels forward with speed V_x, the absolute speed of the tip of the bristle is zero. The bristle tip remains at the same global position, where it first met the road, the bristle remains completely vertical (undeformed) and the wheel rolls over the bristle while moving forward. As the single representative bristle moves backwards in the SAE frame, new bristles gradually enter the contact to cover up all previous positions of the representative bristle. Owing to the lack of bristle deformation in both lateral and longitudinal directions, no tyre forces are generated.

The situation is now considered, where, owing to the application of a **traction torque**, the wheel tends to rotate faster than dictated by its instantaneous forward speed, V_x. Steady state operating conditions are assumed, so the difference between the forward speed and the linear speed of rolling is constant ($\Omega R - V_x$ = constant > 0). A representative bristle enters the contact patch at the leading edge and the bristle tip sticks to the ground due to friction. The bristle base travels backwards with respect to the bristle tip with speed equal to $\Omega R - V_x$. This relative motion produces an equal rate of deflection of the bristle, which, after time Δt will result in a longitudinal deflection equal to $\Delta t (\Omega R - V_x)$. To proceed in detail, it is assumed for simplicity that the contact patch is symmetrical about the zy plane of the SAE moving frame of reference, with a total length of $l = 2 \cdot a$ extending a and $-a$ fore and aft on the x-axis, respectively. Using the SAE frame with its origin coinciding with the centre of the contact, the position of the representative bristle along the contact is denoted x. Since the bristle base moves backwards with speed ΩR in the SAE moving frame of reference, the time Δt elapsed from the moment the bristle enters the contact until it reaches position x can be written as:

$$\Delta t = \frac{a - x}{\Omega R} \qquad (23.31)$$

The total deflection of the representative bristle after time Δt is:

$$u = (\Omega R - V_x) \cdot \Delta t = \frac{\Omega R - V_x}{\Omega R} \cdot (a - x) = \sigma_x \cdot (a - x) \qquad (23.32)$$

The quantity $\sigma_x = (\Omega R - V_x)/\Omega R$ is termed 'theoretical *longitudinal slip-ratio*' and represents the primary kinematic input responsible for tyre force generation. It should be noted that equation (23.32) for the longitudinal bristle deformation holds true only when the bristle tip sticks firmly on the ground and no sliding occurs. This is assumed to be the case for very low slip-ratios and/or high vertical loads. To obtain the total longitudinal force the deformation is multiplied by the stiffness of the bristle and integrated along the contact patch, as follows:

$$F_x = K_x \int_{-a}^{a} u \, dx = -K_x \int_{-a}^{a} \sigma_x (a - x) dx = 2 \cdot K_x \cdot \sigma_x \cdot a^2 \qquad (23.33)$$

where K_x is the longitudinal stiffness of the bristles per unit length of the contact, (in (N/m)/m).

The previous analysis is valid only for small slip-ratios. At higher *longitudinal slip*, bristle deformation builds up faster, up to a point where the longitudinal force developed by the representative bristle exceeds the friction force which keeps the bristle tip stuck to the ground. This is the transition point from sticking to sliding. In a simplified treatment, once sliding has initiated, the longitudinal force becomes equal to the friction force, which in turn is determined by the local vertical load and the coefficient of friction. In order to find the transition point between sticking and sliding, one needs to assume a vertical load distribution (N/m) along the **contact patch**. A parabolic vertical load distribution symmetrical about the yz plane serves as a good first approximation:

$$f_z = \frac{3F_z}{4a} \cdot \left[1 - \left(\frac{x}{a} \right)^2 \right]$$

(23.34)

where f_z denotes the vertical load distribution in (N/m) and F_z is the total vertical load reacted at the tyre contact patch.

In order to find the transition point from stick to slip along the contact patch, a coefficient of static friction, μ_s, is assumed. The transition takes place when the local longitudinal force exceeds the local vertical load multiplied by the coefficient of static friction. Thereafter, the force applied to the bristle tip equals the local vertical load, scaled by the **coefficient of kinetic friction**, μ_k, which is usually assumed smaller than μ_s. Based on the above considerations, the transition point, x_t, along the contact is calculated as follows:

$$K_x |\sigma_x| (a - x_t) = \frac{3}{4} \mu_s F_z \left[1 - \left(\frac{x_t}{a} \right)^2 \right] = \frac{3}{4} \mu_s F_z \frac{(a + x_t)(a - x_t)}{a^3}$$

(23.35)

and, finally:

$$x_t = \frac{K_x |\sigma_x| 4a^3}{3 \mu_s F_z} - a$$

(23.36)

In equation (23.36) the absolute value of the **slip-ratio** is used, to account for the **braking** case where $\sigma_x < 0$. Now, the general expression for the longitudinal force reads:

$$F_x = \mathrm{sgn}\,(\sigma_x) \mu_k \frac{3F_z}{4a^3} \int_{-a}^{x_t} (a^2 - x^2)\,dx - K_x \sigma_x \int_{x_t}^{a} (a - x)\,dx$$ (23.37)

and, finally:

$$F_x = \mathrm{sgn}\,(\sigma_x) \mu_k \frac{3F_z}{4a^3} \left[x_t \left(a^2 - \frac{x_t^2}{3} \right) + \frac{2}{3} a^3 \right] + K_x \sigma_x \left[\frac{a^2}{2} + x_t \left(\frac{x_t}{2} - a \right) \right]$$

(23.38)

A typical traction force curve as predicted by equation (23.38) with the transition point calculated by equation (23.36) is provided in Fig. 23.10. The initial part of the curve is approximately linear. For vanishing slip-ratio, the gradient of the curve can be easily calculated by differentiation of equation (23.33) which expresses the force for $\sigma_x \to 0$:

$$C_t = \frac{\partial F_x}{\partial \sigma_x} = 2K_x a^2 \qquad (23.39)$$

The term C_t can be referred to as 'traction stiffness', in an analogy to the braking, or cornering stiffness (Pacejka, 2002). Equation (23.39) does not directly reveal a vertical load dependency of the traction stiffness. However, both the stiffness of the bristles, K_x, and the length of the contact patch $l = 2 \cdot a$, are vertical load-dependent in a non-linear manner. The general trend is that the traction stiffness increases with vertical load; however, it does so at a decreasing rate.

Depending on the amount of *longitudinal slip*, σ_x, the transition point from stick to slip moves along the contact patch, from the trailing edge at vanishing slip, to the leading edge when sliding takes place throughout the contact. In the latter case, the asymptotic traction force depends only on the vertical load, F_z and the coefficient of kinetic friction, μ_k. The peak traction force shown in Fig. 23.10 is achieved at an intermediate stage, when part of the contact patch sticks to the ground and part of it slides. Thus, the peak force depends on both the coefficients of static and kinetic friction, μ_s, μ_k, respectively. Like the initial gradient of the force curve, C_t, the coefficients

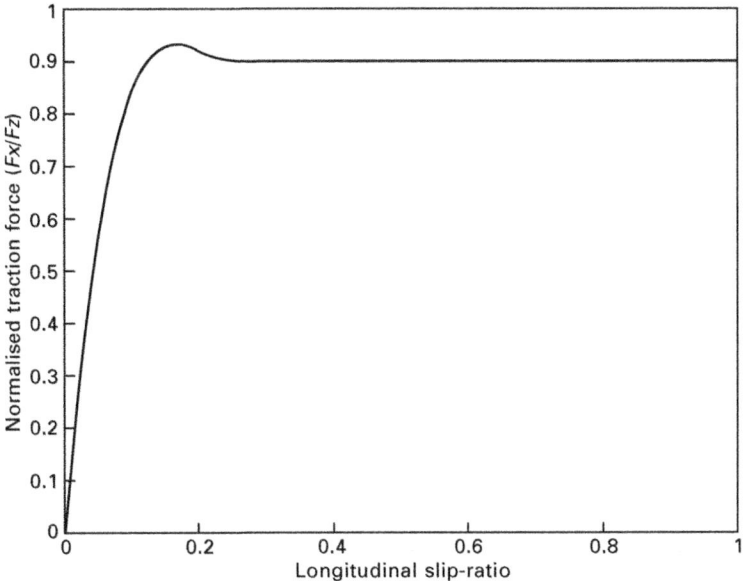

23.10 Typical steady state traction force curve as generated by the brush model.

of friction μ_s, μ_k depend largely on the vertical load. In particular, both coefficients of friction reduce with vertical load so that the total friction force increases with load, but does so at a decreasing rate. It should be emphasised here that the treatment of tyre friction using a couple of friction coefficients is rather simplified. Rubber friction is a field of ongoing research. In reality, the friction force between rubber and a hard substrate depends on a multitude of parameters such as temperature, surface roughness, sliding speed and, of course, vertical load. A more detailed account of the mechanics of friction at the road–rubber interface is given in Moore (1975) and Lemaitre (2001). A frequently used treatment to express the dependency of rubber friction on sliding speed is provided by Savkoor in Lemaitre (2001) and is based on the following expression for the coefficient of kinetic friction:

$$\mu_k(V_s) = \mu_s + (\mu_m - \mu_s)\exp\left(-\frac{h^2}{2}\ln^2\frac{|V_s|}{V_m}\right) \tag{23.40}$$

In the above relation, V_s is the sliding speed, μ_s is the coefficient of static friction and μ_m is the maximum coefficient of friction, experimentally observed at a sliding speed, V_m. The non-dimensional parameter, h reflects the width of the speed range for which equation (23.40) is valid. In order to improve the accuracy of the simple brush model, the constant coefficient of kinetic friction in equation (23.38) could be replaced by the sliding-speed dependent coefficient of friction predicted by equation (23.40), with $V_s = \Omega R - V_x$.

23.8   Relationship between the brush model and the magic formula

The *tyre brush model* presented in the previous section relies on a number of simplifications. For example, the vertical load distribution is assumed to be a perfect parabola and friction is not considered to be load dependent. Accounting for these and other details in a physical tyre model has proven to be a difficult task. The empirical nature of the magic formula allows for the inclusion of such effects in an implicit manner. The full details of one of the latest versions of the formula can be found in Pacejka (2002). Here, a brief description of the formula is given, highlighting the analogies with the brush model.

A general simplified expression of the formula for the *longitudinal tyre force* reads:

$$F_x = D \sin\{C \arctan[Bx - E(Bx - \arctan Bx)]\} \tag{23.41}$$

where x is the input in the form of longitudinal slip-ratio and B, C and D are parameters which primarily depend on vertical load and in some cases on camber-angle and forward speed. Owing to the experimental origins of the formula, the slip-ratio is defined in a slightly different manner from the theoretical slip-ratio appearing in equation (23.32). In particular, the so-called *practical slip-ratio* is used as an input, given by the following equation:

$$s = \frac{\Omega R - V_x}{V_x} \qquad (23.42)$$

The following relationship applies between the two slip-ratios:

$$\sigma_x = \frac{s}{s+1} \qquad (23.43)$$

Apparently, for small slip-ratios ($s \ll 1$), the two slip ratios are interchangeable, i.e. $\sigma_x \cong s$.

Returning to the magic formula, it is easy to see that the parameter, D, equals the peak traction force, while the asymptotic force equals $D \sin [(\pi/2) C]$. Finally, differentiation of relation (23.41) at vanishing slip-ratio, s, yields the initial gradient, or *traction stiffness*, C_t:

$$C_t = BCD \qquad (23.44)$$

Based on the above analysis, some important relationships can be deducted between the magic formula and the brush model presented in Section 23.3.2. In particular, for a given tyre operating at a pre-specified vertical load, equations (23.39) and (23.44) yield, at vanishing slip:

$$C_t = 2 K_x a^2 = BCD \qquad (23.45)$$

Also, at very high slip-ratios where sliding is observed throughout the contact patch, the following relation applies:

$$\mu_k F_z = D \sin[(\pi/2)C] \qquad (23.46)$$

23.9   Transient tyre response: a first approach

The basic *steady state brush model* assumes a rigid disk with the bristles representing the tread compliance in the neighbourhood of the contact patch. In the event of a step-change in the slip-ratio, a shear force is instantly developed according to equation (23.38). This assumes, for example, that all originally undeformed bristles reach their steady state deflections instantly. This is in contrast to the notion that a bristle requires some time in order to build up its deflection, given a rate of deflection equal to $\Omega R - V_x$. Immediately, it appears that the instantaneous generation of tyre shear forces is an invalid assumption. However, when the analysis is confined within the neighbourhood of the contact patch and the compliance of the bristles reflects only that of the outer part of the tyre disk, it is acceptable to neglect the rather fast response of the bristles and assume that the shear force is developed immediately. This assumption can be paralleled to the frequent treatment of dry friction between hard materials using the classical laws of **friction** (see Chapter 3) due to Amontons and Coulomb (Lemaitre, 2001). According to this treatment, the friction force develops instantly as a reaction to a horizontal force attempting to slide one hard object over another. In reality, some deformation needs to take place within the contact, before the friction force can be sensed. Obviously, under transient conditions, the generation of the friction force would depend on the time integral of the rate of contact deformation and hence, on time. Fortunately, such a treatment is only necessary when the time constants of the overall system are comparable to the – usually very small – time constants describing the dynamics of contact deformation. When this argument is projected to tyre-force generation, it is observed that the tread is rather soft compared with typical hard materials (steel, wood, etc.) and requires considerable time, or, alternatively, comparatively large deflections before it can transmit any friction forces. However, the dynamics of tyre force-generation are governed by a much slower process, that is, the build-up of the necessary carcass/sidewall deformation. As a conclusion, the assumption of an instantaneous force generation at contact-patch level is a rather safe one, provided that the slower dynamics of the carcass/sidewall are taken into consideration.

The argument presented thus far allows a first approach to transient tyre force generation, considering the **carcass compliance** and assuming instantaneous force generation at contact patch level. Central to this approach is the concept of the **relaxation length** which has been previously used for the analysis of longitudinal tyre dynamics in Clover and Bernard (1998). A more detailed account of the application of the relaxation length concept for the simulation of transient tyre behaviour in the non-linear range is provided in Higuchi (1997). The main findings from Higuchi's work are also presented in Pacejka (2002).

Figure 23.11 shows a wheel operating in traction. The sliding speed at the neighbourhood of the contact patch is denoted V_{sc}, while the point P on the wheel circumference operates at the nominal slip, $V_s = \Omega R - V_x$. Immediately, the rate of deflection of the tangential/longitudinal spring representing the carcass compliance becomes:

$$\frac{dh}{dt} = V_s - V_{sc} \tag{23.47}$$

where h represents the longitudinal/tangential deflection of the carcass/sidewall.

An approximation of the localised **slip ratio** at the neighbourhood of the contact, is:

$$s_c = \frac{V_{sc}}{V_x} \tag{23.48}$$

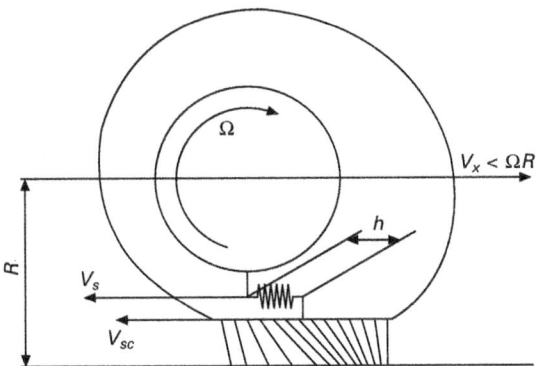

23.11 Schematic of a tyre operating in traction, including the effect of carcass compliance.

Based on the assumption that in the presence of a localised slip-ratio the friction force is generated instantaneously at contact patch level and further assuming a small slip-ratio, ($s_c \ll 1$), the *longitudinal tyre force* is calculated as:

$$F_x = C_t s_c \tag{23.49}$$

The above longitudinal force, F_x, is reacted by the carcass/sidewall so that:

$$F_x = K_c h \tag{23.50}$$

where K_c denotes the equivalent carcass/sidewall longitudinal stiffness at road level in N/m.

Combining equations (23.47) through (23.50) yields the following differential equation:

$$\frac{dh}{dt} + \frac{K_c}{C_t} V_x \cdot h = V_s \tag{23.51}$$

The ratio C_t/K_c which appears inverted in the above relation represents the so-called *longitudinal relaxation length* of the tyre, denoted Sx. With its introduction, equation (23.51) can be written as:

$$\frac{dh}{dt} + \frac{1}{Sx} V_x \cdot h = V_s \tag{23.52}$$

Provided the forward speed, V_x, and a transiently increasing slip-speed $V_s = \Omega R - V_x$, equation (23.52) can be solved in the time domain to provide with the instantaneous value of carcass/sidewall deflection, h. Then, the instantaneous tyre force can be calculated by multiplying the deflection, h, with the corresponding stiffness according to equation (23.50). Alternatively, equation (23.52) can be written in terms of the localised slip-ratio, s_c. Combining equations (23.49) and (23.50) yields:

$$s_c = \frac{h}{Sx} \tag{23.53}$$

Substituting equation (23.53) into (23.52) and considering that $dh/dt = (ds_c/dt)$. Sx and $V_s = \Omega R - V_x = sV_x$, the following relation is obtained:

$$Sx \frac{ds_c}{dt} + V_x \cdot s_c = sV_x \tag{23.54}$$

Now, in the event of a **transient slip-ratio**, $s = (\Omega R - V_x)/V_x$, equation (23.54) can be solved in the time-domain to provide with the instantaneous localised slip, s_c, which, multiplied by the coefficient C_t gives the instantaneous longitudinal force, F_x.

Further appreciation of the significance of the relaxation length, Sx may be achieved by considering the force response of the tyre to a step-increase in the longitudinal slip ratio, s. Solving equation (23.54) for zero initial localised slip, s_c yields:

$$s_c = s \left[1 - \exp \left(-\frac{V_x}{Sx} t \right) \right] \tag{23.55}$$

Equation (23.55) shows that after time $\Delta t = Sx/V_x$ from the application of the step excitation, the localised slip, s_c will have reached $(1 - 1/e) \times 100\% = 63\%$ of its steady state value, s. During the same time, the tyre will have travelled a distance equal to $\Delta t \cdot V_x = (Sx/V_x) V_x = Sx$. Based on this observation, the **longitudinal relaxation length** can be defined as the distance covered by the tyre in order to reach 63% of its steady state force, after the application of a step slip-ratio input. Typically, relaxation length values lie in the range between 0.3 and 0.8 metres. It should be mentioned that the relaxation length for a specific tyre does not remain constant during operation. It has been found experimentally (Higuchi, 1997) that the relaxation length depends both on the vertical loading of the wheel and on the level of slip developed. For example, at higher slip-ratios the tyre tends to respond faster to additional changes in the slip level, hence the equivalent relaxation length reduces (particularly, in a non-linear manner). Such phenomena are relevant to extreme traction manoeuvres, where tyres operate at high slip-ratios. Fortunately, in the case of **driveline shuffle** the levels of longitudinal slip-ratio remain low enough to permit treatment of the problem using equation (23.54) or (23.52).

The assumption that forces develop instantly at contact patch level allows the combination of equation (23.54) with any steady state tyre model. For example, the localised slip, s_c, also referred to as 'transient slip-ratio' can be inserted into the magic formula (eq. (23.41)), or the brush model (eq. (23.38)) for the calculation of the instantaneous tyre force.

23.10   Further tyre modelling possibilities

The tyre modelling approach presented in Sections 23.3.2–23.3.9 is useful to illustrate some of the most important aspects of tyre-force generation. In addition, it can be implemented immediately for the analysis of shuffle, where, due to the low frequencies and low slip-ratios involved, a simplified description of tyre dynamics is deemed adequate. A further advantage of this approach is the minimal requirements in computational power when it comes to implementation in a simulation environment.

On the contrary, in ride analysis, higher-order dynamic tyre behaviour needs to be considered. The in-plane and out-of-plane modes of vibration of the tyre up to a few hundred Hz can be captured by elaborate dynamic models of the tyre belt/carcass, combined with a detailed contact-patch description. Such models are the result of many years of research and can be used for a variety of studies including non-linear handling/ride analyses, ABS (anti-lock braking system), extreme traction simulations, and, of course, shuffle simulations. The requirements of these models in computational power are undoubtedly higher; however, this downside becomes progressively less important as the processing power available increases steadily. Although a number of such models exist in the literature, their complexity prohibits direct application based solely on the existing documentation. Thus, only two of the most popular, well-documented commercially available tyre models are mentioned here.

The *FTire* (*Flexible-ring Tyre*) (Gipser, 2005) is based on the representation of the tyre belt as a series of lumped-mass nodes connected to each other and to the rim using non-linear elastic, damping and friction elements. This structure, including the tension due to air pressure inside the tyre, is primarily responsible for the modal behaviour of the tyre. At contact patch level, forces are calculated using an elaborate brush representation, where the brush elements include stiffness, as well as damping effects. The friction force between the brush tips and the road depends not only on the local vertical load, but also on sliding speed and temperature. In turn, a temperature field is calculated using a thermal model for the prediction of the heat generated due to internal damping and friction and the subsequent transfer of this heat to different parts of the tyre and to the environment.

A different, probably more pragmatic approach has been followed in the development of the *SWIFT* model (*Short-Wavelength – Intermediate Frequency Tyre*) (Pacejka, 2002; Besselink *et al.*, 2005). The main structure of the model responsible for the primary in and out of plane dynamic behaviour of the tyre is a rigid ring with inertia, connected to the rim with elastic elements. Because of the assumption of a rigid ring, the frequency response of the main tyre structure is limited to approximately 60 Hz. The contact forces are treated using an elaborate brush description capable of capturing the effects of finite contact length and width, especially important for the calculation of the self-aligning moment. At higher slip and combined lateral-longitudinal slip conditions, the magic formula model is implemented in order to successfully handle the non-linearities observed in such situations. Whereas the FTire model uses a unified approach for all possible operating conditions, the SWIFT model is based on a combination of physical and semi-empirical treatments resulting in a more-or-less hybrid model.

In recent years, a new breed of tyre models has emerged. These models are mainly based on the ***Dahl hysteretic friction model*** (Dahl, 1976), originally developed for the simulation of machine friction. In its simplest time domain form, the Dahl model reads:

$$\frac{dF}{dt} = K_s \left(u_s - \frac{F}{F_{max}} |u_s| \right)^{\beta} \tag{23.56}$$

where F is the friction force, K_s is the equivalent shear stiffness of the contact, F_{max} is the maximum possible friction force, u_s is the relative sliding speed and β is a parameter used to alter the shape of the hysteretic friction curve (usually $\beta = 1$).

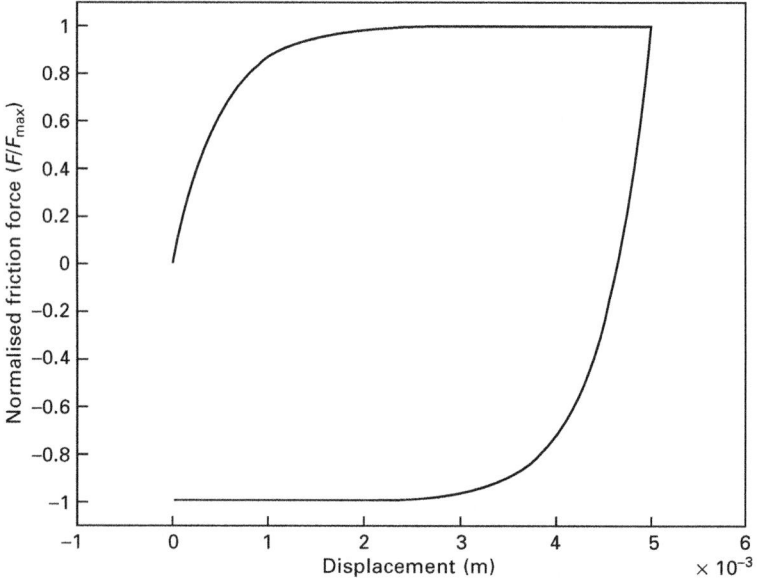

23.12 An example of a hysteretic friction loop as predicted by eq. (23.56).

The resulting friction force versus displacement as predicted by the Dahl model is illustrated in Fig. 23.12. The Dahl friction model does not allow for a distinct friction peak. This is corrected by modification of the Dahl model so as to include the Stribeck effect, resulting in the development of the 'LuGre' friction model (Deur *et al.*, 2002; Canudas-de-Wit *et al.*, 2003). The LuGre friction model has been successfully combined with the brush-modelling concept to produce tyre models capable of generating longitudinal, lateral and combined forces under both steady state and transient operating conditions (Deur *et al.*, 2004, 2005).

23.11   Case study: the influence of transient tyre behaviour on shuffle

In this section, the initial study of the shuffle error state carried out in Sections 23.2.1 and 23.2.2 is enhanced by implementing an appropriate tyre model. In particular, the idea of the tyre operating as a torsional spring is abandoned and a combination of a steady state tyre model with a relaxation-length based transient treatment is adopted. Since shuffle usually involves rather small slip-ratios, the steady state part of the model assumes a linear relationship between the generated force and the longitudinal slip-ratio. Confining the problem within the linear range will also allow a number of interesting comparisons with the initial simplified approach to be made. Implementation of the tyre model requires modification of equation (23.14), as follows:

$$I_w \frac{d^2\theta_6}{dt^2} + C_3\left(\frac{d\theta_6}{dt} - \frac{1}{n_d}\frac{d\theta_4}{dt}\right) + K_3\left(\theta_6 - \frac{\theta_4}{n_d}\right) + RC_t s_c = 0 \quad (23.57)$$

where C_t is the **traction stiffness** as defined by equation (23.39) or (23.45) and s_c is the localised, or transient, slip-ratio as defined by equation (23.48). In addition, equation (23.15) now reads as:

$$m\frac{d^2x}{dt^2} - C_t s_c = 0 \quad (23.58)$$

Finally, the initial system of ten differential equations is augmented by the addition of the law of change of the slip-ratio, s_c, provided by equation (23.54), which is re-written below in terms of the corresponding states of the original system:

$$Sx\frac{ds_c}{dt} + U \cdot s_c = \omega_6 R - U \quad (23.59)$$

Equations (23.57)–(23.59) together with the original equations (23.11)–(23.13) and (23.16)–(23.20) form a system of 11 equations which can be brought into state-space form in order to facilitate eigen-frequency analysis or numerical simulation. Prior to proceeding, it is essential to decide on the exact value of the traction stiffness, C_t, and the **relaxation length**, Sx. The latter will be chosen equal to 0.6 m and this value can be assumed constant, given the anticipated low levels of the slip-ratio. Now, an interesting observation will lead to the choice of the traction stiffness, C_t. For the forward speed, U, approaching zero, equation (23.59), integrated over a period of time, Δt, yields:

$$s_c = \frac{1}{Sx} \int_0^{\Delta t} R\omega_6 dt \tag{23.60}$$

Apparently, the integral in equation (23.60) represents a displacement. The tyre force can be calculated as:

$$F_x = C_t s_c = \frac{C_t}{Sx} \int_0^{\Delta t} R\omega_6 dt \tag{23.61}$$

Taking into account that $Sx = C_t/K_c$, equation (23.61) can be written:

$$F_x = C_t s_c = \frac{C_t}{Sx} \int_0^{\Delta t} R\omega_6 dt = K_c \int_0^{\Delta t} R\omega_6 dt \tag{23.62}$$

This last equation (23.62) reveals that at vanishing forward speed, U, the tyre behaves as a pure longitudinal/tangential spring, with stiffness K_c. This stiffness can be directly related to the torsional stiffness of the tyre, K_t, employed in Section 23.3, using the following relationship:

$$K_c = \frac{K_t}{R^2} \tag{23.63}$$

Replacing the values of K_t, R from Table 23.1 into equation (23.63) yields K_c = 2 × 66 667 N/m, including the influence of both tyres. Finally, $C_t = K_c Sx =$ 0.6 m × 2 × 66 667 N/m = 2 × 40 000 N. To summarise, the following values are chosen for the tyre parameters:

$Sx = 0.6$ m

$C_t = 8 \times 10^4$ N (including the effect of both tyres)

With this choice of parameters and for $U = 0$, it is expected that the updated system of differential equations will be characterised by the same eigen-frequencies as the original system. Following a similar procedure as in Section 23.2.2, Table 23.3 is obtained, listing the eigen-values and the corresponding frequencies of the augmented system. Clearly, the initial hypothesis is confirmed. At higher forward speed, U, a rather modest change is observed in the eigen-frequencies of the system. Obtaining such results is left to the reader. What is most important regarding the behaviour of the system as the forward speed increases, is the change in the rate of attenuation of the shuffle oscillation. In particular, at higher forward speeds, equation (23.55) shows a rapid response of the localised, or transient slip-ratio, s_c, to a step

Table 23.3 Results of the eigen-frequency analysis carried out on the augmented driveline model, including transient tyre behaviour

Roots of char. eq. (23.29)	Eigen-values ($\times 10^3$)	Damped frequencies (Hz)
$\lambda_1 =$	0	0
$\lambda_{2,3} =$	$-0.0581 + 3.1289i$	498.0
	$-0.0581 - 3.1289i$	
$\lambda_{4,5} =$	$-0.0364 + 0.4650i$	74.0
	$-0.0364 - 0.4650i$	
$\lambda_{6,7} =$	$-0.0014 + 0.1939i$	30.9
	$-0.0014 - 0.1939i$	
$\lambda_{8,9} =$	$-0.0000 + 0.0150i$	2.4
	$-0.0000 - 0.0150i$	
$\lambda_{10,11} =$	$0.0000 + 0.0000i$	0
	$0.0000 - 0.0000i$	

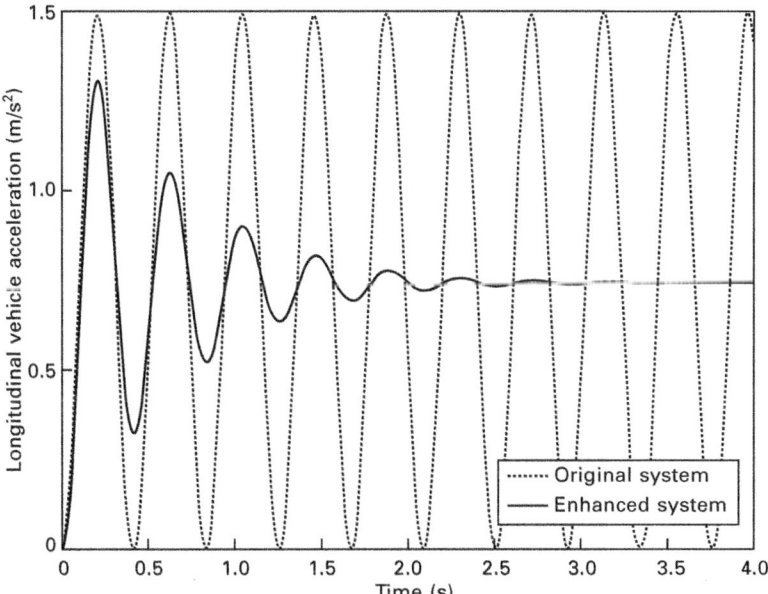

23.13 Shuffle response to a step-torque input of the original and the enhanced driveline models at a forward speed of 2.8 m/s.

slip-ratio input, s. Hence, the tyre force practically depends on the level of slip, $s = (\Omega R - U)/U$, so that:

$$F_x \cong C_t \frac{\Omega R - U}{U} \tag{23.64}$$

Equation (23.64) indicates that since the longitudinal force, F_x, depends on a velocity difference, the tyre can be viewed as a damper. With the increase of forward speed, this mode of operation of the tyre contributes towards a faster

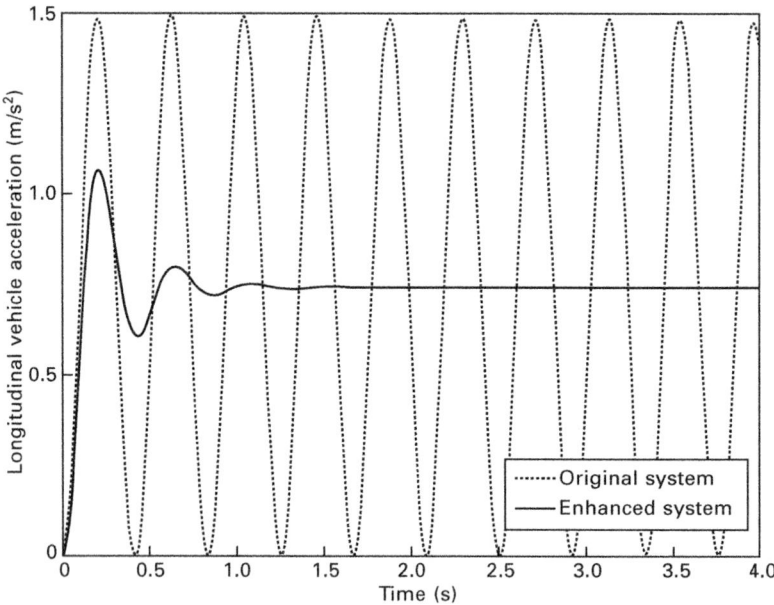

23.14 Shuffle response to a step-torque input of the original and the enhanced driveline models at a forward speed of 8.3 m/s.

attenuation of the shuffle oscillation. This effect is clearly illustrated in Figs 23.13 and 23.14. Figure 23.13 shows the fore–aft acceleration of the vehicle body obtained at a low forward speed of approximately 10 kph (2.8 m/s) by both the original and the enhanced system. Although the oscillation produced by the enhanced system is significant, it attenuates considerably sooner than the oscillation predicted by the initial system. Although the forward speed is rather low, the tyre eventually reaches a steady state where only the longitudinal slip-ratio plays a role in tyre-force generation and, thus, the tyre behaves as a damper. Figure 23.14 shows the response of the two systems at a higher forward speed of 30 kph (8.3 m/s). While the behaviour of the initial system remains the same, that of the enhanced system shows significantly reduced overshoot and a rapid convergence to the corresponding steady state acceleration value.

23.12   References

Bakker, E., Nyborg, L. and Pacejka, H.B., 1987. Tyre modelling for use in vehicle dynamics studies. Papers Presented at the SAE International Congress and Exposition. SAE, Warrendale, PA, USA.

Bakker, E., Pacejka, H.B. and Lidner, L., 1989. A new tire model with an application in vehicle dynamics studies. SAE Paper 890087, pp. 101–113.

Besselink, I.J.M., Pacejka, H.B., Schmeitz, A.J.C. and Jansen, S.T.H., 2005. The MF-Swift tyre model: extending the magic formula with rigid ring dynamics and an enveloping model. *Review of Automotive Engineering*, **26**(2), pp. 245–252.

Biermann, J.W. and Hagerodt, B., 1999. Investigation of the clonk phenomenon in vehicle transmissions – measurement, modelling and simulation. *Proceedings of the Institution of Mechanical Engineers, Part K, Journal of Multi-body Dynamics*, **213**, pp. 53–60.

Canudas-de-Wit, C., Tsiotras, P., Velenis, E., Basset, M. and Gissinger, G., 2003. Dynamic friction models for road/tire longitudinal interaction. *Vehicle System Dynamics*, **39**(3), pp. 189–226.

Clover, C.L. and Bernard, J.E., 1998. Longitudinal tire dynamics. *Vehicle System Dynamics*, **29**(4), pp. 231–259.

Dahl, P.R., 1976. Solid friction damping of mechanical vibrations. *AIAA Journal*, **14**(12), pp. 1675–1682.

Deur, J., Asgari, J. and Hrovat, D., 2002. A dynamic tire friction model for combined longitudinal and lateral motion, 2001 ASME International Mechanical Engineering Congress and Exposition, Nov 11–16 2001, 2002, American Society of Mechanical Engineers, New York, pp. 229–236.

Deur, J., Asgari, J. and Hrovat, D., 2004. A 3D brush-type dynamic tire friction model. *Vehicle System Dynamics*, **42**(3), pp. 133–173.

Deur, J., Ivanovic, V., Troulis, M., Miano, C., Hrovat, D. and Asgari, J., 2005. Extensions of the LuGre tyre friction model related to variable slip speed along the contact patch length. *Vehicle System Dynamics*, **43**, pp. 508–524.

Farshidianfar, A., Ebrahimi, M., Bartlett, H. and Moavenian, M., 2002. Driveline shuffle in rear wheel vehicles. *Heavy Vehicle Systems*, **9**(1), pp. 76–91.

Fredriksson, J., Weiefors, H. and Egardt, B., 2002. Powertrain control for active damping of driveline oscillations. *Vehicle System Dynamics*, **37**(5), pp. 359–376.

Gipser, M., 2005. FTire: A physically based application-oriented tyre model for use with detailed MBS and finite-element suspension models. *Vehicle System Dynamics*, **43**, pp. 76–91.

Higuchi, A., 1997. Transient response of tyres at large wheel slip and camber, TU Delft.

Krenz, R.A., 1985. Vehicle response to throttle tip-in/tip-out. Surface Vehicle Noise and Vibration Conference Proceedings. SAE, Warrendale, PA, pp. 45–52.

Lemaitre, J., ed, 2001. *Handbook of Materials Behavior Models*. San Diego: Academic Press.

Moore, D.F., 1975. *The Friction of Pneumatic Tyres*. Amsterdam, New York: Elsevier Scientific Pub. Co.

Pacejka, H.B., 2002. *Tyre and Vehicle Dynamics*. Oxford: Butterworth and Heinemann (also SAE).

Pacejka, H.B. and Bakker, E., 1993. Magic formula tyre model. *Vehicle System Dynamics*, **21**, pp. 1–18.

Pacejka, H.B. and Besselink, I.J.M., 1997. Magic formula tyre model with transient properties. *Vehicle System Dynamics, Supplement*, **27**(1), pp. 234–249.

Pacejka, H.B. and Sharp, R.S., 1991. Shear force development by pneumatic tyres in steady state conditions. A review of modelling aspects. *Vehicle System Dynamics*, **20**(3), pp. 121–176.

Rahnejat, H., 1998. *Multi-Body Dynamics: Vehicles, Machines and Mechanisms*. London: Professional Engineering Publishing.

Schulz, M., 2005. Low-frequency torsional vibrations of a power split hybrid electric vehicle drivetrain. *JVC/Journal of Vibration and Control*, **11**(6), pp. 749–780.

Stewart, P. and Fleming, P.J., 2004. Drive-by-wire control of automotive driveline oscillations by response surface methodology. *IEEE Transactions on Control Systems Technology*, **12**(5), pp. 737–741.

23.13 Notation

a	half-length of contact patch
B	stiffness factor (magic formula)
C	shape factor (magic formula)
C_i	damping coefficient of the ith driveline component
C_t	traction stiffness
D	peak factor (magic formula)
E	curvature factor (magic formula)
F	friction force
F_{max}	maximum friction force
F_x	longitudinal tyre force
F_z	net vertical load at the contact patch
f_z	vertical load distribution along the contact patch
h	longitudinal/tangential deflection of tyre carcass
I	moment of inertia
K_c	longitudinal/tangential stiffness of tyre carcass
K_i	stiffness coefficient of the ith driveline component
K_s	equivalent shear stiffness of contact
K_t	torsional stiffness of tyre
K_x	longitudinal stiffness of bristle
m	vehicle mass
n_d	differential reduction ratio
n_t	transmission reduction ratio
R	effective wheel radius
R_i	pitch circle radius of the ith gear
s	practical slip ratio
s_c	localised slip ratio
Sx	relaxation length
T_i	torque at the ith gear
U	forward speed of vehicle
u	bristle deflection
u_s	relative sliding speed between bodies
V_m	sliding speed where maximum friction is observed
V_r	linear speed of rolling
V_s	sliding speed
V_x	forward speed at wheel centre
x	position along the contact patch

x_t	transition point between sticking and sliding
β	parameter for the Dahl model
Δt	time interval
θ_i	angular displacement of the ith element
μ_k	coefficient of kinetic friction
μ_m	coefficient of maximum friction
μ_s	coefficient of static friction
ρ	parameter for the Savkoor model
σ_x	theoretical longitudinal slip ratio
Ω	rotational speed of wheel
ω_i	rotational speed of the ith driveline component

24

Tribology of differentials and traction control devices

S. K. MOHAN, Magna Powertrain, USA

Abstract: The tribology of differentials and traction control devices found in modern automobile drivetrains is explored in this chapter. Typical vehicle drivetrain architecture is described to set the context for the discussion of differentials and traction control devices. The fundamentals of vehicle dynamics and the role of different driveline elements are explained. The need for differentials and slip-controlled devices, and how they function and enhance vehicle performance and safety are illustrated. Different types of passive and active, electronically controlled traction devices, and their operating characteristics, are described. The considerations of tribology in the design and development of these devices are explored along with use of modern modelling and simulations tools that allow some predictive engineering capability. A projection of likely future trends in vehicle driving dynamics systems and a number of useful web sites and references are also given.

Key words: tribology, differentials, traction or slip control clutch, viscous coupling, tyre slip, vehicle dynamics.

24.1   Introduction

Modern automotive powertrains provide improved mobility to the driver while maintaining a high degree of traction, directional stability and safety. Much of this performance enhancement is the result of advances in engine, transmission and tyre technologies. The key technology elements that tie all these developments together to give a smooth, transparent and enjoyable vehicle performance are the traction control hardware and software that have become part of the modern powertrain. The central components of such *traction control* systems are the traction and slip control devices based on wet or dry friction elements or viscous shear-based fluid couplings. Many of these devices are passive systems that have inherent, but limited performance characteristics that depend on the initial design factors and the operating environment. To get beyond these inherent limitations, electronically controlled actuation is now very common. Use of electronic control permits multiple external signals and algorithms to determine the most appropriate operating mode and enlarge the operating envelope of the drive system.

This chapter explains the various types of devices currently in use, the

735

fundamental physical principles at work and the tribology of the device where appropriate, and the device's influence on the vehicle's driving dynamics. The devices are broadly classified according to the basic principles used in their design and representative examples are described in greater detail. The fundamentals of drivetrain and vehicle dynamics are explained to provide context to the discussions. While brake-based traction control is described in this chapter for the sake of completeness, discussion of the tribology of brake systems is beyond the scope of this chapter. In the absence of internationally accepted terms to describe various driveline architectures, the nomenclature used in this chapter is guided by the conventional usage in North America as well as the proposed SAE Standard nomenclature. Alternative descriptions are given where appropriate and necessary for clarity.

Where helpful, examples of vehicles commercially sold, with the technology being discussed, are indicated as follows: make-*model* {MY####} (model year of availability) (e.g. Honda-*CRV* {MY1996})

24.2   Vehicle drivetrain architecture

The discussion in this chapter applies by and large to the typical modern automobile that has four wheels arranged in two axles with a steerable front axle. The exact arrangement of the powertrain is a reflection of the designer's attempt to satisfy the styling, marketing, performance, safety and engineering needs for the vehicle. Each variation of the theme results in a combination of strengths and compromises. The following paragraphs describe the major architectures.

24.2.1   Front wheel drive (FWD)

As the name implies, the primary driving wheels are at the front axle (Fig. 24.1). With few exceptions, the engine is transversely mounted with a transaxle that integrates the transmission and the front axle differential into a single housing. Independent drive shafts, often called ***half-shafts***, rotationally connect the differential to the wheels via constant velocity joints that allow the wheels to move in response to steering inputs and suspension demands.

24.2.2   Front wheel drive based all-wheel drive (AWD)

The primary drive remains at the front axle, but as seen in Fig. 24.1, an auxiliary gear box called ***power take-off unit*** (PTO or PTU) connects the drive to the rear axle via a ***propeller shaft***. The output from the transmission may be connected directly to the front axle, as well as the PTO or split up using a centre differential (see Section 24.4.1).

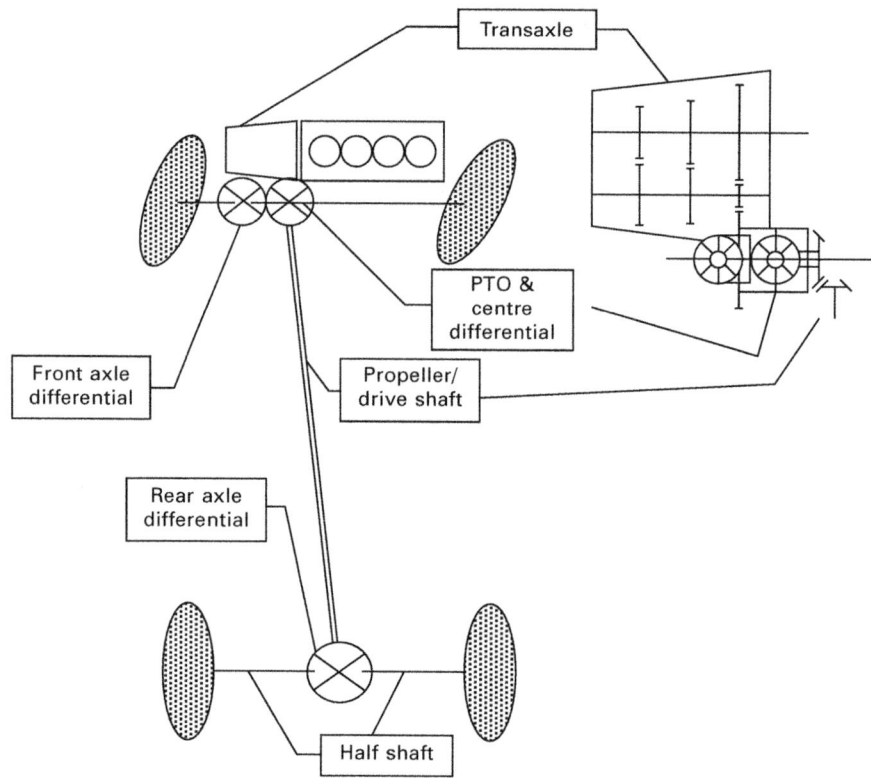

24.1 Front wheel drive based all-wheel drive architecture.

24.2.3 Rear wheel drive (RWD)

The principal driven axle is at the rear. The engine is normally mounted longitudinally with the drive going to the rear axle by means of a propeller shaft attached to the transmission.

24.2.4 Rear wheel drive based all-wheel drive (AWD/4WD)

As shown in Fig. 24.2, special gear-box called the ***transfer case*** or ***transfer box*** is mounted behind the transmission. The transfer case has two outputs – one going to the rear axle and the other going to the front axle. The transfer case might incorporate a centre differential (see Section 24.4.1) or a coupling to permanently or selectively connect the output to the front axle.

In the rest of the chapter, for the sake of simplicity, irrespective of whether the driveline is based on FWD or RWD, the term AWD is used to refer to any driveline architecture, where both axles may be actively driven.

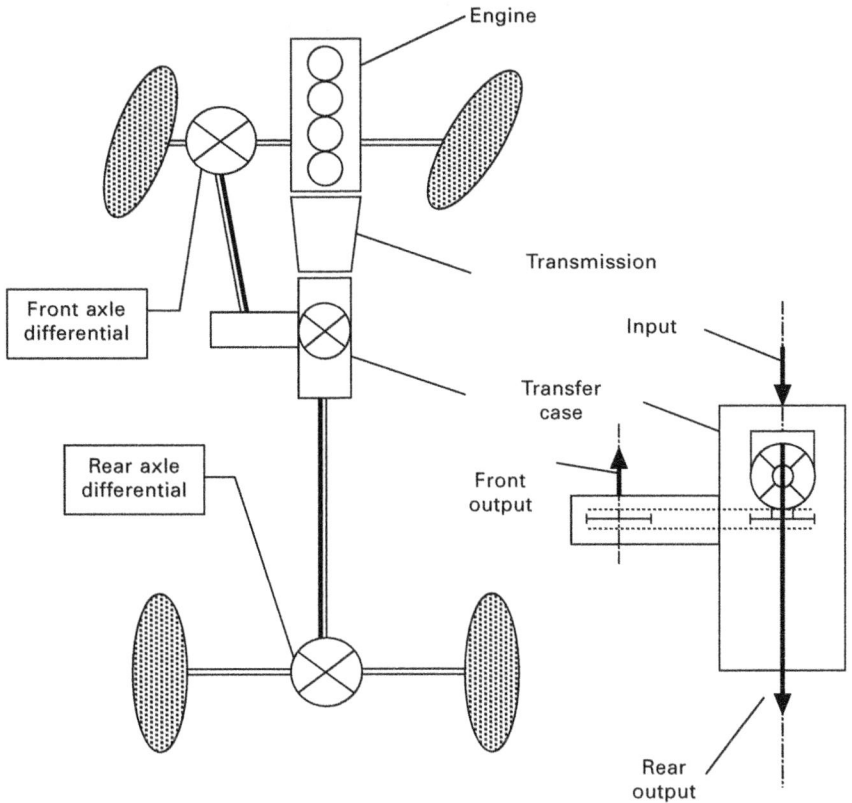

24.2 Rear wheel drive based all-wheel drive architecture.

24.3   The basics of vehicle propulsion and dynamics

The description of vehicle dynamics given in this section is necessarily a simplified one. For a more detailed treatment please refer to Gillespie (1992) or Milliken and Milliken (1995).

24.3.1 Vehicle reference axis system

All references to the left and right side of the vehicle are with respect to the driver sitting in the traditional driver's seat, facing forward in the vehicle. The ISO 8855 axis systems defines X pointing forward, Y pointing to the left of the vehicle and Z pointing up. The SAE J670e has X forward, Y pointing to the right and Z pointing down. In this chapter, the SAE axis system is used (Fig. 24.3).

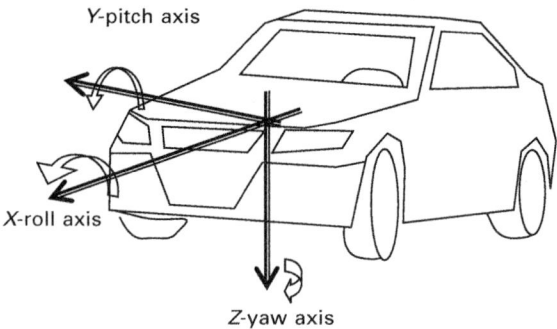

24.3 Vehicle reference axes.

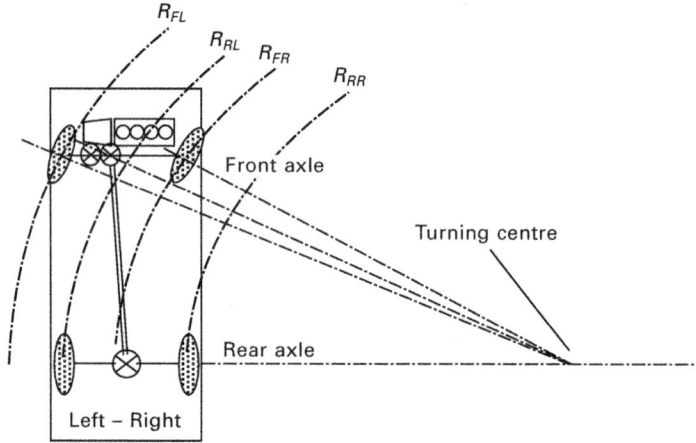

24.4 Vehicle turning radius.

24.3.2 Vehicle motion

The terms longitudinal, lateral and vertical motions are self-explanatory with reference to the X, Y and Z axis system. The rotation about the longitudinal X axis is referred to as **roll**, about the lateral Y axis is called **pitch** and the rotation about the vertical axis Z is called **yaw**. In a vehicle making a turn (Fig. 24.4), the front axle (the steered axle) traverses a longer distance, compared with the rear axle, during the same time interval and so has to turn at a higher speed than the rear axle. Similarly, the outer wheels in both axles have to cover a longer distance compared with the inner wheels and hence have to turn at higher speeds than the inner wheels.

24.3.3 The function of the powertrain

The purpose of the powertrain is to enable the driver to propel the vehicle along the intended path, at the desired speed and yaw-rate and do it in a safe and enjoyable manner. If one ignores the aerodynamic forces, the only external forces on the vehicle come from gravity and the interactions of the tyres with the road surface. In other words, to gain any degree of control over the vehicle's motion, the powertrain has to direct the engine power to the *tyre contact patches* in an appropriate manner. In a hypothetical drive-by-wire system, the powertrain, the steering and the brake systems have to take the throttle, steering and brake signals and interpret and meet the driver's intent in terms of the vehicle's speed, acceleration and yaw rate. The various control systems achieve the desired result by delivering the necessary slip angle at the steering wheels to generate the yaw moment and supplementing it with appropriate traction at each wheel to generate the required acceleration and yaw rate. The introduction of various throttle-by-wire, brake-by-wire and steer-by-wire systems are moving modern vehicles towards this hypothetical model.

24.3.4 Dynamic performance of the vehicle

The dynamic performance of a vehicle may be discussed in terms of its traction performance and its agility and yaw performance. In its simplest form, the inertial equations of motion for a vehicle relate the acceleration and the mass of the vehicle, the traction forces at the four tyre contact patches, the aerodynamic forces, the rolling resistance and the effects of gravity. Similarly, the yaw rate equation for the vehicle relates the rotational inertia of the vehicle to the moment about the centre of mass of all the external forces on the vehicle, especially the lateral and longitudinal forces at the tyre contact patches. It is easy to see that the traction performance is mainly dependent on the ability of the drivetrain to generate the appropriate longitudinal forces at the tyre patches, and the yaw performance is largely dependent on the lateral forces generated.

24.3.5 Tyre patch dynamics

The vertical forces at a tyre patch are related to the static weight distribution and the dynamic effects of vertical acceleration, the rolling and the pitching of the vehicle (see Chapter 23). The instantaneous normal force F_z at each tyre patch controls the available friction forces at the tyre–road interface (Fig. 24.5). This total available force F at the tyre patch can be resolved into orthogonal components on the road surface, in the direction of the plane of the wheel F_x and perpendicular to it F_y (lateral tyre force, see Chapter 23). It is these longitudinal and lateral forces that predominantly determine the acceleration and yaw of the vehicle respectively.

Detailed information on the physics behind the generation of tyre forces may be found elsewhere in this book and in Gillespie (1992) or Milliken

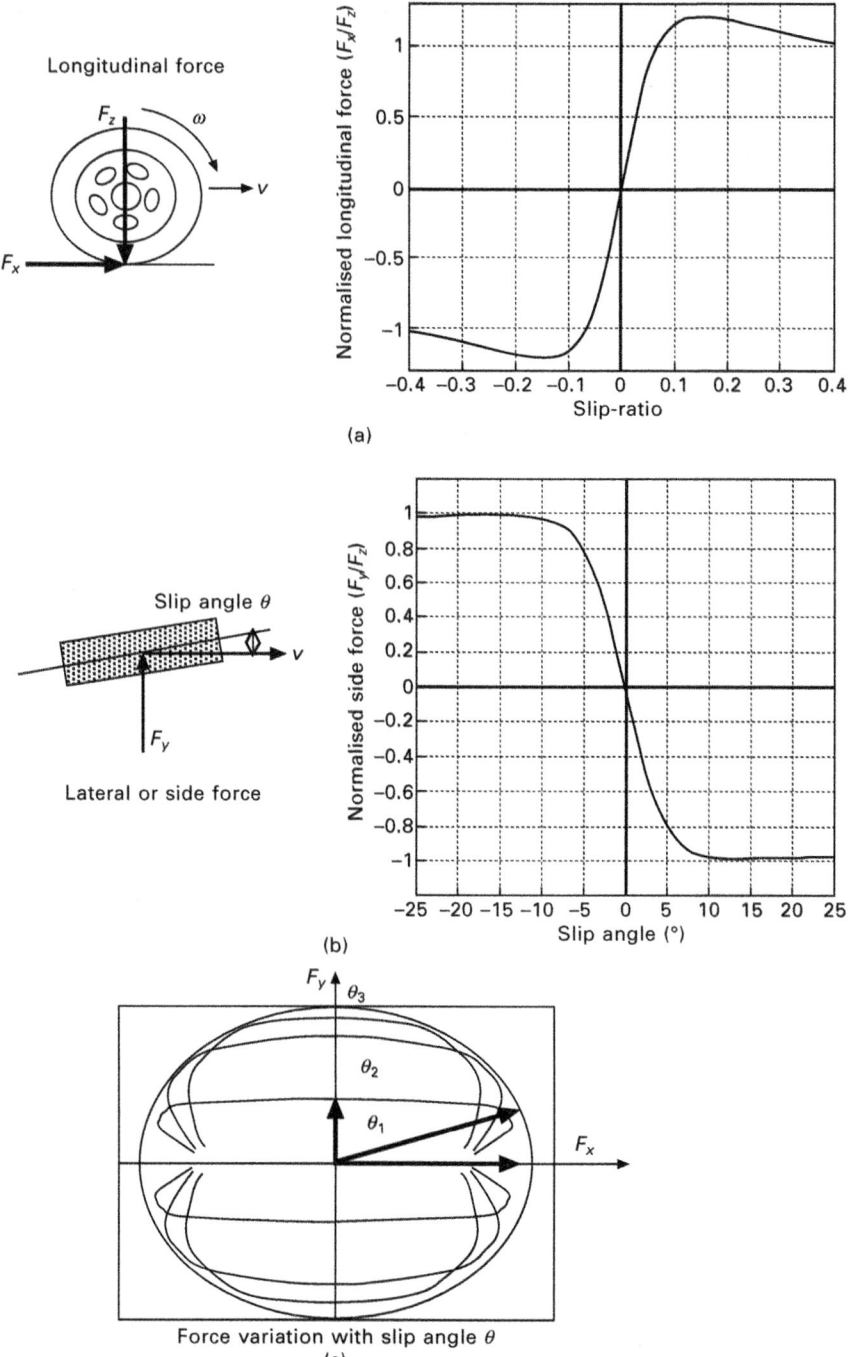

Longitudinal force

Slip angle θ

Lateral or side force

(a)

(b)

(c)

Force variation with slip angle θ

24.5 Tyre forces (a) longitudinal force (b) lateral or side force (c) force variation with slip angle θ.

and Milliken (1995). A much simplified version will be presented here to provide a context to the discussions. The forces F generated at the tyre patch can be modelled as simple friction dependent on the coefficient of friction μ and the vertical force F_z (Fig. 24.5(a)):

$$F = \mu F_z \tag{24.1}$$

The coefficient of friction μ is a characteristic of the nature of the road surface, the properties of the tyre material and geometry, and the slip speed ratio s at the tyre patch. For a tyre moving at a forward translational speed of v, with an effective radius of R and a rotational speed of ω, the slip speed ratio s is defined below. The numerator may be seen as the effective slip at the tyre patch. This is the SAE J670 definition of s.

$$s = \left(\frac{\omega R - v}{v} \right) \tag{24.2}$$

The longitudinal μ rises sharply with the slip ratio, as seen in Fig. 24.5(a), in the *apparent slip* region, up to a maximum value and then drops gradually, in the *real slip* region, as the tyre goes into a run-away slip mode (see Chapter 23).

The μ corresponding to the lateral force increases with the slip angle of the tyre's motion with respect to the wheel heading (Fig. 24.5(b)). The vector sum of the longitudinal and lateral forces at the tyre patch is limited by the available friction. If a tyre patch is required to generate a high longitudinal force, then the available lateral force will be limited. It should also be obvious that to achieve maximum traction, one has to keep at or below the peak μ for the tyre patch. Once the tyre goes beyond the peak μ, it goes into the run-away slip region with constantly decreasing traction value. In other words, limiting the tyre slip increases the available traction at the tyre patch, and this is the rationale for the use of slip limiting devices as traction enhancement systems in the driveline.

24.3.6 Vehicle's yaw dynamics

When a vehicle is going around a curved path, the turning performance varies substantially, depending on whether the front axle, the rear axle or both axles are providing the traction forces. In order for a vehicle to maintain the intended path, the yaw moment generated at the steering front wheels has to be balanced by a counter yaw moment generated at the rear wheels. In a FWD vehicle, since front wheels are required to generate the longitudinal traction forces, the orthogonal steering forces generated are smaller and the vehicle tends to *under-steer* or have smaller yaw rate than desired. In a RWD vehicle, with the rear wheels generating all the traction forces, their ability to generate the balancing counter yaw moment is limited and the vehicle tends to *over-steer* or yaw at a rate higher than desired. An AWD vehicle has a yaw characteristic that is in between the above two cases depending on the traction force distribution among the wheels. As a corollary, one can state that controlling the torque distribution to the wheels has a direct bearing on the yaw behaviour of the vehicle. Since direct measurement and feedback of transmitted torque is difficult and accurate estimation of the instantaneous μ at the tyre patch is almost impossible, the only measurable quantity one can freely depend on is tyre slip. With accurate speed sensors at each wheel becoming standard due to mandatory electronic *traction control systems*, the wheel speed information is readily available on the CAN (controller area network) bus of the vehicle. So, in practice, slip control is the basis by which torque control is indirectly achieved.

24.3.7 Brake-based traction and stability control

With *anti-lock brake systems* (*ABS*) becoming mandatory in most vehicles, the foundations for brake-based *electronic traction control* (ETC) were in place. *ETC* utilises the brakes at each wheel, combined with more sensitive speed sensors and a more powerful ECU (electronic control unit). Unlike an ABS system which attempts to prevent the tyres from locking-up in order to give the driver a measure of continued steering control during heavy braking, the ETC is attempting to limit the slip at the tyre and prevent the tyres from going into a run-away slip mode so as to maintain a degree of traction at the tyre patches.

The *electronic stability program* (ESP) is a lot more complicated in its function and approach. As is described later in this chapter, the directional stability of the automobile is determined by the ECU's control over the vehicle's yaw rate. The turning moment of the net external force about the vehicle's centre of gravity (CG) is a critical element of controlling the yaw rate. The wheel brakes, when applied selectively, can impose sizeable yaw moment on the vehicle. Applying brakes is a dissipative action since it tends to slow the vehicle and could be dangerous if applied incorrectly. So in addition to the wheel speed sensors, *ESP* requires input from the steering system, longitudinal and lateral acceleration sensors as well as yaw rate and roll rate sensors. With the known geometry parameters of the vehicle such as steering ratio, tyre size, wheel track width and wheel base, the ECU of the ESP system computes the turn radius, estimated lateral acceleration at the vehicle's speed and the expected yaw rate. If the actual measured yaw rate varies from the expected yaw rate, the brakes are selectively applied in short spurts to correct the situation by providing additional yaw moment over and above that available from the steering system. Roll-over prevention systems extend this method of correction to the roll rate in addition to the yaw rate.

24.4   The need for differentials and slip control devices

As described in Section 24.3.2, during a turn, the front axle runs faster than the rear axle and the outer wheels of either axle run faster than the inner wheels. If the driveline does not allow for this difference in speed, it could wind-up and even break. The *differential* is a device typically used to permit this speed difference. It could be thought of as a torque balancing element that maintains a fixed torque ratio at the two outputs, while maintaining a corresponding kinematic relationship with the input. If one has to change the torque ratio, it is necessary to introduce additional biasing devices to generate supplementary internal reaction torques. As an extension to this idea, by eliminating the differential and transmitting all the torque through the biasing device or clutch, one can enlarge the range of the torque ratio beyond the threshold set by the inherent torque ratio of the differential. This idea will be explored further in Section 24.6 under electronic active traction control.

24.4.1 Differentials

In the AWD drivetrain, when the transmission drive is split up between the two axles, typically a centre differential is used. Similarly, in an axle, the input drive is directed to the two wheels using an axle differential. Both centre and axle differentials are structurally similar. The differential consists of a set of bevel or spur gears arranged in a way that the two outputs are able to rotate at different speeds, while maintaining the overall kinematic and torque relationship to the input speed and torque (Fig. 24.6). Drivetrains

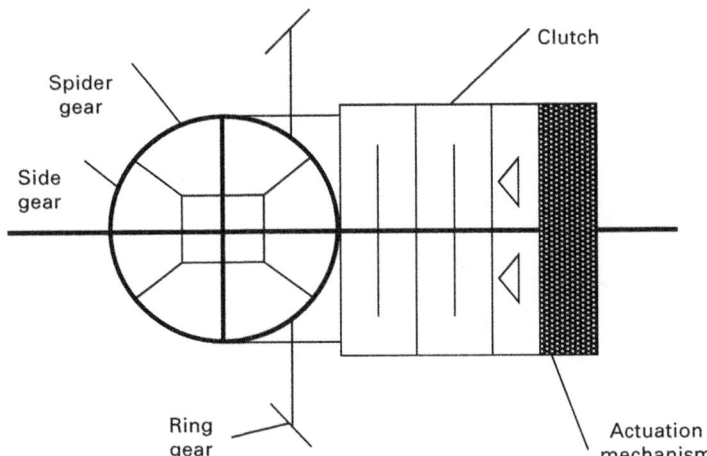

24.6 Bevel gear differential with slip limiting clutch.

with a **centre differential** are sometimes called *'full-time'* systems since they can be used under all conditions irrespective of the nature of the road surface. As shown previously, the centre differential and the axle differentials permit the wheels to rotate at different speeds especially while the vehicle is making a turn. If a centre differential is replaced with a rigid coupling, then such vehicles can be used as a four wheel drive system (**4WD**) only on loose surfaces, such as ice, snow, sand or mud, which will permit the slippage of the tyres and avoid driveline wind-up. Such systems are often called *'part-time'* systems.

In many modern vehicles, the centre differential is supplemented by or even replaced with a slipping clutch which permit controlled amount of slip (Fig. 24.7). Such drivelines are at times called **on-demand AWD systems** as opposed to the 'full-time' AWD. But, since there is no consistency in this nomenclature, the reader must depend on the vehicle manufacturer's description to understand the exact operating nature of the driveline.

Bevel gear differential

Most axle differentials are the **bevel gear** type. This is mainly due to the lower cost of manufacturing the bevel gear sets which can be machined or

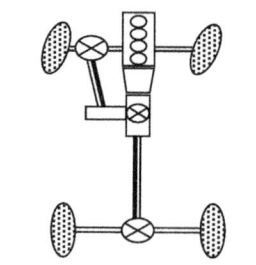

Driveline with transfer case

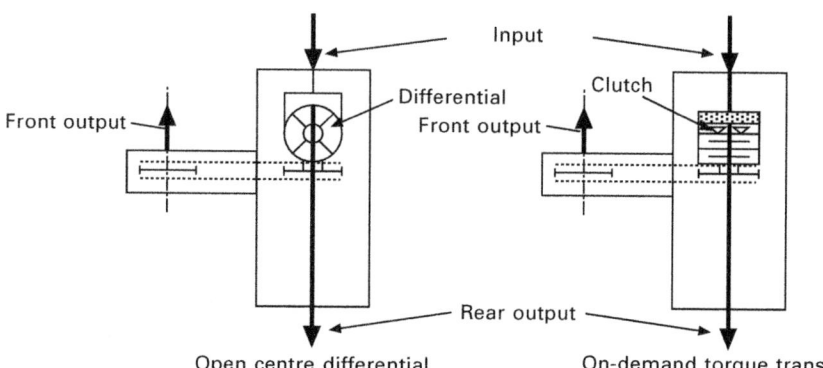

Open centre differential   On-demand torque transfer

24.7 On-demand AWD architecture.

net-formed inexpensively. The bevel gear differential (Fig. 24.6) consists of the pinion carrier acting as the input and the two bevel side gears that mesh with the bevel planetary pinions acting as the two outputs. Typically, the two side gears are identical and hence the two outputs have equal torque ratios. If one ignores the inefficiencies in the system, the input torque equals the sum of the two output torques and the input carrier speed is the average of the two output speeds. Referring to an axle differential, T representing torque and ω representing the speed, the equations for a simple open differential may be written as follows (T_{BR} is the torque bias ratio):

$$T_{left} = T_{right}$$

$$T_{BR} = T_{left}/T_{right} = 1$$

$$T_{input} = T_{left} + T_{right}$$

$$\omega_{input} = \frac{\omega_{left} + \omega_{right}}{2} \tag{24.3}$$

Planetary gear differential

Some axles and most transfer cases utilise the planetary gear architecture for the differential. There is a centrally located **sun gear** and a **ring gear** with internal teeth concentric to the sun gear. A set of **planetary gears** connecting the sun to the ring gear are mounted on a planet carrier, which becomes the third element. Since the ratios of the gearing can be changed within reason, planetary differentials can be designed to be asymmetric with unequal ratios. There are multiple ways to arrange the input and the two outputs. A typical arrangement has a planetary gear carrier as the input element and the sun and the ring gear forming the two outputs. A schematic diagram of a typical planetary differential is shown in Fig. 24.8.

24.4.2 Torque limiting clutches

As the AWD market continues to grow into smaller, more fuel-efficient vehicles, more and more applications use *torque limiting clutches*. These systems provide most of the benefits of the normal lock-up clutch, while allowing for the opportunity for cost reduction, weight reduction and improved fuel economy. In the torque limited system, the clutch capacity is designed to be less than the maximum torque capacity of the tyre patch in the secondary axle. In these situations, friction systems can be subjected to extremely abusive maneuvres. Without proper thermal protection, friction systems can easily be destroyed. One simple event could elevate bulk sump oil temperatures above 200 °C damaging the fluid and failing the system. In early AWD systems without electronic control units (ECU), couplings were

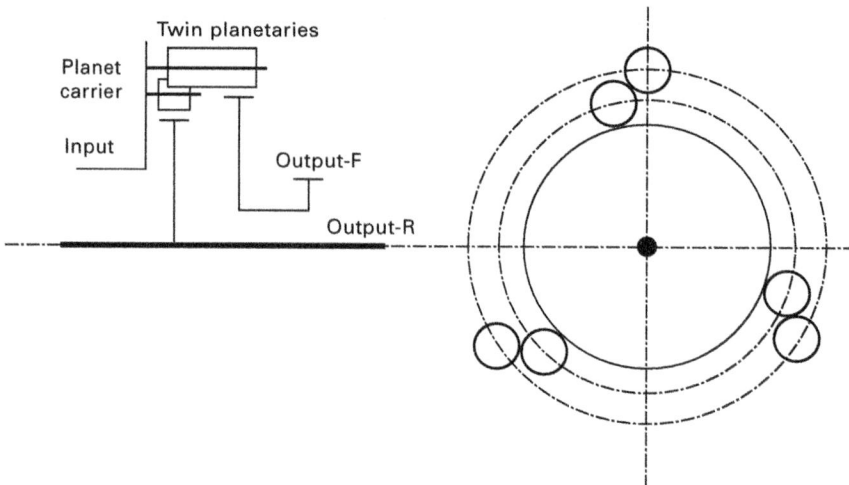

24.8 Planetary differential.

designed to be very robust to deal with abusive situations. Passive thermal protection systems were used which *sensed* the lubricant temperature and caused the system to temporarily shut down or operate at a reduced capacity (Honda-*CRV* {MY 1996} and Saturn-*VUE* {MY2002}). With electronic control available in many modern AWD couplings, thermal protection can easily be integrated into the control module by using a simple temperature sensor. This, however, requires proper understanding of the thermal conditions in the system to ensure effective thermal protection. Be it passive or active control, a well-designed thermal protection system will result in a balance of maintaining pleasability in vehicle performance, while providing component durability under abusive thermal conditions.

24.5   Types of slip control device

There are many different ways to categorise the slip control devices, but not all of them are productive or instructive (Mohan and Williams, 1995). Many traction control devices are purely sensitive to the slip speed across them. There are others that react to the torque being transmitted. Engineers can design into the device both speed and torque sensitivity by selecting the geometric features of the components and their arrangement. Slip control devices form the full spectrum from totally passive devices like the viscous couplings, to semi-active couplings like the Honda-*CRV* twin pump device and to the fully active systems such as the Magna Powertrain ATC (*active torque control*) transfer cases used in the BMW-*xDrive* system (BMW-*X5* {MY2004}) (Magna Powertrain, 2009; BMW, 2009).

In some off-road vehicles, the ability to selectively and completely lock-out the differential function is provided to enable the vehicle to handle extremely slippery conditions at slow speed. This is a special case of slip reduction to zero value and is not described in detail in this chapter. The locking mechanisms are typically some form of dog-clutch or face-clutch with engaging teeth for locking the output and input elements.

In this chapter, instead of describing each major device in the current market, generic devices will be described that capture the essential working principles of a class of devices. The individual devices will be described only to identify the specific manner in which they implement the actuation and control function.

24.5.1 Viscous couplings

Rotary viscous couplings with interleaved, perforated plates and viscous fluids are used in automotive systems to transmit torque (Fig. 24.9). This simple, but effective device is extensively used in small automobiles especially in Asia. They function extremely well during short spurts to provide slip-limiting action that is needed under most road driving scenarios. However, during extended operation as might occur during off-road driving or in very slippery road conditions such as snow or mud, viscous dissipation generates heat and raises fluid temperature, lowers fluid viscosity and causes the torque transmitted to drop monotonically to insignificant levels as seen in Fig. 24.10. Couplings designed with certain plate geometry exhibit a reversal of the torque trend with temperature, and transmit increasingly high torque even under continuous operation. This self-induced torque amplification (STA) is sometimes called *the hump phenomenon*. Such couplings achieve torque amplification factors in excess of 20, compared with earlier couplings. This torque amplification phenomenon had been utilised by industry without fully understanding the mechanisms involved (VW-*Syncro* {MY1986}, VW-*Golf* {MY1988}, Jeep-*Grand Cherokee* {MY1996}). The advantage of such a humping viscous coupling in a drivetrain is the self-protection it affords under continuous operation when an effective lockup replaces the otherwise vanishing fluid shear torque.

A comprehensive theory was proposed by Mohan and Ramarao (2002) and Mohan (2004) to explain the complex sequence of events that results in this *anomalous* but useful phenomenon. Simulation results based on mathematical models compared favourably with experimental findings. The proposed theory identifies, defines and explains the conditions necessary for initiating and sustaining the self-induced torque amplification. A brief explanation of the sequence is given below and differs markedly from the account given in the popular media and even some technical publications. Contrary to some accounts, the *silicone fluid* in the viscous coupling does

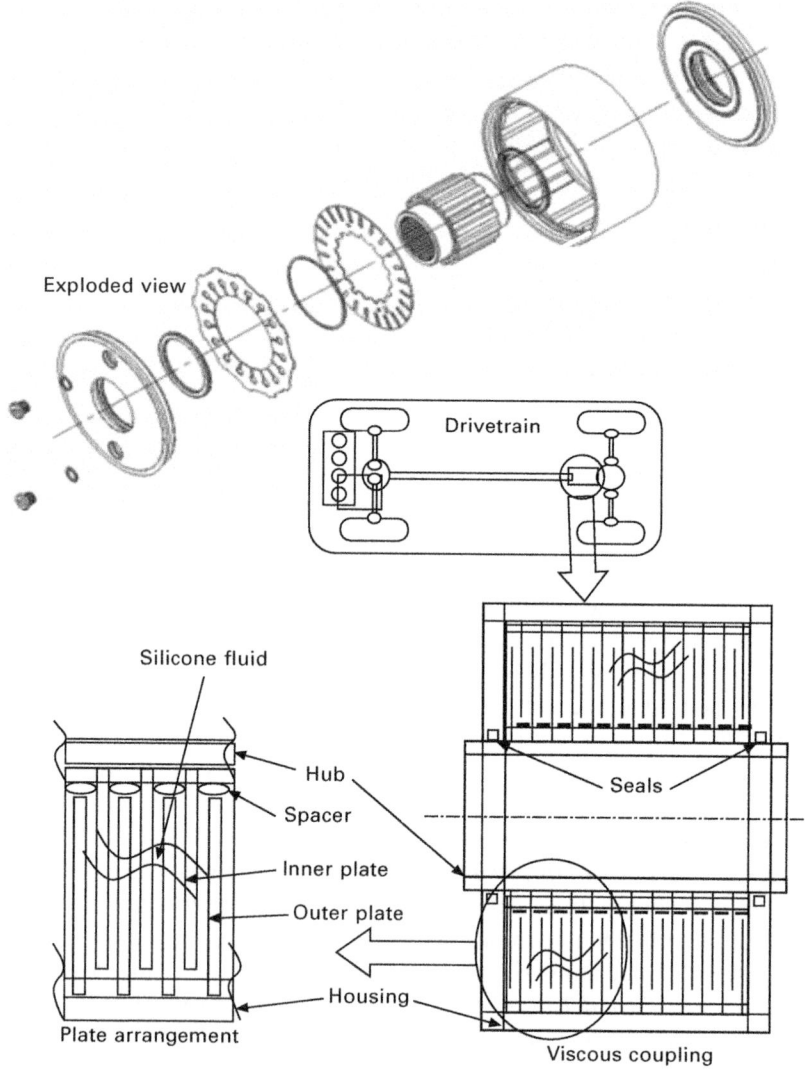

Exploded view

Drivetrain

Silicone fluid

Hub

Seals

Spacer

Inner plate

Outer plate

Housing

Plate arrangement

Viscous coupling

24.9 Viscous coupling.

not become '*almost solid*' under high temperatures. On the contrary, the viscosity of silicones continuously decreases with temperature.

The viscous coupling is filled with silicone fluid of appropriate viscosity to about 80% by volume. As illustrated in Fig. 24.11, when the coupling slips, with the 'inner' plates rotating with reference to the 'outer' plates, the silicone is sheared and the shear work is converted to heat. The coupling temperature increases and the transmitted torque begins to decrease. The tabs in the 'inner' plates deform under the differential shear of the silicone–air

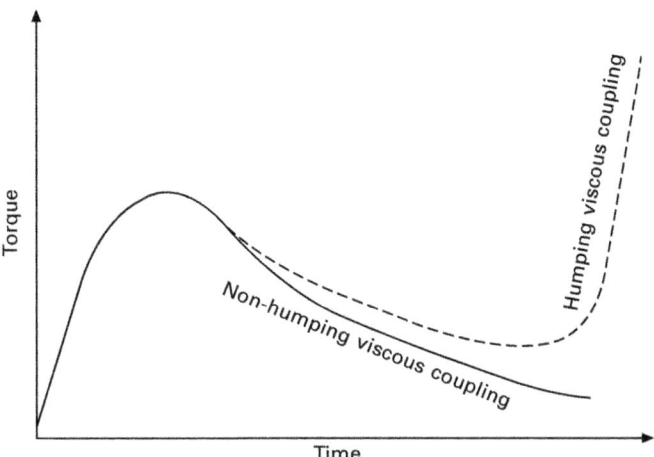

24.10 Viscous coupling torque curve.

mixture created by the small asymmetry in the edge geometry of the tab (Fig. 24.11(b)). The resulting twisting of the tabs causes a differential pressure across the surfaces due to the flow in the wedge-shaped gap between the tabs and the 'outer' plate (Fig. 24.11(c)). This differential pressure pushes the plates towards each other. The pressure tries to equalise due to the leakage flow across the slots in the plates, but this path becomes narrower as the plates approach each other, reduces the leakage flow and increases the differential pressure. Once the leading edge of the tabs make contact with the 'outer' plate, the added Coulomb friction at once increases the twisting moment on the tab and also increases the transmitted torque (Fig. 24.11(d)). The increased twisting of the tabs creates a higher differential pressure and further increase the axial force pressing the plates together. The dissolved air bubbles come out of solution in the low-pressure side and reduce the differential pressure to some extent and at the same time reduce the shear torque in the narrower gap due to the nature of the mixed-phase flow. The transmitted torque stops dropping and levels off under the action of the two opposing trends of decreasing viscosity and increasing differential pressure due to tab twisting. The reduced shear in the two-phase flow created by the air coming out of solution keeps in check the tendency for the torque to monotonically increase and get into the *hump mode*.

The trigger mechanism that creates the torque amplification is the thermal expansion of the silicone that forces all the air to dissolve into the fluid, and, once dissolved, further expansion increases the bulk pressure dramatically in the confined volume of the coupling. The air can no longer come out of solution under the increased bulk pressure even in the 'low'-pressure side and results in single-phase silicone flow on both sides of the tab. Now the pressure

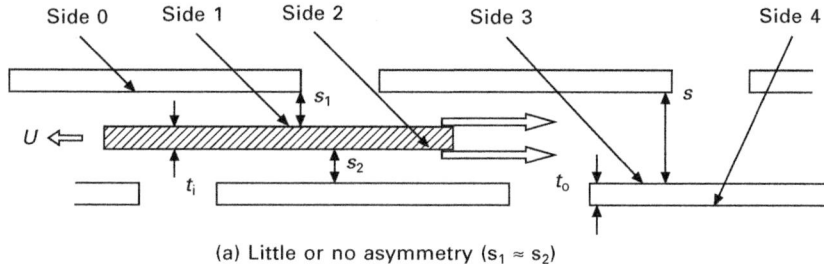

Side 0   Side 1   Side 2   Side 3   Side 4

(a) Little or no asymmetry ($s_1 \approx s_2$)

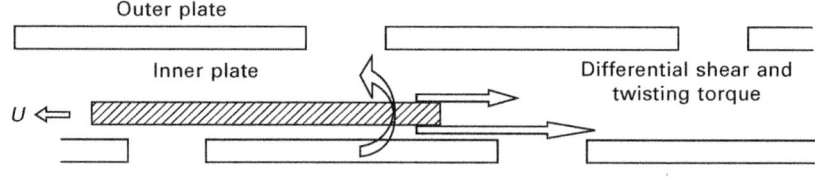

Outer plate

Inner plate   Differential shear and
                                                twisting torque

(b) Increasing asymmetry ($s_1 > s_2$)

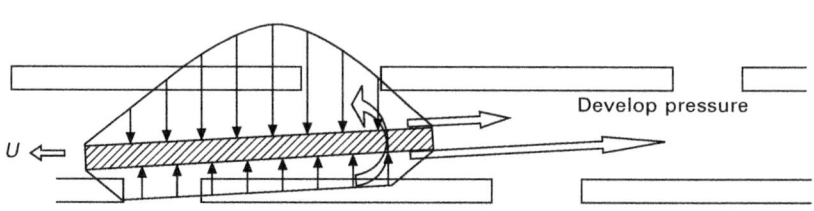

Develop pressure

(c) Tab twist and STA progression

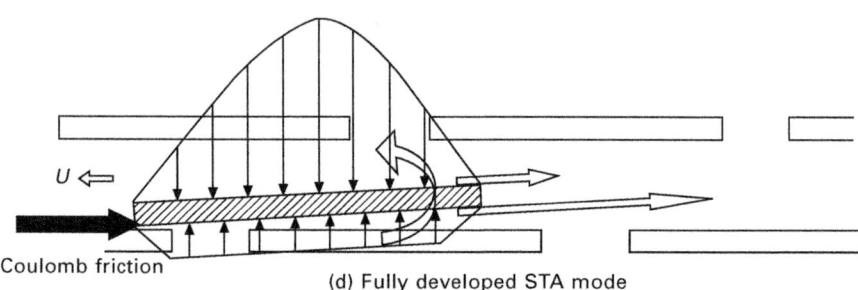

Coulomb friction
(d) Fully developed STA mode

24.11 Viscous coupling STA sequence.

across the plates is self-energised and the torque increases monotonically until
it reaches the maximum Coulomb torque combined with the shear torque
on the side with the narrow gap. Self-energisation is sustained as long as

the slip speed is maintained above a threshold limit and the coupling is not cooled below its *hump* temperature. Otherwise the coupling reverts back to the lower, shear torque mode.

24.5.2 Magneto-rheological (MR) fluid devices

Magneto-rheological (MR) fluid coupling is similar to a viscous coupling, but with a high degree of electronic control. MR coupling uses the strength of an electromagnetic field to regulate the viscosity of the MR fluid, which is filled inside an advanced clutch system. Due to this step-less regulation it becomes possible to create a fully variable, high-precision torque transfer which offers high transmission and traction comfort for AWD applications. With its simple, efficient design the MR fluid coupling offers high control quality, package advantages and is fully compatible with ESP and ABS (Magna Powertrain, 2009; Kieburg *et al.*, 2008).

24.5.3 Wet clutch couplings

Wet clutch, as the name implies and as shown in Fig. 24.12, consists of interleaving friction and separator plates immersed in a lubricant and connected rotationally to the two rotating elements being coupled. The governing equations for a wet clutch are shown in Section 24.8. In simple terms, the torque capacity of a wet clutch is a function of the geometry of the friction surfaces, the number of friction surface pairs, coefficient of friction μ and the axial force F with which the plates are pressed together. The different couplings in the market vary in minor ways with respect to the choice of lubricant, the friction material and its geometry, and in a significant manner in the actuation mechanism that applies the axial force.

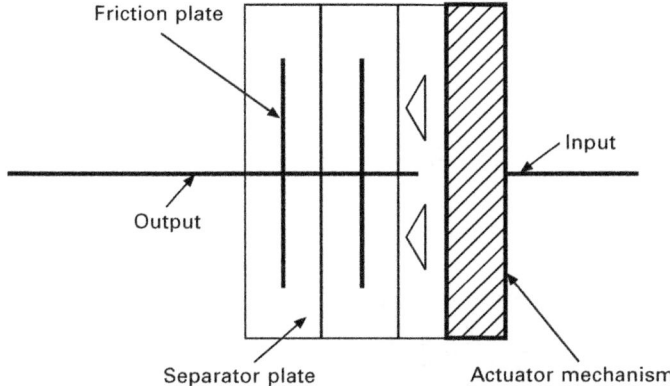

24.12 Representation of a typical wet clutch.

JTEKT's rotary blade coupling (RBC)

The wet clutch in this coupling is applied by a hydraulic piston. The unique feature is the method of generating this hydraulic pressure by means of a rotary blade that utilises the hydrodynamic pressure created by the relative rotation of the tri-blade rotor and the housing (Ashida *et al.*, 1991). This is a strictly passive device, although JTEKT in later years developed an electronically controlled version of this device by using an electric solenoid to trigger the rotation of the tri-blade rotor (JTEKT, 2009). This type of coupling may be seen in Mazda-*MPV* {MY1999}.

Gerodisc device

The gerodisc device was invented by Okcuoglu (1995, 1997) as a hydro-mechanical coupling with a wet friction clutch actuated by a hydraulic piston. Hydraulic pressure is generated by a Gerotor pump that functions only when there is a differential speed between the input and the output (Fig. 24.13).

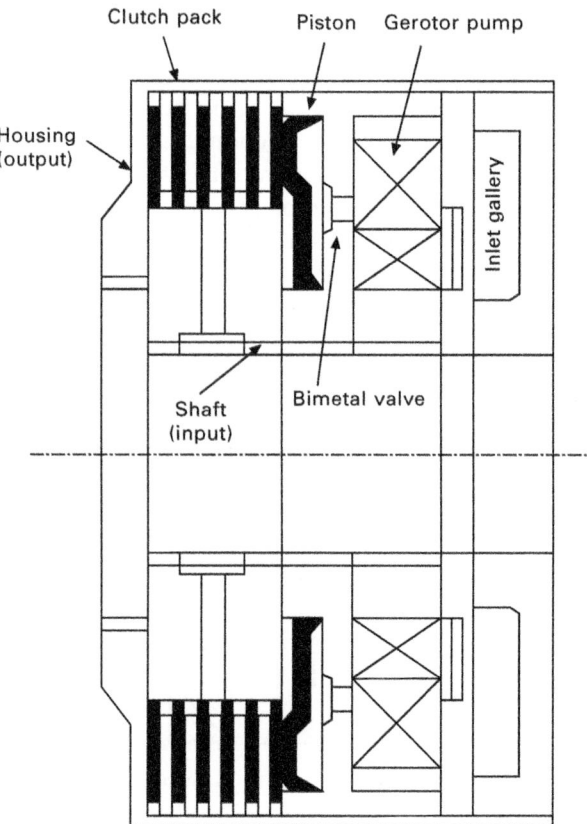

24.13 Gerodisc coupling.

The reaction plates of the clutch and the pump outer ring are rotationally fixed to the housing of the coupling and the clutch friction plates and the inner rotor of the pump rotates with the input shaft. The output pressure from the pump is allowed to act on the apply piston. The pressure at the piston is controlled by the leakage flow permitted by a bimetal valve that compensates for the lowering of the fluid viscosity at elevated temperatures (Zalewski *et al.*, 1999; Jeep-*Grand Cherokee* {MY1999}).

Honda's CRV with the twin pump coupling

In this coupling, the wet clutch is applied by a hydraulic piston connected to the output gallery of a pair of closely coupled pumps, attached to the input and the output elements. When the two elements are rotating at the same speed with no slip, the output from one pump exactly matches the input to the other pump, fluid simply recirculates and no pressure is generated in the output gallery. When there is a speed difference, the excess flow rate accumulates in the gallery, raising the pressure and applying the piston to the clutch. A small unloading solenoid is used to relieve the gallery pressure and establish a small amount of control on the timing and rate of pressure rise in the gallery. Typically during abusive operating conditions, when the oil temperature rises, the solenoid unloads the clutches (Honda-*CRV* {MY1996}).

JTEKT's intelligent torque controlled couplings (ITCC)

JTEKT introduced a series of wet clutch couplings (ITCC) actuated by a ball ramp device, which in turn is rotated by a pilot clutch (JTEKT, 2009). The pilot clutch is made up of a pair of textured and slotted steel plates coated with diamond-like carbon (DLC) layer, resulting in a reliable and stable friction coefficient (Ando *et al.*, 2006). The pilot clutch plates are inserted into a magnetic circuit between a solenoid and an armature plate (Fig. 24.14). When the solenoid is energised, the magnetic circuit is completed through the pilot clutch plates and the armature plate pulls in and applies a clamping force on the pilot clutch plates. When there is relative rotation in the coupling, the small pilot clutch torque is amplified by the ball ramp device into a large axial force against the main wet clutch, providing a reasonably repeatable and predictable coupling torque (Ford-*Escape* {MY 2006}). This device is, in theory, a semi-active system since it requires a small amount of relative rotation in the ball ramp to generate the clutch torque and hence cannot pre-emptively prevent slip. However, in practice, this small rotation is rarely noticed except perhaps by an expert test driver.

Pilot clutch & ball ramp

Main clutch

Solenoid

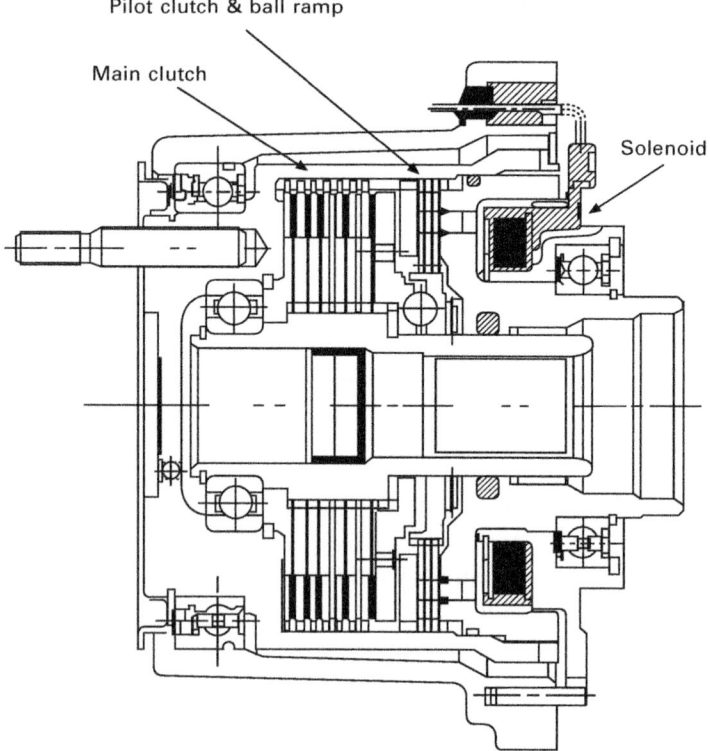

24.14 JTEKT-ITCC coupling.

GKN's driveline torque management coupling

This coupling uses an electric motor actuated wet clutch pack (GKN, 2009). The motor drives a sector gear with a ball ramp device which converts the rotational motion of the sector gear to a translational motion of the ball ramp reaction plate with a greatly amplified axial force, generating the necessary clutch clamping. This is a truly pre-emptive, active clutch actuation independent of the relative slip of the elements. The rear axles of the Porsche-*Cayenne* {MY2004} as well as the VW-*Touareg* {MY2004} have a version of this coupling called ETM.

Haldex coupling

The Swedish company Haldex Traction AB (Haldex, 2009) introduced a unique type of slip reducing coupling in 1998. The basic function is that of an amplified slip sensitive wet clutch. Referring to Fig. 24.15, when both shafts are rotating at the same speed, the pump produces no flow. The slip is *sensed* by an annular piston pump which creates flow proportional to the

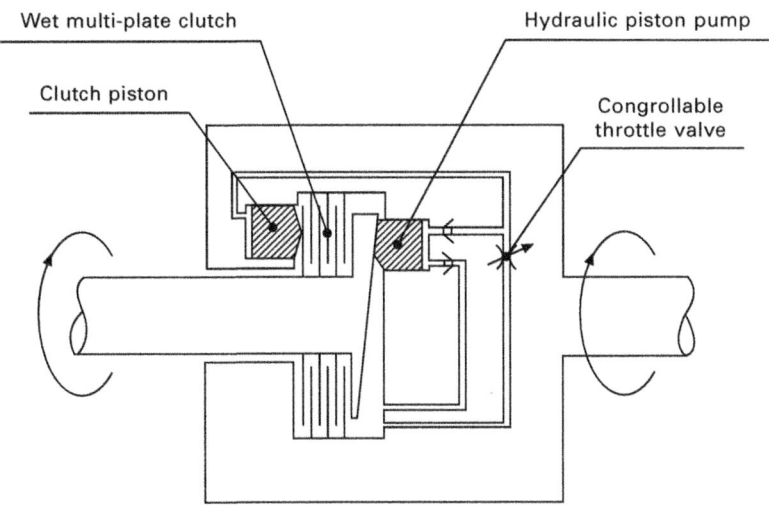

Wet multi-plate clutch   Hydraulic piston pump

Clutch piston   Congrollable
                                       throttle valve

24.15 Haldex coupling.

differential speed, and the force generated by the pressure is amplified by an annular apply piston which compresses the clutch pack. The oil returns to the suction side via a controllable valve which is capable of adjusting the pressure behind the apply piston (Audi-*TT* {MY 1998}). This first generation, semi-passive device was improved in the second generation with electronic sensors and control (Volvo-*XC-90* {MY2002}). The third generation system had enhanced response time by using a high-pressure pre-charge pump, ahead of the slip sensing pump (Land Rover-*Freelander* {MY2006}). The fourth generation Haldex Gen IV eliminated the slip sensitive pump altogether and replaced it with an electric hydraulic pump and an accumulator to store pressurised fluid (Drenth, 2007). Haldex Gen IV can pre-emptively apply pressure on the clutch and control the torque independent of the slip (Saab-*9-3* {MY2007}).

24.5.4 Torsen differential

The Torsen differential, as the name implies is a **Torque sen**sing device. It was invented at the Gleason Corporation using their proprietary gear manufacturing technology. The Torsen differential works just like a conventional differential with one input and two outputs. But, unlike the conventional 'open' differential, which has a torque bias ratio (Eq. 24.3), $(T_{BR}) = 1$, the Torsen differential can have a $T_{BR} > 1$. In practical terms, it implies that even if one of the wheels, say the right one, is on ice and can transmit a low torque of only T_{right}, the Torsen differential can transmit a higher total torque, T_{total}, to the road.

$$T_{\text{total}} = T_{\text{left}} + T_{\text{right}} \qquad\qquad (24.4)$$

For an open differential with $T_{\text{BR}} = 1$, $T_{\text{left}} = T_{\text{right}}$ and $T_{\text{total}} = 2.T_{\text{right}}$.
For a Torsen differential whose $T_{\text{BR}} = 3$, $T_{\text{left}} = 3.T_{\text{right}}$ and $T_{\text{total}} = 4.T_{\text{right}}$.

The basic design concept is based on a version of a helical planetary differential with unsupported or loosely supported planet gears held in pockets in the housing (Fig. 24.16). The radial and axial forces acting on the planets and the sun gear cause the gears to move radially or axially into the housing and the resulting friction creates a reaction torque that is somewhat proportional to the torque going through the gears. It is this reaction torque that allows the 'strong' side to transmit a multiple of the 'weak' side torque (Gleasman, 1958; Gleasman *et al.*, 1980; Chocholek, 1988; Egnaczak, 1994).

There are several variations of the theme by Torsen as well as other manufacturers with cross-axis or parallel-axis planetary gears and different friction conditioning of the surfaces.

Unlike slip sensitive devices, torque sensing devices do not exhibit any reaction lag to the traction loss event. This results in a quicker response to the event, with the caveat that the maximum torque is limited by the design torque bias ratio. In general terms, it is a good characteristic for an on-road vehicle application which requires only a low T_{BR} (Audi-*Quattro* {MY1987}) and an adequate one for an off-road application (AM General-*Hummer* {MY1992}), where a higher torque capability might be more desirable using a design with a higher T_{BR}.

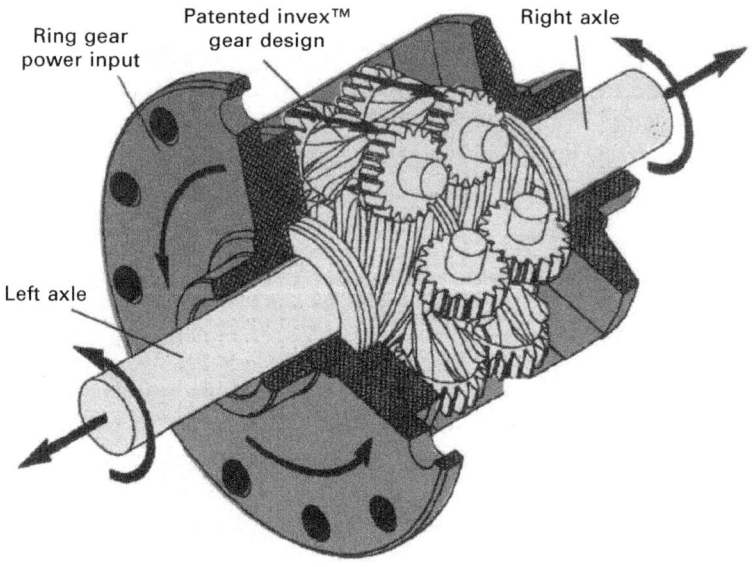

24.16 Torsen differential.

24.6   Advantages of electronically controllable 'active' slip control devices

While the number of 'active' electronically controlled slip-controlled devices are too numerous to list, the following systems are typical of the class of devices and cover the range of simple to quite complex control systems. Nissan introduced the first commercial, production volume 'active' traction control system in the *Skyline* sold in Japan in the 1980s. This was a hydraulically applied wet clutch in a transfer case that could be modulated seamlessly to match the traction requirement. Borg Warner Automotive (2009) introduced an active transfer case in the Ford-*Explorer* {MY1995}. New Venture Gear (later owned by Magna Powertrain, 2009) launched a transfer case with an electro-mechanically applied clutch in the Chevy-*Tahoe* SU {MY 1998}. GKN (2009), Haldex (2009), JTEKT (2009) and Magna Powertrain (2009) have since improved the technology especially in the control strategy and response time. Some of these new systems may be seen in the vehicles BMW-*X5* {MY2008}, Mercedes-Benz-*M-Class* {MY2009}, Porsche-*Cayenne* {MY2008}, VW-*Touareg* {MY 2008} and Volvo-*XC-90* {MY2009}. A direct, electrically actuated clutch pack is shown in Fig. 24.17.

The fundamental advantage of electronic control is the fact that the behaviour of the controlled element is no longer predicated on the inherent nature of the device and its response to the speed, torque or temperature. Instead the controller can command any performance within the device's capability in response to external requirements or even preemptively to avoid an undesirable scenario. The level of authority that the control system has might vary from just minor modification of behaviour to total control and modulation throughout the entire range.

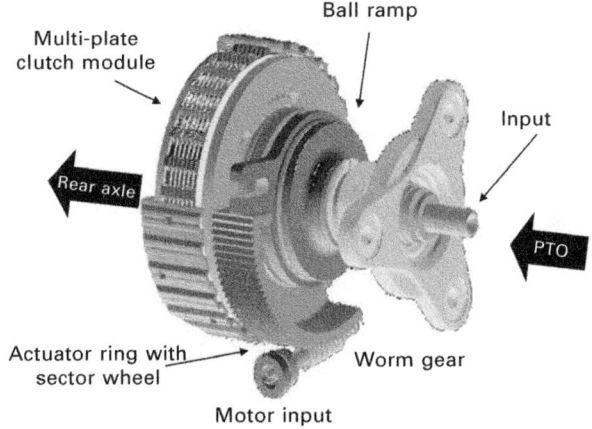

24.17 Direct electric motor actuated clutch.

This capability of the system to enlarge the operating regime of the traction device, and hence the vehicle (Fig. 24.18), has direct bearing on the performance, directional stability and safety of the vehicle. This extension of the vehicle's capability is enabled by technological advances in sensors, actuators, central processing units and control algorithms. A collateral gain is the potential to customise the driving experience by adapting to the driver's habits or changing needs or even the road conditions, traffic or weather.

24.7   Tribological considerations in the design and development of slip control devices

In the application of a friction device, it is important to understand the differences in the operation between a torque transmission clutch and a traction control clutch. These differences dictate the nature of the friction material, the lubricant and the thermal failure mitigation techniques suitable for the application.

24.7.1 Differences between transmission clutch and traction control clutch operation

Transmission clutches are typically designed to lock-up and are expected to slip for 500 ms or less in most applications. In contrast to this, most wet clutches in AWD systems are designed to allow for torque modulation at varying facing pressures and varying relative slip speeds and considerably longer periods of engagement (Fig. 24.19). *AWD clutches* with 1600 N m

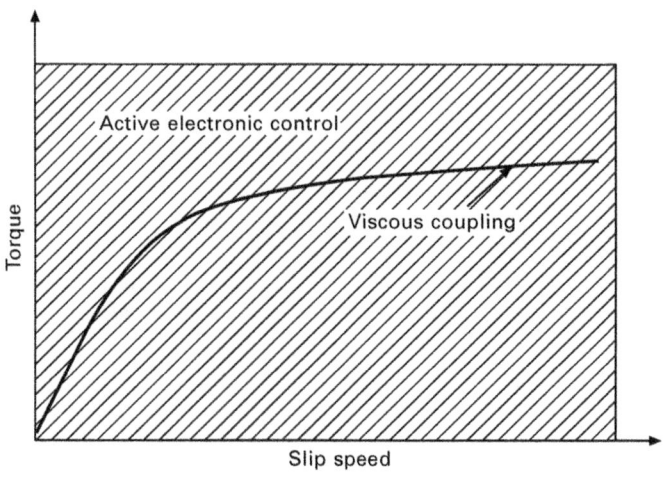

24.18 Enlarged operating regime.

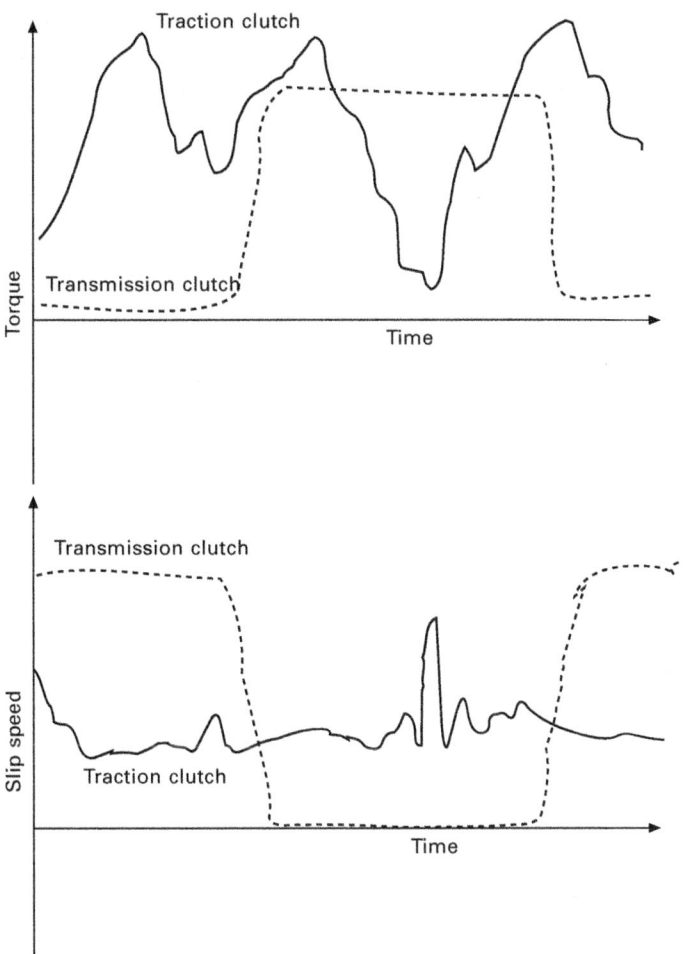

24.19 Clutch engagement cycle.

or more torque capacity are not uncommon, and these systems may see transient heat generation spikes of 20–30 kW or more for short durations. Heat is removed from the wet clutch pack by means of a lubricant, which dissipates the heat to the environment via the housing. AWD clutches influence the directional stability of the vehicle, and therefore need to have good dynamic performance and excellent controllability. Locking a clutch also leads to poor ability to steer in typical automobiles ('crow-hop'), but at the same time, high steering angle manoeuvres also need large amounts of torque transfer (pulling out of a parking lot into moving traffic with maximum acceleration). This means that, based on the demands imposed by vehicle requirements, AWD clutches may be required to slip when under load for a period of several seconds.

Most automatic transmissions have auxiliary cooling systems that AWD systems almost never have. Therefore, AWD clutches have to function with a greater spread in operating temperatures, in addition to wide variations of torque and slip conditions. Coupled with an ever-increasing requirement for clutch torque accuracy, the need to tolerate a wider temperature range makes development of friction material and lubricating oil for AWD systems a challenging proposition.

24.7.2 Friction clutches

As mentioned earlier, a clutch may be used to connect two rotating elements, from a totally disconnected mode to a fully locked mode and anywhere in between (i.e. transmitting the available torque while slipping). A clutch system consists of three tribological elements: the friction lined clutch plate, the reaction plate and the lubricant. The design considerations are very different for wet clutches compared with *dry clutches*. Most traction control clutches are of the wet type and so will be the focus of this chapter.

Since the clutch plates engage with and slide axially along the splined inner and outer elements, the relative hardness of the materials and the geometry of the engaging teeth are selected to minimise the effects of wear and fretting at the contact points. This is a secondary but important tribological consideration since the smooth sliding of the plates on the connected splines is important for predictable torque controllability. The engagement teeth are normally involute spline sections that allow for more uniform pressure distribution and self-centring of the elements under torque.

The clutch plate itself is typically made of a medium carbon steel core plate with the friction material bonded to the core plate on one or both surfaces using a thermosetting organic resin adhesive (Fig. 24.20). The adhesive is selected for good shear and compressive strength, and resistance to delamination under repeated engagement and thermal cycles. Depending on the duty cycle and the operating conditions, the adhesive might experience temperatures of well over 100 °C above the surrounding sump oil temperature.

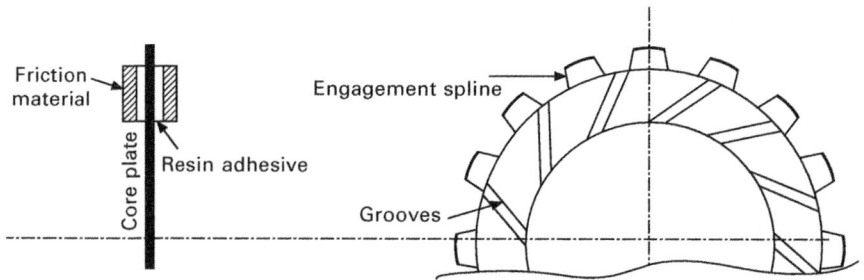

24.20 Typical clutch plate.

The adhesives should also have excellent chemical resistance to the base lubrication oil as well as the other additives and contaminants that might be encountered during operation, especially at elevated temperatures.

The friction material itself is chosen based on the needs for smooth engagement, torque capacity, durability and cost. The organic friction materials are made from fibres, filler material and friction modifying additives held together in a polymer matrix. The resultant material has enough porosity to allow the lubricant to flow through and soak up the heat generated at the friction surfaces. The porous structure also allows the material to conform to the mating surface in spite of small imperfections. The small amount of lubricant carried by the porous friction material and any grooving pattern is important not only to act as a lubricant reservoir, but also to transport the heat away from the interface into the backing core material and the bulk fluid. Other types of friction materials are made of sintered powder metals or strong, thermally resistant fibres of carbon or aromatic polyamide (aramid, Kevlar, Nomex, etc.). Like any material choice, ultimately the decision is a trade-off between competing criteria such as cost, stiffness, durability, good friction coefficient characteristics and desirable thermal properties.

The friction plate pairs are rated based on the axial pressure they can withstand as well as the energy that can be handled during an engagement. The intrinsic characteristic of the friction material and the thermal mass of the system define these two characteristics. In addition, the power that can be handled by the friction system is also a design factor, when considering the lubricant flow and the flow paths in the friction material created by the porosity and the patterns of grooves if present.

Friction systems are in fact just that; a system consisting of the friction material, the mating reaction plate and the lubricant. The characteristics achieved in any one system are highly dependent upon all these three components. It is well known that the friction characteristic can be altered by the use of different lubricants, and/or additives to the same base lubricant. Additionally, the mating surface of a separator plate can be treated, by chemical or mechanical manipulation at the microscopic level (see Chapter 13), to aid in maintaining a boundary layer of fluid under extreme facing pressures. New processes continue to be developed to improve the surface condition, leading to smoother engagement, longer life expectancy and reduced packaging for a given application.

An excellent overview of the tribology of friction clutches is given by Tung and McMillan (2004).

24.7.3 Torque control

The torque transmitted by a clutch system is primarily a function of the coefficient of friction, the mating surface geometry, the axial pressure and the slip speed between the friction plate and the reaction plate. Since the slip speed can be measured and the applied axial pressure can be controlled, it is important to focus on the friction characteristics.

The friction coefficient of the system varies with the slip speed and this pattern of variation could change with surface temperature and with repeated engagements as the interface microstructure and chemical composition change over its life cycle (for example, see Chapters 21 and 22 on the judder phenomenon). Furthermore, the lubricant additives might get depleted over its lifetime and affect the friction coefficient. The dynamic and static friction coefficients of the friction system are typically different from each other. In a locking type clutch, the static friction coefficient must be high enough to have good holding capacity. The nature of the variation of the friction coefficient close to lock-up is critical for smooth engagement (Fig. 24.21). When a clutch is applied and the two rotating elements approach lock-up, the slip speed continually decreases. If the coefficient of friction increases with decreasing slip speed, the interaction of the dynamics of the physical system and the dynamics of the coefficient change with slip will likely set up instabilities and produce undesirable 'stick–slip' shuddering (see Chapter 22). The resulting lock-up would be abrupt and not pleasing to the driver. Therefore, the desirable friction characteristic is for the dynamic coefficient to stay relatively stable over the slip speed and drop in value slightly to the static friction coefficient as it approaches lock-up. Moreover, the friction modifier and other additives should be formulated to provide stability of the coefficient of friction over the life of the clutch and the lubricant.

Some friction materials perform better than others in terms of friction coefficient stability. For example, materials such as carbon composites

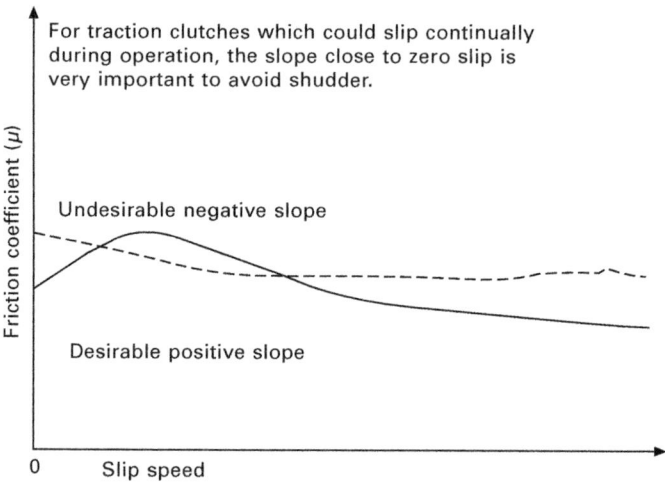

24.21 Traction clutch: μ-slip speed characteristic.

are found to provide a very stable coefficient of friction across a range of operating conditions, and extended slip times. These materials allow for ease of torque capacity control. However, they tend to be of higher cost and present a greater noise, vibration and harshness (NVH) concern during low-speed slip manoeuvres in an aged friction system. On the other hand, lower-cost materials tend to have much greater variation in the coefficient of friction over a range of operating conditions. When torque accuracy is required, a much greater understanding of operating environment (i.e., slip speed, lubricant bulk temperature, interface temperature, apply-force, lubrication flow rate, etc.) is also required. This situation may lead to higher cost in electronic sensing of the condition, and processing of the information, as well as the engineering research and development cost required for proper correlation of models and control algorithms.

24.7.4 Torque control accuracy

The vehicle *traction control system* computes the necessary torque required to be transmitted by the clutch and sends a torque command to the clutch controller. The ideal clutch and control system should react to the torque command within an acceptable response time and accuracy. The response time is typically defined as the delay from the torque command to the time when the torque response has reached an agreed upon percentage of the commanded torque step. Normally the first order system model is used to set the response threshold at ~63%, but some designers arbitrarily set it to 90%.

The control strategy employed assumes that the relationship among friction coefficient, slip-speed and axial pressure is known. The data might be in the form of a look-up table or a set of parametric equations describing the response surface, fitted to test data, under operating conditions. In almost all current commercial automotive traction control systems, there is no direct torque feedback. It is simply too expensive to measure torque with the current state-of-the-art sensors, to the useful accuracy necessary.

Accurate determination of torque or torque capacity, based on applied pressure, or actuator current, or clutch pack compression is a must, since vehicle stability and hence safety is impacted by the clutch performance. So the variation in friction coefficient over the life of the clutch, and the ranges of temperature and facing pressure has to be held at a minimum, or must vary in a predictable fashion.

24.8   Modelling and simulation of traction control devices

The design process of a modern traction control system necessarily involves the multi-physics analysis of the clutch, its control actuator, control software and its interaction with the dynamics of the vehicle and the road conditions. The typical approach is to define distinct modules for the physics-based model of the subsystems and validate them before synthesising the system model from these modules. One critical module is the model for the engagement of the clutch. The model should be complex enough to capture the fine details relevant to the problem, but simple enough to be computationally efficient. As always, it is wise to heed Einstein's maxim for the model to be 'simple, but not simpler' (principle of parsimony). This is especially valid in cases where the simulation model becomes the core of the 'plant model' in the control system and has to run in real-time.

The most basic way of modelling friction is a **Coulomb friction model**. The maximum torque capacity T_m for a pair of friction surfaces subjected to a normal force of F and a friction coefficient of μ is given by:

$$T_m = R_e F \mu \tag{24.5}$$

R_e is the effective radius based on r_o and r_i, the outer and inner radii of the friction material. This equation assumes that the force F is uniformly distributed over the friction surface. Obviously, in a clutch pack with n friction surfaces, the total torque would be T_m multiplied by n.

$$R_e = \left(\frac{2}{3}\right) \frac{r_o^3 - r_i^3}{r_o^2 - r_i^2} \tag{24.6}$$

Computationally, this torque value T_m is equal to the torque transmitted, only when the clutch is slipping. When the clutch is locked up, and there is no slip, T_m simply sets the upper limit of the actual torque transmitted in either direction. So, in case of clutch lock-up, the actual torque will have to be calculated using other equations that impose the equality of the two angular velocities. The sign of the torque is in the direction that opposes slip, but the strongly non-linear nature of the sign function near lock-up often causes numerical problems in simulation. To alleviate the non-linearity, it is easier to use a linear viscous friction model around the lock-up that saturates at the Coulomb friction limit. A very common practice is to use a hyperbolic tangent term for μ, which gives a simplified notation and provides better numerical behaviour (differentiable function):

$$\mu = C_1 \tanh(C_2 v) + C_3 v^k + C_4 \tag{24.7}$$

The constants $C_{1.4}$ and k are fitted to experimental data and v is the sliding velocity between the friction pairs. The main disadvantage of these classes of models is that there is torque only when there is a relative slip speed. Although this is an acceptable compromise for the simulation of slipping clutches, it is not an acceptable model for other simulation tasks (clutch engagement studies, etc). The main advantage is excellent numerical behaviour unless the Coulomb-viscous model is tuned to be extremely stiff.

In cases where a Coulomb-viscous model is not acceptable for behaviour near lock-up, it is common practice to use a microscopic slip model. These models take various forms, but by and large torque is generated only when there is a small amount of relative displacement between the input and the output. An example of such a model is given in Dahl (1977). A properly tuned Coulomb-elastoplastic model can simulate lock-up very gracefully and correlate well with experimental data. The primary disadvantage is that numerical simulations using Coulomb-elastoplastic model is not as well behaved as a simulation using Coulomb-viscous model and stability may not be guaranteed under all conditions. Typically it is not advisable to use these representations for real time simulation models. Further information may be found in Andersson *et al.* (2007).

A number of papers have been published that cover the interaction of the clutch torque with the friction surface features, thermal phenomena taking place in the clutch material and the lubricant in a clutch system: Berger *et al.* (1997), Ito *et al.* (1993), Jang and Khonsari (1999), Lloyd (1974), Mansouri *et al.* (2001), Marklund *et al.* (2007), Miura *et al.* (1998), Natsumeda and Miyoshi (1994), Sharaf *et al.* (2008), Xiang and Kremer (2001) and Yang *et al.* (1998). Others have studied the variation of friction with use; Matsuo and Saeki (1997) and Yesnik (2002). The role of the surface in defining the friction characteristic was studied by Lin *et al.* (2002). Most current proprietary traction control software used in modern vehicles employ algorithms to track the amount of energy going into a clutch, the rise in lubricant temperature and modify the friction and torque calculations to accommodate variations due to temperature and wear. It is assumed with good justification that there is a reasonable correlation among total energy input, time at elevated temperatures and the changes in friction behaviour.

The Mathworks (2009), makers of the Matlab-Simulink suite of simulation software have an adequate model of clutch engagement at their website. Mechanical Simulations (2009) with their CarSim vehicle dynamics simulation software provide an open architecture suite that allows interaction with Simulink models. VeDYNA from TESIS (2009) and CarMaker from IPG Automotive (2009) are other leading vehicle dynamics simulation packages.

Once the clutch control module has been integrated with the vehicle dynamics model, extensive system simulation can be performed under many different driving cycles, vehicle load, road surface and ambient conditions. These simulations will help define the operating envelop of the traction control system and reveal its weaknesses, if any, that need redesign. This iterative process is much less expensive to do in simulation than in hardware. A reliable, compact, computationally efficient model is worth pursuing since it can ultimately become part of the plant model for the control software. In the course of the development process, a complete system simulation model is gradually modified with introduction of an increasing number of hardware elements in-the-loop, replacing the corresponding simulation models, till finally only the system hardware and the control software with the necessary 'plant model' remain.

24.9   Future trends

Safety and efficiency will be the major driving factors in the future. The advances will derive from developments in complex and interactive software, faster processors, innovative sensors and actuators, new material science and especially the paradigm shift to electric drive technology.

Most current torque systems use wet clutches. The lubricant is essential to remove the heat generated in the clutch pack and keep the operation predictable and stable. With the advent of more responsive actuation systems and sensors, and with more powerful computational processing units, it has become feasible to limit the power going into a clutch pack. Such developments in control strategy allow for the use of dry friction clutches which have innately lower parasitic losses. The traction control strategies will work with other vehicle control systems such as the engine, transmission and brake controllers to achieve safety, stability, performance, emission and fuel efficiency, roughly in that order of importance. Hierarchical algorithms may be assigned differing priorities and authorities depending on an adaptive supervisory controller that learns the 'habits' of the driver and interpret the driver's intent as well as watch the external conditions such as the weather, road surface condition and traffic to provide the driver with a safe and enjoyable driving experience.

Developments in the areas of powder metallurgy, friction materials, magnetic materials, fibre and polymer technology, nano-particle coatings and lubricants promise smaller and lighter torque transmission devices with higher energy and power density. While the initial costs may be higher, market forces and high volume adoption should make them more affordable.

The future will also bring more reliance on direct electric drives with the attendant ease of direct torque control, especially when high-torque, lightweight motors enable wheel-end drives. Magnetic inductance-based torque transfer will minimise the bane of parasitic spin losses and torque variation over life, inherent to wet friction or viscous shear based torque transfer. With the introduction of improved magnetic materials, efficient, compact, lightweight, gear-less, lubricant-free, direct-traction drives with built-in regenerative braking capability would become feasible. Work is already moving in this direction at many institutions; e.g. eCorner from Siemens (2007) with concept drivelines that incorporate at least some of these design elements.

24.10   Sources for further information and advice

Many of the topics touched upon in this chapter are developed in much greater detail in the references in Section 24.12. Gillespie (1992) and Milliken and Milliken (1995) are excellent resources for delving deeper into the dynamics of vehicles and understand the need for and effects of traction control. Dowson (1998) still is the most comprehensive book on the history of tribology and when read in conjunction with modern technical literature shows the progress made in understanding of the nature of 'friction' and how much more is yet to be learned. Mohan and Williams (1995) is a somewhat dated (due to the fast pace of technical developments in the last decade) but still useful reference on the classification of different types of traction control systems. Mohan and Ramarao (2002) gives a detailed analysis of viscous couplings, which, in spite of their simplicity of design, are even in 2010, the most widely used traction control device, especially in the Asian market.

Much of the detail on the state-of-the-art technologies on the market is understandably proprietary information and can only be gleaned by referring to the latest patents or articles that the companies choose to publish in technical journals or present at professional societies. With this in mind, the reference section shows a number of papers and patents even though they might not have been directly referenced in the text of the chapter.

The websites of technical organisations such as CAN-CiA, FISITA, IMechE and SAE, as well as institutions that make, use or discuss traction control systems, are given below. Given the ever-changing nature of the Internet and the world-wide web, the URL and the date originally accessed are indicated for each website, with the caution that the reader may have to search the primary website, or use search engines such as Google for archived or cached versions, or access information repositories such as Wikipedia, to get relevant and current information.

24.10.1 Web pages

American Axle & Manufacturing (2009) drivetrain product information available from: http://www.aam.com/ [2009 March]

BMW (2009), xDrive system information available from: http://www.bmw.com/com/en/insights/technology/innovation_lounge/allfacts/phase_3/xdrive.html [2009 March]

BorgWarner Automotive (2009), traction control products information available from: http://www.bwauto.com/products/tts/ [2009 March]

CAN and CiA (CAN in Automation) (2009), information and references available at: http://www.can-cia.de/ [2009 October]

Eaton (2009), differentials and locking differentials information available from: http://www.eaton.com/EatonCom/Markets/Automotive/ProductsandSolutions/Differentials/index.htm [2009 March]

FISITA (2009), International Federation of Automotive Engineering Societies, information available from: http://www.fisita.com/ [2009 March]

Getrag (2009), drivetrain components information available from: http://www.getrag.de/en/205 [2009 March]

GKN (2009), driveline torque management systems information available from: http://www.gkndriveline.com/drivelinecms/opencms/en/products/torque-management/ [2009 March]

Haldex (2009), traction control products available from: http://www.haldex.com/en/North-America/Applications-Products/Product-categories/AWD/ [2009 March]

IMechE (2009), Institution of Mechanical Engineers, Automobile Division information available from: http://www.imeche.org/industries/auto/ [2009 March]

IPG Automotive (2009), information on their CarMaker vehicle dynamics software available from: http://www.ipg.de/428.html?&L=3 [2009 March]

JTEKT (2009), driveline components information available from: http://www.jtekt-na.com/products_driveline.html [2009 March] and http://www.torsen.com/products/products.htm [2009 March]

Magna Powertrain, Inc. (2009), driveline control systems information available from: http://www.magnapowertrain.com/xchg/powertrain_systems/XSL/standard.xsl/-/content/102_1191.htm [2009 March]

Mathworks Inc. (2009), article on clutch lock-up models available from: http://www.mathworks.com/products/simulink/demos.html?file=/products/demos/shipping/simulink/sldemo_clutch.html#1 [2009 March]

Mechancial Simulations (2009), information on their CarSim simulation package available from: http://www.carsim.com/ [2009 March]

Popular Mechanics (2009), article on the ZF torque vectoring system available from: http://www.popularmechanics.com/blogs/automotive_news/4225886.html [2009 March]

SAE (2009), Society of Automotive Engineers International information available from: http://automobile.sae.org/ [2009 March]

Siemens (2007), Siemens Media Summit 2007 eCorner vision information available from: http://w1.siemens.com/press/pool/en/events/media_summit_2007/mediasummit_ sv_vortrag_schelter_final_e_1453124.pdf [2009 October]

TESIS (2009), information on their veDYNA vehicle dynamics simulation package available from: http://www.tesis.de/en/index.php?page=544 [2009 March]

Univance (2009), torque management systems information available from: http://www.uvc.co.jp/english/product/full_4wd.html [2009 March]

ZF (2009), driveline products information available from: http://www.zf.com/corporate/en/products/products.html [2009 March]

24.11   Acknowledgements

The author wishes to acknowledge the help received from many organisations (Magna Powertrain, Borgwarner Automotive, GKN, Haldex, JTEKT), who freely shared technical information and pictures of products, based on which the content for this chapter was created. The author's colleagues at Magna Powertrain, Mr Timothy Burns contributed to Section 24.4 and Mr Anupam Sharma contributed to Section 24.8. Their help is gratefully acknowledged.

24.12   References

Andersson S, Soederburg A and Bjoerklund S (2007), 'Friction models for sliding dry, boundary and mixed lubricated contacts', *Tribology International*, **40**, Issue 4, pp. 580–587.

Ando J, Saito T, Sakai N, Sakai T, Fukami H, Nakanishi K, Mori H, Tachikawa H and Ohmori T (2006), 'Development of compact, high capacity AWD coupling with DLC-Si coated electromagnetic clutch', SAE Technical Paper Series, 2006-01-0820.

Ashida S, Tanigawa Y, Asano H, Yamamoto M, Kojima Y and Yoshida K (1991), 'Development of a rotary tri-blade coupling for four-wheel drive cars', SAE Technical Paper Series, 910806.

Berger E, Sadeghi F and Krousgrill C (1997), 'Analytical and numerical modeling of engagement of rough, permeable, grooved wet clutches', *Journal of Tribology*, **119**, 143–148.

Chocholek S (1988), 'The development of a differential for the improvement of traction control', *C368/88, IMechE* 75–82.

Dahl P (1977), 'Measurement of solid friction parameters of ball bearings', presented at the Sixth Annual Symposium on Incremental Motion, Control System and Devices, University of Illinois.

Dowson D (1998), *History of Tribology*, Second Edition, London, Professional Engineering Publishing.

Drenth E (2007), 'Haldex Cross Wheel Drive', presented at CTI Symposium- Automotive Transmissions North America, Southfield, Michigan.

Egnaczak B (1994), 'The New Torsen II traction technology', SAE Technical Paper Series, 940736.

Gillespie T (1992), *Fundamentals of Vehicle Dynamics*, Warrendale, SAE International.

Gleasman V, (1958), 'Differential Gear Mechanism', US Patent 2,859,641

Gleasman V. et al. (1980), 'Torque Equalizer or Unbalancer for a Cross-Axis Planetary Differential Gear Complex', US Patent 4,191,071

Ito H, Fujimoto K, Egughi M and Yamamoto T (1993), 'Friction characteristics of a paper-based facing for a wet clutch under a variety of sliding conditions', *Tribology Transactions*, **36**, 134–138.

Jang J and Khonsari M (1999), 'Thermal characteristic of a wet clutch', *Journal of Tribology*, **121**, 610–617.

Kieburg C, Oetter G, Laun M, Gabriel C and Steinwender H (2008), 'MR all-wheel-drive prototype car driving tests and durability requirements for the MR fluids used',

Proceedings of the 11th International Conference on Electro-rheological Fluids and Magneto-rheological Suspensions, Technical University Dresden, 2008.

Lin X, Yamamoto Y, Mukai K and Ikawa S (2002), 'The understanding of grooving and surface preparation of high heat resistant materials for heavy-duty vehicles', SAE Technical Paper Series, 2002-01-1482.

Lloyd F (1974), 'Parameters contributing to power loss in disengaged wet clutches', paper presented at the National Combined Farm, Construction and Industrial Machiney and Powerplant Meetings of the Society of Automotive Engineers, 740676.

Mansouri M, Holgerson M, Khonsari M and Aung W (2001), 'Thermal and dynamic characterization of wet clutch engagement with provision for drive torque', *Journal of Tribology*, **123**, 313–322.

Marklund P, Maki R, Larsson R and Hoglund E (2007), 'Thermal influence on torque transfer of wet clutches in limited slip differential applications', *Tribology International*, **40**(5), 876–884.

Matsuo K and Saeki S (1997), 'Study on the change of friction characteristics with use in the wet clutch of automatic transmission', SAE Technical Paper Series, 972928.

Milliken W and Milliken D (1995), *Race Car Vehicle Dynamics*, Warrendale, SAE International.

Miura T, Sekine N, Azegami T and Murakami Y (1998), 'Study on the dynamic property of a paper-based wet clutch', SAE Technical Paper Series, 981102.

Mohan S (2004), 'Comprehensive theory of viscous coupling operation', SAE Technical Paper Series, 2004-01-0867.

Mohan S and Ramarao B (2002), 'A comprehensive study of self-induced torque amplification in rotary viscous couplings', *ASME Journal of Tribology*, **125**(1), 110–121.

Mohan S and Williams R (1995), 'A survey of 4WD traction control systems and strategies', SAE Technical Paper Series, 952644.

Natsumeda S and Miyoshi T (1994), 'Numerical simulation of engagement of paper based wet clutch facing', *Journal of Tribology*, **116**, 232–237.

Okcuoglu M (1995), 'A descriptive analysis of gerodisc type limited slip differentials and all wheel drive couplings', SAE Technical Paper Series, 952642.

Okcuoglu M (1997), 'An all-wheel-drive system utilizing twin hydraulic couplings with gerodisc system', SAE Technical Paper Series, 973235.

Sharaf A, Mavros G, Rahnejat H, King P and Mohan S (2008), 'Optimisation of AWD off-road vehicle performance using visco-lock devices', *International Journal of Heavy Vehicle Systems*, **15**(2–4), 188–207.

Tung S and McMillan M (2004), 'Automotive tribology overview of current advances and challenges for the future', *Tribology International*, **37**, 517–536.

Xiang X and Kremer J (2001), 'A simplified close form approach, for slipping clutch thermal model', SAE Technical Paper Series, 2001-01-1148.

Yang Y, Lam R and Fujii T (1998), 'Prediction of torque response during the engagement of wet friction clutch', SAE Technical Paper Series, 981097.

Yesnik M (2002), 'The influence of material formulation and assembly topography on friction stability for heavy duty clutch applications', SAE Technical Paper Series, 2002-01-1436.

Zalewski J, Durnack M, Burns T and Dober M (1999), 'New technology in passive adaptive traction control for four-wheel-drive vehicles', SAE Technical Paper Series, 1999-01-1262

24.13 Notation

C_x	Constant; subscript x refers to index
F_x	Force; subscript x, if present, refers to direction
k	Constant
r_x	Radius; subscript x, refers to location
R_x	Reference radius; subscript x, if present refers to modifier
s	Slip ratio
T_x	Torque; subscript x, if present, refers to location or modifier
v	Linear velocity
μ	Coefficient of friction
ω_x	Angular velocity; subscript x, if present, refers to location or direction

25
Non-linear dynamics of gear meshing and vibro-impact phenomenon

S. NATSIAVAS and D. GIAGOPOULOS,
Aristotle University, Greece

Abstract: This chapter is devoted to presenting a systematic investigation on the response and stability characteristics of an example gear pair system. Such systems are an important part and affect significantly the response of more complex rotor dynamic systems. The mechanical model examined involves strong non-linear characteristics. In particular, it takes into account gear mesh backlash and static transmission error as well as the essential non-linearities due to the bearing clearance and contact characteristics. The main emphasis is placed on studying long-term dynamics by applying appropriate numerical methodologies. In this way, useful information is obtained for the influence of the loading, the gear mesh and the bearing parameters on the system dynamic response, including periodic, quasi-periodic and chaotic motions. First, classical response diagrams are presented, illustrating the effect of the most important parameters on the system response. Moreover, direct integration of the equations of motion is also performed, demonstrating the existence of quasi-periodic and chaotic long-term response for selected combinations of the system parameters. Finally, the focus is shifted to some related models that have appeared in the literature, as well as to some future trends in the research area examined.

Key words: non-linear gear dynamics, gear backlash, chaotic motion, transmission error, identification.

25.1 Introduction

Geared rotor-bearing systems have found extensive use as power transmission elements in many engineering applications. The continually increasing technological needs for improved performance, compactness, longer life and reduced production costs, require new designs with higher operating speeds and lighter components. In order to satisfy these needs, research in the area of geared systems has been kept active, incorporating new technical advancements and theoretical developments in other related fields. These efforts are also greatly assisted by current rapid enhancements in the level of computing power, which in turn extends the range of applicability of involved numerical algorithms. In particular, the dynamics of systems involving gear mechanisms has long been the epicentre of intensive research efforts. As a result, a large volume of literature exists, dealing with simple

773

gear pair systems as well as with more involved gear trains (see Gregory *et al.*, 1963; Childs *et al.*, 1977; Gunter *et al.*, 1977; Lund, 1978; Yamada and Mitsui, 1979; Adams, 1980; Iida *et al.*, 1980; Simmons and Smalley, 1984; Neriya *et al.*, 1985; Muszynska, 1986; Schwibinger and Nordmann, 1988; Kishor and Gupta, 1989; Choy *et al.*, 1991; Khonsari and Chang, 1993 and references therein). Some related handbooks and review articles have also appeared (Ozguven and Houser, 1988; Ehrich, 1992).

Previous studies on the subject have focused on developing mechanical models of geared systems, ranging from relatively low to high complexity levels, depending on the emphasis and the objectives of the investigation. Consequently, a large variety of important technical topics has already been examined, such as the effect of support and gear-box flexibility, gyroscopics, internal and external damping, shaft shear deformation and coupled torsional-bending vibrations (Gregory *et al.*, 1963; Childs *et al.*, 1977; Gunter *et al.*, 1977; Lund, 1978; Yamada and Mitsui, 1979; Adams, 1980; Iida *et al.*, 1980; Simmons and Smalley, 1984; Neriya *et al.*, 1985; Muszynska, 1986; Schwibinger and Nordmann, 1988; Kishor and Gupta, 1989; Choy *et al.*, 1991; Khonsari and Chang, 1993). The great majority of these studies assume constant average spin speed of the gear shafts. Both response and stability issues have been investigated by means of analytical, numerical and experimental techniques. Among all the technical parameters, those related to the gear backlash and the variable gear meshing stiffness were found to affect the system response in a significant manner. However, gear backlash introduces serious difficulties in the analysis because the equations of motion of such systems become strongly non-linear (Kahraman and Singh, 1991; Padmanabhan and Singh, 1996; Kahraman and Blankenship, 1997; Chen *et al.*, 1997, 1998; Theodossiades and Natsiavas, 2000, 2001a). Moreover, these difficulties are further intensified by the variation in the number of gear teeth pairs which are in contact at a time, causing a variation of the equivalent gear meshing stiffness. On the other hand, some of the earlier studies shifted attention to more fundamental issues and have shown that these complications are responsible for the appearance of complicated and irregular dynamic response (Kahraman and Blankenship, 1997; Theodossiades and Natsiavas, 2001a).

The main focus of the present chapter is directed towards presenting and investigating the dynamics of a simplified but quite representative motor-driven gear pair system. The mechanical model and the accompanying equations of motion are presented in the following section. This model is presented in a way that can easily be adapted as a superelement in a general rotor dynamic configuration. Then, selected numerical results are presented in the form of classical frequency–response diagrams, revealing the influence of various system parameters on the response of the dynamical system examined. Emphasis is placed on examining the effect of the loading parameters as well as the influence of the various non-linearities, including those related to gear backlash and bearing clearance. Some representative results, indicating the presence of irregular long-term response, are also presented. These results are obtained by direct integration of the equations of motion for selected combinations of the system parameters. In the fourth section, some related mechanical models of earlier studies are considered. Moreover, an effort is made in pointing out directions of future research in the area. Finally, the highlights of the work are summarised in the last section.

25.2   Mechanical model and equations of motion of example gear pair

The model of the example mechanical system employed in the present chapter is shown in Fig. 25.1. The components of this model were selected so that most of the essential ingredients of a gear pair system are included (Theodossiades and Natsiavas, 2001a). On the other hand, this model is simple enough to allow a deep and systematic investigation of its dynamics. It consists of a **spur gear pair** with masses m_n, mass moments of inertia I_n and base radii R_n ($n = 1,2$), while both gears are supported on deformable bearings. Neglecting the effect of friction on the gear mesh, the motion transverse to the pressure line is decoupled from the gear motion parallel to the pressure line. Therefore, the essential dynamics of the system is described

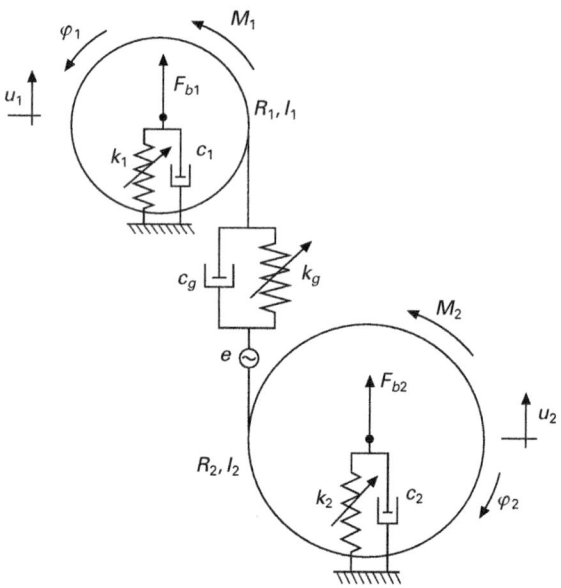

25.1 Mechanical model of a gear pair system on deformable bearings.

by a four degree of freedom system, with co-ordinates φ_1, φ_2, u_1 and u_2. More specifically, application of standard methodologies leads to a set of equations of motion with the following form:

$$I_1 \ddot{\varphi}_1 + R_1 f_g (\varphi_1, \varphi_2, u, \dot{u}) = M_1 (t, \varphi_1, \dot{\varphi}_1) \tag{25.1}$$

$$I_2 \ddot{\varphi}_2 - R_2 f_g (\varphi_1, \varphi_2, u, \dot{u}) = - M_2 (t, \varphi_2, \dot{\varphi}_2) \tag{25.2}$$

$$m_1 \ddot{u}_1 + f_1 (u_1, \dot{u}_1) + f_g (\varphi_1, \varphi_2, u, \dot{u}) = F_{b1} \tag{25.3}$$

$$m_2 \ddot{u}_2 + f_2 (u_2, \dot{u}_2) - f_g (\varphi_1, \varphi_2, u, \dot{u}) = F_{b2} \tag{25.4}$$

The quantities M_1 and M_2 represent the driving and resisting moments, while F_{b1} and F_{b2} refer to pre-tension radial forces applied to the centre of the driving and driven gear, respectively. This model takes into account the **static transmission error**, $e(\varphi_1, \varphi_2)$, representing geometrical errors of the gear teeth profile and spacing. On the other hand, the quantity defined by:

$$u = R_1 \varphi_1 - R_2 \varphi_2 + u_1 - u_2 - e(\varphi_1, \varphi_2) \tag{25.5}$$

is known as the **dynamic transmission error**. In addition, the damping mechanisms in both the gear mesh and the bearings are assumed to have a linear nature, so that the corresponding forces are expressed in the form:

$$f_g (\varphi_1, \varphi_2, u, \dot{u}) = c_g \dot{u} + f_g (u, \varphi_1, \varphi_2)$$

and $f_n (u_n, \dot{u}_n) = c_n \dot{u}_n + f_n (u_n)$

respectively. Furthermore, taking into account the **gear backlash**, the gear meshing restoring force component is expressed in the form:

$$f_g (u, \varphi_1, \varphi_2,) = k_g (\varphi_1, \varphi_2,) h_g (u) \tag{25.6}$$

where the function $k_g (\varphi_1, \varphi_2)$ represents the **variable gear mesh stiffness**, while:

$$h_g (u) = \begin{cases} u - b, & u \geq b \\ 0, & |u| < b \\ u + b, & u \leq - b \end{cases} \tag{25.7}$$

and $2b$ represents the total gear backlash. Likewise, in the special case where bearings with rolling elements are selected, the restoring force developed in the nth rolling element bearing is expressed either in the classical linear form:

$$f_n (u_n) = k_n u_n \tag{25.8}$$

or in the more complicated, but more accurate, form:

$$f_n(u_n) = \begin{cases} \hat{k}_n \sum_{r=1}^{N} (u_n \cos \alpha_{rn} - b_{bn})^{\nu} \cos \alpha_{rn}, & u_n \geq b_{bn} \\ 0, & |u_n| < b_{bn} \\ -\hat{k}_n \sum_{r=1}^{N} (|u_n| \cos \alpha_{rn} - b_{bn})^{\nu} \cos \alpha_{rn}, & u_n \leq -b_{bn} \end{cases} \qquad (25.9)$$

In the last expression, $2b_{bn}$ represents the diametral clearance, α_{rn} is the angular position of the rth rolling element (of the total N elements in contact), ν is a constant (equal to 3/2 for ball bearings and 10/9 for roller bearings), while the coefficient $\hat{k}_n$ is determined from the bearing characteristics and loading conditions (Harris, 1966).

Taking all the above into account, the equations of motion can eventually be put in the form:

$$M\underline{\ddot{x}} + C\underline{\dot{x}} + \underline{k}\,(\underline{x}) = \underline{f}_b + \underline{f}(t) \qquad (25.10)$$

with dynamic co-ordinates:

$$\underline{x}(t) = (\varphi_1 \quad \varphi_2 \quad u_1 \quad u_2)^{T}$$

This represents a strongly non-linear system of coupled ordinary differential equations. In general, the solution of the last set can only be obtained by applying suitable numerical methods. However, a simpler situation is examined frequently, where the total rotation angle of each gear can be decomposed in the form:

$$\varphi_n(t) = \omega_n t + \theta_n(t) \qquad (25.11)$$

Here ω_n are pre-specified mean angular velocity components of the gear shafts and θ_n represent small variations caused by vibration of the mating gear teeth. In such cases, both the gear mesh stiffness and the static transmission error can be considered as time-periodic functions. Neglecting the tooth-to-tooth variations, the fundamental frequency of both of these quantities equals the **gear meshing frequency**:

$$\omega_M = n_1 \omega_1 = n_2 \omega_2 \qquad (25.12)$$

where the integers n_1 and n_2 represent the number of teeth in the driving and the driven gear, respectively. This implies that the aforementioned quantities can be expressed in the Fourier series forms:

$$k_g(t) = k_0 + \sum_{n=1}^{\infty} [k_{cn} \cos (n\omega_M t) + k_{sn} \sin (n\omega_M t)] = k_g(t + T_M) \qquad (25.13)$$

and:

$$e(t) = e_0 + \sum_{n=1}^{\infty} [e_{cn} \cos(n\omega_M t) + e_{sn} \sin(n\omega_M t)] = e(t + T_M) \quad (25.14)$$

respectively, with fundamental period $T_M = 2\pi/\omega_M$. Moreover, the external torques are also assumed to be periodic functions of time. Then, by employing relation (25.5), the rigid body rotation mode can be eliminated from the original set of equations of motion (25.1)–(25.4). In this way, the new set of equations of motion can eventually be put in the form (25.10) with:

$$\underline{x}(t) = \begin{pmatrix} u \\ u_1 \\ u_2 \end{pmatrix}, M = \begin{bmatrix} m_0 & -m_0 & m_0 \\ -m_0 & m_0 + m_1 & -m_0 \\ m_0 & -m_0 & m_0 + m_2 \end{bmatrix}, C = \begin{bmatrix} c_g & 0 & 0 \\ 0 & c_1 & 0 \\ 0 & 0 & c_2 \end{bmatrix},$$

$$\underline{k}(\underline{x}) = \begin{pmatrix} k_g(t) h_g(u) \\ f_1(u_1) \\ f_2(u_2) \end{pmatrix}, \underline{f}_b = \begin{pmatrix} 0 \\ F_{b1} \\ F_{b2} \end{pmatrix},$$

$$\underline{f}(t) = m_0 \left[\frac{R_1}{I_1} M_1(t) + \frac{R_2}{I_2} M_2(t) - \ddot{e}(t) \right] \begin{pmatrix} 1 \\ -1 \\ 1 \end{pmatrix}, m_0 = \frac{I_1 I_2}{I_1 R_2^2 + I_2 R_1^2}$$

Finally, in the special cases where the effect of bearing flexibility is negligible, so that:

$$u_1(t) = u_2(t) = 0,$$

the equations of motion (25.10) are replaced by the single scalar equation:

$$m_0 \ddot{u} + c_g \dot{u} + k_g(t) f_g(u) = f(t) \quad (25.15)$$

where the forcing term:

$$f(t) = m_0 [R_1 M_1(t)/I_1 + R_2 M_2(t)/I_2 - \ddot{e}(t)]$$

is time-periodic and includes contributions from the external torques as well as from the geometric irregularities of the mating gear teeth. In such cases, it is possible to derive approximate analytical solutions by application of appropriate perturbation methods (Natsiavas et al., 2000).

25.3   Typical numerical results from mechanical model and equations of motion of example gear pair

The results presented in this section illustrate effects of selected parameters on regular and irregular dynamics of the mechanical system examined. This provides a basis for capturing effectively and for explaining the results obtained for more complex mechanical models. For convenience in the numerical calculations, the equations of motion are first normalised by introducing the following set of parameters:

$$\zeta_g = \frac{c_g}{2\sqrt{m_0 k_0}}, \; \zeta_n = \frac{c_n}{2m_n \omega_0}, \; b_n = \frac{b_{bn}}{b}, \; r_n = \frac{R_n}{b}, \; v(\tau) = \frac{u(t)}{b},$$

$$v_n(\tau) = \frac{u_n(t)}{b}, \; J_n = \frac{I_n}{m_0 R_n^2}, \; \mu_n = \frac{m_n}{m_0}, \; f_{bn} = \frac{F_{bn}}{k_0 b}, \; \kappa_n = \frac{k_n}{k_0},$$

$$\hat{\kappa}_n = \frac{\hat{k}_n b^{\nu-1}}{k_0}, \; \kappa_g(\psi_1) = \frac{k_g(\varphi_1)}{k_0}, \; \hat{e}(\psi_1) = \frac{e(\varphi_1)}{b}$$

where: $n = 1, 2$, $\omega_0 = \sqrt{(k_0/m_0)}$ and $\tau = \omega_0 t$.

For all the cases considered, the driving and the resisting moments are chosen to have constant magnitude, equal to M_{10} and M_{20}, respectively. Also, the nominal system parameters take the values: $\zeta_g = 0.05$, $\mu_n = 4$, $J_n = 2$ and $b_n = 0.1$. In addition, the bearings possess linear characteristics with $\kappa_n = 1$ and $\zeta_n = 0.01$ and have no pre-tension.

The first sequence of numerical results is presented in the form of classical frequency response diagrams. These diagrams were obtained by applying an appropriate methodology, which is suitable for determining branches of periodic motions of non-linear dynamical systems, together with their stability and bifurcation properties (Doedel, 1986; Nayfeh and Balachandran, 1995; Van de Vorst *et al.*, 1996; Sundararajan and Noah, 1998). The diagrams shown depict the translation amplitude of the driven gear centre as a function of the normalised gear meshing frequency $\Omega = \omega_M/\omega_0$. In all cases, stable and unstable periodic motions are represented by continuous and broken curves, respectively.

First, Fig. 25.2(a) presents results obtained for the default parameter combination and $M_{10} = 0.0375$. Moreover, the gear mesh stiffness was assumed to be constant, while the static transmission error was harmonic with $\hat{e}_{c1} = 0.05$. The results shown indicate that the effects of the gear mesh non-linearities are more pronounced in the vicinity of the second main resonance. This is expected, since for the set of parameters chosen the gear rotation is dominant in that frequency range (Theodossides and Natsiavas,

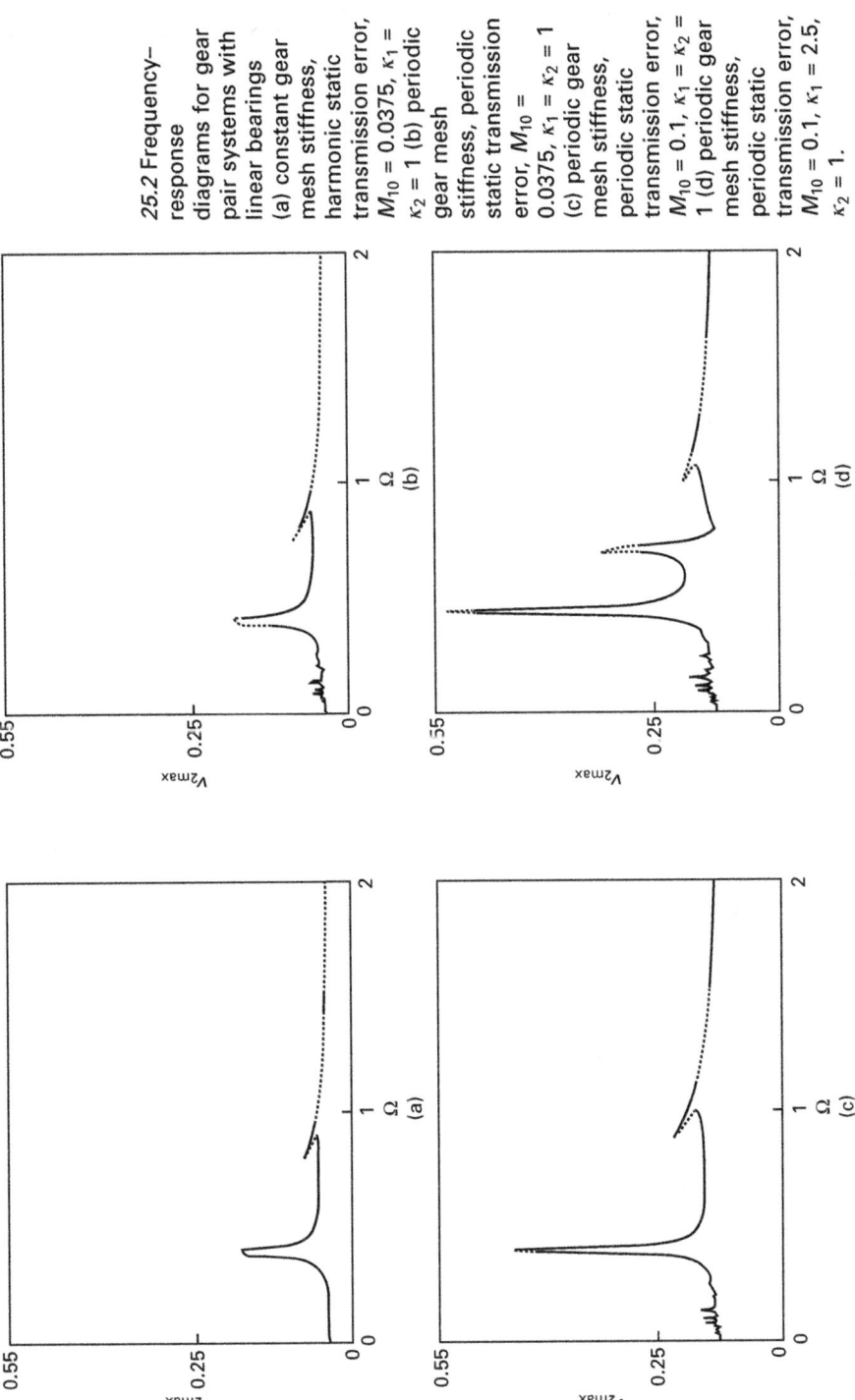

25.2 Frequency–response diagrams for gear pair systems with linear bearings (a) constant gear mesh stiffness, harmonic static transmission error, $\kappa_1 = 0.0375$, $M_{10} = 0.0375$, $\kappa_1 = \kappa_2 = 1$ (b) periodic gear mesh stiffness, periodic static transmission error, $M_{10} = 0.0375$, $\kappa_1 = \kappa_2 = 1$ (c) periodic gear mesh stiffness, periodic static transmission error, $M_{10} = 0.1$, $\kappa_1 = \kappa_2 = 1$ (d) periodic gear mesh stiffness, periodic static transmission error, $M_{10} = 0.1$, $\kappa_1 = 2.5$, $\kappa_2 = 1$.

2001a). Next, the results shown in Fig. 25.2(b) were obtained by expressing both the gear mesh stiffness and the static transmission error as time-periodic functions, according to (25.13) and (25.14). The main effect observed here is confined in the low forcing frequency range, where the response is dominated by multiple super-harmonic resonances. On the other hand, the diagram depicted in Fig. 25.2(c) was determined for exactly the same set of parameters, after increasing the value of the driving moment to $M_{10} = 0.1$. This load increase causes an increase mainly of the response amplitude in the lower main resonance. It also widens the range of the stable motions in the higher frequency range. Finally, Fig. 25.2(d) includes results determined by increasing the support equivalent stiffness parameter of the driving gear to $\kappa_1 = 2.5$. Here, an intermediate resonance branch emerges in the response diagram, due to the symmetry breaking induced by the difference in the values of the bearing stiffness parameters κ_1 and κ_2.

Next, Fig. 25.3 presents another sequence of response diagrams, obtained by including the contact and clearance non-linearities in modelling the behaviour of the bearings. Moreover, in all cases examined here, the gear mesh stiffness and the static transmission error were expressed in a periodic form. First, the response diagram of Fig. 25.3(a) represents results obtained for the nominal case with $M_{10} = 0.0375$ and $\hat{\kappa}_1 = \hat{\kappa}_2 = 0.3594$. Compared with the corresponding system with the linear bearings (shown in Fig. 25.2(b)), besides a considerable shift in the mean value of the gear centre displacement, the major response differences are observed to occur only in the vicinity of the first main resonance range. This is in accordance with the predictions of the linear analysis, since the linear mode with natural frequency close to this resonance is dominated by the gear translation (Theodossides and Natsiavas, 2001a). Likewise, the results shown in Fig. 25.3(b) were determined after increasing the value of the driving moment to $M_{10} = 0.1$. The outcome indicates that this load increase causes qualitatively similar results as those observed for the linear case (Fig. 25.2(c)). The following response diagram, shown in Fig. 25.3(c) was obtained by increasing the support stiffness parameter of the driving gear to $\hat{\kappa}_1 = 2.5\hat{\kappa}_2$. Again, the asymmetry induced in the restoring force of the supporting bearing of the driven gear is responsible for the appearance of a new branch of motions, just after the lower main resonance. Finally, the last diagram of this sequence was captured after introducing a pre-tension applied to the centre of the gears with $f_{b2} = -f_{b1} = 0.05$.

Among other things, the results presented in the response diagrams of Figs 25.2 and 25.3 indicate that there appear extensive frequency ranges where only unstable periodic motions are captured. For instance, consider the branch of Fig. 25.2(a) covering the frequency interval $\Omega \approx 0.97$–1.43. This branch is generated via a ***Hopf bifurcation*** at both of its ends (Nayfeh and Balachandran, 1995). In order to investigate the dynamics of the system

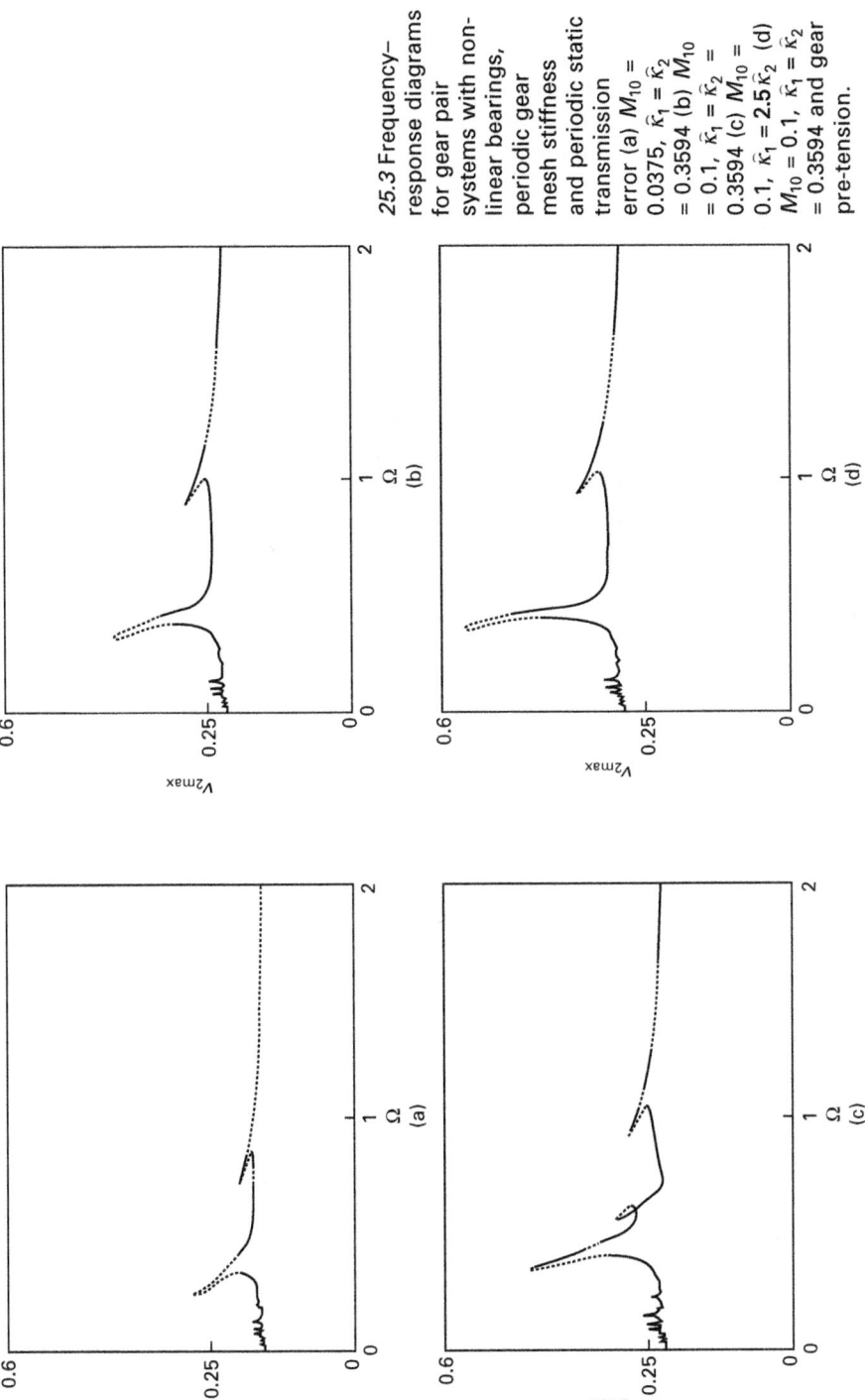

25.3 Frequency–response diagrams for gear pair systems with non-linear bearings, periodic gear mesh stiffness and periodic static transmission error (a) $M_{10} = 0.0375$, $\hat{\kappa}_1 = \hat{\kappa}_2 = 0.3594$ (b) $M_{10} = 0.1$, $\hat{\kappa}_1 = \hat{\kappa}_2 = 0.3594$ (c) $M_{10} = 0.1$, $\hat{\kappa}_1 = 2.5\hat{\kappa}_2$ (d) $M_{10} = 0.1$, $\hat{\kappa}_1 = \hat{\kappa}_2 = 0.3594$ and gear pre-tension.

within this frequency range, direct integration of the equations of motion was performed. A small but typical sample of results obtained is presented in Fig. 25.4, in the form of **Poincaré** sections with the $(v_2, \dot{v}_2)$ plane. According to the information presented by these sections, the periodic motions (Fig. 25.4(a)) lose stability and give their place to quasi-periodic response just after the bifurcation (Fig. 25.4(b)), as expected (Bajaj and Tousi, 1990; Metallidis and Natsiavas, 2000). Moreover, moving away from the bifurcation point these motions were found to repeatedly undergo transitions from quasi-periodic or large order periodic motions (Fig. 25.4(c)) to **chaotic motions** (Fig. 25.4(d)) and vice versa.

A similar situation was found to occur for parameter combinations leading to the response diagram of Fig. 25.3(a). In that case, the roller bearing non-linearity causes a bending to the left even of the first main resonance branch. On this branch, in addition to the **classical saddle-node bifurcations** at the points of vertical tangency, there appears a Hopf bifurcation point. Figure 25.5 depicts a representative sample of histories obtained for the translation of the driven gear, illustrating the evolution of the motions resulting after that bifurcation point. Here, the originally stable periodic motion (Fig. 25.5(a)) loses its stability and turns into a quasi-periodic motion (Fig. 25.5(b)) just after the critical frequency value. Then, this new motion loses its regular characteristics almost immediately and turns quickly into a chaotic motion (Fig. 25.5(c)). The rapid qualitative change observed in the system response is very probably due to the simultaneous loss of contact at all the bearing rollers, which occurs for (normalised) bearing displacement less than 0.1. Moreover, during the **aperiodic motions** shown in Figs 25.5(b)–25.5(e), the driven gear shaft repeatedly loses contact with the rollers on the same side of the bearing, but does not impact with the rolling elements on the opposite bearing side. Note that these motions are characterised by substantially greater amplitudes than the coexisting unstable periodic motions. Also, from the same figures it appears that a gradual decrease in the forcing frequency causes an increase of the intervals of motion with regular form, having amplitude greater than 0.1. Eventually, Fig. 25.5(f) demonstrates that these chaotic motions disappear by getting attracted to the coexisting lower amplitude stable periodic motions, where there is no loss of contact in the bearings.

The response due to the non-linear characteristics of the dynamical system examined causes adverse effects in many areas. For instance, it was observed in a previous study that an identification methodology encountered difficulties for some parameter combinations, leading to either irregular dynamics or to coexistence of multiple motions of the gear-pair system (Giagopoulos *et al.*, 2006). As an example, Fig. 25.6 presents a typical sample of cost functions, obtained by varying the value of the stiffness parameter κ_1 for the case with $M_{10} = 0.1$ and $\kappa_2 = 1$, at four selected values of the meshing frequency Ω. The system parameters and the values of the **meshing frequency** are the

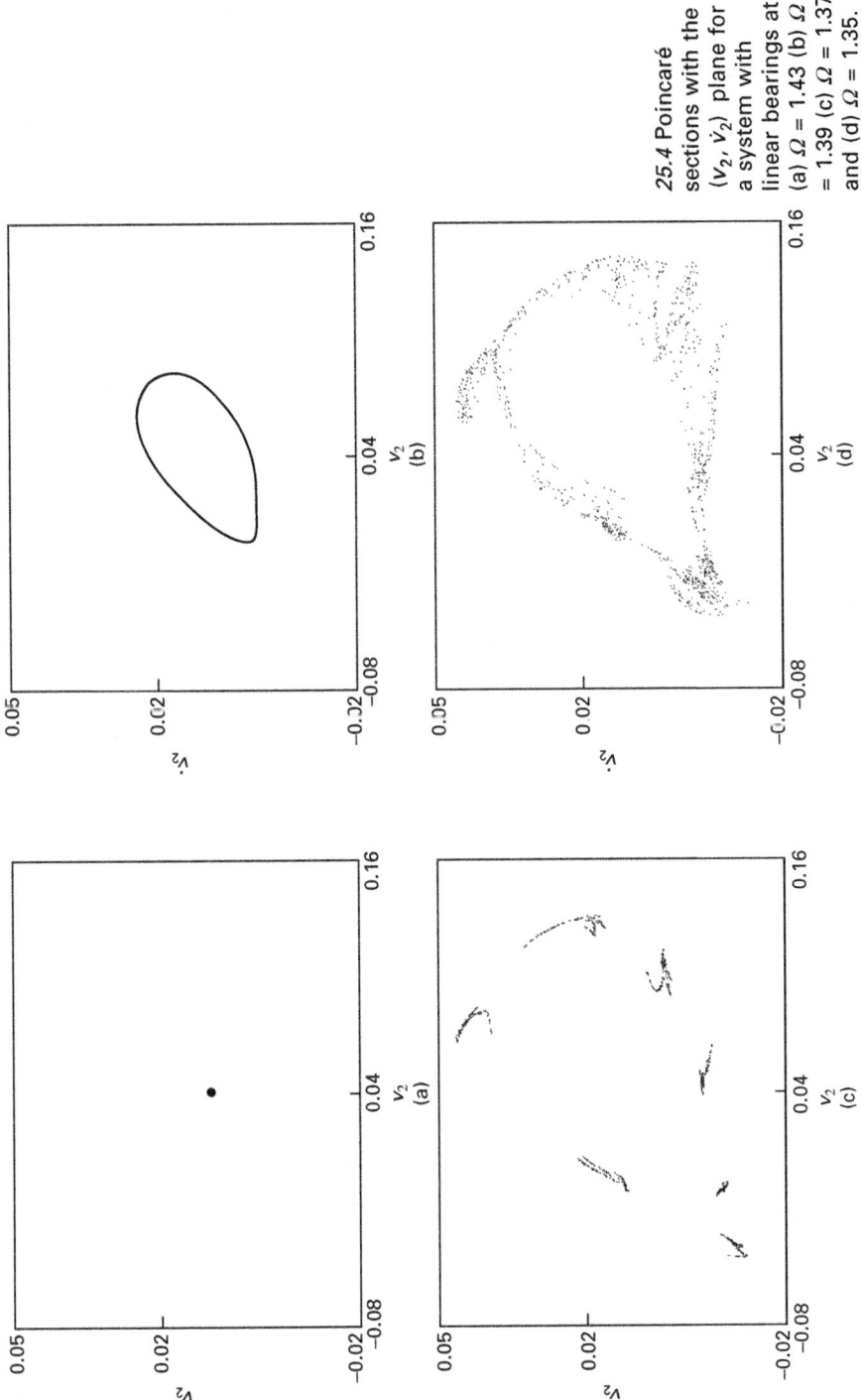

25.4 Poincaré sections with the $(v_2, \dot{v}_2)$ plane for a system with linear bearings at (a) $\Omega = 1.43$ (b) $\Omega = 1.39$ (c) $\Omega = 1.37$ and (d) $\Omega = 1.35$.

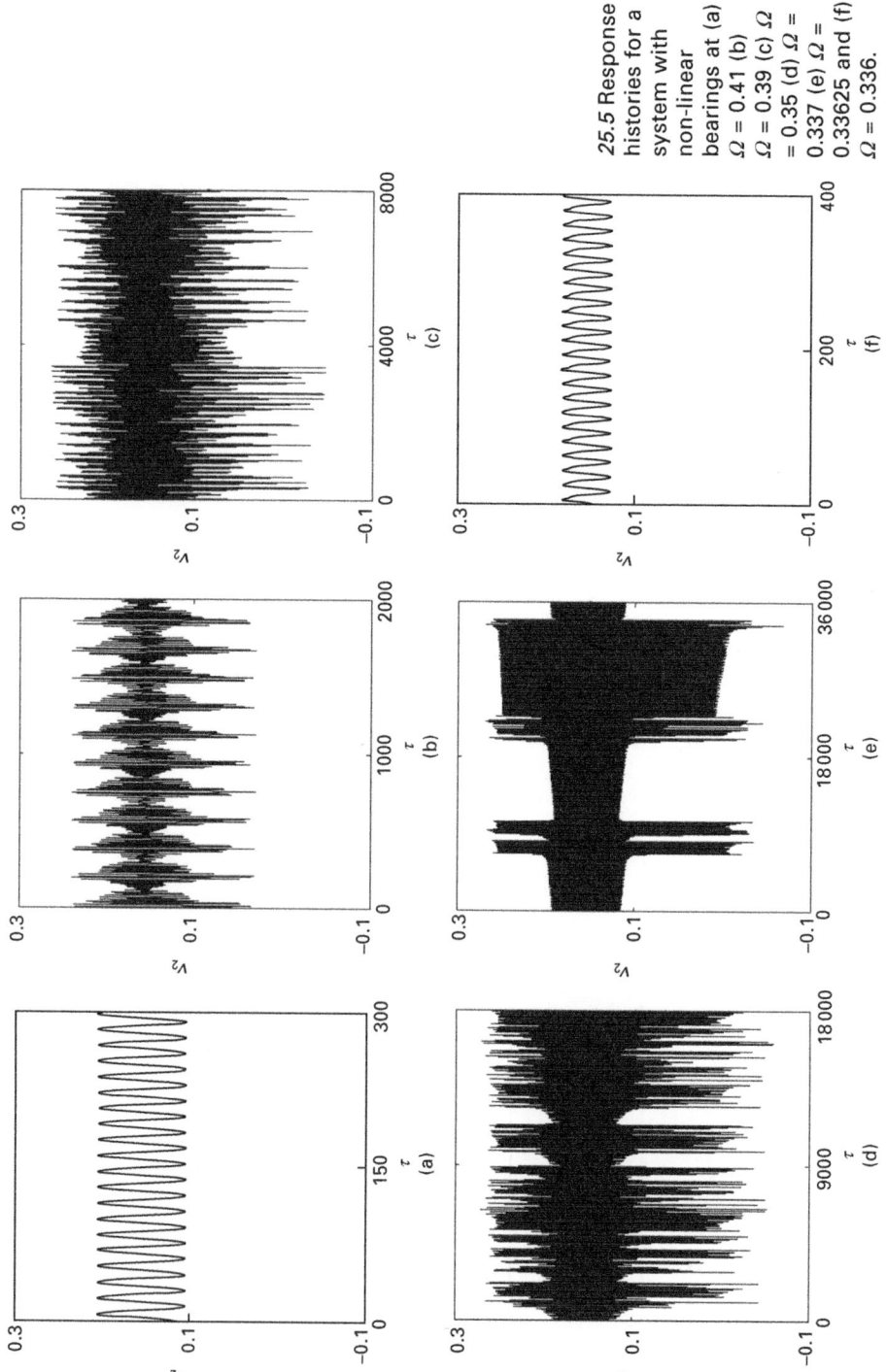

25.5 Response histories for a system with non-linear bearings at (a) $\Omega = 0.41$ (b) $\Omega = 0.39$ (c) $\Omega = 0.35$ (d) $\Omega = 0.337$ (e) $\Omega = 0.33625$ and (f) $\Omega = 0.336$.

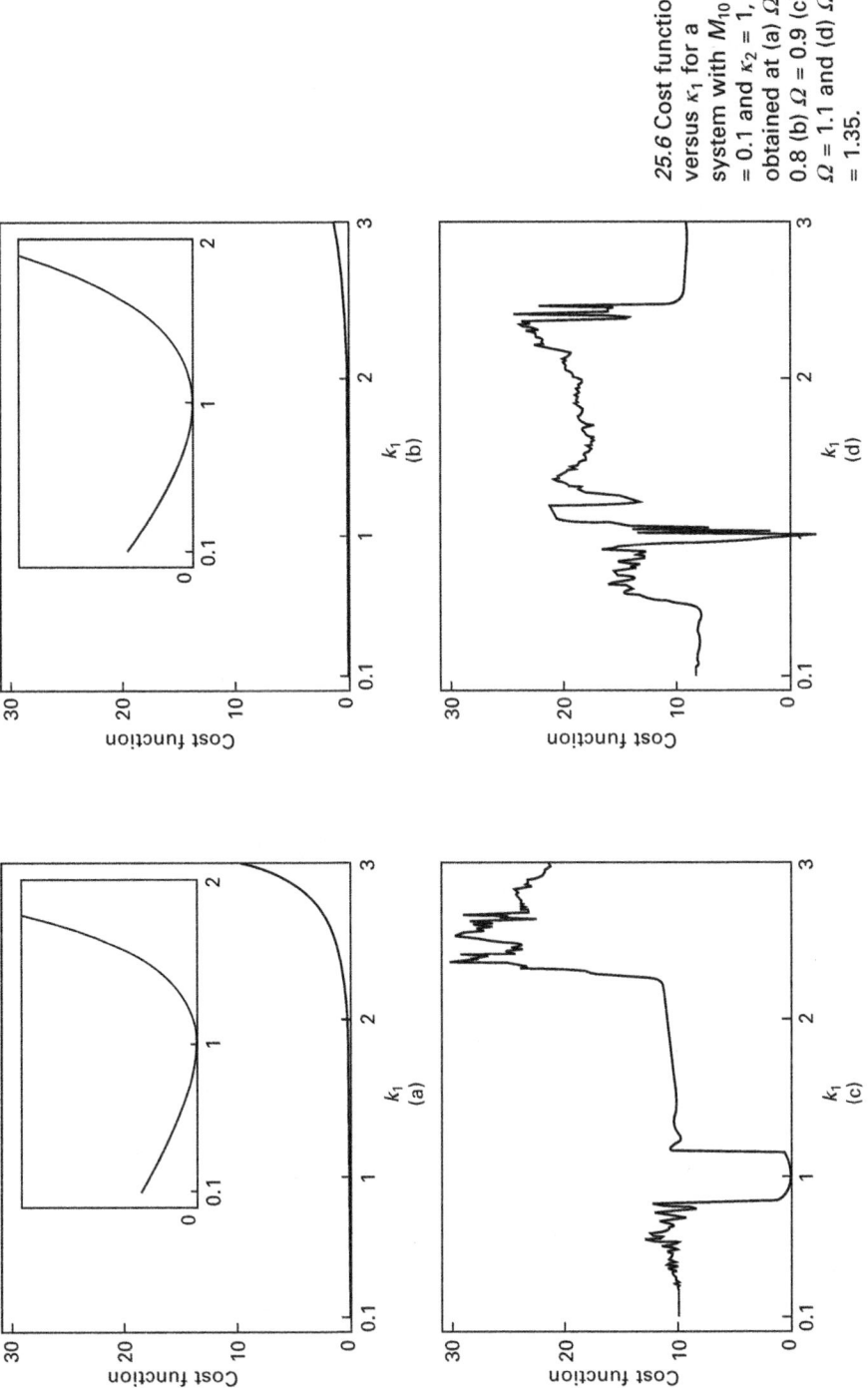

25.6 Cost function versus κ_1 for a system with $M_{10} = 0.1$ and $\kappa_2 = 1$, obtained at (a) $\Omega = 0.8$ (b) $\Omega = 0.9$ (c) $\Omega = 1.1$ and (d) $\Omega = 1.35$.

same as those leading to the results presented in Fig. 25.4. Inspection of Fig. 25.6 shows that the minimum value of the cost function defined in the optimisation part of the identification method applied is reached at the correct value $\kappa_1 = 1$, indeed, for all cases examined. However, the form of the cost function presented for the cases corresponding to $\Omega = 0.8$ and $\Omega = 0.9$ is regular, while the form of the cost function obtained for the other two values of Ω involves multiple minima and jumps and is quite irregular. In an effort to investigate further these peculiar findings, first the emphasis was shifted back into the dynamics of the system examined. It was found that in cases where only one motion occurs, some of these difficulties can be overcome by applying a proper genetic algorithm (Rao, 1996). However, when multiple motions appear in the interval of measurements and calculations, there is no guarantee that the outcome of the calculations will be correct, independently of the optimisation methodology employed (Giagopoulos *et al.*, 2006).

25.4   Future trends

The example mechanical model examined includes most of the traits characterising non-linear dynamics of gear meshing. This or similar models can be employed before more elaborate models are used, in order to reveal and clarify fundamental aspects associated with the related vibro-impact phenomenon taking place between the gear teeth of a gear pair. For instance, similar models have also been developed in earlier studies, where oil journal instead of rolling element bearings were considered (Van de Vorst, *et al.*, 1996; Theodossiades and Natsiavas, 2001b). Since these bearings are also characterised by strongly non-linear properties, similar dynamics was observed to occur. Moreover, some other important effects associated with lubrication of the teeth surfaces were also examined in more recent studies (see also Chapter 29).

In particular, the effect of lubricant film squeeze action between the teeth flanks has been considered in a recent study by introducing non-linear damping due to the ***squeeze film action*** on the adsorbent oil film when the gear teeth are not in direct contact (Brancati *et al.*, 2005). Also, fully flooded conditions between the teeth flanks were assumed in other related studies (Tangasawi *et al.*, 2007; Theodossiades, *et al.*, 2007). This approach is valid for low loads transmitted by ***unselected gears***, where the contact stiffness is governed by the lowest stiffness, which is that of the lubricant rather than that due to Hertzian response. Under high load conditions, where high pressures are generated in the contact area, the elastic deformation of gear teeth also becomes important.

Another related research area with large practical importance is the area of planetary gear dynamics. The existing literature mainly addresses static analysis, natural frequencies and vibration modes. It also estimates dynamic response and predicts cancellation of mesh forces using the planetary gear symmetry through mesh phasing (Cunliffe *et al.*, 1974; Kahraman, 1994). Special issues referring to sensitivity of natural frequencies and modes to operating speeds, natural frequency veering phenomena and mesh stiffness-induced parametric instability were also studied (Lin and Parker, 2002). Among other things, the effectiveness of mesh phasing in suppressing certain harmonics of planetary gear vibration modes based on self-equilibration of the dynamic mesh was also investigated (Parker, 2000).

On the other hand, when the accuracy of the results is of primary concern, more complicated mechanical models are created and examined. For example, more realistic models, including the flexibility of the interconnecting shafts were studied by (Theodossiades and Natsiavas, 2001b). In that study, the equations of motion were first set up by applying the finite element method. The same method was also employed in modelling and determining the exact characteristics of the variable stiffness of the gear pair involved. Due to the large dimension of the resulting dynamical system, the order of the system was first reduced by applying an appropriate co-ordinate reduction method (Craig, 1981; Nelson *et al.*, 1983). This method takes advantage of the fact that the non-linearities are located at the bearings and the gear mesh only. As a result, it was possible to study the dynamics of the reduced rotor-bearing model by employing the methodology that was developed and applied first to the gear pair models. In such studies, the mechanical model examined can be added as a discrete superelement within the overall structural model.

From this point of view, the present gear pair model is a good building block even for more complex rotor dynamic systems, involving planetary gears. In recent years, deformable gear body dynamic models have been developed by employing the finite element method to examine planetary gear dynamics (Parker *et al.*, 2000; Yuksel and Kahraman, 2004). Among other things, the studies performed focused on the effect of ring gear flexibility as well as on tooth wear and its impact on the dynamic behaviour of planetary gears. Recently, some work has also been devoted to characterising the non-linear effects of tooth separation on planetary gear dynamics.

25.5   Conclusions

In the first part of this chapter, a mechanical model of a gear pair system was presented. This model takes into account typical effects, such as gear mesh variable stiffness and backlash, static transmission error and linear or non-linear stiffness effects from bearings with rolling elements. The model selected is realistic and accurate but introduces complications which make more difficult the application of standard analytical methodologies. As a result, the emphasis was first placed on obtaining periodic motions to typical external loading conditions, by applying an appropriate numerical methodology. The results were presented in the form of classical frequency–response diagrams in an effort to demonstrate the effect of system parameters, including the gear mesh stiffness and damping, the bearing stiffness and the gear transmission error. Among other things, it was shown that for small levels of the external moments, small damping and significant bearing non-linearity, there appear frequency intervals where only branches of unstable periodic motions exist. These branches were found to emanate from points where the stability properties of a periodic motion change through a Hopf bifurcation. By performing direct integration of the equations of motion, it was demonstrated that the unstable periodic response inside these branches gives its place to quasi-periodic and chaotic responses, which may possess much higher amplitudes. The adverse effect of this complex dynamics on a parametric identification methodology was also illustrated. Finally, some studies where similar models were employed were also reviewed and possible extensions to more complex rotor dynamic systems were discussed.

25.6   References

Adams, M.L. (1980), 'Nonlinear dynamics of flexible multi-bearing rotors', *Journal of Sound and Vibration*, **71**, 129–144.

Bajaj, A.K. and Tousi, S. (1990), 'Torus doublings and chaotic amplitude modulations in a two degree-of-freedom resonantly forced mechanical system', *International Journal of Non-Linear Mechanics*, **25**, 625–642.

Brancati, R., Rocca, E. and Russo, R. (2005), 'A gear rattle model accounting for oil squeeze between the meshing gear teeth', *Proc. IMechE Part D: J. Automobile Engineering*, **219**, 1075–1083.

Chen, C.-S., Natsiavas, S. and Nelson, H.D. (1997), 'Stability analysis and complex dynamics of a gear-pair system supported by a squeeze film damper', *ASME Journal of Vibration and Acoustics*, **119**, 85–88.

Chen, C.-S., Natsiavas, S. and Nelson, H.D. (1998), 'Coupled lateral-torsional vibration of a gear-pair system supported by a squeeze film damper', *ASME Journal of Vibration and Acoustics*, **120**, 860–867.

Childs, D., Moes, H. and van Leeuwen, H. (1977), 'Journal bearing impedance descriptions for rotordynamic applications', *ASME Journal of Lubrication Technology*, **99**, 198–210.

Choy, F.K., Tu, Y.K., Zakrajsek, J.J. and Townsend, D.P. (1991), 'Effects of gear box vibration and mass imbalance on the dynamics of multistage gear transmission', *ASME Journal of Vibration and Acoustics*, **113**, 333–334.

Craig, R.R. Jr. (1981), *Structural Dynamics – An Introduction to Computer Methods*, New York, J. Wiley & Sons.

Cunliffe, F., Smith, J.D. and Welbourn, D.B. (1974), 'Dynamic tooth loads in epicyclic gears', *ASME Journal of Engineering for Industry*, 578–584.

Doedel, E. (1986), *AUTO: Software for Continuation and Bifurcation Problems in Ordinary Differential Equations*, California Institute of Technology, Pasadena, California.

Ehrich, F.F. (ed.) (1992), *Handbook of Rotordynamics*, New York, McGraw-Hill.

Giagopoulos, D., Salpistis, C. and Natsiavas, S. (2006), 'Effect of nonlinearities in the identification and fault detection of gear-pair systems,' *International Journal of Non-Linear Mechanics*, **41**, 213–230.

Gregory, R.W., Harris, S.L. and Munro, R.G. (1963), 'Dynamic behavior of spur gears', *Proceedings of the Institution of Mechanical Engineers*, **178**, 1–28.

Gunter, E.J., Barrett, L.E. and Allaire, P.E. (1977), 'Design of nonlinear squeeze-film dampers for aircraft engines', *ASME Journal of Lubrication Technology*, **99**, 57–64.

Harris, T.A. (1966), *Rolling Bearing Analysis*, New York, John Wiley.

Iida, H., Tamura, A., Kikuch, K. and Agata, H. (1980), 'Coupled torsional-flexural vibration of a shaft in a geared system of rotors', *Bulletin of the JSME*, **23**, 2111–2117.

Kahraman, A. (1994), 'Planetary gear train dynamics', *Journal of Mechanical Design*, **116**, 713–720.

Kahraman, A. and Blankenship, G.W. (1997), 'Experiments on nonlinear dynamic behavior of an oscillator with clearance and periodically time-varying parameters', *Journal of Applied Mechanics*, **64**, 217–226.

Kahraman, A. and Singh, R. (1991), 'Interactions between time-varying mesh stiffness and clearance nonlinearities in a geared system', *Journal of Sound and Vibration*, **146**, 135–156.

Khonsari, M.M. and Chang, Y.J. (1993), 'Stability boundary of nonlinear orbits within clearance circle of journal bearings', *ASME Journal of Vibration and Acoustics*, **115**, 303–307.

Kishor, B. and Gupta, S.K. (1989), 'On the dynamic analysis of a rigid rotor-gear pair-hydrodynamic bearing system', *ASME Journal of Vibration, Acoustics and Reliability in Design*, **111**, 234–240.

Lin, J. and Parker, R.G. (2002), 'Planetary gear parametric instability caused by mesh stiffness variation', *Journal of Sound and Vibration*, **249**, 129–145.

Lund, J.W. (1978), 'Critical speeds, stability and response of a geared train of rotors', *Journal of Mechanical Design*, **100**, 535–539.

Metallidis, P. and Natsiavas, S. (2000), 'Vibration of a continuous system with clearance and motion constraints', *International Journal of Non-Linear Mechanics*, **35**, 675–690.

Muszynska, A. (1986), 'Whirl and whip – rotor/bearing stability problems', *Journal of Sound and Vibration*, **110**, 443–462.

Natsiavas, S., Theodossiades, S. and Goudas, I. (2000), 'Dynamic analysis of piecewise linear oscillators with time periodic coefficients', *International Journal of Non-Linear Mechanics*, **35**, 53–68.

Nayfeh, A.H. and Balachandran, B. (1995), *Applied Nonlinear Dynamics*, New York, John Wiley & Sons.

Nelson, H.D., Meacham, W.L., Fleming, D.P. and Kascak, A.F. (1983), 'Nonlinear analysis of rotor-bearing systems using component mode synthesis', *ASME Journal of Power Engineering*, **105**, 606–614.

Neriya, S.V., Bhat, R.B. and Sankar, T.S. (1985), 'Coupled torsional-flexural vibration of a geared shaft system using finite element analysis', *The Shock and Vibration Bulletin*, 55, 13–25.

Ozguven, H.N. and Houser, D.R. (1988), 'Mathematical models used in gear dynamics – a review', *Journal of Sound and Vibration*, **121**, 383–411.

Padmanabhan, C. and Singh, R. (1996), 'Analysis of periodically forced nonlinear Hill's oscillator with application to a geared system', *Journal of the Acoustical Society of America*, **99**, 324–334.

Parker, R.G. (2000), 'A physical explanation for the effectiveness of planet phasing to suppress planetary gear vibration', *Journal of Sound and Vibration*, **236**, 561–573.

Parker, R.G., Agashe, V. and Vijayakar, S.M. (2000), 'Dynamic response of a planetary gear system using a finite element/contact mechanics model', *Journal of Mechanical Design*, **122**, 304–310.

Rao, S.S. (1996), *Engineering Optimization: Theory and Practice*, 3rd Edition, New York, John Wiley & Sons.

Simmons, H.R. and Smalley, A.J. (1984), 'Lateral gear shaft dynamics control torsional stresses in turbine-driven compressor train', *ASME Journal of Engineering for Gas Turbine and Power*, **106**, 946–951.

Schwibinger, P. and Nordmann, R. (1988), 'The influence of torsional-lateral coupling on the stability behavior of geared rotor systems', *ASME Journal of Engineering for Gas Turbine and Power*, **110**, 563–571.

Sundararajan, P. and Noah, S.T. (1998), 'An algorithm for response and stability of large order nonlinear systems – application to rotor systems', *Journal of Sound and Vibration*, **214**, 695–723.

Tangasawi, O., Theodossiades, S. and Rahnejat, H. (2007), 'Lightly loaded lubricated impacts: idle gear rattle', *Journal of Sound and Vibration*, **308**, 418–430.

Theodossiades, S. and Natsiavas, S. (2000), 'Nonlinear dynamics of gear-pair systems with periodic stiffness and backlash', *Journal of Sound and Vibration*, **229**, 287–310.

Theodossiades, S. and Natsiavas, S. (2001a), 'Periodic and chaotic dynamics of motor-driven gear-pair systems with backlash', *Chaos, Solitons and Fractals*, **12**, 2427–2440.

Theodossiades, S. and Natsiavas, S. (2001b), 'On geared rotordynamic systems with oil journal bearings', *Journal of Sound and Vibration*, **243**, 721–745.

Theodossiades, S., Tangasawi, O. and Rahnejat, H. (2007), 'Gear teeth impacts in hydrodynamic conjunctions promoting idle gear rattle', *Journal of Sound and Vibration*, **303**, 632–658.

Van de Vorst, E.L.B., Fey, R.H.B., De Kraker, A. and Van Campen, D.H. (1996), 'Steady-state behavior of flexible rotordynamic systems with oil journal bearings', *Nonlinear Dynamics*, **11**, 295–313.

Yamada, T. and Mitsui, J. (1979), 'A study on the unstable vibration phenomena of a reduction gear system, including the lightly loaded journal bearings, for a marine steam turbine', *Bulletin of the JSME*, **22**, 98–106.

Yuksel, C. and Kahraman, A. (2004), 'Dynamic tooth loads of planetary gears sets having tooth profile wear', *Mechanisms and Machine Theory*, **39**, 695–715.

25.7 Nomenclature

m_n	gear masses
I_n	gear mass moments of inertia
R_n	base radii
M_1	driving moment
M_2	resisting moment
F_{b1}	pretension radial force in driving gear
F_{b2}	pretension radial force in driven gear
e	static transmission error
u	dynamic transmission error

c_g	gear mesh damping
c_n	bearing mesh damping
b	Backlash
k_g	gear mesh stiffness
f_g	gear mesh restoring force
f_n	rolling element bearing restoring force
φ	rotation angle
ω_n	mean angular velocity
θ_n	small angle variations due to vibration of mating gear teeth
n_n	gear teeth number
ω_M	fundamental forcing frequency
T_M	fundamental forcing period
t	time

26

Rattle and clatter noise in powertrains – automotive transmissions

S. N. DOĞAN, Daimler AG, USA

Abstract: The chapter discusses rattling and clattering noises in powertrains especially in automotive transmissions, which contribute a considerable proportion of overall vehicle noise. The chapter includes experimental and theoretical studies of production transmissions indicating the cause, type and effect of loose part vibration as well as measures to reduce these noises. In addition modelling is used to simulate the movement patterns of the individual loose parts within the transmission over time and a method to calculate the drag torque of automotive transmissions.

Key words: rattle, clatter, NVH, transmission, powertrain, noise reduction, drag torque.

26.1   Introduction

The development of low noise powertrain requires noise, vibration and harshness (**NVH**) optimisation during the early design phase. This shows that minimising noise is an increasingly important factor in vehicle development. Driveline NVH has always been a critical issue in overall vehicle NVH design. The use of computer-aided engineering (CAE) development tools and methods is a prerequisite for cost and time-effective NVH development, as well as reducing the effort for troubleshooting during the vehicle development process and increasing product quality. For years, numerous efforts have been made in NVH reduction. Rising customer expectations and increasingly stringent legislative curbs on noise emission are placing an ever-greater emphasis on noise consideration. There are numerous sources of noise in motor vehicles, which combine to create a complex acoustic scenario. Apart from the internal combustion engine, the transmission is one of the dominant sources of noise in the driveline. The problem of transmission noise in development is becoming increasingly acute with the growing use of lightweight construction of internal combustion engines and other driveline components, greater concentration on energy saving, increased customer expectations and more rigorous legal restrictions on exhaust and noise emissions. In order to achieve high refinement, NVH development engineers must deliver recommendations prior to the design freeze dates. In the NVH development processes simulation models and techniques are continuously used to lower the powertrain noises.

793

26.2   Significant noises in automotive transmissions

Progressively the overall noise and vibration levels are directly linked to vehicle quality. NVH development engineers are expending considerable effort to eliminate or reduce noise sources in automotive transmissions. These noises in automotive transmissions, in particular, convey an impression of poor quality to the customer (see Fig. 26.1). Addressing these customer complaints in the field can be a significant warranty cost issue. NVH problems have taken on new significance with improvement in overall noise control inside the vehicle.

In recent powertrains, wind and tyre noises have been reduced significantly. Noises in automotive transmission that traditionally were masked by these operating noises have become easily audible for the customers. Automotive transmission noises can be broken down into several groups according to their causes (Lechner and Naunheimer, 1999). The most dominant type of transmission noise is ***rolling contact noise*** from gear pairs under load

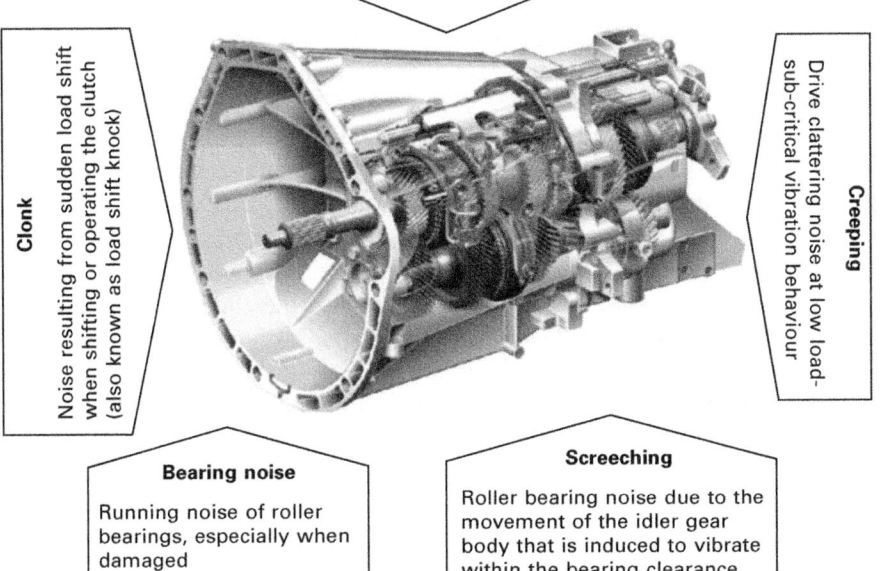

Engagement noise

Scraping and grating of the selector teeth when synchroniser is not functioning correctly

Whining/squealing

Gearwheels vibrating under load
– Meshing impact
– Parametrically excited vibration
– Rolling contact noise

Rattling/clattering

Loose part vibration
– Idler gears
– Synchroniser rings
– Sliding sleeves

Clonk

Noise resulting from sudden load shift when shifting or operating the clutch (also known as load shift knock)

Creeping

Drive clattering noise at low load-sub-critical vibration behaviour

Bearing noise

Running noise of roller bearings, especially when damaged

Screeching

Roller bearing noise due to the movement of the idler gear body that is induced to vibrate within the bearing clearance

26.1 Categorisation of vehicle transmission noises.

known as *whining* and *squealing*, but also as grinding and singing. This type of noise can be caused by meshing impacts, parametrically excited vibration or rolling contact noise due to variations in pitch spacing. Rattling and clattering noises are caused by idle transmission components subject to torsional vibration. This noise is known as *rattle* when the transmission is in neutral, and as *clatter* when a gear is engaged under power or in overrun (pull/push operation). Rattling and clattering noises (Kücükay, 1987; Weidner, 1991; Lang, 1996; Doğan 2001; Ryborz, 2003) are perceived as unpleasant not because of their high airborne sound pressure level, but because of their intrusive characteristic. It, therefore, constitutes a comfort problem (*noise emission*) in the case of passenger cars, and a problem of both comfort and environmental pollution (noise emission) in the case of commercial vehicles. Clattering that arises under power with low load, for example at low road speeds and engine speeds, is called *creeping* (Doğan *et al.*, 1998). Gear shifting noise can also arise in automotive transmissions from scraping and grating of the selector teeth as a result of defective *synchroniser* functioning, and noise can arise from transient load cycle excitation, referred to as *shift clonk* or *load shift knock* (Doğan, 2001). *Bearing noise* can also occur, especially in case of damaged roller bearings, and *screeching* **noise** caused by vibration of the idler body within the bearing clearance (Weidner, 1991).

26.2.1   Causes of rattling and clattering noises

Rattling and clattering noises in automotive transmissions are caused by torsional vibration transmitted from the internal combustion engine to the transmission input shaft (see Fig. 26.2). This is due to discontinuous combustion processes, resulting in periodically fluctuating drive torque at the crankshaft. Unbalanced engine masses also have a sustainable effect on the rotary movement of the engine crankshaft. This rotary oscillation is superimposed on the rotary movement of the crankshaft, resulting in the irregular rotational speed of the internal combustion engine, which is in principle sinusoidal. The relevant portion of the torsional vibration in a four-stroke internal combustion engine is the bisector of the number of cylinders.

The cyclic irregularity of combustion engines is increasing because of efforts to improve fuel consumption and emissions levels, such as the increasing use of turbo-charging, multi-valve technology and direct fuel injection, combined with reducing idling speeds. The engine speed pattern is affected by the operation of additional auxiliaries such as air conditioning, headlights and rear window demister, as well as ignition defects and clutch slip. They increase the irregularity of rotational speed, resulting in larger angular acceleration amplitudes and changes in their characteristics. The

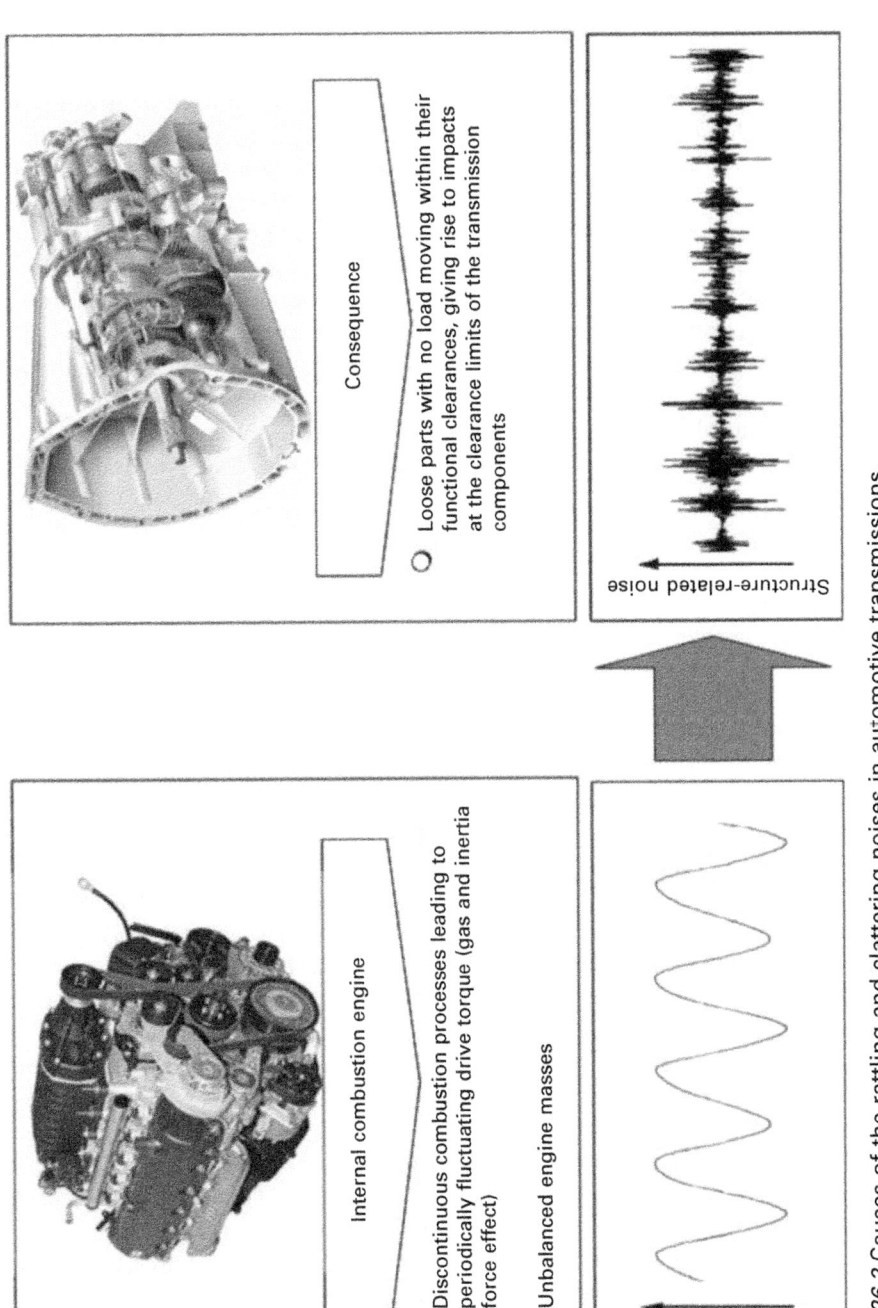

Consequence

○ Loose parts with no load moving within their functional clearances, giving rise to impacts at the clearance limits of the transmission components

Structure-related noise

Internal combustion engine

○ Discontinuous combustion processes leading to periodically fluctuating drive torque (gas and inertia force effect)

○ Unbalanced engine masses

Rotational speed

26.2 Causes of the rattling and clattering noises in automotive transmissions.

torsional vibration transmitted from the internal combustion engine to the transmission excites idle components such as idler gears, synchroniser rings and sliding sleeves to vibrate within their functional clearances. The fixed components encounter idle components at the clearance limits, resulting in impacts perceived as rattling or clattering noises. The intensity and frequency of the impacts are directly related to the *airborne sound* pressure emitted from the gearbox housing (Weidner, 1991; Lang, 1996).

Rattling and clattering noises arise chiefly in manual transmissions and semi-automatic transmissions. Automatic transmissions can also be prone to clattering when the torque converter is locked up. Rattling and clattering noises are perceived as unpleasant, not because of their high airborne sound pressure level, but because of their intrusive characteristic. The rattling and clattering noise of automotive transmissions is in case of passenger cars a comfort problem (noise emission) and in case of commercial vehicles a problem of both comfort and environmental pollution (noise emission).

26.2.2   Movement of loose parts in transmissions

The sinusoidal rotational speed profile of the transmission input shaft (for example that of a four-cylinder four-stroke internal combustion engine) is characterised by acceleration and deceleration phases. Figures 26.3 and 26.4 illustrate the example of an idler gear whose circumferential and axial movement behaviour relates to the rotational irregularity and its impact performance. The sinusoidal rotational speed profile of the transmission input shaft (for example that of a four-cylinder four-stroke internal combustion engine) is characterised by acceleration and deceleration phases (Fig. 26.4(a)). When the rotational speed profile is decelerating (decreasing slope of the curve), the idler gear grips the driven flank of the fixed gear (Fig. 26.4(b)). As soon as an acceleration phase is entered (rising rotational speed profile), the idler gear comes away from the driven flank of the fixed gear with following flying phase and then impacting against the driving flank of the fixed gear, which is perceived as structure-related noise, (Fig. 26.4(d)). After the torsional flank impact, the helical cut idler gear impacts against the drive side thrust collar (Fig. 26.4(c)), which can be clearly recognised in the structure-related noise graph (Fig. 26.2(d)). Each torsional flank impact is followed by an axial impact whose intensity is less than that of the torsional flank impact. Transmission fluid as an engineering design parameter also has a considerable influence on rattling and clattering noises (see Chapter 29). The important factors include the type of oil, the additives used and the viscosity (which is directly related to temperature and pressure (see Chapter 5), and the level of oil in the transmission which together act on a gear pair as drag torque, resulting in a reduction in rattling and clattering noises, especially at low speeds and when cold (Doğan *et al.*, 2003).

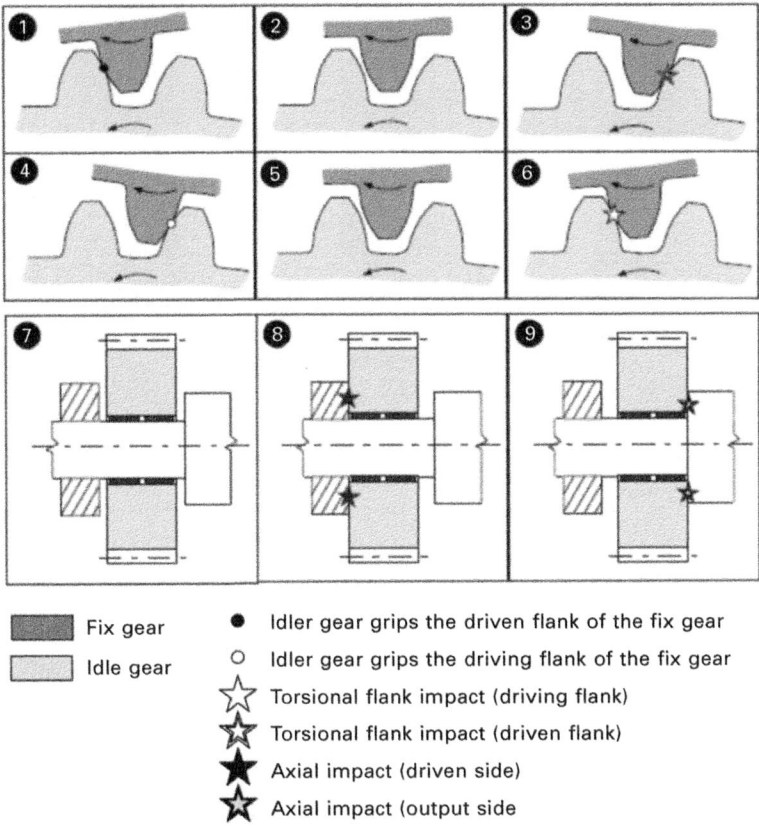

Fix gear

Idle gear

● Idler gear grips the driven flank of the fix gear

○ Idler gear grips the driving flank of the fix gear

☆ Torsional flank impact (driving flank)

★ Torsional flank impact (driven flank)

★ Axial impact (driven side)

☆ Axial impact (output side)

1–9 indicate the running number of possible applications

26.3 Principal movement behaviour of loose parts in transmissions.

26.2.3   Responsible parameters for loose part noise

The parameters responsible for rattle and clatter can be divided into operating parameters and geometrical parameters. The operating parameters include the excitation frequency as the product of rotational speed and engine order, which is responsible for the number of impulses per rotation of the shaft, and the angular acceleration amplitude which is the main factor determining contact between the fixed gear and idler gear flanks. The diameter of the fixed gear and the geometrical parameters of the idler gear such as diameter, moment of inertia, mass, helix angle and the associated reduction ratio are the main parameters that can be influenced at the gear development stage in terms of their rattling and clattering noise behaviour. However, there is some goal conflict between designing the gearwheel stages in respect of power transmission, and their service life and low rattling proneness. The

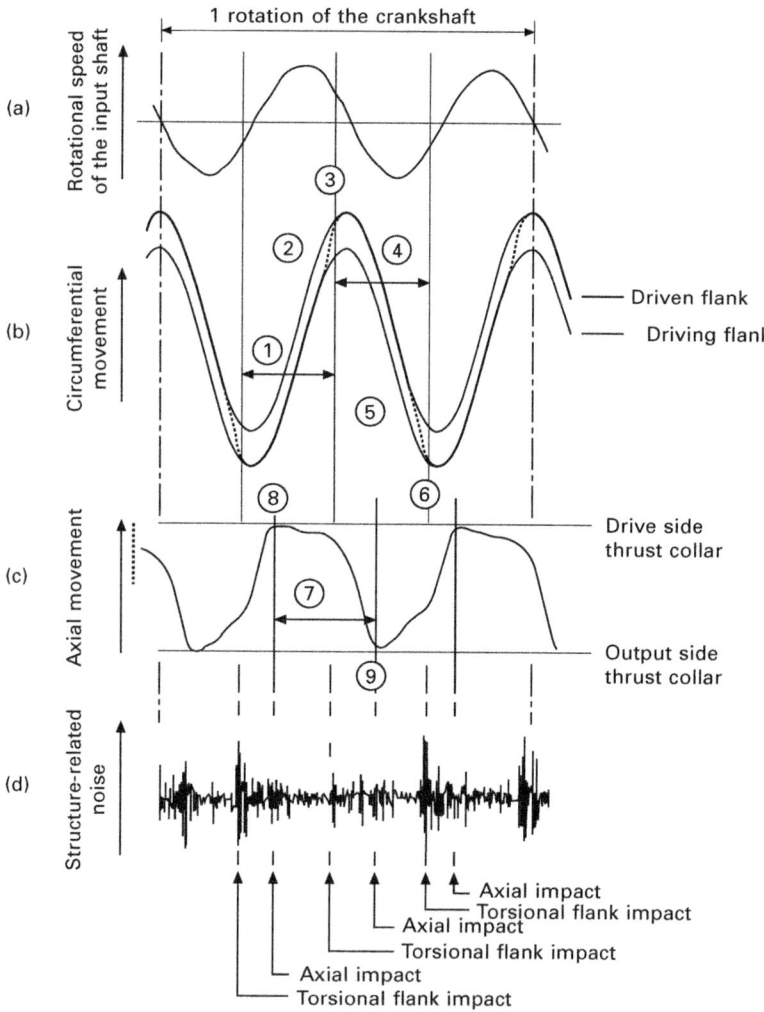

1–9 indicate the running number of possible applications.

26.4 Movement of an idler gear as a result of an irregular rotational speed profile in the case of a fixed/idler gear pair (Doğan, 2001; Ryborz, 2003).

torsional backlash and the axial clearance offer scope for reducing this noise. Reducing the torsional backlash and a defined increase or reduction in the axial clearance result in a reduction of the airborne sound pressure level emitted (Doğan, 2001; Weidner, 1991; Lang, 1996; Ryborz, 2003) (Fig. 26.5).

Transmission fluid as an engineering design parameter also has a considerable influence on rattling and clattering noises as noted above.

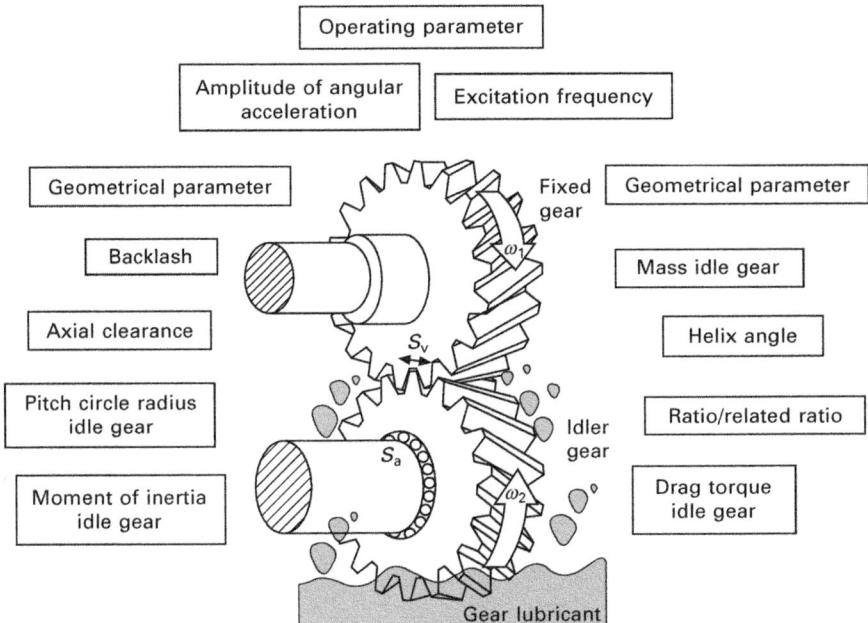

26.5 Parameter influence on rattling and clattering noise.

26.2.4   Assessment of rattling and clattering noises

Assessment of gear rattle behaviour is defined in terms of the airborne noise level profile as a function of angular acceleration, described by the rattle curve in Fig. 26.6. This rattle curve can be divided into three characteristic phases: background noise, rattle limit and profile level.

Background noise occurs up to the *rattle limit*, and is made up of bearing, churning and meshing noise. There is not yet any knocking in the *structure-borne noise* profile 1 shown. The rattle limit 2 marks the point on the rattle curve at which the angular acceleration amplitude has increased to the point where the loose parts start to separate from the driving fixed gears. The rattling noise level starts to rise as angular acceleration amplitudes increase. This is accompanied by the first rattle knocking appearing in the structure-borne noise signal. The profile level (points 3 and 4) shows the noise behaviour at angular acceleration amplitudes above the rattle limit. Both torsional flank knocking and axial knocking are clearly evident.

26.3   Investigation strategy for rattling and clattering noises in automotive transmissions

Throughout the process of developing automotive transmissions with low levels of rattling and clattering noises, suitable equipment is needed to study

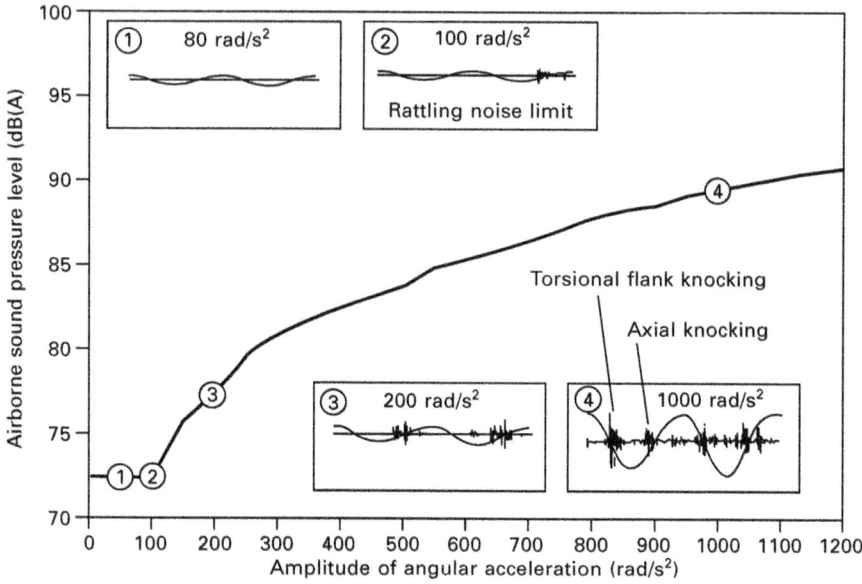

26.6 Rattling noise level curve.

noise behaviour right from the initial stages. This requires suitable models to be developed from the experimental investigation and analysis of rattling and clattering, to determine the noise behaviour of automotive transmissions by simulation. The strategy includes investigating automotive transmissions in test stand trials and by mathematical simulation (Fig. 26.7). It is possible to develop guidelines and recommendations for reducing rattling and clattering noises in automotive transmissions with the aid of fundamental investigations and parameter variation.

In the test stand trials, complete transmissions or their components are investigated in terms of rattling and clattering noise performance. The parameters relevant to rattling and clattering noises are then varied within reasonable limits. The mathematical methods involved can be divided into a simulation and an approximation method. In the case of the simulation method, the pattern of movement of the loose parts and the impact intensities can be determined. The approximation method enables the level of noise to be computed. The simulation models are verified on the basis of the test stand trials.

26.3.1   Experimental investigations

The test bench developed at the Institute of Machine Components at the University of Stuttgart makes it possible to investigate rattling and clattering noises in front-mounted/transverse and standard transmissions, in neutral as

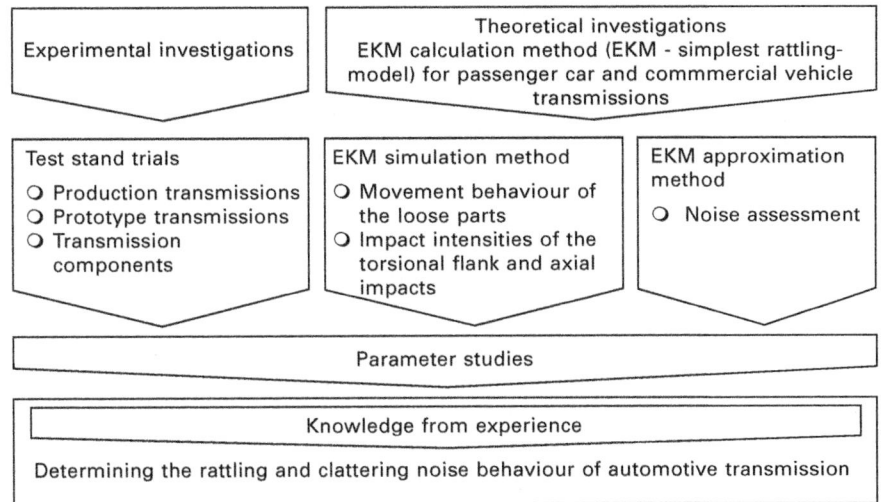

26.7 Investigation strategy for developing low-rattling and low-clattering noise automotive transmissions.

well as under power and in overrun, under realistic operating conditions, excluding extraneous influences. A highly dynamic, brushless, permanently excited three-phase synchronous motor provides the drive unit, capable of simulating the rotational speed profiles of internal combustion engines with different numbers of cylinders. The same motor is available as a braking motor for clattering testing under power and in overrun. Both servomotors are rated at 12 kW with a nominal torque of 30 N m (Figure 26.8). Angular acceleration amplitudes up to 4000 rad/s^2 can be achieved at the input shaft in neutral. To simulate various internal combustion engines in different operating conditions, a PC-controlled function generator is used to specify the corresponding idealised or realistic set values (Doğan, 2001; Ryborz, 2003).

26.3.2   Theoretical investigations

Many years of research into in-gear *rattle* and *clatter* in automotive transmissions at the Institute of Machine Components has resulted in methods of calculation to the point where rattle and clatter can be calculated for complete passenger car and commercial vehicle transmissions (Doğan, 2001; Weidner, 1991; Lang, 1996; Ryborz, 2003). There are two methods available for this, numerical simulation and approximation. The simulation method makes it possible to determine how the loose parts move, and their impact intensity, while the approximation method makes it possible to estimate the noise level of the transmission.

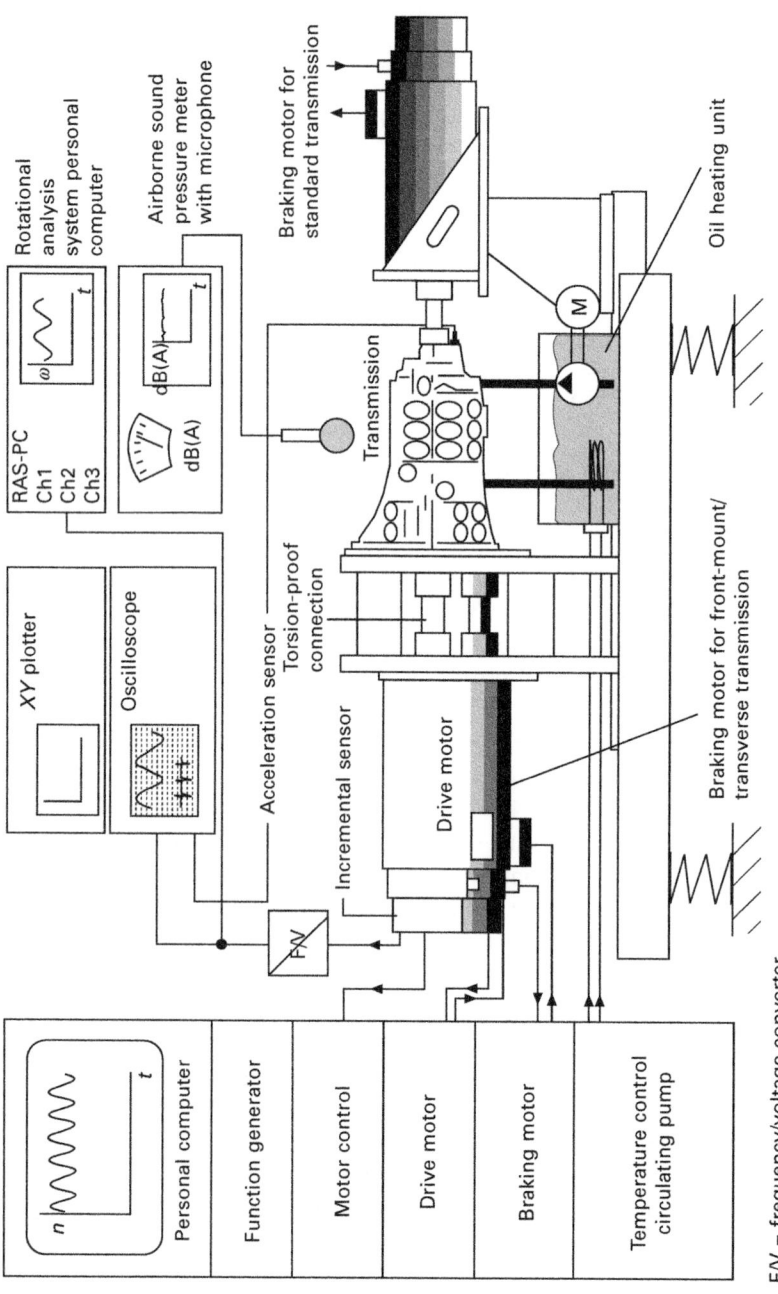

RAS-PC
Ch1
Ch2
Ch3

Rotational analysis system personal computer

Airborne sound pressure meter with microphone

Braking motor for standard transmission

dB(A)

dB(A)

XY plotter

Oscilloscope

Acceleration sensor

Torsion-proof connection

Transmission

Oil heating unit

Incremental sensor

Drive motor

Braking motor for front-mount/ transverse transmission

M

F/V

n

t

Personal computer

Function generator

Motor control

Drive motor

Braking motor

Temperature control circulating pump

F/V = frequency/voltage converter
M = Motor (oil pump)

26.8 Schematic diagram of the rattle and clatter test stand.

The EKM calculation methods (EKM = 'Einfachst-Klapper-Modell' – simplest rattling model) are based on simple models for which the linked rotary and translatory movement relations of rotating loose transmission components can be simulated. For an ***idler gear*** or loose part mounted on the shaft, three more unlinked degrees of freedom (torsional backlash, axial clearance and radial backlash) can be determined. The small effect of the radial backlash, causing wobbling of a loose part in combination with the other degrees of freedom, is ignored. In the case of torsional backlash, the small rotational oscillation amplitudes are described by an equivalent translatory substitution model (Fig. 26.9(a)). Here the external frame corresponds to the driving fixed gear with mass m_1, exciting the loose part (m_2) to vibrate within the torsional backlash s_y. The loose part movements are then opposed by the sum of the individual ***drag torque*** components as an external force R_a. In the case of helical cut gear wheels, impacts are induced in the hub area, because of the axial load and the necessary axial clearance s_a, described by translatory movement (Fig. 26.9(b)). Here the loose part (m_2) impacts against the drive side and output side thrust collar (m_3) as a result of the helical gearing within the axial clearance s_a.

The simplest rattling model (EKM) is derived from the two substitution models to describe the linked rotary and axial vibration of a loose part. The loose part (which can be a no-load idler gear or a cluster gear, e.g. a countershaft or a ***synchroniser ring***) is then modelled as a rigid body, its elastic deformation being negligible.

The EKM simulation method

Using these model parameters, equations of motion can be created for numerical simulation, describing loose part behaviour with defined excitation.

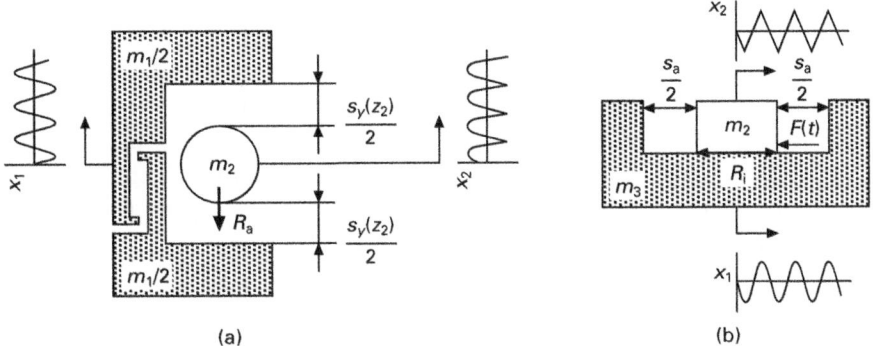

(a)   (b)

26.9 Substitution models for (a) torsional backlash and (b) axial clearance.

The impacts at the clearance limits extend through shafts and bearings both directly as **airborne noise** and also as structure-related noise, thus inducing vibration in the gearbox housing, leading to noise emission. The calculation ignores the complex transmission behaviour, but as a first approximation, the pulse transmitted when impacts arise is proportional to the noise level caused by it. The theorem of momentum and angular momentum is then used for the loose part because of the unknown time profile of the impact forces. This results in a connection between the force and movement profile of the loose part, from which the average impact intensity I_m can be determined, Equation 26.1 (Lang, 1996):

$$I_m = m_2 \cdot \hat{\omega}_1 \cdot r_{b1} \cdot C_{Im} \tag{26.1}$$

The average impact intensity is represented by the average total of all individual shock pulses in a simulation period, where m_2 is the mass of the excited loose part, $\hat{\omega}_1$ is the angular acceleration amplitude, r_{b1} is the pitch circle radius of the exciting part and C_{Im} is the related average **impact intensity** for tooth flank and axial impacts, Equation 26.2 (Lang, 1996):

$$C_{Im} = \frac{1}{\tau_{end} - \tau_{anf}} \sum_{\tau=\tau_{anf}}^{\tau=\tau_{end}} \sqrt{i_{iz}^2 + i_{ix}^2} \cdot i_{ia} \tag{26.2}$$

The related impact intensities i_{iz}, i_{ix} and i_{ia} of the respective co-ordinate directions are determined from the helix angle β, the idler gear mass m_2, the pitch circle radius of the fixed gear r_{b1}, the angular acceleration $\hat{\omega}_1$, the excitation frequency ω_{An} and the two clearances, torsional backlash s_v and axial clearance s_a. The average impact intensity I_m is not an absolute measure of noise and can serve only as a comparative figure between different vibration states and loose parts within a transmission. It is possible to calculate a level of rattling and clattering noises L_p only by correlating between the computed noise value and the measured **noise level**:

$$L_p = 10 \cdot \log (k \cdot I_m + 10^{0.1 \cdot L_{Basic}}) \tag{26.3}$$

The basic noise level L_{Basic} can either be estimated, or is known from measurement. The calibration factor k creates the connection between the real average impact intensity I_m and the acoustic pressure, which is derived from comparing a measured airborne sound pressure level with the average impact intensity from a simulation.

The EKM approximation method

The calculation methods can be used to investigate the rattling and clattering noise behaviour of individual gear steps and also of complete passenger car and commercial vehicle transmissions. An empirical approximation formula seeks to achieve a sufficiently accurate calculation of the **impact intensity**, and thus of the airborne **sound pressure level**, independently of a numerical simulation. In the approximation method, identical standard parameters are used as in case of the numerical simulation method. The related average impact intensity C_{Im} can be determined in the range of small relative axial clearances, using Equation 26.4 (Lang, 1996):

$$C_{Im} = \sqrt{C_{sv}} \cdot \left(1462 - \frac{0.714 \cdot C_{fa} \cdot C_{sa}}{-0.016 \cdot C_{fa} + 0.12 \cdot C_{sv}} \right) \tag{26.4}$$

The related parameters of the clearance in the circumferential direction C_{sv}, of the clearance in the axial direction C_{sa} and the friction force C_{fa} are determined from the excitation frequency ω_{An}, the pitch circle radius of the fixed gear r_{b1}, the angular acceleration $\hat{\omega}_1$, the helix angle β, the idler gear mass m_2, the friction force F_R and the two clearances, torsional backlash s_v and axial clearance s_a.

This makes it possible to expand the average impact intensity I_m for the fixed/loose component pairs under consideration with the related ratio, using Equation 26.1 (Doğan, 2001; Ryborz, 2003):

$$I_m = m_2 \cdot \hat{\omega}_1 \cdot i_b \cdot r_{b1} \cdot C_{Im} \tag{26.5}$$

This makes it possible to determine the rattle and clatter noise level of a complete automotive transmission by logarithmic addition of the individual loose component noise elements using Equation 26.3 as:

$$L_{P Comp.} = 10 \cdot \log \sum_{i=1}^{n} (10^{0.1 \cdot L_{p,i}}) \tag{26.6}$$

Whereas the numerical simulation method can be used without restriction, the approximation method has to be used for validity ranges of the parameters in Equation 26.4. The relative circumferential **drag torque** C_{mv} must be less than 0.3, the relative axial friction force C_{fa} must be less than 0.7 and the axial clearance s_a must be greater than or equal to 0.1 mm. The rattling curve is moreover valid from above an angular acceleration amplitude $\hat{\omega}_1$ of approximately 250 rad/s^2 at the required sinusoidal excitation.

26.4   Automotive transmission lubricants

Transmission lubricants can be divided into **manual transmission fluid** (MTF), **automatic transmission fluid** (ATF) and **synthetic axle fluid** (SAF).

MTF is usually used in manual transmissions, ATF in automatic transmissions because of its special characteristics and SAF in rear axle differential gears. ATF is increasingly used in manual transmissions because of its good friction characteristics and to improve ease of shifting.

The lubricant consists principally of an oil base with additives constituting 2–10% by volume, determining the fluid's characteristics. The functions of the lubricants include building up hydrodynamic lubricant wedges between the contacting surfaces (see Chapter 5) in relative motion. Depending on the state, this creates hydrodynamic areas, mixed friction areas and dry friction areas (see Chapters 3, 5 and 6). The lubricants also serve to dissipate heat, provide adequate lubrication of all transmission components, inhibit corrosion, improve shifting action at low temperatures, enhance dirt removal, reduce losses when idling and when under load, and reduce noise and ageing.

Alternative lubricants, based on organic compounds and having a low coefficient of friction in the hydrodynamic range, have not been considered for use as gear lubricants. They are used in roller bearings and slide bearings for example (see Chapters 5 and 6). In certain pressure and temperature ranges they can take on liquid crystal phases that have directional physical characteristics. The phases can be moved like liquids. The viscosity changes abruptly at the phase transition points. This phase transition characteristic is also known as the ***mesophase***. At the transition from the crystalline state to the liquid crystalline mesophase, the position long-range order (centres of mass of the molecules) is lost, whereas the orientation long-range order (periodic orientation of the molecule axes) is preserved. This causes the crystal-like and anisotropic characteristics of the mesophases, and the neutralised position long-range order causes the liquid-like fluidity (Höhn *et al.*, 1997).

26.4.1   Principles of drag torque calculation

Tribological system

The gear lubricant (oil) in automotive transmissions is an intermediary, forming a tribological system together with the adjacent surfaces in contact (Lechner and Naunheimer, 1999). The function of the lubricant is to ensure separation of the contiguous surfaces for all stress conditions. If this tribological system is transferred to a gear pair, it comprises the gear flanks of the fixed gear and the idler gear and the lubricant between them.

Drag torque components acting on the idler gear

Four drag torque components act on the idler gear as a consequence of the lubricant, and their sum is crucial for loose part vibration. In this section, the theoretical models known from the literature for determining the drag torque components are expanded and compared with test measurements (Doğan,

2001; Weidner, 1991). The drag torque of an automotive transmission is made up of various individual drag torques. The drag torque T_2 acting on an idler gear is divided into the proportions shown in Fig. 26.10 into **compression torque T_{Qu}, synchronisation drag torque T_{Sy}, bearing friction torque T_L and churning torque T_{Pl}**, equation (26.7):

$$T_2 = T_{Qu} + T_{Sy} + T_L + T_{Pl} \qquad (26.7)$$

Theoretical models are known (Doğan, 2001) enabling the individual drag torque components at the idler gear to be determined with sufficient accuracy, which can be used as a basis for calculating the drag torque.

Compression torque

Two areas of flow conditions are distinguished for determining the **compression torque T_{Qu}** between the fixed gear and the idler gear. Relationships are listed in Doğan *et al.* (2003) for laminar and turbulent flow conditions, delimited by a validity range.

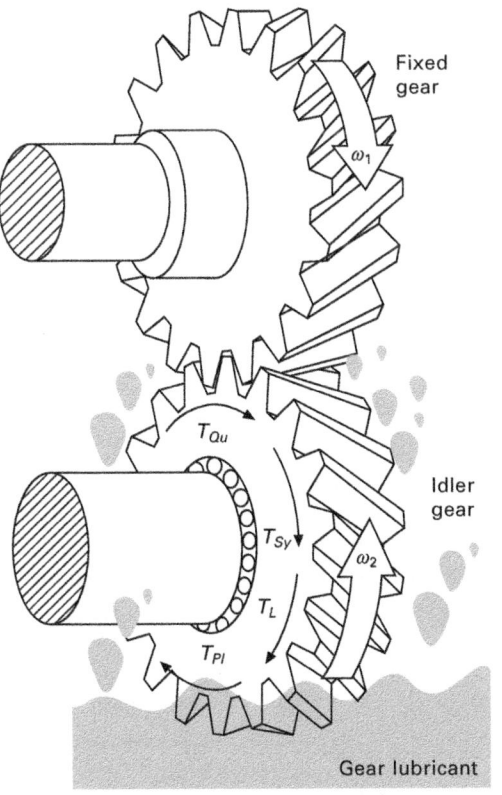

Fixed gear

Drag torque components:

T_{Qu} Compression torque

T_{Sy} Synchronisation drag torque

T_L Bearing friction torque

T_{Pl} Churning torque

Idler gear

Gear lubricant

26.10 Drag torque components at the idler gear.

In the case of laminar flow: $Re_{Qu}^{0.32} \cdot Fr_{Qu}^{-0.23} < 12$

$$T_{Qu\_laminar} = 616.6 \cdot Re_{Qu}^{-0.65} \cdot Fr_{Qu}^{-0.46} \cdot \left(\frac{h}{h_0}\right)^{-1.66} \cdot \left(\frac{b}{h_0}\right)^{-0.46} \cdot \rho \cdot \omega^2 \cdot r_a^3 \cdot b \cdot h$$

(26.8)

In the case of turbulent flow: $Re_{Qu}^{0.32} \cdot Fr_{Qu}^{-0.23} \geq 12$,

$$T_{Qu\_turbulent} = 5623 \cdot Re_{Qu}^{-0.88} \cdot Fr_{Qu}^{-0.78} \cdot \left(\frac{h}{h_0}\right)^{-1.6} \cdot \left(\frac{b}{h_0}\right)^{-0.36} \cdot \rho \cdot \omega^2 \cdot r_a^3 \cdot b \cdot h$$

(26.9)

The **Reynolds number** Re and the **Froude number** Fr are described as:

$$Re_{Qu} = \frac{r_a \cdot h \cdot \omega}{v}, Fr_{Qu} = \frac{r_a^2 \cdot \omega^2}{g \cdot h}$$

(26.10)

With the validity ranges for $h/h_0 = 0.45\text{--}1.8$ and $b/h_0 = 1\text{--}6$, the compression torque can be determined in an approximate manner. Equations (26.8) and (26.9) do not then take into account the location of the gearwheels in the gearbox. One study (Kücükay, 1987) proposes an equation based on trials that take into account the influencing variables as:

$$T_{Qu} = 3.88 \times 10^{-10} \cdot c_{Sp} \cdot r \cdot b^{1.6} \cdot v_t^2 \cdot v^{-0.15} \cdot \rho$$

(26.11)

The oil spray factor c_{Sp} is dimensionless and is affected mainly by the location, immersion depth e, throw-off angle φ and the hydraulic length l_H. These are determined by the size of the gearbox housing.

Synchronisation drag torque

The **synchronisation drag torque** T_{Sy} is based on a simple approach proposed in Doğan (2001) and Weidner (1991), whereby the drag torque is created by the difference in rotational speed between the synchroniser ring and the idler gear or between the synchroniser ring and the sliding sleeve. The assumption is based on the shear stress for a Newtonian fluid, which causes frictional loss due to the shear flow of the gear oil between the friction surfaces:

$$T_{Sy} = \frac{1}{4} \cdot \omega_{Sy,w/2} \cdot v \cdot \rho \cdot \frac{b_{Sy}}{h_{Sy}} \cdot \pi \cdot d_{Sy}^3$$

(26.12)

The geometry of the synchroniser is described by the gap h_{Sy}, the gap width b_{Sy} and the mean friction surface diameter d_{Sy}.

Bearing friction torque

Simplified numerical value equations are given in Doğan (2001) for anti-friction bearing idling losses. The **friction torque** T_L is determined by kinematic viscosity v, the relative movement between idler gear and shaft $n_{2,w}$ and the mean bearing diameter d_m as (see also an alternative suggested in Chapter 29):

$$T_L = f_0 \times 10^{-7} \cdot (v \cdot n_{2,w})^{\frac{2}{3}} \cdot d_m^3 \tag{26.13}$$

The design of the bearing and type of lubrication is reflected in the bearing factor f_0 and is given in the corresponding tables (Doğan, 2001).

Churning torque

There are theoretical models for determining the **churning torque**, developed from various series of trials in Walter (1982), confirmed by measurements on various gear units in Doğan (2001) and Walter (1982). A distinction is made in determining the churning torque T_{Pl} between vertical and horizontal axle orientation, below:

$$T_{Pl} = c_{Pl} \cdot \rho \cdot \omega^2 \cdot r_a^4 \cdot b \tag{26.14}$$

(vertical axle orientation),

$$T_{Pl} = c_W \cdot c_V \cdot c_{Pl} \cdot \rho \cdot \omega^2 \cdot r_a^4 \cdot b \tag{26.15}$$

(horizontal axle orientation).

In the case of vertical axle orientation, it is assumed that only the idler gear is immersed in the gearbox oil sump, whereas in the case of horizontal axle orientation, both the fixed gear and the idler gear run in the oil sump. When two gearwheels are immersed in the oil sump, i.e. in the case of horizontal axle orientation, the individual torque is changed by the fixed gear churning at the same time. Equation (26.15) above applies for various throw-off angles and directions of rotation. The **churning torque factor** c_{Pl} given in equations (26.14) and (26.15) is divided into three areas depending on the flow conditions of the gear oils. For laminar flow conditions greater (equation (26.16)) or less (equation (26.17)) than the value $Re_{Pl}^{-0.6} \cdot Fr_{Pl}^{-0.25} = 8.7 \times 10^{-3}$, are used:

$$c_{Pl} = 4.57 \cdot Re_{Pl}^{-0.6} \cdot Fr_{Pl}^{-0.25} \cdot \left(\frac{e}{r_a}\right)^{1.5} \cdot \left(\frac{b}{r_a}\right)^{-0.4} \cdot \left(\frac{V_z}{V_0}\right)^{-0.3} \cdot \left(\frac{\Sigma V_z}{V_0}\right)^{-0.2}$$

$$\tag{26.16}$$

$$c_{Pl} = 2.63 \cdot Re_{Pl}^{-0.6} \cdot Fr_{Pl}^{-0.25} \cdot \left(\frac{e}{r_a}\right)^{1.5} \cdot \left(\frac{b}{r_a}\right)^{-0.17} \cdot \left(\frac{V_z}{V_0}\right)^{-0.53} \cdot \left(\frac{\Sigma\,V_z}{V_0}\right)^{-0.2}$$

(26.17)

In the turbulent flow area, the following relations should be used to determine the churning torque factor:

$$c_{Pl} = 0.376 \cdot Re_{Pl}^{-0.3} \cdot Fr_{Pl}^{-0.25} \cdot \left(\frac{e}{r_a}\right)^{1.5} \cdot \left(\frac{b}{r_a}\right)^{-0.124}$$

$$\times \left(\frac{V_z}{V_0}\right)^{-0.376} \cdot \left(\frac{\Sigma\,V_z}{V_0}\right)^{-0.2}$$

(26.18)

with $$Re_{Pl} = \frac{\dot{\omega} r_a^2}{v}$$

(26.19)

and $$Fr_{Pl} = \frac{\omega^2 \cdot r_a}{g}$$

(26.20)

The wall gap factor c_W and the oil volume factor c_V were given in equation (26.15) (Doğan, 2001) by approximation formulae, and confirmed on the basis of trials, equations (26.16) and (26.17):

$$c_W = 1 - 0.02 \cdot \left[1 - \frac{s_{zu}}{2 \cdot r_a}\right]^{1.8} \cdot Fr_{Pl}^{0.45}$$

(26.21)

$$c_V = 1 - \left[\frac{V_z}{V_0} - 0.1\right]^{0.4}$$

(26.22)

The density of the gear oils can be calculated using equation (26.23) as a function of ambient pressure and temperature

$$\rho(p, \vartheta) = \rho_0 \cdot (1 + 49.62 \times 10^{-6} \cdot p - 688 \times 10^{-6}$$
$$\times \vartheta + 0.226 \times 10^{-6} \cdot p \cdot \vartheta)$$

(26.23)

The density ρ_0 is given at atmospheric pressure $p_0 = 1.013\,bar$ and at a temperature of $0\,°C$.

26.5   Peripheral instruments for measuring drag torque

The drag torque is measured by a torque measuring shaft with a resolution of $0.004\,N\,m$ which is mounted to the transmission input shaft. The construction and functioning of the torque measuring shaft are shown in Fig. 26.11. It mainly consists of an internal and an external damping tube and a torsion wire connecting the two pipes. Viscous silicon oil is used in the radial gap between the two pipes to dampen vibrations of the motor/torsion-wire/gear-parts system. The ***drag torque*** is determined by the number of counting pulses N of the incremental sensor between the start and stop signal reference points. The stop signal is generated by a pin on the gear unit side by means of a light sensor. The incremental sensor on the motor side also gives another reference signal for each revolution, which is used as a start signal. The counted pulses between the two reference signals are then proportional to the twist angle of the torque measurement shaft. For purposes of zeroing, the transmission shaft is disengaged by slackening the locking screw. The torque measuring shaft is then subjected to load only from the bearing friction

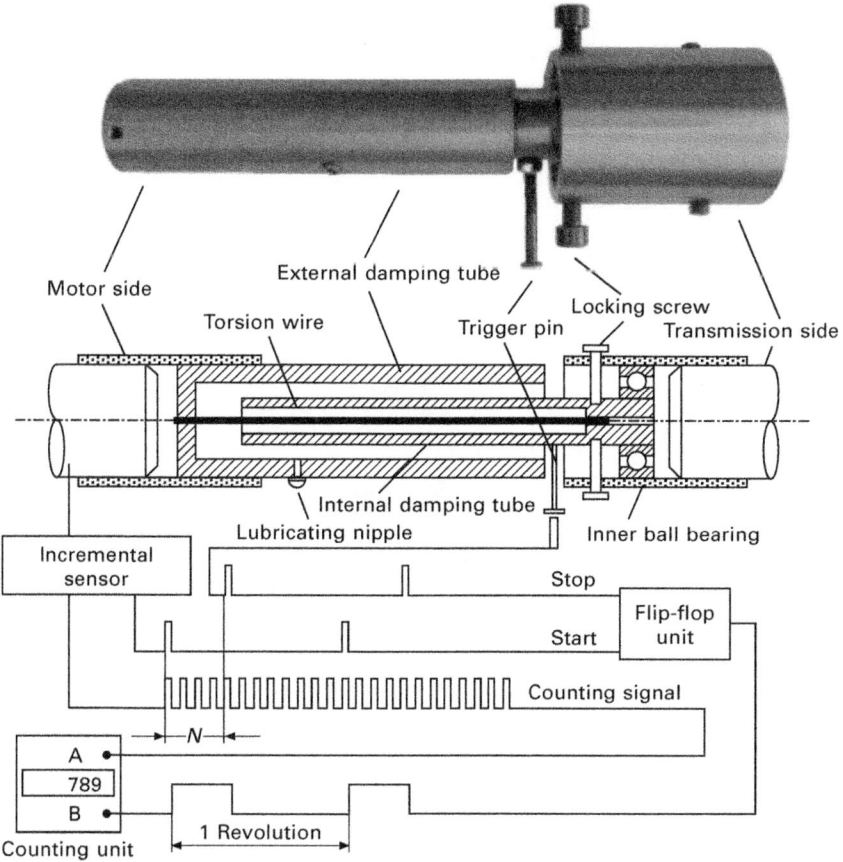

A and B are the counting pulses of the incremental sensor between the start and stop signal reference point.

26.11 Structure and functioning of the torque measuring shaft.

torque, which is negligible. After that the start signal is determined at a low measuring shaft rotational speed.

26.6   Automotive transmissions investigated

The drag torque studies investigated a single gear step of a gearbox, a passenger car coaxial manual gearbox and a coaxial manual gearbox for light commercial vehicles. The gearbox sections with the input torques of the investigated transmissions are shown in Fig. 26.12.

The coaxial transmissions are used in standard drive (rear wheel drive) vehicles. By transmitting the rotational irregularity of the internal combustion

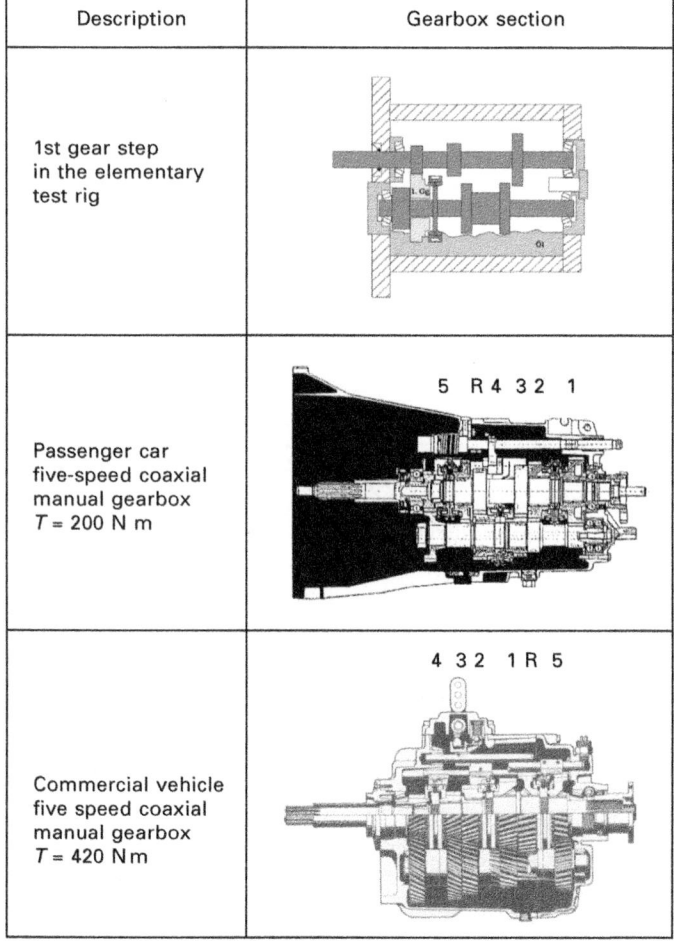

Description	Gearbox section
1st gear step in the elementary test rig	
Passenger car five-speed coaxial manual gearbox $T = 200$ N m	5 R 4 3 2 1
Commercial vehicle five speed coaxial manual gearbox $T = 420$ N m	4 3 2 1 R 5

26.12 Transmissions investigated.

engine via the transmission driveshaft to the countershaft, torsional vibration is induced in the idler gears of the transmission.

The elementary test rig enables both individual gearwheel stages and also several gear pairs and their rattling characteristics to be studied under controlled conditions. In this investigation, the elementary test rig comprises just one gear pair (the 1st gear step), to reveal the effect of the lubricants on one single gear step.

In the case of the passenger car five-speed coaxial manual gearbox, the idler gears of the 1st, 2nd and R gears are located on the output shaft. In neutral, torsional vibration can be induced in the idler gears of 1st, 2nd and R gears and the intermediate gear of the reverse gear, in addition to the head gear, thus contributing to rattling noise tendency. Since the idler gears of the 3rd and 4th gears are mounted on the countershaft, these gear steps do not generate any rattling noises in neutral.

All the fixed gearwheels in the commercial vehicle five-speed coaxial manual gearbox are mounted on the countershaft. This leads to excitation of all the idler gears by the countershaft, both in neutral and with any gear step engaged.

26.7   Automotive transmission lubricants studied

The kinematic viscosity of the lubricants, i.e. the quotient of dynamic viscosity and oil density, is a function of the temperature and is a major factor determining the level of noise generated by transmissions. The viscosity profiles of the lubricants investigated were determined using the Ubbelohde method DIN 51562 (Fig. 26.13).

This principle is based on the different flow times of gear oils through a defined measuring system. The elapsed time recorded is then the measure of viscosity. The viscosity–temperature diagram clearly shows the functional relationship with temperature. The lubricants investigated are made up of different oil bases and additives. At high operating temperatures, the test oils have similar kinematic viscosities between 10 and 18 mm^2/s as at 80 °C.

26.8   Automotive transmission measurement results

Rattling and clattering noise behaviour in automotive transmissions is evaluated using a rattling curve in which the *airborne sound pressure level* is plotted against the angular acceleration amplitude. In the first of four phases of a rattling curve – the constant basic noise phase – only rolling contact noise from the engaged gearwheel pairs, their churning noise and roller bearing noise of the transmission shafts are to be heard. The second phase, called the *rattling limit*, is characterised by the onset of rattling noise, leading to initial impacts resulting from the idler gear flank disengaging from the

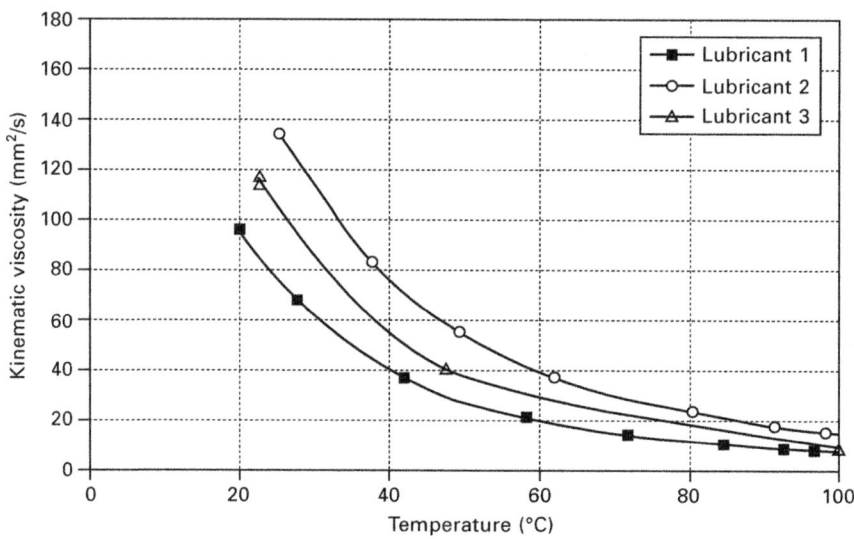

26.13 Kinematic viscosity–temperature diagram for trial lubricants using the Ubbelohde method.

fixed gear flank. During next phase there is a sharp rise in the rattling curve from the rattling limit up to a certain angular acceleration amplitude due in particular to the torsional flank impacts arising. In the fourth and last phase, axial impacts occur in addition to the torsional flank impacts, leading to a further increase in airborne sound pressure level resulting from increasingly large angular acceleration amplitudes.

The effects of the experimental lubricants on rattling noise behaviour were studied for a passenger car five-speed coaxial manual gearbox at an operating temperature of 90 °C. The experimental lubricants were ATF-based (lubricant 1), SAF-based (lubricant 2) and the alternative lubricant shown in Section 26.3 based on organic compounds (lubricant 3). Figure 26.14 shows the rattling curves for the lubricants studied under realistic excitation by a four-cylinder four-stroke internal combustion engine. This reveals a reduction in the airborne sound pressure level when lubricant 3 is used compared with lubricant 1 of up to 3 dB(A), and compared to lubricant 2 of up to 1 dB(A). The rattle limit is also shifted to higher angular acceleration amplitudes.

26.8.1   Comparison of measurement and simulation

Using the example of the 1st gear step in the elementary test rig and of the passenger car five-speed coaxial manual gearbox, measurement results using the approximate calculation of drag torque as set out in Section 26.5 are compared.

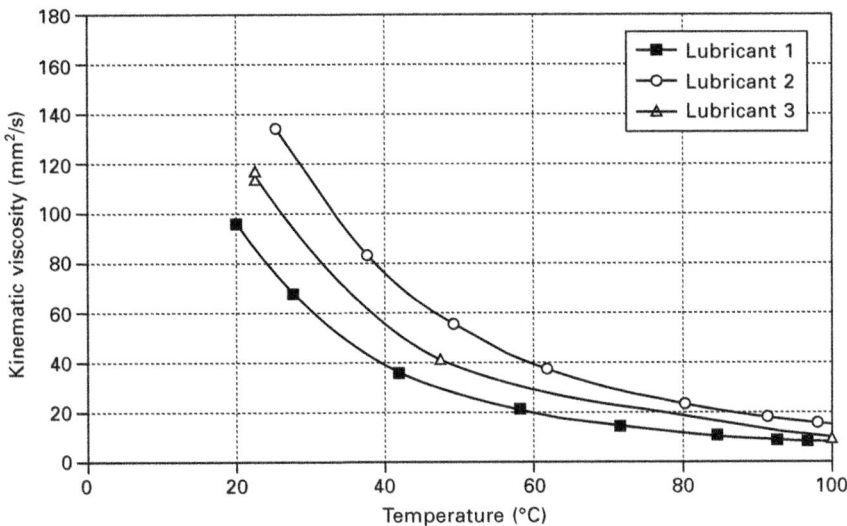

26.14 The effect of different lubricants on rattling noise behaviour in a passenger car five-speed coaxial manual gearbox in neutral.

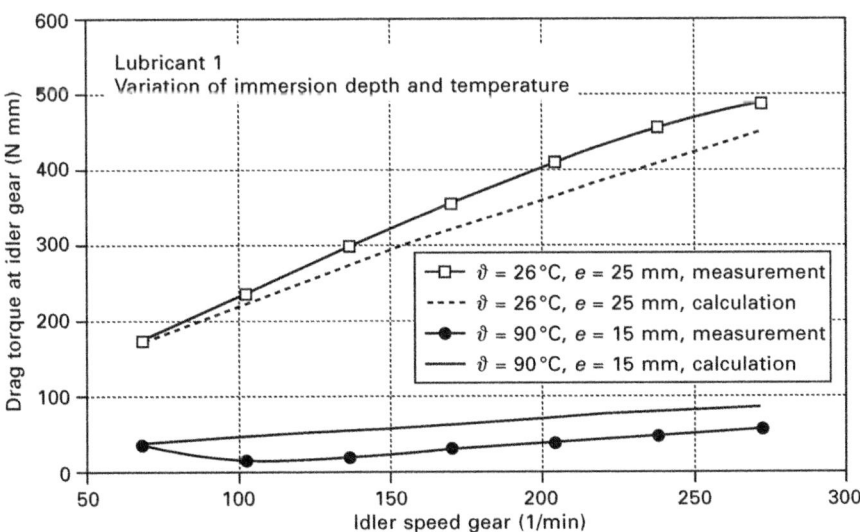

26.15 Measured and calculated drag torques for the 1st gear step in the elementary test rig.

The measured and calculated drag torques at the idler gear of the 1st gear with variation of the immersion depth and the operating temperature of lubricant 1 are compared in Fig. 26.15. It reveals that the approximate simulation of ***drag torque*** provides a good description of the effects of rotational speed, immersion depth and oil temperature. The drag torques differ by up to 10% in the elementary test rig at the operating temperatures studied. This is due in part to the geometry of the gearbox housing that is difficult to define mathematically. The complex flows in the oil sump and the spattering of oil particles in the housing are also difficult to determine mathematically.

The calculated and measured total drag torque curve of the passenger car five-speed manual gearbox in neutral at an operating temperature of 80 °C with lubricant 1 is shown in Fig. 26.16. The total drag torque is calculated from the individual drag torque components of the gearwheels and synchroniser rings involved and then totalled. The approximation of the calculated and the measured curves shows a good correlation up to 2000 rpm. Above this speed the results diverge.

Figure 26.17 shows the simulated drag torques of individual loose components of the light commercial vehicle five-speed manual gearbox in neutral. Each drag torque figure of a loose component contains all four drag torque components. This pattern illustrates the dominant role of the countershaft. The main reason for this is both the design of the transmission (vertical axle orientation – countershaft in the oil sump) and the number of fixed gearwheels all mounted on the countershaft. These fixed gearwheels are immersed in the gear oil with their different diameters, thus significantly contributing to the overall level of drag torque. The small ratio of the head

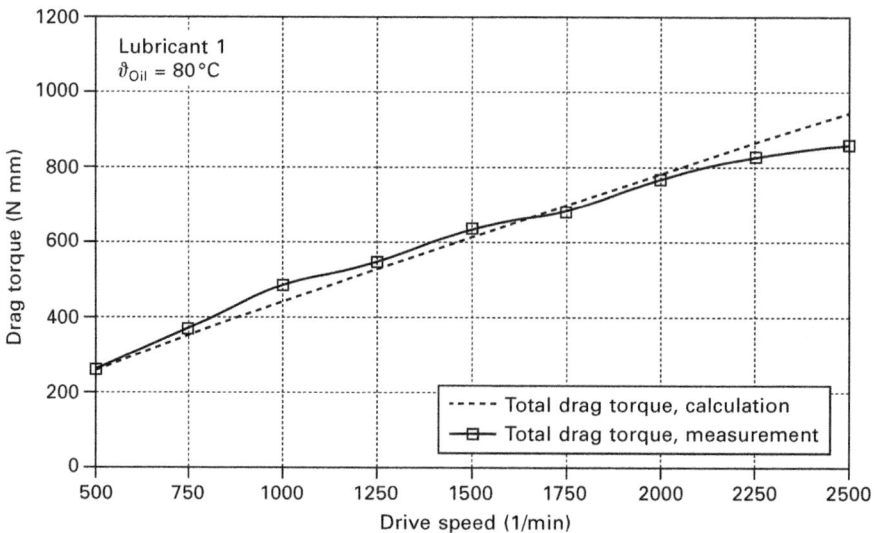

26.16 Measured and calculated overall drag torque based on the example of a passenger car five-speed coaxial manual gearbox in neutral at 80 °C.

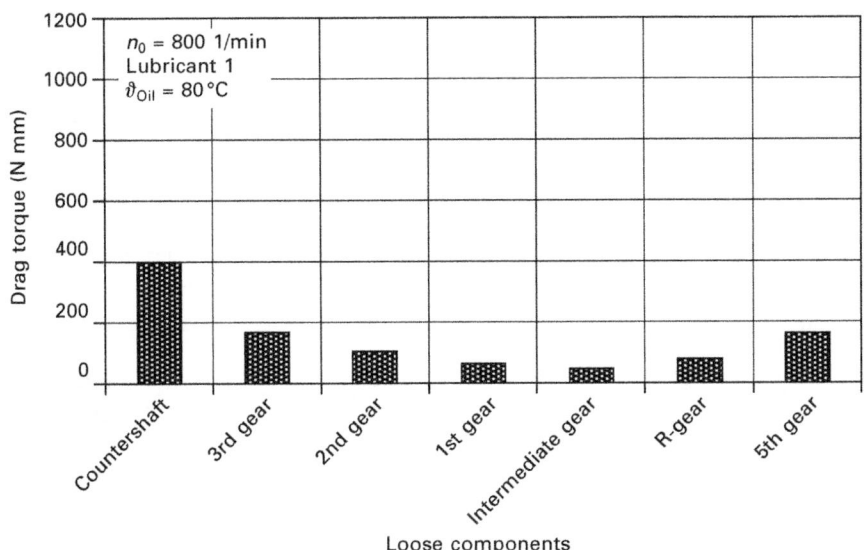

26.17 Drag torques of individual loose components of the commercial vehicle five-speed manual gearbox.

gear is also responsible for the correspondingly high speed of rotation of the countershaft.

The simulated individual drag torques with the compression, synchronisation, bearing friction and churning component, and the overall drag torque of the transmission studied in Section 26.6 are shown in Fig. 26.18. This is based on the same parameters (drive speed 800 rpm, lubricant 1, 80 °C) being used as a basis for modelling the individual drag torques. This illustration shows the weighting of the individual drag torques reflected in the overall drag torque of the gearboxes studied. Whereas the synchronisation drag torque has a small part compared with other drag torque components, the bearing friction torque is not negligible, especially when considering individual gear steps. Churning torque and compression torque are, in principle, dependent on the transmission design, and contribute considerable proportions to the overall drag torque.

It is thus possible to influence the overall drag torque and consequently the onset of rattling (rattling limit) and the rattling and clattering noise proneness of the transmission by considering the individual drag torque components and optimising the parameters on which they are based.

26.8.2   Summary of measurement and simulation

This chapter considers the problem of rattling and clattering noise behaviour in automotive transmissions from a drag torque prospective. The theoretical

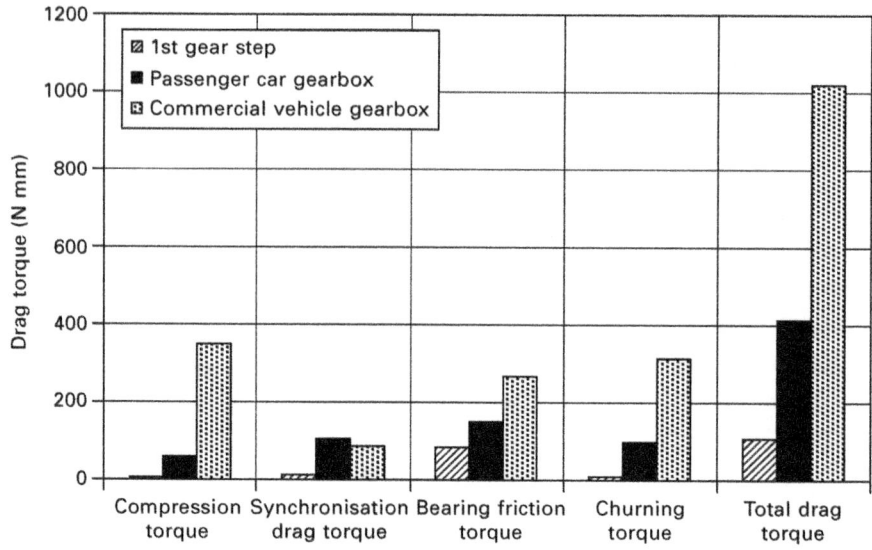

26.18 Drag torque components of a single gear step, a passenger car and a commercial vehicle gearbox.

study shows a simulation method for determining the drag torque of automotive transmissions. Using the example of a single gear step and of a complete passenger car production transmission, various gear lubricants (including an alternative lubricant) were examined for their effectiveness and measurements with different gear oils were analysed and discussed. The theoretical study presents and analyses the drag torques of individual loose components relating to the commercial vehicle transmission.

In the case of coaxial transmissions, the countershaft with its numerous fixed gearwheels applied makes a decisive contribution to the level of overall drag torque in the gear unit as a function of axle orientation. The bearing friction torque, the churning torque and the compression torque are the major drag torque components at the overall automotive transmission drag torque.

Considering the transmission lubricant, it is evident that kinematic viscosity has a major influence on the noise level generated by transmissions. Correct selection of lubricant enables the rattling limit to be moved to higher angular acceleration amplitudes and rattling and clattering noises to be reduced. The findings and the experience gained can thus already contribute to the development and design engineering phase of new automotive transmissions.

26.9   Parameter studies for rattle and clatter noises in automotive transmissions

26.9.1   Influence of torsional backlash and axial clearance

The elementary test transmission was equipped with a single gear step in order to investigate the rattling noise behaviour of a single gear step and the effect of the torsional backlash and axial clearance. Figure 26.19 compares the effect of torsional backlash and axial clearance at constant angular acceleration amplitude of 600 rad/s² with realistic excitation of a four-cylinder four-stroke internal combustion engine (second engine order).

Reducing the production torsional backlash from 0.12 mm by 0.06 mm minimises rattling in this gear step by up to 2 dB(A), whereas increasing it by the same amount results in an increase of 4 dB(A) of the airborne sound pressure level. The *torsional backlash* is a functional value that can be varied within technically permissible limits. Below this minimum threshold, it is not possible to guarantee equalisation of thermal expansion characteristics and

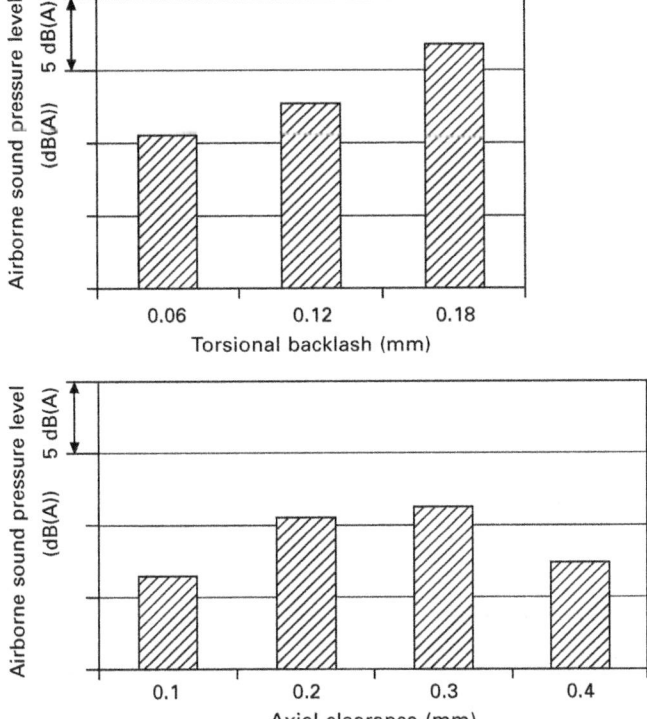

26.19 Effect of torsional backlash and axial clearance at a speed of 800 rpm at an angular acceleration amplitude of 600 rad/s² of a single gear step.

the rolling contact of the gear tooth system. Increasing the axial clearance up to a certain limit of 0.4 mm, at which the idler gear can no longer meet its axial thrust collars, results in a reduction in rattling of up to 3 dB(A) compared with the production axial clearance of 0.2 mm. The idler gear then no longer meets the opposed axial thrust collar, thus reducing the frequency of axial impacts. Reducing the axial clearance also reduces the airborne sound pressure level, but there is a danger of impairing the shifting action and inadequate lubrication of the gear step in question.

26.9.2   Noise of individual gear steps of a front-mounted/ transverse transmission

Figure 26.20 shows the measured rattling curves of individual gear steps of the front-mounted/transverse manual transmission at a rotational speed of

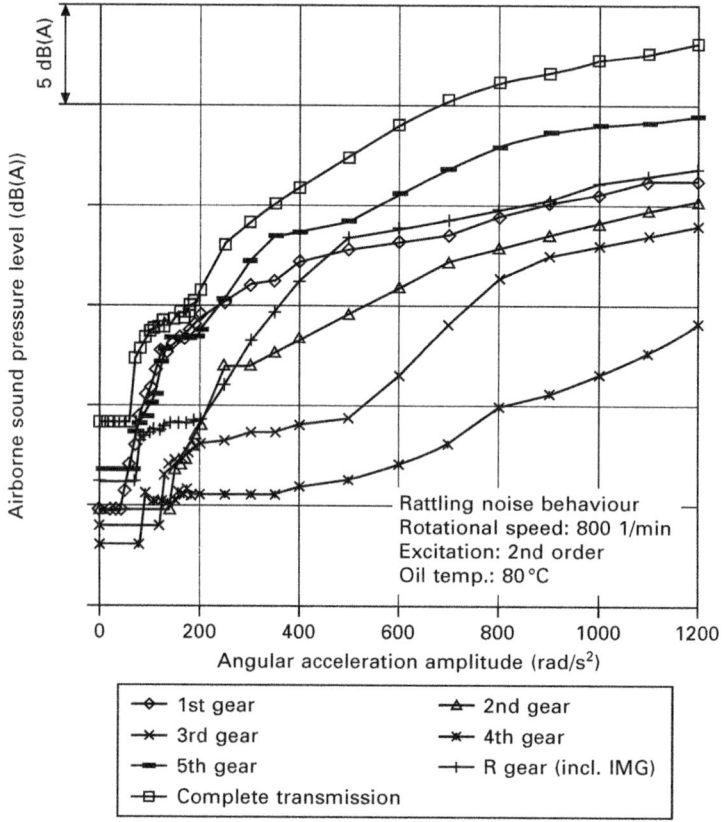

26.20 Five-speed front-mounted/transverse manual transmission: measured rattling curves of individual gear steps and of the complete transmission.

800 rpm, a temperature of 80 °C and second-order sinusoidal excitation. The results relate to the rattling noise behaviour of individual gearwheel stages; all other idler gears were removed. This made it possible to investigate individual rattling positions. The experimentally derived rattling curve of the entire transmission is also shown, running above the individual gear step rattling curves.

In the case of the front-mounted/transverse transmission, in addition to the 5th gear (peculiarity in the design configuration of the idler gear in the transmission), the rattling curves of the R gear (reverse gear) and first gear in particular contribute to the overall level of rattling noises. Two rattling positions between the fixed gear and the intermediate gear (IMG), also being the idler gear of the first gear, and between the intermediate gear and the idler gear of R gear are responsible for the R gear rattling curve. The most influential component for the two gear steps is the first gear idler gear with its large moment of inertia and its large mass, which is excited with the same angular acceleration amplitude as the transmission input shaft. The curves for the 3rd and 4th gears contain the rattling noise of synchroniser rings of these gears with their low moments of inertia and small masses, which have only a subordinate influence on the rattling of the complete transmission because of the logarithmic addition of the airborne sound pressure level.

26.9.3   Noise of individual gear steps of a standard transmission

Figure 26.21 shows the effect of rotational speed and engine order on the rattling noise proneness of the five-speed coaxial manual transmission considered, and its clattering noise behaviour in the engaged gear steps at a transmission fluid temperature of 90 °C. Figure 26.21(a) shows a three-dimensional *rattling curve* of the five-speed standard manual transmission for a rotational speed range between 500 rpm and 1000 rpm in neutral with an idealised third order excitation. The airborne sound pressure level of this three-dimensional rattling curve extends to 7 dB(A) as the rotational speed increases in the range of higher angular acceleration amplitudes. The effect of different engine responses (orders) on the rattling noise proneness of this automotive transmission is shown in Fig. 26.21(b).

Here the transmission is excited in neutral with an idealised four-cylinder, five-cylinder, six-cylinder, eight-cylinder and ten-cylinder internal combustion engines at a rotational speed of 800 rpm, with the bisector of the number of cylinders being the dominant engine response in each case.

As the number of cylinders in an internal combustion engine increases, the *airborne sound pressure levels* of the rattling curve increases. For example in the case of the ten-cylinder engine, the increase in the level of rattling noise is directly related to the number of torsional flank impacts and

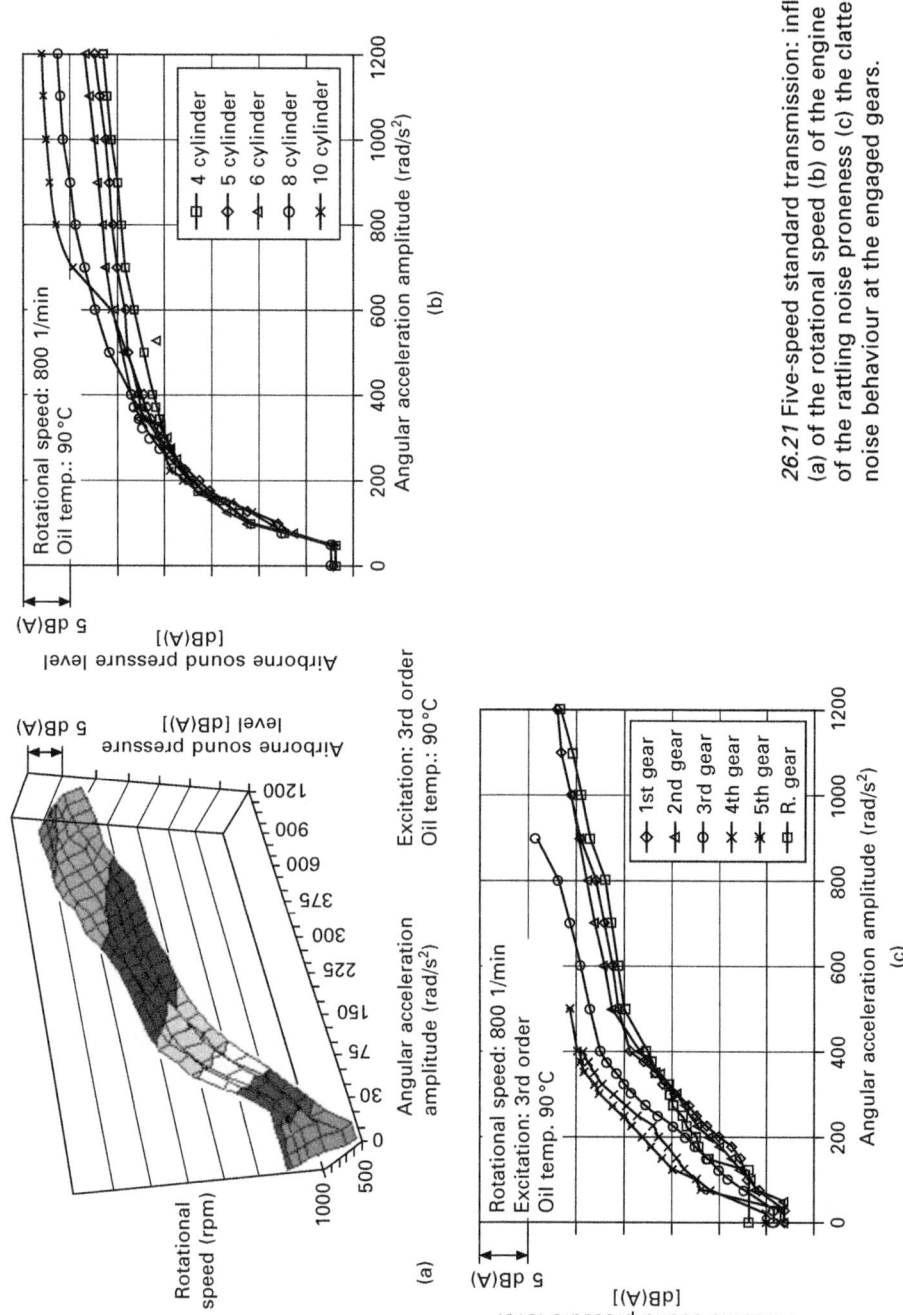

26.21 Five-speed standard transmission: influence (a) of the rotational speed (b) of the engine order of the rattling noise proneness (c) the clattering noise behaviour at the engaged gears.

axial impacts arising in the course of one revolution of the engine shaft. Whereas in the case of the four-cylinder engine there are two sinusoidal oscillations per shaft revolution and therefore only two acceleration phases and two deceleration phases per crankshaft revolution occur, in the case of the ten-cylinder engine there are five sinusoidal oscillations resulting in five acceleration phases and five deceleration phases, and thus a larger number of impacts per shaft rotation. The change in the basic noise level up to the rattling noise limit of the excitation functions considered is insignificant.

For the investigations in case of an engaged gear step of this transmission, a braking torque at the output shaft was selected so that the tooth flanks of the gear step engaged could not lift off at the angular acceleration amplitudes set, and therefore did not lead to any clattering noises. The *clattering noise* investigations were carried out at a speed of 800 rpm and with third order excitation. The largest airborne sound pressure level curve is shown by the rattling curve of the engaged fifth gear (Fig. 26.21(c)). This, in particular, is not only due to the seven rattling positions, but also because of the large mass and moment of inertia of the countershaft, which show a significant noise component. Additionally, the amount of torsional backlash and axial clearance of all loose components considered cannot be ignored. In the other gears, the countershaft with its dominant effect on noise level and the idler gear engaged in each case are in the power flow, resulting in a substantially lower level of the clattering noise.

26.9.4   Influence of the main centre distance of a standard transmission

The influence of the main centre distance a was investigated on a standard transmission. In Fig. 26.22, the main centre distance is infinitely varied by the changing position of the countershaft in 0.2 mm increments. Reducing the main centre distance up to a specific limit of 69.85 mm has the effect of minimising the airborne sound pressure for this standard transmission of up to 2.5 dB(A), whereas increasing it by 0.2 mm (70.15 mm) results in an increase of 7 dB(A).

26.10   Correlation of measurement and calculation results in automotive transmissions

Figure 26.23 shows the correlation between measurements and calculation results for a single-step gear stage, using the example of the first gear of a front transverse transmission. With a single rattling point in the first gear stage, this shows good correlation between the calculated rattling noise level curve and the actual observed level. The rattling noise level curves determined using the EKM simulation method represents a sufficiently

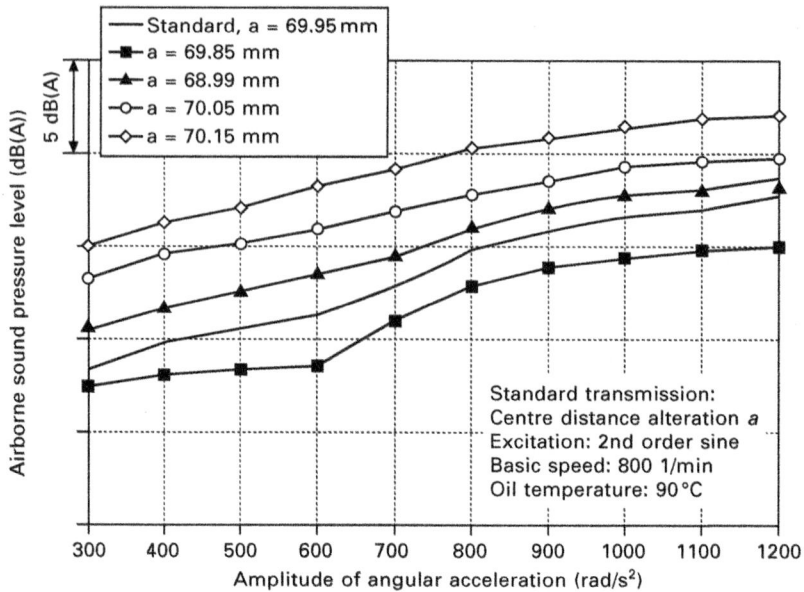

26.22 Effect of centre distance on the rattle noise based on the example of a standard transmission.

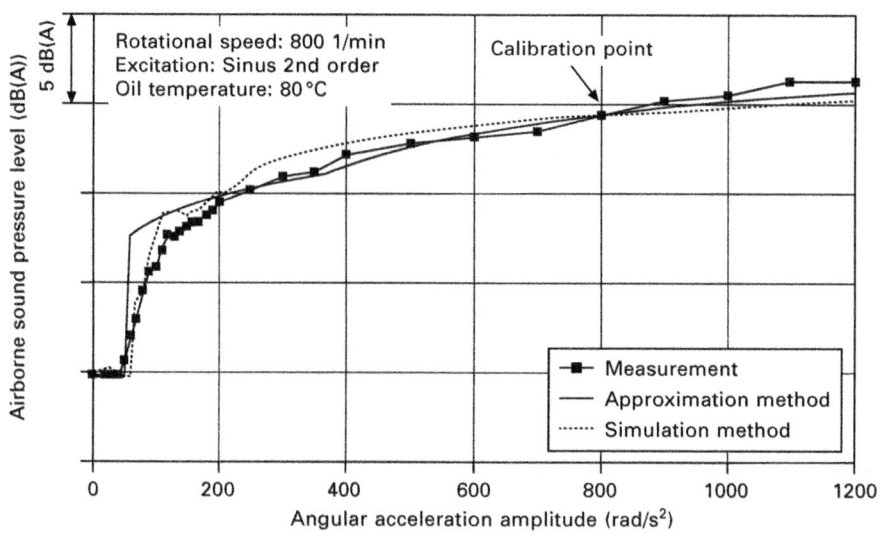

26.23 Correlation between measured and calculated rattling noise level curves with a single-step gear stage.

good result. From an angular acceleration of $200\,\mathrm{rad/s^2}$, the characteristics calculated are approximately equal, with the rattling noise level curve derived by the approximation method coming closer to the measured rattling noise level curve.

There is also satisfactory correlation between the measurement and the EKM calculation method for the standard transmission. Figure 26.24 shows the ***airborne noise pressure levels*** of a five-speed manual transmission with an angular acceleration of $800\,\mathrm{rad/s^2}$. This clearly shows the influence of the countershaft. The 1st gear makes a decisive contribution to the overall rattle diagram profile, as well as the 5th gear.

Figure 26.25 shows the calculated and measured rattling curves of individual gear steps of the standard manual transmission in neutral position at a rotational speed of 800 rpm, a temperature of $90\,°\mathrm{C}$ and with third-order sinusoidal excitation. The dominant influence of the countershaft with its large mass and its large moment of inertia on the overall rattling curve profile is very clearly evident in this case. All other loose components make a subordinate contribution to the overall rattling curve profile of this standard transmission. The clattering noise behaviour for the first gear engaged at the calibration point of $800\,\mathrm{rad/s^2}$ is shown on the left side. The overall airborne sound pressure level is formed here from the individual level value of the 2nd, reverse intermediate, 3rd and 4th gears except for the countershaft and the 1st gear. The large difference in overall airborne sound pressure level

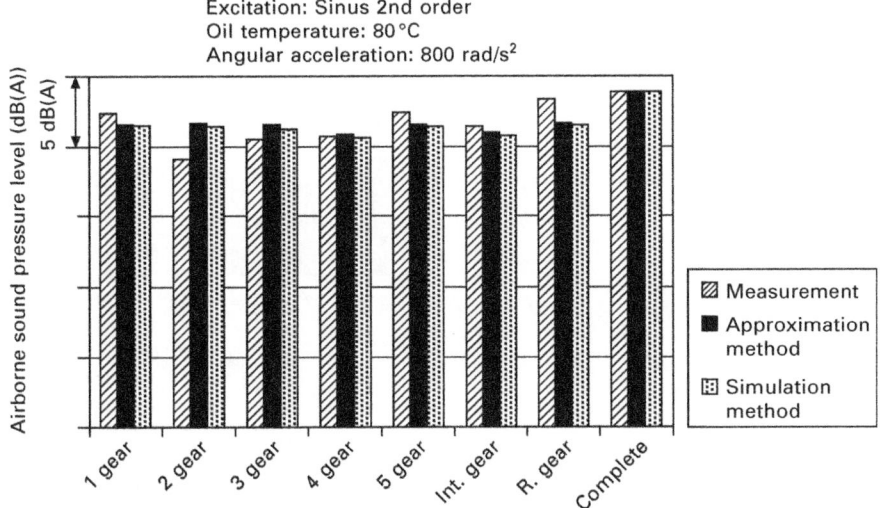

26.24 Airborne noise pressure levels of measured and calculated idling rattling noise level curves at an angular acceleration of $800\,\mathrm{rad/}$ $\mathrm{s^2}$.

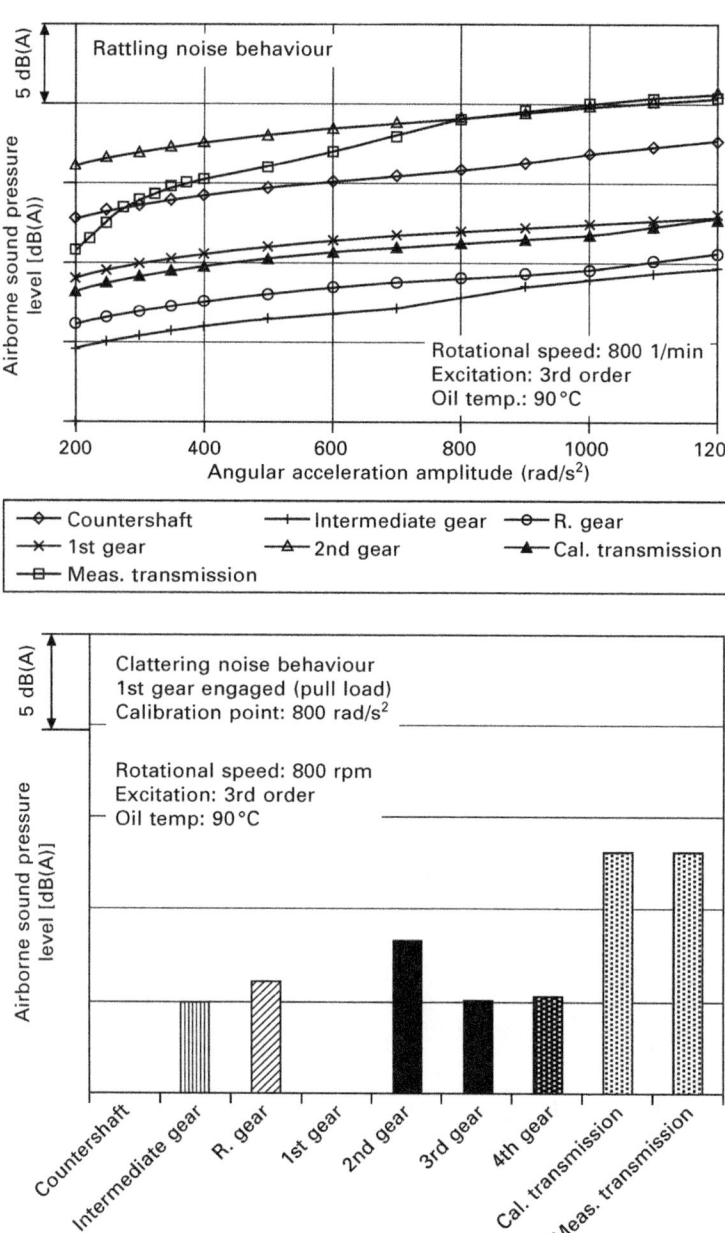

26.25 Five-speed standard manual transmission: calculated and measured rattling curves of individual gear steps and of the complete transmission in neutral position and in gear stage.

between idling and first gear at the calibration point of 6 dB(A), which is also very clearly evident here.

26.11   Noise reduction measures in automotive transmission

Both external and internal measures can be taken to reduce the propensity to *rattle* and clatter in automotive transmissions (Fig. 26.26).

26.11.1 External measures

The purpose of external measures is to tune and optimise the driveline in terms of reducing torsional vibration outside the transmission. A major factor here is decoupling the combustion engine from the driveline.

One notable decoupling system is the *hydrodynamic torque converter* used in automatic transmissions for moving off. However, the clutch can also have a long-term effect on driveline behaviour. For example a hydrodynamic converter clutch, a torsional damper in the *clutch disk*, a *slip-controlled*

External measures
Driveline tuning and optimising
Decoupling of the combustion engine

- Torque converter
- Clutch
- Integral starter–alternator-damper system (ISAD)
- Crankshaft damper

- Clutch pre- and main damper
- Vibration damper
- Encase gearbox

- Mechanical torsion damper (MTD)
- Hydraulic torsion damper (HTD)
- Soundproof bodywork

Combustion engine   Clutch   Transmission   Output

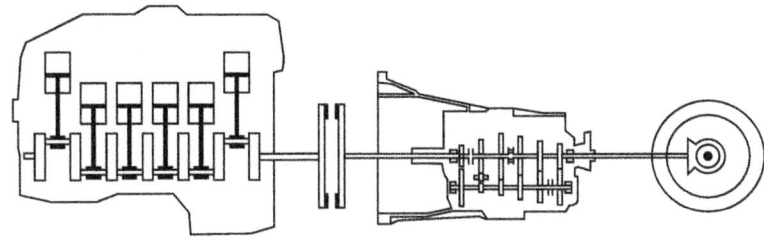

Internal measures
Optimisation of gear parameters

- Moment of inertia
- Mass

- Helix angle
- Gear set configuration

- Restriction movement of loose parts within their functional degrees of freedom
  Torsional backlash   Axial clearance

26.26 External and internal measures to reduce the rattle- and clatter-proneness of automotive transmissions.

clutch or electronic clutch management reduces the irregular rotational movement behaviour of the driveline. Main and ancillary clutch dampers and torsional vibration dampers provide further measures to improve driveline performance.

The *dual-mass flywheel* (DMF) is used between the combustion engine and the transmission as a standard measure for reducing engine torsional vibration. This significantly reduces torsional vibration both in car drivelines with manual or with automatic transmissions, in this case with a locked-up direct-drive clutch. This improves the overall noise and vibration behaviour of the driveline in terms of *clatter* and *hum*, as well as load cycle effect.

Numerous recent developments are concerned with improving the characteristics and optimising the manufacturing costs of the decoupling systems used. The mechanical torsional damper (MTD) represents a cost-optimised refinement of the dual-mass flywheel, which is also based on the principle of spreading the inertia torque of the engine flywheel, but is characterised by simpler design with a smaller number of components. The *hydraulic torsion damper* (HTD) combines the advantages of spreading the flywheel mass inherent in the DMF with a variable damper system. This damper increases its damping effect under critical system conditions, for example in the event of a natural frequency of the engine transmission system being generated, in order to prevent high oscillation amplitudes at the input to the transmission. The *speed-adaptive damper* (SAD) enables resonance-free compensation of the dominant exciting torque at the crankshaft of a combustion engine throughout the speed and load range. This damper principle achieves a significant reduction of cyclic irregularity of the crankshaft and of the vibrations in the belt drive. The *integral starter–alternator–damper* (ISAD) also serves to reduce cyclic irregularity. The primary aim of this system is, however, to provide electrical energy and manage it in motor vehicles. Fitting the electrical machine directly on the crankshaft can generate motive power when starting and generating power to provide energy.

There are also other possible external measures to reduce rattle and clatter that are simple, but involve major disadvantages such as reducing efficiency or high cost; these include encapsulating the transmission and deadening the bodywork. These external measures are, however, subject to physical and economic limits, and they do not always achieve the desired results.

26.11.2 Internal measures

Internal measures to reduce the rattle- and clatter-proneness of a transmission involve judicious design of the geometrical parameters of components that are subject to impact loads, as well as restricting the freedom of movement of loose parts within the degrees of freedom defined by their function. The latter can reduce or prevent impact events occurring or spreading. The aim of restricting the movement of loose parts mainly relates to judiciously impeding freedom of movement in order to reduce or prevent impact events occurring or spreading. Internal measures include for example *idler gear braking*, *tooth gap bracing*, *bracing toothed disks*, auxiliary transmissions, dampers, magnetism, structure-borne noise barriers and axial impact reduction.

With the idler gear brake, the idler gear is braked through friction. With the aid of tangentially acting elastic elements between the idler gear and a thin toothed disk with an equal number of teeth mounted coaxially to the idler gear, space width bracing is achieved by means of the rotary offset. The bracing toothed disk mounted coaxially to the idler gear is pressed against the idler gear by elastic components. An additional tooth on the bracing toothed disk causes the bracing toothed disk to turn more slowly than to the idler gear, resulting in friction that acts in the same direction as the drag torque. In the auxiliary transmission, the fixed gear has an auxiliary transmission subject to friction in addition to the main transmission. This gives rise to a realigning effect for the idler gear, opposite to its direction of rotation. With the damper, the idler gear vibrations are damped with the aid of a secondary mass connected elastically to the loose part.

One means of reducing impact events is developed and patented at the Institute of Machine Components. This is based on using permanent magnets in fixed gears and idler gears. This prevents the idler gear detaching from the fixed gear up to a certain angular acceleration, thus reducing rattle- and clatter-proneness. The structure-borne noise barrier serves to prevent structure-borne noise being transmitted between the point of impact and the hub of the gearwheels using various insulating materials. Axial impact intensities are reduced by ductile thrust collars on the idler gear or the shaft shoulder.

The use of these internal measures depends not only on the degree of the noise reduction achieved, but also on the production engineering and economic parameters. The wear characteristics and the level of power loss and the intrinsic noise involved must also be taken into account. Each transmission system must be individually derived from the large number of various possible solutions.

26.12   Engineering design catalogue for low rattle and clatter automotive transmissions

This section systematically sets out the physical principles of operation in the form of an engineering design catalogue to minimise the rattle- and clatter-proneness of transmissions. This lists possible effects for generating braking forces on idler gears or loose parts, and shows existing solutions (some patented) and new solutions summarised in a convenient form. The physical principles of operation aim to prevent or reduce vibration of loose parts when subjected to excitation. The engineering design catalogue can be used with the advantage of designing low rattle and clatter automotive transmissions since it represents a source of knowledge and contributes to rationalising the engineering design process. It is also a source of new ideas for the design engineer, since the super-ordinate measures enable many existing partial solutions to be combined to find a new solution.

In order to give an overview of measures to reduce rattle and clatter, it is very helpful to summarise the possible principles for generating braking forces at the idler gear or loose part in one solution framework. Possible principles include material locking, frictional locking, elastic power flow, field power flow and momentum change locking (see Table 26.1). In this context, locking or flow (after Roth, 1994) means the reciprocal dynamic effect between two parts (fixed bodies or fluids) is maintained over a particular time period. These possible principles for generating braking forces at the idler gear or loose part are described in Doğan (2001), see Tables 26.2 and 26.3.

The advantage and benefits of this engineering design catalogue are the systematic representation, in tabular form, of known noise reduction methods for automotive transmissions, rapid access to information, ease of use for users, uniform and clear representation of the schematic diagrams, expandability of the engineering design catalogue, and the possibility of finding new approaches by combining existing possibilities.

26.13   Conclusions

Table 26.4 shows the noise reduction levels achievable for the measures synoptically represented according to the influence of the operating parameters, the main geometrical parameters and some internal measures in a five-speed manual transmission. Optimum design configuration of torsional backlash and axial clearance and the selection of a lubricant with high kinematic viscosity are the primary factors leading to low rattle- and clatter-proneness in automotive transmissions.

When designing automotive transmissions from the point of view of minimum rattle and clatter, it is essential to take into account that the airborne noise pressure level output rises as mass and moment of inertia of loose parts increase, and as centre distances increase.

Table 26.1 Overview of measures to reduce rattle and clatter in automotive transmissions

Connection		Mechanism	Engagement	Appendix
Principle of locking	Locking is created by	Example	Braking force is limited by	Comment
Material locking	1 Toothing geometry		Strength	For more examples see Table 26.2
Frictional locking	2 Friction between two touching surfaces		Max. local pressure	For more examples see Table 26.2
Elastic power flow	3 Elastic deformation		Max. local elastic deformation	For more examples see Table 26.3
Field power flow	4 Force field		Max. magnetic force	For more examples see Table 26.3
Momentum change locking	5 Flying component and velocity		Angular acceleration Mass of the flying component Spring force	For more examples see Table 26.3

Shading indicates that there is a cut or different component.

Table 26.2 Typical applications for material and frictional locking

Connection			Mechanism	Engagement	Appendix
Principle of locking	Locking is created by		Example	Braking force is limited by	Comment
Material locking	Changeable gear thickness	1.1		Strength	Toothing geometry
	Tooth flank geometry	1.2		Strength	Toothing geometry
	Tooth flank design	1.3		Strength	Toothing geometry
Frictional locking	Compression spring	2.1		Compression spring force	1 Idler gear 2 Compression spring 3 Toothed disk Patent: DE 3336 669 C2 Date: –.
	Belleville spring	2.2		Belleville spring force	1 Idler gear 2 Belleville spring 3 Toothed disk Patent: 1303-H0/86 Date: 1986
	Elastomer	2.3		Elastic force	1 Idler gear 2 Elastomer 3 Toothed disk Patent: 33336669 Date: 1988

Shading indicates that there is a cut or different component.

Table 26.3 Typical applications for elastic + field power flow and momentum change locking

Connection		Mechanism	Engagement	Appendix
Principle of locking	Locking is created by	Example	Braking force is limited by	Comment
Elastic power flow	Elastomer ring as secondary ratio	3.1		Strength of the elastomer	Patent: DE 2616183 Date: 1979
	Axial fixed elastomer thin toothed disk	3.2		Strength of the elastomer	Patent: DE 3328145 Date: 1983
Field power flow	Magnetic force field of the tooth flanks	4.1		Magnetic force	Patent: 4400874.0 Date: 1994
	Magnetism with coding	4.2		Magnetic force	Patent: 195009355 Date: 1995
Momentum change locking	Braking force at the idler gear Type 1	5.1		Speed	1 Compression spring, 2 Idler gear, 3 disk, 4 Ring, 5 Flying component Patent: DE 2244016 Date: 1972
	Braking force at the idler gear Type 2	5.2		Speed	1 Idle gear, 2 Compression spring, 3 Brake, 4 Shaft Patent: 2320571 Date: 1978

Shading indicates that there is a cut or different component.

Table 26.4 Noise-reducing effect of parameters and internal measures at the passenger car transmissions

Measures to reduce rattle and clatter noise 5-speed manual transmission	Noise-reducing effect compared to initial postion 5-speed manual transmission dB(A)
	No reduction ←— 0 —→ Reduction
Geometrical parameter:	
1. Main distance	
○ Main distance, series *a*=69.95 mm, initial position	**0** -1 -2 -3 -4 -5
○ Reducing main distance up to –0.1 mm (*a*=69.85 mm)	**0** -1 -2 -3 -4 -5
○ Increasing main distance up to +0.1 mm (*a*=70.05 mm)	**+5 +4 +3 +2** +1 0 -1 -2 -3 -4 -5
2. Backlash	
○ Series, 2nd idler gear with 0,12 mm, initial position	+5 **+4 +3 +2 +1** 0 -1 -2 -3 -4 -5
○ Increasing up to 50% (0.18 mm)	+5 +4 +3 +2 +1 **0 1 2** -3 -4 -5
○ Reducing up to 50% (0.06 mm)	
3. Axial clearance	
○ Series, 2nd idler gear with 0.2 mm, initial position	+5 +4 +3 +2 +1 **0** -1 -2 -3 -4 -5
○ Reducing up to 50% (0.1 mm)	+5 +4 +3 +2 +1 **0** -1 -2 -3 -4 -5
○ Increasing up to 50% (0.3 mm)	+5 +4 +3 +2 **+1** 0 -1 -2 -3 -4 -5
○ Increasing up to 100% (0.4 mm)	+5 +4 +3 +2 +1 **0 -1** -2 -3 -4 -5
4. Drag torque	
○ ATF lubricant at 90 °C kinematic viscosity 10 mm^2/s	+5 +4 +3 +2 +1 **0 -1** -2 -3 -4 -5
○ SAF lubricant at 90 °C kinematic viscosity 18 mm^2/s	+5 +4 +3 +2 +1 0 -1 -2 -3 **-4** -5
Operating parameter:	
1. Combustion engine, main order	+5 +4 +3 +2 +1 **0** -1 -2 -3 -4 -5
○ 4-Cylinder (2nd order), initial position	+5 +4 **+3 +2 +1** 0 -1 -2 -3 -4 -5
○ 6-Cylinder (3th order)	
2. Speed	
○ 4-Cylinder (2nd. order, 800 rpm), initial position	+5 +4 +3 +2 +1 **0** -1 -2 -3 -4 -5
○ 4-Cylinder (2nd. order, 500 rpm)	+5 +4 +3 +2 +1 0 **-1** -2 -3 -4 -5
Internal measures:	
1. Elastomer ring as idler gear brake	+5 +4 +3 +2 +1 **0** -1 -2 -3 -4 -5
○ Series, reverse idler gear, initial position	+5 +4 +3 +2 +1 **0 -1** -2 -3 -4 -5
○ Reverse idler gear with elastomer ring	
2. Friction elements at the idler gear	+5 +4 +3 +2 +1 **0** -1 -2 -3 -4 -5
○ Series, 2nd idler gear, initial position	+5 +4 +3 +2 +1 **0 -1** -2 -3 **-4** -5
○ 2nd idler gear with 10 N elastomer friction force	
3. Reducing axial impacts with elastomer components	+5 +4 +3 +2 +1 **0** -1 -2 -3 -4 -5
○ Series, 2nd idler gear, initial position	+5 +4 +3 +2 +1 **0 -1** -2 -3 -4 -5
○ 2nd idler gear with elastomer ring	
4. Idler gear with magnetic elements	+5 +4 +3 +2 +1 **0** -1 -2 -3 -4 -5
○ Series, 2nd idler gear, initial position	+5 +4 +3 +2 +1 **0 -1** -2 -3 -4 -5
○ 2nd idler gear with magnetic elements	

26.14   References

Doğan, S. N.: Zur Minimierung der Losteilgeräusche von Fahrzeuggetrieben. Dissertation, Universität Stuttgart, 2001

Doğan, S. N.; Ryborz, J.; Lechner, G.: Simulation von Losteilschwingungen in Fahrzeuggetrieben, 1998

Doğan, S. N.; Ryborz, J.; Bertsche, B.: Rattling and Clattering Noise in Automotive Transmissions – Simulation of Drag Torque and Noise, Symposium – Transient Processes in Tribology, Leeds–Lyon Symposium, Lyon, September 2003

Höhn, B.-R.; Michaelis, K.; Kopatsch, F.; Eidenschink, R.: Reibungszahlmessungen an mesogenen Flüssigkeiten aus Praxis und Forschung. *Tribologie und Schmierungstechnik*, 44. Jahrgang 3/1997

Kücükay, F.: *Dynamik der Zahnradgetriebe – Modelle, Verfahren, Verhalten*. Springer-Verlag, Berlin, Heidelberg, New York, 1987

Lang, C.-H.: Losteilschwingungen in Fahrzeuggetrieben. Dissertation, Universität Stuttgart, 1996

Lechner, G.; Naunheimer, H.: *Automotive Transmissions*, Springer-Verlag, Berlin, Heidelberg, New York 1999

Roth, K.: *Konstruieren mit Konstruktionskatalogen*. Springer-Verlag, Berlin, Heidelberg, New York, Band II, 2. Auflage, 1994

Ryborz, J.: Klapper- und Rasselgeräuschverhalten von Pkw- und Nkw-Getrieben, Dissertation, Universität Stuttgart, 2003

Walter, P.: Untersuchung zur Tauchschmierung von Stirnrädern bei Umfangsgeschwindigkeiten bis 60 m/s. Dissertation, Universität Stuttgart, 1982

Weidner, G.: Klappern und Rasseln von Fahrzeuggetrieben. Dissertation, Universität Stuttgart, 1991

26.15   Nomenclature

C_{lm}	–	related average impact intensity
C_{mv}	–	related circumferential drag torque
C_{fa}	–	related axial friction force
C_{sa}	–	related parameter in axial direction
C_{sv}	–	related parameter in circumferential direction
$F_R, F[t]$	N	friction force
I_m	N	average impact intensity
k	–	calibration factor
R_a, R_i	N	force
m	kg	mass
L_p	dB	rattling and clattering noise
$L_{p,i}$	dB	noise of one component
L_{Basic}	dB	basic noise level
$L_{p\,Comp.}$	dB	noise level of complete transmission
i_b	–	related ratio
r_b	mm	pitch circle radius
s_a	mm	axial clearance

Symbol	Units	Description
s_v	mm	torsional backlash
β	°	helix angle
τ_{end}, τ_{anf}	s	time (end/start)
ω_{An}	rad s^{-1}	excitation frequency
$\hat{\omega}$	rad s^{-2}	amplitude of angular acceleration
x_1, x_2	–	co-ordinate direction
i_{iz}, i_{ix}, i_{ia}	–	co-ordinate direction
Fr	–	Froude number
N	–	Number of incremental sensor counting pulses
Re	–	Reynolds number
T	N m	Input torque
T_2	N m	Drag torque at the idler gear
T_L	N m	Bearing friction torque
T_{Pl}	N m	Churning torque
T_{Sy}	N m	Synchronisation drag torque
T_{Qu}	N m	Compression torque
V_0	m^3	Total oil volume
V_Z	m^3	Oil volume displaced by the gearwheel
b	mm	Face width
b_{Sy}	mm	Width of the oil gap at the synchroniser ring
c_{Pl}	–	Churning torque factor
c_{Sp}	–	Oil spray factor
c_v	–	Oil volume factor
c_w	–	Wall gap factor
d_m	mm	Mean bearing diameter
d_{Sy}	mm	Mean friction surface diameter
e	mm	Immersion depth
f_0	–	Bearing factor
g	m/s^2	Acceleration due to gravity
h	mm	Tooth depth
h_0	mm	Standardised tooth depth
h_{Sy}	mm	Height of the oil gap at the synchroniser ring
l_H	mm	Hydraulic length
$n_{2, w}$	rpm	Relative speed of rotation between loose part and shaft
p	bar	Pressure
r	mm	Pitch circle diameter
r_a	mm	Tip circle diameter
s_{zu}	mm	Radial wall gap on the oil inlet side
v_t	m/s	Circumferential speed on the pitch circle
φ	rad	Throw-off angle
ϑ_{oil}	°C	Oil temperature
ν	mm^2/s	Kinematic oil viscosity

ρ	kg/m^3	Density
ρ_0	kg/m^3	Density at atmospheric pressure ($p_0 = 1.013\,\text{bar}$, $\vartheta_0 = 0\,^\circ\text{C}$)
ω	rad/s	Angular velocity
$\omega_{Sy,w/2}$	rad/s	Angular velocity between loose part and shaft

Subscripts and superscripts

1	fixed gear
2	idle gear
L	Bearing friction
Laminar	Laminar flow
Pl	Churning
Qu	Compression
Sp	Spray
Sy	Synchroniser ring
Turbulent	Turbulent flow

27

Various forms of transmission rattle in automotive powertrains

P. KELLY, Ford Werke GmbH, Germany and
M. MENDAY, Loughborough University, UK

Abstract: Manual transmissions and their layshaft derivatives, such as double clutch transmissions, have a positive future because of their benefits in fuel economy. The disadvantages regarding noise, such as transmission gear rattle, have to be understood and economically resolved to maintain this dominance. This chapter describes the historical background of powertrain torsional vibration issues, and goes on to explain the gear rattle phenomena and traditional palliatives such as clutch disc dampers and dual mass flywheel. Areas such as testing and modelling are covered, and possible negative future trends on gear rattle are reported. An important point is that the popular dual mass flywheel solution is relatively expensive and technically more challenging to integrate into a powertrain than a solid flywheel, therefore there is always a demand to seek alternatives. This chapter ends with a possible solution in the form of an electronically controlled clutch known as 'eClutch'.

Key words: transmission, gear rattle, dual mass flywheel, clutch disc damper.

27.1 Introduction: history of powertrain torsional vibration issues

The issues of torsional vibration have been recognised and documented since 1890 (Ker Wilson, 1969). One of the early areas of investigation was that related to aircraft powertrain design. The designers of the past did not have the analytical and experimental techniques that are available today. This unscientific, rather trial and error approach relied on intuition and some luck playing an important part in developments. Unfortunately, the torsional vibrations problems occurred early and often. One of the earliest documented in 1901 was the Manley-Balzer five-cylinder radial engine which was the first purpose-built aircraft engine for the Langley Aerodrome project. This experimental aircraft commissioned by the United States Army for US$50 000 was designed by Langley's chief assistant, Charles Manly. He also piloted the numerous failed flying attempts and regrettably did not consider reducing

839

the stiffness of the drive system to alter the natural frequency, but rather added mass, thereby increasing the engine weight.

Another torsional vibration problem occurred in the design of the infamous 1917 British-built ABC Dragonfly nine-cylinder radial engine. Lamentably, it had its rated power at the resonance frequency of the crankshaft, reducing the crankshaft life to a matter of hours (Gunston, 1986). Around 1000 engines were built, but only around ten actually flew (Lumsden, 2003). Fortunately, World War I ended in 1918, otherwise the deficiency would have had serious consequences to the Allied war effort.

A third example of lack of torsional vibration understanding occurred with the LZ127 Graf Zeppelin Air Balloon. It left from Europe for its second trip to the United States in May 1929, but had to return after first two, and then four of its five engines suffered crankshaft failures. These engines were equipped with flexible couplings and vibration dampers, which normally should have been designed to avoid these failures (Baker and Den Hartog, 1931).

The list of issues with torsional vibration, in recognising, understanding and controlling the problems in aircraft engines were numerous. The engine designer, when confronted with a broken crankshaft, in most cases assumed that the issue was lack of capacity and would strengthen the part. In most cases, this resolved the issue, but the possibility that he just moved the crankshaft resonance frequency outside the operating range was not realised, so propagating the incorrect diagnosis. Beginning in the 1920s the natural frequency phenomenon was eventually recognised but analytical tools were inadequate to allow the designer to construct an accurate spring-mass model of the crankshaft–propeller system. Subsequently, late in the 1920s, stiffness tests were carried out on crankshafts to allow the verification of the spring-mass models. However, it was not until the 1930s that the magnitude of the exciting torques and the engine damping values could be both accurately predicted and measured, which proved the turning point in design understanding.

27.2   Noises in the powertrain

Thanks to the understanding of the torsional vibration phenomena from aircraft powertrains, the knowledge led to the vehicle powertrain noise, vibration and harshness (*NVH*) development. For aircraft, the issue related to catastrophic failure; however, in vehicle development this was usually more a noise and vibration issue.

For simplicity the powertrain noises are divided according to the type of noise (see Fig. 27.1). One of the most annoying noises is a high-frequency *whining noise*. This is caused by meshing impacts, or tooth parametric quality variations. A screeching bearing noise can materialise due to bearing

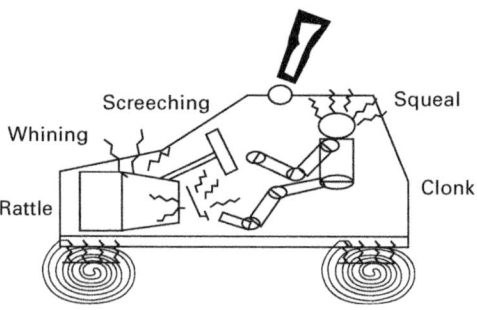

27.1 Typical transmission noises.

damage. The rattling or clattering noise is a broadband frequency caused by the meshing of the transmission loose gears induced by the second engine order (in a four-cylinder engine) (also see Chapters 26 and 29). The rattle noise appears at different drive situations, explained later, and its unpleasant nature gives the perception that something is broken in the driveline. A further group of noises are known as clonk (or clunk in the USA) (see Chapters 1 and 30) and relate to fore and aft shifting events. A scraping or grating noise is also possible when disengagement or synchroniser issues occur. Noises related to the clutch such as whoop (Kelly and Rahnejat, 1997) are outside the scope of this chapter.

27.3   Definition of rattle phenomenon in automotive powertrains

One of the most important *NVH* issues caused by torsional excitation is *transmission gear rattle*. A front wheel drive transmission example, here with cut outs showing internal mechanism, is shown in Fig. 27.2. The engine itself produces numerous noises and some of the most annoying are the ones it transmits to the transmission. The issue of transmission noise during the development process is becoming increasingly important.

On the one side are the lighter and more sensitive bodies, and on the other are the greater excitation inputs coming from the trend of higher-rated downsized powertrains. These smaller engines often have super and turbocharging combined with direct fuel injection to improve fuel economy and emission performance. The downside is that they produce correspondingly more torque and excitations into the drivetrain system. At the same time efforts to improve the engine NVH quality have highlighted the need to attack transmission gear rattle noise.

27.2 A typical front wheel drive transmission.

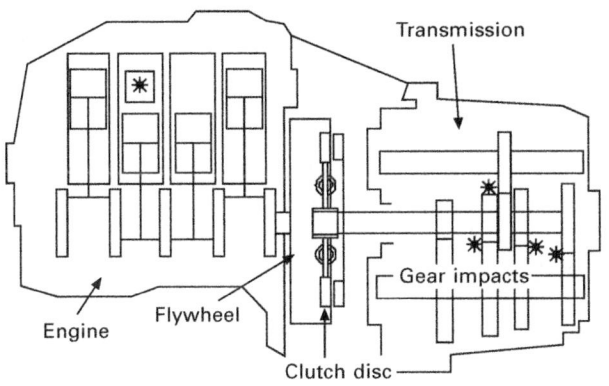

27.3 Engine, clutch and transmission layout.

27.3.1 What is transmission rattle?

Characteristics of impacting gears

Gear rattle is a strong disturbing audible noise from the transmission that can be heard during different drive conditions. The difficulty with gear rattle is that many drivers cannot separate the gear rattle noise from other engine-related noises. This means that many confuse the rattle noise with others such as **diesel knock** and the extent of the particular rattle noise is often underestimated.

To explain gear rattle one must look at it as an issue between the engine output and the transmission reaction (see Fig. 27.3) that shows the main contributing parts of the rattle system. In the powertrain system, it is necessary to understand that the piston engine's power delivery is not smooth. A piston engine makes power pulses rapidly and twice per revolution for a four cylinder engine. These pulses are converted to torsional vibration by rapid torque changes, such as rotational irregularity of the engine caused by the ignition, or alternatively by fast application of either the accelerator or clutch pedals (see Rahnejat, 1998).

The receiving end of this torsional excitation or irregularities is the drivetrain components (transmission or differential) that must have clearance in the gears to function. Therefore, when power pulse engages the transmission, it causes the torsional input through the unloaded or free spinning gear pairs to take up the slack. This effectively causes teeth that were not touching an instant before to touch each other. Figure 27.4 shows a schematic of this rattle phenomenon.

Between each pulse, the components separate again. When a hammer hits a nail, a noise is heard. Similarly when this contact and release cycle is repeated quickly, a staccato sound is produced, which is the characteristic *gear rattle noise*.

To determine the rattle noise portion of the numerous intermingling of noises present in the powertrain a simple way is to disengage the clutch. By repeating the clutch engagement and disengagement process the difference to the background noise can be heard. The rattle noise is perceived as unpleasant in nature, because of its disturbing characteristics. The driver hears a machine gun-type noise that suggests a potentially broken or soon to be broken component in the transmission. This comfort issue together with the noise pollution aspect provide significant loss in quality image of the vehicle.

The rattle noise level is ascertained by two main factors, torsional vibration excitation level at the transmission input shaft and the rattle sensitivity of the transmission at that excitation level.

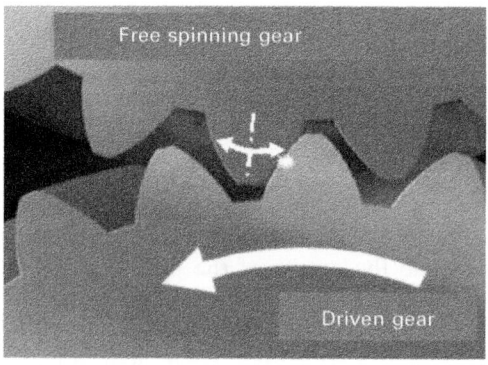

27.4 Schematic representation of gear rattle.

27.3.2 Why does it occur?

Sources of excitation

As already reported, the input causing the transmission to rattle come from the engine. Figure 27.5 shows the excitation speed variation coming from the engine, and in this case, the amplifying the transmission vibration. The excitation is directly related to high-combustion engines. Typically these were diesels engines, but are now also apparent in the trend to turbo-charging downsized petrol engines. These smaller engines often have super and turbo-charging combined with direct fuel injection to improve fuel economy and emission performance. The downside is that they produce correspondingly more torque and excitations into the drivetrain system, resulting in more torsional vibration excitations.

27.3.3 Structure-borne/airborne characteristics

The **structure-borne path** is when the vibration directly transfers from source (i.e. transmission) through driveshaft, axle assembly, suspension and vehicle body path, and the passenger compartment/cavity acts as the receiver. This includes the vibration amplification owing to the resonances of components like mating gears with shafts of transmission, driveshaft and suspension.

The **airborne paths** can also come from one or a multitude of components such as the transmission, drivetrain, axle, suspension and body. These can transfer the vibrations through the air to the passenger's ear. The most significant paths comes directly from the transmission housing directly to the front passenger's/driver's ear. Depending on the sound quality of the engine, this can be masked and lost with the engine's *presence*.

On rear wheel drive vehicles drive shafts are key elements for the transfer of vibrations from the transmission to the suspension. The driveshafts, one or two piece, are long members, around 1 m with at least two constant velocity (CV) joints across their length.

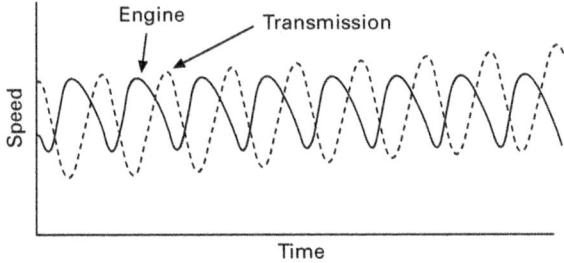

27.5 Engine and transmission speed variation.

27.3.4 Causal factors

There are many causal factors leading to gear rattle. This is not an exhaustive list, but some of the most common ones relate to both the excitation and the sensitivity of the transmission. Examples of excitation are instability at idle speeds, high compression engines, downsized high-powered engines, and thinner engine blocks or sumps.

Examples of increasing transmission sensitivity are thinner transmission case structures, lighter vehicle bodies, inertia of unloaded gears, low driveline stiffness such as rear wheel drive versions, the trend to lower damping using higher transmission oil viscosity, higher normal use oil temperature, specific low-resistance bearing types and tightness, increases in gear teeth free play due to wear, trend to three shaft transmission for package reasons, synchronised reverse gear, longer gear ratios giving higher effective vehicle mass (also see Chapter 30).

Other internal transmission components, such as idler gears, synchroniser rings and sliding sleeves also need to be considered to avoid excitation leading to rattle. The trend to higher viscosity and minimised volumes in transmissions oils to improve shiftability has also a negative impact on gear rattle.

Specifically the rattle sensitivity of transmission architecture is important. The trend from two shafts on a five-speed to three shafts on a six speed has already been mentioned (see Chapters 1 and 29). However, another significant factor on gear rattle is the designed transmission centre distance. The centre distance is directly related to the engine torque capacity required for that transmission. However, the wider the distance between the input and output shaft, the greater the gear rattle sensitivity. The trend of increasing torques over the last 20 years has also been a harmful influence on rattle.

Alignment in the clutch damper system is also a special factor. The idle gear rattle occurs when the clutch disc spline backlash fluctuates by the off-centre between the engine rotation and the transmission rotation, which results in a sudden change in the input shaft angular velocity. Therefore, the larger the off-centre between the engine rotation and the transmission rotation, the maximum value of the clutch disc spline backlash, the angular velocity and angular acceleration of engine rotation become, the greater the excitation force of idle gear rattle becomes.

27.4   System dynamics of automotive powertrans

The rattle quality is dependent on many interactions in the powertrain and vehicle (see Chapters 1, 26 and 29). One example is the system dynamics in the transmission. The sensitivity of the transmission to rattle is affected by the interaction of engine speed, amplitude input from engine, transmission lubrication and the oil temperature. Other factors relating to lubricant force, structural damping and tooth friction forces are also important. The transmission input shaft, in real driving conditions, is excited by main engine order component superimposed to constant rotating speed of the engine.

The two parameters determining the torsional acceleration are:

- the amplitude of engine order vibrations;
- the rotating engine speed.

Figure 27.6 shows an example of transmission sensitivity. The rattle noise map, shown as functions of rotational acceleration and rotating speed at input shaft for one specific transmission oil temperature. Variations in vibrations can lead to gear rattle. Examples include the generation of vibrations owing to the transmission mesh error. This is the result of irregularities of motion transfer due to gear tooth stiffness, mesh stiffness of gears and its alignment. The trend to lower idle speed for fuel economy reasons can result in unstable idle speeds, particularity in diesel engines and again can provoke rattle.

Some rattle issues can be tackled by tuning the system dynamics. For example, changing the system or component structural stiffness, inertia and damping can shift the critical natural frequencies of the system out of the forcing frequency range to reduce the noise and vibration amplitudes. In some cases reducing the transmission input shaft spline backlashes has provided enough improvement.

27.5 Types of rattle and their causes within automotive powertrains

Rattle has been defined by showing the unloaded gears impacting on each other. Unfortunately, it is not possible to eliminate the issue by removing all

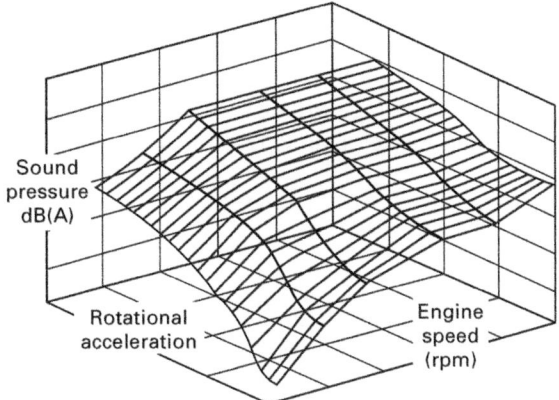

27.6 Rattle sensitivity map.

clearances. These vibrations result from rotational fluctuations of the crankshaft and flywheel induced by combustion and inertia forces in the engine. The angular velocity fluctuations causing the gear rattle are mainly due to the engine firing order. The amplitude of these torsional accelerations depends on the number of cylinders, cycles and the type of fuel (gasoline or diesel). For a four-stroke, four-cylinder engine, the main torsional fluctuations are a second engine order frequency, meaning that they are at twice the crankshaft rotational frequency (see Rahnejat, 1998).

The issue is that these exciting and rattle disturbing resonances are in the normal operating speed range of motor vehicles.

Typical rattle types caused by specific drive conditions that induce significant rattle noise are listed here.

27.5.1 Drive rattle

Drive rattle (see Chapters 1, 21, 26, 30) occurs at low vehicle speed and high engine load, with the clutch engaged, any gear selected, running at partial load or wide open throttle at engine speed around 1500–4000 rpm. The peak rattle is around 1600 rpm for a four-cylinder, and 1100 rpm for a six-cylinder engine.

27.5.2 Overrun rattle

Overrun rattle is the condition when the clutch is engaged, a high gear selected, and running at engine speeds in the range 2000–4000 rpm with negative engine torque demand.

27.5.3 Creep rattle

Creep rattle (see Chapter 1) occurs when moving slowly in 1st or 2nd gear (for example, in a traffic jam) with variations as load input. The conditions are clutch engaged, with and without small throttle input and driving on flat or slightly sloped road. The slope effect is often simulated in a vehicle by lightly applying the hand brake. This is one of the worst conditions for gear rattle.

27.5.4 Idle rattle

Idle rattle (see Chapters 21, 26, 29, 30) is the condition with gear lever in neutral and clutch pedal not applied and the engine at idle rpm. It is normally tested with and without electrical loads applied.

27.6   Traditional rattle palliations in automotive powertrains

There are two traditional ways to tune away gear rattle. Figure 27.7 shows the effect of clutch dampers and *dual mass flywheel* (*DMF*) tuning on reducing the input shaft torsional vibration (Drexl, 1998).

The dual mass flywheel DMF (see Chapter 28) clearly provides a major reduction of engine excitations. The clutch dampers have a positive effect, but need specific application tuning of friction and damping to obtain their optimal potential. The clutch damper is a cheaper solution, but does not completely reduce the engine input excitation over the complete operating range.

The next sections look at these alternatives in some detail.

27.6.1   Palliation by modification of clutch disc characteristics

It is not easy to improve rattle noise by reducing rattle sensitivity of transmission under mass production conditions. Therefore, methods of improving rattle noise by reducing torsional acceleration are more robust.

In the traditional case a *solid mass flywheel* is equipped with a sprung hub clutch. The spring damped clutch disc is mounted between the flywheel and the transmission input shaft. The clutch friction disk is not directly fixed to the clutch hub, but has a spring centrally located in between. The sprung hub clutch has a damper that is tuned to suppress the transmission resonance to

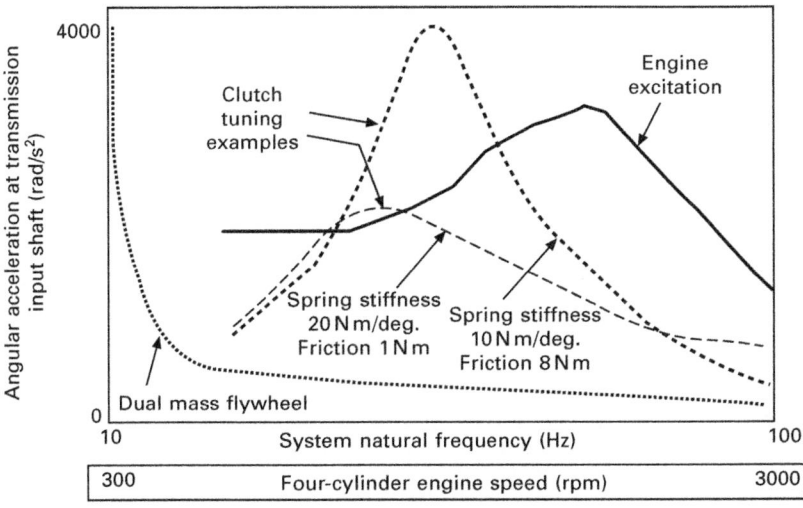

27.7 Clutch and dual mass flywheel tuning affects on transmission input shaft torsional vibration.

minimise the input speed fluctuations, and thus the gear rattle excitation (see Fig. 27.8). Driveline resonance can also be isolated effectively by changing some driveline parameters such as clutch and driveshaft stiffness.

Figure 27.9 shows a typical torsional vibration damper characteristic. The characteristic has a drive and coast region, with stop pins to restrict them. The stop pins limit the maximum torque and prevent overload on the

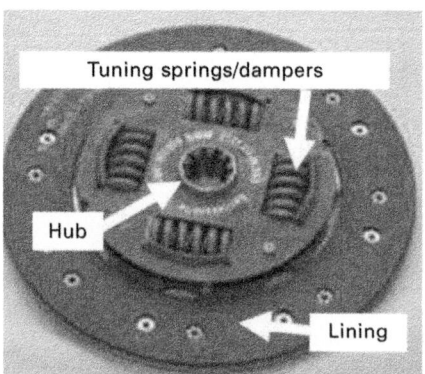

27.8 A traditional damped clutch disc for rattle palliation.

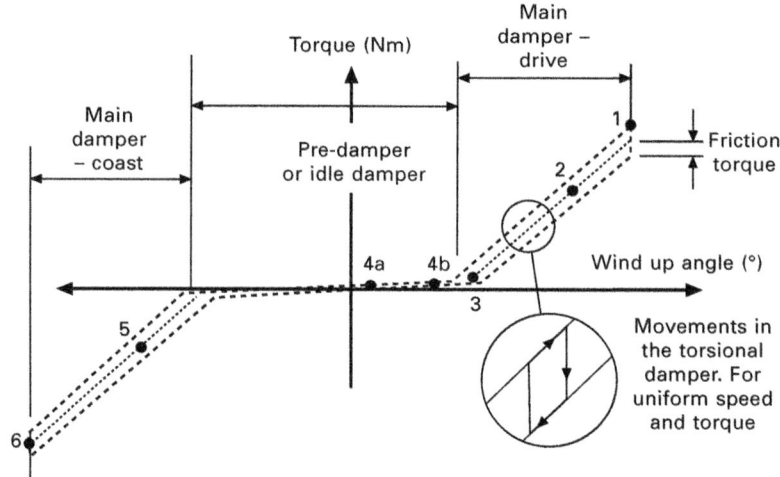

27.9 A typical characteristic curve of the clutch disc damper: 1, maximum damper capacity wide open throttle (WOT), impact shocks; 2, drive 4th gear WOT (drive rattle); 3, idle gear, creep 2nd gear (creep rattle); 4a, idle neutral, low drag torque warm transmission, consumers off (idle rattle); 4b, idle neutral, high drag torque cold transmission, consumers on (idle rattle); 5, coast or overrun, high *gear thrash noise* (coast rattle); 6, coast or overrun, 2nd gear, downhill or transient downshift (coast rattle).

springs. The damper spring is totally used only when a complete torque reversal takes place. Therefore, most of the time, the full operating range is not being used and only small movements occur as shown in the enlarged circle. The engine's rotational irregularity applies a load to the input side of the torsional vibration damper. The lower the spring rate is, the smaller the torque fluctuation at the transmission input shaft become. The objective is to have the lowest spring stiffness derived from the maximum available angle and required maximum stop torque. To reduce the effect of the resonance area a friction hysteresis is added to the damper system to reduce the resulting amplitudes. This is achieved by a device consisting of friction rings and a spring washer that provides the clamping force.

Most passenger cars have four springs around the central plate. The maximum torsional angle is limited by the diameter of the clutch inner lining.

The clutch hysteresis is a compromise between drive and coast conditions. The torsional acceleration under drive decreases for high hysteresis, but in contrast the acceleration under coast decreases for low hysteresis.

Note that gear rattle can still occur at idle with gear in neutral, and with the clutch engaged. This is where the pre-damper stage does not cover the angular excitation range of, say, an unstable idle condition. It isolates at most other times. Above idle (even by a hundred rpm), the increased frequency of power pulses outpaces the speed at which the components couple/de-couple, therefore reducing the rattle.

Conditions where the engine is loaded at idle can also lead to gear rattle being heard. Worse case conditions are hot temperature conditions at idle, where the air conditioning and lights load up the alternator. Headlights and other electrical loads may increase the amount of residual rattle.

27.6.2 Palliation by DMF

For engines with high excitations, such as diesels or high-performance petrol engines, or transmissions that are sensitive to excitation, such as multi-shaft six speeds, the normal solution is the use of a **DMF**. This replaces both the traditional solid flywheel and the tuning springs in the clutch disc. The DMF moves the natural frequency of the system out of the normal operating speed driving range. The lower resonance frequency moves below the idle speed so the resonant condition is overcome in the starting process. With a correctly tuned DMF gear rattle noise is completely resolved.

Figure 27.10 shows the components in the most common design of DMF. The tuning relies on two long arc springs around the circumference of the flywheel that transfer the engine torque. One side of the spring reacts against the primary flywheel that is directly bolted on to the engine crankshaft. The other end of the arc spring reacts against the secondary flywheel that is free

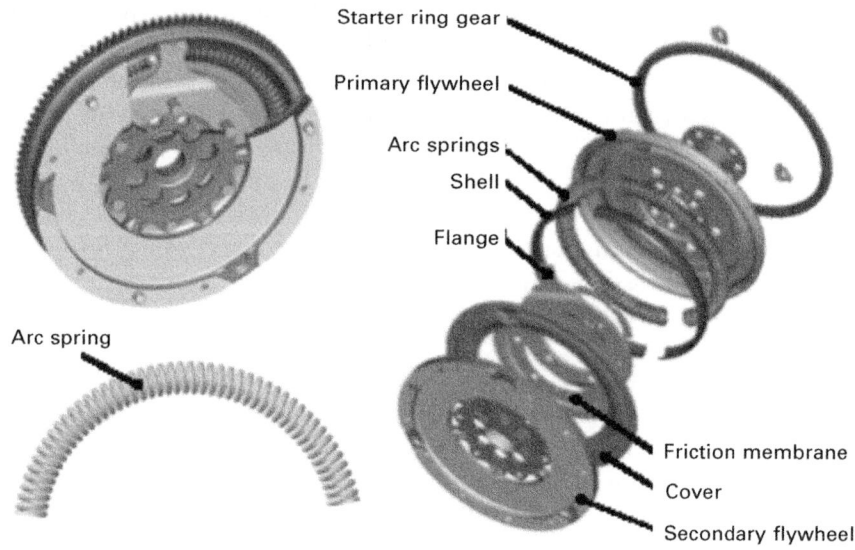

Starter ring gear

Primary flywheel

Arc springs

Shell

Flange

Arc spring

Friction membrane

Cover

Secondary flywheel

27.10 Components of a basic dual mass flywheel.

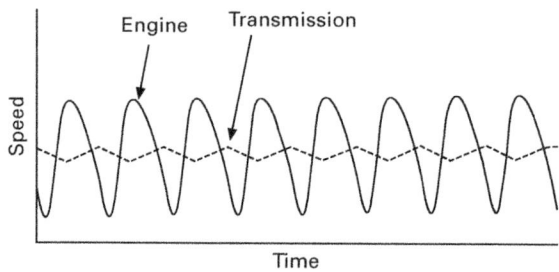

Engine   Transmission

Speed

Time

27.11 The transmission isolation with a DMF.

to rotate relative to the primary. The normal range of movement is around 60° travel, which enables a design to have either a low stiffness single or double stage spring. Additional friction devices are added to the internals to offer further tuning potential.

The clutch is assembled between the secondary flywheel and the input shaft as on the solid flywheel design, but does not normally need any dampers incorporated. The clutch cover is mounted on the secondary flywheel so that one side of the clutch disc facing is clamped directly on the secondary flywheel, enabling torque flow to the transmission if the clutch is engaged.

For drive conditions the DMF proves an unsurpassed isolation (see Fig. 27.11 for the DMF influence). Compare this with Fig. 27.5 to note the improvement. The idle speed isolation in the DMF is given by arc spring lash, which must be larger than normal engine angular fluctuations. With this configuration, the only coupling between primary and secondary flywheels is the so-called basic friction, which is the sum of bearing friction and friction designed into the DMF.

27.7   Experimentation and evaluation method of rattle sensitivity in automotive powertrains

The noise level which is measured close to the transmission cannot be compared directly with the rattling noise level in vehicles because the car passengers do not hear only the gear noise. The other issue is that the direct airborne noise measured on a rig is not possible in a car because in real conditions the engine and transmission are shielded in a car. However, to provide comparative data for a model or to verify it, rattle sensitivity, which is usually investigated experimentally on a rig, is very valuable.

Figure 27.12 shows a transmission prepared for NVH measurements. A transmission installed on a rig in semi-anechoic chamber will then provide the possibility to assess the influencing components such as gears, synchronisers, bearings, case, oil and non-linear properties such as meshing stiffness and viscosity.

To measure the rattle sensitivity, the transmission is excited directly to the transmission input shaft by the main engine order acceleration. In the case of a four cylinder engine this is second order. The torque carrying gear pairs of the transmission are loaded so that the gear teeth come into contact on the drive side.

Gear rattle sensitivity can be measured as a function of engine speed fluctuation and the radiated rattle noise and input shaft acceleration are also measured. The sound pressure level is measured at four ends of the

27.12 A transmission prepared for an NVH test.

transmission at a set distance. Noise signals are dB (A) weighted with calculated mean noise levels of the microphone.

The input shaft speed is measured at a fixed gear fixed on input shaft. The torsional acceleration is calculated from which the peak to peak angular acceleration in rad/s^2 is determined.

Rattle sensitivity can be described using a curve of rattling noise versus rotational acceleration of transmission input shaft The underlining noise occurs from start of rattle and consists of bearing noises, plunging and tooth meshing noises. As the engine speed increases, the torsional accelerations result in the sliding gears separating from their pairs which is the disturbing gear rattle case.

With increasing acceleration amplitudes, the sound pressure level changes non-linearly with torsional acceleration. This depends on the effect of gear teeth bending. Each transmission type has its own signature as the resultant sensitivity curve depends on the dimensions and components used. It is important in any comparison to compare like for like at constant oil temperature which has a significant effect on *drag torque* and gear sliding (also see Chapters 1, 26 and 29).

27.8   Simulation of rattle phenomenon in automotive powertrains

With major reductions in vehicle development times, it is not always possible to wait for expensive prototype parts to be manufactured and physically tested. In the automobile industry, most parts are first designed and evaluated well before a prototype is produced. This also reduces the increasing cost of developing new parts.

The building of a model is relatively simple and the design data to run it are also readily available. The most difficult aspect now is determining the so-called *noise factors* to run the model. Customer usage and system to system interactions are the critical elements to ensure that the final design is acceptable.

NVH is an important part of the development process and many simulation tools are available. Figure 27.13 shows the mass and spring system schematic for the model of a DMF and a conversional system together with the prediction of the torsional vibration from the engine and the resulting torsionals after passing the respective damper systems.

27.9   Future trends

An important point is that DMF is relatively expensive and technically more challenging to integrate into a powertrain than a solid flywheel, therefore there is always a demand to seek alternatives.

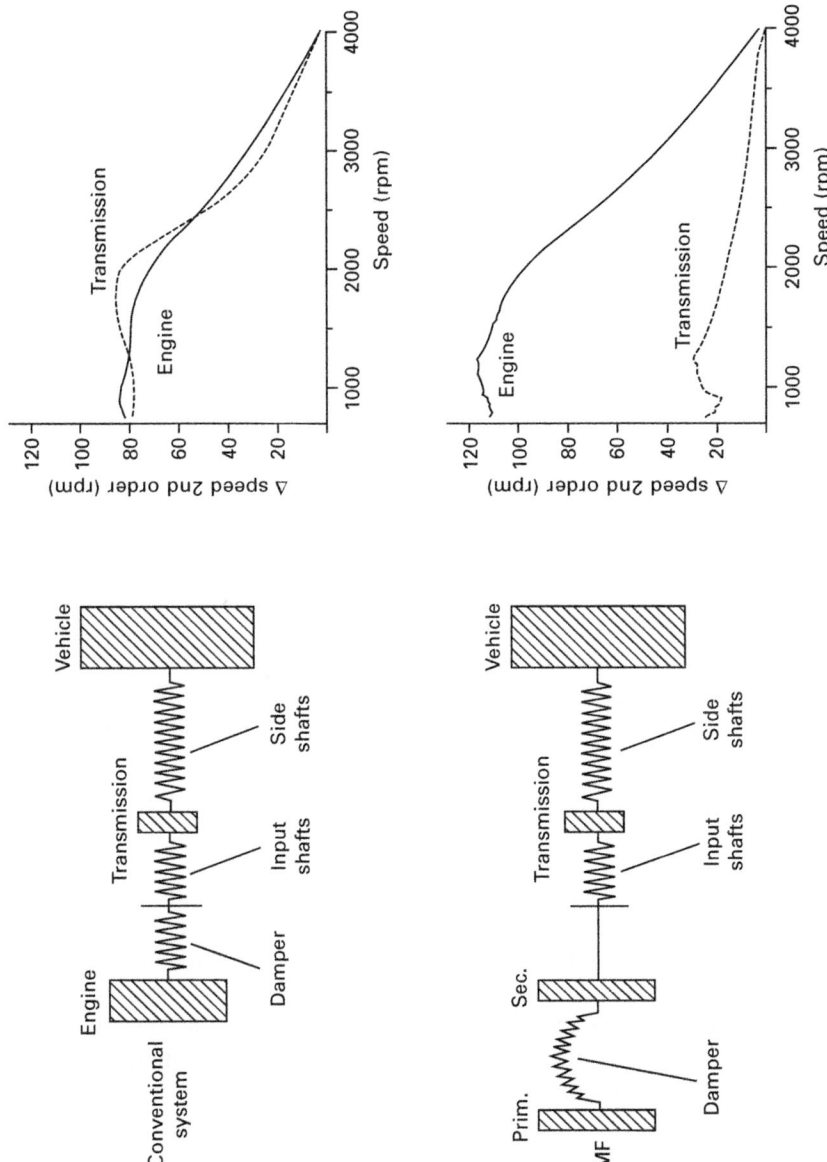

27.13 Principle and effect of the dual mass flywheel, in comparison with a conventional clutch disc.

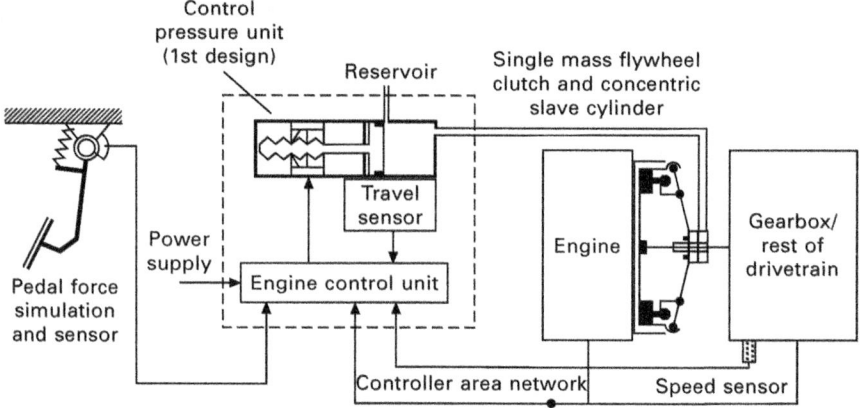

27.14 A schematic of an eClutch.

Manual transmissions are the most popular form of transmission even today. Although there have not been many major developments except the move to six speeds, one area that has seen developments is clutch operation. The traditional clutch control of disengaging and engaging to allow torque transfer has changed from cable actuation to hydraulic. This clutch control is easy to use, and robust in coping with the torque demands. With the increasing trend of higher engine torques, the traditional system is being evaluated again. The next development is electrical control. Figure 27.14 shows an electronic clutch (*eClutch*) schematic.

With the knowledge obtained from slip control of the ***double clutch transmissions***, it is feasible to use this technology on the single clutch manual transmissions to control clutch slip. This gives the opportunity of a clutch-by-wire or eClutch system to filter the torsional vibrations flow (also see Chapter 24). One possible benefit is the revaluation of the filter function with today's conventional or DMF solution.

In the future, an electrical sensor will be able to measure the clutch pedal position and the data read by an electronic module that would control an actuator to disengage and engage the clutch. Consequently, pedal travel and effort curve can be tailored to suit an optimised ergonomically and comfortable position for the driver. The integration of the system to the engine, transmission and other vehicle systems enables many function features that were either not possible before or too costly.

27.10 References

Baker and Den Hartog., 'A complete description and analysis of the Zeppelin failures are in "Zeitschrift für Flugtechnik und Motorluftschiffahrt" for September 1929', *SAE Journal, Feb.* 1931. p. 465.

Drexl, H-J., *Motor Vehicle Clutches*, Landsberg: Verlag Moderne Industrie, 1998.

Gunston, W., *World Encyclopedia of Aero Engines*. London: Guild Publishing, 1986. p. 8.

Kelly, P. and Rahnejat, H., 'Clutch pedal dynamic noise and vibration investigation' in Rahnejat, H. and Whalley, R. (eds.) *Multi-body Dynamics: Monitoring and Simulation Techniques*, London: Mechanical Engineering Publications (IMechE), 1997.

Ker Wilson, W., *Practical Solution of Torsional Vibration Problems*, 5 volumes, 3rd edition, London: Chapman and Hall Ltd, 1969.

Lumsden, A., *British Piston Engines and their Aircraft*. Marlborough: Airlife Publishing, 2003, p. 53.

Rahnejat, H., *Multi-body Dynamics: Vehicles, Machines and Mechanisms*, London: PEP and Troy: SAE, 1998.

28
Dual mass flywheel as a means of attenuating rattle

P. KELLY, Ford Werke GmbH, Germany and B. PENNEC,
R. SEEBACHER, B. TLATLIK and M. MUELLER,
LuK GmbH & Co. oHG, Germany

Abstract: With the ever-increasing demand to improve noise, vibration and harshness (NVH) in motor vehicles one of the most significant powertrain contributions comes from the *dual mass flywheel* (DMF). Its primary aim is to reduce rattle and boom in full and part load drive situations. This chapter illustrates the range of development effort aimed at reducing engine-induced rotational vibrations in the drivetrain.

Key words: dual mass flywheel, drivetrain rattle, full and part load drive.

28.1   Basic dual mass flywheel (DMF)

The primary aim of the dual mass flywheel (DMF) is to reduce *rattle* (see Chapters 21, 26, 27 and 29) and *boom* in full and part load drive situations. Its location in a rear wheel drive application can be seen in Fig. 28.1 and its main components are shown in Fig. 28.2.

The primary flywheel, which is bolted onto the engine crankshaft, is free

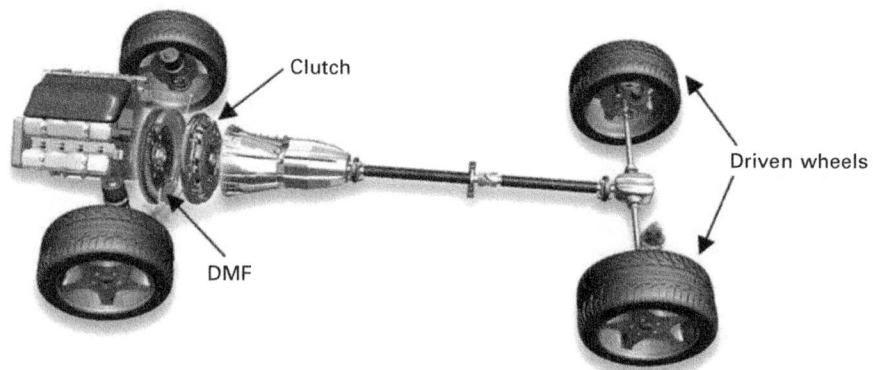

28.1 Location of the DMF in a standard drivetrain.

857

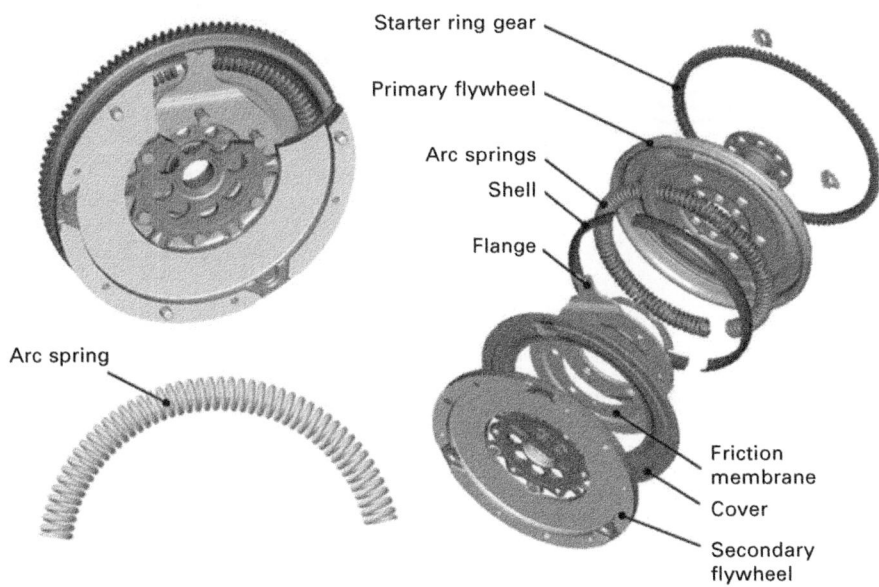

28.2 Components of a basic DMF.

to rotate relative to the secondary flywheel. DMF arc springs located on the outer edge of the flywheel are used to transmit the torque between the primary flywheel and a flange, which is riveted to the secondary flywheel. With a positive load, the arc springs are pushed by the end stop on the primary side against the flange on the secondary side. With a negative load, the arc springs are pushed the other way round by the flange against the end stop. Furthermore, the damping characteristic of the DMF is equivalent to a mechanical low-pass vibration filter.

The clutch cover is mounted on the secondary flywheel so that one side of the clutch disc facing is clamped directly onto the secondary flywheel, enabling torque flow to the transmission if the clutch is engaged.

28.1.1 Physics of arc spring

Lash within the **DMF arc spring** channel and Coulomb friction due to the typical behaviour of the arc spring generate non-linear terms in the equations of motion, which can subsequently affect the rest of the drivetrain behaviour. This part describes basic physical properties of the arc springs.

The basic principle of the force equilibrium conditions within the operating arc spring is sketched in Fig. 28.3. A force F_L, acting at one end of the spring, is transferred along the curved line. Assuming a homogeneous mass distribution this line represents the centre of mass of the spring with respect to its arc angle. For the equilibrium at coil i inside the spring a radial force

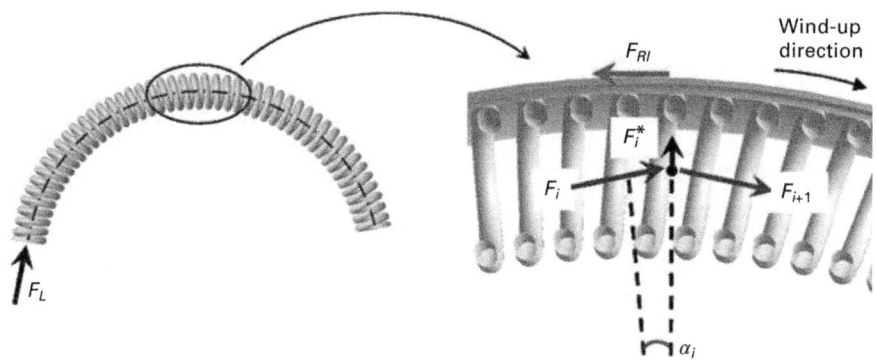

28.3 Force equilibrium schematics for the arc spring.

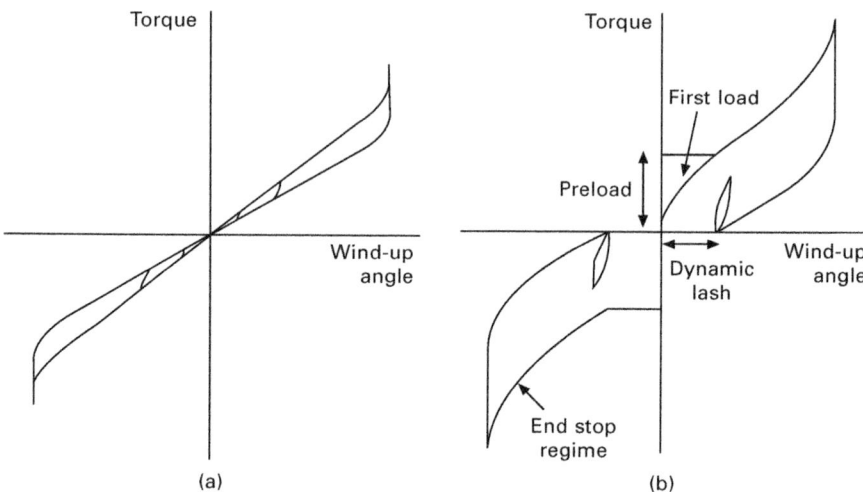

28.4 Homogeneous arc spring characteristics (a) static case and (b) at 3000 rpm.

F_i^* results from the deviation of the elastic forces F_i and F_{i+1} at both sides of the coil, thus:

$$\vec{F}_i^* = \vec{F}_i - \vec{F}_{i+1}, \; F_i^* = F_i \cdot \sin(\alpha_i), \; F_{Ri} = \mu \cdot F_i^* \qquad (28.1)$$

The dependency of the resisting friction F_{Ri} on the preload of the neighbouring coil F_i determines the current deviation angle α_i and leads to the typical trumpet shape of the arc spring characteristics, until closed to the end stop as shown in Fig. 28.4(a)). When the coils move under increasing load, the friction between coil wire and shell results in a subsequent decrease of the elastic force within the individual coils and vice versa for the unloaded case. Consequently, in the end stop regime, the spring becomes stepwise blocked, starting from the introduction side of the force and the resulting stiffness increases (Fig. 28.4).

In addition to the so-called *deviation friction*, the contribution originating from the centrifugal force of the spring in the DMF device increasingly dominates the friction torque above about 1000 rpm by the square of the rotational speed ω of the rotating crankshaft. This friction force does not depend on the applied load. Expressing the situation in terms of torques the total friction at coil i yields:

$$T_{Ri} = \mu \cdot T_i \cdot \sin(\alpha_i) + \mu \cdot m_i \cdot \left(r - \frac{d}{2}\right) \cdot \omega^2 \cdot r \qquad (28.2)$$

with μ being the coefficient of friction, m_i the mass of coil i, r the outer spring radius and d the coil diameter.

If the theoretical friction torque T_{Ri} prevails, then the difference of the elastic torques T_{Ri} and T_{Ri+1} at both sides of the coil i will stick and the following coils also follow suit. Under load the resulting stiffness subsequently decreases with an increasing number of active coils as depicted in Fig. 28.4(b)) for the first application of load to the unloaded spring and in the beginning of the unloading process due to the reversion of sign of the friction torque.

Friction due to centrifugal force will result in a large dynamic lash of the arc springs (additionally due to a probable geometrical lash), as well as a significant preload torque (see Fig. 28.4(b)). Obviously, the arc springs will then only be completely pushed through the channel after the preload torque is exceeded by the applied torque on the DMF.

When operating in a certain region of the characteristics (e.g. constant drive with mean torque T_{eng}) typically only a few coils n_{act} are active. This results in a larger stiffness as compared with the nominal value C of the spring and a low hysteresis which is sketched in Fig. 28.5(a) as the slope

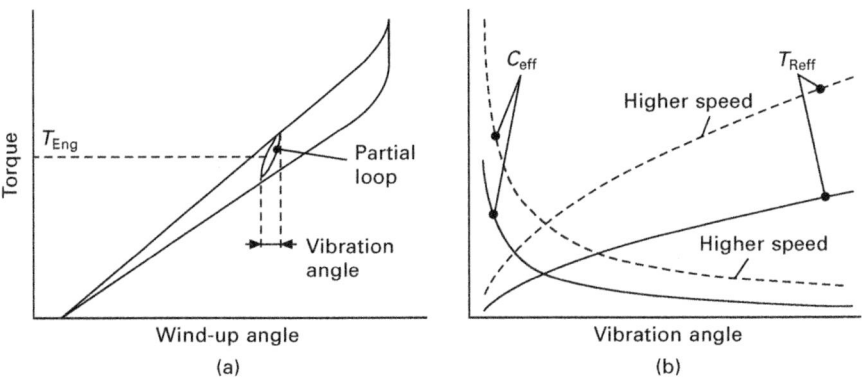

28.5 (a) Partial loop at operating point (b) angle and speed dependence of effective stiffness and friction.

and the area of a partial loop. The higher the speed, the smaller and steeper the partial loop becomes, even for the same mean engine torques. According to Eq. 28.2 the complete instantaneous friction torque adds to:

$$T_{Reff} = \sum_{i}^{n_{act}} T_{Ri}$$

(28.3)

and the effective stiffness results in:

$$C_{eff} = \sum_{i}^{n_{act}} C_{i_{homogeneous\ spring}}\ n_{act} \cdot C$$

(28.4)

In summary (Fig. 28.5(b)) the effective stiffness C_{eff} of the arc spring increases with increasing friction and decreasing number of active coils or width of the vibration angle respectively. The effective friction T_{Reff} increases with the vibration angle and the speed of the crankshaft.

Passing the natural frequency of the DMF at the start condition with large vibration angles is no problem, because of the very large damping of the arc spring, whereas under permanent load in a driving cycle it hardly incurs any frictional losses (shaded areas in Fig. 28.5(a)).

28.2   Dual mass flywheel (DMF) interactions at different operating points

28.2.1 Operating condition drive

The engine irregularity increases when a conventional flywheel is replaced with a DMF. However, this usually does not affect transmission rattle negatively because the vibration isolation of a DMF-equipped system is superior to that of a system fitted with a conventional flywheel. This becomes obvious when comparing the amplitudes of the transmission signals (see Fig. 28.6).

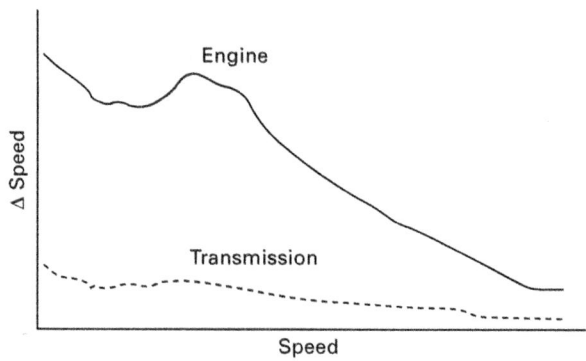

28.6 DMF vibration isolation during full-load drive.

Since the DMF resonance is at sub-idle speeds, the vibration isolation range covers all engine speeds encountered during normal driving conditions. The isolation of the transmission and the differential speeds, which is only a fraction of the engine irregularity, can be seen in Fig. 28.6 as a function of engine speed.

During drive, the arc spring angles are small because only a few coils are active. This also means that the effective spring rate of the partial loops during drive is stiffer than the nominal spring rate of the total DMF characteristic curve (as mentioned previously). The stiffer spring rate increases the DMF's resonance speed, and the increased frictional hysteresis due to centrifugal forces increases damping. In most vehicles, this does not cause any acoustic problems because engine irregularity also decreases at higher speeds, so trade-offs can be made with respect to the DMF's filtering function.

Even though the main purpose of the dual mass flywheel is to raise the comfort level during drive situations, the DMF design is influenced by other vehicle operating points. In fact, every DMF design represents a compromise for different operating conditions. Therefore, two other important operating points will be taken into account later: engine start and idle.

Although the DMF is not rigid at high speeds, acoustic problems may occur. In this case, the DMF can be equipped with an inner damper (Fig. 28.7). The inner damper is integrated into the flange of the DMF in such a way to connect serially to the arc springs. Its main advantage is the nearly frictionless damping, which compensates the effects of centrifugal forces on the arc springs at higher engine speeds and leads to an improvement in drive isolation. This is achieved by employing straight coil springs, on which, contrary to the arc spring, only the end coils of the springs make contact and the actuating forces are oriented in the same direction.

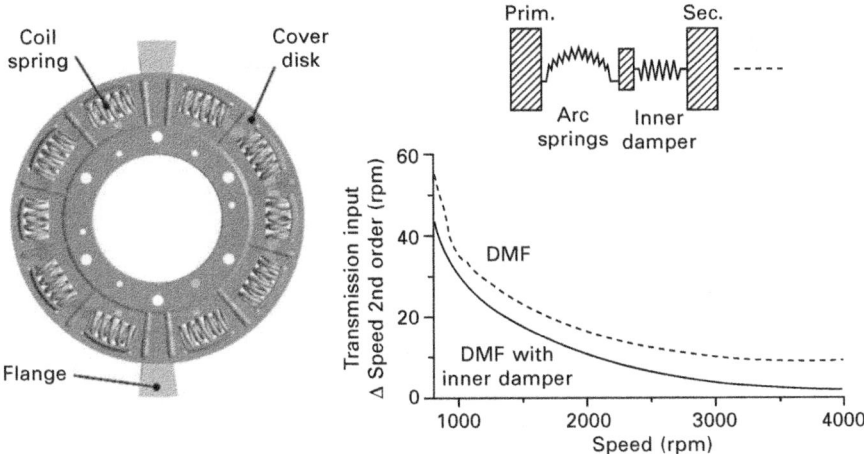

28.7 Inner damper.

28.2.2 Operating condition start

Every time the engine is started, the DMF has to be accelerated to speeds above its resonance frequency, meaning that it has to pass through a critical speed range. A typical engine start can be seen in Fig. 28.8. During the compression and expansion phase of the engine, when no fuel is injected, the starter motor accelerates the DMF up to a speed of about 200 rpm. The first fuel injection with a subsequent combustion occurs at about $t = 0.35$ seconds. From then on, the energy used to accelerate the engine and the DMF up to the idle speed is obtained mainly from the combustion process instead of the starter motor.

Naturally, the arc springs wind up to their maximum angle while the DMF passes through its resonance speed of about 350 rpm. Fortunately, the characteristic curve of the arc springs produces a maximum damping effect when large relative angles occur, as during an engine start. When the engine speed is high enough, the idle speed controller is activated and the engine operating point changes to the idle.

Nevertheless, resonance problems may still occur during engine starting. A low DMF resonance frequency combined with a strong starter motor usually resolves this conflict. A strong starter motor can provide enough energy for the DMF to clear the critical resonance speed. Additionally, decreasing the spring rate decreases the DMF's resonance frequency accordingly.

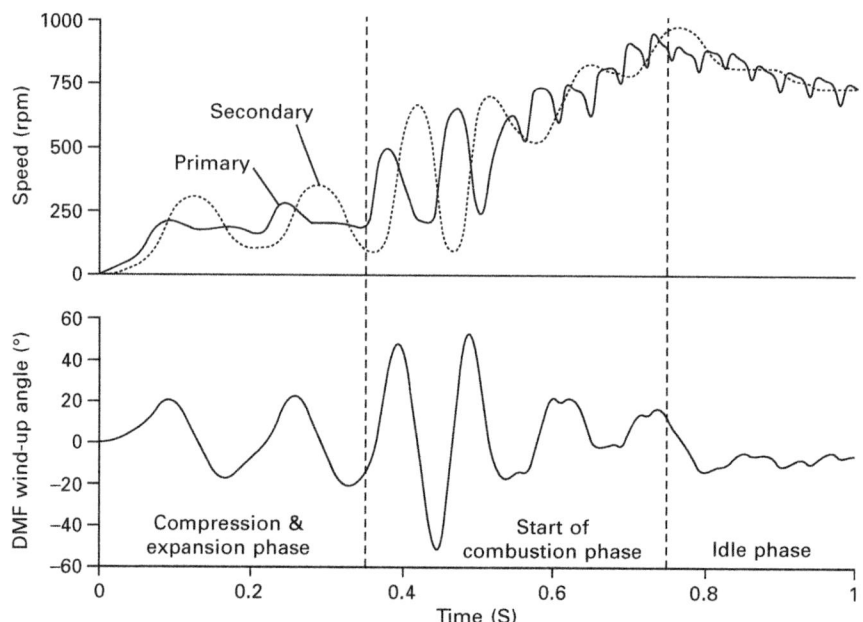

28.8 Normal engine starting with a DMF.

When lowering the *arc spring rate*, one must consider the maximum torque which the DMF should be able to transmit, since the engine torque remains constant. Furthermore, the maximum wind-up angle is fixed due to design constraints. Thus, lowering spring rates for engine starting can be achieved only by applying a multi-stage characteristic as shown in Fig. 28.9. However, this leads to degraded drive isolation, since a multi-stage DMF characteristic curve is far stiffer at high engine torques than that of a comparable single-stage arc spring characteristic.

In critical applications (usually diesel engines with few cylinders and high displacement per cylinder), the mentioned countermeasures may still not render an adequate starting behaviour. Durability issues and acoustic problems occur if the arc springs are fully compressed, which may occur during DMF resonance. In such cases, starting behaviour can be improved by using a *friction control plate* (FCP). Figure 28.10 shows engine starts with and without an FCP. As can be seen, vibrations of both DMF are very different during the DMF resonance between 0.35 and 0.5 seconds.

An FCP provides additional Coulomb friction for large wind-up angles of the arc spring. An overview of basic FCP functionality is shown in Fig. 28.11. For small arc spring angles, no additional friction is generated because the FCP oscillates in the free angle of the flange. When the arc springs' wind-up angles become large enough to clear the lash of the free angle in the flange, the FCP is rotated relative to the *primary flywheel*. An axial preload of the FCP (see Fig. 28.11) induces Coulomb friction at this point, which results in a torque between the two components.

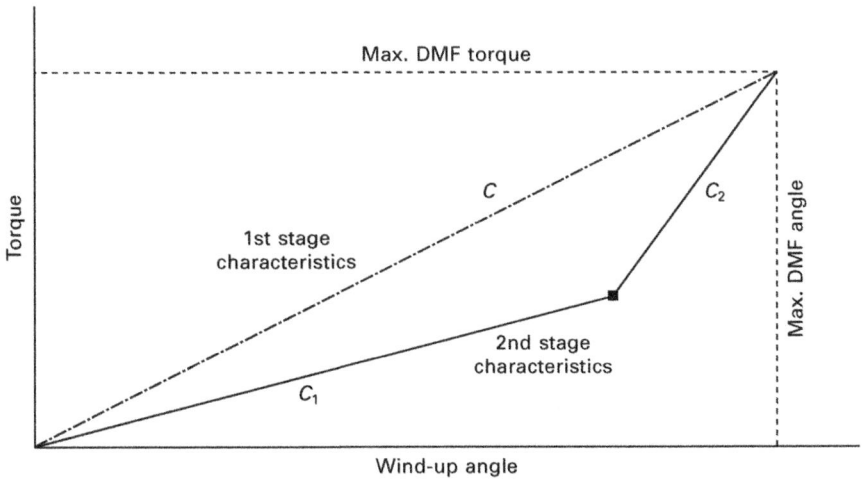

28.9 Effect of multiple stages on arc spring characteristic curves.

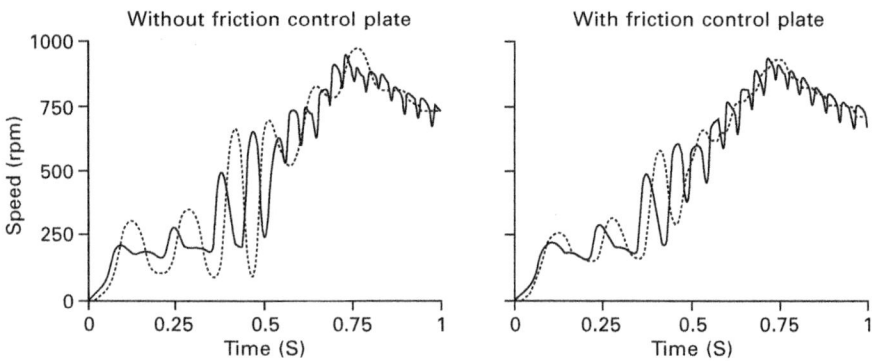

28.10 Improvement in starting behaviour due to a friction control plate.

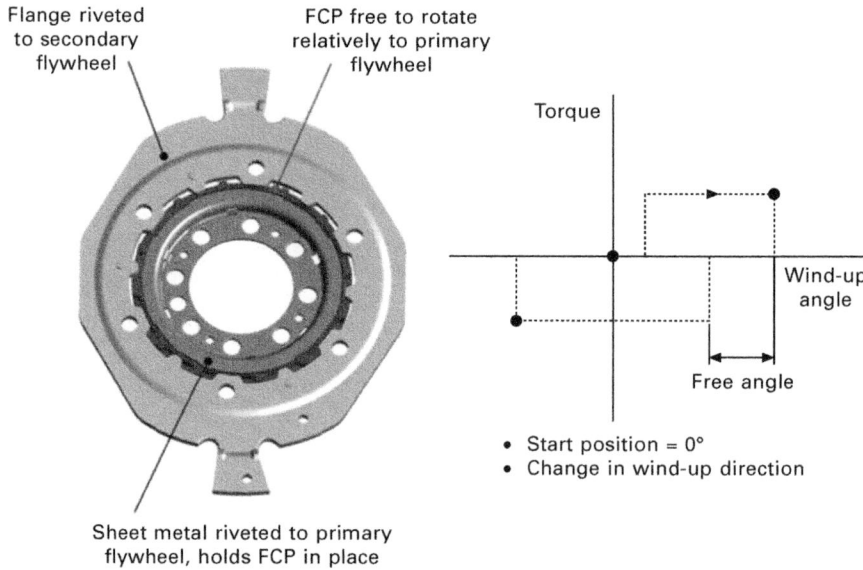

28.11 Friction control plate design and torque characteristics.

28.2.3 Operating idle condition

The arc springs are usually short enough to allow a certain amount of lash in the DMF. During idle, the engine speed remains constant at the set point of the idle speed controller. The DMF should oscillate within the arc spring lash (i.e. the flange should not activate the arc springs). Isolation is provided by the basic friction of the DMF, owing to the drag torque of the bearing between the flywheels and a friction membrane (see Fig. 28.2).

In Fig. 28.12, the graph on the left shows a DMF with enough lash to

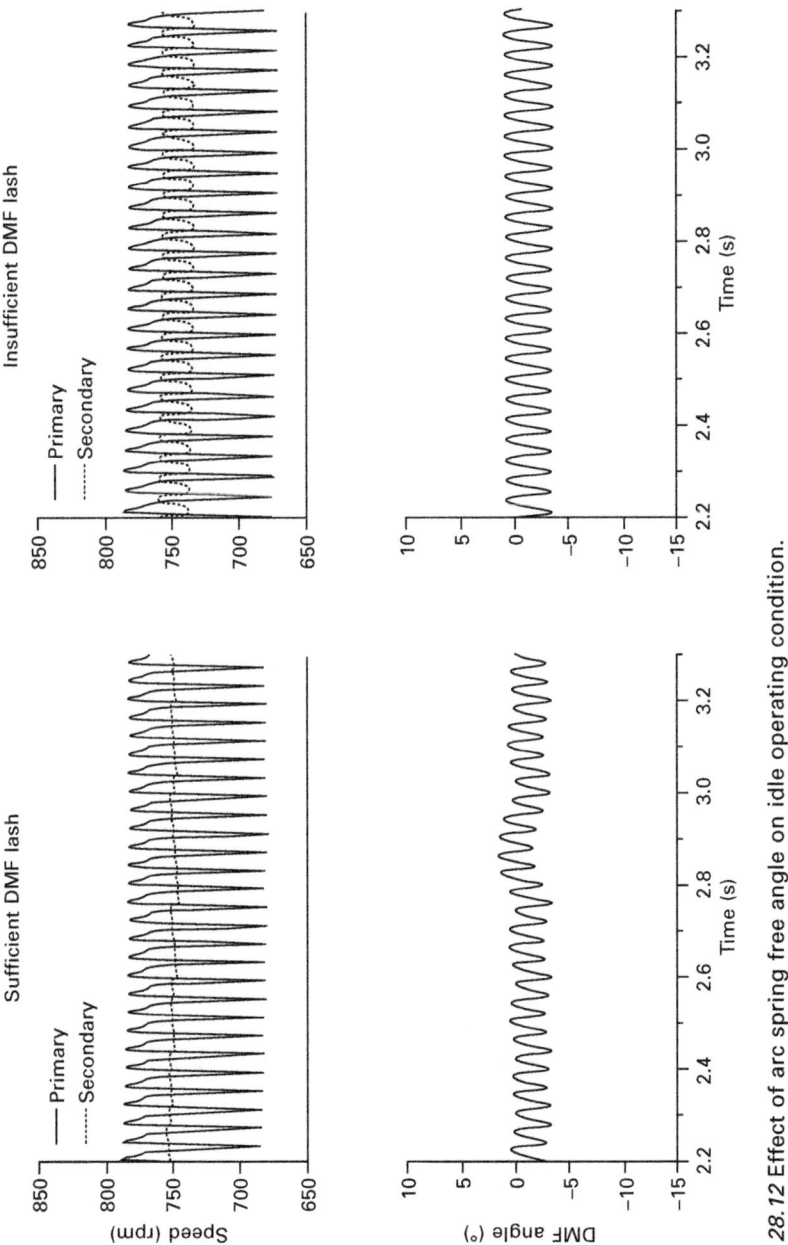

28.12 Effect of arc spring free angle on idle operating condition.

provide adequate idle isolation. Accordingly, the ***DMF wind-up*** angles are smaller than the lash, and the secondary speed exhibits almost no irregularity. In contrast, the graph on the right shows the effect of a poorly defined DMF lash during engine idle. The free angle is too small for this application. Instead of being transmitted solely by friction elements, torque is also transferred by the arc springs, resulting in a high irregularity of the secondary speed. Associated acoustic problems are likely to occur.

28.3   Dual mass flywheel (DMF) interactions with engine environment

In some operating points, the behaviour of the DMF can be strongly influenced by other components like the engine, the engine control unit (ECU) and the starter system. This section describes some key examples during engine starting, idle and engine stop.

28.3.1 Engine starting

The engine starting quality is strongly influenced by the starter motor and the injection strategy. Indeed, an insufficient amount of kinetic energy initially provided to the system will not compensate for the energy loss caused by the DMF resonance. An adequate starter motor must be carefully chosen to fulfil this requirement, even under critical conditions with low battery voltage or corroded components of the starter system. Moreover, the engine should not be fired too soon during the starting phase before the starter motor reaches a stationary speed.

This is illustrated in Fig. 28.13, where the initially satisfactory vibrations during an engine start (top diagram) are strongly worsened due to a much earlier injection (middle diagram). Combined with a less powerful starter motor, this unfavourable strategy would even lead to a long DMF resonance (bottom diagram), which could damage the DMF.

28.3.2 Engine idle

As explained in Section 28.5, idle isolation in the DMF is given by arc spring lash, which must be larger than the normal engine angular fluctuations. With this configuration, the only coupling between the primary and secondary flywheels is the so-called *basic friction*; the sum of the bearing friction and the friction generated on purpose in the DMF.

However, ***idle stability*** can be disturbed by such events as an additional torque demand on the engine (air-conditioning, power-steering, consumers, etc.) or a clutch engagement in neutral gear. The engine speed can subsequently deviate from the idle target over a short time. During this process, the DMF

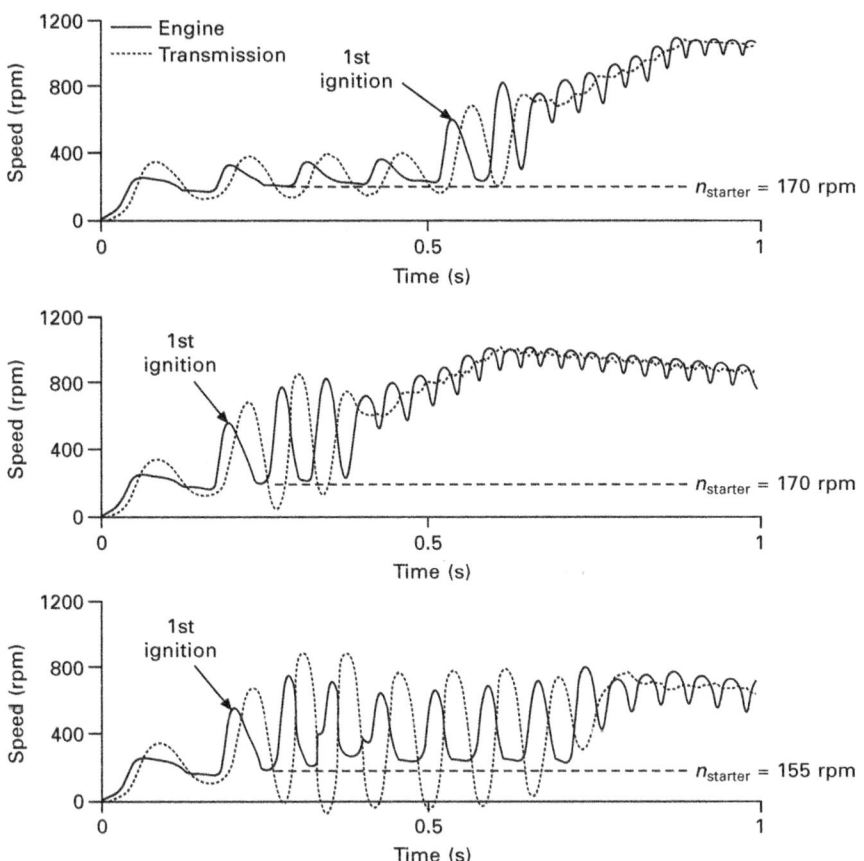

28.13 Influence of injection strategy and starter motor on engine start with DMF.

flange may hit the normally inactive arc springs. This state no longer matches the ideal DMF state in idle without spring contact between the primary and secondary flywheel. Owing to their large torsional excitation at idle, diesel engines are mostly of concern in this regard.

Under certain circumstances, the so-called sub-harmonic vibrations appear, in which the flange does not return to its initial position after hitting the arc spring, and would oscillate with a frequency lower than the *ignition frequency*. For example, the movement may have a period of three ignitions: one arc spring being hit at first ignition, flange oscillation within lash at second ignition, and the other arc spring being hit at the third ignition. The graphs in Fig. 28.14 illustrate the engine and transmission speed signals with and without sub-harmonic vibrations for a six-cylinder diesel engine.

This state can remain perfectly stable until another disturbance is brought to the system. Unfortunately, sub-harmonic vibrations can generate body shaking

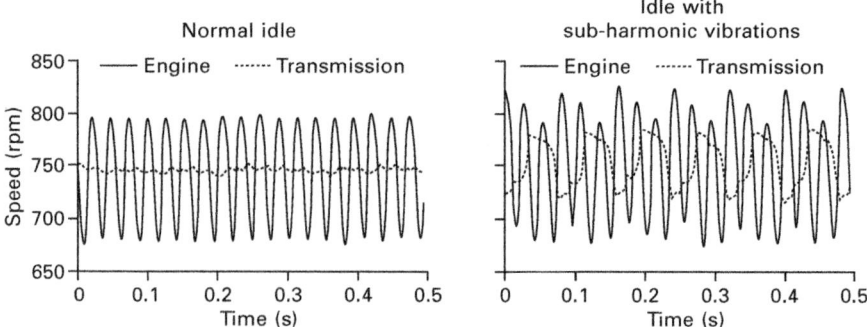

28.14 Time domain comparison of normal engine idle and idle with sub-harmonic vibrations.

or gearbox rattle, and must therefore be avoided. The (DMF) parameters that most influence this phenomenon are the lash and the *basic friction*, which cannot be easily restricted to certain values due to production tolerances.

When sub-harmonic vibrations occur in idle, the ECU can be disturbed. Indeed, the torque demand calculation of the idle controller uses the average engine speed as an input signal, determined at each ignition cycle. Since sub-harmonic vibrations generate a fluctuation of the average engine speed, the engine ECU can misinterpret this speed variation, and attempt to compensate for it. In this process the proportional controller of the idle governor plays a dominant part, for its corrections are applied directly on the ignition cycle following the speed variation.

On the other hand, it is also possible with a calibration change of the idle controller, to turn a normal idle state into a critical state with sub-harmonic vibrations. Different sub-harmonics may even be generated in this process, especially when the target idle speed is modified. In the example in Fig. 28.15, doubling the proportional and integral gains from a base calibration excites a 2/3 order, while a 100 rpm reduction of the idle speed excites a 1/2 order.

A total system consideration is necessary to optimise its behaviour. The use of simulation is very helpful, and can avoid time-consuming vehicle tests with different DMF parameter sets, which would be necessary to ensure satisfactory vehicle behaviour over the whole production range. Variations of main DMF and idle governor parameters can, with a reasonable computational time, provide a good overview. It is thus even possible to determine the stability limit of the idle controller, beyond which it can no longer sustain the desired idle speed (Fig. 28.16).

28.3.3 Engine stop

Unlike engine starting, engine stop normally does not represent an issue regarding arc spring and DMF durability. Nevertheless, the excitation of

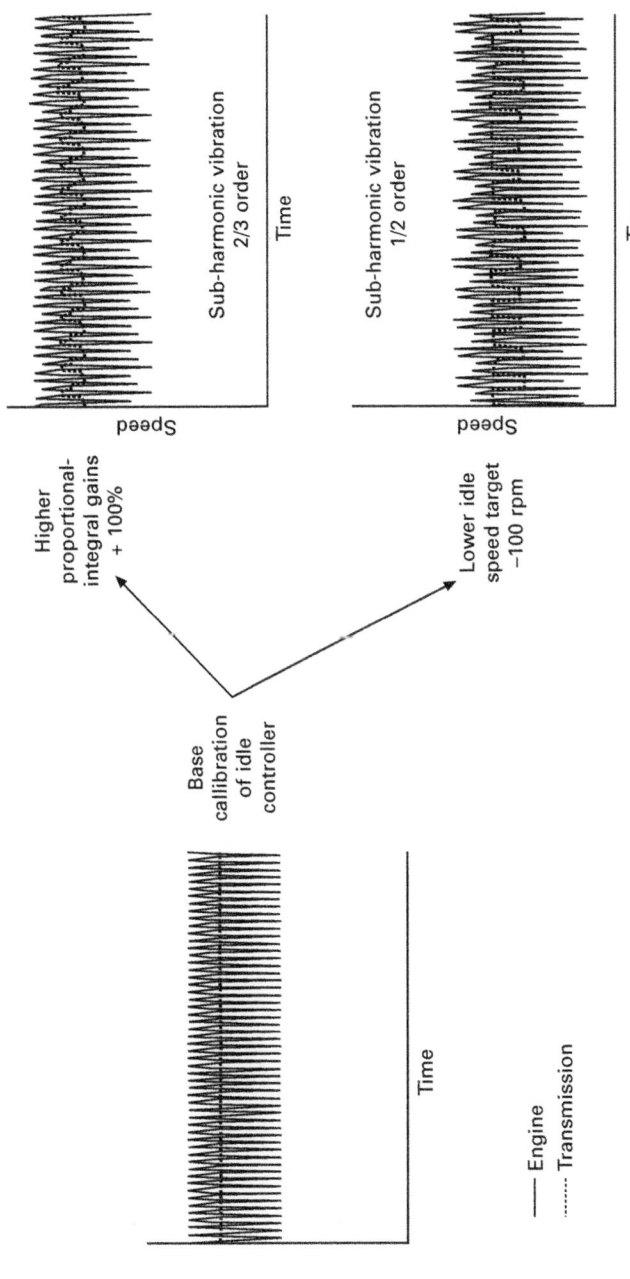

28.15 Effect of calibration changes of idle controller on engine idle for a four-cylinder diesel.

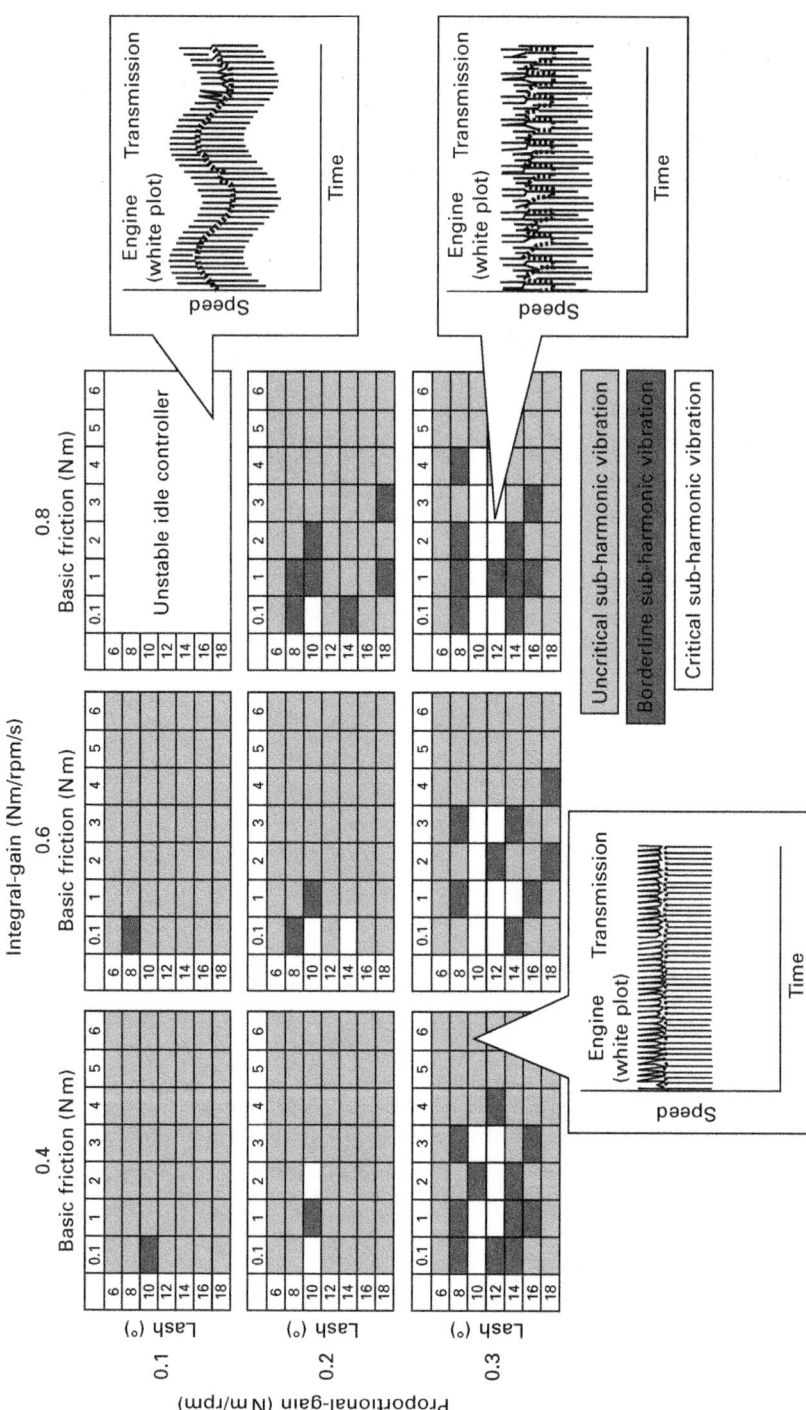

28.16 Example of a four-dimensional variation of DMF and ECU parameters with simulation.

the DMF resonance at low speeds can typically generate a ***clatter noise*** just before the engine stands still. This situation can be relevant in particular in diesel engines, which deliver large torsional vibrations at no load due to their higher compression and expansion forces.

In this case, high torsional accelerations in the transmission may be prevented by closing a throttle valve during the stop phase. The reduced amount of air in the intake manifold will result in a much lower engine and transmission excitation (Fig. 28.17).

28.3.4 Other operating points

Possible negative interactions between the DMF and its environment are not limited to the prior mentioned cases. Issues involving other ECU subsystems (e.g. launchability, false misfire detection, DMF impacts) may also occur.

Improvement potentials may then be identified within the ECU or the DMF. Nevertheless, changes in the DMF characteristics may improve one operating situation, while worsening another one. In those cases, unsatisfactory functional trade-offs must be made if no acceptable ECU calibration changes are to be found. The total vehicle and damper system should therefore be investigated for all relevant operating points with simulation support, as mentioned by Balashov *et al.* (2006).

28.4   Future trends

28.4.1 Global vehicle trends

Engine torque

Especially starting from the mid-1990s, engine torques available for a given engine displacement have risen considerably. This trend was first driven by diesel engines, with improved, high-pressure direct injection systems and better turbochargers. These have allowed reduction in fuel consumption, while improving driving dynamics and engine acoustic behaviour. For a series production diesel engine, a specific torque of 200 N m/l and power of 75 kW/l was achieved in 2007, thanks to two-stage turbocharging (see Fig. 28.18) (Anon, 2007).

In a similar manner, direct injection systems have started to become standard on gasoline engines on certain vehicle types. Turbo- and supercharged gasoline engines combined with direct injection and homogeneous mixture filled the gap between turbo diesel engines and naturally aspirated gasoline engines, leading to increased torque outputs, and thus to larger engine excitations at full load.

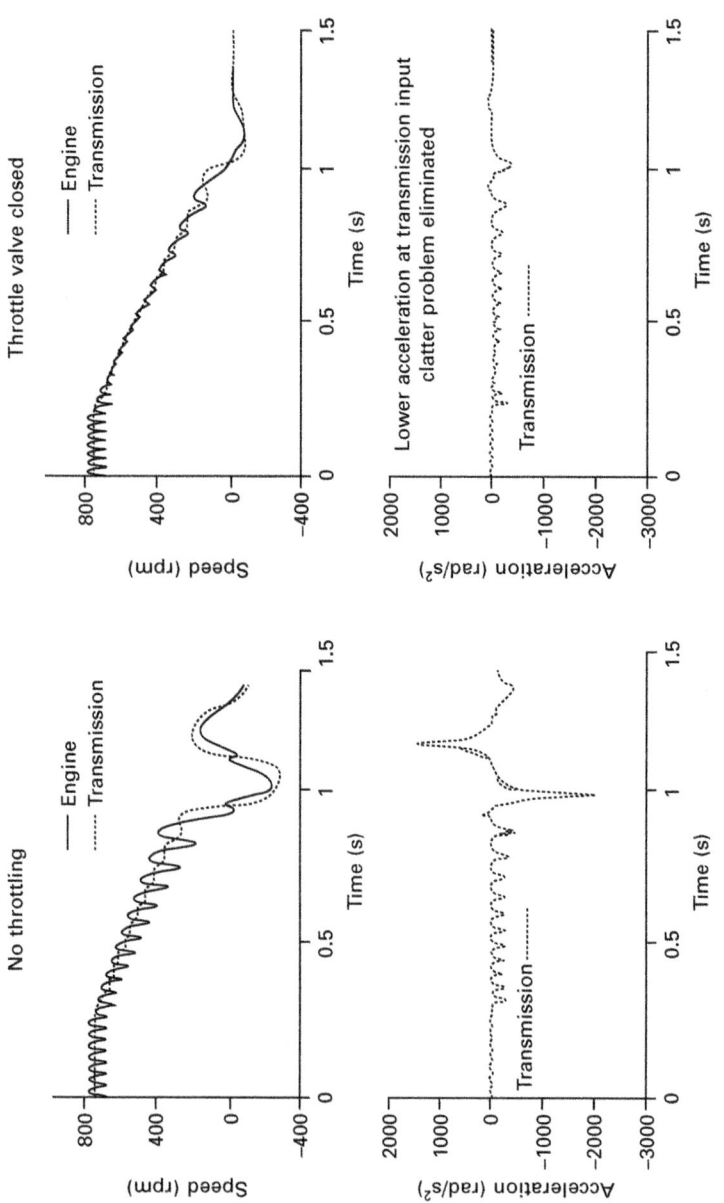

28.17 Effect of throttle valve closing on engine stop behaviour.

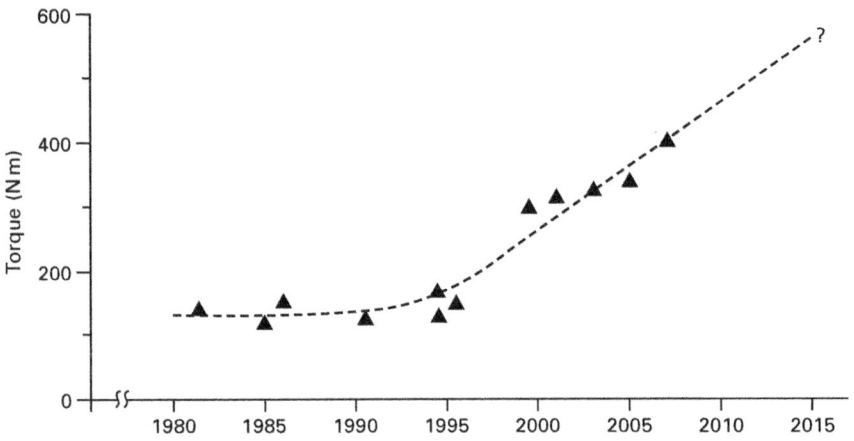

28.18 Maximum torque delivered by a four-cylinder diesel engine for different model years.

Engine efficiency

In naturally aspirated direct injection gasoline engines, the so-called 'stratified charge' combustion with globally lean mixture allows improvement in efficiency at low speeds and low loads. However, since the engine is not throttled in these operating points, the air-excess results in higher torque fluctuations due to compression forces, creating potential NVH issues at idle or part load.

Thanks to supercharging devices, car manufacturers have also used the possibility to maintain a given torque output while reducing the engine displacement and/or the number of cylinders. This method known as ***engine downsizing*** is beneficial to fuel consumption and emissions. A good example for this is a 1.4 l engine fitted with supercharger and turbocharger, which delivers 240 N m and 125 kW (170 N m/l and 90 kW/l). This four-cylinder engine can be compared to a 2.5 l naturally aspirated six-cylinder engine in terms of performance, with an average fuel consumption of 7.2 l/100 km in the new european driving cycle (NEDC) (Krebs *et al.*, 2005).

Indeed, apart from the weight gain, such an engine has less drag torque than an engine with larger cubic capacity. On the other hand, both reducing the number of cylinders and supercharging the engine cause the engine torsional excitation to increase. Damper devices developed for these downsized engines must fulfil both requirements of light weight and transmit a high full load torque; vibration isolation must therefore be investigated carefully in these cases.

Vehicle sensitivity

In parallel to the engine improvements, vehicle drivetrains have also been optimised to achieve better efficiency (e.g. by using components with less friction or damping: oil, bearings, etc.), which consequently are more sensitive to vibrations.

With the average number of gears rising from five to six and more, manual transmissions have become increasingly complex. New structures such as double-clutch transmissions have made a significant breakthrough on the automotive market. These new transmission concepts are characterised by more components being able to oscillate outside the normal power flow in the gearbox. As a result, loose gears will generate more rattle issues than in traditional transmissions (see Chapters 1, 27 and 29). Last but not least, lightweight design and the introduction of thinner transmission housings have a negative impact on sound propagation from the gearbox to the vehicle interior.

28.5   Consequences on damper development

Despite the excellent filtering quality of the DMF, it is no longer possible to compensate for the increasing engine excitations. On the contrary, higher engine torques have caused spring rates to rise and vibration isolation to worsen. The sum of the aggravating factors, mentioned in the previous section, has made it necessary to introduce innovative damper systems, which will be able to fulfil more severe isolation requirements.

28.19 DMF flange with mounted CAP.

One interesting solution in combination with the DMF is the so-called *centrifugal pendulum absorber* or CPA. This does not rely on a spring system like the DMF, but on additional relatively small masses (ca. 1 kg in total) fixed on the flange of the DMF, as shown in Fig. 28.19.

Driven by the remaining excitation on the DMF's secondary side (after the filtering effect of the arc springs), these masses have an oscillating movement in opposite phase to the excitation. A vibration in opposite phase is consequently superimposed on the secondary oscillation, resulting in an additional vibration filtering. (Fig. 28.20).

The effect of this device can be rather spectacular, especially at low engine speeds, where vibrations at the transmission input shaft can be halved. The pendulum absorbers function is indeed optimal in a range where most vehicles have rattle or boom issues, allowing the driver to stay in a high gear even at low speeds with far less acoustic disturbances, and taking advantage of a lower fuel consumption. Figure 28.21 represents the benefits of a centrifugal pendulum absorber in comparison with a standard DMF.

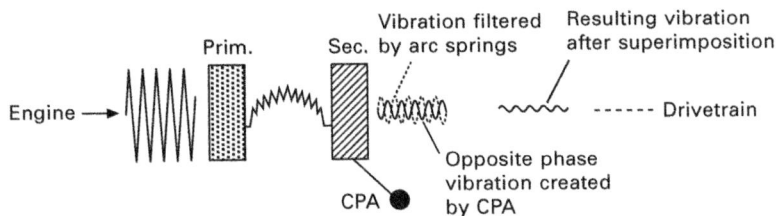

28.20 CPA on DMF secondary flywheel.

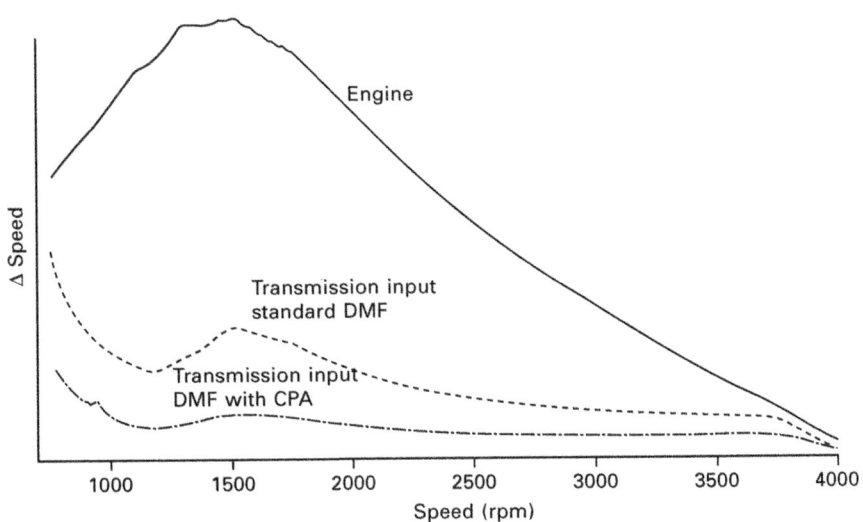

28.21 Speed amplitude at flywheel and transmission input. Comparison of standard DMF against DMF with CPA.

Thanks to innovative damper systems, individual comfort demands can still be met, matching more global requirements of vehicle development as driving dynamics and fuel economy.

28.6   References

Anon, 'Der neue Vierzylinder-Dieselmotor von BMW', *MTZ* December 2007, Vieweg Verlag

Balashov, D., Burkovski, L., Ferderer, F., Fidlin, A., Kremer, A., Pennec, B., Seebacher, R., 'Simulation of Torsional Vibration Dampers' in *ATZ* December. 2006, Vieweg Verlag

Krebs, R., Szengel, R., Middendorf, H., Sperling, H., Siebert, W., Theobald, J., Michels, K., 'Neuer Ottomotor mit Direkteinspritzung und Doppelaufladung von Volkswagen', *MTZ* December 2005, Vieweg Verlag

Multi-physics approach for analysis of transmission rattle

S. THEODOSSIADES, O. TANGASAWI and
H. RAHNEJAT, Loughborough University, UK

Abstract: Transmission rattle is a major concern in the automotive industry. It occurs as the direct result of system downsizing (reducing the gap between the transmission input, output and lay shafts) and increased engine order torsionals. The latter has been the result of a growing trend towards higher output torque as a drive towards greater use of diesel powered engines. The torsional oscillations onboard the transmission input shaft cause repetitive impacts between the mounted pinions with the loose unselected gear wheels due to their proximity. These events cause airborne noise radiation at the impact sites as well as transmission of vibration through the output shafts and their support bearings to the transmission casing, as structure-borne noise. The phenomenon does not pose any structural integrity issues, but is nevertheless quite disconcerting and, thus, has become a build quality issue for vehicle customers. The chapter shows that in the presence of backlash any form of contact separation causes rattling condition. With light loads and in the presence of a film of lubricant entrained into the contact domain, repetitive squeeze and separation effect is a key contributor to the rattle phenomenon.

Key words: transmission rattle, lightly loaded hydrodynamic impact, lubricated Petrov drag torque, squeeze film action, rattle ratio.

29.1 Introduction

The internal combustion engine produces a fluctuating torque due to combustion as well as induced inertial imbalance in the reciprocating motion of pistons. When the clutch is engaged, the resulting torsional oscillations of the crankshaft are transmitted to the transmission. In the transmission the fixed driving gears (pinions) transfer the motion to the driven *unselected gears* (loose gears) through impacting teeth pairs in close proximity due to the compact nature of modern transmission systems. The driven gears in turn rotate freely on their bearings, as they are unselected (referred to as *loose gears*). As a result, the idle (loose) gears oscillate within their backlash limits, leading to impacts with the driving gears. The vibrations caused by these impacts are transferred through the transmission shafts and their support bearings to the transmission bell housing, and is mostly radiated as sound that resembles the noise produced when a marble rolls inside a metallic can. This sound is onomatopoeically referred to as *rattle*.

878

Idle rattle (neutral rattle) occurs when the engine runs at its idling speed, the transmission is set to neutral (no gear is selected), and the clutch is engaged (Rust *et al.*, 1990). This produces the highest rattle intensity compared with other forms of rattle, since the engine noise level is at its minimum (Wang *et al.*, 2001). Furthermore, under idling condition the vehicle is usually stationary, for example at a traffic light (or near other stationary vehicles) and, thus, the radiated noise from the transmission is reflected from the neighbouring surfaces. This accentuates the passengers' noise perception, as well as other road users. Kim and Singh (2001) pointed out that rattle favours loose (idle) gears, because they are unconstrained, thus move freely when teeth pair separation occurs. The condition for separation to occur is when the inertial torque of the idle gear is larger than the resisting drag torque (Comparin and Singh, 1990).

Gear contacts have been modelled by a number of investigators. The **backlash** zone has usually been regarded as a dead-space function with a constant teeth pair contact stiffness assumed (Sakai *et al.*, 1981; Singh *et al.*, 1989; Comparin and Singh, 1990; Couderc *et al.*, 1998; Kim and Singh, 2001). Other contact models have been considered as time-varying piecewise linear functions (Wang *et al.*, 2001), or in accord with the classical Hertzian theory (Bellomo *et al.*, 2000) or other time-varying functions (Yakoub *et al.*, 2004; Brancati *et al.*, 2005). In the aforementioned models, damping has mainly been included as bearing drag torque in the system (Sakai *et al.*, 1981; Seaman *et al.*, 1984; Ohnuma *et al.*, 1985; Kim and Singh, 2001), using coefficients that may be regarded as constant or proportional to the penetration velocity, referring to the mutual convergence of gear teeth flanks (Couderc *et al.*, 1998; Bellomo *et al.*, 2000; Yakoub *et al.*, 2004). The effect of lubricant *squeeze film action* has been considered by Brancati *et al.* (2005) and Gnanakumarr *et al.* (2002). Brancati *et al.* (2005) assumed non-linear damping due to squeeze film action of the adsorbent oil film when teeth are not in direct contact. On the other hand, in Gnanakumarr *et al.* (2002), metal to metal contact was considered not to take place. Instead, a hydrodynamic lubricant reaction was assumed to be present at all times between the approaching and rolling convergent **wedge effect** of teeth pairs. The latter approach is valid for low loads transmitted by unselected gears, where the contact stiffness is governed by the lowest stiffness, which is that of the lubricant rather than that due to a Hertzian response, requiring localised deformation, which is unlikely at such lightly loaded impacts.

Sakai *et al.* (1981) found experimentally that rattle occurs when the inertial torque of the idle gear exceeds the drag torque acting on it. Therefore, when the ratio RR = *inertia torque/drag torque* exceeds unity, rattle occurs. This ratio is called the **rattle ratio**. Seaman *et al.* (1984) defined the rattle threshold as the angular acceleration at which the inertial torque at the unloaded teeth mesh exceeds the drag torque at the same unloaded teeth mesh. A similar approach was also undertaken by Smith (1999).

Engine speed fluctuations have been found to be an important parameter, affecting gear vibrations and noise levels. The larger the fluctuations, the higher the rattle response becomes according to Sakai *et al.* (1981). Gear backlash has a similar effect when double-sided impacts take place between the pinion and gear teeth (Sakai *et al.*, 1981; Seaman *et al.*, 1984; Doğan, 2001; also see Chapter 26 and the appendix in Chapter 1). Furthermore, when the gears' centre distance is decreased, the noise level is also decreased (Doğan, 2001).

It has been found that drag torque is higher at lower temperatures (Sakai *et al.*, 1981; Seaman *et al.*, 1984; Fujimoto *et al.*, 1987) and, therefore, rattle noise is somewhat attenuated. It is, therefore, surmised that ***lubricant viscosity*** plays an important role. It introduces a drag torque between a loose gear and its retaining shaft which dampens the impacts between the gear teeth, thus attenuating the noise produced. It has been shown by Seaman *et al.* (1984) that at the same oil sump temperature, drag torque is higher at higher rotational speeds (due to a thicker lubricant film); but as temperature increases, the variation between different values of drag torque at different rotational speeds decreases. Therefore, theoretically, there could be a temperature at which drag torque is the same for all rotational speeds. Fujimoto and Kizuka (2001) reported that there is an optimum temperature range, at which rattle is at its minimum.

This chapter presents numerical models for a gear pair contact, as well as for a complete transmission in order to investigate idle rattle conditions. The method is validated through comparison with experimental measurements taken from a vehicle under the same conditions. Parametric studies are also presented for both the aforementioned models in order to explain experimental observations reported by various researchers and hopefully shed some light on the physical behaviour of a rattling system.

29.2   Theoretical formulation for analysis of transmission rattle

The single degree of freedom model for a gear pair comprises a driving gear (pinion) and an idling driven gear (Fig. 29.1) mounted on a retaining shaft via a conformal bearing surface. The angular displacement $\varphi_p(t)$, velocity $\dot{\varphi}_p(t)$ and acceleration $\ddot{\varphi}_p(t)$ time histories of the pinion are known kinematic quantities for a given engine running condition. The equation of motion of the loose gear wheel is obtained as:

$$I_w \ddot{\varphi}_w = \sum_{}^{N} (W \cos \beta_b \, r_{bw}) - \sum_{}^{N} (F_f \rho_w) - F_p r_s \qquad (29.1)$$

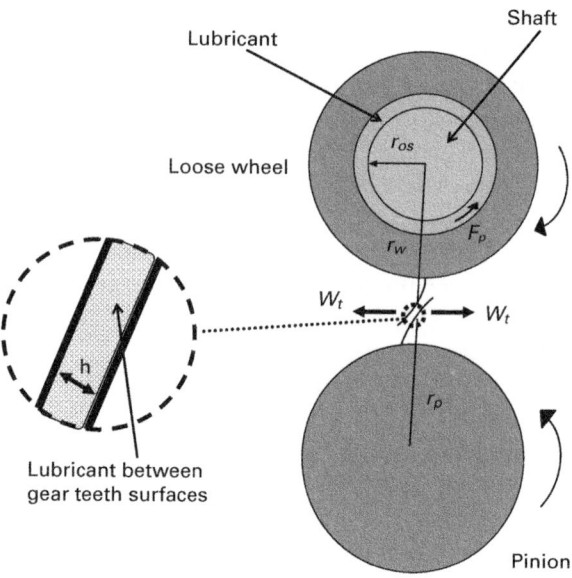

29.1 The gear pair model.

The wheel is driven by a **hydrodynamic (contact) force**, W, applied by the normally approaching and rolling pinion teeth. The motion is resisted by the **hydrodynamic viscous friction force**, F_f, acting on the teeth flanks, and a tractive force, F_p, known as **Petrov friction** due to the action of the lubricant between the loose wheel and its supporting shaft. The hydrodynamic reaction force is calculated for line contact condition as (Rahnejat, 1985):

$$
\left.
\begin{aligned}
W_j &= \frac{L_j \eta_0 r_{eqj}}{h} \left(2u_j - \frac{3\pi}{\sqrt{\dfrac{2h}{r_{eqj}}}} \frac{\partial h}{\partial t} \right), \quad \text{if } \frac{\partial h}{\partial t} < 0 \\[2em]
W_j &= \frac{L_j \eta_0 r_{eqj}}{h} (2u_j), \qquad\qquad\qquad \text{if } \frac{\partial h}{\partial t} \geq 0
\end{aligned}
\right\}
\tag{29.2}
$$

The rigid film thickness (rigid hydrodynamic contact) is given by:

$$
h = C_b - \frac{|(r_{bw}\varphi_w - r_{bp}\varphi_p)|}{\cos \alpha_n \cos \beta}
\tag{29.3}
$$

The viscous friction force acting on the tooth flank surface as a result of the lubricant film shearing is given by Gohar (2001) as:

$$F_{fj} = \frac{\pi \eta_0 L_j |u_{sj}| \sqrt{r_{eqj}}}{\sqrt{2h}} \qquad (29.4)$$

The friction between the gear wheel and the supporting shaft is due to Petrov friction in a conforming hydrodynamic contact with an assumed eccentricity ratio of zero as:

$$F_p = \frac{\pi \eta_0 v l_1 r_s}{C} \qquad (29.5)$$

More details on equations (29.1–29.5) are provided in Tangasawi *et al.* (2007) and Theodossiades *et al.* (2007).

The seven-speed (including reverse) ***transaxle transmission*** was modelled using known angular displacement $\varphi_{in}(t)$, velocity $\dot{\varphi}_{in}(t)$ and acceleration $\ddot{\varphi}_{in}(t)$ time histories of the transmission input shaft from experimental measurements in a vehicle. The seven torsional degrees of freedom, representing the idle gears are shown in Fig. 29.2. Each gear $i\,(i = 1...7)$ is driven by the hydrodynamic contact force, W_i, due to contact with its pinion. Each idle gear is also acted upon by hydrodynamic friction force, F_{fi}, along the teeth flanks, and the tractive force, F_{pi}, due to the lubricant action between the gear wheel and the supporting shaft. The motion of the first gear, however, is also resisted by the transverse component of the reaction force due to the contact between the reverse pinion and its idling gear, W_7. Therefore, the equation of motion for the 1st speed idle gear is obtained as:

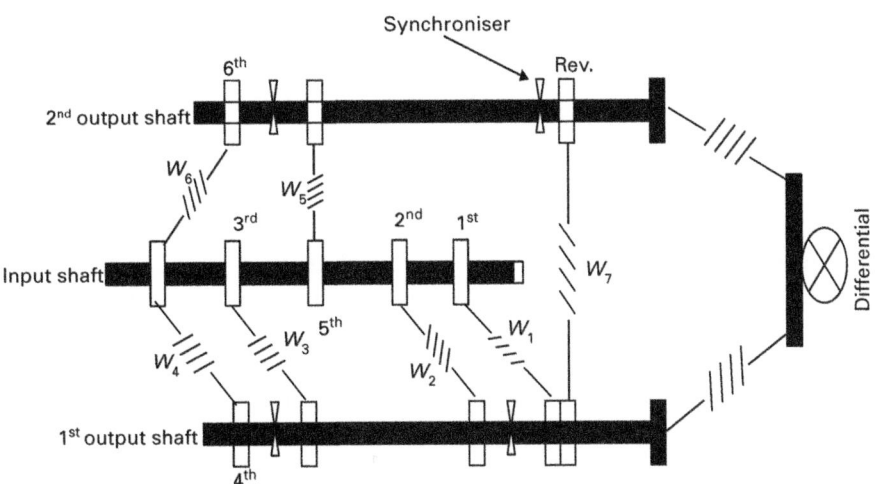

29.2 The idle gear pairs of the manual transmission.

$$I_1\ddot{\varphi}_1 = \overset{N_1}{\Sigma} (W_1 \cos \beta_{b1} r_{b1}) - \overset{N_7}{\Sigma} (W_7 \cos \beta_{b7} r_{bp7}) - \overset{N_1}{\Sigma} (F_{f1}\rho_1) - F_{p1} r_{s1}$$

$$(29.6)$$

The remaining idle gear wheels ($i = 2, \ldots, 7$) are represented by the following equation:

$$I_i\ddot{\varphi}_i = \overset{N_i}{\Sigma} (W_i \cos \beta_{bi} r_{bi}) - \overset{N_i}{\Sigma} (F_{fi}\rho_i) - F_{pi} r_{si} \qquad (29.7)$$

The radii $r_{bi=1\ldots7}$ are the base radii of the idle gears; and r_{bp7} is the base radius of the reverse pinion, while $r_{si=1\ldots7}$ are the internal radii of the idle gears, which represent the outside radii of the conforming bearings, on which the unselected gears rest. The components of the hydrodynamic reaction force W_i and the flank friction $F_{fi}(i = 1, 2, \ldots, 7)$ are, respectively, the sum of the hydrodynamic and friction forces acting on all teeth pairs in simultaneous contact.

The lateral (transverse) degrees of freedom of the supporting shafts are determined along two arbitrary orthogonal coordinates for each shaft. The equations of motion for the two shafts are obtained as follows.

For the 1st shaft (Fig. 29.3, views A and C) in the x_1 and y_1 directions, respectively:

$$M_1\ddot{x}_1 = F_{x1} - \Lambda_{ax} - \Lambda_{bx} \qquad (29.8a)$$

$$M_1\ddot{y}_1 = F_{y1} - \Lambda_{ay} - \Lambda_{by} \qquad (29.8b)$$

And for the 2nd shaft (Fig. 29.3, views B and D) in the x_2 and y_2 directions, respectively:

$$M_2\ddot{x}_2 = F_{x2} - \Lambda_{cx} - \Lambda_{dx} \qquad (29.9a)$$

$$M_2\ddot{x}_2 = F_{x2} - \Lambda_{cx} - \Lambda_{dx} \qquad (29.9b)$$

The rectilinear forces F_{x1}, F_{y1}, F_{x2} and F_{y2} are the resultant forces acting on the shafts due to the radial and transverse components of the hydrodynamic film reactions taken in the x_1, y_1, x_2 and y_2, directions, respectively. These forces couple the translational degrees of freedom to the torsional ones. The forces denoted by Λ_{ax}, Λ_{ay}, Λ_{bx}, Λ_{by}, Λ_{cx}, Λ_{cy}, Λ_{dx} and Λ_{dy} are the bearing reactions. They are calculated based on the *bearing stiffness coefficients* K_{ax}, K_{ay}, K_{bx}, K_{by}, K_{cx}, K_{cy}, K_{dx} and K_{dy} as shown in Fig. 29.3.

For simplicity one may assume that the supporting *taper roller bearings* behave as non-linear springs, acting in the x and y directions. Assuming no radial clearance, the total restoring force Λ_k of the kth bearing can be calculated as follows (Harris, 2001):

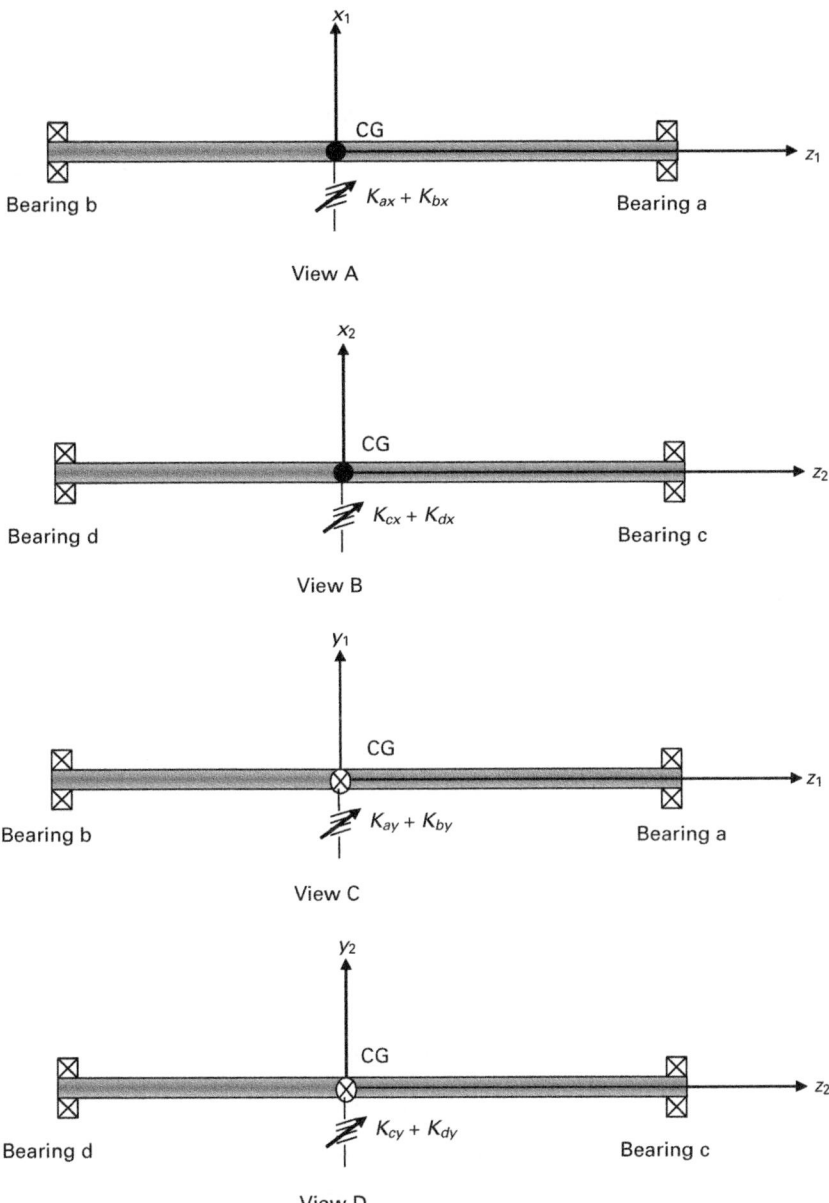

29.3 The lateral degrees of freedom of the seven-speed manual transmission.

$$\Lambda_k = 0.2453 \ m_k Q_k \ \delta_k^{1.11} \cos \gamma_k, \ (k = a, b, c, d) \qquad (29.10)$$

The coefficient Q_k is the reduced coefficient of the inner and outer raceways contacts with a taper roller and it can be determined as follows:

$$Q_k = \frac{7.86 \times 10^4 l_k^{8/9}}{2^{1.11}} \tag{29.11}$$

Combining the above equations yields the following expression for the radial restoring force Λ_k:

$$\Lambda_k = \frac{l_k^{8/9} m_k \cos \gamma_k}{1.12 \times 10^{-4}} \delta_k^{1.11} = K_k \delta_k^{1.11} \tag{29.12}$$

where γ_k is the contact angle, l_k is the roller length and m_k is the number of rollers per row of the kth bearing. The radial deformation (pseudo-penetration), δ_k, can be estimated using a suitably small time step from the instantaneous displacement d_k of the shaft at the relevant bearing support as:

$$\delta_k = \begin{cases} d_k - \Delta_k, & d_k > \Delta_k \\ 0, & -\Delta_k \le d_k \le \Delta_k \\ d_k + \Delta_k, & d_k < -\Delta_k \end{cases} \tag{29.13}$$

where Δk is the radial clearance of bearing k and the displacement d_k is the lateral motion of the shaft at the location of bearings a, b, c and d. Since the infinitesimal rocking motions of the shafts are not considered, it can be concluded that:

$$\left. \begin{array}{l} x_{1a} = x_{1b} = x_1 \\ y_{1a} = y_{1b} = y_1 \\ x_{2c} = x_{2d} = x_2 \\ y_{2c} = y_{2d} = y_2 \end{array} \right\} \tag{29.14}$$

Hence, with the existence of clearance, the radial reaction force is a piecewise non-linear function and it can be expressed as:

$$\Lambda_k = \begin{cases} -\text{sign}\,(d_k)\, K_k |\delta_k|^{1.11}, & d_k > \Delta_k, d_k < -\Delta_k \\ 0, & |d_k| \le \Delta_k \end{cases} \tag{29.15}$$

The main assumptions of the methodology are summarised as follows:

- During engine idling condition light teeth-pair impact loads and sufficient loose gear speeds permit the formation of a hydrodynamic lubricant film between the mating teeth-pairs.
- Owing to the low impact forces (of the order of a few newtons), shaft bending (0.1–0.3 μm) has been neglected. These are in fact orders of magnitude lower than the bearing deformations. Shaft elasticity effects are more important when high loads are transmitted to the drivetrain.

- The rocking motion of the shafts are not included as this would significantly complicate the calculation of lubricant film thickness between the teeth flanks, an assumption which is justified because of the light impact loads during idling conditions.
- The gear tooth profile is considered to follow a perfect involute shape.
- No misalignments are considered.
- No tooth-to-tooth geometrical variations are considered.
  The loads on the teeth are too low to cause any local or global deformation of the teeth (local being due to Hertzian condition and global being due to tooth bending or rocking).
- Changes of viscosity due to generated pressures within the contact zone are considered as negligible because of low hydrodynamic pressures (i.e. iso-viscous assumption).
- No other forms of external excitations other than engine torsional excitations are considered.

The equations of motion are solved in a series of suitably small time steps using Newmark's linear acceleration method (Timoshenko *et al.*, 1974; Rahnejat, 1985). The parameters of Newmark integration scheme were carefully selected so as to minimise numerical damping and improve its accuracy albeit at the expense of conditional stability (Timoshenko *et al.*, 1974). To cater for problem of numerical stability care had to be taken in the selection of a suitable time step size (1 μs) and definition of initial conditions. Sufficiently long simulation run times were undertaken in order to exceed well beyond any initial transience.

29.3   Experimental set-up for analysis of transmission rattle

Vehicle tests were carried out in a semi-anechoic chamber environment under engine idling condition (with the transmission in the neutral position). Tests were carried out at various oil sump temperatures. Piezo-electric accelerometers were used to capture vibrations of the gearbox housing, following the response path from the bearing housings to the transmission wall. A total of five accelerometers were used in the vehicle measurements. The three accelerometers, whose results are analysed in this work, were arranged as follows:

- Two accelerometers were mounted on the transmission wall at the bearing locations of the input shaft (indicated by (1) in Fig. 29.4) and the second (upper) output shaft (indicated by (2) in Fig. 29.4).
- An accelerometer was mounted on the under-side of the transmission bell housing (indicated by (3) in Fig. 29.4), where the wall is relatively more compliant (away from the bell housing ribs).

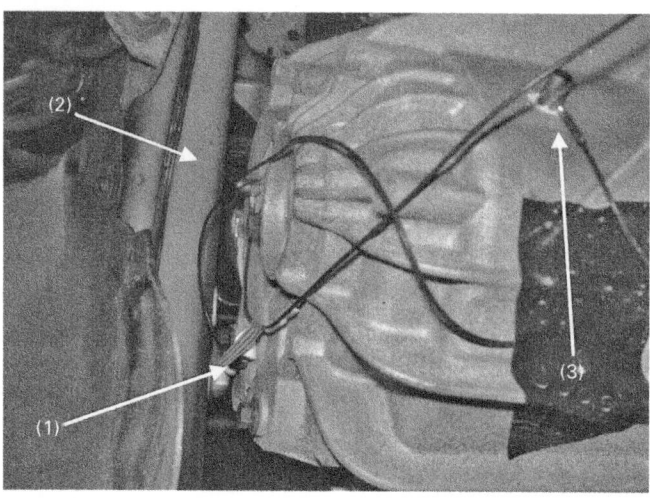

29.4 The accelerometers mounted on the gearbox surface.

Additionally, a thermocouple was inserted into the oil filling hole, measuring the transmission oil spray temperature. The data acquisition sampling rate was set to 12000 Hz; hence, giving a Nyquist frequency of 6000 Hz, which is reasonable for gear rattle measurements.

29.4 Parametric studies for analysis of transmission rattle: discussion

Parametric studies were conducted for the gear pair model and for the full seven-speed transmission model. The former was used as a simple model to demonstrate the basic gear vibration phenomena, before proceeding to a more complex full transmission model. The latter is used to examine the interactions between the torsional and lateral degrees of freedom, as well as for comparison purposes and validation with the experimental measurements.

29.4.1 The gear pair model

The gear pair in the first set of parametric studies has the geometric properties of the 5th gear pair of the manual transmission used in the experimental measurements. The bearing clearance (between the loose gear and its retaining shaft) has been taken as 35 μm, while the gear backlash value is 79 μm.

Figure 29.5 shows that as the lubricant viscosity drops due to temperature rise, the *rattle ratio* increases and crosses the threshold ($RR = 1$) at 80 °C – Fig. 29.5(b) – where the separation between the meshing teeth markedly increases. This can be seen in the corresponding film thickness variation

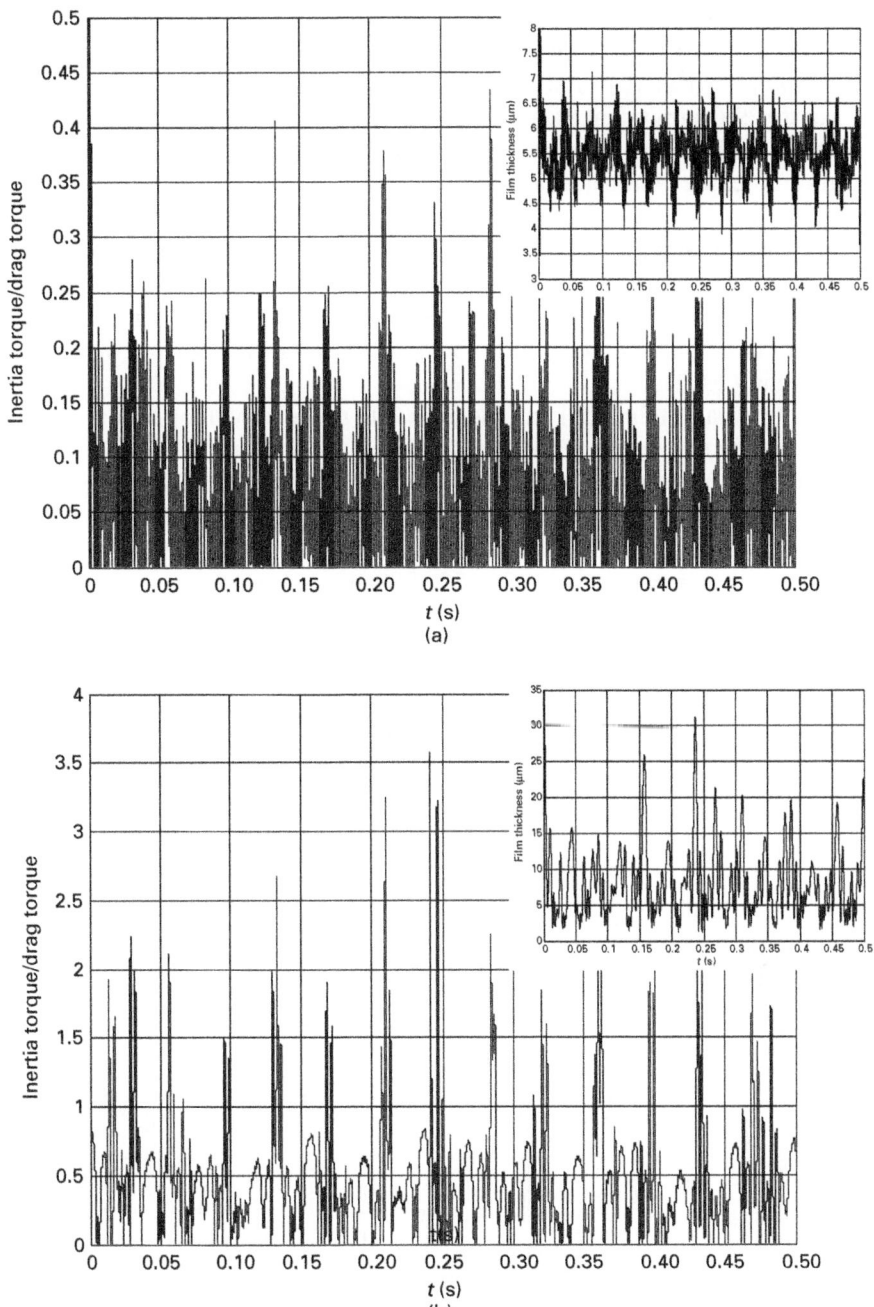

29.5 Rattle ratio and teeth separation time history at (a) 20 °C and (b) 80 °C.

(upper right corner figures). It is noted that teeth-pair separation increases with temperature rise (from approximately $3\,\mu m$ peak-to-peak variation at $20\,°C$ to more than $10\,\mu m$ at $80\,°C$).

Figure 29.6 shows that the threshold for *onset of rattle* is mainly breached by the wheel, when the pinion motion undergoes a rise in acceleration. Furthermore, the pinion acceleration shows that the peaks tend to be stronger at a spacing of $13\,Hz$ (the engine rotational fundamental frequency; *first engine order*) rather than at the second engine order. The effect of backlash for one-sided impacts – shown in Fig. 29.7 – shows a negligible influence on the rattle ratio (comparing Figs 29.7 and 29.5(a)). This is in line with the findings of Sakai *et al.* (1981), Seaman *et al.* (1984), Rust *et al.* (1990) and Doğan (2001).

The dominant influential parameter in idle rattle is the lubricant viscosity variation with temperature (it decreases as temperature increases). This shifts the response frequencies to a lower spectral range (reduced hydrodynamic stiffness with an increasing film thickness), and lowers the amplitude at the higher spectral contributions. The gear response contains the engine order excitations (approximately at 26, 52, 78 and $106\,Hz$), the gear meshing frequency ($465\,Hz$) and a band of frequencies related to contact stiffness. As the bulk oil temperature increases this band of frequencies shifts towards

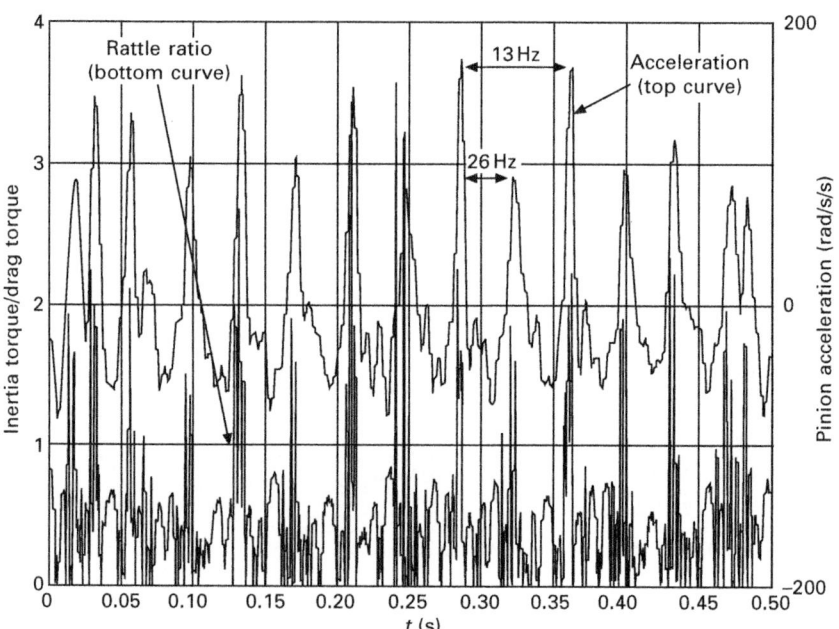

29.6 Rattle ratio (lower curve/left label) and pinion acceleration (upper curve/right label) time histories at $80\,°C$.

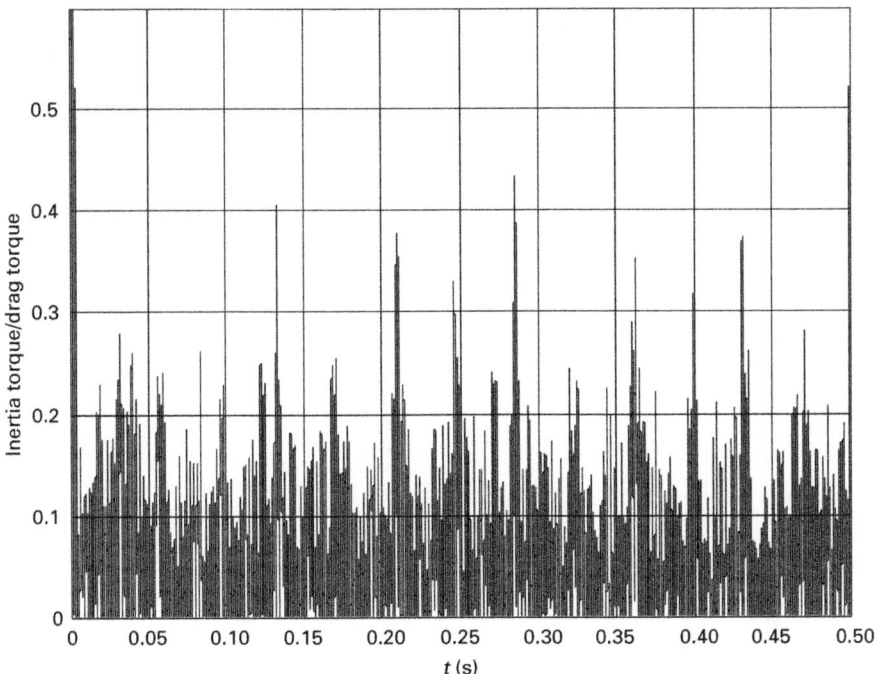

29.7 Rattle ratio at 20 °C and 158 μm backlash.

the lower spectral region (see Fig. 29.8). Modulation is observed in the spectra at the pinion rotational frequency of 13 Hz. At 20 °C (Fig. 29.8(a)) the *gear meshing frequency* is dominant. As the temperature rises to 39.4 °C (Fig. 29.8(b)) the engine orders become dominant and the *frequency due to lubricated impacts* makes its appearance (261 Hz). At a lower temperature the drag torque dominates (higher system effective torsional stiffness). Therefore, the spectrum of vibration is dominated by the meshing frequency, similar to a pair of load transmitting gears. With decreasing drag torque due to reduced Petrov friction, the impact energy is dissipated by vibrations within lubricated contacts and damping of engine torsional vibration is less effective. Finally, in Fig. 29.8(c) (80 °C), the band of impact-related frequencies shifts further to lower spectral region due to a further reduction in the hydrodynamic stiffness. The moving band of frequencies can be associated with gear rattling by correlating the change in rattle ratio with lowered viscosity, reduced Petrov friction and larger contact separations (the trend is shown in Fig. 29.5).

With a 79 μm backlash at 20 °C, the frequency response shifts to the lower spectral region as the clearance between the gear and the retaining shaft increases (Fig. 29.9). This is because thicker hydrodynamic films have lower stiffness. With much reduced clearance, Fig. 29.9(a) shows vibration at high frequency with a greater share of dissipation power than with higher clearance

(Fig. 29.9(b)). Although engine orders still dominate the spectrum, a band of frequencies initially centred on 1565 Hz, when the bearing clearance is 10 μm (Fig. 29.9(a)) moves to 249 Hz (Fig. 29.9(b)) as the bearing clearance is increased. Furthermore, the gear meshing frequency's 3rd order (1393 Hz)

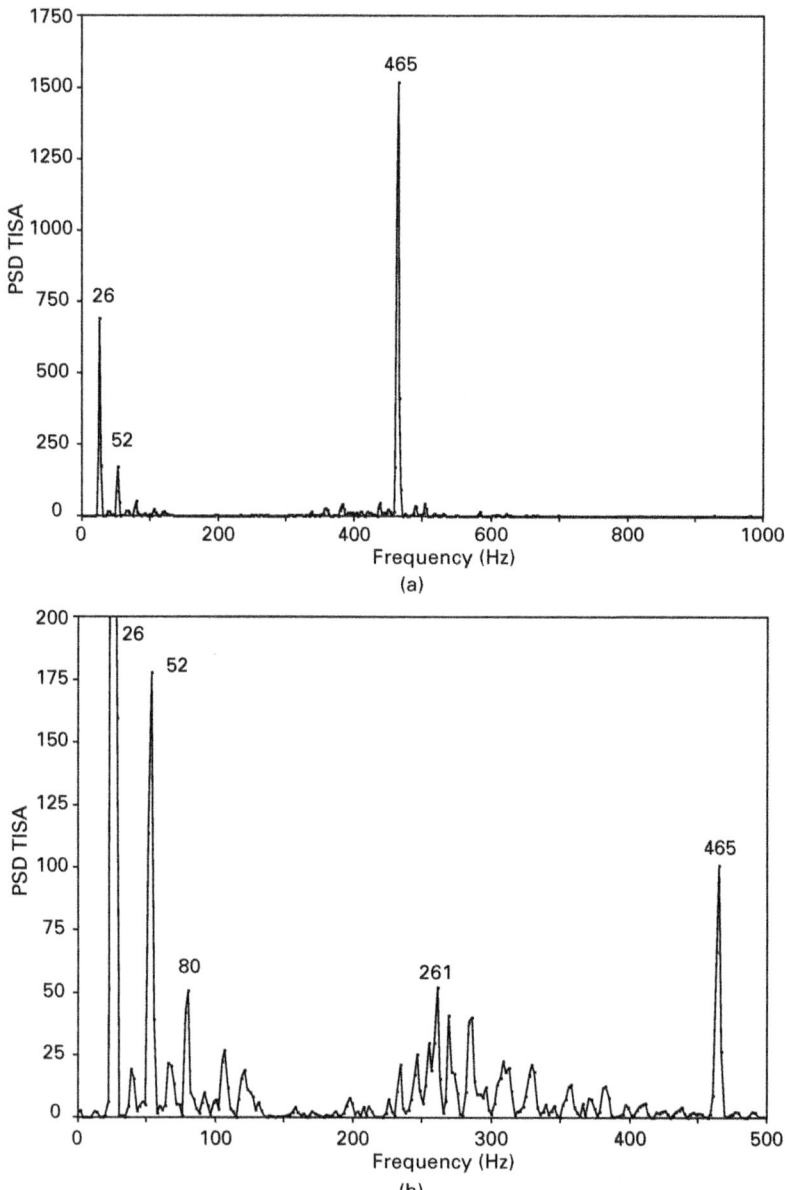

29.8 Fast Fourier transform (FFT) spectra of the gear wheel response at (a) 20 °C (b) 39.4 °C and (c) 80 °C.

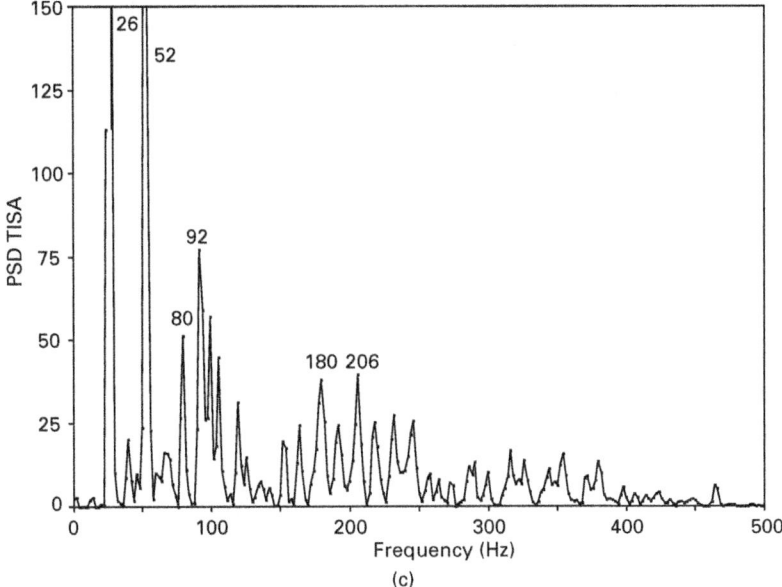

29.8 Continued

has a strong presence when bearing clearance is low while the 1st order gets stronger with an increased clearance. A similar trend is observed at the higher temperatures of 80 °C, where the system response crosses the rattle threshold. This confirms that the clearance between the gear and the shaft can be effective in constraining the system from crossing the rattle threshold even when the temperature is relatively high.

The above case studies show that rattling occurs as a band of frequencies, which shift towards the lower spectral contributions as drag, resisting the wheel motion decreases either due to an increase in the lubricant temperature, thus the lowering of its viscosity or an increase in the bearing clearance.

29.4.2 The complete transmission model

Figure 29.10 shows the FFT spectra of the 2nd and 4th loose gears at the bulk oil temperature of 39.4 °C with an assumed 10 µm clearance in the gear–shaft conjunctions. The overall response of an idling gear consists of two particular motions: a carrier-type motion, and superimposed infinitesimal fluctuations, containing different frequency characteristics (the former modulates the latter). The lower engine order responses dominate in the carrier-region, while the higher vibro-impact frequencies are expected to govern the lower amplitude fluctuations (see Theodossiades *et al.*, 2007). Meshing frequencies of various gear pairs are listed in Table 29.1.

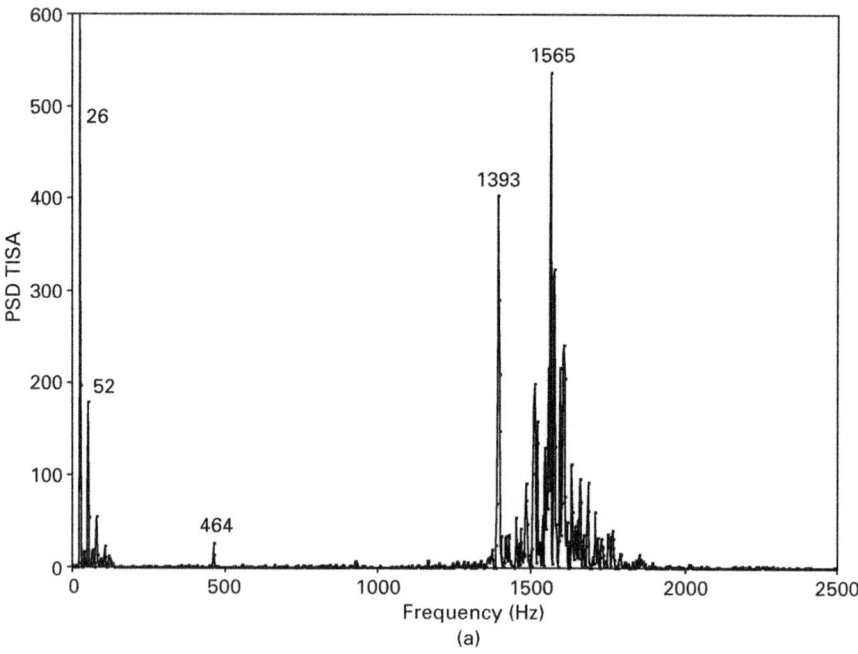

(a)

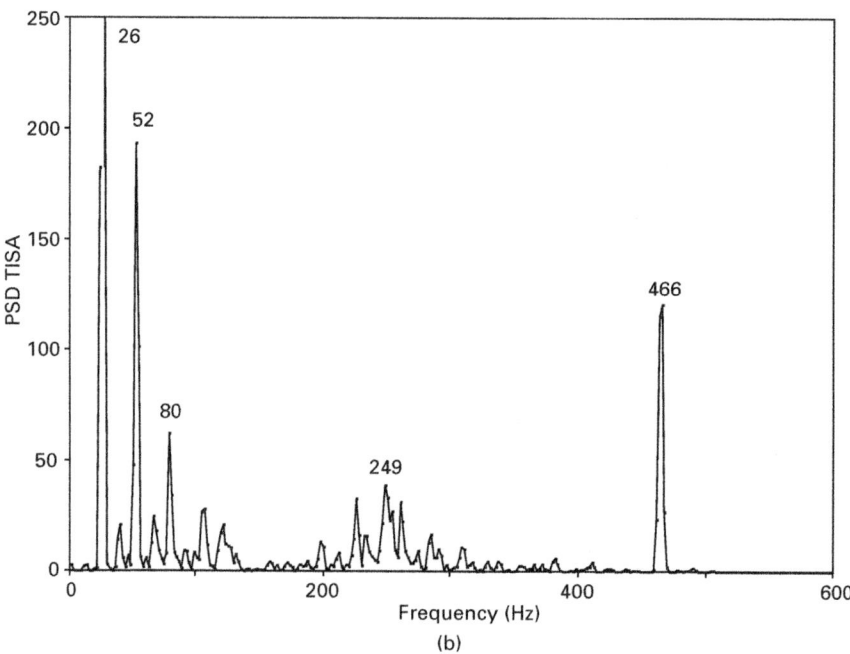

(b)

29.9 Gear response FFT spectra for bearing clearance (a) 10 μm and (b) 60 μm.

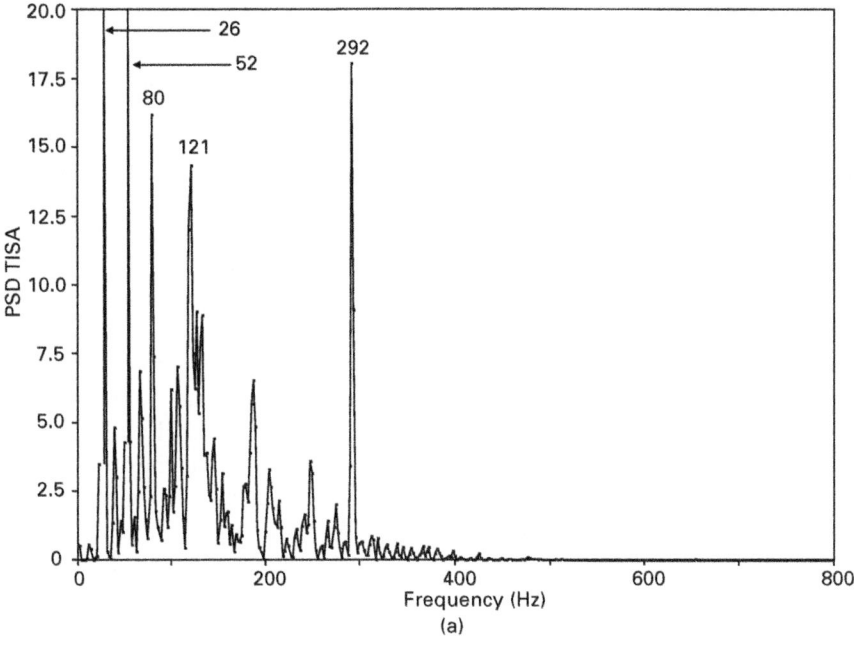

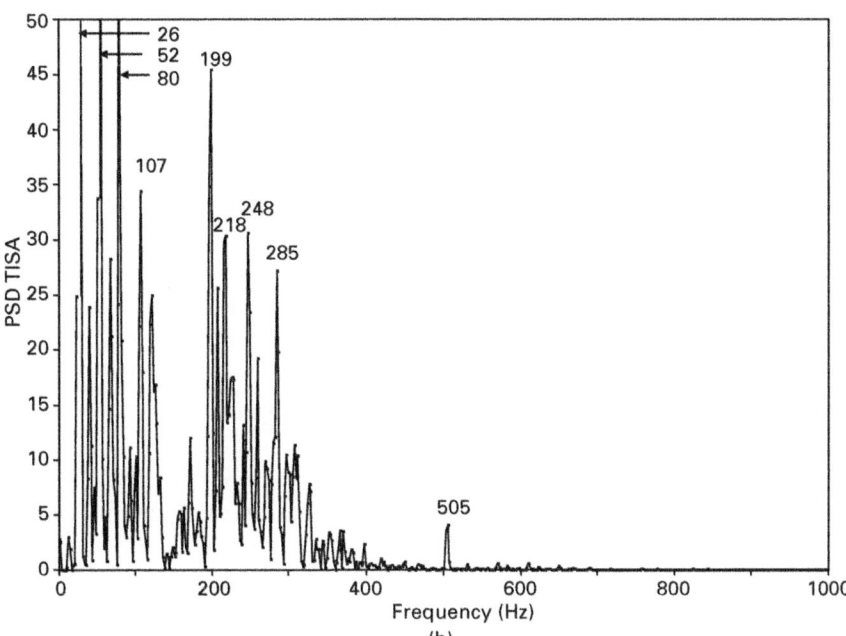

29.10 Simulated response FFT spectra of (a) 2nd and (b) 4th idle gear wheels (at 39.4 °C).

Table 29.1 Meshing and rolling frequencies of the idle gear pairs at 800 rpm engine speed

Gear set	1st	2nd	3rd	4th	5th	6th	Reverse
Meshing frequency (Hz)	172	291	384	504	464	504	107

To better understand these higher-frequency contributions and the lubricant effect on the system dynamics, the undamped natural frequencies of the model arising through linearisation of equations (29.6)–(29.9) are calculated. Since the lubricant film is the main factor determining the interactions between the gear flanks – behaving as a non-linear spring, which follows the meshing cycle variation – it is additionally important to calculate the natural frequencies of the mechanical model and observe any interactions indicated by the normal modes of the linearised system. The equation of motion of the input shaft is also included in this process. The lubricant stiffness is determined by differentiating equation (29.2) with respect to the film thickness, considering the rolling term only. This assumption is reasonable, because firstly the differentiation of the squeeze film term is only valid in normal approach (i.e. $\partial h/\partial t < 0$) and secondly that its contribution is negligible as $\partial h/\partial t \lll u$ in eq. (29.2). Thus, the torsional gear stiffness is given as:

$$K(\varphi_i) = \left| \frac{\partial W_i}{\partial h_i} \right| \qquad (29.16)$$

The lubricant stiffness between the various pairs of meshing gear teeth is a function of time, since it depends on the film thickness and contact kinematics as a function of the rolling/sliding geometry. If the tooth-to-tooth variations are neglected, the fundamental period of the impacting stiffness is equal to $\varphi_{iT} \equiv 2\pi/n_i$, where n_i represents the tooth number of the driving gear (Theodossiades and Natsiavas, 2001). Therefore, the lubricant stiffness can be approximately expressed in a Fourier series as follows:

$$K(\varphi_i) = K_{0i} + \sum_{p=1}^{\infty} K_{cp} \cos (pn_i \varphi_i) + K_{sp} \sin (pn_i \varphi_i) \qquad (29.17)$$

The linearised model occurs by only retaining the constant term for the lubricant film stiffness in the Fourier series expansion of equation (29.17). Here, the aim is to identify the undamped natural frequencies of the lubricated system. Therefore, the friction terms have not been included, since these are terms depending on the velocity of motion and, consequently, they are regarded as damping terms. It is now well established that fluid film damping contribution is quite insignificant (Dareing and Johnson, 1975; Mehdigoli et al., 1990; Gohar and Rahnejat, 2008). Therefore, the equations of motion for the linearised model are formulated as follows.

For the input shaft:

$$I_{in}\ddot{\varphi}_{in} + \sum_{p=1}^{\infty} K_{0i}r_{pi}\left(r_{pi}\varphi_{in} - r_{wi}\varphi_i - y_1\right)$$

$$+ \sum_{i=5}^{6} K_{0i}r_{pi}\left(r_{pi}\varphi_{in} - r_{wi}\varphi_i - y_2\right) = 0 \tag{29.18}$$

For the 1st speed idle gear (combined with the reverse gear pinion):

$$\left(I_1 + I_{prev}\right)\ddot{\varphi}_{1,prev} + K_{01}r_{w1}\left(r_{w1}\varphi_{1,prev} + y_1 - r_{p1}\varphi_{in}\right)$$

$$+ K_{0(rev)}r_{prev}\left(r_{prev}\varphi_{1,prev} + x_1 \sin \varepsilon_1 + y_1 \cos \varepsilon_1 - r_{wrev} - \varphi_{wrev}\right.$$

$$\left. - x_2 \sin \varepsilon_2 + y_2 \cos \varepsilon_2\right) = 0 \tag{29.19}$$

For the 2nd speed idle gear:

$$I_2\ddot{\varphi}_2 + K_{02}r_{w2}\left(r_{w2}\varphi_2 + y_1 - r_{p2}\varphi_{in}\right) = 0 \tag{29.20}$$

For the 3rd speed idle gear:

$$I_3\ddot{\varphi}_3 + K_{03}r_{w3}\left(r_{w3}\varphi_3 + y_1 - r_{p3}\varphi_{in}\right) = 0 \tag{29.21}$$

For the 4th speed idle gear:

$$I_4\ddot{\varphi}_4 + K_{04}r_{w4}\left(r_{w4}\varphi_4 + y_1 - r_{p4}\varphi_{in}\right) = 0 \tag{29.22}$$

For the 5th speed idle gear:

$$I_5\ddot{\varphi}_5 + K_{05}r_{w5}\left(r_{w5}\varphi_5 + y_2 - r_{p5}\varphi_{in}\right) = 0 \tag{29.23}$$

For the 6th speed idle gear:

$$I_6\ddot{\varphi}_6 + K_{06}r_{w6}\left(r_{w6}\varphi_6 + y_2 - r_{p6}\varphi_{in}\right) = 0 \tag{29.24}$$

For the reverse speed idle gear

$$I_{wrev}\ddot{\varphi}_{wrev} + K_{0rev}r_{wrev}\left(r_{wrev}\varphi_{wrev} + x_2 \sin \varepsilon_2 - y_2 \cos \varepsilon_2\right.$$

$$\left. - r_{prev}\varphi_{1,prev} - x_1 \sin \varepsilon_1 - y_1 \cos \varepsilon_1\right) = 0 \tag{29.25}$$

For the 1st shaft:

$$M_1\ddot{x}_1 + \sum_{i=1}^{4} K_{x1i}\varphi_i + K_{x1rev}\varphi_{rev} + K_{x1in}\varphi_{in} + K_{x1x1}x_1 + K_{x1y1}y_1 = 0 \tag{29.26}$$

$$M_1\ddot{y}_1 + \sum_{i=1}^{4} K_{y1i}\varphi_i + K_{y1rev}\varphi_{rev} + K_{y1in}\varphi_{in} + K_{y1x1}x_1 + K_{y1y1}y_1 = 0 \tag{29.27}$$

For the 2nd shaft:

$$M_2 \ddot{x}_2 + K_{x21}\varphi_1 + \sum_{i=5}^{6} K_{x2i}\varphi_i + K_{x2rev}\varphi_{rev} + K_{x2in}\varphi_{in}$$

$$+ K_{x2x1}x_1 + K_{x2y1}y_1 = 0 \tag{29.28}$$

$$M_2 \ddot{y}_2 + K_{y21}\varphi_1 + \sum_{i=5}^{6} K_{y2i}\varphi_i + K_{y2rev}\varphi_{rev} + K_{y2in}\varphi_{in}$$

$$+ K_{y2x1}x_1 + K_{y2y1}y_1 = 0 \tag{29.29}$$

The solution of the eigen-value problem determines the natural frequencies and normal modes of the system, which are shown graphically in Fig. 29.11. Generally, the lower the hydrodynamic film thickness, the higher the natural frequency of the corresponding gear pair (Tangasawi *et al.*, 2007). In fact, the film stiffness is proportional to $1/h^2$. The natural frequencies of the shafts' rectilinear motions clearly dominate the corresponding fast fourier transform (FFT) spectra, as it can be seen from the graphs of Fig. 29.11. The normal modes also reveal strong coupling between the motions of the 1st, 4th and reverse gears and between the motions of the 5th gear and second output shaft in the *y* direction, while a weaker form of coupling exists between the 3rd and 6th idle gears.

The 2nd idling gear pair (Fig. 29.10(a)), possessing a higher inertia, exhibits lower spectral components when compared with the 4th pair. The frequencies 292 and 505 Hz are the *meshing frequencies* of the 2nd and 4th gear pairs, respectively. On the other hand, the frequencies 198 and 266 Hz are the *natural frequencies* (those due to the balance of lubricated conjunctional characteristics) of the 2nd and 4th gear pairs. The additional frequencies observed in the spectra (218, 248 and 285 Hz are natural frequencies of the other gear pairs that are transmitted via the shafts and also excite the 4th idling gear).

Two simulated cases are compared with experimental measurements taken from a vehicle equipped with the same transmission type. The engine idling speed is 800 rpm. Typical values for gear backlash are in the range 80–150 μm, with the larger value corresponding to the higher inertial gear pairs. The first comparison is made for the bulk oil temperature of 39.4 °C with the lubricant dynamic viscosity being 0.0512 Pa s.

The responses of the housing, captured by the accelerometers mounted on its wall and at the input shaft bearing position (Fig. 29.12(a) and (b) respectively), contain contributions at 240 and 238 Hz, respectively. These contributions correspond to the 4th wheel acceleration response (Fig. 29.10(b)) with an error of 4%. The contribution at 350 Hz is the second harmonic of the first gear meshing frequency, while the lower frequencies, 26, 52 and 80 Hz, are engine order multiples. Additionally, the 1st gear's natural frequency is

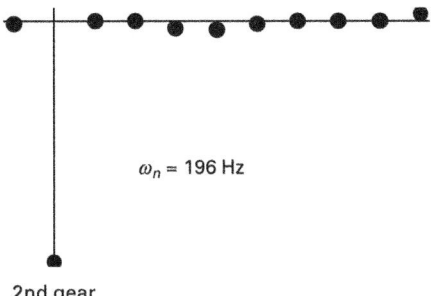

$\omega_n = 196\,\text{Hz}$

2nd gear

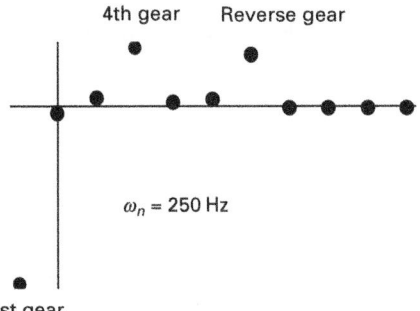

4th gear   Reverse gear

$\omega_n = 250\,\text{Hz}$

1st gear

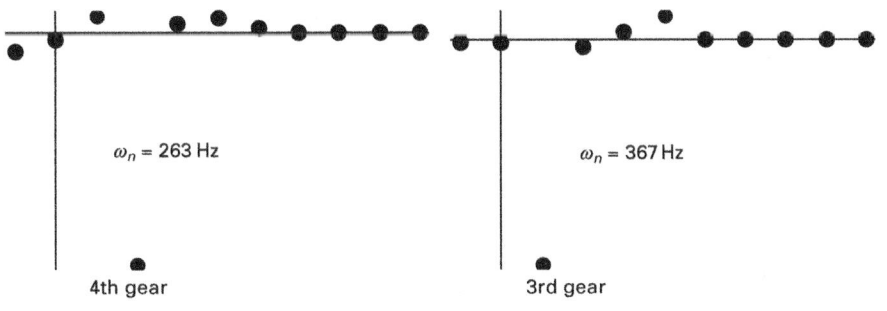

$\omega_n = 263\,\text{Hz}$

4th gear

$\omega_n = 367\,\text{Hz}$

3rd gear

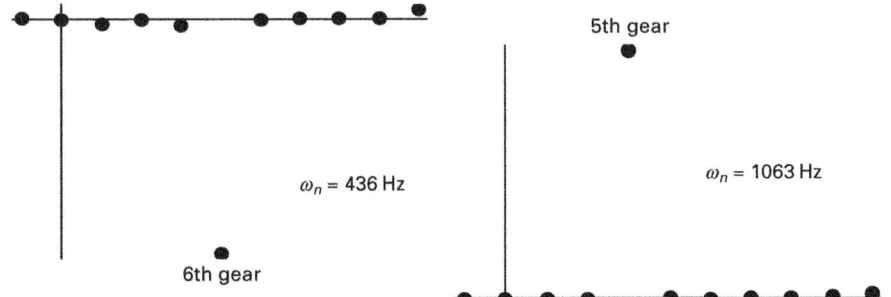

$\omega_n = 436\,\text{Hz}$

6th gear

5th gear

$\omega_n = 1063\,\text{Hz}$

29.11 Natural frequencies and mode shapes of the linearised system.

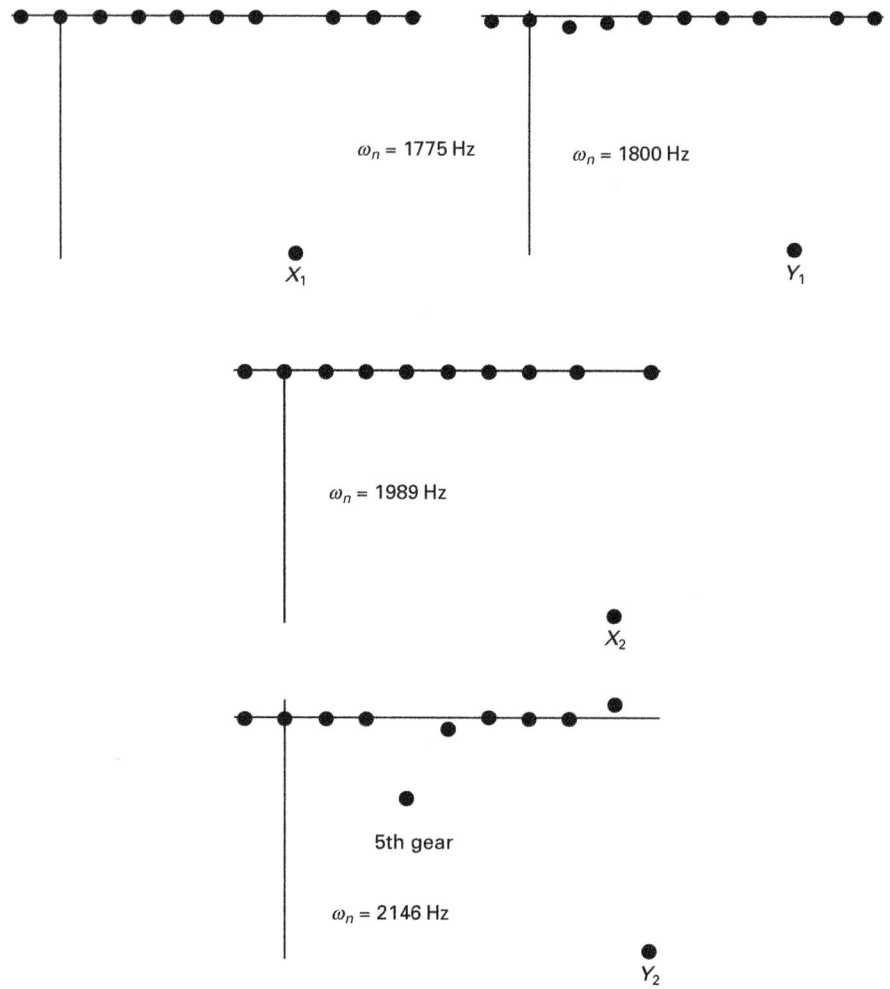

$\omega_n = 1775\,\text{Hz}$ $\omega_n = 1800\,\text{Hz}$

X_1 Y_1

$\omega_n = 1989\,\text{Hz}$

X_2

5th gear

$\omega_n = 2146\,\text{Hz}$

Y_2

29.11 Continued

at 208 Hz, which was obtained by solving the eigen-value problem. This is also visible in the spectra of Fig. 29.10(a) and (b). The frequency component at 286 Hz for the 4th loose gear response in Fig. 29.10(b)) is within a margin of error of 5% from the 273 Hz component captured by the transmission wall accelerometer in Fig. 29.12(a). The contributions around 540 Hz in Fig. 29.12 represent approximately the second harmonic of 273 Hz. The meshing frequencies of the 4th, 5th and 6th gear pairs are also present in Fig. 29.12 (in the regions of 466 and 506 Hz). The frequencies above 1000 Hz in the same figure correspond to natural frequencies of the output/input shafts and of the gearbox casing (Tangasawi *et al.*, 2007).

Figures 29.13 and 29.14 show the rattle and *squeeze-roll ratio* for the

1st, 4th and 6th gear pairs respectively. Only the 1st gear crosses the rattle threshold, mainly due to the higher inertia of its loose wheel. The lubricant film reaction due to rolling action is clearly dominant in all the three gears, since their corresponding squeeze-to-rolling force ratios are less than unity. The squeeze-to-rolling ratio assesses the contribution of squeeze action (approach leading to impact) to the system dynamic and it is obtained by

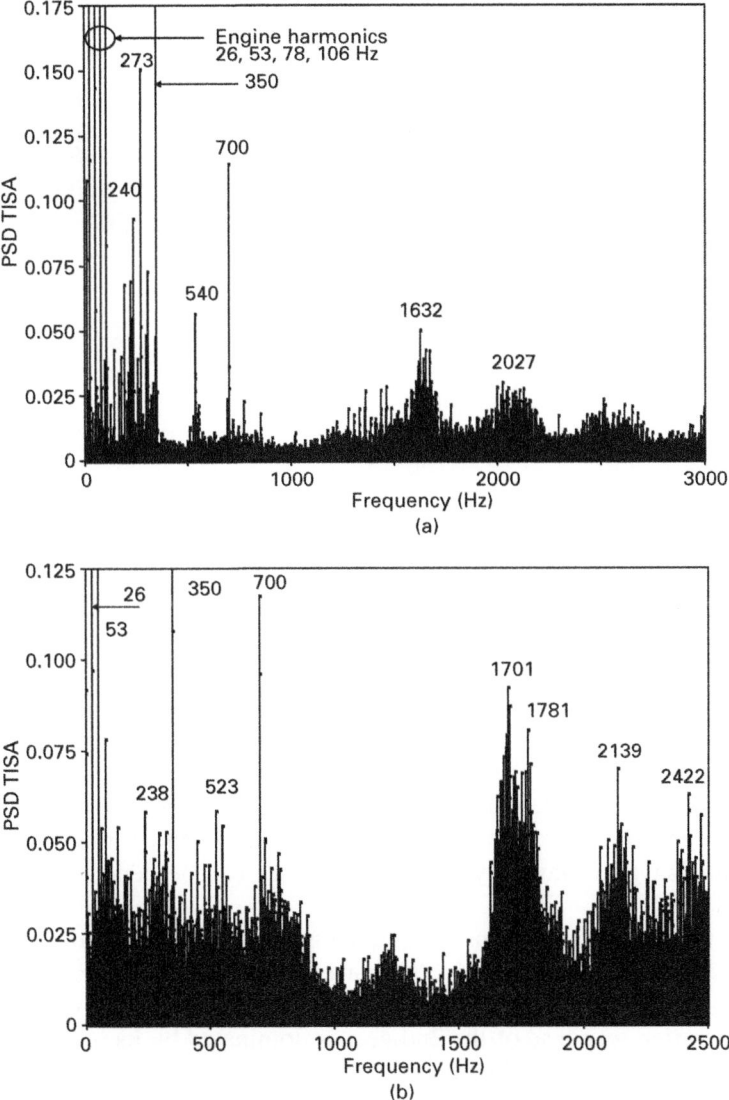

29.12 FFT spectra of the (a) transmission wall response (b) input bearing housing response and (c) upper output bearing housing response (at 39.4 °C).

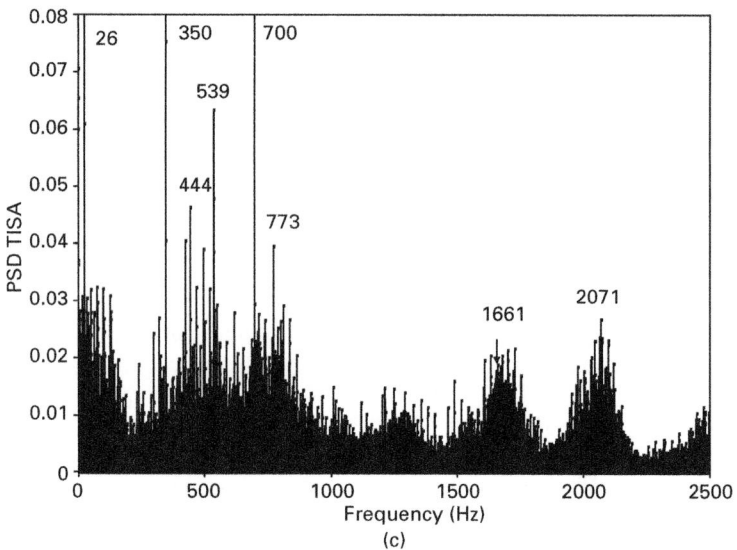

29.12 Continued

dividing the squeeze term (second term) by the rolling term (first term) in equation (29.2):

$$\text{Squeeze-to-roll force ratio} = \frac{3\pi}{2\sqrt{2}}\ \frac{\left|\dfrac{\partial h}{\partial t}\right|}{u\sqrt{\dfrac{h}{r_{eq}}}},\ \text{if}\ \frac{\partial h}{\partial t} < 0$$

$$\text{Squeeze-to-roll force ratio} = 0.0,\qquad \text{if}\ \frac{\partial h}{\partial t} \geq 0$$

(29.30)

Equation (29.30) indicates that a combination of low squeeze action, as well as the equivalent radius of curvature with high entraining velocity and larger film thickness would attenuate rattle. Thus, for design purposes, optimising gear geometry in such a manner can be beneficial in terms of NVH behaviour.

In the second predictions–measurements comparison case, the transmission lubricant temperature is 24.2 °C, corresponding to a dynamic viscosity of 0.0951 Pa s (a stiffer spring/damper lubricant element). In Fig. 29.15(a) the acceleration response of the 2nd gear is dominated by its lubricant reaction contribution due to rolling action (engine orders and meshing frequency at 291 Hz), while the squeeze action has a weaker presence at 266 Hz, compared with that of Fig. 29.10(a) (corresponding to a higher temperature). Similar trend can be seen when comparing Figs 29.15(b) and 29.10(b). In Fig.

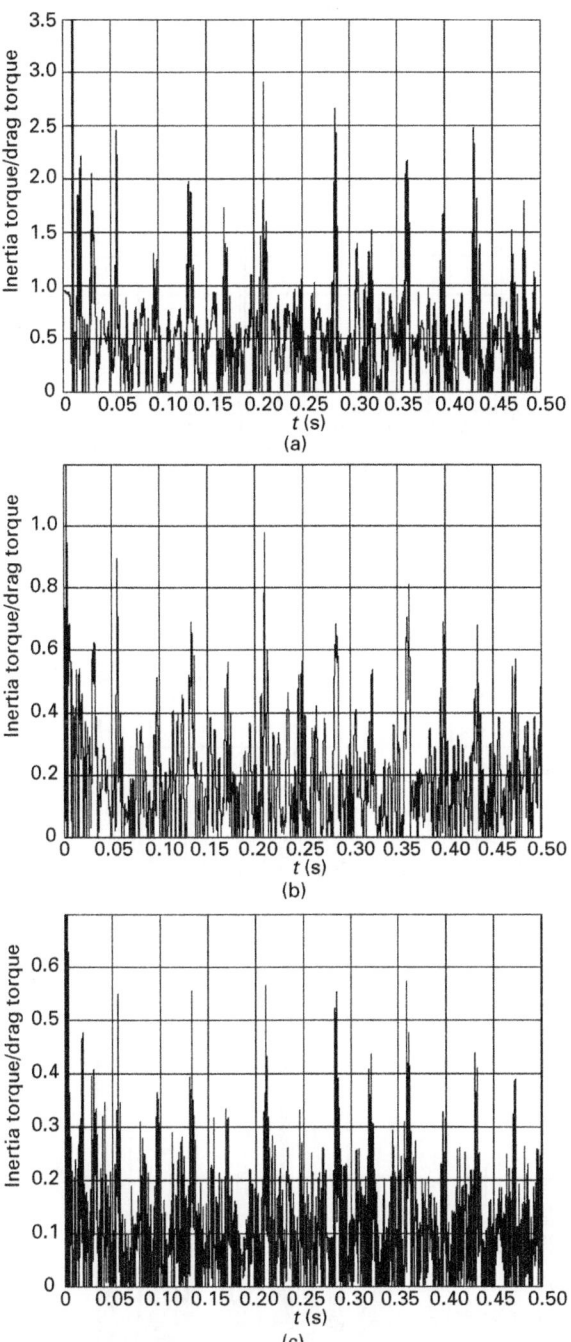

29.13 Simulated rattle ratios of (a) 1st (b) 4th and (c) 6th gear wheels (at 39.4 °C).

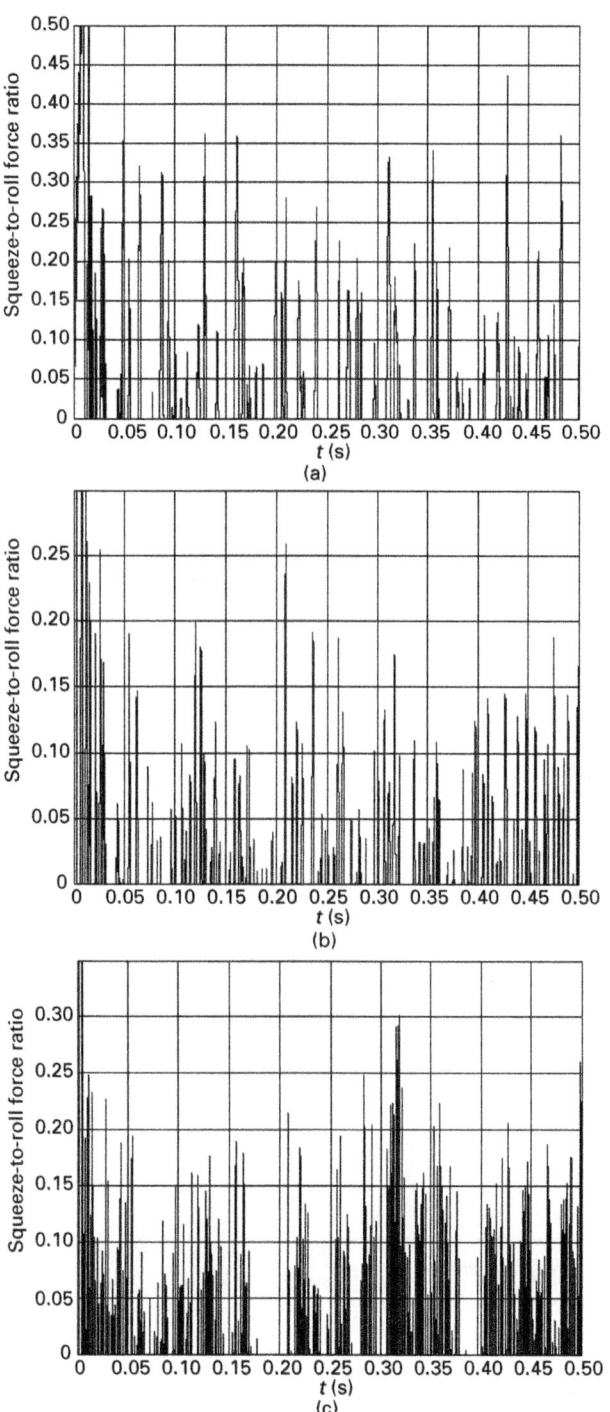

29.14 Simulated squeeze-to-rolling force ratios of (a) 1st (b) 4th and (c) 6th gear wheels (at 39.4 °C).

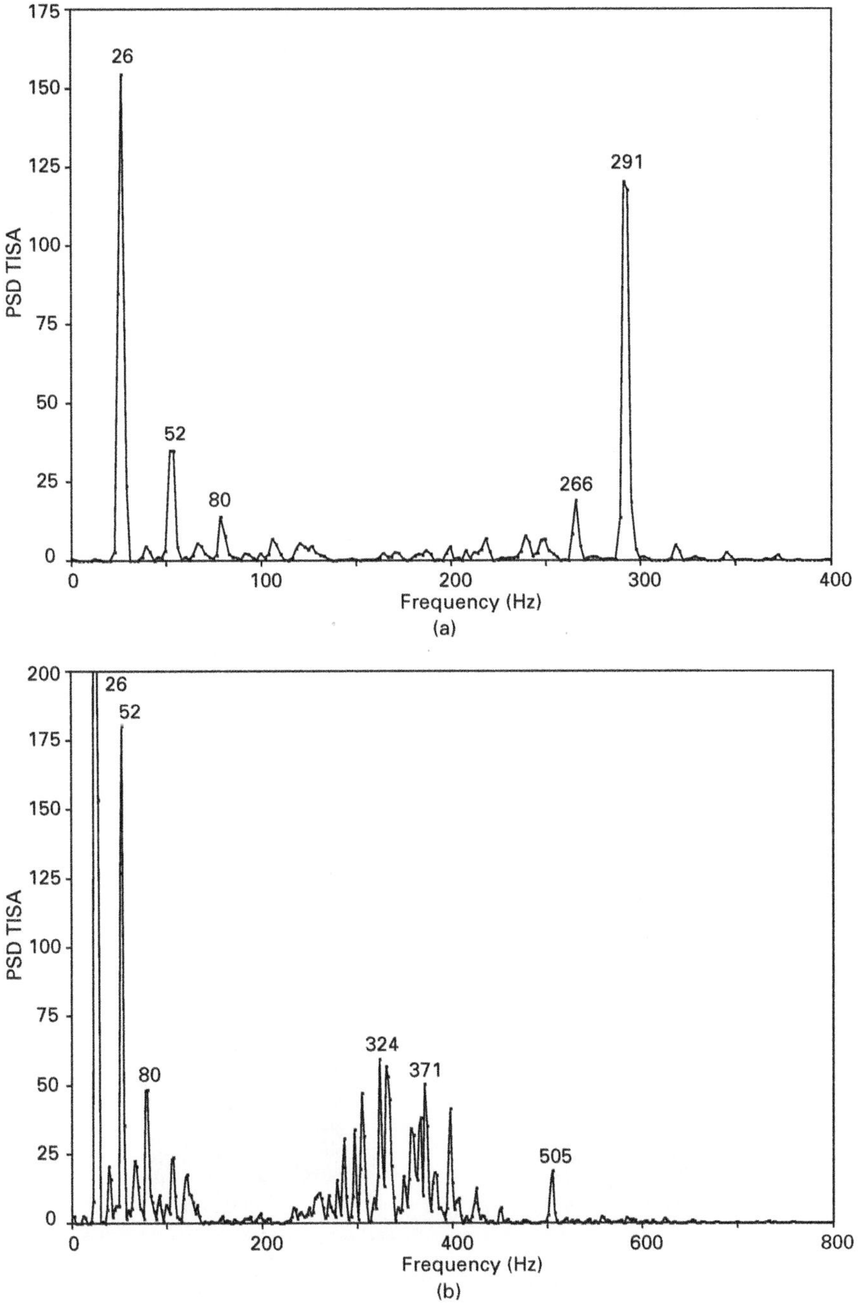

29.15 Simulated response FFT spectra of (a) 2nd and (b) 4th idle gear wheels (at 24.2 °C).

29.15(b) the natural frequency of the 4th gear pair (in the region of 352 Hz) has a stronger presence than its meshing frequency (505 Hz).

The lower contributions in Fig. 29.16 (26, 52 and 80 Hz) are ***engine order harmonics***. The natural frequencies of the 1st and 2nd gear pairs (258 and 256 Hz, respectively) are close to the spectral contribution at 233 Hz,

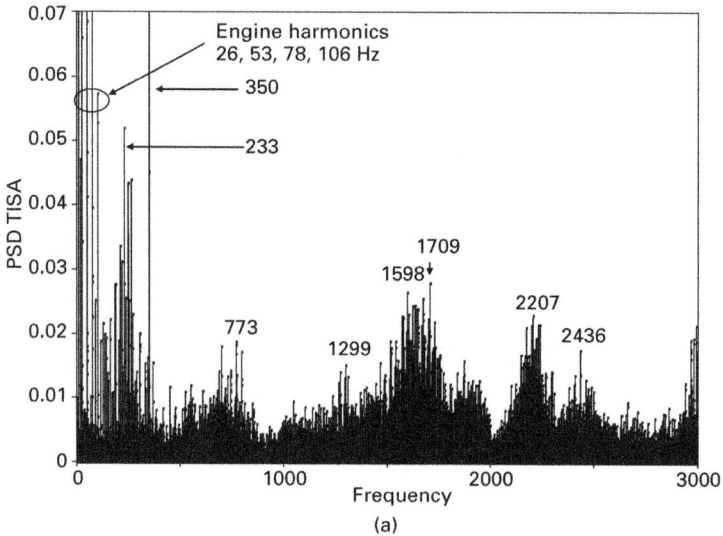

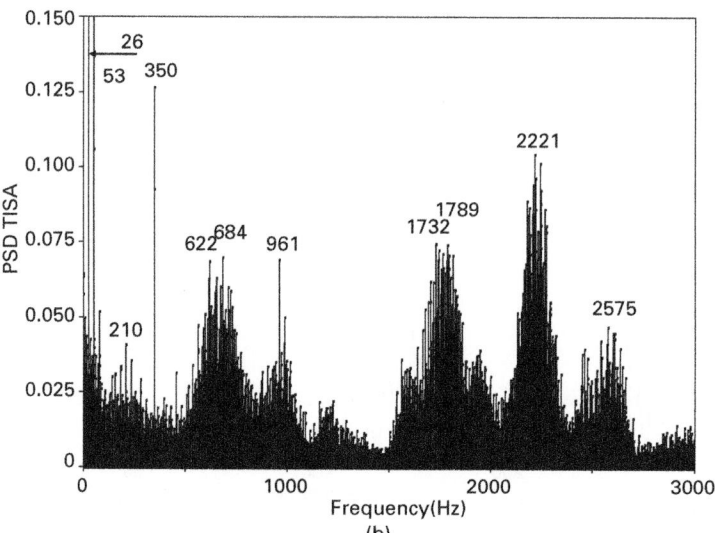

29.16 FFT spectra of the (a) transmission wall response (b) input bearing housing response and (c) upper output bearing housing response (at 24.2 °C).

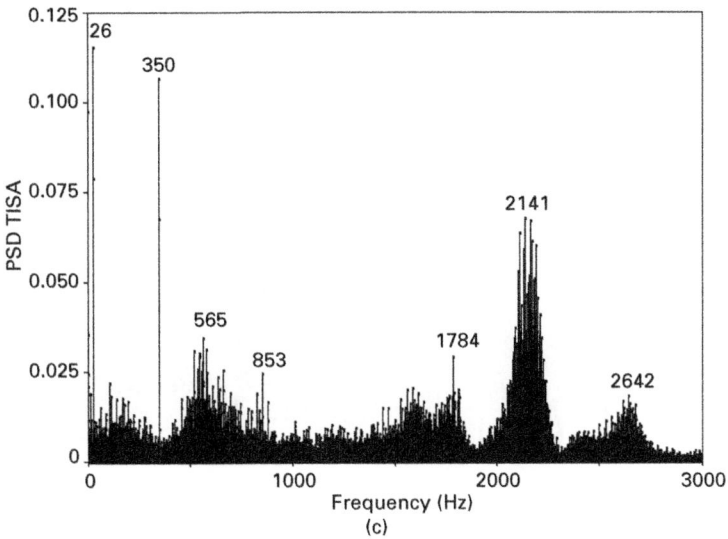

29.16 Continued

which appears in the transmission wall spectrum (Fig. 29.16(a)). The 350 Hz contribution in the same figure is close to the natural frequency of the 4th gear pair, while that at 565 Hz corresponds to the 6th gear pair natural frequency. The 684 Hz contribution is associated with the 4th order of the 1st gear meshing frequency and modulations due to the engine order excitations. Again, the frequency components above 1000 Hz are associated with the shafts' natural frequencies and with the frequencies of the bell housing (Tangasawi *et al.*, 2007). When inspecting the rattle ratios (Fig. 29.17), the responses of the 1st and reverse gears cross the rattle ratio threshold of unity. However, the squeeze-to-rolling ratio (Fig. 29.18) reveals that only the reverse gear has a squeeze-to-roll force ratio that also crosses unity. This suggests that rattle condition is more severe in the case of reverse gear than for the 1st gear pair.

29.5   Conclusions

Idle rattle conditions have been studied using gear pair and full transmission lumped parameter models. Comparisons with experimental measurements taken from a vehicle under similar conditions have shown good agreement with the numerical predictions. Most of the observed frequencies correspond to the model predictions. It has been concluded that gear rattle manifests itself as a band of frequencies, which shift towards lower spectral regions as the lubricant temperature rises. This band includes frequencies induced by the rolling and squeezing motions of lightly loaded hydrodynamic films.

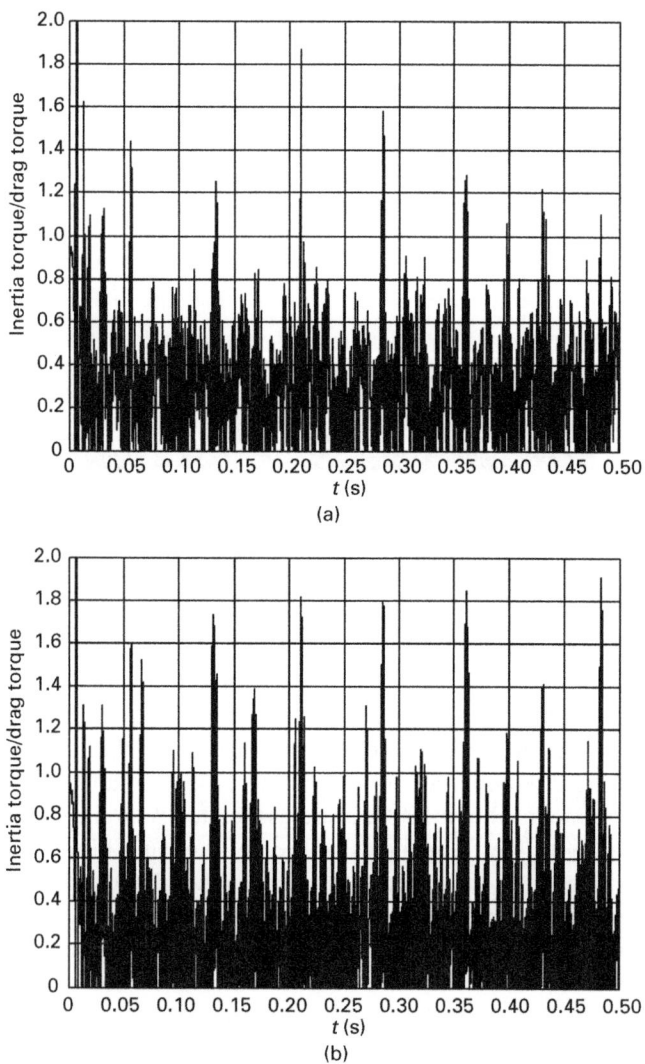

29.17 Simulated rattle ratios of (a) 1st and (b) reverse gear wheels (at 24.2 °C).

The effect of these contributions becomes clear by the squeeze-roll ratio, such that an emphatic squeeze action accompanies rattle behaviour. The gear design parameters determine the contact radii of curvature and lubricant entraining motion, which can be carefully selected in order to favourably influence the severity of rattle.

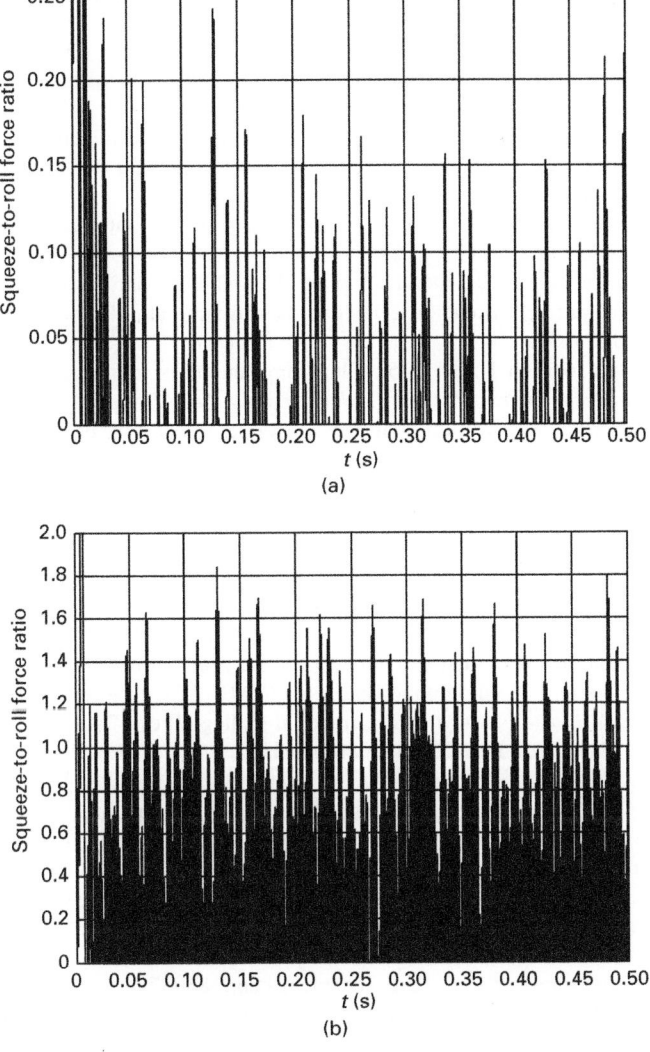

29.18 Simulated squeeze-to-rolling ratios of (a) 1st and (b) reverse gear wheels (at 24.2 °C).

29.6   Acknowledgements

The authors gratefully acknowledge the financial support extended to this investigation by the Engineering and Physical Sciences Research Council (EPSRC) and Ford Motor Company. Technical support of Getrag Ford (GFT) is also appreciated.

29.7   References

Bellomo, P., Cricenti, F., Vito, N. De, Lang, C., and Minervini, D., (2000), 'Innovative vehicle powertrain systems engineering: beating the noisy offenders in vehicle transmissions', SAE Technical Paper 2000-01-0033.

Brancati, R., Rocca, E., and Russo, R., (2005), 'A gear rattle model accounting for oil squeeze between the meshing gear teeth', *Proc. IMechE Part D: J. Automobile Engineering*, **219**, pp. 1075–1083.

Comparin, R., and Singh, R., (1990), 'An analytical study of automotive neutral gear rattle', *J. Mech. Design*, **112**, pp. 237–245.

Couderc, P.H., Callenaere, J., Der Hagopian, J., and Ferraris, G., (1998), 'Vehicle driveline dynamic behaviour: experimentation and simulation', *J. Sound Vibration*, **218**(1), pp. 133–157.

Dareing, D.W., and Johnson, K.L., (1975), 'Fluid film damping of rolling contact vibrations', *Proc. Instn. Mech. Engrs., Part C: J. Mech. Engng. Sci.* **17**, pp. 214–218.

Doğan, S.N., (2001), 'Zur Minimierung der Losteilgeräusche von Fahrzeuggetrieben', Berichte Nr. 91, Institut für Maschinenelemente, Universität Stuttgart, PhD Thesis.

Fujimoto, T., and Kizuka, T., (2001), 'An improvement of the prediction method of the idling rattle in manual transmission – in the case of the manual transmission with backlash eliminator', SAE Technical Paper 2001-01-1164.

Fujimoto, T., Chikatani, Y., and Kojima, J., (1987), 'Reduction of idling rattle in manual transmission', SAE Technical Paper 870395, pp. 2.99–2.109.

Gnanakumarr, M., Theodossiades, S., and Rahnejat, H., (2002), 'The tribo-contact dynamics phenomenon in torsional impact of loose gears promoting gear rattle', SAE 02ATT-138, Society of Automotive Engineers (SAE)-ATT Congress, Paris.

Gohar, R., (2001), *Elastohydrodynamics*, Imperial College Press, 2nd edition.

Gohar, R. and Rahnejat, H., (2008), *Fundamentals of Tribology*, Imperial College Press.

Harris, T., (2001), *Rolling Bearing Analysis*, John Wiley & Sons, Inc., 4th edition.

Kim, T., and Singh, R., (2001), 'Dynamic interactions between loaded and unloaded gear pairs under rattle conditions', SAE Technical Paper 2001-01-1553, pp. 1934–1943.

Mehdigoli, H., Rahnejat, H., and Gohar, R., (1990), 'Vibration response of wavy surfaced disc in elasohydrodynamic rolling contact', *Wear*, **139**, pp. 1–15.

Ohnuma, S., Yahata, S., Inagawa, M., and Fujimoto, T., (1985), 'Research on idling rattle of manual transmission', SAE Technical Paper 850979.

Rahnejat, H., (1985), 'Computational modelling of problems in contact dynamics', *Engineering Analysis, Computational Mechanics*, **2**(4), pp. 192–197.

Rust, A., Brandl, F.K., and Thien, G.E., (1990), 'Investigations into gear rattle phenomena – key parameters and their influence on gearbox noise', IMechE, C404/001, pp. 113–120.

Sakai, T., Doi, Y., Yamamoto, K., Ogasawara, T., and Narita, M., (1981), 'Theoretical and experimental analysis of rattling noise of automotive gearbox', SAE Technical Paper 810773, pp. 1–10.

Seaman, R., Johnson, C., and Hamilton, R., (1984), 'Component inertial effects on transmission design', SAE 841686, pp. 6.990-6.1008.

Singh, R., Xie, H., and Comparin, R., (1989), 'Analysis of automotive neutral gear rattle', *J. Sound Vibration*, **131**(2), pp. 177–196.

Smith, J.D., (1999), *Gear Noise and Vibration*, Marcel Dekker, Inc.

Tangasawi, O., Theodossiades, S., and Rahnejat, H., (2007), 'Lightly loaded lubricated impacts: idle gear rattle', *J. Sound Vibration*, **308**, 418–430.

Theodossiades, S., and Natsiavas, S., (2001), 'On geared rotordynamic systems with oil journal bearings, *J. Sound and Vibration*, **243**(4), pp. 721–745.

Theodossiades, S., Tangasawi, O., and Rahnejat, H., (2007), 'Gear teeth impacts in hydrodynamic conjunctions promoting idle gear rattle', *J. Sound Vibration*, **303**, 632–658.

Timoshenko, S., Young, D.H., and Weaver Jr. W., (1974), '*Vibration problems in engineering* 2, 4th edition, John Wiley & Sons, New York, London, Sydney, Toronto.

Wang, M., Manoj, R., and Zhao, W., (2001), 'Gear rattle modelling and analysis for automotive manual transmissions', *Proc. Instn. Mech. Engrs., Part D.*, **215**, pp. 241–258.

Yakoub, R., Corrado, M., Forcelli, A., Pappalardo, T., and Dutre, S., (2004), 'Prediction of system-level gear rattle using multibody and vibro-acoustic techniques', SAE Technical Paper 2004-32-0063/20044350.

29.8   Nomenclature

C Clearance between gear wheel and shaft (m).

C_b Normal backlash (m).

d_k Shafts' lateral displacement at bearing k ($k = a, b, c, d$) (m).

F_f Viscous flank friction force acting on the loose gear wheel (N).

F_{fi} Viscous flank friction force acting on the ith loose gear wheel (N).

F_{fj} Viscous flank friction force acting on the jth tooth in contact (N).

F_p Petrov's hydrodynamic tractive force (N).

F_{pi} Petrov's hydrodynamic tractive force acting on the ith ($i = 1, 2 \cdots 7$) loose gear wheel (N).

F_{xi} The resultant force acting on the shaft in the x-direction due to the radial and transverse components of the hydrodynamic film reaction ($i = 1, 2$) (N).

F_{yi} The resultant force acting on the shaft in the y-direction due to the radial and transverse components of the hydrodynamic film reaction ($i = 1, 2$) (N).

h Lubricant film thickness between the gear pair teeth surfaces (m).

$\dfrac{\partial h}{\partial t}$ Squeeze film velocity (m/s).

I_i Mass moment of inertia of the ith gear wheel (kg m^2).

I_{in} Mass moment of inertia of the input shaft (kg m^2).

I_{prev} Mass moment of inertia of the reverse pinion (kg m^2).

I_w Mass moment of inertia of the gear wheel (kg m^2).

I_{wrev} Mass moment of inertia of the reverse gear wheel (kg m^2).

K Linearised lubricant stiffness (N/m).

$K_{0i(i=1,\dots,6)}$ Mean lubricant stiffness of the ith gear pair (N/m).

$K_{0(rev)}$ Mean lubricant stiffness of the reverse gear pair (N/m).

K_{cp} Amplitude of the alternating lubricant stiffness component (pth cosine term) (N/m).

K_{sp} Amplitude of the alternating lubricant stiffness component (pth sine term) (N/m).

K_k Coefficient of the bearing $k(k = a, b, c, d)$ non-linear stiffness (N/m$^{1.11}$).

K_{kr} Coefficient of the bearing $k(k = a, b, c, d)$ non-linear stiffness in the $r(r = x, y)$ direction (N/m$^{1.11}$).

L_j Length of contact line on the meshing teeth flank of the jth teeth pair (m).

l_1 Length of contact line in the conformal contact of gear wheel and retaining shaft (m).

l_k Length of roller bearings $k(k = a, b, c, d)$ (m).

m_k The number of rollers in supporting bearing $k(k = a, b, c, d)$.

M_i Mass of the output shaft including supported gears ($i = 1,2$) (kg).

n_i Teeth number of the ith gear wheel.

N Total number of teeth in simultaneous contact.

N_i Total number of teeth in simultaneous contact of the ith ($i = 1, 2 \ldots 7$) gear wheel.

Q_k The reduced non-linear coefficient of the inner and outer raceways' contact with the bearing $k(k = a, b, c, d)$ roller.

r_{bp} Base radius of the pinion (m).

r_{bw} Base radius of the gear wheel (m).

r_{bi} Base radius of the ith ($i = 1, 2 \ldots 7$) gear wheel (m).

r_{bp7} Base radius of the reverse pinion (m).

r_{eqj} The reduced (equivalent) curvature radius of the jth meshing teeth pair in normal plane (m).

r_{pi} Contact radius of the ith pinion ($i = 1,2\ldots6, rev$) (m).

r_s Internal radius of the loose gear wheel (m).

r_{si} Internal radius of the ith loose gear wheel (m).

r_{wi} Contact radius of the ith gear wheel ($i = 1, 2\ldots6, rev$) (m).

u_j Lubricant entraining velocity associated with the jth teeth pair (m/s).

u_{sj} Sliding velocity associated with the jth teeth pair (m/s).

v Lubricant entraining velocity in the conformal contact between gear and the retaining shaft (m/s).

W Hydrodynamic reaction force (N).

W_i Hydrodynamic reaction force acting on the ith ($i = 1,2\ldots7$) gear wheel (N).

W_j Hydrodynamic reaction force acting on the jth tooth in contact taken in the normal plane (N).

x_i Output shaft rigid body translational displacement in the direction along the gear pair line of centres ($i = 1,2$) (m).

$\ddot{x}_i$	Output shaft rigid body translational acceleration in the direction along the gear pair line of centres ($i = 1, 2$) (m/s^2).
x_{ik}	Displacement of the ith ($i = 1, 2$) shaft at bearing k ($k = a, b, c, d$) along the x-direction (m).
y_i	Output shaft rigid body translational displacement in the direction normal to the gear pair line of centres ($i = 1, 2$) (m).
$\ddot{y}_i$	Output shaft rigid body translational acceleration in the direction normal to the gear pair line of centres ($i = 1, 2$) (m/s^2).
y_{ik}	Displacement of the ith ($i = 1, 2$) shaft at bearing k ($k = a, b, c, d$) along the y-direction (m).
α_n	Normal pressure angle (rad).
β	Helix angle (at pitch circle) (rad).
β_b	Helix angle (at base circle) (rad).
β_{bi}	Helix angle (at base circle) of the ith ($i = 1, 2...7$) gear wheel (rad).
γ_k	Contact angle of supporting bearing k ($k = a, b, c, d$) (rad).
Δ_k	Radial clearance of bearing k ($k = a, b, c, d$) (m).
δ_k	Radial deformation of bearing k ($k = a, b, c, d$) (m).
ε_i	The angular distance between the reverse pair centre line and the axis x_i ($i = 1, 2$) (rad).
η_0	Lubricant dynamic viscosity (Pa s).
Λ_k	The restoring force of bearing k ($k = a, b, c, d$) (N).
ρ_w	Tooth surface radius of curvature (m).
ρ_i	Radius of curvature of the ith ($i = 1, 2...7$) gear wheel surface (m).
$\varphi_{1,prev}$	Angular displacement of the 1st speed gear and reverse pinion (rad).
φ_i	Angular displacement of the gear wheels ($i = 2, 3...6$) (rad).
$\ddot{\varphi}_i$	Angular acceleration of the ith ($i = 1, 2...7$) gear wheel (rad/s^2).
φ_{in}	Angular displacement of the input shaft (rad).
$\dot{\varphi}_{in}$	Angular velocity of the input shaft (rad/s).
$\ddot{\varphi}_{in}$	Angular acceleration of the input shaft (rad/s^2).
φ_{iT}	Fundamental period of the gear teeth contact stiffness (s).
φ_p	Angular displacement of the pinion (rad).
$\dot{\varphi}_p$	Angular velocity of the pinion (rad/s).
$\ddot{\varphi}_p$	Angular acceleration of the pinion (rad/s^2).
φ_w	Angular displacement of the wheel (rad).
$\dot{\varphi}_w$	Angular velocity of the gear wheel (rad/s).
$\ddot{\varphi}_w$	Angular acceleration of the gear wheel (rad/s^2).
φ_{wprev}	Angular displacement of the reverse speed gear (rad).

Subscripts

xji	Related to the x_j axis and ith gear ($j = 1, 2$).

xjin   Related to the x_j axis and input shaft ($j = 1,2$).
xjrev Related to the x_j axis and reverse gear ($j = 1,2$).
xjyk   Related to the x_j axis and y_k displacement ($j = 1,2; k = 1,2$).
xjxk   Related to the x_j axis and x_k displacement ($j = 1,2; k = 1,2$).
yji   Related to the y_j axis and *i*th gear ($j = 1,2$).
yjin   Related to the y_j axis and input shaft ($j = 1,2$).
yjrev Related to the y_j axis and reverse gear ($j = 1,2$).
yixk   Related to the y_j axis and x_k displacement ($j = 1,2; k = 1,2$).
yjyk   Related to the y_j axis and y_k displacement ($j = 1,2; k = 1,2$).

30

High-energy impact-induced phenomena in driveline clonk

M. GNANAKUMARR, Loughborough University, UK

Abstract: Recent years have witnessed increased output power, particularly with modern diesel engines and reduced powertrain inertia. Increased efficiency has been accompanied by certain drawbacks. Among these, noise and vibration have been particularly notable. Lightweight powertrain systems are poorly damped and any significant torsional impulse can lead to metallic noise output. A significant preoccupation for the drivetrain engineers is the clonk phenomenon. This is an unacceptable high-frequency metallic noise, which is accompanied by a tactile drivetrain response. It occurs under several driving conditions, resulting in impulsive action. Clonk is induced by a power torque surge, which can cause abrupt load change reactions, such as throttle tip-in or rapid clutch actuation. It is shown that structural modal responses of the powertrain components can lead to radiated noise through elasto-acoustic coupling. The coincidence of the structure-bore waves with the acoustic modes of sub-system thin-walled components causes these audible responses.

Key words: powertrain NVH, structure-borne vibration, elasto-acoustic coupling, clonk noise.

30.1 Introduction

Recent years have witnessed increased output power, particularly with modern diesel engines and reduced powertrain inertia. Increased efficiency has been accompanied by certain drawbacks. Among these, noise and vibration have been particularly notable. Lightweight powertrain systems are poorly damped and any significant torsional impulse leads to metallic noise output.

There exist many sources of torsional input, including those induced by the power torque. For a four-stroke engine the signature of power torque comprises the *half engine order response* and all its multiples (Rahnejat, 1998). A major source of torsional impulse is *impact* in the lash zones; in the transmission, splines and the differential. The response is broad band, from a few Hz to a few kHz. Reduction of inertial imbalance has lessened the influence of the lower spectral content. However, the converse is true of the higher spectral contributions due to *elastodynamics* (see Chapter 1).

Thin-walled components, such as the bell housing and driveshaft pieces are efficient noise radiators, with energy input induced by an impulsive torque.

914

These are usually impact-induced. When the speed of wave propagation in the structure coincides with that in the adjacent fluidic medium, the acoustic modal behaviour is referred to as **supersonic** (Wang and Lai, 2000). In the case of hollow structures, the efficient noise radiating modes are a combination of high-frequency **elasto-acoustic** coupled waves. These are referred to as the **breathing modes**.

The outward bound acoustic waves reflect from nearby obstacles, such as parked vehicles or the kerbside. If the vehicle windows are lowered the echo can be heard by the vehicle occupants as metallic noise, referred to as **clonk** in the automobile industry (Kelly *et al.*, 1999). No structural damage is usually noted but the event is quite disconcerting for both the vehicle occupants and other nearby road users.

This chapter discusses some experimental and numerical investigations of high-frequency behaviour of light truck drivetrain systems, when subjected to sudden impulsive actions, which can also be due to driver behaviour. The problem is treated as a **multi-physics** interactive phenomenon under transient conditions, because it is as the result of interactions of system dynamics, structural modal responses of elastic members, impact dynamics in lash zones and acoustic wave propagation (theoretical treatment is provided in Chapter 1). A representative powertrain system rig was designed and implemented (Theodossiades *et al.*, 2004), and controlled tests simulating driver behaviour, such as side-slipping from the clutch pedal were undertaken. This action results in a torsional input which causes impact in the aforementioned lash zones. In fact many noise and vibration phenomena are impact induced.

30.2   Impact-induced noise and vibration

30.2.1 Repetitive steady state impacts in transmission system: rattle

Various forms of rattle have been cited, owing to the mechanism of manifestation and operating conditions (see also Chapters 26–29):

- **Idle rattle**: with an engaged clutch and transmission in neutral and with the engine at idle rpm. This is very audible owing to the lack of engine noise and is switched off with clutch disengagement.
- **Creep rattle**: occurs between 1200–2000 rpm in gear and is strongly related to the torsional modes of the drivetrain.
- **Over run rattle**: with throttle off and coasting between 1500 and 4000 rpm. It occurs at low engine speeds and high loads and when the inertial torque exceeds the **drag torque**. It is thought that this mode of rattle would be sensitive to bulk lubricant properties such as viscosity.

Some remedial actions are currently undertaken to palliate for these modes of rattle. These are:

- clutch first stage is used for idle rattle;
- clutch second stage is used for creep-in-gear rattle;
- impact forces from non-torque transmitting teeth are transferred to neighbouring housing.

The following additional observations are also made:

- Rattle is aggravated by ancillaries, when switched on.
- Needle-type torque pulses result in wide band frequency spectra, thus a connection between rattle and clonk may be initiated.

30.2.2 Transient impulsive action: clonk

Drivetrain *clonk* is an audible sound, which is accompanied by a tactile drivetrain response. This may occur under several different driving conditions. Many drivetrain noise, vibration and harshness (*NVH*) concerns are related to impact loading of subsystems down-line of engine. These concerns are induced by power torque surge through engagement and disengagement processes, which may propagate through various transmission paths as structural waves. The coincidence of these waves with the acoustic modes of subsystem components leads to audible high frequency responses, referred to as clonk.

Krenz (1985) was one of the first who carried out a detailed investigation of clonk. He described clonk as 'A short duration audible transient response, usually the result of a load reversal and in the presence of backlash'. Clonk may occur under several different driving conditions:

- Rapid throttle applied (*throttle tip-in*) from coast or released (*throttle back-out*) from drive condition.
- Rapid engagement/disengagement of clutch at low road speeds. It may also occur after gear selection, if the clutch is rapidly engaged.

In either case, the resulting torsional impulse to the drivetrain causes a short duration vehicle jerk action and an accompanying clonk noise. It is more noticeable in low gear and low vehicle speeds. It is thought that clonk is as a result of a high-energy impact (frequency range: 1000–5000 Hz).

There are two basic types of clonk.

1. *Throttle-induced clonk*: rapid application or release of throttle.
2. *Clutch clonk*: Rapid engagement or disengagement of clutch at low road speeds and in low gear (Menday, 2003).

Often torque reversals take place through the lash zones in the drivetrain system, which are accompanied by coupled torsional oscillations of the drivetrain system at its first rigid body fundamental mode, referred to as *shuffle*, and the longitudinal fore and aft motion of the vehicle body, referred to as *shunt* (Rahnejat, 1998; Farshidianfar *et al.*, 2000; Capitani *et al.*, 2001).

The high-frequency clonk noise may be heard, coincident with cycles of shuffle. In the presence of inevitable backlash in the drivetrain the torque surge over a very short time period leads to impact conditions. This is why the clonk noise may be heard with each cycle of shuffle (Arrundale *et al.*, 1998), although it has been found that this is not the necessary condition for clonk to occur, for example in the case of clutch clonk (Theodossiades *et al.*, 2004).

Given the importance of this problem in industry surprisingly very few controlled rig-based investigations have been reported. They include those by Krenz (1985) and Biermann and Hagerodt (1999) for a front wheel drive vehicle and Menday (2003), Kelly *et al.* (1999) and Theodossiades *et al.* (2004) for a rear wheel drive light truck. Other vehicle-based investigations include that of Petri and Heidingsfeld (1989), who observed that a sudden throttle change transforms into vehicle acceleration/deceleration, which may excite the system *shuffle*, accompanied by a hard metallic clonk.

30.3   Fundamentals of impact-induced noise

The difference between clonk and the usual rattle of meshing pairs is in the severity of the impulsive action, caused by driver behaviour, although both phenomena are in fact impact induced. A sudden demand in torque caused by driver action, as in throttle tip-in or abrupt clutch engagement yields high impact velocities through *backlash*. The greater the impact energy, the higher is the spectral modal composition of the response.

Localised impact of gear teeth pair may be regarded as Hertzian and are related to the *crushing stiffness* of teeth surfaces and their geometry. With sufficient energy, global-type deformation and wave propagation would result. Global deformation refers to tooth bending and rocking modes at much higher frequencies than those due to crushing stiffness. Torsional impulsive action that initiates clonk condition are thought to induce modal behaviour of impacting teeth rather than the lower impact conditions that are responsible for rattle. The modal behaviour may be regarded as St Venant-type deformation. Therefore, the distinction between rattle and clonk can be made with respect to the local versus global nature of deformation. In the case of rattle, the deformation remains local.

Sound is propagated by two distinct processes, when two elastic bodies impact. Firstly, when bodies are solid an *accelerative noise* would result as in the case of common Newton trolley. When the solids are hollow, elastic waves propagate through their structures. Therefore, the initial, accelerative noise is followed by a *ringing noise*. The number of natural modes increases with reduced thickness of hollow thin-walled structures.

30.3.1 Accelerative noise

Rapid changes in velocity of the moving parts during impact gives rise to a pressure perturbation, which in turn emits accelerative noise (Richards *et al.*, 1979a). This is not dependent on damping or vibration isolation. Accelerative noise arising from impact is a function of impact duration, size factor, area in contact and the impact velocity. The dominating accelerative noise mechanisms are those which relate to very short impact times and in association with backlash. The shorter the ***impact duration***, the greater the radiated noise energy becomes.

30.3.2 Ringing noise

Ringing noise is due to sound radiation from vibrating modes of the attached structures. This continues until all the impact energy has been radiated as sound or has been absorbed in the structural damping system (Richards *et al.*, 1979b). ***Ringing noise*** is a function of structural damping, the rate of propagation of vibration energy, Young's modulus of the structure, radiation efficiency of the structural panels liable to vibrate flexurally, surface areas and velocities and vibration amplitudes. Richards *et al.* (1979b) have developed empirical formulae for the prediction of noise generated by impact, so that design action can be undertaken early in a programme to avoid costly palliation later on.

For bodies with sufficiently compact dimensions, it has been found that acceleration noise and structural ringing noise are of similar order. However, if local resonant flexural modes are excited by an impact, the radiation energy can be significantly higher than the energy radiated during the time of impact. The reason for this is the longer time taken for lightly damped structures to radiate the noise energy by ringing.

30.4   Lashes in vehicular drivetrain

Drivetrain lash may be described as the summation of lashes at each contact zone in the drivetrain, including the major lash effect of the clutch disc's low stiffness pre-damper springs. There are many lash points in a geared system and each lash point has an associated pair of impacting inertias. In each lash zone the impact velocity is a function of the lash or the gap that an impacting inertia accelerates through. The larger this gap, the higher the velocity reached prior to an impact, if one ignores the damping effect of the usually present lubricant film. In rattle, the velocity is a function of engine order vibration, which is most prominent as the ***second engine order*** (i.e. twice the crankshaft angular velocity) in a four-stroke, four-cylinder engine (Rahnejat, 1998). Therefore, palliative action can be undertaken to reduce the second engine order, such as the use of a ***dual mass flywheel*** (***DMF***) (see Chapter 28). Reduction of lash is the optimum solution to the problem. However, the costs associated are not usually permissible in mass manufactured vehicles. Piece-to-piece variations also have an important effect in this respect. Dogan *et al.* (2004) discuss the importance of tuning the amount of driving inertia up line of the impact zone to reduce the effect of impact, particularly in the case of transmission rattle (see also Chapter 26). Littlefair (2004) justifies the use of dual mass flywheel for attenuation of rattle, based upon the reduction of second engine order due to damping effect of grease drag, when incorporated between the primary and secondary inertias.

Other palliations can include control of:

- clutch disc hub spline lash clearance to transmission input shaft;
- clutch disc low rate pre-damper springs: the largest single lash source;
- transmission free play between gears in mesh;
- universal joint's spline free play;
- final drive lash between the pinion and the crown wheel;
- half-shaft spline lash.

30.5   An experimental rig for driveline clonk

An experimental rig was devised, shown in Fig. 30.1 (Theodossiades *et al.*, 2004). It comprises all the components of a light truck's powertrain system, from flywheel to the rear axle road wheels. The system is driven by a four-pole electric motor, delivering 145 N m torque at 1500 rpm (22.7 kW), typical of baseline conditions, with transmission engaged in the 2nd gear, which can lead to clonk, when the clutch is suddenly disengaged. It is important that experimentation should be carried out in a repeatable manner and under

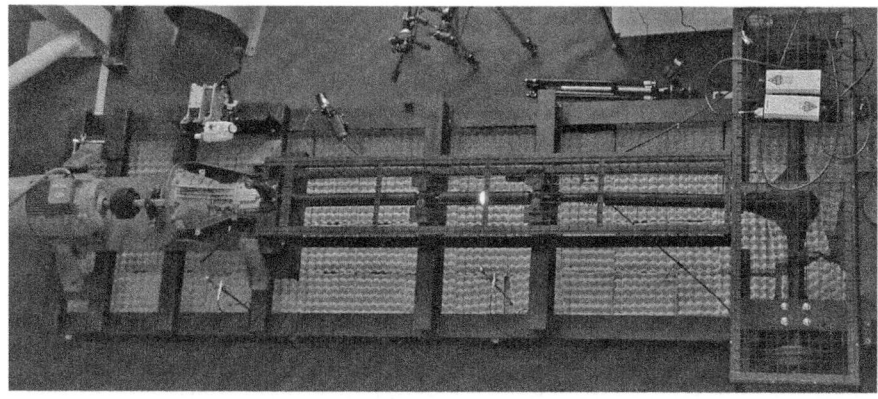

30.1 The experimental rig.

controlled conditions. This gives the major advantage, of being able to examine the behaviour of subsystems and components. Thus, a motorised rig rather than a vehicle is preferred for this investigation. Good correlation has been found between the rig and vehicle behaviour.

Since the rear axle of the rig is rigidly mounted onto structural cross-beams, clonk induced by throttle action, though possible by sudden reversal of motor action, is not very representative of vehicle conditions, where for the aforementioned actions the vehicle would normally undergo fore and aft motions, referred to as **shunt**, with associated weight transfer forward or backward in dive or squat. The rig does not allow shunt, thus best represents **clutch-induced clonk**. Various laden states of the vehicle in **shuffle** are nevertheless allowed for by controlling the rear axle resistance by an adjustable hydraulic handbrake system. This also enables the introduction of representative rolling resistance in the very low-speed parking manoeuvres, which are typical of circumstances leading to clutch clonk.

The instrumentation included **laser Doppler vibrometers** (LDV), placed normal to the axis of the driveshaft tubes and 400 mm away, microphones positioned directly opposite to the lasers in order to capture the corresponding airborne signal, and various accelerometers on the driveline support needle bearings and flanges of transmission and the differential.

LDV relies on the detection of a **Doppler shift** in the frequency of the coherent light scattered by a moving target, from which a time-resolved measurement of the target velocity is obtained (Bell and Rothberg, 2000). There usually exists a problem with LDV measurement of members that rotate about an axis normal to the beam's direction. This is caused by a fictitious velocity component due to the rigid body rotation of the shaft, being: $v = \omega r$, where ω is the angular velocity of the driveshaft and r is its outer radius. This corrupts the lateral vibration of the shaft at the point of measurement, given by $\dot{y}$, y being the direction of the beam. Thus: $v_y = \dot{y} \pm \omega r$. This is a known problem with LDV. However, it is not significant with high-frequency measurements from a vibrating structure, which is rotating at a relatively low speed. The speed of rotation in the experiments was 1500 rpm, equivalent to 25 Hz, which is considerably below the clonk frequencies: 1000–5000 Hz. Thus, the use of LDV is justified.

The directed beams along the surface normal to any given cross-section of the tubes capture their transient behaviour during the passage of the **clonk wave**. Although the system is operating while the spectral information is being captured, mode shape reconstruction is not attempted under such fast transient conditions due to the high modulating modal density of the tubes.

Microphones are used to measure **sound pressure level** during testing. Using $V = f\lambda$, and with the lowest frequency taken as 1000 Hz, these were placed 340 mm away from the tube surface and 570 mm from the laboratory floor (noting that velocity of sound is: $V = 340\,\text{m/s}$).

The sound pressure recorded by a microphone can be used to obtain the noise level (in dB), using the following expression:

$$L_P = 20 \log_{10}\left(\frac{P}{P_r}\right) \tag{30.1}$$

where P is the recorded pressure in pascal, and P_r is the reference pressure: $P_r = 20\,\mu\text{Pa}$

All data were collected simultaneously by a multi-channel data acquisition system.

30.6   Results and discussion of driveline clonk experiment

Abrupt clutch actuation as a hasty driver action can cause severe load-change reaction in the presence of lash zones in vehicle drivetrain system. The resulting impact(s) of sufficient energy propagates as structural waves along the driveline system, exciting many modes of vibration of lightly damped thin-walled structures, such as the *driveshaft tubes*. Some of these *structural modes* coincide with the *acoustic modes* of the same components, which are efficient noise radiators, referred to as the breathing modes. The radiated noise is at high frequency and has a metallic nature, onomatopoeically referred to as *clonk*.

Figure 30.2 shows a typical clonk signal, obtained by a microphone, positioned normal to the surface of the first driveshaft tube at its mid-span. Note that the signal commences with a ramp-up period of approximately 100 ms. Then a very short impulse is recorded of the duration 0.25–5 ms, corresponding to the *accelerative noise* (due to impact), followed by a long decaying period of 80–150 ms, corresponding to the ringing noise response. These characteristics are typical of clonk signals. In this case, the peak pressure is obtained as 2.91 Pa. The noise level is obtained using equation (30.1) as 103 dB in this case. The measured noise level here is rather typical of the annoying noises measured from vehicles on the road (e.g. at 80–85 dB reported by Hachenbroich, 1994, for heavy trucks, even for low-noise transmission with engine encapsulation).

The response in Fig. 30.2 corresponds to the rig configuration, employing a single *solid mass flywheel* (SMF) and an aggressive clutch actuation of duration approximately 80 ms (this represents hurried side-slipping off the pedal). The proper operating range for clutch actuation is 200–300 ms.

It is necessary to decompose the recorded signal in order to obtain its spectral content. This enables identification of the troublesome contributions with high output power. When compared with those obtained by LDV for

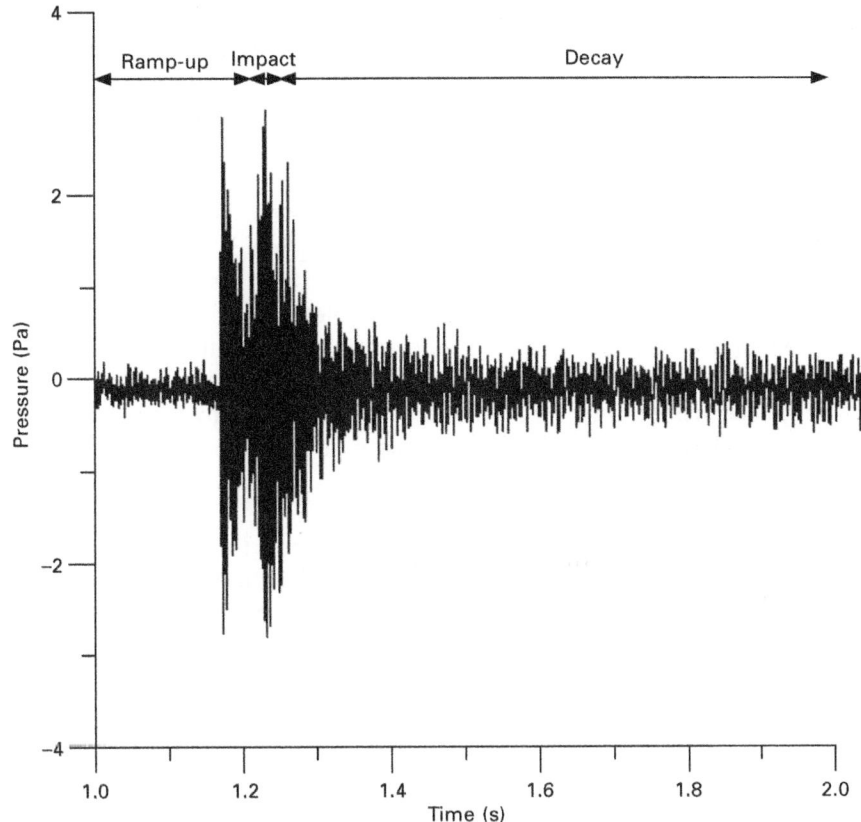

30.2 A typical clonk signal.

structural vibration, one can then pinpoint coincidence between structure and acoustic modes.

Fourier transformation of the signal requires a sampling rate that captures the highest spectral composition of interest. Furthermore, to avoid aliaising, one needs a sampling rate at least twice that of the highest frequency of interest, according to the *Nyquist criterion*. Since the highest frequency of interest according to the observation of vehicle data is 5000 Hz, it is safe to gather 10 000 samples per second, but prudent to take a larger record. Thus, a sampling rate of 16 384 samples per second (or $n = 14$ for integer multiples of 2^n) was used.

It is clear that the percentage of the samples related to the actual accelerative response (i.e. the *impact time*) represents less than 10% of the acquired record. The problem with Fourier transformation is its averaging nature. Thus, if a fast Fourier transform (*FFT*) or power spectral density (*PSD*) analysis is carried out over the entire sample, the result will be unrepresentative of the actual frequency content of the signal. Therefore, it is necessary to window around areas of interest in the signal (Vafaei *et al.*, 2001). Nevertheless, one can employ windowing with Fourier analysis, so long as a 2^n sample size is used with the appropriate sampling rate for a ***short fast Fourier transform*** (SFFT), leading to a spectrogram.

The regions of interest are: the initial ramp-up, prior to the second region, being the accelerative response, followed by the final region of exponential decay, where hollow structures would ring. This approach not only provides the power sources in the overall signal, but it also gives an indication of the timing of each contribution, similar to wavelet analysis. The importance of combined time–frequency analysis becomes clear in attributing spectral contributions to given causes, thus pointing to palliative actions.

The ***spectrogram*** in Fig. 30.3 is the accelerative noise (the impact region) in Fig. 30.2. This takes place approximately at the time 1.21 s for 40 ms. There are 1024 samples here, which enable spectral analysis to be carried out.

Wavelet of the same signal provides combined time and frequency information, which is very useful to understand the effect of various contributions in a transient event (Daubechies, 1990; Newland, 1993). Plate VIII (between pages 514 and 515) shows the wavelet spectrum for the same condition and for the impact region. It is clear that the high-frequency content of the signal occurs in the same 30–40 ms region as that obtained by the spectrogram of Fig. 30.3. The wavelet results pinpoint the 5 ms duration of the main accelerative noise.

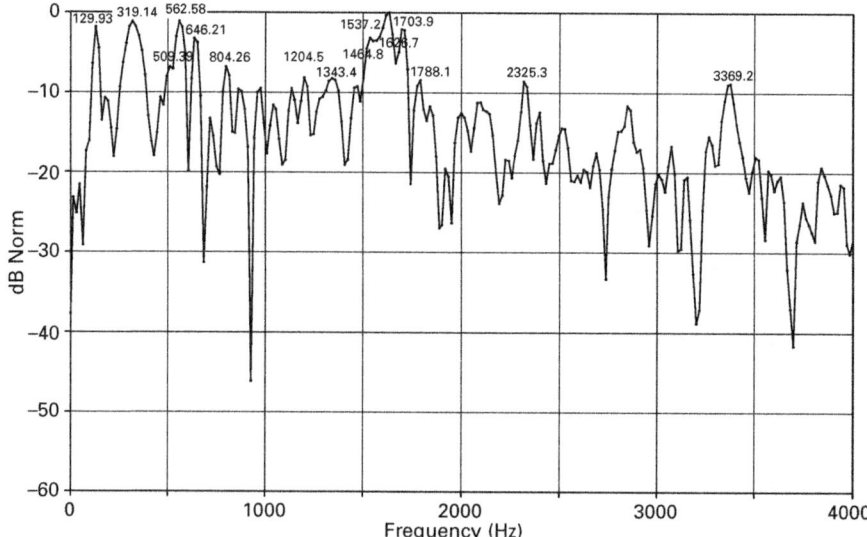

30.3 A spectrogram of noise monitored from the first tube.

It is important to identify the possible underlying causes for all the significant spectral contents. Here the spectrogram of Fig. 30.3 is more useful, because it 'quantifies' the contribution of the various spectral contents. The main contributions are at: 129, 315, 563, 650, 800, 1631, 1700, 2325 and 3375 Hz. The contributions in the band: 1630–1700 Hz and at 2454 represent the *breathing modes* of the driveshaft tubes, being the efficient noise radiators, obtained by numerical analysis, reported elsewhere (Theodossiades *et al.*, 2004) (see Plate IX (between pages 514 and 515), breathing modes at 1830 and 2554 Hz).

To confirm elasto-acoustic coupling one needs to show coincidence of structure-borne vibration with the radiated noise in Fig. 30.3, as well as the breathing nature of these modes. The second condition is corroborated by Plate IX.

Figure 30.4 shows the spectrum of structural modal responses, obtained by an LDV. Note the significant contributions at 1631 and 2456 Hz. Therefore, coincidence of these with the acoustic modes has been established. An interesting feature is the existence of a high-frequency response at 3375 Hz in Fig. 30.3, with low amplitude in Fig. 30.4. This indicates that insufficient impact energy exists to excite this mode under clutch clonk conditions, and with the experimental rig. There is a much greater inertial effect with throttle-induced clonk, where tip-in and back-out action with the throttle causes dive and squat of the vehicle. These actions cause vehicle shunt, thus a greater inertial action, more severe impacts and higher modal responses.

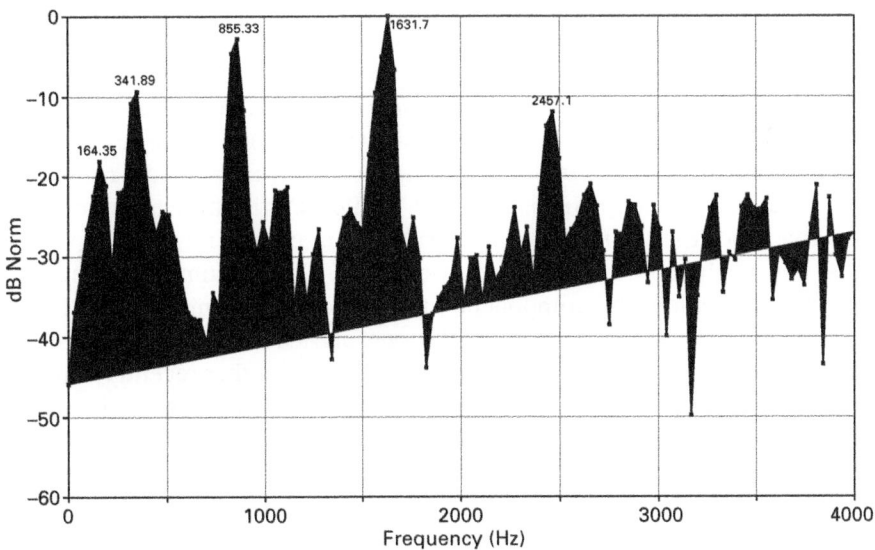

30.4 Structural modes of first drive shaft tube obtained by LDV.

30.7   Some methods of palliation of driveline clonk

As in all other investigations, one could have set other aims of scientific or technical importance with the benefit of hindsight. A number of enhancing features could have been included in the current investigation. On the numerical side, the current methods include the use of finite element analysis and *component mode synthesis* to represent component flexibility into the multi-body model of the drivetrain system (see Chapter 1). This approach is quite suitable for combined rigid body and structural dynamics, but precludes the determination of acoustic modes of the thin hollow structures. These were obtained using an analytical model or by experimentation. Alternatively, boundary element method can be used to obtain both structural and the acoustic modes.

Additionally, the impact models are for dry meshing solid pairs. Lubricated contacts tend to increase the impact time, as well as alter the nature of deformation (see for example Dowson and Wang, 1994; Al-Samieh and Rahnejat, 2002). Although this effect has been found to be not very significant for small backlash, its inclusion will improve the accuracy of predictions. However, the effort required warrants a specific research programme of its own.

30.7.1 Construction of driveshaft tubes

On the experimentation side, the existing rig can be used to test various driveshaft tubes with changes to their material of construction, geometric alterations or insertion of *sound-absorbing media*. These may be regarded as alternative palliation methods, with varying degrees of success. For instance altering the geometry of the tubes would change their modal behaviour, and the clonk-type response may become more significant at different spectral contributions. Filling the tubes with *industrial foams* can be hazardous, as some of these have already been banned under health and safety regulations. Foam filling can lead to attenuation of some of the spectral modal responses and not all, as found by Arrundale *et al.* (1998). Partial filling is more cost effective, and it is claimed that if carried out in certain locations it would most benefit the absorption of acoustic waves. These claims and others in industry, and chiefly by component suppliers do not seem to be based on any fundamental study, similar to the one carried out here. Insertion of *card-board inserts* into the driveshaft tubes is yet another recently claimed palliation method. One would expect that some attenuation of the clonk response would take place, but equally certain undesired effects may also ensue. One such undesired effect would be increased inertia of the tubes, as they are quite light and any significant cardboard thickness can add to their inertia. This has the adverse effect of increased rigid body inertial dynamics, which is mainly due to unavoidable imbalances in assembly and alignment of driveshaft pieces. Therefore, increased mechanical losses will decrease drivetrain efficiency. One should note that hollow driveshaft pieces have been progressively introduced in the past two decades in order to reduce the unbalanced problems with solid shafts of earlier constructions.

These palliation methods should form the basis of future investigations in a fundamental manner. This can involve the determination of absorption coefficients and inclusion of these into boundary element analysis. In general, a root-cause solution may be found in modification of impacting geometries, which would require an in-depth contact mechanics analysis.

30.7.2 Dual mass flywheel

A *dual mass flywheel* is a device which incorporates a conventional flywheel, referred to as the primary inertia, constituting the main inertial element. It also comprises a secondary member, which has a much lower inertia, being essentially a shell (see also Chapter 28). A torsional spring conforms to the shape of this secondary member and sits inside it. The contact between this spring and the shell is achieved via a number of mounted members, referred to as shoes. The cavity formed by the arrangement is filled with grease, often doped with metallic compounds, which improve grease stability at high generated temperatures. The primary inertia is attached to the spring and compresses it, when subjected to torque and the arrangement resists torsional input. The spring stiffness is chosen to dampen the second engine order in four-stroke engine, this being the main torsional vibratory signal. The drag introduced by the grease also plays a significant role (Littlefair, 2004).

30.8   Acknowledgements

The author wishes to express his gratitude to Ford Motor Company and the Vehicle Foresight Directorate (EPSRC and DTI) for sponsorship and financial support to this research project.

30.9   References

Al-Samieh, M.F. and Rahnejat, H. (2002): Physics of lubricated impact of a sphere on a plate in a narrow continuum to gaps of molecular dimensions, *J. Phys. D: Appl. Phys.* **35**, 2311–2326.
Arrundale, D.P., Hussain, K., Rahnejat, H. and Menday, M.T. (1998): 'Acoustic response of driveline pieces under impacting loads (clonk)', Proc. 31st ISATA, Dusseldorf, pp. 319–331.
Bell, J.R. and Rothberg, S.J. (2000): 'Rotational vibration measurements using laser doppler vibrometry: comprehensive theory and practical application', *J. Sound Vibration*, **238**(4), 245–261.
Biermann, J.W. and Hagerodt, B. (1999): 'Investigation into the clonk phenomenon in

vehicle transmission – measurement, modelling and simulation', *Proc. Instn. Mech. Engrs., Part K: J. Multi-body Dyn.*, **213**(K1), 53–60.

Capitani, R., Delogu, M. and Pilo, L. (2001): 'Analysis of the influence of a vehicle's driveline dynamic behaviour regarding the performance perception at low frequencies', SAE-ATT Conference, Barcelona, Pap.: 01ATT-328.

Daubechies, I. (1990): 'The wavelet transform time-frequency localization and signal analysis', *IEEE Trans. on Magnetics* **32**, 1715–1720.

Dogan, S.N., Ryborz, J. and Bertsche, B. (2004): 'Low-noise automotive transmissions investigations of rattling and clattering', in *Multi-body Dynamics: Monitoring and simulation techniques – III*, Rahnejat, H and Rothberg, S. (Eds.), Professional Engineering Publishing (IMechE), pp. 323–338.

Dowson, D. and Wang, D. (1994): 'An analysis of the normal bouncing of a solid elastic ball on an oily plate', *Wear*, **179**, 29–37.

Farshidianfar, A., Ebrahimi, M., Rahnejat, H. and Menday, M.T. (2000): 'Low frequency torsional vibration of vehicular driveline systems in shuffle', in *Multi-body Dynamics Monitoring & Simulation Techniques – II*, H. Rahnejat, M. Ebrahimi and R. Whalley (Eds.), Professional Engineering Publishing (IMechE), pp. 269–282.

Hackenbroich, D. (1994): 'Reduction of the commercial vehicle noise level using numerical optimisation of the structure-borne sound and sound radiation behaviour of a drive shaft', *VDI Berichte*, No. **1153**, 1–22.

Kelly, P., Menday, M.T., Rahnejat, H. and Ebrahimi, M. (1999): 'Powertrain refinement: a combined experimental and multi-body dynamics analysis approach', 8th Aachener Kolloquium, Aachen, Germany.

Krenz, R.A. (1985): 'Vehicle response to throttle tip-in/tip-out', Proc. Conf. Surface Vehicle Noise and Vibration, SAE Transactions, paper no. 850967, pp. 45–51.

Littlefair, G.P. (2004): 'Arc spring sliding friction within the dual mass flywheel', MSc thesis, Loughborough University.

Menday, M.T. (2003): 'Multi-body dynamics analysis and experimental investigation for the determination of the physics of drivetrain vibro-impact induced elasto-acoustic coupling', PhD Thesis, Loughborough University.

Newland, D.E. (1993): *An Introduction to Random Vibrations and Spectral and Wavelet Analysis*, 3rd edition, Longman Scientific & Technical.

Petri, H. and Heidingsfeld, D. (1989): 'The hydraulic torsion damper-a new concept for vibration damping in power train', SAE Trans. Pap. 892477.

Rahnejat, H., (1998) *Multi-body Dynamics: Vehicles, Machines and Mechanisms*, Professional Engineering Publications (IMechE)/Society of Automotive Engineers (SAE).

Richards, E.J., Westcott, M. and Jeyapalan, R. (1979a): 'On the prediction of impact noise: Part 1: Acceleration Noise', *J. Sound Vibration*, **162**, pp. 547–575.

Richards, E.J., Westcott, M. and Jeyapalan, R. (1976b): 'On the prediction of impact noise: Part 2: Ringing Noise', *J. Sound Vibration*, **162**, pp. 419–451.

Theodossiades, S., Gnanakumarr, M., Rahnejat, H. and Menday, M. (2004): 'Mode identification in impact-induced high-frequency vehicular driveline vibrations using an elasto-multi-body dynamics approach', *Proc. Inst. Mech. Eng. Part K: J. Multi-body Dynamics*, **218**, 81–94.

Vafaei, S., Menday, M.T. and Rahnejat, H. (2001): 'Transient high-frequency elasto-acoustic response of a vehicular drivetrain to sudden throttle demand', *Proc. Instn Mech. Engrs., Part K: J. Multi-body Dyn.*, **215**, pp. 35–52.

Wang, C. and Lai, J.C.S. (2000): 'The sound radiation efficiency of finite length acoustically thick circular cylindrical shells under mechanical excitation I: theoretical analysis', *J. Sound Vibration*, **232**, pp 431–447.

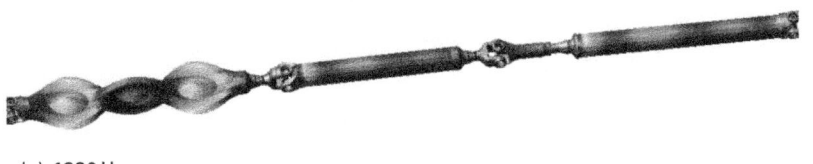

(a) 1830 Hz

(b) 2554 Hz

Plate IX Breathing modes of the first driveshaft tube.

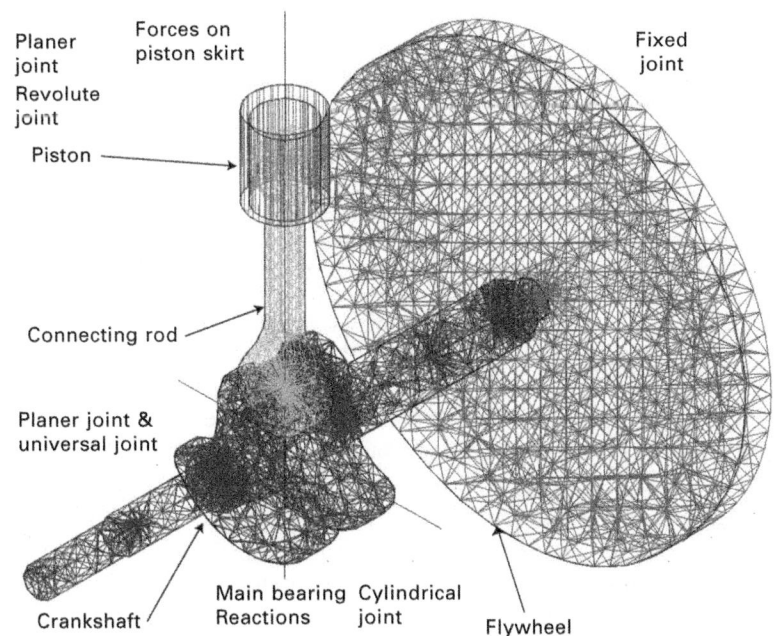

Planer joint

Forces on piston skirt

Fixed joint

Revolute joint

Piston

Connecting rod

Planer joint & universal joint

Crankshaft

Main bearing Reactions

Cylindrical joint

Flywheel

Plate X The multi-body dynamic model of a single cylinder engine.

31
Tribo-elasto-multi-body dynamics of a single cylinder engine under fired condition

M. S. M. PERERA, S. THEODOSSIADES and
H. RAHNEJAT, Loughborough University, UK

Abstract: The use of multi-body dynamics within the multi-physics, multi-scale environment of an internal combustion (IC) engine is described. It is shown that realistic analysis should necessarily include the physical interactions at all scales. In the case of IC engines these include tribological behaviour of load bearing and transmitting conjunctions, such as those of piston–cylinder and main crankshaft support bearings in the case of the highlighted study in this chapter, concerning piston–connecting rod–crank subsystem. Such analyses take into account many requirements in engine development, which are concurrent and driven by market economy, legislation or socio-economic trends and aspirations. The multi-physics multi-scale approach recognises the hierarchical nature of engine development; what is now termed as down-cascading to root causes and palliation, and up-cascading to ascertain their effects and repercussions. It also recognises the highly interactive nature of modern engineering practice, referred to as simultaneous engineering, by creating multi-physics investigations within a unique analysis platform. There is a long way to go in this approach, but the chapter makes an initial contribution in this regard.

Key words: IC engines, multi-body dynamics, component flexibility, thermo-hydrodynamics, piston skirt–cylinder conjunction, piston ring pack, crankshaft support bearings, crankshaft offset, frictional losses.

31.1   Introduction

Theoretical basis for multi-body dynamics is discussed in the main body of this chapter, including the inclusion of component flexibility. The theory for various tribological contacts used in this chapter can be found in various chapters, for example, Chapters 5, 11, 15 and 18. In the example provided in this chapter these approaches are combined in a multi-physics, multi-scale analysis. It includes multi-body inertial dynamics, component flexibility, tribology and contact thermal analysis.

The internal combustion (IC) engine as a whole is a very complex system in terms of its dynamic behaviour. There are a number of subsystems within the engine itself. However, major excitations that transfer into the driveline come through the piston, connecting rod, crankshaft and flywheel subsystem. Therefore, the case study here is confined to this subsystem.

928

The aforementioned engine sub-system is modelled in a multi-body environment (in this case the MSC Software's ADAMS). The model is shown in Plate X (between pages 514 and 515). In a multi-body model *parts* or components are assembled together by means of joints, which are made of one or a number of constraint functions. This is to ensure a model with correct functional performance, representing the real system. In the above single cylinder engine model, the crankshaft is connected to the flywheel by a *fixed joint*, which removes all the relative degrees of freedom. The crankshaft is connected to the ground (here representing the engine block) by a combination of a *planar joint* perpendicular to the crank axis and a *universal joint* that removes only the translational motion along the crankshaft axis. Combinations of joints are chosen so that only the correct degrees of freedom remain, also avoiding redundant or repeat constraints (see Chapter 1; Boysal and Rahnejat, 1997; Zeischka *et al.*, 1994; Rahnejat, 2000; Kushwaha *et al.*, 2002).

The crankshaft is balanced by the reaction forces acting upon it by the main engine bearing supports at the correct locations. The crankshaft is attached to the connecting rod using a *cylindrical joint*. The connecting rod and the piston are connected through a *revolute joint* so that the piston can rotate with respect to the connecting rod along the axial direction of the crankshaft. The piston is connected to the ground (the engine block) using a planar joint, allowing its secondary motions to take place, which are resisted by contact forces, described later. A number of dummy parts are used in order to enable the use of a combination of constraint functions. Detailed description of the model may be found in Perera *et al.* (2007). Here Table 31.1 provides a summary of joint types used, with the resulting constrained degrees of freedom. The engine modelled is Ricardo E6, widely available, particularly as a teaching tool in the academe in the UK.

The *combustion gas force* (Fig. 31.1) is measured using a Kistler spark-plug type pressure transducer inserted into the combustion chamber of an E6 engine. Cycle-to-cycle variations are ignored. The resisting torque or the load torque is given as an external torque applied at the flywheel end

Table 31.1 Constraints in the single cylinder engine model

No	Description	Joint type	Constraints						Total
			dx	dy	dz	rx	ry	rz	
1	Crankshaft–ground	Planar & Universal			*				1
2	Crankshaft–flywheel	Fixed	*	*	*	*	*	*	6
3	Crankshaft–connecting rod	Cylindrical		*	*		*	*	4
4	Piston–connecting rod	Revolute	*	*	*	*	*		5
5	Piston–ground	Planar		*		*	*	*	4
					Total constraints				23

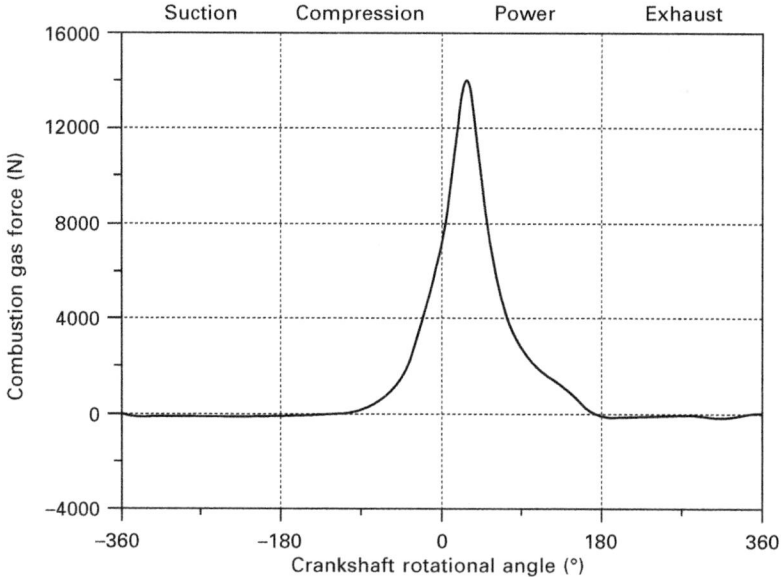

31.1 Combustion gas force variation with crank angle.

Table 31.2 Resistive forces on engine components

No.	Type	Position
1	Journal bearing reaction	Left main journal bearing
2	Journal bearing reaction	Right main journal bearing
3	Combustion gas force	Perpendicular to piston top surface
4	Load torque	Crankshaft end
5	Friction torque	Journal bearings
6	Friction at piston	Piston top corners (1, 2) as shown in Fig. 31.3
7	Normal forces at piston edge	Piston top and bottom corners (1, 2, 3, 4) as shown in Fig. 31.3

of the crankshaft. In general, load torque multiplied by the angular velocity gives the power output of an engine, in the same way as the gas pressure force multiplied by the piston velocity gives the power input to the engine. If there are no losses within the system the input power should equal the output power. In practice there exist many losses in an engine in the form of heat and parasitic losses, comprising inertial out-of-balance and friction, the latter being the main source of mechanical losses. All the forces acting on the model *parts* are listed in Table 31.2. These are described later.

31.2   Engine model with flexible components

With the *parts* shown in Plate X and joints/constraints indicated in Table 31.1, a rigid multi-body system results. However, component flexibility plays an important role in noise, vibration and harshness (NVH) issues (see Chapter 1) and should be included in the model, where appropriate. Flexibility of a component is introduced via use of finite element analysis (FEA), converting the component into a finite number of nodes, connected by Euler beams. The discretisation of a flexible component into a series of finite elements represents an infinite number of degrees of freedom (DOF), with a finite but large number of modes. For a better representation of a component, there should be a large number of elements in the finite element model that would be sufficient to capture all the major modes that can be excited during its operation. However, a finite element model with such an accuracy needs a significant computational memory, which can render it almost impractical. Therefore, it is necessary to use a representation method to condense this large finite element model into an acceptable size, which can work with the available computing resource. Component mode synthesis (CMS) is used as described in Chapter 1. Referring to Plate X, note that for the engine concerned the crankshaft/flywheel assembly and the connecting rod are considered to be elastic. In many cases, certainly more modern engines, the piston is also considered as elastic. The same would be true of advanced cylinder liners with thin wall thickness.

Component mode synthesis (CMS) is a method that can be used to transform large number of DOFs of a component into a smaller number of modal DOFs, using the modal truncation method (Craig and Bampton, 1968). Essentially, it condenses the mass and stiffness matrices of the chosen elastic components. The most important step in this transformation is matching the properties such as mass, location of centre of mass, mass moment of inertia and natural frequencies of an actual component with the created flexible element(s). During this transformation few boundary nodes can be selected through which the component is excited. Then, the vector summation of these boundary nodes transfers to another single node, which is the connecting node of one component to another. These types of nodes are created in places such as the main bearing positions of the crankshaft and the external reaction forces are applied to them during the formation of equations of motion.

31.3   Tribological conjunctions in the models

31.3.1 Main bearing reactions

As noted above, the big-end bearing is represented by constraint functions. This is implemented for simplicity of the model and the fact that tribological behaviour of the bearing there is not of interest in this analysis. However, it can easily be included in a more complex model such as those described in Chapters 19 or 20. In the single cylinder engine described here the crankshaft is supported by two main journal bearings. These carry the applied loads due to combustion and inertial imbalances (see Chapter 18). The combustion force has already been described above, as that of the combustion pressure at any instant of time acting over the piston crown's active surface area. The inertial forces are determined through multi-body dynamics analysis at any instant of time, as described in Chapter 18. The applied loads and moments induce bearing reactions, which are caused by the hydrodynamic oil film formed in the wedge between the journal and its bushing. These are described in Chapters 5 and 18 (also see Pinkus and Sternlicht, 1961; Gohar and Rahnejat, 2008). Thus:

$$W_z = \frac{u_b \eta_0 L^3}{4c^2} \frac{\pi \varepsilon}{(1 - \varepsilon^2)^{3/2}} \tag{31.1}$$

$$W_x = \frac{u_b \eta_0 L^3}{c^2} \frac{\varepsilon^2}{(1 - \varepsilon^2)^2} \tag{31.2}$$

Therefore, the main bearings in the model are represented by the resultant of the above forces (for the front and rear main bearings) and shown in Fig. 31.2. When the combustion force is at its maximum the bearing reaction forces also reach their maximum values.

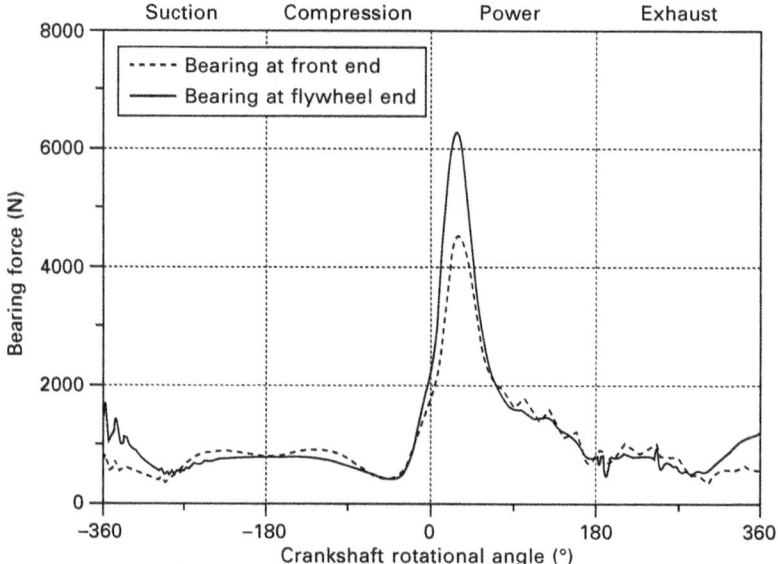

31.2 Variation of bearing reaction forces against crankshaft rotation.

31.3.2 Piston interactions with the cylinder wall

Owing to the relatively high stiffness of the piston and the clearance between the cylinder and the piston to compensate for thermal expansion, the piston can be considered as a rigid body moving within the confines of the cylinder. Therefore, **piston secondary motions** in lateral and tilting motions are represented by the four piston edges, touching the cylinder wall (Fig. 31.3). All these edges are considered to form lubricant wedges, where hydrodynamic pressures can be obtained by solution of Reynolds equation with appropriate boundary conditions (in this case Swift–Steiber or Reynolds boundary conditions are used, see Chapter 5). The contacts in these locations are considered as line contacts since the circumferential length is much larger than their width. It is also assumed that no side leakage of the lubricant occurs along the circumference. Therefore, dimensionless pressure distribution can be obtained as (Rahnejat, 1984):

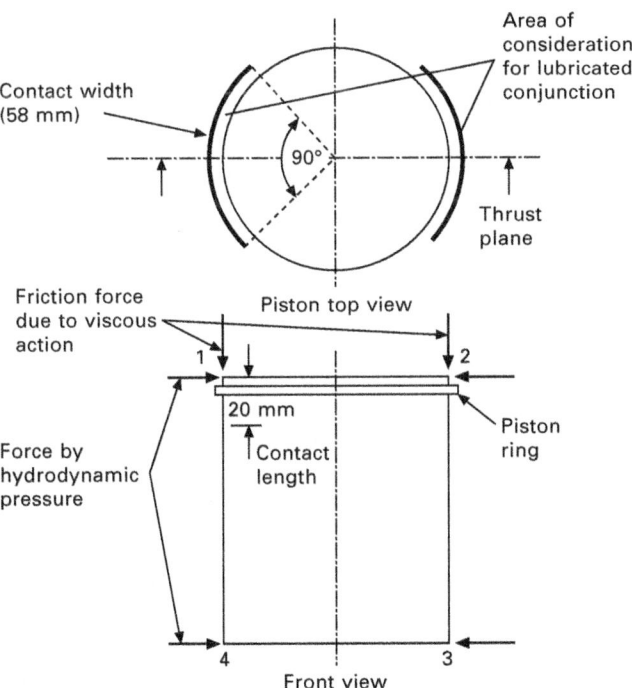

31.3 Conjunctions considered between the piston skirt and cylinder liner.

$$P^* = \frac{1}{8}\bar{x} - \frac{1}{32}\sin 4\bar{x} - \tan^2 \bar{x}_a \left(\frac{3}{8}\bar{x} + \frac{1}{4}\sin 2\bar{x} + \frac{1}{32}\sin 4\bar{x} \right)$$

$$+ \frac{4w_s^*}{\sqrt{2h_0^*}} \left[\begin{array}{l} -\frac{3}{32} - \frac{1}{8}\cos 2\bar{x} - \frac{1}{32}\cos 4\bar{x} \\ + \tan \bar{x}_a \left(\frac{3}{8}\bar{x} + \frac{1}{4}\sin 2\bar{x} + \frac{1}{32}\sin 4\bar{x} \right) \end{array} \right]$$

$$+ \frac{\pi}{16}\left(1 - 3\tan^2 \bar{x}_a + \frac{12w_s^*}{\sqrt{2h_0^*}}\tan \bar{x}_a \right) \tag{31.3}$$

31.3.3 Piston ring interaction with cylinder wall

In this case only the compression ring-liner/bore conjunction is considered. When the piston moves upwards the pressure at the back of the ring is considered to be the combustion chamber pressure. When it moves downwards the pressure is considered to be that of the crankcase. The ring is pre-tensioned so that it always presses against the cylinder, sealing against the combustion gases. The same approach as the previous section is again used to described the piston ring–cylinder wall conjunction and the pressure distribution can be sought from the same equation (31.3). As such the ring is considered in a state of equilibrium with lubricant hydrodynamic reaction forces F_{lr} (the integrated pressure distribution resulting from equation 31.3) and asperity force P_a (when a proportion of contact force is supported by the solid surfaces, for example in mixed regime of lubrication), radial gas force F_{rg} and the ring tension force F_{rt} as (Fig. 31.4):

$$F_{lr} + P_a - F_{rg} - F_{rt} = 0 \tag{31.4}$$

Also note that due to the clearance between the piston groove and the ring, the ring does not follow the exact motion of the piston itself. There is a good chance of exciting the various mode shapes of the ring as the result of the forces acting on it. These forces also include friction (viscous and boundary, see Chapter 18) and contact reactions with the ring groove. At any instant of time there is a net force between friction and these contact forces in the axial direction, which is balanced by the inertial force of the ring. This, for example, causes ring flutter.

31.4   Conjunctional friction in the engine model

Frictional forces are associated with sliding motions. However, not all contribute equally to the total frictional losses at any instant of time.

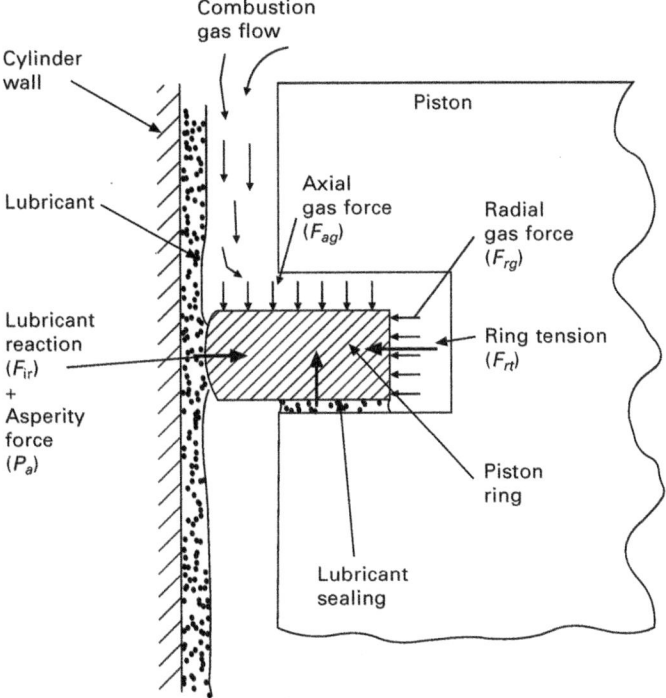

31.4 Forces acting on the piston ring.

Therefore, in this model the frictional forces generated at main journal bearings, piston skirt to cylinder wall and the piston ring to cylinder wall contacts are considered.

Main journal bearings always enjoy a hydrodynamic regime of lubrication (see Chapters 3, 5 and 6) due to the large surface area of the bearing and thickness of the journal shell. Friction in a lubricated conjunction is generated due to the relative sliding motion of the two contacting surfaces. With short bearing approximation friction at the bearing surface can be simplified to (Cameron, 1966; Gohar and Rahnejat, 2008):

$$F = \frac{-2\pi\eta_0 u_j r_j L}{c} \cdot \frac{1}{\sqrt{1 - \varepsilon^2}} \tag{31.5}$$

The first quotient is known as the *Petrov friction*, which is present where the journal and the bearing shell are concentric (i.e. rotating cylinder friction). The second quotient is termed the *Petrov multiplier*. Note also the higher the eccentricity ratio ε, the greater is the friction and hence more heat is generated. Figure 31.5 shows the variation of the frictional torque at the main bearings during a combustion cycle. The frictional torque is high at

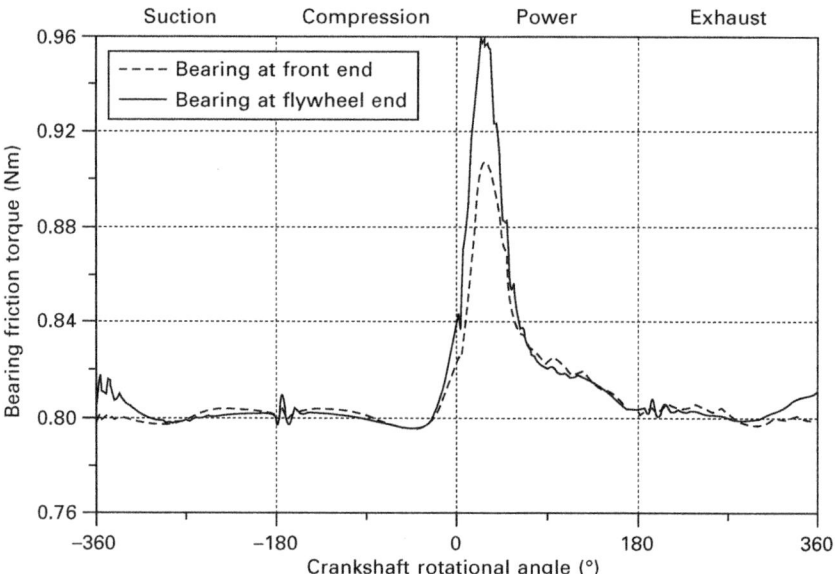

31.5 Bearing friction torque variation with crank angle.

the beginning of the power stroke. This is when the combustion gas force is at its maximum. Also the friction torque at the flywheel end bearing is higher compared with the front end bearing. This is obvious as the load due to flywheel is taken by the bearing at the flywheel end.

At the dead centres the motion of the piston and the ring pack ceases instantaneously and reverses direction. This constitutes diminishing of lubricant entrainment into the contact, thus a mixed or boundary regime of lubrication results. Therefore, fluid film lubrication (viscous friction) is the dominant source during most of the time, while Coulomb friction may be present under **boundary regime of lubrication** particularly at dead centres. These two phenomena contribute to the total friction in **mixed regime of lubrication**. Thus, it is reasonable to consider friction force between two surfaces to be made up of two contributions: a boundary friction F_b and a viscous friction F_v:

$$F_f = F_b + F_b \tag{31.6}$$

Boundary lubrication occurs at the asperities in contact, while viscous lubrication conditions prevail when a film of lubricant is sheared, also when such a film is trapped between valleys of asperity pairs. Depending on the oil film thickness and rate of shearing of the trapped lubricant, it can behave as either a Newtonian fluid or non-Newtonian fluid (see Rahnejat and Gohar, 2008; and Chapters 15 and 20). The distribution of peaks and valleys on a surface cannot be calculated precisely due to their rather random nature.

Thus, a statistical approach is used to predict the tribological aspects of the lubricated contact (Greenwood and Tripp, 1971):

$$F_b = \tau_e A_a + c_{pb} P_a \qquad (31.7)$$

$$F_v = \tau(A - A_a) \qquad (31.8)$$

where the shear stress τ depends on the Newtonian or non-Newtonian behaviour of the lubricant (see Perera *et al.*, 2007).

Figure 31.6 shows the variation of friction at the above-mentioned four edges of the piston. Owing to the long piston length and relatively low speed of lubricant entrainment, the regime of lubrication remains hydrodynamic throughout the cycle except close to the dead centres. As can be observed, friction is at its maximum at the mid-span of the power stroke, where the sliding speed attains its highest value and viscous shear dominates the mechanism of friction generation in this case. This trend prevails in all the four strokes of the engine under consideration, indicating the fact that viscous friction is dominating piston skirt to cylinder wall friction. This is not always the case, particularly in more modern engines, with higher side forces and tighter clearances. Figure 31.7 shows the breakdown of viscous and boundary frictional contributions for the piston ring to cylinder wall contact. In this case boundary friction is dominant at dead centres and its value is quite high compared with the viscous friction both in piston skirt as well as that in piston ring conjunction.

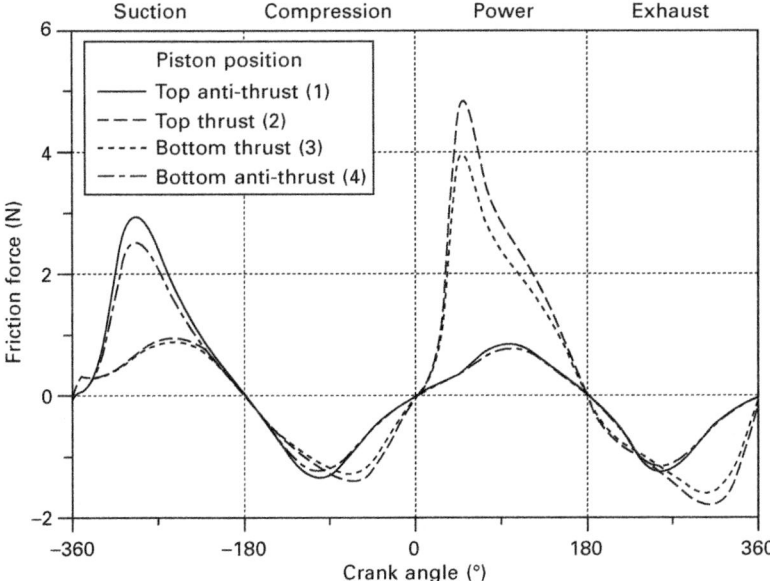

31.6 Piston skirt to cylinder wall friction.

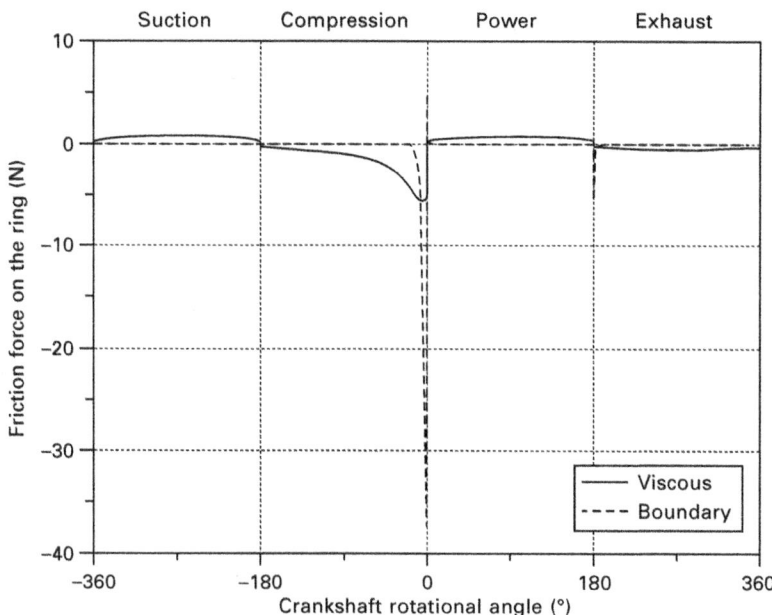

31.7 Viscous and boundary friction forces in piston rings to cylinder wall contact.

31.5   Temperature effects in lubricated contacts

Thus far the analytical methods described are based on the assumption that isothermal conditions prevail in a lubricated contact. However, in all contacts the temperature of the lubricant rises from its bulk oil temperature. This reduces its viscosity, which in turn reduces the film thickness, sometimes quite considerably. In conjunctions with lower lubricant pressure and larger gaps (film thickness) the temperature rise is less pronounced. However, contacts such as piston ring to cylinder wall, where the load is concentrated on a very small contact area the temperature rise cannot be neglected. In a lubricated contact the heat generated due to compressive and viscous heating is dissipated to the environment through conduction and convection (see the energy equation in Chapter 5). However under hydrodynamic conditions heat generated due to compression can be neglected as this is usually quite small compared with *viscous shear heating*. As such the temperature rise in a contact can be given by the following equation (Perera *et al.*, 2007):

$$\frac{\eta u_s^2 B}{h} = \frac{\rho u_s C_p h \Delta\theta}{4} + \frac{k_c (\Delta\theta) B}{2h} \tag{31.9}$$

The term on the left-hand side of the equation above is the heat generated through viscous shear. The terms on the right-hand side of the equation are cooling terms due to **convection cooling** and **conduction cooling** respectively. An analytical approach is undertaken here, where the heat loss due to convection by the flow of a lubricant film is superimposed upon heat conducted through the bounding surfaces. This approach is described in some detail by Gohar and Rahnejat (2008). Here it is applied to the piston ring pack to cylinder liner conjunction. The same approach can be applied for the case of journal bearings, as indicated in Chapter 18. In both cases heat generated by asperity interactions is ignored.

Once the new temperature is calculated, based on the temperature rise, the new viscosity (**effective viscosity**) for the new temperature can be calculated using Vogel's equation (Vogel, 1921):

$$\ln \eta_e = -1.845 + \left(\frac{700.81}{\Theta_e - 203} \right) \tag{31.10}$$

where $\Theta_e = \theta_e + 273$ and $\theta_e = \theta_i + \Delta\theta$.

Figure 31.8 shows the heat dissipation due to convection and conduction for a one combustion cycle in the ring–bore conjunction. As expected, the heat loss due to convection is higher at the mid-span, where the sliding speed is at its maximum, yielding a thicker lubricant film. However, heat dissipation due to conduction is higher near to dead centres, because of relatively thin films.

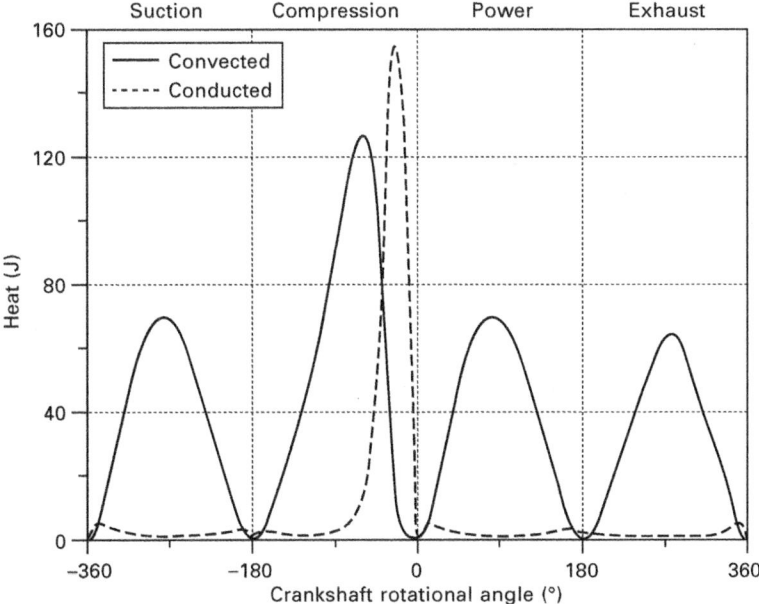

31.8 Heat dissipated due to conduction and convection in piston ring to cylinder wall contact.

31.6   Parametric study – crankshaft offset

Once a model of the kind described here is developed it can be used as a design tool for *what-if* scenario building simulations, potentially reducing the cost of physical prototyping and tests, particularly at the conceptual design stage. One particular current question of interest is the ***crankshaft offset*** in order to ensure adherence of the piston to the major thrust side, thus reducing the side to side motion of the piston and any subsequent problems due to loss of lubrication and piston slapping action and, thus, radiated noise (also see Wakabayashi *et al.*, 2003).

The side force acting on the piston exactly at the top dead centre (TDC) is zero, if the crankshaft is in-line with the piston axis. However, a few degrees before the TDC a net resultant force F_s acts towards the anti-thrust side of the cylinder wall, while this reverses in direction a few degrees after the TDC, as shown in Fig. 31.9. Such a sudden force variation results in ***piston slap***. The sudden rise in the side force can also lead to the depletion of lubricant film, which is an undesirable outcome. Offsetting the crankshaft towards the major thrust side is expected to smoothen this force transformation.

Figure 31.10 shows frictional variation at piston skirt-to-cylinder wall contact with crankshaft offset. The absolute maximum friction occurs during the power stroke and it decreases with an increase in crankshaft offset.

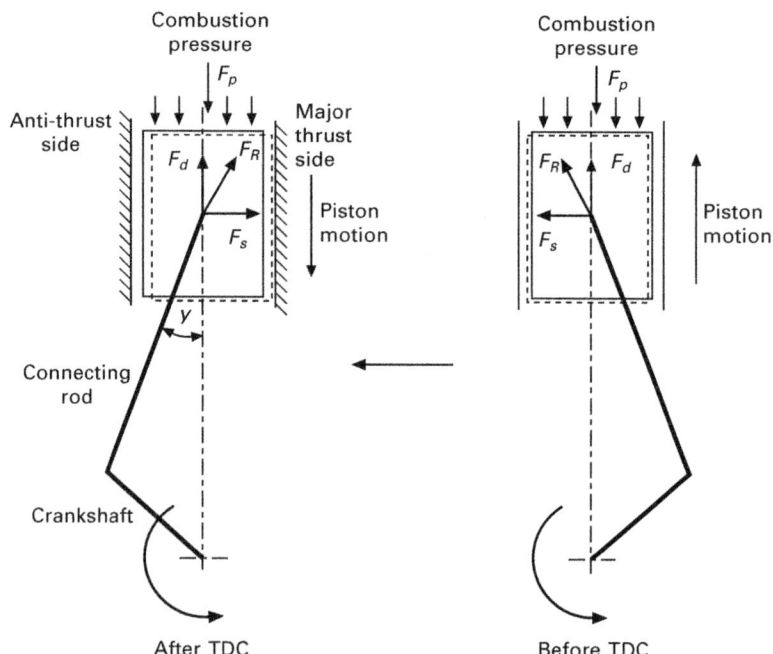

31.9 Piston orientation and forces acting before and after the TDC.

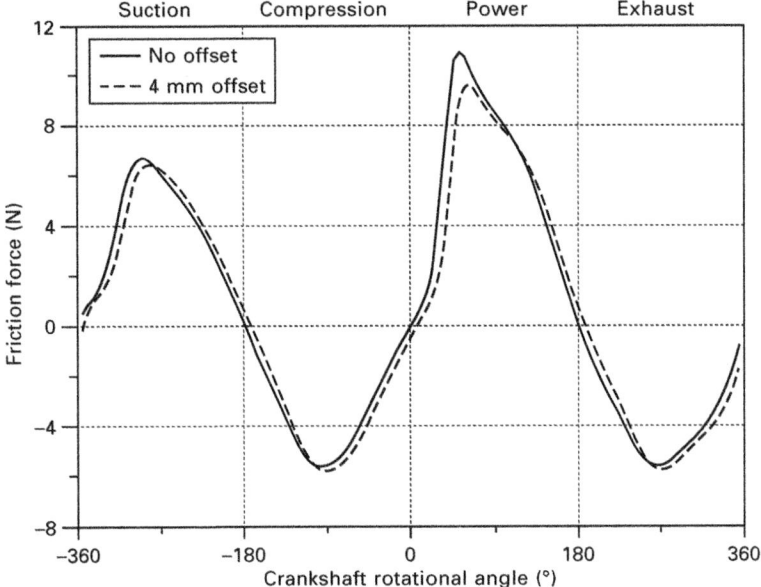

31.10 Friction at piston skirt to cylinder wall contact.

Figure 31.11 shows the variation of major thrust side force with crankshaft offset for a four-stroke engine cycle. This shows that an increase in the crankshaft offset progressively decreases the peak side force during the suction and power strokes, while it slightly increases the side force, particularly during the exhaust stroke. This force variation results in the piston slap noise. Therefore, the reduction in peak to peak force due to crankshaft offset should reduce the generated slap noise. However, certain crankshaft offset may cancel the reduction in piston friction during the expansion stroke because of the increased piston friction during the compression stroke.

31.7   Concluding remarks

This chapter shows the multiplicity of interacting phenomena in a complex system such as the piston-connecting rod–crank subsystem of an engine. These include interactions across the physics of scale; from micro-scale tribology to sub-millimetre deformation of elastic members through to constrained inertial multi-body dynamics. The analysis shows that increased computing power and better methods of integration of phenomena enables realistic models to be developed. This is one of the trends in future numerical analysis, increasingly termed as multi-scale, multi-physics analysis. The approach enables in-depth investigation of phenomena, observed at the system level, by down-cascading into component and interaction levels. It also enables

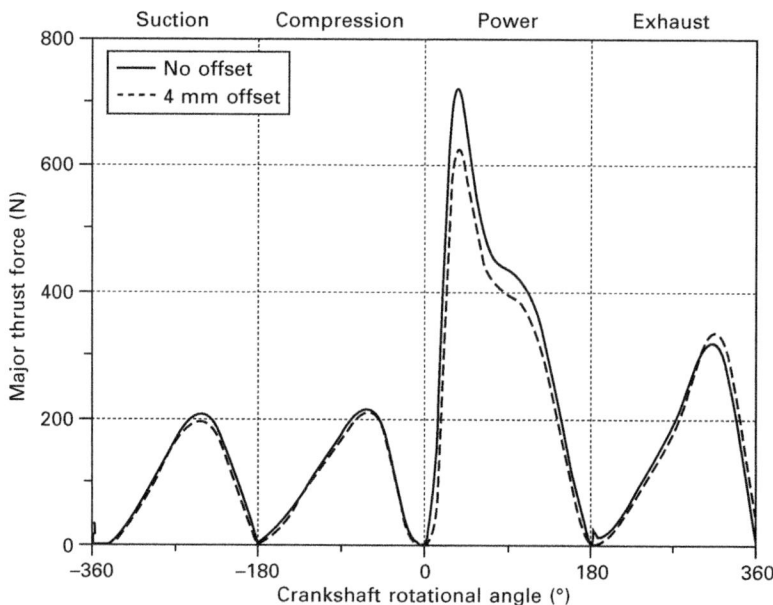

31.11 Major thrust side force variation during combustion cycle.

prediction of performance of a system as an outcome of choices that are made at component or interaction levels through up-cascading. It also puts aside many compartmentalised distinctive disciplinary approaches to analysis. Hence, issues such as frictional performance and NVH refinement can be addressed in a unified analysis framework.

31.8   References

Boysal, A. and Rahnejat, H., 'Torsional vibration analysis of a multi-body single cylinder internal combustion engine model,' *Appl. Math. Modelling*, **21**(8), 1997, pp. 481–493.

Cameron, A., *The principles of lubrication* Longmans, 1966.

Craig, R.R. and Bampton, M.C.C., 'Coupling of substructures for dynamics analysis' *AIAA J.*, **6**, 1968, pp. 1313–1319.

Gohar, R. and Rahnejat, H., *Fundamentals of Tribology*, Imperial College Press, London, 2008.

Greenwood, J.A. and Tripp, J.H., 'The contact of two nominally flat rough surfaces', *Proc. Instn. Mech. Engrs., Part C: J. Mech. Engng. Sci.*, **185**, 1971, pp. 625–633.

Kushwaha, M., Gupta, S., Kelly, P. and Rahnejat, H., 'Elasto-multi body dynamics of a multi-cylinder internal combustion engine', *Proc. Instn. Mech. Engrs., Part K: J. Multi-body Dyn.*, **216** (4), 2002, pp. 281–293.

Perera, M., Theodossiades, S. and Rahnejat, H., 'A multi-physics multi-scale approach in engine design analysis', *Proc. Instn. Mech. Engrs., Part K: J. Multi-body Dyn.*, **221**, 2007, pp 335–348.

Pinkus, O. and Sternlicht, B., *Theory of Hydrodynamic Lubrication*, McGraw-Hill, 1961.

Rahnejat, H., *Influence of Vibration on the Oil film in Concentrated Contacts'*, PhD Thesis, Imperial College, University of London, London, 1984.

Rahnejat, H., 'Multi-body dynamics: historical evolution and application', *Proc. Instn. Mech. Engrs., Part C: J. Mech. Engng. Sci.*, **214** (1), 2000, pp. 149–173.

Vogel, H., 'The law of the relation between the vis costing of liquids and the temperature', *PHYSIK Z*, **22**, 1921, pp. 645–646.

Wakabayashi, R., Takiguchi, M., Shimada, T., Mizuno, Y. and Yamauchi, T., 'The effects of crank ratio and crankshaft offset on piston friction losses', SAE Technical Paper, Pap. No.2003-01-0983 Detroit, Michigan, 2003.

Zeischka, J., Mayor, L.S., Schersen, M. and Maessen, F., 'Multi-body dynamics with deformable bodies applied to the flexible rotating crankshaft and the engine block', Proc. ASME Fall Tech. Conf., Laffayette, USA, 1994.

31.9   Nomenclature

A_a	Asperity contact area	(m^2)
B	Contact width	(m)
c	Clearance	(m)
C_p	Specific heat at constant pressure	$(m^2/s^2\,^{\circ}C)$
c_{pb}	Pressure coefficient of boundary shear strength	(m^2)
F_b	Boundary friction force	(N)
F_v	Viscous friction force	(N)
F_{lr}	Lubricant reaction acting on ring	(N)
F_{rg}	Gas pressure force acting on ring	(N)
F_{rt}	Radially outward ring tension	(N)
h	Film thickness	(m)
h_0	Minimum film thickness	(m)
k_c	Thermal conductivity	$(N/s\,^{\circ}C)$
L	Length or width of the bearing	(m)
p	Local pressure	(Pa)
P_a	Pressure at asperities	(Pa)
R	Equivalent radius of the contact	(m)
r_j	Radius of the jth journal	(m)
u_b	Tangential velocity at journal surface	(m/s^2)
u_j	Tangential velocity at journal surface	(m/s^2)
u_s	Sliding velocity	(m/s)
w_s	Squeeze velocity	(m/s)
W_x	Load along the line of centres in bearing	(N)
W_z	Load perpendicular to the line of centres in bearing	(N)
x	Length along the contact	(m)
x_a	Length from the centre of the contact to maximum pressure point	(m)

η_e	Effective dynamic viscosity	(Pa s)
η_0	Atmospheric dynamic viscosity	(Pa s)
ε	Eccentricity ratio	(–)
τ_e	Eyring shear stress	(Pa)
θ_i	Inlet lubricant temperature (bulk)	(°C)
θ_e	Effective average lubricant temperature in the contact	(°C)
$\Delta\theta$	Temperature rise	(°C)

Non-dimensional quantities

$$\tan \bar{x} = \frac{x}{\sqrt{2Rh_0}}$$

$$\tan \bar{x}_a = \frac{x_a}{\sqrt{2Rh_0}}$$

$$p^* = \frac{h_0^{3/2} p}{6u_s \eta_0 (2R)^{1/2}}$$

$$w_s^* = \frac{w_s}{u_s}$$

$$h_0^* = \frac{h_0}{R}$$

Part III

Micro-systems and nano-conjunctions

32
Microengines and microgears

M. TEODORESCU, Cranfield University, UK and
S. THEODOSSIADES and H. RAHNEJAT,
Loughborough University, UK

Abstract: This chapter considers some aspects of the developments in the past few decades arising from miniaturisation of many electromechanical systems. These mechanisms are progressively developed and used as microactuactors, switches and trigger devices, all of which have load-bearing conjunctions in the scale of minutiae. Consequently, there is a plethora of as-yet not fundamentally understood prevalent kinetics. The chapter describes the current knowledge about these interactions and uses this to predict the behaviour of some nanoscale conjunctions. It is found that overall, the predictions are in accord with observed phenomena in the case of microelectromechanical gears, but concern remains as to the fundamental and generic nature of the employed force laws.

Key words: microelectromechanical systems (MEMS), microgears, nanoscale kinetics, adhesion, meniscus action, hydration.

32.1   Introduction

A later development of the second half of the twentieth century and still ongoing has been the miniaturisation of components and devices. Aside from the perceived advantage of personalisation of consumer devices, downsizing has the significant advantage of reducing inertial effects such that out-of-balance would progressively play a less important role in the moving mechanisms, thus improving their efficiency and useful life. Some would argue that the evolutionary processes in nature themselves appear to follow the same objectives. The process seems to favour the smaller of the species, requiring fewer resources for survival. Smaller planetary systems are also subject to lesser ferocious forces.

As it stands, either due to a lack of understanding or for some inexplicable reason(s), the world of very small things seems to be quite different from our everyday experiences, which may be regarded as a Newtonian world. The world of minutiae has its upper limit in the nanoscale. The quest to exploit this vanishingly small world seems to have been initiated by Richard Feynman's question, posed at the annual meeting of the American Physical Society (Feynman, 1959): 'Why cannot we write the entire 24 volumes of

947

the *Encyclopaedia Britannica* on the head of a pin?'. Feynman proceeded to demonstrate that there would not only be a problem in creating such very small writing and reading devices, but also one would need to redefine some key accepted principles, which would govern the interactions at very small scale. Feynman predicted that one of the *most interesting points* would be the mechanism of lubrication: 'the effect of viscosity of oil would be higher and higher in proportion as we went down (and if we increase the speed as much as we can)'. Therefore, the solution he proposed was to 'let the bearing run dry'. Fifty years on, this is precisely what the scientists are proposing! Namely, in the scale of minutiae, the bulk properties of matter do not appear to play any significant role, instead individual interactions between molecules and particles dominate (Chan and Horn, 1984; Matsuoka and Kato, 1997; Al-Samieh and Rahnejat, 2001; Teodorescu *et al.*, 2006, also see Chapter 3). This largely discrete behaviour is rather alien to the comfort of a continuum enjoyed by the traditional tribology. In nature itself some tiny and resilient entities like the flagellar bacteria such as *Escherichia-coli* propel themselves forward with speeds in excess of 100 000 rpm, supported by dry bearings!

Feynman also predicted that 'in the year 2000, when they will look back at this age, they will wonder why it was not until the year 1960 that anybody begun seriously to move in this direction' (referring to the idea of miniaturisation of devices by understanding and harnessing of intermolecular and surface forces in the vanishing scale). After all, insects and geckos exploit these forces skilfully to cling to surfaces rather effortlessly. Feynman was right, perhaps because he was a keen observer of nature.

The quest for bringing technology to individuals as a business proposition rather than an understanding of nature belies the advances made in science. As a result the implied Feynman's pessimism has not quite withstood the test of time. Downsizing has seen one of the greatest improvements in technology in the past couple of decades. However, as it seems usual in the history of science, the entrepreneurship of the technologist has run ahead of methodical scrutiny of the scientist. As a result, although microengineering and nanotechnology exist, many questions remain unanswered in nanoscience, upon which they should logically reside. Nevertheless, the overall scientific understanding of phenomena at the scale of minutiae is progressively improving.

The advent of the computer age has increased the demand for miniaturisation of electronics and mechanical systems at a greater pace than that perhaps expected by Feynman. This, however, has been because of the earlier developments in the sub-micrometre science and technology which were primarily intended for manufacture of components for the electronic industry. As a result, techniques for depositing and cutting successive nanoscale layers of silicon have advanced much quicker than was expected. Techniques that resemble carving and printing rather than the traditional manufacturing methods have meant that micro- and nanoscale interconnected structures and mechanisms could then be created (Tanner *et al.*, 2000; Kim *et al.*, 2006; Barlian *et al.*, 2007).

There is a vast range of applications where microscale mechanisms can be used, which can be loosely grouped into two categories. Firstly, there are a large variety of sensors, including miniaturised piezo-accelerometers and even microscale gyroscopes (Lee *et al.*, 2000) and almost any other type of sensor, all of which were traditionally made at macroscale (Bao and Middelhoek, 2000). Beside the obvious advantage of small package space, the other advantage of microscale sensors is their appeal for cheap mass manufacture (once the difficult hurdle of creating the precise cutting technique was overcome). These sensors are ubiquitous in the modern day equipment, from mobile phones to automotive applications (Spangler and Kemp, 1996). The second class of *microelectromechanical systems* (MEMS) consists of a large variety of actuators, which have a small integrated power source. One of the most commonly cited MEMS has a small size *electrostatic comb engine*, usually integrated with a miniature mechanism (e.g. a small size gearbox) (Tanner *et al.*, 2000). These can be deposited/carved together with an electronic circuit, which can provide the source of power and control for a *microengine*. Therefore, there are several disadvantages; the most obvious being that the mechanism cannot operate without the electrical control of an embedded chip. However, since these small mechanisms can potentially be used for a large variety of applications (e.g. microbatteries or microflying machines), alternative power sources have been devised (Chigier and Gemci, 2003; Jacobson and Epstein, 2003). Some of the most promising alternatives are the highly efficient microturbines manufactured by the Gas Turbine Laboratory and Microsystems Technology Laboratory at MIT (Epstein *et al.*, 1997), microscale rotary combustion engines and MEMS internal combustion engine (Fernandez-Pello, 2002).

Irrespective of the technology used to power MEMS devices, all these microscale mechanisms have a large number of contacting surfaces in nanoscale, which transmit motion and power. As Feynman predicted, most of these surfaces *ideally* run dry. It is implicit in Feynman's words his appreciation that such mechanisms should operate at significant speeds. Should they desist, the considerable adhesion at nanoscale would be their nemesis. Thus, to emulate natural control of adhesion in *gecko-fashion* would be the aim. This has become the preponderant scientific quest (see Chapters 3 and 33). Therefore, if the inertial forces cannot exceed adhesion, the counterfaces in contact can stick, with catastrophic consequences (Tanner *et al.*, 2000). Therefore, the advantages of reduced inertia in MEMS has its limits. To prevent stiction, two potential preventative measures can be undertaken. Firstly, the speed of the mechanisms should be increased to the levels where the inertial dynamics can overtake adhesion of components (Teodorescu *et al.*, 2009a). Secondly, the surfaces must be protected with materials, which have very low surface free energies (see Chapter 3). Considering the diminutive scale of the entire mechanism (contact footprints in sub-micrometre or nanoscale) the most effective approach is seen as the introduction of *self-assembled monolayers* (SAM) of long chain molecules (Maboudian and Carraro, 2004; see also Chapter 33).

One of the complexities in the nanoscale is the plethora of interaction potentials, which in itself points to a lack of fundamental understanding (see the discussions in the introduction part of Chapter 3). Thus, the situation is more complex than that described in the preceding paragraph. In nanoscale conjunctions molecular interactions cannot be ignored. Stiction by the wetting action of any ingressed moisture into the contact can also play an important role. The protective SAM layers should, therefore, be ideally hydrophobic. This is the case with beta-carotene, nature's choice for the *spatulae* at the extremities of the gecko feet (see Chapter 3). However, hydrophobic layers introduce a repulsive potential very near surfaces, generally termed as *hydration*. This is one of the least understood near-surface effects, but nevertheless a crucial one.

Irrespective of the advances made in the understanding of engineering surface characteristics at nanoscale, most of the design principles used for small-scale devices are directly adopted from their macroscale brethren almost by a process of extrapolation. However, as already mentioned, most established theories do not immediately lend themselves to such presumptuous extrapolation, particularly with respect to contact/impact dynamics in gaps of molecular dimensions (Al-Samieh and Rahnejat, 2002; Teodorescu and Rahnejat, 2008). Consequently, from the outset it has become clear that fabrication of such components would rely, at least in part, on an implicit acceptance of some empirical approach, which has clearly manifested itself in an inherent level of unreliability (Tanner *et al.*, 2000). To improve the reliability of small-scale devices, once again science is faced with the challenge of friction, this time at the scale where large number of interaction types poses a significant barrier, not hitherto encountered. Then, multi-physics as described by Perera *et al.* in Chapter 31 assumes a whole new meaning altogether.

Besides skilful engineering of surface properties (as previously mentioned), smooth operation of MEMS can only be guaranteed if a high operation speed is maintained. This can be controlled if the energy lost in frictional impacts between load-bearing surfaces is minimised.

32.2   Impact dynamics in microelectromechanical systems (MEMS) gears

Macroscopic gears are machined and gear teeth are carefully treated for improved tribological performance. In contrast, *MEMS gears* are roughly (in relative terms) cut *in situ*. Therefore, their surface roughness and minimum separations (backlash) depend on the cutting technique used. Hence, contact between meshing teeth is actually relatively rough. This means that adhesion between individual asperities can play a significant role (Teodorescu *et al.*, 2009a).

As an example of the type of problems faced in nanoscale mechanisms and the correlation between surface properties and overall mechanism behaviour, this chapter considers two examples. The frictional impact between a silicon roller (with surface and inertial characteristics representative of MEMS gear conditions) and a flat plane is considered. Then, the microdynamics of a pair of silicon gears (a pinion and a wheel), incorporating detailed nanoscale impact dynamics of rough meshing gear teeth, protected with a self-assembled mono-layer of hydrophobic ***octadecyltrichlorosilane*** (OTS) long chain molecules is taken into account. The mathematical model accounts for an assumed progressive degradation of the SAM layer (emulating gradual wear) and its effect on the dynamics of a microgear assembly. Figure 32.1 gives schematic representations of the contacts investigated. The current analysis is confined to small strain localised contact mechanics of asperity tips' interactions at nanoscopic level, when they are treated as hemispheres.

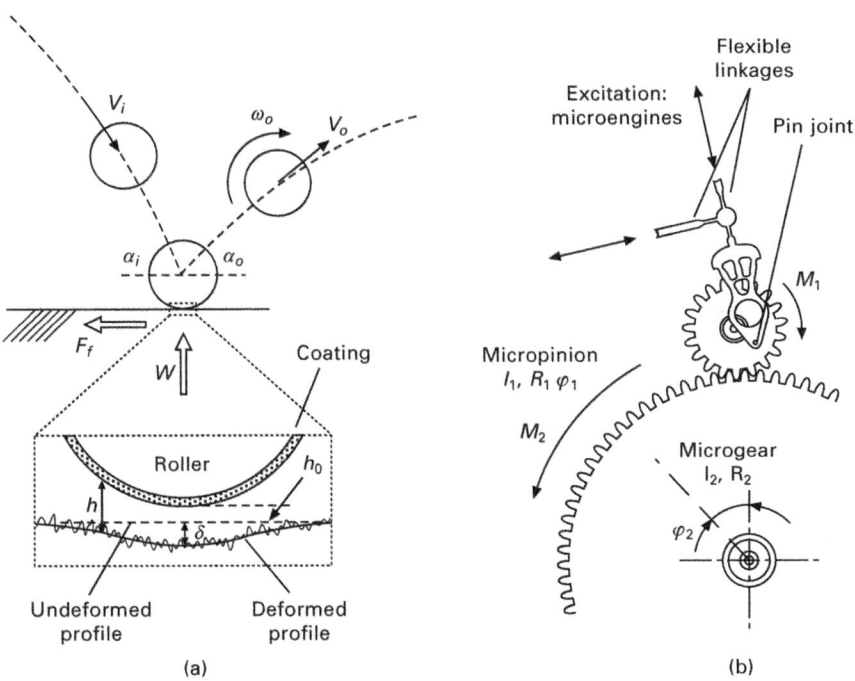

32.1 Generic representation of MEMS frictional impact conditions (a) roller against flat plane (b) two MEMS gear.

Microscale mechanisms usually operate at very high speeds (frequencies of the order of kHz). To understand progressive contact degradation (SAM wear), these contacts are subjected to step changes in contact kinematics, where the lower operational frequency can be considered to be as little as 1 Hz. At high speeds, the gears operate smoothly even if the surface/ protective layer is somewhat damaged. At low speeds, however, increased adhesion (and therefore friction) changes the operation significantly and can eventually cause seizure.

The contact/impact between two meshing teeth can be modelled as an equivalent rigid roller, impacting a flat semi-infinite elastic plane with an equivalent elastic modulus of the pair of impacting bounding surfaces (see Chapter 4). An oblique impact disregarding any pre-impact roller misalignment or post-impact lateral or tilting dynamics can be described in two degrees of freedom as (see Fig. 32.1):

$$\begin{cases} W = m(\ddot{z} + g) \\ I\alpha_0 = M_f \end{cases} \tag{32.1}$$

where $\alpha_0 = \dot{\omega}_0$ and $M_f = F_f R$ and W is the balance of forces acting during impact and rebound; these being the accumulated localised Hertzian impact force between pairs of opposing asperities and their adhesion. Furthermore, some asperity pairs form nanoscale menisci, which add to the work of adhesion. At very close separation, hydrophobic hydration potential acts as a repulsive force as described in Chapter 3.

The normal impact force W and friction F_f are computed as the summation of the interactions between individual asperities (Teodorescu *et al.*, 2008, 2009a,b). These interactions are as outlined above and are chiefly due to adhesion. The formulation for adhesion of asperities is given in Chapter 3 and not repeated here. However, with any ingression of water moisture tiny menisci bridges can also form between the opposing asperity pairs (see Fig. 3.6 in Chapter 3). The effect is increased work of adhesion, because these menisci have to be broken for contact separation and continuance of motion. Therefore, under normal atmosphere, ***meniscus action*** would play an important role as the result of ***condensation***. The thickness of water condensate on a surface depends on the prevailing atmospheric and saturation pressures as (Riedo *et al.*, 2002):

$$h_{mv} = \ln(t/t_a)[\ln(p_s/p_{act})A_m\rho]^{-1} \tag{32.2}$$

where the cross-sectional area of a meniscus bridge between a pair of opposing asperities is $A_m = \pi r^2$ (see Fig. 3.6 in Chapter 3). The radius of meniscus bridge is approximated to:

$$r \approx 2\gamma \ V/R'T \ \log(p/p_s) \tag{32.3}$$

where $\gamma \in 10 - 50 \, \text{mJ/m}^2$.

The condensation activation time, t_a in equation (32.2) is usually quite small; of the order of a few tens of microseconds. Once a monolayer is formed, the depth of the condensation film, h_{mv} increases fairly rapidly and sufficient water can accumulate, leading to the formation of menisci bridges. These bridges are in the nanoscopic range. In time further accumulation of water merges these menisci together and forms a microscale one (some asperities submerge). If a sufficient coherent film is formed, then possibility of hydrodynamics may arise with sufficient speed of entraining motion. However, this is unlikely because of a number of reasons. Firstly, MEMS gears are actually quite rough (roughness in the order of sub-micrometre), thus an uninterrupted film is not envisaged. Secondly, the use of hydrophobic SAM and maintenance of an inert atmosphere is expected to guard against ingression of moisture, and if condensation occurs there would still be an insufficient supply of it to form a coherent film, particularly at very high operating speeds. These high speeds would require a sufficient supply of fluid in order not to avoid starvation. Therefore, to some extent Feynman's wish for a dry contact is almost naturally met.

Within a nanomenisci, a meniscus force tends to bring the opposing asperity pairs together, whilst the hydrophobic *hydration potential* strives to keep them apart. Thus (refer to Chapter 3):

$$p_h = -2 \, \pi R_e \gamma (\cos \theta_1 + \cos \theta_2) + 2 A_m \gamma e^{-(h-z)/\lambda_0}/\lambda_0 \tag{32.4}$$

where $\lambda_0 = 1.5 \, \text{nm}$ is an assumed gap equivalent of six molecules of water deep, a molecule of water being $0.25 \, \text{nm}$. This makes the hydration force monotonic. In closer gaps an oscillatory behaviour is superimposed on the monotonic variation (Israelachvili, 1992), which is similar to the *solvation effect*, described by Al-Samieh and Rahnejat (2001, 2002) and Gohar and Rahnejat (2008). This is unlikely with rough cut surfaces of MEMS. z is the approach of the bounding surfaces and h is the instantaneous gap between the surfaces' average profile (excluding individual opposing asperity pairs) (see Fig. 32.1(a)):

$$h = x^2/2R + h_0 + \delta \tag{32.5}$$

where $x^2/2R$ is the assumed parabolic profile of the roller, h_0 the rigid gap and δ the Hertzian deflection of the roller at any locality. Thus, for a meniscus to form at any instant of time: $h - z \leq h_{mv}$ for any pair of opposing asperities, within which the hydration potential would act.

Now the normal forces p_h due to meniscus action and hydration and p_a for deformation and adhesion of asperity pairs (see Chapter 3) in the various states; deformed, stretched, with a meniscus or completely submerged are obtained with a Gaussian distribution $\phi(z)$, such as that in Chapter 3 as:

$$W = N \int (p_a + p_h)\phi(z) \, dz \tag{32.6}$$

where N is the number of asperities. The friction force F_f, required for equation (32.1), is also obtained as:

$$F_f = N \int (f_a + f_h)\phi(z) \, dz \qquad (32.7)$$

For a pair of asperities the adhesive friction is given as: $f_a = \pi \tau a^3$, where a is the Hertzian contact footprint radius of a pair of assumed hemispherical asperities of equivalent radius R_e and τ is the interfacial shear stress, $\tau \approx$ 10 MPa for silicon surfaces. Assuming that a film of fluid $h–z$ is formed, constituting a meniscus bridge, the viscous shear required to break it is $f_h = A_m \eta U / (h - z)$, where U is the speed of sliding and η the effective dynamic viscosity of water forming a meniscus at the prevailing thermodynamic balance; $\eta \propto p_s, T$.

For the roller impacting a semi-infinite flat plane (see Fig. 32.1(a)), the impact velocity, V_i is chosen to be representative of a pair of MEMS gear teeth. The rebound velocity is clearly obtained by solution of equations of motion (32.1). Therefore, the ***coefficient of restitution*** can be obtained as: $c = V_0/V_i$, this being a measure of damping property of the surface; deviation from totally elastic Hertzian impact. Clearly, c is affected by the work of adhesion which arises from adhesion of asperities and the meniscus action by the formed bridges as described above. The analysis shows that the effect of hydration is small compared with this work of adhesion. When SAM is used, no moisture is expected as the contact angle that water makes with SAM-covered surfaces is obtuse (***hydrophobic***), thus diminishing the meniscus action. Also adhesion is reduced due to decreased free surface energy. However, as SAM is worn a larger percentage of the surface area is exposed and consequently both the meniscus action and adhesion become significant. Hence, it is interesting to note the behaviour of the roller in rebound with different assumed percentage coverage of SAM.

Figure 32.2(a) shows the coefficient of restitution as a function of the impact and rebound velocities with different levels of SAM coverage. It can be noted that for a full protective layer (100% SAM), where adhesion and friction are quite small, the coefficient of restitution remains close to unity, regardless of the impact velocity. Low work of adhesion results. This is true for all impact velocities.

When the level of SAM coverage is reduced, at high impact velocity, the impact force remains high and inertial dynamics dominates. On the other hand, at low impact velocities, the impact force is reduced, and the rebound characteristics is dominated by surface adhesion. Therefore, predictions are in accord with observations of the behaviour of MEMS devices (Tanner *et al.*, 2000). This indicates that better understanding of kinetics action in nanoscale is gradually emerging.

For a pair of MEMS gears in contact (Fig. 32.1(b)), the equations of

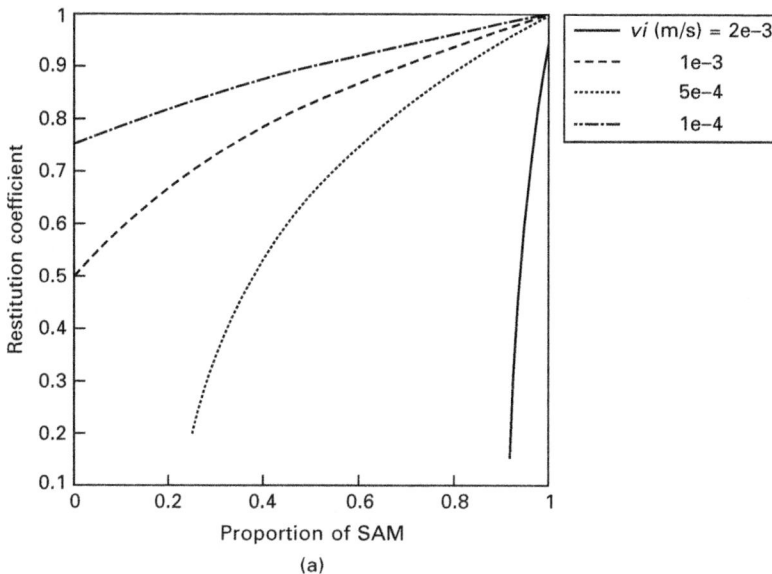

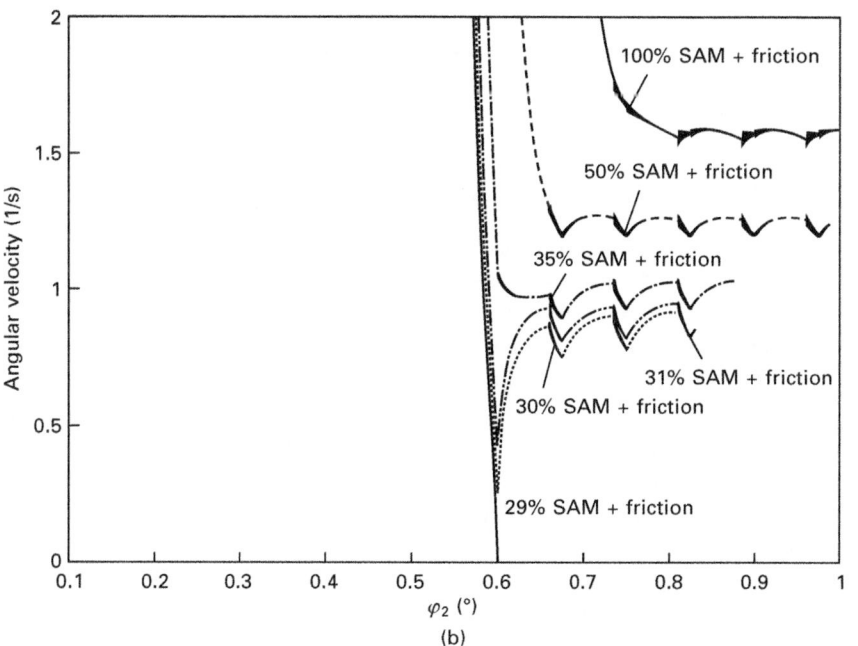

32.2 Impact characteristics at the small scale (a) coefficient of restitution for the impact between a roller and a rough plane (b) angular velocity as a function of wheel rotation.

motion for the two degrees of freedom torsional vibration system $\begin{Bmatrix} \varphi_1 \\ \varphi_2 \end{Bmatrix}$ are (Teodorescu *et al.*, 2009a):

$$\begin{cases} I_1\ddot{\varphi}_1 + R_1 f_g(\varphi_1) = M_1(t,\,\varphi_1,\,\dot{\varphi}_1) \\ I_1\ddot{\varphi}_2 + R_1 f_g(\varphi_1) = -M_2(t,\,\varphi_2,\,\dot{\varphi}_2) \end{cases} \tag{32.8}$$

where $M_1(t,\varphi_1,\dot{\varphi}_1)$ represents the excitation or driving moment on the pinion and is $M_2(t,\,\varphi_2,\,\dot{\varphi}_2)$ s the resisting moment developed at the driven gear. These are described in some detail in Teodorescu *et al.* (2009a). $f_g(\varphi_1)$ is the net interaction force between pairs of gear teeth due to adhesion, meniscus action, deformation and hydration, as described previously. It can be seen that a step reduction in the excitation moment M_1 would result in a sharp drop in gear angular velocity $\mathrm{d}\varphi_2/\mathrm{d}t$. If the protective SAM layer wears out, the substrate becomes exposed. The reduced inertial force is at the same time accompanied by an increase due to the forces contributing to the work of adhesion. Figure 32.2(b) shows that progressively degraded SAM layer demands a higher angular velocity for safe operation of the MEMS gears. Hence, as the SAM layer is worn steadily, reduced impact force fails to overcome the sources of adhesion and after a small angle of rotation the gear pair stick and presumably damaged. On the other hand, with sufficient coverage of SAM, the work of adhesion is overcome even at fairly low rotational speeds.

32.3   Concluding remarks

The opportunities arising from component downsizing have become apparent and have resulted in almost a revolution in certain commercial sectors; mainly in personalised products driven by the microelectronic industry. Although some evolutionary gains have been made in micromechanical counterparts, the plethora of assumed force laws in nanoscale conjunctions is rather troublesome and represent a serious obstacle to rapid progress. Thus, tribology will remain the key area of research in the future. Feynman had the foresight to note the complexities of tribology in nanoscale, a half century ago. What he was actually hinting at was the need to understand the interaction potentials in nanoconjunctions. When and if finally a fundamental understanding is reached, one would note that a generic potential belies all facets of interaction and thus the so-called *theory of everything* is presumably all, but found.

32.4   References

Al-Samieh, M.F. and Rahnejat, H., 'Ultra-thin lubricating films under transient conditions', *J. Phys., Part D: Appl. Phys.*, **32**, 2001, pp. 2610–2621.

Al-Samieh, M.F. and Rahnejat, H., 'Physics of lubricated impact of a sphere on a plate in a narrow continuum to gaps of molecular dimensions', *J. Phys., D: Appl. Phys.*, **35**, 2002, pp. 2311–2326.

Barlian, A.A., Park, S.-J., Mukundan, V. and Pruitta, B.L., 'Design and characterization of microfabricated piezoresistive floating element-based shear stress sensors', *Sensors and Actuators A: Physical*, **134**(1), 2007, pp. 77–87.

Bao, M.-H. and Middelhoek, S., *Micro Mechanical Transducers: Pressure Sensors, Accelerometers, and Gyroscopes*, Elsevier Science, 2000.

Chan, D.Y.C. and Horn, R.G., 'The drainage of thin liquid films between solid surfaces', *J. Chem. Phys.*, **83**, 1984, pp. 5311–5324.

Chigier, N. and Gemci, T., 'A review of micro propulsion technology', 41st Aerospace Sciences Meeting and Exhibition, 6–9 January 2003, Reno, Nevada, USA.

Epstein, A.H., Senturia, S.D., *et al.*, 'Micro-heat engines, gas turbines and rocket engines-The MIT microengine project', AIAA 97–1773, 28th AIAA Fluid Dynamics Conf. Snowmass Village, CO, USA, 1997.

Fernandez-Pello, A.C., 'Micropower generation using combustion: issues and approaches', *Proc. Combustion Institute*, **29**(1), 2002, pp. 883–899.

Feynman R.P., 'Plenty of room at the bottom', Transcript of a Talk to the American Physical Society, Pasadena, USA, December 1959.

Gohar, R. and Rahnejat, H., *Fundamentals of Tribology*, Imperial College Press, London, 2008.

Israelachvili, J.N., *Intermolecular and Surface Forces*, Academic Press, New York, 1992.

Jacobson, S.A. and Epstein, A.H., 'An informal survey of power MEMS', Int. Sympos. on Micro-Mechanical Engng., December 1–3, 2003.

Kim, J.B., Choo, H., Lin, L. and Muller, R.S., 'Microfabricated torsional actuators using self-aligned plastic deformation of silicon', *IEEE/ASME, J. Microelectromechanical Sys.*, **15**(3), 2006, pp. 553–562.

Lee, S., Park, S., Kim, J. and Cho, D, 'Surface/bulk micromachined single-crystalline-silicon micro-gyroscope', *J. Microelectromechanical Sys.*, **9**(4), 2000, pp. 557–567.

Maboudian, R. and Carraro, C., 'Surface chemistry and tribology of MEMS', *Annu. Rev. Phys. Chem.*, **55**, 2004, pp. 35–54.

Matsuoka, H. and Kato, T., 'An ultra-thin liquid film lubrication theory: calculation method of solvation pressure and its applications to EHL problem', *Trans. ASME, J. Trib.*, **119**, 1997, pp. 217–226.

Riedo, E., Lévy, F. and Brune, H., 'Kinetics of capillary condensation in nanoscopic sliding friction', *Phys. Rev. Lett.*, **88**(18), 2002, 4 pp.

Spangler, C. and Kemp, C., 'A smart automotive accelerometer with on-chip airbag deployment circuits,' in Tech. Dig. IEEE Solid-State Sensor and Actuator Workshop, Hilton Head Island, SC, 1996, pp. 211–214.

Tanner, D.M., Smith, N.F., Irwin, L.W., Eaton, W.P., Helgesen, K.S., Clement, J.J., Miller, W.M., Walraven, J.A., Peterson, K.A., Tangyunyong P., Dugger, M.T. and Miller, S.L., *MEMS Reliability: Infrastructure, Test Structures, Experiments, and Failure Modes*, Sandia Report, Sand 2000–0091, 2000.

Teodorescu, M. and Rahnejat, H., 'Dry and wet nano-scale impact dynamics of rough surfaces with or without a self-assembled monolayer', *Proc. Instn. Mech. Engrs., Part N: J. Nanoengng. Nanosys.*, **221**, 2008, pp. 49–58.

Teodorescu, M., Balakrishnan, S. and Rahnejat, H., 'Physics of ultra-thin surface films on molecularly smooth surfaces', *Proc. Instn. Mech. Engrs., Part N: J. Nanoengng. Nanosys.*, **220**, 2006, pp. 7–19.

Teodorescu, M., Majidi, C., Fearing, R. and Rahnejat, H., 'Effect of surface roughness on adhesion and friction of microfibers in side contact', STLE/ASME Int. Joint Trib. Conf., Miami, Florida, 2008.

Teodorescu, M., Theodossiades, S. and Rahnejat, H., 'Impact dynamics of rough and surface protected MEMS gears', *Trib. Int.*, **42**, 2009a, pp. 197–205.

Teodorescu, M., Theodossiades, S. and Rahnejat, H., 'Prediction for MEMS micro-gears frictional impact degradation', Proc. World Trib. Congress, Kyoto, Japan, September 6–11, 2009b.

32.5   Nomenclature

A_m Cross-sectional area of a meniscus
c Coefficient of restitution
g Gravitational acceleration
f_a Adhesive friction per asperity pair contact
f_h Viscous friction per asperity pair contact
F_f Overall friction
h Local elastic gap between the surfaces
h_0 Rigid gap between the surfaces
h_{mv} Cummulative thickness of condensation film
I Polar moment of inertia of the roller
I_1 Polar moment of inertia of the pinion gear
I_2 Polar moment of inertia of the gear wheel
m Mass of the roller
M_1 Driving moment
M_2 Resistive moment
M_f Frictional moment acting on the roller
N Number of opposing asperity pairs
p Pressure
p_a Adhesive force per asperity pair
p_h Net meniscus-hydration force per asperity pair
p_{act} Condensation activation pressure
p_s Saturation pressure
r Radius of a meniscus
R Roller radius
R' Vapour/gas constant
R_e Equivalent radius of a pair of hemispherical opposing asperities
R_p Base radius of the pinion gear
R_w Base radius of the gear wheel
t Time
t_a Condensation activation time
T Absolute temperature
U Sliding velocity
V Volume of a meniscus

V_i	Impact velocity
V_0	Rebound velocity
W	Impact force acting on the roller
x	Contact distance from minimum rigid separation
z	Approach of the surfaces (vertical motion)
α_0	Rebound angular acceleration of the roller
δ	Hertzian deflection
γ	Free surface energy
η	Dynamic viscosity
φ_1	Angular motion of the pinion gear
φ_2	Angular motion of the gear wheel
θ_1, θ_2	Contact angles
ρ	Density
τ	Interfacial shear stress of solid surfaces
ω_0	Rebound angular velocity of the roller

33
Small-scale surface engineering problems

F.W. DELRIO, National Institute of Standards and Technology,
USA and C. CARRARO and R. MABOUDIAN,
University of California, Berkeley, USA

Abstract: This chapter begins by providing a theoretical framework for
the various interfacial forces that lead to adhesion of micromechanical
structures. Next, a number of test structures used to quantify adhesion
between micromachined surfaces are reviewed. Adhesion results for both
horizontal and vertical surfaces are presented, with particular emphasis
placed on the role of surface roughness, relative humidity, surface chemistry
and contact pressure. Finally, several techniques for reducing the work of
adhesion are examined. Initial attempts to decrease adhesion have focused
on reducing the real contact area by physically altering the surfaces via
dimples, spacers and surface roughening techniques, while more recent
work has focused on chemically modifying the surfaces to control roughness
and surface energy via hydrogen-terminating treatments, self-assembled
monolayers and wear-resistant coatings.

Key words: adhesion, microelectromechanical systems (MEMS), self-
assembled monolayers, surface engineering.

33.1   Introduction

The fabrication and integration of miniaturised mechanical components
with microelectronic components have spawned the technology known as
microelectromechanical systems (MEMS). This technology is enabling the
benefits of microelectronic fabrication to be extended to sensing and actuating
functions. A number of fabrication techniques have been developed for this
technology and have been reviewed elsewhere (Kovacs, 1998; Gad-el-Hak,
2002; Madou, 2002). In this chapter, we focus on surface micromachining
technology and adhesion problems related to surface-micromachined
polycrystalline silicon (polysilicon) structures, although many of the principles
discussed here will also apply to other material systems and fabrication
techniques.

Surface micromachining is defined as the fabrication of micromechanical
structures from deposited thin films. The basic steps in a surface micromachining
process are illustrated in Fig. 33.1. First, a substrate is typically coated with
an isolation layer that protects it during subsequent etching steps, followed
by a ground plane structural layer that can be patterned to form electrical

960

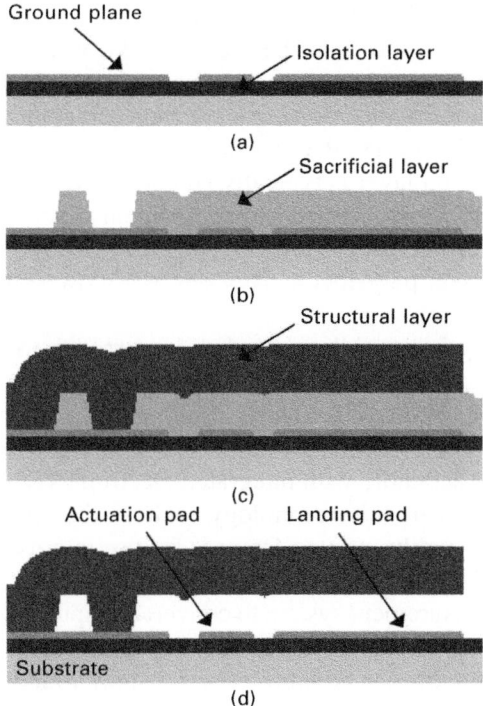

33.1 Basic steps employed in surface micromachining (commonly used materials are shown in parentheses) (a) deposit and pattern isolation layer (silicon nitride) and ground plane structural layer (polysilicon) on substrate (crystalline silicon) (b) deposit and pattern sacrificial spacer layer (silicon oxide) to define anchor and dimple locations (c) deposit and pattern structural device layer (polysilicon) to create mechanical components (d) etch the sacrificial layer to create freestanding structures.

interconnects and landing pads for the devices. A sacrificial spacer layer is then deposited and patterned to define anchor (i.e. electrical and mechanical connections between structural layers) locations. A structural film is then deposited and etched to create the individual mechanical components. The process is repeated to construct multiple layer microstructures. Finally, the sacrificial layers are removed to create freestanding structures such as the cantilever beam shown in Fig. 33.1(d). Typically, surface microstructures have lateral dimensions of 50–500 µm, with thicknesses of 0.1–2.5 µm, and are offset 0.1–2 µm from the substrate.

Surface micromachining is quite flexible in that it can be applied to a variety of material systems, such as semiconductors, metals, ceramics, piezoelectrics and plastics. Single crystal silicon has very good mechanical properties (e.g. in comparison to stainless steel, yield strength of 4.8 vs.

1.0 GPa; Knoop hardness of 9.5 vs. 6.5 GPa; Young's modulus of 190 (111) vs. 220 GPa) (Madou, 2002). It is also the basis for the well-established integrated circuit (IC) technology. As a consequence, **polysilicon** is the most commonly used structural material in surface micromachining. With polysilicon as the structural material, the sacrificial layer is typically silicon oxide (SiO_2), and the isolation layer is typically silicon nitride (Si_3N_4). The microstructures are 'released' by removing the sacrificial oxide layer in a hydrofluoric acid (HF) solution, which etches silicon nitride relatively slowly and has a negligible effect on polysilicon. This step is followed by rinsing and drying.

Surface micromachining has rapidly expanded, in part, due to the well-established infrastructure for depositing, patterning and etching of thin films using IC microfabrication technology. Another reason for this rapid expansion has been the potential of integrated microsystems, which incorporate surface micromachined sensors or actuators with integrated electronics on the same substrate. Early applications of this technology include the ADXL-50, a $\pm 50\,g$ accelerometer developed by Analog Devices for use in airbag control systems (Core et al., 1993). The ADXL-50, shown in Fig. 33.2(a), employs a differential capacitive measurement system to convert an input acceleration into an output voltage, which is used to trigger the airbag deployment. More recently, Texas Instruments introduced a projection display founded on the digital micromirror device (DMD), an array of reflective micromachined mirrors that digitally adjust light to produce a high-quality image on a screen (van Kessel et al., 1998). A scanning electron microscope (SEM) image of the DMD is shown in Fig. 33.2(b). Each aluminium mirror ($16\,\mu m \times 16\,\mu m$) is rotated by an electrostatic force from the underlying complementary metal oxide semiconductor (CMOS) memory cell, which is the result of a voltage difference between the two surfaces.

One of the most common MEMS failure mechanisms is adhesion, also known as stiction, which prevents relative motion between structures. MEMS are especially vulnerable to adhesion as a result of the large surface-to-volume ratio, small surface separations and highly compliant components. Kendall (1994) illustrated the transition from the engineering (macroscale) regime to the adhesive (microscale) regime by comparing the adhesion force to the force of gravity for a ball bearing with various diameters. For a ball bearing with a diameter of 10 mm, the adhesion force is 1.2 mN, while the force of gravity is about 41 mN. In contrast, a ball bearing with a 1 μm diameter experiences an adhesion force more than six orders of magnitude greater than the force from gravity. In this case, the 'crossover' into the adhesive regime occurs somewhere between a diameter of 1 μm and 1 mm. As a result of this transition, the design of microscopic devices must be based on different principles from those of macroscopic components. To establish this new set of design principles, it becomes necessary to gain a thorough

(a)

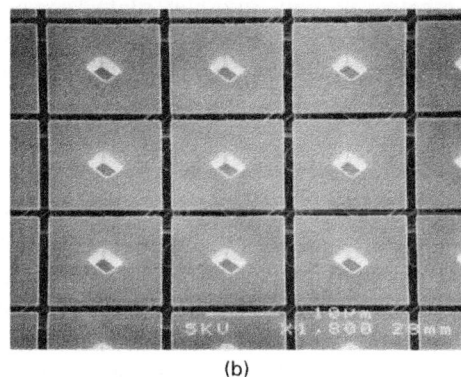

(b)

33.2 Examples of commercially available surface micromachined devices (a) the Analog Devices ADXL-50 is a ±50*g* accelerometer developed for use in airbag control systems (reprinted with permission from Analog Devices) (b) the TI DMD is an array of reflective micromachined mirrors that digitally adjust light to produce a high quality image on a screen (reprinted with permission from Texas Instruments).

understanding of factors contributing to adhesion via novel theoretical and experimental studies.

This chapter begins by providing a theoretical framework for the various interfacial forces that lead to adhesion of micromechanical structures. Parameters appropriate for silicon are used to compare the forces. Next, we review a number of test structures that have been developed to quantify adhesion between micromachined surfaces. Adhesion results for both horizontal and vertical surfaces are presented, with particular emphasis placed on the role of surface roughness, relative humidity, surface chemistry and contact pressure. Finally, we examine several techniques for reducing the work of adhesion. Initial attempts to decrease adhesion have focused on reducing the real contact area by physically altering the surfaces via dimples, spacers and surface roughening techniques. More recent work has focused on chemically modifying the surfaces to control the roughness and surface energy. Chemical passivation schemes that are reviewed include hydrogen-terminating treatments, self-assembled monolayers, and hard, wear-resistant coatings.

33.2   Interfacial forces between two flat plates

The sources of strong adhesion can be traced to the interfacial forces existing at the dimensions of microstructures. These include capillary, van der Waals and electrostatic forces. In this section, we examine each of these interfacial forces for a simplified geometry, namely two flat plates separated by a distance d. Chemical forces such as hydrogen bonding also contribute to the observed adhesion (Legtenberg *et al.*, 1994), but owing to their short-range nature, they are not discussed in this section.

33.2.1 Capillary meniscus forces

In a humid environment, a liquid **meniscus** spontaneously condenses from vapour at an interface (see Chapter 3). The first equation of capillarity, also known as the Young–Laplace equation, is derived by analysing a small section of an arbitrarily curved surface in mechanical equilibrium (Adamson, 1990). Using this technique, the pressure difference across the surface ΔP is shown to be

$$\Delta P = \frac{\gamma}{r_e} \tag{33.1}$$

where γ is the liquid–vapour surface energy and r_e is the mean radius of curvature. r_e is related to the principal radii of curvature r_a and r_m by

$$\frac{1}{r_e} = \frac{1}{r_a} + \frac{1}{r_m} \tag{33.2}$$

The azimuthal radius, r_a, and the meridional radius, r_m, exist on orthogonal planes and are positive when the centre is inside the meniscus (de Boer and de Boer, 2007). In addition, r_a and r_m vary locally, while r_e remains constant along the meniscus surface, assuming that the influence of gravity is negligible (Orr *et al.*, 1975).

The second fundamental relationship of capillarity, or the **Kelvin equation**, utilises thermodynamics and the aforementioned Young–Laplace equation to demonstrate the effect of surface curvature on the free energy (and thus the vapour pressure) of a substance (Adamson, 1990). At constant temperature, the free energy of a substance is related to the mechanical pressure via

$$\Delta G = \int V \mathrm{d}P \qquad (33.3)$$

where V is the molar volume. Assuming constant V and using Equation 33.1 for ΔP, we find that

$$\Delta G = \frac{\gamma V}{r_e} \qquad (33.4)$$

By means of the ideal-gas law, we can also relate the free energy to the relative vapour pressure of the liquid P/P_0 by

$$\Delta G = RT \ln(P/P_0) \qquad (33.5)$$

where R is the gas constant and T is the temperature. In the case of water, P/P_0 is equivalent to the **relative humidity** (RH). Combining Equations 33.4 and 33.5, the Kelvin equation is found, which is given by

$$r_K = \frac{\gamma V}{RT \ \ln(P/P_0)} \qquad (33.6)$$

where r_K is the Kelvin radius, or the mean radius of curvature at equilibrium. For water at 300 K, $\gamma V/RT = 0.53$ nm. Thus, the values for r_K are negative, indicating that the **Laplace pressure** ΔP across the liquid meniscus is negative and the **capillary meniscus** forces act to pull the surfaces together.

The **adhesion energy**, or the work required to separate unit area of the surfaces (also see Chapter 3), depends on the amount of evaporation or condensation during the separation process (de Boer and de Boer, 2007). In the case of slow separation of the two surfaces, the liquid at the interface is allowed to reach thermodynamic equilibrium via evaporation or condensation. As a result, r_K (and thus ΔP) remains constant as d increases from an initial separation d_0 to a rupture separation $2|r_K|\cos\theta$, where θ is the contact angle between the meniscus and the surface. For constant r_K, the **capillary meniscus** force per unit area F_{cap} between two flat plates is

$$F_{cap} = \frac{\gamma}{r_K} \qquad (33.7)$$

If no evaporation or condensation is allowed, the volume of the liquid remains constant. This corresponds to a rapid separation of the interface, which prevents the capillary menisci from reaching thermodynamic equilibrium. Here, the liquid bridge breaks at separations much larger than $2|r_K|\cos\theta$

(Sirghi *et al.*, 2006). For constant volume, the capillary meniscus force per unit area F_{cap} between two flat plates is related to the separation d by

$$F_{cap} = \frac{2\gamma d_o \cos\theta}{d^2}$$
(33.8)

33.2.2 van der Waals dispersion forces

Unlike capillary meniscus forces, van der Waals dispersion forces cannot be eliminated and pose a fundamental limit to the adhesion between micromachined surfaces. London showed that *van der Waals* forces arise from a fluctuation in the molecular dipole moment produced by a virtual fluctuation of the electronic cloud in a molecule (London, 1937; see also Chapter 3 and Gohar and Rahnejat, 2008). This fluctuating dipole polarizes the electron distribution of a nearby molecule, creating an attractive dispersion energy proportional to $1/r^6$, where r is the distance between the molecules. London theory, however, assumes the nearby molecule responds instantaneously to the fluctuating dipole and therefore only applies for separations less than a so-called '*retardation length*', set by the ratio of the speed of light c to the frequency spacing between electronic levels ω (typically, $c/\omega \approx 10$ nm) (Dzyaloshinskii *et al.*, 1961). In other words, information regarding the electron distribution travels at the speed of light with wavelengths corresponding to the emission spectrum of the molecule. Casimir and Polder demonstrated that the interaction energy between molecules becomes proportional to $1/r^7$ for distances larger than these wavelengths (Casimir and Polder, 1948). As a result, the *Casimir force* (or retarded van der Waals) force governs at separations greater than a few tens of nm (Israelachvili and Tabor, 1972). Assuming the forces are additive, the normal and retarded van der Waals forces per unit area between two smooth flat surfaces are

$$F_{vdW} = \frac{A}{6\pi d^3}$$
(33.9)

and

$$F_{rvdW} = \frac{B}{d^4}$$
(33.10)

respectively, where A is the *Hamaker constant* and B is the retarded van der Waals constant (Tabor and Winterton, 1969). A gradual transition from normal to retarded van der Waals forces occurs between these separations (Israelachvili and Tabor, 1972; Anandarajah and Chen, 1995). Undoubtedly, these expressions neglect the repulsive portion of the surface potential energy, which results from overlapping electron orbitals at extremely small separations. However, using the Lennard-Jones 6-12 potential (Lennard-Jones, 1931), it is possible to show that the work of adhesion is overestimated by less than a factor of 2 by ignoring the repulsive term (Maboudian and Howe, 1997).

33.2.3 Electrostatic forces

Electrostatic forces are known to arise from charge trapping in a dielectric layer and a work function difference between the contacting materials. Dielectric charging is the process by which charges tunnel into the dielectric layer and become trapped, screening the applied potential and hindering device operation. The exact physics of dielectric charging are not fully understood. Goldsmith *et al.* (2001) propose that the charges travel through the layer according to the *Frenkel–Poole relationship* for insulating films (Sze, 1981):

$$J \sim V e^{+2a\sqrt{V}/T - q\phi_B/kT} \tag{33.11}$$

where J is the current density due to Frankel–Poole emissions, V is the applied voltage, T is the temperature, ϕ_B is the barrier height, k is *Boltzman's constant* and a is a constant based on material and geometric considerations. This equation indicates an exponential relationship between current density (i.e. rate of trapped charge) and the applied voltage, which was observed experimentally in capacitive radio-frequency (RF) MEMS switches (Goldsmith *et al.*, 2001). Eventually, the trapped charge becomes large enough to keep the switches pulled down.

The *electrostatic force* per unit area acting between surfaces with potential difference V can be written as

$$F_{el} = \frac{\varepsilon_o V^2}{2d^2} \tag{33.12}$$

where $\varepsilon_o = 8.8542 \times 10^{-12}$ F/m is the *permittivity* of free space (see also Chapter 3). It is important to note, however, that this expression does not take into account the thickness of the dielectric layer, deformation of the top surface or fringing-field effects. These effects will not be considered here, but details on each can be found elsewhere (Osterberg and Senturia, 1997; Rebeiz, 2003).

33.2.4 Comparison

The capillary, van der Waals and electrostatic forces acting across a $1\,\mu m^2$ area are plotted as a function of surface separation in Fig. 33.3. For capillary meniscus forces (constant volume case), we assume a contact angle of $\theta = 0°$, *surface energy* of $\gamma = 72$ mJ/m^2 for water, relative humidity of 50%, and an initial separation of $d_o = 2|r_K|\cos\theta$. For van der Waals forces, we use

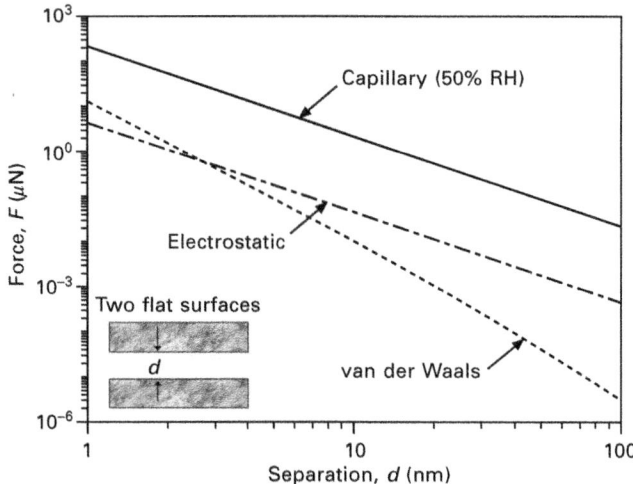

33.3 Capillary, van der Waals and electrostatic forces between two flat surfaces with a 1 μm^2 area separated by a distance *d*. Even at relatively low RH values (50% RH), capillary meniscus forces are shown to dominate van der Waals and electrostatic forces by several orders of magnitude.

d_{co} = 0.2 nm as the cut-off separation and $A = 2.6 \times 10^{-19}$ J as the Hamaker constant (Israelachvili, 1992). A potential difference of $V = 1$ V is assumed for electrostatic attraction. In comparison, the restoring force for spring-suspended MEMS devices deflected by 1 μm in the transverse direction is between 10^{-3} μN and 10 μN (Maboudian and Howe, 1997). Even at relatively low RH values (50% RH), capillary meniscus forces are shown to dominate van der Waals and electrostatic forces by several orders of magnitude. However, it is important to note that even van der Waals and electrostatic forces may exceed restoring forces at small surface separations.

33.3   Experimental methods

While the flat plate relationships provide insight into the magnitude of the attractive forces as a function of surface separation, they neglect the actual surface topography. Interfacial forces are reduced, in some cases by several orders of magnitude, as a result of surface roughness, which reduces the interfacial area to the contacting asperities. Standard rough surface adhesion models account for surface roughness by assuming a statistical distribution of summit heights in contact with a rigid flat plane (Fuller and Tabor, 1975; Maugis, 1996; see also Chapters 2 and 3). These models account for surface forces at or around the contact areas, which is a valid assumption for micron-scale spheres or surface roughness (see also Chapter 32). As a result of the planar deposition technology in MEMS, however, surfaces normally exhibit nanometer-scale roughness. Thus, the points along the interface are separated by less than 100 nm and the adhesion contribution from these non-contacting areas can no longer be neglected (Houston *et al.*, 1997; DelRio *et al.*, 2005). The size and separation of these surfaces is further reduced in nanoelectromechanical systems (NEMS) (see also Chapter 32). In this regime, the surface forces across non-contacting areas will require even more attention to prevent adhesion-induced failures. Therefore, it becomes necessary to develop new theoretical and experimental techniques that take into account adhesion contributions from both contacting and non-contacting areas. A number of test structures have been developed to quantify adhesion in MEMS technology. We will next review the ones that have been most extensively used.

33.3.1 In-plane adhesion

The cantilever beam geometry has long been used to measure ***adhesion*** of mica and glass (Obreimoff, 1930; Bailey, 1961; Wiederhorn, 1967; Michalske and Freiman, 1983; Roach *et al.*, 1988). In these experiments, a crack is propagated along a cleavage plane using a wedge or universal testing machine and the resulting equilibrium crack geometry is used to calculate the surface energy. A similar technique, which involves an array of micromachined cantilever beams, can be used to measure the adhesion energy between micromachined surfaces. The structures are fabricated using standard surface micromachining techniques, which consist of depositing and patterning alternating layers of structural and sacrificial materials as discussed above. At the conclusion of this process, the sacrificial layers are selectively etched and the resulting structures are rendered freestanding using a variety of drying and coating techniques. A schematic representation of a cantilever beam in contact with the substrate is shown in Fig. 33.4(a); an ***SEM*** image of a cantilever beam array is shown in Fig. 33.4(b). The critical dimensions include gap height h, thickness t, width w and length L.

The cantilever beam is forced into contact with the substrate via electrostatic (Houston *et al.*, 1997; de Boer *et al.*, 2000) or mechanical (Jones *et al.*, 2003) loading, and the shape of the deformed structure can be related to the adhesion energy at the interface (Mastrangelo and Hsu, 1992). Using fracture mechanics terminology, the non-adhered length from the support post to the point where the beam meets the substrate is defined as the crack length s. Longer beams are adhered over a large portion of the length and bent into an S-shape. In this configuration, the crack length s is significantly shorter than the length of the beam L and the ***adhesion energy*** Γ is given by

$$\Gamma = \frac{3}{2}\left(\frac{Et^3h^2}{s^4}\right) \tag{33.13}$$

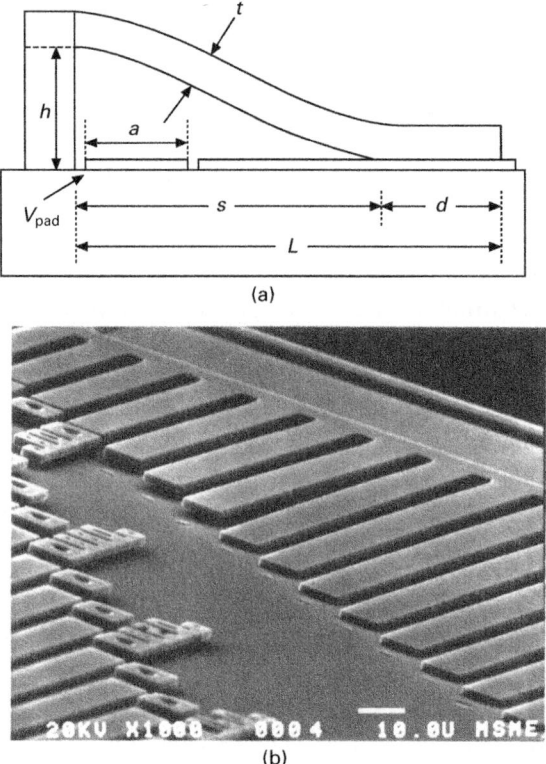

(a)

(b)

33.4 (a) Schematic representation of a cantilever beam in contact with the substrate (reprinted with permission from DelRio F W, Dunn M L, Phinney L M, Bourdon C J, and de Boer M P (2007), *Appl Phys Lett*, **90**, 163104) (b) SEM image of a cantilever beam array (reprinted with permission from Houston M R, Howe R T, and Maboudian R (1997), *J Appl Phys*, **81**(8), 3474–3483).

where E is the Young's modulus. Shorter beams come into contact with the substrate only at their tip, bending the beam into an arc-shape. In this case, the crack length is approximately equal to the length of the beam and the adhesion energy is given by

$$\Gamma = \frac{3}{8}\left(\frac{Et^3h^2}{L^4}\right)$$
(33.14)

De Boer and Michalske (1999) later modelled the two configurations from a linear elastic fracture mechanics perspective. Interestingly enough, the apparent adhesion energy calculated from the experimentally measured shortest arc-shaped beam can be different from that for the S-shaped beam. The discrepancy was resolved by analysing the total system energy of the cantilever in the adhered state as a function of the crack length. For the S-shaped geometry, there is a deep energy well at the equilibrium position, resulting in an accurate adhesion energy measurement (immune to small disturbances that may temporarily force the system out of equilibrium). In the arc-shaped configuration, however, the energy well approaches zero for the shortest beam. As a result, the shortest beam will often pop off the surface and a longer beam will be used in the calculations, yielding an inaccurate measure of the adhesion energy. It is only fair to mention, however, that the analyses based on S-shaped beams require an out-of-plane measurement technique (e.g. interferometry) to measure the crack length s, whereas the arc-shaped technique only requires a high-powered objective on an optical microscope to observe the shortest adhered beam within an array of different lengths.

Mastrangelo and Hsu (1992, 1993a,b) fabricated an array of cantilever beams with various lengths to investigate *release-induced adhesion*, or adhesion that occurs during the final stages of the fabrication process. On the wet etching of the sacrificial layer and the ensuing drying process, capillary menisci formed between the cantilever beams and the underlying substrate, resulting in strong attractive forces that pulled the surfaces together. The transition from adhered to free beams was detected by means of an optical microscope with a Michelson interferometeric attachment. Using Equation 33.14, the adhesion energy for both hydrophobic (HF etch, deionised water rinse) and hydrophilic (HF etch, H_2SO_4:H_2O_2 clean, deionised water rinse) surfaces was evaluated (see also Chapter 3). It was found that the adhesion energy is almost equal for the two systems, suggesting that a thin hydrophilic oxide may be present on both samples. This is consistent with results from Gräf et al. (1989), which indicate that HF-treated silicon surfaces quickly oxidise in water due to (1) the development of surface OH groups, (2) the rupture of Si–Si bonds and (3) the formation of Si–O–Si bridges. In addition, the adhesion energy values were found to be about $140 \, mJ/m^2$, which is roughly equivalent to twice the liquid–vapour surface energy for water, again suggesting that both surfaces thus produced were hydrophilic. Experimentally, this issue was resolved by Houston et al. (1997), who showed that by a careful release process, H-termination of silicon surfaces can be preserved, leading to a large reduction in adhesion in comparison to hydrophilic oxide-passivated surfaces. This will be further discussed in Section 33.5.

A number of techniques have been developed to minimise release-induced adhesion, such as freeze–dry sublimation (Guckel et al., 1989), supercritical CO_2 drying (Mulhern et al., 1993), vapour-phase HF etching (Lee et al., 1997), self-assembled monolayers (Srinivasan et al., 1998) and polystyrene microspheres (Mantiziba et al., 2005). After successful drying of the cantilever beams, these test structures can also be used to quantify *in-use adhesion*, or adhesion that occurs during device operation when the surfaces may come into contact. Houston *et al.* (1997) noted that the surface roughness leads to statistical variation in contact area, and hence a statistical variation in the measured adhesion. To take this into account, they used a statistical method to determine an average crack length and hence an average work of adhesion. For a SiO_2-coated polysilicon surface with an rms roughness of 14 nm, they reported a work of adhesion of 20 mJ/m^2 at 50% RH. At RH approaching 100%, the work of adhesion was found to increase to 140 mJ/m^2, which, as expected, is close to twice the surface tension of water.

DelRio *et al.* (2007a) performed cantilever experiments on hydrophilic SiO_2-coated polysilicon surfaces as a function of surface roughness *and* relative humidity. Prior to adhesion testing, the samples were cleaned using a glass DC plasma generator with oxygen gas and water vapour, which resulted in a water contact angle of <10°, and moved to an environmental interferometer without intermediate exposure to the ambient. The cantilevers were brought into contact with the substrate at 0% RH by applying a voltage to the actuation pad. The RH within the environmental chamber was then increased from 0% RH to 95% RH in 5% RH increments. Using a long-working distance interference microscope (Sinclair *et al.*, 2005) with a phase shifting algorithm (Hariharan *et al.*, 1987), the deflection profiles of the cantilevers (surface roughnesses ranging from 2.6 nm rms to 10.3 nm rms) were recorded as a function of RH as shown in Fig. 33.5(a). The experimental data indicated a strong correlation between surface roughness and capillary condensation. As the landing pad roughness increases, the RH at which the adhesion initially jumps due to capillary condensation also increases. Once the initial jump occurs, the adhesion increases towards the upper limit of 144 mJ/m^2. A detailed model based on the measured surface topography qualitatively agrees with the experimental data only when the topographic correlations between the upper and lower surfaces are considered as illustrated in Fig. 33.5(b).

33.3.2 Sidewall adhesion

In the cantilever beam geometry, the contacting surfaces consist of the top of the landing pad and the bottom of the **structural layer**, also known as in-plane surfaces. However, micromachined devices often contact on vertical rather than horizontal surfaces (these surfaces are referred to as sidewalls). For instance, the ratchet-driven micromotor developed at Sandia National Laboratories requires a significant amount of contact between sidewalls to provide the unidirectional rotation of the drive gear (Sniegowski and de Boer, 2000). The chemical composition and surface texture of sidewalls are expected to be drastically different from in-plane surfaces (Maboudian and Carraro, 2004). The chemical differences between sidewalls and in-

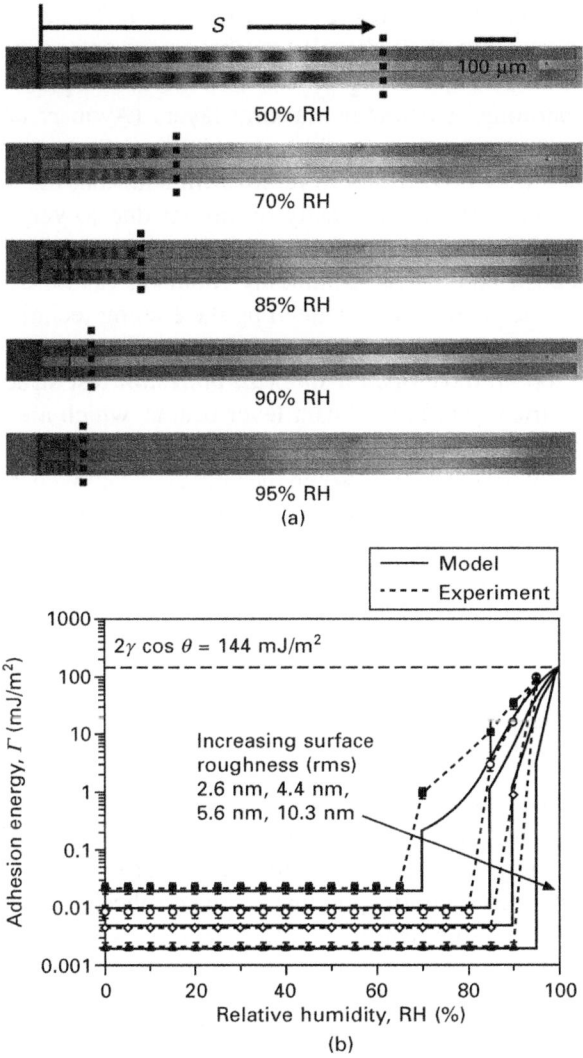

33.5 (a) Interferograms of cantilever beams with a landing pad roughness of 2.6 nm rms as a function of relative humidity (b) adhesion results from the experiments (data points with dashed lines to guide the eye) and model (solid lines) for landing pad roughnesses ranging from 2.6 nm rms to 10.3 nm rms. Solid vertical lines indicate RH at which capillary forces are included in the model. The maximum adhesion energy due to capillary condensation $2\gamma\cos\theta = 144\,\text{mJ/m}^2$ is shown for reference (reprinted with permission from DelRio F W, Dunn M L, Phinney L M, Bourdon C J, and de Boer M P (2007), *Appl Phys Lett,* **90**, 163104).

plane surfaces are a result of exposing the surfaces to different processing environments. For example, the sidewalls can be exposed to reactive gases and high-energy ions during etching processes, whereas in-plane surfaces are protected by photoresist or other masking layers (Ayon *et al.*, 1999). As a result of the exposure, a polymer-like material sometimes develops on the sidewalls, which may be difficult to remove. In addition, the texture of sidewalls and in-plane surfaces are vastly dissimilar due to very different formation techniques. The roughness of the in-plane surfaces is controlled by the thin film deposition process, commonly chemical vapour deposition, while the texture of the sidewalls is dictated by the etching technique, often reactive ion etching.

Ashurst *et al.* (2003a) developed a *sidewall adhesion* test structure that consists of two electrically grounded cantilever beams, which are designed to contact along the sidewall surfaces when a voltage is applied to a centre electrode located between the beams. When the voltage is removed, the beams peel apart to an equilibrium position and the apparent work of adhesion can be calculated (see Fig. 33.6 for schematic diagram and SEM image). In principle, the test structure is similar to the cantilever beam structure used for measuring adhesion between in-plane surfaces. In detail, however, the *work of adhesion* for this two-beam system changes to

$$\Gamma = 3\left(\frac{Et^3h^2}{s^4}\right) \tag{33.15}$$

as a result of the additional stored elastic energy in the beams. The sidewall structures were fabricated alongside their in-plane counterparts using surface micromachining techniques. Interestingly, the adhesion for sidewall surfaces was found to be similar to the corresponding in-plane adhesion. This was not to be expected due to the aforementioned differences in texture and chemical composition. The discrepancy was attributed to the fact that the adhered beams were not perfectly aligned due to strain gradients in the structural material, an event defined as 'scissoring', which reduces the apparent contact area and the resulting adhesion energy values.

Timpe and Komvopoulos (2005) designed and fabricated another *sidewall adhesion* test structure, which consisted of two shuttles supported by double-folded flexure suspensions as illustrated in Fig. 33.7(a). One set of comb-drive actuators was used to bring the shuttles into contact, while another set of comb-drive actuators was used to separate the surfaces and calculate the *adhesion force* in dry and wet environments. The authors draw the following conclusions from the investigation on SiO_2-passivated polysilicon surfaces. First, both van der Waals and capillary meniscus forces contribute to the adhesion (pull-off) force. For apparent contact pressure up to 10 kPa, the adhesion of sidewall surfaces is controlled by van der Waals

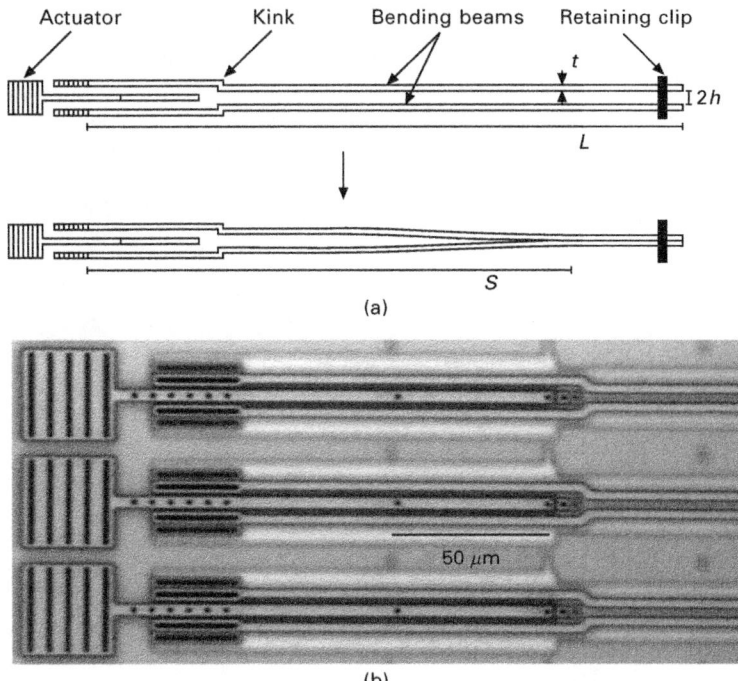

33.6 (a) Schematic representation of the sidewall test structure (before and after actuation). Hashed regions are anchored to the substrate (b) SEM image of an array of the sidewall test structures (reprinted with permission from Ashurst W R, de Boer M P, Carraro C, and Maboudian R (2003), *Appl Surf Sci*, **212–213**, 735–741).

dispersion forces, whereas capillary meniscus forces are shown to play a dominant role at contact pressures well above 10 kPa. Second, the pull-off force increases as the relative humidity increases as depicted in Fig. 33.7(b), which agrees with results for in-plane surfaces (DelRio *et al.*, 2007a). Third, the contribution of non-contacting asperities to the adhesion force is greater than that of contacting asperities. Again, these results agree with those for in-plane surfaces (Houston *et al.*, 1997; DelRio *et al.*, 2005), which further highlights the significance of adhesion models that take non-contacting portions of the interface into account.

33.4   Physical modification of surfaces

By examining the flat-plate relationships and the subsequent rough surface adhesion experiments, it becomes clear that one way to reduce the adhesion energy between in-plane and sidewall surfaces is to increase the average surface separation. Fan *et al.* (1988) fabricated a micromechanical bushing,

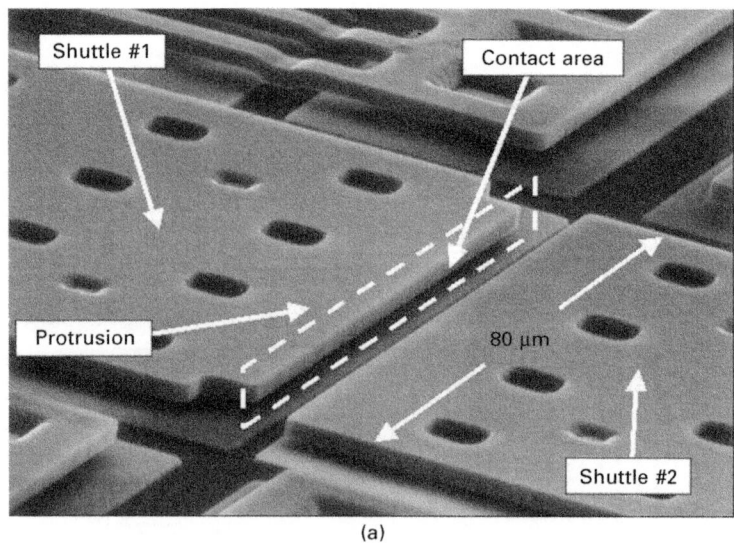

(a)

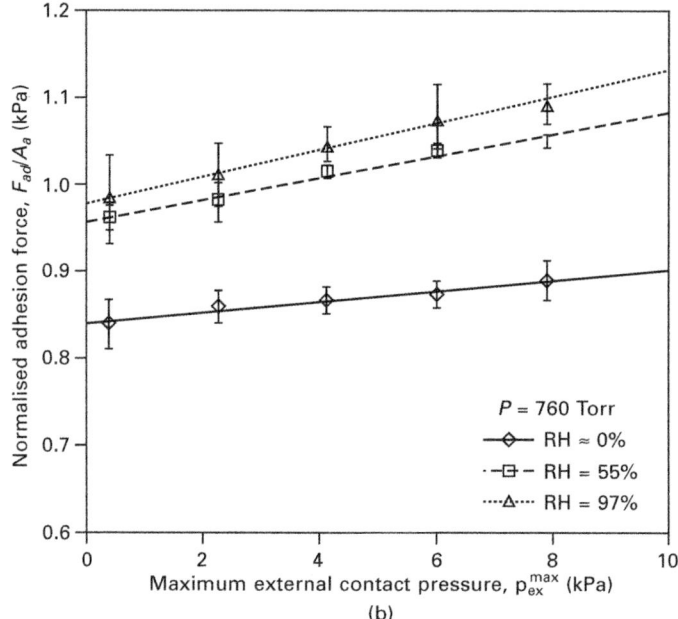

(b)

33.7 (a) SEM image of the contacting shuttles in the sidewall adhesion test structure (b) adhesion force as a function of maximum external contact pressure for ambient pressure and various RH values (reprinted with permission from Timpe S J and Komvopoulos K (2005), *J Microelectromech Syst,* **14**, 1356–1363).

also known as a 'dimple', as part of the test structure using the so-called undercut-and-refill technique. In the process, the sacrificial layer is etched and refilled in a select area prior to depositing the structural device layer, which results in a small protrusion that defines, and reduces, the *real contact area*. Another technique for reducing the real contact area is to modify the surface roughness. By increasing the surface roughness, the average separation between the surfaces is increased, thereby reducing the adhesion energy at the interface. Alley *et al.* (1993) have used texture etchback techniques to obtain very rough polysilicon surfaces. Initially, an undoped polysilicon layer was deposited at 630 °C by low-pressure chemical vapour deposition on a Si (100) substrate. Next, the polysilicon was thermally oxidised at 800 °C for various times and textured via $CCl_4/He/O_2$ plasma etching. The non-uniform removal of the SiO_2 layer in addition to the high polysilicon-to-oxide selectivity resulted in modifications to the surface roughness. This texturing process reduced the release-based adhesion of cantilever beam test structures by about a factor of five. Houston *et al.* (1997) used NH_4F etching of Si to increase the surface roughness, followed by the growth of a chemical oxide using H_2O_2, to vary the surface roughness of the as-deposited polysilicon during the release process. The rms roughness was varied from 14 nm for as-deposited polysilicon to 40 nm following NH_4F roughening as shown in Fig. 33.8.

Yee *et al.* (1995) used a two-step dry etch process to modify the surface roughness of heavily-doped polysilicon layers. In *phosphorus-doped polysilicon*, the phosphorus preferentially segregates to the grain boundaries at elevated temperatures. As a result, a thicker oxide forms at the grain boundaries due to phosphorus-enhanced oxidation. With the thicker oxide at grain boundaries as mask for etching polysilicon, grain holes are formed without additional photolithographic steps. The authors used cantilever test structures to quantify the effect of the roughening technique on the adhesion energy. The crack length was found to increase as the etch time was increased (corresponding to deeper grain holes, and hence a rougher texture). For samples with the deepest grain holes (300 nm deep), there was nearly a factor of 20 reduction in the work of adhesion.

DelRio *et al.* (2006, 2007b) took a slightly different approach to reduce *in-use adhesion*. Instead of incorporating a physical spacer into the design or roughening the surfaces, which may degrade properties such as reflectivity, fracture strength and material wear resistance, they demonstrated the growth of silicon carbide (SiC) nanoparticles using SiO_2 films deposited from tetraethylorthosilicate (TEOS, $Si(OC_2H_5)_4$). Several SiC nanoparticles on a polysilicon surface are shown in Fig. 33.9. High-temperature annealing allowed residual carbon in the TEOS film to diffuse to and react with the silicon substrate to form SiC nanoparticles. Given that SiO_2 from TEOS is extensively used in MEMS as a sacrificial layer (Sniegowski and de Boer,

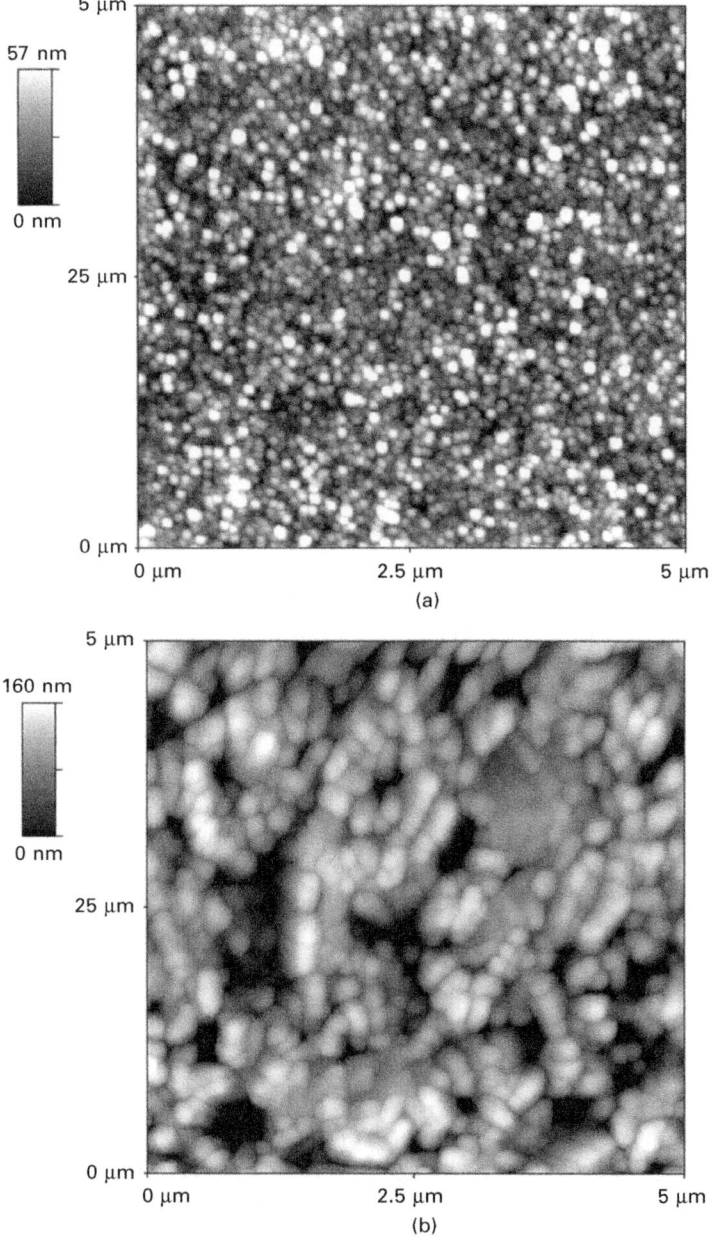

33.8 Atomic force microscope (AFM) images of the polysilicon surface (a) before and (b) after NH_4F roughening. The rms roughness was varied from 14 nm for as-deposited polysilicon to 40 nm for the NH_4F-treated polysilicon (reprinted with permission from Houston M R, Howe R T, and Maboudian R (1997), *J Appl Phys,* **81**(8), 3474–3483).

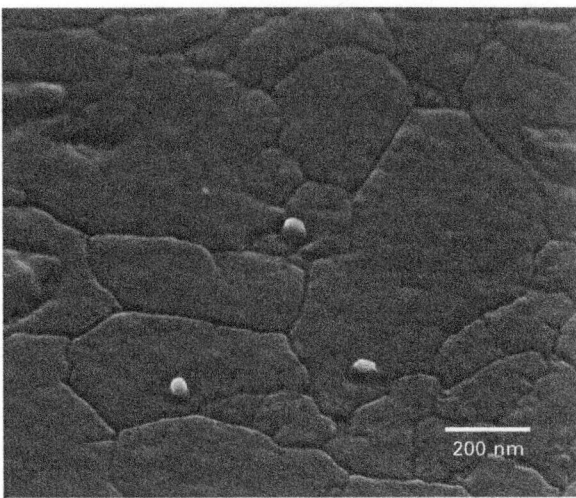

33.9 SEM image of several SiC nanoparticles on a polysilicon surface (reprinted with permission from DelRio F W, Dunn M L, Boyce B L, Corwin A D, and de Boer M P (2006), *J Appl Phys*, **99**, 104304).

2000), these particles provided a new method to reduce adhesion between micromachined surfaces. Above a threshold density, the particles introduced a topography that is more significant than the intrinsic surface roughness. As a result, the adhesion energies were independent of the surface roughness with values as low as $0.4\,\mu J/m^2$. Furthermore, the particles provided high immunity to capillary adhesion. The adhesion of microcantilever beams increased at RH as low as 30% in the absence of particles (de Boer, 2007). In contrast, there was no increase in adhesion until 90% RH to 95% RH with particles on the polysilicon surface. Unfortunately, the size, density and location of the particles were not easily controlled, making the adhesion energy smaller, but stochastic in nature.

33.5   Chemical modification of surfaces

As described above, initial attempts to alleviate in-use adhesion focused on reducing the contact area between adjacent surfaces via dimples (Fan *et al.*, 1988) and surface roughening (Alley *et al.*, 1993; Yee *et al.*, 1995; see also Chapter 13). Another approach involves *chemical modification* of the polysilicon surface which allows one to manipulate the surface energy and roughness. Below, we shall briefly review three approaches to chemically modify the surface of silicon in order to reduce adhesion in microstructures. The first technique involves etching the surface oxide and terminating the surface with hydrogen. The second approach involves coating the surface with thin organic films such as self-assembled monolayers (see also Chapter 32).

Last, we shall explore the potential of hard coatings for reducing adhesion in microdevices.

Houston *et al.* (1997) explored the effect of H-termination on adhesion using the cantilever beam array technique. In order to prevent the rapid reoxidation of the hydrogen-terminated surfaces seen during water rinsing, after HF treatment, structures were rinsed with a 4:1 methanol:water mixture instead of pure water, followed by supercritical CO_2 drying. The work of adhesion was observed to decrease significantly from $20\,mJ/m^2$ for the oxidised surfaces to $30\,\mu J/m^2$ for the hydrogen-terminated surfaces. Further reduction was reported by using NH_4F not only to H-terminate but also to further roughen the surfaces. The large reduction is traced to the combined effect of the surface chemistry (to reduce capillary meniscus forces) and topography (to reduce effective area) (see also Chapters 2 and 3). The measured values highlighted that the interactions between parts of the surface that are nearly in contact may be comparable to or even larger than the interactions at the actual contact points. However, the H-terminated surfaces reoxidise in air, leading the adhesion to increase as a function of time.

Another class of chemical modification involves the use of *self-assembled monolayer* (SAM) coatings (Maboudian, 1998) (see also Chapter 32). SAM precursor molecules include octadecyltrichlorosilane (OTS, $CH_3(CH_2)_{17}SiCl_3$), dichlorodimethylsilane (DDMS, $Cl_2Si(CH_3)_2$), 1H,1H,2H,2H, perfluorodecyltrichlorosilane (FDTS, $CF_3(CF_2)_7(CH_2)_2SiCl_3$) and 1H,1H,2H,2H, *perfluorodecyldimethylchlorosilane* (FDDMCS, $CF_3(CF_2)_7(CH_2)_2(CH_3)_2SiCl$); some of the physical properties for these various coating schemes are shown in Table 33.1. The self-limiting nature of *SAM* growth ensures a conformal coating, including undersides and sidewalls of microstructures. When applied to the microstructures in a liquid-based process during the release, the water capillary forces responsible for release-related adhesion are eliminated, and hence no drying technique is needed (Srinivasan *et al.*, 1998). Furthermore, the surface coatings significantly reduce in-use adhesion, with work of adhesion values often in the few $\mu J/m^2$ range (Ashurst *et al.*, 2001a,b). However, unlike hydrogen-terminated surfaces, the *SAM*-coated surfaces are much more stable under ambient conditions. Owing to difficulties with scaling up the liquid-based process, the SAM deposition has also been demonstrated from vapour phase (Ashurst *et al.*, 2003b).

One of the key parameters affecting the utility of a surface modification for MEMS is thermal stability, since micromachines may be exposed, during packaging or in use, to extreme temperatures. Moreover, studying adhesive behaviour at high temperature may prove useful in developing accelerated testing protocols for in-use adhesion. The effect of annealing on the anti-adhesion performance of chlorosilane-based SAMs has been characterised using polysilicon cantilever beam arrays. FDTS and FDDMCS monolayers

Table 33.1 Physical property data for various surface treatments (repr nted with permission from Maboudian *et al.* (2002), Tribological challenges in micromechanical systems, *Trib Lett*, **12**(2), 95–100)

Surface treatment	Contact angle		Work of adhesion (mJ/m²)	Coefficient of static friction	Thermal stability in air (°C)	Particulate formation	Selective to Si	References
	Water	Hexadecane						
OTS	110°	38°	0.012	0.07	225	High	No	Ashurst *et al.* (2001a)
FDTS	115°	68°	0.005	0.10	400	Very high	No	Srinivasan *et al.* (1998)
DDMS	103°	38°	0.045	0.28	400	Low	No	Ashurst *et al.* (2001a)
Octadecene	104°	35°	0.009	0.05	200	Negligible	Yes	Ashurst *et al.* (2001b)
Oxide	0–30°	0–20°	20	1.1	–	–	–	Ashurst *et al.* (2001a,b)

deposited from both liquid and vapour phases were investigated (Frechette *et al.*, 2006a,b). It was observed that adhesion *decreases* upon annealing for both monolayers and for both types of deposition. FDTS, however, displays greater temperature stability than FDDMCS regardless of the mode of deposition. The higher thermal resistance of the FDTS underscores the importance of **monolayer crosslinking** since, unlike FDDMCS, FDTS forms a siloxane network on the surface. Based on X-ray photoelectron spectroscopy experiments, incipient monolayer degradation was observed, with loss of the whole fluorinated monolayer chain. This process appears drastically different from the decomposition mechanism of **hydrogenated alkylsiloxane monolayers** such as **OTS** (Kluth *et al.*, 1998).

One of the limitations of SAMs is that they wear under repetitive contact (Ashurst *et al.*, 2001a,b). A number of **hard coatings** have thus been explored (see also Chapter 3). Given that often contacting surfaces in MEMS are not in the line of sight, deposition methods such as chemical vapour deposition (CVD) and atomic layer deposition (ALD) are particularly well suited for MEMS. For example, CVD can be used to deposit silicon carbide (SiC) thin films from 1,3-disilabutane (DSB, $SiH_3CH_2SiH_2CH_3$); the ball and stick model for the DSB precursor molecule is illustrated in Fig. 33.10(a). The single source precursor CVD method permitted conformal, pin-hole free coatings at temperatures as low as 650 °C (Stoldt *et al.*, 2002) as shown in Fig. 33.10(b). The in-use adhesion characteristics of SiC films when used as a substrate material in MEMS applications were investigated using polysilicon cantilever beams (Gao *et al.*, 2006). Apparent work of adhesion of less than $6 \mu J/m^2$ was obtained to be compared to a value greater than $20 mJ/m^2$ when a polysilicon substrate is used. It was suggested that the topography as well as the slower oxidation rate of SiC films are responsible for the observed adhesion reduction. Furthermore, SiC coatings have been shown to significantly reduce friction and wear in released polysilicon microstructures (Ashurst *et al.*, 2004a).

ALD is a vapour phase thin film growth technique based on a series of two self-limiting reactions between gas phase precursor molecules and a solid surface as shown in Fig. 33.11(a). ALD Al_2O_3 films are deposited using alternating trimethylaluminium (TMA) and H_2O exposures. The A and B surface reactions, which define the AB cycle for ALD Al_2O_3, are given by (Elam *et al.*, 2003):

$$AlOH^* + Al(CH_3)_3 \rightarrow AlOAl(CH_3)_2^* + CH_4 \qquad (33.16a)$$

$$AlCH_3^* + H_2O \rightarrow AlOH^* + CH_4 \qquad (33.16b)$$

where the asterisks designate the surface species. Similarly, ALD ZnO films are deposited using alternating diethylzinc (DEZ) and H_2O exposures. The A and B surface reactions, which define the AB cycle for ALD ZnO, are described by (Elam *et al.*, 2003):

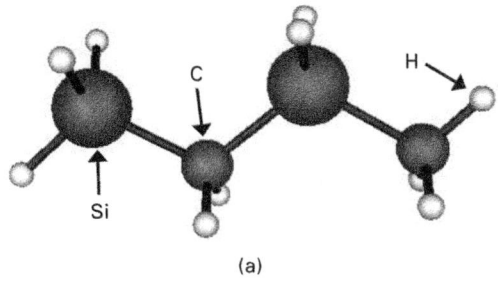

(a)

33.10 (a) Ball and stick model for the DSB precursor molecule (b) SEM image of a thin SiC film deposited on a Si cantilever beam at 800 °C (reprinted with permission from Stoldt C R, Carraro C, Ashurst W R, Gao D, Howe R T, and Maboudian R (2002), *Sens Actuators A*, **97–98**, 410–415).

$$ZnOH^* + Zn(CH_2CH_3)_2 \rightarrow ZnOZn(CH_2CH_3)^* + CH_3CH_3 \quad (33.17a)$$

$$Zn(CH_2CH_3)^* + H_2O \rightarrow ZnOH^* + CH_3CH_3 \quad\quad\quad (33.17b)$$

where the asterisks again designate the surface species. Hoivik *et al.* (2003) used ALD Al_2O_3 as a protective coating for released polysilicon cantilever beams as shown in Fig. 33.11(b). Owing to the self-limiting nature of the ALD process, a highly uniform coating was obtained. Electrostatic testing of the coated cantilever beams revealed that the ALD prevented electrical shorting and failure when the devices were activated beyond the pull-in voltage. Herrmann *et al.* (2007) expanded upon this idea by investigating ALD Al_2O_3/ZnO alloys as charge dissipative dielectric layers in RF MEMS. ALD Al_2O_3/ZnO alloys are formed by alternating the TMA/H_2O and DEZ/H_2O exposure sequence. Using this technique, the resistivity can be tuned over 18 orders of magnitude from 10^{-2} Ωcm for pure ZnO to 10^{16} Ωcm for pure Al_2O_3 (Elam *et al.*, 2003). The number of cycles prior to failure for the RF MEMS switches with ALD Al_2O_3 and ALD Al_2O_3/ZnO was 661 million and 1057 million, respectively. The results suggest that device lifetime increases with increasing ALD ZnO content, which is attributed to

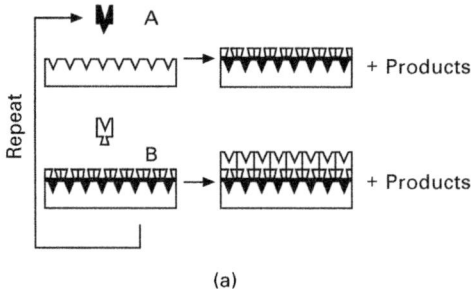

(a)

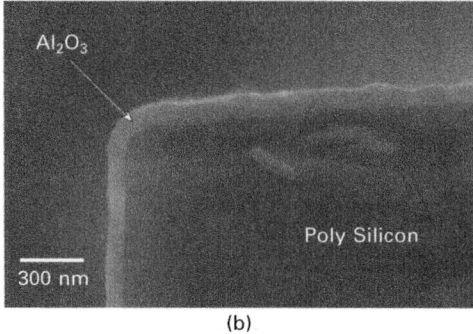

(b)

33.11 (a) Schematic drawing of the AB reaction sequence in the ALD process (b) SEM image of the tip of a cantilever beam coated with 80 nm Al_2O_3 (reprinted with permission from Hoivik N D, Elam J W, Linderman R J, Bright V M, George S M, and Lee Y C (2003), *Sens Actuators A,* **103**, 100–108).

the conductivity of the ALD ZnO. In other studies, a significant reduction in wear, and consequently, a much increased time-to-failure were observed for microdevices whose sliding contacts had been coated by ALD deposited titania (Ashurst *et al.*, 2004b) and carbon-doped alumina (Carraro *et al.*, 2007).

33.6   Future trends

Adhesion forces scale unfavourably with size in small-scale mechanical systems, and as a consequence, the design of microscopic machinery must be based on different principles from that of macroscopic mechanical components. In MEMS, surface contact should be avoided at all cost at the design level. But when this cannot be done, the surfaces can be engineered in such a way as to minimise the deleterious consequences of strong surface forces. The use of rough surfaces made out of, or coated with, hard and/or hydrophobic materials, contributes to alleviating adhesion substantially. Similarly, coating a surface with a molecular lubricant or with a self-lubricating film reduces friction and wear, enhancing the number of cycles-to-failure of devices with sliding contacts (see also Chapters 3, 8 and 14).

The increasing surface-to-volume ratio of microsystems also enhances the adverse consequences of surface chemical reactivity, especially when the microsystem cannot be encapsulated in a completely inert ambient (see also Chapter 32). This is certainly the case for a chemical sensor that has to interact with the atmosphere and, more generally, for any small component fabricated to function at high temperature or in harsh environments, such as a combustion *microengine* (Fernandez-Pello *et al.*, 2003). In this case, the choice of materials becomes crucial. Unfortunately, the materials that have been best characterised, from a tribological standpoint, for use in micromechanical systems – namely silicon and aluminium (and polymers to a lesser extent) – fail the criterion of resilience to harsh environments or to extreme operating conditions (such as high temperature, voltage, radiation and relative humidity). In contrast, materials that are slated to perform well in harsh environments (such as ceramics, refractory metals, carbides and diamond) have not been tested adequately in microscale applications, partly because technologies have yet to be developed to process and to integrate these materials in microscale applications. The pursuit of such technologies will surely become a fertile field for research both in microscale materials properties and in microsystems engineering.

33.7   References

Adamson A W (1990), *Physical Chemistry of Surfaces*, New York, John Wiley and Sons.

Alley R L, Mai P, Komvopolous K, and Howe R T (1993), Surface roughness modification of interfacial contacts in polysilicon microstructures, *Proceedings of the 7th International Conference on Solid-State Sensors and Actuators (Transducers 93)*, 288–291.

Anandarajah A and Chen J (1995), Single correction function for computing retarded van der Waals attraction, *J Colloid Interface Sci*, **176**, 293–300.

Ashurst W R, Yau C, Carraro C, Maboudian R, and Dugger M T (2001a), Dimethyldichlorosilane as an anti-stiction coating for MEMS: a comparison to the octadecyltrichlorosilane self-assembled monolayer, *J. Microelectromech Syst*, **10**(1), 41–49.

Ashurst W R, Yau C, Carraro C, Lee C, Kluth G J, Howe R T, and Maboudian R (2001b), Alkene based monolayer films as anti-stiction coatings for polysilicon MEMS, *Sens Actuators A*, **91**, 239–248.

Ashurst W R, de Boer M P, Carraro C, and Maboudian R (2003a), An investigation of sidewall adhesion in MEMS, *Appl Surf Sci*, **212–213**, 735–741.

Ashurst W R, Carraro C, and Maboudian R (2003b), Vapor phase anti-stiction coatings for MEMS, *IEEE Trans Dev Mat Reliability*, **3**, 173–178.

Ashurst W R, Wijesundara M B J, Carraro C, and Maboudian R (2004a), Tribological impact of SiC encapsulation of released polycrystalline silicon microstructures, *Trib Lett*, **17**(2), 195–198.

Ashurst W R, Jang Y J, Magagnin L, Carraro C, Sung M M, and Maboudian R (2004b), Nanometer-thin titania films with SAM-level stiction and superior wear resistance for reliable MEMS performance, *Proceedings of IEEE MEMS Conference*, Maastricht, Netherlands, 153–156.

Ayon A A, Braff R, Lin C C, Sawin H H, and Schmidt M A (1999), Characterization of a time multiplexed inductively coupled plasma etcher, *J Electrochem Soc*, **146**(1), 339–349.

Bailey A I (1961), Friction and adhesion of clean and contaminated mica surfaces, *J Appl Phys*, **32**, 1407–1412.

Carraro C, Chinn J, and Kobrin B (2007), Exceptional wear resistance of MEMS devices coated with carbon-doped alumina films, *Proceedings of Solid-State Sensors, Actuators and Microsystems Conference, Transducers 2007*, Lyon, France, 1319–1320.

Casimir H B G and Polder D (1948), The influence of retardation on the London–van der Waals forces, *Phys Rev*, **73**(4), 360–372.

Core T A, Tsang W K, and Sherman S J (1993), Fabrication technology for an integrated surface-micromachined sensor, *Solid State Tech*, **36**(10), 39–47.

de Boer M P (2007), Capillary adhesion between elastically hard rough surfaces, *Experimental Mechanics*, **47**, 171–183.

de Boer M P, and de Boer P C T (2007), Thermodynamics of capillary adhesion between rough surfaces, *J Colloid Interface Sci*, **311**, 171–185.

de Boer M P, and Michalske T A (1999), Accurate method for determining adhesion of cantilever beams, *J Appl Phys*, **86**(2), 817–827.

de Boer M P, Knapp J A, Michalske T A, Srinivasan U, and Maboudian R (2000), Adhesion hysteresis of silane coated microcantilevers, *Acta Mater*, **48**(18–19), 4531–4541.

DelRio F W, de Boer M P, Knapp J A, Reedy E D, Clews P J, and Dunn M L (2005), The role of van der Waals force in adhesion of micromachined surfaces, *Nature Materials*, **4**, 629–634.

DelRio F W, Dunn M L, Boyce B L, Corwin A D, and de Boer M P (2006), The effect of nanoparticles on rough surface adhesion, *J Appl Phys*, **99**, 104304.

DelRio F W, Dunn M L, Phinney L M, Bourdon C J, and de Boer M P (2007a), Rough surface adhesion in the presence of capillary condensation, *Appl Phys Lett*, **90**, 163104.

DelRio F W, Dunn M L, and de Boer M P (2007b), Growth of silicon carbide nanoparticles using tetraethylorthosilicate for microelectromechanical systems, *Electrochem Solid-State Lett*, **10**(1), H27–H30.

Dzyaloshinskii I E, Lifshitz E M, and Pitaevskii L P (1961), General theory of van der Waals forces, *Sov Phys Uspekhi*, **4**(2), 153–176.

Elam J W, Routkevitch, and George S M (2003), Properties of ZnO/Al$_2$O$_3$ alloy films grown using atomic layer deposition techniques, *J Electrochem Soc*, **150**(6), G339–G347.

Fan L S, Tai Y C, and Muller R S (1988), Integrated movable micromechanical structures for sensors and actuators, *IEEE Trans Electron Devices*, **35**, 724–730.

Fernandez-Pello A C, Pisano A P, Fu K, Walther D C, Knobloch A, Martinez F, Senesky M, Stoldt C, Maboudian R, Sanders S, and Liepmann D (2003), MEMS rotary engine power system, *Transactions of the Institute of Electrical Engineers of Japan, Part E*, **123**-E, 326–330.

Frechette J, Maboudian R, and Carraro C (2006a), Effect of temperature on in-use stiction of cantilever beams coated with perflorinated alkysiloxane monolayers, *J Microelectromech Syst*, **15**, 737–744.

Frechette J, Maboudian R, and Carraro C (2006b), Thermal behavior of perfluoroalkylsiloxane monolayers on the oxidized Si(100) surface, *Langmuir*, **22**, 2726–2730.

Fuller K N G, and Tabor D (1975), The effect of surface roughness on the adhesion of elastic solids, *Proc Roy Soc Lond A*, **345**, 327–342.

Gad-el-Hak M (2002), *The MEMS Handbook*, Boca Raton, FL, CRC Press.

Gao D, Carraro C, Howe R T, and Maboudian R (2006), Polycrystalline silicon carbide as a substrate material for reducing adhesion in MEMS, *Trib Lett*, **21**, 226–232.

Gohar R, and Rahnejat H (2008), *Fundamentals of Tribology*, London, Imperial College Press.

Goldsmith C, Ehmke J, Malczewski A, Pillans B, Eshelman S, Yao Z, Brank J, and Eberly M (2001), Lifetime characterization of capacitive RF MEMS switches, *IEEE MTT-S International Microwave Symposium Digest*, **1**, 227–230.

Gräf D, Grundner M, and Schulz R (1989), Reaction of water with hydrofluoric acid treated silicon (111) and (100) surfaces, *J Vac Sci Technol A*, **7**, 808–813.

Guckel H, Sniegowski J J, and Christenson T R (1989), Advances in processing techniques for silicon micromechanical devices with smooth surfaces, *IEEE Microelectromechanical Systems Workshop*, Salt Lake City, UT, USA, 71–75.

Hariharan P, Oreb B F, and Eiju T (1987), Digital phase-shifting interferometry: a simple error-compensating phase calculation algorithm, *Appl Opt*, **26**, 2504–2506.

Herrmann C F, DelRio F W, Miller D C, George S M, Bright V M, Ebel J L, Strawser R E, Cortez R, and Leedy K D (2007), Alternative dielectric films for RF MEMS capacitive switches deposited using atomic layer deposited Al_2O_3/ZnO alloys, *Sens Actuators A*, **135**, 262–272.

Hoivik N D, Elam J W, Linderman R J, Bright V M, George S M, and Lee Y C (2003), Atomic layer deposited protective coatings for micro-electromechanical systems, *Sens Actuators A*, **103**, 100–108.

Houston M R, Howe R T, and Maboudian R (1997), Effect of hydrogen termination on the work of adhesion between rough polycrystalline silicon surfaces, *J Appl Phys*, **81**(8), 3474–3483.

Israelachvili J (1992), *Intermolecular and Surface Forces*, New York, Academic Press.

Israelachvili J N and Tabor D (1972), The measurement of van der Waals dispersion forces in the range 1.5 to 130 nm, *Proc Roy Soc Lond A*, **331**, 19–38.

Jones E E, Begley M R, and Murphy K D (2003), Adhesion of micro-cantilevers subjected to mechanical point loading: modeling and experiments, *J Mech Phys Solids*, **51**, 1601–1622.

Kendall K (1994), Adhesion: molecules and mechanics, *Science*, **263**, 1720–1725.

Kluth G J, Sander M, Sung M M, and Maboudian R (1998), Study of the desorption mechanism of alkylsiloxane self-assembled monolayers through isotopic labeling and high resolution electron energy-loss spectroscopy experiments, *J Vac Sci Technol A*, **16**, 932–936.

Kovacs G T A (1998), *Micromachined Transducers Sourcebook*, Boston, MA, McGraw-Hill.

Lee Y-I, Park K-H, Lee J, Lee C-S, Yoo H J, Kim C-J, and Yoon Y-S (1997), Dry release for surface micromachining with HF vapor-phase etching, *J Microelectromech Syst*, **6**, 226–233.

Legtenberg R, Tilmans H A C, Elders J, and Elwenspoek M (1994), Stiction of surface micromachined structures after rinsing and drying: model and investigation of adhesion mechanisms, *Sens Actuators A*, **43**, 230–238.

Lennard-Jones J E (1931), Cohesion, *Proc Phys Soc*, **43**, 461–482.

London F (1937), The general theory of molecular forces, *Trans Faraday Soc*, **33**, 8–26.

Maboudian R (1998), Surface processes in MEMS technology, *Surface Science Reports*, **30**, 207–269.

Maboudian R and Carraro C (2004), Surface chemistry and tribology of MEMS, *Annu Rev Phys Chem*, **55**, 35–54.

Maboudian R, and Howe R T (1997), Critical review: adhesion in surface micromechanical structures, *J Vac Sci Technol B*, **15**(1), 1–20.

Maboudian R, Ashurst W R, and Carraro C (2002), Tribological challenges in micromechanical systems, *Trib Lett*, **12**(2), 95–100.

Madou M (2002), *Fundamentals of Microfabrication*, Boca Raton, FL, CRC Press.

Mantiziba F, Gory I, Skidmore G, and Gnade B (2005), Wet-etch release process for silicon-micromachined structures using polystyrene microspheres for improved yield, *J Microelectromech Syst*, **14**(3), 598–602.

Mastrangelo C H and Hsu C H (1992), A simple experimental technique for the measurement of the work of adhesion of microstructures, *Solid State Sensor and Actuator Workshop (Hilton Head 1992)*, 208–212.

Mastrangelo C H and Hsu C H (1993a), Mechanical stability and adhesion of microstructures under capillary forces – part I: basic theory, *J Microelectromech Syst*, **2**(1), 33–43.

Mastrangelo C H and Hsu C H (1993b), Mechanical stability and adhesion of microstructures under capillary forces – part II: experiments, *J Microelectromech Syst*, **2**(1), 44–55.

Maugis D (1996), On the contact and adhesion of rough surfaces, *J Adh Sci Tech*, **10**(2), 161–175.

Michalske T A and Freiman S W (1983), A molecular mechanism for stress corrosion in vitreous silica, *J Am Ceram Soc*, **66**(4), 284–288.

Mulhern G T, Soane D S, and Howe R T (1993), Supercritical carbon dioxide drying of microstructures, *Proc. 7th Int. Conf. Solid-State Sensors and Actuators (Transducers 93)*, Yokohama, Japan, June 7–10, 296–299.

Obreimoff J W (1930), The splitting strength of mica, *Proc Roy Soc London A*, **127**, 290–297.

Orr F M, Scriven L E, and Rivas A P (1975), Pendular rings between solids: meniscus properties and capillary force, *J Fluid Mech*, **67**(4), 723–742.

Osterberg P M and Senturia S D (1997), M-TEST: a test chip for MEMS material property measurement using electrostatically actuated test structures, *J Microelectromech Syst*, **6**(2), 107–118.

Rebeiz G M (2003), *RF MEMS Theory, Design, and Technology*, Hoboken, NJ, John Wiley and Sons.

Roach D H, Lathabai S, and Lawn B R (1988), Interfacial layers in brittle cracks, *J Am Ceram Soc*, **71**(2), 97–105.

Sinclair M B, de Boer M P, and Corwin A D (2005), Long working-distance, incoherent light interference microscope, *Appl Opt*, **44**, 7714–7721.

Sirghi L, Szoszkiewicz R, and Riedo E (2006), Volume of a nanoscale water bridge, *Langmuir*, **22**, 1093–1098.

Sniegowski J J, and de Boer M P (2000), IC-compatible polysilicon surface micromachining, *Annu Rev Mater Sci*, **30**, 299–333.

Srinivasan U, Houston M R, Howe R T, and Maboudian R (1998), Alkyltrichlorosilane-based self-assembled monolayer films for stiction reduction in silicon micromachines, *J Microelectromech Syst*, **7**(2), 252–260.

Stoldt C R, Carraro C, Ashurst W R, Gao D, Howe R T, and Maboudian R (2002), A low-temperature CVD process for silicon carbide MEMS, *Sens Actuators A*, **97–98**, 410–415.

Sze S M (1981), *Physics of Semiconductor Devices*, New York, John Wiley and Sons.

Tabor D, and Winterton R H S (1969), The direct measurement of normal and retarded van der Waals forces, *Proc Roy Soc Lond A*, **312**, 435–450.

Timpe S J, and Komvopoulos K (2005), An experimental study of sidewall adhesion in microelectromechanical systems, *J Microelectromech Syst*, **14**, 1356–1363.

van Kessel P F, Hornbeck L J, Meier R E, and Douglass M R (1998), MEMS-based projection display, *Proc IEEE*, **86**(8), 1687–1704.

Wiederhorn S M (1967), Influence of water vapor on crack propagation in soda–lime glass, *J Am Ceram Soc*, **50**(8), 407–414.

Yee Y, Chun K, and Lee J D (1995), Polysilicon surface modification technique to reduce sticking of microstructures, *Proc. 8th Int. Conf. Solid-State Sensors and Actuators (Transducers 95)*, Stockholm, Sweden, June 25–29, 206–209.

Index

CPI Antony Rowe
Eastbourne, UK
November 22, 2022